# Global Congress on Construction, Material and Structural Engineering (GCoMSE 2017)



IOP Conference Series: Materials Science and Engineering
Volume 271

Johor Bahru, Malaysia
28 - 29 August 2017

**Editors:**

**Norashidah Abd Rahman**
**Faisal Sheikh Khalid**
**Nor Hazurina Othman**
**Tong Yean Ghing**
**Lee Yee Yong**
**Noorwirdawati Ali**

ISBN: 978-1-5108-5337-9
ISSN: 1757-8981

**Printed from e-media with permission by:**

Curran Associates, Inc.
57 Morehouse Lane
Red Hook, NY 12571

CURRAN ASSOCIATES INC. proceedings .com

**Some format issues inherent in the e-media version may also appear in this print version.**



Printed by Curran Associates, Inc. (2018)

For permission requests, please contact the Institute of Physics
at the address below.

Institute of Physics
Dirac House, Temple Back
Bristol BS1 6BE UK

Phone: 44 1 17 929 7481
Fax: 44 1 17 920 0979

techtracking@iop.org

**Additional copies of this publication are available from:**

Curran Associates, Inc.
57 Morehouse Lane
Red Hook, NY 12571 USA
Phone: 845-758-0400
Fax: 845-758-2633
Email: curran@proceedings.com
Web: www.proceedings.com

# Table of contents

## Volume 271

**Global Congress on Construction, Material and Structural Engineering 2017 (GCoMSE2017)**

**28–29 August 2017, Johor Bahru, Malaysia**

**Accepted papers received: 15 November 2017**
**Published online: 7 December 2017**

### Preface



### Papers

#### Advanced Construction Materials

**Build environment**

**Green Materials and Technology**

**Structural Health Monitoring System**

**Structure**

IOP Conf. Series: Materials Science and Engineering **271** (2017) 011001 doi:10.1088/1757-899X/271/1/011001

**PREFACE**

We are very grateful that this global congress is convened today to gather more than one hundred speakers presenting papers pertaining to the field of Construction, Material and Structural Engineering. GCoMSE 2017 jointly organised by UTHM Commercial and Jamilus Research Center for Sustainable Construction (JRC) as a platform for scientists, researchers and engineers to share their knowledge and experiences in creating sustainable and environmental friendly world for the present and the future. We all know that this world is borrowed from the next generation and in other for us to ensure that it is well looked-after, we have to maintain the three important pillars of sustainability i.e. economics, environment and society. These three pillars are the catalyst to ensure that the world is livable, continue to exist and at the same time is developed to enhance the socio-culture of mankind.

The GCoMSE 2017 theme, 'Green Technology for Global Awareness and a Better Future' indicates that the needs to prepare the world for future generation to continue the responsibility to make the world a better place to live. It is the desire of all population and generations that mankind will continue to exist and lead the world. It has been portrayed by many in science fiction movies that the apes will rule the world or the world will be destroyed because of our negligence and too engross in the development of artificial intelligence (AI). It is always possible if we do not care to ensure that what exists can be de-existed and what is impossible is always possible. The best example is the telephone evolution. From wired to wireless and from a typical phone to a smart phone. Constructing buildings was considered difficult was limited to certain people. But now, everybody can construct a house with some skills with the application of industrialized building system (IBS). With research works on reducing the usage of natural materials and increasing the use of recycled materials, pollution and negative environmental impacts can be reduced. Thus, it is our duty to ensure that we conduct the necessary steps towards sustainability of the world in whatever discipline and role that we have.

Technically Above the Rest

Sincerely,

**DR SHAHIRON SHAHIDAN**
Executive Chairman of GCoMSE 2017

IOP Conf. Series: Materials Science and Engineering **271** (2017) 011002 doi:10.1088/1757-899X/271/1/011002

## Conference Committees

### Organizing Committees

| | | |
|---|---|---|
| Chairman | : | Dr. Shahiron Shahidan |
| Secretary | : | Dr. Noorwirdawati Ali |
| Website | : | Dr. Basil David Daniel |
| Scopus & Open Acces | : | Dr. Norashidah Abd Rahman |
| | | Dr. Nor Haslinda Abas |
| | | Mr. Muhamad Hanafi Rahmat |
| | | Mrs. Nur Nasuha Abd. Salam |
| Promotion | : | Dr. Nor Hayati Abd Ghafar |
| | | Ir. Shamrul-Mar Shamsuddin |
| | | Dr. Norwati Jamaluddin |
| | | Dr. Siti Radziah Abdullah |
| Paperwork & Proceeding | : | Dr. Faisal Sheikh Khalid |
| | | Dr. Nor Hazurina Othman |
| | | Dr. Tong Yean Ghing |
| | | Dr. Lee Yee Yong |
| Technical | : | Dr. Shahrul Niza Mokhatar |
| | | Dr. Muhammad Fikri Hasmori |
| | | Mr. Afandi Abu Bakar |
| | | Mr. Suhaimi Harun |
| | | Mr. Amran Abd Rahman |
| | | Mr. Mohd Khairi Zainal |

### Managing Editor

Dr. Norashidah Abd Rahman
Dr. Faisal Sheikh Khalid
Dr. Nor Hazurina Othman
Dr. Tong Yean Ghing
Dr. Lee Yee Yong
Dr. Noorwirdawati Ali

### Scientific Committee

Prof. Dr. Badorul Hisham Abu Bakar (Universiti Sains Malaysia, Malaysia)
Assoc. Prof. Dr. Mohd Wira Mohd Shafiei (Universiti Sains Malaysia, Malaysia)
Assoc. Prof. Dr. Abdul Rahman Mohd Sam (Universiti Teknologi Malaysia, Malaysia)
Assoc. Prof. Dr. Norhazilan Bin Md. Noor (Universiti Teknologi Malaysia, Malaysia)
Assoc. Prof. Dr. Zainah binti Ibrahim (Universiti Malaya, Malaysia)
Assoc. Prof. Dr. Bashar S. Mohammed (Universiti Teknologi PETRONAS, Malaysia)
Assoc. Prof. Dr. Redzuan Abdullah (Universiti Teknologi Malaysia, Malaysia)
Assoc. Prof. Dr. Norazura Muhamad Bunnori (Universiti Sains Malaysia, Malaysia)

Assoc. Prof. Dr. Rozana Binti Zakaria (Universiti Teknologi Malaysia, Malaysia)
Assoc. Prof. Dr. Mohd Suffian Yusoff (Universiti Sains Malaysia, Malaysia)
Assoc. Prof. Dr. Ilias Said (Universiti Sains Malaysia, Malaysia)
Assoc. Prof. Sr. Dr. Mohamad Sukeri bin Khalid (Universiti Utara Malaysia, Malaysia)
Assoc. Prof. Dr. Mohd Fairullazi Ayob (International Islamic University Malaysia, Malaysia)
Assoc. Prof. Dr. Zaiton Haron (Universiti Teknologi Malaysia, Malaysia)
Dr. Sharifah Akmam Syed Zakaria (Universiti Sains Malaysia, Malaysia)
Sr. Dr. Mohd Nasrun Mohd Nawi (Universiti Utara Malaysia, Malaysia)
Ir. Dr. Daniel Muhammad Edzhan (JK Bersatu Perunding, Malaysia)
Dr. Hamed Bolandi (Michigan State University, United States)
Dr. Shaker S. Bitar (Universiti Teknologi Brunei, Brunei)
Dr. Rendy Thamrin (Andalas University, Indonesia)
Dr. Mochamad Solikin (Universitas Muhammadiyah Surakarta, Indonesia)
Dr. Rita Irmawaty (Hasanuddin University, Indonesia)
Dr. Samer Al Martini (Abu Dhabi University, UAE)
Dr. Rahmita Sari Rafdinal (Japan Development Sector, Japan)
Dr. Amged O. Abdelatif (University of Khartoum, Sudan)
Dr. Ahmed Elamin (University of Nottingham Ningbo China, China)
Dr. Mazizah Ezdiani binti Mohamad (University College of Technology Sarawak, Malaysia)
Dr. Mustaqqim Abdul Rahim (Universiti Malaysia Perlis, Malaysia)
Dr. Khairunisa Muthusamy (Universiti Malaysia Pahang, Malaysia)
Dr. Nurul Nadrah Aqilah Tukimat (Universiti Malaysia Pahang, Malaysia)
Dr. Doris Toe Hooi Chyee (Universiti Teknologi Malaysia, Malaysia)
Dr. Noorsuhada Md Nor (Universiti Teknologi MARA Pulang Pinang, Malaysia)
Dr. Noorhazlinda Abd Rahman (Universiti Sains Malaysia, Malaysia)
Eng. Dr. Mia Wimala (Universitas Katolik Parahyangan, Indonesia)
Ir. Dr. Syuhaida Ismail (Universiti Teknologi Malaysia, Malaysia)
Dr. Chai Chang Sa'ar (Universiti Teknologi Malaysia, Malaysia)
Dr. Fazdliel Aswad Ibrahim (Universiti Malaysia Perlis, Malaysia)
Dr. Lee Yeong Huei (Universiti Teknologi Malaysia, Malaysia)
Dr. Raudhah Ahmadi (Universiti Malaysia Sarawak, Malaysia)
Dr. Muhamad Azry Khoiry (The National University of Malaysia, Malaysia)
Dr. Tan Ooi Kuan (Universiti Tunku Abdul Rahman, Malaysia)
Dr. Suhailah Mohamed Noor (Universiti Teknologi MARA, Malaysia)
Dr. Fadzli Mohamed Nazri (Universiti Sains Malaysia, Malaysia)
Dr. Lee Yee Ling (Universiti Tunku Abdul Rahman, Malaysia)
Dr. Nur Farhayu Ariffin (Universiti Malaysia Pahang, Malaysia)
Dr. Eeydzah Aminudin (Universiti Teknologi Malaysia, Malaysia)
Dr. Farid Ezanee Mohamed Ghazali (Universiti Sains Malaysia, Malaysia)
Dr. Noor Nabilah Sarbini (Universiti Teknologi Malaysia, Malaysia)
Dr. Mohamad Anuar Kamaruddin (Universiti Sains Malaysia, Malaysia)
Dr. Izian Abd. Karim (Universiti Putra Malaysia, Malaysia)

Dr. Chow Ming Fai (Universiti Tenaga Nasional, Malaysia)
Dr. Puziah Ahmad (Universiti Teknologi MARA, Malaysia)
Dr. Salmia Beddu (Univesiti Tenaga Nasional, Malaysia)
Dr. Nur Hafizah Abd Khalid (Universiti Teknologi Malaysia, Malaysia)
Dr. Norhaiza Nordin (Infrastructure University Kuala Lumpur, Malaysia)
Dr. Sharifah Maszura Syed Mohsin (Universiti Malaysia Pahang, Malaysia)
Noor Aina Misnon (Universiti Pertahanan Nasional Malaysia, Malaysia)
Soffian Noor Mat Saliah (Universiti Teknologi MARA, Malaysia)
Aizam Yusoff (Universiti Sains Malaysia, Malaysia)
Norliana Manap (Public Work Department, Malaysia)
Assoc. Prof. Dr. Mohd Haziman Wan Ibrahim (Universiti Tun Hussein Onn Malaysia, Malaysia)
Assoc. Prof. Dr. Noor Yasmin Zainun (Universiti Tun Hussein Onn Malaysia, Malaysia)
Assoc. Prof. Dr. Mohd Irwan Juki (Universiti Tun Hussein Onn Malaysia, Malaysia)
Dr. Riduan Yunus (Universiti Tun Hussein Onn Malaysia, Malaysia)
Dr. Goh Wan Inn (Universiti Tun Hussein Onn Malaysia, Malaysia)
Dr. Azeanita Suratkon (Universiti Tun Hussein Onn Malaysia, Malaysia)
Dr. Nor Hayati Abd Ghafar (Universiti Tun Hussein Onn Malaysia, Malaysia)
Dr. Nor Haslinda Abas (Universiti Tun Hussein Onn Malaysia, Malaysia)
Dr. Nickholas Anting Anak Guntor (Universiti Tun Hussein Onn Malaysia, Malaysia)
Dr. Mohd Hanif Ismail (Universiti Tun Hussein Onn Malaysia, Malaysia)
Dr. Sallehuddin Shah Ayop (Universiti Tun Hussein Onn Malaysia, Malaysia)
Dr. Sharifah Salwa Mohd Zuki (Universiti Tun Hussein Onn Malaysia, Malaysia)
Dr. Muhammad Nizam Zakaria (Universiti Tun Hussein Onn Malaysia, Malaysia)
Dr. Shahrul Niza Mokhatar (Universiti Tun Hussein Onn Malaysia, Malaysia)
Dr. Junaidah Jailani (Universiti Tun Hussein Onn Malaysia, Malaysia)
Dr. Muhammad Fikri Hasmori (Universiti Tun Hussein Onn Malaysia, Malaysia)
Dr. Mohd Hilton Ahmad (Universiti Tun Hussein Onn Malaysia, Malaysia)
Dr. Rafikullah Bin Deraman (Universiti Tun Hussein Onn Malaysia, Malaysia)
Dr. Alvin John Lim Meng Siang (Universiti Tun Hussein Onn Malaysia, Malaysia)
Ir. Mohd Norazam Yasin (Universiti Tun Hussein Onn Malaysia, Malaysia)
Masni A Majid (Universiti Tun Hussein Onn Malaysia, Malaysia)
Norhafizah Salleh (Universiti Tun Hussein Onn Malaysia, Malaysia)
Hasniza Abu Bakar (Universiti Tun Hussein Onn Malaysia, Malaysia)
Mohamad Hanafi Rahmat (Universiti Tun Hussein Onn Malaysia, Malaysia)
Syed Burhanuddin Hilmi Syed Mohamad (Universiti Tun Hussein Onn Malaysia, Malaysia)

# Peer review statement

All papers published in this volume of *IOP Conference Series: Materials Science and Engineering* have been peer reviewed through processes administered by the proceedings Editors. Reviews were conducted by expert referees to the professional and scientific standards expected of a proceedings journal published by IOP Publishing.

IOP Conf. Series: Materials Science and Engineering **271** (2017) 012001 doi:10.1088/1757-899X/271/1/012001

# Utilization of sugarcane bagasse ash in concrete as partial replacement of cement

**Sajjad Ali Mangi[1], Jamaluddin N[1], Wan Ibrahim M H[1], Abd Halid Abdullah[1], A S M Abdul Awal[1], Samiullah Sohu[1] and Nizakat Ali[2]**

[1]Jamilus Research Centre, Faculty of Civil and Environmental Engineering, Universiti Tun Hussein Onn Malaysia, 86400 Batu Pahat, Johor, Malaysia.
[2] Graduate Student, Department of Civil Engineering, MUET, Campus, Khairpur, Pakistan.
Corresponding author: sajjad.nec@gmail.com

**Abstract.** This research addresses the suitability of sugarcane bagasse ash (SCBA) in concrete used as partial cement replacement. Two grades of concrete M15 and M20 were used for the experimental analysis. The cement was partially replaced by SCBA at 0%, 5%, and 10%, by weight in normal strength concrete (NSC). The innovative part of this study is to consider two grades of concrete mixes to evaluate the performance of concrete while cement is replaced by sugarcane bagasse ash. The cylindrical specimens having size 150 mm x 300 mm were used and tested after curing period of 7, 14 and 28 days. It was observed through the experimental work that the compressive strength increases with incorporating SCBA in concrete. Results indicated that the use of SCBA in concrete (M20) at 5% increased the average amount of compressive strength by 12% as compared to the normal strength concrete. The outcome of this work indicates that maximum strength of concrete could be attained at 5% replacement of cement with SCBA. Furthermore, the SCBA also gives compatible slump values which increase the workability of concrete.

## 1. Introduction

Sugarcane is main food crop in tropical and subtropical countries. It is the major resource for the sugar production. Sugarcane bagasse (SCB) is the waste created after juice extraction from sugarcane. The Sugarcane bagasse ash (SCBA) is acquired through the control burning of sugarcane bagasse. The SCB creates the environmental nuisance due to direct disposal on the open lands and forms garbage heaps in that area [1]. According to Barroso [2] that one ton sugarcane generates the 280 kg of bagasse waste. It generates economics as well as environmental related issues, to solving this issues, enormous efforts have been global towards the bagasse waste management i.e. handling, disposed-off and application. To reduce the environmental burden, the usage of waste materials in concrete is a significant aspect, the sugarcane bagasse ash (SCBA) is a waste material of sugar industry, which has a good potential to utilize in concrete as cement replacement. Sugarcane is main crop and besides that high worth crop of south Asia normally termed as cash-crop. It is important to sugar and production related to it. The sugar industry plays an essential part in the countrywide economy of Pakistan. According to the report of Pakistan sugar mills association, Pakistan produced 65.45 million tons of sugarcane in the year 2015-16 and production of year 2016-17 is forecasted at 71.371 million tons [3]. Furthermore, globally, the researchers are concentrating on usage of both industrial and agricultural waste, as raw material sources for the industry, the appropriate usage of this waste material would not only be economical benefits, but may also result in foreign exchange wages and environmental pollution control [4].

Cement is the most widely consumable material in the infrastructure development works. It is considered as a durable material of construction. However, the environmental issue of cement has become a rising concern, as cement industries are accountable around 2.5% of total worldwide waste emissions from industrial sources [5, 6]. It is need of time to rise the use of cement replacement materials in the concrete which can reduce the significant amount of cement consumption, because the production of cement required huge energy and conferring to Asma [7] it is also accountable for 5% of global anthropogenic $CO_2$ release (every ton of cement produces around 01 ton of $CO_2$), and their usage can also improve the properties of concrete. The burning of organic waste of sugar industry known as bagasse, produces the considerable amount of ash named as sugarcane bagasse ash (SCBA). SCBA is freshly acknowledged as a pozzolanic material; though, there is partial research statistics accessible to the effects of SCBA on the behavior of concrete.

Therefore, it was highly recommended to conducting research on the bagasse and their impact on concrete behavior. Generally, the bagasse waste is disposed to the landfills or disposal sites where ever present in the country and rare studies has been conveyed yet. The bagasse ash can be used as partial cement replacement in concrete. Meanwhile, in the present era there is a huge rise in the production of sugar worldwide, and almost 1500 Million tons of sugarcane are yearly produced in all over the world, which leaves around 40 - 45% bagasse afterward juice removal. So, a normal yearly production of bagasse is projected as 600Million tons, which is a bulky waste from sugar industry [8]. For the construction industry the concrete is one of the most important item which is prepared for mixing of cement, fine aggregates and coarse aggregates and within the concrete the role of cement is very vital. Without cement, one cannot build reinforced structures. However, the high used of cement are an important concern of world environmental professionals. Considering the facts, one of the effective way to reduce the environmental impact is to use mineral admixtures, as a partial cement replacement in concrete, which will have the possible to cost reduction, energy conservation, and waste emission minimization.

Furthermore, Malyadri [9] has carried out the research on SCBA with 5% cement replacement in M25 (1:1:2) grade of concrete and found that the increase in the strength of concrete with use of SCBA but he never produce the optimum amount of SCBA that proposed to be utilized in sustainable concrete production. In addition to that Kawade [10] has carried out the research work on sugarcane bagasse ash as a partial replacement of cement with 10%, 15%, 20%, 25% & 30% cement replacement in M20, M30 & M40 grades of concrete and found that the OPC could favorably be replaced with SCBA up to maximum replacement of 15%. Although, the optimal replacement value of SCBA content was attained as 15.0%.

Therefore, realizing the significance of the issue, this research work is carried out to find out the optimum percentage cement replacement of SCBA in M15 & M20 grades of concrete, because grades M15 or M20 are widely used for reinforced concrete works. Grades M40 used for very heavy reinforced concrete/pre-cast/pre-stressed and M30 used for heavy reinforced concrete/pre-cast [11]. Hence, the object of this research is to evaluate the performance of concrete while incorporating sugarcane bagasse ash as cement replacement in different mixes proportions.

## 2. Methodology

### *2.1 Material used for the research*

*2.1.1 Cement.* The ordinary Portland cement (ASTM C150 Type-I) was used in this research to prepare the control specimens.

*2.1.2 Fine aggregate.* Fine aggregate (commonly known as hill sand) free from debris were brought from nearby having 2.61 of specific gravity and size below 4.75 mm were used.

*2.1.3 Coarse aggregate.* Coarse aggregate commonly known as crushed aggregates were also brought from nearby having 2.65 of specific gravity and nominal maximum size of 20 mm were used.

*2.1.4 Water.* Water (palatable liquid) accessible within the campus laboratory were utilized for the mixing and curing of concrete cylinders.

*2.1.5 Sugarcane bagasse ash.* Sugarcane bagasse brought from vicinity of Khairpur Mir's, Pakistan and was burnt in a closed drum (uncontrolled burning), SCBA was obtained after passing through 300μm standard sieve used for experimental study. Figure 1 and 2 shows the sugarcane bagasse ash after passing from sieved size 300μm and sugarcane bagasse ash retained materials on sieved 300μm respectively after the process of sieving.

**Figure 1.** Bagasse ash after passing from sieve 300μm.

**Figure 2.** Bagasse ash retained materials on sieve 300μm.

*2.1.6 Density & microstructure of SCBA.* The density basically defines as the mass per unit volume and Aigbodion [12] mentioned that the sugarcane bagasse ash is very light weight material and its density normally about 1.95g/cm$^3$. The obtained density value fall within the category of carbon and silica which is 1.8 and 2.2 g/cm$^3$ respectively [13, 14]. Furthermore, figure 3 represents the structure, size and shape of the particles varies; however, it can be classified into three main groups; prismatic, spherical and fibrous and SCBA sample analysis under XRD analysis, which indicated that bagasse ash, having substantial amount of silica particles. The SEM image also demonstrated that $SiO_2$ and carbon have a fine structure, the latter having a finer one. This could be associated with pore size of 4.29 Å against carbon which has substantial surface area in the range of hundreds of square meters per gram and microspores with pore size maxima around 3.36Å [15]. This is comparable to SCBA acquired in previous findings [2, 12, 15].

**Figure 3.** SEM structure of the SCBA [12].

Subsequently, the typical chemical composition of bagasse ash as demonstrated by Idris [4] are provided in table 1.

IOP Conf. Series: Materials Science and Engineering **271** (2017) 012001 doi:10.1088/1757-899X/271/1/012001

**Table 1.** Typical chemical composition of bagasse ash [4].

| S.No. | Component | Percentage |
|---|---|---|
| 1. | $SiO_2$ | 65 |
| 2. | $Al_2O_3$ | 3.95 |
| 3. | $Fe_2O_3$ | 9.17 |
| 4. | CaO | 12.6 |
| 5. | MgO | 0.6 |
| 6. | $SO_3$ | 0.1 |
| 7. | L.O.I | 9.02 |

*2.2 Casting of specimen*

For the experimental analysis total 36 concrete specimens were casted (refer table 2). The size of each specimen was taken as 150 mm diameter and 300 mm length cylindrical in shape. Concrete mixes were prepared as per grades given in the table 2, then, concrete cylinders were kept for the curing purpose as shown in figure 4, for the period of 7, 14 and 28 days. While preparing the concrete cylinders, the slump test was also checked for the each grade for NSC and after replacing the SCBA in concrete. Afterward, cement is replaced by SCBA as per percentage given in the table 2 and same shall be cured for the mentioned period of time. For the each grade of concrete the pair of cylinders were casted to check the average results. After the completion of the required curing period every specimen was checked for the compressive strength in the compressive strength machine as shown in figure 5. After the necessary investigations tested samples will be placed outside of the laboratory as shown in figure 6.

**Table 2.** Quantity of specimens.

| % of SCBA | Mix Proportions | |
|---|---|---|
| | M20 (1:1 ½:3) | M15 (1:2:4) |
| 0 | 2 x 3 = 6 | 2 x 3 = 6 |
| 5 | 2 x 3 = 6 | 2 x 3 = 6 |
| 10 | 2 x 3 = 6 | 2 x 3 = 6 |
| Sub Total: | 18 | 18 |
| Total specimens | | 36 |

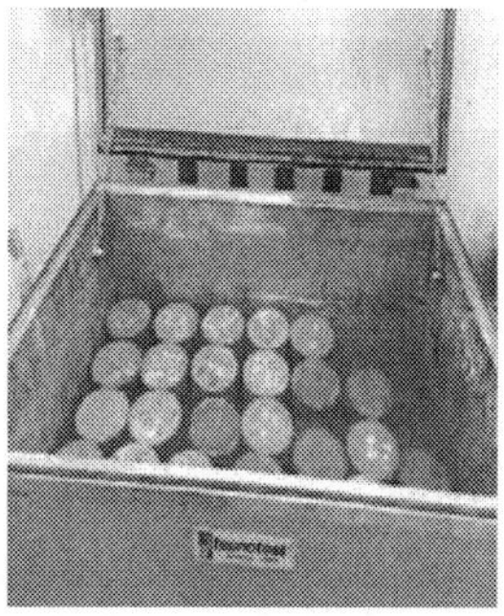

**Figure 4.** Specimen in curing tank.

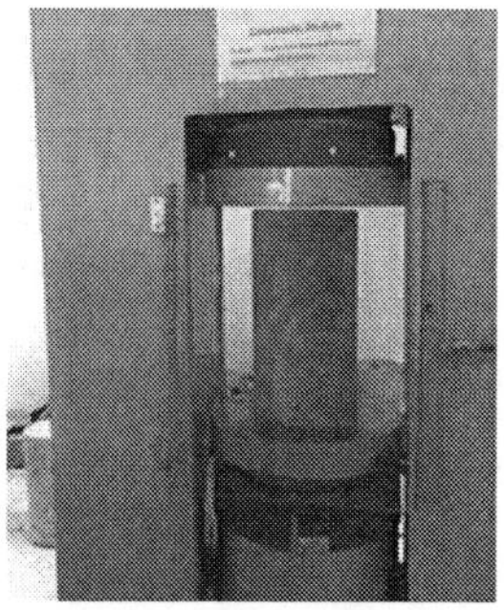

**Figure 5.** Specimen testing for compressive strength.

**Figure 6.** Specimen after testing.

## 3. Results and discussion

The slump test was performed for the concrete with and without SCBA to check the consistency of NMC and the observed result are reported in table 3.

**Table 3.** Slump values of different grades of concrete.

| Sr.No. | Mix Proportion | OPC % Replacement with SCBA | Average slump value (mm) | Increase |
|---|---|---|---|---|
| 1 | | 0% | 27.89 | - |
| 2 | M20 (1:1 ½:3) | 5% | 42.50 | 34% |
| 3 | | 10% | 51.12 | 45% |
| 4 | | 0% | 31.94 | - |
| 5 | M15 (1:2:4) | 5% | 37.48 | 15% |
| 6 | | 10% | 44.64 | 28% |

After the execution of slump test for concrete grade M15 & M20, the continuous rise in the slump value were recorded, when cement replaced by SCBA. The range for slump test was targeted in between 25 mm to 75mm (as recommended by ACI). It is prominent that slump value represent the workability of fresh mix concrete and the observed values fall in the category of low and medium degree of workability. Therefore, the usage of super plasticizer (SP) is not essential in this case.

### *3.1 Compressive strength of concrete*

The compressive testing machine was used to test the entire concrete cylinders for crushing strength at 7, 14 and 28 days respectively. The compressive strength for concrete grade M20 (1:1 ½:3) and M15 (1:2:4) were investigated for the control mix and while cement was partially replaced by SCBA. The results of compressive strength test at different curing periods are provided in table 4 and 5. Through the laboratory observations it was perceived that early age strength is lower than the later age strength, because concrete gain its strength with the passage of time.

**Table 4.** Compressive strength of concrete for M20 (1:1 ½:3).

| OPC % replacement with SCBA | Compressive Strength ($N/mm^2$) | | | Avg. Compressive Strength ($N/mm^2$) | | |
|---|---|---|---|---|---|---|
| | 7 days | 14 days | 28 days | 7 days | 14 days | 28 days |
| 0 | 18.07 | 20.68 | 26.60 | 17.16 | 20.33 | 25.525 |
| | 16.25 | 19.98 | 24.45 | | | |
| 5 | 21.39 | 26.78 | 29.92 | 20.92 | 25.11 | 28.500 |
| | 20.45 | 23.44 | 27.10 | | | |
| 10 | 19.52 | 22.84 | 5.81 | 19.225 | 21.09 | 26.435 |
| | 18.93 | 19.34 | 24.06 | | | |

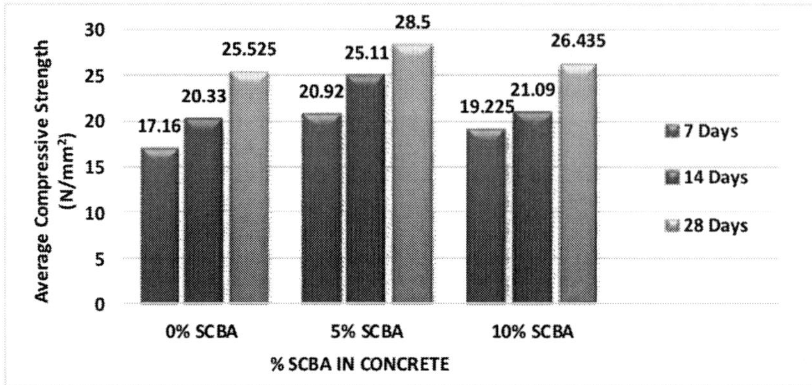


**Figure 7.** Average compressive strength of concrete grade M20 containing different percentage of SCBA at different curing periods.

The compressive strength test results has been presented in figure 7 and it was observed that if cement is replaced by SCBA in concrete at 5% then the average amount of compressive strength will concrete for M20 grade at 28 days curing period around 12% strength increased as compared to the normal mix concrete.

**Table 5.** Compressive strength of concrete for M15 (1:2:4).

| OPC % replacement with SCBA | Compressive Strength ($N/mm^2$) | | | Avg. Compressive Strength ($N/mm^2$) | | |
|---|---|---|---|---|---|---|
| | 7 days | 14 days | 28 days | 7 days | 14 days | 28 days |
| 0 % | 17.30 | 18.50 | 24.11 | 16.025 | 18.19 | 23.805 |
| | 14.75 | 17.88 | 23.50 | | | |
| 5 % | 18.89 | 21.95 | 27.20 | 17.645 | 20.56 | 26.775 |
| | 16.40 | 19.17 | 26.35 | | | |
| 10 % | 17.54 | 20.62 | 24.67 | 16.805 | 19.35 | 24.16 |
| | 16.07 | 18.08 | 23.65 | | | |

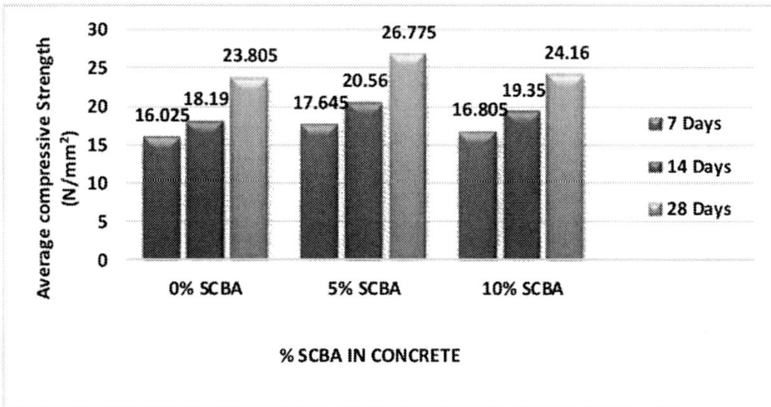


**Figure 8.** Average compressive strength of concrete grade M15 containing different percentage of SCBA at different curing periods.

Subsequently, experimental analysis has done for concrete grade of M15. The result of compressive strength test for grad M15 has been presented in figure 8. The higher results were recorded at 28 days curing period, when cement is replaced with SCBA in concrete at 5% then the average amount of compressive strength will increase about 11.50% as compared to the normal strength concrete. Furthermore, for the both grade (M20 and M15) of concrete while containing SCBA, the compressive strength were detected as lower at the early ages but improves significantly by increasing the curing period. This phenomena of gaining strength with increasing curing time is comparable with previous findings while using cement replacement materials [16].

## 4. Conclusion

This research was successfully carried out, to the establishment of SCBA as an alternative cement replacement material in concrete. After the detailed investigation the following conclusions have been drawn:

- SCBA in concrete gives the higher compressive strength as compared to the normal strength concrete, hence optimal results were found at the 5% replacement of cement with SCBA.
- The usage of SCBA in concrete is not only a waste-minimizing technique, also it saves the amount of cement.
- The replacement of cement with SCBA increases the workability of fresh concrete; therefore, use of super-plasticizer is not essential.
- It is recommended that future research should be performed to assess the use of SCBA in concrete for several properties of concrete for example modulus of elasticity, flexure test, split tensile test, drying shrinkage etc.

## 5. References

[1] Abdulkadir T S, Oyejobi D O and Lawal A A 2014 *Evaluation of Sugarcane Bagasse Ash as a Replacement for Cement in Concrete Works* Acta Tehnica Corviniensis – Bulletin of Engineering (Ilorin: University of Ilorin) pp 71–76

[2] Barroso J, Barreras F, Amaveda H and Lozano A 2003 On the optimization of boiler efficiency using bagasse as fuel, *Fuel* **82** 1451–63

[3] Iskander M K 2016 *Annual review status by Pakistan Sugar Mills Association Centre,* Retrieved on October 20, 2016 from http://www.psmacentre.com/aboutus.php?id=6&type=annual_review&status=1

[4] Idris M K, Eldin K and Yassin E 2015 Determination of the effects of bagasse ash on the properties of portland cement, *Journal of Appl. and Industr. Sci.* **3** 6–11

[5] Jayminkumar A P and Raijiwala D B 2015 Experimental studies on strength of RC concrete by partially replacing cement with sugar cane bagasse ash, *Int. J. of Innovative Research in Science, Engineering and Technology* **4** 2228–2232
[6] Caldarone M A, Gruber K A and Burg R G 1994 High reactivity metakaolin: A new generation mineral admixt for high performance concrete, *Concrete International* **16** 37–40
[7] Asma Abd E H, Shafiq N, Nuruddin M F and Memon F A 2014 Compressive strength and microstructure of sugar cane bagasse ash concrete, *Research J. of Applied Sciences, Engineering and Technology* **7** 2569–2577
[8] Srinivasan R and Sathiya K 2010 Experimental study on bagasse ash in concrete, *Int. J. for Serv. Learn. in Engineering* **5** 60–66
[9] Malyadri T and Supriya J 2015 Experimental study on bagasse ash in concrete by partially replacement with cement, *Int. J. of Comput. Engineering in Research Tre.* **2** 995–1001
[10] Kawade U R, Rathi V R and Vaishali D G 2013 Effect of sugarcane bagasse ash on strength properties of concrete, *Int. J. of Innovative Resarch. in Science, Engineering and Techology* **2** 159–164
[11] Anum I, Williams F N, Adole A M and Haruna A C 2014 Properties of different grades of concrete using mix design method, *Int. J. of Geology, Agriculture and Environmental Sciences* **2** 6–10
[12] Aigbodion V S, Hassan S B, Ause T and Nyior G B 2010 Potential utilization of solid waste (bagasse ash), *J. of Minerals and Materials Characterization and Engineering* **9** 67–77
[13] Osinubi K J and Alhassan M 2007 Bagasse ash modification of Shika laterite, *Nigerian Journal of Engineering* **14** 96-74104
[14] Kevorkijan V 2003 *Functionally graded aluminum-matrix composites* American Ceramic Society Bulletin (Ohio: American Ceramic Society) pp 60
[15] Paya J, Monzo J, Borrachero M V, Diaz Pinzon L and Ordonez L M 2007 Sugar-cane bagasse ash (SCBA): Studies on its properties for reusing in concrete production, *Journal of Chemical Techology and Biotechnology* **77** 21 – 32
[16] Mangi S A, Jamaluddin N, Wan Ibrahim M H, Noridah M and Sohu S 2017 Utilization of sawdust ash as cement replacement for the concrete production: A review, *Engineering Science and Technology Int. Research J.* **1** 11-15

**Acknowledgment**
The authors of this research work are gratefully acknowledging the technical support of Mehran UET, SZAB, Campus Khairpur, Pakistan regarding accessories and equipment provided in the laboratory for the experimental work and authors are also thankful to Center of Graduate Studies, Universiti Tun Hussein Onn Malaysia (UTHM) for their financial support and guidance.

# Strength development of pervious concrete containing engineered biomass aggregate

**A A M Sharif[1], S Shahidan[1], H B Koh[1], A Kandash[1] and S S Mohd Zuki[1]**

[1]Jamilus Research Center, Faculty of Civil and Environmental Engineering, Universiti Tun Hussein Onn Malaysia, Johor Bahru Batu Pahat.
Corresponding author: shahiron@uthm.edu.my

**Abstract.** Pervious concrete with high porosity has good permeability and low mechanical strengths are commonly used in controlling storm water management. It is different from normal concrete. It is only containing single size of coarse aggregate and has lower density compared with normal concrete. This study was focused on the effect of Engineered Biomass Aggregate (EBA) on the compressive strength, void ratio and water permeability of pervious concrete. EBA was prepared by coating the biomass aggregate with epoxy resin. EBA was used to replace natural coarse aggregate ranging from 0% to 25%. 150 mm cube specimens were prepared and used to study the compressive strength, void ratio and water permeability. Compressive strength was tested at 7, 14 and 28 days. Meanwhile, void ratio and permeability tests were carried out on 28 days. The experimental results showed that pervious concrete containing EBA gained lower compressive strength. The compressive strength was reduced gradually by increasing the percentage of EBA. Overall, Pervious concrete containing EBA achieved higher void ratio and permeability.

## 1. Introduction

Concrete is considered as major cause of natural materials deployment. However there are many by-product materials which are being problematic to the environment as being waste material. Biomass aggregate (BA) is conceded as by-product material from the industrial effluents therefore, it is being dumped into the landfills apart from that, and biomass can be recycled to be used in the production of concrete as biomass aggregate [1-5]. Biomass aggregate is used in this study. It was estimated in 1998 that more than 390 million tons of biomass was produced annually but its use was less than 15%. Production of biomass will increase especially in countries such as China and India. It is now estimated that about 780 million tons of biomass was produced in 2010 [3, 5-8]. Biomass aggregate (BA) is produced from the biomass wastes which used to generate heat in the furnace at around 800 C□ for the manufacturing process. After the heat generation procedure, the remaining biomass waste ranging from 10% to 20% and the wastes are left and it becomes aggregate after cool down. This kind of aggregate can define as biomass aggregate (BA) and it is potentially become the sustainable material for concrete [6, 9, 10].

Pervious concrete is a type of concrete with a high porosity used for concrete flatwork applications that allows water from precipitation and other sources to pass through it, thereby reducing the runoff from a site and recharging ground water levels [5, 11-13]. It is also known as "no-fines concrete" composed with Portland cement, coarse aggregate, water, admixtures, and with a small amount or no sand. In the past of 30 years, pervious concrete has been increasingly used in the United States, and among the Best Management Practices (BMP) recommended by the Environmental Protection Agency (EPA). By capturing storm water and allowing it to seep into the ground, pervious concrete is instrumental in recharging groundwater and reducing storm water runoff. The use of pervious concrete in building site design can also aid in the process of qualifying the building for Leadership in Energy and Environmental Design (LEED) Green Building Rating System credits [7, 8, 12, 14]. Innovative development of engineered biomass aggregate pervious concrete for sustainable construction is the main aim of this research.

## 2. Material and procedure

All the procedure and method preparation have been comply with the ASTM and EN BS standard [15-18].

### *2.1. Materials*

The constituent materials used in this research were composite cement, crushed granite aggregate with a nominal size of (8-20) mm, water, superplasticizers (SP) and engineered biomass aggregate (EBA).Engineered biomass aggregate (EBA) was produced by immersing the biomass aggregate with epoxy resins and hardener. Once epoxy resins are hardened, it will produce a coating layer on the surface of biomass aggregate. The coated biomass aggregate is named as engineered biomass aggregate (EBA). The epoxy resins, hardener, biomass aggregate and engineered biomass aggregate (EBA) are shown in figure 1 and figure 2 respectively.

**Figure 1.** Epoxy resin and hardener.

**Figure 2.** Biomass and engineered biomass aggregate.

*2.2. Producing engineered biomass aggregate*

The EBA is produced by mixing the epoxy resin (Sikafloor-161 HC) with the hardener [7]. Before mingling the hardener with the epoxy resin, the epoxy resin needs to be mixed up. A 0.42 kg of hardener was mixed with 1.52 kg epoxy resin inside a small pan and mingles it very well for a short time. 8 kg of biomass aggregate was mixed with the 2 kg of epoxy resin and hardener inside a pan to produce the EBA. After mixing, the EBA are taken out of the pan for drying, and underlid by plastic to avoid sticking to the plate surface. The EBA should be left on the room temperature for 24 hours to ensure well drying, and then collect the EBA after 24 hours drying period.

*2.3. Mix design*

Six pervious concrete mixes, as shown in table 1 were prepared. Since, there is no existing standard method of designing concrete mixes incorporating engineered biomass aggregate, the method of mix design proposed by [6]. The previous research [6, 19] was employed to design a pervious concrete mix with engineered biomass aggregate coated with plastic bag. The mix was designed to have a 28-day target compressive strength of 10 MPa. Superplasticizer of 1 liter/100 kg cement was used to increase the workability of the low water cement ratio pervious concrete mixtures.

**Table 1.** Mix design for pervious concrete by adding EBA.

| Mix | Percentage | Water | Cement | C.A | EBA | SP1% |
|---|---|---|---|---|---|---|
| M1 | 100% N.A | 114 | 380 | 1620 | ------ | 3.8 |
| M2 | 95 NA,5EBA | 114 | 380 | 1539 | 81 | 3.8 |
| M3 | 90 NA,10EBA | 114 | 380 | 1385 | 154 | 3.8 |
| M4 | 85 NA,15EBA | 114 | 380 | 1177 | 208 | 3.8 |
| M5 | 80 NA,20EBA | 114 | 380 | 942 | 236 | 3.8 |
| M6 | 75NA,25EBA | 114 | 380 | 706 | 236 | 3.8 |

*2.4 Specimens and tests*

Concrete cubes of 150 x 150 x 150 mm were used to evaluate the compressive strength, water permeability and void ratio. All hardened concrete cube specimens will be subjected to water curing. The number of specimens used in the tests and the distribution is tabulated in table 2.

**Table 2.** Number of specimens.

| Specimen | Test | Number of specimens | | |
|---|---|---|---|---|
| | | 7-day | 14-day | 28-day |
| 150 x 150 x 150 mm cube | compressive strength | 3 | 3 | 3 |
| 150 x 150 x 150 mm cube | water permeability | - | - | 3 |
| 150 x 150 x 150 mm cube | void ratio | - | - | 3 |

Compressive strength test was conducted based on BS EN 12390-3: 2009 [9]. Three cubes were tested for each age. Concrete cubes used for void ratio are the same cubes used for the 28 day compressive strength test. Void ratio test (figure3) was carried out at 28 day accordance to BS EN 12390-7: 2009 [10]. Void ratio of pervious concrete is defined as the percentage of the total volume of void to the total volume of specimen. It was calculated based on the following equation,

$$\text{Void Ratio } (\%) = 1 - \left(\frac{(W2 - W1)}{V\rho}\right) x100 \qquad (1)$$

Where,as ;
W1 is the weight of specimen submerged under water (kg)
W2 is the weight of specimen at a saturated surface dry condition (kg)
V is the volume of specimen ($m^3$)
ρ is the density of water in ($kg/m^3$)

Permeability test was carried out according to JCI-SPO3-1 (Japan Concrete Institute) standard. Permeability test was conducted on the 150 mm pervious concrete cubes at 28 days. Figure 4 shows the water permeability test equipment. It was calculated based on the following equation:

$$K = \frac{L\,x\,Q}{A\,(L+H)} \tag{2}$$

Where, Q: Amount of water ($mm^3/s$).
A: Cross-section area of pavement ($mm^2$).
H: Water level.
L: Thickness of specimen (mm).
K: Coefficient of water permeability (mm/s).

**Figure 3.** Void ratio test.

**Figure 4.** Water permeability test.

## 3. Result and discussion

Table 3 shown below summarizes the tests conducted on this study, which are, density, void ratio, water permeability, flow rate and compressive strength test. Compressive strength was tested at 7, 14 and 28 days. Meanwhile, void ratio and permeability tests were carried out on 28 days.

**Table 3.** Values of all tests.

| Type of Mix | Percentage | Density (kg/ m³) | Void ratio % | Flow rate (mm/s) | Compressive Strength (MPa) | | |
|---|---|---|---|---|---|---|---|
| | | | | | 7days | 14days | 28days |
| M1 | 100% N.A | 1956 | 21.8 | 8.87 | 9.4 | 8.4 | 13.6 |
| M2 | 95NA,5EBA | 1985 | 15.3 | 9.62 | 4.5 | 6.3 | 2.6 |
| M3 | 90NA,10EBA | 1828 | 25.3 | 10 | 10.7 | 9.7 | 6 |
| M4 | 85NA,15EBA | 1843 | 24.7 | 11 | 7.4 | 6.4 | 6.8 |
| M5 | 80NA,20EBA | 1766 | 26.8 | 11.8 | 4.7 | 3.3 | 8.1 |
| M6 | 75NA,25EBA | 1778 | 22.4 | 12.15 | 2.9 | 5.1 | 2.9 |

*3.1. Density*

Table 3 shows the results of density when adding different percentages of EBA. Decreases of densities were recorded when adding EBA to the mix, and the reason behind this decrease is the lower density of EBA comparing to the normal aggregate. Only the second mix M2 recorded an increase of density, though it included EBA. The maximum density recorded were the M2 mix which included 5% of EBA, while the lowest density were recorded for the M5 mix, which included 20% of EBA.

*3.2. Compressive strength*

Figure 5, summarized the values of compressive strengths for all mixes at 7, 14 and 28 days of curing. From the Figure, it can be observed that the increase of percentage of EBA resulted in decreasing the values of compressive strength, since the EBA has low strength comparing to the normal aggregate. Only M3 showed an increase of the compressive strength, though it included 10% of EBA. The highest value were recorded for M1 mix that includes 0% engineered biomass aggregate at 28 days of curing time, while the lowest value were also recorded for the mixture with 5% of engineered biomass aggregate at 28 days of curing time, with the value of 2.6 MPa. 2.6 MPa, which were recorded for the 5% engineered biomass aggregate mixture at 28days of curing time.

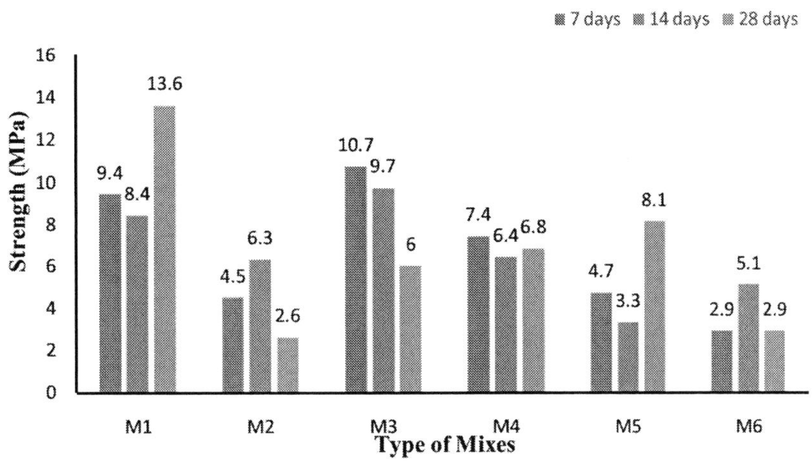


**Figure 5.** Compressive strength of EBA for 7, 14 and 28 days.

Based on the previous study conducted by [6] which was conducted about the biomass aggregates wrapped with plastic, and recorded an increase of the compressive strength with the increase of engineered biomass aggregates values. It can be observed that the biomass aggregates wrapped with plastic resist solving with water and helped in increasing the compressive strength, while the epoxy resin with hardened need to be studied to observe the effect of water to it, since the compressive strength decreased with the increase of biomass aggregates immersed with epoxy resin and hardened and also with the increase of curing time [6].

*3.3. Total void ratio*

Table 3 shows the results of void ratio for various mixes including different percentages of EBA. Basically, the increase of EBA will lead to increase of void ratio. The reason behind the increase of void ratio is the absence of the aggregate in the mixes, giving a space for voids. The highest values of void ratio were obtained at M5 which included 20%, while the lowest values were obtained at M2 which included 5% of EBA.

### *3.4. Water permeability test*

Determining the time for the water to pass through the cubes is called the water permeability. The test was conducted using a tube with specific dimensions to calculate discharge time. From table 3, it can be observed the higher the percentage of EBA, the higher the water permeability. M6 which included 25% of EBA recorded the highest value of water permeability, while the M1, which include 0% of EBA recorded the lowest value of water permeability. The behind the increase of water permeability with the increase of EBA percentage is the smooth surface of the EBA, which allow the water to move in a faster rate.

## 4. Conclusion

To sum up, the addition of Engineered Biomass Aggregate (EBA) lead to an increase of void ratio and water permeability, and thus decrease the compressive strength of the pervious concrete. Engineered Biomass Aggregate (EBA), as a partial replacement of conventional biomass aggregates, tends to increase the water permeability, and this increase of water permeability results in decreasing the compressive strength, except for the M3 that contains 10% of EBA which results in increasing the compressive strength. Void ratio was affected by the replacement of EBA. Void ratio tended to increase by the addition of EBA, except for the M2 mix that contains 5% of EBA. In terms of compressive strength, the optimum percentage was 10% EBA, which recorded the highest value of compressive strength. The mix that included 20% EBA recorded the highest value of void ratio, while the optimum mix that recorded the highest value of water permeability was the mix included 25% EBA.

## 5. References

[1] Montes F and Haselbach L 2006 Measuring hydraulic conductivity in pervious concrete, *Environ. Eng. Sci.* **23** 960-969

[2] Shahidan S, Isham I and Jamaluddin N 2016 A review on waste minimization by adopting in self-compacting concrete *MATEC Web of Conf.* vol 47 (Paris: EDP Sciences) 01003

[3] Lee Y L and Krishna R N 2014 Biomass aggregate geopolymer concrete–Malaysia experience, *Int. J. Civ. Eng. Technol. (IJCIET)* **5** 340-356

[4] Teraiya D, Doshi U, Viradiya P, Yagnik A and Joshi T 2015 To develop method to find out permeability and void ratio for pervious concrete, *Int. J. Res. Eng. Technol.(IJRET)* **4** 177–182

[5] Ali N, Zainal N A, Burhanudin M K, Abdul Samad A A, Mohamad N, Shahidan S and Abdullah S R 2016 Physical and mechanical properties of compressed earth brick ( CEB ) containing sugarcane bagasse ash *MATEC Web of Conf.* vol 47 (Paris: EDP Sciences) 01018

[6] Shahidan S, Koh H B, Sharif Alansi A M and Loon L Y 2016 Strength development and water permeability of engineered biomass aggregate pervious concrete *MATEC Web of Conf.* vol 47 (Paris: EDP Sciences) 01007

[7] Unnikrishnan K P and Thachil E T 2006 *Studies on the Toughening of Epoxy Resins* PhD Thesis (India: Cochin University of Science and Technology) pp 1 - 126

[8] Shahidan S, Mohd Zuki S S, Rahim M A, Ali N and Mohamad N 2016 Amplitude distribution of emission wave for cracking process *MATEC Web of Conf.* vol 47 (Paris: EDP Sciences) 02013

[9] Lee Y L, Goh K S, Koh H B and Bakar I 2009 Foamed aggregate pervious concrete – an option for road on peat *Proc. of MUCET 2009, Malaysian Technical Universities Conf. on Engineering and Technology (Pahang, Malaysia)* (Malaysia: Malaysian Technical Universities Network (MTUN)) pp 1 - 4

[10] Shahidan S, Leman A S, Senin M S and Hannan N I R R 2017 Suitability of coconut shell concrete for precast cool wall panel-A review *MATEC Web of Conf.* vol 87 (Paris: EDP Sciences) 1005

[11] Khalid F S, Herman H S and Azmi N B 2017 Properties of sugarcane fiber on the strength of the normal and lightweight concrete *MATEC Web of Conf.* vol 103 (Paris: EDP Sciences) 01021
[12] Luckett K 2009 *Green Roof Construction and Maintenance* (USA: McGraw-Hill Companies, Inc.) pp 1-175
[13] Zaetang Y, Sata V, Wongsa A and Chindaprasirt P 2016 Properties of pervious concrete containing recycled concrete block aggregate and recycled concrete aggregate, *Constr. Build. Mater.* **111** 15-21
[14] Ramzi N I R, Shahidan S, Maarof M Z and Ali N 2016 Physical and Chemical Properties of Coal Bottom Ash (CBA) from Tanjung Bin Power Plant *IOP Conference Series: Materials Science and Engineering* vol 160 (United Kingdom: IOP Publishing) 12056
[15] ASTM-C128-97 2007 *Standard Test Method for Relative Density (Specific Gravity) and Absorption of Fine Aggregate* (USA: ASTM International) vol. 04 pp. 1–5
[16] ASTM: C33-02 2010 *Standard Specification for Concrete Aggregates* (USA: ASTM International) pp 1–11
[17] ASTM: C29/C29M-09 2009 *Standard Test Method for Bulk Density ('Unit Weight') and Voids in Aggregate* (USA: ASTM International) pp 1–5
[18] BS EN 12390-3 2009 *Testing Hardened Concrete: Compressive Strength of Test Specimens* (London: British Standard Institution) pp 1-22
[19] Senin M S, Shahidan S, Leman A S and Hannan N I R R 2016 Analysis of physical properties and mineralogical of pyrolysis tires rubber ash compared natural sand in concrete material *IOP Conference Series: Materials Science and Engineering* vol 160 (United Kingdom: IOP Publishing) 12053

**Acknowledgement**
The authors acknowledge the technical and financial support provided by Universiti Tun Hussein Onn Malaysia and Grant Vot (U523).

IOP Conf. Series: Materials Science and Engineering **271** (2017) 012003 doi:10.1088/1757-899X/271/1/012003

# Utilization of fly ash as partial sand replacement in oil palm shell lightweight aggregate concrete

**A Z Muhammad Nazrin Akmal[1], K Muthusamy[1], F Mat Yahaya[1], H Mohd Hanafi[1] and Z Nur Azzimah[1]**

[1] Faculty of Civil Engineering and Earth Resources Universiti Malaysia Pahang, Lebuhraya Tun Razak, 26300, Gambang, Pahang
Corresponding author: nazrinakmal89@gmail.com

**Abstract.** Realization on the increasing demand for river sand supply in construction sector has inspired the current research to find alternative material to reduce the use of natural sand in oil palm shell lightweight aggregate concrete (OPS LWAC) production. The existence of fly ash, a by-product generated from coal power plant, which pose negative impact to the environment when it is disposed as waste, were used in this research. The effect of fly ash content as partial sand replacement towards workability and compressive strength of OPS lightweight aggregate concrete were investigated. Four concrete mixes containing various percentage of fly ash that are 0%, 10%, 20% and 30% by weight of sand were used in the experimental work. All mixes were cast in form of cubes before subjected to water curing until the testing age. Compressive strength test were conducted at 1, 3, 7 and 28 days. The finding shows that the workability of the OPS LWAC decreases when more fly ash are used as sand replacement. It was found that adding of 10% fly ash as sand replacement content resulted in better compressive strength of OPS LWAC, which is higher than the control mix.

## 1. Introduction

Fly ash is a type of by-product resulting from the burning of pulverized coal in power plant. Approximately 8.5 million tonnes of coal ash consist of fly ash and bottom ash are produced in Malaysia annually [1]. Due to large production of fly ash, a large landfill area is required for disposal of this solid waste. In addition, the coal industry needs to spend in managing the waste disposal. Dumping of waste in abundance without any treatment would cause negative effect to environment such as land pollution, water pollution, air pollution as well as destruction to flora and fauna. However, these problems are possible to be avoided by utilizing the fly ash in production of construction material that would be benefit to the industry as well as community.

At the same time, the demand for sand supply, which used in construction activity and concrete production is increasing along with the growing construction sector. Natural sand usually obtained from river mining activity. Uncontrolled and excessive sand mining to meet the increasing demand of the construction industry would cause negative impact to the river environment. Excessive river sand mining would cause destruction of flora and fauna as well as bad water quality [2]. Due to the rising demand of sand in construction, government has increased the price of sand in order to restrict the extraction of sand thus increase the overall cost of construction project [3- 4]. It is seen that one of the alternative way to solve this problem is to use another materials to replace or reduce the normal sand use in concrete. Looking at the research in this area, there are researchers [5-7] that looked on the possible use of waste materials to replace fine aggregate in concrete. However, the properties of OPS lightweight aggregate concrete containing fly ash as partial sand replacement has not received much attention. Therefore, this paper investigates the effect of various percentage of fly ash as partial sand replacement towards workability and compressive strength of OPS LWAC.

## 2. Experimental programme

### *2.1 Materials*

The oil palm shell lightweight aggregate concrete (OPS LWAC) prepared for this experimental work were produced using cement, water, river sand, superplasticizer, oil palm shell, and fly ash. Type 1 ordinary Portland cement and Type A water-reducing superplasticizer as stated in ASTM C494-05 [8]. was used in all the concrete mixes. All the cement were stacked well away from damp environment to prevent it from hardened. Tap water were used for concrete mixing and curing purpose. A fixed water to cement ratio of 0.45 was introduced in all the OPS LWAC mixes. A superplasticiser of 1.0% by cement weight was added into the fresh concrete to improve the workability.

The fly ash used in this research was obtained from a coal fire plant located at West Malaysia. Fly ash of 0%, 10%, 20% and 30% by sand weight was added as fine aggregate replacement. Oil palm shell (OPS) was use to fully replace conventional coarse aggregate to produce oil palm shell lightweight aggregate concrete (OPS LWAC). The OPS was used in this research was procured from a local palm oil extraction factory as illustrated in figure 1. At the laboratory, the shells shown in Figure 2 were cleaned from impurities before using. OPS having a moisture content, water absorption, fineness modulus and specific gravity value of 12.45%, 12.47%, 6.53 and 1.37 kg/m$^3$ were used respectively. The bulk density of OPS used in this research is 568 kg/m$^3$ which is in the range of 500-600 kg/$^3$. This value is in line with the OPS bulk density done by previous researchers [9,10]. The physical properties of the OPS used in this study are tabulated in Table 1.

**Figure 1.** OPS at the palm oil mill.

**Figure 2.** Oil palm shell (after cleaning).

**Table 1**. Physical properties of oil palm shell.

| Physical properties | **OPS** |
|---|---|
| Moisture content (%) | 12.45 |
| Water absorption (%) | 12.47 |
| Fineness modulus | 6.53 |
| Specific gravity (kg/m$^3$) | 1.37 |
| Bulk density (compacted) | 568 |
| Los Angeles abrasion (%) | 7.6 |
| Aggregate impact value (%) | 18.18 |
| Aggregate crushing value (%) | 14.85 |

### *2.2 Mix proportions*
The amount of constituent materials such as ordinary Portland cement (OPC), OPS, water and superplasticizer (SP) is kept constant except for the sand and fly ash. The effect of sand replacement with 0%, 10%, 20% and 30% of fly ash were studied in this research. Four concrete mixes proportions of OPS LWAC containing fly ash as partial sand replacement are shown in Table 2.

**Table 2.** Mix proportion of OPS LWAC mixes.

| Mix design | Cement ($kg/m^3$) | Sand ($kg/m^3$) | Fly Ash ($kg/m^3$) | OPS ($kg/m^3$) | Water ($kg/m^3$) | Superplasticizer ($kg/m^3$) |
|---|---|---|---|---|---|---|
| FA0 | 500 | 800 | 0 | 360 | 225 | 5 |
| FA10 | 500 | 720 | 80 | 360 | 225 | 5 |
| FA20 | 500 | 640 | 160 | 360 | 225 | 5 |
| FA30 | 500 | 560 | 240 | 360 | 225 | 5 |

### *2.3 Specimen preparation and testing method*
For mixing process, firstly, the OPC, OPS, sand and fly ash were blended in mixer for a while to make sure that the materials is well distributed. Then, half of water and superplasticizer were added to the mixture. After few minutes, remaining water and superplasticizer were added and mixed until the fresh concrete become a uniform mixture. Fresh concrete workability was investigated by conducting slump test in accordance to BS EN 12350:2 [11]. For the determination of compressive strength, 100 mm cubes were prepared as illustrated in figure 3. Then, the specimens were demoulded after 24 hours. Then, all the specimens were immersed in water for curing purpose until the testing date. The compressive strength test as shown in figure 4 was conducted at 1, 3, 7 and 28 days following the procedure in BS EN 12390:3 [12].

**Figure 3.** Concrete mixes was filled in cube mould.

**Figure 4.** Compressive strength test in progress.

## 3. Result and discussion

### *3.1 Workability*
Figure 5 shows that the workability of concrete mix decreases as the amount of fly ash used becomes higher. The reduction of concrete mix workability when fly ash is added is attributed to the physical properties of fly ash which is porous and tends to absorb water. The characteristic of fly ash being porous and causes difficulties during mixing process has been highlighted by researcher elsewhere [13]. In this experimental work, the slump is still in good working range for lightweight concrete when

10% fly ash was added. However, when the percentage of replacement is increased up to 20 and 30%, the slump value dropped significantly. Inclusion of 30% produces a very dry concrete mixture making it difficult to be mix and compacted.

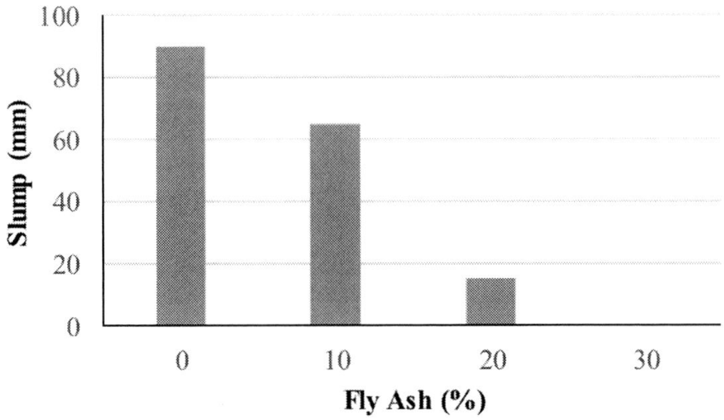


**Figure 5.** Slump value vs fly ash percentage.

*3.2 Compressive strength*

The compressive strength of OPS LWAC concrete containing fly ash as partial sand replacement is shown in Figure 6. Generally, all concrete mixes exhibit strength increment as the curing time become longer. Application of continuous water curing has allowed undisturbed hydration process which is important for formation of calcium silicate hydrate gel that contribute to densification of concrete internal structure resulting in strength increment [6]. As for the effect of using fly ash as partial sand replacement on the compressive strength of concrete, it is observed that replacing river sand with fly ash with adequate amount has resulted in higher compressive strength. Excessive use of fly ash although appears more environmental friendly but it causes significant drop in strength of concrete.

It is interesting to note that, fly ash replacement of 10% by sand produces strength higher than control mixes of about 7% at 28 days. This increase probably attributed to filling effect of fly ash making the concrete denser and capable of resisting higher load as compared to control specimen. Past researcher [14] noted that use of fine fly ash particles contribute to better packing of concrete ingredients when the particles fills the voids between the cement particles. However, it is noticed that the presence of fly ash beyond 20% replacement of sand, evidently reduces the concrete strength till 7% at 28 days. This shows that replacement of more than 10% did not significantly improve the mechanical properties. This probably due to the addition of more fly ash makes the mix become drier causing difficulties during compaction. As a result, a non-homogenous concrete possessing lower strength were produced.

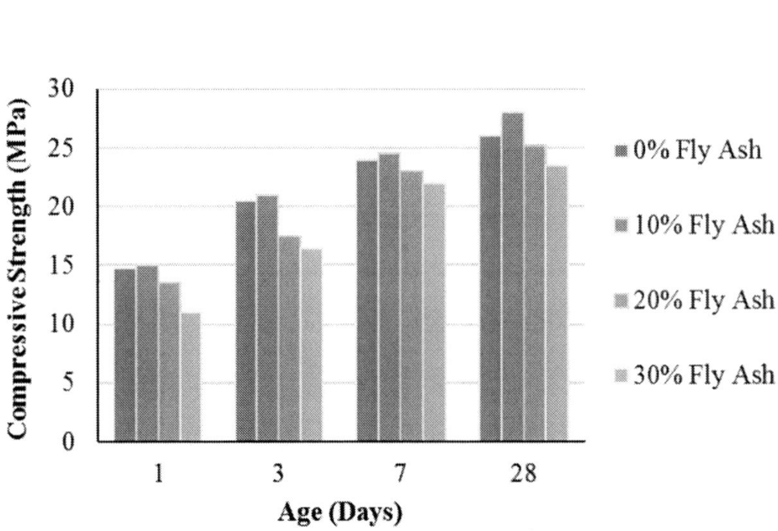


**Figure 6.** Compressive strength vs fly ash percentage.

## 4. Conclusion

The effect of fly ash as partial sand replacement in OPS LWAC at various percentages was investigated. Based on the investigation, the used of fly ash decreases the workability of OPS LWAC containing fly ash due to porous fly ash properties that tends to absorb water during mixing process. However, the slump workability is still in the range of good workability for lightweight concrete when 10% fly ash is used as partial sand replacement. Interestingly, it was found that adding of 10% fly as partial sand replacement increases the compressive strength of oil palm shell lightweight aggregate concrete (OPS LWAC) as compared to control specimen. Fly ash that act as filler enhanced the compressive strength with more than 7% higher than the control mix. The increment of fly ash of more than 10% in the mixes however lowered the strength by 3% and 9% for fly ash replacement of 20% and 30%.

## 5. References

[1] Rafieizonooz M, Mirza J, Salim M R, Hussin M W, Khankhaje E 2016 Investigation of coal bottom ash and fly ash in concrete as replacement for sand and cement *Construction and Building Materials* **116** 15-24

[2] Wang Z F, Ding J Y and Yang G S 2012 Risk analysis of slope instability of levees under river sand mining conditions *Water Science and Engineering* **5** (3) 340-349

[3] Ganesh P G, Hyun J H and Kim Y Y 2014 Effect of foundry sand as a fine aggregate in concrete production *Construction and Building Materials* **70** 514-521

[4] Ahmad Zawawi M N A, Muthusamy K and Mohd Ali N 2016 Compressive strength of oil palm shell lightweight aggregate concrete containing clinker as sand replacement *Proc. of The National Conference for Postgraduate Research 2016* (Pahang: Universiti Malaysia Pahang) pp 165-168

[5] Singh M and Siddique R 2013 Effect of coal bottom ash as partial replacement of sand on properties of concrete Resources *Conservation and Recycling* **72** 20-32

[6] Mat Yahaya F, Muthusamy K and Hussin M W 2015 Long term investigation of sulphate resistance of aerated concrete containing palm oil fuel ash *ARPN Journal of Engineering and Applied Sciences* **11**(4) 2406-2411

[7] Muthusamy K, Tukimat N N A, Sarbini N N and Zamri N A 2016 Exploratory study on the use of crushed cockle shell as partial sand replacement in concrete *International Journal of Research in Engineering and Science* **4** 67-71

[8] ASTM C618 2005 Standard Specification for Fly Ash and Raw Material or Calcined Natural Pozzolan for Use as a Mineral Admixture in Portland Cement Concrete (Pennsylvania: American Society of Testing and Materials)
[9] Shafigh P, Jumaat M Z and Mahmud H 2010 Mix design and mechanical properties of oil palm shell lightweight aggregate concrete: a review *International Journal of Physical Sciences* **5** (14) 2127-2134
[10] Abdullahi M, Al-Mattarneh H M A, Hassan A H A, Hassan M H and Mohammed B S 2008 Trial mix design methodology for Palm Oil Clinker (POC) concrete *The International Conference on Construction and Building Technology* (Kuala Lumpur, Malaysia)
[11] BS EN 12350-2 2000 Testing Fresh Concrete: Slump test (London: British Standard)
[12] BS EN 12390-3 2002 Testing Hardened Concrete: Compressive strength of test specimens (London British Standard)
[13] Wang and Wu 2006 Environmental-benign utilisation of fly ash as low-cost adsorbents *Journal of Hazardous Materials* **136** (3) 482-501
[14] Shirish and Arun 2015 Parametric study for replacement of sand by fly ash for better packing and internal curing *Open Journal of Civil Engineering* **5** pp 118-130

**Acknowledgement**

The authors would like to acknowledge the help and co-operation received from the technical staff of the Structural and Materials Laboratory of Universiti Malaysia Pahang (UMP) in conducting the experimental work. The financial support received from Universiti Malaysia Pahang through Graduate Research Grant was also gratefully acknowledge.

IOP Conf. Series: Materials Science and Engineering **271** (2017) 012004 doi:10.1088/1757-899X/271/1/012004

# Physical and mechanical properties of self-compacting concrete containing superplasticizer and metakaolin

**Shahiron Shahidan[1], Bassam A Tayeh[2], A A Jamaludin[3], N A A S Bahari[1], S S Mohd Zuki[1], N Ali[1] and F S Khalid[1]**

[1] Jamilus Research Center, Faculty of Civil and Environmental Engineering, Universiti Tun Hussein Onn Malaysia, 86400 Batu Pahat, Johor.
[2]Civil Engineering Department, Islamic University of GAZA, P.O Box 157, Gaza Palestine.
[3]Choice Approach Sdn. Bhd., No3-10-B Jln Desa 2/2 Desa Aman Puri 5210 Kepong Kuala Lumpur.
Corresponding author:shahironshahidan@gmail.com

**Abstract.** The development of concrete technology shows a variety of admixtures in concrete to produce special concrete. This includes the production of self-compacting concrete which is able to fill up all spaces, take formwork shapes and pass through congested reinforcement bars without vibrating or needing any external energy. In this study, the main objective is to compare the physical and mechanical properties of self-compacting concrete containing metakaolin with normal concrete. Four types of samples were produced to study the effect of metakaolin towards the physical and mechanical properties of self-compacting concrete where 0%, 5%, 10% and 15% of metakaolin were used as cement replacement. The physical properties were investigated using slump test for normal concrete and slump flow test for self-compacting concrete. The mechanical properties were tested for compressive strength and tensile strength. The findings of this study show that the inclusion of metakaolin as cement replacement can increase both compressive and tensile strength compared to normal concrete. The highest compressive strength was found in self-compacting concrete with 15% metakaolin replacement at 53.3MPa while self-compacting concrete with 10% metakaolin replacement showed the highest tensile strength at 3.6MPa. On top of that, the finishing or concrete surface of both cube and cylinder samples made of self-compacting concrete produced a smooth surface with the appearance of less honeycombs compared to normal concrete.

## 1. Introduction

Due to the high demand and increasing applications of concrete, concrete technology is developing rapidly and many types of concrete have been studied to increase the quality and its properties. One of the achievements of concrete technology is the production of self-compacting concrete. This type of concrete is considered to have a very high workability due to its flow able properties. As a result, the concrete can be placed and compacted without needing any vibration medium as well as external energy. This vibration or compaction primarily aims to minimize the entrapped air in fresh concrete in order to obtain a homogenous mix with no cavities or honeycombs [9]. Due to its flowable properties, this type of concrete can also slip through congested reinforcement bars and take any shape of the formwork [13].

In terms of strength, the inclusion of metakaolin was introduced as a cement replacement and it is believed to increase the strength of concrete in both compress and tensile state [1]. Metakaolin is a type of mineral admixture and is also known as calcined clay. It is a very fine material and is produced by heating up kaolin until 800°C [12]. It has been reported that the use of metakaolin in concrete can increase the strength of mixtures especially during the early ages [1].

## 2. Self-Compacting concrete

Self-compacting concrete (SCC) can be defined as concrete which can flow and be compacted with its own weight, able to pass the space in between the reinforcement bars to fill up the formwork and stabilize its material composition at the same time [12,18-22]. It was first developed in Japan during the late 1980's [2]. In Japan, SCC has become the standard concrete mixture [16].

Various tests associated with self-compacting concrete include slump flow test, J-ring test, L-box test etc. The slump flow test is used to assess the flow ability and the flow rate of self-compacting concrete in the absence of obstructions [5,19,23-24]. The J-ring test is used to assess the passing ability of self-compacting concrete to flow through tight openings including spaces between reinforcing bars and other obstructions without segregation or blocking [7,25]. The L box test is used to assess the passing ability of self-compacting concrete to flow through tight openings including spaces between reinforcing bars and other obstructions without segregation or blocking. There are two variations; the two-bar test and the three-bar test. The three-bar test simulates more congested reinforcement [6].

There are 3 properties that differentiate self-compacting concrete from normal concrete which are flowing ability, passing ability and resistance to segregation [2]. Three properties which make self-compacting concrete different from normal concrete are as follow:

### *2.1. Flowing ability*

A higher flowing ability of concrete shows its ability to fill up every space in the formwork.

**Table 1.** Classes of slump flow.

| Class | Slump flow (mm) |
|---|---|
| SLF 1 | 550 – 650 |
| SLF 2 | 660 – 750 |
| SLF 3 | 760 – 850 |

### *2.2. Passing ability*

The ability of concrete to fill up the formwork and pass through the congested reinforcing bars.

**Table 2.** Classes of passing ability.

| Class | Step height (mm) |
|---|---|
| PJ 1 | ≤ 10 to 12 bars |
| PJ 2 | ≤ 10 to 16 bars |

### *2.3. Resistance to segregation*

The ability of material inside the concrete to remain mixed and not separated between each material.

**Table 3.** Classes of resistance to segregation.

| Class | Sieve segregation (%) |
|---|---|
| SR 1 | ≤ 20 |
| SR 2 | ≤ 15 |

## 3. Experimental program

The experimental program in this study consist of metakoalin, materials preparation and testing method.

### *3.1 Metakoalin*

Many efforts have been made to minimize the use of cement as a binder in concrete production. Concrete containing metakaolin has been claimed to possess enhanced engineering properties that are comparable to silica fume concrete [7].

Metakaolin is a type of mineral admixture and is generally known as calcined clay. It is an ultrafine material produced by the dehydroxylation of a kaolin precursor upon heating in the temperature range of 700° - 800°C [15]. Metakaolin enhances the strength and durability of concrete through three primary actions which are the filler effect, the acceleration of ordinary Portland cement hydration and the pozzolanic reaction with calcium hydroxide [11]. In addition, the inclusion of metakaolin can increase the resistance to acids and sulphates, reduce porosity, reduce oxygen permeability, reduce chloride ion diffusivity, prevent or minimize the risk of alkali-silica reaction and reduce the unsightly effect of efflorescence [4].

The main characteristics of metakaolin are its high reactivity with calcium hydroxide, Ca(OH)2, and its ability to accelerate cement hydration [3]. Due to these characteristics, the addition of metakaolin in concrete can increase its strength especially during the early ages [14].

**Table 4.** Chemical composition comparison between metakaolin and OPC.

| Chemical | Metakaolin | OPC |
|---|---|---|
| $SiO_2$ | 20.69 | 51.6 |
| $Al_2O_3$ | 4.72 | 41.3 |
| $Fe_2O_3$ | 3.06 | 4.64 |
| CaO | 63.76 | 0.09 |
| MgO | 2.08 | 0.16 |
| $TiO_2$ | - | 0.083 |
| $SO_3$ | 2.92 | - |
| $K_2O$ | 0.61 | 0.62 |
| $Na_2O$ | 0.26 | 0.01 |

### 3.2 *Material preparation*

The cement used in producing the concrete was Ordinary Portland Cement (OPC) from Tasek Cement brand. The aggregate used was uncrushed aggregate type with nominal size of 20mm and should be oven dried first to avoid any changes in water cement ratio. The sand used should also be dried first. The water used is normal pipe water.

In terms of admixture and material to be studied, superplasticizer was used to increase the workability of concrete. The superplasticizer used was Darex Super 20 as shown in Figure 1(b). Metakaolin shown in Figure 1(a) was used marked by the company as KM5CL which is the finest particle size available.

Four batches of self-compacting concretes were produced and another sample of normal concrete should be used as a control sample. For each batch of self-compacting concrete, different percentages of metakaolin were used as cement replacement (0%, 5%, 10% and 15%). Cube mould size of 150 x 150 x 150mm was used for the compressive strength test [7-10] whereas cylinder mould size of 150mm x 300mm was used for tensile strength test [11] in accordance to British Standards.

(a) (b)

**Figure 1.** (a) Metakoalin (b) Superplasticizer (Darex Super 20).

*3.3 Testing method*

In the testing method, all tests were conducted according to British Standards. The slump flow test was used to assess the flowability and the flow rate of self-compacting concrete in the absence of obstructions shown in figure 2(c). It is based on the slump test described in BS EN 12350-2. The result is an indication of the filling ability of self-compacting concrete. T500 time is a measure of the speed of flow and an indication of the relative viscosity of the self-compacting concrete [4].When the cone is withdrawn upwards, the time from commencing upward movement of the cone to when the concrete has flowed to a diameter of 500 mm is measured; this is known as t500 time. The largest diameter of the flow spread of the concrete and the diameter of the spread at right angles to it are then measured and the mean is the slump-flow [4]. In addition, the compressive strength and tensile strength tests should follow as described in BS EN 12390-3 and 6.

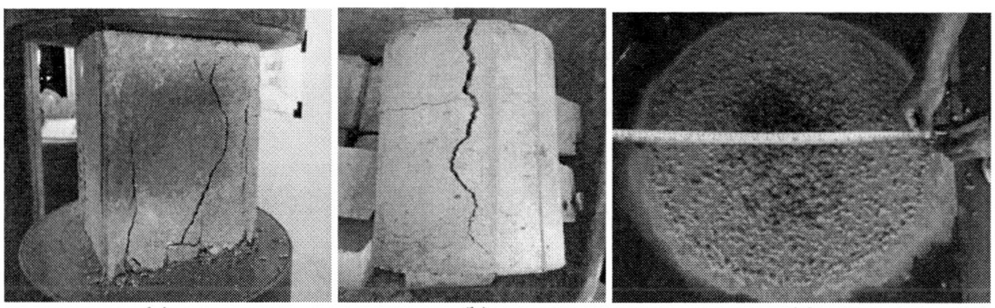

(a) (b) (c)

**Figure 2.** Testing method of compressive test (a), tensile test (b), slump flow test (c).

## 4. Results and discussion

*4.1. Slump flow test*

Four type of samples of self-compacting concrete with differ percentages of metakoalin have been tested using this method.

**Table 5.** Diameter and SF value for each sample.

| Type of sample | D1 (mm) | D2 (mm) | SF (mm) |
|---|---|---|---|
| SCC + 0% Metakaolin | 600 | 596 | 598 |
| SCC + 5% Metakaolin | 589 | 586 | 588 |
| SCC + 10%Metakaolin | 571 | 575 | 573 |
| SCC + 15%Metakaolin | 560 | 567 | 564 |

*4.2. Compressive test*

From the figure 3, shown the results of slump flow with differ percentages of metakoalin for self-compacting concrete produced, the slump flow recorded are in class SLF 1 where the flow is between 550-650mm.

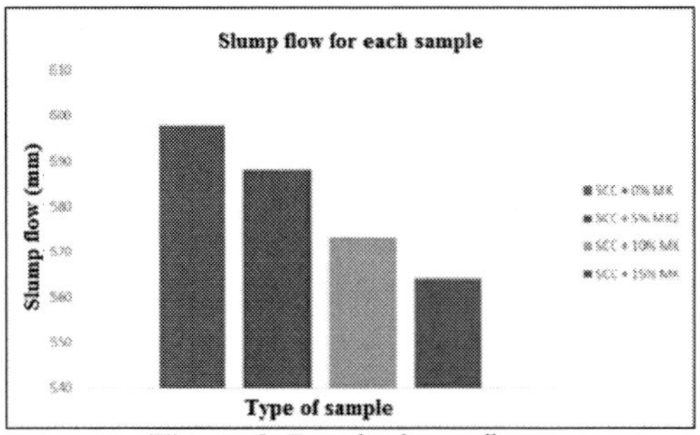


**Figure 3.** Result slump flow.

From figure 4, there is no significant difference between 2 types of normal concrete and self-compacting concrete in terms of compressive strength at 28 days. The difference recorded is only 3.1% where the recorded strength of normal concrete was 44.8MPa whereas the recorded strength of SCC was 43.8MPa.

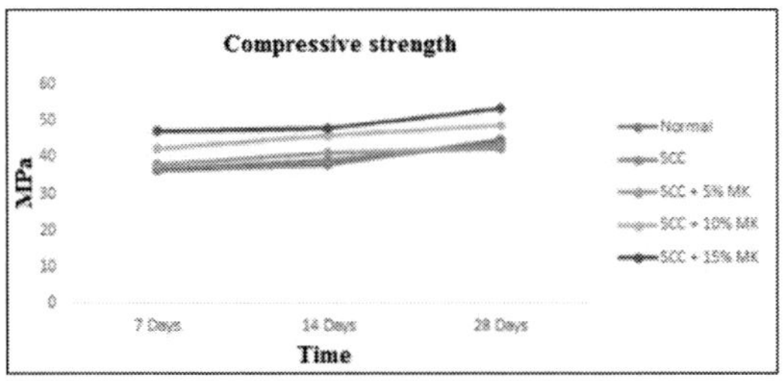


**Figure 4.** Development of compressive strength for each sample until 28 days.

When the effect of metakaolin in SCC was compared, the highest compressive strength recorded was SCC + 15% metakaolin sample where its strength at 28 days was 53.3MPa while the lowest strength recorded was SCC + 5% metakaolin. However, the difference between normal concrete and SCC + 5% was low, that is a difference of only 5% (2.3MPa). The 28 days strength recorded for SCC + 10% metakaolin was 48.6MPa.

In terms of early strength development, it can be said that the inclusion of metakaolin in concrete increases the early strength of the concrete as all 3 samples of concrete with metakaolin showed high early strength until day 14 as compared to normal concrete and SCC without metakaolin.

*4. 3 Tensile strength*

From figure 5, shown the significant difference between normal concrete and self-compacting concrete. Normal concrete recorded the lowest tensile strength and its value was nowhere near any of the samples.

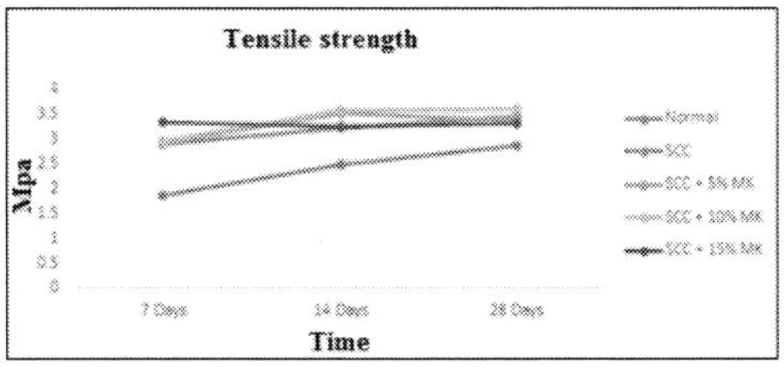


**Figure 5.** Development of tensile strength for each sample until 28 days.

The biggest difference between normal concrete and self-compacting concrete was recorded when the age of concrete was 7 days with a difference of 35.2% (1.02MPa) and the difference decreasesover time. In terms of the usage of metakaolin, self-compacting concrete + 10% metakaolin produced the highest tensile strength of 3.6MPa at 28 days and the data plotted for this sample is consistent with increasing tensile strength over time.

Overall, the inclusion of metakaolin increases the tensile strength of concrete with 10% of metakaolin showing the optimum result. However, the addition of superplasticizer also increases the tensile strength compared to normal concrete.

## 5. Conclusion

In conclusion, the addition of metakaolin in concrete helps increase its strength. In terms of compressive strength, the higher inclusion of metakaolin increases the strength linearly. However, the inclusion of metakaolin in concrete reduces the workability of concrete. In order to counter the problem, superplasticizer is used in concrete to produce self-compacting concrete. The usage of superplasticizer in concrete doesn't have any effect on compressive strength. It does, however, help to increase the tensile strength of concrete.

## 6. References

[1] Ahmad S, Adekunle S. K, Maslehuddin M, Azad A. K 2014 Properties of self-consolidating concrete made utilizing alternative mineral fillers, *Constr. Build. Mater.* **68** 268-276

[2] Alawneh M R 2001 Overview on self-compacted concrete: Innovations and emerging technology in concrete industry

[3] Leman A S, Shahidan S, Senin M S, Ramzi Hannan N I R 2016 A preliminary study on chemical and physical properties of coconut shell powder as a filler in concrete *IOP Conf. Ser. Mater. Sci. Eng.* vol 160 (London: IOP Publishing) p 12059

[4] Ali N, Zainal N A, Burhanudin M K, Abdul Samad A A, Shahidan S 2016 Physical and mechanical properties of compressed earth brick (CEB) containing sugarcane bagasse ash vol 47 (Paris: MATEC Web Conf.) pp 1–7

[5] Badogiannis E G, Sfikas I P, Voukia D V, Trezos K G, Tsivillis S G 2015 Durability of metakaolin self-compacting concrete, *Constr. Build. Mater.* **82** 133-141

[6] Brooks J J, Johari M A 2001 Effect of metakaolin on creep and shrinkage of concrete, *Cement & Concrete Composites* **23** 495-502

[7] BS EN 12350-8 2010 *Testing fresh concrete Part 8:Self-compacting concrete – slump flow test* (London: British Standard)

[8] BS EN 12350-10 2010 *Testing fresh concrete Part 10:Self-compacting concrete – L-box test* (London: British Standard)

[9] BS EN 12350-12 2010 BSI *Testing fresh concrete Part 12:Self-compacting concrete – J-ring test* (London: British Standard)
[10] BS EN 12390-3 2009 *Testing hardened concrete. Compressive strength of test specimens* (London: British Standard)
[11] BS EN 12390-6 2009 *Testing hardened concrete. Tensile splitting strength of test specimens* (London: British Standard)
[12] Demie S, Nuruddin M F, Ahmeed F M, Shafiq N 2011 Effects of curing temperature and superplasticizer on workability and compressive strength of self-compacting geopolymer concrete vol 11 (USA:IEEE)
[13] Ismail I, Shahidan S 2017 A preliminary modified volumetric mix design for self-compacting concrete (SCC) by utilizing asphalt dust waste (ADW) vol 103 (Paris: MATEC Web Conference) p 1008
[14] Ismail I, Jamaluddin N, Shahidan S 2016 A review on performance of waste materials in self compacting concrete (SCC), *J. Teknol.* **5(78)** 29–35
[15] Karahan O, Hossain MA, Ozbay E, Lachemi M, Emre S 2012 Effect of metakaolin content on the properties of self-consolidating lightweight concrete, *Constr. Build. Mater.* **31** 320-325
[16] Kovarik T, Belsky P, Novotny P, Riha J, Savkova, J, Medlin R, Rieger D, Holba P 2015 Structural and physical changes of re-calcined metakaolin regarding its reactivity, *Constr. Build. Mater.* **80** 98-104
[17] Le H T, Muller M, Siewart K, Ludwig H M 2015 The mix design for self-compacting high performance concrete containing various mineral admixtures, *Materials and Design* **72** 51-62
[18] Madandoust R, Mousavi S Y 2012 Fresh and hardened properties of self-compacting concrete containing metakaolin, *Constr. Build. Mater.* **35** 752-760
[19] Senin M S, Shahidan S, Leman A S, Hannan N I I R 2016 Properties of cement mortar containing rubber ash as sand replacement vol 160 (London: IOP Conf. Ser. Mater. Sci. Eng.) pp 12055
[20] Qian X, Li Z 2001 The relationship between stress and strain for high performance concrete with metakaolin, *Cement and Concrete Research* **31** 1607-1611
[21] Seng V, Shima H and Ouchi M 2004 Possibility of Self-Compacting Concrete in Cambodia *Proceedings of the Seventh International Conference on Concrete Technology in Developing Countries* (Kuala Lumpur)
[22] The Self-Compacting Concrete European Project Group 2005 *The European Guidelines for Self-compacting Concrete* (United Kingdom: EFNARC)
[23] Shahidan S, Mohd Zuki S S, Rahim M A, Ali N 2016 Amplitude distribution of emission wave for cracking process vol 47 (Paris: MATEC Web Conference) pp 2–8
[24] Shahidan S, Isham I, Jamaluddin N 2016 A review on waste minimization by adopting in self compacting concrete vol 47 (Paris: MATEC Web Conference) pp 1–7
[25] Shahidan S, Ismail I, Zulkarnaian M S S, Abd Rahman N 2017 Influence of asphalt dust waste material in mix design for self-compacting concrete *Key Engineering Materials* **730** 473-478

**Acknowledgments**
The author would like to thank to Jamilus Research Center, Universiti Tun Hussien Onn Malaysia and Grant VOT U523, Ministry of Higher Education, Malaysia

# Stress distribution study on multi-holes configurations in woven fabric kenaf composite plates

**K Supar[1] and H Ahmad[1]**

[1] Faculty of Civil and Environmental Engineering, Universiti Tun Hussein Onn Malaysia, 86400 Parit Raja, Batu Pahat, Johor Darul Ta'zim.
Corresponding author:khairisupar@gmail.com

**Abstract.** In structures applications, composite components are joined by bolted connection is commonplace requires introduction of drilling multi-holes. In order to study the structures response of these material properly, a parametric study on stress analysis is required to take into account arrays of variation including lay-up types, multi-hole configurations, intensity of applied stress etc. The failure and fracture of composite plates is somewhat complex, stress analysis study from numerical modelling able to predict the location of crack exhibited. Current study implemented 2-D finite element analysis (FEA) modelling framework on various multi-holes configurations woven fabric kenaf composite (WFKC) plates, the findings are then compared with conducted experiment. It was found that in non-staggered holes, maximum stress occurred at the outer holes suggesting net-sectional failure to initiate and self-similar crack behaviour through plate thickness. On the contrary, net-sectional failure in staggered hole configurations are dependent upon smallest net-sectional area in any line perpendicular to plane axis. The stress analysis conducted provides good agreement with conducted experiment.

## 1. Introduction

Reported works available in the literatures are limited to single hole plate, where the crack initiation and crack propagations occurred along the net-section plane perpendicularly to the applied loading. However, in most construction design, connection between structure parts involved more than single-bolts leading to more complicated stress field as by-pass stress able to be transferred to the neighbouring bolts. Net-tension mode is associated with stress raisers at the vicinity of the hole edge, susceptible in multi-hole configurations due to by-pass stress transfer [1]. Net-tension failure is dependant upon bearing stress applied, by-pass stress transfer from neighbouring bolts increases linearly and bearing stress decreased concurrently prone to net-tension failure to occur. Many researchers studying the optimization of composite joint design [2-4], a study on multi-holes problem need to be carried out prior to multi-bolted design of composite materials in order to optimize joint efficiency during composite joining design phase. Eugene *et al.* [5] found the notched strength of double-column plate is greater than single-column plate counterparts. This implies that the presence of another holes in adjacent column along the loading direction in a plate would cause to maximum stress reduction as a results of stress relief. He also concludes that lower notched strength were observed in staggered configurations than non-staggered configurations counterpart. This is because extensive fibre breakages due to high maximum stress value in staggered configurations as given in finite element analysis (FEA) contours. Jia *et al.* [6] explained that the stress concentrations becomes more concentrated and higher with hole size increment, where the larger hole size may reduce stiffness within hole locality in the form of separated length or the broken width [associated with damage zone length]. A classical approach to study stress distribution in quasi-isotropic composites of pin-loaded hole is carried out by deJong [2] using superposition method. Three stress distribution expression was developed, i.e., radial stress, $\sigma_r$, tangential stress, $\sigma_\theta$ and shear stress, $\tau_{xy}$ by assuming rigid pin and cosinusoidal radial stress distribution. The work is later extended to incorporate effect from clamp-up [3]. A parametric study on pin-loaded hole of six CFRP plate with different orthotropy and normalised plate width, *W/d* ranged from 2 to 20 were reported by Crews *et al.* [4]. They found that maximum tangential stress, $\sigma_\theta$ in low *W/d* and maximum radial stress in larger *W/d* suggesting prone of net-tension and bearing failure occur respectively. Owing to discontinuities in composite plate, stress concentration occurred at the notch tip, higher stress concentration factor, $K_T^\infty$ in highly orthotropic lay-ups. Recently, with the excellence computing technology leading to aggressive implementation of finite element method in stress analysis study. Ahmad *et. al.* [7, 8] compares stress distribution of single-bolted CFRP plates from analytical expression [2] and 3D FEA modelling in double-lap joints [7] and single-lap joints [8]. It was found that finite element modelling able to capture more physically realistic results than analytical expression due to simplistic assumption made in pin contact interactions. They also found that relatively higher tangential stress from FEA modelling due to analytical approach has not accounted for plate width effect. There is less study on stress distribution multi-hole problem available in the literature, however there are experimental observations [1] and stress concentration in multi-square holes using finite element method (FEM) [5] were reported.

Current work investigates stress distribution in multi-holes configurations of various staggered and non-staggered arrangement in different lay-ups, plate thickness and holes configurations. Finite elements analysis (FEA) were carried out by using ABAQUS CAE 6.13 and estimation of crack path in multi-holes configurations was investigated.

## 2. FEA modelling

2-D FEA modelling framework were carried out and description on idealizations and discretization, and testing series and boundary conditions are given in the following section.

### *2.1. Idealizations and discretization*

Finite element modelling is carried out by implementing two-dimensional modelling framework using ABAQUS CAE Version 6.13 software package. Stress distributions of multi-holes configurations was carried out to include staggered and non-staggered multi-holes configurations as given in testing series as given in table 1. Three lay-up types were investigated in current work, i.e., cross-ply lay-up (PX2, PX4) and quasi-isotropic lay-up (PQ4). Elastic properties includes in-plane tensile modulus, ($E_x/E_y$), shear modulus, $G_{xy}$ and Poisson's ratio, $v_{xy}$ of WFRP used in the current work and were considered as "smeared-out" properties. These elastic properties were carried out experimentally using UTM machine following ASTM standard D3039.

**Table 1.** Smeared-out elastic properties used in current work.

| Orientation | Lay-up | Thickness (mm) | In-plane elastic properties | | | |
|---|---|---|---|---|---|---|
| | | | $E_x$ (MPa) | $E_y$ (MPa) | $v_{xy}$ | $G_{xy}$ (GPa) |
| Cross-ply | PX2 $(0°/90°)_s$ | 1.8 | 2260.33 | 2260.33 | 0.07 | 198.23 |
| Cross-ply | PX4 $(0°/90°)_{2s}$ | 3.8 | 2291.04 | 2291.04 | 0.09 | 200.92 |
| Quasi-isotropic | PQ4 $(0°/90°/\pm45°)_s$ | 3.8 | 2008.93 | 2008.93 | 0.33 | 755.23 |

Figure 1 shows the meshing refinement at vicinity of hole edge and remaining plate regions were made coarser to reduce computational effort. It is important to refine the meshing at the vicinity of the hole is sufficiently small to make sure the post-processing output is free from mesh refinement. Second-order plane stress element with element designation code of CPS8 was used as element type due to the plate is infinitely thin (less than 4 mm). A simplified 2-D modelling able to achieve faster convergence without sacrificing physical representation as compared to 3-D modelling framework which is highly time- and cost-consuming that lead to convergence difficulties.

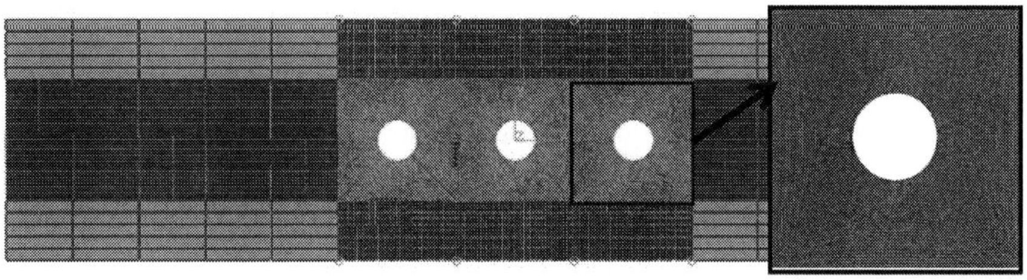

**Figure 1.** Meshing of multi-hole FEA model.

*2.2. Testing series and boundary conditions*

Woven fabric kenaf composite plates has a constant normalised pitch, $p/d = 3$ and constant plate width , $W = 30$ mm with drilling open holes size of 5mm diameter, $d$ according to specified geometric dimension as given in figure 2. Five testing series were carried out in current works with various lay-up types, plate thickness, and spacing between adjacent holes. There are three non-staggered and two staggered multi-holes configurations investigated in current study as shown in Figure 2. In current work, a constant value of three in normalised pitch ($p/d$), normalised gage length ($g/d$) and edge distance ($e_2/d$) were implemented to assure that stress distributions possible compare with all testing series geometry.

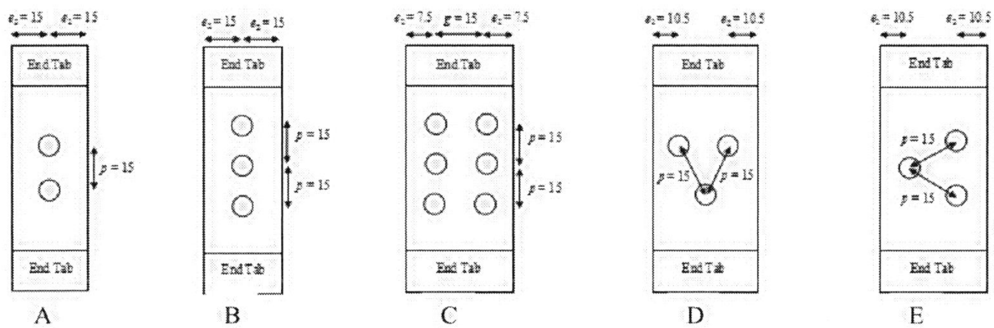


**Figure 2.** Testing plates geometry of all testing series tested.

The boundary condition and applied loading were properly assigned according to mechanical testing condition, where applied traction pressure was assigned on the right edge and left edge was assigned as fixed boundary as shown in figure 3. Multi-holes configurations are made full model as no existence of symmetrical axis due to some models (particularly in staggered-holes configurations) impossible to apply half-models to save time and computational effort.

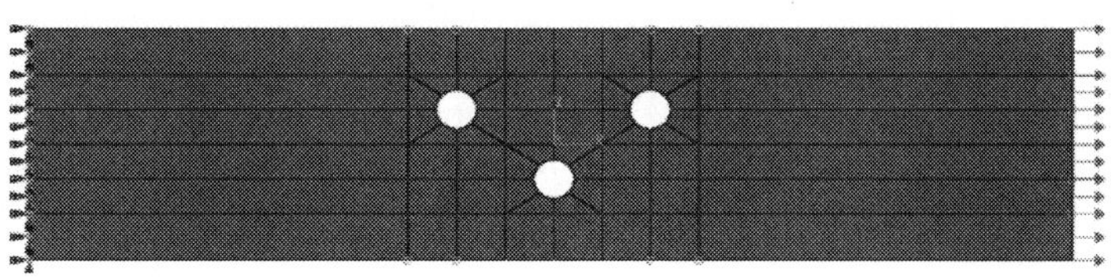

**Figure 3.** Boundary conditions of multi-hole FEA model.

IOP Conf. Series: Materials Science and Engineering **271** (2017) 012005 doi:10.1088/1757-899X/271/1/012005

## 3. Results and discussions

The stress distribution study has been carried out and the effects of various parameters were explained in the following section. Direction of angle along the boundary is given in figure 4.

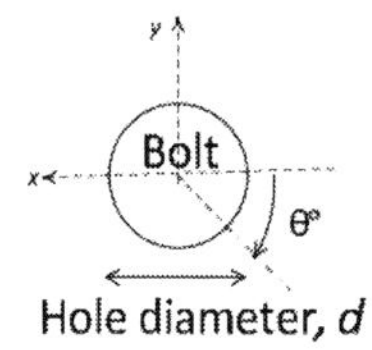


**Figure 4.** Direction of angle along boundary.

### *3.1. Effect of various applied stress*

Current work studied the tangential (or hoop) stress with various applied stresses between 20 - 40 MPa in configuration B (single-row three hole) of cross-ply lay-up (PX4) as illustrated in figure 2. Figure 5 shows the tangential stress, $\sigma_\theta$ distribution (a) along hole boundary and (b) across net-tension plane in hole 1 of outer hole, where outer hole demonstrated largest tangential stress compared to middle hole (hole 2). As expected, highest tangential stress occurred at net-sectional plane (90° to the direction of applied loading) and tangential stress is associated with net-tension failure across the net-sectional plane as given in figure 5(a). Various applied traction gives similar pattern on tangential stress distributions, where the tangential stress increased with increased traction magnitude. Owing to hole elongation due to applied loading, circular hole is now giving almost elliptical hole shape, therefore there is compression stresses exhibited at an angle of 0°-25°.

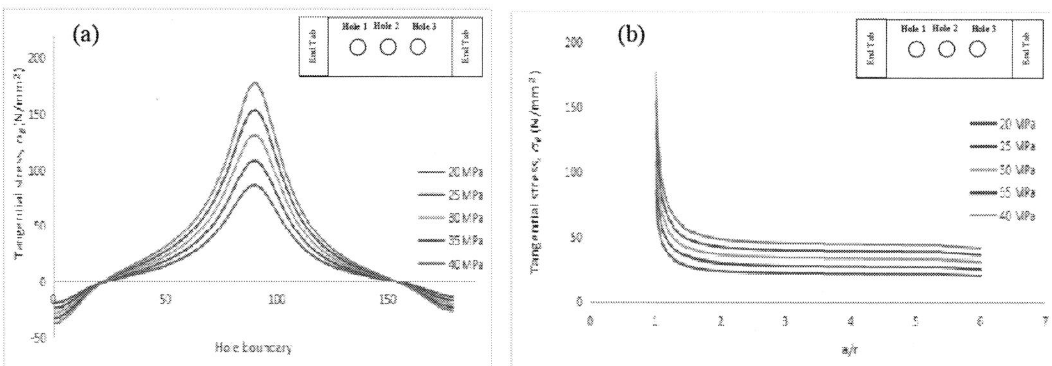


**Figure 5.** Stress distribution in configuration B of PX4 lay-ups (a) along hole boundary and (b) along net-sectional plane.

Figure 5(b) gives tangential stress along net-sectional plane from notch tip where largest stress is given at the notch tip and gradually reduced to applied stress at a certain distance. This distance is associated with damage zone length where extensive micro-damage events occurred. Calculation of normalized maximum tangential stress at the notch tip to far-field stress gives stress concentration factor, $K_T^\infty$ gives a value of about 4.25 in PX4 with applied stress of 40 MPa (theoretical value for $K_T^\infty$ in unidirectional cross-ply lay-up is 4.75). Similar finding were found in PQ4 with applied stress 40 MPa where stress concentration factor, $K_T^\infty$ of 2.95 (against theoretical $K_T^\infty$ in quasi-isotropic lay-up is 3).

### *3.2. Effect of adjacent holes*

Figure 6 shows tangential stress (a) along hole boundary and (b) along net-tension plane in configuration B of PX4 lay-ups under uniform applied stress of 40 MPa. Curve of both outer holes in configuration A is almost overlapped, but slightly lower curve exhibited in mid-hole. Hole 1 and 3 in configuration B exhibited largest tangential stress where the net-sectional failure prone to occur within the respective hole. Similarly maximum tangential stress at notch tip occurred in outer hole as shown in figure 6 (b).

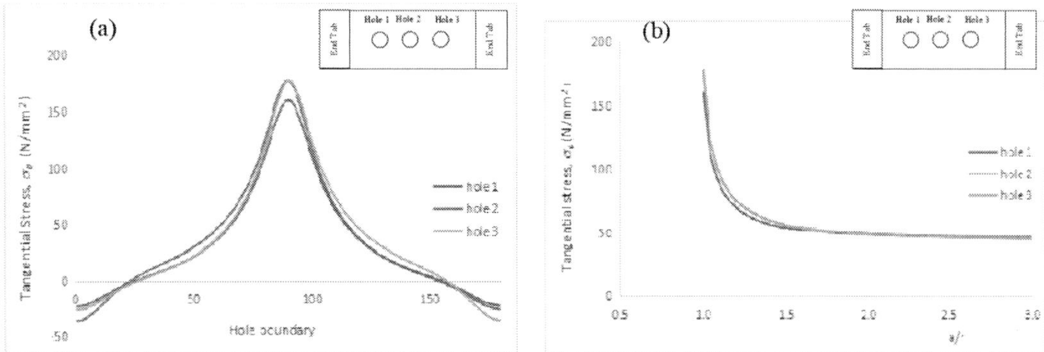


**Figure 6.** Tangential stress of PX4 lay-up of configuration B (a) along hole boundary and (b) along net-sectional plane with uniform applied traction of 40 MPa.

Figure 7 displayed the stress contours of configuration A and B in PX4 lay-up, where the maximum stress occurred at the notch tip of outer hole as described above. In configuration A, there is no middle hole exist and both outer holes in configuration A has similar hole elongation due to linear elastic exhibited in composite plate. Therefore, it is expected that the failure occurred in one of the outer holes. On the other hand, hole elongation in middle hole (hole 2) is relatively small than the two outer holes and lower tangential stress were exhibited at the middle hole (from figure 6) leading to less likely the failure and fracture is initiated from middle hole. The crack is occurred in the outer hole as given in figure 7 (b). It is tally with conducted experiment where the failure is susceptible to occur at outer hole as shown in photograph of figure 7.

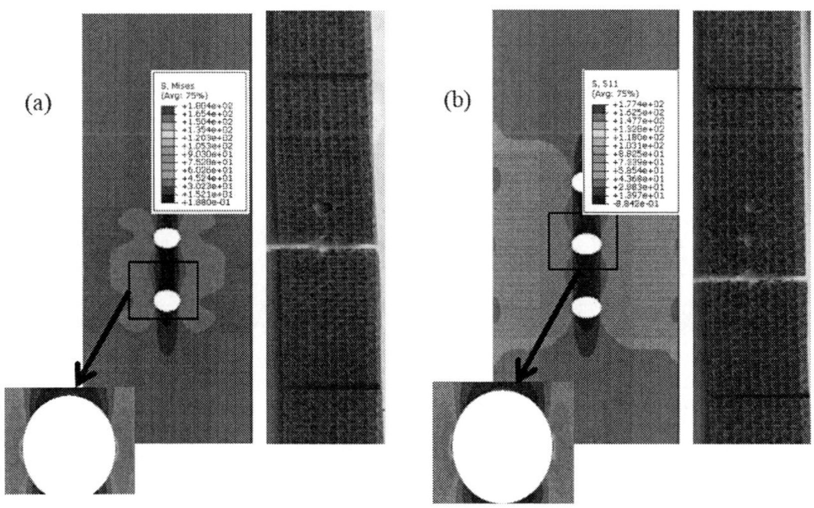


**Figure 7.** Stress contours in PX4 lay-up (a) configuration A (b) configuration B (small pictures is the enlargement at the vicinity of outer hole as specified for visual clarity).

### *3.3. Effect of lay-up types*

Figure 8 showed stress distribution on configuration B between cross-ply lay-up (PX4) and quasi-isotropic lay-up (PQ4), both has similar nominal plate thickness of 4 mm. The stress distribution along hole boundary in cross-ply and quasi-isotropic lay-up exhibited different curve profiles. In cross-ply lay-up, the 0°/90° fiber orientation gives high tangential stress at 90° due to more fiber volume fraction of 0° fiber directions than quasi-isotropic lay-up counterpart, and due to absence of ±45° layer, more shallow curve (especially at angle 45°) is exhibited. In quasi-isotropic lay-up, stress distribution are more evenly distributed at angle 0° to 180° along the hole boundary.

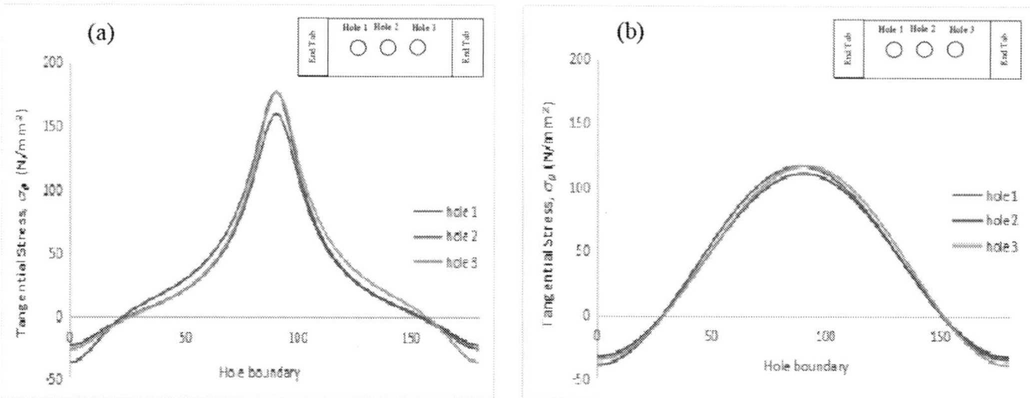


**Figure 8.** Tangential stress of configuration B (a) cross-ply (PX4) lay-up and (b) quasi-isotropic (PQ4) lay-up.

### *3.4. Effect of various rows in non-staggered configurations*

Figure 9(a) and 9(b) displays tangential stress distribution along hole boundary in configuration B and C respectively. In Figure 9(b), the plot given taken at the upper hole due to symmetrical stress distribution in lower holes. Larger tangential stress is exhibited in configuration C than configuration B, this is expected due to smaller net-sectional area in configuration C. Similar findings were reported by Eugene *et al.* [5], where larger stress occurred within pitch area in plates with larger holes number in double-row configurations. They also found from stress analysis results that the maximum stress along the fiber direction of 0° ply is reduced as additional holes are placed in a plate row.

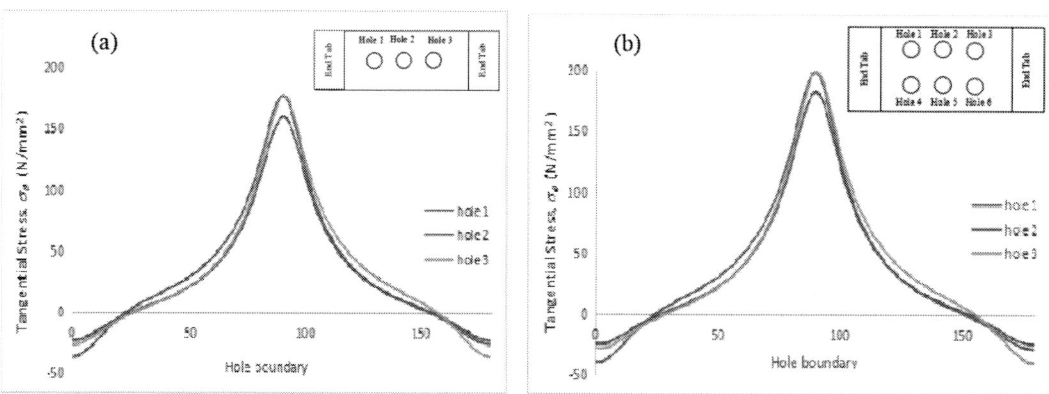


**Figure 9.** Tangential stress of PX4 lay-up in (a) configuration B and (b) configuration C.

### 3.5. *Effect of hole arrangement in staggered configurations*

Figure 10(a) and 10(b) shows stress analysis on hole arrangement in staggered configuration, i.e., configuration D and E respectively. From figure 10(a), tangential stress along hole boundary at configuration D are similar in every holes but in configuration E (given in figure 10(b)), middle hole (hole 2) exhibited larger tangential stress than two outer holes (hole 1 and 3). In configuration D, it was found that the hole elongations are approximately equal in all open hole leading to similar stress distributions. On the other hand, in configuration E, the lower hole (hole 2) exhibited larger elongations leading to higher stress than two upper holes (hole 1 and 3). From stress-strain relationship, increasing hole elongation of composite plate (associated to plate displacement) exhibited higher stress at the notch tip.

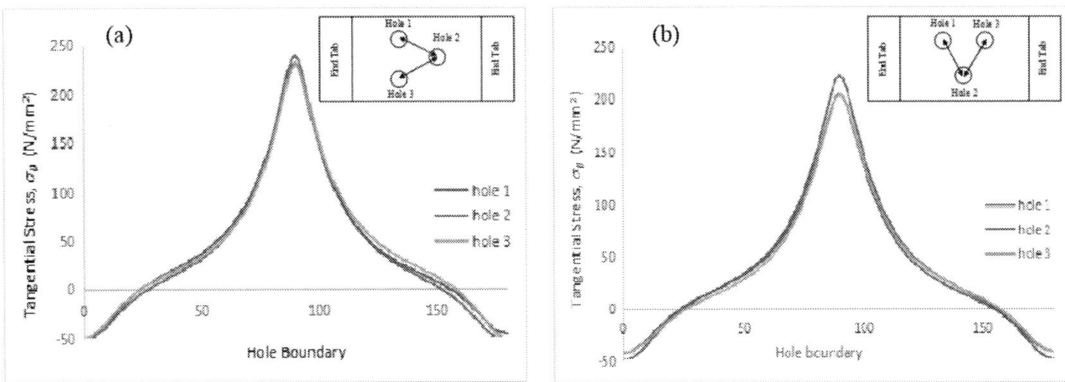


**Figure 10.** Stress distribution along hole boundary (a) configuration D and (b) configuration E.

It was found that the stress contours in configuration D at the vicinity of the hole is similar in all holes, but due to low net-sectional area, it is expected that failure and fracture is initiated and propagated along plane A-A. In-line with largest stress exhibited in hole 2 compared to other two holes as shown in figure 10, crack exhibited in hole 2 is perhaps not surprising. Similar observations were seen from experimental works (photograph) in configuration D and E as given in figure 11(a) and 11(b) respectively.

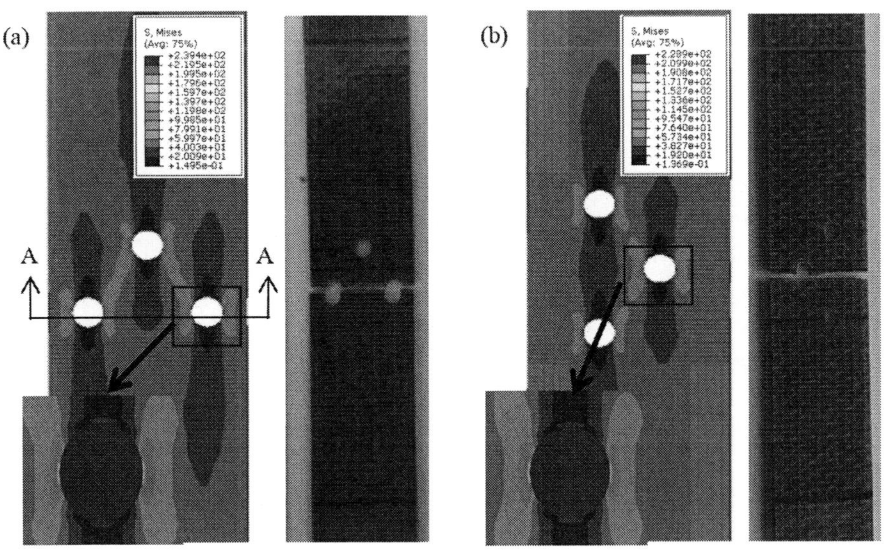


**Figure 11.** Stress contours in PX4 lay-up of (a) configuration D (b) configuration E.

## 4. Conclusions
A 2-D FEA modelling of staggered and non-staggered multi-holes configurations has been implemented within FEA framework. The parametric study of stress analysis includes various applied stresses, effect of adjacent holes, lay-up types, hole arrangement in staggered and non-staggered configurations. Fracture and failure prone to occur at the outer holes for non-staggered arrangement due to higher tangential stress exhibited. On the other hand, net-sectional failure is demonstrated within lowest net-sectional area of any plane perpendicular to loading direction. This arguments is supported by experimental observations in both configuration types. Due to existing of bearing stress behind the bolt shank may alter the stress distribution in bolted joints problem. Therefore, further work in multi-bolted joints problem need to be conducted to explore the applicability of woven fabric kenaf composite in structures engineering works.

## 5. References
[1] Cunningham D, Harries K A and Bell A J 2015 Open-hole tension capacity of pultruded GFRP having staggered hole arrangemen, *Engr. Struc.* **95** 8-15
[2] deJong T 1977 Stresses around pin-loaded holes in elastically orthotropic and isotropic plate, *J. Comp. Mater.* **11** 313-331
[3] Smith P A, Ashby M F and Pascoe K J 1987 Modelling clamp-up effects in composite bolted joints. *J. of Comp. Mater.* **21** 879-897
[4] Crews J H, Hong C S and Raju I S, 1981 *Stress Concentration Factors for Finite Orthotropic Laminates with a Pin-loaded Hole* NASA Technology Paper *1982* (Virgiilia: Larlgley Research Certter, Hampton) pp 1-40
[5] Eugene D J, Russel K, Wen S C and Selvaraj S 2009 *Strength of composite laminate with multiple holes* (Arlington: The Boeing Company) pp 1-9
[6] Jia Y, Liao D, Cui H, Ji A, Bai X and Yasir M 2016 Modelling the needling effect on the stress concentrations of laminated C/C composites, *Mater. & Desg.* **104** 19–26
[7] Ahmad H, Crocombe A D and Smith P A 2014 Strength Prediction in CFRP woven laminate bolted single-lap joints under quasi-static loading using XFEM, *App. Scie. & Manufac.* **56** pp 192-202
[8] Ahmad H, Crocombe A D and Smith P A 2014 Strength prediction in CFRP woven laminate bolted double-lap joints under quasi-static loading using XFEM, *App. Scie. & Manufac.* **66** pp 82-93

**Acknowledgements**
This paper was partly sponsored by the Centre for Graduate Studies Universiti Tun Hussein Onn Malaysia (UTHM).

IOP Conf. Series: Materials Science and Engineering **271** (2017) 012006 doi:10.1088/1757-899X/271/1/012006

# Properties of concrete containing coconut shell powder (CSP) as a filler

**A SLeman[1], S Shahidan[1], A J Nasir[1], M S Senin[1], S S Mohd Zuki[1], M H Wan Ibrahim[1], R Deraman[1], F S Khalid[1]and A T S Azhar[2]**

[1]Jamilus Research Centre, Faculty of Environmental and Civil Engineering, Parit Raja, 86400, Batu Pahat, Johor Malaysia.
[1]Research Center for Soft Soil, Universiti Tun Hussein Onn Malaysia, 86400 Batu Pahat Malaysia
Corresponding author:shahironshahidan@gmail.com

**Abstract.** Coconut shellsare a type of agricultural waste which can be converted into useful material. Therefore,this study was conducted to investigate the properties of concrete which uses coconut shell powder (CSP)filler material and to define the optimum percentage of CSP which can be used asfiller material in concrete. Comparisons have been made between normal concrete mixes andconcrete containing CSP. In this study, CSP was added into concrete mixes invaryingpercentages(0%, 2%, 4%, 6%, 8% and 10%). The coconut shell was grounded into afine powder before use. Experimental tests which have been conducted in this study include theslump test, compressive test and splitting tensile strength test. CSP have the potential to be used as a concrete filler and thus the findings of this study may be applied to the construction industry. The use of CSP as a filler in concrete can help make the earth a more sustainable and greener place to live in.

## 1. Introduction

In recent years, the world has been facing the problem of increasing solid waste disposal, some of which include agricultural waste. Agricultural waste has negatively affected the environment [1-2]. It is the main source of pollution in water and may lead to health problems among humans. Various methods can be used to reduce pollution caused by the disposal of solid waste produced from coconut shells. One of the ways is to use coconut shells as an admixture in the production of concrete.

Annually, 40–50 million tons of coconuts are grown worldwide in countries such as Indonesia, the Philippines and India which make up75% of the world's coconut production [3,4]. Furthermore, coconuts are also used for religious purposes and traditional celebrations. In Malaysia, for example, coconuts are important for the Indians during the Diwali festival. For the Malays, coconut milk is widely used in traditional cuisine and food production. However, coconuts shells are also a type of solid waste which can contribute to the pollution of the environment. The annual production of coconut shells amount to approximately 3.18 million tonnes [5-7].

The disposal of agricultural waste materials such as rice husks, groundnut husks, corn cobs and coconut shells have constituted an environmental challenge. Hence, there is a need to convert them into useful materials to minimise their negative effects on the environment [8]. In recent times, researchers have begun to utilise these wastes as partial replacement for conventional concrete-making materials and came up with interesting findings [9].

Moreover, the reuse of agricultural waste in the production of concrete may reduce the cost of production and harmful diseases caused by the disposal of solid waste. In short, both the cost of construction materials and disposal of waste can be reduced if agricultural waste can be reused [5]. In previous studies, most researchers have focused on the use of other natural fillers in composites. Coconut shell as filler material is a potential candidate for the development of new composites because of their high strength and modulus properties [10]. There is a lack of research on the use of CSP in concrete material. Therefore, this study emphasises on the effectiveness of using CSP as filler material in normal concrete. This study focuses on the mechanical properties of CSP as a filler in concrete due to the gap identified in the literature.

## 2. Coconut shell properties

A coconut shell refers to the non-food part of a coconut which is considered as a type of hard, lignocellulosic agro-waste. Comparative studies on concrete properties using coconut shells and oil palm shells show that concrete containing coconut shells exhibits higher compressive strength compared to concrete containing oil palm shells [3]. The ultimate bond strength of this type of concrete was much higher compared to theoretical bond strength [5]. Thus, coconut shells have shown great potential to be used as admixtures in the production of concrete due to its excellent properties.

Previous studies have been conducted on the utilisation of coconut shells as a composite material in concrete mixes. Coconut shell concrete is able to provide ample warning before failure compared to conventional concrete; this shows that coconut shells play a vital role in the ductility of concrete. Concrete containing coconut shells can be useful for consumers as it provides warning signs before failure [11]. The average specific gravity and the apparent specific gravity were 1.05 – 1.20 and 1.40 – 1.50 respectively and these values were comparatively less than the specific gravity of conventional aggregates [12]. Thus, further studies need to be conducted to improve or increase the use of coconut shells as a composite material or admixture in the production of concrete.

## 3. Material and laboratory work

### *3.1. Material preparation*

Firstly, the coconut shells were cleaned and manually broken into small pieces and then fed into a pulveriser. The powder obtained from the pulveriser was then fed into a rotor lift, coiled and passed through a dresser so that the required mesh size can be obtained. Efficient pulverising and screening are critical aspects of material preparation. All of these processes were conducted at a coconut shell factory. In this research, the percentages of CSP used in concrete were 0%, 2%, 4%, 6%, 8%, and 10% respectively. The percentage of CSP was calculated based on the weight of concrete. The cement used as the binding agent was Ordinary Portland Cement. The coarse aggregates used were those which could pass through a 20mm sieve but retained on a 5mm sieve whereas the fine aggregates used were those which could pass through a 150µm sieve and were retained on a 300µm sieve.

### *3.2. Laboratory work*

*3.2.1. Workability test.* The workability test was performed as per ASTM C143 [13]. The apparatus for the slump test consists primarily of a hollow mould in the form of a frustum of a cone which is 305mm in height. The base of the cone which measures 203 mm in diameter is placed on a smooth surface with a smaller diameter measuring 102 mm at the top. The mould is filled with concrete in three layers of equal volume; each layer is rodded 25 times with a steel rod measuring 16mm in diameter. The mould is then lifted away vertically and the slump is measured by determining the difference between the height of the mould and the height of concrete over the original centre of the base of the specimen.

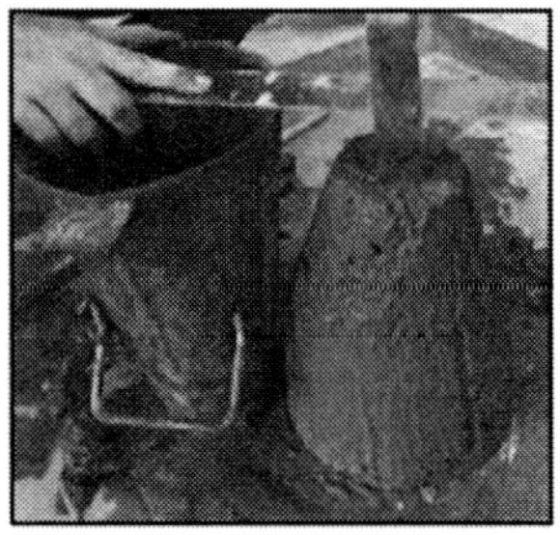

**Figure 1.**Workability test.

*3.2.2. Compressive strength test*

Cylindrical (150mm in diameter x 300mm in height) specimens of concrete were tested using digital compression according to ASTM C39 [14]. The load was applied such that the stress increases at a given constant rate until failure. The maximum load was recorded and the value was divided by the cross-sectional area of the specimen to obtain the compressive strength of concrete. The average compressive strength of three cylinders for each batch was taken.

**Figure 2.** Compressive strength test.

*3.2.3. Splitting tensile strength test.* The compressive splitting tensile strength test was done according to ASTM C496 [15]. The specimens used were 150mmin diameter and300mm in height. The test specimen was placed in the centering jig with packing stripped carefully along a diametrically vertical plane at the top and bottom of the specimen. The splitting tensile strength $f_{sp}$ was calculated using the following formula:

$$f_{sp} = 2W/(\pi DL) \quad (1)$$

where W = maximum load applied to the specimen (N). D = cross-sectional diameter of the specimen (mm) and L = length of the specimen (mm).

**Figure 3.** Splitting tensile strength test.

*3.2.4. Modulus of Elasticity test.* Cylinders measuring 150mm in diameter and 300mm in height were used to find the modulus of elasticity of concrete. The specimens were loaded up to 40% of the ultimate load as per ASTM469 [16].The modulus of elasticity was calculated using formula Eq. 2:

$$E = (S_2 - S_1) / (\epsilon_2 - 0.000050) \tag{2}$$

where E = chord modulus of elasticity (MPa). $S_2$ = stress corresponding to 40% of the ultimate load (MPa). $S_1$ = stress corresponding to the longitudinal strain of 50 x $10^{-6}$ (MPa). $\epsilon_2$ = longitudinal strain produced by stress $S_2$.

**Figure 4.** Modulus of Elasticity test

## 4. Results and discussion

*4.1. Sieve analysis*

A sieve analysis was conducted and the results complied with the particle size distribution with applicable specification requirements. In this study, the sieve analysis has been used for fine aggregates and coconut shell powder. Based on the results of the sieve analysis test, the cumulative percentage passing for CSP measuring 600 $\mu$m, 300 $\mu$m and 150 $\mu$m was higher than fine aggregates. It can be concluded that CSP has smaller particle size compared to fine aggregates. Figure 5 indicates the size distribution of fine aggregates and CSP.

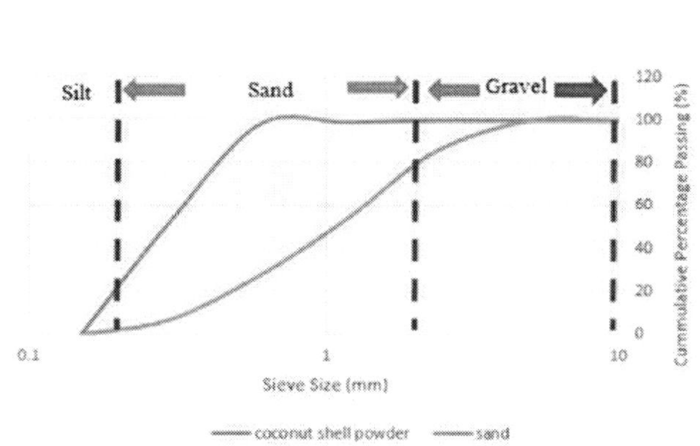


**Figure 5.** Percentage passing versus sieve size.

### 4.2. Workability

The workability test was conducted for each mix of concrete produced namely normal concrete and concrete containing 2%, 4%, 6%, 8%, and 10% of CSP. All specimens used the same water-cement ratio which was 0.45. The concrete mixes containing 0%, 2%, 4% and 6% of CSP had a slump range between54mm to 65mm.These mixtures had a medium degree of workability and a compacting factor of 0.92. Concrete mixes with 8% and 10% of CSP had a slump value of 46mm.This mixture had a low degree of workability and a compacting factor of 0.78. From the results, it can be seen that if there is an increase in the percentage of CSP added to the concrete mix, the degree of workability of the concrete mix will decrease. This may be because coconut shell has relatively high ability to absorb water compared to conventional aggregates [5]. Figure 6 shows the data of the workability test.

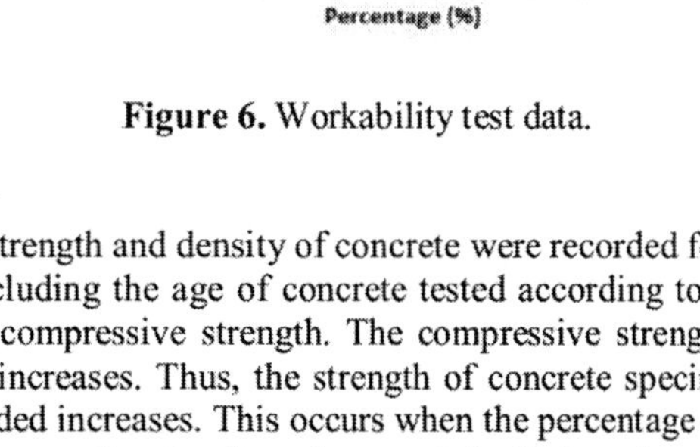


**Figure 6.** Workability test data.

### 4.3. Compressive strength test

The data for the compressive strength and density of concrete were recorded for each specimen. Figure 7 shows the data collected including the age of concrete tested according to the percentages of CSP added in concrete as well as compressive strength. The compressive strength of concrete increases when the density of concrete increases. Thus, the strength of concrete specimen decreases when the percentage of coconut shell added increases. This occurs when the percentage of coconut shell powder added is higher. This is because the density of cement is higher than coconut shell [8]. The compressive strength of normal concrete cured for a period of 7 days was 38.9MPa whereas the compressive strength for 2% CSP concrete was 29.7MPa which was 23.6% lower than the strength of normal concrete. The compressive strength for 4% CSP concrete and 6% CSP concrete was 26.5MPa and 25.6MPa or 31.9% and 34.2% lower than normal concrete respectively. Concrete containing 8% CSP recorded with lowest compressive which was 23MPa or 40.1% lower than normal concrete.

However, the compressive strength of concrete containing 10% CSP recorded a compressive strength of 24.4MPa or 37.3% lower than the compressive strength of normal concrete. Therefore, the compressive strength results for a curing period of 28 days differ. The compressive strength for normal concrete was 41.8MPa while the average strength for concrete containing 2% CSP was 33.1MPa or 20.6% lower than the strength of normal concrete (the optimum percentage of CSP as filler material in concrete). The compressive strength of 4% CSP concrete was 32Mpa or 23.45% lower than the strength of normal concrete. The results for 6% and 8% of CSP concrete were lower than 4% CSP concrete which were 28.5MPa and 26.1MPa respectively. However, when 10% of CSP was added into the concrete mix, the strength of concrete obtained was more than that obtained by 8% CSP concrete.

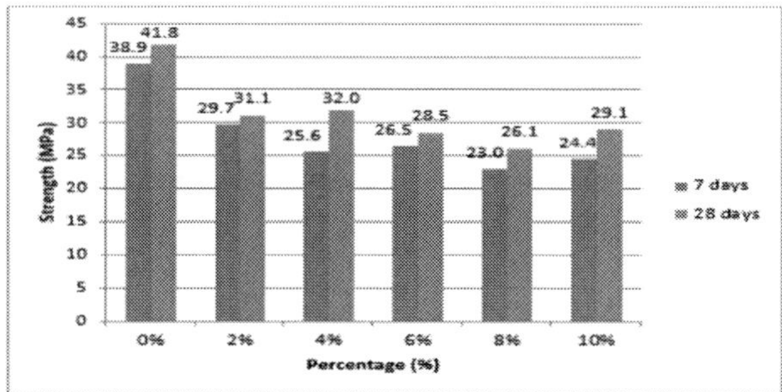


**Figure 7.** Compressive strength for a curing period of 7 days and 28 days.

*4.4. Density of concrete*

Figure 8 shows of the density of normal concrete and concrete containing different percentages of coconut shell powder as filler material. The results indicated that the density of normal concrete was the highest at 2438 kg/m$^3$. From the results shown, it can be concluded that the density of concrete containing coconut shell powder decreases as the percentage of coconut shell powder used in concrete increases.

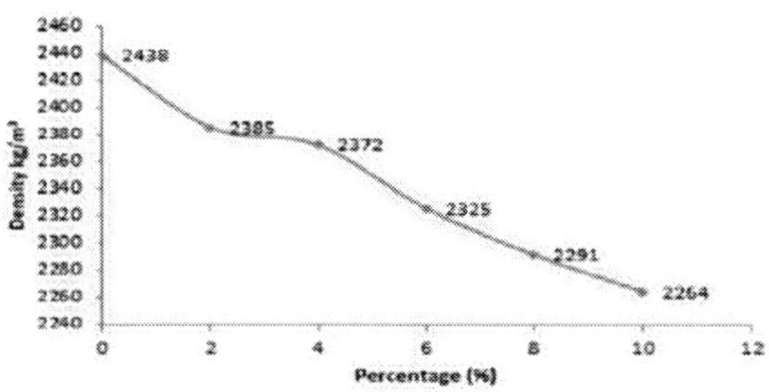


**Figure 8.** Density of concrete

*4.5. Splitting tensile strength*

Figure 9 shows the graph percentage versus splitting tensile strength. The highest value of splitting tensile strength recorded for normal concrete was 6.58MPa (15.74% of compressive strength). Concrete containing 2% of CSP concrete had the lowest splitting tensile strength which was 4.63MPa (13.98% of compressive strength). However, the value of splitting tensile strength increased when4% up to 10% of CSP was used in concrete. Thus, the highest splitting tensile strength was recorded for concrete containing 10% CSP at 6.25MPa (21.48% of compressive strength). It can be concluded that the use of CSP enhances the tensile strength of concrete. The tensile strength increases with the increase of filler content [17].

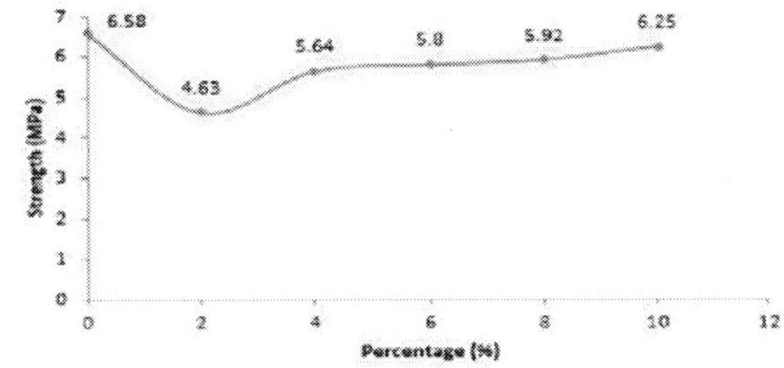


**Figure 9.** Splitting tensile strength.

*4.6. Modulus of Elasticity*

Based on results tabulated in table 1, the modulus of elasticity varied between 20.47 GPa to 27.82 GPa. The highest modulus of elasticity of 27.82 GPa was achieved by concrete containing 4% of coconut shell powder while the lowest modulus of elasticity of 20.47 GPa was achieved by concrete containing 8% of coconut shell powder. The typical range of the modulus of elasticity for concrete (cured for a period of 28 days) with a compressive strength of 30 MPa is 20 GPa – 32 GPa [18-19].

**Table 1.**Compressive strength and modulus of elasticity of concrete containing different percentages of CSP.

| Percentage (%) | Compressive Strength, $\sigma$ (MPa) | 40% of ultimate load, $\sigma$ (MPa) | Strain, ∈ x$10^{-4}$ (mm) | Modulus of Elasticity, $E_c$ (GPa) |
|---|---|---|---|---|
| 0 CSP | 39.30 | 15.72 | 6.6 | 25.77 |
| 2 CSP | 33.10 | 13.24 | 5.9 | 24.51 |
| 4 CSP | 32.00 | 12.80 | 5.1 | 27.82 |
| 6 CSP | 28.50 | 11.40 | 4.7 | 27.14 |
| 8 CSP | 26.10 | 10.44 | 5.6 | 20.47 |
| 10 CSP | 29.10 | 11.64 | 5.4 | 23.76 |

## 5. Conclusion

This early study found that the addition of CSP as filler material in concrete reduces the workability of concrete because coconut shell has a higher level of water absorption. Hence, the compressive strength of concrete decreases when an increasing percentage of CSP is added into concrete. The highest compressive strength obtained fornormal concrete was 38.9 MPa and the highest compressive strength for CSP concrete was recorded at 33.1 MPa which is 2% of CSP in concrete. However, the result shown CSP concrete was good in splitting tensile strength where the strength was increasing by the percentage of CSP is increasing. The modulus of elasticity of normal concrete and CSP concrete was varied in range of 20.47 GPa to 27.82 GPa.

## 6. References

[1] Aprianti E 2016 A huge number of artificial material can be supplementary cementitious material (SCM) for concrete production, *Constr. Building Mater.***74**176–187

[2] Shahidan S, Zuki S S M and Jamaluddin N 2016 Damage grading system for severity assessment on concrete structure, *Case Stud. Constr. Mater.* **5** 79–86

[3] Shafigh P, Mahmuda H, Jumaata M Z and Majid Z 2014 Agricultural wastes as aggregate in

Concrete mixtures - A review, *Constr. Build. Mater.* **53** 110–117
[4] Leman A S, Shahidan S, Senin M S and Ramzi Hannan N I R2016 A preliminary study on chemical and physical properties of coconut shell powder as a filler in concrete *IOP Conf. Series: Materials, Science and Engineering* vol 160 (London: IOP Publishing)12059
[5] Gunasekaran K, Annadurai Rand Kumar PS 2012 Long term study on compressive and bond strength of coconut shell aggregate concrete, *Constr. Build. Mater.* **28(1)** 208–215
[6] Shahidan S, Leman A S, Senin M Sand Hannan N I R R 2017 Suitability of coconut shell concrete for precast cool wall panel-a review *MATEC Web Conf.* vol 87 (Paris: EDP Sciences)1005 pp1–6
[7] Ismail I, Jamaluddin N and Shahidan S2016 A review on performance of waste materials in self compacting concrete (SCC) *J. Teknologi* **5**(78) 29–35
[8] Utsev and Taku J K 2012 Coconut shell ash as partial replacement of ordinary portland cement in concrete production *Int. J. of Scientific & Technology Research* vol 1(8) pp 86–89
[9] Hung Mo K, Alengaram J, Jumaat M Z, Poh Yap S and Cheng Lee S 2016 Green concrete partially comprised of farming waste residues: A review, *Journal of Cleaner Production* **117** 122–138
[10] Salleh Z, Islam M M, Yusopa M Y M, Mun'aim M A and Idrusa M 2014 Mechanical properties of activated carbon (ac) coconut shell reinforced polypropylene composites encapsulated with epoxy resin. *Science Direct,* pp 92-96
[11] Khalid F S, Azmi N B, Sumandi K A S M and Mazenan P N 2017 Mechanical properties of concrete containing recycled concrete aggregate (RCA) and ceramic waste as coarse aggregate replacement *IOP Conf. Proc.* vol 1891 (London: IOP Publishing)020079
[12] Gunasekaran K, Annadurai R and Kumar PS 2013b. Study on reinforced lightweight coconut shell concrete beam behavior under shear, *Materials and Design* **50** 293–301.
[13] ASTM C 143/C 143M *Standard Test Method for Slump of Hydraulic Cement Concrete* Annual Book of ASTM standards(USA: ASTM)
[14] ASTM C 39-04a *Standard Test Method for Compressive Strength of Cylindrical Concrete Specimens* Annual Book of ASTM standards (USA: ASTM)
[15] ASTM C 496/C 496M-04 *Standard Test Method for Splitting Tensile Strength of Cylindrical Concrete Specimens* Annual Book of ASTM standards (USA: ASTM)
[16] ASTM C 469-02 *Standards Test Method for Static Modulus of Elasticity and Poisson's Ratio of Concrete in Compression* Annual Book of ASTM standards(USA: ASTM)
[17] Sarki J, Hasan S B, Aigbodiona V S and Oghenevwetaa J E 2011 Potential of using coconut shellparticle fillers in eco-composite materials, *J. of Alloys and Compounds* **509(5)** 2381–2385
[18] Neville A and Brooks J 2010 *Concrete Technology Second Edition* (England: Pearson Education limited)
[19] Senin M S, Shahidan S, Leman A S and Hannan N I R R 2016 Analysis of physical properties and mineralogical of pyrolysis tires rubber ash compared natural sand in concrete material *IOP Conf. Proc.* vol 160 (London: IOP Publishing)12053

**Acknowledgments**
This paper was partially sponsored by the centre for Graduate Study UTHM.The authors would like to thank Universiti Tun Hussein Onn Malaysia for the GPPS grant U449, Grant Vot U523 and the Ministry of Higher Education (MOHE) (MyPhD) for the approved funding which made this important research viable and effective.

# Durability of coconut shell powder (CSP) concrete

**A S Leman[1], S Shahidan[1], M S Senin[1], S M Shamsuddin[1], N A Anak Guntor[1], S S Mohd Zuki[1], F S Khalid[1], A T S Azhar[2] and N H S Razak[1]**

[1]Jamilus Research Centre, Faculty of Environmental and Civil Engineering, Parit Raja, 86400, Batu Pahat, Johor, Malaysia.
[2]Research Center for Soft Soil, Universiti Tun Hussein Onn Malaysia, 86400 Batu Pahat Malaysia
Corresponding author:shahironshahidan@gmail.com

**Abstract.** The rising cost of construction in developing countries like Malaysia has led concrete experts to explore alternative materials such as coconut shells which are renewable and possess high potential to be used as construction material. Coconut shell powder in varying percentages of1%, 3% and 5% was used as filler material in concrete grade 30 and evaluated after a curing period of 7 days and 28days respectively. Compressive strength, water absorption and carbonation tests were conducted to evaluate the strength and durability of CSP concrete in comparison with normal concrete. The test results revealed that 1%, 3% and 5% of CSP concrete achieved a compressive strength of 47.65MPa, 45.6MPa and 40.55% respectively. The rate of water absorption of CSP concrete was recorded as 3.21%, 2.47%, and 2.73% for 1%, 3% and 5% of CSP concrete respectively. Although CSP contained a carbon composition of 47%, the carbonation test showed that CSP no signs of carbon were detected inside the concrete. To conclude, CSP offers great prospects as it demonstrated relatively high durability as a construction material.

## 1. Introduction

The use of fillers in concrete is common in the construction industry nowadays. In many developing countries, concrete mixes which contain no fillers tend to be an exception. Fillers can be quite costly. That is the main reason why it is so important to explore alternatives in order to reduce the cost of fillers. On the other hand, waste generated by industrial and agricultural processes has created disposal and management problems which threaten the conservation of the environment [1-2]. Hence, using waste materials as a filler in concrete not only helps in reducing disposal and management problems, but also helps in protecting the environment from possible effects of pollution [3].

Malaysia was among the top coconut producers in the world back in 2005 [4-6]. Coconut trees are also famous as all parts of the tree can be used in multiple ways. After the coconut flesh is scraped out, the shell is usually discarded as waste. Due to the increasing production of agricultural waste, researchers have attempted to use coconut shells as filler material in concrete. Previous studies have shown that by using coconut shells as filler material, a lightweight concrete can be produced. As time passes, concrete structures tend to deteriorate when they come into contact with carbolic acid especially for buildings in the city where vehicles release carbon dioxide which reacts with water and form carbolic acid. Thus, the demand for high durability concrete is growing. In this paper, concrete which contain filler material in the form of coconut shell powder is tested in terms of durability to meet the demand of highly durable concrete. In order to conserve the environment, it is vital to solve the problem of the excessive production of agricultural waste.

In a nut shell, reduction of cost in terms of transportation [7-8], improved landfill management [9] and higher durability in concrete are among the benefits of using coconut shell as s filler material in concrete mix design. Many studies have been carried out are using coconut shell as a partial replacement for cement or aggregates. Therefore, this study mainly emphasizes on the durability of coconut shell powder (CSP) concrete in comparison with normal concrete.

## 2. Coconut shell powder properties

Coconut shell in the form of powder was sourced from a local coconut miller. It is a light brown material which has a bulk density of 0.7gcm-3 [10]. Also, coconut shells are found to have varying thickness between 2-8 mm and have an approximate density value of 1.60g/cm$^3$ [11]. Also, previous studies have shown that coconut shell in the form of powder has a complex structure [10]. A wide range of particle sizes with a mean value of 34.2 microns was observed. Results from the scanning electron micrograph revealed under a magnification of 300 showed a range of shapes and sizes from very small pebble-shaped particles to much larger particles measuring 75 microns which have a distinctly uneven rough surface and irregular shape.

## 3. Material and laboratory work

Basically, concrete is made up of Portland cement, aggregates and water. In this study, CSP was added into the concrete mix as a filler material by the percentage of volume of 0%, 1%, 3% and 5%. The total number of samples used in this experiment was54samples.

### *3.1. Material preparation*

The coconut shells were crushed using a machine. The size of the coconut shells used was about the same as coarse aggregates which measured about 20mm. Next, the coconut shells were grounded until a fine powder was obtained. Now, the size of the coconut shell powder was about the same as the size of cement particles. After that, the coconut shell powder was sieved again to separate fiber-like material from the powder.

(a) (b) (c)

**Figure 1.**(a) Coconut shell, (b) grinding process of coconut shell, and (c) coconut shell powder.

### *3.2. Laboratory work*

*3.2.1. Compressive strength test.* Concrete cubes measuring 100 mm x 100 mm x 100 mm according to ASTM C 192 standards were used. All concrete cubes were cured for a period of 7 days and 28days respectively.

*3.2.2. Water absorption test.* The water absorption test was conducted according to BS 1881: Part 122. 15 samples were weighed and placed inside an oven for 24hours at 105$^0$C. Next, the specimens were submerged in a tank full of water provided at the lab for another 24hours as shown in figure 2. After that, the specimens were taken out and weighed once again after the surface water was removed. The percentage of water absorption was calculated as according to equation 1

$$\text{Water Absorption (\%)} = \frac{m2-m1}{m1}\ x\ 100\% \qquad (1)$$

Where:

$m_1$ = the dry weight of concrete sample (kg)
$m_2$ = the wet weight of concrete sample (kg)

**Figure 2.**Water absorption process.

*3.2.3. Carbonation test.* This test based on EN 14630:2006 (E) was carried out to determine the depth of concrete affected due to the combined attack of atmosphere, carbon dioxide and moisture causing a reduction in the level of alkalinity of concrete. A solution of 0.2% phenolphthalein was used as the pH indicator of concrete. Figure 3 shows the phenolphthalein solution used.

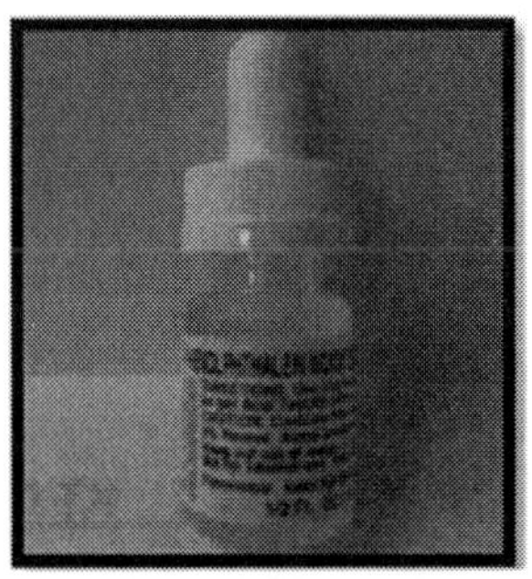

**Figure 3.** Phenolphthalein solution.

## 4. Results and discussion

*4.1. Compressive strength*

The compressive strength development of concrete is shown in figure 4. Normal concrete has an average compressive strength of 42.3Mpa after a curing period of 28days. Meanwhile, concrete that contained 1% of CSP achieved a compressive strength of 47.65MPa while concrete containing 3% and 5% of CSP achieved a compressive strength of 45.6MPa and 40.55MPa. This result was good as all specimens achieved a compressive strength above the minimum strength of grade 30 concrete. Furthermore, 1% and 3% of CSP concrete displayed higher compressive strength compared to normal concrete. Unfortunately, 5% CSP concrete recorded a value lower than normal concrete. However, the value was still higher than the minimum grade strength. It is clear that 5% of CSP was not suitable as it reduces the strength of concrete.

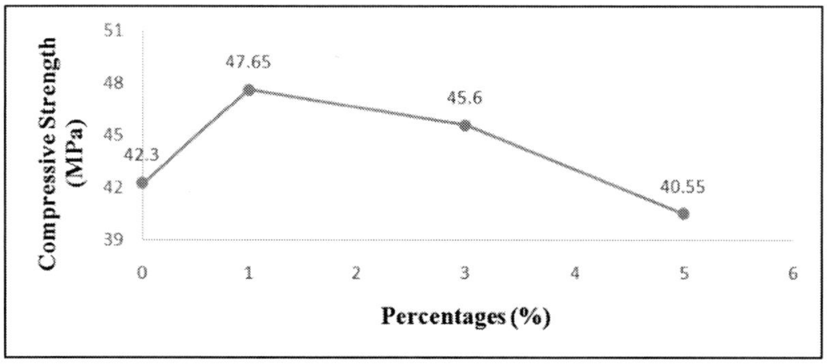


**Figure 4.** Compressive strength varies with percentages of CSP used.

The results on concrete density are shown in figure 5. Concrete with 1% of CSP had the highest density at 2500kg/m$^3$while concrete containing 3% of CSP achieved a density of 2470kg/m$^3$.Concrete with 5% of CSP had the lowest density of 2350 kg/m$^3$ compared to normal concrete which had a density of 2430kg/m$^3$. The data recorded in this study is related to a previous study which stated that if the density of concrete is lower, it will cause the strength of concrete to decrease [1].

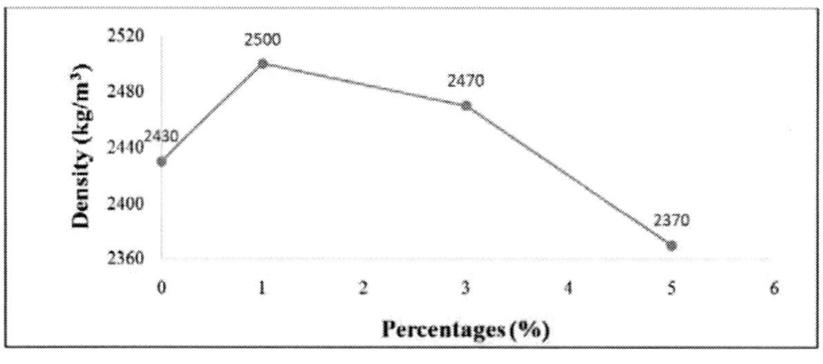


**Figure 5.** Density versus percentages of CSP.

*4.2. Water absorption test*

The results of the water absorption test are shown in figure 6. The highest water absorption was achieved by normal concrete (4.46), followed by concrete containing 1% of CSP (3.21). Concrete containing 3% of CSP had the lowest water absorption whereas the second lowest water absorption was achieved by concrete containing 5% of CSP. This shows that using CSP as a filler material in concrete is a good decision as it controls the ability of concrete to resist water absorption. Moreover, the control sample had the weakest rate of water absorption hence the use of CSP may improve the rate of water absorption of concrete.

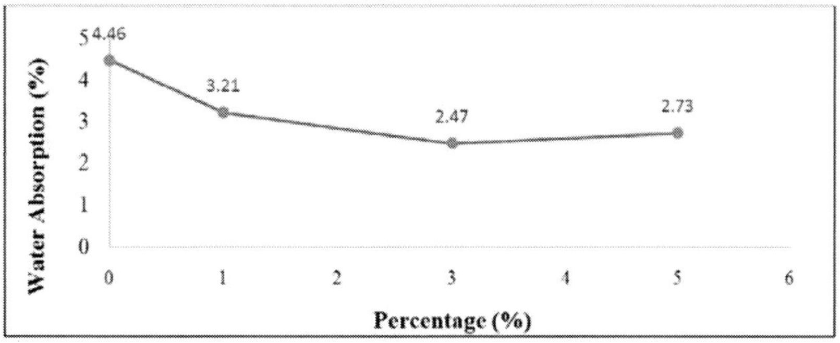


**Figure 6.** Different percentages of CSP in concrete vs water absorption rate.

*4.3. Carbonation test*
After a curing period of 28 days, the concrete was left at the lab for 60 days. The carbonation test results are shown in figures 7(a)-(d).

(a) (b) (c) (d)

**Figure 7.** (a) 0% after 60days, (b) 1% of CSP after 60days, (c) 3% of CSP after 60days, (d) 5% of CSP after 60 days.

The colour purple shows that there is no carbonation detected inside the concrete. The average depth for all samples was 1mm. This shows that 1%, 3% and 5% of CSP concrete are able to resist carbonation. This results also proved that although CSP itself contained47% of carbon[8], it is able to blend and produce good concrete that can resist carbonation attacks.

*4.4. Relationship between compressive strength, density and percentage of CSP*
Results in figure 8 show the relationship between compressive strength, density and concrete added with different percentages of CSP. Concrete containing 1% of coconut shell powder had the highest density. This proves that coconut shell powder plays its role as a filler in concrete hence increasing the strength as well as the density of the concrete. Meanwhile, as the percentage of CSP increases, the compressive strength and density of CSP concrete will decrease. Hence, adding more than 1% of coconut shell powder to a concrete mixture is not suitable as it causes concrete to lose its strength as well as its density.

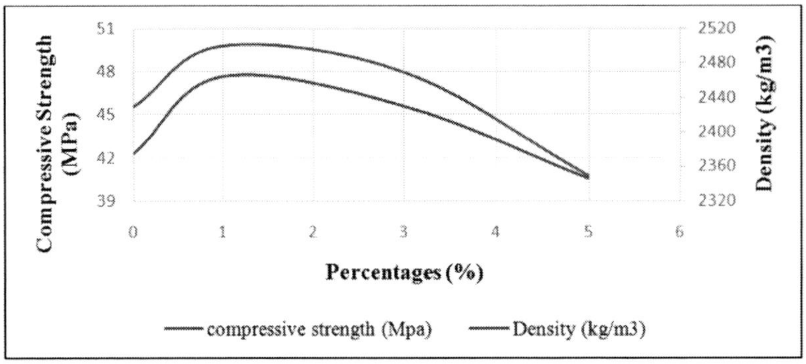


**Figure 8.** Relationship between compressive strength, density and percentages of CSP.

*4.5. Relationship between compressive strength, water absorption and percentages*

Figure 9 shows the relationship between compressive strength and water absorption which varies with different percentages of CSP in concrete. Again, 3% of CSP has been found to be optimal for obtaining concrete with high strength and low water absorption ability. The increase in water absorption maybe due to the development of micro cracks in the weak transition zone at an early stage which may reduce the strength of the concrete.

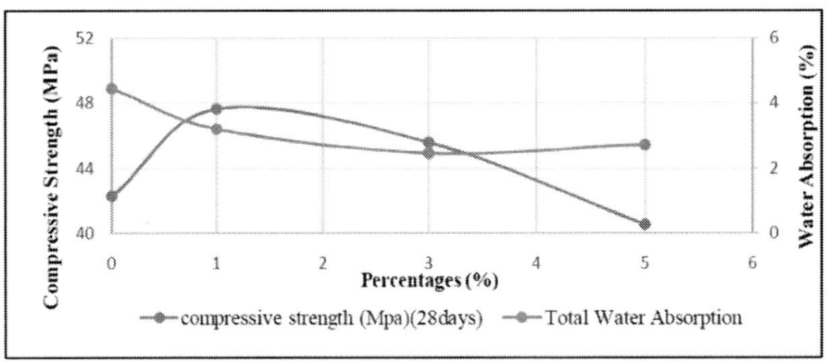


**Figure 9.** Relationship between compressive strength, water absorption and different percentages of CSP added to concrete.

## 5. Conclusion

Concrete added with CSP produce good durability and results from the water absorption test show that concrete added with CSP also can resist water absorption which proves its suitability to be used as an admixture. Also, 3% of CSP concrete has been shown to be better than normal concrete because of its relatively high compressive strength. On the other hand, the correlation between strength and durability of concrete was demonstrated by 3% of CSP concrete which has been found to be optimal for obtaining higher strength and durability. All in all, further studies on the durability of CSP concrete must be conducted as CSP has the potential to be used as filler material in concrete.

## 6. References

[1] Osei D Y, Engineering C and Coast C2013 Experimental assessment on coconut shells as aggregate in concrete *Int. J. Eng. Sci. Invent.* **2(5)** 7–11

[2] Ismail I, Jamaluddin N and Shahidan S 2016 A review on performance of wastematerials in self compacting concrete (SCC), *J. Teknol.* **5(78)** 29–35

[3] Akshay S P G, Shelke S, Kalyani R. Ninghot, Pooja P and Kunjekar 2014 Coconut shell as partial replacement of coarse aggregate in concrete *Int. J. of Civil Engineering Research* **5**(3) pp 211–214
[4] Gunasekaran K, Kumar P S and Lakshmipathy M 2011 Mechanical and bond properties of coconut shell concrete,*Constr. Build. Mater.* **25** (1) 92–98
[5] Leman A S, Shahidan S, Senin M S and Ramzi Hannan N I R 2016 A preliminary study on chemical and physical properties of coconut shell powder as a filler in concrete *IOP Conf. Series: Material Science Engineering* vol 160 (London: IOP Publising) 12059
[6] Senin M S, Shahidan S, Leman A S and Hannan N I R R 2016 Analysis of physical properties and mineralogical of pyrolysis tires rubber ash compared natural sand in concrete material *IOP Conf. Series: Material Science Engineering* vol 160 (London: IOP Publising)12053
[7] Shahidan S, Mohd Zuki S S, Rahim M A and Ali N 2016 Amplitude distribution of emission wave for cracking process *MATEC Web of Conf.* vol 47 (Paris: EDP Sciences) pp 2–8
[8] Lee J 2003 Utilization of solid wastes as aggregates in concrete, *J. of Waste Glass and Rubber Particle* **3** 123-134
[9] Mamlouk M S and Zaniewski J P 2006 Laboratory Manual for Mansory (Manila: National Bookstore) pp 487-523
[10] Kumar S, Vishwas B G and Kukarni P 2013 Comparative study on coconut shell aggregate with conventional concrete *Int. J. Eng. Innov. Technol.* **2**(12) pp 67–70
[11] Tomas U Ganiron Jr 2013 Sustainable management of waste coconut shells as aggregates in concrete mixture *Journal of Engineering Science and Technology Review* vol 5 (4) pp 48–55

**Acknowledgments**
The authors would like to thank Universiti Tun Hussein Onn Malaysia for theGPPS U449 grant as well as Vot U523.The authors are also grateful for the funds granted by theMinistry of Higher Education (MOHE) (MyPhD) which made this research possible.

IOP Conf. Series: Materials Science and Engineering **271** (2017) 012008 doi:10.1088/1757-899X/271/1/012008

# Dynamic modulus of nanosilica modified porous asphalt

**A K Arshad[1,2], K A Masri[2], J Ahmad[2] and M S Samsudin[2]**

[1] Institute for Infrastructure Engineering and Sustainability Management (IIESM), Universiti Teknologi MARA, Shah Alam, Selangor, Malaysia.
[2] Faculty of Civil Engineering, Universiti Teknologi MARA, Uitm Shah Alam, Selangor, Malaysia.
Corresponding author:khairilazmanmasri@yahoo.com

**Abstract.** Porous asphalt (PA) is a flexible pavement layer with high interconnected air void contents and constructed using open-graded aggregates. Due to high temperature environment and increased traffic volume in Malaysia, PA may have deficiencies particularly in rutting and stiffness of the mix. A possible way to improve these deficiencies is to improve the asphalt binder used. Binder is normally modified using polymer materials to improve its properties. However, nanotechnology presently is being gradually used for asphalt modification. Nanosilica (NS), a byproduct of rice husk and palm oil fuel ash is used as additive in this study. The aim of this study is to enhance the rutting resistance and stiffness performance of PA using NS. This study focused on the performance of PA in terms of dynamic modulus with the addition of NS modified binder to produce better and more durable PA. From the result of Dynamic SPT Test, it shows that the addition of NS was capable in enhancing the stiffness and rutting resistance of PA. The addition of NS also increase the dynamic modulus value of PA by 50%.

## 1. Introduction

Malaysia is a tropical country which have a rainy season along the year. Rainfall is well known as a main contributor to the occurrence of various problems for asphalt pavement. Such problem include water ponding above the pavement, which cause splashing and skidding for the road users, leading to accidents and traffic congestion. These problems are closely related to traffic safety which is a highly concerned issue to the people and governments all over the world. Thus, porous asphalt (PA) is introduced to overcome these problems. PA is well-known for its advantages in improving skid resistance of pavement during rain, reducing splashing effects, and producing lower riding noise [1]. These advantages are due to the high porosity possessed by the pavement layer which allows for high drainage capability of surface run-off. However, due to the high porosity, water tends to flow inside the pavement, which may cause moisture penetration, raveling, loss of stability and load bearing capacity of PA [1]. This will result in the loss of adhesive bonding between aggregate and binder which. Besides that, rutting easily develops due to the lack of strength of PA layer [1]. Thus, the binder used should be strong enough to provide adequate bonding to the aggregate especially for PA. In addition, although PA is considered as a non-structural layer in flexible pavement, porous asphalt is still the uppermost layer of pavement. Thus, it deteriorates and defects such as rutting and fatigue cracking occur due to the imposed traffic loadings and effects of the environment. Main parameter that usually used to measure rutting resistance of PA is dynamic modulus.

Dynamic modulus measures the stiffness and viscoelastic properties of the asphalt concrete mixes. The protocol of dynamic modulus defines as a linear viscoelastic test for asphaltic materials. Typically, mixtures with higher dynamic modulus values tend to have better rutting resistance [2,3]. Besides that, the stiffness value (dynamic modulus) is a key property of the material to evaluate strains and displacements in any pavement or road structures [2-3]. In addition, the mixtures with a softer asphalt binder revealed a lower dynamic modulus compared to those with a stiffer asphalt binder [4]. Brown *et al.* [5] stated that the evaluation of dynamic modulus of asphalt mixture has significant relationship with various aggregate structures. A study by Apeagyei [6] also reported that there was a good correlation between dynamic modulus and rutting.

One of the way to enhance dynamic modulus of PA is by binder modification. Previously, binder is modified using polymers such as SBS, SBR, EVA and EGA. However, these types of binder modifiers have several weaknesses such as low ageing resistance and poor storage stability [7]. Therefore, there is a need to improve binder modification to overcome those problems. Other than polymer modifier, nanomaterials is also can be used to improve the properties of asphalt binder. In this regard, this study focused on the usage of the nanoparticle technology as an additive for the binder to overcome the permanent deformation and at the same time preserve the drainage capability of PA.

## 2. Experimental Plan

### *2.1. Material selection*

Porous Asphalt grading A and B as per PWD Malaysian Specification for Road Works [8] were used for this study. The weight of aggregate for each sample was approximately 1100 g. PEN 60-70 type of binder was used and the binder was mixed with different proportions of nanosilica (NS) ranging from 1 to 6 % by weight of binder. NS used was in colloidal form with the average size within 10 to 15 nanometer (nm). Figure 1 below shows the aggregate gradation for PA Grading A and B in accordance to PWD Malaysia Standard Specification while table 1 shows the properties of NS used in this study.

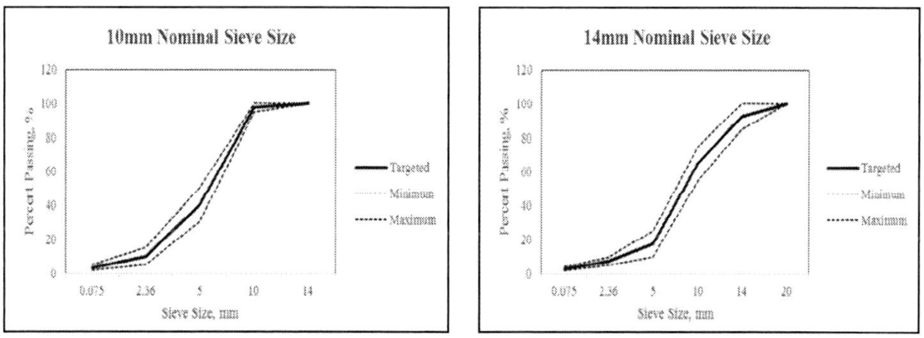


**Figure 1.** Aggregate gradation for PA.

**Table 1.** Specification of nanosilica.

| Properties | Value |
|---|---|
| Appearance | Slight Milky Transparent |
| $SiO_2$ (%) | 30% |
| $Na_2O$ (%) | 0.5% |
| pH | 8.5-10.5 |
| Density | 1.19-1.22 g/cm3 |
| Particle Size | 10-15 nm |

### *2.2. Binder modification*

During modification process, NS was mixed in a reactor with continuous stirring. In this study, a control binder (Penetration Grade Pen 60-70) was used to prepare the modified binder. The binder then was heated at 160°C. The NS blends were then be transferred to the hot-melted asphalt and mixed with continuous stirring at 1800 rpm. This process was continued for one hour to obtain nano-modified asphalt binder. The asphalt binder was blended with 0% to 6% NS in increments of 2% by weight of the binders.

### *2.3. Design binder content*

There are three main processes to obtain Design Binder Content (DBC) for porous asphalt, which are air void determination, lower limit determination and upper limit determination [8]. Air void determination was calculated based on the volumetric properties of PA. Among the volumetric properties are Theoretical Maximum Density ($G_{mm}$) and Bulk Specific Gravity ($G_{mb}$), Void in total Mix (VTM), Void in Mineral Aggregate (VMA) and Void filled with Asphalt (VFA). Then, Lower Limit and Upper Limit value of DBC are based on Cantabro Loss Test and Binder Draindown Test.

45 samples of PA (1100g per sample) for each grading were prepared for DBC determination. Fifteen samples were used for Cantabro Test, fifteen samples were used for Binder Draindown and another 15 samples were used for the determination of $G_{mm}$ & $G_{mb}$. Samples for Cantabro Test and Gmb were compacted with 50 blows of Marshall hammer at every side based on Marshall mix design in accordance with [8] while loose samples were prepared for Binder Draindown test and $G_{mm}$.

The lower limit for design binder content is the average loss of mass in Cantabro test, where it should not more than 15%. In addition, the average air voids shall be less than 25% but shall not less than 18%. For the upper limit of design binder content, the average binder draindown shall not exceed 0.3% of total weight of the specimen [8].

The mean value from these limits were taken as initial design binder content. Each of design binder contents was tested and the binder content that meets all criteria was selected as Design Binder Content (DBC). Then, the same process were repeated to obtain DBC for 2%, 4% and 6% NS-PA.

### *2.4. Dynamic SPT test*

Dynamic Modulus Test is crucial to indicate the engineering properties of porous asphalt, especially to evaluate the resistance of porous asphalt towards permanent deformation. Simple Performance Tester (SPT) equipment was used to evaluate dynamic modulus of PA incorporated with various amount of NS (0%, 2%, 4% and 6% NS-PA). A total of 24 Superpave samples with two porous asphalt gradation (GA and GB) were prepared for this test. Each sample was approximately 5000 g in weight with 150 mm in diameter and 165 mm in height. After compaction and cooled to room temperature for 24 hours, the specimens were cored and trimmed from the centre of gyratory compacted specimen. The axial deformations were measured with displacement transducers referenced to gauge points connected to the specimens.

Prior to testing, the specimens must be placed in the testing chamber until the effective temperature, contact stress and confining pressure were achieved. To begin the test, the LVDTs were zeroed, and minimal contact load was applied to specimen. A sinusoidal compressive load was applied on the specimen in a cyclic manner. Three Linear Variable Displacement Transducers (LVDT) were placed at 120 degrees on the specimen surface to capture full range accumulation of the compressive permanent deformation. A continuous uniaxial sinusoidal (haversine) compressive stress at a specified test frequency is applied to the unconfined cylindrical test specimen in a cyclic manner. The dynamic load was properly adjusted during the tests to keep the strain level between 75 to 125 microstrain. The strain level should be checked after completion of the test, otherwise the dynamic pulse load is increased or decreased to adjust the strain to be within the limits. Each specimen was tested at five different temperatures and six frequencies for each test temperature as shown in table 2 while figure 2 shows the SPT machine.

**Table 2.** Dynamic SPT test condition.

| Parameters | Values |
|---|---|
| Temperature (°C) | 25, 30, 35, 40 and 45 |
| Frequency (Hz) | 25, 10, 5, 1, 0.5 and 0.1 |
| Contact Load (kPa) | 15 |
| Axial Strains | 75 to 150 microstrains |
| Dynamic Stress (kPa) | 100 |

**Figure 2.** Dynamic SPT Machine.

## 3. Result and discussion

### *3.1. DBC determination*
There are three main criteria used for DBC determination which are air void determination, lower limit determination and upper limit determination.

*3.1.1. Air void determination* .Air void content of bituminous material is significant to control the quality of asphalt binder for being laid and compacted in the field especially in constructing porous asphalt pavement. $G_{mm}$ and $G_{mb}$ were calculated to determine value of air void. For PA Grading A (PA-GA), the air voids for 4.0, 4.5, 5.0, 5.5 and 6.0 % of binder were 19.54, 19.40, 19.91, 21.13 and 20.97 % respectively. For PA-GB, the air void values were 19, 18, 16, 18 and 18 % respectively. Thus, all the design mixtures in this study met the requirement stated by PWD Malaysian Specification for PA. Table 3 shows the summary of air void value for both PA-GA and PA-GB. From the result, air void value for each binder percentages were in the range of the requirement stated in PWD Malaysia's Specification for Road Works which was in between 18 to 25 %.

**Table 3.** Air void values.

| Percentage of Binder (%) | Air Void (%) | |
|---|---|---|
| | GA | GB |
| 4 | 19 | 19 |
| 4.5 | 19 | 18 |
| 5 | 20 | 16 |
| 5.5 | 21 | 18 |
| 6 | 21 | 18 |

*3.1.2. Cantabro loss test.* Fifteen numbers of samples per grading were prepared and tested using LA drum without steel balls subjected to 300 revolutions. The weight losses after the test were calculated in terms of percentages as Cantabro Loss value. The highest value of Cantabro loss for PA-GA and PA-GB were obtained at 4.0% binder content with 40.37% and 36% respectively. PWD Malaysian Specification stated that the average loss of mass in the Cantabro Test should not exceed 15%. Based on the result recorded, only 5.0% to 6.0% binder content met the requirement for both grading. The result pattern for PA-GB was also almost the same with PA-GA where the value of Cantabro Loss drop rapidly due the increment of binder percentage. In addition, at 5%, 5.5% and 6% binder content, the Cantabro loss values passed the minimum requirement.

The lower amount of losses indicates the stronger bond between aggregate and binder. Thus, this test was conducted to indicate the lower limit of binder content to obtain design binder content (DBC) of PA based on PWD Malaysian Specification. To obtain the lower limit of DBC using graph, the maximum value of average loss of Cantabro Test which is 15% was used by intersecting with graph plotted down to the binder content. From figure 3, the lower limit of DBC for PA-GA and PA-GB were 5.3% and 5.0% respectively.

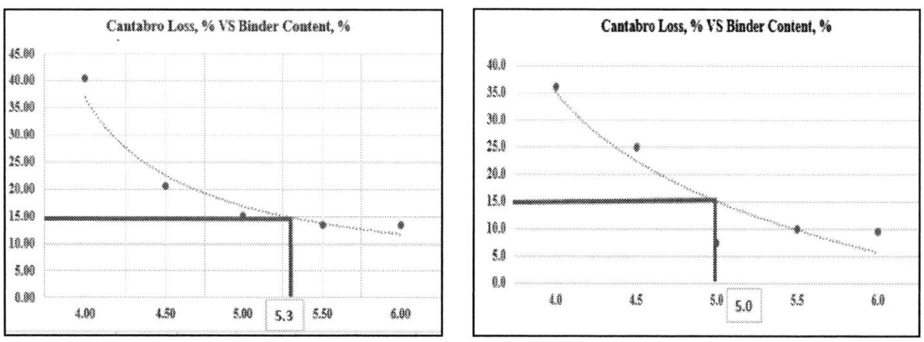


**Figure 3.** Cantabro loss (PA-GA & PA-GB).

*3.1.3. Binder draindown test*. For this test, loose mixes of both grading were prepared and placed into wire basket in an oven for 3 hours. According to PWD Malaysian Specification, binder draindown value of porous asphalt should not exceed 0.3% by weight of the total mix. From the result shown in table 4, the binder draindown obtained for 4.0, 4.5, 5.0, 5.5 and 6.0 % of binder content for PA-GA were 0.114, 0.157, 0.228, 0.415 and 0.326 % respectively, while 0.155, 0.123, 0.265, 0.486 and 0.323 % respectively for PA-GB. The occurrence of binder draindown would reduce the permeability of PA mix. A maximum draindown of 0.3 % by weight of total PA mix is typically the maximum allowable value for PA mix draindown. Thus, only 5.5% and 6% binder content exceeded the limit for both grading. This maximum value was used to determine the upper limit of DBC as shown in figure 4.

Based on this figure, the upper limit of DBC for PA-GA and PA-GB were 5.4 % and 5.1% respectively. From both tests, DBC values for PA-GA and PA-GB were 5.35% and 5.05% respectively. Cantabro Loss Test and Binder Draindown Test were repeated to obtained the DBC values for 2%, 4% and 6% NS-PA. Table 5 summarises the DBC values for all the NS-PA mixes that were used in this study.

**Table 4.** Binder draindown values.

| Percentage of Binder (%) | Binder Draindown (%) | |
|---|---|---|
| | GA | GB |
| **4** | 0.114 | 0.155 |
| **4.5** | 0.157 | 0.123 |
| **5** | 0.228 | 0.265 |
| **5.5** | 0.415 | 0.486 |
| **6** | 0.326 | 0.323 |

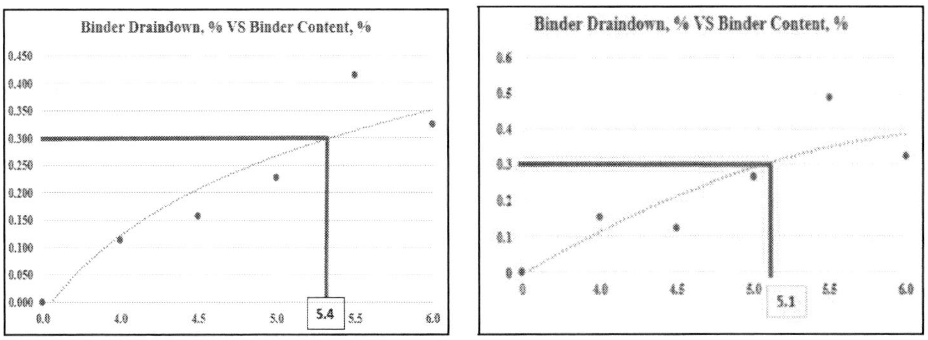


**Figure 4.** Binder Draindown (PA-GA & PA-GB).

**Table 5.** Summary of DBC values.

| NS-PA | Design Binder Content (%) | |
|---|---|---|
| | GA | GB |
| **0** | 5.35 | 5.05 |
| **2** | 5.22 | 4.93 |
| **4** | 5.16 | 4.87 |

*3.2. Dynamic SPT test*

The dynamic modulus evaluation of NS-PA with various frequencies and temperatures was conducted using Simple Performance Test (SPT) machine. This test is essential to evaluate the effect of stiffness characteristic for NS-PA at different temperatures. Dynamic modulus is a mechanically based laboratory test that is carried out to characterise the strength and load resistance of hot-mix asphalt (HMA) mixes including PA mixtures. This test was carried to further analyse the performance, PA mixture characteristics and also identify any possible relationship between the other performance tests of PA with the dynamic modulus values of NS-PA mix.

In this test, five test temperatures (25°C, 30°C, 35°C, 40°C and 45°C) at six different frequencies for each temperature level (20 Hz, 10 Hz, 5 Hz, 1 Hz, 0.5 Hz and 0.1 Hz) were performed for both NS-PA grading A and B with various amount of NS (0%, 2%, 4% and 6% NS).

Figure 5 and 6 illustrate the Dynamic Modulus graph versus frequency for all NS-PA at 25°C and 45°C for NS-PA Grading A and NS-PA Grading B respectively. The dynamic modulus values for all mixes show a similar trend of decreasing value as the frequency decrease at increasing test temperature. In terms of different grading, both NS-PA-GA and NS-PA-GB produced almost the same maximum dynamic modulus values which were 7442 MPa and 7102 MPa respectively. In addition, all modified NS-PA (2%, 4% and 6% NS-PA) produced higher dynamic modulus values compared to unmodified specimens (0% NS-PA). This shows that the addition of NS was capable in enhancing the resistance of PA towards distresses and permanent deformation such as rutting and fatigue cracking.

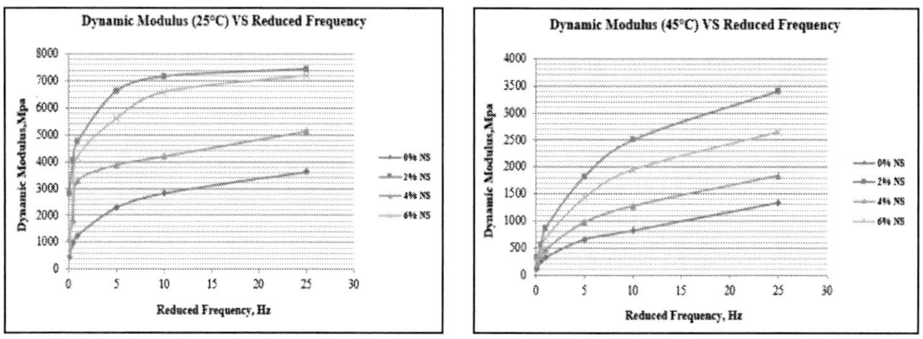


**Figure 5.** Dynamic modulus for NS-PA-GA.

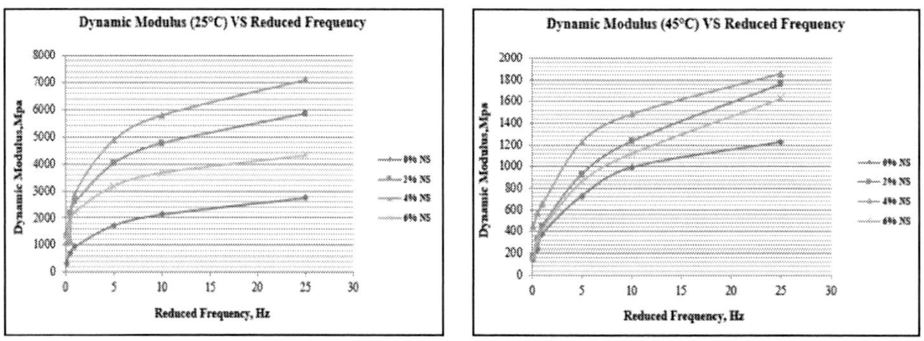


**Figure 6.** Dynamic modulus for NS-PA-GB.

The values of dynamic modulus for all NS-PA were decreased at high temperatures (35°C to 45°C) especially for NS-PA-GB where the values of dynamic modulus decreased more than 50%. This finding proved that PA mixture was very prone to high temperature. This may be due to the softening of the asphalt binder that cause the reduction of stiffness for NS-PA mixtures. When the stiffness decreases, the PA structure become unstable and reduce the resistance towards distresses. Hence, the dynamic modulus value will drop. Figure 7 shows the dynamic modulus values with different test temperatures.

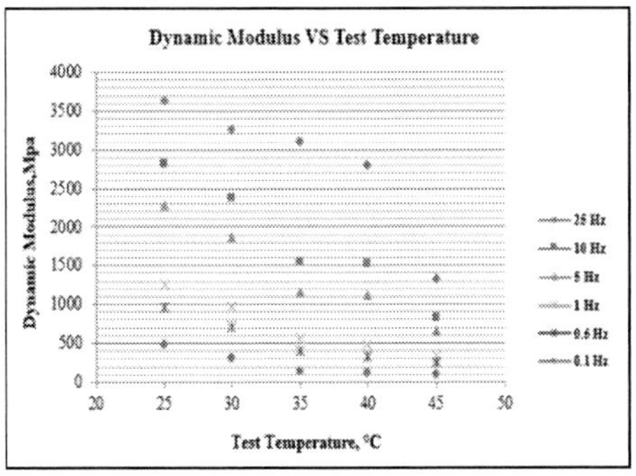


**Figure 7.** Dynamic modulus vs test temperatures.

## 4. Conclusion

Based on Dynamic SPT Test, the existence of NS was capable in enhancing the stiffness properties of PA, thus increasing the resistance towards permanent deformation. The results showed that the addition of NS was capable in increasing the dynamic modulus value for about 50% for both gradation of PA. This shows that NS was capable in providing PA with adequate rutting resistance.

## 5. References

[1] Liu Q and Cao D 2009 Research on material composition and performance of porous asphalt pavement, *J. of Mater. in Civil Engineering* **21(4)** 135–140

[2] Xie J, Wu S, Lin J, Cai J, Chen Z and Wei W 2012 Recycling of basic oxygen furnace slag in asphalt mixture: Material characterization & moisture damage investigation, *Constr. Build. Mater.* **36** 467–474

[3] Goh S W and You Z 2012 Mechanical properties of porous asphalt pavement materials with warm mix asphalt and RAP, *J. of Transportation Engineering* **138(1)** 90–97

[4] Marasteanu M O and Clyne T R 2006 Rheological characterization of asphalt emulsions residues, *J. of Mater. in Civil Engineering* **18(3)** 398–407

[5] Brown B E R, Kandhal P S K, Lee D Y and Lee K W 1996 Significance of tests for highway materials, *J. of Mater. in Civil Engineering* **8(1)** 26–40

[6] Apeagyei A K 2011 Rutting as a function of dynamic modulus and gradation, *J. of Mater. In Civil Engineering* **23** 1302-1310

[7] Zhu J, Birgisson B and Kringos N 2014 Polymer modification of bitumen: Advances and challenges, *European Polymer Journal* **54** 18-38

[8] JKR/SPJ/2008 2008 *Standard Specification for Road Works* (Kuala Lumpur: Public Works Department)

**Acknowledgments**

The authors expressed utmost gratitude to FRGS Research Grant: FRGS/1/2015/TK08/UITM/02/3 from Ministry of Higher Education, Malaysia for financial support which enables this paper to be written.

IOP Conf. Series: Materials Science and Engineering **271** (2017) 012009 doi:10.1088/1757-899X/271/1/012009

# The optimum content of rubber ash in concrete: flexural strength

**M S Senin[1], S Shahidan[1], S M Shamsuddin[1], S F A Ariffin[1], N H Othman[1], R Rahman[2], F S Khalid[1] and F M Nazri[3]**

[1] , Jamilus Research Center, Faculty of Civil and Environmental Engineering, Universiti Tun Hussein Onn Malaysia, Johor Bahru Batu Pahat, Johor, Malaysia
[2]Smart Driving Research Center, Faculty of Civil and Environmental Engineering, Universiti Tun Hussein Onn Malaysia, Batu Pahat, Johor, Malaysia
[3] School of Civil Engineering, Engineering Campus, Universiti Sains Malaysia, 14300 Nibong Tebal, Penang, Malaysia
Corresponding author: shahironshahidan@gmail.com

**Abstract.** Discarded scrap tyres have become one of the major environmental problems nowadays. Several studies have been carried out to reuse waste tires as an additive or sand replacement in concrete with appropriate percentages of tire rubber, called as rubberized concrete to solve this problem. The main objectives of this study are to investigate the flexural strength performance of concrete when adding the rubber ash and also to analyse the optimum content of rubber ash in concrete prisms. The performance total of 30 number of concrete prisms in size of 100mm x 100mm x 500 mm were investigated, by partially replacement of rubber ash with percentage of 0%, 3%, 5%, 7% and 9% from the volume of the sand. The flexural strength is increased when percentage of rubber ash is added 3% from control concrete prism, RA 0 for both concrete prism age, 7 days and 28 days with value 1.21% and 0.976% respectively. However, for RA 5, RA 7 and RA 9, the flexural strength was decreased compared to the control for both age, 7 days and 28 days. In conclusion, 3% is the optimum content of rubber ash in concrete prism for both concrete age

## 1. Introduction

The accumulation of waste tyres is one of major environmental problems worldwide including Malaysia. Abdullah [1] stated the fact that one scrap tyre per person is generated constantly from all cars in the world every year. As scrap tyres are non-degradable materials, they can cause serious problems to the environment and public health. For example, they can become breeding sites for mosquitoes and rodents as the water becomes stagnant inside the tyres. It also causes diseases, accidental fires and takes up landfill space. The situation can worsen if it is illegally dumped in cities, disposed in lakes, streams, forest and not in the right places such as landfills. Other than that, there are also some people who use the easiest way to get rid of waste rubber tyres through open burning and using them as a source of fuel. These processes can lead to very serious health problems [2-9].

One way to solve the increasing amount of waste tyres is by incorporating waste rubber in concrete mixtures to produce rubberised concrete. Rubber can replace fine aggregates in the concrete mixture. This method can also create landfill avoidance [10-14]. A lot of previous research have been done regarding on rubberized concrete. Koreňová stated that concrete which contain rubber show excellent flexibility, ductility and energy absorbency as compared with controlled concrete [11]. Other researchers were used ground rubber powder and the crushed tyre chips particles range in size from about 15 to 4 mm as a replacement for coarse aggregate in normal concrete in various ratios. These group research are tested on the effect of rubber type and rubber content on strength and also modulus of elasticity [12,15-18]. Bakri worked on rubberized concrete by using waste tyre rubber as replacement of coarse aggregate to produce early age concrete and founded the decrease in compressive strength [8]. The compressive strength and density of concrete was affected on the amount of rubber added in the concrete [19]. Investigations on waste rubber in concrete applications are still in the process for better used. The previous researchers are mostly focusing in additive material and coarse aggregate replacement. Due to that matter, there need more detail investigation on the fundamental usage of waste rubber in concrete. Therefore, this study will be mainly emphasized on the use of rubber ash as partially replacement of fine aggregate concrete mixtures. The objectives for this study are to investigate the flexural strength performance of concrete when adding the rubber ash and also to analyse the optimum content of rubber ash in concrete prisms in term of flexural strength.

## 2. Rubber ash

Pyrolysis is a word basically based on the Greek words "pyr" which means fire and "lysis" which means separation. This means that pyrolysis is a process of decomposing material and usually the process operates at higher temperatures above 430 °C (800 °F). This process is done in the absence of oxygen otherwise the organic material is burned. Pyrolysis is also defined as one way of reprocessing scrap tyres through thermal degradation in the absence of oxygen. The products are fuel gas, oils, and a solid residue (char) which contains appreciable quantities of mineral matter and low-grade carbon black [20]. Products of tyre pyrolysis experiments are gases, liquid oils and solid carbon residues. The typical weight percentage of tyre pyrolysis are 33-38 wt. % char, 38-55 wt. % oil, and 10-30 wt. % gas which are affected by conditions such as temperature and heating rate. Gases produced from tyre pyrolysis are mainly hydrogen, carbon dioxide, carbon monoxide, methane, ethane and butadiene, with lower concentrations of propane, butane and other hydrocarbon gases. If water is present during the process, the production of hydrogen and carbon monoxide can increase from the occurrence of carbon gasification by steam. The yield of oil from tyre pyrolysis is high (50% of initial tyre rubber), reflecting the potential of tyre rubber as a substitute for fossil fuel and chemical feedstock. It is a potential source of energy and chemicals and can be used directly as fuel or added to petroleum refinery feedstock.

Furthermore, tyre pyrolysis can also be performed in an inert environment which can produce 33-38 wt. % of carbon residue and can become a marketable product if the properties are similar to those of manufactured carbons. Chars, 15 wt. % of ash, with the majority of this ash being zinc oxide are parts of the products from tyre pyrolysis. Factors that increase the char yield include a decrease in pyrolysis temperature and an increase in heating rate. Pyrolysis should also be performed under mildly oxidizing conditions, especially in the case of granulated tyres [18-20].

## 3. Laboratory work

### *3.1. Material preparations*

*3.1.1 Rubber ash.* Rubber ash was used as partial sand replacement by using a different percentage of rubber which is 0%, 3%, 5% 7%, 9% from the volume of the sand. Figure 1 shows the rubber ash that is used in this study. The rubber ash has been obtained from rubber factory process at Gopeng, Perak.

**Figure 1.** Rubber ash.

*3.1.2 Cement.* Ordinary Portland Cement (OPC: TYPE I) was used in this study because it has medium rate of hardening and suitable for most type of concrete work and easily available in market and it follow Standards ASTM C150 [21]

*3.1.3 Coarse and fine aggregate.* Crushed stone with maximum size of 20mm as coarse aggregate and natural sand with maximum size of 4.75 mm as fine aggregate has been used in this study.

*3.2. Specimen preparations*
There were 30 numbers of specimens (Prims size: 100mm x 100mm x 500mm) had been prepared in this study by using concrete mix design grade 30.

**4. Results and discussions**
This part discussed the result and analysis that was conducted in the laboratory. Flexural strength test was conducted using Flexural Testing Machine based on standard ASTM C1161-02C [22] at Jamilus Research Centre Laboratory, UTHM.

*4.1 Flexural strength*
Figure 2 illustrates the flexural strength results of different types of concrete prisms containing different percentages of rubber ash. RA 0, RA 3, RA 5, RA 7 and RA 9 underwent a curing process for 7 days. The maximum load was achieved by RA 3 with a value of 20.126kN. The second highest reading was achieved by the control concrete prism, RA 0, followed by RA 5, RA 7 and RA 9 which achieved readings of 19.886kN, 18.566kN, 18.406kN and 17.325kN respectively. When analysing the bar chart from figure 2, the difference in percentage for flexural strength can be determined by calculating the percentage ratio compared to the value achieved by the control concrete prism, RA 0. For RA 3, the flexural strength increased by 1.21% compared to RA 0. The flexural strength of RA 5 decreased by 6.64% compared to RA 0. Furthermore, the flexural strength of RA 7 decreased by 7.74% compared to RA 0. Lastly, RA 9 showed the highest percentage in terms of the reduction of flexural strength when compared to RA 0, which is 12.87%.

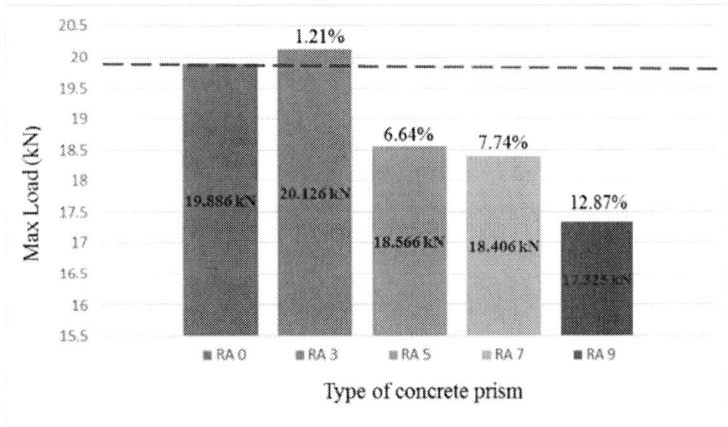


**Figure 2.** Flexural strength result for 7 days.

Figure 3 shows the maximum deflection of concrete prism for all types of concrete with different percentages of rubber ash. The highest value was 5.261mm for RA 3 followed by concrete prisms RA 0, RA 5, RA 7 and RA 9 with maximum deflection values of 4.877mm, 3.635mm, 0.648mm and 0.5099mm respectively. The difference in percentage in terms of maximum deflection for each concrete prism was compared with the control concrete prism, RA 0. RA 3 showed an increase of about 7.87% while RA 5, RA 7 and RA 9 showed a decrease in value of 25.5%, 86.7% and 89.54% respectively.

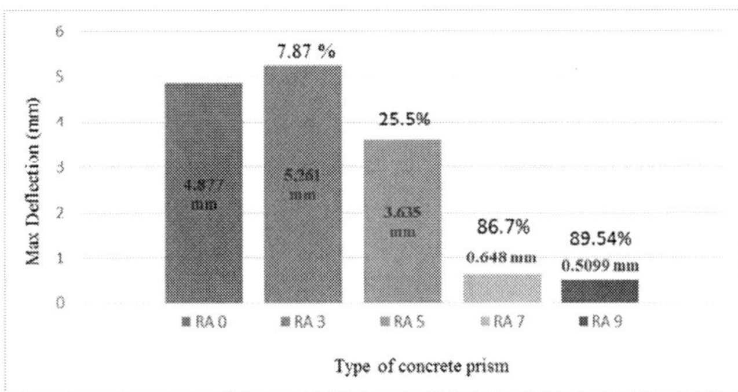


**Figure 3.** Deflection result for 7 days.

From both graphs in figure 2 and figure 3, RA 3 was found to have the optimum content for concrete prisms because it showed the highest values for both maximum load and deflection for the flexural test. For RA 5, RA 7, and RA 9, the flexural strength decreased due to increase in rubber ash content in the concrete. This is because of low bonding between the rubber and the particles in the concrete. Besides that, rubber also has lower bonding abilities compared to fine aggregates. The graph shows that flexural strength decreases when particles of rubber ash are added into the concrete prism. This is similar to a study by Grdic′ et al (2014) [23] where there was a reduction of about 14.6%, 30.4% and 51% with the inclusion of 10%, 20% and 30% of rubber sand respectively for the flexural strength test. It showed a reduction in flexural strength increased with increasing rubber sand content. This is due to the increase in voids and weaker bonding as the rubber content increases

Figure 4 shows the maximum load of all concrete prisms which underwent 28 days of the curing process. RA 3 shows highest value which was 26.168kN, followed by RA 0, RA 5, RA 7 and RA 9 valued at 21.527kN, 21.647kN, 20.606kN and 18.125kN respectively. The different in percentage for the maximum load for each concrete prism was compared with the control concrete prism RA 0. RA 3 showed an increased value in flexural strength of about 0.976% while RA5, RA7 and RA9 showed a decrease in flexural strength of about 0.006%, 0.043% and 0.158% respectively when compared with RA 0.

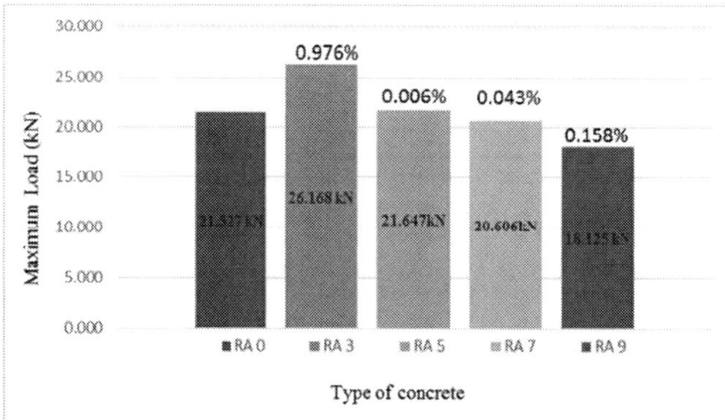


**Figure 4.** Flexural strength result for 28 days.

Figure 5 shows the maximum deflection of concrete prisms after a curing period of 28 days during the flexural test. This graph also showed that RA 3 had the highest value among the rest which is 3.923mm. This is followed by RA 0, RA 5, RA 7 and RA 9 with values of 1.842mm, 1.584mm, 0.930mm and 0.780 mm respectively. When compared with the control concrete prism RA 0, RA 3 showed an increase in percentage of about 1.13%. While the others, RA 5, RA 7 and RA 9 showed a decrease in terms of percentage difference of about 0.14%, 0.5% and 0.58%.

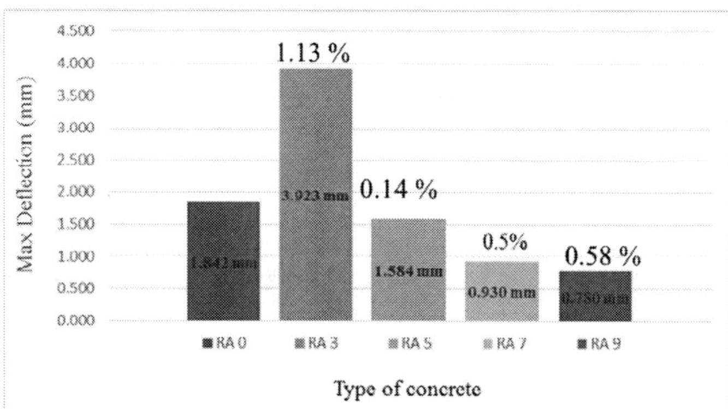


**Figure 5.** Deflection result for 28 days.

Overall, from both graphs in figure 4 and figure 5, increasing values were observed for concrete prisms cured for 28 days compared to those cured for 7 days. This is because the concrete prisms gained 99% in strength in 28 days compared to only 65% in strength for those cured for 7 days. Flexural tensile strength increases when the age of the concrete increases [18]. RA 3 has highest values for both graphs which means that RA 3 is the optimum content for concrete prisms. This trend is similar with a study by Parveen *et al.* [24] where the result of the flexural strength test for concrete cured for 28 days is higher than the ones cured for 7 days, with the inclusion of 0%, 5%, 10%, 15%, 20%, and 25% of rubber as a replacement for sand in concrete.

## 5. Crack pattern of concrete prism

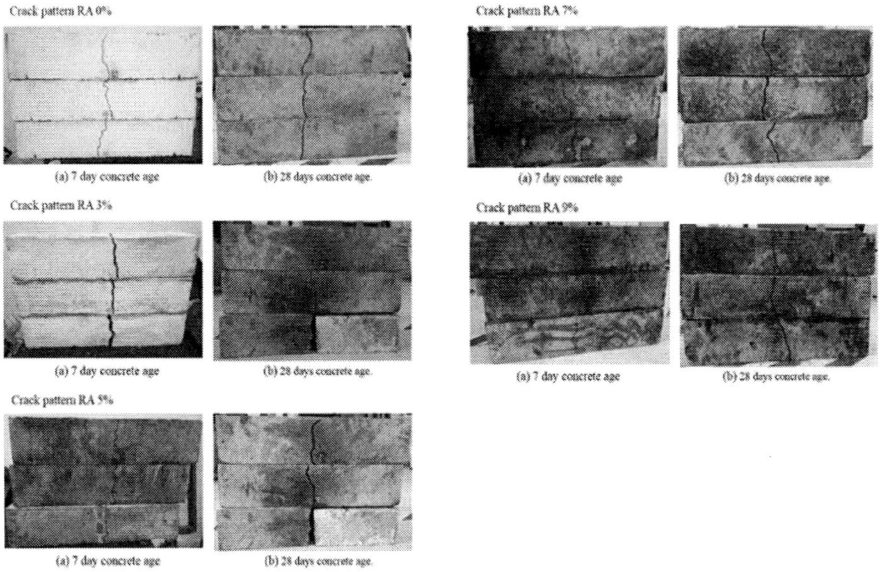


**Figure 6.** Crack pattern for each concrete.

Starting from RA 0 to RA 9, the concrete prisms showed the same crack pattern in the middle. The crack happened due to the brittle nature of the concrete. This is because all of these concrete prisms did not provide reinforcement, causing the prism to have no stiffness. Therefore, the concrete prism cannot bear the increasing load capacity. The crack pattern of the concrete prisms only occurred in the middle due to the increasing bending stress in the prism during the flexural test. When compression failure happened, maximum deflection took place and all the concrete prisms became brittle and started to crack. There were no shear cracks at the support point because the concrete prisms are not resistant against shear forces [20,25].

## 6. Conclusions

RA 3 showed the highest flexural strength, 1.21% and 0.976% increased compared to the normal concrete for concrete aged 7 days and 28 days respectively. RA 3 was found to have the optimum content and the best strength for both concrete aged. However the used of rubber ash more than 3% decreased the flexural strength of the concrete. This concrete could be improved by adding silica fume or fly ash.

## 7. References

[1] Abdullah S R, Abidin W R W Z and Shahidan S 2016 Strength of Concrete Containing Rubber Particle As Partial Cement Replacement *MATEC Web of Conf.* vol 47 (Paris: EDP Sciences) pp 2–5

[2] Ali N, Zainal N A, Burhanudin M K, Abdul Samad A A, Mohamad N, Shahidan S, Abdullah S R 2016 Physical and Mechanical Properties of Compressed Earth Brick ( CEB ) Containing Sugarcane Bagasse Ash *MATEC Web of Conf.* vol 47 (Paris: EDP Sciences) pp 1–7

[3] Zheng L, Huo X S and Yuan Y 2008 Strength, Modulus of Elasticity, and Brittleness. Index of Rubberised concrete, *J. Mater. Civil Eng.* **20**(11) 692-699

[4] Leman A S, Shahidan S, Senin M S and Hannan N I R R 2016 A Preliminary Study On Chemical And Physical Properties Of Coconut Shell Powder As A Filler In Concrete *IOP*

*Conference Series: Materials Science and Engineering* vol 160 (London: IOP Publishing) 12059

[5] Senin M S, Shahidan S, Leman A S and Hannan N I R R 2016 Properties of Cement Mortar Containing Rubber Ash as Sand Replacement *IOP Conference Series: Materials Science and Engineering* vol 160 (London: IOP Publishing) 12055

[6] Ramzi N I R, Shahidan S, Maarof M Z and Ali N 2016 Physical and Chemical Properties of Coal Bottom Ash (CBA) from Tanjung Bin Power Plant *IOP Conference Series: Materials Science and Engineering* vol 160 (London: IOP Publishing) 12056

[7] Senin M S, Shahidan S, Leman A S and Hannan N I R R 2016 Analysis of Physical Properties and Mineralogical of Pyrolysis Tires Rubber Ash Compared Natural Sand in Concrete material *IOP Conference Series: Materials Science and Engineering* vol 160 (London: IOP Publishing) 12053

[8] Bakri A M M A, Fadli S A S N, Bakar M D A and Leong K W 2007 Comparison of rubber as aggregate and rubber as filler in concrete *1st Int. Conf. on Sustainable Materials (Penang)* (Perlis: Universiti Malaysia Perlis) pp 1-9

[9] Wójtowicz M A, Bassilakis R and Serio M A 2004 *Carbon Black Derived From Waste Tire Pyrolysis Oil* Retrieved on April 20, 2017 fom https://www.researchgate.net/profile/Michael_Serio/publication/238102979_CARBON_BLACK_DERIVED_FROM_WASTE_TIRE_PYROLYSIS_OIL/links/00b7d-52ec1d7b067f3000000.pdf?inViewer=0&pdfJsDownload=0&origin=publication_detail.

[10] Serio M A, Teng H, Bassilakis R, Marrison P W, Solomon P R 1993 Reprocessing Of Used Tires into Activated Carbon and Other Products, *Ind. Eng. Chem. Res.* **1995** *34* (9), 3102-3111

[11] Juma M, Koreňová Z, Markoš J, Annus J and Jelemenský L 2006 Pyrolysis and Combustion of Scrap Tire, *Petroleum & Coal* **48**(1) 15-26

[12] Khalid F S, Herman H S and Azmi N B 2017 Properties of Sugarcane Fiber on the Strength of the Normal and Lightweight Concrete ) *MATEC Web of Conf.* vol 103 (Paris: EDP Sciences) 01021

[13] Khalid F S, Herman S H, Azmi N B and Juki M I 2017 Sand Cement Brick Containing Recycled Concrete Aggregate as Fine-Aggregate Replacement *MATEC Web of Conf.* vol 103 (Paris: EDP Sciences) 01016

[14] Shahidan S, Senin M S, Kadir A A, Yee L H and Ali N 2017 Properties of Concrete Mixes with Carwash Wastewater *MATEC Web of Conf.* vol 87 (Paris: EDP Sciences) 1018

[15] Shahidan S, Leman A S, Senin M S and Hannan N I R R 2017 Suitability of Coconut Shell Concrete for Precast Cool Wall Panel-A Review *MATEC Web of Conf.* vol 87 (Paris: EDP Sciences) 1005

[16] Leman A S, Shahidan S, Yusuf M Y, Mohd Zuki M M and Misnon N A 2017 Workability and Compressive Strength for Concrete With Coconut Shell Aggregate 3 Behaviour Of Concrete With Coconut *MATEC Web of Conf.* vol 87 (Paris: EDP Sciences) 1017

[17] Jing L, Zhou T, Du Q and Wu H 2015 Effects of rubber particles on mechanical properties of lightweight aggregate concrete, *Constr. Build. Mater.* **91** 145–14

[18] Antil Y, Verma V and Singh B 2014 Rubberized concrete with crumb rubber, *Int. J. Sci. Res. (IJSR)* **3**(5) 1481–1483

[19] Shahidan S, Ismail I, Zulkarnaian M S S and Abd Rahman N 2017 Influence of Asphalt Dust Waste Material in Mix Design for Self-Compacting Concrete, *Key Eng. Mat.* **730** 473-478

[20] Shahidan S, Pullin R, Bunnori N M and Zuki S S M 2017 Active crack evaluation in concrete beams using statistical analysis of acoustic emission data, *Insight - Non-Destructive Test. Cond. Monit.* **59**(1) 24–31

[21] ASTM International 2017 *ASTM C150 / C150M - 17 Standard Specification for Portland Cement* (West Conshohocken: ASTM International)

[22] ASTM International 2017 *ASTM C1161 – 02C Standard Test Method for Flexural Strength of Advanced Ceramics at Ambient Temperature* (West Conshohocken: ASTM International)
[23] Grdic´Z, Topličić-Curčić G, Ristic´N, Grdic´D and Mitkovic P 2014 Hydro-abrasive resistance and mechanical properties of rubberized concrete, *GRAĐEINAR* **66** (1) 11–20
[24] Parveen, DassS and Sharma A 2013 Rubberized concrete: needs of good environment (overview), *Int. J. Emerg. Technol. Adv. Eng.* **3**(3) 192–196
[25] McGraw-Hill *Encyclopedia of Science & Technology 10th Edition* (New York: McGraw-Hill) pp 557-558

**Acknowledgements**
The author would like to thank to Jamilus Research Center, Universiti Tun Hussien Onn Malaysia and Grant VOT U523 and VOT 572 (GPPS).

# Comparative study of modified bitumen binder properties collected from mixing plant and quarry.

**M Mustafa Kamal[1], R Abu Bakar[1] and K A Hadithon[1]**

[1]Malaysian Rubber Board
Corresponding author: mazlina@lgm.gov.my

**Abstract.** Quality control and assurance are essential in pavement construction. In general, the properties of bitumen change as it ages in bulk storage, transport, and storage on site. The minimization of bituminous hardening during storing, transportation and mixing depends on careful control of binder temperature. Hence therefore, bitumen should always be stored and handled at the lowest temperature possible, consistent with efficient use. The objective of the work is to monitor the quality of bitumen samples collected from mixing plant and quarry. Results showed that, samples modified bitumen which collected from quarry showed some adverse effects on rheological properties and physical properties after subjecting to high temperature storage within a period of time. The dynamic stiffness, elastic properties and other common binder properties were deteriorated too. The chemical changes that occurred during storage were analysed using Fourier transform infra-red spectroscopy (FTIR). Thus studies developed an understanding of bitumen ageing in storage.

## 1. Introduction

Bitumen may be rigid and friable at low temperature, liquid and fluid at high temperature and semi-rigid or rigid at medium temperature. Bitumen dynamic modulus due to temperature fluctuations from −30 °C to 50 °C change even 37 000 times [1]. When asphalt mixture is produced of too cool and, therefore, too viscous (consistent) bitumen, it hardly turns into a thin film, which should coat all mineral particles with solid layers of appropriate thickness and make a strong compacted structure of the mixture. When too hot or overheated mineral aggregates are used, bitumen may burn: it degrades chemically, volatilizes, its colloidal composition changes and rheological properties deteriorate.

Bitumen ageing is one of the main causes of the degradation of asphalt pavement. In bitumen ageing, two mechanisms are involved: chemical changes in binder (irreversible) and physical hardening (reversible). The most important bitumen ageing mechanisms are as follows: oxidation, volatilisation, steric or physical factors and exudative hardening. Ageing due to oxidation is the key factor if compared with other factors impacting on ageing [2, 3]. Oxidation rate mostly depends on temperature [4] duration and intensity of interaction with oxygen as well as the thickness of bitumen film, which coats aggregate particles [5]. Heating duration and temperature have a significant impact on such rheological properties of bitumen as viscosity or complex modulus [6]. In general, a higher viscosity is contributed by modified bitumen. As therefore, the temperature is raised up during mixing and storage at the quarry site unlike normal procedure. However, the concern with elevated temperatures is that they may further degrade the asphalt binder to a level beyond that caused by normal production temperatures; thus increasing the degree of short term aging to a level that may adversely affect the performance of the asphalt pavement.

Several research studies have shown that the volatilization and oxidation are the biggest contributors to the hardening (aging) process of asphaltic binders. The combined effect of elevated temperature and longer exposure time may cause excessive short term aging and adversely affect the performance of the pavement. This research study aimed on quantifying how much further aging occurs beyond normal mixing temperatures and exposure times of asphalt mixes. The quantification is achieved by measuring viscosities of aged binders as well as measuring the dynamic modulus, softening point, penetration and other properties of bitumen at production and quarry plant. The properties of the binder from the tanker were also tested for a comparison. The results were compared to establish correlations between the properties of the aged samples and their laboratory performance, upon which conclusions was drawn.

## 2. Experimental

Samples of modified bitumen supplied were collected from various locations, production and quarry plant. In general sample from tanker remained almost the same with the one collected from the production plant. As therefore samples from production plant and quarry will only be discussed in this work. The temperature recorded at the production and quarry plant was 160°C and 200°C respectively. Conventional bitumen tests such as softening point (ASTM D36) and penetration grade (ASTM D5) are performed to characterize the bitumen. The rheological properties of modified bitumen at high and intermediate temperatures are measured by using Dynamic Shear Rheometer (DSR) (ASTM D7175) and the viscosity of the bitumen is evaluated by application of Brookfield rotational viscometer (ASTM D4402).

Nicolet 6700 FTIR spectrometer (Thermo Scientific) was used to record the infrared spectra with 32 scans at a resolution of 4 $cm^{-1}$ in the range of 4000 – 600 $cm^{-1}$.

## 3. Results and discussion

Table 1 shows the viscosity, penetration and softening results of the rubberised bitumen tested in this study from those two locations. In general, the viscosity of bitumen binders at high temperatures is an important property as it is a good indicator of the binder's ability of pumping through bitumen plant. Results showed that, sample collected from quarry exhibited lesser viscosity and softening point with higher penetration point as compared to fresh binder from mixing plant, which indicating possible ageing of the asphalts.

**Table 1.** Bitumen properties collected from Mixing plant and Quarry.

| | Bitumen from Mixing Plant | Bitumen from Quarry |
|---|---|---|
| Viscosity at 135 °C (cp) | 3000 | 1500 |
| Penetration at 25°C (mm) | 58 | 80 |
| Softening Point °C | 60 | 50 |

The rheological properties of the binders with various temperatures are obtained from the Dynamic Shear Rheometer (DSR). Complex shear modulus (G*) is determined as a result of the DSR experiments by applying stress at different frequencies on bitumen samples. The relationship between the complex shear modulus and the temperature is shown in figure 1. Results showed that the complex modulus decreased with temperatures. However, the complex shear modulus of bitumen collected from mixing plant is greater than the one collected from quarry which indicates stiffer binder and greater resistance to deformation respectively.

Figure 2 illustrated changes in values of phase angles to different test temperatures. Lower phase angles value data of bitumen collected from mixing plant than bitumen from quarry. A smaller phase angle values obtained from mixing plant is an indicator greater binder elastic property and of lesser in binder permanent deformation occurred as a result of applied stress.

It is known that the temperature of the bitumen binders when G*/sin δ is equal to 1 kPa is defined as a criterion for the high temperature (good visco- elastic) performance of bitumen. It is also known that higher G*/sin δ values are found to correlate with higher rutting resistance. Changes in rutting parameters (G*/sin δ) to temperature decreases is shown in figure 3. It is clear seen that rutting parameters of bitumen collected from mixing plant is lesser than bitumen collected from quarry. These results show that bitumen from mixing plant has more considerable effects on rutting resistance than the one collected from quarry.

Considering the results obtained by all DSR tests it is seen that bitumen from mixing plant exhibits greater complex shear modulus (G*) values and rutting parameters (G/sinδ), and has lesser phase angle values compared to binder from quarry plant.

Decreased in complex shear modulus (G*) values and viscosity might be attributed to the arrangement /degradation of NR and bitumen into lower molecular weight fragments after long hot storage of the modified bitumen sample collected from quarry. At this elevated temperature there is an increase in phase angle indicating more viscous response. This is could due presence of degradation of unsaturated polymer as in the case of NR. The oxidation of the binder also could attribute to diminishing of physical and chemical crosslink or polymer network decomposed under the influenced of heat and oxygen and rearranged in smaller fragments. Therefore chain scission of the polymer could cause the binder properties to change. Collin [7] observed penetration decreased at a much great rate at high temperature. This behaviour has also been observed in SBR modified bitumen binders. Indeed during the mixing of the modifier with bitumen, there is some polymer breakdown due to thermal and mechanical action of the mixing process, but this is minimal (4-5%) [8].

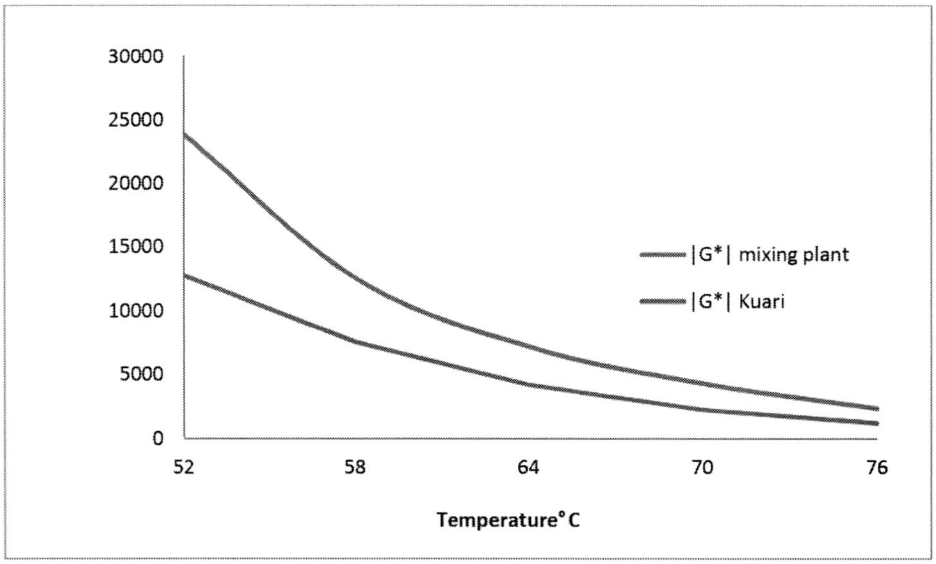


**Figure 1.** Rheological Behaviour of Modified Bitumen collected from Quarry and Mixing Plant.

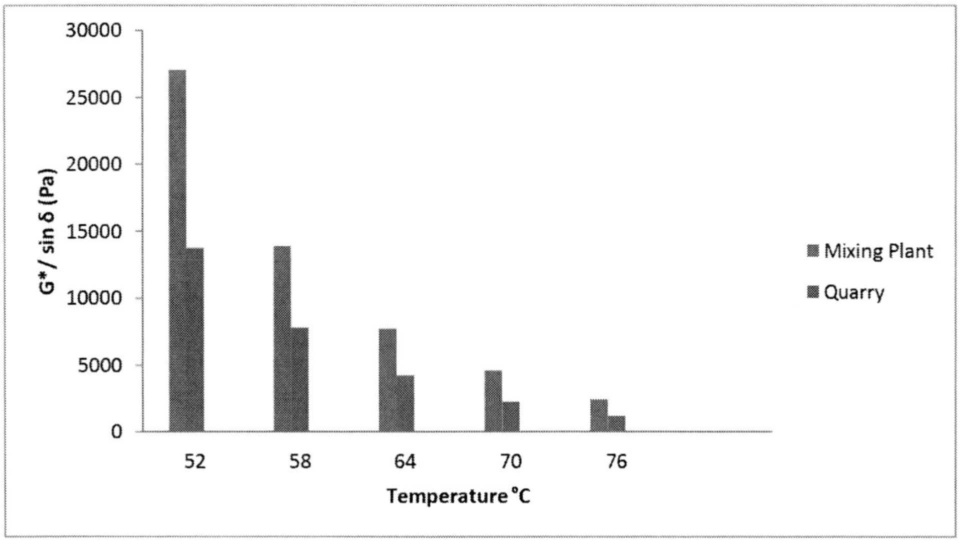


**Figure 2.** Rheological Behaviour of Modified Bitumen collected from Quarry and Mixing Plant.

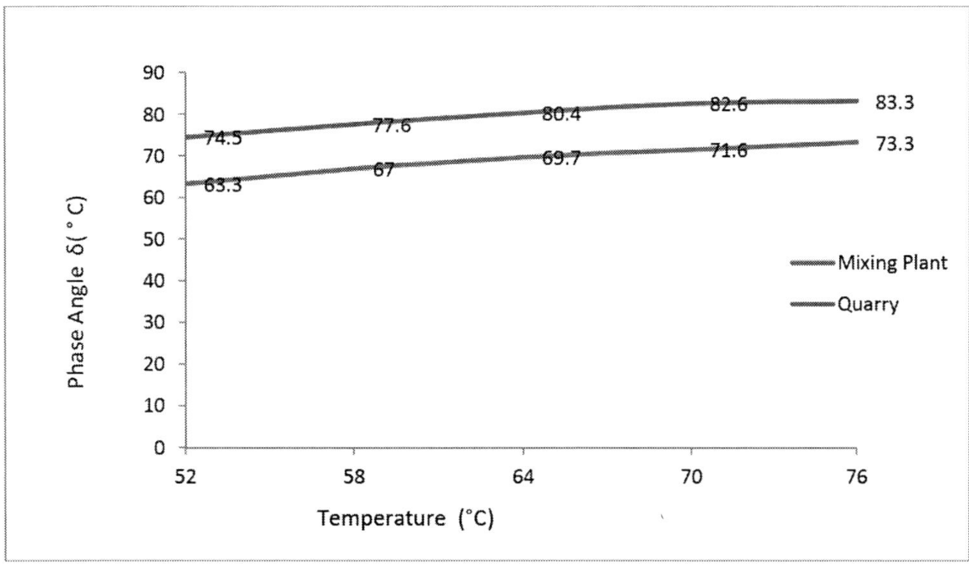


**Figure 3.** Rheological Behaviour of Modified Bitumen collected from Quarry and Mixing Plant.

FTIR spectra of pure bitumen (60/70 grade), samples modified bitumen from mixing plant and quarry are shown in figure 4. The spectra are similar to each other and the typical absorbance peaks are listed in Table 2. However, the new absorbance peak of 1109 $cm^{-1}$ attributed to the C-O-C is seen for sample collected from quarry. This could be due to some chemical changes in the binder as discussed earlier. Araujo [9] found that a molecular reorganisation and hydrogen/carbon ratio decreased as the bitumen was subjected to weathering chamber test after 150h. Hence, as ageing proceed, ether bonds can be formed and the content of carbonyl decreased as indicated in this work. Therefore further analysis is necessary in order to confirm the ageing mechanism.

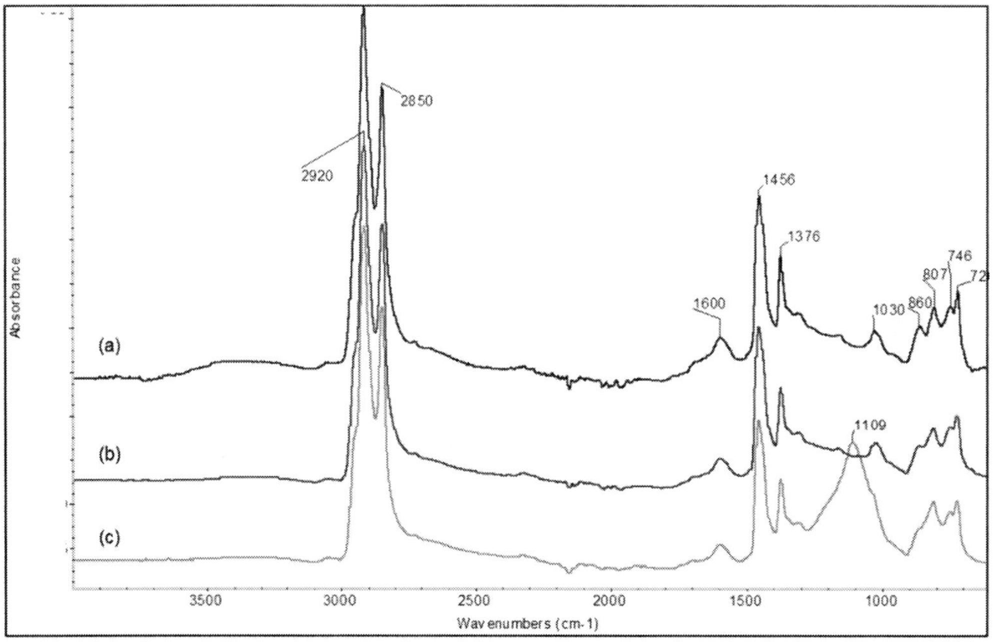


**Figure 4.** FTIR spectra of a) PG60/70 and modified bitumen collected from (b) mixing plant and (c) quarry.

**Table 2.** Absorbance peaks of bitumen [8].

| Wavenumbers ($cm^{-1}$) | Functional group |
|---|---|
| 2920, 2850 | C-H aliphatic stretching |
| 1600 | C=C aromatic stretching |
| 1456 | C-H of –($CH_2$)- bending |
| 1376 | C-H of $CH_3$ bending |
| 1030 | S=O sulfoxide stretching |
| 862, 807 | C=C of alkene stretching |
| 746, 720 | C-H or C-S bending |

## 4. Conclusion

Results showed adverse effects on modified bitumen properties collected from quarry. The properties such as penetration, viscosity and softening point changed drastically after subjecting to high temperature storage at quarry in contrast to modified bitumen collected from mixing plant. The IR analysis indicated that a new peak formation of ether peak at 1109 $cm^{-1}$ which highly possible due to oxidation of the binder.

IOP Conf. Series: Materials Science and Engineering **271** (2017) 012010 doi:10.1088/1757-899X/271/1/012010

## 5. References


[1] Jenks C W, Jenks C F, Harrigan, E T, Adcock M, Delaney E P and Freer H 2011 *A Manual for Design of Hot Mix Asphalt with Commentary* NCHRP Report 673 (Washington, D.C.: Transportation Research Board) pp 273

[2] Petersen J C and Glaser R 2011 Asphalt oxidation mechanisms and the role of oxidation products on age hardening revisited, *Road Mater. Pavement* **12** 795–819

[3] Morian N, Hajj E Y, Glover C J and Sebaaly P E 2011 Oxidative aging of asphalt binders in hot-mix asphalt mixtures, *Transport. Res. Rec.: J. Transport. Res. Board.* **2207** 107–116

[4] De Moraes M B, Pereira R B, Simão R A and Leite L F M 2010 High temperature AFM study of CAP 30/45 pen grade bitumen, *J. Microsc.* **239** 46–53

[5] Herrington Ph R 2012 Diffusion and reaction of oxygen in bitumen films, *Fuel* **94** 86–92

[6] Filippova A G, Kirillova L G, Okhotina N A, Dvoyashkin N K, Filippov A V, Vol'fson S I, Liakumovich A G and Samuilov Ya D 2000 Viscosity of polymer–bitumen binders, *Colloid J.* **62** 755–758

[7] Collins J H and Bouldin M G 1992 Long and short term stability of straight and polymer modified asphalts, *Rubber World* **205** 32 - 38

[8] Parvez M A, AL-Mehthel M, Al-Abdul Wahhab H I and Hussein I A 2014 Utilization of sulfur and crumb rubber in asphalt modification, *J. Appl. Polym. Sci.* **131** 40046

[9] Araujo M.FA, Lins V F C, Pasa V M D and Fonseca C G 2012 Infrared spectroscopy study of photodegradation of polymer modified asphalt binder, *J. Appl. Polym. Sci.* **125** 3275-3281

# Compressive and flexural strength of concrete containing palm oil biomass clinker and polypropylene fibres

**Wan Ibrahim M H[1*], Sajjad Ali Mangi[1], Burhanudin M K[1], Ridzuan M B[1], Jamaluddin N[1], Shahidan S[1], Wong YH[1], Faisal SK[1]., Fadzil M A[2], Ramadhansyah P J[3], Ayop S S[1] and Othman N H[1]**

[1]Jamilus Research Centre, Faculty of Civil & Environmental Engineering, Universiti Tun Hussein Onn Malaysia, 86400 Batu Pahat, Johor, Malaysia
[2]Faculty of Civil Engineering, Universiti Teknologi MARA, Shah Alam, Selangor, Malaysia
[3]Faculty of Civil Engineering, Universiti Teknologi Malaysia, Skudai, Johor, Malaysia
Corresponding author:haziman@uthm.edu.my

**Abstract.** This paper presents the effects of using palm oil biomass (POB) clinker with polypropylene (PP) fibres in concrete on its compressive and flexural strength performances. Due to infrastructural development works, the use of concrete in the construction industry has been increased. Simultaneously, it raises the demand natural sand, which causes depletion of natural resources. While considering the environmental and economic benefits, the utilization of industrial waste by-products in concrete will be the alternative solution of the problem. Among the waste products, one of such waste by-product is the palm oil biomass clinker, which is a waste product from burning processes of palm oil fibres. Therefore, it is important to utilize palm oil biomass clinker as partial replacement of fine aggregates in concrete. Considering the facts, an experimental study was conducted to find out the potential usage of palm oil fibres in concrete. In this study, total 48 number of specimens were cast to evaluate the compressive and flexural strength performances. Polypropylene fibre was added in concrete at the rate of 0.2%, 0.4% and 0.6%, and sand was replaced at a constant rate of 10% with palm oil biomass clinker. The flexural strength of concrete was noticed in the range of 2.25 MPa and 2.29 MPa, whereas, the higher value of flexural strength was recorded with 0.4% polypropylene fibre addition. Hence, these results show that the strength performances of concrete containing POB clinker could be improved with the addition of polypropylene fibre.

## 1. Introduction

Concrete is versatile and widely being used as construction material due to its desirable engineering properties. In the twenty-first century, it has emerged as the dominant construction material for infrastructure needs. However, it is strong in compression and weak in tension. Therefore, the innovative materials such as fibres and admixtures is being introduced for better performances of concrete [1, 2]. Considering the environmental sustainability, the waste materials need to be utilized in the concrete production. One of such waste is the palm oil waste, which is very abundantly available in Malaysia. While, comparing the globally palm oil production, the Malaysia is one of the largest palm oil producer in the world. It creates large amount of waste in the form of fibres, nutshells, kernels and empty fruit bunches. The wastes such as kernels, nutshell and fibres are used to generate steam by burning them in boilers to run a turbine to generate electricity. Palm oil biomass clinker is obtained when these wastes are incinerated. It was reported by Zarina [3] that palm oil mill industry in Malaysia has been producing solid waste in large quantities around 4 million tons per year. Now a days

Malaysia also considered as second largest producer of palm oil in the world, the production of 18.6 million metric tons has been reported in the year 2010 [3]. Ultimately, the wastes are disposed-off and do not serve any purpose. According to the particle size the palm oil biomass clinker are very similar to sand and having low specific gravity which makes it suitable to be utilized as sand replacement material. Therefore, the wastes that are produced can be utilized as construction materials to act as partial replacement of sand in concrete.

The term Fibre Reinforced Concrete (FRC) can be used to describe concrete with the addition of fibre. Addition of polypropylene fibre which is one type of synthetic fibres is being used due to its high ductility. According to Saeed [4], lower dosage rate of polypropylene fibres from 0.18% to 0.40% increases the compressive strength of concrete by about 5%. While the addition of 0.20% Polypropylene fibres can rise the flexural strength about 80% [4]. Hence, the aim of this study is to investigate the effect of polypropylene fibre on the flexural strength of concrete containing palm oil biomass clinker as a sand replacement.

## 2. Materials and Method

### *2.1. Fine and coarse aggregate*

Fine and coarse aggregates conforming to the standards BS EN 12620 [5] was used for this study. The coarse aggregates were sieved passing from size 20 mm and retained in size 10 mm and passing in 5 mm for fine aggregate. The aggregates that are sieved and collected were kept in safe place.

### *2.2. Palm Oil Biomass (POB) Clinker*

Palm oil biomass clinker used in this study was a waste by-product from the steam generation process of Tanjung Langsat Biomass Steam Plant. Palm oil biomass (POB) clinker was first being washed to wash off the impurities of palm oil biomass clinker and dried before sieving to pass the size of 5mm. The percentage of palm oil biomass (POB) clinker as partial replacement of sand in concrete used are 10%, 20% and 30%. The figure 1 shows that the particle size distribution of fine aggregates and palm oil biomass clinker. The gradation of palm oil biomass clinker appeared to be similar to sand.

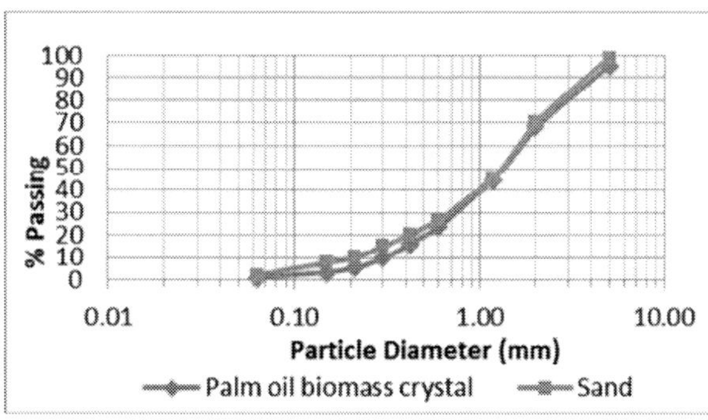


**Figure 1:** Particle size distribution of sand and palm oil biomass clinker

### *2.3. Specific Gravity*

Specific gravity is a dimensionless parameter that relates the density of the soil particles to the density of water. It was carried out to determine the weight-volume relationships. Table 1, shows the specific gravity of palm oil biomass (POB) clinker and sand. It was experimentally perceived that the specific gravity of palm oil biomass (POB) clinker is about 21% lesser than natural sand, which indicate that palm oil biomass (POB) clinker is porous material.

**Table 1:** Specific gravity of sand and palm oil biomass clinker

| | Sand | | | Palm oil biomass (POB) clinker | | |
|---|---|---|---|---|---|---|
| Specific Gravity | 2.54 | 2.66 | 2.68 | 2.07 | 1.98 | 2.17 |
| Average | | 2.63 | | | 2.07 | |

*2.4. Polypropylene (PP) Fibres*

The type of synthetic fibre used is polypropylene macro synthetic fibre as shown in figure 2. Different proportions of fibres such as 0%, 0.20%, 0.40% and 0.60% were used in this research with replacement of optimum percentage of palm oil biomass clinker to replace sand. Table 2 shows the detailed properties of polypropylene fibres.

**Table 2:** Properties of polypropylene fibre

| Description | Result |
|---|---|
| Length | 55mm |
| Diameter | 0.85mm |
| Aspect ratio | 65-70 |
| Specific Gravity | $0.90 kg/dm^3$ |
| Tensile strength | $425 N/mm^2$ |
| Young Modulus | $7\ 288 N/mm^2$ |

From the above mentioned result, the properties of polypropylene macro synthetic fibre considering the length of 55 mm, diameter of 0.85mm with aspect ratio of 65-70 and tensile strength of 425 MPa.

**Figure 2:** Macro synthetic polypropylene fibres

*2.5. Water and Cement*

There are no specific requirements according to BS EN 1008 [6] for the water used to produce concrete as long as it is potable water, therefore water tap was used as long as the water is not severely contaminated. In this study, the Ordinary Portland Cement complying with standard of BS EN 197-1 [7] was used and water to cement ration were used as 0.50 for all the mix proportion. Whereas, Table 3 shows the chemical composition of Ordinary Portland Cement.

**Table 3:** Chemical composition of ordinary Portland cement [2]

| Material | Chemical Composition | Percentage Limit |
|---|---|---|
| Lime | $CaO$ | 60-67 |
| Silica | $SiO_2$ | 17-25 |
| Alumina | $Al_2O_3$ | 3-8 |
| Iron Oxide | $Fe_2O_3$ | 0.5-6 |
| Magnesium | $MgO$ | 0.1-4 |
| Alkalis | $Na_2O$, $K_2O$ | 0.2-1.3 |
| Sulphur Trioxide | $SO_3$ | 1-3 |

*2.6. Mix Proportions*

The concrete mix design was based on Design of Experimental (DoE) British standard method. There were four different percentage of polypropylene fibre with labels as PP0, PP0.20, PP0.40 and PP0.60 for 0%, 0.20%, 0.40% and 0.60% of replacement percentage respectively were prepared. The amount of palm oil bottom clinker was chosen as optimum percentage of 10% and water-cement ratio was kept contact as 0.50 for all the mixes. Table 4 shows the mix proportions for one batch with six concrete cube of size of 100 mm x 100 mm x 100 mm and another six concrete prisms of size 100 mm x 100 mm x 500 mm.

**Table 4:** Mix proportions of concrete for one batch

| Mix notation | Cement ($kg/m^3$) | Sand ($kg/m^3$) | Palm oil biomass clinker ($kg/m^3$) | Water ($kg/m^3$) | PP Fibre (kg) | Water-cement ratio |
|---|---|---|---|---|---|---|
| PP0 | 18.22 | 34.13 | 3.79 | 9.11 | 0 | 0.5 |
| PP0.20 | 18.22 | 34.13 | 3.79 | 9.11 | 0.07 | 0.5 |
| PP0.40 | 18.22 | 34.13 | 3.79 | 9.11 | 0.14 | 0.5 |
| PP0.60 | 18.22 | 34.13 | 3.79 | 9.11 | 0.21 | 0.5 |

The samples preparation started with the 24 concrete cubes samples of size 100 mm x 100 mm x 100 mm were casted to find the optimum percentage of palm oil biomass clinker as partial replacement of sand in concrete. The overall 48 specimens were prepared, 24 for compressive strength test and 24 for flexural strength test. After getting the optimum percentage of palm oil biomass clinker, the other specimen was prepared for the addition of polypropylene fibre for the curing period of 7 and 28 days. The experimental work was divided into sets, one is the fresh concrete and second is the hardened concrete. For fresh mix concrete the workability was evaluated with the help of slump test fallowing the BS EN 12350-2 [8]. While, for the determination of compressive and Flexural strength test the standard BS EN12390-3 [9] and BS EN 12390-5 [10] fallowed respectively.

## 3. Results and Discussions

*3.1. Workability of concrete*

It was observed that the slump value tends to decreases from 20mm to 10mm with the addition of polypropylene fibre. The figure 3, represents the slump value of concrete, which indicates the slump values were decreases with the increase of polypropylene fibre. This phenomena expresses that the workability of concrete decreases with addition of polypropylene fibre. It was also reported by Payam [11] that palm oil biomass clinker while used as fine aggregates in concrete can absorb more water with even three times to the normal weight of sand. The addition of palm oil biomass clinker decreases the workability of concrete due to the porosity of palm oil biomass clinker grain which absorbs more water.

The decreasing values of slump can be associated with the formation of network structure in concrete due to high volume of polypropylene fibres as stated by Machine [12], the high content and large surface area of fibres also causes more cement paste to wrap around the fibres thus increasing the viscosity and reducing the slump value. Whereas, the non-homogenous distribution of fibre also leads to decrease in the workability of concrete. In addition, in research conducted by Mohod [13], it is shown that with usage of polypropylene fibres in concrete decreases the workability of concrete.

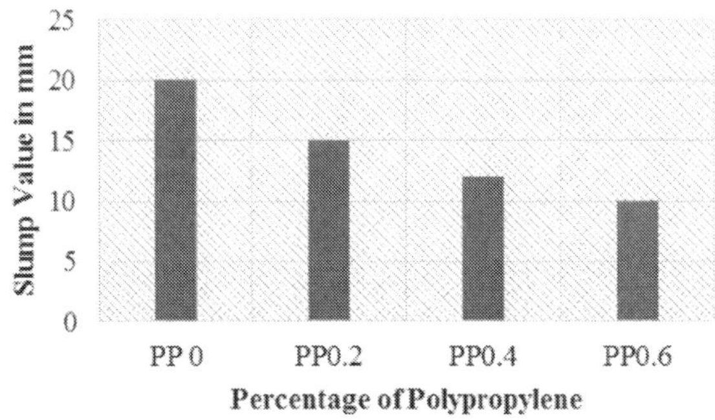


**Figure 3:** Slump value of concrete

*3.2. Compressive Strength*

Generally, it was perceived that the compressive strength of concrete tends to decrease when sand is replaced with palm oil biomass (POB) clinker. The figure 4, represents the compressive strength values at 7 and 28 days. Whereas, at the age of 28 days, the compressive strength of concrete shows decreasing as compared to the control samples of 10% replacement of palm oil biomass clinker is 7.44%, for 20% replacement is 9.94% and for 30% replacement is 37.43%. It was observed that there is no any considerable difference in the strength for the 10% and 20% replacement percentage, which is 27MPa and 26.27MPa respectively with difference of 0.73MPa. On the basis of performances, it is concluded that the percentage of sand replacement increase the concrete strength decrease with significant reduction especially for 30% palm oil biomass clinker replacement for sand.

This is mostly due to the high water absorption and low specific gravity of palm oil biomass clinker as compared to sand. The porosity nature of palm oil biomass clinker causes high water absorption, which eventually reduces the strength of concrete.

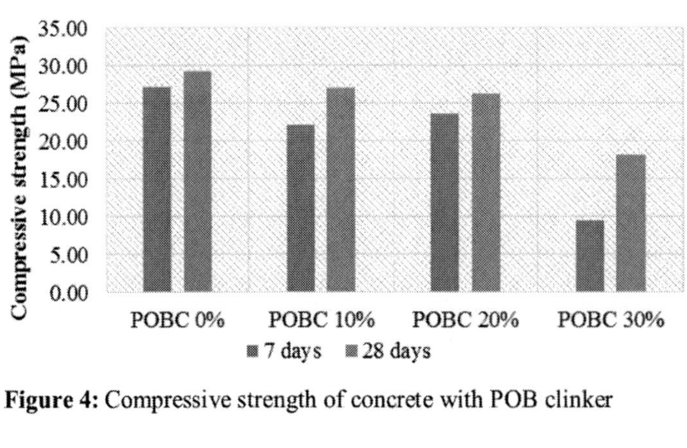


**Figure 4:** Compressive strength of concrete with POB clinker

The compressive strength shows increase from 7 days curing age to 28 days curing ages with 0.6% polypropylene fibres addition the lowest compressive strength as compared to other percentage of fibres. As observed from the figure 5, the compressive strength of concrete with 0.4% polypropylene fibres addition shows highest strength of 27.71MPa as compared to 0% polypropylene fibres addition of 27.23MPa. However, as compared to the control the compressive strength of 0.4% addition of polypropylene fibres is still lower.

According to Saeed [4], the addition of polypropylene (PP) fibre at low values such as 0.18% to 0.40% the compressive strength increases but when the values are high like 0.55 % to 0.60%, compressive strength decreases by 3 to 5%. Higher dosage rate of polypropylene (PP) fibre can also decrease the strength of concrete due to the high volumes of polypropylene fibres interfering with the cohesiveness of the concrete matrix.

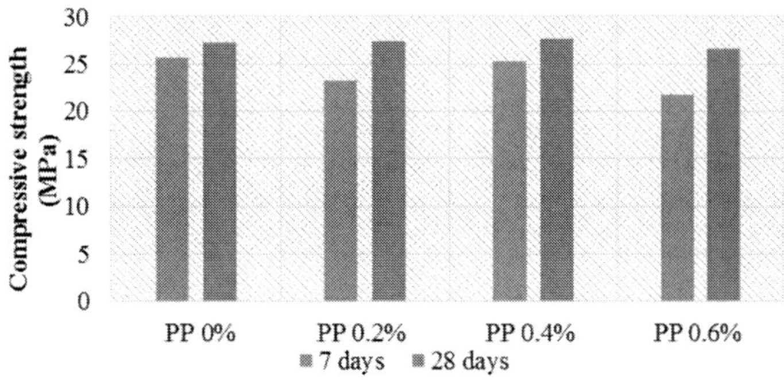


**Figure 5:** Compressive strength of concrete with POB clinker and PP fibres

### *3.3. Flexural Strength*

The flexural strength of concrete or control without any replacement of sand and polypropylene fibres is the highest as compared to the concrete after 10% replacement of palm oil biomass clinker and addition of polypropylene fibre with different percentages of 0%, 0.2%, 0.4% and 0.6%. While the value of flexural strength of 10% replacement of palm oil biomass clinker without addition of polypropylene fibre is 21.50% which shows reduction as compared to the control samples. However, when the addition of 0.2% of polypropylene fibre concrete with palm oil biomass clinker replacement were tested, the flexural strength value shows an increase to 2.265MPa and continue to increase to 2.297MPa for 0.4% of addition of polypropylene fibre. However, when the polypropylene fibre addition reached to 0.6%, the flexural strength started to decrease to 2.251MPa. Fibres reinforcement tends to increase the ductility of concrete. This is because the capability of fibres to transfer tensile stress across a cracked section. Fibres tend to decrease the crack depth [11]. It was indorsed by Wan Ibrahim [14] that the flexural strength of foamed concrete is not much rise with the addition of polyolefm fibers.

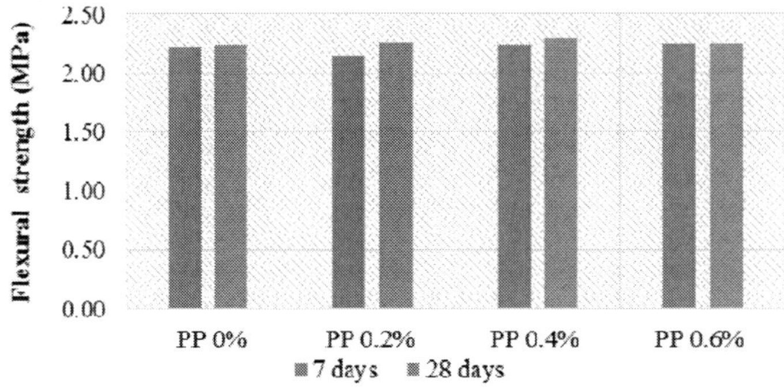


**Figure 6:** Flexural strength of concrete with POB clinker and PP fibres

## 4. Conclusion and Recommendations

Based on the results obtained, the palm oil biomass clinker considered as a potential material of construction. It was observed that the addition of polypropylene fibre can considerably be improved the flexural strength performances of concrete, about 8% raise in flexural strength was noticed at 0.4% fibre addition. Although the increase of the flexural strength is not in high percentage, but it is still proven that addition of polypropylene (PP) fibres can increase the flexural strength of concrete. Other than that, the concrete containing palm oil biomass (POB) clinker as sand replacement causes decrease in compressive strength due to the porosity of the clinker. Even though the target strength of concrete has been achieved and flowing conclusions has been drawn:

i. The concrete containing palm oil biomass (POB) clinker as sand replacement along with polypropylene (PP) fibres, decreases the workability of the concrete.
ii. The compressive strength of concrete at 28days was recorded as 27.40MPa, 27.71MPa and 26.61MPa for 0.2%, 0.4% and 0.6% respectively, with 10% of palm oil biomass (POB) clinker as sand replacement along with addition of polypropylene (PP) fibres.
iii. The highest flexural strength of concrete was noticed with 10% oil biomass (POB) clinker as sand replacement and 0.4% addition of polypropylene (PP) fibres, which about 8% higher than the control specimen at the curing period of 28days.
iv. According to the compressive and flexural strength performances, the 0.4% addition of polypropylene (PP) fibres in concrete with 10% palm oil biomass (POB) clinker as sand replacement could be considered as optimum content.

Hence, it is recommended for the future works that the in-depth research is required to investigate the strength and durability performances of concrete containing ground palm oil biomass (POB) clinker with polypropylene (PP) fibres.

## 5. References

[1] Ahsana Fathima K M, Shibi V 2014Behavioral study of steel fiber and polypropylene fiber reinforced concrete*Int. Journal of Res. in Eng. & Tech.* **2(10)** 17-24
[2] Neville A M, Brooks J J 2010 Concrete Technology 2nd Edition (England: Pearson)
[3] Zarina Y, Mustafa A M, Kamarudin H, Khairul N I, Rafiza A R 2013 Review on the various ash from palm oil waste as geopolymer material *Adv. Mater. Sci.* **34** 37-43.
[4] Saeed A, Bukhari I A, Sidiqui J I, Qureshi S A2006 A Studies on properties of polypropylene fibre reinforced concrete *Proc. Int. Conf.On our world in concrete & structures,* 16 - 17 Aug. 2006,*Singapore.* 1-10
[5] BS EN 12620 2002 *Specification for aggregate for concrete*
[6] BS EN 1008 2002*Specification for Mixing water of concrete*
[7] BS EN 197-1 2000 *Specification for cement composition*

[8] BSEN 12350-2 2009 *Testing of Fresh Concrete,Slump Test*
[9] BS EN 12390-3 2009 *Testing Hardened Concrete, Compressive Strength of Test Specimens*
[10] BS EN 12390-5 2009 *Testing Hardened Concrete. Flexural Strength of Test Specimens*
[11] Payam S, Halim B M, Mohd Zain B J, Rasel A, Syamsul Bahri 2014 Structural Lightweight aggregate concrete using two types of waste from the palm oil industry as aggregate *Journal of Cleaner Production.* **80** 187-196
[12] Machine H, Chijen T, Song P S 2008 Mechanical properties of polypropylene hybrid fiber-reinforced concrete *Mat. Sci. and Engineering.* **494** 153-157
[13] Milind V M 2015 Performance of Polypropylene Fibre Reinforced Concrete *IOSR Journal of Mech. and Civil Engg.* **12** 28-36
[14] Wan Ibrahim M H, Jamaluddin N, Irwan J M, Suraya hani A *2014* Compressive and Flexural Strength of Foamed Concrete Containing Polyolefin Fibers *Adv. Materials Res.* **911** 489-493

**Acknowledgment**
The authors thankfully acknowledge the UniversitiTun Hussein Onn Malaysia (UTHM) for the funds and facilities provided for this research.

# Foamed concrete containing rice husk ash as sand replacement: an experimental study on compressive strength

**R H M Rum[1], Z M Jaini[2,*], K H Boon[2], S A A Khairuddin[1] and N A Rahman[2]**

[1]Faculty of Civil and Environmental Engineering, Universiti Tun Hussein Onn Malaysia, 86400 Parit Raja, Johor, Malaysia
[2]Jamilus Research Center, Universiti Tun Hussein Onn Malaysia, 86400 Parit Raja, Johor, Malaysia
Corresponding author: rizuan@uthm.edu.my

**Abstract.** This study presents the utilization of rice husk ash (RHA) as sand replacement in foamed concrete. The study focuses on the effect of RHA on the compressive strength of foamed concrete. RHA contains high pozzolanic material that reacts with cementitious to enhance the strength and durability of foamed concrete. RHA also acts as filler causing the foamed concrete to become denser while retaining its unique low density. A total 243 cube specimens was prepared for the compression test. Two sets of mix design were employed at water-cement (W/C) ratio of 0.55, 0.60 and cement-sand ratio of 0.50, 0.33. The results revealed that the presence of RHA as sand replacement resulted in an increase in the compressive strength of foamed concrete. Moreover, 30% to 40% RHA was the optimum content level, contributing to the compressive strength of 18.1MPa to 22.4MPa. The W/C ratio and superplasticiser dosage play small roles in improving workability. In contrast, density governs the compressive strength of foamed concrete.

## 1. Introduction

Foamed concrete is a lightweight and versatile material. The basic components of foamed concrete are cement, fine aggregate, stable foam, water and other type of admixtures. The non-presence of coarse aggregate causes foamed concrete to have low density (400kg/m$^3$ to 1600kg/m$^3$). Cement consumption is also high due to the need for decent quality and adequate strength in foamed concrete. However, the compressive strength of foamed concrete is deficient when in the range of 10MPa to 15MPa [1-4]. The corresponding fracture energy can be reached at around 20N/m to 35 N/m [5, 6]. These characteristics limit the application of foamed concrete and the prohibition of its use as a structural component. According to Eurocode 2 [7], the minimum compressive strength employed in reinforced concrete structures should be 22MPa.

Previous studies have attempted to improve the compressive strength of foamed concrete. The modifications of cementitious using fly ash [8] and palm oil fuel ash [9] have yielded the most favourable results in terms of producing good foamed concrete. Another approach in enhancing the material properties of foamed concrete is by exploiting lightweight aggregates, such as pelletized fibre [3] and oil palm shell [10], where promising results (as high as 50MPa) can be achieved. This study, however, intends to utilize rice husk ash (RHA) as sand replacement in foamed concrete to fulfil the requirement of characteristics strength for structural purposes. More focus is given to the effect of volume fraction, the role of water-cement (W/C) ratio, superplasticiser (SP) dosage and the density of foamed concrete toward the optimum contain of RHA.

## 2. Rice Husk Ash

Rice husk ash (RHA) is an agricultural by-product obtained from the process of burning rice husk under controlled temperatures below 800 °C. The by-product is black, grey, or pinkish-white, depending on the burning process [11]. The burning process produces about 25% of ash containing 85% to 90% amorphous silica as well as about 5% alumina, which makes it highly pozzolanic [12]. Studies have thoroughly investigated the mechanical properties and behaviour of concrete incorporating RHA. Kishore et al. [13], Xu *et al.* [14], Jamil et al. [15], Sathawane *et al.* [16] and Abalaka [17] applied RHA as partial replacement for cement. To some extent, RHA can be used to partially substitute for cement, but the addition of RHA exhibits better performance in flexure strength.

RHA with volume fraction of more than 5% to 10% has exhibited a decrease in compressive strength Compressive strength has been found to decrease linearly with the increase in the increments of RHA. However, RHA can enhance the durability of concrete particularly in terms of environmental exposure and chemical attacks. The addition of around 20% to 30% RHA to cement has been found to be advantageous and does not adversely affect the durability of concrete [18]. Che Wan [19] determined that the incorporation of RHA as cement replacement significantly improved its resistance toward sodium sulfate. Ganesan *et al.* [20] and Raman *et al.* [21] revealed that the blending of RHA with cement resulted in approximately 35% decrease in water permeability, 28% reduction in chloride diffusion and 75% reduction in chloride permeation of concrete.

In foamed concrete, Kawabata [22] and Bayuaji [23] performed a series of experimental studies on the influence of RHA as cementitious base and cement replacement and confirmed that RHA was capable of enhancing the strength and workability of foamed concrete. RHA caused an increase in setting times, which improved the early strength of foamed concrete. However, replacing large volumes of cement with RHA would decrease its strength and workability. RHA tends to contribute up to 80% and above to the compressive strength of foamed concrete if the percentage replacement does not exceed 20% and is line with the proper mix proportion, water-cement (W/C) ratio and fineness of RHA.

Until recently, literature reviews on foamed concrete incorporating RHA as sand replacement have been very limited. Hadipramana *et al.* [24] investigated the utilization of RHA in foamed concrete as partial replacement for fine aggregate. Several mixes of RHA with different ratios and densities have been used and the results showed that 25% RHA yielded a higher value of compressive strength. This study suggested that the presence of RHA acts simultaneously as a pozzolanic reaction and a filler, thereby contributing to the good compressive and tensile strengths of foamed concrete. In contrast, Kunchariiyakun *et al.* [25] and Qu & Zhao [26] demonstrated that RHA from the burning process at 180 °C for 8 to 18 hours reduced compressive strength. However, in terms of the microstructure, the highly reactive silica in RHA had a significant effect on the tobermorite transformation.

## 3. Experimental study

### *3.1. Material preparation*

Material preparation involved the provision of mix design and required quantity of cement, sand, RHA, water, pre-foamed+ foam agent and SP to produce high quality of foamed concrete. Two sets of mix design were considered. In Set 1, cement-sand (C/S) and W/C ratios were 0.50 and 0.55, respectively. The mix design for Set 2 used C/S ratio of 0.33 and W/C ratio of 0.60. The details of the required quantity of materials are specified in tables 1 and 2. In this experimental study, RHA was used as partial sand replacement at the range of 0% to 50%. RHA was obtained from controlled burning at 700 °C for 6 hours. The effects of W/C ratio, SP dosage and density were examined at the optimum RHA content for Set 1 only.

In this experimental study, foamed concrete was produced using pre-foaming method. Foam agent was diluted with water in an LCM foam generator and compressed to create the pre-formed foam. The pre-formed foam is a stable substance added to the foamed concrete mixture to produce air-void. The pre-formed foam functions as an expansion factor and as control for the desired density of foamed concrete. In general, pre-foaming method is better in producing foamed concrete because the pre-formed foam is in a liquid state and reacts effectively with the mixture. Foam agent Sika AER 50/50, which is a synthetic surfactant and polymer-based with brown colour and specific gravity of 1.01, was used with foam-cement (F/C) ratio of 0.07, while foam-water (F/W) ratio of 0.05 for Sets 1 and 2. The concentrated of foam agent with water was kept constant for all cube specimens. The stabilising component of the foam agent promotes uniformity of the air-void with even sizes and shapes.

**Table 1.** Quantity of foamed concrete for 1m$^3$.

| Material | Set 1 (kg) | Set 2 (kg) | Remarks |
|---|---|---|---|
| Cement | 500 | 375 | BS EN 197-1, Ordinary Portland Cement Type 1 |
| Water | 250 | 225 | BS EN 1008:2002, water, free from any impurities |
| Foam agent | 35 | 27 | Sika AER 50/50, the ratio of foam to water is 0.05 |
| SP | 4 | 3 | ESTOP Admix AP, consistence content along the mixing |

**Table 2.** Quantity of sand and RHA in proportion to the volume fraction for 1m$^3$.

| Volume fraction of Sand (%) | Volume fraction of RHA (%) | Set 1 | | Set 2 | |
|---|---|---|---|---|---|
| | | Mass of Sand (kg) | Mass of RHA (kg) | Mass of Sand (kg) | Mass of RHA (kg) |
| 100 | 0 | 920 | - | 1035 | - |
| 90 | 10 | 828 | 14 | 932 | 15.8 |
| 80 | 20 | 736 | 28 | 828 | 32 |
| 70 | 30 | 644 | 42 | 725 | 48 |
| 60 | 40 | 552 | 56 | 621 | 64 |
| 50 | 50 | 460 | 70 | 518 | 79 |

### *3.2. Specimen preparation*

A total of 243 cube specimens were prepared using mould size of 100mm x 100mm x 100mm. Thus, 108 cube specimens were prepared for volume fraction of RHA (Set 1), while 135 cube specimens were prepared for the parametric investigations on W/C, SP dosage and density at the optimum RHA content (Set 2). The casting process was conducted in batches because the preparation involves a huge number of cube specimens. The cube specimens underwent air curing at 7, 14 and 28 days. In total, this experimental study consumed 0.243m$^3$ foamed concrete.

### *3.3. Test programme*

The test programme of the cube specimens involved a compression test, which was conducted according to BS EN12390-3:2009 [27]. Before the compression test, the cube specimens were visually inspected, labelled and sized. The cube specimens were also weighed to determine the density of foamed concrete. The compression test was performed using a calibrated Ele-Campact Machine 1500. Load increased continuously at a normal rate (0.14MPa/min to 0.34MPa/min) was applied to the cube specimens until failure.

## 4. Results and discussion

### *4.1. Compressive strength*

The results of the compression test on cube specimens indicated the resulting compressive strengths of foamed concrete, which are shown in figures 1 and 2. Compressive strength was observed to increase slightly according to the volume fraction of RHA. Plain foamed concrete developed compressive strength at 12.5MPa and 15.1MPa for Set 1 and Set 2, respectively. The values are in accordance with that determined by the British Cement Association [28]. Jaini *et al.* [3] and Abd Rahman *et al.* [5]

showed that plain foamed concrete with W/C ratio of 0.55 and C/S ratio of 0.55 had compressive strength around 12MPa to 18MPa. Therefore, this experimental study successfully reproduced plain foamed concrete with appropriate compressive strength. Although the mix design has a significant role in producing good foamed concrete, the quality of pre-formed foam also contributed to the uniformity of air voids, which had a direct effect on compressive strength.

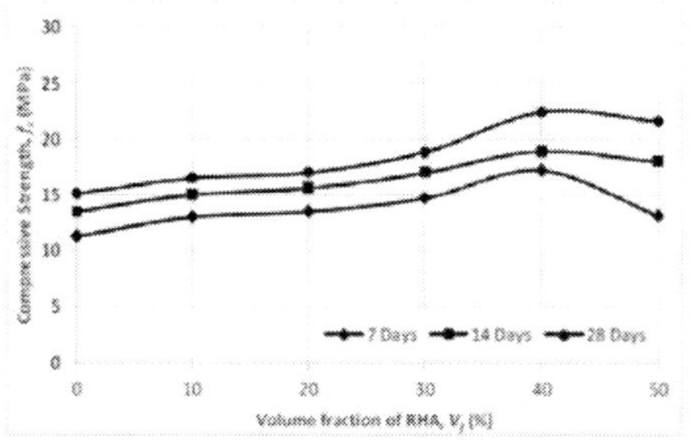


**Figure 1.** Compressive strength of foamed concrete for mix design Set 1.

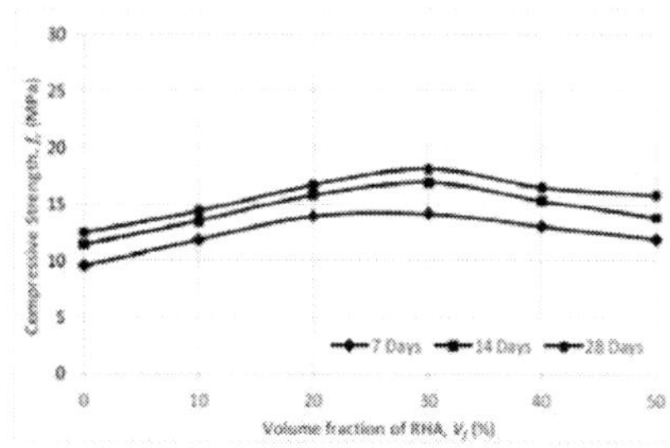


**Figure 2.** Compressive strength of foamed concrete for mix design Set 2.

Utilization of RHA as sand replacement in foamed concrete yielded significant improvement in compressive strength. Boost phenomenon on the compressive strength could be clearly observed for both sets. Foamed concrete in Set 1 had optimum RHA content at 40%, whereas Set 2 had 30% RHA. Compressive strength increased by up to 22.4MPa and 18.1MPa for Set 1 and Set 2, respectively. However, although the compressive strength increased the decay, the value obtained was still higher than plain foamed concrete. The addition of RHA resulted in higher compressive strength compared to plain foamed concrete by up to 48.34%, which also contributed to the satisfactory characteristic strength of the foamed concrete for the mixed design. If the density of foamed concrete is set at1800kg/m$^3$, the compressive strength will probably be higher than that obtained in this experiment.

The increase in compressive strength can be attributed to the pozzolanic reaction and the presence of reactive silica in RHA as reported by Rashid *et al.* [29]. The pozzolanic material also improved the impermeability of the foamed concrete, leading to higher durability against environmental exposures and chemical attacks. However, the presence of large RHA and coarse particles were found to delay the pozzolanic reaction, thereby resulting in lower compressive strength [30, 31]. Unlike its utilization as cement replacement that advantageously improved the workability of foamed concrete similar to the findings of Bayuaji [23], the workability of foamed concrete containing RHA as sand replacement indicated negative behaviour. Workability decreased as the volume fraction of RHA increased. Although the workability is not measured directly, it was also observed during the mixing process and specimen preparation.

*4.2. Effect of water-cement ratio*

The W/C ratio plays an important role in the mix design of foamed concrete and is the main factor affecting the performance of foamed concrete. W/C ratio is dependent purely on the cement content and other admixtures. Nambiar & Ramamurthy [32] reported that low water content caused the mixture to become too stiff and to have broken air-voids during the mixing process, which resulted in increased density. Similarly, at high water content, the mixture was too thin to hold the air-voids, causing segregation of the pre-formed foam and sand from the mixture, causing an increase in the density. The range of W/C ratio is within 0.40 to 1.25 [33]. However, the British Cement Association [28] suggested that the optimum W/C ratio should be limited to between 0.5 to 0.6 only. Moreover, high W/C ratio affected the porosity of foamed concrete, which also contributed to low compressive strength.

At 40% RHA, the amount of water needs to be high enough to maintain workability of the foamed concrete; however, increased water content has an adverse effect on compressive strength. In addition, cement and RHA would absorb water and cause rapid degeneration of pre-formed foam. Figure 3 shows the compressive strength of foamed concrete corresponding to W/C ratio. The increase in W/C ratio caused the compressive strength to decrease. However, the decrease was still too small at around 2.5% to 12% when evaluated based on W/C ratio of 0.50. At 28 days, only W/C ratios of 0.50, 0.55 and 0.60 achieved compressive strength equal or higher than 22MPa. The compressive strength of foamed concrete using W/C ratios of 0.50, 0.55, 0.60, 0.65 and 0.70 is 24.7MPa, 24.1MPa, 22.2MPa, 21.6MPa and 20.9MPa respectively. Blending during casting and drying shrinkage were observed in foamed concrete with W/C ratios of 0.65 and 0.70.

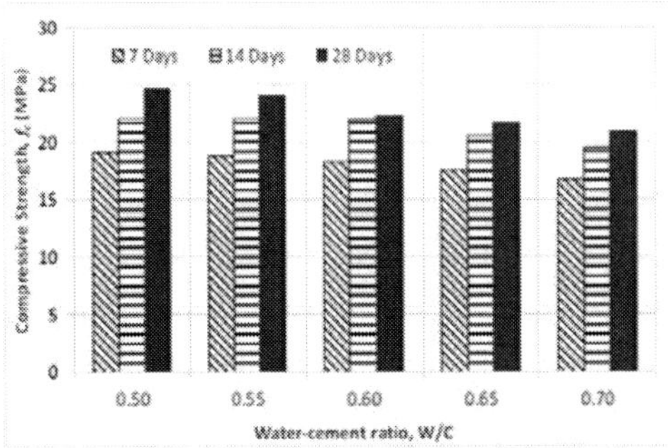


**Figure 3.** Compressive strength of foamed concrete due to water-cement ratio.

*4.3. Effect of superplasticizer*

The effect of SP was analyzed using the ratios of W/C and C/S at 0.55 and 0.50, respectively. At constant water content and 40% RHA, the dosage of SP was 0.4% to 0.8% of the cement content. The results of compressive strength could be attributed to the dosage of SP as shown in figure 4. The SP dosage had no effect on the compressive strength of foamed concrete instead it caused an increase in its workability. Basically, foamed concrete containing RHA requires higher water content than plain foamed concrete because of the higher specific surface area and higher carbon composition in RHA. In fact, RHA absorbs water during the mixing process. Increasing the volume fraction of RHA in foamed concrete requires more water, which in turn caused an increase in the SP dosage as the volume fraction of RHA increases. Hence, treating SP as a water reducing agent is important to ensure that the bond and stability in foamed concrete can be enhanced.

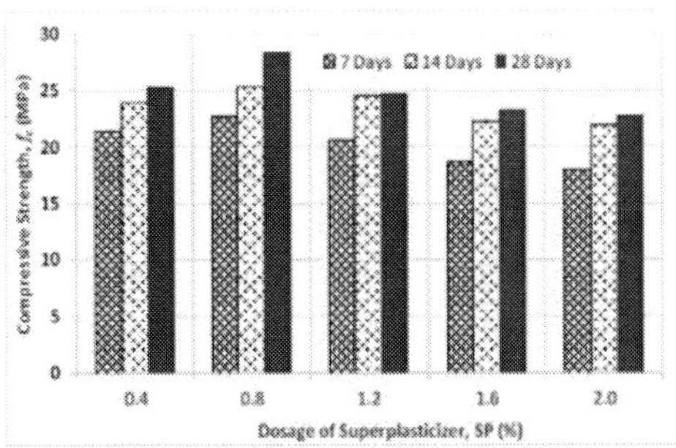


**Figure 4.** Compressive strength of foamed concrete due to dosage of SP.

However, it can be observed that 0.8% SP resulted in an improvement in compressive strength by approximately 24% as compared with foamed concrete with no SP. Increasing the dosage of SP at a constant volume fraction of RHA led to the decrease in compressive strength, which is a similar trend to that observed by Memon *et al.* [34]. The compressive strengths of foamed concrete containing 40% RHA + 0% SP is 17.2MPa, 18.9MPa and 22.4MPa at 7, 14 and 28 days. The addition of SP accelerated the early strength of foamed concrete. As a comparison, at 7 days, foamed concrete with 40% RHA + 0.8% SP surpassed the compressive strength of foamed concrete with 40% RHA + 0% SP. Foamed concrete with 40% RHA + 0.8% attained compressive strength of 22.8MPa, which is 32.5% above foamed concrete with 40% RHA + 0% SP. In the fresh mixture, no significant effect on segregation was observed at any dosage of SP.

### *4.4. Effect of density*

Two density behaviours were observed in foamed concrete containing RHA as shown in figure 5. The first behaviour pertains to the effect of density due to the volume fraction of RHA. The density of foamed concrete was reduced with the increase in volume fraction of RHA. In Set 1, the density of foamed concrete with 0%, 10%, 20%, 30%, 40% and 50% RHA was 1800kg/m$^3$, 1740kg/m$^3$, 1717kg/m$^3$, 1663kg/m$^3$, 1633kg/m$^3$ and 1600kg/m$^3$, respectively. Sparse reduction was observed around 1.5% to 3.5%. Although the density decreased with the addition of RHA, the compressive strength significantly increased by 40% RHA. RHA is low in density, which can also be attributed to RHA having a filler effect with air-void in foamed concrete. As a comparison, the density of sand is 1380kg/m$^3$ while RHA has density of 210kg/m$^3$. If sand is replaced partially by 40% to 50% RHA, the decrease in density would be approximately 10% to 15%.

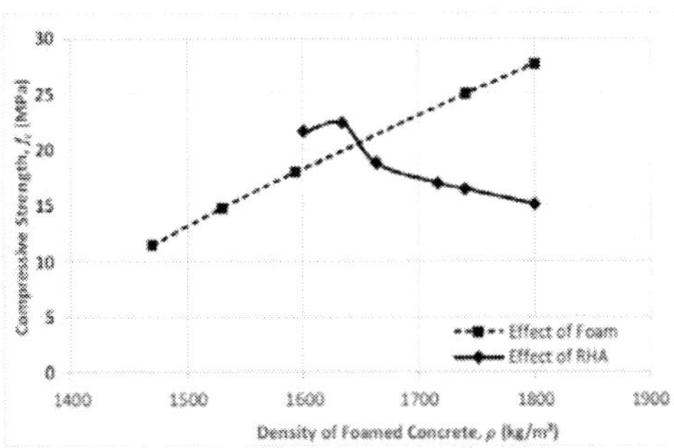


**Figure 5.** Density of foamed concrete due to the volume fraction of RHA and amount of preformed foam.

The second behaviour was observed from the effect of pre-formed foam in foamed concrete at 40% RHA. Adding a larger amount of pre-formed foam in foamed concrete contributed to higher porosity and larger air-void size. Therefore, ultimately, the density of the foamed concrete will be reduced, subsequently lessening its compressive strength. A pronounced relationship can observed between compressive strength, porosity and density. In this experimental study, the decrease in density caused the compressive strength to decrease linearly. At the same W/C ratio and volume fraction of RHA, the higher amount of pre-formed foam used, the higher the porosity and the lower the density. Narayanan & Ramamurthy [35] found that the main factor affecting the density of the foamed concrete is the pre-formed foam, which can be controlled to obtain the desired compressive strength.

## 5. Conclusion

An investigation on the effect of RHA as sand replacement in foamed concrete was conducted using two sets of W/C and C/S ratios. The results indicated that RHA had a significant effect on the compressive strength of foamed concrete. Increasing the volume fraction of RHA contributed the significant increments in compressive strength. However, the optimum RHA content depended on the W/C and C/S ratios. The highest value of compressive strength at 22.4MPa was obtained at 40% RHA, which is 48.34% higher than plain foamed concrete. The improvement on compressive strength of foamed concrete can be credited to the presence of pozzolanic material and reactive silica in RHA. In addition, RHA also acts as filler, which caused the foamed concrete to become denser. However, the presence of large volume fraction of RHA dramatically reduced compressive strength.

The effects of W/C ratio, SP dosage and density on the compressive strength of foamed concrete were also examined at 40% RHA. The W/C ratio was the main factor that influenced the compressive strength of foamed concrete. When W/C ratio was increased from 0.50 to 0.70, the compressive strength decreased at a limited rate from 2.5% to 12%. The results indicated that the W/C ratios of 0.50, 0.55 and 0.60 were the only ones suitable for use in foamed concrete incorporating RHA. The suitable SP dosage is 0.8% of the cement content, which increased the compressive strength to approximately 24% and improved the workability. An increase in SP dosage could result in unsatisfactory performance of the foamed concrete. The investigation on the effect of density indicated that presence of RHA contributed to the low density but high compressive strength. However, the portion of pre-formed foam significantly reduced density and compressive strength.

## 6. References

[1] McCarthy M R and Jones A 2005 Preliminary views on the potential of foamed concrete as a structural materials, *Mag. Concr. Res.* **7(1)** 21-31

[2] Nambiar E K and Ramamurthy K 2006 Models relating mixture composition to the density and strength of foamed concrete using response surface methodology, *Cem. Concr. Compos.* **28(9)** 752-760

[3] Jaini Z M, Mokhatar S N, Mohd Yusof A S, Zulkiply S and Abd Rahman M H 2016 Effect of pelletized coconut fiber on the compressive strength of foamed concrete, *MATEC Web of Conf.* **4** 01013

[4] Mugahed Amran Y H, Farzadnia N and Abang Ali A A 2015 Properties and application of foamed concrete: A review, *Constr. Build. Mater.* **101** 990-1005

[5] Abd Rahman N, Jaini Z M and Mohd Zahir N N 2015 Fracture energy of foamed concrete by means of the three-point bending test on nocthed beam specimens, *ARPN J. of Eng. Appl. Sci.* **10(15)** 6562-6570

[6] Kozlowski M, Kadela M and Kukielka A 2015 Fracture energy of foamed concrete based on three-point bending test on notched beams, *Proceedia Eng.* **108** 349-354

[7] EN 1992-1-1:2004 Design of Concrete Structure *Part 1-1*: *General Rules and Rules for Buildings*, (British Standard Institution, London)

[8] Kearsley E P and Wainwright P J 2002 Ash content for optimum strength of foamed concrete, *Cem. Concr. Res.* **32(2)** 241-246

[9] Lim S K, Tan C S, Lim O Y and Lee Y L 2013 Fresh and hardened properties of lightweight foamed concrete with palm oil fuel ash as filler, *Constr. Build. Mater.* **46** 39-47

[10] Jumaat M Z, Alengaram U J and Mahmud H 2009 Shear strength of oil palm shell foamed concrete beams, *Mater. Des.* **30(6)** 2227-2236

[11] Juma A 2012 A review on experimental behaviour of self compaction concrete incorporated with rice husk ash, *Int. J. of Sci Adv. Technol.* **2(3)** 75-80

[12] Atan N O R and Awang H 2011 The compressive and flexural strengths of selft-compacting concrete using raw rice husk ash, *Mater. J.* **6(6)** 720-732

[13] Kishore R, Bhikshma V and Prakash P J 2011 Study on strength characteristics of high strength rice husk ash concrete, *Procedia Eng.* **14** 2666–2672

[14] Xu W, Lo T Y and Memon S A 2012 Microstructure and reactivity if rice husk ash, *Constr. Build. Mater.* **29** 541-547

[15] Jamil M, Kaish A B M A, Raman S N and Zain M F M 2013 Pozzolanic contribution of rice husk ash in cementitious system, *Constr. Build. Mater.* **47** 588-593

[16] Sathawanea S H, Vairagadeb V S and Kenec K S 2013 Combine effect of rice husk ash and fly ash on concrete by 30% cement replacement, *Procedia Engineering*, **51** 35-44

[17] Abalaka A E 2013 strength and some durability properties of concrete containing rice husk ash produced in a charcoal incinerator at low specific surface, *Int. J. of Concr. Struc. Mater.* **7** 287-293

[18] Lung H C, Tuan B L A and Tsun C C 2011 Effect of rice husk ash on the strength and durability characteristics of concrete, *Constr. Build. Mater.* **25** 3768-3772

[19] Che Wan C N, Jaya R P, Abu Bakar B H and Arshad M F 2014 Strength of concrete containing rice husk ash subjected to sodium sulfate solution via wetting and drying cylic, *Appl. Mech. Mater.* **534** 3-8
[20] Ganesan K, Rajagopal K and Thangavel K 2008 Rice husk ash cement: Assessment of optimal level of replacement for strength and permeability properties of concrete, *Constr. Build. Mater.* **22(8)** 1675-1683
[21] Raman S N, Ngo T, Mendis P and Mahmud H B 2011 High-strength rice husk ash concrete incorporating quarry dust as a partial substitute for sand, *Constr. Build. Mater.* **25** 3123-3130
[22] Kawabata C Y, Junior H S and Coutinho J S 2012 Rice husk derived waste materials as partial cement replacement in lightweight concrete, *Sci. Agrotechnol.* **36(5)** 567-578
[23] Bayuaji R 2015 The influence of microwave incinerated rice husk ash on foamed concrete workability and compressive strength using Taguchi method, *J. Teknologi*, **75(1)** 265-274
[24] Hadipramana J, Samad A A A, Zaidi A M A, Mohammad N and Riza F V 2013 Effect of uncontrolled burning rice husk ash in foamed concrete, *Adv. Mater. Res.* **626** 769-775
[25] Kunchariyakun K, Asavapisit S and Sombatsompop K 2015 Properties of autoclaved aerated concrete incorporating rice husk ash as partial replacement for fine aggregate, *Cem. Concr. Compos.* **55** 11-16
[26] Qu X and Zhao X 2017 Previous and present investigations on the components, microstructure and main properties of autoclaved aerated concrete - A review, *Constr. Build. Mater.* **135** 505-516
[27] BS EN 12390-3:2009, *Testing Hardened Concrete - Compressive Strength of Test Specimens* (British Standard Institution, London, 2011).
[28] British Cement Association 1994 *Foamed concrete: Composition and properties*, Report 46.042, (Slough, United Kingdom)
[29] Rashid N H, Molla M K A and Ahmed T U 2010 Durability of mortar in presence of rice husk ash, *World Acad. Sci. Eng. Technol.* **43** 736-739
[30] Nehdi M, Duquette J and Damatty A 2003 Performance of rice husk ash produced using a new technology as a mineral admixture in concrete, *Cem. Concr. Res.* **33** 1203-1210
[31] Venkatanarayanan H K and Rangaraju P R 2013 Material characterization studies on low-and high-carbon rice husk ash and their performance in portland cement mixtures, *Adv. Civil Eng. Mater.* **2(1)** 266-287
[32] Nambiar E K and Ramamurthy K 2006 Influence of filler type on the properties of foam concrete, *Cem. Concr. Compos.* **28(5)** 475-480
[33] Kearsley E P and Visagie M 1999 *Micro-Properties of Foamed Concrete. Specialist Techniques and Materials for Construction* (Thomas Telford, London)
[34] Memon S A, Muhammad A S and Hassan A 2010 Utilization of rice husk ash as velocity modifying agent in self compacting concrete, *Constr. Build. Mater.* **25(2)** 1044-1048
[35] Narayanan N and Ramamurthy K 2000 Structure and properties of aerated concrete: A review, *Cem. Concr. Compos.* **22(5)** 321-329

**Acknowledgments**
Authors would like to thank Universiti Tun Hussein Onn Malaysia (UTHM) and Ministry of Education Malaysia for the continuous supports in term of facilities and financial. This experimental study was conducted under the Research Acculturation Grant Scheme (RAGS), Vot. No. R068. The paper also was sponsor by centre graduate study. This paper also was partly sponsored by the Centre for Graduate Studies Universiti Tun Hussein Onn Malaysia (UTHM).

IOP Conf. Series: Materials Science and Engineering **271** (2017) 012013 doi:10.1088/1757-899X/271/1/012013

# Effects of heating durations on normal concrete residual properties: compressive strength and mass loss

**Fadzli Mohamed Nazri[1*], ShahironShahidan[2], Nur Khaida Baharuddin[1], Salmia Beddu[3] and Badorul Hisyam Abu Bakar[1]**

[1]School of Civil Engineering, Engineering Campus, Universiti Sains Malaysia, 14300 Nibong Tebal, Penang, Malaysia.
[2]Jamilus Research Center, Faculty of Civil and Environmental Engineering, Universiti Tun Hussein Onn Malaysia,86400, Pt Raja Johor
[3]Universiti Tenaga Nasional, Jalan IKRAM-UNITEN, 43000 Kajang, Selangor
*Corresponding author: cefmn@usm.my

**Abstract.** This study investigates the effects of high temperature with five different heating durations on residual properties of 30 MPa normal concrete. Concrete cubes were being heated up to 600°C for 30, 60, 90, 120 and 150 minutes. The temperature will keep constant for 30, 60, 90, 120 and 150 minutes. The standard temperature-time curve ISO 834 is referred to. After heating the specimen were left to cool in the furnace and removed. After cooling down to ambient temperature, the residual mass and residual compressive strength were observed. The obtained result shows that, the compressive strength of concrete decrease as the heating duration increases. This heating duration influence, might affects the loss of free water present and decomposition of hydration products in concrete. As the heating duration increases, the amount of water evaporated also increases led to loss in concrete mass .Conclusively, the percentage of mass and compressive strength loss increased as the heating duration increased.

## 1. Introduction

The properties of concrete will deteriorated when it was subjected to high temperature. This is due to the changes of physical properties and chemical composition of concrete. Concrete will undergo different type of damages depending on level of temperature. The effects of concrete component to high temperature were reduction in compressive strength, micro-cracking within the concrete microstructure, colour changes consistent with strength reduction, increase in pore structure, and various degree of spalling. There was certain reduction in strength of concrete when it is exposed to high temperature. Based on Ma Q*et al.* [1] residual compressive strength of concrete after being exposed to high temperature occurs in three stages which at room temperature-300°Cthe compressive strength keeps constant. Then at 300-800°C, compressive strength of concrete drop drastically and at more than 800°C temperature, concrete lost its compressive strength.

Prior to technical note by Phaedonos[2] concrete lose its strength up to 30-40% once the temperature hit 300°C. This is because of thermal expansion, internal cracking, and weak adhesion between paste and steel reinforcement within the concrete. At 500-600 °C, more than 70% of concrete strength reduced. In previous research by Haddad& Al-Rousan[3] the mechanical performance of beams after fire was scaled down due to reduction in compressive strength, weak concrete to steel bond and reinforcement steel deteriorated. The load capacity of beam, after heated at 300, 500, and 600°C was reduced to 32%, 79% and 80%.Up to now, previous study by Park *et al.*[4] has established that post fire curing condition surely enhanced the residual properties of fire-damaged concrete. The mass and concrete density gradually decreases after heating (300°C, 500°C and 700°C). In addition, after post fire curing also recovers the density, mass and also the concrete tensile strength. When concrete exposed to 800°C temperature, there was significant reduction of strength [5]. As the temperature increase, the strength loss become smaller and at 1000°C the value become null [6]. The high temperature surely gave negative effect to the concrete strength [7].

The factors that affect post fire concrete's compressive strength are the rate of heating, the duration of heating, whether the specimen was loaded or not, the type and size of aggregate, the water/cement ratio, and the percentage of cement paste. In general, concrete heated by a building fire always loses some compressive strength and continues to lose it on cooling. Previous research done by Mazza[8] stated that concrete damaged zone comes bigger with longer heating duration. Concrete structures that exposed to high temperature for 60 minutes experienced more damaged compared to concrete that heated for 45 minutes. Bikhiet [9] compared the experimental and theoretical analysis for column heated at constant 600°C temperature with different durations. The findings stated that, for column sample as the period of fire increased from 10 min to 15 or 20 min, the strain of column increased by 45% and 55%, respectively. Both experimental and theoretical analysis reported that as the period of fire increases, the column stiffness decreases. The analysis also showed that with increase in the period of fire (10, 15, 20 min), the corresponding column failure load compared with unheated column decreased by nearly (12%, 15%, 25%) respectively.

## 2. Experimental programme

### *2.1. Sample preparation*

The normal concrete (NC) with design strength of 30 MPa was prepared. The specimens were in cube (100 x 100 x 100mm). The normal concrete used Ordinary Portland cement, sand, coarse aggregates with maximum size 20mm, water cement ratio of 0.55 and slump value between 150mm to 180 mm. The mix proportion of normal concrete substrate is presented in Table1.

**Table 1.** NC substrate mix design

| Item | Mass (kg/m$^3$) | Remark |
|---|---|---|
| OPC | 380 | Type-1 |
| Coarse Aggregates | 1115 | Max. 20mm |
| Fine Aggregates | 685 | F.M 2.7 |
| Water | 205 | |
| W/C ratio | 0.55 | |
| Super plasticiser | 0.5% | |

As the objective is to investigate the effect of heating duration subjected to residual properties of normal concrete, the specimen will be exposed to 600°C temperature for five different durations. The NC was tested for 7 and 28 days strength and then it was heated at 600°C for 30, 60, 90, 120 and 150 minutes. The samples will be heated up using a furnace owns by Concrete Laboratory, University Sains Malaysia. The specification and photo of this furnace were shown in Table 2 and Figure 1.

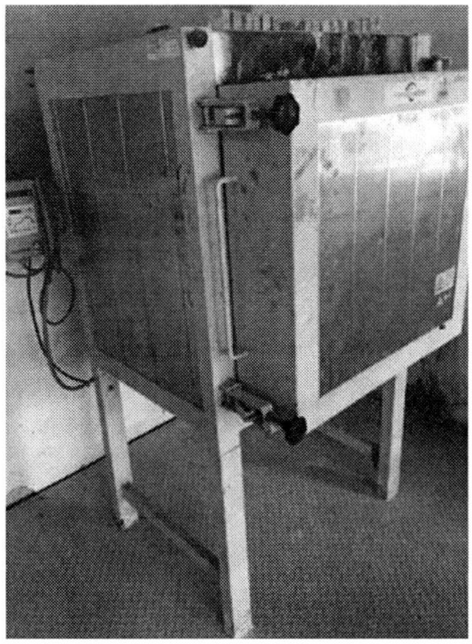

**Figure 1.** KC 80/15 furnace

**Table 2.** KC 80/14 furnace specification.

| Model | Temperature max (°C) | Inside dimension (w x d x h) | Litre (L) | External Dimension (w x d x h) | Power (kW) | Weight (kg) | Voltage (V) |
|---|---|---|---|---|---|---|---|
| KC 80/14 | 1400 | 430 x 430 x 430 | 80 | 985 x 1066 x 1540 | 13 | 280 | 4003/N |

Samples are then placed in the furnace as shown in Figure 3.3. Temperature exposed to the substrate was adopted from previous research by [9-12]. All of this works were preparing their fire damaged substrate by exposing it to 600 °C temperature. The 600 °C also considered as critical temperature for concrete as at this point its strength will lose about 50%. In addition, second phase of C-S-H decomposition occur at this stage and concrete can be consider as thermally damaged [11]. All substrates sample will be placed into furnace for firing. The firing durations specified according to [13]. The temperature will keep constant for 30, 60, 90, 120 and 150 minutes. The standard temperature-time curve ISO 834, also known as the Cellulosic curve / standard nominal fire curve is referred to. After heating the specimen were left to cool in the furnace and removed. The residual mass and residual compressive strength were observed.

*2.2. Test method*

The Compressive strength test is often considered as the most vital property of concrete. The compressive strength of NC substrate (before and after fire) and UHPFC were determined using 100 x 100 x 100 mm cubes as per BS 1881-116: 1983(Method for determination of compressive strength of concrete cubes)[14].A compression machine with maximum capacity of 3000kN was used to compress the cube specimens at loading rate of 0.30 MPa/s. The cubes of the NC were tested in compression at, 7 days, and 28 days. The sample was demoulded after 24 hours and cured until the age of test in the water tank. Standard curing in water follows until testing. For weight loss assessment, the weight of the concrete cubes was measured before and after exposure to high temperature. The mass loss of the entire investigated specimen is expressed as a percentage of the original mass at the ambient temperature to the mass after exposure to high temperature.

## 3. Result and discussion

### *3.1. Compressive strength*

According to Kodur, V[15]the compressive strength of concrete can be reduced if it is exposed to high temperature. In this study, the effects of heating durations to normal concrete (NC) compressive strength can be determined. The average compressive strength of normal concrete before and after fire is illustrated in Table 3 and Table 4. The lowest compressive strength recorded is 17.7 MPa. The highest compressive strength achieved is 40.3 MPa, for concrete at 28 days. The concrete's compressive strength decreases as the heating duration increases.

**Table 3.** Compressive strength of NC after fire (7 days).

| Heating Durations (Min) | F1 (N) | F2 (N) | F3 (N) | Average,F (N) | Compressive strength (MPa) |
|---|---|---|---|---|---|
| 0 | 320 | 300 | 310 | 310 | 31.0 |
| 30 | 270 | 275 | 280 | 275 | 27.5 |
| 60 | 270 | 240 | 250 | 253 | 25.3 |
| 90 | 230 | 240 | 235 | 235 | 23.5 |
| 120 | 230 | 210 | 215 | 218 | 21.8 |
| 150 | 180 | 170 | 180 | 177 | 17.7 |

**Table 4.** Compressive strength of NC after fire (28 days).

| Heating Durations (Min) | F1 (N) | F2 (N) | F3 (N) | Average,F (N) | Compressive strength (MPa) |
|---|---|---|---|---|---|
| 0 | 410 | 420 | 380 | 403 | 40.3 |
| 30 | 250 | 335 | 300 | 295 | 29.5 |
| 60 | 270 | 300 | 275 | 282 | 28.2 |
| 90 | 210 | 250 | 260 | 240 | 24.0 |
| 120 | 250 | 220 | 235 | 235 | 23.5 |
| 150 | 205 | 240 | 220 | 222 | 22.2 |

The effect of heating duration on compressive strength of normal concrete at 7 and 28 days ages are shown in Figure 2.

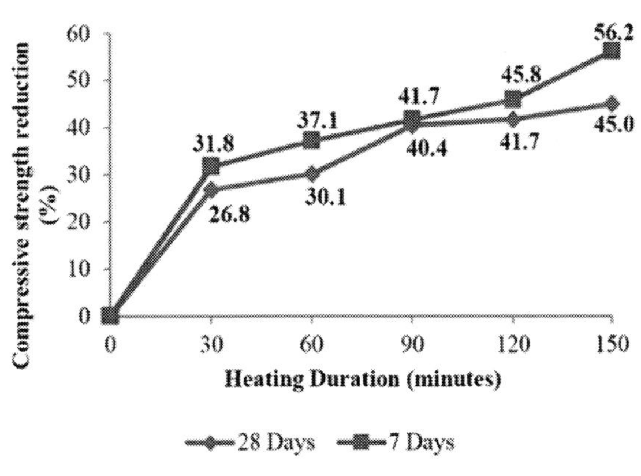


**Figure 2.** Compressive strength percentage of reduction.

The compressive strength of 7 days ages concrete reduces up to 31.8% after being heated for 30 minutes. Normal concrete loose about 56% of their compressive strength after being exposed to 600 °C temperature in 150 minutes duration. As for 28 days, the compressive strength reduces 26.8%. After being heated for 60, 90 and 120 minutes, the compressive strength values drop to 30.1%, 40.4% and 41.7 %. The reduction of compressive strength is directly proportional with the heating duration. As the heating duration increases, the reductions in compressive strength of concrete become larger. This might be caused by the driving out of moisture and concrete hydration due to high temperature. According to Hager [11] the strength of concrete reduced after being heatedmainly cause of the evaporation of water. There is a sequence of water removal and itdepends on the energy that straps the water and the solid. In short, free waterevaporates first, capillary water next and lastly by the physical bound water. Theprocess of releasing water that is chemically bound with cement hydrates is the last tobe initiated. Longer fire period makes more moisture hydrated and lead to the increase in concrete porosity which makes the concrete becomes weaker.

*3.2. Mass loss*
For mass loss assessment, the mass of concrete cubes were measured before and after fire. The impact of 600°C temperature with different heating duration on normal concrete is shown in Figure 4.2.

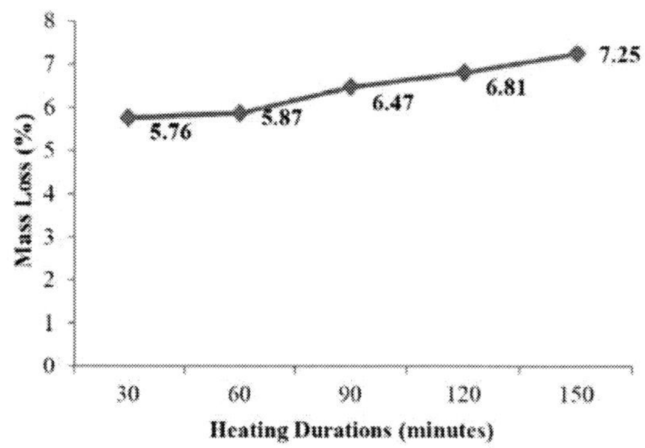


**Figure 3.** Mass loss of different heating duration.

The mass loss of all samples expressed as a percentage of the original mass at room temperature to the mass of concrete cubes after heated. Figure 3 displays that the longer the heating period, the mass loss of the concrete had the tendency to increase. At first 60 minutes of heating, small percentage of mass loss is observed. When concrete were heated for 90 minutes, its mass reduced about 6.47%. This is due to the loss ofwater content in the concrete. Plus, the concrete shows some spalling, cracking and the piecesof concrete start to breaks away. After being heated up for 120 and 150 minutes sample loss, about 6.81% and 7.25% of its original mass. This heating duration influence, might affects the loss of free water present and decomposition of hydration products in concrete. When concrete is exposed to high temperature, free water in the concrete will firstly be removed through evaporation. As the heating duration increases, the amount of water evaporated also increases. Later, the calcareous aggregates will decomposed, carbon dioxide ($CO_2$) liberated and altered the mechanical properties of concrete [16] and[1]. Conclusively, the percentage of mass loss increased as the heating duration increased.

The reduction of compressive strength is directly proportional to heating duration. As the heating duration increases, there are more drive out of moisture and concrete hydration [11]. This loss of moisture will lead to reduction in concrete mass too. The evaporation of water and decomposition of hydration product will increase the concrete's porosity. The longer the heating duration make the amount of evaporated water increases, followed by increases in void volume. The expansion in void volume of concrete will reduce its quality, thus there is reduction in their UPV and rebound number values and mass.

**4. Conclusion**

The current paper investigates the residual properties of normal concrete exposed to 600°C temperature for five different heating durations. The compressive strength of concrete decrease as the heating duration increases. The lowest compressive strength recorded when sample being heated for 150minuteswhich is 17.7Mpa. At this stage the compressive strength loss about half of it's original value. This heating duration influence, might affects the loss of free water present and decomposition of hydration products in concrete. As the heating duration increases, the amount of water evaporated also increases leads loss in concrete mass. The highest loss of mass recorded when concrete heated for 150 minutes. The lowest percentage of mass loss recorded is5.76%. This occurs when sample heated up to 30 minutes only. Conclusively, the percentage of mass loss increased as the heating duration increased.

**5. References**

[1] Ma Q, Guo R, Zhao Z, Lin Z and He K 2015 Mechanical properties of concrete at high temperature: A review, *Constr. Build. Mater.* **93** 371-383
[2] Phaedonos F A 2011 *Fire Damaged reinforced Concrete – investigation, Assessment And repair* Vic Roads Technical Note No.102 (Melbourne: Victoria Road) pp 1-4
[3] Haddad R H and Al-Rousan R Z 2016 An anchorage system for CFRP strips bonded to thermally shocked concrete, *International Journal of Adhesion and Adhesives*. **71** 10-22
[4] Park S J, Yim H J and Kwak H G 2015 Effects of post-fire curing conditions on the restoration of material properties of fire-damaged concrete, *Constr. Build. Mater.* **99** 90-98
[5] Li M, Qian C and Sun W 2004 Mechanical properties of high-strength concrete after fire,*Cement and Concrete Research*. **34** 1001-1005
[6] Husem M 2006 The effects of high temperature on compressive and flexural strengths of ordinary and high-performance concrete, *Fire Safety Journal*. **41** 155-163
[7] Dos Santos C C and Rodrigues J P C 2016 Calcareous and granite aggregate concretes after fire, *Journal of Building Engineering*. **8** 231-242
[8] Mazza F 2015 Seismic vulnerability and retrofitting by damped braces of fire-damaged r.c. framed buildings. *Engineering Structures*. **101** 179-192
[9] Bikhiet M M, Nasser F E S and El-Hashimy H M 2014 Behavior of reinforced concrete short columns exposed to fire, *Alexandria Engineering Journal*. **53** 643-653
[10] Haddad R, Al-Mekhlafy N and Ashteyat A 2011 Repair of heat-damaged reinforced concrete slabs using fibrous composite materials, *Constr. Build. Mater.* **25** 1213-1221
[11] Hager I 2013 Behaviour of cement concrete at high temperature. Bulletin of the Polish Academy of Sciences: *Technical Sciences*. **61** 145-154
[12] Mohammad hosseini H and Yatim J M 2017 Microstructure and residual properties of green concrete composites incorporating waste carpet fibers and palm oil fuel ash at elevated temperatures. *Journal of Cleaner Production*. **144** 8-21
[13] BS 476 1987 *Method for determination of the fire resistance of elements of construction* (London: British Standard) Part 20 pp 1 -22
[14] BS 1881-116 1983Method for determination of compressive strength of concrete cubes (London: British Standard) page 1-8.
[15] Kodur V 2014 Properties of Concrete at Elevated Temperatures *ISRN Civil Engineering* vol 2014 Article ID 468510
[16] Düğenci, O, HaktaniR T and Altun F 2015 Experimental research for the effect of high temperature on the mechanical properties of steel fiber-reinforced concrete, *Constr. Build. Mater* **75** 82-88

**Acknowledgments**
The authors acknowledge the technical and financial support provided by UniversitiSains Malaysia under Research University (Individual) Grant (number 814190) and UniversitiTun Hussein Onn Malaysia Vot (U523).

IOP Conf. Series: Materials Science and Engineering **271** (2017) 012014 doi:10.1088/1757-899X/271/1/012014

# Vibration behaviour of foamed concrete floor with polypropylene and rise husk ash fibre

**N A Mohd Azaman[1], N H Abd Ghafar[1], N Ayub[1] and M Z Ibrahim[1]**

[1]Department of Structural and Material Engineering, Faculty of Civil and Environmental Engineering, Universiti Tun Hussein Onn Malaysia, Malaysia
Corresponding author: noryati@uthm.edu.my

**Abstract.** In the history of the construction industry, lightweight concrete or foamed concrete is a special concrete which can very useful in the construction sector because it is very lightweight and it can compact by itself at each angle of foamwork. Foamed concrete is one of lightweight concrete which widely used for floor construction due to its light weight and economic. The significant challenges in the floor design process are considering the vibration that needs improvements for the poor dynamic behaviour insulation. An alternative material to replace sand with certain amount of rice husk ash (RHA) and polypropylene was introduced. Research was determine the dynamic behavior of foam-polypropylene and foam-RHA concrete by using impact hammer test. The natural frequency for normal foamed concrete, 0.5 % of Polypropylene and 15% of RHA is 29.8 Hz, 29.3 Hz and 29.5 Hz respectively.

## 1. Introduction

Lightweight concrete has becoming a demand in a construction development due to its numerous benefits especially with the attempts at achieving weight reduction along with the standard of comfort that was established by the light weight floor system. Nowadays, many construction industries have been using foamed concrete floor system in construction. This is because the foamed concrete floor has many advantages such as low density, low construction cost and low material disposal [1, 2].

The practical range of concrete density for lightweight concrete is between 300 kg/m$^3$ and 1850 kg/m$^3$. Compared with normal weight concrete which has the density in the range of 2250 to 2400kg/m$^3$, lightweight concrete structure has the density of 1400 to 1840kg/m$^3$ which generally 20% to 40% lighter than normal weight concrete. For application to structural, partition, insulation and filling grades, a wide ranges of densities (400-1600 kg/m$^3$) of foamed concrete can be obtained by proper controlled in dosage of foam [3].

Research had been done by Zegers on the lightweight floor system, with the focus on the low frequency comfort, he stated that in designing light weight floor systems specific problems are encountered. Particularly comfort demands of such a lightweight floor system are harder to meet, as has been reviewed on the existing floor systems; especially vibrations related demands in combination with lightweight floor systems are not yet successfully combined [4].

Modern technologies preferred lighter weight and long floor system which may bring trouble to vibration. A floor system is one of the most complex and important parts of a building structure. The demands for the improvement of floor vibrations have been brought up since years. Since excessive floor vibration may lead to discomfort and is inconvenience to the people especially to the building occupants as heavy-weight impact noise causes by any rhythmic activities such as foot walk, rearranging furniture and rattling of boards and windows in which becomes a main source of indoor noises [5].

In order in improvise the vibration behaviour of light-weight concrete, this research was continued with the vibration behaviour of foamed concrete floor with RHA and Polypropylene. Impact hammer test was used to determine the frequency estimation, damping ratio and mode shape of the floor structure. According to Murray, limits for vibration frequencies between 4Hz to 8Hz. Outside this frequency range, people accept higher vibration acceleration [6].

Many researchers has investigated to use other composite material to incorporated with the concrete to enhance the floor performance such as by replacing certain amount of cement or sand with organic or synthetic waste material such as rice husk ash and polypropylene.

Rice Husk Ash (RHA) which is an agricultural by-product has been reported to be a good pozzolan by numerous researchers. Investigation that have done by Mehta and Pirth on used of RHA to reduce temperature in high strength mass concrete and got result showing that RHA is very effective in reducing the temperature of mass concrete compared to OPC concrete[7]. Mehta again reported that ground RHA with finer particle size than OPC improves concrete properties, including that higher substitution amounts results in lower water absorption values and the addition of RHA causes an increment in the compressive strength[8]. There were many research on using varying percentages of RHA for concrete strength and its workability however lack of research on dynamic behaviour or RHA in foamed concrete. Therefore, a certain percentage of RHA was used in foamed concrete to study its dynamic behaviour on floor performances.

In order to increase strength and reduce brittleness of foam concrete Polypropylene fibre is introduced. Polypropylene (PP) fibre is introduced to increase the flexural strength and impact resistance of the foam concrete as well as can help in shrinkage failure problem of the foam concrete [9]. The usage of these fibres has increased tremendously in construction of structures because addition of fibres in concrete improves the toughness, flexural strength, tensile strength and impact strength as well as failure mode of concrete [12].

## 2. Experimental of RHA and Polypropylene Concrete

The slab design is 1.6m long and 0.75m width with the thickness of 0.125m as shown in figure 1. Additional material is 0.5% and 15% for poplypropylene and RHA repectively added into the mix of foam concrete.Table 1 illustrated the material characteristic of the floor system.

**Figure 1.** Actual size of slab.

**Table 1.**Characteristic of floor system.

| Material characteristic | Foamed concrete with RHA | Foamed concrete with Polypropylene |
|---|---|---|
| Density (kg/m$^3$) | 1600 | 1600 |
| RHA | 0.15 to sand replacement | - |
| Polypropylene | - | 0.05 to sand replacement |
| Cement-sand ratio (C/S) | 1:3 | 1:3 |
| Water-cement ratio (W/C) | 0.55 | 0.55 |
| Foam Agent-Water ratio (FA/W) | 1:20 | 1:20 |

## 3. Vibration due to Impact Hammer

Impact hammer test were used to determine the vibration behaviour of the slabs. Data were obtained through vibration impact from the hammer strike and accelerometer were recorded the data. Before the testing conducted, all slabs have a gridline to make testing process easier. Hammer was strike on location A as shown in figure 2. The impact hammer with weighted about 5 kg was been strike using super-soft tip. The hammer have been imparted in about 90° angle in relation to the object surface and applied constant force. For each tests of impact hammer, it were been take ten times of data collection. Two accelerometer were been placed on the grid point and it was been roving during the test until the test done. Accelerometer is used to measuring acceleration and it is very sensitive and must be placed on plane surface to avoid from noise.

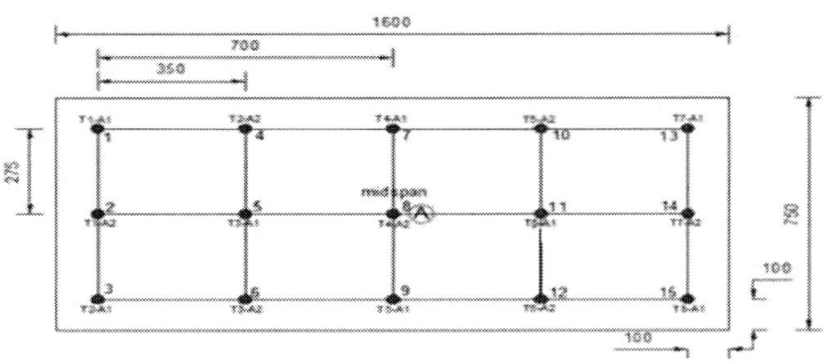


**Figure 2.** Gridline of slab for testing.

### *3.1. Result and Discussion*

From the analysis of dynamic behaviour, natural frequency are identified by peaks in the Fast Fourier Transform (FFT) graph. FFT resolves a time waveform into its sinusoidal components. The FFT takes a block of time-domain data and return the frequency spectrum of the data. Natural frequency and of normal foamed-concrete, 0.5% Polypropylene and 15% RHA is 29.8 Hz, 29.3 Hz and 29.5 Hz, respectively. Damping ratio is defined using ratio of two peaks and calculated using Equation (1)[10].

$$\delta = \frac{1}{m} ln \frac{y_n}{y_{n+m}} \tag{1}$$

where, $m$ is cycle apart, $y_n$ is the amplitude on $n^{th}$ and $y_{n+m}$is the amplitude of the $n+m^{th}$ cycle. Damping ratio for normal foamed-concrete, 0.5% Polypropylene and 15% RHA is 3.37%, 4.79% and 3.98%, respectively.

Figure 3 and figure 4 illustrated the mode shapes for each type of slab. The measured mode shapes and frequencies presented were calculated from the data obtained. The analysis showed that, the first mode shape and second mode shape for each slab were approaching the similarity feature shape between all slabs.

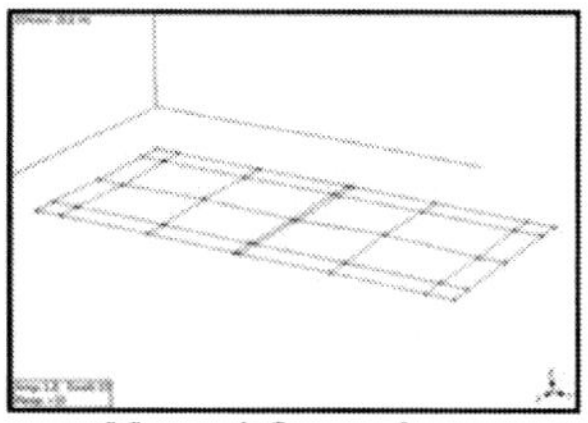

a. Normal foamed-concrete

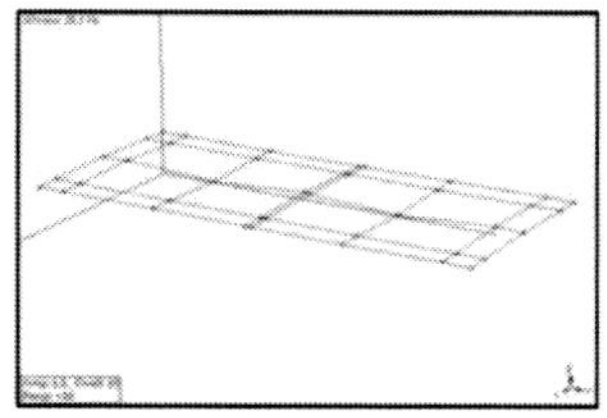

b. 0.5% of Polypropylene

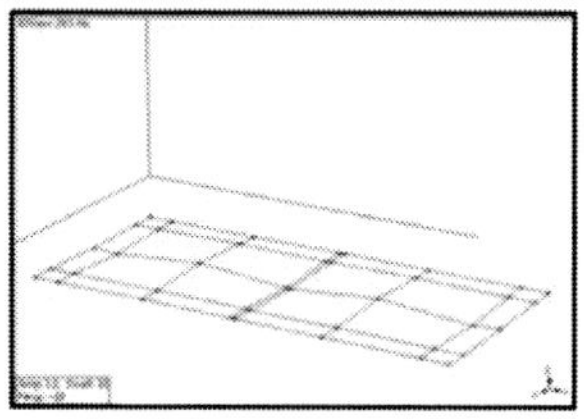

c. 15 % of RHA

**Figure 3**. First mode shape.

The first mode shape is shown in figure 3 for all types of slab. For normal foamed concrete, its frequency is 29.8Hz, 0.5% of Polypropylene is 29.3 Hz and 15% of RHA is 29.5 Hz. The mode shape is quite similar to the deflected mode caused by uniform gravity loading. The highest displacement is located at the center of mode shape.

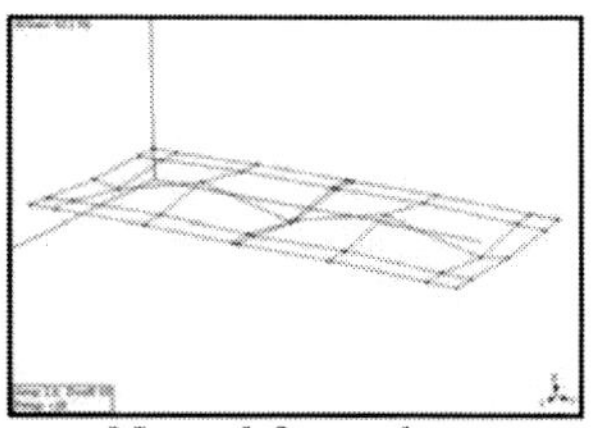

a. Normal foamed-concrete

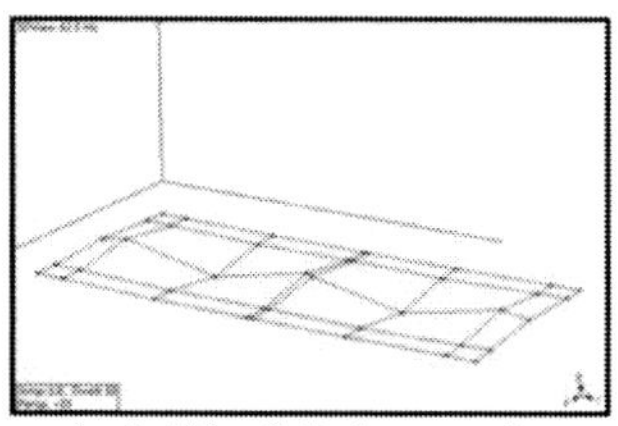

b. 0.5% of Polypropylene

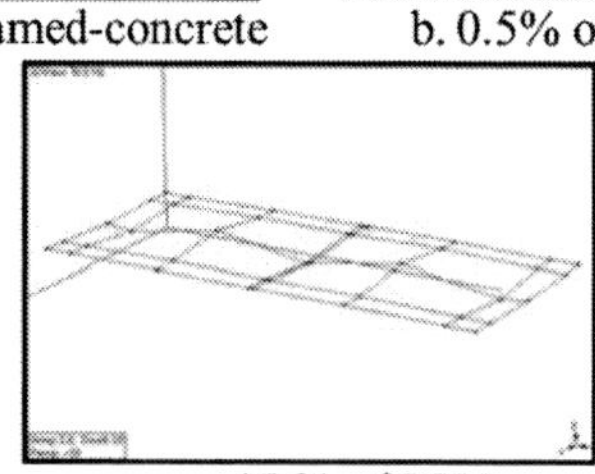

c. 15 % of RHA

**Figure 4**. Second mode shape.

IOP Conf. Series: Materials Science and Engineering **271** (2017) 012014 doi:10.1088/1757-899X/271/1/012014

The second mode shape is shown in figure 4 for all types of slab. For normal foamed concrete, its frequency is 62.1 Hz, 0.5% of Polypropylene is 62.5 Hz and 15% of RHA is 50.6 Hz. The mode shape is quite similar to the deflected mode caused by uniform gravity loading.
Natural frequency is the frequency at which a structure will vibrate when displaced and then quickly released. Overall, the experimental modal analysis gives the value of natural frequencies in range 29.3Hz to 29.8 Hz for all type of slabs. These frequencies are acceptable because its above limit for vibration frequencies between 4 Hz to 8Hz based on Design Guide. Damping usually expressed as the percent of critical damping or as the ratio of actual damping to critical damping. For damping that determined from the experimental data, it gives high value of 3.37%, 4.79% and 3.98%. Although it high but still recommended based on previous reference by Aalami [11] which is from 2% -3% for bare concrete floor and 5% -8% with full height partitions.

Elastic modulus plays an important part in the vibration of wave. Since the disturbance from surrounding cannot perfectly eliminate, it was important to obtain the vibration reactions characteristics. Wave propagation was occurred by excitation and by bouncing. Figure 5 demonstrated the wave propagation from the analysis. It shown by using wave propagation, vibration interference can be easily recognized.

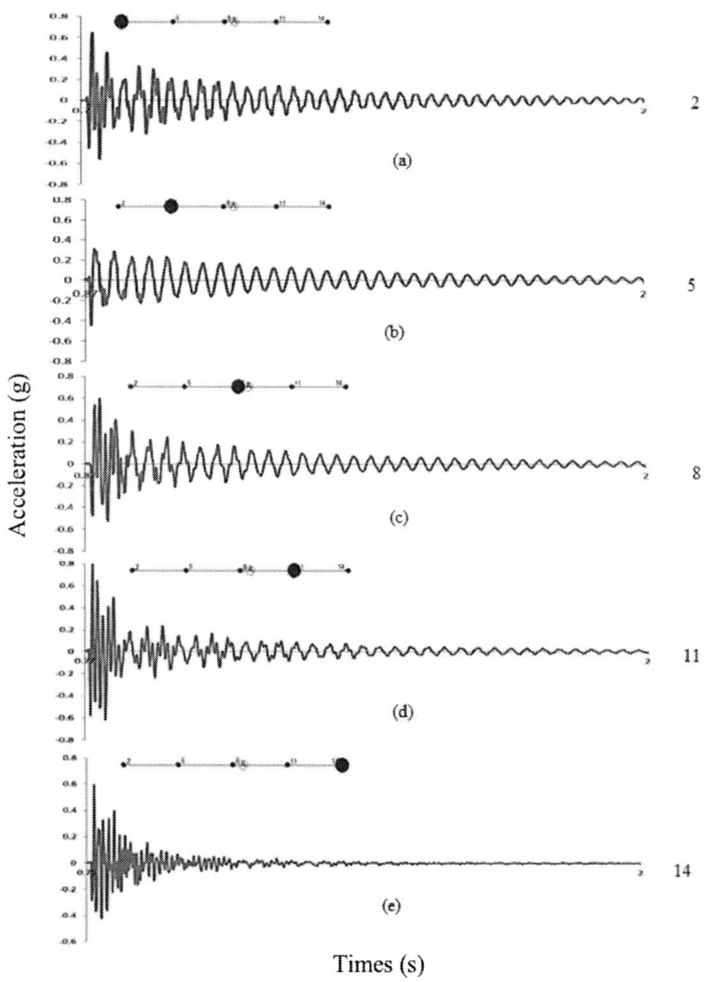


**Figure 5**. Wave propagation.

## 4. Conclusion

The main conclusion that can be made from the case study are, the experimental natural frequency for normal foamed, 0.5% of Polypropylene and 15% of RHA is 29.8 Hz, 29.3 Hz and 29.5 Hz, respectively. Therefore, it is accepted based on previous research [6], which is stated the limit frequency between 4-8 Hz.

For damping, the experimental damping ratio for first mode foamed-concrete slab is 3.37%, 0.5 % of Polypropylene is 4.79% and for 15% of RHA is 3.98%. Recommended damping ratio for bare concrete floor based on previous research, Aalami [11] is 2.0-3.0%. Therefore, the damping ratio are accepted based on recommended from previous study.

## 5. References

[1] Bala A, Sehgal V K and Saini B 2014 Effect of Fly ash and waste rubber on properties of concrete composite, *Concrete Research Letters*. *5*(3) 842-857

[2] Weigler H and Karl S 1980 Structural lightweight aggregate concrete with reduced density-lightweight aggregate foamed concrete. *International Journal of Cement Composites and Lightweight Concrete*.**2**(2) 101-104.

[3] Ramamurthy K, Kunhanandan Nambiar E K and Indu Siva Ranjani G 2009 A classification of studies on properties of foam concrete. *Cement and Concrete Composites*. **31**(6) 388–396

[4] Zegers S 2011 *Lightweight floor system for vibration comfort*. Eindhoven University

[5] Chik T N T, Yabi S, Yusoff N A and Ghazali M I2015Analysis of the Effect of Vibration from Footfalls on Office Building. *Proceedings of the International Civil and Infrastructure Engineering Conference(InCIEC)* pp. 3-15

[6] Murray T M and Allen D E and Ungar E E 1997 Floor Vibration. Vibration Due to Human Activity, *AISC/CISC Steel Design Guide 11*, Chicago : American Institute of Steel Construction.

[7] Mehta P K and Pirtz, D 2000 Use of rice husk ash to reduce temperature in high strength mass concrete. *ACI Journal Proceedings*. **75** 60-63

[8] Malhotra V M and Mehta P K 2004 Pozzolanic and Cementitious Materials. *CRC Press 2003* (London: Taylor & Francis)

[9] Alhozaimy A M, Soroushian P and Mirza F 1996 Mechanical properties of polypropylene fiber reinforced concrete and the effects of pozzolanic materials. *Cement and Concrete Composites*. **18(2)** 85-92

[10] Saidi I, Haritos N, Gad E F and Wilson J L 2006 Floor vibrations due to human excitation - damping perspective, Earthquake Engineering in Australia , 257–264.

[11] Aalami B O 2008 Technical Note. Vibration Design of Concrete Floors for Serviceability, ADAPT Corporation, Redwood City, CaliforniaUSA**650** 1–20

[12] Zollo R F 1997 Fiber-reinforced concrete: an overview after 30 years of development. *Cement and Concrete Composites* **19(2)** 107–12

## Acknowledgement

This work was supported by Universiti Tun Hussein Onn Malaysia. The author also wishes to acknowledge Centre of Graduate Studies, UTHM.

# Sound absorption and morphology characteristic of porous concrete paving blocks

**N H Abd Halim[1], H Md Nor[1], P J Ramadhansyah[1], A Mohamed[1], N Abdul Hassan[1], M H Wan Ibrahim[2], N I Ramli[3] and F Mohamed Nazri[4]**

[1] Faculty of Civil Engineering, Department of Geotechnics and Transportation, Universiti Teknologi Malaysia, 81310 Johor Bahru, Malaysia
[2] Faculty of Civil and Environmental Engineering, Universiti Tun Hussein Onn Malaysia, 86400 Batu Pahat, Johor Bahru, Malaysia
[3] Faculty of Civil Engineering and Earth Resources, Universiti Malaysia Pahang, 26300 Gambang, Pahang, Malaysia
[4] School of Civil Engineering, Universiti Sains Malaysia, Engineering Campus, 14300 Nibong Tebal, Penang, Malaysia
Corresponding author: ramadhansyah@utm.my

**Abstract.** In this study, sound absorption and morphology characteristic of Porous Concrete Paving Blocks (PCPB) at different sizes of coarse aggregate were presented. Three different sizes of coarse aggregate were used; passing 10 mm retained 5 mm (as Control), passing 8 mm retained 5 mm (8 – 5) and passing 10 mm retained 8 mm (10 – 8). The sound absorption test was conducted through the impedance tube at different frequency. It was found that the size of coarse aggregate affects the level of absorption of the specimens. It also shows that PCPB 10 – 8 resulted in high sound absorption compared to the other blocks. On the other hand, microstructure morphology of PCPB shows a clearer version of existing micro-cracks and voids inside the specimens which affecting the results of sound absorption.

## 1. Introduction

An increase in volume of road traffic has resulted in an increase in road-generated disturbance to nearby residents as well as contributing to environmental noise pollution [1,2]. If the traffic noise sources are considered in more detail it turns out that vehicle noise is related by the engine, the exhaust opening, vehicle speed and the tyres [3]. Among all, tyres noise is certainly a very important problem in noise control. Tyres noise is generated by tyres vibration and the movement of air particles in the tread tyre pattern [4]. Porous surface has been introduced to reduce the tyre noise. A dense road surface reflects the sound energy, while a porous road surface absorbs the sound energy [5-7]. Ferguson also states that a porous surface absorb sound energy and allow some of the air around the tires to be pressed into the voids, wipe out air pressure before any noise is generated. Nowadays, porous surface has been applied in concrete block shape used for pavement; known as Porous Concrete Paving Blocks (PCPB) or Permeable Blocks. The use of open-graded aggregate with little or no fine aggregates can achieve a continuous void network inside the block; which allowed water to infiltrate directly through the pavement and also help in reducing the noise generate from interconnected between tyres and road surface [8]. Yu et al. [9] reported that noise level generated from PCPB is found to be very similar to those of the ordinary asphalt pavement.

IOP Conf. Series: Materials Science and Engineering **271** (2017) 012015 doi:10.1088/1757-899X/271/1/012015

## 2. Materials and Methods

### *2.1. Ordinary Portland Cement and Coarse Aggregate*
The three sizes of crushed granite as coarse aggregate were used; passing 10 mm retained 5 mm (Control), passing 8 mm retained 5 mm (8 - 5) and passing 10 mm retained 8 mm (10 – 8) and mixed with Ordinary Portland cement (OPC) as a major binder. No fine aggregate used, to allow voids inside the block. The water-cement ratio of 0.35 was used and for compaction process; the mixed were vibrated using vibrating table for less than 3 seconds. The blocks were air cured for 7 days and 28 days at room temperature.

### *2.2. Sound Absorption Test*
The absorption coefficient is commonly reported as a measure of a material ability to absorb sound [10]. In this study, the acoustic absorption coefficient was measured using an impedance tube as per ASTM E 1050 [11]. The specimens with diameter 100 mm and thickness of 80 mm was placed in a thin cylindrical and placed against a rigid backing at one end of the impedance tube (as in Figure 1a and Figure 1b). Microphones placed along the length of the tube were used to detect the sound pressure level (P), which were translated into the reflection (R) and absorption coefficient (α)1.

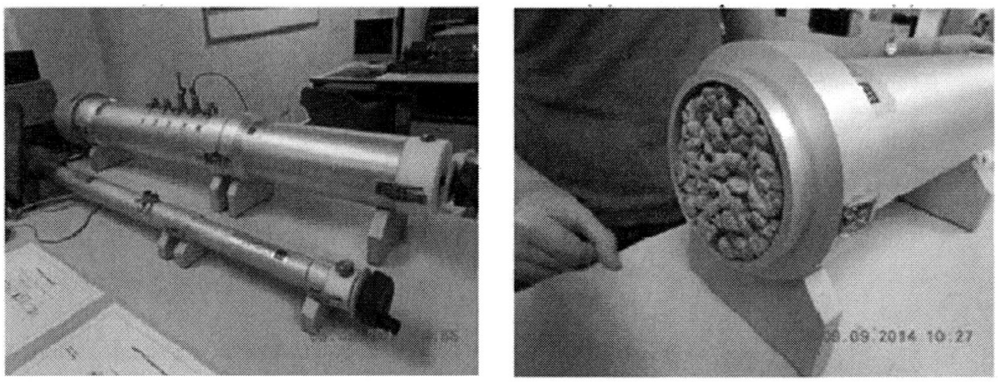

**Figure 1**. (a) Sound absorption test instrument, left and (b) Position of a specimen, right.

### *2.3. Field Emission Scanning Electron Microscopy*
Field emission scanning electron microscopy (FESEM) studies were carried out on platinum-coated fractured surface PCPB samples using Hitachi Scanning Electron Microscope SU8020. In general, FESEM is microscopies techniques are used to magnify images of surface morphology for example the shape and size of the topographic features of the specimen surface.

## 3. Results and Discussion

### *3.1. Sound Absorption of PCPB*
In order to analyse the sound absorption characteristics of PCPB, the specimens were measured for each frequency using the impedance tube. The measurement result is shown in figure 2. It can be seen that the results of Jusli *et al.* [12] the PCPB of control specimens, the absorption coefficient became highest in the frequency range of 500 – 550 Hz; Nur Hidayah *et al.* [13] for the PCPB 8 – 5, it becomes highest in the frequency range of 550 – 600 Hz; and Jusli *et al.* [14] for the PCPB 10 – 8, become highest in the frequency range of 600 – 650 Hz. This signifies that the frequency of the highest absorption coefficient was obtained as the coarse aggregate size increased. This is because absorption became higher as the porosity and voids inside the blocks increased. This result has been supported by previous findings by Park et al. [15] where the absorption becomes higher as the surface area of voids increased and resulting in the highest sound absorption coefficient.

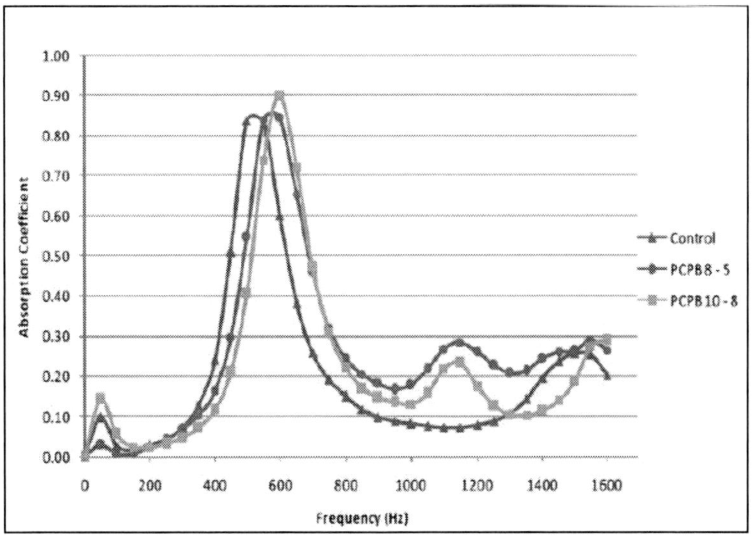


**Figure 2**. Sound absorption of PCPB specimens.

*3.2. Sound Absorption of CPB*

By comparing the results of PCPB with the CPB in sound absorption, it proved that by eliminating the fine aggregate in the mix will contribute to high sound absorption. In general, the results of CPB specimen (Figure 3) with all sizes of coarse aggregate in sound coefficient is less than 0.1 compared to PCPB specimens which is more than 0.8. This shows that, PCPB specimens suggested was applied at road with high noise generated.

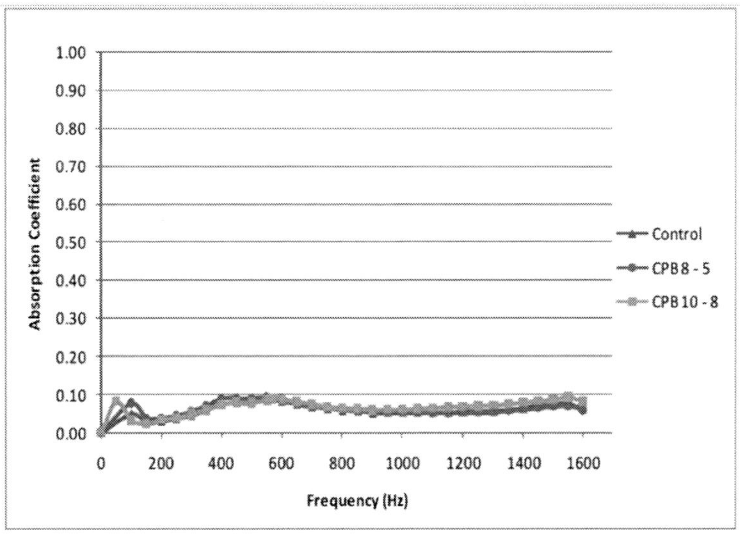


**Figure 3**. Sound absorption of CPB specimens.

### *3.3. Morphology Image*

FESEM investigations on hardened PCPB show distinct changes in the morphology after 7 days curing process at room temperature. The samples drawn from the fractured surface of the PCPB specimens were examined. A number of FESEM micrographs illustrating various characteristic of PCPB as in Figure 4. FESEM micrographs of hardened PCPB at room temperature, respectively, reveal well developed hydrated phases such as Portlandite, $Ca(OH)_2$ crystals (marked "CH") intermixed with C-S-H (marked "CSH"), voids (marked "V") and deformation of ettringite, $\{Ca_6[Al(OH)_6]2\cdot 24H_2O\}[(SO_4)3\cdot 1.5H_2O]$ (marked "E"). In addition, due to no fine aggregate in the PCPB, it leads to the predominance of micro-cracks and voids as shown in Figure 4.

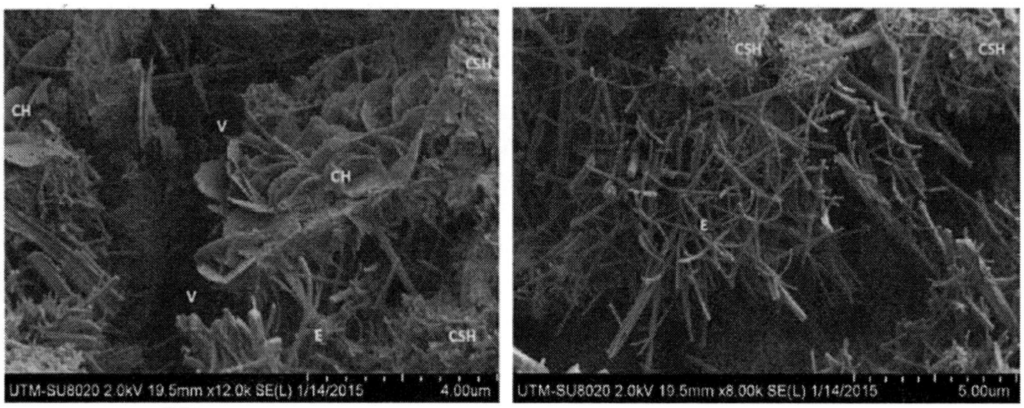


**Figure 4**. Morphology in the porous concrete paving blocks specimen.

## 4. Conclusions

a) By eliminating the fine aggregate in mixed, results in an increasing absorption coefficient. In this study, it shows that approximately 80 % to 90 % improves in sound absorption with PCPB and it's also suggested as noise reducing concrete block for pavement.
b) An increase in coarse aggregate size results in high sound absorption, it is due to the increased of the pore size and porosity. In this study, PCPB with CA 10 – 8 give in the result of high absorption coefficient at the highest frequency compared to other specimens.

## 5. References

[1] Nur Hidayah A H, Nor H M, Ramadhansyah P J 2016 Effect of open-graded coarse aggregate on concrete paving blocks properties for pavement, *ARPN J. Eng. Appl. Sci.* **11** 11836-11839
[2] Graf R A G, Kuo C Y, Dowling A P, Graham W R 2002 On the horn effect of a tyre/road interface, part I: experiment and computation, *J. Sound Vibration* **256** 417-431
[3] Heckl M 1986 Tyre noise generation, *Wear* **113** 157-170
[4] Gołebiewski R, Makarewicz R, Nowak M, Preis A 2003 Traffic Noise reduction due to the porous road surface, *Appl. Acoustics* **64** 481-494
[5] Jusli E, Nor H M, Putra Jaya R, Haron Z 2014 Double layer concrete paving blocks using waste tyre rubber as aggregate replacement, *Appl. Mech. Mater.* 554 128-132
[6] Yamaguchi M, Nakagawa H, Mizuno T 1999 Sound absorption mechanism of porous a sphalt pavement, *J. Acoustical Society Japan* **20** 29-43
[7] Bérengier M C, Stinson M R, Daigle G A, Hamet J F 1997 Porous road pavements: Acoustical characterization and propagation effects, *J. Acoustical Society America* **101** 155-162.
[8] Nur Hidayah A H, Nor H M, Ramadhansyah P J 2014 Porous concrete paving blocks using coarse aggregate, *Appl. Mech. Mater.* **554** 111-115
[9] Yu C, Ke-Jin W, Di L 2012 Mechanical properties of pervious cement concrete, *J. Cent. South. Univ.* **19** 3329–3334

IOP Conf. Series: Materials Science and Engineering **271** (2017) 012015 doi:10.1088/1757-899X/271/1/012015

[10] Marolf A, Neithalath N, Sell E, Wegner K, Weiss J, Olek J 2004 Influence of aggregate size and gradation on acoustic absorption of enhanced porosity concrete, *ACI Mater. J.* **101** 82-91

[11] ASTM E1050 2012 *Standard test method for impedance and absorption of acoustical materials using a tube, two microphones and a digital frequency analysis system* (United States: American Standard) pp 1-12

[12] Jusli E, Nor H M, Putra Jaya R, Haron Z 2014 Mechanical properties of double layer rubberized concrete paving blocks, *Adv. Mater. Res.* **911** 463-467

[13] Nur Hidayah A H, Nor H M, Azman M 2014 Effect of jointing sand sizes and width on horizontal displacement of concrete block pavement, *J. Teknologi* **71** 83-86

[14] Jusli E, Nor H M, Jaya R P, Haron Z, Azman M 2014 Effect of using waste tyre rubber on the properties of double layer rubberized concrete paving blocks, *J. Teknologi* **71** 1-5

[15] Park S B, Seo D S, Lee J 2005 Studies on the sound absorption characteristics of porous concrete based on the content of recycled aggregate and target void ratio, *Cement Concrete Res.* **35** 1846-1854

**Acknowledgments**

The support provided by Malaysian Ministry of Higher Education and Universiti Teknologi Malaysia in the form of a research grant vote number Q.J130000.2522.18H05 and Q.J130000.2522.11H76 for this study is highly appreciated.

IOP Conf. Series: Materials Science and Engineering **271** (2017) 012016 doi:10.1088/1757-899X/271/1/012016

# Study on effects of different patterns and cracking for wastes FRP (used banner) wrapping on compressive strength of confined concrete

**Alif Syazani Leman[1], Shahiron Shahidan[1],M. A. M Azmi[1], Mohamad Syamir Senin[1], N Ali[1], S.R. Abdullah[1], S.S. Mohd Zuki[1], M. H. Wan Ibrahim[1] and Fadzli Mohamed Nazri[2]**

[1]Jamilus Research Centre, Faculty Of Environment And Civil Engineering, Parit Raja, 86400, Batu Pahat, Johor Malaysia.
[2]School of Civil Engineering, USM, Engineering Campus, 14300 Nibong Tebal, Seberang Perai Selatan, P. Pinang, Malaysia
Corresponding author:shahironshahidan@gmail.com

Abstract.Previous researches have shown that FRP are being introduce into wide variety of civil engineering applications. Fibre Reinforce Concrete (FRP) are also used as repairing method in concrete structures. FRP such as S-glass, AR-glass, E-glass, C-glass, and Aramid Fibre are the common material used in industry. The FRP strips provide the necessary longitudinal and hoop reinforcement. However, there are lots waste materials that can be form as fibre and used in repairing. Banner is a type of waste material fibre that can be used in repairing. In this study, banner will be used as the replacement of the common FRP. The confined concrete (cylinder) of 300mm height and 150mm diameter were cast with M35 grade concrete and tested until it is crack. Next banner are used as the wrapping along the cracking of the concrete with three different pattern that are full wrapping, two band wrapping and cross wrapping using epoxy. Epoxy is a common name for a type of strong adhesive used for sticking things together and covering surface. The objective of this study is to determine the maximum strength and the effect of different patterns wrapping of FRP (banner) on the compressive strength of confined concrete. The results are shows that banner are suitable as a replacement of material for FRP.

## 1. Introduction

Recently, Waste material or unused material is one of the biggest contributions to environmental pollution. Examples of waste material include municipal solid waste (household trash/refuse), hazardous waste, waste water, radioactive waste and others. According to [1-3], wastes materials are not the prime product for which the initial user has no further use in terms of purposes of production, transformation or consumption of which it will be dispose. Waste material may be generated during the extraction of raw materials, the processing of raw materials into intermediate and final products, the consumption of final products and other human activities.

Banner is an example of waste or unused material. Banner is a flag or other piece of cloth bearing a symbol, logo, slogan or other message as shown in Figure 1.3. It is used in many business ventures, marketing to their potential audience. In Malaysia, banner is widely used to inform the people about the upcoming event. All the details such as name of program, date and venue are printed on the banner.

When the event or program is already underway, the banner become a useless item and will be thrown away or leave it hanging without been used by people.

## 2. Literature Review

Banner is a type of fibre that is strong enough to withstand any type of temperature. Banner can be used in substituting existing FRP because of the similar properties between each other. In Malaysia, banner is used as an information part to show the upcoming event as shown in Figure 1 Therefore it becomes waste material when the event has been completed.

According to P. Sadeghian[4], many researchers studied the behaviour of concrete members wrapped with external bonded fibre reinforced polymer (FRP) composites and proposed a variety of confined models in the past two decades.This FRP has two different field of application in repair and strengthening of existing reinforced concrete column[5-7]. The first field is to repair and upgrading for earthquake resistance and the second field is to repair of corrosion damaged members where the transverse and the longitudinal reinforced are not dependable.

In a past few decade, FRP were used to strengthen reinforced concrete [8]. FRP is made up by combining two or more materials that is to be form as composite materials. Sadeghian and Fam [4] and Gunasekaran *et al.* [9] stated that FRP is made up of fibre which will provide composite strength and matrix that gives rigidity and environmental protection as shown in figure 1.

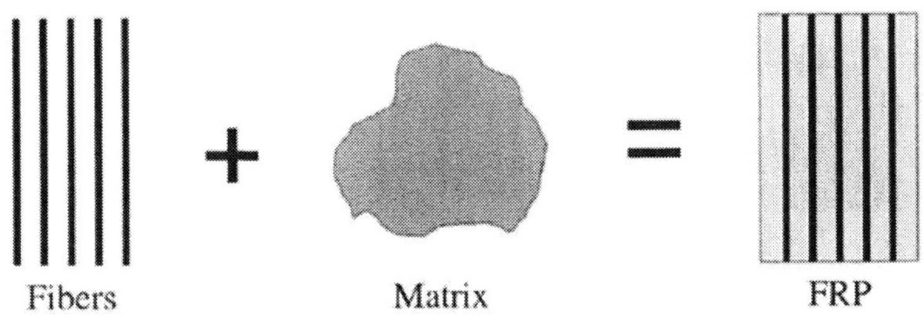


**Figure 1**. How Fibre Reinforce Polymer Was Formed [10].

## 3. Type of FRP

Typical fibre reinforcement used in the composite industry is glass (E-glass and S-glass), carbon and aramid.

### *3.1. Glass Fibre*

Glass fibres are produced by fusing silicate with silica or with potash, lime, or various metallic oxides. The molten mass is passed through micro-fine bushings and rapidly cooled to produce glass fibre filaments ranging in diameter from 5 to 24μm. These filaments are then drawn together into closely packed strands or loosely packed roving. During this process, the fibres are frequently covered with a coating, know as sizing, to minimize abrasion-related degradation of the filaments [10-12].

### *3.2. Carbon Fibre*

Commercial quantities of carbon fibres are derived from three major feedstock or precursors sources: rayon, polyacrylonitrile, and petroleum pitch. Rayon precursors, derived from cellulose materials, were one of the earliest sources used to make carbon fibres. Their primary advantage was their widespread availability. The most important drawback was the relatively high weight loss, or low conversion yield to carbon fibre, during carbonization. Carbonization is the process by which the precursor material is chemically changed into carbon fibre by the action of heat. On average, only 25% of the initial fibre mass remains after carbonization. Therefore, carbon fibre made from rayon precursors is more expansive than carbon fibres made from other materials [13-14].

*3.3. Aramid Fibre*

This fibre has the lowest specific gravity and the highest tensile strength-to-weight ratio among the reinforcing fibres used today. It is 43% lighter than glass approximately 20% lighter than most carbon fibres. In addition to high strength, the fibres also offer good resistance to abrasion and impact, as well as chemical and thermal degradation. Major drawbacks of these fibres include low compressive strength, degradation when exposed to ultraviolet light and considerable difficulty in machining and cutting [11-12 ,15].

## 4. Waste Material As Fibre Reinforced Polymer (FRP)

According to Vermaet al [16], the abundant availability of natural fibre such as banana, sugarcane, bamboo etc. gives attention on the development of natural fibre composites primarily to explore value-added applications avenues. Such natural fibre composites are well suited as wood substitutes in the housing and construction sector. Reinforcement with natural fibre in composites has recently gained attention due to low cost, low density acceptable specific properties, ease of separation, enhanced energy recovery.

*4.1Sugarcane Bagasse*

According to Hajihaand Sain [17],sugarcane bagasse has a reasonable tensile strength and modulus, thus it has the potential to be used as reinforcement in composites. It has a tensile strength in the range of 170-290 MPa and a modulus elasticity in the range of 15-19 GPa. [18], state that bagasse is a fibrous residue obtained from sugarcane during extraction of sugar juice. As a rough estimate, for every 10 tonnes of sugarcane which is crushed, 3 tonnes will become wet bagasse.

*4.2 Banana Fibre*

Chandramohan and Marimuthu [19], stated that banana fibre obtained from the pseudo-stem of banana plant (Musa sepientum), is a bast fibre with relatively good mechanical properties. The "pseudo-stem" is clustered, cylindrical aggregation of leaf stalk bases. The extraction of fibre from the pseudo-stem is not common practice and much of the stem is not used for production of fibres.

*4.3 Used Banner*

Banner is flag or other piece of cloth bearing a symbol, logo, slogan or other message. It is use as a promotion for the upcoming event or any other details that are relevant for a program. Figure 2 below shows the example of a typical banner that is mostly found in Malaysia. The most commonly used material is a heavy weight vinyl technically known as PVC. The weights of the different banner substrates from as light as 255 gram to 623 gram per square yard, and may be double or single sided. Grommets can also be added in order to facilitate hanging of the banner. Banner tapes are also used to protect the lamination of the banner, fasten the hems neatly and adhere to various surfaces. Large banners (which can be so large that they cover the side of a building) are printed on a special mesh material so that the wind can pass through them.

IOP Conf. Series: Materials Science and Engineering **271** (2017) 012016 doi:10.1088/1757-899X/271/1/012016

**Figure 2**. Typical banner.

## 5. Material
Good and quality materials used in this study will ensure a good result or building.

### *5.1 Concrete*
Concrete mix design of grade 35 is used in the preparation of concrete cylinder with the height of 300mm and diameter of 150mm. There are total of 21 concrete cylinder need to be cast. Table 1 shows the summary of concrete mix design.

**Table 1**. Summary of Concrete Mix Design.

| | |
|---|---|
| Water Cement Ratio | 180 kg/m3 |
| Water | 180 kg/L |
| Cement | 450 kg |
| Fine Aggregate | 530 kg |
| Course Aggregate | 1240 kg |

### *5.2 Used Banner As FRP*
The banner is then cut to several lengths to make it easier to be patch with epoxy for this study. There were 3 type of cutting with different dimension for this research as shown in figure 3.

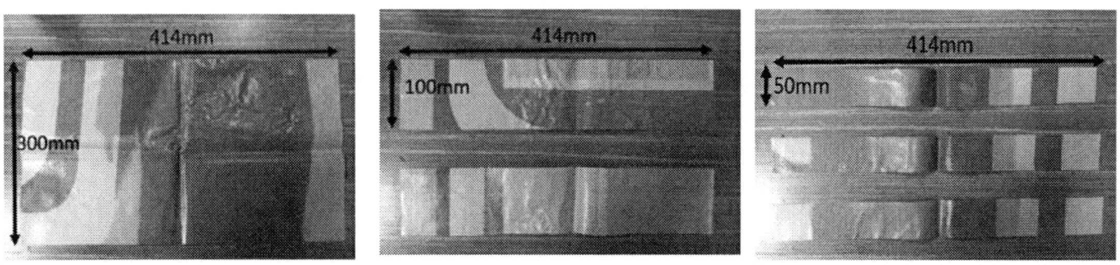


**Figure 3**. Shows the three type of cutting used banner.

## 6. Testing Method
This study had conducted 3 testing which is slump test, compression test and tensile test.

*6.1 Slump Test*
Slump test was conducted to determine the workability of the concrete mix. This test must be conducted before concrete is poured into the moulds. This test was conducted for each mixture to ensure that the concrete quantities of materials used, and designed in accordance with the ruins.

*6.2 Compression Test*
The concrete cylinder ware cast and cure for 28 days. It then were taken out from the curing tank. Excess water was wiped out from the specimen. The specimen is then placed vertically on the platform of compression testing machine. The load were apply continuously and uniformly without shock at the rate of 315 KN/min and continue until the load is half of the maximum load. The concrete cylinder is then taken out and repair by using FRP method with 3 different pattern. It then were placed again after being repaired and testing until it fails. The final load that it can withstand is then recorded. Figure 4 shows the concrete cylinder test after being repair.

**Figure 4**. Concrete cylinder testing after being repaired

*6.3 Repair Pattern*
After the concrete cylinder is being loaded with half loading it then were taken out and the cracking was check. Then the specimen were repaired using FRP. 3 batch of FRP wrap pattern crack repaired were used to see which is the most effective to be use worldwide. It then being test again to see which batch can withstand more loads. Figure 5 shows the type of wrapping pattern that will be considered to use in the study.

**Figure 5**. Wrapping pattern (A) Full wrapping pattern, (B) Cross wrapping pattern, (C) 2 band wrapping.

*6.4 Tensile Test*

Tensile strength were conducted in this study to know the tensile between CFRP and Banner. There were four specimen that is shaped look like alphabet ' I ' were used for this test. One layer of CFRP were used to get it maximum tensile. Next, banner with 1 layer, 2 layer and 3 layer were used to determine the layer that can archive the same maximum value as one layer CFRP. The machine is then stretched the specimen as in Figure 6 and the data were recorded in graph mode.

**Figure 6**. Tensile test were conducted.

## 7. Result and Analysis

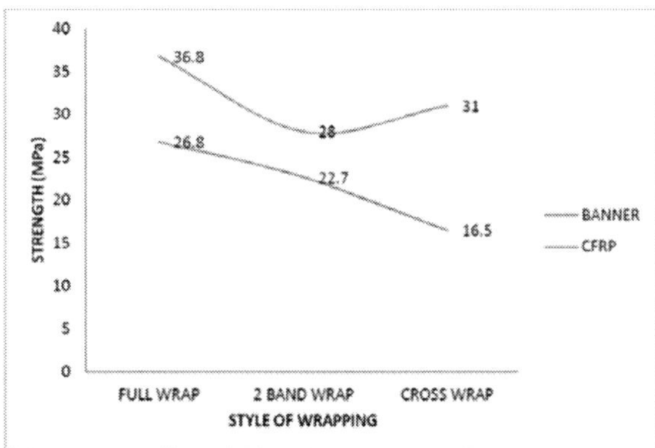


**Figure 7**. Wrapping pattern between banner and CFRP.

By referring to the Figure 7 it shows three type of wrapping style which is full wrap, 2 band wrap and cross wrap. For full wrap pattern, the value for used banner is 26.8 MPa while CFRP is 36.8 MPa. There are difference by 10 Mpa for this type of wrap. For 2 band wrap pattern, used banner value are 22.7 MPa and CFRP is 28 MPa. the difference value are 5.3 MPa. Lastly for cross wrap pattern, used banner value are 16.5 MPa and CFRP are 31 MPa. The difference between this two specimen are 14.5MPa.

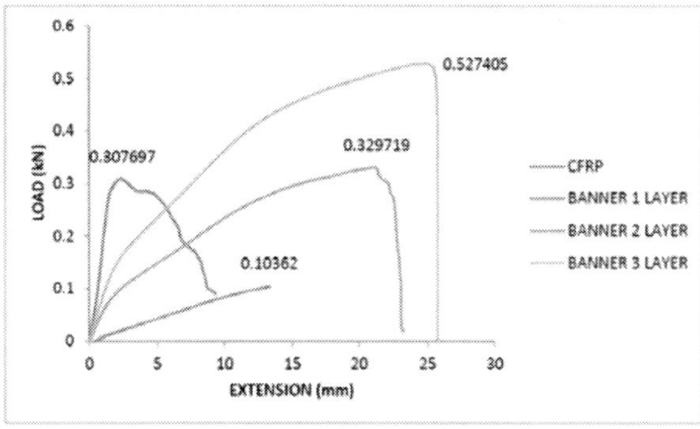


**Figure 8**. Graph of load vs extension for CFRP and used banner.

From the Figure 8, it shows that one layer of CFRP have the maximum value of load that it can withstand is 0.31kN with the extension of 9.32mm. Next for one layer of used banner, it can only withstand the maximum load of 0.10kN with the highest extension of 13.33mm. This reflect the result in the Figure 5.1 of this study when the used banner were attached to the concrete cylinder have a lower strength than the CFRP. The tensile test had been proved that one layer of used banner is weaker than one layer of CFRP.

From this graph, the 2 layer used banner specimen can withstand maximum load of 0.33kN with maximum extension of 23.20mm. This shows that 2 layer of used banner can have the same strength of one layer of CFRP. Lastly this graph also gives the result of three layer used banner which is 0.53kN with maximum extension of 25.78mm. This specimen gives a better result than one layer of CFRP. It can be conclude that 2 layers is enough to repair an ordinary crack and 3 layer can be used to repair a major crack.

## 8. Conclusions

Banner is a suitable material to replace CFRP for concrete repairing based on the data receive. However a deeper study needs to be conducted to know the properties of banner before it can be widely used in the industry.

## 9. References

[1] Shahidan S, Zuki S S M and Jamaluddin N 2016 Damage Grading System for Severity Assessment on Concrete Structure *Case Stud. Constr. Mater.* **5** 79–86

[2] Shahidan S, Mohd Zuki S S, Rahim M A and Ali N 2016 Amplitude Distribution of Emission Wave for Cracking Process vol 47(Paris: MATEC Web of Conference) pp 2–8

[3] Ismail I, Jamaluddin N and Shahidan S 2016 A Review on Performance of Waste Materials in Self Compacting Concrete (SCC*) J. Teknol.* **5** (78) 29–35

[4] Sadeghian P and Fam A 2015 Improved design-oriented confinement models for FRP-wrapped concrete cylinders based on statistical analyses, *Engineering Structures* **87** 162–182

[5] Pantazopoulou S J, Bonacci J F, Sheikh S, Thomas M D A and Hearn N 2001 Repair of Corrosion-Damaged Columns with FRP Wraps, *Journal of Composites for Construction* **5**(1) 3-11

[6] Shahidan S, Abdullah S R and Ismail I 2016 Relationship between AE Signal Strength and Absolute Energy in Determining Damage Classification of Concrete Structures *J. Teknol.* **78** (5) 1–6

[7] Leman A S, Shahidan S, Senin M S and Ramzi Hannan N I R 2016 A Preliminary Study On Chemical And Physical Properties Of Coconut Shell Powder As A Filler In Concrete **160** (London: IOP Conference Series: Materials Science and Engineering) 012059

[8] Shahidan S, Mohd Zuki S S, Rahim M A and Ali N 2016 Amplitude Distribution of Emission Wave for Cracking Process vol 47 (Paris: MATEC Web of Conference) pp 2–8
[9] Gunasekaran K, Annadurai R and Kumar P S 2012 Long Term Study on Compressive and Bond Strength of Coconut Shell Aggregate Concrete *Construction and Building Materials* **28**(1) 208–215
[10] Ranolia K V, Thakkar B K and Rathod J D 2013 Effect of Different Patterns and Cracking in FRP Wrapping on Compressive Strength of Confined Concrete, *Procedia Eng* **51** 169–175
[11] Abdul Rahim M, Ibrahim N M, Idris Z, Ghazaly Z M, Shahidan S, Rahim N L, Sofri L A and Isa N F 2014 Properties of Concrete with Different Percentange of the Rice Husk Ash (RHA) as Partial Cement Replacement *Mater. Sci. Forum* **803** 288–293
[12] Shahidan S, Koh H B, Alansi A M S and Loon L Y 2016 Strength Development and Water Permeability of Engineered Biomass Aggregate Pervious Concrete vol 47 (Paris: MATEC Web of Conference) pp 2–7
[13] Tajudin S A A, Azmi M A M, Shahidan S, Abidin M H Z and Madun A 2016 Relationship of physical parameters in Pb-contaminated by stabilization/solidification method vol 47 (Paris: MATEC Web of Conference) 03015
[14] Azmi M A M, Tajudin S A A, Shahidan S Masrom M A N and Nabila A T A 2016 Preliminary Study on Remediation of Contaminated Clay Soil Using Cement and Sugarcane Bagasse vol 103 (London: IOP Conference Series: Materials Science and Engineering) 07001
[15] Kambli P S and Mathapati S R 2014 Application of Coconut Shell as Coarse Aggregate in Concrete : A Technical Review **4**(3) 498–501
[16] Verma D, Gope P C, Maheshwari M K and Sharma R K 2012 Bagasse Fiber Composites-A Review, *J. Mater. Environ. Sci.* **3**(6) 1079–1092
[17] Hajiha H and Sain M 2015 The Use Of Sugarcane Bagasse Fibres as Reinforcements in Composites,Biofiber Reinforcements in Composites (Woodhead Publishing Ltd) pp 525-549
[18] Ghazali M J, Azhari C H, Abdullah S and Omar M Z 2008 Characterisation of Natural Fibres (Sugarcane Bagasse) in Cement Composites *Proc. of the World Cong. on Eng.* pp 23–5
[19] Chandramohan D and Marimuthu K 2011 A Review On Natural Fibers *Inter. J. of Research and Reviews in Applied Sciences* **8** 194–206

**Acknowledgments**
The author would like to thank Universiti Tun Hussein Onn Malaysia for Grant GPPS U449, Grant VOT U523and Ministry of Higher Education (MOHE) (MyPhD) for the approved fund which makes this important research viable and effective

IOP Conf. Series: Materials Science and Engineering **271** (2017) 012017 doi:10.1088/1757-899X/271/1/012017

# The comparison of properties and cost of material use of natural rubber and sand in manufacturing cement mortar for construction sub-base layer

**R Rahman[1], M S Nemmang[1] , Nor Hazurina[2], S Shahidan[2], Raden Khairul Tajuddin Jemain[1], Abdullah M E[1], Hassan M F[3]**

[1] Department of Infrastructure & Geomatic (JKIG) Faculty of Civil and Environmental Engineering University Tun Hussein Onn Malaysia, Batu Pahat
[2] Jamilus Research Center, Faculty of Civil and Environmental Engineering University Tun Hussein Onn Malaysia, Batu Pahat
[3] Batu Pahat Municipal Council, Johor
Corresponding author: raha@uthm.edu.my

**Abstract.** The main issue related to this research was to examine the feasibility of natural rubber SMR 20 in the manufacturing of cement mortar for sub-base layer construction. Sub-base layers have certain functions that need to be fulfilled in order to assure strong and adequate permeability of pavement performance. In a pavement structure, sub-base is below the base and serves as the foundation for the overall pavement structure, transmitting traffic loads to the sub-grade and providing drainage. Based on this research, the natural rubber, SMR 20 was with the percentages of 0%, 5%, 10% and 15% to mix with sand in the manufacture of the cement mortar. This research describes some of the properties and cost of the materials for the natural rubber and sand in cement mortar manufacturing by laboratory testing. Effects of the natural rubber replacement on mechanical properties of mortar were investigated by laboratory testing such as compressive strength test and density. This study obtained the 5% of natural rubber replaced in sand can achieved the strength of normal mortar after 7 days and 28 days. The strength of cement mortar depends on the density of cement mortar. According to the cost of both materials, sand shows the lower cost in material for the cement mortar manufacturing than the uses of natural rubber. Thus, the convectional cement mortar which used sand need lower cost than the modified rubber cement mortar and the most economical to apply in industrial. As conclusion, the percentage of 5% natural rubber in the cement mortar would have the same with normal cement mortar in terms of the strength. However, in terms of the cost of the construction, it will increase higher than cost of normal cement mortar production. So that, this modified cement mortar is not economical for the road sub-base construction.

## 1. Introduction

In the highway engineering, the composition of road structure consists of several types of cross section layers which are sub-grade, sub-base, base and bituminous materials which is wearing course. Sub-base is the layer of aggregate material which is on the sub-grade located under the base course layer. Sub-base may be constructed of granular materials, cement mortar materials, lean concrete or open-graded, highly-permeable materials which may be stable or unstable [1]. Sub-base will be the main issue to be discussed in this research based on the cement mortar producing with different material in the sub-base construction. Cement mortar is widely used for construction of sub-base in the roadway, but financial and environmental concerns are causing concern, leading to the quest for alternative and efficient stabilizers to provide a better modified sub-base layer. The road pavement that occur distress and failure that commonly happen in foundation of the road which involve the sub-grade and sub-base need the improvement to upgrade the condition of road pavement for better and long lasting use. The reason that may lead to these failure road condition due to inadequate maintenance, excessive loads, climatic and environmental conditions, poor drainage leading to poor subgrade conditions, non-uniform support of the surface layer, poor subgrade soil, and disintegration of the component materials [2]. According to this research, the alternative material which is natural rubber will be used as the replacement of sand in the manufacturing of the modified cement mortar to produce a better of the sub-base construction plus involved the factor of cost-efficiency to ensure the economical method in the construction.

## 2. Literature review

The assessment of the road pavement would be identified by the research on the material in the road construction in sub-base layer to fulfill the requirements for good performances of road pavement. According to this research, status of rubber properties to perform in the cement mortar to apply in the road sub-base construction would be identified based on the review the knowledge on past and current study. Based on the Table 1, there are many discussions about the uses rubber in the highway construction in Malaysia. The uses of the rubber in the highway construction also had been long time established for the rubberized road in Malaysia especially for the surface layer that involved the bituminuous of the wearing course of the road structure. But the future study of the lower part of the road structure had not been established especially for the foundation of the road pavement. The utilization of rubber in cement mortar have widely applied in the industry based on the previous study in its capability to perform in the cement mortar. The previous study in Table 2 has shown the uses of the alternatives of rubber material in the with cement mortar.

**Table 1.** Rubberized road in malaysia.

| Year | Location | Rubber used (Wet process) |
|---|---|---|
| 1950 | Kota Bharu- Kuala Krai (90m) | Crumb rubber and latex (5%) |
| 1968 | KL-Seremban, KL-Bentong | 1.5 % & 3% latex |
| 1988 | Klang | 2% latex |
| 1993 | Rembau-Tampin (1km) | Crumb rubber and latex |
| 1996 | Sungai Buluh (3 km) | Crumb rubber |
| 1996 | KLIA (50 km) | Crumb rubber |
| 2003 | Jalan kuantan-Gambang (4 km) | Crumb rubber |
| 2010 | Bukit Kuantan (0.6 km) | Crumb rubber |
| 2015 | Kota Tinggi, Johor (1km) | Cup lump (5%) |
| 2016-2017 | Rubberized asphalt pilot projects in Negeri Sembilan, Pahang, Kedah, Kelantan and Johor | Cup lump (5%) |

**Table 2.** The previous study on utilization of rubber in cement mortar

| Title/Publisher/Year | Scope of research | Result |
|---|---|---|
| Investigation of mechanical and durability properties of cement mortar and concrete with varying replacement levels of crumb rubber as fine aggregate. Md Salamah Meherier, [3]. | Mechanical properties of rubber-based cement mortar were investigated and compared to conventional cement mortar without crumb rubber | Compressive strength tests showed a sharp decline in strength when crumb rubber was used as a replacement of sand in all three batches. |
| The use of recycled rubber tire crumbs for waterproofing of concrete Anjerick J. Topacio *et al.* [4]. | To find a rubber-concrete mixture using recycled rubber tire crumbs that can enhance the waterproofing capacity of concrete. | Recycle rubber tire crumbs be used as partial replacement of aggregated which can reduce the water absorption, permeability and sorpitivity of concrete which can make it waterproof. |
| A comprehensive review on the applications of waste tire rubber in cement concrete. Blessen Skariah Thomas *et al.*[5]. | Use of waste tire rubber as a partial substitute for aggregate in cement concrete. | It was clear from the density of rubberized concrete decreases with increasing amount of rubber. |

## 3. Materials and methods

Based on this study, it discussed about the materials that used in cement mortar producing where the replacement of sand with natural rubber SMR 20 by 0% (as a control), 5%, 10% and 15% volume of the sand. Before that, the testing on properties of the natural rubber would be conducted to identify its performances with cement mortar by the laboratory testing. The ratio used in this cement mortar producing is 1:3 (BS 5628-1:1992) [6] and Ordinary Portland Cement (OPC) which representing the ratio of cement and sand. Other material used to produce mortar is water where the use is fixed due to water cement ratio of 0.5 [7]. This modified cement mortar produced following the flowchart below:

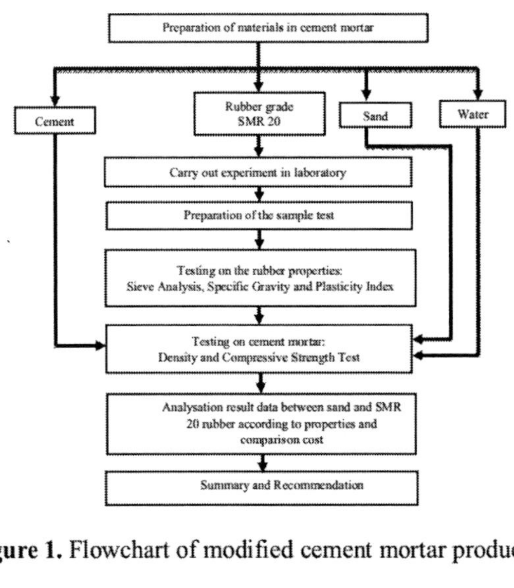


**Figure 1.** Flowchart of modified cement mortar production

## 4. Materials and methods

### *4.1. The comparison properties of natural rubber and sand*

#### *4.1.1 The comparison properties of natural rubber and sand*

**Table 3.** Sieve test for fine aggregate

| Sieve size BS (mm) | Retain weight (g) | Cumulative retain weight (g) | Cumulative percent retain weight (%) | Cumulative passing weight (%) |
|---|---|---|---|---|
| 10 | 0 | 0 | 0 | 100 |
| 5 | 0 | 0 | 0 | 100 |
| 2.36 | 50 | 50 | 10 | 90 |
| 1.18 | 50 | 100 | 20 | 80 |
| 0.600 | 53 | 153 | 31 | 69 |
| 0.300 | 98 | 251 | 50 | 50 |
| 0.150 | 129 | 380 | 76 | 24 |
| 0.075 | 85 | 465 | 93 | 7 |
| Pan | 35 | 500 | 100 | 0 |
| Total | 500 | | 280 | |
| **Fineness modulus = 2.80** | | | | |

**Table 4.** Sieve test for rubber

| Sieve size BS (mm) | Retain weight (g) | Cumulative retain weight (g) | Cumulative percent retain weight (%) | Cumulative passing weight (%) |
|---|---|---|---|---|
| 10 | 0 | 0 | 0 | 100 |
| 5 | 0 | 0 | 0 | 100 |
| 2.36 | 15 | 15 | 3 | 97 |
| 1.18 | 30 | 45 | 9 | 91 |
| 0.600 | 170 | 215 | 43 | 57 |
| 0.300 | 165 | 380 | 76 | 24 |
| 0.150 | 75 | 455 | 91 | 9 |
| 0.075 | 35 | 490 | 98 | 2 |
| Pan | 10 | 500 | 100 | 0 |
| Total | 500 | | 320 | |
| **Fineness modulus = 3.20** | | | | |

The fineness modulus of rubber is higher than sand, signifying the existence of more coarser particles in rubber aggregate than sand aggregates [3]. Based on this research, the fineness modulus of natural rubber SMR 20 rubber is higher than sand had been proven. Notice that most of the fine aggregate pass through a No.4 (4.75mm) sieve, but a very large percentage of the coarse aggregates retained on a No. 4 (4.75mm) sieve. From the table as shown above, we noticed that most of the fine aggregates and crushed rubber pass through a 4.75 sieve which with the 90% and 97% not less than 45% as specified in JKR/SPJ/2008-S4. So both the data verified for Specification Standards for Road Work JKR/SPJ/2008-S4 [6].

*4.1.2 Specify gravity*

**Table 5.** Specific gravity test (fine aggregate)

| **Bottle no** | **1** | **2** | **3** | **4** |
|---|---|---|---|---|
| Mass of density bottle and stopper (ml) | 29.18 | 29.88 | 29.54 | 27.88 |
| Mass of density bottle plus oven dried soil (m2) | 34.32 | 34.9 | 34.67 | 32.97 |
| Mass of density bottle plus stopper plus soil plus distilled water (m3) | 81.43 | 82.20 | 82.25 | 80.19 |
| Mass of density bottle plus stopper plus distilled water (m4) | 78.06 | 79.26 | 79.19 | 77.08 |
| Mass of sample (m2-ml) | 5.14 | 5.02 | 5.13 | 5.09 |
| Mass of water contained by the density bottle (m4-ml) | 48.88 | 49.38 | 49.65 | 49.20 |
| Mass of water occupying the volume not occupied by the soil (m3-m2) | 47.11 | 47.30 | 47.58 | 47.22 |
| (m4-ml) – (m3-m2) | 1.77 | 2.08 | 2.07 | 1.98 |
| $Gs = \frac{(m2-m1)}{(m4-m1)-(m3-m2)}$ | 2.90 | 2.41 | 2.48 | 2.57 |
| Gs | 2.60 | | | |

According to this study, the specific gravity of fine aggregates and rubber were analysed according to ASTM C 127 for the rubber and ASTM C 128 for fine aggregates. The average specific gravity for both materials used in this research for fine aggregates and the natural rubber were 2.60 and 0.96 respectively. Both the specific gravity in the range as specified which 2.4 - 2.9 for fine aggregate and 0.9 – 1.2 for rubber.

**Table 6.** Specific gravity test (rubber)

| **Container** | **A** |
|---|---|
| Mass of the mesh containing natural rubber in the water after the soak 24 hours at room temperature (ml) | 76.66 |
| Mass of the mesh containing before the soak (m2) | 23.10 |
| Mass of natural rubber with burner container (after been wiped) (m3) | 131.95 |
| Mass of the burner container and natural rubber (after drying in the sun 3-6 hours) (m4) | 120.6 |
| Mass of the burner container (m5) | 100.74 |
| Specific Gravity (drying oven) $Gs = \frac{(m4-m5)}{(m3-m5)-(m1-m2)}$ | $\frac{(120.6 - 100.74)}{(131.95 - 100.74) - (76.66 - 23.10)}$ = 0.88 |
| Specific Gravity (humidity) $Gs = \frac{(m3-m5)}{(m3-m5)-(m1-m2)}$ | $\frac{(131.95 - 100.74)}{(131.95 - 100.74) - (76.66 - 23.10)}$ = 1.4 |
| Specific Gravity (significantly) $Gs = \frac{(m4-m5)}{(m4-m5)-(m1-m2)}$ | $\frac{(120.6 - 100.74)}{(120.6 - 100.74) - (76.66 - 23.10)}$ = 0.6 |
| Averege Specific Gravity (Gs) | 0.88 + 0.6 + 1.4 = 0.96 |

*4.1.3 Plasticity index*

Plasticity Index (Ip) of a soil is the numerical difference between the liquid limit and the plastic limit. Related to this study, the plasticity index cannot be defined because natural rubber SMR 20 in solid state. So it can be defined as non-plastic. So the plastic limit would be considered as 0. Based on Manual Pavement Design JKR, either the liquid limit or plastic limit cannot be determine, the plasticity index will be as non-plastic. The value of plasticity index for the fine aggregates would be 0.218 which from difference between the liquid limit and the plastic limit and dividing by 100. According to the standard JKR JKR/SPJ/2008-S4 for road sub-base work specification, the plasticity index when tested in accordance with BS 1377 shall be not more than 12 %. The value of plasticity index for the fine aggregates for this research is verified as it not more than 12 % [6].

**Table 7.** Liquid limit test (fine aggregate)

| Can number | 1 | 2 | 3 | 4 |
|---|---|---|---|---|
| Mass of can + moist soil (Mcws) | 46 | 39 | 42 | 40 |
| Mass of can + dry soil (Mcs) | 43 | 37 | 39 | 37 |
| Mass of can (Mc) | 34 | 29 | 29 | 29 |
| Mass of dry soil (Ms) | 9 | 8 | 10 | 8 |
| Mass of water(Mw) | 3 | 2 | 3 | 3 |
| Water content | 33.33 | 25 | 30 | 37.5 |
| Average Water content | 31.46 | | | |

However, based on the plasticity properties for the rubber, the Wallace plasticity test reports have two measures:

a) Plasticity (Po), which is a measure of the compression of a sample after a load has been applied for a defined time. For the value of plasticity of rubber properties for SMR 20 that used for this research is 30 [10].

b) Plasticity retention index (PRI), which measures recovery after a sample has been compressed, heated and subsequently cooled. PRI % is defined as ([ ]/Po) x 100 and where Po is the plasticity and [ ] is the plasticity after aging for 30 minutes at typically 140°C [9]. During processing in, for example, a tire factory, natural rubber with low PRI values tends to break down more rapidly than those with high values. For the value of plasticity retention index (PRI) of rubber properties for SMR 20 that used for this research is 40 [10]. Based on the standard JKR/SPJ/2008-S4, the value of plasticity retention index (PRI) of SMR 20 failed to fulfil the standard by JKR which stated the plasticity index when tested in accordance with BS 1377 shall be not more than 12. So the uses of crushed rubber does not verified in the standard JKR/SPJ/2008-S4 [6].

*4.1.4 Properties of the cement mortar*

**Table 8.** Summary of data

| **No.** | **Density** ($kg/m^2$) | | **Compressive Strength Test** ($N/mm^2$) | |
|---|---|---|---|---|
| | **7 days** | **28 days** | **7 days** | **28 days** |
| A (Control) | 2000 | 1947 | 17.87 | 31.29 |
| B (5%) | 1920 | 1867 | 9.13 | 15.99 |
| C (10%) | 1840 | 1787 | 6.88 | 12.04 |
| (15%) | 1760 | 1760 | 3.16 | 5.52 |

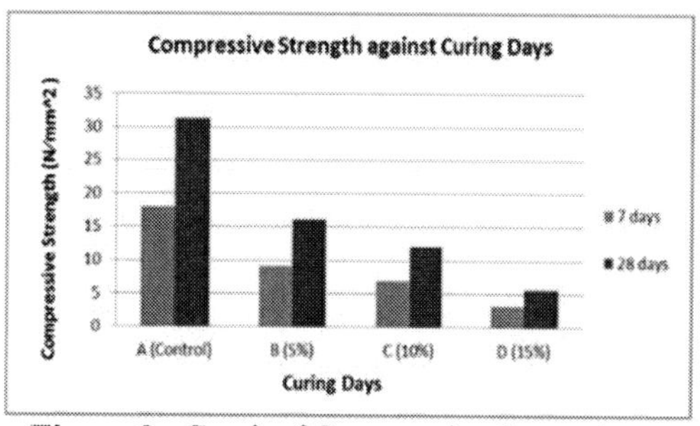


**Figure 2.** Graph of Compressive Strength

Based on the Table 8, it shows the differential of mechanical properties of mortar by using the vary percentage of natural rubber SMR 20. The value of the compressive strength of cement mortar shows the increasing of curing days, which provide more strength of the cement mortar. From the result obtained, the vary percentage of natural rubber SMR 20 provides the different strength of mortar. But when compared between the normal cement with the cement mortar with the percentage of natural rubber, show the decreasing the compressive strength value between normal cement mortar and cement mortar with the percentage of natural rubber. The result shows that at 7 days, the normal mortar achieved the early strength where greater than 16 N/mm$^2$ as specified in BS 4551:2005 [11]. The highest value of compressive strength is indicated by 5% of natural rubber SMR 20 as a sand replacement. The strength is increased a little until 28 days curing. The control sample gave the higher strength at 7 days and 28 days curing higher than the replacement of sand with percentage of natural rubber SMR 20. Based on graphs in Figure 2, it shows the compressive strength of cement mortar in different percentage of rubber in the cement mortar against curing days. For the control sample, it shows the higher strength at 7 days and 28 days than other cement mortar. It differs with the cement mortar containing 5% of natural rubber where it increases and almost achieved to the near strength for normal mortar at 28 days. Meanwhile, cement mortar in C and D show the relation between compressive strength with the curing days from 7 days until 28 days with increasing strength with the increasing curing days even though cement mortar C and D show the decreasing strength than the mortar A and B. Based on the graph obtained, it was concluded the strength of each type of mortar increased with the increasing of curing days and achieve the maximum strength at 28 days. In Figure 3, the graph shows the density of mortar decreased by the increasing of percentage replacement of natural rubber in cement. The density of rubberized concrete decreases with increasing amount of rubber [12]. While Figure 4 shows the compressive strength of mortar decreased during the increasing of percentage replacement of natural rubber. This may be attributed to the decrease in the adhesive strength between the surface of the rubber and cement paste [13].

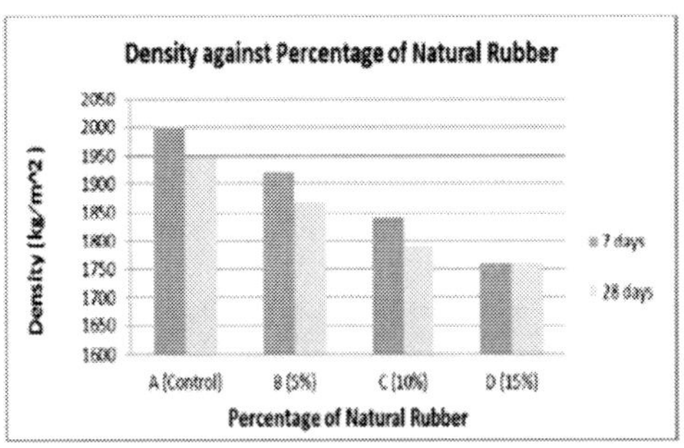


**Figure 3.** Graph of Density against Percentage of Natural Rubber

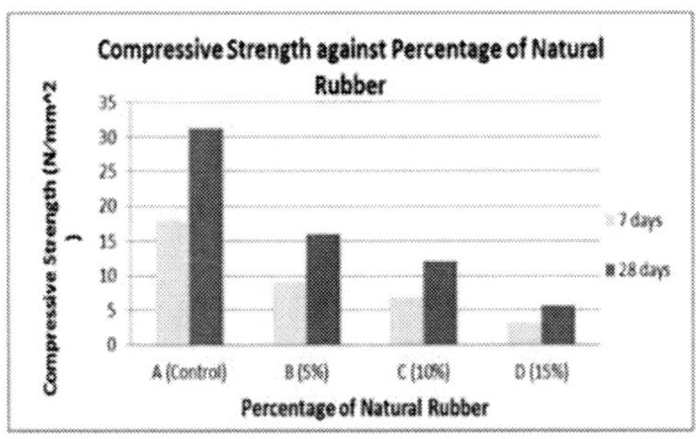


**Figure 4.** Graph of Compressive Strength against Percentage of Natural Rubber

**Table 9.** Cost of production in cement mortar for sand and rubber smr 20.

| Rubber percentage (%) | Natural rubber content in cement mortar per mould (kg) | Cost of natural rubber(RM) | Sand content in cement Mortar per mould (kg) | Cost of sand (RM) | Cost of cement per mould | Total cost in cement mortar (RM) |
|---|---|---|---|---|---|---|
| 0 (Normal) | 0 | - | 0.203 | 0.0164 | 0.026 | 0.04 |
| 5 | 0.011 | 0.08 | 0.192 | 0.0156 | 0.026 | 0.12 |
| 10 | 0.021 | 0.15 | 0.182 | 0.0147 | 0.026 | 0.19 |
| 15 | 0.030 | 0.21 | 0.173 | 0.0140 | 0.026 | 0.25 |

**Table 10.** Cost production in conventional and modified cement mortar in the sub-base.

| Rubber percentage in cement mortar (%) | Total cost in cement mortar (RM) | Amount of cement mortar to occupy 7 meter width along 1 kilometer sub-base | Total cost of cement mortar for the road sub-base (RM) |
|---|---|---|---|
| 0 (Normal) | 0.04 | 2,800,000 | 112,000 |
| 5 | 0.12 | 2,800,000 | 336,000 |
| 10 | 0.19 | 2,800,000 | 532,000 |
| 15 | 0.25 | 2,800,000 | 700,000 |

The type of sub-base design should be selected while considering the purpose of the sub-base, locally available materials, and their cost-effectiveness. At the early stage, the study would compared the cost of both materials which are sand and rubber in the construction of road sub-base for 1 kilometers with the width of 7 meters with the normally thickness of 200 mm. Related to this research, the uses of 50 mm thickness of the cement mortar would be replaced from the 200 mm thickness layer of road sub-base structures as the modified road sub-base with the identification to its cost-effectiveness of the materials in the cement mortar manufacturing. From the Table 9, it shows that the normal cement mortar shows the lowest cost of material in the production of the cement mortar than the modified cement mortar. Besides that, it shows that the increasing the percentage of the natural rubber in the cement mortar provides the increasing the cost of the materials in the manufacturing of the cement mortar. Meanwhile, Table 10 shows that the normal cement mortar shows the lowest cost of the road sub-base construction than the modified cement mortar with different percentage of natural rubber. Based on the objectives that have been defined, we concluded that the most economical method for this research was convectional method mortar because it provides higher strength and have a reasonable market value in industrial than the modified cement mortar even though the modified cement mortar for the percentages 5% almost achieved to the strength for the normal cement mortar. Obviously, based on the comparison between both materials, the replacement of natural rubber with sand in the cement mortar production provide the higher cost than the conventional cement mortar due to the additional material in the mixture and cost in the market price.

## 5. Conclusions

The potential of mortars was identified through the laboratory testing where the mechanical properties are investigated such as density and compressive strength. By replacing some percentage of natural rubber SMR 20, it indicated the mechanical of properties of cement mortar achieved low strength and low density compared to the control sample. Based on this study, it found that 5% of natural rubber SMR 20 replaced in sand material got the initial strength higher than others percentage of natural rubber uses after it cured in 7 days. The mechanical strength becomes increased and almost achieved strength of control sample after 28 days. Besides that, the strength of cement mortar increased with the increasing of curing days. It also could be concluded that the higher of mortar density will be performed in higher compressive strength. This study shows that the percentage of 5% of natural rubber would be achieved strength of normal cement mortar and the most optimum compared to other percentage of rubber. The increasing of the percentage of cement mortar would be decreasing the density and compressive strength of the cement mortar because there is lack of the workability of the natural rubber to perform in the cement mortar. As conclusion, the percentage of 5% natural rubber in the cement mortar would have the same with normal cement mortar in terms of the strength. However, in terms of the cost of the construction, it will increase higher than cost of normal cement mortar production. So that, this modified cement mortar is not economical for the road sub-base construction. But, this study could be applied for the future research with the new improvement to upgrade the road sub-base in highway construction for better and long lasting use.

## 6. References

[1] Association, A C P *1995 Sub-grade-Sub-base.* Pdf. Retrieved on Mei 5, 2017 from http://www.dot.state.mn.us/stateaid/projectdelivery/pdp/tools/subgrade subbase.pdf.

[2] IHRB Report, F 2008 Design Guide for Improved Quality of Roadway Subgrades and Subbase, (September).

[3] Meherier Md S 2016 *Investigation Of Mechanical And Durability Properties Of Cement Mortar And Concrete With Varying Replacement Levels Of Crumb Rubber As Fine Aggregate* Retrieved on May 5, 2017 from https://open.library.ubc.ca/cIRcle/collections/ubctheses/24/items/1.0305047.

[4] Topacio A J, Gozo, N C, Marquez S R V, Poblete, R V M and Ramos H S 2016 Use of Recycled Rubber Tire Crumbs for Waterproofing of Concrete *International Journal of Civil and Structural Engineering Research* **3(2)** 95–132.

[5] Skariah B and Gupta R C 2016 A comprehensive review on the applications of waste tire rubber in cement concrete. *Renewable and Sustainable Energy Reviews* **54** 1323–1333

[6] British Standard B.S. 1992 *Code of practice for use of masonry* (1st ed., 1-2) [1]. BS 5628-1:1992 Retrieved May 5, 2017, from xa.yimg.com/kq/groups/21948400/1287076697/name/5628.

[7] Toledo Filho R D, Goncalves J P, Americano B B and Fairbairn E M R 2007 Potential for use of Crushed Waste Calcined-clay Brick as Supplementary Cementitious Material in Brazil *Science Direct: Cement and Concrete Composites* **37**(9) 1357 – 1365

[8] Ketua Pengarah Kerja Raya J K R 2008 *Standard Specification for Road Work* (2008 ed., Ser. 4)

[9] Standard Malaysian Rubber S M R 2012 Astlett Rubber Inc., (905), 2701.

[10] Sambhi, M S 1989 An analysis of the plasticity retention index of the standard Malaysian rubber scheme *Journal of the Natural Rubber Research* **4(2)** 133-140

[11] BS 4551: 2005 *Methods for testing mortar, chemical analysis and physical testing.*

[12] Skariah B and Gupta R C 2016 A comprehensive review on the applications of waste tire rubber in cement concrete *Renewable and Sustainable Energy Reviews* **54** 1323–1333

[13] Abas F O, Ghafoor, E A A, Abass M U and Abd T K 2015 Re-Use of Waste Tires Rubber As *International Journal Of Engineering Sciences & Research Technology* **4(3)** 110–120

IOP Conf. Series: Materials Science and Engineering **271** (2017) 012018 doi:10.1088/1757-899X/271/1/012018

# Effect of water curing duration on strength behaviour of portland composite cement (PCC) mortar

**M A Caronge[1], M W Tjaronge[1], H Hamada[2] and R Irmawaty[1]**

[1] Department of Environmental Engineering, Faculty of Engineering, Hasanuddin University, Gowa, Indonesia
[2] Department of Civil and Structural Engineering, Faculty of Engineering, Kyushu University, Fukuoka, Japan
Corresponding author: caronge_eng@yahoo.co.id

**Abstract.** Cement manufacturing of Indonesia has been introduced Portland Composite Cement (PCC) to minimize the rising production cost of cement which contains 80% clinker and 20% mineral admixture. A proper curing is very important when the cement contains mineral admixture materials. This paper reports the results of an experimental study conducted to evaluate the effect of water curing duration on strength behaviour of PCC mortar. Mortar specimens with water to cement ratio of (W/C) 0.5 were casted. Compressive strength, flexural strength and concrete resistance were tested at 7, 28 and 91 days cured water. The results indicated that water curing duration is essential to continue the pozzolanic reaction in mortar which contributes to the development of strength of mortar made with PCC.

## 1. Introduction

Cement manufacturing companies in Indonesia has been introduced Portland Composite Cement (PCC) to minimize the rising production cost of cement. The PCC equivalent with CEM Type II/A-M cement which contains 80% clinker and 20% mineral admixture including ground granulated blast furnace slag, silica fume, fly ash and gypsum. The use of the PCC on concrete production requires a proper curing due the presence of mineral admixture. Although it is very simple and inexpensive procedure, proper initial curing has an important influence on improving the concrete quality [1-2]. The objective of curing is considered by the duration of providing concrete with sufficient humidity and appropriate temperature conditions to reduce the loss of moisture to ensure the progress of hydration reactions causing the filling and segmentation of capillary voids by hydrated compounds [3-4]. The matter would be more critical in the case of the cement containing supplementary cementitious material because the pozzolanic reaction is very sensitive to the curing procedure. According to the ACI Committee 308 [5], the curing period should be extended to 14 days when the cement contains supplementary cementitious materials, owing to the slow hydration reactions between supplementary cementitious materials and the calcium hydroxide. Furthermore, curing condition also could be an important parameter in controlling durability of the reinforced concrete exposed to the marine environment [6].

In this study, the effect of water curing duration on strength behaviour of mortar made from PCC was evaluated. Mortar specimens with water to cement ratio of (W/C) 0.5 were casted. Compressive strength, flexural strength and concrete resistance were tested at 7, 28 and 91 days cured water.

## 2. Experimental procedure

### *2.1. Material and mix proportion*
Portland Composite Cement (PCC), Ordinary Portland Cement (OPC), washed sea sand were used and physical properties of material is shown in table 1. Further, the chemical compounds of the PCC and OPC is presented in table 2. For all specimens a mortar mix with water to cement ratio (W/C) 0.5 was casted. The mix proportion of mortar used is described in table 3.

**Table 1.** Physical properties of materials.

| Material | Description |
|---|---|
| Portland Composite Cement (PCC) | SSD density = 3.08 g/cm$^3$<br>Specific surface area (SSA)= 3410 cm$^2$/g |
| Ordinary Portland Cement (OPC) | SSD density = 3.16 g/cm$^3$<br>Specific surface area (SSA) = 3390 cm$^2$/g |
| Washed sea sand | SSD density = 2.58 g/cm$^3$<br>Fineness modulus (F.M) = 2.77 |

**Table 2.** Chemical compounds of cement.

| Chemical compounds (%) | PCC | OPC |
|---|---|---|
| Magnesium oxide (MgO) | 0.99 | 1.2 |
| Sulfur trioxide ($SO_3$) | 1.81 | 2.23 |
| Silicon dioxide ($SiO_2$) | 18.39 | 19.71 |
| Aluminium oxide ($Al_2O_3$) | 5.15 | 5.20 |
| Ferric oxide ($Fe_2O_3$) | 3.14 | 3.73 |
| Calcium oxide (CaO) | 61.79 | 62.91 |
| Loss of ignition (LOI) | 4.61 | 2.15 |

**Table 3.** Mix proportion of mortar.

| W/C | Water (kg/m$^3$) | Cement (kg/m$^3$) | Sand (kg/m$^3$) |
|---|---|---|---|
| 0.5 | 255 | 510 | 1508 |

### *2.2. Mixing and curing*
Water and cement were first placed in a standard mortar mixer and mixed for about 1 min. Then, the sands were added and mixed for 2 min. Finally, the mixture was mixed for an additional 2 min to complete the whole mixing process. The fresh mortar samples were then put into the steel moulds in two layers of similar depths. After filling up each layer, compaction was achieved by placing the moulds on a mechanical vibrating table. Thereafter, the mortar specimens were covered with a plastic sheet and allow to cure in the laboratory environment at 20 ± 2 °C for 24h. After 24h, the samples were demoulded and cured in water until further testing.

### *2.3. Specimens and testing*
The flow table test was used for determining the fluidity of fresh mortar mix as described by ASTM C1437 [7]. A three-point flexural strength test in conformity with ASTM C348 [8] was performed at 7, 28 and 91 days on prism specimens with a size of 40 x 40 x 160 mm (figure 1). The remaining portions of the broken prisms in the flexure strength test were used for determining the equivalent compressive strength according to ASTM C349 [9] as shown in figure 2. In addition, the electric resistance of mortar was measured on prism specimen (40 x 40 x 160 mm) with steel bar of 10 mm in

diameter centrally located with a length of 120 mm. Mortar resistance was performed by immerse method using portable rebar corrosion meter as shown in figure 3.

**Figure 1.** Three-point flexural strength test

**Figure 2.** Compressive strength test

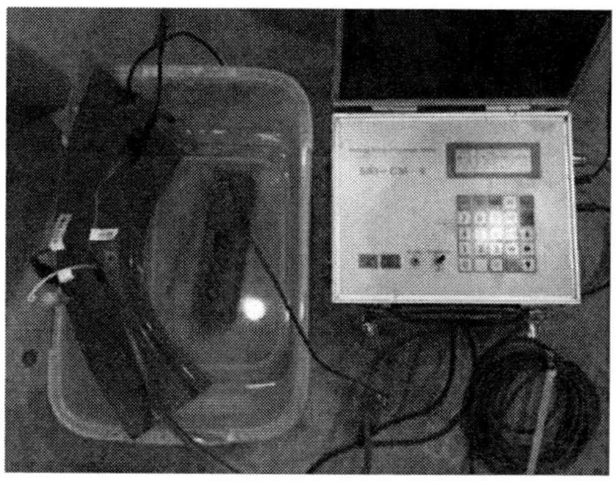

**Figure 3.** Mortar resistance measurement

## 3. Results and discussion

### *3.1. Slump flow*

The slump flow results of the fresh mortar mixes are shown in figure 4. The results show that the PCC decreased the fluidity of the mortar due to the higher surface area of PCC than OPC which cause low workability. The slump flow of PCC and OPC mortar were 160 mm and 165 mm, respectively.

### *3.2. Flexural strength*

The flexural strength of PCC and OPC mortar was determined at the ages of 7, 28 and 91 days. Figure 5 shows the development of flexural strength with age for PCC and OPC mortar. The flexural strength of PCC mortar is found lesser than of OPC mortar at all the ages, which could be attributed for slow hydration process of PCC. The percentage increase in flexural strength from 7 to 28 days and 28 days to 91 days has been 4.95% and 6.32%, respectively for OPC mortar. The increase in strength for similar comparison is observed as 12.37% and 14.36%, respectively for PCC mortar.

**Figure 4.** Slump flow of PCC and OPC mortar

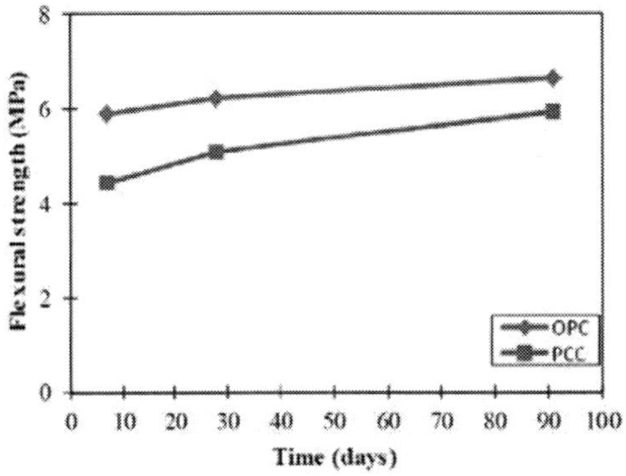


**Figure 5.** Flexural strength of mortar specimens

*3.3. Compressive strength*

Figure 6 show the development of compressive strength of PCC and OPC mortar at 7, 28 and 91 days cured water. The compressive strength of mortar increased with increasing curing time. Test results showed that the 7 days compressive strength for OPC mortar is 25.05% higher than PCC mortar. At 28 days, PCC and OPC mortar reached the compressive strength about 34.38 MPa and 42.19 MP. A significant increasing in compressive strength of PCC mortar was noticed at 91 days. The PCC mortar shows that compressive strength is almost identical with that of OPC mortar after 91 days cured water. This indicate that PCC mortar have lower strength than OPC mortar at early ages (7 and 28 days), but achieved similar compressive strength at 91 days.

*3.4. Resistivity measurement*

The resistivity of PCC and OPC mortars specimens at 7, 28 and 91 days are shown in Figure 7. Similar trends to that flexural and compressive strength can be observed. The resistivity of mortar increased with increasing curing time. The resistivity of PCC and OPC mortar is found similar at 7 days. At 28 days, OPC mortar showed higher resistivity value than PCC mortar due to the slow hydration process in the presence of mineral admixture of PCC. But at later ages, the resistivity of PCC and OPC mortar is almost same.

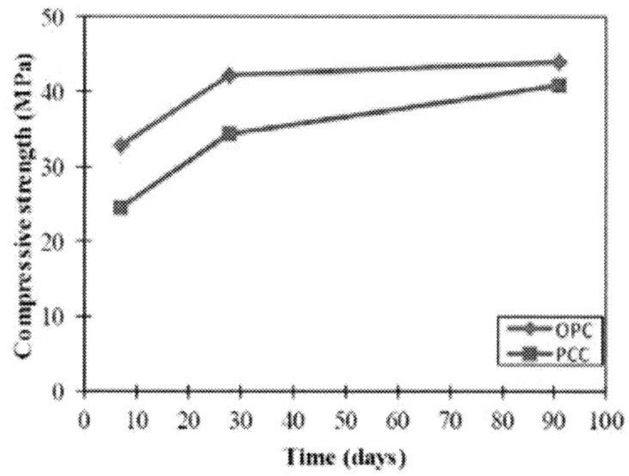


**Figure 6.** Compressive strength of mortar specimens

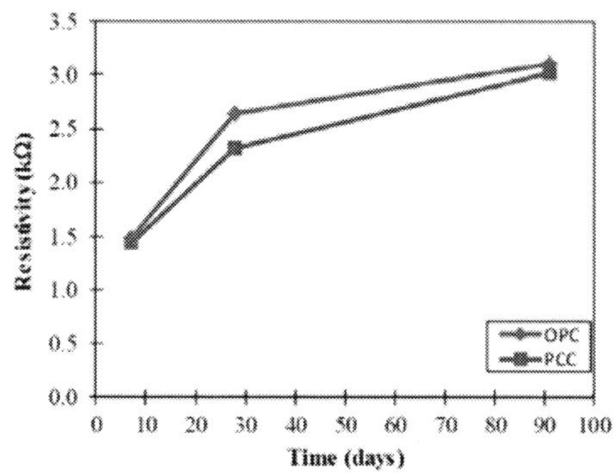


**Figure 7.** Resistivity of mortar specimens

The properties of mortar are significantly influenced by duration of curing since it greatly effects the hydration of cement. A proper curing maintains a suitably warm and moist environment for the development of hydration products and thus reduces the porosity in hydrated cement paste and increases the density of microstructure in mortar [10]. The chemical composition of the calcium silicate formed in hardened blended cement paste was different from that of Portland cement hydration products. For this reason, mortar made with PCC will have lower strength than OPC at early ages and reached almost similar strength at longer ages of curing (91 days).

## 4. Concluding remarks

This experimental study has investigated the effect of water curing duration on strength behaviour of Portland Composite Cement (PCC). The strength at early ages of PCC mortar is lower than that of OPC mortar whereas, at later ages both of the mortar is providing approximately similar strength due to continuous curing condition and well performed pozzolanic reaction activities. It can be concluded that adequate curing at early ages as well as later ages is essential to continue the pozzolanic reaction in mortar which contribute to the development of strength of mortar made with PCC.

## 5. References

[1] Alizadeh R, Ghods P, Chini P, Hoseini M, Ghalibafian M and Shekarchi M 2008 Effect of curing condition on the service life design of RC structures in the Persian Gulf Region *Journal of Materials in Civil Engineering* **20** 2-8.

[2] Radlinski M and Olek J 2015 Effects of curing conditions on the properties of ternary (ordinary portland cement/fly ash/silica fume) concrete, *ACI Material Journal* **112(1)** 49-58.

[3] Guneyisi E, Ozturan T and Gesoglu M 2005 A study on reinforcement corrosion and related properties of plain and blended cement concretes under different curing conditions, *Cement and Concrete Composites* **27** 449-461.

[4] Guneyisi E, Gesoglu M, Ozturan T and Ozbay E 2009 Estimation of chloride permeability of concretes by empirical modelling: Considering effects of cement type, curing condition and age *Construction and Building Materials* **23** 469-481.

[5] ACI Committee 308 2009 *Recommended practice for curing concrete* (MI: MCP, American Concrete Institute) Farmington Hills

[6] Caronge M A, Hamada H, Irmawaty R, Sagawa Y and Yamamoto D 2015 Application of sacrificial anode for prevention of steel corrosion in cracked concrete *Journal of Advanced Concrete Technology* **13** 479-488.

[7] ASTM C1437 2007 *Standard test method for flow hydraulic cement mortars* (ASTM International) West Conshohocken PA

[8] ASTM C348 2008 *Standard test method for flexural strength of hydraulic cement mortars* (ASTM International) West Conshohocken PA

[9] ASTM C349 2008 *Standard test method for compressive strength of hydraulic cement mortars (using portions of prisms broken in flexure)* (ASTM International) West Conshohocken, PA.

[10] Uddin Md A, Jameel M, Sobuz H R, Islam Md S and Hasan N Md S 2013 Experimental study on strength gaining characteristic of concrete using portland composite cement, KSCE *Journal of Civil Engineering* **17(4)** 789-796.

IOP Conf. Series: Materials Science and Engineering **271** (2017) 012019 doi:10.1088/1757-899X/271/1/012019

# Effect of notch position on fracture energy for foamed concrete

**Mohd Naqiuddin Zamri[1], Norashidah Abd Rahman[1,*], Zainorizuan Mohd Jaini[1] and Nurul Shamila Bahador[1]**

[1] Jamilus Research Centre, Faculty of Civil and Environmental Engineering, Universiti Tun Hussein Onn Malaysia, 86400 Parit Raja, Malaysia
Corresponding author: nrashida@uthm.edu.my

**Abstract.** Foamed concrete is one of the lightweight concrete used to replace normal concrete. Foamed concrete has potential as a building construction material in Malaysia due to low density range. However, the behavior of fracture energy on foamed concrete still under investigation. Therefore, a study to determine the fracture energy of foamed concrete was conducted. In this study, foamed concrete fracture energy was obtained using the three-point bending test methods develop by RILEM and Hillerborg. A total of 12 beams with different types of notch and positions of notch were tested on the load–deflection condition. In addition, a total of 9 cube samples were cast to support the result of fracture energy by using model from Bazant and Becq-Giraudon and Comite Euro-International du Beton (CEB). Results showed the far the position of the notch from midpoint, the higher the value of fracture energy. In this study, the value of fracture energy ranges between 15 N/m and 40 N/m.

## 1. Introduction

In concrete construction, permanent action that included self-weight contributes a very large proportion of the total load of structures. To reduce the dead weight, the use of foamed concrete is one of the solution. Typically the density of foam concrete are range around 1000 kg/m$^3$ to 1800kg/m$^3$ while the compressive strength can be achieved up to 12 MPa [1]. Therefore, using lightweight concrete, such as foamed concrete, can reduce the self-weight of concrete structures. Lightweight concrete has unique benefits in construction to increase productivity such as low density and thermal conductivity.

From previous studies, researchers have focused more on the behavior of foamed concrete in terms of density, permeability, compressive strength, and other factors. The behavior of fracture energy on foamed concrete is still under investigation. Fracture energy can be defined as energy required to open a unit area of crack surface. A current study showed that fracture energy was influenced by the properties of aggregate, with high strength aggregate producing high fracture energy [2-4].

Fracture energy, $G_f$, is the energy required to form a unit area of crack. Fracture energy on concrete is important when concrete fails in tension. This behavior is characterized by the peak stress and the energy required to open the crack fully. Conventional concrete have a fracture energy approximately 75- 100N/m depending on the grade of concrete [5]. Meanwhile fracture energy of foamed concrete is roughly between 15 – 35N/m [6-10].

Three empirical formulas from RILEM and Hillerborg [4], Bazant and Becq-Giraudon [2], and Comite Euro-International du Beton (CEB) [11] is used to determine fracture energy.

The Hillerborg model is assumed that crack bridging stress in the fracture zone is a function of the crack opening displacement of materials [3] and known as a load–deflection method. The fracture energy using Hillerborg model can be obtained from equation 1:

$$G_f = (W_o + m_g\delta_f)/A \quad (1)$$

where : $W_o$ is area under the load-deflection curve, $m_g$ is mass of the specimens, $\delta_f$ is maximum deflection and A is the cross section area of the specimen.

According to Rahman and Jaini [10], the value of fracture energy was influenced by the size of the specimen and the load-deflection curve.

Therefore to support the fracture energy value proposed by RILEM and Hillerborg model's, Bazant and Becq-Giraudon and Comite Euro-International du Beton (CEB) were used as a comparison for fracture energy determination. Bazant and CEB also known as size effect method where we can see in equation 2 and 3 respectively. From this model, fracture energy will be influenced directly by compressive strength ($f_c$), aggregate size ($D_{max}$), and water–cement ratio (w/c).

$$G_f = 2.5\alpha_o\,(f_c/0.058)^{0.4}\,(1 + D_{max}/1.94)^{0.43}\,(w/c)^{-0.18} \quad (2)$$

$$G_f = [0.0469(D_{max})^2 - 0.5D_{max} + 26] \times (f_c/10)^{0.7} \quad (3)$$

The aim of this study is to obtain the value of fracture energy based on the mortar formulation of fracture energy. Previous studies showed the value of fracture energy for foamed concrete with a density of 1400 kg/m$^3$ to 1,600 kg/m$^3$ ranges between 18 N/m and 25 N/m [11].

## 2. Materials and method

Materials used in this study are fine aggregate, cement, water, and foaming agent and foamed concrete with a density of 1400 kg/m$^3$. The methodology will be explained in following sub section to achieve the aim of this study. Several tests were conducted to investigate the fracture energy of foamed concrete.

### *2.1. Specimen*

All beams were designed with the dimension of 150 mm × 150 mm × 700 mm and with different distances of U-notch and V-notch, which are at the midpoint of the beam, 100 mm from the midpoint, and 200 mm from the midpoint. In this study, two beam specimens were prepared for different notch positions. The design of notched beam specimens is based on RILEM [12] and ASTM E1820-13 [13].

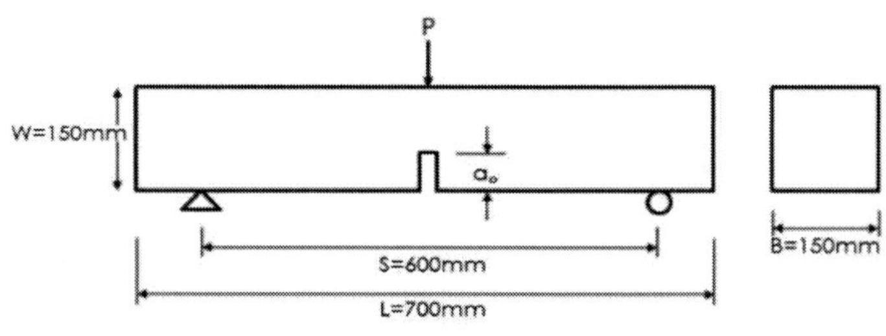


a) U-notch beam specimens

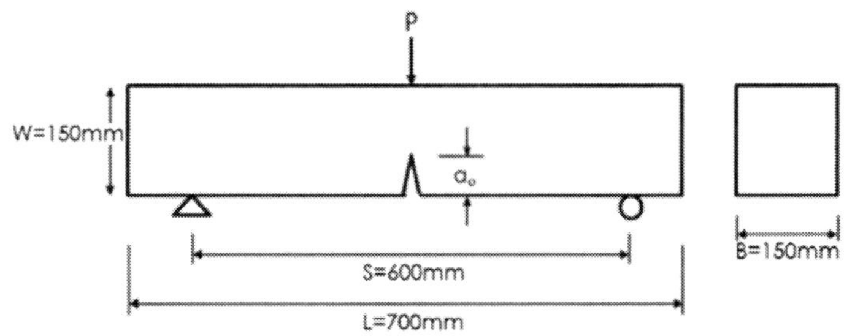


b) V-notch beam specimens

**Figure 1.** Schematic design of the beam specimens, with depth and width of beam (B and W) = 150mm

*2.2. Compression test*

Compression test was conducted based on BSEN12390: Part 3 [14] to determine the strength of the material. The results of compressive strength are shown in figure 2. Result showed the compressive strength for this study is 4.3 MPa at 28 days, where the strength is lower than propose strength by British Cement Association [15] due to the selection of formwork made by wood that used in this experiment. Wood formwork absorb more water and caused the composition of foamed concrete disrupted and at a time made the strength of foamed concrete decreased. The compressive strength proposed by British Cement Association for density 1400kg/m$^3$ ranges between 6.0 MPa and 8.0 MPa.

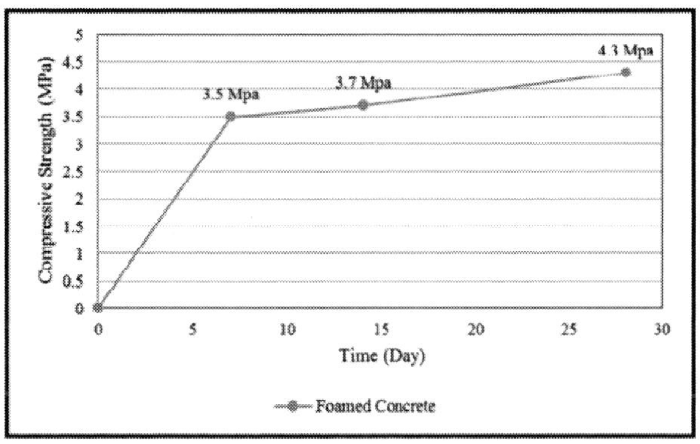


**Figure 2.** Graph of compressive strength (MPa) against time (day)

*2.3. Experimental test*

Three-point bending test (figure 3) was conducted on beam specimens to determine maximum force applied, deflection, and failure behavior after the specimen failed. From the test, the maximum load and maximum deflection at U-notch and V-notch were determined.

## 3. Result and discussion

*3.1. Load–Displacement*

Table 1 shows the results of the maximum load of beam with maximum deflection. The graph of load against deflection is plotted and shown in figure 4.

**Figure 3.** Three-point bending test

**Table 1.** Results of maximum load and maximum deflection

| Beam specimen | Notch type | Notch position from midpoint (mm) | Maximum load (N) | Maximum deflection (mm) |
|---|---|---|---|---|
| FCB-U0 | U | 0 | 2,087.50 | 0.688 |
| FCB-V0 | V | 0 | 1,081.00 | 0.640 |
| FCB-U100 | U | 100 | 3,518.75 | 0.789 |
| FCB-V100 | V | 100 | 1,894.00 | 0.870 |
| FCB-U200 | U | 200 | 4,337.50 | 1.298 |
| FCB-V200 | V | 200 | 3,844.00 | 0.925 |

Figure 4 shows the variations of load against deflection for two types of notch which are U-notch (U) and V-notch (V) at different distances. Figure 4 and table 1 showed that the further the position of the notch from the midpoint, the higher the values of load and deflection. The high value of load of 4.3 kN occurred on the sample with U-notch for all the distance of notch from the midpoint. In addition, the load of the sample with V-notch is lower than that of the sample with U-notch at the distance of 200 mm from the midpoint. Thus, the load that occurred on the sample with U-notch is higher than that on the sample with V-notch because the specimen needs to develop a crack between two edges before the crack appeared on either edge and resulted in failure. By contrast, for V-notch, the crack directly occurred on the edge of the notch, resulting in the lower value of load. Thus, the development of deflection can also be observed when the position of the notch is far from the midpoint, which will increase the value of deflection itself.

### *3.2. Fracture Energy*

Fracture energy, $G_f$, is determined using data collected from the three-point bending test. The value of fracture energy is computed using three empirical equations derived from three reference equations, as follows:

i. Hillerborg [4] model;
ii. Bazant and Becq-Giraudon [2] model;
iii. CEB [11] model.

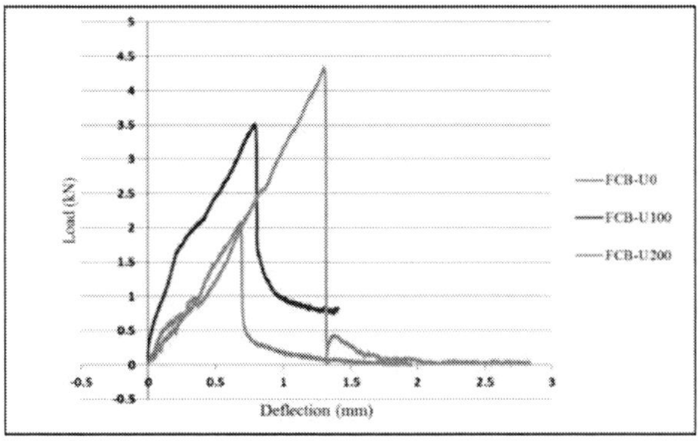


(a)

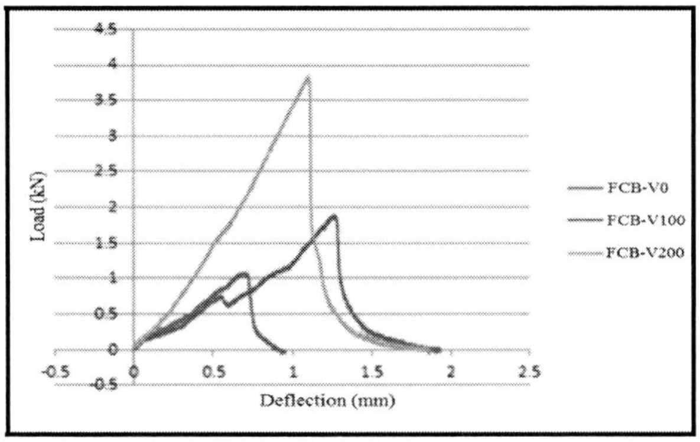


(b)

**Figure 4.** Graph of load versus deflection $G_f$,
a) U-notch beam b) V-notch beam

**Table 2.** Analysis of fracture energy, $G_f$, by Hillerborg [4] model

| Notch position | $W_0$ (Nm) | Mass (kg) | $g$ (N) | $\delta_f$ (mm) | $A$ ($m^2$) | $G_f$ (N/m) |
|---|---|---|---|---|---|---|
| FCB-U0 | 0.17 | 21.83 | 9.81 | 0.69 | 0.0158 | 20.280 |
| FCB-V0 | 0.07 | 22.34 | 9.81 | 0.71 | 0.0158 | 14.170 |
| FCB-U100 | 0.22 | 22.00 | 9.81 | 0.79 | 0.0158 | 24.800 |
| FCB-V100 | 0.20 | 21.71 | 9.81 | 1.26 | 0.0158 | 29.270 |
| FCB-U200 | 0.22 | 21.91 | 9.81 | 1.16 | 0.0158 | 30.360 |
| FCB-V200 | 0.28 | 21.90 | 9.81 | 1.10 | 0.0158 | 32.300 |

Results in table 2 shows that the value of fracture energy varies with different equations, where the value of fracture energy for foamed concrete ranges between 15 N/m and 40 N/m [8]. The values obtained from the equations derived from Hillerborg [4] model (table 2) and from Bazant and Becq-Giraudon model (table 3) are 15 N/m to 35 N/m and 15.595 N/m, respectively. Using Hillerborg model, the lowest value of $G_f$ is on the midpoint with V-notch (14.170 N/m), whereas the highest value of $G_f$ is on 200 mm from the midpoint with V-notch (32.300 N/m). The calculation of $G_f$ from Hillerborg model depends on the area under the graph, which will result in a high value of fracture energy when the area is large. The notch shape does not significantly affect the fracture energy where the percentage is 10% in average. The test done by Rahman [7] was shown that there are 15% of average for different notch shape and different density.

The formula proposed by Bazant and Becq-Giraudon model and CEB model was used to support the calculation of fracture energy using the formula derived from Hillerborg model. Table 3 and Figure 5 shows that fracture energy, $G_f$, from Bazant and Becq-Giraudon model is 15.595 N/m, whereas that from CEB model is slightly lower at 14.400 N/m. This is due to the model proposed by CEB does not take into account the w/c ratio meanwhile the Bazant and Becq-Giraudan model does take into account the w/c ratio. Fracture energy obtained using this formula is slightly lower because of the lower compressive strength. Thus, it still supports the value of fracture energy and can be accepted when the value lies in the range previously stated.

**Table 3.** Analysis of fracture energy, $G_f$, using model Bazant and Becq-Giraudon and CEB

| Equation | $D_{max}$ (m) | $w/c$ | Compressive strength, $f_c$ (MPa) | $G_f$ (N/m) |
|---|---|---|---|---|
| (2) | 0.003 | 0.55 | 4.3 | 15.595 |
| (3) | 0.003 | Nil | 4.3 | 14.400 |

## 4. Conclusion

The study was conducted to determine the effect of distance on notch position under the three-point test. The Results using the Hillerborg model shows that different positions of the notch will directly affect the value of fracture energy. Meanwhile fracture energy using model Bazant and Becq-Giraudon and CEB does not show an effect to the fracture energy. This is due to the both of model only affected by the diameter of aggregate and the compressive strength.

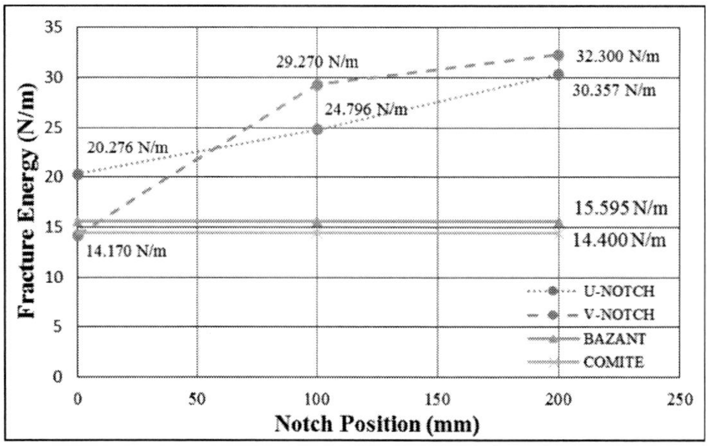


**Figure 5.** Overall fracture energy, $G_f$

Overall, study shows that, the further the position of the notch from the midpoint, the higher the value of fracture energy. The overall fracture energy for U-notch and V-notch ranges between 15 N/m and 35 N/m.

## 5. References

[1] Ramamurthy K and Nambiar E K K 2009 A classification of studies on properties of foam concrete *Cement and Concrete Composites* **31(6)** 388-396
[2] Darwin D, Barham S, Kozul R and Luan S 2001 Fracture Energy of High-Strength Concrete *ACI Materials Journal* **98(5)** 410-417
[3] Byung-Wan J and Ghi-Ho T 2001 Experimental Study On Fracture Energy Of Low-Heat Concrete By Three Point Bend Test *Russian Journal of Nondestructive Testing* **37(12)** 907-915
[4] Wittmann F H 2002 Crack formation and fracture energy of normal and high strength concrete *Sadhana* **27(4)** 413–423
[5] Muralidhara, S, Prasad, B K R, Karihaloo, B L and Singh, R K 2011 Size independent fracture energy in plain concrete beams using tri-linear model *Journal of Construction and Buildings Materials* **25(7)** 3051-3058
[6] Mier J G M V 1997 Fracture Processes of Concrete *CRP Press* (Florida, United States)
[7] Rahman N A, Jaini Z M and Zahir N M 2015 Fracture energy of Foam Concrete by Means of the Three Point Bending Test on Notched Beam Specimens *ARPN Journal of Engineering and Applied Sciences* **10(15)** 6562 – 6570
[8] Abd Rahman N, Jaini Z M, Abd Rahim, NA and Abd Razak, SA 2015 An experimental study on the fracture energy of foamed concrete using v-notched beams *Proc of the Int Civil and Infrastructure Engineering Conference* **InCIEC 2014** 97-108
[9] Kozlowski M, Kadela M and Kukielka A 2015 Fracture energy of foamed concrete based on threepoint bending test on notched beams *Procedia Engineering* **108** 349-354
[10] Rahman M H A and Jaini Z M 2013 The combined finite-discrete element analysis of precast lightweight foamed concrete sandwich panel (PLFP) under axial load *Proc of the Int Conf on Advances in Structural Civil and Environmental Engineering*
[11] FIP Model Code 1990 *Design Code, Euro-International du Beton and Federation International de la Precontrainte* (ISBN 0 7277 1696 4) (Telford, London) pp 437
[12] RILEM 1990 Fracture mechanics of concrete-test methods, determination of fracture parameters of plain concrete using three-point bend tests *Materials and Structures* **23** 457-460
[13] ASTM E1820-13 2014 *Standard test method for measurement of fracture toughness* (ASTM International) (West Conshohocken, United States)
[14] British Standard Institution 2009 *Testing Hardened Concrete, Compressive Strength of Specimen* (BSEN 12390 Part 3)
[15] British Cement Association 1994 *Foamed Concrete; Composition and Properties* (London, UK)

## Acknowledgments

The author would like to thank the Ministry of Education Malaysia, Universiti Tun Hussein Onn Malaysia and Centre for Graduate Studies for the funding and facilities. This study was conducted under FRGS Vot 1219 and grant scheme code U-573

# Thermal repellent properties of surface coating using silica

**Y Y Lee[1], M S Halim[1], E Aminudin[2] and N A Guntor[1]**

[1] Jamilus Research Centre (JRC), Faculty of Civil and Environmental Engineering, Universiti Tun Hussein Onn Malaysia, 86400 Parit Raja Batu Pahat, Johor, Malaysia.
[2] Faculty of Civil Engineering, Universiti Teknologi Malaysia, 81310 Johor Bahru, Johor, Malaysia.
Corresponding author: yylee@uthm.edu.my

**Abstract.** Extensive land development in urban areas is completely altering the surface profile of human living environment. As cities growing rapidly, impervious building and paved surfaces are replacing the natural landscape. In the developing countries with tropical climate, large masses of building elements, such as brick wall and concrete members, absorb and store large amount of heat, which in turn radiate back to the surrounding air during the night time. This bubble of heat is known as urban heat island (UHI). The use of high albedo urban surfaces is an inexpensive measure that can reduce surrounded temperature. Thus, the main focus of this study is to investigate the ability of silica, $SiO_2$, with high albedo value, to be used as a thermal-repelled surface coating for brick wall. Three different silica coatings were used, namely silicone resin, silicone wax and rain repellent and one exterior commercial paint (jota shield paint) that commercially available in the market were applied on small-scale brick wall models. An uncoated sample also had been fabricated as a control sample for comparison. These models were placed at the outdoor space for solar exposure. Outdoor environment measurement was carried out where the ambient temperature, surface temperature, relative humidity and UV reflectance were recorded. The effect of different type of surface coating on temperature variation of the surface brick wall and the thermal performance of coatings as potential of heat reduction for brick wall have been studied. Based on the results, model with silicone resin achieved the lowest surface temperature which indicated that $SiO_2$ can be potentially used to reduce heat absorption on the brick wall and further retains indoor passive thermal comfortability.

## 1. Introduction

Malaysia is a tropical country where sunlight shines upon the faces of buildings in a normal sunny day. Malaysia is experiencing rapid economic development since the last 2 decades [1]. The high exposure to solar heating in tropical climate is causing high ambient temperature occurred when the sunlight is blazing. When the wall surface is exposed to excessive solar heating, there is a portion of heat energy being absorbed by the wall. The brick wall will store the heat and then re-emitted back especially at nightfall whenever the ambient temperature is lower than the heated brick wall. This is because of the high density nature of brick makes itself an effective thermal mass to absorb and release high amounts of heat energy before experiencing the temperature increase [2]. The rise of ambient temperature at the night can be described as the urban heat island (UHI) phenomena and may magnify by these construction materials.

IOP Conf. Series: Materials Science and Engineering **271** (2017) 012020 doi:10.1088/1757-899X/271/1/012020

The heat dissipation during night time is one of the key problems faced by the tropical countries. Brick walls absorb heat from sunlight during day time and release to the surrounded environment at night when the ambient temperature is lower than that surface temperature of brick walls. The high density of brick wall makes it act as a thermal mass where it has the capacity to absorb heat energy [3] and the heat will be dissipated into the surroundings when the surrounding temperature is lower than that of the brick wall. The excess heat is released into indoors causes higher room temperature at night than usual. It also poses higher energy consumption for the air-conditioning system to cool down the room.

Due to the changes of environmental feature, the attention is drawn within thermal comfort study in order to reduce the heat transfer to the building by using high reflectivity surface coating material. The amount of energy that is reflected by a surface is determined by the reflectivity of that surface called albedo. A high albedo represents a surface which can reflect majority and absorb minority of the radiations that reach the surface, whereas, a low albedo represents a surface which can reflect a small amount and absorb the rest of the incoming radiation [3]. It is generally agreed that buildings with high reflectance surface can improve the indoor thermal effectively and serve to provide comfortability for applicants and reduce costs for heating and cooling.

There are many types of thermal insulation systems. At present, the common materials of exterior insulation in Malaysia are external heat-insulation wall of polystyrene panel, aerated concrete blocks and thermal insulation coating. The choice of the proper insulation materials type and form depends on the application as well as the desired materials physical, thermal and other properties. Thermal insulation coating is found beneficial to both new and existing buildings without modification of existing building. The use of high reflective surface coatings that content hiding pigments, can protect subtract from the harmful effect of ultraviolet light, and further reduce indoor temperature of a building. Surface reflectance can be optimized by using the materials that are characterized by a high solar reflectance. The use of high reflectivity material as a surface coating in the urban environment may contribute to a lower surface temperature which can minimize the thermal exchanges with the ambient [4].

Therefore, this paper investigates the effect of different type of surface coating on temperature variation of the surface brick wall and the thermal performance of coatings as a potential of heat reduction for brick walls. There are five small-scaled brick wall with different coatings of Silicon Resin, Silicone Wax, Rain Repellent and Jota shield paint, including an uncoated control specimen, were studied with the data collection of ambient temperature, surface temperature, relative humidity and UV reflectance for a week period. Thermal performance analysis of the specimens is conducted using the collected data.

## 2. Materials and methods

This study was conducted to determine the effect of coating materials on building thermal performance. In this experiment, three different silica coatings which were silicone resin, silicone wax and rain repellent, and one commercial paint were applied to the casted small-scaled brick wall models, with dimensions of 260 × 360 mm, as shown in figure 1, by spraying and roller/brush method. An uncoated brick wall model also has been studied for comparison. The experimental setup area were away from the building wall and above the pavement to avoid addition heat radiation to the brick wall models.

Ambient temperature, relative humidity and surface temperature were measured for a week period from 9:00 am to 9:00 pm. A continuous measurement of surface temperature on brick wall models was recorded at the interval 5 minutes by using GL200 midi data logger (GRAPHTEC). The thermal performances of the insulated models were compared with the control specimen. The ambient and surface temperatures of the specimens with and without coating were analysed to determine the effect of different type of surface coatings on temperature variation. Meanwhile, comparison was developed to investigate the thermal performance of coating as potential of heat reduction for brick wall. The

IOP Conf. Series: Materials Science and Engineering **271** (2017) 012020 doi:10.1088/1757-899X/271/1/012020

infrared thermal camera was used to capture the 'heat contour' at brick wall surface. The images were captured during the highest ambient temperature on the first day at the time of 2:15 pm.

**Figure 1.** Experimental setup.

## 3. Results and discussions

Thermal insulation is one of key factor for assessing the material composition of building envelops [5] which is important for urban planning of thermal balance. Basically, one of the sustainable approaches to cooling buildings by natural means is the passive cooling strategy. The flow of energy is based on natural means where the heat from solar radiation will be absorbed and the infrared will radiate and dissipate part of the accumulated heat through convection, conduction and radiation [6]. Throughout the experiment, all specimens were placed in outdoor environment of tropical climate.

### *3.1. Surface temperature variation*

Control specimen always record with the highest among the specimens, as shown in figure 2, which indicated coating is necessary in order to reduce the surface temperature of building materials. Moreover, ambient temperature is always the lowest values as the building materials are denser than air which can absorb more heat within its medium.

Day 1 showed the highest surface temperature of 45°C and ambient temperature of 35.9°C, recorded in figure 2. This is because there was no rainfall or cloudy sky while day 2 to day 7 were cloudy and even rain that caused the drop in surface and ambient temperatures. Therefore, only the day 1 data were critically discussed.

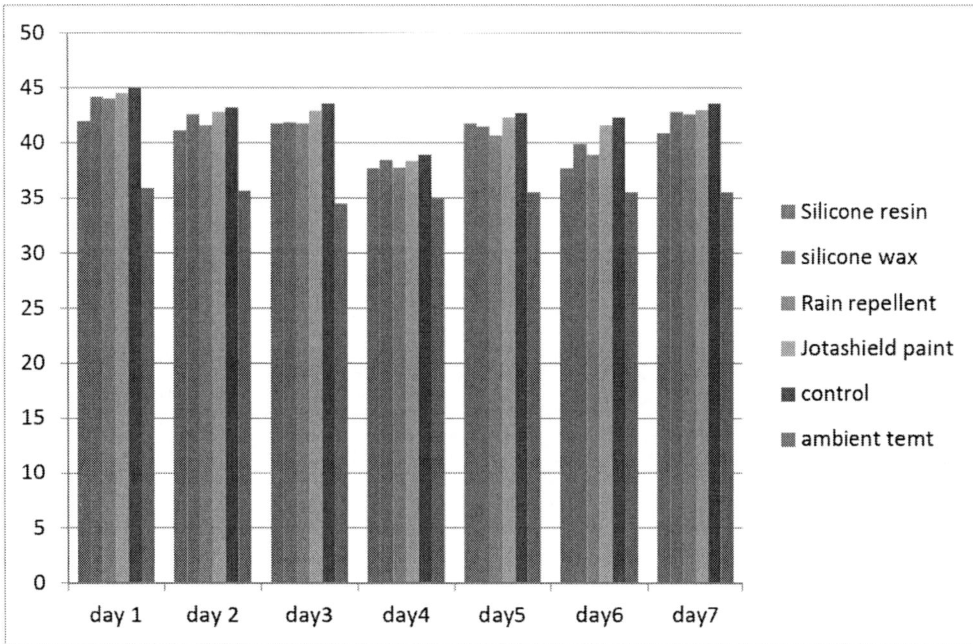


**Figure 2.** Maximum surface and ambient temperature in 7 days.

*3.1.1. Effect of ambient temperature on surface temperature.* According to figure 3, surface temperature was fluctuated from 9:00 am until 3:00 pm. At 2:15 pm, surface temperature rose to 45°C (control sample) higher than the ambient temperature by 9.1°C. It is due to solar heat absorbed by the building materials [7]. The temperature difference was caused by their different thermal inertia. This situation corresponding to [8] where the brick began to increase the temperature faster because it had larger capacity to store heat. At 5:00 pm, the surface temperature of silicon resin was lower than ambient temperature this is due to thermal mass, brick wall have high thermal mass which can act like a battery for storing thermal energy, it can store the heat during midday and release that heat to the environment [9].

The surface temperature of silicone resin coating was the fastest to reach the thermal equilibrium stated, where it was placed in contact with heat transferred until it reached the same temperature of both surface and ambient temperatures, followed by rain repellent, silicone wax and Jota shield paint. Meanwhile, for control sample, the surface temperature is the highest. It is an uncoated sample that no insulator to minimize heat flow through the brick wall and caused the absorption of heat much more than the other specimens. Thus, make it the slowest to release the heat back to environment [5]. In this study, therefore, silicon resin obtained better potential of heat reduction for brick wall because it absorbed the least amount of heat and recorded the fastest period to release the stored heat.

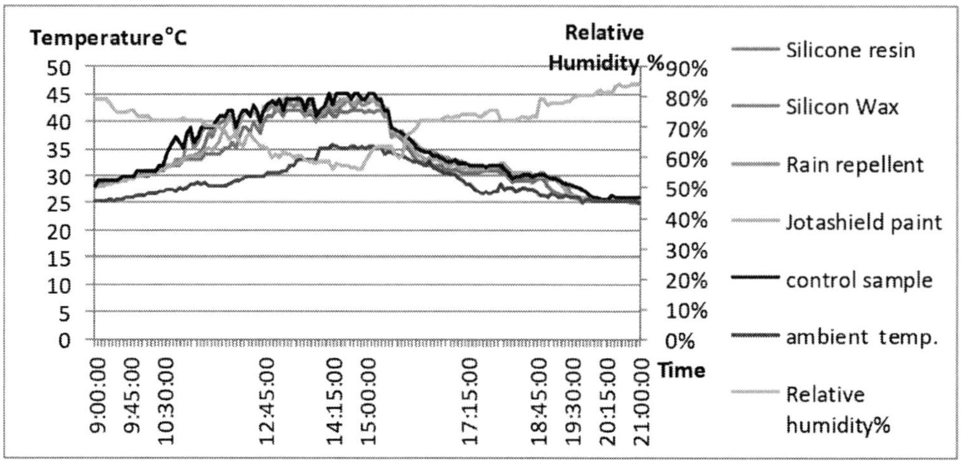


**Figure 3.** Observation of temperature variation of day 1.

The coolest surface temperature were achieved by silicone resin followed by rain repellent, silicone wax and Jotashield paint. Meanwhile, the control sample depicted the warmest surface temperature. The difference of surface temperature between silicone resin and control sample was 3°C and 2.5°C between silicone resin and Jotashield Paint. Silicone resin is a silica $SiO_2$ based product where $SiO_2$ can be a heat reflector by reflecting the exceeded radiation to a brick wall. This will create a heat-insulating layer on the brick wall surface. The $SiO_2$ also consists of excellent UV reflectance [10].

On the other hand, control specimen absorbed more solar heat radiation and has low albedo of 0.20 – 0.40 [11] which caused to absorb heat more than the coated specimens. The rising temperature will affect the process of energy transfer between a building and its surroundings, also known as thermal performance of a building. Solar radiation will be absorbed by the vertical wall and the radiation will be converted into heat energy that will increased the wall's surface temperature. Some of the heat will reradiate to the atmosphere and surroundings. This reradiated heat would caused urban heat island (UHI) phenomenon.

### *3.2. Thermal infrared image*

Figure 4 showed the thermal image for all brick wall surface with different coated surface and uncoated surface. The test was recorded during the highest ambient temperature at 2:15 pm of day 1. Based on the obtained thermal images, it can be observed that the silicone resin have the weakest heat absorption since the colour on the surface was still consisted of more blue colour even it started to change to green and red colour. Thus, it was proved that the silica resin coating was the most effective in solar heat reduction as compared with other specimens. The control sample was fully red colour due to its capability to absorb more heat radiation since there was no insulator to act as barrier to minimize heat absorption [7].

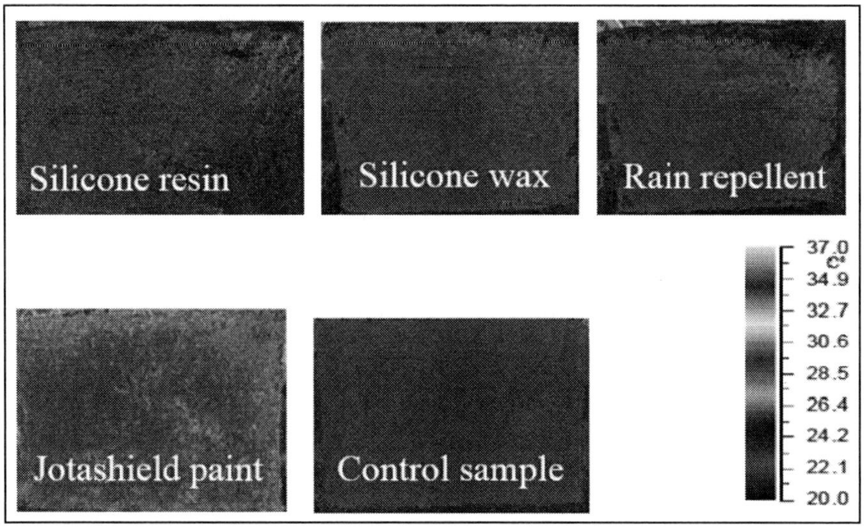


**Figure 4.** Thermal image for different type of coating on brick wall surface.

## 4. Conclusion

In this study, five small-scaled brick wall were developed for thermal performance investigation. Three different silica coating (silicone resin, silicone wax, rain repellent), one exterior commercial paint ( Jota shield paint) and one uncoated control sample were tested to determine the effect of different types of surface coating on temperature variation for the surface of brick wall. Further analysis on the thermal performance of coatings as a potential product of heat reduction for brick wall also has been carried out.

From the analysis, all the surface temperature of the samples were affected by the surrounding environment where the ambient temperature and relative humidity was measured to demonstrate the thermal conditions. Several conclusions have been drawn from the experimental study.

- The surface temperature of brick wall is directly proportional to the ambient temperature.
- The surface temperature is inversely proportional to the relative humidity.
- The lowest surface temperature was coated with silicone resin followed by rain repellent, silicone wax and Jota shield paint.
- The warmest surface temperature was the uncoated control specimen.

## 5. References

[1] Elsayed I S M 2009 A study on the urban heat island of the city of kuala lumpur, Malaysia, *JKAU: Met.Env. & Arid Land Agric. Sci.* **23** 121 - 134

[2] Steven V, Kotol M, Grunau B and Craven C 2016 The effect of thermal mass on annual heat load and thermal comfort in cold climate construction, *J. Cold Reg. Eng.* **30** 1 - 13

[3] Akbari H, Matthews H D and Seto D 2012 The long-term effect of increasing the albedo of urban areas, *Environ. Res. Lett.* **7** 1 - 10

[4] Md Din M F, Dzinun H, Ponraj M, Chelliapan S, Noor Z Z, Ossen D S and Iwao K 2012 Investigation of thermal effect on exterior wall surface of building material at urban city area, *Int. J. Energy Environ.* **3** 531 - 540

[5] Aktacir M A, Büyükalaca O and Yilmaz T 2010 A case study for influence of building thermal insulation on cooling load and air-conditioning system in the hot and humid regions, *Applied Energy* **87** 599 - 607

[6] Al-Obaidi K, Ismail M and Abdul Rahman A M 2014 Passive cooling techniques through reflective and radiative roofs in tropical houses in Southeast Asia: A literature review, *Front. Archit. Res.* **3** 283 - 297
[7] Md Din M F, Dzinun H, Ponraj M, Chelliapan S, Noor Z Z, Ossen D S and Iwao K 2012 Investigation of Heat Impact Behavior on Exterior Wall Surface of Building Material at Urban City, *J. Civil Environ. Eng.* **2** 1 - 6
[8] Guan K K 2011 Surface and ambient air temperatures associated with different ground material: a case study at the University of California, Berkeley, *Environ. Sci.* **196** 1 - 14
[9] Reardon C, McGee C and Milne G 2013 *Thermal mass* Retrieved on October 12, 2017 from http://www.yourhome.gov.au/sites/prod.yourhome.gov.au/files/pdf/YOURHOME-PassiveDesign-ThermalMass.pdf
[10] Rezaei E and Moghaddas J 2016 Thermal conductivities of silica aerogel composite insulating material, *Adv. Mater. Lett.* **7** 296 – 301
[11] Taha H 1997 Urban climates and heat islands: albedo, evapotranspiration, and anthropogenic heat, *Energ. Buildings* **25** 99 - 103

**Acknowledgments**
Authors wish to acknowledge assistance from Universiti Tun Hussein Onn (UTHM) for the technical supports. The author also would like to extend their gratitude to financial support from Universiti Teknologi Malaysia under vot 12J53.

IOP Conf. Series: Materials Science and Engineering **271** (2017) 012021 doi:10.1088/1757-899X/271/1/012021

# The potential of net zero energy buildings (NZEBs) concept at design stage for healthcare buildings towards sustainable development

**Roy Hazli Abdellah[1], Md Asrul Nasid Masrom[1], Goh Kai Chen[1], Sulzakimin Mohamed[1] and Roshartini Omar[1]**

[1] Faculty of Technology Management and Business, University Tun Hussein Malaysia
Corresponding author: asruln@uthm.edu.my

**Abstract.** The focus on net-zero energy buildings (NZEBs) has been widely analysed and discussed particularly when European Union Parliament are progressively moving towards regulation that promotes the improvement of energy efficiency (EE). Additionally, it also to reduce energy consumption through the recast of the EU Directive on Energy Performance of Buildings (EPBD) in which all new buildings to be "nearly Zero-Energy" Buildings by 2020. Broadly, there is a growing trend to explore the feasibility of net zero energy in healthcare sector as the level energy consumption for healthcare sector is found significantly high. Besides that, healthcare buildings energy consumption also exceeds of many other non-domestic building types, and this shortcoming is still undetermined yet especially for developing countries. This paper aims to review the potential of NZEBs in healthcare buildings by considering its concept in design features. Data are gathered through a comprehensive energy management literature review from previous studies. The review is vital to encourage construction players to increase their awareness, practices, and implementation of NZEBs in healthcare buildings. It suggests that NZEBs concept has a potential to be adapted in healthcare buildings through emphasizing of passive approach as well as the utilization of energy efficiency systems and renewable energy systems in buildings. This paper will provide a basis knowledge for construction key players mainly architects to promote NZEBs concept at design stage for healthcare buildings development.

## 1. Introduction

Over the years, the concern of energy efficiency along with the attention on net-zero energy buildings (NZEBs) has received incrementing attention by most countries. NZEBs concept is developed to represents a zero or low energy building that using renewable energy systems installed to meet the buildings energy consumption demands [1]. In the context of Malaysia's construction industry, the government has recognised the important concern of energy consumption and negative impacts of the inefficient use of energy in developing the country. It can be clearly seen through Construction Industry Transformation Programme (CITP) 2016-2020 which one of strategic thrust that seeks to transform the industry is environmental sustainability [2]. Generally, between 1978 and 2014, the residential and commercial sector's final energy demand in Malaysia had increased rapidly to reach 7458 ktoe [3]. As a result of uncontrollable energy consumption of resources and energy, the sector has been greatly responsible for problems related to environmental sustainability. However, despite today's buildings energy consumption demand increasing significantly, NZEBs concept practices enable to develop a network of clean domestic energy assets.

Healthcare buildings is a complex building since these buildings involve several of service area which requires different working environment [4]. For instance, in the current Malaysia healthcare built environment, with healthcare building as the main energy users, reducing $CO^2$ emission and improving energy efficiency are becoming more important. Based on Kamaludin *et al.* [5], healthcare building built environment in Malaysia associated with great energy consumption and $CO^2$ emissions. It is important to realise that the accumulation of carbon dioxide $CO^2$ emission may result in global warming [6].

Besides that, according to the study that has been conducted by Shaza & Rozana [7] Amanjaya Specialist Center Green Hospital and Columbia Hospital, Petaling Jaya are the only two numbers of private healthcare buildings have been evaluated by using GBI NRNC rating system and achieved as 'certified' category. Contrastingly, none public healthcare buildings in Malaysia certify under the GBI system. With the attention to this issue, it clearly showed that the implementation of healthcare sustainability as well as energy efficiency awareness is still at a low level. Therefore, NZEBs should become a core concept that needs to follow by all construction players as well as government agencies.

## 2. Net Zero Energy Buildings (NZEBs) concept

This section presents a fundamental aspects of the net-zero energy buildings (NZEBs) concept that used currently. An overview of NZEBs as highlighted by Patiño *et al.* [8] are buildings that achieve the benchmark of efficiency by following or overcoming a list of technical fundamentals. Furthermore, NZEBs concept is developed to represents a zero or low energy building that using renewable energy systems installed to meet the buildings energy consumption demands [9]. The concept was analysed in several perspectives. The detail discussion of NZEBs definitions is explained in the following subsection 2.1.

### *2.1. What is NZEBs?*

The earliest definition and classification of net-zero energy buildings (NZEBs) can be found in the study by Paul *et al.*, [9], that the authors addressed four main primary definitions of NZEBs. The four main definitions for various types of NZEBs according to Paul *et al.* [9] are net zero site energy, net zero source energy, net zero energy costs, and net zero emissions as illustrated in Figure 1. According to the four definitions, it pointed out that the NZEBs is a grid connected building where the annual energy generation from a renewable source is equally with energy consumption during the term of one year. Similarly, Sartori *et al.*, [10] emphasized that NZEBs is conceptually a building with a lower of energy demand and are balanced by the on-site generation of electricity, or other energy carriers, from renewable sources. Moreover, the authors also mentioned that the term NZEBs refer to buildings that are connected to the energy infrastructure. Furthermore, another study by Kurnitzki *et al.* [11] and Karsten *et al.* [12] highlighted that NZEBs are typically a grid-connected building with a very high energy performance.

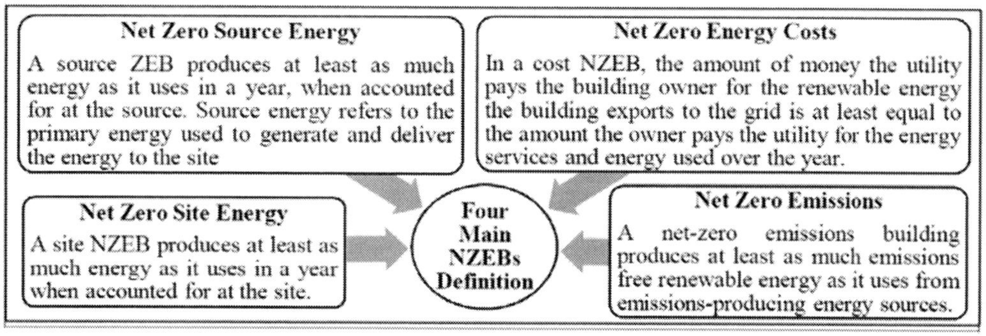


**Figure 1.** Four main definitions of NZEBs [9].

However in 2009, The Massachusetts Zero Net Energy Buildings Taskforce which comprised of more than 70 leaders has been worked together to establish key solutions that will guide to the worldwide adoption of NZEBs and deep energy reduction retrofits throughout the Commonwealth by 2030. The taskforce defined NZEBs as a building that is capable to generate energy onsite with a quantity that is equal to or greater than the total quantity of energy consumed onsite within a period of a year, and the building uses clean renewable resources [13].

### *2.2. What are NZEBs criteria?*

The continuous improvement on the NZEBs definition can be seen through the study of Sartori *et al.* [10] which to introduces a harmonised framework for describing the important elements of NZEBs in a series of criteria. Based on their study, the authors have listed several criteria that should be questioned and addressed in order to strengthen the overall NZEB "goodness of design" as well as to give a sound NZEB definitions in a formal, systematic and comprehensive way [10]. The summary of the scope of questioned criteria among the task activities is illustrated as in Figure 2.

| First Criteria | **Boundary Condition** |
|---|---|
| •System boundary (Is the boundary on a single building or on a group of buildings?)<br>•Functionality and effectiveness (What type of building is it?)<br>•Climate and Comfort (What is the reference climate?) | |
| Second Criteria | **Crediting System** |
| •Credit metrics (What metrics is used?)<br>•Credit accounting (How are the credits accounted for?) | |
| Third Criteria | **Net Zero Balance** |
| •Item of balance (What loads are included in the balance?)<br>•Balancing period (What is the basis for calculating the balance?)<br>•Energy effiency (Are there mandatory minimum requirements on energy efficency?)<br>•Energy supply (Is there an explicit hierarchy of supply options?) | |
| Fourth Criteria | **Temporal Energy Match** |
| •As grid connected buildings are investigated, it is useful to define some indicators which can evaluate the impact on the energy carrier exchange between the building and the energy infrastructures. | |
| Fifth Criteria | **Monitoring Procedure** |
| •Is the definition based only on design data and simulations or a monitoring procedure is mandatory to check the effective balance? | |

**Figure 2.** NZEBs definition criteria [10].

IOP Conf. Series: Materials Science and Engineering **271** (2017) 012021 doi:10.1088/1757-899X/271/1/012021

Additionally in 2012, a consistent framework for setting NZEB definitions has been developed through the study of Sartori *et al.,* [14] in which evaluation of the criteria in the definition framework and selection of the related options becomes a methodology to set NZEB definitions in a systematic way. According to the study, the authors stated that there is the difference between the terms of net-zero energy buildings (NZEBs) and zero energy buildings (ZEBs) as the terms of ZEBs is broader and it could include autonomous buildings. The wording 'Net' in NZEBs indicates that the energy obtained from and provided back to the energy grids nominally within a period of a year is a balance [14]. Moreover, the study showed that the core concept of NZEBs is where there is a balance between weighted supply and weighted demand. In other words, the sum of all generated energy is equal to the sum of all delivered energy.

Therefore, since 2006 to 2012, much improvement on NZEBs definition had been made in order to give a sound and comprehensive definition of this concept. Hence, it is understood that NZEBs represent buildings that connected to the electricity grid and the total energy usage is equal to the total 'onsite' generation of energy from a renewable source over the course of a year.

## 3. Adapting NZEBs design features for healthcare buildings

There are numerous different methods to reach net zero energy balance within a year. Based on Matt *et al.,* [15] there is a growing interest in NZEBs as many NZEBs demonstration projects showed that the goal of this concept is achievable. However, the need to identify NZEBs approaches is crucial in order enable broad replication of NZEBs. Furthermore, according to Aelenei *et al.,* [16] in the study on close inspection of the relevant design strategies and relative performance indicators of the eight case studies, the authors stated that there is no standard approach for designing NZEBs. Similarly, Gandhi *et al.,* [17] highlighted that it is important for architects or engineers to have specific design guidelines and strategies in order to popularize NZEBs. Even though there are still no specific design strategies in achieving NZEBs design, but based on Yuehong *et al.* [18] there are several common design elements and some consensus in designing NZEBs.

Generally, according to Aelenei *et al.* [19] and Yuehong *et al.* [18], there are three main steps of design approaches that can be applied in designing NZEBs which are the first design approach is passive approach that focused on reducing energy demand. The second design approach is the use energy efficiency system, and the third design approach is renewable energy system which is needed to generate renewable energy as well as to offset in large measure the energy demand. The combination of this three design approach able to succeed in reaching the desired energy performance as well as NZEBs target. Therefore, this section intends to analyse the current design practices of NZEBs and the detail explanation of the first design approach will be explained in subsection 3.1, subsection 3.2 will explain on the second design approach and the third design approach will be explained in subsection 3.3.

### *3.1. First design approach: passive approach*

Passive approaches are one of the key element when designing NZEBs as this approach will affect directly to the energy needs for the buildings mechanical and electrical systems, and also indirectly affect the renewable energy generation [16]. Similarly, Alessandra *et al.,* [20] also highlighted that the passive approach plays a significant part in the NZEBs design. Besides that, the improvement of energy consumption of a building can be achieved through passive design strategies as buildings are constructed that react to the environment, hence it is possible for a building to achieve high environmental quality [21].

Furthermore, Gandhi *et al.,* [17] stated that architects have an important role in controlling aspects of passive design when designing NZEBs. One of the most significant aspects of passive design strategies is maximizing the amount of natural daylight into a building and reduce the amount of energy requires for artificial lighting in a building [22]. Besides that, it is important to control passive design strategies in a building as this approach will influence the energy performance of a building such as natural lighting, heat gain, shading, and envelope conduction.[17,23]. In addition, Aleinei *et al.* [16] stated that the passive design strategies in buildings should be appropriately orientating building towards solar heating maximization when buildings are dealing with heating challenges. Moreover, according to Thalfeldt *et al.* [24] when designing NZEBs, façades has a strong impact on heating, cooling and artificial light in the building needs as well as on daylight. Thus, it is important to realise that the passive heating solutions are essential to be studied along with passive cooling solutions in order to prevent overheating in a building [16].

This indicates that it is important to consider passive approach such as orientation, shading devices, ventilation and thermal insulation while designing NZEBs as it will enable the reductions of energy needs in buildings and this approach able to provide significant benefits for healthcare buildings.

### *3.2. Second design approach: energy efficiency system*

Energy efficiency has a significant part to act as a parameter that indicates the reduction level of energy consumption in carrying out a related task [25]. Besides that, it is important to realise that the pathway towards NZEBs is reducing energy demand by means of energy efficiency measures [10]. Therefore, energy efficiency measure plays a significant role towards NZEBs as it is one of the key actions in achieving NZEBs target. Moreover, the government of Poland also has supported the energy efficiency measure in order to successfully implementing NZEBs [26]. The main purpose to enhance energy efficiency for the reason that it will able to protect scarce energy resources as well as to minimise the costs of energy by means of preventing wastefulness, and possibly to minimise carbon dioxide ($CO^2$) emissions which will effect to climate change [27].

Generally, energy efficiency can be improved by several design strategies including airtightness to avoid infiltration and mechanical ventilation systems with heat recovery to provide air conditioning and indoor air quality (IAQ) [28]. However, the selection of high-efficiency technologies is also crucial in the improvement of energy efficiency in a building. For instance, the use of low power lighting, energy efficient electrical equipment such as washing machines and dishwashers with a warm water connection are strategies in planning an equated energy balance [29].

Similarly, based on Maassen [30], healthcare building's energy demand can significantly reduced by considering energy efficiency as a factor in buying medical equipment, besides the proper use of electricity. Furthermore, many researchers have demonstrated that by implementing different energy efficiency measures in a building, the variation of energy savings can be achieved. According to Erica *et al.*, [31] that carried out an investigation of the effectiveness of energy efficiency measures for different occupancy patterns. They concluded that similar savings to passive approach and conversion devices can be achieved through the combinations of less expensive and less invasive energy efficiency measures. As a result, these energy efficiency measures options should be greatly promoted.

In addition, based on Rahmi [32] in the investigation of energy consumption reduction by utilizing high-efficiency office equipment and HVAC system by using DesignBuilder as simulation tool in this research. The simulation findings showed that there are several factors generally will give significant impacts to the reduction of cooling energy demand in buildings. The factors involved are the application of high-efficiency HVAC system, the use of high performance glazing material (low SHGC value and high light transmission value) for building envelope with majority glass façade and involving task lighting which includes the lighting sensor and indoor set point temperature.

This shows that the adoption of energy efficiency measure such as energy efficient mechanical system and energy efficient electrical system in a building is vital towards NZEBs as energy reduction, as well as energy savings in healthcare buildings, can be achieved. Moreover, by improving the energy efficiency of the various incorporated building systems, it will help to reduce building's energy demand [33].

### *3.3. Third design approach: renewable energy system*

The necessity of renewable energy systems such as solar electric PV and solar water heating is vital towards NZEBs since these systems are required to reduce and as well as to offset the thermal energy need in buildings. Moreover, designing NZEBs is not only achieved by reducing the energy consumption of the building with passive design methods or the application of energy efficient system in buildings, but NZEBs should also be designed with the balances energy requirements between active energy production techniques and renewable technologies in order to achieve the NZEBs objective [34]. In this regard, innovative renewable energy system strategies have to be taken into consideration. Besides that, the utilization of renewable energy systems in buildings can be a critical need and solutions against global warming and environmental pollution [35].

With respect to renewable energy systems, there are a lot of benefits that can be gain from these system and several studies have demonstrated the benefits of these systems. For instance, Phuangpornpitak & Tia [36] studying the integration of renewable energy in a smart grid system and they concluded that renewable energy system is one of incomparable option to generate energy since these systems able to provide a clean energy resource. However, they suggested that some issues such as the design, sizing, and the suitability of the system in terms cost for energy generation need to be addressed in order to ensure that it can be used thoroughly and commercially.

In addition, Rishi *et al.* [37] presented a comparative analysis of four prevalent solar lighting technologies by surveying selected photovoltaic (PV) programs in India. Authors found that a better operation of solar systems can be achieved by a strong management and monitoring mechanisms of solar PV programs. It also highlighted that the key contributors for extensive use of PV in India are strong technical, institutional and financial aspects since developing a strong framework and institutional policies vital towards achieving higher success rates in PV application. Based on Erni *et al.*[38] studied the green concept of healthcare facilities in tropics stated that utilization of natural energy and by using solar cell and the creation of a simple alternative energy is vital to minimise the energy dependency on the electricity power plant and generators. Additionally, according to Shanti & Paul [39], there are four classifications of NZEBs that based on their renewable energy sources:

- NZEB A: indicates a building uses a combination of energy efficiency and renewable sources gained within the building footprint.
- NZEB B: indicates a building uses a combination of energy efficiency and renewable sources gained within the building footprint and within the site.
- NZEB C: indicates a building uses a combination of energy efficiency and renewable sources gained from off-site to generate energy on-site.
- NZEB D: indicates building purchases certified off-site renewable energy sources such as utility-scale wind.

As a summary, there are three key areas that need to focus in order to support the implementation of NZEBs concept for healthcare buildings into action which is namely, passive design strategies, the application of energy efficiency system, and renewable energy system. These design approaches are crucial towards the reduction of energy demands in healthcare buildings as well as the generation of clean energy sources for healthcare buildings. Hence, Figure 3 shows a summary of the main design approach for NZEBs in healthcare buildings.

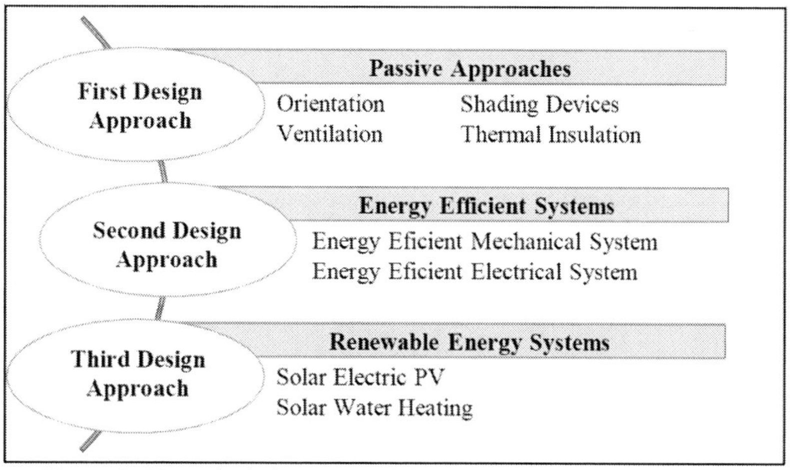


**Figure 3.** The main design approach for NZEBs in healthcare buildings.

## 4. Conclusion

Healthcare buildings is a complex building and these buildings involve several of service area which requires different working environment and the aspect of low energy and environmental control should be more focus. It is crucial to determine the principle design of healthcare buildings to achieve net zero energy buildings (NZEBs) requirement. The focus on the NZEBs in healthcare building had become significant especially in developing country. The conclusion is drawn that three principles were used to identify possibilities to reach towards NZEBs requirements for healthcare building. This principle consisted of passive approach, energy efficiency, and renewable energy systems. By considering to this context, it is expected that NZEBs concept is potential to be adopted in healthcare building for ensuring the energy improvement as well as to achieve a maximum of energy efficiency. Besides that, the adaptation of NZEBs design features into healthcare buildings will enable the healthcare sustainable development.

## 5. References

[1] Paul T, Shanti P and Chad L 2010 Main street net-zero energy buildings: the zero energy method in concept and practice *Paper ASME 2010 4th Int., Conf., on Energy Sustainability* (Phoenix, Arizona: ASME) 1009-17

[2] Construction Industry Development Board (CIDB) 2015 *Construction industry transformation programmes 2016-2020* (Kuala Lumpur: Construction Industry Development Board) p 19

[3] Suruhanjaya Tenaga 2015 *Statistical handbook malaysia* (Putrajaya: Suruhanjaya Tenaga) p 45

[4] Norwina M N, Abdul R S, Noor H A M and Srazali A 2013 Hospital designs in tropical Malaysia - towards a green agenda *The uia/phg 2013 annual healthcare forum + Gupha meeting at IIDEX* (Canada, Toronto: IIUM Repository) pp 1-47

[5] Kamaluddin K A, Imran M S and Yang S S 2016 Development of energy benchmarking of Malaysian government hospitals and analysis of energy savings opportunities *J. of Build. Perf.* **7**(1) 72-87

[6] Michaelides and Stathis E E 2012 *Alternative energy sources* (Berlin: Springer) p 35

[7] Shaza R S and Rozana Z 2014 Green assessment criteria for public hospital building development in Malaysia *Proc. Envir. Sc.* **20** 106-15

[8] Patiño-Cambeiro F, Armesto J, Patiño-Barbeito F and Bastos G 2016 Perspectives on near zeb renovation projects for residential buildings: the Spanish case *Energies* **9**(8) 628

[9] Paul T, Shanti P and Michael D 2006 Zero energy buildings: a critical look at the definition *Preprint No. NREL/CP-550-39833*

[10] Sartori I, Napolitano A, Marszal A J, Pless S, Torcellini P and Voss K 2010 Criteria for definition of net zero energy buildings *In Proc., of EuroSun 2010: Int., Conf., on Solar Heating, Cooling and Buildings* (Graz, Austria: EuroSun) 1-8
[11] Kurnitzki J, Allard F, Braham D, Goeders G, Heiselberg P, Jagemar L, Koso-nen R, Lebrun J, Mazzarella L, Railio J, Seppänen O, Schmidt M and Virta M 2011 How to define nearly zero energy buildings *REHVA J.* **48**(3) 6–12
[12] Karsten V, Eike M and Markus L 2011 From low-energy to net zero-energy buildings: status and perspectives *J. of Green Build.* **6**(1) 46-57https://doi.org/10.3992/jgb.6.1.46
[13] Massachusetts Zero Net Energy Buildings Task Force 2009 Getting to zero: final report of the massachusetts zero energy buildings task force (Boston, USA: Mass.gov) **25** 46
[14] Sartori I, Assunta N and Karsten V 2012 Net zero energy buildings: a consistent definition framework *Ener. and Build.* **48** 220-32
[15] Matt L, Shanti P and Paul T 2014 Cost control best practices for net zero energy building projects *Preprint No. NREL/CP-5500-61365.*
[16] Aelenei D, Aelenei L, Musall E, Cubi E, Ayoub J and Belleri A 2013 Design strategies for non-residential zero-energy buildings: lessons learned from Task40/Annex 52: towards net zero-energy solar buildings. *In: CLIMA 2013 - 11th REHVA World Congress & 8th Int., Conf., on IAQVEC* (Prague, Czech Republic: Laboratorio Nacional de Energia e Geologia I.P) 1-10
[17] Gandhi H, Daniel C, Peter F, Alec R, Riadh H and Will N 2014 Sustainable design of a nearly zero energy building facilitated by a smart microgrid *J. of Renew. Ener.* **2014** 1–11
[18] Yuehong L, Shengwei W and Kui S 2015 Design optimization and optimal control of grid-connected and standalone nearly/net zero energy buildings *App. Ener.* **155** 463-77
[19] Aelenei L, Ayoub J and Aelenei D 2012 Net zero energy residential building case studies *Ener. Conserv. in Build. & Comm. Sys.* **56** 10-2
[20] Alessandra S, François G and Harald N R 2014 How net zero energy buildings and cities might look like? new challenges for passive design and renewables design *Ener. Proc.* **61** 1163–66.
[21] Hootman T 2012 *Net zero energy design: a guide for commercial architecture* (Hoboken, NJ: John Wiley) p 186
[22] Ajla A 2015 Regenerative design of existing buildings for net-zero energy use *Proc. Engin.* **118** 72–80
[23] Hossein O and Abdul K M 2015 Optimization of building energy performance through passive design strategies *Brit. J. of App. Sc. & Tech.* **13**(6) 1-16
[24] Thalfeldt M, Pikas E, Kurnitski J and Voll H 2013 Facade design principles for nearly zero energy buildings in a cold climate *Ener. and Build.* **67** 309-21
[25] Parameshwaran R, Kalaiselvam S, Harikrishnan S and Elayaperumal A 2012 Sustainable thermal energy storage technologies for buildings: a review *Renew. Sustain. Ener. Rev.* **16**(5) 2394–433
[26] Buildings Performance Institute Europe (BPIE) 2012 *Implementing nearly zero-energy buildings (nzeb) in Poland - towards a definition and roadmap (publication)* Buildings Performance Institute Europe.
[27] Jonathan M C and Julian M A 2010 Theoretical efficiency limits for energy conversion devices *Ener.* **35**(5) 2059-69
[28] Fausto B, Paolo C and Luca G 2017 A design strategy to reach nZEB standards integrating energy efficiency measures and passive energy use *Ener. Proc.* **111** 205-14
[29] Musall E and Voss K 2012 "The passive house concept as suitable basis towards net zero energy buildings" *Proc.* in *Conf.: Passive House Conference* (Hannover: Passiv Haus Institut) 1-6
[30] Maassen W 2017 (Nearly) zero energy healthcare buildings (Report No. I&BBE3112R001F0.1). Netherland, Haskoning DHV Nederland B.V.

[31] Erica M, Julia K S, Valerie D and Timothy J F 2016 Combining energy efficiency measure approaches and occupancy patterns in building modelling in the UK residential context *Ener. and Build.* **111** 98–108
[32] Rahmi A 2014 The role of building thermal simulation for energy efficient building design *Ener. Proc.* **47** 217-26
[33] Sartori I, Assunta N and Karsten V 2012 Net zero energy buildings: a consistent definition framework *Ener. and Build.* **48** 220-32
[34] Kolokotsa D, Rovas D, Kosmatopoulos E and Kalaitzakis K 2011 A roadmap towards intelligent net zero- and positive-energy buildings *Solar Ener.* **85**(12) 3067-84
[35] Koo C, Hong T, Park H and Yun G 2013 Framework for the analysis of the potential of the rooftop photovoltaic system to achieve the net-zero energy solar buildings. Progress in photovoltaic *Res. and App.* **22**(4) 462-78
[36] Phuangpornpitaka N and Tiab S 2013 Opportunities and challenges of integrating renewable energy in smart grid system *Ener Proc.* **34** 282–90
[37] Rishi R B, Debajit P and Sadhan M 2014 Comparative analysis of solar photovoltaic lighting systems in India *Ener. Proc.* **54** 680–89
[38] Erni S, Arnis R H and Yasmina N F 2013 Green building design concept of healthcare facilities on the orthopaedic hospital in the tropics *Proc. Soc. and Nat. Envir.* **101** 189–99
[39] Shanti P and Paul T 2010 Net-zero energy buildings: a classification system based on renewable energy supply options (Tech. No. NREL/TP-550-44586). National Renewable Energy Laboratory.

**Acknowledgement**
The authors would like to thank to Registrar Office, Universiti Tun Hussein Onn Malaysia and Faculty of Technology Management and Business (FPTP) for sponsoring this research.

# Passive wall cooling panel with phase change material as a cooling agent

**Masni A. Majid [1] ,Rasyidah Ahmad Tajudin [1], Norhafizah Salleh [1], Noor Azlina Abd Hamid [1]**
[1]Jamilus Research Center, Faculty of Civil and Environmental Engineering, University of Tun Hussein Onn Malaysia, Batu Pahat
Corresponding E-mail: masni@uthm.edu.my

**Abstract**. The study was carried out to the determine performance of passive wall cooling panels by using Phase Change Materials as a cooling agent. This passive cooling system used cooling agent as natural energy storage without using any HVAC system. Eight full scale passive wall cooling panels were developed with the size 1500 mm (L) x 500 mm (W) x 100 mm (T). The cooling agent such as glycerine were filled in the tube with horizontal and vertical arrangement. The passive wall cooling panels were casting by using foamed concrete with density between 1200 kg/m$^3$ – 1500 kg/m$^3$. The passive wall cooling panels were tested in a small house and the differences of indoor and outdoor temperature was recorded. Passive wall cooling panels with glycerine as cooling agent in vertical arrangement showed the best performance with dropped of indoor air temperature within 3°C compared to outdoor air temperature. The lowest indoor air temperature recorded was 25°C from passive wall cooling panels with glycerine in vertical arrangement. From this study, the passive wall cooling system could be applied as it was environmental friendly and less maintenance.

## 1. Introduction

Nowadays, we were facing the global warming problem. Global warming was a serious threat to our planet and would result in a wide variety of harsh environmental impacts. Greenhouse gas (GHG) emissions (especially $CO_2$ emission) were the underlying cause of this warning. Due to this problem, the people in world feel uncomfortable. For providing comfortable environment, they would find the alternative way such as using air conditioner and using fan. All of this needed to consume the electrical energy [1].

Energy used and supplied current trends were clearly unsustainable from social, economic and environmental perspective. The increase in energy demand results in increasing level of greenhouse gas emission that contributes to the increased global warming and cost of energy. Without decisive steps, carbon dioxide ($CO_2$) emission associated with energy consumption expected double in 2050 which leads to growing demand of fossil energy and intensity concerns over the supplies security [1].

One of the major challenges for administration and government is energy demand to satisfy thermal comfort in building. In energy systems where temporal differences exist between the supply of energy and its utilization, thermal storage is necessary to ensure the continuity of a thermal process [2].

Consuming to much electrical energy gave the big problem for future generation. This was because for the future development, most of their life would use all the technology which was needed electrical energy. Besides, consuming more electrical energy and air conditioner was high in cost. Most of people could not afford for it especially for rural population. So that, they would not find the comfortable environment for their life. Figure 1 showed the breakdown of energy consumption in Malaysia. This research was carried out to find the best way and solution to solve the thermal problem without using electrical energy and environmental friendly by using passive cooling system in order to save the future generation's life.

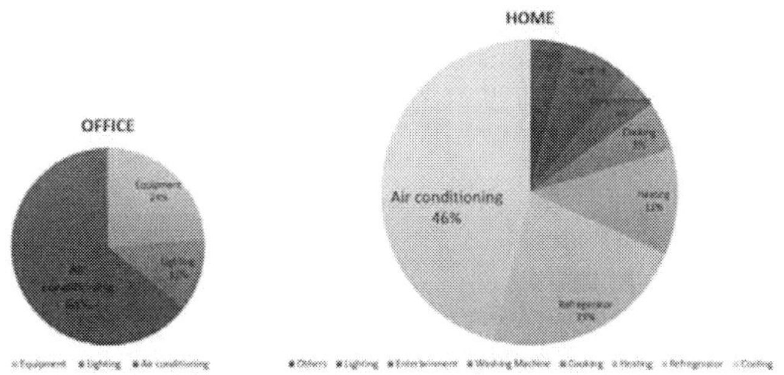


**Figure 1.** Breakdown of energy consumption
Source: (Malaysia Energy Commission)

## 2. Literature Review

### *2.1. Phase Change Materials*

Phase change materials (PCMs) could be defined as unique materials that store and release heat by means of latent heat. PCMs were used as thermal energy storage (TES) for building to control temperature over a cycle period of 24 hours to change and discharged the heat. PCMs were one of the best management methods used to control building temperature within the daily cycle.

A first classification of substances used for thermal energy storage materials stating those utilized as PCM which was useful and widely used. The basic mechanism of phase change materials (PCMs) application in free cooling was for reduction of overheating inside buildings, as PCMs had enormous thermal capacity to absorb large amounts of heat and large ranges of melting temperatures [3]. As such, the used of PCMs helped to reduce the need for regular heating and cooling and worked as an alternative to air-conditioning to control the temperature inside the space [4]. For the selection of PCM certain specifications must be taken into account [5], melting point of PCM must be very close to the application temperature, high specific heat must be considered for the PCM, and high thermal conductivity in solid and liquid phases must be taken to support charging and discharging process in storage system used.

Changing of material phase had been classified into four states, solid-solid, solid-liquid, gas-solid and gas-liquid. For practical purposes, only the solid-liquid variety had been used for building cooling or heating because the other varieties had technical limitations [6]. There was a wide variety of PCMs on the market with different melting point ranges. The most common classification of PCMs are organic, in organic and eutectic as presented in figure 2 [7].

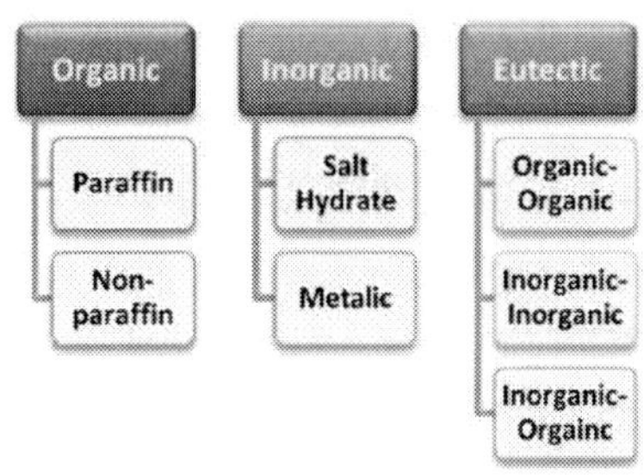


**Figure 2**. Different types of PCMs [7]

## 3. Methodology

### *3.1. Preparation of materials*

The materials were used in this research were foam agent, Ordinary Portland Cement, aggregate, water, cooling agent (glycerine & tap water), tube and formwork. The volume of foaming agent used for panel with the size 1500 mm x 500 mm x 100 mm was 500 ml. with the mixing of water ten times the volume of foam agent. The Ordinary Portland Cement used was 27.72 kg for each panels. The aggregate used was fine aggregate with the size of 5 mm, quantity used was 50 kg and the aggregate was in saturated and dry surface condition. The quantity of water and glycerine used were 12.73 kg and 1000 ml. The type of tube used was reinforcement tube because of its high durability and could withstand the glycerine. Size of formwork was used same as the size of wall cooling panels, 1500 mm (L) x 500 mm (W) x 100 mm (T) and the formwork was made up of high quality of plywood and waterproofing.

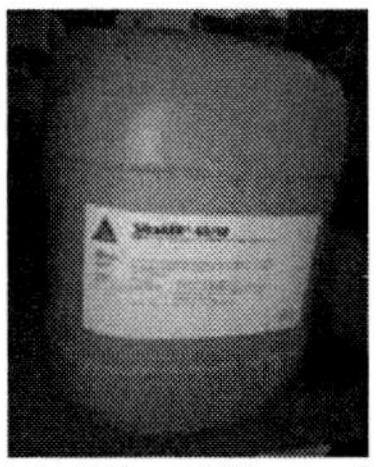

(a) SikaAER synthetic foaming agent

(b) Portland cement

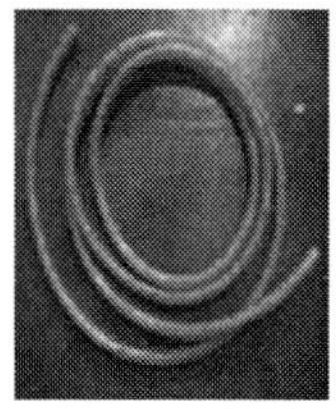

(c)Tube

(d) Sand

(e) Glycerine

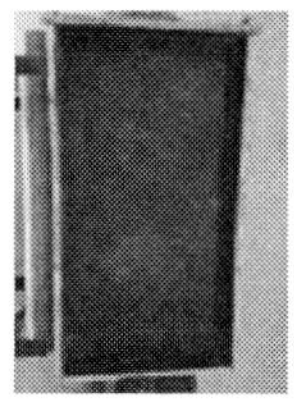

(f) Formwork

**Figure 3.** (a) – (f) : Materials of passive wall cooling panels

### *3.2. Casting of wall cooling panels*

The wall cooling panel with dimension 1500 mm x 500 mm x 100 mm had been cast. The preparation of foamed concrete with foaming agent and installation of reinforcement tube and cooling agent in wall cooling panel had been done before using them to the field measurement. The targeted wet density for foam concrete was 1300-1400 kg/m$^3$. Table 1 showed, list of sample, cooling panel with horizontal arrangement (CPWH) with water, cooling panel with vertical arrangement (CPWV) with water, cooling panel with horizontal arrangement (CPWH) with glycerine, cooling panel with vertical arrangement (CPWV) with glycerine, and brick panel (BP). Table 2 showed the mix design for each of the panels.

**Table 1.** List of samples

| Type of wall panel | Cooling agent | No of samples |
|---|---|---|
| CPWH | Water | 2 |
| CPWV | Water | 2 |
| CPWH | Glycerine | 2 |
| CPWV | Glycerine | 2 |
| BP | - | 2 |
| Total | | 10 |

**Table 2.** Mix Design

| Material | Sand | OPC | Water | Foam Agent | Glycerine |
|---|---|---|---|---|---|
| Mix design (kg/m$^{3}$) | 666.67 | 369.60 | 169.73 | 6.67 | 13.33 |

Eight wall cooling panel was developed to further the next phase, which was field measurement. The wall cooling panel with 1500 mm x 500 mm x 100 mm was developed. The wall cooling panels were the different arrangement of tube which are in horizontal and vertical.

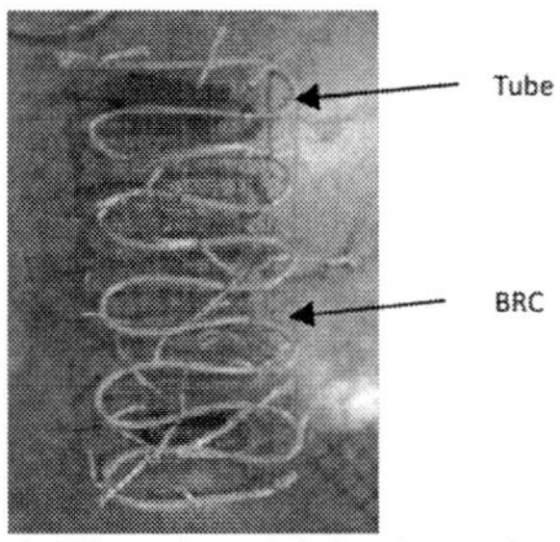


**Figure 4.** The tube with horizontal arrangement

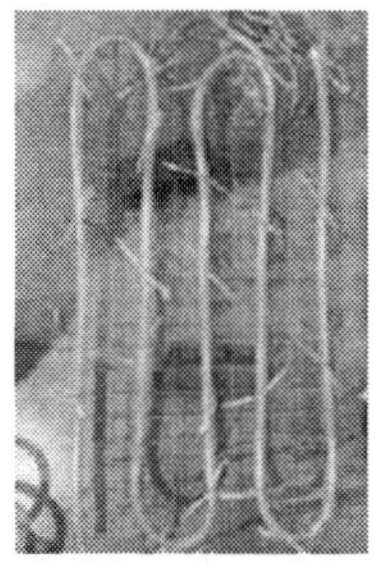

**Figure 5.** The tube with vertical arrangement

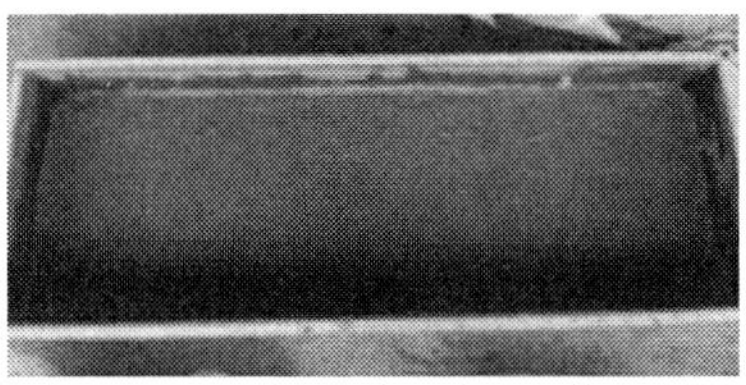

**Figure 6.** Casting of wall cooling panel

**Figure 7.** Aluminium foil as thermal conductivity

**Figure 8.** Casting of wall cooling panel with vertical arrangement

*3.3. Testing on foamed concrete*

For fresh properties, the slump flow test was carried out according to ASTM C1611/C1611 to accesses the properties of the fresh foam concrete. The primary characteristics of fresh foamed concrete was evaluated by this two tests and was carried out directly after concrete mixing.

The flowability of self-compacting concrete was evaluated by slump flow test without any obstruction. The diameter require was shown in figure below was the measurement required for all self-compacting lightweight concrete. The self-compacting concrete should had the value between 500 mm to 850 mm for a slump value, while $T_{50}$ time are the time spend for the concrete to reach the 500 mm spread circle [8]. Foamed concrete workability performance was visually evaluated, which aims to be achieved an appropriate viscosity of the mix. The workability by using spreadability method. This test was done on a fresh mix of low-strength materials by measuring the spread in two directions of a sample placed in a 75 mm diameter and 150 mm long open-ended cylinder, after cylinder was raised vertically. The average of two measured diameters was calculated and reported to the nearest 5 mm.

For hardened concrete properties, compression test was carried out. The test was conducted according to BS 1881: Part 116: 1983 [9]. According to ASTM C 796, at least three days before proceeding the testing, foamed concrete cube that will be conducted the compression test should be cured in a moist room with 100% relative humidity. The sample need to take out from the curing room and be oven-dried for 72 hours at 60° [10]. The desired compressive strength will be obtained when the sample have been cured in a normal moist air for one day and the temperature must increase at 20°C/h for steam state to preserve at 65°C for 4 hours and then cooled in air [11].

**Figure 9**. Air curing

**Figure 10.** Curing of wall cooling panel

### *3.4. Field Measurement Test*

For field measurement test, a small house had been developed for the testing of the wall cooling panel. The size of the small house was 2 m x 2 m x 2 m. The area of the floor was 4.0 m$^2$ The test for temperature performances of passive wall cooling panels were tested in the small house for 3 days. The air temperature reading were taken inside and outside of the small house. Due on Ibrahim and M. Hazrin, the thermal comfort parameter should be measure at least one meter from the floor to represent the height of occupant at seated level. The point of sampling was selected at 8 point according to the Industry Code of Practice On Indoor Air Quality 2010. However, it will change based on the condition of the nature of the building. Since the area of small house 4m$^2$ only 1 point had consider when doing field measurement. The parameter such as air velocity, air temperature and relative humidity was used as thermal comfort parameter. The measurement tool used for this test was an anemometer.

**Figure 11.** Small house

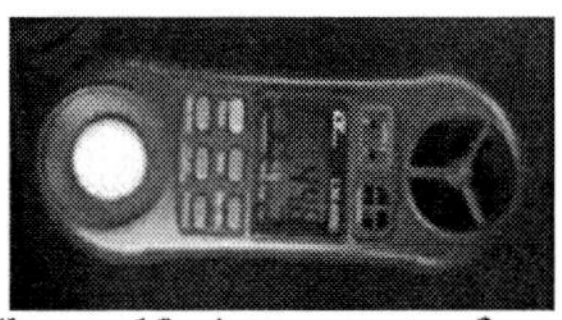

**Figure 12.** Anemometer for temperature

## 4. Results and Discussion

### *4.1. Slump Flow Test*

According to the ASTM C1611/C1611M, all of the result showed the pass result for slump flow test and it was acceptable. From the testing, the spreadability of fresh foamed concrete was shown in the table 3 below. This result was acceptable because the spreadability of fresh concrete mixes was in arranged of the limited, between 40% and 60% of 20 seconds for 500 mm.

**Table 3.** Result of Slump Flow Test

| Panels | Cooling Agent | Time (s) | Result |
|---|---|---|---|
| CPWH | Water | 10 | Pass |
| CPWV | Water | 12 | Pass |
| CPWH | Glycerine | 11 | Pass |
| CPWH | Glycerine | 10 | Pass |

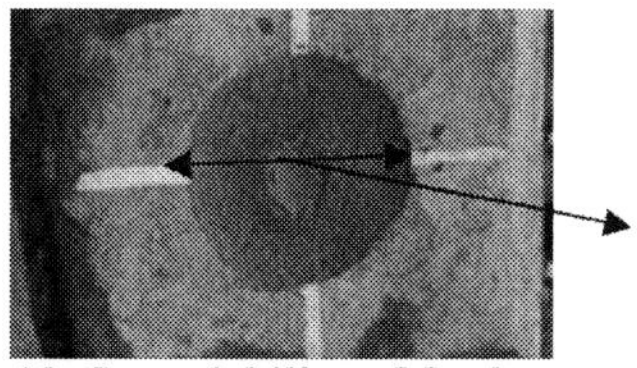


**Figure 13.** Spreadability of fresh concrete

*4.2. Space considerations*

Compressive test was conducted to test the strength of harden concrete. It was the capacity of a concrete sample to withstand axially the direct force until the limit of failure load was achieved, the sample crushed. Six sample of cubes were tested to get the value of compressive strength. Figure 14 shown the compressive strength result. Based on the result below, the strength of foamed concrete increase with the day. The result of compressive strength according to BS 1881: Part 116: 1983 passed because of correct technique of rate of foam agent, w/c ratio, sand particle type, the curing method and cement-sand ratio.

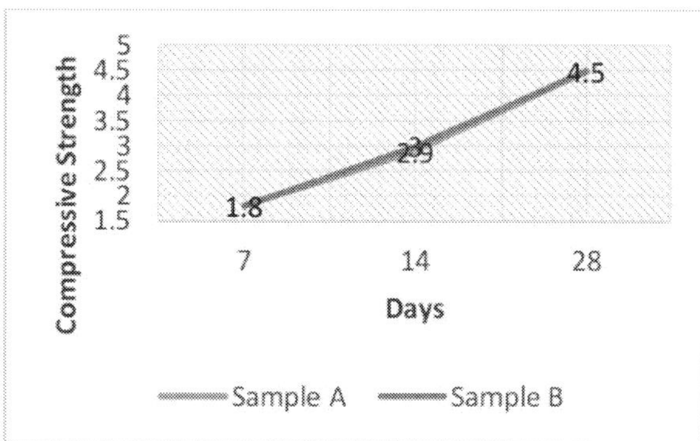


**Figure 14**. Compressive strength result

*4.3. Space considerations*

The results were analysed for difference arrangement of tube and different type of cooling agent. For difference arrangement, the results were analysed for glycerin and water. According to figure 15, the result showed the differences of the temperature between horizontal and vertical arrangement of tube by using glycerine as cooling agent. The result showed 8 out of 12 times, the vertical arrangement had biggest value of difference temperature and the biggest difference was on time 10 which vertical was 5.5°C and horizontal was 1.1°C. This differences happened due to wall cooling panels with vertical arrangement had a large surface area of the exposed tube with the cooling agent compared to horizontal arrangement. Then, the cooling agent reacted quickly and decreased more temperature in the small house.

As conclusion, the wall cooling panels with vertical arrangement had better performance in decreasing air temperature in the small house compared to horizontal arrangement.

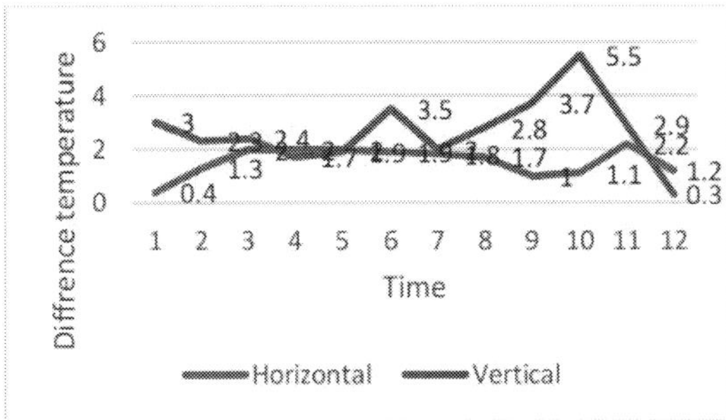


**Figure 15.** Temperature of different arrangement of wall cooling panel with glycerine

According to figure 16, the result showed the differences of the temperature between horizontal and vertical arrangement of tube by using glycerine as cooling agent. The result showed 8 out of 12 times, the vertical arrangement had biggest value of difference temperature and the biggest difference was on time 8 which vertical was 2.2°C and horizontal was 0°C. This differences happened due to wall cooling panels with vertical arrangement had a big surface area of the exposed tube with the cooling agent compared to horizontal arrangement. Then, the cooling agent reacted quickly and can decrease more temperature in the small house. As conclusion, the wall cooling panels with vertical arrangement had good performance in decreasing temperature in the small house compared to horizontal arrangement.

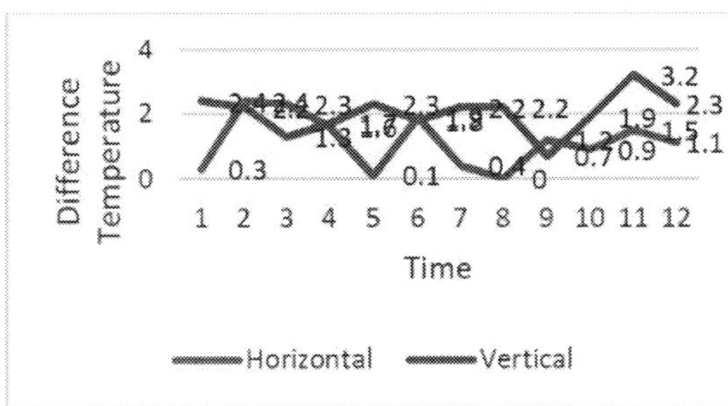


**Figure 16.** Temperature of different arrangement of wall cooling panel with tap water

For different cooling agent, indoor temperature was analysed. Based on figure 17, the temperature performance was analysed with different cooling agent. The graph showed wall cooling panels in vertical with glycerine had the best performance compared to others wall cooling panels. The cooling agent of glycerine could reduce the temperature up to 25°C. The reason because, glycerine could accelerate the process of cooling because of it properties. Besides, the higher temperature was 36.3°C for wall cooling panel in vertical with glycerine because the climate changing. For others panels, the result showed the changes but wall cooling panel in vertical arrangement with glycerine was more effective to use as wall cooling panel.

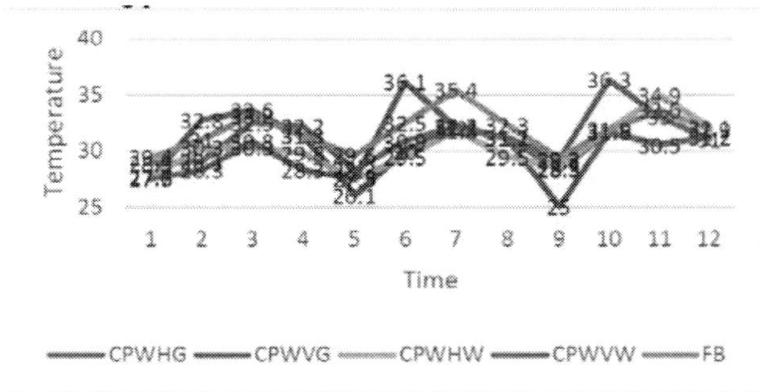


**Figure 17.** Indoor air temperature

## 5. Conclusion

Based on the study that had been carried out, the result obtained was to determine the temperature performance of wall cooling panel by using difference arrangement of the tube and to analyse temperature performance of wall cooling panel by using difference type of cooling agent. Comparison between the brick panel as control and all the wall cooling panel had been analysed. Hence, the conclusions were concluded, the wall cooling panel with vertical arrangement had better temperature performance because they had big area exposed to tube contained glycerine. The wall cooling panel with PCMs which was glycerine as passive cooling agent have the lowest temperature such as 25°C compared to others panel with more than 25°C. The Phase Change Materials, glycerine was one of the suitable cooling agent that showed lowest air temperature compared to other panels. Based on the comparison between all the cooling agent and medium with the brick panel as control, it was proved that vertical arrangement with glycerine were the suitable medium for passive cooling.

## 6. References

[1] International Energy Agency 2002 *Technology Roadmap Energy-efficient Building.* (France: Heating and Cooling Equipment).

[2] Alizadeh M and Sadrameli S M 2016 Development of free cooling based ventilation technology for buildings: Thermal energy storage (TES) unit, performance enhancement techniques and design considerations- A review *Renewable and Sustainable Energy Reviews* **58** 619-645.

[3] Baetens R, Jelle B P and Gustaven A 2010 Phase change materials for building application: A state review. *Energy Building.* **42(9)** 1361-1368.

[4] Biswas K, Lu J, Soroushian P and Shrestha S 2014 Combined experimental and numerical evaluation of a prototype nano-PCM enhanced wallboard. *Applied Energy* **131** 517-529.

[5] Tyagi V V and Buddhi D 2007 PCM thermal storage in buildings: A state of art. *Renewable and Sustainable Energy Reviews* **11(6)** 1146-1166.

[6] Sun X, Zhang Q, Medina M A and Lee K O 2016 Experimental observation on the heat transfer enhancement caused by natural convection during melting of solid-liquid phase change materials (PCMs). *Applied Energy, 162,* pp 1453-1461.

[7] Cao L, Su D, Tang Y, Fang G and Tang F 2015 Properties evaluation and application of thermal energy storage in building. *Renewable and Sustainable Energy Reviews* **48** 500-522.

[8] Biswas K, Lu J, Soroushian P and Shrestha S 2014 Combined experimental and numerical evaluation of a prototype nano-PCM enhanced wallboard. *Applied Energy* **131** 517-529.

[9] Barreneche K, Navarro L, De Gracia A, Fernandez A I and Cabeza L F 2016 In situ thermal and acoustic performance and environmental impact of the introduction of a shape-stabilized PCM layer for building applications *Renewable Energy* **85** 281-286.

[10] Amran Y H M Farzadnia N and Ali A A A 2015 Properties and applications of foamed concete: A review *Construction and Building Materials* **101(1)** 990-1005.

[11] Fujiwara H, Sawada E and Ishikawa Y 1995 Manufacture of High-Strength Aerated Concrete Containing Silica, vol 153. *USA.* ACI Special Publication

[12] Barreneche K, Navarro L, De Gracia A, Fernandez A I and Cabeza L F 2016 In situ thermal and acoustic performance and environmental impact of the introduction of a shape-stabilized PCM layer for building applications *Renewable Energy* **85** 281-286.

**Acknowledgment**
The authors would like to thank the Faculty of Civil and Environmental Engineering, UTHM for the given supports towards this study.

IOP Conf. Series: Materials Science and Engineering **271** (2017) 012023 doi:10.1088/1757-899X/271/1/012023

# Preliminary analytical study on the feasibility of using reinforced concrete pile foundations for renewable energy storage by compressed air energy storage technology

**S Tulebekova[1], D Saliyev[1], D Zhang[1], J R Kim[1], A Karabay[1], A Turlybek[1] and L Kazybayeva[1]**

[1] Department of Civil Engineering, Nazarbayev University, 53 Kabanbay batyr ave., Astana 010000, Kazakhstan
Corresponding author:dichuan.zhang@nu.edu.kz

**Abstract.** Compressed air energy storage technology is one of the promising methods that have high reliability, economic feasibility and low environmental impact. Current applications of the technology are mainly limited to energy storage for power plants using large scale underground caverns. This paper explores the possibility of making use of reinforced concrete pile foundations to store renewable energy generated from solar panels or windmills attached to building structures. The energy will be stored inside the pile foundation with hollow sections via compressed air. Given the relatively small volume of storage provided by the foundation, the required storage pressure is expected to be higher than that in the large-scale underground cavern. The high air pressure typically associated with large temperature increase, combined with structural loads, will make the pile foundation in a complicated loading condition, which might cause issues in the structural and geotechnical safety. This paper presents a preliminary analytical study on the performance of the pile foundation subjected to high pressure, large temperature increase and structural loads. Finite element analyses on pile foundation models, which are built from selected prototype structures, have been conducted. The analytical study identifies maximum stresses in the concrete of the pile foundation under combined pressure, temperature change and structural loads. Recommendations have been made for the use of reinforced concrete pile foundations for renewable energy storage.

## 1. Introduction

Over the past decades, a huge number of technologies for the production of so-called "green" energy have been developed, but none of them has reached the proper level of development for a full large-scale industrial application [1]. The main reason for that is the high cost of renewable energy in comparison with the financial and environmental benefits of its use. However, taking into account the inevitable depletion of traditional energy sources, developed countries continue to invest in further research in the field of renewable energy, which leads to a gradual reduction in the cost price of green energy production.

Another serious obstacle for the full-scale use of renewable energy is the lack of reliable and effective means for its storage as energy received from the sun, wind or water should be used immediately, otherwise it will be wasted [2]. Nowadays, the new methods for the partial accumulation of renewable energy by using various technologies are being developed. One of them is the conversion of electricity into compressed air through the pressurization of the reservoir for the subsequent generation of electricity [3]. This technology is widely known as Compressed Air Energy Storage (CAES).

Currently, the application of CAES technology is limited to pressurization of large scale underground caverns using compressors driven by power plants [4]. This paper explores the possibility to apply CAES into small scales. Renewable energy, generated from the solar panels or windmills attached to building structures, is proposed to be stored inside the pile foundation with hollow sections via compressed air. Given the relatively small volume of storage provided by the foundation, the required storage pressure is expected to be higher than that in the large-scale underground cavern. The high air pressure typically associated with large temperature increase, combined with structural loads, will make the pile foundation in a complicated loading condition, which might cause issues in the structural and geotechnical safety. Therefore this paper presents a preliminary analytical study on the performance of the pile foundation subjected to high pressure, large temperature increase and structural loads. The study is conducted using finite element analyses on pile foundation models which are built from selected prototype structures. The analytical study identifies maximum stresses in the concrete of the pile foundation under combined pressure, temperature change and structural loads. Recommendations have been made for the use of reinforced concrete pile foundations for renewable energy storage.

## 2. Background

The proposed compressed air energy storage system shown on simplified scheme (See figure 1) is designed to store compressed air inside reinforced concrete pile foundations with hollow sections. The advanced-adiabatic (AA-CAES) process is used for thermodynamic cycles of energy storage [5]. The process includes compression of air, extraction of heat from it, separate storage of heat and compressed air, and expansion with preceding heating.

Heat storage is critical for CAES as it directly affects the total efficiency of the system. The air temperature increases during compression and drops down while expanding. Hence, the temperature has to be controlled at compression, storage and expansion as extreme temperatures may damage equipment and pile foundation.

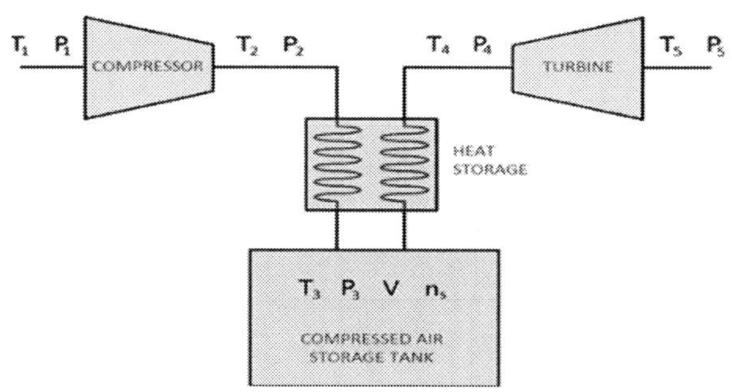


**Figure 1.** Compressed air storage simplified scheme.

### *2.1. Compression process*

The compression process model is based on the energy conservation principle. For simplification purposes, the isentropic adiabatic compression is assumed as in equation(1) [6].

$$\dot{w} = \frac{7}{2\eta_1}\dot{n}R(T_2 - T_1) \quad (1)$$

where $\dot{w}$ is the electric power consumed by the compressor; $\dot{n}$ is the rate of air flow through the compressor; $\eta_1$ is the compressor efficiency; $T_1$ is the ambient air temperature; $T_2$ is the compressed air temperature; $R$ is the universal gas constant. The adiabatic temperature to pressure relation for air is shown in equation(2) [6].

$$T = CP^{\frac{2}{7}} \tag{2}$$

where $C$ is the constant. Considering the volume of the storage tank, $V$, and initial amount of air stored, $n_{s,i}$, the time required to charge the tank to a pressure $P_2$, can be obtained as given in equation(3) [6].

$$t = \frac{3.5}{\dot{w}\eta_1}\left(VP_2 - n_{s,i}RCP_2^{\frac{2}{7}} - \frac{T_1V}{C}P_2^{\frac{5}{7}} + n_{s,i}RT_1\right) \tag{3}$$

*2.2. Storage*

During compression, the extra heat is extracted from the air with certain efficiency,$\eta_2$, hence, temperature and pressure decrease. The heat is accumulated in a separate storage for subsequent air heating at an expansion stage. Pressure and temperature at heating and cooling processes may be assumed to have the isochoric relation as shown in equation(4) [6].

$$\frac{P}{T} = \frac{n_sR}{V} = const \tag{4}$$

Also, heat extracted with zero work is directly proportional to temperature change. Then, storage temperature and pressure can be determined as in equation(5) and equation(6) [6].

$$T_3 = T_2 - \eta_2(T_2 - T_1) \tag{5}$$

$$P_3 = \frac{n_sRT_3}{V} \tag{6}$$

*2.3 Expansion process*

The expansion process can be assumed to be a reverse of compression with power generation with efficiency$\eta_3$. Considering the heat losses inside the air storage tank and heat storage tank with heat storage efficiencies $\eta_4$ and$\eta_5$ respectively, the total efficiency of the system is given as in equation(7) [6].

$$\eta_T = \eta_1\eta_3\frac{\eta_4 + \eta_5}{2} \tag{7}$$

*2.4. Energy generation and consumption*

This paper uses solar energy as an example for study the energy storage. Accumulation of solar energy is of a high importance as energy generated at daylight is to cover the day and night consumption. The data acquired from National Laboratory of Astana for solar panel performance was used to obtain different electricity generation patterns in 24 hours period per square meter of solar panel [7]. Also, for the same time period, the electricity consumption patterns were obtained per square meter of area based on different building occupation types. Figure 2 shows the example of electricity generation and consumption patterns for 100 $m^2$ of prototype residential area with 20 $m^2$ of solar panels. The area of solar panels was adjusted based on the total efficiency of the energy storage system.

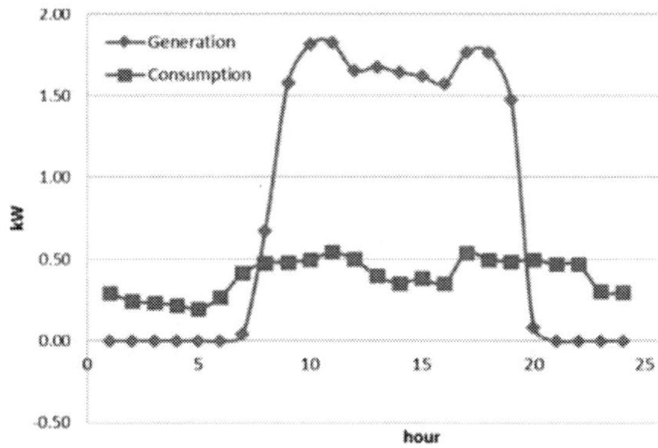


**Figure 2.** Electricity generation and consumption.

### *2.5. Required storage pressure and temperature*

One of the prototype residential buildings was assumed to have 4 stores and 5m x 5m column grid, which implies 25 $m^2$ of tributary area for the column and 100 $m^2$ of total tributary consumption area. As will be shown further, the column was designed to be supported by four 20 m long piles with cap. The internal hollow region in each pile has 200 mm in diameter and 18 m in length, resulting in the total volume of 2.26$m^3$ available for storage. Using the energy surplus (the difference between generation and consumption shown in figure 2 and the given storage area, the thermal cycle processes are calculated for the CAES via equation (1) to equation (7). The CAES processes results in maximum storage pressure of $P_3$=3.125 MPa and maximum temperature increase inside the pile of$T_3$=60.27$^0$C. The change of the storage pressure and temperature inside the pile for a 24-hour cycle are shown in figure 3.

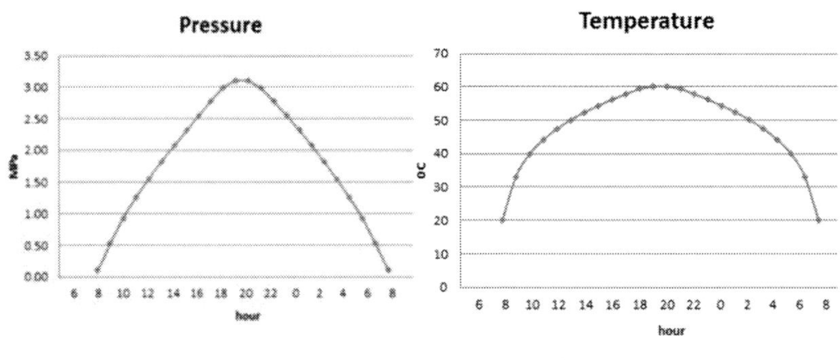


**Figure 3.**Pressure and temperature variations in a pile.

## 3. Analytical Study

The main objective of this paper is to conduct the preliminary study on the behavior of the pile foundation subjected to combined air pressure, temperature change and structural loads through finite element analyses.

### *3.1. Structural and geotechnical design*

Superstructure above the pile foundation is a 4-story moment frame building with column grid of 5m x 5m. The loads are assumed to be as shown in table 1. The pile dimensions have been determined based on the factored structural load and soil conditions shown in Sec.3.2. Table 2 shows the summary of the factored structural load and pile dimensions.

IOP Conf. Series: Materials Science and Engineering **271** (2017) 012023 doi:10.1088/1757-899X/271/1/012023

**Table 1.** Loads.

| Load type | Value | |
|---|---|---|
| Dead Load | 10 | kN/m$^2$ |
| Live Load | 3 | kN/m$^2$ |
| Total Gravity Load | 18 | kN/m$^2$ |

**Table 2.** Pile Parameters.

| Parameter | Value | |
|---|---|---|
| Structural vertical load | 900 | kN |
| Length | 20 | m |
| Outer diameter | 0.5 | m |
| Inner diameter | 0.2 | m |

*3.2. Soil-structure interaction*

The modeling of the soil-structure interaction involves formulating the soil-pile resistance through the set of uniformly distributed independent springs with nonlinear properties [8]. Figure 4 shows the diagram of the soil-pile interaction model. The spring properties in the model, such as normal and lateral stiffness, are governed by load-displacement curves, which are obtained from the empirical tests and are discussed further (See figure 5). Skin friction on the pile is governed by T-Z curve springs; horizontal soil pressure is determined by P-Y curve; end-bearing is governed by Q-Z curve.

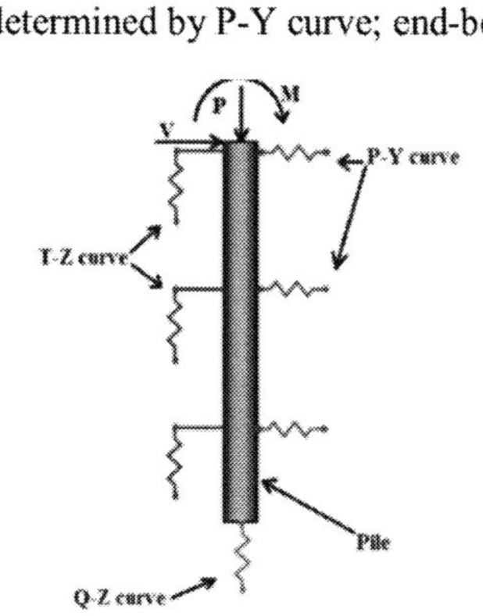


**Figure 4.** Spring model of a pile.

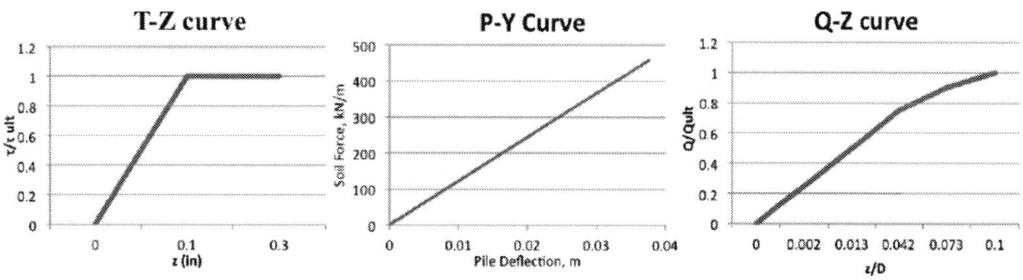


**Figure 5.** Load-displacement curves [8, 9]

Soil model for calculating vertical load-displacement curve (skin friction and end-bearing) has been obtained through American Petroleum Institute (API) recommendations [9]. In this study, sandy soil that is homogeneous across the depth is considered. API recommendation for skin-friction in the sand is linear elastic perfectly plastic transfer curve from which shear stiffness can be obtained. Shear stiffness coefficient value depends on the depth of the pile. End-bearing stiffness can be determined from API load transfer curve presented in [9]. For lateral load-displacement curve the provisions presented by Reese and Van Impe [10] have been applied. From this curve the normal stiffness of soil can be obtained, which is constant throughout depth of the pile. Table 3 presents the summary of the soil properties for the model.

Concrete foundation pile has been modeled as a cylinder with the hollow section with linear elastic properties. Since the study is to check the maximum stress in the concrete, at this stage the nonlinear concrete model is not needed. The nonlinear concrete model will be used for further analytical investigation. The amount of the concrete that can be removed from the inner section of the pile has been determined based on the axial load capacity of the reinforced concrete section. Table 4 shows the summary of the properties of concrete material used for the pile.

**Table 3.** Soil Properties.

| Soil Properties | Values |
|---|---|
| Soil Type | Medium Sand |
| Unit Weight | 18 kN/m$^3$ |
| Internal Friction Angle, $\varphi$ | 30° |
| Soil-Pile Friction Angle, $\delta$ | 25° |
| Normal stiffness, $K_N$ | 1220 kN/m |

**Table 4.** Concrete Properties.

| Concrete Properties | Values |
|---|---|
| Grade of Concrete | C-40 |
| Young's Modulus, $E_c$ | 3.6 GPa |
| Poisson's Ratio, $\mu$ | 0.15 |
| Density of Concrete, $\rho$ | 2307 m$^3$ |

*3.3. Finite element model*

In this study, finite element models for the pile and surrounding soil have been developed in ANSYS. The pile has been modeled as a 3D solid element using SOLID45 element and material properties of elastic concrete have been assigned to it. The interface between structure and soil has been modeled using CONTAC52. Predefined negative gaps were used to model the lateral static earth pressure, and they increased linearly with the depth of the pile. The skin friction resistance has been defined through shear stiffness $K_s$ and coefficient of friction $\mu$ using Coulomb friction model. The end-bearing of the soil has been modeled using nonlinear springs. Figure 6 shows the model of the cylindrical hollow pile cross-section with CONTAC52 elements uniformly distributed along the circumference of the outer layer. Figure 7 shows the perspective view of the pile with spring elements uniformly distributed at the bottom of pile.

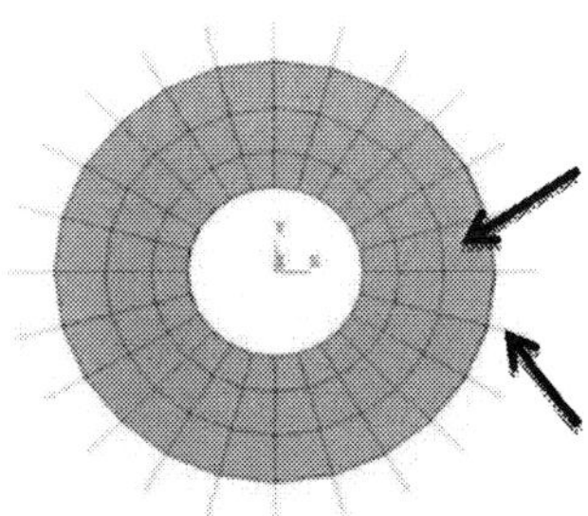


**Figure 6.** Pile cross-section with contacts.

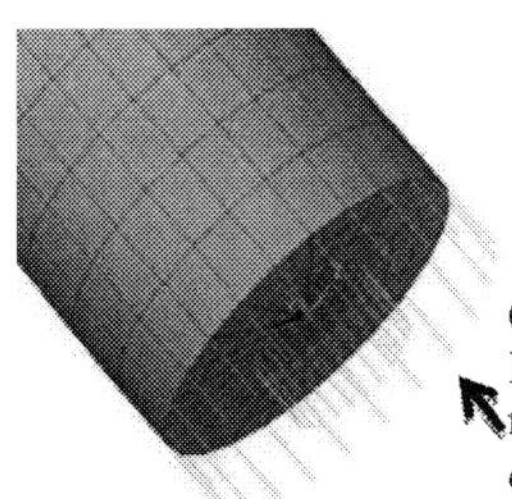


**Figure 7.** Pile with end-bearing elements.

## 4. Analytical results

In this study, the response of the pile structure to the air pressure and temperature change combined with structural loads is investigated. For the purpose of investigating the effect of each load on the pile foundation separately, three different load combinations are considered: case (a)structural load (refer to Table 2) + pressure load (refer to Figure 3a); case (b) structural load+ temperature load (refer to Figure 3b); case (c) structural load+ pressure load+ temperature load.

The results obtained from the finite element analysis show that significant stress in the concrete of the pile foundation is caused by high air pressure and temperature change. Figure 8 shows the circular tensile stress diagrams for the three cases. As it can be seen, the combination of the temperature change and pressure loading causes significant circular tensile stress (about 6.1MPa) on the pile.

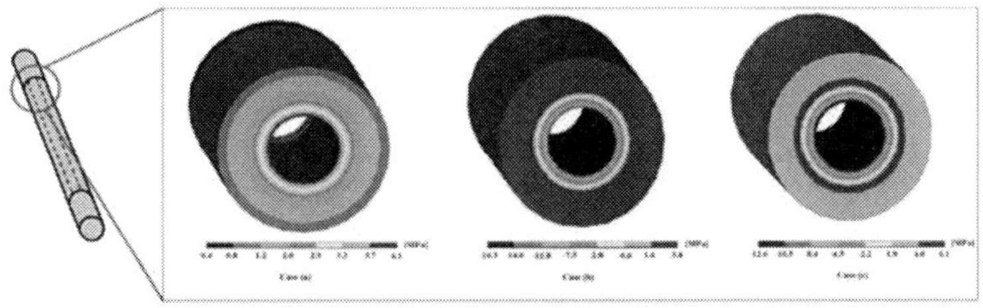

**Figure 8.** Circular tensile stress diagrams for three cases.

Since the pile is modeled with elastic elements, it is important to check the critical stresses in the pile. The maximum compression, tension and shear stresses have been obtained and are shown in the figures 9, 10, and 11 respectively. The summary of the critical stresses on the pile foundation is shown in Table 5.

**Table 5.** Summary of critical stresses.

| Stress type | Value | |
|---|---|---|
| Maximum compression | 21 | MPa |
| Maximum tension | 6.1 | MPa |
| Maximum shear | 0.6 | MPa |

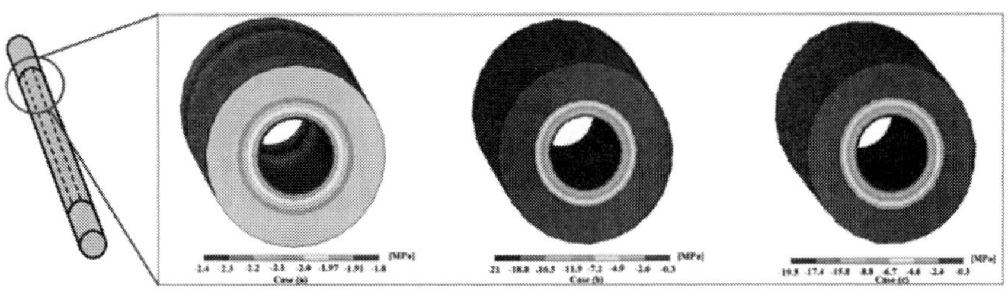

**Figure 9.**Maximum compression stress for three cases.

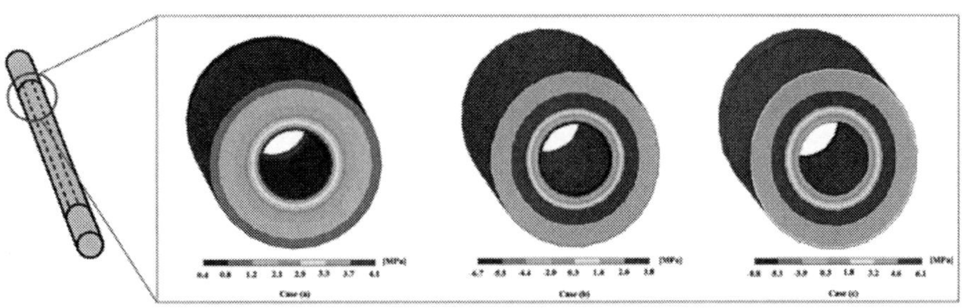

**Figure 10.**Maximum tension stress for three cases.

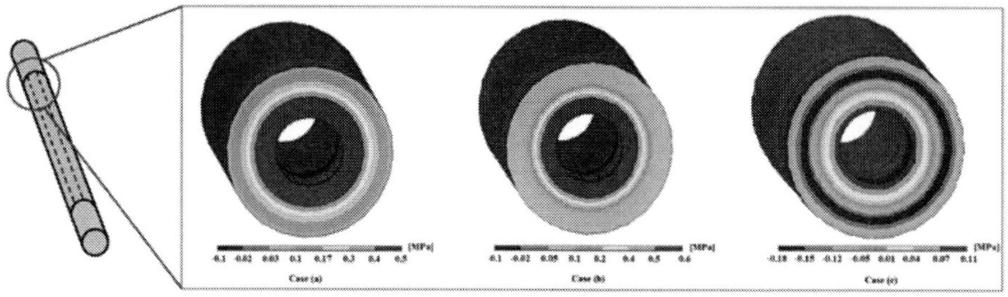

**Figure 11.**Maximum shear stress for three cases.

As it can be seen from the comparison of three loading cases, the temperature increase causes larger stresses in the concrete than the air pressure. However, the simulation for the temperature load does not consider the thermal conductivity of the concrete. This assumption creates a conservative structural demand of the pile foundation due to the temperature change. Coupled thermal and structural analyses will be performed in detailed studies.

As seen in table 5, the maximum compression stress (21MPa) and shear stress (0.6MPa) is within the capacity of the normal concrete used in this study. However, the maximum tensile stress (6.1MPa) exceeds the tensile strength of normal concrete and will cause cracks in the concrete, which might not be suitable for retain the air pressure and heat inside the pile. Thus it might require higher strength concrete for the construction of the pile foundation. The nonlinear inelastic behavior as well as the effects of reinforcement cannot be investigated using the elastic pile foundation model. Further studies will be performed using the concrete model to consider the concrete cracking and effects of reinforcement.

## 5. Conclusion and future work

A new pile foundation system is being proposed to storage the renewable energy using compressed air. This paper presents the preliminary study on the structural behavior of the pile foundation subjected to combined high air pressure, large temperature increase and structural loads. The maximum compression, tension and shear stresses inside the concrete caused by these combined loadings are obtained through finite element analyses. The analytical results indicate: (1) the temperature increase leads to a larger structural stress demand than the high air pressure does; (2) the compression and shear stresses caused by the combined loading can be resisted by the normal strength concrete; and (3) the tensile stress might cause cracking in the normal concrete. The cracking might cause the leakage of the pressure and heat. Thus it is recommended to adopt high strength concrete and high insulation material for the proposed pile foundation system.

The preliminary study has several limitations: (1) the concrete material for the pile foundation is modeled as elastic material; (2) the reinforcement inside the concrete is ignored; (3) thermal conductivity of concrete is not considered; and (4) the soil-structure interaction is modeled as simple springs. All these limitations will be addressed in a more detailed analysis to fully understand the behavior of the new pile foundation system.

## 6. References

[1] Center for Resource Solutions 2013*About Green-e* Retrieved on June 10, 2017 fromhttps://www.green-e.org/about.

[2] Delony J 2017 *Storage and Solar Finance: Programs for the Rising Behind-the-meter Markets* Renewable Energy World (May 24)

[3] Energy Storage Association 2017*Compressed Air Energy Storage (CAES)* Retrieved on June 5, 2017 fromhttp://energystorage.org/compressed-air-energy-storage-caes.

[4] Zhang L and Ahmari S 2008*Solar energy storage using compressed air in geologic repositories*

Progress Report, September-October (Arizona: Department of Civil Engineering and Engineering Mechanics, University of Arizona)

[5] Energy Storage Association 2017*Advanced Adiabatic Compressed Air Energy Storage* Retrieved on June 10, 2017 from *http://energystorage.org/advanced-adiabatic-compressed-air-energy-storage-aa-caes*

[6] Borgnakke C and Sonntag R 2013 *Fundamentals of Thermodynamics8th ed* (United States: John Wiley & Sons, Inc.) pp 51-159

[7] National Laboratory of Astana 2016*Report on solar panel performance* (Kazakhstan: Nazarbayev University) pp 1-5

[8] Küçükarslan S and Banerjee P 2002 Behavior of axially loaded pile group under cyclic loading, *Eng. Struct. J.* **25** 303-311

[9] American Petroleum Institute 2002*Recommended Practice for Planning, Designing and Constructing Fixed Offshore Platforms- Working Stress Design, 21st ed.* (United States: API) pp 55-71

[10] Reese C and Van Impe W 2001*Single Piles and pile groups under lateral loading* (London: Taylor & Francis Group plc.) pp 49-89

**Acknowledgements**

This research was supported by the Nazarbayev University Research Fund under Grant "Development of a Renewable Energy Storage System using Reinforced Concrete Foundations". The authors are grateful for this support. Any opinions, findings, and conclusions or recommendations expressed in this material are those of the author(s) and do not necessarily reflect the views of the Nazarbayev University.

IOP Conf. Series: Materials Science and Engineering **271** (2017) 012024 doi:10.1088/1757-899X/271/1/012024

# Alternative design of pipe sleeve for liquid removal mechanism in mortar slab layer

**W M H Wan Nazri[1], N Anting[1], A J M S Lim[2], J Prasetijo[3], S Shahidan[1], M F Md Din[4] and M A Mohd Anuar[1]**

[1]Jamilus Research Center, Department of Structural and Material Engineering, Faculty of Civil and Environmental Engineering, Universiti Tun Hussein Onn Malaysia, 86400 Parit Raja Batu Pahat, Malaysia
[2] Research Centre of Soft Soil, Department of Infrastructure and Geomatic Engineering, Faculty of Civil and Environmental Engineering, Universiti Tun Hussein Onn Malaysia, 86400 Parit Raja Batu Pahat, Malaysia
[3] Smart Driving Research Centre, Department of Infrastructure and Geomatic Engineering, Faculty of Civil and Environmental Engineering, Universiti Tun Hussein Onn Malaysia, 86400 Parit Raja Batu Pahat, Malaysia
[4] Centre for Environmental Sustainability and Water Security (IPASA), Research Institute for Environmental Sustainability, Universiti Teknologi Malaysia, 81310, Johor, Malaysia
Corresponding author:nickholas@uthm.edu.my

**Abstract.** Porosity is one of the mortar's characteristics that can cause problems, especially in the room space that used high amount of water, such as bathrooms. Waterproofing is one of the technology that normally used to minimize this problem which is preventing deep penetration of liquid water or moisture into underlying concrete layers. However, without the proper mechanism to remove liquid water and moisture from mortar system, waterproofing layer tends to be damaged after a long period of time by the static formation of liquid water and moisture at mortar layer. Thus, a solution has been proposed to drain out water that penetrated into the mortar layer. This paper introduces a new solution using a Modified Pipe Sleeve (MPS) that installed at the mortar layer. The MPS has been designed considering the percentage surface area of the pipe sleeve that having contact with mortar layer (2%, 4%, 6%, 8% and 10%) with angle of holes of 60°. Infiltration test and flow rate test have been conducted to identify the effectiveness of the MPS in order to drain out liquid water or moisture from the mortar layer. In this study shows that, MPS surface area 10%, angled 60°, function effectively as a water removal compared to other design.

## 1. Introduction

Penetration of water into the concrete layer that exists in the form of moisture can significantly affect the durability of the building structure, such as wall and floor structure. High levels of moisture caused the corrosion of reinforcement bar inside the concrete, promoting growth of microorganisms due to migration of moisture through the wall and also shorten concrete structure's life [1 - 3]. This moisture penetration phenomenon is called leakage.

The term of leakage is defined as a way of fluid to escape outside containers, tanks or also pipes. While leakage in the context of structure is a way in concrete that permit the fluid or gases permeate across the concrete structure [4, 5]. Concrete leakage (especially used for hydraulic-concrete) is a key factor that deteriorating the concrete structures since leakage can aggravate the chemical corrosion and frost damage [6, 7]. Concrete leakage usually occurred at part of the building that significantly expose to water, normally observed in bathroom area.

Concrete leakage happens when the water able to penetrate the concrete layer through seepage, due to crack or porous design of the concrete or mortar. The main point source of this problem is from the bathroom, where most of the water (bath/shower) been used and dropped onto the floor surface. Although a slab finishing is a water resistant and waterproofing liquid has been applied to the slab surface, some portion of water still penetrate into the mortar layer via tiny pores due to the porosity of the mortar itself or the existence of crack. The hydraulic transport process of the water (in the form of moisture) from the mortar to the concrete layer could be characterized using three fundamental mechanisms, which are absorption, condensation and capillarity [8].

The applying of waterproofing coating between the concrete and mortar layer will prevent the fluid to seep into the concrete layer or in other words absorbs not more than 2.5% of moisture [9]. However, this will cause another problem. Fluid will get trapped into the mortar layer. This phenomenon will cause other serious problem such as corrosion of embedded metals, fungus at the structure surface, deterioration of concrete and so on [10].

Without proper treatment and immediate action, floor or slab leakage will cause more serious problems to the building in the future, such as corrosion of steel reinforcement, fungus attack on the wall surface, formation of stalactites from the concrete slab and disturbance of electrical wiring system. Therefore, the structures and infrastructures need to be maintained and inspected all the time.

As a result, the cost of maintenance will increase due to the needs of major repair to the affected building. A lot of resources have been employed in the rehabilitation of infrastructure. As an example, in Italy, the rehabilitation work total cost amounted to half of the expenditure invested in construction [11]. The repair cost for Algiers Airport was estimated about 3 Millions USD [12]. While in UK, corrosion-related cost in building and construction industry was estimated to be 250 Million GBP [13].

Based on the problems stated, new technology or innovation is needed to enhance the existing product and technology. Preventing the water penetration into deep layers of the concrete is not enough since the behaviour of water keeps searching the occurrence of seepage and flow through it. Thus, more investigation of hydraulic behaviour of the concrete and mortar is needed as fundamental knowledge to design the suitable product that could remove the liquid and moisture from the concrete system.

## 2. Methodology

### *2.1. Sample preparation*

Five mortar specimens installed with Modified Pipe Sleeve (MPS) have been cast with dimension of 300mm x 300mm x 50mm. The weep-holes surface area were applied on the pipe sleeve as percentage shown in table 1. Table 1 tabulates the varieties design of the MPS and their abbreviations.

**Table 1**. Designs of MPS and abbreviation.

| Angle | Weep-Holes Surface Area (%) | Abbreviation |
|---|---|---|
| 60° | 0 | Conventional |
| 60° | 2 | A |
| 60° | 4 | B |
| 60° | 6 | C |
| 60° | 8 | D |
| 60° | 10 | E |

*2.1.1 Modified pipe sleeve design.* The detail dimensions of the MPS are as can be seen in figure 1.

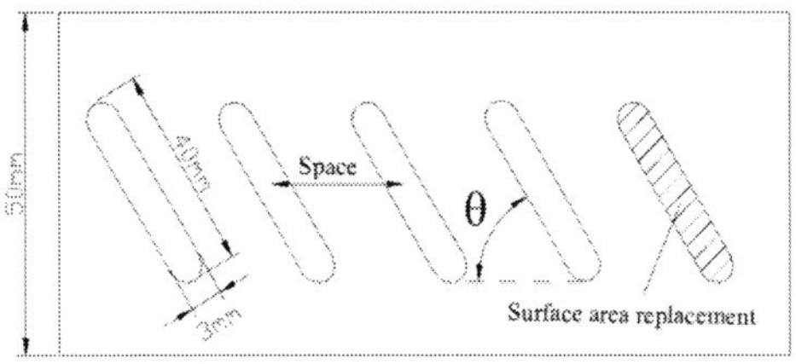


**Figure 1.** Detail dimension of MPS.

Table 2 and table 3 show the number of holes and the spaces between holes for every type of MPS.

**Table 2.** Number of holes.

| Weep-Holes Surface Area (%) | No. of holes |
|---|---|
| 0 | 0 |
| 2 | 3 |
| 4 | 6 |
| 6 | 9 |
| 8 | 12 |
| 10 | 15 |

**Table 3.** Space between holes.

| No. of holes | Space between holes (mm) |
|---|---|
| 0 | 0 |
| 3 | 115 |
| 6 | 58 |
| 9 | 38 |
| 12 | 29 |
| 15 | 23 |

*2.1.2 Model preparation.* As usual, Ordinary Portland Cement (OPC) was used as a binder. River sand was used as fine aggregate must passing 3mm sieve size. Table 4 indicates the mix design composition for mortar based on [14] with fixed water-cement ratio (w/c) at 0.5.

**Table 4.** Mortar mix design composition.

| Materials | Mix Composition (kg) |
|---|---|
| Ordinary Portland cement | 18.9 |
| Fine aggregate | 37.8 |
| Water (0.5 w/c) | 9.45 |

*2.2. Experimental work*

There are two main parameters have been conducted in this study which are infiltration and flow rate. The infiltration test was conducted to obtain the infiltration rate of water that penetrates in mortar layer that installed with 5 varieties design of the MPS. Three liters of water were pond on the mortar surface in a period of 12 hours and the result was recorded for every 30 minutes. Figure 2 reveals the infiltration rate test.

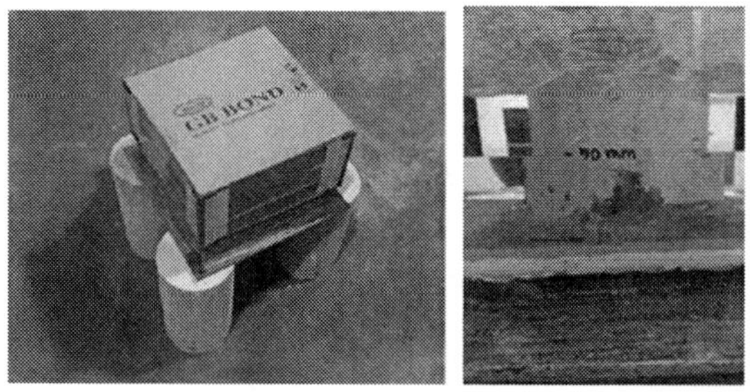

**Figure 2**. Infiltration test.

The flow rate test was conducted to determine the flow rate of water that drain out from the mortar layer through MPS. It is conducted simultaneously with infiltration test. Water pond that penetrates into the mortar layer will be drained out through the MPS. The drained out water then will be collected and measured. The test was carried out for 12 hours and the result was recorded for every 30 minutes. Figure 3 demonstrates the flow rate test that has been conducted.

**Figure 3**. Flow rate test.

## 3. Results and discussions

### *3.1. Infiltration test to the mortar*

Table 5 represents the cumulative volume of water infiltrate for the 10% surface area design pipe sleeve for every 30 minutes.

Figure 4 illustrates, the cumulative volume of water infiltrates in 6 hours for each surface area percentage pipe sleeve design. For pipe sleeve E, the cumulative volume of water infiltrate was increased rapidly in 6 hours. This is because the higher percentage of surface area will design a pipe sleeve with more holes to drain out water. So, the number of holes will influence the infiltration rate of water in the mortar. The total volume of water infiltrates for 6 hours for pipe sleeve E is 945 ml. Same goes to pipe sleeve D, the cumulative volume of water was increased rapidly but the total volume infiltrates for the last 6 hours was less than pipe sleeve E which is 810 ml. Next, the cumulative volume of water infiltrates for the pipe sleeve C was increased gradually in 6 hours with the total volume of water infiltrates 720 ml. While for pipe sleeve A, B and Conventional, the cumulative of water infiltrates was increased gradually with the total volume of water infiltrates 270 ml, 225 ml and 180 ml. Based on the data collected for infiltration rate test, it shows that the higher the percentage area of holes of the pipe sleeve design, the higher the total volume of water infiltrates in 6 hours. So that, the design holes can affect the result of the volume of water infiltrates.

**Table 5.** Cumulative volume of water infiltrates for pipe sleeve E.

| Time (min) | Height of water (mm) | Volume of water (ml) | Volume of water infiltrates (ml) | Cumulative volume of water infiltrates (ml) | Percentage of infiltration of water (%) |
|---|---|---|---|---|---|
| 0 | 40.0 | 3000 | 0 | 0 | 0 |
| 30 | 39.0 | 2910 | 90 | 90 | 2.50 |
| 60 | 38.0 | 2820 | 90 | 180 | 5.00 |
| 90 | 37.5 | 2775 | 45 | 225 | 6.25 |
| 120 | 37.0 | 2730 | 45 | 270 | 7.50 |
| 150 | 36.0 | 2640 | 90 | 360 | 10.00 |
| 180 | 35.0 | 2550 | 90 | 450 | 12.50 |
| 210 | 34.0 | 2460 | 90 | 540 | 15.00 |
| 240 | 33.5 | 2415 | 45 | 585 | 16.25 |
| 270 | 32.0 | 2280 | 135 | 720 | 20.00 |
| 300 | 31.0 | 2190 | 90 | 810 | 22.50 |
| 330 | 30.0 | 2100 | 90 | 900 | 25.00 |
| 360 | 29.5 | 2055 | 45 | 945 | 26.25 |

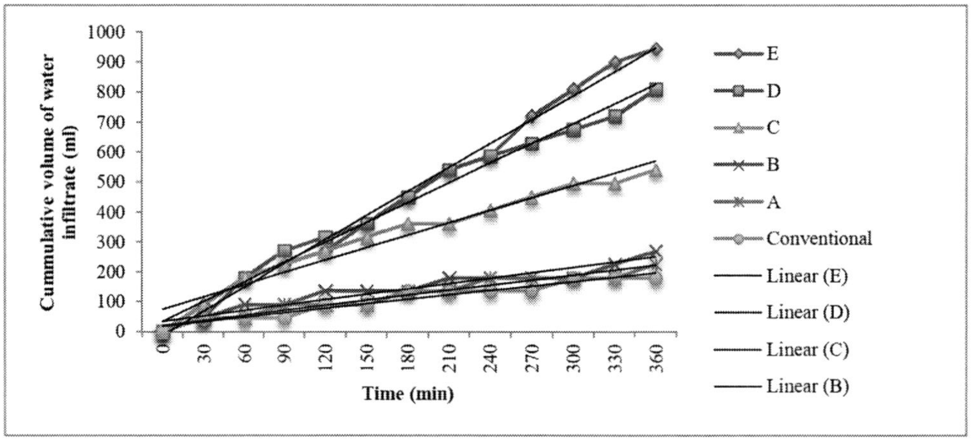


**Figure 4**. Cumulative volume of water infiltrates for each surface area percentage pipe sleeve design.

Figure 5 demonstrates the mortar layer that using pipe sleeve E design gives the highest infiltration result which is 79.863 ml/min. This is because of the water have an alternative way to drain out from the mortar layer which is the designated number of holes. Then, the infiltration rate of mortar for pipe sleeve D is less than pipe sleeve E which is 66.016 ml/min. The infiltration rate of mortar for pipe sleeve C was 41.291 ml/min. While the infiltration rate of the mortar layer of B, A and Conventional pipe sleeve design were 18.049 ml/min, 16.813 ml/min and 14.835 ml/min respectively. It is proved that the infiltration rate of mortar was influenced by the surface area percentage of the pipe sleeve design as the increase of surface area percentage of the pipe sleeve design, the increase the infiltration rate of mortar layer. The most effective design of pipe sleeve is pipe sleeve E as the infiltration rate of mortar layer was the highest among others.

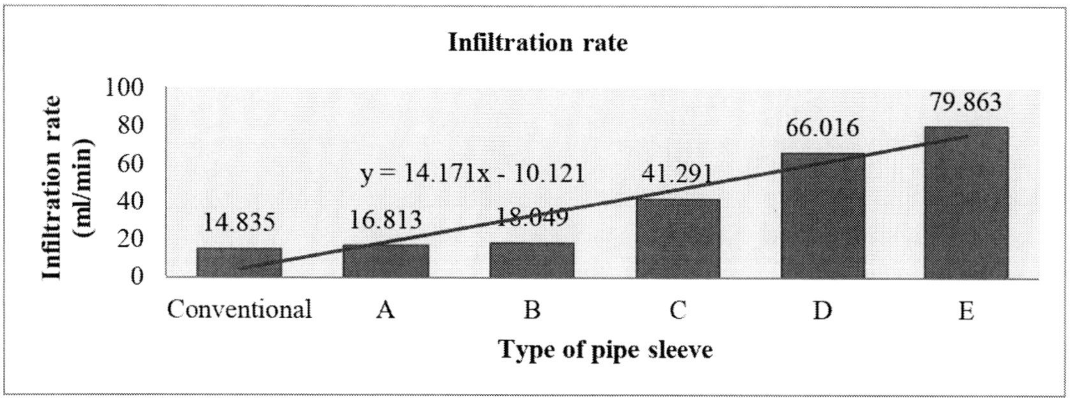


**Figure 5**. Infiltration rate of water.

*3.2. Flow rate test to the mortar*

Table 6 reveals the volume of water that flow out through the 10% surface area design pipe sleeve for every 30 minutes.

**Table 6**. Volume of water that flow out through pipe sleeve E.

| Time (min) | Volume of water collected (ml) |
|---|---|
| 0 | 0 |
| 30 | 40 |
| 60 | 70 |
| 90 | 100 |
| 120 | 125 |
| 150 | 170 |
| 180 | 210 |
| 210 | 250 |
| 240 | 290 |
| 270 | 340 |
| 300 | 360 |
| 330 | 400 |
| 360 | 460 |

Figure 6 shows the volume of water collected in six hours for every design pipe sleeve. For pipe sleeve E, it shows that the volume of water that flow out was increased rapidly in six hours in time. While for pipe sleeve D, the volume of water collected was a bit slow for the first hour as it remains at zero value and then for the next hour, it was gradually increased for the next five hours. For C, the volume of water flow out was increased slowly but in constant mode for the six hours. Lastly, the volume of water collected for B, A and Conventional pipe sleeve were in the same state which was constantly in zero as there is no water collected for the six hours period of time. Pipe sleeve E shows the highest volume of water flow out in the period of six hours because it covers more surface area percentage resulting the higher number of holes to flow out the water from the mortar.

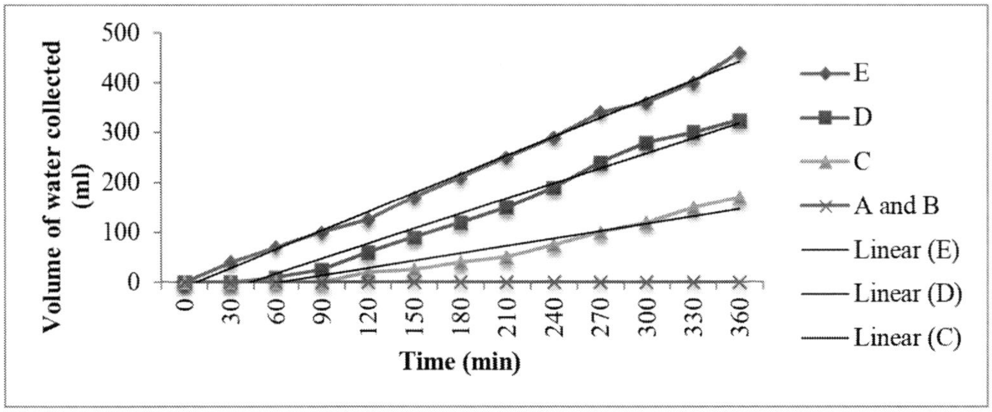


**Figure 6**. Volume of water collected in six hours for every design pipe sleeve.

As depicted in figure 7, the flow rate of water in the mortar layer that using pipe sleeve E was the highest among all of them all. It is clearly shown that the alternative holes provided by pipe sleeve design at mortar layer were able to drain out the water from mortar layer. The flow rate that remains at zero value shows that the surface area covered was too small and it required more time to drain out water. Then, the flow rate of water in a mortar layer only started to increase when the pipe sleeve C has been used. The flow rate of water for C is 14.753 ml/min. Then, the flow rate of water increased rapidly for the D and E which is 30.192 ml/min and 37.637 ml/min respectively. The flow rate of water for pipe sleeve E recorded the highest rate and it shows that it was the best design compared to the other design.

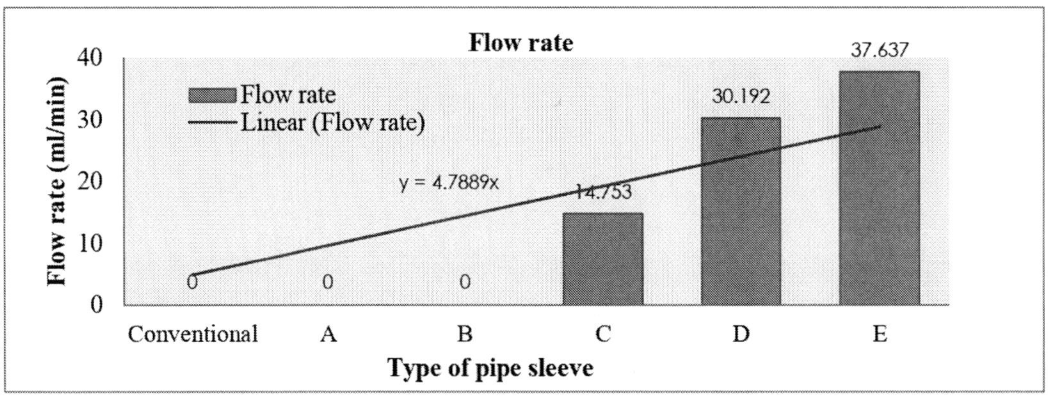


**Figure 7**. Flow rate result.

## 4. Conclusion

The pipe sleeve design was to enhance innovative product for problem solving in Malaysia, thus to help in reducing the moisture content of the building structure problem. Refer to the problem of slab leakage, this study can be a solution to reducing this problem. This study has achieved the objective to identify the performance of the ordinary pipe sleeve system and the MPS in order to remove water in the mortar layer where the performance of MPS is better compared to ordinary pipe sleeve.

Based on this study also, the MPS system absolutely can drain out water effectively that trapped in the mortar. This will help in reducing the moisture content of the slab structures, thus decreased the structure failures probability in future. This MPS can be an alternative solution for existing solution of leakage problem such as waterproofing and other solution.

## 5. References

[1] Lu X 2002 Modelling of heat and moisture transfer in buildings I . Model program, *Energy Build.* **34** 1033–1043
[2] Piaia J C Z, Cheriaf M and Rocha J C 2013 Measurements of water penetration and leakage in masonry wall : Experimental results and numerical simulation, *Build. Environ.* **61** 18–26,
[3] Muhammad N Z, Keyvanfar A, Abd. Majid, M Z, Shafaghat A and Mirza J 2015 Waterproof performance of concrete: A critical review on implemented approaches, *Constr. Build. Mater.* **101** 80–90
[4] Al-Kadi T, Al-Tuwaijri Z and Al-Omran A 2013 Wireless sensor networks for leakage detection in underground pipelines: A survey paper *Procedia Computer Science* vol 21 (Netherland: Elsevier) pp 491–498
[5] Shahidan S, Koh H B, Alansi A M S and Lee L Y 2016 Strength development and water permeability of engineered biomass aggregate pervious concrete *MATEC Web of Conf.* vol 47 (Paris: EDP Sciences) 01007
[6] Turkmen S 2003 Treatment of the seepage problems at the Kalecik Dam (Turkey), *Eng. Geol.* **68** 159–169
[7] Song X F, Wei J F and He T S H 2009 A method to repair concrete leakage through cracks by synthesizing super-absorbent resin in situ, *Constr. Build. Mater.* **23** 386–391
[8] Freitas V P D E and Abrantes V 1996 Moisture Migration in Building Walls- Analysis of the Interface Phenomena, *Build. Environ.* **31** 99–108
[9] BS EN 14695 2010 *Flexible Sheets for Waterproofing. Reinforced Bitumen Sheets for Waterproofing of Concrete Bridge Decks and Other Trafficked Areas of Concrete. Definitions and Characteristics* (London: British Standard) pp 9-11
[10] Jamaluddin N, Ayop S S, Wan Ibrahim M H, Boon K H, Yeoh D, Shahidan S, Mohamad N, Tuan Chik T N, Abd. Ghafar N H, Abdul Ghani A H and Shamrul-Mar S 2017 Forensic building: deterioration and defect in concrete structures *MATEC Web of Conf.* vol. 103 (Paris: EDP Sciences) 02016
[11] Jumaat M Z, Kabir M and Obaydullah M 2006 A review of the repair of reinforced concrete beams, *J. Appl. Sci. Res.* **2** 317–326
[12] Kenai S and Bahar R 2003 Evaluation and repair of Algiers new airport building, *Cem. Concr. Compos.* **25** 633–641
[13] Bahaskaran R, Palaniswamy N and Rengaswamy N S 2013 A review of differing approaches used to estimate the cost of corrosion, anti-corrosion method and materials, *J. Appl. Sci. Res.* **52** 29–41
[14] BS 7542 1992 *Method of Test for Curing Compound for Concrete* (London: British Standard) pp 2-4

### Acknowledgments

This paper was partly sponsored by the Centre for Graduate Studies UTHM. The author would like to thank Universiti Tun Hussein Onn Malaysia for their financial support in term of grant, Short Term Grant (STG) for vote number of U344 and Geran Pembangunan Produk (GPP) for vote number of B064. Special thanks to Research, Innovation, Commercialization and Consultancy Management Office UTHM and Department of Structural and Material Engineering (FKAAS), for supporting this research.

# Impacts of flood on children and adults' health and ways to sustainable development

**S Mohamed[1], I Y Ebenehi[1,3], A Adaji[1,2], T W Seow[1], N W Chan[4], K C Goh[1] and M H I Abd Rahim[1]**

[1]Department of Construction Management, Faculty of Technology Management and Business, Universiti Tun Hussein Onn Malaysia, 86400 Parit Raja, Batu Pahat, Johor, Malaysia
[2]Department of Building Technology, Kogi State Polytechnic Lokoja, Nigeria
[3]Department of Building Technology, The Federal Polytechnic Bauchi, Nigeria
[4]School of Humanities, Universiti Sains Malaysia, Pulau Pinang, Malaysia
Corresponding author: zakimin@uthm.edu.my

**Abstract**: One of the events that will remain fresh in the minds of Kelantanese is the 2014 massive flood that occurred at the end of that year. Heavy rains fell initiating vast flooding in most areas of Kelantan leading to great destruction of livelihood of local communities. Natural hazard such as floods are not only caused by natural processes but also by human activities. The flood recorded severe fatalities, injuries and exposed many vulnerable to diseases. This paper through critical review of literature considered the long-term impact of floods on human's health as the effects could meaningfully contribute to the worldwide burden of disease. Also, its outcomes are relentless hence need to be adequately comprehended and addressed through sustainability. This study revealed vulnerability to flood inclined ailments as psychological distress in the survivors is liable for a quota of all physical ailments. Hence, sustainable approach to flood preparedness and prevention is instantly needed. Accolades should be given to the Malaysian government for taken bold steps in that direction in recent time but success will be achieved if implementation is in compliance with sustainability agenda spelt in the New Sendai Framework for Disaster Risk Reduction 2015-2030.

## 1. Introduction

The flooding experience of 2014 cannot be forgotten by Malaysians especially the Kelantanese. Precisely, on the 26th of December 2014 to January 2, 2015, there was a massive flood in the state of Kelantan, Malaysia. Heavy rains fell initiating great flooding in most regions of Kelantan resulting into great disruption of livelihood of residents [1].

Report had it that across the country, quite a large population of residents were affected by the ugly visitor (flood) and Kelantan state was identified as the most affected among the affected states [2]. In an attempt to describe the scenario, it is not out of place for one to say that the speed of the flood water in the affected regions flowed so fast with vitality equivalent to that of Tsunami, displacing anything that obstruct its channel of flow including buildings and other infrastructures as displayed in figure 1 and figure 2 respectively.

**Figure 1**. Property damaged-House the cars.

**Figure 2**. Tsunami-like disaster damaged another house.

Record has the flood as the worst that hit the country of Malaysia in two centuries (20 decades) with the assumption that surge of such magnitude happens seldom perhaps once in 1000 years. Specifically, for the state of Kelantan, the catastrophe was exceptional in the trace of its existence. The property that was ravaged by the floods comprehensively in parts of peninsular Malaysia is anticipated to cost Putrajaya and the Kelantan state governments more than RM 1billion [3]. Considering the assessment of destruction caused by the natural hazard, it is appropriate to call it a national disaster because of the huge economic loss of approximately USD 0.7 billion that Malaysia as a nation suffered.

It is knowledge widening and truthful that natural hazard such as floods are not caused by natural processes only but also by human activities. A number of investigators have revealed that human activities contribute largely to natural menace like the one under context. For example, Azuhan [4] and Sathiamurthy & Kong [5] unwaveringly blamed the happening of the flood on the unusual rainfall within short interval and illegal exploitation of natural vegetation such as logging activities. The flood caused acute fatalities, injuries and left many residents to diseases.

Great and astonishing floods such as the one under context, subject the elderly, who are in need of assistance relating evacuation and admission to medical amenities, and who perhaps hesitate to abandon their houses, at bigger risk of injury and death [6, 7]. Giving clear proofs that residents of the affected locations actually suffered these discomforts can be observed in the work of Baharuddin *et al* [8] captioned "The Record-Setting Flood of 2014 in Kelantan". Baharuddin *et al* [8] acknowledged the inflow of flood victims (patients) of Kelantan to Hospital University Sains Malaysia (HUSM) to add to the numbers transferred to HUSM by Hospital Raja Perempuan Zainab 2 (HRPZ2) after the vacation of their health centre on the 25th day of December 2014. The transferred patients' situations were largely critical issues from protracted medical situations namely; acute exacerbation of chronic obstructive airway disease, acute coronary syndrome, sepsis, heart failure, among others. This paper will ponder on the long-term impact of floods on human's health as the effects could meaningfully contribute to the worldwide burden of disease.

A few researchers such as Ramin & McMichael [9] and Schwartz *et al.* [10] opined that floods may be directly responsible for long term mortality such as increase in diarrheal deaths in developing nations or indirectly, affecting health, food and economic settings, aggravating poverty, starvation and noncommunicable diseases [11]. Furthermore, its consequences are unrelenting therefore need to be better understood and addressed through sustainable approach. Sustainable innovations or ideas like preventive measures and adaptation planning to decline post-flood mortality and morbidity should be holistically pursued.

## 2. Literature review

Flooding is the most popular kind of natural disaster globally, and has so become an important point of concern within the developed and developing countries. In Europe, floods have been reported to have affected more than 3.4 million people over the last decade, with number of death extending into the thousands [12]. The prediction of the WHO Regional Office for Europe [13], is that by the year 2085 flooding could directly affect about 5.5 million people in Europe yearly, if sustainable measures regarding climate change are not taken serious. With tenacious rainfall causing various types of flooding, this affected over 7000 households in the United Kingdom while incurring major disruption to personal livelihoods, property and public transport systems [14]. Flooding affects numerous aspects of man's environment and these include his economic activities; settlements and health.

Flooding is a momentous rise of water level in a stream, lake, reservoir or coastal system that overflows the banks. Otomofa, Okafor, and Obienusi [15] uncovered that there is increasing susceptibility or vulnerability of populations and infrastructure to flooding and flood related hazards. ProVention Consortium [16] expressed that normal floods are anticipated and welcomed as they offer rich soil, water and means of transport, but flooding at a big scale may causes damage to life, built and natural environment.

Floods are complex events by nature, caused by a collection of human exposures, incorrect development planning and climate change [16, 17]. Hence, attempting to describe any form of floods gives prominence to central factors such as weather, climate and climate change. These three words (weather, climate and climate change) are unified. In trying to figure out their meaning, we will examine from the International Strategy for Disaster Reduction (ISDR) perspectives. International Strategy for Disaster Reduction [18] views weather as a set of meteorological settings like rain, wind, snow, temperature and sunshine at a specific time and place. While by disparity, the term climate refers to the overall long-term features of the weather experienced at a certain place. In a more concise expression, Othman [19] addressed climate change as a long term changing tendency of climate. Climate change is a stern issue because its consequences are serious and devastating [1].

Climate change is becoming a global interest in the recent times therefore gaining prominence for recognition. Climate change is affecting the world at large and Asia is of no exception as natural disasters consequences are frequently experienced in this continent. To buttress this statement, Sakar *et al.* [20] confirm that for the past few decades, Malaysia like other nations of the world has experienced climate change through changing rainfall patterns, increasing temperature, and life-threatening weather conditions. These factors lead to growing incidence of climate related disasters which are primarily floods that promotes regular global warming episode.

Worthy of mentioning is the 2014 flood in Malaysia where three states namely; Kelantan, Terengganu, and Pahang experienced unusual and heavy rains that caused great flooding. Among the affected states, Kelantan was the most severe and left with the greatest ravage. The Kelantan flood of 2014 can be regarded as unpredictable and unprecedented. Going by the level of devastation, it is marked the biggest and worst flood in the state in the past 10 decades (century).

Many people were homeless and discomforted. Both social and emotional costs were involved in flooded areas. These costs comprise displacement from homes, the loss of private valuables and the continuing fear and insecurity caused by the ugly experience. Portable water supplies were polluted and lost during episodes of flood. Borneo Post [21] presented that, Kelantan recorded the highest number of evacuees, followed by Terengganu and Pahang respectively as summarized in table 1.

**Table 1.** Summation of evacuees in states hit by 2014 flood.

| S/N | States | Total Number of Evacuees |
|---|---|---|
| 1 | Kelantan | 20,468-24,765 |
| 2 | Terengganu | 21,606 |
| 3 | Pahang | 10,825 |

As vulnerability to floods is inescapable, relentless attention through sustainable development or approach has to be dedicated to lessening hazard connected with its happening especially the long-term effects. This can be achieved through the introduction or adoption of adequate planning to enhance operational capabilities and preventive steps that are focused at reducing the impacts there of.

This paper examines the impact of floods on the built environment with emphasis on human's health in Kelantan State, Malaysia.

**3. Consequences of flood on sustainable development**

Floods may be responsible for unsmiling variations in environmental cleanliness and individual health. Huang *et al.* [22] said pollutants over the flooded scenes can be spread by the action of flood with shared characteristics of abating drinkable water supply and conditions of living or place of abode. The shocking consequence of flood on the environment and infrastructure, poor diet, overcrowded accommodations and unhygienic conditions perhaps give rise to the risk of contagious ailments. Vachiramon *et al.* [23] and Balato *et al.* [24] concurred that diseases such as bacterial and fungal skin infections, eczema, urticaria (nettle-rash) and scabies are said to be powered by floods. In another development, Wade *et al.* [25] and Chen *et al.* [26] claimed that floods and heavy precipitation are affiliated to increased gastrointestinal (GI) sicknesses namely; enterovirus infection and bacillary dysentery.

Floods can be viewed as a facilitator or booster to the quick or fast spread of some silent killing health irregularity. Developing countries are said to suffer severe threats of infections particularly residents that are residing in floods-prone areas and elderly and children are the high ranked on vulnerability than any other class within the range of the flooded locations [27]. As water turbidity increases, the elderly and children increase their chances of contacting gastrointestinal (GI) tract infection. For it is a common or usual practice to see children using flooded water as playground without the understanding or knowledge that they are widening their susceptibility of infections [28, 29]. Stating clear need for sustainable development in the management of epidemics in the now and the afterward bearing in mind that the source of the ailments (floods) is unavoidable or unstoppable or inseparable from man. Hence, need to consider impact of flood on health for both children who are the future generation and adult becomes non-substitutable.

*3.1 Children's health*

It is no longer false but true that the destruction caused by natural disaster on individual is extensive and Raj & Sekar [30] have ranked children as the most exposed class, followed by women, elderly and people with disabilities respectively. Factors such as dislodge from familiar surroundings, loss of parents, witnessing the passing unto glory of their beloved ones, horror of disaster repetitions among others, may be responsible for the high number of vulnerability among children.

Pine and Cohen [31] acknowledged that distress undergone by children in disaster leads to several psychological problems. Despite being prone to this awkward and unavoidable situation, research on children's health in flood occurrence receives little attention. Vranda & Sekar [32] stated that poor recognition and underestimation of the post-disaster psychological effects on children by mental health professionals in flood dominated locations is obvious. This is one of the motivations for this research in the direction of sustainability bearing in mind the popular slogan that "better safe is better than sorry".

Disaster preparedness with distinct orientation to psychosocial aspects through care givers is one of the sure measures to ameliorate the distress of both children and adult affected by any disaster (on-set or slow set) because it decreases the susceptibility factors that curtails the pain of any hazard on the affected person or family. There is need for more research to be conducted on the most vulnerable cohort to flood-related diseases as this can help in preserving them from further contact with both the discovered and yet to be discovered diseases.

*3.2* Adult's *health*

A disaster whichever form it manifests, is the misfortune of a natural or human induced hazard that depressingly shakes people and environment. According to the findings of Vranda & Sekar [32] the wrath of disaster rests on the psychological, social, physical and economic capacities of the victims' i.e. individual, family and community.

It is evident that investigation on the impact of flooding on human's health is not gaining popularity in Malaysia. Even the few that exist regarding methods to diminish the impact of flooding suffer little assessment. This study finds it imperative to reflect on flooding since it has the ability to harm anything regardless of settings (from single house to many houses) and locations (particular or several locations).

According to Alderman *et al.* [11], the predominance of mental disorderliness involving downheartedness, concern and even post-traumatic stress disorder (PSTD) is approximately between 9 to 53 per cent for as long as two years after the flood. When home is flooded, there is possibility for dominance of signs among those with health and financial challenges to increase triple times [33]. This shows that, people with mental conditions before the flood stand a greater risk if anxiety is not controlled. Been affected by flood in the past and the constant thought that it will still surface in the nearest future; largely contribute to poor health [11, 33-35]. Stanke *et al.* [35] proves the existence of some elevating household fury among males and females in the aftermath of flooding. Interruption of daily activities, congested momentary shelter, reflection on owned belongings but lost to flooding, living condition that is so demanding and financial incapability among others, may be affiliated to the violence rising.

According to Reacher *et al.* [36] physical ailments developed following flood is rooted in emotional agony and these have a prolong effects on the behaviour of the survivors. From the above explanations, it can be assumed that feelings after shock such as flooding have negative results on human's well-being. As such, prevention is absolutely non-negotiable. The record of health impairments after flooding is influenced by a number of factors namely:

- The level of devastation caused to properties.
- The record of death and injury sustained.
- The duration of normalcy restoration or recovery.
- Accessibility to insurance and response rate from the insurance managers.
- The state of preparedness prior to flooding, community resilience and social supports network.

### *3.3 Malnutrition*

Floods do not only affect human health but also cause malnutrition and absence of morale. Floods affect the agricultural sector by destroying crops, livelihood of the people, and also destroy homes and infrastructures [37]. Thus, it has the capability to reduce agricultural produce thereby causing food shortages in the stricken regions.

As witnessed in neighbouring countries to Malaysia such as People's Republic of Bangladesh, selected parts of Australia and the continent of Africa, floods overwhelm land and destroy agricultural harvests [11]. Kelantan in particular was not spared in the recent strike of flood in the year 2014. As Shamshuddin *et al.* [2] confirmed that this actually happened in Kelantan during the 2014 flood. As their report shows devastation of agricultural plots regularly used for cultivation thereby causing immeasurable damage to farmers. Furthermore, natural grasses in the environment were absolutely destroyed leaving nothing for the surviving livestock to graze. With the report, one would say that the lucky and innocent livestock who might have survived the disaster had no means of feeding enough thereby reducing their economic value if they are to be sold or exchanged for money.

The recent work of Shamshuddin *et al.* [2] corresponds with the assertion of Austin and Baharudin [38] that Agriculture which contributes approximately 10 per cent of Malaysia's Gross Domestic Products and gives sustenance of livelihoods to minimum one- third population of Malaysia is been threatened by flood [38]. This implies that, the agricultural sector can be affected in terms of production and the people employed under this would have their socio-economic life depressed as well. It is appropriate to say that flooding in Malaysia is a potential threat to food security and agriculture which is capable of causing malnutrition if not appropriately controlled. Ambu [39] expressed that increased rate of malnutrition is a function of floods that affect food safety. From the foregoing, it is eminent for natural disaster specifically flood which is predominant in Malaysia, to be incorporated into the existing crucial framework for malnourishment. Urgent steps should be taking in the direction of sustainability in order to salvage the affected regions from subsequent or future hardships or difficulties.

**4. Sustainable development as the way forward**

With increasing dependency on fossil-fuels and a lessening supply of resources, people need innovative, cutting-edge approaches from appropriate stakeholders to make a real difference for sustainable development. The challenge is to solve matter regarding the needs of citizen in a sustainable manner, so as to generate continuing development and activities that meet the needs of the people and protecting, sustaining and enhancing the human and natural resources for the future [40, 41]. Apart from that, the recognition of sustainable development as a current trend and global pursuit has driven this study to map out some strategies to reduce the negative outcomes of flooding on human health in the study area. These strategies are highlighted as follow:

- Develop a comprehensive list of the vulnerable class as identified in this paper that may need help and medical assistance.
- Mark top risk geographical areas and key buildings such as health care facilities.
- Creation of online data for sharing flood risk data with relevant organisation.
- Provide prior news of flooding through the design and implementation of early warning system.
- Facilitate community engagement and supports in the recovery phase.
- Recovery needs should be freely accessed by flood victims and advice in that direction.
- Educate stakeholders to understand the frustration of flood victims and should work in tolerance with them in the quest to better their living conditions.
- Empower the affected people to guaranty sustenance of livelihoods.

**5. Conclusion**

The flood of 2014 had unforgettable marks on the residents of Kelantan states. It created multiple risks to all stakeholders with the flood victims carrying the brunt of it. Its impact cut across the physical, psychosocial and economic lives in a significant way. Report reveals vulnerability to flood diseases as psychological distress in the survivors is guilty for a quota of all physical ailments. Sustainability plan has the potential to reduce vulnerability and realize opportunities associated with flood effects and menaces. Thus, the strategies above should be looked into. Accolades should be given to the Malaysian government who has taken bold steps in recent time, to uncover the long-term impacts of 2014 flood as well as to fashion out safety measures for the future. Government should ensure strict implementation of reports in compliance with the Sendai Framework for Disaster Risk Reduction 2015-2030 targets.

## 6. References

[1] Ahmad W I W and Abdurahman S M 2015 Kelantan flood 2014: Reflections from relief aid mission to Kampung Kemubu, Kelantan, *Med. J. Soc. Sc.* **6** 340-44

[2] Shamshuddin J, Panhwar Q A, Othman R and Ismail R 2016 Effects of December 2014 great flood on the physico-chemical properties of the soils in the Kelantan plains Malaysia, *J. Wat. Res. and Prot.* **8** 263-76

[3] Berita Harian 2014 *Amaran Hujan Lebat Peringkat Jingga di Kelantan, Terengganu* Retrieved on December 24, 2016 from http://www.utusan.com.my/berita/nasional/amaran-hujan-lebat-di-kelantan-terengganu-1.36974

[4] Azuhan M 2015 Kelantan flood-divine vs anthropogenic causes *Proc. Nat. Geoscience Conf. (Kota Bharu)* (Kuala Lumpur: Geological Society of Malaysia) pp 2-5

[5] Sathiamurthy E and Kong Y H 2015 Kelantan state great flood event in December 2014-hydrology perspective *Proc.Nat. Geoscience Conf. (Kota Bharu)* (Kuala Lumpur: Geological Society of Malaysia) pp 41-42

[6] Jonkman S N and Kelman I 2005 An analysis of the causes and circumstances of flood disaster deaths, *Disasters* **29** 75-97

[7] Yeo S W and Blong R J 2010 Fiji's worst natural disaster: The 1931 hurricane and flood, *Disasters* **34** 657-83

[8] Baharuddin K A, Wahab S F A, AB Rahman N H N, Mohamad N A N, Kamauzaman T H T, Noh A Y M D and Majod M R A 2015 The record-setting flood of 2014 in Kelantan: Challenges and recommendations from an emergency medicine perspective and why the medical campus stood dry, *Malaysian J. of Med. Sc.* **22** 1-7

[9] Ramin B and McMichael A 2009 Climate change and health in sub-Saharan Africa: A case-based perspective, *EcoHealth* **6** 52-57

[10] Schwartz B, Harris J, Khan A, Larocque R, Sack D and Malek M 2006 Diarrheal epidemics in Dhaka, Bangladesh, during three consecutive floods: 1988, 1998, and 2004, *American. J. Trop. Med. Hyg.* **74** 1067-73

[11] Alderman K, Turner L R and Tong S 2012 Floods and human health: A systematic review, *Environ. Int.* **4** 37-47

[12] Jakubicka T, Vos F, Phalkey R and Marx M 2010 *Health impacts of floods in Europe: Data gaps and information needs from a spatial perspective* A MICRODIS report Brussels Centre for Research on the Epidemiology of Disasters (CRED) Retrived on April 20, 2017 from http://www.cred.be/sites/default/files/Health_impacts_of_floods_in_Europe.pdf

[13] WHO Regional Office for Europe 2017 *Climate change: Data and statistics* Retrived on April 14, 2017 from http://www.euro.who.int/en/health-topics/environment-and-health/climate-change/data-and-statistics

[14] Muchan K, Lewis M, Hannaford J and Parry S 2015 The winter storms of 2013/2014 in the UK: hydrological responses and impacts, *Weather* **70** 55-61

[15] Otomofa J O, Okafor B N and Obienusi E A 2015 Evaluation of the impacts of flooding on socio-economic activities in Oleh, Isoko South Local Government Area, Delta State, *J. Environ. and Earth Sc.* **5** 155-71

[16] ProVention Consortium 2008 *Flood Disasters: Learning from Various Relief and Recovery Operations ALNAP* Retrieved from www.proventionconsortium.org

[17] Jha A K, Bloch R and Lamond J 2012 *Cities and flooding: a guide to integrated urban flood risk management for the 21st century* (Washington DC: World Bank Publications) pp 582-620

[18] International Strategy for Disaster Reduction 2008 *Climate change and disaster reduction* Geneva ISDR Retrieved on May 13, 2017 from http://eird.org/publicaciones/Climate-Change-DRR.pdf

[19] Othman S 2011 *Adaptation to Climate Change and Reducing Natural Disaster Risk: A Study on Country Practices and Lesson Between Malaysia and Japan* Final Research Report (Tokyo: Asian Disaster Reduction Centre (ADRC) pp 1-67

[20] Sakar S K, Begum R A, Periera J J and Abdul H J 2013 Addressing disaster risk reduction in Malaysia: Mechanisms and responds *2nd Int. Conf. Env. Agric & Food Sc (ICEAFS 2013) (Kuala Lumpur)* pp 81-85

[21] Borneo Post 2014 *Government to Carry Out Post Mortem on Floods* Retrieved on December 30, 2016 from http://www.theborneopost.com/2014/12/31/government-to-carry-out-post-mortem-on-floods/

[22] Huang L-Y, Wang Y-C, Wu C-C, Chen Y-C and Huang Y-L 2016 Risk of flood-related diseases of eyes, skin and gastrointestinal tract in Taiwan: A retrospective cohort study, *Plos One* **11** 1-11

[23] Vachiramon V, Busaracome P, Chongtrakool P and Puavilai S 2008 Skin diseases during floods in Thailand, *J. Med. Assoc. Thai.* **91** 479-84

[24] Balato N, Megna M, Ayala F, Balato A, Napolitano M and Patruno C 2014 Effects of climate changes on skin diseases, *Expert Rev. Anti Infect. Ther.* **12** 171-81

[25] Wade T J, Sandhu S K, Levy D, Lee S, LeChevallier M W, Katz L and Colford J M 2004 Did a severe flood in the Midwest cause an increase in the incidence of gastrointestinal symptoms?, *American J. Epidemiol* **159** 398-405
[26] Chen M J, Lin C Y, Wu Y T, Wu P C, Lung S C and Su H J 2012 Effects of extreme precipitation to the distribution of infectious diseases in Taiwan, *Plos One* **7** 1994-2008
[27] Ahern M, Kovats R S, Wilkinson P, Few R and Matthies F 2005 Global health impacts of floods: epidemiologic evidence, *Epidemiol. Rev.* **27** 36-46
[28] Gertle M, Dürr M, Renner P, Poppert S, Askar M, Breidenbach J, Frank C, Preußel K, Schielke A, Werber D, Chalmers R, Robinson G, Feuerpfeil I, Tannich E, Gröger C, Stark K and Wilking H 2015 Outbreak of cryptosporidium hominis following river flooding in the city of Halle (Saale), Germany, August 2013, *BMC Infectious Diseases* **15** 1-10
[29] Schwartz J, Levin R and Goldstein R 2000 Drinking water turbidity and gastrointestinal illness in the elderly of Philadelphia, *J. of Epidemiol.* **54** 45-51
[30] Raj E A and Sekar K 2011 Psychosocial disaster preparedness program for school children, *Prehosp. Disaster Med.* **26** 83
[31] Pine D S and Cohen J A 2002 Trauma in children and adolescents: risk and treatment of psychiatric sequelae, *Biological psychiatry* **51** 519-531
[32] Vranda M N and Sekar K 2011 Assessment of psychosocial impact of flood on children-Indian experience, *Prehosp. Disaster Med.* **26** 83-84
[33] Murray V, Caldin H, Amlôt R, Stanke C, Lock S, Rowlatt H and Williams R 2011 *The Effects of Flooding on Mental Health* Health Protection Agency report (London: Health Protection Agency)
[34] Tapsell S M and Tunstall S M 2008 "I wish I'd never heard of banbury": The relationship between 'place' and the health impacts from flooding, *Health Place* **14** 133-54
[35] Stanke C, Murray V, Amlot R, Nurse J and Williams R 2012 The effects of flooding on mental health: Outcomes and recommendation from a review of the literature, *Plos Current Disasters* **1** 1-17
[36] Reacher M, McKenzie K, Lane C, Nichols T, Kedge I, Iversen A, Hepple P, Walter T, Laxton C and Simpson J 2004 Health impacts of flooding in Lewes: a comparison of reported gastrointestinal and other illness and mental health in flooded and non-flooded households, *Communicable Disease and Public Health* **7** 1-18
[37] Padli J, Habibullah M S and Abdul Hamid B 2013 Determinants of flood fatalities: Evidence from a panel data of 79 countries, *Pertanika J. Soc. Sci. and Humanities* **21** 81-98
[38] Austin O C and Baharuddin A H 2012 Risk in Malaysian agriculture: The need for a strategic approach and a policy refocus, *Kajian Malaysia* **30** 21-50
[39] Ambu S 2015 Floods-the consequence of human intrusion into nature, *Int. E-J. Sci. Med. and Education* **9** 1-2
[40] Clémençon R 2012 From Rio 1992 to Rio 2012 and beyond: Revisiting the role of trade rules and financial transfers for sustainable development, *The J. of Environment & Development* **21** 5-14
[41] Masrom M A N, Abd Rahim M H I, Ann S C, Mohamed S and Goh K C 2017 A preliminary exploration of the barriers of sustainable refurbishment for commercial building projects in Malaysia *Procedia Engineering* vol 180 (United Kingdom: Elsevier) pp 1363-1371

**Acknowledgements**
The authors are unreservedly grateful to Ministry of Education (MOE), Pejabat Pendaftar UTHM, ORICC UTHM, under the Research Acculturation Collaborative Effort (RACE: No.1517) Grant Scheme, Centre for Graduate Studies (UTHM) and UTHM for the support of this study.

IOP Conf. Series: Materials Science and Engineering **271** (2017) 012026 doi:10.1088/1757-899X/271/1/012026

# The impact of floods in hospital and mitigation measures: A literature review

**N A Yusoff[1], H Shafii[1] and R Omar[1]**

[1]Faculty of Technology Management and Business, Universiti Tun Hussein Onn Malaysia, 86400 Parit Raja, Batu Pahat, Johor, Malaysia
Corresponding author: arein1804@yahoo.com

**Abstract.** In late December 2014, the flood was most significant and largest recorded specifically in the Kelantan, Malaysia. It was considered to be a "tsunami like disaster" in which 202,000 victims were displaced and causing widespread collapse of public infrastructure. Flooding of hospital results in interruption of business, loss of infrastructure, such as electrical power and water supplies, increased difficulty in providing routine medical and increased patient admissions and nursing care for patients with chronic diseases, such as renal failure, diabetes, cancer, cystic fibrosis and mental illness. The aimed of this paper to identify the best of measures for reduce the risk of flood in hospital. Method of this paper uses the previous study result. Several related previous study can be used as measures to mitigation flood risk in Malaysian hospitals. Early stage research of related studies hope to help add more information to assist researchers in reducing the risk of flooding in hospital. The findings with proper pre-event preparation framework for mitigation flood risk of hospitals, the continuing medical services can be provided to patient especially during emergency.

## 1. Introduction

Floods are natural disaster resulting from rainfall in certain seasons because water levels of rivers and lakes overflowed and went into the surroundings [1][2]. Floods can create various of damage to property and loss of life is significant. Scenario flooding and destruction from flooding not only recorded in Malaysia but also throughout the world. According to the World Meteorological Organisation (WMO), the flooding is three worst natural disasters in the world and has claimed thousands of lives and destruction of property valuation hundred thousand million [3]. A record number of floods in Malaysia, such as in year 1926, 1931, 1947, 1954, 1957, 1963, 1965, 1967, 1969, 1971, 1973, 1983, 1988, 1993, 1998, 2001, 2006, 2007 and 2010. Reports from Department of Irrigation and Drainage stated that about 29,000 square kilometers, or 9% of the total area and more than 4.82 million people (22%) of the population affected by floods every year [4].

Floods are an annual occurrence of varying in state on the east coast of Peninsular Malaysia such as Pahang, Terengganu and Kelantan. In late December 2014, the flood was most significant and largest recorded specifically in the Kelantan. It was considered to be a "tsunami like disaster" in which 202,000 victims were displaced and causing widespread collapse of public infrastructure [5][6]. Kelantan faces Northeast monsoon from November to March each year and compounded by unplanned urbanization, geographical characteristic and land use planning, triggered the severe flood [7]. Healthcare facilities were severely affected by the flood and patients had difficulties receiving medical treatment in a timely and effective manner as most of the healthcare facilities and public amenities were located on the flood plain [5].

Hospital is among most important in which it gives a medical facility and treatment of the highest quality to in and out of patients[8]. In addition to being a major referral centre that provides diagnostic and treatment services. Hospital building equipped with modern facilities and building of sophisticated medical equipment as well as involving professional staff to provide medical services and care to patients [9][10]. Therefore, this study was to review the issue of the impact of floods in the hospital and measures to reduce its impact.

## 2. Impact of floods on health services

Flooding can either damage hospital facilities directly or disrupt access to them [11][12]. Damage may be direct costs such as losses in infrastructure, expensive medical equipment, hospital furniture, lifeline installations and medical supplies. Indirect costs are unforeseen expenses after emergencies such as increased risk of outbreaks due to loss of laboratory and diagnostic support, temporary solutions like field hospitals and the loss income normally generated by health care services [8][13].

### *2.1. In Malaysia*

On December 2014, a massive flood happened in the state of Kelantan, Malaysia. It was recorded as the most significant and disastrous flood in Kelantan. The governments estimated around 200,000 people were affected [14]. Hospital are also affected by the floods 2014 in Kelantan, such as Hospital Raja Perempuan Zainab II (HRPZ II), Kuala Krai, Tumpat, Pasir Mas, Gua Musang and Tanah Merah [15]. Hospital Raja Perempuan Zainab II (HPRZ II) is the worst hospital affected due to water entered and had to stop 80 percent of their services due to flooding and left more 200 patients had to be transferred to the Hospital Universiti Sains Malaysia (HUSM) [16]. The situation became chaotic in Hospital Universiti Sains Malaysia (HUSM) in Kubang Kerian due to increasing the number of patients because of the closure Raja Perempuan Zainab II (HPRZ II) hospital.

Another challenge that was encountered during flood was a shortage of staff. Almost 50% of the staff was affected by the floods, including the head of the Emergency Department. Furthermore, there was no disaster declaration from the top-down at HUSM on day 1 of the disaster, meaning that Emergency Department staff had to handle the situation themselves. With the increasing number of patients, this became a disaster for them [6]. About five of patients were evacuated to Hospital Besut, Terengganu to receive more critical treatment due to insufficient number of care beds. There are many cases referred from HPRZ II involving victims who suffer from chronic diseases such as acute respiratory syndrome, chronic respiratory disease and heart disease jammed [17]. While the average amount of losses incurred by the Kuala Krai hospital due to floods in 2014 is estimated at RM1 millions of which including damage to non-drug items such as medical equipment, non-medical and other facilities is estimated at RM600 million. While items involving drugs is estimated at RM500 thousand [18]. In addition, the fuel shortage is causing disruption to the electricity supply in Kuala Krai hospital. This is because there is no storage fuel for use in emergencies.Increasing the number of flood victims in Kuala Krai hospital has caused shortages of food, drugs, oxygen gas and clean water. Because of that, the critical patients should be transferred for getting further treatment [19].

### *2.2. In Overseas*

During the past 10 years, in the European Region, 1000 persons are reported to have been killed by floods and more than 3.4 million affected. Flooding of health facilities results in interruption of business, loss of infrastructure, such as electrical power and water supplies, increased difficulty in providing routine medical and increased patient admissions and nursing care for patients with chronic diseases, such as renal failure, diabetes, cancer, cystic fibrosis and mental illness. Ukraine reported damage to 122 hospital due to flooding [13]. Typhoon Nari hit the coastal region of northern Taiwan on the night of September 16, 2001. The health care system in Taipei was seriously damaged. There were 11 hospital damaged by storm induced flooding, which left them in vulnerable condition requiring evacuation. Storm water filled the basements and first floor of many hospitals and caused the backup generators to be inoperable. Power was limited to critical care units, operating rooms and to important laboratory equipment. Therefore, providing medical service at full capacity was nearly impossible in the aftermath of the typhoon. Some emergency services were closed to the public for days [10]. In 2007, floods in the New South Wales Hunter Valley has caused the Morriset Hospital was without power for almost two days [20]. Some hospital were completely destroyed during 2004 Hurricane Katrina in the USA and in Australia many hospital have been damaged and cut-off from surrounding communities during severe cyclones, storms and major flooding events in the recent past [21]. Flooding occurred in the Gloucestrershire Hospital UK has caused water contamination and had disruption inoperability of medical services and cease of activities [22].

## 3. Flood mitigation measurement for hospital

The paper is written based on an overview of literature from journal, articles, newspaper, book and report. The review focus about local and other country experiences regarding flood effect in hospital issue and identify the best of measures for reduce the risk of flood in hospital. Given growing international concern about the resilience of health facilities in the face of climate change, many countries have strengthened and developed legal frameworks, research and funding schemes for disaster reduction and climate change adaptation [23]. Floods also caused a large damage of property and life threatening population. Therefore, these publications mainly focus on floods[24]. Flooding can either damage health care facilities directly or disrupt access to them [13]. Federal Emergency Management Agency (FEMA) design guide for improving hospital safety provides important insights into variety of hazards mitigation for health facilities [25]. The FloodProBe project aims to provide cost effective solution for flood risk reduction in urban UK areas [26]. The World Health Organization (WHO) Regional Office for Europe and the United Kingdom Health Protection Agency (HPA) collaborated to assess the health effects of floods as well as to identify measures to prevent or minimize their health effects. This Hyogo Framework for Action (2005-2015) adopted by the World Conference on Disaster Reduction highlights the need to integrate planning for disaster risk reduction into the health sector. This framework addresses primary, secondary and tertiary prevention for managing flood risk with a range of interventions and measures to reduce the impact of human health (refer table 1) [13].

In studies Plate and Chan said, both of structural and non-structural method needs identified for disaster reduction planning. Selection the best of measures to be able solve or reduce the impact of disaster problems. All activities the both management method of structural aimed "controlling" and non-structural is to reduce the floods impact [4, 27]. FEMA recommends to reducing flood effects in critical facilities must relocating the facilities outside of the floodplain [28].Having listed some study recommended to mitigate the impacts of flood, it is important to point out that multi method mitigation flood is considered the best practice.

**Table 1.** Preventive measures for flood risk reductionin hospital.[13]

| | |
|---|---|
| Primary | Structural (physically engineered interventions) including tree planting, control of water source and reconstruction of the defense to the beach / river.<br>Nonstructural (policy and organization) including land used management, stewardship and government, resource generation, health financing and service delivery. |
| Secondary | Flood forecasting, warning system, moving belongings and assets, Evacuation shelters and temporary and mobile structures, maintaining health services. |
| Tertiary | Include ensuring clean water and food supplies, close relationship with another agency(military, fire and rescue, police, water organizations, gas organization, human resources) |

## 4. Result/discussion

Sudden and unexpected flood often lead to severe damage and dramatic disruptions to the affected area especially in hospital. According to Sam and Pesigan the destructive effects of natural disasters on infrastructure and operations due to inadequate planning and disaster management [8]. Because prevention is more cost-effective than dealing with problems caused by the calamity, several improvements can be made in the initial preparation for hospitals in a flood. A comprehensive preparation should include a systemic process of planning, which includes structural, nonstructural and administrative issues [10]. The common organisational objectives to across all case studies which emerged in hospital were [23]:

- Continuity of service delivery;
- Preserving the building structure's integrity along with its building services;
- Having effective communication both externally and internally;
- Maintaining access to and from the site and;
- Ensuring availability and safety of relevant staff on hand to respond to the crises.

Hospital located near riverbeds, storm drains and areas of lower altitudes sustained even more damage. Thus, hospital must be designed and built to prevent potential disruption of function caused by flooding [29]. Structural elements for hospital building such as safe location, design, and structure are important considerations for allowing buildings to withstand the recent events. Tree planting also important, it is because a canopy of tree can slow precipitation, intercept rainfall and slow its flow into natural and engineered drainage system [13].

While floods can also cause damage to the equipment or drugs. Previously, hospital design focused on the optimum use of space to prepare the most effective interaction and action among difference department [8]. Thus, heavy equipment such as the telephone lines, electric generators, heavy medical equipment and medical supply storage are usually placed in basement. Therefore, when occur flooding, emergency response is interrupted immediately. Several measures can prevent this damage, such as regular inspections of the drainage system and sewage system, relocating the important medical equipment to above grade and convenient maintenance of the equipment to keep it in best working order [10]. In hospital located in flood prone areas transformer, emergency generator and electrical room must be relocated above grade or relocating the hospital outside of the floodplain, so that they can safe from flooding [28]. This is non-structural suggestions, if applied this suggestion will prevent costly damage where interferes with the efficiency and functionally of a hospital.

Features Administration also needs to be considered in preparation for the floods. This includes preparation and training is not limited the emergency staff only, but also involves the management of hospital staff [17]. Besides, the effective communication through top-down management is needed. This ensures the flood information flow can be channeled properly. Effective communications from other parties that cooperated with hospitals to help flood victims are also required [6]. The information chain in communication system should be monitored every moment. Effective human resource management is important to ensure that the hospital staff is required during flooding sufficient to ensure continuity of operation. The list of contact numbers of hospital staff should be updated [30].

Increasing the number of patients caused the beds care inadequate [6]. Sufficient number of critical care beds must be available for the transfer of patients from disaster impacted hospitals or mass causalities [10]. Hospitals often rely on electric generators during electric supply disruption. However, inadequate fuel supply and cause an electric generator are not operating. Therefore, adequate the fuel supply should be available during disaster and the fuel storage is needed [19]. Hospital should be ensuring that they can prepare water and food for the first 72 hours of the flood [13].Due to methods of communication depend on a power supply, communication hospital are interrupted during power outages. Both external and internal communication must have other, prearranged methods of communication. Hospital should be having backup equipment to support their communication such as radios, battery-run speakerphones and walkie-talkie [31].

The main reasons for evacuation are the presence of water in building and loss of power [13]. Critical patients should be transferred to other hospital for further treatment. Because roads were flooded, rafts and boats were use as alternative method for patient evacuation during this time. Same also with supplies such as fuel, food and water were brought to hospital by the same method [10].

Several related previous study can be used as measures to reduce flood risk in Malaysian hospitals affected by floods. For example, location of the hospital is clearly giving potential damage when flooding occurs. Hazard assessment should be started before construction of a hospital. Flood can affect the internal hospital facilities, such as communications, electricity, medical equipment and supplies also transportation to evacuated patients. Therefore, proper preparation is needed before an event occurs.

## 5. Conclusion

Floods catastrophic can cause severe damage and can affect the national economy. The variety of damage includes structural, non-structural and administrative. Thus, the risk assessment of flood hazards is conducted to assess the weaknesses the exiting to ensure the best elements needed to reduce the effects the flood hazards in hospital. Selection the best measures through structural and non-structural methods to solve or reduce the problems that faced the impact of the floods. Although we cannot prevent natural disasters from occurring, early warning systems and a well-organized disaster management plan can lessen the disruption. The findings with proper pre-event preparation guidelines for mitigation flood risk of hospitals, the continuing medical services can be provided to patient especially during emergency.

## 6. References

[1] Lin G F, Chou Y C and Wu M C 2013 Typhoon flood forecasting using integrated two-stage support vector machine approach *Journal of Hydrology* **486** 334-342

[2] Abidin A F Z 2010 *Analysis of Rainfall Event and Bridge on Galing River* Doctoral Dissertation (University Malaysia Pahang)

[3] Hussain T P R S, Nor A R M, and Ismail H 2014 The Level of Satisfication Towards Flood Management System in Kelantan, Malaysia *Pertanika Journal of Social Sciences & Humanities* **22**(1) 257-269

[4] Chan N W 2015 Impacts of disasters and disaster risk management in Malaysia: The case of floods *Resilience and Recovery in Asian Disasters* (Japan: Springer ) pp 239-265

[5] Abdul Ghani M N, Tuan Kamauzaman T H,Arifin W N, MohdFudzi M F M and Chew K 2016 Development and psychometric evaluation of flood disaster management questionnaire-(Flooddmq-Bm): Exploratory factor analysis and item response theory analysis *International Journal of Public Health and Clinical Sciences* **3**(3) 59-70

[6] Baharuddin K A, Wahab S FA, Ab Rahman N H N, Mohamad N A N, Kamauzaman T H T, Noh A Y M and Majod M R A 2015 The record-setting flood of 2014 in Kelantan: challenges and recommendations from an emergency medicine perspective and why the medical campus stood dry *The Malaysian Journal of Medical Sciences: MJMS* **22**(2) 1

[7] Khan M M A, Shaari N A B, Bahar A M A, Baten M A and Nazaruddin D A B2014 Flood impact assessment in Kota Bharu, Malaysia: A statistical analysis *World Applied Sciences Journal* **32**(4) 626-634

[8] Sam A G L and Pesigan A M 2011 Disaster risk reduction for health facilities in the Western Pacific Region *International Journal of Disaster Resilience in the Built Environment* **2**(3) 268-277

[9] Labarda C, Labarda M P and Lamberte E E 2017 Hospital resilience in the aftermath Typhoon Haiyan in the Philippines *Disaster Prevention and Management: An International Journal* **26**(4) 424-436

[10] Lai T I, Shih F Y, Chiang W C, Shen S T and Chen W J 2003 Strategies of disaster response in the health care system for tropical cyclones: Experience following Typhoon Nari in Taipei City *Academic emergency medicine* **10**(10) 1109-1112

[11] Loosemore M, Carthey J, Chandra V andMirti A 2010 Risk management of extreme weather events: A case study of Coffs Harbour Base Hospital, Australia *Management* 1239-1246

[12] Meusel D and Kirch W 2005 Lessons to be learned from the 2002 floods in Dresden, Germany *Extreme Weather Events and Public Health Responses* (Berlin Heidelberg:Springer) pp 175-183

[13] Menne B, Murray V and World Health Organization 2013 Floods in the WHO European Region: health effects and their prevention *WHO IRIS Technical Document* (Copenhagen: WHO Regional Office for Europe) p 146

[14] Chong Z X 2015 Experience of a medical student: Volunteering in the emergency department during massive flood in Kelantan, Malaysia *International Journal of Medical Students* **3**(3) 163-164

[15] Bernama 2014 *Hospital Kuala Krai Bergelap* Berita Harian Retrieved on May 8, 2017 from http://www.bharian.com.my/node/25291

[16] Sinar Harian 2014 *HRPZ II Hentikan 80 Peratus Perkhidmatan* Sinar Harian Retrieved on May 8, 2017 from http://www.sinarharian.com.my/hrpz-ii-hentikan-80-peratus-perkhidmatan-1.346139

[17] Kamarul Aryffin B, Shaikh Farid A W, Nik Hisamudin N A R, Nik Arif NM, Tuan Hairul nizam T K, Abu Yazid M N and Mohd Roslani A M 2015 Therecord-setting flood of 2014 in Kelantan: Challenges and recommendations from an emergency medice perspective and why the medical campus stood dry *Malaysia J. Med. Sci.* **22(2)** 1-7

[18] Hospital Kuala Krai 2016 *Laporan Kerosakan, Kerugiandan Gambar-Gambar Berkaitan Banjir 2014* (Malaysia:Hospital Kuala Krai, Kelantan)

[19] Adibah Abdullah 2015 *An Island Of Hope: Hospital Kuala Krai, A Refuge During The Floods* Medical Officer, Anaesthesiology Department Hospital Kuala Krai Retrieved on May 8, 2017 from http://www.ihm.moh.gov.my/images/files/MOHAMM/slide/12082015_Lunch%20Symposium/DR_ADIBAH_AN-ISLAND_OF_HOPE.pdf

[20] Carthey J, Chandra V and Loosemore M 2008 Assessing the adaptive capacity of hospital facilities to cope with climate-related extreme weather events a risk management approach *Procs 24th Annual ARCOM Conference* pp 1-3

[21] Chand A M and Loosemore M 2016 Hospital disaster management's understanding of built environment impacts on healthcare services during extreme weather events *Engineering, Construction and Architectural Management* **23**(3) 385-402

[22] Achour N, Miyajima M, Pascale F and D F Price A 2014 Hospital resilience to natural hazards: Classification and performance of utilities*Disaster Prevention And Management* **23**(1) 40-52

[23] Loosemore M, Chow V and Mc George D 2014 Managing the health risks of extreme weather events by managing hospital infrastructure *Engineering, Construction and Architectural Management* **21**(1) 4-32

[24] Said M. Z, Gapor S A, Samian M N and Abd Aziz A M 2013 Konflik di Pusat Pemindahan Banjir: Kajian Kes di Daerah Padang Terap, Kedah *Journal of Society and Space* **9**(1) 69-78

[25] Federal Emergency Management Agency 2007*Design Guide for Improving Hospital Safety in Earthquakes, Floods and High Winds: Providing Protection to People and Buildings,* Federal Emergency Management Agency (New York)

[26] Van Ree C C D F, Van M A, Heilemann K, Morris M W, Royet P and Zevenbergen C2011 FloodProBE: Technologies for improved safety of the built environment in relation to flood events *Environmental Science & Policy* **14**(7) 874-883

[27] Plate E J 2002 Flood risk and flood management *Journal of Hydrology* **267**(1) 2-11

[28] Federal Emergency Management A agency 2013 *Reducing Flood Effects in Critical Facilities,* Federal Emergency Management Agency (New York)

[29] World Health Organisation 2010 *Safe hospitals in emergencies and disasters: Structural, non-structural and functional indicators* (Manila: World Health Organization Regional Office for the Western Pacific)

[30] Kamarulazizi I, Kanayathu C K, Noor Adelyna M A, Radeah M N, Normaliza A M, Sharifah Nurlaili Farhana S A and Marlinah M 2015 *Resolusi Persidangan Pengurusan Bencana Banjir Kelantan 2015* Pusat Kajian Kelestarian Global Universiti Sains Malaysia

[31] Da Costa S and Williams J 2008 Surviving flood and contamination: How team work kept our diabetes centre a float *Journal of Diabetes Nursing* **12**(8) 305-309

**Acknowledgments**

The authors wish to acknowledge the contribution of our partners in this research: Ministry of Health Malaysia, Hospital Raja Perempuan Zainab II, Hospital Kuala Krai, Hospital Pasir Mas, Ministry of Higher Education (MyBrain15),Universiti Tun Hussein Onn Malaysia (UTHM) and facilities. This paper also was partly sponsored by the Centre for Graduate Studies UTHM.

# Thermal characteristics of a residential house in a new township in Johor Bahru

**Y Y Lee**[1], **Y H Lee**[2], **S Mohammad**[2], **P N Shek**[3] and **C K Ma**[2]

[1] Jamilus Research Centre (JRC), Faculty of Civil and Environmental Engineering, Universiti Tun Hussein Onn Malaysia, 86400 Parit Raja Batu Pahat, Johor, Malaysia.
[2] Faculty of Civil Engineering, Universiti Teknologi Malaysia, 81310 Johor Bahru, Johor, Malaysia.
[3] Construction Research Centre (UTM-CRC), Universiti Teknologi Malaysia, 81310 Johor Bahru, Johor, Malaysia.
Corresponding author: yhlee@civil.my

**Abstract.** As one of the densest population occurred in Johor Bahru, development of a new township become necessary in order to cater this growth. Rapid urbanization is a factor of degradation of environmental quality which may increase surrounded temperature through urban heat island (UHI) phenomena. Therefore, this paper is investigating thermal comfort for a new township, Bandar Dato Onn in Johor Bahru, in order to understand the thermal characteristics of a new township and for further mitigation of UHI phenomena. Surface temperature, ambient temperature and relative humidity have been collected from a terrace house of Bandar Dato Onn. Thermal characteristics with the features of time lag and decrement factors were discussed based on the collected data. The time lag of 5 to 8 hours and the decrement factors of 0.02 to 0.06 were recorded during a five-day measurement. It was concluded that human thermal comfort may not be achieved as the heating, ventilation and air conditioning (HVAC) system needed for bedroom and living room. Passive energy design is suggested to be included in the house in order to have human thermal comfort while reducing the electricity consumption towards sustainable future.

## 1. Introduction

As population of Johor in Malaysia increased from 3.46 to 3.66 million [1], between 2012 and 2016, development of new township should be launched to cater the population growth. Johor Bahru is the capital of Johor which consists of the highest population density among districts of Johor.

Urban heat island (UHI) has been identified to be likely occurred in dense developing countries where most of the South-east Asia countries are facing rapid change of natural resources and environment. This change is associated with urbanization, industrialization and economic development [2]. A dramatic growth of urban population is associated with serious degradation of environmental quality [3] which may link to UHI phenomena, increasing surrounded temperature.

Human thermal comfort is one of the considered features in the selection of living criteria. Outdoor thermal comfort may induce good social relationship among the residents where social activities can be conducted frequently. Moreover, the usage of electricity can be reduced if the indoor thermal comfort is achieved. Global warming and rapid urbanization are the main sources of increment of temperature for urban thermal environment [4]. Some mitigations have been suggested to improve human thermal comfort, such as increasing vegetation surface coverage and shading effects for Dutch city, Rotterdam [4].

Thermal bioclimate and its influence as urban design factor have been discussed in Cuba [5]. The obtained results may provide information to create space for tourism and usable in the reconfiguration of cities [5]. Thermal comfort surveys have been conducted in the city of Belo Horizonte, Brazil to evaluate thermal comfort conditions and identify potential thermal adaption processes [6]. Behavioural adaptation is suggested to be considered in designing strategies for urban planning [6].

Thermal comfort is one of the considered factors in urban planning and development by avoiding UHI phenomena for developing countries. Therefore, this paper investigates the thermal characteristic of a terrace house in Bandar Dato Onn Township where climatic data will be collected for further analysis. Suggestions have been made in order to achieve human thermal comfort in this new city development.

## 2. Data collection

Surface and ambient temperatures have been recorded inside and outside the house. The surface temperature was collected through thermocouple wire type-T with Graphtec Midi data logger, whereas, the ambient temperature was obtained by using HOBO data logger.

### *2.1. Location selection*

The data collection was performed in a residential terrace house of Bandar Dato Onn. Bandar Dato Onn is one of the latest township development in Johor Bahru to cater the population growth. The location of the site investigation was set at a residential house of Bandar Dato Onn.

### *2.2. Site setup*

The experimental setup is shown in figure 1 where the red line was the location of the measurement of surface temperature. The red lines or measured walls were located at Bedroom 1 and living room where direct sunlight was radiated on the morning and afternoon respectively. The data have been collected for five day period for further thermal performance analysis.

**Figure 1.** Floor plan of the site.

## 3. Results and discussion

### *3.1. Ambient temperature and relative humidity*

Surface temperature, ambient temperature and relative humidity have been recorded and analysed for human thermal comfort study. Figure 2 showed the ambient temperature of living room, bedroom 1, bedroom 2 and upstairs foyer. Bedroom 1 has a constant ambient temperature as it was not occupied and the door was closed all the time. Although it was around 29°C to 30°C, the room was isolated from the temperature variation from nearby space. For bedroom 2, it was occupied and air-conditioning was operated on the night time where the temperature was kept under 25°C. The temperature could rise up to 31°C at night time unless air-conditioning was set on. Ambient temperature at living room was recorded high in first two days where air-conditioning system was switched on for human thermal comfort. For next three days, the temperature was expected to be achieved the highest during evening time. Upstairs foyer had a normal temperature variation according to the testing period.

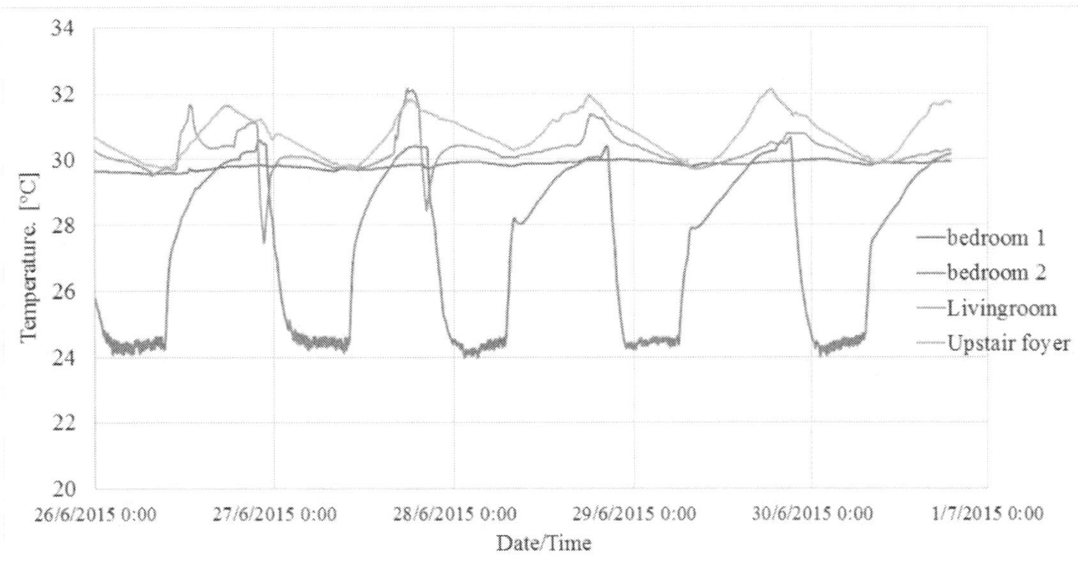


**Figure 2.** Ambient temperature for all locations.

For figure 3, relative humidity graph has been plotted. There was a drastic drop of relative humidity for living room and bedroom 2 when the air-conditioning system was operating. Other spaces were recording the indoor relative humidity ranged 65 to 80% during the test period. Rain may not affect much on indoor relative humidity.

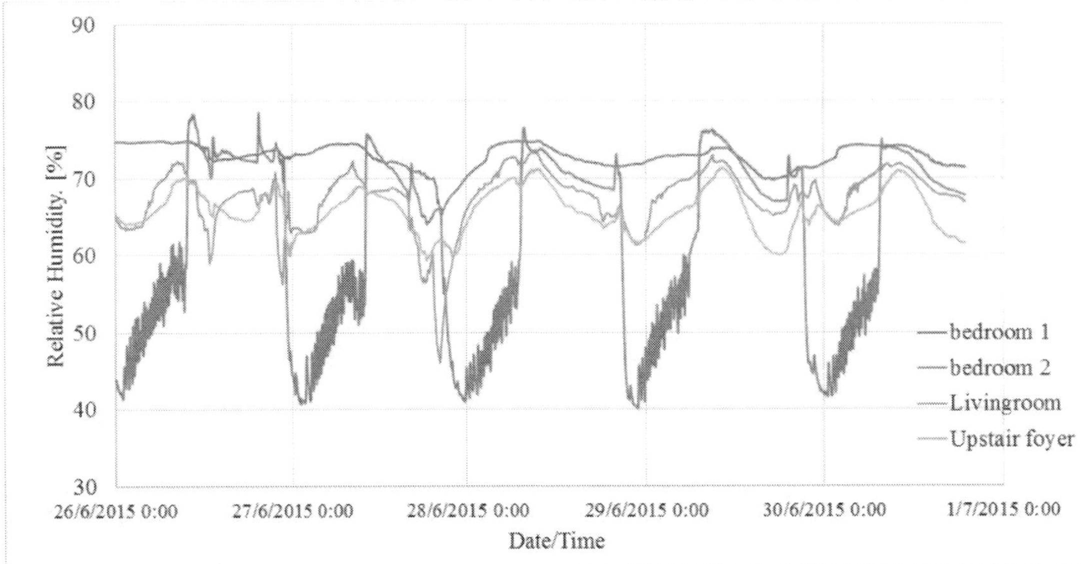


**Figure 3.** Relative humidity for all locations.

### *3.2. Surface Temperature*

The surface temperature and relative humidity of bedroom 1 were shown in figure 4. The outside wall surface temperature obviously fluctuated as compared to inside wall surface temperature and ambient temperature. The relative humidity was inversely varied with outdoor surface temperature. The outdoor temperature fluctuation was doubtedly influenced to the indoor environment. For the inside walls, when facing to the East, left indicated beside with bathroom, whereas, right indicated the wall shared with next house. Both surface temperatures have almost constant variation throughout the testing period.

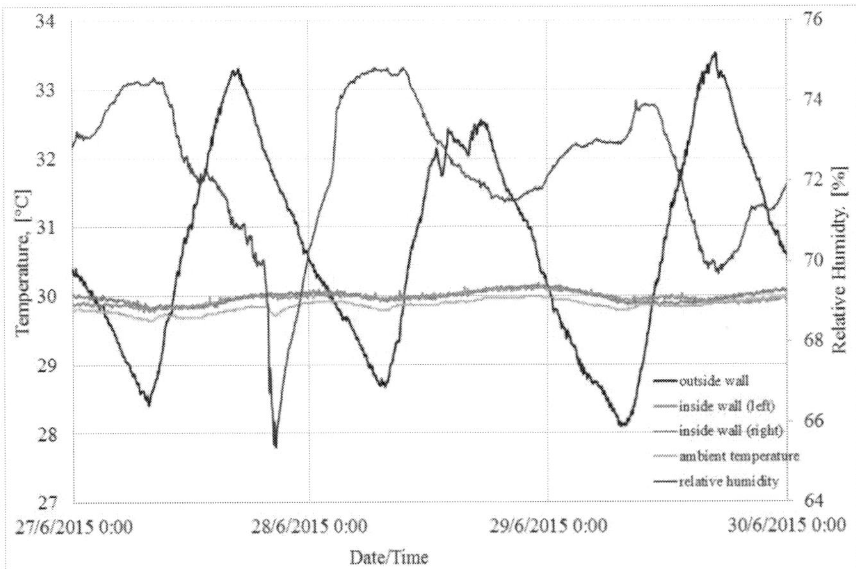


**Figure 4.** Recorded temperatures and relative humidity for bedroom 1.

According to figure 5, the surface temperature and relative humidity are shown for the living room. The investigated wall at outdoor environment consists of two, where "far" indicated wall at car porch (refer Figure 1) and near indicated wall at main entrance. Higher temperature was recorded at outer wall at car porch as there was no shading and exposure to direct sunlight in the afternoon. On the other hand, the wall at main entrance showed slightly lower temperature than at car porch at the evening, but exhibited higher temperature at midnight until the next morning. This correspond to the location of the measurement taken where the wall at car porch re-emitted the heat to the surrounding environment by convection process due to lower temperature at outdoor to achieve thermal equilibrium. However, the area of the main entrance was enclosed by the car porch that might trap the heat within the area causing less heat released to the outer environment and obtained a higher surface temperature. Meanwhile, ambient temperature and inside wall surface temperature behaved constantly during the testing observation.

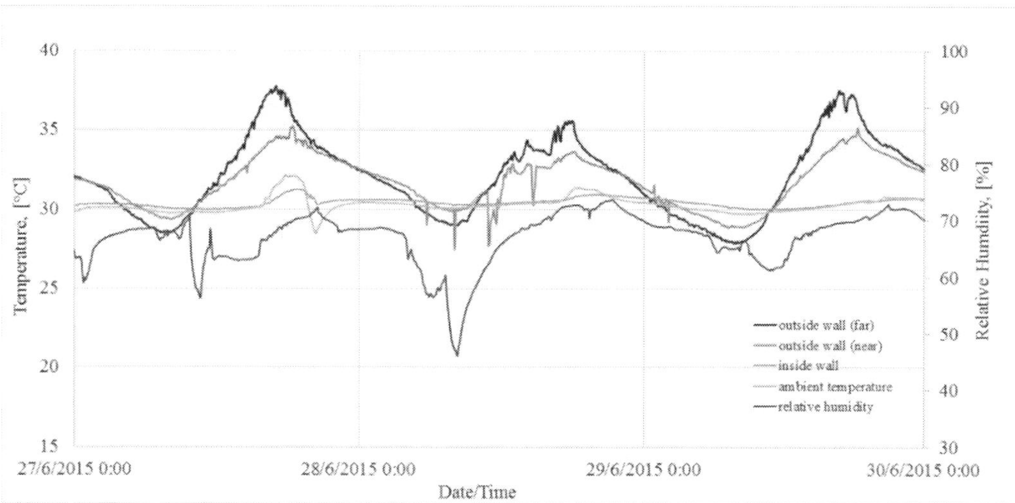


**Figure 5.** Recorded temperatures and relative humidity for living room.

## 4. Discussion

Time lag and decrement factor are two characteristics to determine the heat storage capabilities of any materials [7]. Time lag is referred as the times used for heat wave to propagate from outer to inner surfaces, whereas, decrement factor is decreasing ratio of its amplitude during the heat propagation [7].

The thermal variations between indoor and outdoor surface temperatures were ranged 2.37°C and 3.4°C for bedroom 1 as shown in figure 6. Living room was not included in the analysis due to the non-appropriate temperature distribution of indoor wall surface temperature. Time lag was measured the time taken between outdoor maximum temperature to indoor maximum temperature. For bedroom 1, the recorded time lag was ranged 355 to 530 minutes. The time lag showed a 5 to 8 hours of new amplitude to be occurred.

Table 1 shows decrement factor for bedroom 1. The decrement factor included the materials of paint, concrete cover and brick. The small indoor amplitude variation contributed to lower decrement factors. Another factor that contributing to the decrement factor was the applied HVAC system at the testing period.

From the analysis, HVAC system was still needed in day 1 and 2 for the space of bedroom where the heat from the outer surface was released to inner space at night time. Although the decrement factor was low, however, human thermal comfort may not be achieved. Therefore, it is suggested that the passive energy design should be applied in order to achieve thermal comfort and reduce the electricity consumption. Further detailing on the passive design should be investigated in the next stage of study.

**Table 1**: Temperature variation and decrement factor for bedroom 1.

| Exp. Day | Outer | | Inner | | Temp. amplitude | | $f = A_i / A_o$ |
|---|---|---|---|---|---|---|---|
| | Max. | Min. | Max. | Min. | $A_o$ | $A_i$ | |
| 2 | 33.07 | 28.47 | 30.04 | 29.81 | 4.60 | 0.23 | 0.05 |
| 3 | 33.28 | 28.67 | 30.01 | 29.92 | 4.61 | 0.09 | 0.02 |
| 4 | 32.51 | 28.09 | 30.13 | 29.86 | 4.42 | 0.27 | 0.06 |
| 5 | 33.48 | 28.51 | 30.07 | 29.91 | 4.97 | 0.16 | 0.03 |

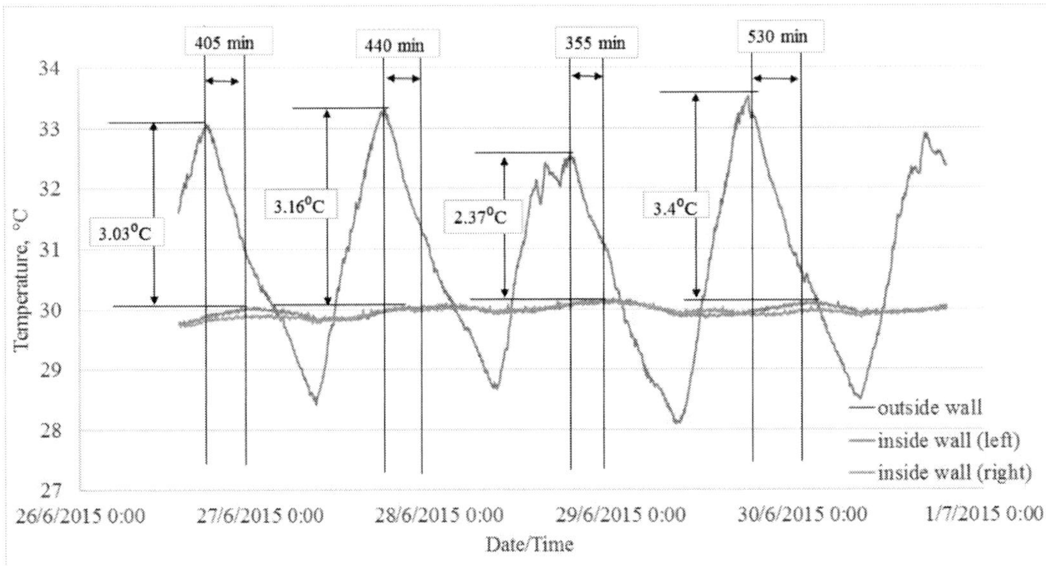


**Figure 6.** Thermal variations of indoor and outdoor surface temperatures and the time lag.

## 5. Conclusion

A thermal characteristic of a terrace house of a newly developed township in Johor Bahru, Bandar Dato Onn, has been performed. Surface temperatures, ambient temperatures and relative humidity were collected. From the collected data, the time lag of 5 to 8 hours and decrement factors of 0.02 to 0.06 have been recorded for bedroom 1. Although low decrement factors have been found, the human thermal comfort may not be achieved for day 1 and 2, as HVAC system was still needed for living room and bedroom 2. Passive energy design is recommended for better human thermal comfort and reduction of electricity consumption. This study was carried out in one of the residential house where larger data collection around Bandar Dato Onn area (residential and commercial areas) is suggested for better representation of the thermal characteristics of this new township.

## 6. References

[1] Department of Statistics Malaysia 2017 *Johor @ a Glance* Retrieved on May 27, 2017 from https://www.dosm.gov.my/v1/index.php?r=column/cone&menu_id=d1dTR0JMK2hUUUFnTnp5WUR2d3VBQT09

[2] Das M K and Karmakar S 2015 *Urban Heat Island Assessment for a Tropical Urban Air-Shed in Bangladesh* Retrieved on May 27, 2017 from http://www.meteo.fr/icuc9/LongAbstracts/poster_1-12-7881645_a.pdf

[3] Roth M 2007 Review of urban climate research in (sub) tropical regions, *Int. J. Climatol.* **27** 1859-1873
[4] Dou Y 2014 *The Influence of Urban Planning on Urban Thermal Comfort* Master thesis. (Netherlands: Wageningen University) pp 1 - 41
[5] Algeciras J A R, Coch H, Perez G D P, Years M C and Matzarakis A 2016 Human thermal comfort conditions and urban planning in hot-humid climates – The case of Cuba, *Int. J. Biometeorol.* **60** 1151-1164
[6] Hirashima S Q S, Assis E S and Nikolopoulou M 2016 Daytime thermal comfort in urban spaces: A field study in Brazil, *Build. Environ.* **107** 245-253
[7] Asan H 2000 Investigation of wall's optimum insulation position from maximum time lag and minimum decrement factor point of view, *Energy Build.* **32** 197-203

**Acknowledgments**
The supports from Ministry of Higher Education Malaysia (MOHE) and Universiti Teknologi Malaysia (UTM) through grant 4F763 and 17H53 are gratefully acknowledged.

# Influence of children pedestrian behaviour on pedestrian space usage

**F P Makalew[1], S A Adisasmita[1], S Wunas[2] and S Hamid[1]**

[1] Civil Engineering Department, Hasanuddin University, Poros Malino Street KM 6 Gowa, South Sulawesi, Indonesia
[2] Architecture Department, Hasanuddin University, Poros Malino Street KM 6 Gowa, South Sulawesi, Indonesia
Corresponding author: fmakalew@yahoo.com

**Abstract.** School children pedestrian behaviour can be seen along their journey to and from school. Pedestrian spaces used by children are places available in urban and rural areas including streets with and without pedestrian pathways. Samples data are collected from 23 elementary schools in urban and rural areas in North Sulawesi, Indonesia in the form of video records and photos taken. The aim of this research is to analyse children pedestrian behaviour and its influence on the space usage on pedestrian areas. Method of analysis is a comparative study on urban and rural areas. Results of this research are types of behaviour, factors that influence the behaviour, physical condition of pedestrian areas and space usage by children. The behaviours are duck-line walking, running, playing, walking backward and walking with bare foot in which running is the main behaviour. These behaviours are influenced by factors including following friends and responding to acts. There are similarities and differences between pedestrian space usage in urban and rural areas. Space use by children pedestrian demonstrates the way pedestrian areas should be planned. Space usage by children pedestrian indicates that there is a need of evaluation of the space available considering pedestrian children behaviour.

## 1. Introduction

Pedestrians require access to their destination. Children pedestrians use pathways available when walking to and from school. Easy access is important in helping children to arrive at their destination safely and comfortably. An adequate pedestrian pathway is important including its supporting facilities. Previous studies show that existing pedestrian pathways in city of Manado, Indonesia are not well maintained and most streets lack adequate facilities for pedestrian [1, 2]. On the other hand, children use pedestrian spaces as places for more than just walking areas. Their behaviour can be seen along their walking journey to and from school. They use any available access on the street with or without pedestrian pathways. Space usage is important in planning urban and rural public spaces and access to places. This condition raises question on the influence of children behaviour on the pedestrian space usage. This research focuses on children pedestrian behaviour in school areas and their surroundings. The types of children pedestrian behaviour and material use on pedestrian pathways are revealed in this research. Moreover, factors that influence their behaviour and the pattern of space usage are analysed using comparative analysis. Pedestrian areas in elementary schools in urban and rural areas are chosen based on preliminary study on potential children pedestrian available and fair distribution of data in study area.

The location for this research includes 23 elementary schools in North Sulawesi province in Indonesia. Samples distribution can be seen on figure 1 and 2.

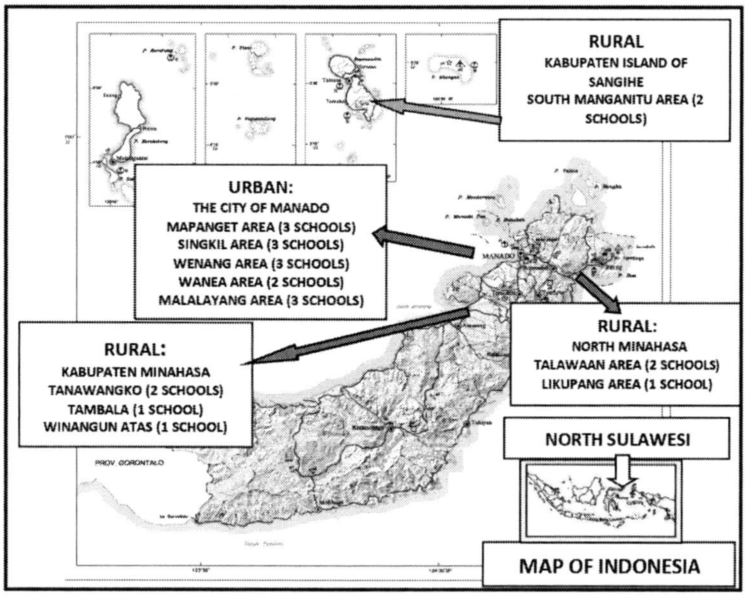


**Figure 1**. Map of sample distribution in North Sulawesi.

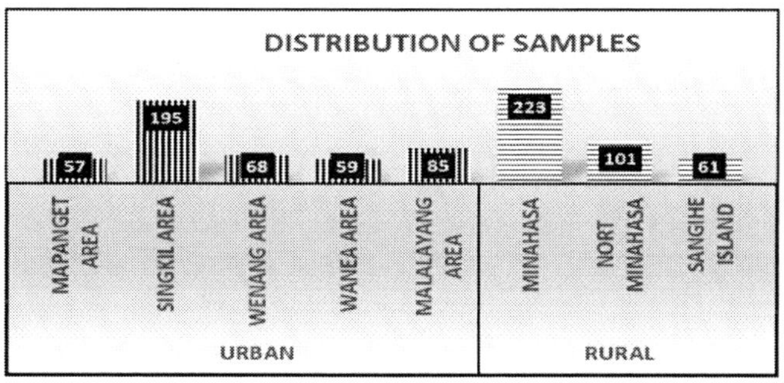


**Figure 2**. Distribution of samples.

## 2. Children pedestrian behavior

Previously researches on pedestrian behaviour are including adult pedestrian and children pedestrian behaviour. Study on the influence of pedestrian behaviour to space usage in recently is related to adult pedestrian. Pedestrian behaviour is any act and decision by pedestrian when walking through pedestrian areas. Space usage is the area used by children and the way they use. Research on model of pedestrian behaviour can be seen in model of children behaviour in urban settlements [3], special character of modifying micro model of social force, evaluation of pedestrian behaviour model based on human factor considering relation between pedestrian and the environment [4]. Evaluation on pedestrian behaviour in rural and urban areas is the evaluation on environment and its close relation with area capacity as pedestrian pathway [5]. Moreover, pedestrian behaviour is an important factor in measuring the level of walking in the environment in urban and rural areas [5]. Behaviour based on factors influence walking areas including location and access [6].

Pedestrian behaviour in different pedestrian areas such as in affordable housing and medium housing has been compared by Azmi et al. [6] in terms of the use of facilities by the community. Research shows that facility is used more often in pedestrian areas in affordable housing than those in medium housing. Recommendation for this study is to create flexible access, maximise community facilities and access standard. Public facilities can be accessed by walking are schools, playgrounds and local shops [6]. School is considered as the characteristic of environment in micro system that should be varied [7].

Children behaviour can be seen in the way they use natural elements [7, 8]. Research on children sense of place in New Mexico involving 8 to 9 years old children in three communities show gradient between city and village [8]. For children in cities and villages, exploration has similar functions. In rural areas, children with more responsibility in collecting wood for fire reveal their trip to the mountain and find special places. Children in the urban areas find their special places near their house. Children who explore, expose their freedom, control and personal needs. Playing is one of the children' behaviours which has benefit [9] and function approach [10]. Children's behaviour can be seen in their activities in rural areas [11, 12], cities and suburbs [11]. Children interaction with the environment includes playing to prepare children to be adult and apply their roles [13]. Behaviour as one of children pedestrian characteristics in urban and rural areas in relation to the environment shows differences and similarities [8, 11].

Crossing area is one of children pedestrian areas where we can see their behaviour [14, 15]. The examples of this behaviour are children taking risk [16] and safe time selected when crossing [17]. Study on children taking risk in the neighbourhood crossing area used simulator test where the sample is 10-year-old children [16]. Other study found that children aged 5 to 9 years old are the riskiest as pedestrians in crossing [17]. In the pedestrian space usage, Monteiro et al (2012) study the factors to understand and measure pedestrian space. In this study, indicators in measuring pedestrian space usage include facilities and accessibility [18].

## 3. Research method

Source of data is from survey on 23 elementary schools in urban and rural areas in North Sulawesi province, Indonesia. Videos and photos recorded from 14 schools in urban areas and 9 schools in rural areas. Sample data is elementary students from grade 1 to 6 aged between 6 to 12 years. Preliminary research has been done to find potential pedestrian available in schools selected. Areas of research is walking area at school and its surrounding area. Walking distance is between the boundary of school and its adjunct street to maximum 1000m from school area. Average walking distance for this research is 500m. Urban area is the capital city of North Sulawesi province, the city of Manado. Five areas in Manado have been the research location namely Mapanget, Singkil, Wenang, Wanea and Malalayang with fair distribution of school's location. Rural area is chosen based on government regulation for rural areas in North Sulawesi. Rural area is defined as area of villages or perdesaan under the area of Regency (Kabupaten). Locations for Kabupaten are chosen after preliminary research for potential children pedestrian available in the area. From many rural areas, three Regencies have been chosen namely Minahasa Regency, North Minahasa Regency and Island of Sangihe Regency. Number of sample is counted based on Krejcie Norman formula. Based on this formula and its table the number of sample for 183,167 students in rural area and 98,686 students in urban area [19], there is a need of minimum 383 sample data for each area. Based on survey data available are 464 students in urban areas and 385 students in rural areas. Method of analysis used is a comparative analysis of pedestrian behavior in urban and rural areas.

## 4. Result and discussion

Results of this research are similarities and differences between children pedestrian behaviours in urban and rural areas. The comparison of these areas is analysed in terms of number of children pedestrian groups, types of children pedestrian behaviours, factors that influence children behaviour, types of pedestrian surface material and pattern of pedestrian usage areas. From sample data in urban areas there are 194 students or 41.81% from the total sample and from sample data in rural areas there are 207 students or 53.77% from the total sample.

The lowest number of children walking is 1 child and the highest is group of 7 children. Figure 3 shows the difference between number of children pedestrian behaviour in urban and rural areas. Children pedestrian behaviours in both areas are duck-line walking, running, playing and backward walking. Duck-line walking is an act where a child walks followed by one or more children and shapes a line. This name of behaviour is taken from the way how group of duck walk.

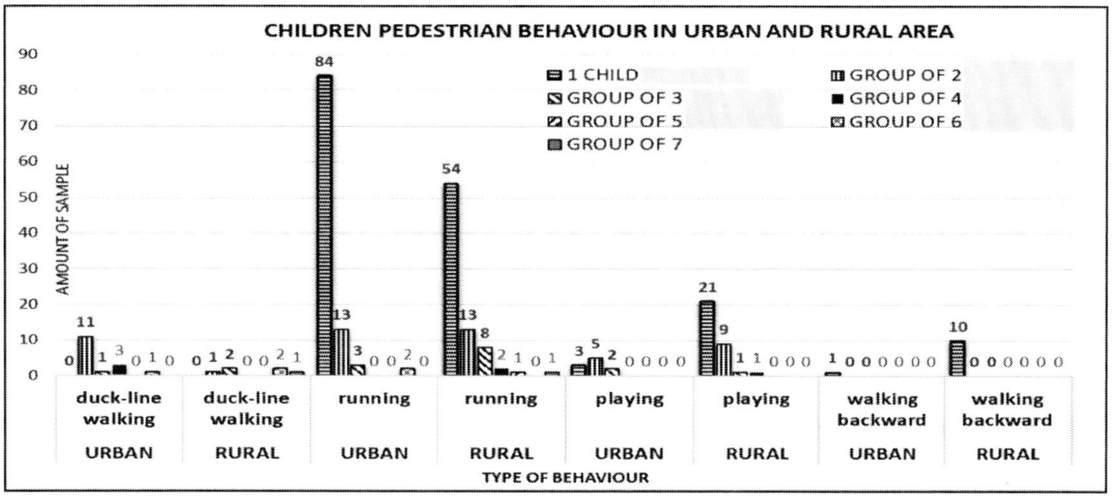


**Figure 3**. Children pedestrian behaviours in urban and rural areas.

Factors that influence children pedestrian behaviours include responding to do the behaviour by himself or herself, following friends' behaviours, avoiding vehicle traffic and crossing streets.
The main children pedestrian behaviour in urban and rural areas is running which can be seen in figure 4. Responding to running is an act to run by a child without being influenced by other children. This act is the main reason for children to run in both areas.

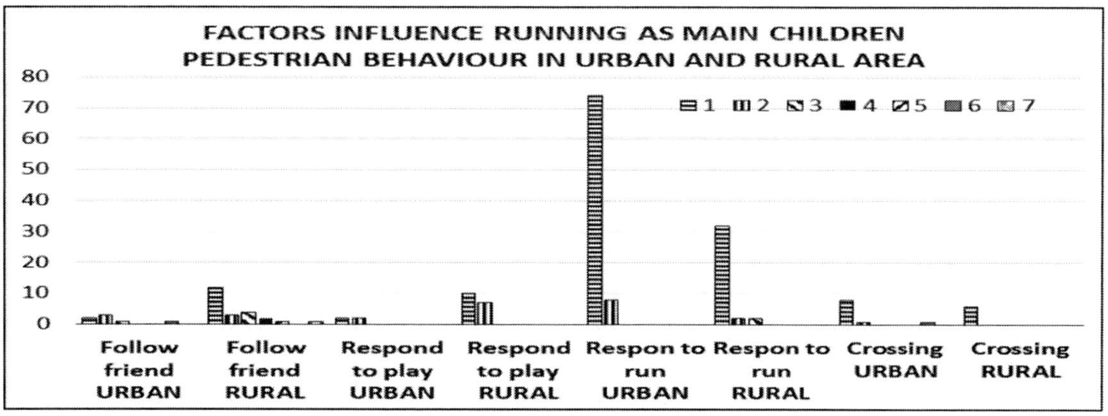


**Figure 4**. Running as main children pedestrian behaviour in urban and rural areas.

The physical condition of pedestrian area includes material type of pathway surface such as asphalt, concrete, grass and soil. The number of children pedestrian using pathway based on type of material on surface can be seen from figure 5.

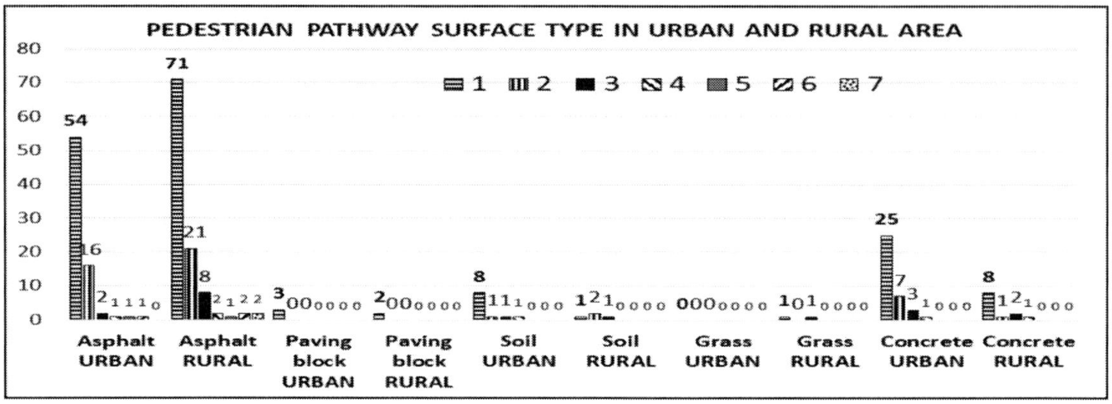


**Figure 5**. Pedestrian pathway surface type in urban and rural area.

Based on data provided, there are similarities and differences between children pedestrian behaviours in urban and rural areas. The comparison is based on the types of behaviour, pathway surface type and obstacle along the journey as shown in table 1.

**Table 1**. Similarities and differences of children pedestrian behaviour in urban and rural area.

| NO | ITEM | SIMILARITIES | DIFFERENCES | |
|---|---|---|---|---|
| | | | URBAN | RURAL |
| 1 | PATHWAY SURFACE TYPE | | | |
| a | Asphalt | V | | |
| b | Paving block | V | | |
| c | Soil | V | | |
| d | Grass | | X | V |
| e | Concrete | V | | |
| 2 | OBSTACLE ON PATHWAY | | | |
| a | Puddle | | V | X |
| b | Tree roots | | X | V |
| c | Broken concrete | | X | V |
| d | Open Concrete drains | V | | |

Notes: V = Available. X = Not Available

The similarity of behaviour type in urban and rural areas is in both areas all behaviour types are available although the number of children is different. In terms of pathway surface material type and the similarity of both areas are most types available. The difference is that grass is not available in urban areas. Material asphalt is the highest due to the use of street as pedestrian access. Children pedestrians use streets as pedestrian pathways because the pathways for pedestrians are unavailable or inaccessible. Inaccessibility can be caused by the pedestrian pathway that is used for planting trees, tree pots and other materials. The similarity of the obstacle in pedestrian pathways is that both areas have open concrete drain.

Factors that influence children pedestrian behaviour are varied based on different behaviours which can be seen from table 2. Children behaviour of running is influenced by responding to running. Responding to running is an act by children themselves to run. The cause of children to run is a response to reach the school faster in the morning due to time limitation, a response to reach places such as shops and homes faster in the afternoon and no other friends who influence them to run.

Responding to running is not necessarily one child's act but also two or more children without intending to follow friends. Following friends is an act to do the same activity such as duck-line walking, running, playing and walking backward.

Local behaviour is the behaviour which occurs considering its uniqueness and limitation in a certain place yet have a great influence on the children pedestrian space use. Local behaviour found in this research area is the use of object to play during their walking to and from school and the use of any available facilities along the street. Objects use can be divided into manmade objects and objects from natural environments. Manmade includes objects, toys, and instruments such as flutes and used objects such as bike wheels and mineral water bottles. Natural objects include parts of tree such as leaves, branches and roots and other objects such as stones. Facilities provided along the street are used by children to play such as bridge and water tap in rural areas. The example of local behaviour is a child in a school in a rural area in Winangun Atas running home from school while playing bike wheel and roll it with stick. This is a response to play during his journey home. He uses the local street available as there is no pedestrian pathway. Other example is the use of tree leaf as an umbrella to cover his head during walking. More varieties of local behaviour during walking can be found in rural areas. Children in these areas tend to play more during their journey with larger number of students comparing with those in urban areas.

**Table 2**. Similarities and differences of factors children pedestrian behaviour in urban and rural area.

| NO | ITEM | | SIMILARITIES | DIFFERENCES | |
|---|---|---|---|---|---|
| | BEHAVIOUR TYPE | FACTORS | | URBAN | RURAL |
| 1 | Duck-line walking | Following friend | V | | |
| | | Avoiding vehicle | | X | V |
| 2 | Running | Following friend | V | | |
| | | Responding to play | V | | |
| | | Responding to run | V | | |
| | | Crossing | V | | |
| 3 | Playing | Responding to play | V | | |
| | | Following friend | | X | V |
| 4 | Walking backward | Responding to play | V | | |
| | | Following friend | | X | V |
| 5 | Walking Barefoot | Responding to play | V | | |

Notes: V = Available. X = Not Available

Space use based on children pedestrian behaviour shows similarities and differences in urban and rural areas as can be seen on figure 6 and 7. Space usage in this research is a micro area of pedestrian pathway. The areas include pathways, road side and streets. The similarities of space usage in urban and rural areas are the behaviour of duck-line walking, running and playing on streets without pathway and playing on local streets. The areas used are roadsides of local and secondary streets for duck-line walking, the area of local and secondary streets for running and playing and the area of street for local street without roadside.

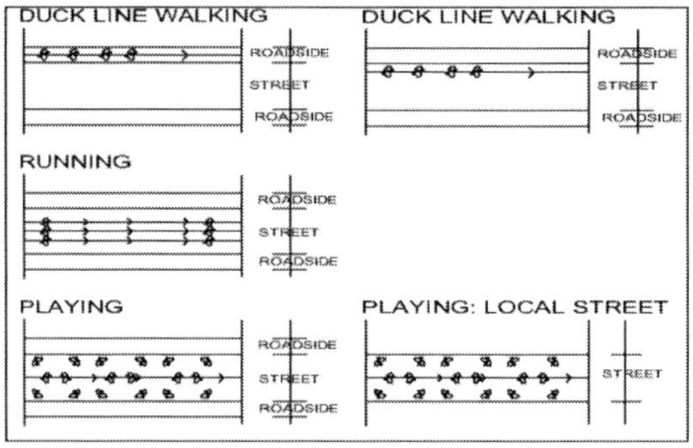


**Figure 6**. Similarities of space usage based on children behaviour in urban and rural area.

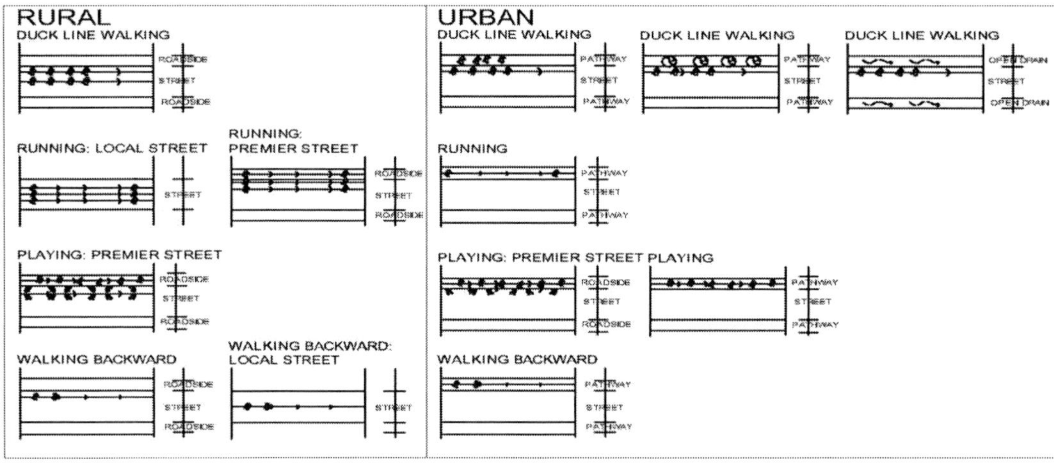


**Figure 7**. Differences of space usage based on children behaviour in urban and rural areas.

The differences of space usage in duck-line walking are in rural areas children pedestrian walk in two lines while for those in urban areas this act is not available. In urban areas, children walk duck-line at the edge street with pathways. They do not use the pathway because the area is occupied by large trees. Children also use the edge of street with pathway due to limited space of pathway for sharing with another pedestrian. Moreover, the edge of street is used by children for streets with open drain.

In the behaviour of running, the differences of space usage between urban and rural areas are that in rural areas children use streets in local streets without roadsides and the roadsides and half of the street for premier streets while in urban area children use pathways for premier streets.

The differences of space usage in playing are that in rural area children use roadside areas and up to half of the street in premier streets while in urban area children use roadside areas and the edge of the street for premier streets and use pathway for streets with pathways.

The differences of space usage in walking backward are that children use the edge of the street for streets with roadsides and the middle of street for local streets without roadsides or pathways while in urban areas children use pathways for streets with pathways.

From the evaluation of similarities and differences of space usage by children pedestrians, different types of behaviour influence the area used by children on different types of streets. On behaviour of running and playing, the use of space is larger due to the use of any available space on streets. On this type of street, there is no pathway therefore children use street as their space for running and playing. Children pedestrian behaviour must be considered in planning streets and pedestrian areas. Spaces used by children are influenced by type of behaviour, number of group pedestrians, the way children face the obstacles on pedestrian areas and surface material. From all obstacles faced by pedestrian children, the similarity is children use open drain as area of walking by stepping on the edge of it as part of playing. In the areas where there is no pathway for pedestrians, children use streets as wide as possible in all types of streets from local to premier.

## 5. Conclusion

Children pedestrian behaviours in urban and rural areas in North Sulawesi Indonesia show differences and similarities and their influence on the space usage. All types of behaviours are walking duck-line, running, playing and walking backward can be seen in both areas. Spaces in pedestrian pathways are not just for children to walk but also for their activities. The activities are more varied for those in rural areas. Based on the children pedestrian behaviour and how they use the spaces available, there is a need for planning a good pedestrian space and its facility to accommodate children pedestrian behaviour in urban and rural areas. Pedestrian pathways should be planned to accommodate children activities. Planning for pedestrian area in urban and rural area should consider the differences between urban and rural areas. From the children behaviour differences and similarities, areas required for children pedestrians in urban and rural areas, there is a need further investigation considering space usage with different numbers of children in a group.

## 6. References

[1] Makalew F P 2014 Toward great streets using urban design approach, the case study of Manado *The Int. Conf. on Environmentally Friendly Civil Engineering Construction and Materials (Manado, Indonesia)* (Indonesia: Civil Engineering Department Sam Ratulangi University) pp 385 - 394

[2] Makalew F P 2014 Identifikasi fasilitas jalan perkotaan: tantangan jalan kaki sebagai transportasi ramah lingkungan, studi kasus kota Manado Urban street facility identification: a challenge for walking as environment friendly, case study of Manado *Pros. Hasil Penelitian Terapan (Manado State Polytechnic)* vol 2 (Indonesia: Unit Penelitian Politeknik Negeri Manado)

[3] Rezasoltani M, Mostafa B and Said I 2015 A model development for children's walking in neighborhood *Procedia - Social and Behavioral Sciences* vol 201 (United Kingdom: Elsevier) pp 30 – 38

[4] Papadimitriou E, George Y and John G 2009 A critical assessment of pedestrian behaviour models *Transp. Res. F-Traf.* **12** 242–255

[5] Azmi D I, Karim H A and Amin M Z M 2012 Comparing the walking behaviour between urban and rural residents *Procedia - Social and Behavioral Sciences* vol 68 (United Kingdom: Elsevier) pp 406 – 416

[6] Azmi D I and Karim H A 2012 A comparative study of walking behaviour to community facilities in low-cost and medium cost housing *Procedia - Social and Behavioural Sciences* vol 35 (United Kingdom: Elsevier) pp 619 – 628

[7] Kytta M 2006 Environmental child-friendliness in the light of the bullerby model *Children and Their Environments, Learning, Using and Designing Spaces* ed C Spencer and M Blades (Cambridge: Cambridge University Press) chapter 9 pp 141 - 158

[8] Derr T 2006 Sometimes birds sound like fish': perspectives on children experiences *Children and Their Environments, Learning, Using and Designing Spaces* ed C Spencer and M Blades (Cambridge: Cambridge University Press) chapter 7 pp 108 - 123

[9] Lester S and Maudsley M 2007 The values and benefits of children's play in natural settings *Play, Naturally, A Review of Children's Natural Play* (England: National Children's Bureau Enterprises Ltd) chapter 3 pp 23 - 47

[10] Heft H 1988 Affordances of children's environments: a functional approach to environmental description *Children's Environments Quarterly*, **5** 29-37

[11] Kytta M 2002 Affordances of children's environments in the context of cities, small towns, suburbs and rural villages in Finland and Belarus *J. Environ. Psychol.* **22** 109 - 123
[12] Yatiman N A, Aziz N F and Said I 2012 Affordances of homeschool journey in ruralenvironment for children's performances *Procedia-Social and Behavioral Sciences* vol 68 (United Kingdom: Elsevier) pp 395 - 405
[13] Mrnjaus K 2014 The child's right to play?!, *Croat. J. Educ.* **16** 217-233
[14] Geßler A L, Peter H, Sara K, Rene K, Xue H L, Qi H S, Johanna B, Zhi J and Guan Z Y 2013 Crowd research at school: crossing flows, *Traffic and Granular Flow* **13** 137-144
[15] Kadali B R and Vedagiri P 2013 Modelling pedestrian road crossing behaviour under mixed traffic condition, *European Transport Trasporti Europei* **55** 1 - 17
[16] Charron C, Aurélie F and Nicolas G 2015 Do child pedestrians deliberately take risks when they are in a hurry? An experimental study on a simulator, *Transp. Res. F-Traf.* **15** 635–643
[17] Meir A, Parmet Y and Oron-Gilad T 2013 Towards understanding child-pedestrians' hazard perception abilities in a mixed reality dynamic environment, *Transp. Res. F-Traf.* **20** 90–107
[18] Monteiro F B and Campo V B G 2012 A proposal of indicators for evaluation of the urban space for pedestrians and cyclists in access to mass transit station *Procedia - Social and Behavioral Sciences* vol 54 (United Kingdom: Elsevier) pp 637 – 645
[19] Biro Pusat Statistik (BPS) Statistic of North Sulawesi Province 2016 Social *North Sulawesi in figures* (Indonesia: BPS-Statistics of North Sulawesi Province) chapter 4 pp 81 - 142

IOP Conf. Series: Materials Science and Engineering **271** (2017) 012029 doi:10.1088/1757-899X/271/1/012029

# Effect of urban noise to the acoustical performance of the secondary school's learning spaces – A case study in Batu Pahat.

**Y G Tong[1],Abu Bakar H[1],Mohd. Sari K A[2], Ewon U[3], Labeni M N [1] and Fauzan N F A[1]**

[1]JamilusResearch Centre, Faculty of Civil and Environmental Engineering, University Tun Hussein Onn Malaysia, Batu Pahat, Johor, Malaysia
[2]Department of Civil Engineering Technology, Faculty of Technical Engineering, University Tun Hussein Onn Malaysia, Batu Pahat, Johor, Malaysia
[3]Kota Bharu Polytechnic, Ketereh, Kelantan, Malaysia
Corresponding author:ygtong@uthm.edu.my

**Abstract.** Classrooms and laboratories are important spaces that use for teaching and learning process in the school. Therefore, good acoustical performances of these spaces are essential to ensure the speech or message from the teacher can be delivered to the students effectively and clearly. The aims of this study is to determine the acoustical performance of the teaching and learning spaces in public school that situated near to the traffic roads. The acoustical performance of the classrooms and laboratories at Sekolah Menengah Kebangsaan Convent Batu Pahat was evaluated in this study. The reverberation time and ambient noise of these learning spaces which are the main parameters for classroom design criteria were evaluated. Field measurements were carried out inside six classrooms and four laboratories unoccupied furnished according to the international standards. The acoustical performances of the tested learning spaces were poor where the noise criteria and reverberation times inside the measured classrooms and laboratories were higher than recommended values.

## 1. Introduction

Urban noise especially that generated from road traffic is the major noise pollutant that has affected surrounding populations in many cities. This environmental noise annoyance bring outs various effects to the human daily activities and health [1]. Many social-acoustic surveys were carried out over the century to evaluate the effects of traffic noise exposure to the nearby residents. In several studies, traffic noise is found to be the most critical effects on human sleep disturbances [2–5] and could lead to deterioration of subsequent daytime life quality such as tiredness, sleepy and low working performance [6]. Excess exposure to long-term high level of traffic noise can increase the risk to health diseases [7–9] as well as health problems.

From the social survey in London, the critical threshold for acceptable noise level based on the residents live near the traffic roads was range between 65 to 75 dB(A) $L_{10}$[10]. WHO has recommended the limit of noise level ($L_{eq(A)}$) outside dwellings for daytime to be 55dB(A) while 45 dB(A) is suggested as the limit during the night time to protect people from being seriously annoyed [11]. In Malaysia, the guideline for limiting sound level from road traffic has been set for different categories of receiving areas in order to protect the population from this environmental noise [12]. For the noise sensitive areas such as institutional or schools, Malaysian Department of Environment [12] had limit the permissible sound levels outside these areas to the 50 dBA during the daytime from 7am to 10pm. However, there were still many institutions in Malaysia are facing with this urban noise problem [13,14]. Mohmadisa *et. al.*[14] had studied traffic noise levels outside five selected school environments in Batu Pahat. The results from their study showed that the noise level outside all studied school areas were exceed the permitted level that set by Department of Environment.

Although traffic noise measurement outside school environment can be used for assessing the urban noise impact to the children in the school, but it cannot be used to evaluate the efficiency of teaching and learning processes. The efficiency of the communication in classroom is depends on the acoustic conditions of the classrooms. High levels of noise in the classrooms will affect the concentration of the students. Excessive noise levels in the school area may affect the psychology and physiology of both teachers and students [15]. It can affect communications and more extreme can default hearing. Students will feel difficulty in paying attention and understanding the teaching information that given by their teachers [16]. Better acoustical design of learning spaces will ensure the efficient distribution of desirable sound inside the rooms. That's mean when teachers are giving instructions to the students in the classrooms, the messages can be delivered to the students effectively and clearly. A good quality of sound in the classroom is importance for students' audio comfort. Recent survey showed that higher physiological stress reaction happened on teachers and students who were exposed to noisy classrooms [17].

The acoustical performance of the classrooms very depends on the background noise and reverberation time of the spaces [18]. Both parameters are very importance in evaluate the speech intelligibility of the classroom. When the background noise of the classroom is high, teachers have to rise up their voice in order to increase the clarity of their speech. The contributions of background noise of a classroom include mechanical ventilation system, lighting systems and outdoor noises such as transportation, construction works, student activities at school field, neighboring classrooms and others. The reverberation time value of the classrooms should not be too long in order to avoid a masking effect.

Noise criteria (NC) curves are common noise evaluation curves that used for rating the noise inside the indoor environment. The recommended NC level for a classroom is NC 25 to NC 30 [18]. For the reverberation time, the recommendation values were different based on difference country or standards with difference derivations [19–21]. In overall, these standards or guidelines recommended the reverberation time of unoccupied classroom should be below 0.8 second.

A recent study showed that the traffic noise levels outside the school environment of Sekolah Menengah Kebangsaan (SMK) Convent, Batu Pahat was about 70 dBA during day time for both working and non-working days [14]. The noise levels about 20 dBA exceed the recommended value set by Malaysian Department of Environment. Thus, investigation on the effects of this urban noise to the acoustical performance of the teaching and learning spaces of this school is needed. This study aims to determine acoustical performance of the learning spaces in high school that located next to heavy traffic roads. The background noise and reverberation time inside the 6 classrooms and 4 laboratories of this secondary school were investigated.The acoustical performance of tested classrooms and laboratories were compared with the recommended values.

## 2. Learning spaces and measurements

Field measurement was conducted at SMK Convent, Batu Pahat during the school holiday on weekday and office hours. The location of this school was next to heavy traffic road of Jalan Tan Swee Hoe as shown in Figure 1 with the latitude of 1°51'45" U and longitude of 102°57'0" T. A total of ten learning spaces in this school were surveyed in the present study. Six unoccupied furnished classrooms and four laboratories at different location of the school environment were measured as shown in Figure 2. The example of indoor environment of the tested learning spaces are illustrated in Figure 3.All the classrooms that chosen for the present study were located at the first floor while all laboratories were located at the ground floor of the buildings. Wooden desks and plastic chairs were used in all classrooms in this school while tables and chairs in laboratories were made of wood and finished with paint. The classroom can occupied about 30 students while each laboratory can occupied about 40 students. Table 1 summarizes the measured learning spaces conditions and information which related to the acoustical performance of the spaces. Natural ventilation with ceiling fans were used in all classrooms and laboratories in this school.Since the layouts of the learning spaces in this school were similar, the materials, sizes and number of windows and doors were not present in details.

**Figure 1.**Outside view of the studied area.

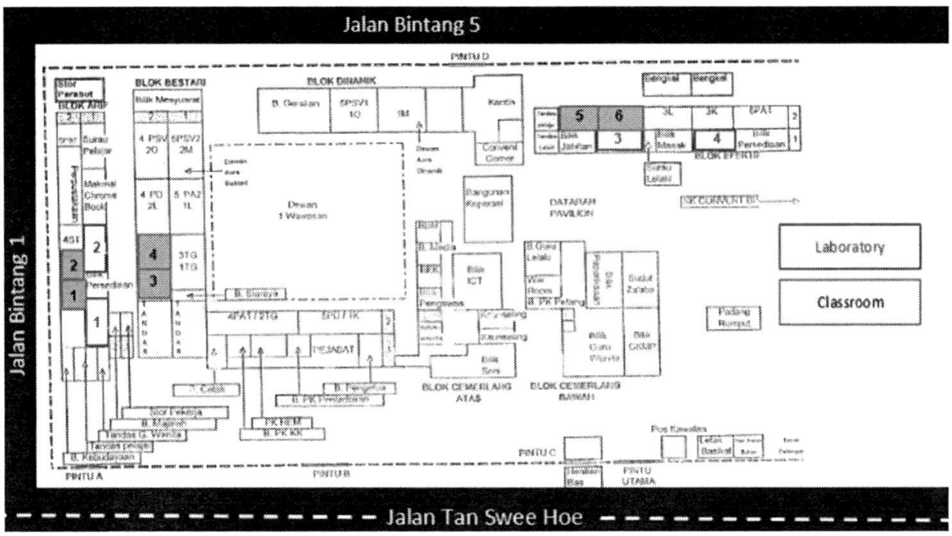


**Figure 2.** Measured learning spaces (classrooms and laboratories).

(a) (b)

**Figure 3.** Indoor environment of the (a) classroom and (b) laboratory.

**Table 1.** Summary of learning spaces information and conditions.

| Learning spaces | Dimension (length X width X height) | Desk/ Table | Chairs | Partition/ wall between the spaces | External landscape |
|---|---|---|---|---|---|
| **Classroom 1** | 9.0 m X 7.0 m X 3.4 m | Wooden | Plastic | Wooden | Light traffic and heavy traffic |
| **Classroom 2** | 9.0 m X 7.0 m X 3.4 m | Wooden | Plastic | Wooden | Light traffic |
| **Classroom 3** | 9.0 m X 7.5 m X 3.6 m | Wooden | Plastic | Concrete | Garden |
| **Classroom 4** | 9.0 m X 7.5 m X 3.6 m | Wooden | Plastic | Concrete | Garden |
| **Classroom 5** | 9.0 m X 7.0 m X 3.4 m | Wooden | Plastic | Wooden | School's field and light traffic |
| **Classroom 6** | 9.0 m X 7.0 m X 3.4 m | Wooden | Plastic | Wooden | School's field and light traffic |
| **Laboratory 1** | 12.0 m X 7.5 m X 3.0 m | Wooden | Wooden | Concrete | Light traffic and heavy traffic |
| **Laboratory 2** | 12.0 m X 7.5 m X 3.0 m | Wooden | Wooden | Concrete | Light traffic |
| **Laboratory 3** | 12.0 m X 8.8 m X 3.6 m | Wooden | Wooden | Concrete | School's field and light traffic |
| **Laboratory 4** | 12.0 m X 8.8 m X 3.6 m | Wooden | Wooden | Concrete | School's field and light traffic |

A sound level meter was used in the study to capture the background noise levels and reverberation time inside the classrooms and laboratories. For the background noise measurement, a measurement point was sufficient to capture the ambient noise of the enclosed space. Reverberation time is depend on the volume and sound absorption of the room. Since the room layout and dimension for some classrooms were same, only 3 classrooms were tested to examine the sound decays of the rooms. Reverberation time measurements were carried out according to the ISO 3382 [22] using Brüel&Kjær Type 2250 sound level meter, Brüel&Kjær Type 4260 omni-directional sound source and Brüel&Kjær Type 2734 amplifier. In classrooms, a total of 4 measurement points were chosen to capture the reverberation times of the room while 5 measurement locations were set in the laboratories to record the sound decays. The sound level meter was put at different heights, 1.2 meter and 1.6 meter above from the floor, and at least 1 meter away from all walls in order to capture the reverberation time of whole enclosed space and avoid the reflections sound from the walls.

## 3. Results and Discussions

### *3.1. Background noise / Noise Criteria (NC)*

Figure 4 shows the comparison of background noise between six classrooms at one-third octave band central frequencies. From the graph, the background noise at classroom 1 and classroom 2 were slightly higher than other classrooms. This may due to the facts that these classrooms were located next to the residential road and main road of Jalan Tan Swee Hoe. Combination of the traffic noise from both roads made the background noise at classroom 1 and classroom 2 get higher value around 8 dB to 14 dB compared to the classroom 3 and classroom 4 which is only exposed to the traffic noise from the main road only as shown in Figure 2. From the graph, the lowest frequency in each of the classroom give the higher value of the background noise. This is affect from the traffic noise which produce lower frequencies of sound.

The background noise of all tested laboratories was shown in Figure 5. From the graph, the data obtained from laboratory 4 were slightly higher at range of frequency 800Hz to 1.25 kHz octave bands. Higher value of background noise obtained in this laboratory may due to the facts that the distance between the laboratories 4 with the school field was very near which was approximately 4.6 m from laboratory wall. Besides, there were some students playing softball at that school field during the field measurement was conducted. One can conclude that, the background noise of the spaces inside the school not only depend on the traffic noise but also depends on the noise from the school field. Thus, the teaching and learning process in the laboratories can be interrupted by the noise produced from the school field.

For the spaces like laboratories or classrooms, the recommended Noise Criteria (NC) value is between NC 25 – NC 30 [18]. However, the results obtained in this secondary school showed that all measured learning spaces were exceed the recommended value. This indicated the urban traffic noise and surrounding noise gave the great impact to the learning spaces indoor environment.

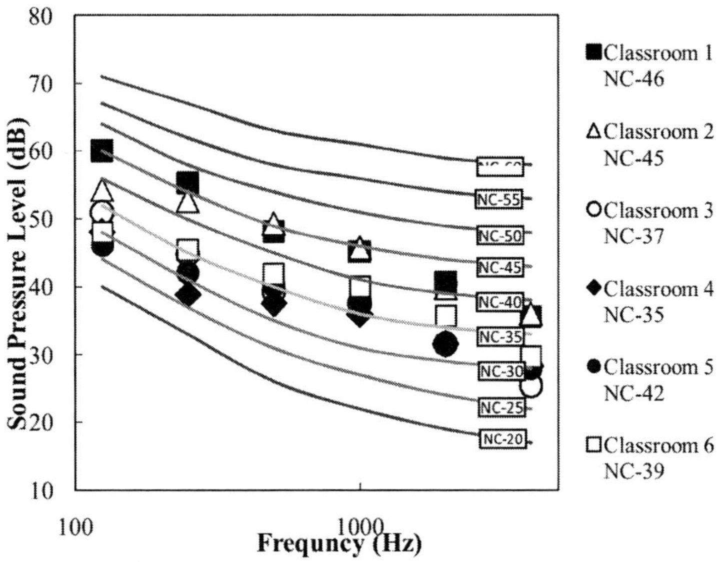


**Figure 4.** Background noise (NC curves) of measured classrooms.

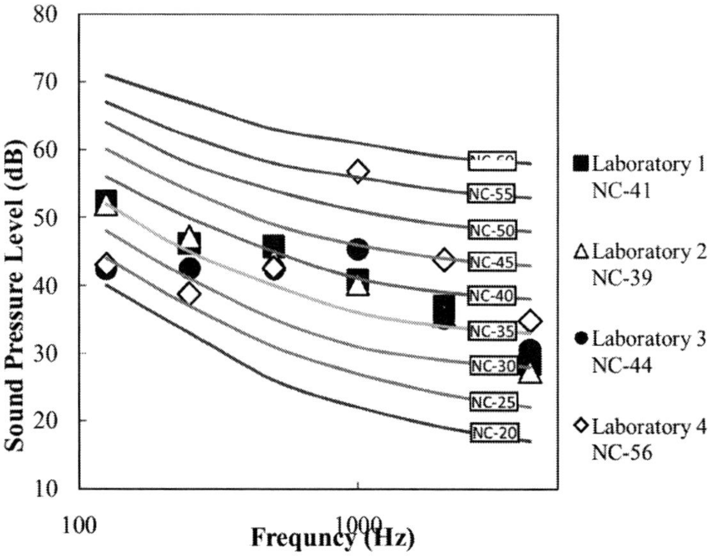


**Figure 5.** Background noise (NC curves) of measured classrooms.

### *3.2. Reverberation time*

Figure 6 shows the average of reverberation time for classroom 2, classroom 4 and classroom 6. From the graph, classroom 4 obtained higher reverberation time compared to other classrooms. This may due to facts that volume of this classroom was bigger than other two classrooms where classroom 2 and classroom 6 have the same design and space volume of 214.20 $m^3$ while classroom 4 has the space volume of 244.35 $m^3$. Besides, the higher value of sound decays recorded in classroom 4 probably because of the condition inside the classroom which had poor sound absorption. Based on the field observation, there was no curtain and notice board inside this classroom.

Based on recommended value, reverberation time for room with volume of $200m^3$ - $300m^3$ should be below 0.8 second, but almost all measured classrooms had the longer sound decay than it should be except classroom 2 at 4 kHz and 5 kHz. The speech intelligibility in these spaces can be said was very poor especially classroom 4 which obtained longer reverberation time than a classroom should be.

Figure 7 illustrates the average sound decays for all measured laboratories. At the lower frequencies, the value of reverberation time for laboratory 1, 2 and 3 were equivalents which are in range 1.0 s to 1.7 s and the time for sound to decays for laboratory 1, 3 and 4 were rose sharply between 125Hz to 200Hz. This may due to time decays at those octave band frequencies influenced by the multi reflections inside the laboratories. Based on 4 tested laboratories, laboratory 1 and 2 obtained lower reverberation time because the volume of these two laboratories were smaller than that volume of laboratory 3 and 4. Besides, there were some wooden furniture that placed inside these two laboratories which can increase the total absorb of the spaces and this made the sound decays become shorter. According to the results obtained from this study, the overall reverberation time values for all laboratories were higher than recommended value.

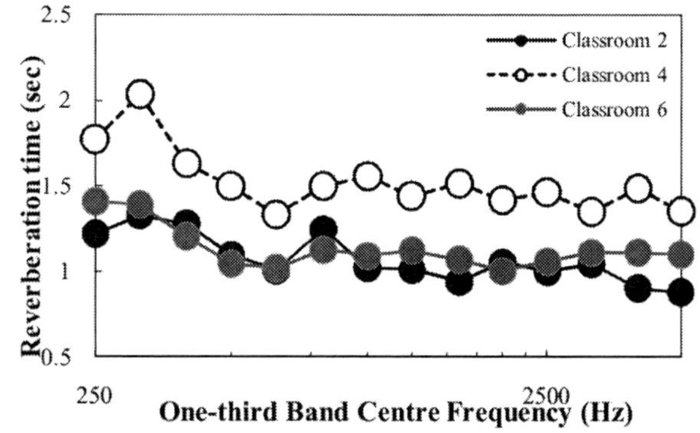


**Figure 6.** Average reverberation time in tested classrooms.

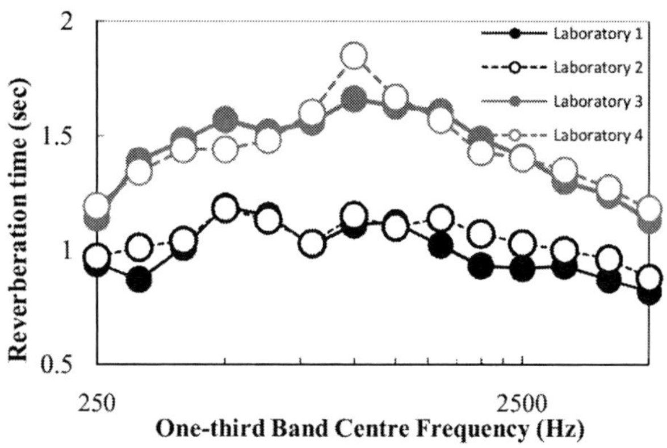


**Figure 7.** Average reverberation time in tested laboratories.

## 4. Conclusions

Field measurement on acoustical performance of classrooms and laboratories in a public high school in Batu Pahat was carried out. Two importance parameters (reverberation time and background noise) for evaluating the audio comfort of the classroom was tested. A total of 6 classrooms and 4 laboratories were tested for its ambient noise. The background noise inside the learning spaces is influenced by the outdoor traffic noise and school's recreation activities from other students. All measured learning spaces obtained higher noise criteria (NC) levels than recommended value of NC 25 to NC30. This will create poor classroom acoustic condition when the background noise in the rooms are so high that they would interfere with learning and teaching process.

The evaluation of spaces in term of speech performance (reverberation time) for 3 tested classrooms and 4 laboratories was done in this study. The results showed that all tested learning spaces except classroom 2 exceed the recommended value of 0.8 second. This implied that the level of speech intelligibility of the spaces was low. Students may not be able to hear the instructions or massages from the teachers clearly.

As a conclusion, the acoustical performances of classrooms and laboratories in SMK Convent Batu Pahat were unsatisfied as places for teaching and learning process. This may affect the speech understanding, attention, concentration, academic achievement and also student's behavior in classroom.

## 5. References

[1] Ouis D 2001Annoyance from road traffic noise: A review *J. Environ. Psychol.* **21**(1) 101–120

[2] Robinson D W 1970 *An outline guide to criteria noise for the limitation of urban* Current Papers No. 1112 (London: Ministry of Technology)17-20

[3] Skånberg A andÖhrström E 2006Sleep disturbances from road traffic noise: A comparison between laboratory and field settings *J. Sound Vib.* **290**(1-2)3–16

[4] Öhrström E, Skånberg A, Svensson H and Gidlöf-Gunnarsson A 2006Effects of road traffic noise and the benefit of access to quietness*J. Sound Vib.* **295**(1–2) 40–59

[5] Öhrström E, Hadzibajramovic E, Holmes M and Svensson H 2006 Effects of road traffic noise on sleep: Studies on children and adults *J. Environ. Psychol* **26**(2) 116–126

[6] Muzet A 2007 Environmental noise, sleep and health*Sleep Med. Rev.***11**(2)135–42

[7] Sørensen M, Hvidberg M, Andersen Z J, Nordsborg R B, Lillelund K G, Jakobsen J, Tjønneland A, Overvad K and Raaschou-Nielsen O2011Road traffic noise and stroke: a prospective cohort study *Eur. Heart J.* **32**(6) 737–44

[8] Babisch W 2008 Road traffic noise and cardiovascular risk *Noise Health* **10**(38) 27–33

[9] Wagner J, Cik M, Marth E, Santner B I, Gallasch E, Lackner A and Raggam R B 2010 Feasibility of testing three salivary stress biomarkers in relation to naturalistic traffic noise exposure *Int. J. Hyg. Environ. Health* **213**(2) 153–5

[10] Langdon F 1976 Noise Nuisance caused by road traffic in residnetial area: part III*J. Sound* **49**(2)241–256

[11] Berglund B and Lindvall T 1995 Community noise. Archives of the Centre for Sensory Research **2**(1) 1–195

[12] Department of Environment 2004 *The Planning Guidelines for Environmental Noise Limit and Control* (Putrajaya, Malaysia: Ministry of Natural Resources and Environment) pp 12

[13] Rahim L A, Hashim M and Nayan N 2011Road Traffic Noise Pollution and its Management in Tanjong Malim, Perak *Journal of Techno-Social* **3(2)** 1-12

[14] Mohmadisa H, Hairul F M, Yazid S, Nasir N, and Mohamad S Y C N2014Analisis Bunyi Bising Trafik Persekitaran Sekolah di Bandar Batu Pahat, Johor, Malaysia *Geografi* **2(2)**66–79

[15] Yuen F K2014A vision of the environmental and occupational noise pollution in Malaysia **16(73)** 427–436

[16] Ismail M, Abdullah S andYuen F S2015Study on environmental noise pollution at three different primary schools in Kuala Terengganu, Terengganu State *J. Sustain. Sci. Manag.***10(2)** 103–111

[17] Tiesler G, Machner R and Brokmann H 2015Classroom acoustics and impact on health and social behaviour *6th Int. Building Physics Conf. (Eindhoven, Netherlands)*vol 78 (United Kingdom: Energy Procedia)3108–3113

[18] Raichel D R 2006 *The science and application of acoustics*2nd Ed. (New York: Springer)

[19] Zannin P H T and Zwirtes D P Z 2009Evaluation of the acoustic performance of classrooms in public schools *Appl. Acoust.***70**(4)626–635

[20] Sala E and Rantala L 2016Acoustics and activity noise in school classrooms in Finland *Applied Acoustics* **114** 252–259

[21] Fukuchi T and Ueno K 2004 Guidelines on acoustic treatments for school buildings proposed by the Architectural Institute of Japan *Proc. Int. Congr. Acoust.* 909–910

[22] British Standard BS EN ISO 3382-2: 2008 *Acoustics — Measurement of room acoustic parameters —Part 2: Reverberation time in ordinary rooms* (London: British Standard Institution)

# The application, benefits and challenges of retrofitting the existing buildings

**Muhammad Khairi[1], Aini Jaapar[1] and Zaharah Yahya[1]**

[1]Faculty of Architecture, Planning & Surveying, Kompleks Tahir Majid, Universiti Teknologi MARA, 40450 Shah Alam, Selangor, Malaysia
Corresponding author: mckhairi@yahoo.com

**Abstract.** Sustainable development has been the main topic of debate for years in some countries such as United Kingdom, United State of America and Malaysia. Depletion of natural resources, global warming, economics uncertainty and health issues are some of the reasons behind sustainable development movements, it is not just a political debate in the parliament but more towards collective works among sectors in order to minimizing the negative impact of development to the environment and other living organism. Retrofit an existing building is one of the solutions to reduce the dependency on constructing new buildings. There are huge numbers of existing building stocks that suitable to be retrofitted such as historical buildings, offices, residential, warehouse, factories, vacant buildings and other historical buildings. Therefore, the aim of this research is to provide information on the application, benefits and challenges of retrofitting an existing building. Two buildings were chosen as case studies following by site visits and observation to the buildings. The data were then compared in a table form. Primary and secondary sources were also used for this research. The application of retrofit should be promoted across the construction and conservation industries since it has significant tangible and intangible benefits. It is one of the most environmentally friendly and efficient solutions to optimize the energy performance and could also helps to extend the life of the existing building or historical buildings while ensuring optimum thermal comfort for the occupants which leads to higher productivity.

## 1. Introduction

The depletion of natural resources, global warming, economic uncertainty and health issues leads to sustainable development movements across the world [1]. There's a need for mechanisms or approach that can reduce the negative effects of development, construction and urbanisation to the environment. Retrofitting an existing building is one of the most environmentally friendly, sustainable and efficient solutions to optimize the energy performance of building [2-4]. As a matter of fact, when compared to new buildings construction, this kind of intervention reduces the consumption of land energy and could be applied to a large building stock [5].An in-depth interviews with 37 key players, Dixon et al., [2] found that many instances a distinction was indeed made between retrofit, where a building(s) could be refitted with relatively "light touch" energy efficiency measures, for example, whilst a tenant was still in occupation, as opposed to the case of "refurbishment" which entails a much "deeper" level of refit with changes to the internal and external fabric of the building, with the latter frequently occurring at lease renewal. However, in other cases refurbishment was used rather than retrofit [2].

National Refurbishment Centre (NRC) United Kingdom referred green refurbishments as the use of multiple sustainable fabric measures applications which are then applied to a specific space, room or even to the whole existing building, such as insulation mechanism or micro-generation solutions. Retrofit in the other hand, refers specifically to the installation of micro-generation solutions to buildings or a single measure, such as fitting insulation to an external wall, while 'renovate' is more commonly used in Europe including in the UK to specifically describe the improvement activities or repair works on dilapidated heritage buildings, which may also include the installation of new energy-efficient system or to a whole house [6].

## 2. Methods

Two buildings were chosen as case studies, which is Kuala Lumpur performing art centre (KLPac) and Shah Alam Royal Theater. Site visit and observation to the two buildings were carried out. The data were then compared in a table form followed by a discussion. Primary and secondary sources were also used for this research.

## 3. Retrofitting Mechanism and Application

Dixon *et al.*, [2] stated that retrofit measures not only focusing on energy, but also look into the usage of water and the production of waste as well. As an example, up to 30% - 40%energy cost per annum could be reduced by light-touch retrofit, such as installation of energy-efficient lighting and controls, building services and management systems and controls. In addition to that, recycling water and waste (e.g. in shopping centre, office, school and public buildings) could also bring significant and positive sustainability and cost impacts. Furthermore, research from UK by Dixon *et al.*, [2] and from USA Institute for Building Efficiency [7] also suggests the following retrofit measures are the most common in the commercial property sector:

- energy-efficient lighting and controls;
- management systems and controls; and
- building services.

However, the research studies done by Nils Kok *et al.*, [4] found out that there are other more expensive improvements and retrofitting measures such as replacing roofs, installing PV solar cells, changing floors, insulation, and operable windows or better glazing system that could be done to an existing buildings as shown in Table 1. This associate to thermal comfort of the buildings as well as making the space more usable and to improved productivity of the space.

**Table 1**. Major improvement during Retrofitting [4].

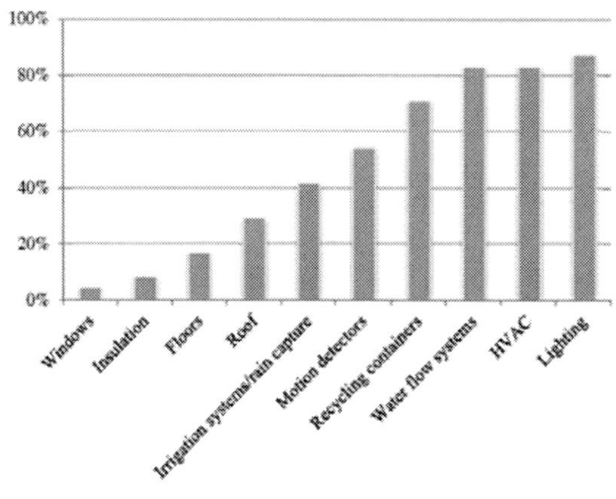

From the Table 1, window is one of the elements that were improvised during retrofitting. Estimated USD 20 billion was spent in 1990 to offset heat losses through windows in the commercial and residential buildings in United State [8]. United State Energy Information Administration, United State Department of Energy [9] estimated, 33% of the average heat losses arise through window and doors, which caused by poorly fitted and air leak. Resulting the heating system to work more in order to maintain the desire temperature in the winter, thus increasing the power consumption of the heater. Caulking and weather stripping are the easiest way to overcome air leak in windows and doors, while draperies works as thermal barrier and could help to reduce heat loss by 25%. Installing insulated window panels, windows films panels or Low-Emissivity (Low-E) windows may also help to reduce infrared (IR) heat transfer into the buildings, thus ensuring the thermal comfort level is achievable without causing extra power consumption from the heater or air-condition [9]. Windows can help minimize the cost of cooling, heating and lighting if it is selected properly based on the building typology.

In order for the carbon targets to be successfully achieved, an estimated 28 million building which includes 25 million homes from almost 30 millions buildings (domestic and non-domestic) in the UK are required to be retrofitted by 2050 [10-11]. According to UK Green Building Council, the existing buildings may requires retrofit measures to enable the occupants to live and work more sustainably. The council furthermore noted that refurbishing the existing buildings will be one of the major solutions to reduce carbon emissions [12]. The UK government has proposed an incentive for businesses (non-domestic) owners and households (domestic) owners to take up the energy efficiency measures, and offers variable stamp duty, variable council tax and an energy efficiency feed-in-tariff to create an attractive market opportunity for the industry [12].

## 4. Retrofitting Matrix

NRC also produced Retrofitting Matrix. The matrix was first developed by the East Midlands Development Agency. It is a self assessment tool that was created to help any individual or organisation to develop their own retrofit strategy or integrated and holistic approach on energy efficient projects that meet the local demands or requirement and achieving the goals of the projects. It also aimed to help the users to assess their current situation or activities and what is needed to solve problems or enhance their retrofit strategy. It highlights eight critical themes and each one of it sets out a series of suggestion for the users. The eight critical themes in the retrofit strategic matrix tools are; partnership, governance and leadership; evidence and monitoring base; finance and investment; planning, development and regulatory control; skills and knowledge; industry and business development; procument; cultural and social factors which then cross-check with five maturity level [6]. In order to develop a strategy based on the prompts, the user need to use one of the theme as a Chapter heading and the prompts in the matrix itself to both assess the current scenario of authority or partners, which then brings to critically plan what is needed to enhance and improve the current scenario for each themes.

## 5. The Benefits & Challenges of Retrofitting

The contribution of existing buildings in regards to sustainability should not be underestimate. Through retrofitting, existing buildings can be benefited for sustainability purposes hence lessening the wastage [13]. The time has come to begin concentrating on retrofitting the existing buildings [14]. Tangible and intangible benefits from greening the existing buildings could be achieved through retrofitting. The tangible benefits are any benefits or advantages that can be measured in financial terms such as cost of construction and intangible benefits are any benefits or advantages that cannot be measured by financial term but it still has significant impact to business or project such as comfort level and satisfactory level [15].

According to Abdul Qayyum *et al.*, [13] one of the advantages of executing retrofit to an existing building compared to constructing new green buildings is that, the production of associated wastes could be prevented while conserving the embodied energy. It is also a cost-effective way to greening the existing buildings by cutting down the energy consumption, cost of operation and the lifecycle cost thus creates higher investment returns to the client. As a result, retrofitting the existing buildings is a highly significant approach in achieving the aim of sustainability [13].

Another outcome of retrofit is related to productivity of building occupants. The productivity of the employee plays an important role in any organization and each company would want their employees to be productive. Surprisingly over 50% occupants who work in environmentally certified buildings found their employees to be more productive [16-17]. A survey done by Corenet Global [18] members in 2012 suggested that the tenants wanted more natural light, better ventilation and better temperature control in their buildings. These features are consistent with more sustainable and greener space [4]. Furthermore, greening the existing building through retrofitting could provide better indoor environment quality for the end users and the construction period could be completed in a shorter time compared to building up a new green buildings which takes longer time to be completed [13]. In term of equipment, retrofitting existing equipment can be a cost effective, timesaving and technically sound solution, as opposed to purchasing new equipment. In addition to that, by reusing back the existing equipment to the maximum limit, may results in lower financial expenditure compared to the installation of new equipment [19].

Energy efficiency in buildings is often reported as one of the direct economic benefits for real estate investment companies when considering energy efficiency and sustainability in their portfolios [4]. Jones Lang LaSalle [20] reports stated that, 115 office properties in its portfolio managed to improve the building energy efficiency in 2006, which leads to an average savings of USD 2.24 Mil in 2007 and USD 3.0 Mil in 2008.

CIOB through its Carbon Action 2050 [21] has listed down the benefits and the challenges of retrofitting as follows:

The benefit:

- Retrofitted buildings are more adaptable and suitable to existing activities or future activities if required, more comfort;
- Retrofitted buildings are more energy efficient, lower carbon emissions from the building operations;
- Greater sustainable use of embodied-carbon investment (capital carbon).

The challenges of retrofitting:

- Expensive and inconvenient;
- Internal spaces may reduce upon installation of internal wall-insulation;
- Might cause negative impact to heritage and archaeological assets caused by usage of unproven methods, technologies or instruments;
- Further research is needed especially on insulation mechanism on walls and the effect on retrofit on buildings fabrics;
- More education, training and activities on maintaining and preserving the buildings need to be taught to address issues and to create awareness;
- The risk of retrofitting needs to be highlighted, not just focusing on the benefits of retrofitting and discussion between retrofit and refurbishment.

Other benefits of retrofit existing buildings includes; cost saving in long run by reducing the usage of energy and water by incorporating new technology, services or equipment; increasing the comfort level in a building by redesign the façade and interior to improve end users productivity and satisfaction through improving interior thermal comfort; lowering the greenhouse emission and optimizing the water usage in building; future-proof buildings which means the building will be use and last for many years when it is equipped with current technology based; and preserving cultural and heritage significance of an existing buildings by enhancing the building exterior and interiors to meet current standards and design based on the needs of end users [22].

Although CIOB highlighted cost of retrofit as one of the challenges, it is still open for more research to be done on this. Retrofit could be a cheaper solution to some existing building; if it is compare between constructing new buildings that has same function and capacity (e.g floor area or number seats) with retrofitting the existing building. Two buildings were compared in this research, between Kuala Lumpur Performance Art Centre (KLPac) and Shah Alam Royal Theater, followed by site visit and observation. The two buildings were selected due to the same typology, total number of seats, its usage and the location which both are situated in Klang Valley. Data collected were compared as shown in the Table 2 below, noted that the research was unable to get detail data on the cost and pricing of construction phases such as earthwork, machineries and fittings due to private and confidential information by the developer.

**Table 2**. Comparison between KLPac and Shah Alam Royal Theater.

| Building | KLPac | Shah Alam Royal Theater |
|---|---|---|
| Construction Type | Retrofit existing railway warehouse | New |
| Typology | Theater | Theater |
| Seats | 794 [23] | 800 [25][26] |
| Construction Cost (MYR) | 30 Mil [23] | 44 Mil [25][26] |
| Construction Period | 15 months [23] | 44 months [25][26] |
| Spaces<br><br>KLPAc<br><br>Shah Alam Royal Theater | *Pentas 1* [24]*:*<br>A 504-seat proscenium theatre (fixed seating)<br>*Pentas 2:*<br>A 190-seat experimental black box theatre (modular seating)<br>*Indicine:*<br>A 100-seat flexible space<br>*The Actors Studio Academy @ klpac:*<br>9 studios for rehearsals & training<br>*Supporting facilitie:*<br>Holding Room, Resource Centre, Box Office, Pentas 1 Foyer Bar, Pentas 2 Foyer Art Gallery, Surau, Merchandise Counters, Workshop, Green Room for Artistes, Wardrobe & Laundry Room, Costume & Props Store, Technical, Training Studio, Security Office, Conference Room, Administration Office & Production Offices | Main stage with 800 seats, easy-to-move stage, conference rooms, multipurpose hall, cafeteria, exhibition hall, management office and surau [26] |

There is a significant cost saving of MYR 14 Million on retrofitting the existing building based on the 2 buildings above. KLPac project also managed to significantly saves construction time by 29 months earlier as compared to construct a new building, which has identical same numbers of seating; KLPac with 794 seats and 800 seats for Shah Alam Royal theater.

The reason behind the time and cost saving factor was because, KLPac reused back some of its original structure sand fittings from the existing railway warehouse which includes the original brick walls and large doors that can still be seen clearly until today. The original structures were carefully chosen based on its rigidity, strength and suitability to be integrated with the new design and additional spaces. The new 'add-on' spaces were carefully design by the in-house architect from YTL Sdn Bhd in order to intergrade with the existing buildings while maintaining the architectural balance of design which includes material selections and the proportionate scale between the original structures and additional structures. New column were added to support the additional new floors, new roof, new electrical-mechanical equipment, fittings and finishes. Natural lighting and natural ventilation were used as the main source of lighting and ventilation in KLPac but the Pentas area, toilet, surau remain using the electrical-power lighting and only the Pentas were equipped with centralised air-condition which the unit located at the back of KLPac while the other active spaces like the office and studios was equipped with the split unit air-condition which can be control accordingly. The optimisation of natural lights and natural ventilation in KLPac was possible due to the unique design features from the architect and the usage of large uv-coated swing windows that can be open from inside, thus permits natural air to enter the buildings. This will helps to minimize the operational cost of KLPac in long run. Interestingly, railway woodblock and bricks from the original railway warehouse were also being reused for KLPac landscape, thus creates unity between the buildings and the surrounding landscape.

Different scenario with Shah Alam Royal Theater where it was built from ground up from its first project inception resulting higher embodied energy through out the construction period of 44 months. The whole building was equipped with centralized air-condition unit and electrical lighting units while only the lobby area used large uv-coated window that permit natural lighting. With this it may result in a higher operational cost of the buildings in term of electrical power usage since the building depending on the air-condition unit to ensure the thermal comfort for the occupants.

The cost-effective approach to turn existing buildings into green buildings requires consideration of various factors and aspect of complexities such as technology, funding, law and knowledge [13, 27, 28, 39]. Although there are available support from the government such as Green Building Index certification for existing building, National Policy on Climate Change, Green Technology Financing Scheme for retrofitting, and MS1525:2007; a Malaysian code of practice on energy efficiency (EE) and renewal energy (RE) [30] to encourage retrofitting the existing buildings in Malaysia, but still not many clients had taken an extra mile to grab the opportunity and experience the benefits of retrofitting to their buildings. This unfortunate event caused by the lack of awareness among professionals and developers in the construction industries, thus resulting in the slow progress and reluctance of retrofitting [31].

Dixon et al., [2] has found that the inflexible regulations, complicated validating and approval process for new technologies, selective authorized lists of technologies are all seen as aspects that can reduce the entry of new retrofit products into the market. Furthermore the insignificant payback periods in funding, contracting arrangements, difficulties with the process of validating and approving technologies were also identified as problem that hinder retrofitting technology [2]. Therefore, the vast stock of existing buildings in Malaysia need to be retrofitted if the Malaysia building sector wanted to achieved sustainability by sustaining the building energy consumption and finding ways to generate and sustain more energy from renewable energy sources that later could use for the buildings [13].

Even though retrofit bring benefits to the buildings, environment and end user of the buildings, the planning of making good the existing buildings such as the historical buildings need to be done with manner and respect to its authentic building elements and the surrounding community. Fail to do so may result on negative impacts such as damaging the most authentic building elements of the buildings or creating nuisance to the community. Nuisance to the local community may occur if the newly retrofitted buildings crossed the line of private privacy. As an example, the residents of residential building, Neo Bankside which situated besides Tate Modern Museum London, launched a legal bid in London High Court to force the Tate Modern shut its newly open Switch House viewing platform due to constant surveillance from the visitors and which some then post photographs film of their homes on social media which breaching the European Convention of Human right, respect for private and family lives and their homes [32].

Thus it is very important for the architects, planner and conservator to work closely with each other and understand the needs of local community and gets feedback from them before construction started. Any information will then help the designer to design a building that has consideration and respect to the surround community especially on humans privacy, avoiding any negative impact socially and economically in future to both sides.

## 6. Suggestion

Retrofitting the existing buildings is one of the most environmentally friendly, economical competent and proven as an efficient solutions to optimize the energy performance and could also helps to prolong the life of the existing building especially to the historical buildings. Thus, the application of retrofit should be promoted across the construction and conservation industries. More research need to be done in order to have complete sets of detail data on the direct and indirect impacts of retrofit to the environment, cost differences between retrofit with the normal construction of a building, cost of maintenances as well as, the impacts to the end users and to the surround area of retrofitted buildings.

## 7. References

[1] Oh T H , Pang S Y and Chua S C 2010 Energy Policy And Alternative Energy In Malaysia: Issues And Challenges For Sustainable Growth *Renewable and Sustainable Energy Reviews* **14** 1241-1252
[2] Tim Dixon Commercial 2014 Commercial Property retrofitting: What does "retrofit" mean, and how can we scale up action in the UK sector*? J. of Property Investment and Finance* **32(4)** 443 – 452
[3] Kamarudin K, Harun R, Jaapar A and Yahya Z 2013 Retrofitting As Environmental Hybrid Approach (EHA) In Conservation Works On Historical Buildings In Malaysia *Computational Methods in Applied Sciences* (Germany: Springer) pp 153-158
[4] Kok N, Miller N and Morris P 2012 The Economics of Green Retrofits *J. of Sustainable Real Estate* **4(1)** 4-22
[5] Boeri A, Gabrielli L and Longo D 2011 Evaluation and Feasibility Study of Retrofitting Interventions on Social Housing in Italy *Procedia Engineering* (United Kingdom: Elsevier) **21** pp 1161–1168
[6] National Refurbishment Centre (NRC) 2011 Retrieved on June 15, 2017 from *http://www.rethinkingrefurbishment.com*
[7] Institute for Building Efficiency 2013 *Energy Efficiency Indicator Survey* (Washington: Institute for Building Efficiency)
[8] Pisupati S 2017 *Windows and Heat loss, EGEE102* (The Pennsylvania State University: Energy Conservation and Environmental Protection. Penn State College of Earth and Mineral Sciences)
[9] United State Department of Energy 2008 *Improving the energy efficiency of existing windows* Retrieved on November 1, 2017 from https://energy.gov/energysaver/articles/improving-energy-efficiency-existing-windows
[10] EHA 2010 *The Existing Homes Alliance 2010 Manifesto* Retrieved on November 1 2017 from http://assets.wwf.org.uk/downloads/existing_homes_alliance_manifesto_2010.pdf
[11] CIOB 2011 *CIOB Buildings under refurbishment and retrofit* (UK:The Chartered Institute Of Building) pg 1
[12] United Kingdom Green Building Council (UKGBC) 2015 Retrieved on Mei 15, 2017 from http://www.ukgbc.org/resources/key-topics/new-build-and-retrofit/retrofit-domestic-buildings
[13] Nazria A Q, Mohammadb I S, Babab M, Zainol N N, Lokman M A A, Woon N B and Ramli N A 2015 The Need For Retrofitting To Achieve Sustainability of Malaysian Buildings *Journal Teknologi* **75(10)** 171-176
[14] Yu S M, Tu Y and Chenxi L 2011 *Green Retrofitting Costs and Benefits: New Research Agenda* Working Paper Series (Singapore: National University of Singapore)

[15] Ludwig Kleynhans, 2016 *What are the tangible and intangible benefits of ERP system?* Retreived on Jun 15, 2017 from www.linkedin.com/pulse/what-tangible-benefits-erp-system-ludwig-kleynhans
[16] Miller N G, Pogue D, Gough Q D and Davis S M 2009 Green Buildings and Productivity *J. of Sustainable Real Estate* **1** 65–89
[17] Miller E and Buys L 2008 Retrofitting Commercial Office Buildings for Sustainability: Tenants' Perspectives *Journal of Property Investment & Finance* **26**(6) 552-561
[18] Corenet Global 2012 *Corporate Occupier Sustainability Perspectives 2012* (San Diego: CBRE and The University of San Diego)
[19] Phoenix Electric Corporation (PEC) 2015 Retrieved on June 150 2007 from *http://www.pec-usa.biz/page144.html*
[20] Jordan M, McCarty T and Velo B 2009 The Performance Measurement Challenge *Journal of Corporate Real Estate* **11(2)** 106-114
[21] Chartered Institute of Building (CIOB) UK 2015. *Carbon Action 2050 (CA 2050): Building Under Refurbishment & Retrofit* Retrieved on November 1, 2017 from *http://www.carbonaction2050.com/*
[22] Raniga U 2017 *Five benefits of retrofitting buildings for sustainable future* Thesis (Australia: RMIT University)
[23] The Star 2007 *Salvage Art Sunday* Retrieved on October 28, 2007 from http://www.malaysiapropertynews.com/2007_10_28_archive.html
[24] KLPac 2017 *Facilities* Retrieved on Mei 10, 2017 from at *http://www.klpac.org/about-klpac/facilities/*
[25] Selangor Symphony Orchestra 2017 Retrieved on June 10, 2017 from http://sso.org.my/index.php/2016/03/19/theatre-shah-alam/
[26] The Star 2008 *Shah Alam Royal Theater.* Retrieved at *http://www.thestar.com.my/news/community/2008/10/24/rm43mil-stateoftheart-theatrecomplex-opens-in-state-capital/*
[27] Kamarudin K, Yahya Z, Jaapar A and Harun R 2015 The Options Of Project Financing And Funding For Retrofitting The Historical Building In Malaysia *Australian J. of Basic and Applied Science* **9(7)** 179-181
[28] Dixon T, Britnell J and Watson G B 2014 *City-Wide' Or 'City-Blind?' An Analysis Of Emergent Retrofit Practices In The UK Commercial Property Sector* Project report EPSRC Retrofit2050 (UK: University of Reading)
[29] Cantell S F and Huxtable A L 2005 *The Adaptive Reuse of Historic Industrial Buildings : Regulation Barriers , Best Practices and Case Studies* Master Thesis (Virginia: Virginia Polytechnic Institute and State U)
[30] Green Building Index 2009 *Ms1525:2007 Clause 5: Building Envelope.* Retrieved on November 1, 2017 from *http://www.greenbuildingindex.org/Resources/20090214%20-%20GBI%20MS1525-2007%20Seminar/20090214%20-%20GBI%20MS1525-2007%20Seminar%20(VKL)%20Presentation.pdf*
[31] Mohd R E, Mohd A M, Rostam Y, Ahmad A H, N H N R and Hamimah A 2011 Obstacles in Implementing Green Building Projects in Malaysia *Australian Journal of Basic and Applied Science* **5(12)** 1806-1812
[32] Colin Marrs, Architect Journal 2017 *Neo Bankside residents head to court over Tate Modern extension*. Retrieved on June 15, 2017 from https://www.architectsjournal.co.uk/news/neo-bankside-residents-head-to-court-over-tate-modern-extension-nuisance/10019120.article

**Acknowledgement**
Acknowledgment goes to Ministry of Education Malaysia through Fundamental Research Grant Scheme (FRGS) and Universiti Teknologi MARA for their financial supports.

# A review on the current issues and barriers of Industrialised Building System (IBS) adoption in Malaysia's construction industry

**M A Mohd Amin[1], N H Abas[1], S Shahidan[1], M H Rahmat[1], N A Suhaini[1], S Nagapan[1] and R Abdul Rahim[2]**

[1] Jamilus Research Centre (JRC), Faculty of Civil and Environment Engineering, Universiti Tun Hussein Onn Malaysia, 86400 Parit Raja, Batu Pahat, Johor, Malaysia
[2] Faculty of Electric and Electronic Engineering, Universiti Tun Hussein Onn Malaysia, 86400 Parit Raja, Batu Pahat, Johor, Malaysia
Corresponding author:nhaslin@gmail.com

**Abstract.** Malaysia considers the construction industry as one of the main contributors to its Gross Domestic Product (GDP). However, there are some unresolved issues arising from the ongoing and widespread adoption of the conventional method of construction such as the resultant fragmentation of the industry itself; delays in production and delivery time of unnecessary wastages and lack of sustainability practice. Malaysian Government has been continuously encouraging the industry to use, partly or if not wholly, the Industrialized Building System (IBS), which is considered to be an important part of sustainable construction initiative. IBS was introduced to Malaysia as the solution to issues related to dependencies of foreign workers, raising demand for affordable accommodations and improving image, quality and productivity of construction industry. However, the IBS adoption in Malaysia remains low. This paper presents the review of the current issues and barriers of IBS adoption in Malaysian construction industry.

## 1. Overview of Malaysian construction industry

The construction industry in Malaysia is generally divided into two areas, namely general construction, which comprises residential construction, non-residential construction and civil engineering construction; and special trade works, which comprises activities of metal works, electrical works, plumbing, sewerage and sanitary works, refrigeration and air-conditioning works, planting works, carpentry, tiling and flooring works and glass works. Malaysia's construction industries have a strong influence on the country's economy and accounted for about 4.6% of economic growth in 2007 [1]. In 2009 where the global economy had to pace itself against slow economic growth, the construction sector was the only sector that plotted positive growth during every quarter of the year in Malaysia. The Construction Sector registered a strong growth of 5.8% in 2009, and subsequently 8.7% for the first quarter of 2010 as against the overall GDP growth of 10.1% during the first quarter of that year [2].

The strong economic growth has created a high demand for construction activities in Malaysia. For example, the Malaysian government has targeted 4,964,560 unit houses to be built between the years

1995-2020 [3]. This figure does not include the number of other building construction such as schools and infrastructure. The demand and consequent shortfall in labour and skills attracted a huge number of foreign workers into the country to take up employment on site as unskilled labour doing manual jobs [4]. According to the Construction Industry Development Board (CIDB) of Malaysia, 69% (552,000) out of a total of 800,000 registered workers in the construction industry (as at June 2007) were the foreign workers [5]. Regardless of the over-dependency on foreign labours, the industry is still saddled with serious problems such as low product quality, low productivity, poor image, economic volatility, bureaucratic delays, lack of ethics, shortage of skilled manpower and lack of data and information [6]. Moreover, the OHS performance of the Malaysian construction industry is poor as evidenced by its high accident and injury rates, lagging behind other industries such as manufacturing.

In addition, the huge demand for construction projects, especially building construction, has fostered a new interest in Industrialised Building Systems (IBS) to replace the traditional approaches. These approaches could save on labour, cost and time of construction, and confer quality and durability [4, 7]. However, the main aim of IBS in Malaysia is to reduce the dependency on foreign labour and save the country's loss in foreign exchange [8].

The importance of IBS implementation is highlighted in the *Construction Industry Master Plan* (CIMP 2006-2015), under the *Strategic Thrust 5* [9]. The Government of Malaysia has emphasized the full utilization of IBS for government projects by the inclusion of not less than 70% IBS component (IBS Roadmap *2003-2010*, 2003). Further to this, the IBS Roadmap 2011-2015 aims to sustain the existing momentum of 70% IBS content for public sector building projects through to 2015. Further, this new Roadmap will also impact the private sector involved in "public-private- partnership" (PPP), through which the average 50% IBS score for private projects could be achieved by 2015 [10]. Further, in a recent Construction Industry Transformation Plan (CITP) 2016- 2020, the significance of IBS is highlighted under 'productivity' key component.

## 2. Current IBS adoption in Malaysia

The construction industry has led to embrace IBS as a method of attaining better construction quality and productivity, reducing risks related to occupational safety and health (OSH), alleviating issues for skilled workers and dependency on manual foreign labour and achieving the ultimate goal of reducing the overall cost of construction [11]. Apart from this, it offers minimal wastage, fewer site materials, a cleaner and neater environment, controlled quality, and lower total construction costs [12-14]. Already utilized in Malaysia since the 1960s, IBS is the way forward for the industry stakeholders to make leaps and bounds progress in the Malaysian construction industry [11].

Despite government's initiatives in promoting and encouraging the use of IBS, the IBS adoption in Malaysia remains low. In a press statement by the CIDB General Manager of the IBS & Mechanization Division, Technology Development Sector, Noraini Bahri stated that, as at 31st December 2015, only about 42% of public projects and 70% of private projects in Malaysia use modern Industrialised Building Systems (IBS) technology [14]. Furthermore, when the target of the IBS Roadmap 2011- 2015 is to have all public projects to obtain an IBS score of 70 or more, the IBS usage in public projects fell below expectations, with only 24% of public projects valued at above RM10 million have an IBS score of 70 or more. To date, as at May 2016, Works Minister, Datuk Fadillah Yusof said that about 69% of government projects used IBS, while the adoption rate by the private sector is still low around 14%, according to CIDB's study in 2014 [15].

## 3. Issues and barriers of IBS adoption

The issues and challenges that impede the adoption of IBS in Malaysia can be found in abundance from the existing literatures. The followings describe the barriers of IBS implementation that lead to low adoption of IBS in Malaysia.

*3.1 Transition from conventional to IBS*

Despite numerous IBS benefits, Malaysian contractors were perceived as not willing to make the transition to IBS and find it easier to stick to conventional construction method [14,16,17]. Changes from conventional construction method to IBS will have an influence on the design, manufacturing and site work as the result from an increased number of prefabricated elements and other innovative solutions used. This requires changes in the overall process such as synchronization of construction, manufacturing and design [18].

Factors like project management, rationalization, standardization, repetition, collaboration, IT, lean management, integration, supply chain partnering, planning, skills and training would be essential [19]. Apart from that, utilizing IBS means there was a necessity of specialized equipment and machineries requirements. Based on previous studies, lack of these equipment and machineries proves to be a major hurdle that hinders work in IBS-based projects [4,16].

Apart from that, the construction companies were exposed and trained in the traditional method of construction for decades and there was an availability of cheap foreign workers in Malaysia. Shifting to IBS seemed to be too complicated, unless the legislators impose a policy or strict governmental requirement on the use of IBS, or redefine the market by setting a mandatory quota for IBS projects. Therefore, a long term comprehensive policy and directive statement towards the industrialization of the building and construction sector should be pursued by the government for future business model in Malaysian's construction industry [20]. Small contractors were not willing to change to mechanized based on the IBS system because the conventional method perceived by them suited most of the small scale projects as usually undertaken by them [21].

*3.2 Limited enforcement of IBS requirements*

Despite massive government's initiatives to increase the utilization of IBS for every construction projects in Malaysia, the execution process was believed not strongly enforced. In 2008, in order to create sufficient momentum for the demand of IBS components, the use of IBS had been made mandatory for public buildings construction under Treasury Circular 7/2008 [18]. The significances of IBS inclusion for construction projects were highlighted in both IBS Roadmaps 2003-2010 and 2011-2015, however the adoption was still low.

Another requirement for mandatory IBS utilization for construction projects had been spoken up by the Works Minister, Datuk Fadillah Yusof recently, in which the government intends to make the adoption of IBS in construction projects compulsory in 2018. Datuk Fadillah Yusof quoted, "Ministry of Work intend to make the usage of IBS compulsory in 2018, with the assistance of local authorities, so that any proposed plans will use IBS in its construction." However, this particular initiative could only be realized if one of the strategies to IBS adoption is to include the IBS score as part of satisfactory conditions before the granting of development orders by local authorities. This is in line with the current CITP 2016-2020 main components, i.e. productivity, which will be driven by the adoption of new technologies, not only IBS but also BIM (building information modeling) to reduce waste and improve productivity [15].

*3.3 Lack of IBS expertise in the market*

Kamar *et al.*, [18] revealed that the main attributes to the low IBS adoption by contractors mostly related to the organizational strategy and soft issues which underpin the capability of organization to implement IBS, for example higher capital cost, high capital investment, difficulties in achieving economies of scale and lack of knowledge in IBS . Technical factors such as level of IT, building regulation and code of standard were perceived as not significant.

IBS requires high construction precision [22]. Studies had indicated that the availability of cheap foreign labour which offsets the cost benefit of using IBS was the root cause of the slow adoption in the past. As long as it was easy for the industry to find foreign workers, labour rates would remain low and builders would find it unattractive to change into simplified solutions such as IBS [11]. Furthermore, many local authorities were not fully familiar with modular co-ordination and standardization concept related with IBS design and assembling procedure. This had slowed down building approval and caused unnecessary delay in the development process. Due to the lack of knowledge and awareness, these local authorities tend to misunderstand IBS guidelines adding to further delays in approval [23].

According to Sasitharan *et al.* [22], IBS required high skilled workers compared to the 139 conventional construction methods. Under this system, the demand for on-site manual labor particularly carpenters, bar benders and concreters becomes less. The system demands more machine-oriented skills, both on sites and in factories. Thus, this leads to a transformation requiring the restructuring of human resource in an organization in terms of training and education

Generally, Malaysia is still lacking of skilled workers. As such, more intensive training programs are needed in the specialized-IBS-skills-training like integrating or assembling system. However, these special needs require more time and investment [17].

In an IBS project, the role of the contractor is shifted from being a 'builder' to becoming an 'assembler' on site [22]. This requires contractors to be equipped technologically with IBS knowledge and skills. The needs are made more imperative if the contractors were to promote their IBS products and compete in the industry. From July to September 2002, the situation was suddenly worsened when many trained foreign workers were forced to leave the country after a wide spread crackdown on illegal foreign workers [17]. The 'new batches' of foreign workers did not possess the required skill in IBS and had to be retrained [17].

*3.4 Cash flow problems for contractors*

In IBS approach, payment and procurement mechanism needs to be reviewed to tailor to IBS activities which are reliable and for safer payment and procurement [23]. Change in construction methods and processes from conventional to IBS will affect the change in the mode of payment and any related clauses in the contract. At this time, there is no security of payment designed for IBS projects and contractors need to deposit to manufacturers, which resulting in cash flows issues [9, 24]. The existing procurement structures had constrained the contractors by requiring them to have a high upfront financial capital to be able to procure IBS components from the suppliers.

Besides, Nawi *et al.* [25] discussed about the problematic payment issues that plagued with IBS implementation in Malaysia. Current practice of payment for IBS project is perceived as not suitable to the nature of IBS project itself, which affects the productivity and development progress of the project. According to Nawi, Mydin, Nursal, Nifa, & Bahaudin (2015), IBS system was based on prefabrication process where the contractors need to purchase construction component from the manufacturers and this concept required them to have a strong cash flow [25]. In the current practice of IBS project, the awarded contractor would be paid by the client an initial payment between 10% and 25% of contract value [25]. However, the manufacturing company would require 75% of the capital in order to manufacture the IBS components before delivering the components to the construction sites [24].

Unfortunately, small contractors do not have strong financial funding to follow this concept. To deal with this situation, some of the contractors required a bond from a financial institution as a guarantee [25]. Besides that, some contractor would only pay 75% of the cost to the manufacturer after the components were delivered on site [24]. Both of the situations were unfair to both parties.

This is in agreement with a study by Fateh *et al.* [26] who conducted a series of interviews with several clients, contractors and manufacturers, and found that the main issue that led to low adoption of IBS was due to no standard form of contract to tailor IBS construction in Malaysia. Therefore, it is recommended for the construction industry to have a standard form of contract for IBS which is different from traditional design and build project, and suit IBS specific activities and construction processes.

A study by Meynagh *et al.* [20] revealed that most contractors strongly agree that supply chain and method of procurement as primary factor contributes to the success of IBS project. Therefore, it is crucial for legislators and construction actors, to establish a good relationship and collaboration among the project members from the moment when the project starts up to the completion to ensure the smooth running of supply chain.

A study by Dzulkalninea *et al.* [26] revealed that there were five (5) issues related to payment for IBS contractors, as below:

- Cost
  The cost incurred for the implementation of IBS was to set up the factory. The IBS contractor needed more than 30% of the up-front payment. The payment was based on work progress for the conventional and it did not applicable when it came to IBS project. Besides that, there was an issue of high prices in terms of production cost which comprised of buying new equipment, machinery, technology and training for the manpower [27].
- Difficulty in securing timely and adequate financing
  The difficulty of financial funding for the IBS contractor was because of the progress work payment. The payment for the IBS contractor could not be treated the same way as the conventional project because the initial cost was not the same. It was essential for the IBS contractor to receive the initial payment from the client to sustain their business [27].
- Lack of integration at design stage
  In terms of the integration at the design stage, respondents mentioned that sometimes it was related to the payment issue to the IBS contractor. At this stage, it was important for each party to be responsible and had good cooperation to ensure that the payment could be made on time based on the right design [27].
- Difficulty to get loans from the financial institution
  Some of the respondents mentioned that dealing with the financial institutions for loan application ware not the main problem for the IBS contractor. As long as the documents were fully-completed, the financial institution would be able to approve the loans. As for now, there were few financial institutions would be able to provide loans for the construction industry such as EXIM Bank, SME Bank and others [27].
- Increment of material prices
  Most of the respondents explained that the increment of the material prices would directly affecting the production cost. The high cost of material would increase the initial cost of production. One of the respondents revealed that the material prices increase in every three (3) months that had made them difficult to sustain the equilibrium of the profit [27].

These factors had caused the IBS business to be much more risky as it needs extra capital for investment as compared to the conventional method. Therefore, it was very hard for new companies to compete for opportunities with international competitors which were stronger in terms of financial capability, technology and specialization [23]. In addition, contractors prefer not to use IBS as they find it was easier to stick to the traditional construction method [17, 25]. Adopting a new system means that more time needed to be allocated for training of human resources, specialized equipment and machineries, also the need to have a substantial and sustained budget. Based on previous studies, lack of these superior equipment and machineries proved to be the major hurdle that hinders work in IBS-based projects [23, 25].

*3.5 Industrialization objective*

In Malaysia, the introduction of IBS was primarily to reduce over-dependency on foreign labour. It differed from other developed countries such as Australia, where the main objective of IBS was to cater the housing demand due to the increasing number of migrants each year. The idea of having an offsite housing production was considered a success as the housing-assembling productivity had been increased and the issues were excellently solved. It was also reported that the gap between demand and supply for housing in Australia had reached up to the crisis proportions. This had led to more houses to be prefabricated in the modular unit in the factory. As in 2014, the demand for prefabricated houses was increased from less than 1% in 2004 (Sarah Backhouse, Chief Executive Officer at prefabAUS). It was predicted that the demand of prefabricated houses would rise between 10% and 20% in the next 10 years [28].

IBS in Malaysia had been established, introduced and applied in the construction industry to deal with a growing demand for affordable housing, solving issues associated with foreign workers and improving quality, efficiency and productivity of construction industry. Nowadays, IBS had evolved and implemented in hybrid construction projects to build national landmark as significant valuable national assets [29].

Malaysian IBS contractors needed to benchmark IBS technologies, learn from the lesson and adapting the best practices from other countries. Construction industrialisation is a worldwide agenda. IBS was already successfully adopted in Finland, Sweden, Japan, Germany and Singapore where offsite technologies had eventually modernised and improved the construction industry [11]. The government should launch a forum on a regular basis for the academicians and associated practitioners who are actively involved with the IBS technology for exchange of information and experiences, development of new techniques and advice on promotion and implementation of IBS. An online portal was also suggested to disseminate international trends, products and processes associated with the IBS [23].

## 4. Conclusion

This paper had identified and discussed the barriers in IBS adoption in the Malaysian construction industry. The level of IBS adoption in the construction industry was still low even though the government had seriously promoted and encouraged the use of IBS in order to have a better practice in the construction industry. The issues and barriers that impeded the adoption of IBS had been discussed and mainly revolves around the lack of expertise, the transitional shift from conventional construction to IBS, cash flow problems for contractors, limited enforcement of IBS requirements and industrialization objective. Despite the negative perspective concerning IBS method, this does not mean the future for adoption of IBS method by the prospects would be low. Future research is sought for more investigation to be done for the qualitative stakeholders' perspectives on the factors that impede the IBS adoption in the construction industry.

## 5. References

[1] Bank Negara Malaysia 2008 *Monthly Statistically Bulletin* (Kuala Lumpur: Bank Negara Malaysia)

[2] Mansor S A 2010 *Review of construction sector The 7th Malaysia Construction Sector Review and Outlook Seminar* Retrived on August 3, 2017 from http://www.kkr.gov.my/files/PRESS CIDB

[3] Alaghbari W, Kadir M R A, Salim A and Ernawati 2007 The significant factors causing delay of building construction projects in Malaysia, *Engineering, Construction and Architectural Management* **14** 192-206

[4] Hamid A H A, Kamar Z, Zain K A M, Ghani and Rahim 2008 Industrialized Building System (IBS) in Malaysia: The Current State and R&D Initiatives, *Malaysia Constr. Res. J* **2** 1–13

[5] CIDB 2007 *IBS Digest at Malbex in IBS Digest* Special Issues on 24th Malaysian International Building Exposition pp 19-22

[6] CIDB 2008 *Master Plan for Occupational Safety and Health in the Construction Industry 2005-2010* Retrieved on July 20, 2017 http://www.cidb.gov.my/cidbv4/images/pdf/OHSAS

[7] Ismail F, Mohamad Yusuwan N and Baharuddin H E A 2012 Management factors for successful IBS projects implementation *Procedia - Social and Behavioral Sciences (Egypt)* vol 68 (United Kingdom: Elsevier) pp 99-107

[8] Mohamad Kamar K A 2010 The critical success factors and readiness of contractors in adopting Industrialised Building System (IBS) construction *IBS Construction in Malaysia* (Machester: University of Salford) chapter 1 pp 10-23

[9] CIDB 2008 *Construction Industry Master Plan (CIMP 2006-2015)* (Kuala Lumpur Malalysia: Construction Industry Development Board) pp 20-21
[10] IBS Centre 2012 *IBS Roadmap 2011-2015* Retrieved July 20, 2017 from http://www.vertilite.asia/f/2012/09/IBS-Roadmap-2011-2015.pdf
[11] Din M I, Bahri N, Dzulkifly M A, Norman M R, Kamar A M and Hamid Z A 2012 The adoption of Industrialised Building System ( IBS ) construction in Malaysia : The history, policies, experiences and lesson learned *Proceedings of the 2012 29th International Symposium of Automation and Robotics in Construction (ISARC) (Eindhoven, The Netherlands)* pp 26-29
[12] Pan W, Gibb A G F, and Dainty A R J 2008 Leading UK Housebuilders' Utilization of Offsite Construction Methods, *Build. Res. Inf* **36** 56–67
[13] Pan W, Gibb A G F, and Dainty A R J 2007 Perspectives of UK housebuilders on the use of offsite modern methods of construction, *Constr. Manag. Econ* **25** 183–194
[14] Lee R 2015 *IBS adoption in Malaysia remains low* Retrieved on July 20, 2017 from http://www.theedgeproperty.com.my/content/ibs-adoption-malaysia-remains-low-cidb
[15] Idris A 2016 M*inister: IBS to be compulsory for construction projects in 2018 The Edge Markets* Retrieved on July 22, 2017 from http://www.theedgeproperty.com.my /content/1099599/minister-ibs-be-compulsory-construction-projects-2018
[16] Nasrun M, Nawi M and Utara U 2015 Suitability of IBS formwork system in Malaysian construction industry, *Australian Journal of Basic and Applied Sciences* **9** 89-91
[17] Thanoon W A M, Peng K W, Kadir W R A, Jaafar M S and Salit M S 2003 The essential characteristics of Industrialised Building System *Proceeding of International Conference Industrialized Building Systems* vol 10 pp 11
[18] Mohamad Kamar K A, Hamid Z A and Dzulkalninea N 2012 Industrialised Building System (IBS) construction: measuring the perception of contractors in Malaysia *2012 IEEE Business, Eng. Ind. Appl. Colloq (Kuala Lumpur, Malaysia)* (United States: IEEE) pp 328–333
[19] Goodier C and Gibb A G F 2007 Future opportunities for offsite in the UK, *Construction Management and Economics* **25** 211-224
[20] Meynagh M H Z, Marsono M M, Kama A K and Isaabadi 2014 Main Factors of contractor reluctancy to adopt IBS projects in Malaysia, *Aust. J. Basic Appl. Sceinces* **8** 251–254
[21] Mohamad D, Ramli M Z, Danuri H N, and Sapuan W K 2006 Main factors of contractor reluctancy to adopt IBS projects in Malaysia, *J. Sci. Res. Dev* **3** 77–82
[22] Sashitharan S, Jusoh M S, Amlus M H, Rahimi A, Ibrahim A and Ismail M S 2014 Industrialized Building System (IBS) performance in Malaysian construction industry : Analysing the cost, training and construction policies, *J. Appl. Sci. Agric* **9** 6–11
[23] Mohamad Kamar K A, Alshawi M and Abd Hamid Z 2009 Barriers to Industrialized Building System (IBS): The case of Malaysia *Built Hum. Environ. 9th Int. Postgrad. Res. Confrence (Salford, United Kingdom)* vol 24 pp 1–16
[24] Shukor A S A, Mohammad M F, Mahbub R and Ismail F 2011 Supply chain integration in Industralised Building System in the Malaysian construction industry, *The Built & Human Environment Review* **4** 108-121
[25] Nawi M N M, Mydin M A O, Nursal A T, Nifa F A A and Bahaudin A Y 2015 Payment issues in Malaysia Industrialised Building System (IBS), *Adv. in Environmental Biol* **9** 185–188
[26] Fateh M A M, Mohammad M F and Shukor A S A 2016 Review in formulating the standard form of contract for Industrialised Building system (IBS) construction approach in Malaysia *MATEC Web of Conference* vol 87 (Paris: EDP Science) 01001
[27] Dzulkalninea N, Azman M N A, Khing W B, Habidin N F and Ayub A R 2016 Issues of payment provurement process for Industrialised Building System (IBS) project, *J. Teknol. Engineering Sci* **78** 1–15

[28] James C 2014 *What's happening in the world of prefab housing?* Retrieved on July 25, 2017 from http://www.realestate.com.au/news/whats-happening-world-prefab-housing/
[29] Abedi M, Fathi M S and Mirasa A K 2011 Establishment and development of IBS in Malaysia *Proceedings of International Building and Infrastructure Technology Conference 2011 (Penang, Malaysia)* (Penang: Universiti Sains Malaysia) pp 405-412

**Acknowledgement**
This paper was partly sponsored by the Centre for Graduate Studies UTHM.

# Theoretical framework of the causes of construction time and cost overruns

**K Ullah[1], A H Abdullah[1], S Nagapan[1], S Suhoo[1] and M S Khan[2]**

[1]Jamilus Research Centre, Faculty of Civil and Environmental Engineering, Universiti Tun Hussein Onn Malaysia, 86400 Batu Pahat, Johor, Malaysia
[2]Faculty of Business and Management Sciences, Limkokwing University of Creative Technology, Cyberjaya, Selangor, Malaysia
Corresponding author: gf150051@siswa.uthm.edu.my

**Abstract.** Any construction practitioner fundamental goal is to complete the projects within estimated duration and budgets, and expected quality targets. However, time and cost overruns are regular and universal phenomenon in construction projects and the construction projects in Malaysia has no exemption from the problems of time overrun and cost overrun. In order to accomplish the successful completion of construction projects on specified time and within planned cost, there are various factors that should be given serious attention so that issues such as time and cost overrun can be addressed. This paper aims to construct a framework for the causes of time overrun and cost overrun in construction projects of Malaysia. Based on the relevant literature review, causative factors of time overrun and cost overrun in Malaysian construction projects are summarized and the theoretical frameworks of the causes of construction time overrun and cost overrun is constructed. The developed frameworks for construction time and cost overruns based on the existing literature will assist the construction practitioners to plan the efficient approaches for achieving successful completion of the projects.

## 1. Introduction

Construction industry performs a fundamental role in the economic growth of any country as it stimulates the development of other industries too [1]. Worldwide large amount of money are spent on construction development works, which provides essential infrastructures and large number of employment opportunities. The input of the construction industry in growth of national economy requires better efficiency in the industry by means of successful project delivery. Timely completion of construction projects within anticipated cost and to the level of quality standards settled by the owner is an index of successful project delivery [2]. Despite the importance of time and cost in construction projects, majority of the construction projects both in developed and developing countries are frequently experiencing the severe issue of time and cost overruns. Many studies are carried out on construction time and cost overruns in various countries such as in United State [3, 4], Nigeria [5, 6], India [7, 8], Australia [9, 10], United kingdom [11], Germany [12], Korea [13] and Kuwait [14].

In Malaysia, construction industry is one of the growing industries contributing in the country's socioeconomic development, though it contributes less than the services, manufacturing and agriculture sectors. Construction projects in Malaysia are suffering from a major problems of delay or time overrun and cost overrun constantly since many years [11]. A study carried out by Shehu et al.

[15] in Malaysia indicated that more than 50% projects experienced the problem of cost overrun while 79.95% of the public sector construction projects and 66.65% of the private sector construction projects were finished after the scheduled time. While, Rahman et al. [16] carried out a study using questionnaire survey amongst construction stakeholders of central and southern region of Peninsular Malaysia, the results of the survey showed that 92% of survey respondents reported that they are facing with the issue of time overrun and 89% of the projects experienced cost overrun. This confirms that time and cost overruns are severe problem in Malaysian construction industry, which needs to be minimized or controlled. For efficient control of the time and cost overruns, the understanding of causative factors of these issues is very essential. Hence, this paper aiming on constructing theoretical frameworks for the causes of time overrun and cost overrun in Malaysian construction projects.

## 2. Construction time and cost overruns

Time overrun and cost overrun in construction projects can also be called is "slippage of project-schedules" [17]. Time overrun can also be defined as the time increased to finish the construction project after scheduled date which is affected by internal and external causes surrounded the construction project [18].

Cost overrun can also be termed as budget overrun, cost increase, or cost escalation [19]. Cost overrun is a variance between initially estimated or projected cost and final cost at the completion of the project [20]. Final costs are described as the total costs actually spent on construction project as determined at the project completion time while, projected or initial costs is known as the planned or predicted costs at the project approval time [13].

## 3. Objective and research method

The aim of this study is to construct a theoretical frameworks for the causes of time and cost overruns. The factors influencing time and cost overruns are collected from the previous research studies. During literature review for theoretical frameworks only those research papers are considered which were conducted to study the issues of time overrun and cost overrun in Malaysian construction projects.

## 4. Previous studies on causative factors of time overrun

In Malaysia, various studies were conducted to figure out the main causes of time overrun in construction projects. Memon [21] examined the major causative factors of time overrun in building construction projects by conducting a questionnaire survey. Results of the survey revealed that most significant causative factors of time overrun were repeated changes in design, modification in the scope of the project, financial problems of client, delays in process of decisions making, unexpected site conditions, delay in the progress of payment by client, shortage of site workers, inaccuracies and errors in design, very slow process of preparation and approval of drawing documents, and incapable subcontractors.

Mydin et al. [22] investigated the influential causes of time overrun in Malaysian private housing projects through a questionnaire survey. Top 10 common and highly severe factors of cost overrun were unpredictable weather conditions, poor management at the site by contractor, incomplete design documents, and lack of contractor's experience, financial difficulties, slow process of approval of major changes, changes in contract agreement, lack of contractor coordination with other construction stockholders, mistakes in construction, and poor quality works.

A questionnaire survey was performed by Azlan et al. [23] among construction practitioners to determine main factors contributing to time overrun construction projects in state of Kedah, Malaysia. Results of the survey showed that top ten root causes of time overruns as observed by the three key construction parties comprises delay in sub contractor's work, improper arrangement and scheduling of project, problems in financing project, shortage of labors, delay in process of decision making, slowness in progress payment by owner, delay in material delivery to site, late procurement of materials, escalation in raw material prices, and delay in process of approving major variations in scope of work. Shehu et al. [24] found out that the main causative factors of time overrun were cash flow and financial problems of contractor, delay in payments by owner, delay in payment from contractor to sub-contractors and materials suppliers, late permits by local government authorities, unproductive planning and scheduling of the project, improper control of the project progress by the contractor, bureaucracy in government organizations and delay in decision making process by the owner.

Ramanathan et al [17] examined the factors that cause time overrun in construction projects and identified that key factors of cost overruns were rain effect on construction activities, shortage of labors, contractors' poor site management and control, unqualified workforce, lack of contractor experience, late progress payments by client, lack of communication and coordination of contractor with other stakeholders, low productivity level of labors, and delay in decision making process by client.

Rahman et al. [25] investigated the causative factors of time overrun or schedule overrun in MARA construction projects from perspective of "project management consultant" and found that dominate factors were poor site management by contractor, lack of contractor experience, lack of site labors, escalation of material prices, practice of awarding contract to lowest bidder, problems with subcontractors, lack of communication among parties, and change management. To identify the factors that affects timely completion of building construction projects, a questionnaire survey was conducted by Alaghbari et al [26] in Klang Valley, Malaysia. The study found that financial problems and lake of coordination among construction parties are main factor initiating time overrun in construction projects. The theoretical frame of the causes construction time overrun is develop from seven scholars research papers [17, 21-26] as shown in figure 1.

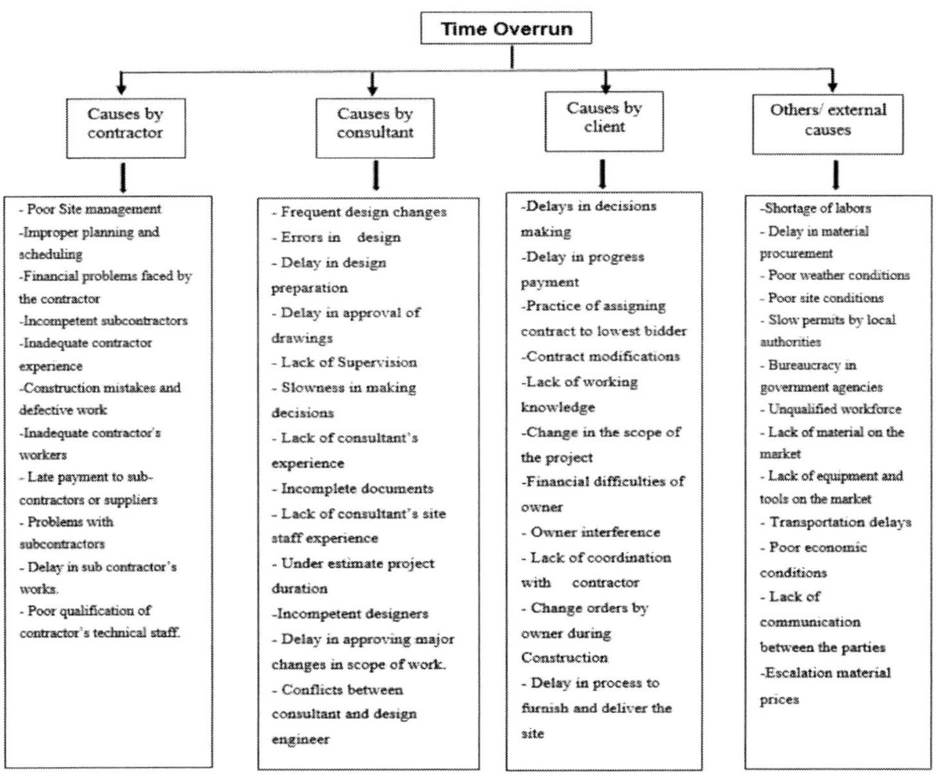


**Figure 1.** Theoretical framework for construction time overrun.

**5. Previous studies on causative factors of cost overrun**

A Study conducted by Ali and Kamaruzzaman [27] identified 13 major causes that influenced cost overruns in building construction projects in Klang valley, Malaysia. A questionnaire survey was conducted among construction project manager, quantity surveyor, Civil & Structure Engineer, and other related respondents. The survey discovered that the causes of overrun in budget of building construction projects were attributed to incorrect / poor estimate of original cost, underestimation of construction projects duration, inappropriate planning, contractor poor project management, poor contract management, lack of contractor experience, increase in project costs, fluctuation in cost of heavy machineries, variation in price of materials, unexpected site situations, inadequate funds for project financing, obsolete construction equipments. /unsuitable construction methods, mistakes in design.

Memon et al. [28] determined 15 main factors effecting cost performance in large construction projects of MARA. Based on the project management consultants surveyed the most important cost overrun factors were: financial problems experienced by contractors, inadequate site management, lack of monitoring of work progress by contractor, incapable contractors, unavailability of site labors, indelicate planning and arrangement by contractors, instability in costs of construction materials, practice of allocating contract to lowest bidder, lack of communication and coordination among key construction stakeholders, underestimate project duration, and late material procurement. Further, Memon et al. [29] conducted another study on cost overrun in Malaysian building construction projects. The result of his research specified that the core factors of cost overrun in building construction projects were errors in design, slowdown in design preparation, impractical contract duration, incapable contractor, delay in delivery process of materials & equipments to construction site, poor relationship between top management & labors, delay in preparation of drawings, slowness in approval of drawing documents, insufficient planning and scheduling, and errors during construction process resulting reworks.

A quantitative research carried out by Ramanathan et al. [17] on cost increase in construction projects in Malaysia identified 18 causes of construction cost overrun. The findings of his research showed that the main causes of cost overrun were extension of time, fluctuation in cost of raw materials, design changes, unpredictable weather conditions, insufficient project preparations and planning, delay in delivery of raw materials and equipment to site, lack of cost plan / improper monitoring of pre and post contract stages, monopoly by some suppliers in project materials, deficiency of coordination at design stage, and re-measurement of provisional sum.

To investigate the factors that contribute to increase in scheduled budget of large construction projects in Malaysia, a study was conducted by Rahman et al. [30]. A questionnaire with 35 common causes of cost overrun was precisely designed from initial investigations. The questionnaire form was directed towards three groups: clients of the large construction projects, consulting offices, and contractors handling large construction projects. The questionnaire form was circulated to a random sample of 150 client firms, 150 contractor firms and 30 consultant personals. The results of the survey revealed the key factors of cost overruns or budget overrun in large construction projects: variation of cost of raw material, ineffective site management and supervision by contractor, lake of contractor experience in handling large construction projects, construction mistake resulting schedule delay, improper planning and scheduling, unskilled subcontractors, errors in design, regular variations in design, , financial difficulties faced of owner, poor financial control on site unavailability of construction materials, inaccurate cost estimate, and underestimate of project duration.

A survey was carried out by Jamaludin et al. [31] on the factors inducing the cost variance during the construction stage amongst the building contractors registered under Class 'A' within the Klang Valley, Malaysia. The results of the survey showed that the important causes of cost increase were imperfect design drawings and specifications at tendering phase, alterations in client requirements, cash flow and financial problems faced by the contractors, fluctuation of material prices, poor planning, scheduling and monitoring, increasing of labour salaries, fluctuation of plant and heavy machineries cost, lack of coordination and communication among the stakeholders, scrape and rework, lack project team's experience.

Research conducted by Ismail et al. [32] studied risk level of several factors which cause cost overrun throughout the life cycle of a construction project. The study revealed that 6 causes have a high risk on cost overrun, which were ineffective site management & monitoring, unskilled subcontractors, construction mistakes, incomplete design documents at the tendering stage, and late payment.

The theoretical frame of the causes of construction cost overrun is develop from seven scholars research papers [17, 27-32] as presented in the figure 2.

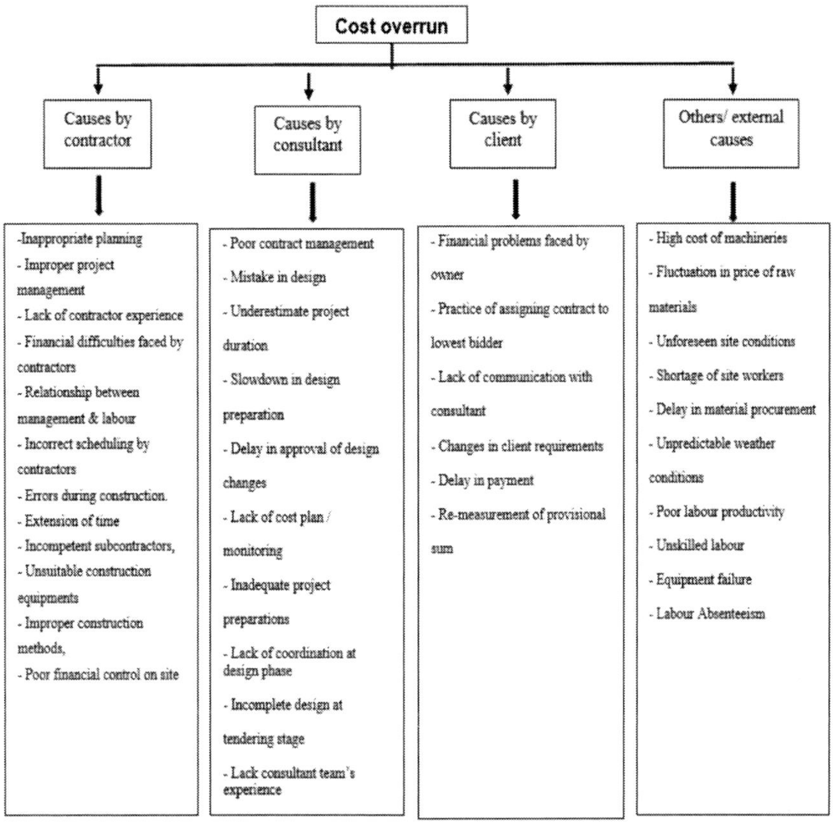


**Figure 2.** Theoretical framework for construction cost overrun.

## 6. Conclusion

The problem of construction cost and time overruns has been remarkable in Malaysian construction projects for many years. There are various causes of construction time and cost overrun that have been determined by various researches. The developed frame work for time overrun results in identification of 48 causative factors for time overrun in construction projects in Malaysia, while the framework for cost overrun showed 38 cusses of cost overrun in Malaysian construction projects. There is need of proper depth studies on the problems of time and cost overrun in construction industry, which can identify possible solution and measuring the level of possible solutions to insure the successful completion of construction projects.

## 7. References

[1] Rum N A and Akasah Z A 2011 Implementing life cycle costing in Malaysia construction industry : A review *Proc. of International Building & Infrastructure Technology Conference (Penang, Malaysia)* vol 7 (Penang: Universiti Sains Malaysia) **1** 272–280
[2] Chan W and Kumaraswamy M M 2002 A survey of time-cost relationships in Hong Kong construction projects, *Build. Technol. Manag. J.* **20** 54–72
[3] Chang A S 2002 Reasons for Cost and schedule increase for engineering design projects, *J. Manag. Eng.* **18** 29–36
[4] Shrestha P P, Burns L A and Shields D R 2013 Magnitude of construction cost and schedule overruns in public work projects, *J. Constr. Eng.* **20** 1-9
[5] Amu O O and Adesanya D A 2011 Mathematical expressions for explaining project delays in South Western Nigeria, *Singapore J. Sci. Res.* **1** 59–67
[6] Ameh O J, Soyingbe A A, and Odusami K T 2010 Significant factors causing cost overruns in telecommunication projects in Nigeria, *J. Constr. Dev. Ctries.* **15** 49–67
[7] Rajan T A, Gopinath G and Behera M 2014 PPPs and project overruns : Evidence from road projects in India, *J. Constr. Eng. Manag.* **140** 1 – 10
[8] Salunkhe A A and Patil R S 2014 Effect of construction delays on project time overrun : Indian scenario, *Int. J. Res. Eng. Technol.* **3** 543–547
[9] Love P E D, Wang X, Sing C, and Tiong R L K 2012 Determining the probability of project cost overruns, *J. Constr. Eng. Manag.* **139** 321–330
[10] Wong K and Vimonsatit V 2012 A Study of the factors affecting construction time in Western Australia, *Sci. Res. Essays* **7** 3390–3398
[11] Olawale Y A and Sun M 2010 Cost and time control of construction projects: Inhibiting factors and mitigating measures in practice, *Constr. Manag. Econ.* **28** 509–526
[12] Kostka G and Anzinger N 2016 *Large Infrastructure Projects in Germany: Between Ambition and Realities* (Berlin: Palgrave Macmillan Publishers Limited)
[13] Lee J K 2008 Cost overrun and cause in Korean social overhead capital projects: Roads, rails, airports, and ports, *J. Urban Plan. Dev.* **134** 59–62
[14] Koushki P A, Rashid K and Kartam N 2005 Delays and cost increases in the construction of private residential projects in Kuwait, *Constr. Manag. Econ.* **23** 285–294
[15] Shehu Z, Endut I R, Akintoye A and Holt G 2014 Cost overrun in the Malaysian construction industry projects: A deeper insight, *Int. J. Proj. Manag.* **32** 1471–1480
[16] Rahman I A, Memon A H, Nagapan S, Alias Q B, Latif I, Asmi A and Azis A 2012 Time and cost performance of costruction projects in southern and cenrtal regions of Penisular Malaysia *IEEE Colloquium on Humanities, Science & Engineering Research* (United States: IEEE) pp 52-57
[17] Ramanathan C, Potty N S and Idrus A B 2012 Analysis of time and cost overrun in Malaysian construction, *Adv. Mater. Res.* **45** 1002–1008
[18] Najjar J 2008 *Factors Influencing Time and Cost Overruns on Construction Projects in The Gaza Strip*, Master Thesis (Gaza: The Islamic University of Gaza)
[19] Zhu K and Liu L 2004 Structuring finance to meet client's needs - A stage-by-stage factor control framework for cost estimation of construction projects., *Proc. of Clients Driving Innovation International Conference (Surfers Paradise, Australia)* (Brisbane: CRC for Construction Innovation)
[20] Amoa K and Allotey S 2014 Cost overruns in building construction projects : A case study of a government of Ghana project in Accra, *Dev. Ctry. Stud.* **4** 54–65
[21] Memon A H 2014 Contractor perspective on time overrun factors in Malaysian construction projects, *Int. J. Sci. Environ. Technol.* **3** 1184–1192
[22] Mydin M A, Sani N M, Salim N and Alias M 2014 Assessment of influential causes of construction project delay in Malaysian private housing from developer's viewpoint E3S Web of Conf. vol 3 (Paris: EDP Sciences) 01027

[23] Azlan O and Syuhada I 2014 Delay in government project delivery in Kedah, Malaysia *Proc. of Recent Advances in Civil Engineering and Mechanics* (*Florence, Italy)* WSEAS Press pp 248–254
[24] Shehu Z, Endut I R and Akintoye A 2014 Factors contributing to project time and hence cost overrun in the Malaysian construction industry, *J. Financ. Manag. Prop. Constr.* **19** 55–75
[25] Rahman I, Memon A H, Abdullah M R and Aziz A A A 2011 Time overrun in construction projects from the perspective of project management consultant (PMC), *J. Surv. Constr. Prop.* **2** 54–66
[26] Alaghbari W, Razali A K, Azizah S and Ernawati 2007 The significant factors causing delay of building construction projects in Malaysia, *Eng. Constr. Archit. Manag.* **14** 192–206
[27] Ali A S and Kamaruzzaman S N 2010 Cost performance for building construction projects in Klang Valley, *J. Build. Perform.* **1** 110–118
[28] Memon A H, Rahman I R, Abdullah M R and Azis A A A 2010 Factors affecting construction cost performance in project management projects :Perspective of project management consultant, *Int. J. Sustain. Constr. Eng. Technol.* **1** 30–35
[29] Memon A H, Rahman I R, Asmi A and Azis A A A 2011 Preliminary study on causative factors leading to construction cost overrun, *Int. J. Sustain. Constr. Eng. Technol.* **2** 57–71
[30] Rahman I A, Memon A H and Karim A T A 2013 Significant factors causing cost overruns in large construction projects in Malaysia, *J. Appl. Sci.* **13** 286–293
[31] Jamaludin S H, Mohammad M F and Ahmad K 2014 Enhancing the quality of construction environment by minimizing the cost variance *Procedia - Social and Behavioral Sciences* vol 153 (United Kingdom: Elsevier) pp 70–78
[32] Ismail I, Memon A H, and Rahman I R 2014 Expert opinion on risk level for factors affecting time and cost overrun along the project lifecycle in Malaysian construction projects, *Int. J. Constr. Technoogy Manag.* **1** 10–15

## Acknowledgement
The authors would like to thank Faculty of Civil and Environmental Engineering, and Centre for Graduate Studies (CGS), Universiti Tun Hussein Onn Malaysia, (UTHM) for support and providing facilities to accomplish the study. We are also thankful to all individuals for their help during this study.

# An initial investigation on the challenges of managing construction workforce in Saudi Arabia

**N Al-Emad[1] and I A Rahman[1]**

[1]Jamilus Research Centre, Faculty of Civil and Environmental Engineering, Universiti Tun Hussein Onn Malaysia, 86400 Parit Raja, Batu Pahat, Johor, Malaysia.
Corresponding author: df080116@gmail.com

**Abstract.** Most of the construction workers in Saudi Arabia are foreigners from several countries having different knowledge, skills and cultures. These create challenges to administer the workers in ensuring project success. This paper presents an initial investigation to uncover the challenges faced by construction professionals in managing construction workforce in Saudi Arabia. It describes insight processes of handling the workforce during planning stage, recruitment procedures and construction stage based on interview with senior manager who are well experienced in handling mega construction projects in Saudi Arabia. The interview was carried out in semi structured mode where the interviewee was given ample time to express the experiences encountered in dealing the workforce issue. This preliminary work able to identify among important issues related to construction workers are restrictions to non-Muslim skilled workers, limited visa quota, being away from family, delay in salary payment, cheating of workers skill's status, safety issues, communication barriers and living conditions. Hence, these issues require quality leadership attributes such as continuous empathy with workers, respectful, trustful, sincere, reliable, good communication skills and problem solving skills. These findings are useful to construction practitioners and also research work related to construction leadership in handling worker's issues.

## 1. Introduction

Saudi Arabia is considered amongst the fastest growing economies of the Middle East countries [1-3]. It has a robust growth due to more government spending on construction, infrastructure and hospitality to promote tourism especially religious tourism in order to host the people who are continuously coming for Hajj and Umrah in the holy cities Makkah Al Mukarramah and Madinah [4]. Saudi Arabia population is approximately 20 million with 10 million are foreigners mainly from India, Pakistan, Bangladesh, Egypt, Yemen, Philippine, Sri Lanka, Nepal, Turkey and others. Most of these foreigners obtained working permit (visa) to work in Saudi Arabia with majority are workforces for construction industry as Saudi Arabia is now experiencing rapid construction growth in all aspects of construction projects both in urban and rural areas [5]. The total number of foreigner's workforces that are working in construction industry is 4.4 million [6]. In holy cities like Makkah and Madinah only Muslim workers are allowed to work in; which restricts the choice of the contractors and narrows the band for the best choice. Hence construction professionals faced challenges in managing workforces' issues specifically in Makkah city. This article uncovers on how construction professionals able to handle issues related to construction workers. It is mainly based on the first author observation while working in Makkah construction industry as civil engineer and also based on interview session with senior manager who has involved working for more than 15 years in mega construction projects in different part of Saudi Arabia.

## 2. Literature review

Literatures show that worker is considered an essential component to all industries including the construction industry. Works will not be able to be carried out without the involvement of skilled, semi-skilled or unskilled workforces [7-9]. With the engagement of these workforces many related issues are created which engulfing construction industry. An example of these issues are availability of skilled and unskilled construction workers, low productivity level, lack of experience, safety issues, long working hours, racial discrimination based on nationality and others [10-12]. According to Yankov & Kleiner [10], among important issues is selection of quality workforce because of a growing shortage of skilled workers in that field. Pramanik & Chackrabarti [13] stated that most of the construction workers are illiterate and ignorant about market conditions and labour market variations. In Saudi Arabia there are enormous problems and challenges surrounding construction workforces such as shortage of skilled workforce, poor living condition, low wages, delaying in wages payment, poor living condition, safety issues and others [12, 14]. These problems have adverse impacts to construction project performance which requires serious attention by construction professionals to handle them.

Addition to that, construction workforce face many problems in their daily life, these problems not only related to construction demanding tasks but also related to social, cultural and climate adoption [14].

## 3. Methodology

This initial study was conducted in an investigative and qualitative mode to identify challenges faced by construction professionals in managing construction workforce in Saudi Arabia focussing to Makkah city. It was conducted through semi structured interview with a senior manager who has 15 years experiences in handling mega construction projects in Saudi Arabia construction industry. Prior to the interview session, the objectives and questions to be asked were designed to ensure that the expert shared all his experiences in handling the workers. At the beginning of the interview, the expert was requested to describe his involvement in construction industry. Subsequently, requested to share his approaches and leadership qualities in handling the issues related to construction workforce. Interview session was conducted in face to face mode where all the given information was captured manually and also electronically. Expert was given ample time to voice his opinions on the given topic and gently guided to avoid deviation from the interview's objectives. The information gathered from the interview was converted into transcription. The issues faced by construction workers highlighted by the expert were analysed based on priority of the issues being mentioned by the expert. The issues brought up by the expert were extracted, sorted and tabulated as in table 1.

**Table 1.** Workers' issues.

| Nos. | Worker's Issue |
|---|---|
| 1 | Restrictions to non-Muslim skilled workers |
| 2 | Limited visa quota |
| 3 | Being away from family |
| 4 | Delay in salary payment |
| 5 | Cheating of workers skill's status |
| 6 | Safety issues |
| 7 | Communication barriers |
| 8 | Living conditions |

Table 1 shows the identified issues faced by Saudi construction workers from the views of construction expert who is well experienced in handling mega construction projects in whole Saudi Arabia especially in Makkah city. Based on these issues, the following section presented as a discussion on the issues as challenges to construction professionals.

## 4. Workforce planning and recruitment challenges

Planning and recruitment of Saudi Arabia construction industry workforces is crucial to ensure that the workers are available according to the need of work schedule. According to the expert, construction professional needs to plan the manpower for the whole construction project duration by determining man-hours for each activity based on bill of quantity work. Then it should be multiplied by a coefficient 1.2, due to some losses like vacations, absentees, pray times, weather conditions, accidents, training periods for semiskilled labours and possible increase in the BOQ quantities. The manpower schedule is prepared in line with construction schedule for each month of the project duration. The expert mentioned that to apply for visa quota, turn-over of the manpower should be considered where it is roughly 1.3 to 2 times more than the required manpower based on the nature of the works (length of project, difficulty factors and others). Professionals should invite for discussion with several sub-contractors in order to overcome lack of manpower at construction site.

Besides planning, recruitment of workforce is also a challenge to the construction professionals. The recruitment process is either can be done by labour agency of the workers home country or those workers who have been working in previous projects in Saudi Arabia. For new workers, a company representative should assess the capacities and eligibilities of the workers. While for the experienced workers, their skills are known and recruitment process can be carried out with proper documentation. According to the expert, activities for unskilled workers such as rebar and concreting can be easily performed by Indian and Pakistanis workforces. However activities such as formwork, ceramic, marble and paint preparation cannot be performed efficiently by these workers. Professional should structure the workforce skeleton of these labours' activities for countries like Turkey, Egypt and others even though the salary is high. Prior recruitment, the contractor should arrange accommodation, transportations and food services for their workforce. According to the expert, these services should not be compensated with the workers' salary as they will choose the cheapest alternatives and this consequently will affect their productivity negatively. The provided accommodation should be with comfortable space area of approximately 8 $m^2$ for each worker. While, the food provided should preferably comply with the workers' taste and healthy to ensure they are fit to work and the transportations to construction site should be proper.

## 5. Challenges in handling workforce

Upon joining the company, welcome meeting for all new workforces should be conducted with the presence of all top site management. Workers are presented and explained about site's general rules in a friendly atmosphere. Subsequently, safety orientation should be conducted by explaining all safety rules that must be strictly followed during carrying out construction tasks. Once the workforce has been structured for the construction site, then the managers' role is to ensure maximum progress with these workforces in-order to deliver construction project on time with stipulated budget and accident free. While handling this workforce at site, there are two levels of managements involved which are top level management including project director, project manager and construction manager and middle level management which also known as first manager's level in Saudi Arabia construction industry. The challenges related to each level of these two managements are elaborated in more details as in the following subsections.

### *5.1. Top management challenges*

Under normal circumstantial, the contract duration between company and foreign worker is valid between one to two years. Workers from Turkey and Egypt are usually given one year contract while from India and Pakistan normally given two years contract. This is considered long duration in human life and have tremendous impacts in their social life. Being away from families may affect their behaviours and consequently their working productivity. Hence, the construction professionals should establish good relations with their workers. Based on the expert experiences, the most important compensation for being away from home is the sincere relations of the workers with their managers. If a resemblance is required, top managements should act like father while the Forman should be their big brother. Also professional should encourage workers to treat each other as close friends and thus, lessen the psychological pain that they may suffer. A good professional should make continuous empathy with the workers and treat them as human being regardless of their country or position. Careful considerations and respect should be shown by construction professionals towards workers which make them less stressful. The workers will feel themselves in more friendly and less threatening environment. Once this is established and maintained, the trust to management will increase and resulted to high productivity. Hence, Trust, Sincerity and Reliability are important characteristics which should be possessed by top management.

### *5.2. Middle management challenges*

Officer of middle management is called first manager, they are either foreman or site engineer. The expert mentioned that the first managers should be close to the workers as they are in a direct contact on daily basis. Construction tasks carried out by workers need to be checked and supervised well by first managers. Hence, the first manager plays important roles to harmonise member of the worker team. The first managers should know the individual characters of each team member and harmonize the team members for the targets and success of the construction project. Based on the expert, harmonization should be accompanied by continuous training during execution at construction site. Unskilled workers should work closely with skilled workers and their mistakes are to be correctly routed to better results. The expert has pointed out that organizing good construction works is really difficult and largely depends on the managerial and leadership skills of the first manager. The first managers should possess almost all the main skills of top management but in a smaller scale such as good communication, empathy, trust, sincerity and reliability. First manager should evaluate the workers' performance continuously and report to top management accordingly. Also first manager should ensure the importance of safety rules to be followed strictly at site. Additionally, first managers should use medium language which is widely used by most workers and also a translator for special circumstances. This will lessen miscommunication between workers and their managers. Besides that, first manager should immediately try to solve small problems by himself without any delay as to avoid the problems getting bigger. Any problem that cannot be resolved should be brought up to higher management's attention immediately for immediate actions.

## 6. Conclusion

Success of construction project largely depends on management of human resources. Interactions between top management with first managers and also workers are crucial to ensure that the project can be carried out in harmonise manner and then the success of the project will be unavoidable. This paper has demonstrated an initial investigation to uncover the challenges faced by construction professionals in handling construction workforces. The findings of this study mainly based on interview session with very well experienced expert who has more than 15 years of working experiences in the construction industry of Saudi Arabia. Based on the expert's experiences, among important issues related to construction workers in Makkah construction industry are restrictions to non-Muslim skilled workers, limited visa quota, being away from family, delay salary payment, cheating of workers skill's status, safety issues, communication barriers and living conditions. However, these issues need quality leadership attributes such as continuous empathy with workers, respectful, trustful, sincere, reliable, good communication skills and problem solving skills. By identifying these challenges and issues, it will be helpful to the related construction parties around Makkah city and also whole Saudi Arabia in dealing with their workforces.

## 7. References

[1] Alrashed A, Taj S, Phillips M and Kantamaneni K 2014 Risk assessment for construction projects in Saudi Arabia, *Research J. of Management Sciences*. **3** 1-6

[2] Rahman I A, Al-Emad N and Nagapan S 2016 Projects delay factors of Saudi Arabia construction industry using PLS-SEM path modelling approach. *MATEC Web of Conf.* vol 81 (Paris: EDP Sciences) 07001

[3] Rahman I A, Al-Emad N and Nagapan S 2016 Categorization of Saudi Arabia's construction delay factors using factor analysis technique *Engineering Challenges for Sustainable Future: Proc. of the 3rd Int. Conf. on Civil, Offshore and Environmental Engineering (Kuala Lumpur, Malaysia),* vol 138 (London: CRC Press Taylor & Francis Group) pp 177-181

[4] Al-Emad N H and Nagapan S 2015 Identification of delay factors from Mecca's construction experts perspective, *Int. J. of Sustainable Construction Engineering and Technology* **6** 16-25

[5] Al-Emad N, Rahman I A, Nagapan S and Gamil Y 2017 Ranking of delay factors for Makkah's construction industry *MATEC Web of Conf.* vol 103 (Paris: EDP Sciences) 03001

[6] Jeddah Chamber 2016 *Saudi Arabia –Manpower & Employment, Talent Management, and Compensation* Retrieved on July 2017 from http://docplayer.net/36358782-Saudi-arabia-manpower-employment-talent-management-and-compensation-may-2016.html

[7] Loosemore M, Dainty A and Lingard H 2003 *Human Resource Management in Construction Projects: Strategic and Operational Approaches* (London: Spon Press Taylor & Francis Group) pp 50-90

[8] Dutta A B 2014 Study of labour factor in construction: A review, *Int. J. of Management and Social Sciences* **3** 147-52

[9] Ngwenya L and Aigbavboa C 2017 Improvement of productivity and employee performance through an efficient human resource management practices, *In Advances in Human Factors, Business Management, Training and Education* vol 498 (Switzerland: Springer International Publishing) pp 727-737

[10] Yankov L and Kleiner B H 2001 Human resources issues in the construction industry, *Management Research News* **24** 101-105

[11] Memon A H, Rahman I A, Abdullah M R and Azis A A 2011 Factors affecting construction cost in Mara large construction project: Perspective of project management consultant, *Int. J. of Sustainable Construction Engineering and Technology* **1** 41-54

[12] Almannaee, R 2014 *Migrant workers in the construction industry in United Arab Emirates (UAE): The role of the Ministries of Labour and Foreign Affairs in enforcing fair payments and minimum wages* M.Sc. Thesis (The Hague: Int. Institute of Social Studies) pp 11-64

[13] Pramanik S and Chackrabarti S N 2013 A study on problems of construction workers in West Bengal based on neutrosophic cognitive maps, *Int. J. of Innovative Research in Science, Engineering and Technology*. **2** 6387-6394

[14] Andrieu J B, Ucla A and Lee M 2016 *Addressing workers' rights in the engineering and construction sector – Opportunities for collaboration* BSR Working Paper (San Francisco: Humanity United) pp 12-38

## Acknowledgments

Authors wishing to acknowledge Universiti Tun Hussein Onn Malaysia for the financial support, also authors would like to thank the construction expert for his significant inputs for this research.

IOP Conf. Series: Materials Science and Engineering **271** (2017) 012034 doi:10.1088/1757-899X/271/1/012034

# The management of building fire safety towards the sustainability of Malaysian public universities

**I Y Ebenehi[1,2], S Mohamed[1], N Sarpin[1], M A N Masrom[1], R Zainal[1] and M A Mohd Azmi[3]**

[1] Department of Construction Management, Faculty of Technology Management and Business, Universiti Tun Hussein Onn Malaysia, 86400 Parit Raja, Batu Pahat, Johor, Malaysia
[2] Department of Building Technology, The Federal Polytechnic Bauchi, Nigeria
[3]Faculty of Civil and Environmental Engineering, Universiti Tun Hussein Onn Malaysia, 86400 Parit Raja, Batu Pahat, Johor, Malaysia
Corresponding author: zakimin@uthm.edu.my

**Abstract.** Recently, there had been reduction in annual budgetary allocations to public universities in Malaysia due to some economic tensions. This situation had left many institutions in question with the options of scaling down their expenses as well as sourcing for other means of meeting up with the shortfalls in allocated funds. Hence, it affects the sustainability of the building itself. This paper is an attempt to look at the possibility of reducing incidents that could lead to expending unbudgeted fund to rehabilitating property unfortunately destroyed by fire on campus, in addition to limiting risk to life and interruption of academic and business activities. Several research had been conducted on FSM, nevertheless very few consider Higher Education Institutions (HEI)s holistically. Hence this research intends to fill that gap.

## 1. Introduction

Buildings are among the most important investments for any Higher Educational Institution (HEI) which serve as meeting point for students, teachers and the communities [1]. Malaysian government and private organisations made huge investments running into billions of Ringgits in constructing these structures and the facilities within them. This is evident from the number of HEIs including Universities and Colleges that are publicly and privately funded [2]. For instance, located in Malaysia are over 500 tertiary institutions comprising 20 public universities, over 70 private universities, 33 polytechnics, 72 community colleges, 27 teachers' education institutes and about 403 active colleges [2, 3]. HEIs provide students and the entire institutions' community with an environment that is attractive, conducive to learning and academic success [4]. It is therefore important to adequately protect buildings from fire destruction, as well as the life of the users through appropriate FSM implementation [1]. Fire safety management (FSM) encompasses the combination of or coordination of some activities or programs to avert destruction from fire [1]. Such programs include escape routes provisions and maintenance, fire prevention measures, staff training, and fire drill training, etc. [5] described FSM as "the application by a manager of policy, standards, tools, information and practices to the task of analysing, evaluating and controlling fire safety". Pickard [6] asserted that fire safety strategy for a specific building requires management policies and procedures for the strategy to function effectively. Fire safety strategy should be a continuing process such that fire safety systems are regularly checked and maintained. Normally, fire takes place without warning allowing building occupants limited time to react either to extinguish the fire or to escape [7]. Effective FSM requires identifying all the potential risk associated with the premises and effectively carrying out an assessment of the adequacy of the measures provided or needed to resist the risk [8]. In addition, the concept of sustainable development was firmly embedded within the environmental movement, but still has social and economy values in sustainable use of natural resources. This implies that sustainability is a wide area covering topics from economy to social and the issues relating to ecological sustainability. While all business enterprises can make contributions towards its sustainability attainment, the ability to make a difference varies by sector and organisation size where the role of business in contributing to sustainable development remains unlimited [9, 10]. In view of that, sustainable development is good business in itself. It creates opportunities for suppliers of 'green consumers', developers of environmentally safer materials and processes. Hence, firms that invest in 'green' are eco-efficiency and engage themselves in social well-being.

## 2. Background of study

Furthermore, FSM is concerned with lessening the potential of harm to life and damage to properties resulting from the occurrence of fire in buildings [11, 12]. The importance of FSM in relation to modern large or complex buildings, has been identified for several years, though, the challenges are yet to be completely addressed. The growing complexity of buildings as well the fire safety systems' sophistication place more responsibility on management than has been traditionally essential [13]. Protecting buildings from fire can be tackled from two perspectives. Namely Building design, and Building operation /management [14]. Although, fire safety priority is very low during the building design process, it is generally included to satisfy minimum requirements for building regulations and recommendation of the insurance company [15]. Therefore, if an efficient management team is put in place to operate a property, such property would be safe from fire even if poorly designed. However, a well-designed building that is not operated by efficient management team may likely have issues with fire safety during the building lifecycle. Many of the major disasters including fires which happened in recent decades were characterised by a failure of management, whether before or during those incidences. Example was the 1987 fire at King's Cross underground station, London. Consequently, great emphasis has been placed on the importance of management in safety as well as focusing on both corporate and individual liability in this respect [13]. Apart from that, sustainable development is a process of change in which exploitation of resources, the direction of investments, the orientation of technological developments and institutional are all in harmony and enhance current and future potential to meet human needs and aspirations. It is reflected on many Governments around the world when they take on sustainable development on the agenda for the development project [16, 17]. Organisations around the world that incorporate sustainable practices to strengthen their organisation goals can increase shareholders' values and build better global market share.

## 3. Problem statement

Campus Firewatch [18] stated that no fewer than 146 people perished in campus fire in the United states between 2000 and 2011. These comprised of both on-campus and off-campus fire incidences. Several injuries were sustained and property loss worth $9 million were reported [19]. The impact of fire fatality on HEIs campuses could be very serious due to the homogenous nature of students' population with respect to age and experience [20]. Furthermore, substantial fire loads such as books, papers, and other document in lecturer's offices in High education institutions (HEI)s could contribute significantly to fire severity [21].

Research had shown that a number of fire incidence happen in tertiary institutions' buildings [22]. The cost of the incidents to the HEIs from both financial and public image perspectives could be substantial [23]. These incidents take place especially among students in hostels while they are attending classes or during siesta [4]. Numerous fire incidents mainly of small scale occur in campuses without being reported. Report from the United States Fire Administration (USFA) indicated that an

IOP Conf. Series: Materials Science and Engineering **271** (2017) 012034 doi:10.1088/1757-899X/271/1/012034

average 1700 fires were recorded annually in the United States of America (USA) [20]. The abovementioned figure increased by over 100% in the USFA's report between 2007 and 2009 as 3,800 fires were recounted in university houses in 2009. The campus fires usually occur during the months of September and October when schools are starting fresh academic session which lead to estimated fatalities of five people annually [19].

In the United Kingdom, two campus fire incidents were recorded in 2001 [25, 26]. The first one occurred in the main campus of the City University London and destroyed College Building on 25 of May. Result of the fire investigation revealed that the fire started from an office of a member of staff. The affected building was grade II listed building accommodating five academic departments, the school of journalism and computing as well as the vice-chancellor's office. The second incident took place on November 2, 2001. The fire attacked a 100 years old Bower Building at the Glasgow University, Scotland destroying PhD work of ten Botany students. It took two years to carryout restoration work on the destroyed building at the cost GBP 6.5 pound and GBP 3.5 million for the equipment. The building was opened for use in 2005 [26, 27]. Furthermore, on September 12, 2014, the GlaxoSmithKline Carbon Neutral Laboratory for Sustainable Chemistry at the Nottingham University, UK was destroyed by fire attributed to electrical fault. The GBP 20 million structure which was at advanced stage of completion was completely razed down [28, 29]. Other examples of fire incident in HEIs include a clubhouse fire at Nelson Mandela Metropolitan University South Africa in October, 2016, the University of Jos Nigeria Library fire on October 10, 2016, the hostel fire at International Islamic university of Malaysia (IIUM), 2014, Dewan Tunku Canselor (DTC), of Universiti Malaya (UM) among others [29–32]. The need for HEIs to protect students, employees, and physical facilities cannot be overstated. This is because if a disastrous loss ensues, media coverage may affect the institutions' reputation, posing a risk to future admissions, financial strengths, and endowments [21].

Report from the Fire and Rescue Department of Malaysia (FRDM) indicated a steady increase in fire incidents from 2000 to 2015 [7, 33, 34]. For instance, in 2006 FRDM attended to 18,913 fire calls, and 20,225 in 2007. Fire occurrences in 2012 was 29,848 which was 11% higher than the relatively stable statistics from 2009 to 2012. From 2013 to 2015, fire incidents recorded were 33,640 for 2013, 54,517 for 2014 and 80,183 for 2015 [7, 34]. Abdul Rahim [34] also conveyed that fire related death doubled between 2011 to 2012 that is, from 72 deaths in 2011 to 152 in 2012 and then to 165 in 2013. The cost implication of damages from fire in 2013 is about 20 billion Malaysia Ringgit (MYR 20 billion), which was an increase of MYR 874.31 million from 2012 [35, 36]. The main causes of fire include arson, cooking, smoking, electrical sources, and other unknown sources, to this Abdul Rahim [34] called for more investigative efforts to obtain useful information for developing precautionary strategies in Malaysia.

No doubt enormous investments made in the education especially in the higher education sector is in line with fulfilling a set target of creating an attractive environment, conducive for learning and academic excellence [4]. It is therefore duty bound on all stakeholders to zealously guard various infrastructure from fire destruction. In accordance with building world class higher education institutions, Malaysian government spent more on education compared to other sectors, for instance [37] reported that about 27% of the National budget was spent on education. Such gesture allows for training of human resources that shall tackle future challenges in various sectors of economy [38]. However, in recent time, the fund allocation to public universities is dwindling due to some challenging economic situation [39, 40]. Therefore, necessitating paying great attention to protecting buildings from all form of disasters including fire.

## 4. Discussion

Substantial effort had also been made at internationalizing Malaysian higher education with a resultant saving of about RM4 billion annually. RM2.5billion is the amount that could be saved if Malaysian students decide not to study abroad and take advantage of provisions made in Malaysian universities and other higher educations, whereas RM1.5 billion in net revenue realised from over 50000 students from various countries [38]. Malaysia had also set the target of becoming by 2020, the world's sixth-biggest education exporting country, and therefore making plans to accommodating about 200,000 international students [41]. This projection doubled the aspiration in the 9th Malaysia plan for attracting 100,000 students from overseas by 2010 [38]. In view of these strides, research relating to the managing fire safety in Malaysian higher education's buildings becomes imperative especially as no building have total immunity against fire occurrences [34, 44]. Previous research concerning management of fire safety in buildings had dealt with other types of building/structure such as karaoke establishment [43], oil and gas [44], non-residential high rise buildings [45, 46], residential high rise buildings [43], nursing homes [44], hotel buildings [45], heritage buildings [7], passenger terminals [50] airport terminals [51], enclosed shopping centres [52], cruise vessel construction [53] and hospital [50]. However, few of these studies had specifically looked at tertiary education buildings holistically. Though, Kong [21] studied possible implementation of performance-based design for fire safety provisions in higher education institutes, no emphasis was placed on fire safety management strategies adoption. In addition, there are research covering students' housing [4, 12, 55, 56], and other publications dealing with cafeteria [57] and library [58, 59]. This research seeks to investigate the implementation of fire safety management implementation strategies in Malaysian Higher Educational Institutions (HEI) with a view to developing a framework for effective fire safety management for all building types and occupancies in the institutions.

Statistical surveys in different parts of the world established that fires occur regularly in buildings with costly consequences [12, 60]. In the United Kingdom (UK), the direct cost of fire was estimated as 8 billion GBP in 2003. In addition, the health and safety commission (HSC) approximated the annual cost of health and safety failure as 18billion GBP [61]. The United States Fire Administration reported that about 118 deaths related to campus fire had been witnessed from year 2000 to 2015 in the United States of America (USA). 80% of those incidents occurred in off-campus housing due to the following factors [62]:

- Careless disposal of smoking materials.
- Missing or disabled smoke alarms.
- Lack of automatic fire sprinklers.
- Alcohol consumption.

According to the guidance jointly produced by [8], the obligation of an organisation's management to fire safety is essential to achieving suitable fire safety standards in buildings and in upholding staff's culture with respect to fire safety. Dublin Fire Brigade [63] consider FSM as a key factor which must be present to forestall fire disaster in buildings such as large shopping malls. According to the Dublin Fire Brigade [63] and the British Standards Institution [64] FSM structure should make provisions for the following:

- Clear lines of responsibility, authority, accountability, and resources.
- Replacements of absentees with specific responsibilities.
- An emergency services liaison officers to call for supplying information to the fire and rescue services.

FSM encompasses the whole lifecycle of the building and include the following constituents [63]:

- Day to day operation of the building.
- Changes of building.
- Change of use.
- Unit in disuse.

Similarly, Tsui and Chow [65] outlined the objectives of FSM components to include maintenance of fire safety measures and fire prevention, staff training, emergency action plan, and assessment on building alternative. FSM programmes consist of inspection, education and training, fire suppression, emergency service, evaluation of fire probability, fire prevention, report and record keeping, as well as communication [66, 67]. The following benefit can be accomplished when and effective FSM is developed and implemented [67]:

- Reduction of property insurance premiums.
- Continuity of business operation.
- Enhancement of public images and customer service.
- Promote an efficient work environment.
- Quality gain realization.
- Influence an organisation profitability.

For the above stated benefits to be realised, the fire safety managers should up to his responsibilities as stated by the Dublin fire brigade, thus:

- Awareness of fire safety features and their functions.
- To monitor general maintenance and building/refurbishment work.
- To maintain register of fire certificates and compliance for all buildings/ refurbishment works.
- Fire safety risk assessment.
- To be present at the commencement of building occupation.
- To liaise with fire authority and seek advice.

The person appointed as a fire safety manager is responsible for total control and daily building safety management. The appointed fire safety manager should be practically knowledgeable to direct firefighters to affected areas in buildings during emergency especially in complex buildings [67]. Hence, Fire safety practices and awareness is very necessary as adequate knowledge of fire, cause, prevention, suppression and the provision of adequate firefighting equipment are important to be available and handled adequately by the appointed person in any building.

## 5. Conclusion

This research seeks to investigate the implementation of fire safety management strategies in Malaysian Higher Educational Institutions (HEI) with a view to have sustainability aspects. The work creates a space for research in to detail understanding of fire safety management practices, and current implementation strategies of FSM principles among Malaysian universities. Therefore, a framework for effective fire safety management for all building types and occupancies in the institutions shall be developed and validated on completion of the research.

## 6. References

[1] Nadzim N and Taib M 2014 Appraisal of fire safety management systems at educational buildings *SHS Web of Conf.* vol 11 (Paris: EDP Sciences) 01005

[2] StudyMalaysia.com 2015 *The Malaysian Higher Education System – An Overview* Retrieved from https://www.studymalaysia.com/education/higher-education-in-malaysia/the-malaysian-higher-education-system-an-overview

[3] Ahmad A R 2013 *The Impact of Government Funding Reforms on The Strategic Planning of Malaysian Public Universities* PhD Thesis (Victoria University, Australia)

[4] Hassanain M A 2008 Fire safety in the design and operation of student housing facilities, *Struct. Surv.* **26** 55–62

[5] Howard DJ and Kara-Zaitri C 1999 Fire safety management at passenger terminals, *J. Disaster Prev. Manag.* **8** 362–369

[6] Pickard R 1994 Fire safety and protection in historic buildings in England and Ireland–Part I, *Struct. Surv.* **12** 27–31

[7] Salleh NH and Ahmad A G 2009 Fire Safety Management in Heritage Buildings: the Current Scenario in Malaysia *22nd CIPA Symp. (Kyoto Japan)* (Kyoto: CIPA) pp 1-6

[8] The Scottish Government's Police, Community Safety Directorate, HM Fire Service Inspectorate for Scotland and the Scottish Building Standards Agency and the Health and Safety Executive (2008) Practical fire safety guidance for educational and day care for children premises Retrieved on January 31, 2017 from http://www.firesafetyfirst.co.uk /publications/Scotland%20%20Educational%20&%20Day%20Care%20for%20Children%20Premises%20Guide.pdf

[9] Mohamed S and Wee S T 2016 Initial findings on knowledge transfer practices towards sustainable development at Malaysian local government, *Int. J. of Trend in Research and Development* **2** 1-3

[10] Mohamed S, Wee S T and Masrom M A N 2016 Sustainable development at Malaysian local governments from management view of the knowledge transfer practices, *IOSR Journal of Humanities and Social Science* **21** 11-17

[11] Canadian Wood Council (CWC) 2000 *Fire Safety in Residential Building* (Ontario: Canadian Wood Council)

[12] Sanni-Anibire M O and Hassanain MA 2015 An integrated fire safety assessment of a student housing facility, *Struct. Surv.* **33** 354–371

[13] Porter A 1990 Management of fire safety, *Prop. Manag.* **8** 154–158

[14] Ramli A, Akasah Z A and Masirin M I M 2013 Factors contributing building safety and health performance of low cost housing in Malaysia, *J. of Safety Engineering* **2** 1–9

[15] Mydin M A O 2014 Human Factors in Fire Safety Management and Prevention, *Analele Universităii "Eftimie Murgu"* **21** 213–219

[16] Masrom M A N, Abd Rahim M H I, Seow C A, Mohamed S and Goh K C 2017 A preliminary exploration of the barriers of sustainable refurbishment for commercial building projects in Malaysia, *Procedia Engineering* vol 180 (United Kingdom: Elsevier) pp 1363-1371

[17] Mohamed S, Abd Rahim, M H I, Wee, S T and Goh K C 2017 Preliminary concept of knowledge transfer practices, *The Social Science* **12** 502-508

[18] Campus Firewatch 2011 Fatal Campus Fires Retrieved on February 20, 2017 from http://www.campusfirewatch.com/resources/resource center/fatalcampusfires.html

[19] United States Fire Administration 2011 University Housing Fires (2007-2009), *Top. Fire Rep. Ser.* **12** pp 1–12

[20] Taub, D J and Servaty-Seib 2008 Developmental and contextual perspectives on bereaved college students, *New Directions for Student Services* **121** 15–26

[21] Kong S K 2009 A Study of Implementing Performance-Based Design for Fire Safety Provisions in Higher Education Institutes, PhD Thesis (Hong Kong: The Hong Kong Polytechnic University)

[22] Mowrer F W 1999 *Fire Safe Student Housing: A Guide for Campus Housing Administrators* Retrieved on December 5, 2016 from https://www.hsdl.org/?view&did=446415

[23] Bubka M A and Coderre P 2010 *Best Practices in Risk Management for Higher Education Addressing the 'What If' Scenarios* Retrieved on December 6, 2016 from https://www.pmacompanies.com/pdf/.../PMA_Education_BestPractices_WhitePaper.pdf

[24] United States Fire Administration 2001 Dormitory Fires, *Trop. Fire Res. Ser.* **1** 14

[25] Times Higher Education 2001 *City University's Fire* Retrieved on November 30, 2016 from https://www.timeshighereducation.com/news/city-universitys-fire/160570.article

[26] Times Higher Education 2001 *Fire Destroys PhD Work* Retrieved on November 30, 2016 from https://www.timeshighereducation.com/news/city-universitys-fire/160570.article

[27] University of Glasgow 2017 *Bower Building* Retrieved on January 17, 2017 from http://www.universitystory.gla.ac.uk/building/?id=90

[28] British Broadcasting Corporation (BBC) 2015 *University of Nottingham Laboratory Fire Caused by Electrical Fault, Says Report* Retrieved on November 29, 2016 from

www.bbc.co.uk/news/uk-england-nottinghamshire-3075143
[29] Nottinghamshire and City of Nottingham Fire and Rescue Authority Community Safety Committee 2015 *Fire Investigation for the University of Nottingham Jubilee Campus* Report of the Chief Fire Officer (United Kingdom: Nottinghamshire Fire and Rescue Service)
[30] Sadiq L 2016 *Fire Ravages Unijos Library, 6 Departments* Retrieved on October 15, 2016 from http://www.dailytrust.com.ng/news/general/fire-ravages-unijos-library-departments/165984.html#F2wXguuQ7fviSDVp.99
[31] Spies D 2016 *NMMU Clubhouse Torched Overnight* Retrieved on December 01, 2016 http://www.news24.com/SouthAfrica/News/pics-nmmu-clubhouse-torched-overnight-20161017
[32] Ali Z M 2003 *Rehabilitation of the Tunku Canselor Hall, University of Malaya, Kuala Lumpur, Malaysia* Retrieved on October 15, 2016 from https://www.researchgate.net/publication/272420112_REHABILITATION_OF_THE_TUNKU_CANSELOR_HALL_UNIVERSITY_OF_MALAYA_KUALA_LUMPUR_MALAYSIA
[33] Sulaiman A 2008 *Exploring Fire Safety Awareness Among the Malaysian Public MSc Building Technology* MSc Thesis (Penang: Universiti Sains Malaysia)
[34] Abdul Rahim M S N 2015 The current trends and challenging situations of fire incident statistics, *Malaysian J. Forensic Sci.* **6** 63–78
[35] Department of Statistics Malaysia 2013 *State/District Social Statistics Malaysia 2013* (Putrajaya: Department of Statistics Malaysia)
[36] FRDM 2013 *Statistik Kebakaran mengikut Jenis Kebakaran Tahun* (Putrajaya: Fire and Rescue Department of Malaysia)
[37] Musa B M 2003 An Education System Worthy of Malaysia (Bloomington: iUniverse)
[38] Kaur S, Sirat M and Azman N 2008 Globalisation and internationalisation of higher education in Malaysia (Penang: National Higher Education Research Institute, Gelugor, Penang, Malaysia)
[39] Malay Mail 2016 *Budget 2017: Public Universities Suffer Almost 20pc Spending Cut* Retrieved on January 17, 2017 from http://www.themalaymailonline.com/malaysia/article/budget-2017- public-universities-suffer-almost-20pc-spending-cut.
[40] Else H 2017 *Malaysia Cuts Public University Funding* Retrieved on January 17, 2017 from https://www.timeshighereducation.com/news/malaysia-cuts-public-university-funding
[41] Hughes J 2015 *Malaysia Seeks 200,000 International Students by 2020* Retrieved on May 11, 2017 from https://www.masterstudies.com/news/Malaysia-Seeks-200-000-International-Students-by-2020-495/.
[42] Ramachandran G 1999 Fire safety management and risk assessment, *Facilities* **17** 363–377
[43] Lui C H G and Chow W K 2000 A preliminary proposal on fire safety management for karaoke establishments *Conf. Proc. of the 18th Int. System Safety Conference (Fort Worth, Texas, USA)* (Virginia: The International System Safety Society) pp 76-84
[44] Santos-reyes J and Beard A N 2001 A systemic approach to fire safety management, *Fire Safety Journal* **36** 359–390
[45] Chow W K 2002 Proposed fire safety ranking system EB-FSRS for existing high-rise nonresidential buildings in Hong Kong, *J. Archit. Eng.* **8** 116–124
[46] Tharmarajan P A 2007 *The Essential Aspects of Fire Safety Management in High Rise Building* MSc Thesis (Johor: Universiti Teknologi Malaysia)
[47] Yatim Y M 2009 *Fire Safety Models for High-Rise Residential Buildings in Malaysia* PhD Thesis (Edinburgh: Heriot-Watt University)
[48] Moore J 2012 *Assessment of Fire Safety and Evacuation Management in Nursing Homes* MPhil Thesis (Dublin: Dublin Institute of Technology)
[49] Chen Y Y, Chuang Y J, Huang C H, Lin CY and Chien S W 2012 The adoption of fire safety management for upgrading the fire safety level of existing hotel buildings, *Build. Environ.* **51** 311–319

[50] Howarth D J and Kara-Zaitri C 1999 Fire safety management at passenger terminals, *Disaster Prev. Manag. An Int. J.* **8** 362–369
[51] Chow W K and Ng M Y 2003 Review on fire safety objectives and application for airport terminals, *J. Archit. Eng.* **9** 47–54
[52] Freeman M 2010 *How Fire Risks Should Be Managed in Enclosed Australian Shopping Centres* PhD Thesis (Victoria: Victoria University)
[53] Räisänen P 2014 *Fire Risk and its Management in Cruise Vessel Construction Projects* PhD Thesis (Greater Helsinki: Aalto University)
[54] Ong C and Suleiman M Z 2015 Problems in implementation of fire safety management in Malaysia government hospital, *Adv. Environ. Biol.* **9** 47–50
[55] Griffin B D 2011 *Effective Fire Safety Education for Residential Students* PhD Thesis (Greensboro: The University of North Carolina)
[56] Agyekum K, Ayarkwa J and Amoah P 2016 Challenges to fire safety management in multi-storey students' hostels, *Mod. Manag. Sci. Eng.* **4** 53–61
[57] Hassanain M A and Hafeez M A 2005 Fire safety evaluation of restaurant facilities, *Struct. Surv.* **23** 298–309
[58] Khalid S J B M 2015 Disaster preparedness in academic libraries in Malaysia: An exploratory study, *Int. J. of Humanities and Soc. Sc.* **9** 3413-4319
[59] Hassanain M A and Al Ashwal N 2005 An approach to assess fire safety requirements in library facilities, *Facilities* **23** 239–252
[60] Chen T, Pan L H, Zhang S, Narayanan and Soldner N 2013 Experimental study of evacuation from a 4-storey building *Procedia Eng.* vol 62 (United Kingdom: Elsevier) pp 538–547
[61] Furness A and Muckett M 2007 *Introduction to Fire Safety Management* (Oxford: Butterworth-Heinemann) pp 2
[62] U.S. Fire Administration 2015 *Campus Fire Fatalities in Residential Buildings (2000-2015)* (Maryland: US Fire Administration)
[63] Dublin Fire Brigade 2012 *Fire Safety Management of Shopping Centres with Covered Malls* (Dublin: Dublin City Council)
[64] The British Standards Institution 2017 *Section 9: Managing Occupied Buildings* (London: British Standard) pp 220–250
[65] Tsui SC and Chow W K 2004 Legislation aspects of fire safety management in Hong Kong, *Facilities* **22** 149–164
[66] Malhotra H 1993 Proposed Code for Fire Safety in Buildings for the State of Sao Paulo (United Kingdom: AGNICONSULT) pp 1-98
[67] Chow W K 2001 Review on fire safety management and application to Hong Kong, *International Journal on Engineering Performance-Based Fire Codes* **3** 52–58

**Acknowledgements**
The authors are unreservedly grateful to Ministry of Education (MOE), Pejabat Pendaftar UTHM, Office of Research, Innovation, Commercialisation and Consultancy (ORICC), UTHM, under the Graduate Postgraduate Incentive Scheme (U729), RACE (1517), Centre for Graduate Studies (UTHM) and FPTP, Universiti Tun Hussein Onn Malaysia for the support of this study.

IOP Conf. Series: Materials Science and Engineering **271** (2017) 012035 doi:10.1088/1757-899X/271/1/012035

# Need of risk management practice amongst bumiputera contractors in Malaysia construction industries

**Nur Syuhada Fadzil [1], Nurazuwa Md Noor [1] and Ismail Abdul Rahman[1]**

[1]Jamilus Research Centre, Faculty of Civil and Environmental Engineering, Universiti Tun Hussein Onn Malaysia, 86400 Parit Raja, Batu Pahat, Johor, Malaysia.
Corresponding author: nursyuhada_nsf@yahoo.com

**Abstract.** Malaysia construction industry has been growing continuously with double-digit growth for the past two decades. There are many issues surrounded the industry such as delay in delivery, cost overrun, quality and safety. In bumiputera contractor's context, the literature review found that 32 issues were encountered in the construction project. These issues can be handled properly with the application of risk management. According to previous researchers, application of risk management in Malaysia construction industry is at the low level where it is implemented in traditional ways that are brainstorming and checklist. These were due to lack of knowledge, the high cost of hiring experts and avoiding extra cost. Besides that, this study also intentions to analyse the risk categories in the construction industry as well as identify risk management process to resolve the construction issues. Hence, this paper presents issues engulfed by Bumiputera contractors which can partly be resolved by applying risk management practice in carrying out the construction activity. This may inspire the contractors to apply the risk management practice in ensuring the success of their construction project.

## 1. Introduction

Risk management plays a vital role in the construction industry to remove or reduce the negative impact because there are many risks encountered in construction project particularly [1]. In Malaysia construction industry, only a few companies and industrial practitioners apply risk management during handling project [2].

According to Alkaf [3], the construction industry is usually riskier because of the complexity in coordinating various activities compared to other industry activity such as manufacturing, agricultural, and mining. This can be proved by the data prepared via Malaysia Department of Occupational Safety and Health [4] shows that five years of average mortality for this industry is 99 employees per year (from 2011 to 2015). Death rate per 100,000 construction workers by 2015 is 10.94, higher than the average death rate for five years (2011-2015) which is 8.17. The trend of the deadly injury of construction workers shows an increase since 2012, and 140 construction workers have died on the year 2015, the highest recorded since 2001 and the highest in the 21st century. Thus, the effective risk management should be part of the construction project organizational culture for allow project going competently, enhancing productivity and bring greater rewards to project performance [5, 6]. Besides that, stakeholder and project team of the construction projects will gain benefits from the effective risk management [7]. In the construction projects, define contractors are the most powerful and carry ultimate responsibility amongst stakeholder [8]. Contractors also are considered as one of the main parties to ensure project success [9, 10].

In Malaysia context, contractors are separated into two groups which are bumiputera and non-bumiputera. It is mandatory for all construction companies in Malaysia, regardless of whether they are bumiputera or non-bumiputera, local or foreign contractor to register with the Construction Industry Development Board of Malaysia (CIDB) before they undertook any tendering capacity and paid up capital [11]. Registering as bumiputera contractor must require the majority of shares (at least 51%) owned by Malays/bumiputera and the majority (at least 51%) of the workforce in the company are Malays/bumiputera and also awarded bumiputera status by Contractor Service Centre (PKK) [10]. Study by Rahman dan Rahmat [10] indicates that bumiputera contractors are predominantly facing failure because of the inadequate planning, poor site management, inadequate contractor experience, lack of financial management, problems with subcontractors, lack of construction materials and machineries, labour supply, lack of communication among parties and mistakes during the construction phase. These issues may arise due to the lack of implementing a systematic risk management because finding by Abdul-Rahman states that implementation of risk management process in Malaysian construction industry is still at a low level [12]. Therefore, contractors have to reconsider their approach to risk within their project and organization via implementing a good risk management practice amongst bumiputera contractors during handling the construction projects as it will able to anticipate the occurrence and impact of negative events which may affect the overall performance of the projects [13].

## 2. Category of risk in construction industry

Risks associated with construction industry can be categorized in numerous methods and this study classified the risk categories into four (4) main groups which are financial risks, logistical risks, contractual risks, and managerial risks as shown in table 1.

**Table 1.** Risk main groups in construction industry.

| Risk Category | References* | | | | | | | | | |
|---|---|---|---|---|---|---|---|---|---|---|
| | A | B | C | D | E | F | G | H | 1 | J |
| 1. Financial | * | * | * | * | * | * | * | * | * | * |
| 2. Logistical | * | * | * | * | * | * | * | * | * | * |
| 3. Contractual | | * | * | * | * | * | * | | * | * |
| 4. Managerial | * | * | * | * | * | * | * | * | * | * |

* A[2], B[3], C[5], D[8], E[14], F[15], G[16], H[17], I[18], J[19]

### *2.1 Financial Risks*

A literature review conducted in this study found ten (10) articles related to financial risk in the construction industry as in Table 1. In any organization, financial is one of the major risks which is very important and always need to be aware. This factor could have a substantial impact on the profit or loss [13]. Financial risks are due to lack of financial management, inadequate cash flow, mismanagement of cash flow, inaccurate project estimation , difficulty in securing the bank loan, fluctuation in foreign exchange, or late of the progress payment [3, 5, 14, 16, 19]. The financial risk can give impact to construction project such as poor of quality, workmanship, environment, safety and health that leads to project failure [18].

### *2.2 Logistical Risks*

Logistics risks involve the planning and organizing process to ensure that resources are always available when they are needed, so that an activity or process happens effectively. Effective logistics planning is essential for any construction project as it could provide significant benefits in efficiency and productivity that significantly reduce the overall cost of the project. An example of resources needed is transportation facilities, construction equipment spare parts, material, fuel and labour [2, 3, 5, 8, 14-19].

*2.3 Contractual Risks*
Based on the finding by Adnan et al., contractual risks caused by disagreements arising from flawed contract documents, inappropriate types of contract, improper tendering procedure or improper contractual clauses [13]. This contractual risk also including with design risk. For instance of design risks are incomplete design, design changes, design failure, and appropriateness of specifications [8, 15, 16, 19]. Three (3) main components must be considered and managed effectively to avoid contractual risks are quality, timeliness, and cost-effectiveness as well as sufficient knowledge on a standard of procedure.

*2.4 Managerial Risks*
Managerial risk refers to the failure of the organization or company in managing their management which eventually contributes to the negative impact of the company's reputation. Concurrently, it induces industrial relation problems [19]. Referred to Reddy, human resource management is one of the factors contributes to managerial risk [15]. This is due to lack of communication and attitude of participant [16]. Other contributors to managerial risks include natural disasters such as earthquakes, floods, cyclones and typhoons [20]. As a result, these natural disasters cause destruction of facilities, equipment, material and labour.

## 3. Construction issues faced by bumiputera contractors
Contractors are independent business organizations who have been awarded the project to produce the required end product as stipulated in the contract documents [21]. Also, Rahman and Rahmat defines contractors are someone who is legally given responsibilities to execute the awarded construction project as in the contract [10]. Alkaf et al. and Siang et al. previously reported that different project phases in the company resulted in different risk management practice and approach. Hence it exposed to different types of risks. In bumiputera contractor context, problem encountered by bumiputera contractor was summarized and clustered according to the suitability of the identified risk categories are shown in table 2.

## 4. Risk management in construction project
Risk management is important in managing a project being able to anticipate occurrence of events that may affect construction projects and define actions that can minimise their effects [22]. In PMBOK [23], risk management is one of the ten (10) components should ponder to ensure project success. The risk management comprises of six (6) elements which are plan risk management, identify risk, perform qualitative risk analysis, perform quantitative risk analysis, plan risk response and control risks as shown in figure 1. This process adopted from the PMBoK 2013.

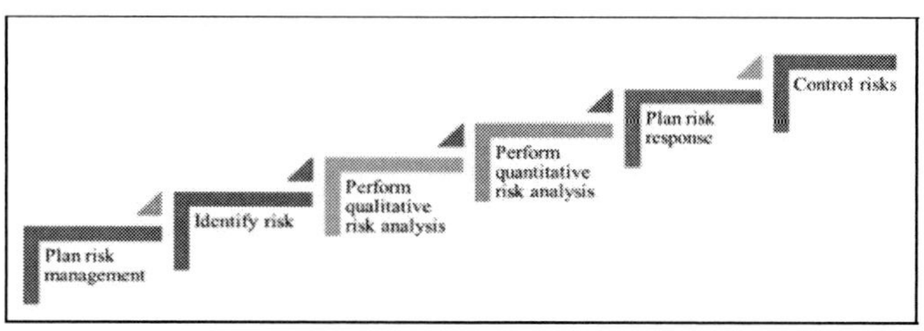


**Figure 1**. Risk management process [23].

*4.1 Plan risk management*

Plan of risk management is the planning process to conduct risk management activities via the approach of risk management implementation to ensure it is commensurate with the project requirement [5, 23]. It also involved of the past problems, which is a trigger change to future work as basic practices before work [24].

*4.2 Risk identification*

Risk identification is a process for identifying the type and effect of the risk that can affect the project and document their characteristics [2, 23]. The purpose of this process for uncovering any risks that could potentially affect a process [14]. According to PMBoK, it stated that stakeholders and experts on risk management are a participant in risk identification activities. This stakeholder involving project manager, project team members, risk management team, customers and other.

*4.3 Perform qualitative risk analysis*

Qualitative risk analysis is prioritizing risks by appraising and combining likelihood event and effect for further analysis [23]. Moreover, qualitative risk analysis is to select the most significant techniques for easier applied [24]. The purpose of this method is to enable project managers to decrease the level of uncertainty and to focus on high-priority risk as well as to reveal the risk attitude to a project team and other stakeholders.

*4.4 Perform quantitative risk analysis*

Perform quantitative risk analysis is analysing numerical process about the impact of identified risk on project goals [23]. Also measuring the probability and consequences of risk and estimating their implications for project objective [25]. On most projects, risk does not happen one at a time. Instead, they interact in groups, with some risk causing others to be more likely and some risks making others impossible. For the most part, qualitative risk analysis considers risks individually and allows development of a good understanding of each one (although grouping risks into categories can give some insights into patterns of risk exposure). However, it is sometimes necessary to analyse the combined effects of risks on project outcomes, regarding how they might affect overall time and cost particularly.

*4.5 Plan risk responses*

Plan of risk responses is an options and actions to increase chances and to decrease threats against project goals [23]. Necessary to identify the available approaches when responding on risks. Four possible approaches when risks respond are risk avoidance, risk transfer, risk retention and risk reduction [2]. Risk avoidance is known as risk elimination and it includes not undertaking an activity, which may carry risk. Risk transfer is assigned to other parties by using contracts or insurances. Risk reduction is a search for methods to reduce the likelihood of the risk. Risk retention refers to the loss/benefit of the gain of a risk is accepted when it occurs. Hence, this method giving benefit to addressing risks according to priority, inserting activities and resources into the financial plan, schedule and project management plan as necessary as identified by [23].

IOP Conf. Series: Materials Science and Engineering **271** (2017) 012035 doi:10.1088/1757-899X/271/1/012035

**Table 2.** The list of construction issues which can affect the performance of the Bumiputera contractors and these issues are clustered into four (4) risk categories which are identified as financial, workforce, contractual, and managerial.

| Issues | References* | | | | | | | | | | |
|---|---|---|---|---|---|---|---|---|---|---|---|
| | A | B | C | D | E | F | G | H | I | J | K |
| **Financial Category** | | | | | | | | | | | |
| 1. Difficult to retain skilled staff because financial issues | | | | * | * | | | | | | |
| 2. Challenge in obtain equipment at reasonable price | | | | * | | * | * | | | | |
| 3. Lack of financial management | | | * | * | * | | | | | | |
| 4. Lack of financial capital to start new project | | | | * | | | | | | | |
| 5. Inadequate cash flow to run the construction project | * | | * | * | * | | * | | | | |
| 6. Mismanagement of cash flow | | | | | | * | | * | | | |
| 7. Difficult to purchase construction material | | * | | | | | * | | | | |
| 8. Fluctuation of construction materials cost | | | | | * | | | | | | |
| 9. Improper or inaccurate project estimation | | | * | | | | | | | | |
| 10. Difficulty to securing bank loan | | | * | * | * | | | | | | |
| 11. Late of progress payment | | | | | | | * | | | | |
| **Workforce Category** | | | | | | | | | | | |
| 1. Lack of knowledge and skill on new technology approach | | | * | | | | | | | | |
| 2. Lack of entrepreneurial skills in managing construction projects | | | | | | | | * | * | | |
| 3. Lack of knowledge and experience on construction | | | | | * | | * | | | | |
| 4. Lack of apprenticeship to resolve problem | | | | * | * | | | | | | |
| 5. Lack of local construction workers | | | * | * | * | | | | | | |
| 6. Difficulty to getting skilled workers | | | | * | * | | | | | | |
| 7. Lack of machineries and construction material | * | * | * | | | | | | | | |
| **Managerial** | | | | | | | | | | | |
| 1. Lack of teamwork amongst parties | | | | | * | | | | | | |
| 2. Lack of communication skill | | | * | | | | * | | | | |
| 3. Conflict between contractor and other parties | * | | | | | | | * | | | |
| 4. Lack of opportunity to participate and getting new project | | | | | | | | | | * | * |
| 5. Lack of networking regarding project procurement information | | | | | | | | * | | | |
| 6. Lack of handling workers using systematic approach | * | | * | | | | | | | | |
| 7. Lack of good reputation based on the previous work record | | | | | | | | * | | | |
| 8. Inefficient of planning and management | | | * | | | | | * | | | |
| 9. Lack of management | | | | | | | | * | | | |
| 10. Poor site conditions | * | | * | * | | | | | | | |
| **Contractual** | | | | | | | | | | | |
| 1. Changes in requirements and specifications during construction phase | * | | | | | | | | | | |
| 2. Lack of tendering technique | | | | * | | | | | | | |
| 3. Dependent on government allocated project | | | | | | * | | * | | | |
| 4. Insufficient instruction and information in the contract specification, drawing and design | | | * | | | | | | | | |

* A[8], B[10], C[21], D[26], E[27], F[28], G[29], H[30], I[31], J[32], K[33]

## 5. Conclusion

There are many issues/challenges in ensuring the success of construction project which depends on the given time and cost with safety and quality as stipulated in the contract document. Hence, it is important to apply risk management practices in navigating the implementation of the project by identifying the real risk which leads to the project failure. Irrespective of the size of the project, by adopting the risk management the assurance of project success is quite certain. Bumiputera contractors should consider risk management practice because it provides guidance and awareness to control any risks in their construction project. Besides that, implementing risk management practice will increase the likelihood and impact of positive events and restrain the negative events in the project. This can be achieved by implement the process in project risk management. Therefore, the contractor that applied the recommended approach such as from PMBOK (project management book of knowledge) will be able to manage project better and offer assistance to bumiputera contractors in managing the awarded projects successfully.

## 6. References

[1] Haseeb M 2011 Hazard risk analysis and management in construction sector of Pakistan, *International Journal of Engineering Research* **2** 35–42
[2] Kang B G, Fazlie M A, Goh B H, Song M K, and Zhang C 2015 Current practice of risk management in the Malaysia construction industry–The process and tools/techniques, *Int. J. of Structural and Civil Engineering Research* **4** 371–377
[3] Alkaf N, Karim A, Rahman I A, Memmon A H and Jamil N 2012 Significant risk factors in construction projects: contractor ' s perception *IEEE Colloqium on Humanities, Science & Engineering Research* (United States: IEEE) pp 351–354
[4] Ab-Rahman M F 2016 *Kecederaan Maut Kepada Pekerja Pembinaan Malaysia* Retrieved on January 30, 2017 from http://www.dosh.gov.my/index.php/en/construction-safety/e-buletin/2016-2/2258-bil-3-2016-accident-statistics/file
[5] Goh S and Abdul-rahman H 2013 The identification and management of major risks in the Malaysian construction industry, *J. of Construction in Developing Countries* **18** 19–32
[6] Yusuwan N M, Adnan H and Omar A F 2008 Clients' perspectives of risk management practice in Malaysian construction industry, *J. Politic and Law* **1** 121–130
[7] Chun Siang L and Ali A S 2012 Implemention of risk management in the Malaysian construction industry, *J. of Surveying, Construction & Property* **3** 1–15
[8] Ahmad Zaini A, Adnan H and Che Haron R 2010 Contractors' approaches to risk management at the construction phase in Malaysia *Proc. of the International Conference on Construction Project Management (Chengdu, China)* (Germany: MPRA) pp 330–334
[9] Alaghbari W, Mohd. Razali A K, Azizah S and Ernawati 2007 The significant factors causing delay of building construction projects in Malaysia, *Engineering, Construction and Architectural Management* **14** 192–206
[10] Rahman I A and Rahmat N I 2016 Causes of failure among Bumiputra contractors, *ARPN J. of Engineering and Applied Sciences* **11** 9882-9886
[11] Ernawati M K, Syarmila H H, Norhidayah M U and Faizal B 2012 The critical review on the Malaysian construction industry, *J. of Economics and Sustainable Development* **3** 81–87
[12] Abdul-Rahman F, Wang H C and Sheikh-Mohamad 2015 Implementation of risk management in the Malaysian construction industry, *J. of Construction Engineering* **3** 1–6
[13] Adnan H, Rahmat M N, Fatanah N and Mazali N 2008 Risk management assessment for partnering projects in the Malaysian construction industry, *J. of Politics and Law* **1** 76–81
[14] Tipili L G and Yakubu I 2016 Identification and assessment of risk factors affecting construction projects, *Int. J. of Engineering and Advanced Technology Studies* **4** 20–32
[15] Reddy A S 2015 Risk Management in construction industry, *Int. J. of Innovative Research in Science, Engineering and Technology* **4** 10058–10067

[16] Mahendra P A, Pitroda J R and Bhavsar J J 2013 A study of risk management techniques for construction projects in developing countries, *Int. J. of Innovative Technology and Exploring Engineering* **3** 139–142

[17] Ahmad-Zaini Z, Endut A and Shehu I R 2012 A preliminary study of the subcontractor's risk identification for the construction projects *IEEE Symposium on Business, Engineering and Industrial Applications* (United States: IEEE) pp 878-883

[18] Ahmad-Zaini R and Takim A 2011 Contractors' strategic approaches to risk assessment techniques at project planning stage *IEEE Symposium on Business, Engineering and Industrial Applications* (United States: IEEE) pp 320–325

[19] Subramaniyan V S and Veerakumar K 2017 Risk management in construction industry, *Int. J. for Risk Assesment* **2** 1–4

[20] Kamane S K and Mahadik S A 2014 Risk management in construction industry, *IOSR J. of Mechanical and Civil Engineering (IOSR-JMCE)* **2** 59–65

[21] Abdul Rahman Ayub J E 2006 *Identification of Challenges Faced by Bumiputra Contractors & Roles of Local Government in Ensuring a Successful Completion of a Project* Retrieved on December 28, 2016 from http://www.oocities.org/icolg2006/Identification_of_Challenges.htm

[22] Serpell A, Ferrada X, Rubio L and Arauzo S 2015 Evaluating risk management practices in construction organizations *Procedia - Social and Behavioral Sciences* vol 194 (United Kingdom: Elsevier) pp 201–210

[23] Project Management Institute 2013 *A guide to the project management body of knowledge (PMBOK ® guide)* (Pennsylvania: Project Management Institute, Inc.)

[24] Kendrick T 2003 *Identifying and Managing Project Risk* (New York: American Management Association)

[25] Barkley B T 2004 *Project Risk Management* (United States of America: McGraw-Hill)

[26] Yin K Y 2005 *Why are There So Few Successful Bumiputera Contractors?* Retrieved on February 22, 2017 from https://aliran.com/archives/monthly/2005a/5k.html

[27] Yin K Y 2013 *Bumiputera Contractors: Producing Competitive Bumiputra Contractors* Retrieved on January 11, 2017 from https://www.malaysiakini.com/letters/223247

[28] Mohamed N J 2011 *Kontraktor Bumiputera Perlu Jadi Trend Setter* Harian Metro pp 20 (August 11)

[29] Hanif M I M 2011 *Problems and Dilemmas of Class F Contractors' Participation in Large Projects* Master's Thesis (Johor: Universiti Teknologi Malaysia)

[30] Thomas Menkhoff C Y H, Evers H-D and Chay Y W 2013 *Catalyst for Change: Chinese Business in Asia* (Singapore: World Scientific Publishing Co. Pte. Ltd.)

[31] Muhammad N 2011 *Kontraktor Bumiputera Lemah Dalam Praktikal* Retrieved on March 23, 2017 from http://www.theborneopost.com/2011/12/12/kontraktor-bumiputera-lemah-dalam praktikal/

[32] Bernama 2012 *Persatuan Kontraktor: Beri Bumiputera Lebih Ruang Sertai Projek Johor* Retrieved on March 23, 2017 from http://www.sinarharian.com.my/persatuan-kontraktor-beri-bumiputera-lebih-ruang-sertai-projek-johor-1.84407.

[33] Hasli F A 2014 *Beri Peluang Pada Kontraktor Bumiputera* Retrieved on from March 23, 2017 from http://www.sinarharian.com.my/edisi/sabah-sarawak/beri-peluang-pada-kontraktor-bumiputera-1.292101.

**Acknowledgements**
This paper is partly sponsored by BP Renal Care and the Centre for Graduate Studies, Universiti Tun Hussein Onn Malaysia.

IOP Conf. Series: Materials Science and Engineering **271** (2017) 012036 doi:10.1088/1757-899X/271/1/012036

# Causative factors of cost overrun in highway projects of Sindh province of Pakistan

**S Sohu[1], A Halid[1], S Nagapan[1], A Fattah[2], I Latif[3] and K Ullah[1]**

[1]Jamilus Research Centre, Faculty of Civil and Environmental Engineering, Universiti Tun Hussein Onn Malaysia, 86400 Parit Raja, Batu Pahat, Johor, Malaysia
[2]University of Engineering Science and Technology Nawabshah, Pakistan
[3]Department of Civil and Environment Engineering, University of Nizwa, Initial Campus, Oman
Corresponding author: sohoosamiullah@gmail.com

**Abstract**. Cost overrun is an increase of cost of project from approved budget which was signed by parties at the time of tender. Cost overrun in construction of highway projects is a common problem worldwide and construction industry of Pakistan is also facing this crucial problem of cost overrun in highway projects of Pakistan. The main objective of this research is to identify the causative factors of cost overrun in highway projects of Sindh province of Pakistan. A well designed questionnaire was developed based on 64 common factors of cost overrun from literature review. Developed questionnaire was distributed among selected 30 experts from owner/client, designer/consultant and contractor who have experience more than 20 years' experience in highway projects. The collected data was statistical analyzed. After analysis results showed that delay process in payment by client, inadequate planning , client interference, poor contract management, delay of decision making, change of scope of project and financial problems faced by client were most causative factors of cost overrun in highway projects. This research will provide alertness to stakeholders of highway projects of Sindh province to avoid cost overrun in projects.

## 1. Introduction

Construction industry is developing in developed and developing countries all over the world is [1]. Construction industry plays an important role in the growth of economy of the any country, furthermore it is proved that it is facing many challenges and issues, from others cost overrun or budget overrun is one of the critical issue [2]. Through the project management life cycle of construction projects, cost is known as most important consideration as well as the main one of the most significant parameters for measuring achievement of construction projects [3]. According to Agren *et al* [4] that cost overrun is a crucial problem globally and that many projects are not completed within the approved cost and budget of the project. Cost overrun is a critical issue in construction industry which causes disputes between client, contractor and consultant [5]. Like developing countries Pakistan is also facing the critical cost overrun problem where the cost of the projects exceeds up to 100 % of actual cost of the project. Many factors like cash flow problem faced by contractor and inadequate planning leads to delay and stopping of construction activities at site which causes cost overrun [6].

## 2. Cost overrun

Cost overrun can be defined as cost of any project is increased from the actual and approved cost of the project [7]. Cost overrun can be known as budget overrun where the final cost of the project exceeds the approved and original cost of the project [8]. Cost overrun can also be defined as the change between the original budget and estimated budget as a percentage of the estimated budget, with all budget calculated in constant prices [9]. Cost overrun is a change of initially approved cost and the final cost of the completed project [5].

## 3. Previous Research

In Pakistan, many researches have been conducted to identify major and critical factors cost overrun or budget overrun in construction industry. A study carried by Chouhadary *et al* [10] in which major factors of cost overrun or cost increase in construction industry of Pakistan were found. From his study most causative factors of cost increase and cost overrun. From his study most causative factors of cost increase were errors and mistakes in design and drawings, inaccurate cost and time estimates, cash flow problem faced by client and contractor, variations in materials price and inadequate planning. Another study carried by Azhar *et al* [6] in which most critical factors which causes cost overrun were identified by using quantitative method from stakeholders of construction industry. Most critical factors of cost increase were fluctuation or change in material prices, machinery maintenance cost, bidding process at lower rate, delay in approval of drawings, mistakes in cost estimates, addition works by contractors and sudden changes and adequate planning by government. Furthermore, another study carried Ejaz *et al* [11] regarding critical and causative factors of cost increase or cost overrun. Results of study showed that bidding on lower rate, delay process of payment to contractor, shortage of materials, poor cost control, delay in approvals and rework at site were the causative and critical factors of cost increase or cost overrun.

## 4. Research procedure

In this research quantitative method is adopted to understand the opinion of construction stakeholders towards factors contributing cost overrun in construction of highway projects in Sindh province of Pakistan. In this regard, investigation was made up in two phases. First phase included literature work and conducting interviews. From this phase 64 factors of cost overrun were identified. In second phase part a questionnaire was developed which comprises of two stages. Stage A includes demography of the respondents (experience, biodata, qualification, contact details), stage B includes factors of cost overrun in construction industry. Each respondents were asked to rank and score factor according to their knowledge.

## 5. Data collection and analysis

In this research a questionnaire was established and designed on likert scale. Likert scale indicate the statement of the respondent is asked and to calculate according to subjective or objective criteria; mostly level of disagreement and agreement is measured. The designed questionnaire of five ordinal actions of Likert scale from one (1) to five (5) according to level of a contribution. Each scale indicates the resulting score:

(5) = Very high Effective
(4) = High effective;
(3) = Medium effective;
(2) = Low effective; and
(1) = Very low effective

Designed questionnaire was distributed hand to hand among 30 respondents of having more than 20 years' experience in construction of highway projects in Sindh province. Reliability test was conducted to check the reliability of gathered data by using SPSS 20. For this Cronbach's alpha was calculated and value of Cronbach's alpha was 0.874. According to Yuan *et al* [12] that if the value of

Cronbach's alpha is more than 0.7 to 1 is acceptable and less than 0.7 is not acceptable. Ranking of the factors was accord to mean value from the response of each respondents. Higher value indicates the maximum value of cost overrun factor. In order to find the causative factor of cost overrun mean value of each factor is calculated.

## 6. Results and discussions

Table 1 shows the results of analysis of each factor with score. Based on results of survey for each factor, the factors whose mean score above than 4.6 were selected as causative factors of cost overrun in construction of highway projects. after analysis of results shows that financial delay in process payment by owner/client, inadequate planning, owner/client interference, poor contract management delay in decision making, laws and regulations frame work, change in scope of project, financial difficulties faced by Owner/Client and policy bidding tender to the lowest price with their mean value 4.90, 4.668, 4.667, 4.651, 4.644, 4.633, 4.621, and 4.618 respectively are the causative factors of cost overrun in construction of highway projects in Sindh province of Pakistan as shown in table 2.

### *6.1 Delay in process payment by owner/client*

Delay in progress payment by owner/client was most causative factor with mean value 4.90. Mostly owner/client delay the payment process to the contractors and with this the construction activities at site mostly stop and causes the cost overrun and time overrun respectively. For the small, medium and large scale owners delay in payment may cause unpredicted events. Construction of highway projects is too expensive for one owner to complete the project without continuous flow of payment [13].

### *6.2 Inadequate planning*

Inadequate planning was found common problem in construction of highway projects which causes cost overrun problem. Overlooking the ground conditions leads to cost overrun of project. Inadequate planning by client and contractor in project also leads the cost overrun in project [14]. Inadequate planning of the project creates problem of cost overrun because so many activities are involved in construction phase. Due to inadequate planning, project comes with cost overrun problem [14].

### *6.3 Owner/client interference*

Owner/client is the main player of the project and interference of owner/client can changes many things in construction of project. Survey results showed that factor interference by owner/ client causes cost overrun in highway. Owner/client interference like changes in key posts and appointment of contractor on his own selection causes cost overrun or sometimes changes in policies and approval and delays in construction activities also causes cost overrun [15].

### *6.4 Poor contract management*

Poor contract management is also one of causative factor of cost overruns in highway projects which is mostly causes cost overruns when project is given on not clear and well understood. Mostly projects are not clear and contractor perspective is not clear about contract document which mostly causes cost overrun [16].

### *6.5 Delay in decision making*

Delay in decision making is one common factor with high mean value which causes cost overrun in construction of highway projects. Delay in taking decision is mostly on part of client and consultant, many construction activities remain stopped until final decision. Because of not frequent meetings and other parameters causes factor delay in decision making [17].

### *6.6 Laws and regulations frame work*

Law and regulations frame work is a causative and critical factor of cost overrun in construction of highway projects. Because of not clear and general policies of projects contractor faces problem is execution which causes cost overrun in every stage of execution [18].

**Table 1.** Critical factors with score.

| S.No | Factor | Score | S.No | Factor | Score |
|---|---|---|---|---|---|
| 01 | Delay in process payment by owner/Client | 4.9000 | 31 | Additional works | 3.6200 |
| 02 | Inadequate planning | 4.6681 | 32 | Schedule delay | 3.6100 |
| 03 | Owner/Client interference | 4.6673 | 33 | Mistakes and discrepancies in contract document | 3.6010 |
| 04 | Poor contract management | 4.6510 | 34 | Contractual procedure and type of contract | 3.6000 |
| 05 | Delay in decision making | 4.6443 | 35 | Omissions and errors in the bills of quantities | 3.5667 |
| 06 | Laws and regulations frame work | 4.6333 | 36 | Impractical and complicated design | 3.5666 |
| 07 | Change in scope of project | 4.6213 | 37 | Social and cultural impacts | 3.5333 |
| 08 | Financial difficulties faced by Owner/Client | 4.6180 | 38 | Rework | 3.5233 |
| 09 | Policy bidding tender to the lowest price | 4.596 | 39 | Delay payment to supplier/subcontractor | 3.5313 |
| 10 | Lack of experience | 4.5900 | 40 | Limited range of supplier | 3.5033 |
| 11 | Shortage of skilled labour | 4.5667 | 41 | Poor project management | 3.5003 |
| 12 | Unsuitable construction methods | 4.5333 | 42 | Incompetent subcontractor | 3.5133 |
| 13 | Late delivery of equipment's | 4.5000 | 43 | High cost of machinery and its maintenance | 3.5033 |
| 14 | Shortage of workers | 4.5000 | 44 | Fraudulent procedures and kickboards | 3.5011 |
| 15 | Poor site management | 4.4667 | 45 | Problem with neighbors | 3.4667 |
| 16 | Cash flow and financial difficulties faced by contractor | 4.4667 | 46 | Waste on sites | 3.4657 |
| 17 | Frequent design changes | 4.4667 | 47 | Relationship between labour and management | 3.4333 |
| 18 | Inaccurate time and cost estimates | 4.3667 | 48 | Inadequate modern equipment | 3.4113 |
| 19 | Quality assurance control | 4.3667 | 49 | Contractual claims, such as, extension of time with cost claims | 3.4100 |
| 20 | Late delivery of materials | 4.3333 | 50 | Inaccurate quantity take-off | 3.4000 |
| 21 | Shortage of materials | 4.2667 | 51 | Problem with neighbors | 3.3667 |
| 22 | Fluctuations of price of materials | 4.2333 | 52 | Waste on site | 3.3657 |
| 23 | Slow information between parties | 4.1667 | 53 | High cost of labor | 3.3467 |
| 24 | Inadequate monitoring and control | 4.0000 | 54 | Disputes between management | 3.335 |
| 25 | Approval of drawings | 3.9000 | 55 | Delay in approval of Completed works | 3.333 |
| 26 | Delay in material procurement | 3.8667 | 56 | In adequate monitoring | 3.333 |
| 27 | Quality of materials | 3.8333 | 57 | Inadequate new equipment | 3.267 |
| 28 | Bureaucracy in tendering document | 3.7333 | 58 | Incompetent subcontractors | 3.267 |
| 29 | Number of construction projects going on at same time | 3.7000 | 59 | Low labour productivity | 3.267 |
| 30 | Disputes on site | 3.6667 | 60 | Poor cost control at site | 3.067 |
| 61 | Poor project management | 3.033 | 62 | Lack of communication | 3.033 |
| 63 | Inaccurate site investigation | 2.900 | 64 | Complicated design | 2.871 |

**Table 2**. Causative factors of cost overrun in construction of projects in Sindh province of Pakistan.

| Factor | Mean Value | Rank |
|---|---|---|
| Delay in process payment by owner/client | 4.900 | 1 |
| Inadequate planning | 4.668 | 2 |
| Owner/Client interference | 4.667 | 3 |
| Poor contract management | 4.651 | 4 |
| Delay in decision making | 4.644 | 5 |
| Laws and regulations frame work | 4.633 | 6 |
| Change in scope of project | 4.621 | 7 |
| Financial difficulties faced by owner/client | 4.618 | 8 |

*6.7 Change in scope of project*

Change in scope of project can change many construction activities. Mostly client or owner changes the scope of project by influencing of politicians or sometimes initial stage the scope of project is not clear. If the scope of the construction project is changed which leads to increase in cost of the project [19].

*6.8 Financial difficulties faced by owner/client*

Financial difficulties faced by owner are most common and causative factor found in construction of highway projects in Sindh province. Mostly funds of projects are transferred to other projects. Mostly projects are facing problem of cost overrun because of financial difficulties faced by owner/client [20].

## 7. Conclusion

This study found that cost overrun is a major and important issue in construction of highway projects of Sindh province. Total 64 factors were found from deep literature review. After analysis the factors whose mean value was above 4.6 were found as causative factors. Most causative factors of cost overrun were Delay in Process payment by owner/client, inadequate planning, Owner/Client interference, Poor contract management, Delay in decision making, Laws and regulations frame work, Change in scope of project and financial difficulties faced by owner/client in construction of highway projects in Sindh province. The results of this research will not only be beneficial in controlling cost overrun in construction of highway projects but it will also help in the improve the environment in construction industry of Pakistan.

## 8. References

[1] Nagapan S, Rahman I A, Asmi A, Hameed Memon A and Zin R M 2012 Identifying causes of construction waste -case of central region of Peninsula Malaysia, *Int. J. of Integrated Engineering* **4** 22–28

[2] Mahamid I and Dmaidi N 2013 Risks leading to cost overrun in building construction from consultants' perspective, *Organization, Technology & Management in Construction: An Int. J.* **5** 860–873

[3] Tejale D S, Khandekar D S and Patil D J 2015 Analysis of construction project cost overrun by statistical method, *Int. J. of Advance Research in Computer Science and Management Studies* **3** 348-355

[4] Agren R, Jingmond M and Landin A 2011 Cost and time overrun in construction projects in a multicultural setting *Proc. of the 2nd Int. Conf. on Advances in Engineering and Technol.(India)* vol 4 (Noida: IDES Conference Publishing System) pp 291–297

[5] Amoa-abban K and Allotey S 2014 Cost overruns in building construction projects : A case study of a government of Ghana project in Accra, *Developing Country Studies* **4** 54–65

[6] Azhar N, Farooqui R U and Ahmed S M 2008 Cost overrun factors in construction industry of Pakistan *1st Int. Conf. on Constr. in Developing Countries "Advancing and Integrating*

*Construction Education, Research & Practice" (Karachi)* vol 1 ( Karachi: NED University) pp 499–508

[7] Liu L and Zhu K 2007 A stage-by-stage factor control framework for cost estimation of construction projects, *J. Constr. Eng. Manage* **133** 91-95

[8] Avotos I 1983 Cost-relevance analysis for overrun control, *Int. J. of Project Management* **1** 142–148

[9] Lee J K 2008 Cost overrun and cause in Korean social overhead capital projects: Roads, rails, airports, and ports, *J. of Urban Planning and Development* **134** 59–62

[10] Chouhadary M R, Nasir A R and Gabriel H M 2012 Cost and time overruns in highway projects in Pakistan, *Journal of Pakistan Engineering Congress* **2** 353 369

[11] Ejaz N, Ali I and Tahir M F 2013 Assessment of delays and cost overruns during construction projects in Pakistan *Int. Conf. on Structural Engineering Construction and Management* (Sri Lanka: University of Moratuwa)

[12] Yuan W, Deng C, Zhu H and Li J 2013 The statistical analysis and evaluation of examination results of materials research methods course, *Creative Education* **3** 162-164

[13] Ayudhya B I N 2011 Evaluation of common delay causes of construction projects in Singapore, *J. of Civil Engineering and Architecture* **5** 101-107

[14] Sunjka B P and Jacob U 2013 Significant causes and effects of project delays in the Niger delta region, Nigeria *Southern African Institute of Industrial Engineering (Stellenbosch, South Africa)* (South Africa: The Southern African Institute of Industrial Engineers) 641

[15] Ismail I 2014 *Risk Assessment of Time and Cost Overrun Factors Throughout Construction Project Lifecycle* PhD Thesis (Johor: Universiti Tun Hussein Onn Malaysia)

[16] Haseeb M, Lu X, Hoosen A B and Rabbani W 2011 Causes and effects of delays in large construction projects of Pakistan, *Kuwait Chapter of Arabian Journal of Business and Management Review* **1** 18-42

[17] Ambituuni A 2011 *Five Causes of Project Delay and Cost Overrun, and Their Mitigation Measures* PhD Thesis (Aberdeen: The Robert Gordon University)

[18] Larsen J K, Shen G Q, Lindhard S M and Brunoe T D 2015 Factors affecting schedule delay, cost overrun, and quality level in public construction projects, *J. of Management in Engineering* **32** 1-10

[19] Mohammad N, Ani A I C and Rakmat R A O 2017 Causes and effects of variation orders in the construction of terrace housing projects: A case study in the state of Selangor, Malaysia, *Int. J. of Supply Chain Management* **6** 226-232

[20] Alaryan A 2014 Causes and effects of change orders on construction projects in Kuwait, *Int. J. of Engineering Research and Applications* **1** 1-8

[21] Mukuka M J, Aigbavboa C O and Thwala W D 2014 A theoretical review of the causes and effects of construction projects cost and schedule overruns *Int. Conf. on Emerging Trends in Computer and Image Processing (Pattaya, Thailand)* vol 1 (Planetary Scientific Research Center) pp 112-115

# Causes of shortage and delay in material supply: a preliminary study

**M M Rahman[1], Y H Yap[1], N R Ramli[1], M A Dullah[1] and M S W Shamsuddin[1]**

[1]Civil Engineering Programme Area, Faculty of Engineering, Universiti Teknologi Brunei, Tungku Highway, Gadong BE1410, Brunei Darussalam
Corresponding author: motiar.rahman@utb.edu.bn

**Abstract.** Shortage and delay in materials supply is argued to be one of the most important factors that lead to delay in construction project delivery globally. However, the relevant underlying reasons vary from country to country. As such, this paper summarises the outcomes of a study that targeted identifying causes of shortage and delay in materials supply in Brunei Darussalam. The study was conducted through fifteen semi-structured interviews of contractors and materials suppliers in Brunei. The study identified six causes of shortageof materials and nine causes of delay in materials supply in Brunei.The most importantcausefor shortage of materials relates to the origin or availability of construction materials. On the other hand, the most influential cause of delay in material supply was found to be poor materials procurement and inventory management system, which has other underlying reasons such as late identification of the type of materials needed. The observations are expected to help in formulating or reviewing relevant policies, in order to ensure on-time project delivery.

## 1. Introduction

Delays are common in construction projects. For example, Morris and Hough [1] examined more than four thousand construction projects from UK and Europe and observed that projects were rarely completed on schedule time. Similar outcomes were observed in many other countries as well[2-6]. There are many causes for such project delays, of which shortage and delay in materials supply are among the most notable in many studies. For example, Assaf *et al.* [4] studied the causes of delay in large building projects in Saudi Arabia and identified a group of factors relating to materials, which included causes related to shortages and delivery of materials. Abd Majid and McCaffer [7] observed that late delivery and slow mobilization of materials ranked 1 among 25 factors contributing to causes of non-excusable construction delays in United Kingdom. Koushki and Kartam[8] studied 450 small, medium and large private residential projects in Kuwait and found that nearly one-fourth of the total project delays were due to the late delivery of materials. Similarly, studies in Nigeria, Egypt, Thailand and India [9-12] observed that delay in supply, and/or shortage of, materials caused the project delay. It was also observed in Brunei that shortage and delay of construction materials were the leading causes of project delay [13].Whilestudies in many other countries attempted to identify the causes of delay or shortage of materials, no such study was previously conducted in Brunei, at least to the knowledge of the authors. As such, the present study was undertaken to identify the causes of delays insupplyandshortageofmaterialsinBruneiconstruction.

## 2. Research methodology

The study was undertaken in three phases. The first phase was a literature review, to extract the causes of project delays elsewhere (i.e. in other studies or countries) due to delay in supply, and shortage of, materials. This also led to the identification of a number of focus areas on which information for the present study needed to be collected. The second phase was to understand how theconstruction materials of both local and overseas origin, are sourced, procured and transported to construction sites. This was done by informal discussion with two industry experts. The third phase of the study identified the Brunei-specific causes of delay in supply and shortage of construction materials through face-to-face semi-structured interviews of materials suppliers and contractors. A total of 15 interviews were conducted: ten suppliers and five contractors; only two were contractors' quantity surveyors and other thirteen were holding managerial position, such as contract manager, operation manager, marketing manager and general manager. Their experience in their 'current' positions varied from a minimum of three years to a total of forty years in the construction industry. Construction projects require a variety of materials. Five contractors were interviewed to cover the use all commonly used construction materials. On the other hand, not all suppliers supply all kinds of commonly used construction materials. Therefore, more (i.e. 10) suppliers were interviewed to cover the supply of all commonly used construction materials. Different organisations were approached for interview who import materials from overseas, and either use those materials in their own projects (e.g. by contractors), or supply to constructors (e.g. by suppliers). Data was analysed manually, by comparing opinions from individual interviews, and gradually compiling the causes in to different groups.The study collected a diverse range of information, but the present paper reports the summary of the key outcomes only.

## 3. Summary of outcomes

The study identified a total of fifteen causes of delay in supply and shortage of materials. Those are summarised in the following subsections.

### *3.1. Causes of shortage of materials*

Six causes of shortage of materials were identified. They are summarised in the following subsections.

*3.1.1 .Origin or availability of materials.* Only very few construction materials are locallyavailable in Brunei, so it relies on imported materials in general. The only available materials locally are: ordinary Portland cement, sand, aggregates, timber, bricks and glass. However, their local supplies cannot wholly meet the demand of the local construction industry. Therefore, these materials, along with other construction materials, are imported to meet the demand, mainly from China, Indonesia and Malaysia.This takeslonger time for sourcing, procuring and transportingof the materials. Moreover, land transport is the only delivery method available within Brunei that results in to a long lead time for materials. With higher demand of materials, for example during the construction of national housing projects, which involve the construction of hundreds or thousands of houses, local suppliers experience shortages of their stock.Saudi Arabia also experienced project delays due to similar causes during their booming construction industry in 1990s [4, 15]. On the other hand, locallyavailable materials may also encounter shortage in production. For example, one of the Brunei industrial areashad suffered power failure for several days a few years back, resulting in to temporary stoppage of cement production and the factory ran out of stock. During the times of higher demand (i.e. increased number of ongoing projects), Brunei also suffers from shortage of bricks at times and contractors need to import bricks from neighbouring provinces of Malaysia. Moreover, local authority limits timber logging toonly 17% by the year 2045, although 75% of Brunei is covered with forest [14]. This may cause shortage in timber supply and Brunei may need to import it from neighbouring countries to cater for current needs.

*3.1.2. Poor estimation of materials quantity.* Poor estimation causes either shortage of supply or oversupply. Shortageof supply disrupts construction progress on site,as reordering can cause prolonged delay and additional costs, especially if the material needs to be imported from overseas. Reordering might cause other problems too, such as inconsistency in colour and size tolerance. On the other hand, oversupply incurs additional costs / losses to contractor.

*3.1.3. Poor workmanship.* This occurs due to lack of care by the contractor while installing the material (e.g. imported fittings and fixtures, and modular parts),resulting in poor finishing and the product (i.e. material) being non-functional. Similar to Brunei, rework of bad quality performancewas found to be one of the most significant factor causing project delays in India [16]. However, the problems in Brunei come from contractors' installers, by not installing the product in the proper way(i.e. according to the instructions of suppliers), as they want to complete the work faster. Some materials need certain techniques to install, but some contractorsignore these and use their own way to complete the job faster. When problems arise, such as cracking or malfunction, an investigation by the supplier then takes place. The outcomes of the investigation decide whether the supplier should replace the materials or the contractor needs to compensate for the losses. In either case, such problems delaythe project activities asthe investigation process first needs to be done, followed by negotiations(between the contractor and the supplier), and the ordering of new product results in more waiting time.

*3.1.4. Quality of material.* The quality of materials can delay their usage on site if the materials do not meet the standard or contain defects. The defect can be due to improper handling during packaging, shipment or delivery. This was also observed in a study in Iran [17]. Some materials may have minor defects that can be repaired, but some materials may have major defects that cannot be repaired and therefore cannot be used at all, e.g. broken sanitary appliances. This happened with one of the interviewed companies, when they ordered a manufactured door from Malaysia. The procurement/ordering process was made online, where the items were viewed through the manufacturer's website, and appeared to be nice products with good quality. However, when the doors arrived on site, they were found to be of poor quality with holes throughout the door panel, possibly due to insects or termites. The interviewed company made a report to the supplier within three days of receiving the doors, with proof of defect,to arrange for compensation. This consumed more time due to the need to wait for the new doors to arrive. If it is already the time to install the door, then there is no choice other than to wait for the next delivery.

*3.1.5. Inconsistent demand.* Inconsistent demand occurs when the sales of a material or product do notdepend on the demand of the local market. Paint is one such material. It is hard to know which colour is saleable, because sometimes clients look for a certain colour of paint that is not in stock at that time. If the quantity/volume required by the client is large and the local distributor is also without stock, the material needs to be imported from overseas, usually from Singapore. The manufacturer starts producing the paint once they receive the order, and it normally takes one month to arrive in Brunei. But by the time the ordered colour is available, some different colour that is unavailable isthen demanded by the client. Therefore,the paint of unwanted colouris kept in stock. However, once there is demand, the paint will be sold out quickly and a new order then has to be made. This seems to be a unique problem to Brunei, probably due to smaller volume of its construction industry.

*3.1.6. Special materials.* An example of such materials is special appliances, fittings and lightings that are used in buildings that have differential needs, e.g. prison, hospital or for diplomaticoffices. One of the interviewees is the sole distributor of specialisedand sophisticated lighting appliances in Brunei. Therefore, they should ensure the availability of this material to cater the local demand. However, the technology of such appliances change very fast, so they cannot take the risk of keeping a large stock due to the fear that they cannot sell those after a few months. Therefore, they only order the products when demanded by certain projects. Moreover, this special material/ appliance is imported from all over the world, as the factories of these lighting appliances are in places as diverse as Hungary, Holland, China, Thailand and Malaysia. Each factory produces only certain kind of appliances. Therefore, there involves very long lead times. If orders are not placedsufficiently long before by certain projects, delay in supply is likely to occur, which then leads to project delay. In this regard, Seboru [18] observed that the nature of demand for materials is both project specific and client specific.

*3.2. Causes of delay in material supply*
Nine causes of delay in supply of materials were identified.

*3.2.1. Labour productivity.* The Brunei construction industry largely relies on overseas labour, mainly from India, Indonesia and Bangladesh. The labourers work in every sphere of construction related activities: materials production, transportation, loading-unloading and on construction sites. Most of them are unskilled, especially when they first come to work in the construction sector. Some time is needed to trainthem, but even after training, many of them remain less motivated, especially in materials production activities. They seem to be idle for prolonged time, resulting in reduced productivity. The situation worsens after servicing, cleaning and adjusting the machineries; the older the machines, the more episodes of servicing and adjustment, therefore higher frequencies of prolonged idle time. Furthermore, in Brunei, renewable labour work permits are limited up to two years and therefore labourers are usually appointed on two-year contract terms. Many labourers go back to their home country after one or two contracts, but not all of them come back to Brunei. Hence, many new and unskilled labourers are required to make up the shortfall, meaning a continuous need for training and reduced productivity. Similar situation was observed in Saudi Arabia and Turkey [15, 19].

*3.2.2. Inclement weather.* Inclement weather affect materials supply in two ways: while importing materials from overseas via sea freight, and during delivery to the project site. Shipment via sea freight depends on the weather condition of source country of the materials, as it influences the loading, and of Brunei as it influences unloading. Moreover, it is not safe for ships to sail in the sea during bad weather. As a result, delays are common during monsoon season. On the other hand, Brunei is a tropical country with rain throughout the whole year, and the rain comes with little notice. Rain can prevent the supply of some materials to construction sites. For example, ready-mixed concretesuppliersdo not produce concrete during the spells of rains, because the water ingress cannot be controlled. If heavy rain starts when concrete carrying vehicles reach the site, the concrete cannot be placed, because the concrete will have excessive water content, and hence be too workable or too weak (once hardened), or the concrete may even be washed away.As fresh concrete is perishable, there is no turning back once the cement in the concrete has been mixed with water.Such negative impact on schedule from undesired events like rainy weather seems to be common in tropical or semi-tropical countries, e.g. in Malaysia [20] and Sri Lanka [21].

*3.2.3 .Government restrictions.* Certain materials (e.g. timber) need permit approvals from government departments before placing import orders, which may be time-consuming if not planned well in advance. Permit approvals mainly deal with the declaration of type, source, quantity and quality of materials to be ordered. This is to ensure that only certain materials are imported, and exactly the declared type and quantity of materials are imported. The economic planning control policy of the country also prohibits importing cheaper materials and selling the same at higher prices. Therefore, the importing companies have to specify correctly the type of material in order to receive approval of their permits. If the permit is not granted, more time is needed to get a new permit approval, resulting in to delayed delivery of materials to site.

*3.2.4. Slow decision-making.* Slow decision-making delays the project completion as the contractor waits for confirmation to carry out certain works. Similar observation was made in a study in Kenya: slow decision-making was identified as one of the most important factors for delay [18].However, this usually happens in Brunei while selecting finishing products and patterns, such as the tiles. An owner can simply decide to change the scope of work in a project and also can dramatically impact project schedules by failing to respond to requests for information or change orders, or making decisions on material selections. Any lastmoment decision leads to delayed order for materials from overseas and consequently delayed supply / delivery of materials on site. This certainly takes more time, but allows much lower prices than local market. Some materials at local market cost higher, butcan make up project time to some extent. However, those materials may not be available in the local market, and if by any chance they are available, may not have sufficient stock, in which case import from overseas is the only option, resulting in to prolonged delay.

*3.2.5. Shortage of raw materials.* Not all the materials imported from the overseas manufacturers are 100% produced by them. The raw materials (i.e. ingredients or components) of certain materials are sometimes ordered and collected from some other countries. For instance, paintis imported from Singapore, and its manufacturer importsrelevant raw materials from other countries. The manufacturers usually maintain a certain amount of storage of the raw materials, but they may also suffer from shortage. If manufacturers receive orders from Brunei when they are experiencing shortage of raw materials, they will have to delay manufacturing of the paint. This will in turn lead to delayed supply of paint to Brunei.

*3.2.6. Logistics.* There are instances that the quantity of ordered materials/products isinsufficient to be placed in a standard shipping container. Such orders are delivered as Loose Cargo Load (LCL), which means that the shipped volume will either be palletized or crated using heat-treated lumber following customs regulations. Local suppliers in Brunei manage such pallets. Once they order some materials, they send those pallets to the forwarder, who then sends the order to Brunei through Singapore (or the source country). Thus, thetravelling time for the shipment from source country to Brunei doubles, plus the packaging time. So, the overall time taken would be,for example, roughly about ten days, assuming that the manufacturer has the stock. If the manufacturer is out of stock, more time is taken. This problem of logistics can be of particular relevance to Brunei, as it is a relatively small country with a population of 411,900 [14], and notwithstanding its high gross domestic product per capita, it also has a commensurately small construction industry and construction materials supply market.

*3.2.7. Poor planning and scheduling.* Planning and scheduling (P&S) of activities plays an important role in time management of a project. It is normally done before the construction begins, and should entail sourcing and procuring/ordering (i.e. P&S) of relevant materials. Otherwise, construction delays are likely to befall, along withcorrespondingimpact on cost. It is therefore important to ensure that materials are available on site when they are to be used. In order to achieve that, one of the interviewees suggested toordering materials as early as six months before the actual startof relevant activities. This was due to uncertainty over what could happen when the materials are about to be used. For example,materials may be out of stock or no production in both local and overseas markets (e.g. for lack of supply order from agents or users), or manufacturers may be unable to supply (e.g. due to excess order than their production capacity), hence demanding a search for alternate sources, and the need to procure materials usually with higher prices. The interviewee recalled that they experienced a shortage of rebar supply from China during the Beijing Olympic 2008 construction works. China, which is usually the cheapest supplier, was unable to export rebar due to their internal high demand at that time. Early P&S of rebar would have spotted the low chance of getting sufficient supplies in time, and allowed procuring rebar from other neighbouring countries.Although this would have come with a slightly higher price, this could have saved them from delays that came with higherfinancial consequences.However, some materials cannot be ordered too early, e.g. cement, as they need to be used within days or a few weeks. Early order of such materials would make them unduly aged after delivery and therefore unusable, requiring a new order for the same materials, with additional cost and possible delays. Therefore, P&S of materials has to be both project and materials-specific [18], while appreciating their availability both in local and overseas sources, trends of market demands, and continuously monitoring alternate sources. Any deviation to such practice, i.e. poor P&S of materials, may delay their delivery and thereby lead to project delay and incur extra cost. P&S is seen as an important issue not only in Brunei, but also in many other countries, including India [12] and Malaysia [20].

*3.2.8. Unrealistic construction duration.* The duration of construction project is normally set by clients or their representatives, which refers to the work programme of the project. The contractor has to follow the set time in order to prevent any penaltydue to time overruns. While taking into account of such project duration, the contractor should ensure during tendering that all materials are ordered in good time and that they can be readily available when needed to be used. However, occasionally some projects are set with unrealistically short duration, due to the urgent cases such as the preparation for international meeting, or to meet any emergency situations. Contractors have to face those challenges, by exploring sources for urgent supply of materials and ensuring their installations / construction, by some special arrangements, and certainly with higher costs, since such projects must not be delayed.

*3.2.9. Variations and changes.* Variation orders and changes to tendered scope of works are common to construction projects in manycountries, e.g. in India [12], Iran [17], Kenya [18] and Malaysia [20]. In Brunei, they emerge due to a range of reasons, including missing information, changes in client decisions, inability to build as specified, changes to the design, unforeseen conditions, or procurement issues in sourcing a required product. Whatever the reason is, changes have impact on materials. The change may be an addition to any existing item of work that will require previously ordered / used materials, or a new work that will require different materials, new orders may need to be placed in any case. However, if the materials are not available in local market, or available stock is limited, theyhave to be imported from overseas. This will lead to delays in materials supply and eventually project execution. Many interviewees mentioned this issue, which may be considered as indicative of the severity of the problem.

## 4. Concluding observations

The reported study attempted to identify the causes of delays in supply and shortage of materials in Brunei construction industry.The semi-structured interview-based study identified a total of fifteen causes. Lack of locally available materials, that compels import from overseas, seems to be at the core of shortage of materials. On the other hand, inability to timely procure and maintain a suitable inventory system seems to be the most important cause of delay in materials supply. This is due to the late identification of what materials to be required, which in turn originates from change orders and very late changes in decisions. However, not all the causes were present in all the projects, but some projects shared only a few causes, while a few causes may be unique. The outcomes are expected to assist policy makers for taking effective measures to improve the timely delivery and available stock of materials at local market. The next step in this study will be to quantify and prioritise the causes of shortage and delay in materials supply, which will further help policy makers to more specific measures for improved supply of materials and timely delivery of construction project in Brunei Darussalam.

## 5. References

[1] Morris P and Hough G 1998 *The Anatomy of Major Projects* (New York: Wiley)
[2] Al-Hammad A 1993 Factors affecting the relationship between contractors and their sub-contractors in Saudi Arabia, *Building Research and Information* **21**(5) 269-273

[3] Arditi D, Akan G and Gurdamar S 1985 Reasons for delays in public projects in Turkey *Journal of Construction Management and Economics* **3**(2) 171-181
[4] Assaf S, Al-Khalil M and Al-Hazmi M 1995Causes of delay in large building construction projects *Journal of Management in Engineering***1 1**(2) 45-50
[5] Rad P 1979 Delays in construction of nuclear power plants *Journal of Energy Division* ASCE**105**(1) 34-46
[6] Taha J, Thorpe A and McCaffer R1993 Contractor project risks contingency allocation using linguistic approximation *Computing Systems in Engineering* **2**(4) 281-293
[7] Abd-Majid M Z and McCaffer R 1998 Actors of non-excusable delays that influence contractors' performance *Journal of Management in Engineering* **14**(3) 42-49
[8] Koushki P A and Kartam N 2004Impact of construction materials on project time and cost in Kuwait *Engineering, Construction and Architectural Management* **11**(2) 126-132
[9] Okpala D C and Aniekwu A N1988Causes of high costs of construction in Nigeria, *Journal of Construction Engineering and Management* **114**(2) 233–244
[10] Abd El-Razek M E, Bassioni H A and Mobara A M2008 Causes of Delay in Building Construction Projects in Egypt *Journal of Construction Engineering and Management* **134**(11) 831-841
[11] Toor S R and Ogunlana S O 2008 Problems causing delays in major construction projects in Thailand *Construction Management and Economics* **26**(4) 395-408.
[12] Doloi H, Sawhney A, Iyer K C and Rentala S 2012 Analysing factors affecting delays in Indian construction projects *International Journal of Project Management* **30** 479–489
[13] Salleh R 2009 *Critical success factors of project management for Brunei Construction Projects: improving project performance* PhD Thesis (Brisbane, Australia: Queensland University of Technology)
[14] Department of Statistics 2015 *Brunei Darussalam Key Indicators 2015* (Bandar Seri Begawan: Department of Economic Planning and Development, Prime Minister's Office, Brunei)
[15] Al-Kharashi A and Skitmore M 2009 Causes of delays in Saudi Arabian public sector construction projects *Construction Management and Economics* **27**(1) 3-23
[16] Shanmugapriya S and Subramanian K 2013 Investigation of significant factors influencing time and cost overruns in Indian construction projects *International Journal of Emerging Technology and Advanced Engineering* **3**(10) 734–740
[17] Asnaashari E, Knight A, Hurst A and Farahani S S 2009 Causes of construction delays in Iran: project management, logistics, technology and environment.Dainty A (Ed) *Proc. 25th Annual ARCOM (Association of Researchers in Construction Management) Conference* (Nottingham UK) 897-906
[18] Seboru M A 2015 An investigation into factors causing delays in road construction projects in Kenya *American Journal of Civil Engineering* **3**(3) 51-63
[19] Kazaz A, Ulubeyli S and Tuncbilekli NA 2012 Causes of delays in construction projects in Turkey *Journal of Civil Engineering and Management* **18**(3) 426-435
[20] Mydin M A O, Sani N M, Taib M and Alias N M 2014 Imperative causes of delays in construction projects from developers' outlook *MATEC Web of Conferences* vol 10 (Paris: EDP Sciences)
[21] Dolage D A R and Rathnamali D L G 2013 Causes of time overrun in construction phase of building projects *Engineer Journal of the Institution of Engineers of Sri Lanka* **46**(3) 9-18

# Best practices of Building Information Modelling (BIM) implementation in design phase for construction project

**N Kasim[1], N A Zainal Abidin[1], R Zainal[1], N Sarpin[1], M H I Abd Rahim[1] and M Saikah[1]**

[1] Department of Construction Management, Faculty of Technology Management and Business, Universiti Tun Hussein Onn Malaysia
Corresponding author: narimah@uthm.edu.my

**Abstract.** Implementation of Building Information Modelling (BIM) was expected to bring improvement in current practices of Malaysian construction industry. In the design phase, there is a lack of a ready pool of skilled workers who are able to develop BIM strategic plan and effectively utilise it. These create boundaries for BIM nature in Malaysian construction industry specifically in the design phase to achieve its best practices. Therefore, the objectives of this research are to investigate the current practices of BIM implementation in the design phase as well as the best practices factors of BIM implementation in the design phase. The qualitative research approach is carried out through semi-structured interviews with the designers of different organisations which adopt BIM in the design phase. Data collection is analysed by executing content analysis method. From the findings, the best practices factors of BIM implementation in design phase such as the incentive for BIM training, formal approach to monitoring automated Level of Detailing (LOD), run a virtual meeting and improve Industry Foundation Class (IFC). Thus, best practices factors which lead to practices improvements in the design phase of project development which subsequently improves the implementation of BIM in the design phase of Malaysian construction industry.

## 1. Introduction

Malaysian construction industry's project development is divided into three major stages, which are pre-construction stage, construction stage and post-construction stage [1]. These three stages also being called as a 'project life cycle' involving numerous workers, a list of activities, costs, and stakeholders. According to National Centre for Construction Education and Research (2008), there are three activities in pre-construction stage, which are project design, cost estimates and project scheduling while the actual construction work is carried out in construction stage and as in post-construction stage, final inspection, project administration, project close out and operating and facilities management are the activities involved. This research is focusing on pre-construction stage since it plays a critical role to commence the construction project. The first phase of pre-construction stage which is a design phase, its necessity to verify the design so that the project meets client's requirements. Design challenges must have been resolved before the end of the final design stage. Furthermore, sufficient details must be provided with the drawings as well to allow accurate construction estimates and operating costs together with the construction schedule. On the other hand, in order to achieve success in the critical phase of the project which is a design phase, best practices of technology implementation is vital. Moreover, Malaysian Construction Industry Transformation Programme 2016-2020 (CITP) encourages the adoption of Building Information Modelling (BIM) technology as a productivity enhancing tool which will change in the way the industries goes about designing a building and infrastructure as a third strategic thrust [2].

Building Information Modelling (BIM) is one of the technologies which represents the process of development and the use of a computer generated model to execute construction activities such as planning, designing, constructing and operation of a building in a virtual environment [3]. Thus, BIM plays its own roles in the design phase to bring the basic traditional method into the new emerging method by the use of 3D and 4D parametric authoring tools as object-based modelling software. BIM was introduced in Malaysian construction industry in 2007 by Public Work Department (PWD) and the implementation of BIM was expected to bring improvement in current practices of Malaysian construction industry. Switching from Computer-Aided Drafting (CAD) to Building Information Modelling (BIM) by the construction parties creates several challenges and benefits in the design phase, but yet, the benefits from BIM implementation outweigh the challenges [4]. BIM in Malaysian construction industry is still emerging but yet, it is currently practiced by developers. In addition, there are several challenges that limit the implementation of BIM in Malaysian construction industry focusing on the design phase which was driven by cost, market trends and organizational behaviours [5]. Even though BIM drives to benefits, there are still barriers towards implementing it and achieving those benefits, for example, 'lack of competent staff to operate the software' is the most significant barrier in contractor's perspective [6]. Thus, the awareness of potential benefits of BIM implementation is vital and need to be strengthened by spreading the reason on why it should be implemented at its very best, overcoming the obstacles and optimizing its adoption in the design phase of project development.

There is a lack of skilled workers who are capable of developing BIM implementation strategic plan and effectively utilize BIM in the design phase [2]. The strategic plan involves the selection of BIM tools to carry out design development by considering the strengths and weaknesses. Besides, BIM usage guideline is also limited regarding Malaysian ways of working. Here, BIM can be challenging and there is no remedy to treat those challenges and transform it into potential benefits that the adopters might get. This will then give skepticism to the construction firms in implementing it at a very first phase of project development as well. Therefore, to ensure the best practices of BIM in Malaysian construction industry, it is crucial to spread the awareness of BIM advantages and inducements among all parties in this industry in order to improve the current practices of BIM implementation in the design phase. Several best practices are already identified, BIM training for example, but there is a need to find the key factor to ensure the continuous success of BIM training. Hence, this research is carried out to investigate the current practices of BIM implementation in the design phase and the best practices factors of BIM implementation in the design phase.

## 2. BIM implementation in design phase

Building Information Modelling (BIM) is used in diverse ways in project development by the key players of the construction industry. The architect, for example, implements BIM to produce architectural drawings, followed by engineers producing structural or energy data management and contractor develop coordination model of the building. Concept design does not only provides a framework of design in terms of its structure, spatial layout, and environmental condition, but also the cost estimates, building practices and the aesthetic considerations as well [7]. Furthermore, BIM does provide several tools to assist designers that involve in design phase which is Autodesk Revit that composed of Revit Architectural, Revit Structural, Revit MEP as well as Tekla and Bentley that currently practices in Malaysian construction industry [8]. In brief, there are various types of BIM tools available for designers to develop project designs. However, it is not necessarily for companies or firms to practice only one type of BIM tools for their designers. This is due to the differences involving the design team in terms of skills, preferences and desired outputs [9]. Hence, monitoring is needed to ensure the designers achieves what has been expected from them in term of work quality and productivity if the whole team design practices different BIM tools.

In order to increase the practices of Building Information Modelling (BIM) in the design phase, it is important to address the potential challenges of its implementation. From challenges identification, solutions and initiatives could be addressed afterward. Eventually, from the solutions of those challenges, the benefits of BIM implementation including its awareness will efficiently spread among the parties in the construction industry. In brief, there are several challenges highlighted by various researchers focusing on BIM practices in the design phase. The BIM implementation challenges in design phase such as changes in practices, BIM disrupts established design workflow, requires skilled design team, designers preferences, time constraints, and integrated use of expert models [7][10][5][11][12]. Generally, it involves design teams with other stakeholders such as contractor and sub-contractors, this is where BIM implementation comes to its limits. However, those challenges are quite needed for the government, Malaysian government, for instance, to enhance initiatives in order to encourage the use of BIM implementation in Malaysian construction industry. Without challenges and problems, there will be no lessons to learn and improve the current practices of BIM implementation in the design phase.

Building Information Modelling (BIM) offers huge benefits to the adopters, especially in architectural, engineering and construction (AEC) sector. However, those benefits only could be fully achieved if there are proper practices implements by the adopters. Without best practices of BIM, still, it is hard to exploit the benefits from the implementation. Hence, it is important to execute BIM implementation best practices strategic plan to ensure the works are clearly defined and understood which makes BIM as a useful tool to develop construction projects [13]. Also, the best practices factors of BIM implementation in the design phase is crucial. Best practices factors are basically the idea of how to improve current practices. Hence, in the design phase of construction project lifecycle, it is beneficial for the design team as well as other stakeholders to formulate the best practices factors of BIM implementation to obtain the benefits that BIM might offers. From the designers' perspectives, more '3D Building Product Manufacturer-specific content' is favourable as an improvement [14]. Thus, regardless of how beneficial BIM is for designers, without a proper plan to its best practices, there are still several limitations to earn those benefits. The best practices of BIM will then drive to a great success of construction project development and brings the construction industry to the new high level.

## 3. Methodology

The research design refers to the strategy to integrate the different components of the research in a coherent and logical way. This will ensure the effectiveness of addressing research issues towards achieving research objectives. According to [15], a good research design is when all the components work harmoniously together, promotes efficient and successful operation towards achieving objectives, while poor research design, on the other hand, leads to poor operation or failure. Meanwhile, according to [16], research designs are types of inquiry within the qualitative analysis, quantitative analysis and mixed methods approaches that provide specific direction for research procedures. Hence, this research is conducted by using qualitative approach by using interview questions. The collection of primary data involves the interviews from key players in the construction industry that experienced BIM implementation in the design phase. The secondary data for this research involved literature review covers the current practices of BIM implementation in the design phase and best practices factors. The interview is conducted during respondent's working hours. It is from 10am to 4pm during weekdays. The duration of each interview is basically varied among the respondents. In brief, the interviews are between 40 minutes minimum until 2 hours maximum depends on the respondent's answers, elaborations and also personal thoughts and judgments. After all, the findings from interviews with the respondents do satisfied data collection requirements for this research.

The target population for this research is the company that implements Building Information Modelling (BIM) in the design phase for their past project, current projects and upcoming projects as well. According to the Chief Executive of Construction Industry Development Board in 2014, only 10% of construction industry key players adopt BIM for their projects, and until 2014 only 20 projects were reported in accordance with the implementation of BIM [17]. Apart from that, several developers, contractor, and consultant that implements BIM has been identified as part of this research population which is located in Selangor and Wilayah Persekutuan Kuala Lumpur. A semi-structured interview is the main instrument for this research where the respondents have to answer open-ended questions. The set of question for the interview is the composition of the core question with other associated questions [18]. The target respondents are among designers which are engineer and architect based on their contribution in implementing BIM in the design phase. As a matter of fact, semi-structured interviews is suitable in order to fully understand the respondent's point of view regarding the current issues related to BIM implementation in the construction industry where it is not just limited to the questions itself. Thus, broad information and justification can be obtained from the respondents. In brief, the method in data analysis for this research is content analysis. Content analysis is a research technique used to interpret textual information or it is called qualitative-interpretative since this research performs interviews as primary data [19]. Meanwhile, according to [20], content analysis is the method of subjective interpretation through a systematic coding process. Thus, content analysis helps in structuring the information from various sources and perspectives as well. Hence, by executing content analysis, it will interpret the information from recorded interviews for further analysis with secondary data which is a literature review.

## 4. Results and discussion

### *4.1 Respondent's background*

In this sub-topic, all the respondents' background in construction industry specifically on their contribution in BIM is identified. The selection of respondents involves consultant firm, developer, contractor and local authority by targeting the designer in that organization. Hence, Table 4.1 indicates the respondents' background which participated in this research. In brief, the selection of respondents for this research is based on their academic qualification, experiences in the construction industry as well as experiences in BIM during design development phase. Apart from that, the collected data and information given by the respondents are precise and complementary with the objectives of this research. Hence, the selection of respondents for this research is suitable in order to achieve the research aim and objectives.

**Table 4.1.** Respondent's background.

| Respondent | Types of organisation | Position | Experience in construction industry | Experience in implementing BIM |
|---|---|---|---|---|
| R1 | Developer | Architect | 8 years | 4 years |
| R2 | Local Authority | Chief Assistant Director Architect | 15 years | 9 years |
| R3 | Consultant | BIM Coordinator (Structural-based) | 10 years | 6 years |
| R4 | Contractor | BIM Executive (Structural-based) | 2 years | 2 years |
| R5 | Developer | Senior Executive Building Technology | 20 years | 8 years |

### *4.2 Current practices of BIM implementation in design stage*

Table 4.2 below shows the current practices of BIM implementation in design phase based on the summary of findings from collected data through semi-structured interviews with selected respondents. It consists of strengths and weaknesses of BIM tools, BIM challenges in the design phase as well as benefits of BIM implementation in the design phase.

**Table 4.2.** Current practices of BIM implementation in design phase.

| Current practices of BIM implementation in design phase | Description | |
|---|---|---|
| **(a) Strengths and Weaknesses of BIM Tools** | <u>*Strengths*</u> | <u>*Weaknesses*</u> |
| • *Autodesk Revit; Revit Architecture, Revit Structure and Revit MEP* | • Lowest initial cost<br>• Concurrent operation<br>• Great coordination | • System lagging |
| • *ArchiCAD* | • Large object libraries or families | • Poor sharing capability |
| • *Bentley* | • Great coordination | • Poor sharing capability |
| • *Tekla Structure* | • Great coordination | - |
| **(b) BIM Challenges in Design Phase** | | |
| • *Before designing* | • Collaboration and teaming<br>• Unenthusiastic stakeholders<br>• Unclear BIM standards | |
| • *During designing* | • Time constraint<br>• Risk increase | |
| • After designing | • Unclear benefits | |
| **(c) Benefits of BIM Practices in Design Phase**<br>• *Clash detection*<br>• *High quality of BIM model output*<br>• *Contributes to high productivity* | • Reduce error<br>• Fast fabrication drawing<br>• Great coordination<br>• LEED compliance<br>• Predict environmental performance<br>• More client's input<br>• Consistent view<br>• Faster decision making<br>• Reduce manpower<br>• Less time and cost | |

According to the respondents, the strengths of BIM tools can be divided into four (4) categories; lowest initial cost for Autodesk Revit; concurrent operation for Autodesk Revit; large object libraries (families) for ArchiCAD; and great coordination for Autodesk Revit, Bentley, and Tekla Structure. Meanwhile, the weaknesses of BIM tools are system lagging for Autodesk Revit and poor sharing capability for ArchiCAD and Bentley. All those categories are complementary with previous research [6][7]. It is important to discover each BIM tools strengths and weaknesses in order to give a clear figure to the designers in a construction project team of what they actually embrace. The respondents agreed that it helps to improve the decision making of which BIM tool is appropriate to be implemented by the designers of all trades. Equally important, the information gathered can be used as a result of a survey for the BIM tools' developers as well in order to develop a plan to improve their products.

Next, the challenges of BIM implementation in the design phase are also being identified. The respondents for this research claimed that challenges before designing in BIM process is collaboration and teaming, unenthusiastic stakeholders and unclear BIM standards. Meanwhile, the challenges of designing in BIM process are time constraint to develop a model and increase the risk of disputes. Lastly, unclear benefits are the challenge of BIM implementation after designing. All of the respondents agreed that those challenges will affect the performance of design team in a construction project. Hence, the design team needs to develop their own strategic plan to solve the challenges when it comes to BIM implementation in design phase because after all, designing is a crucial phase before commencing actual construction works. Any disputes during designing will affect the whole project performance and decrease the effectiveness of BIM implementation itself. Once the challenges are fully overcome by the construction key players, the benefits of BIM implementation in the design phase can be fully exploited.

Lastly, according to respondents for this research, there are three (3) classifications of BIM benefits in design phase; clash detection which to contributes to the reduction of error, fast fabrication drawing, and great coordination among designers; high quality of BIM model output since it is certified with LEED compliance and capable of predicting environmental performance; and contributes to high productivity since it allows more client outputs during designing, consistent models view, faster decision making, reduce manpower and require less time and cost. The first research objective is also achieved since the respondents agreed that the advantages gained in the use of BIM in the design phase of the construction project will be able to increase the implementation awareness of BIM in Malaysian construction industry. Also, there is a need to identify the best practices factors of BIM implementation in the design phase in order to promote its benefits and spread the awareness of BIM potentials to the construction industry.

*4.3 Best practices of Building Information Modelling (BIM) implementation in design phase*

From current practices of Building Information Modelling (BIM) implementation, there are several improper techniques which create boundaries to achieve its best practices in the design phase. In order to make BIM as a useful tool to develop construction projects, it is important to execute BIM implementation's best practices strategic plan which ensures the works to be clearly defined and understood by the adopters [13]. When it comes to the formulation of the strategic plan to achieve best practices in BIM implementation during design development, all of the respondents in this research claimed that it is more appropriate if all the stakeholders in project development are involved. 'Top-down and bottom-up management' would be the keys to encourage the complicity of the stakeholders to develop a strategic plan towards achieving best practices in BIM implementation. Table 4.3 below shows the best practices factors of BIM implementation in design phase based on the summary of findings which consider findings from literature as well as from collected data through semi-structured interviews with selected respondents. It consists of BIM training among designers, tactical design development, enhances effective collaboration in design review and capabilities of maintaining exchanged data.

**Table 4.3.** Best practices' factors of BIM implementation in design phase.

| Best practices' factors of BIM implementation in design phase | Description |
|---|---|
| 1 | • <u>*BIM training among designers*</u><br>Best practice's factor: Provide incentive for designers to undergo training process |
| 2 | • <u>*Tactical design development*</u><br>Best practice's factor: Provide formal approach to monitor level of detailing (LOD) in BIM feature |
| 3 | • <u>*Enhance effective collaboration in design review process*</u><br>Best practice's factor: Carry out virtual meeting in conducting design review process |
| 4 | • <u>*Capabilities of maintaining exchanged data*</u><br>Best practice's factor: Improve Industry Foundation Class (IFC) compatibility |

From the findings of this research, there are four (4) best practices factors of BIM implementation in design phase; provide incentive for the designers to go to a BIM training, provide a formal approach to monitor automated Level of Detailing (LOD) in BIM feature during tactical design development, run a virtual meeting for enhancing effective collaboration in design review process; and lastly, improve Industry Foundation Class (IFC) to support the capabilities of maintaining exchanged data among designers of different trades. From the interviews, the best practices factors that have been identified covers the construction team management and planning, designer's scope of work and the technology as well. People, process, and technology are interrelated in order to develop a strategic plan towards achieving best practices of BIM in design phase through its identified best practices factors. In brief, BIM urges the adopters to develop a strategic plan on how to make it beneficial for all stakeholders to ensure it is best practice. The strategic plan needs to include people, process, and technology, since BIM changes the whole design process through its broad types of technology and people (construction stakeholders), needs to deal with the changes. It is basically the factors on how to make the best practices effective. Certainly, those best practices are also supported by previous research [6][8][9][14].

## 5. Conclusion

In conclusion, the current practices of BIM tools implementation in the design stage are Autodesk Revit (Revit Architecture, Revit structure and Revit MEP), ArchiCAD, Bentley and Tekla Structure. All the tools have specific challenges and weakness in the implementation of BIM in the design stage. The challenges of BIM implementation in the design phase are also being identified such as before designing (collaboration and teaming, unenthusiastic stakeholders and unclear BIM standards), during designing (time constraint to develop a model and increase the risk of disputes) and after designing is unclear benefits. Furthermore, there are three (3) classifications of BIM benefits in design phase; clash detection which to contributes to the reduction of error, fast fabrication drawing, and great coordination among designers; high quality of BIM model output since it is certified with LEED compliance and capable of predicting environmental performance; and contributes to high productivity since it allows more client outputs during designing, consistent models view, faster decision making, reduce manpower and require less time and cost. There are four (4) best practice's factors of BIM implementation in design phase; provide incentive for the designers to go to a BIM training, provide a formal approach to monitor automated Level of Detailing (LOD) in BIM feature during tactical design development, run a virtual meeting for enhancing effective collaboration in design review process; and lastly, improve Industry Foundation Class (IFC) to support the capabilities of maintaining exchanged data among designers of different trades. Thus, by identification of best practices factors of BIM implementation in the design phase will bring improvements in BIM current practices during design development. This will then accelerate the percentage of BIM implementation in Malaysian construction industry through the awareness of how BIM is beneficial to the adopters once it is practiced at its very best.

## 6. References

[1] National Centre for Construction Education and Research (NCCER) 2008 *Project management* (New Jersey: Pearson Education, Inc.)

[2] Construction Industry Development Board (CIDB) 2015 Construction industry transformation programme 2016-2020 (Malaysia: Construction Industry Development Board)

[3] Gardezi S S S, Shafiq N and Khamidi M F 2013 Prospects of Building Information Modeling (BIM) in Malaysian construction industry as conflict resolution tool *Journal of Energy Technologies and Policy* **3** (11) 346-350

[4] Hopple M 2015 *Real Life Perspective: Transitioning from AutoCAD to BIM* Retrieved April 8, 2016 from https://www.augi.com/

[5] Gardezi S S S, Shafiq N, Nurudinn M F, Farhan S A and Umar U A 2014 Challenges for implementation of Building Information Modelling (BIM) in Malaysian construction industry *Applied Mechanics and Materials* **567** 559-564

[6] Memon A H, Abdul Rahman I, Memon I and Azman N I A 2014 BIM in Malaysian construction industry: Status, advantages, barriers and strategies to enhance the implementation level *Research Journal of Applied Sciences, Engineering and Technology* **8** (5) 606-614

[7] Eastman C, Teicholz P, Sacks R and Liston K 2011 *BIM Handbook: A Guide to Building Information Modeling for Owners, Managers, Designers, Engineers and Contractors* 2nd Ed. (New Jersey: John Wiley & Sons, Inc.)

[8] Ahmad Latiffi A, Mohd. S, Kasim N and Fathi M S 2013 Building Information Modelling (BIM) applications in Malaysian construction industry *International Journal of Construction Engineering and Management* **2** (4A) 1-6
[9] Mohd Nor M F I and Grant P M 2014 Building Information Modeling (BIM) in Malaysian architecture industry *WSEAS Transaction on Environment and Development* **10** 264-273
[10] Deutsch R 2011 *BIM and Integrated Design* (New Jersey: John Wiley and Sons, Inc.)
[11] Navendren D, Manu P, Shelbourn M and Mahamadu A 2014 Challenges to BIM implementation in UK: Designer's perspectives *Procs. 30th Annual ARCOM Conference* (Portsmouth, UK: Association of Research in Construction Management) pp 733-742
[12] Kerusuo H, Miettinen R, Paavola S, Maki T and Korpela J 2015 Challenges of the expansive use of Building Information Modelling (BIM) in construction projects *Production* **25** (2) 289-297
[13] Smith A 2012 *Best Practices: First Steps with BIM* Retrieved April 23, 2016 from The Architect's Journal: http://www.architectsjournal.co.uk/
[14] Mitchell D 2014 *BIM Adoption: Why Can't It Just Be About Best Practice?* Retrieved April 23, 2016 from: https://sourceable.net/
[15] Maxwell J A 2013 *Qualitative Research Design: An Interactive Approach* Third Edition (United States of America: SAGE Publications, Inc.)
[16] Creswell J W 2014 *Research Design: Qualitative, Quantitative, and Mixed Methods Approaches* 4th Edition (United States of America: SAGE Publications, Inc.)
[17] Construction Industry Development Board (CIDB) 2014 BIM Project *Construction Industry Development Board (CIDB) Malaysia*
[18] Jamshed S 2014 Qualitative research method-interviewing and observation *Journal of Basic and Clinical Pharmacy* **5** (4) 87-88
[19] Mayring P 2002 Qualitative Content Analysis *Qualitative Social Research 1(2)* Retrieved April 21, 2016 from http://www.qualitative-research.net/
[20] Hsieh H and Shannon S E 2005 Three approaches to qualitative content analysis *Qualitative Health Research* **15** (9) 1277-1288

**Acknowledgement**
The author would like to thank the Universiti Tun Hussein Onn Malaysia (UTHM) for supporting financial aid for this research.

IOP Conf. Series: Materials Science and Engineering **271** (2017) 012039 doi:10.1088/1757-899X/271/1/012039

# Current practices of construction waste reduction through 3R practice among contractors in malaysia: Case study in penang

**L S Ng[1], L W Tan[1] and T W Seow[2]**

[1] Department of Water and Environmental Engineering, Faculty of Civil and Environmental Engineering, University of Tun Hussein Onn Malaysia (UTHM), Batu Pahat, Johor, Malaysia.
[2] Department of Construction Management, Faculty of Technology Management and Business, University of Tun Hussein Onn Malaysia (UTHM), Batu Pahat, Johor, Malaysia.
Corresponding author:leeshan3122@gmail.com

**Abstract.** The effectiveness of the implementation of construction waste reduction through 3R reflects the sustainability in construction waste management. Weak implementation of construction waste reduction through 3R among contractors will lead to unsustainable construction waste management. Increase in construction waste on landfills is critical especially on islands where land is very limited for solid waste disposal. This aim of this paper is to investigate current practice of construction waste reduction through 3R practice among contractors in Penang, Malaysia. The findings reported herein is based on feedbacks from 143 construction contractors of grade CIDB G7, G6 and G5 in Penang and experts from Penang Local Authority, CIDB in Penang and its Headquarters, National Solid Waste Management Department, and Headquarters of Solid Waste and Public Cleansing Management Corporation. Interviews and questionnaire surveys have been found that 3R practice is not mandatory in construction waste management in Penang. Only 39.8% construction contractors practiced 3R in managing their waste. Therefore, 3R practices should be emphasized in construction industry. Reducing wastes through 3R practices in construction industry is a way forward towards sustainable construction waste management especially in expanding the lifetime of landfill.

## 1. Introduction

Construction industry consumes a huge amount of natural resources and generates large quantities of construction waste [1]. Construction waste is defined as waste generated from construction industry during construction activities, building renovation, civil construction and building, construction site cleaning, road construction and demolition activities, including soil excavation. Some of construction wastes are recyclable and reusable; however most of the wastes are usually dumped on landfill [2]. Basic construction waste such as dirt, brick, concrete, asphalt, glass, wood, plastic and metal [3]. According to [4], construction waste management is a tool to control the cost of construction waste disposal and to facilitate other alternative than disposal method such as reuse and recycling to reduce construction waste disposed at landfills at the last stage. Construction waste should be recognized as a valuable resource because many types of construction waste can be reused or recycled [5].

### *1.1. Construction waste reduction through 3R in Malaysia*

Solid waste reduction through 3R is one of the thrusts of National Solid Waste Management (NSWM) Policy. Construction waste is one of the controlled solid wastes [6]. 3R practices represent the concept of reduce, reuse and recycle [7].

- **Reduce:** Reduction is considered as the most effective and efficient method to manage construction waste. Reduction does not only reduce construction waste generation, it also can reduce the cost for transportation, waste disposal and waste recycling [1,8-10].
- **Reuse:** Reuse is usually a favourite option because some construction waste can be reused in other construction project. Reuse is most beneficial and contractors can save money since disposal involved cost [11]. Reuse is using the same material more than once for the same function such as formwork in construction [12]. Any material which cannot be reused but can be recyled will be sent to recyling center [13].
- **Recycle:** When reduction and reuse become difficult, recycling is desired. Some new materials can be made out through recycling [14]. Recycling construction waste can be categorized into on-site and off-site. On-site recycling is defined as segregation of construction waste for subsequent use as the raw materials in construction project. Meanwhile, off-site recycling is segregation of construction waste which are then transported to other organizations or locations and the waste is used as raw materials [15].

3R practice has become more popular as it has been included in policy and option among other alternative of waste hierarchy concept and it is based on the idea of fully utilizing the resources before disposal [16]. Efforts have been placed in focusing on 3R towards reducing the solid waste [17].

Malaysia like other developing countries are facing problem of high construction waste generation due to rapidly growth of construction industry. Landfill dumping is common practices in handling construction waste in Malaysia [18]. At much 70% to 80% recyclable materials can be found on landfills in Malaysia [19]. This causes landfill to reach their full capacity ahead of time [18]. Sustainable waste management is remains a low priority among majority of contractors in Malaysia. The 3R practices in construction industry are limited since it is not a mandatory requirement for construction companies [20]; and roadside illegal dumping is still an issue for the authorities [21]. Majority of contractors do not practice source separation, reduction, reuse or recycling at construction sites in Malaysia [20].

Moreover, practice of reuse and recycling construction waste has always been neglected because most of the efforts were focused on handling domestic waste [5]. Municipal Council only provides disposal services at landfill but does not interfere in construction waste management since it is the responsibility of contractors to manage construction waste on-site [13]. Currently, there is no systematic information record on the volume and type of construction waste being generated, amount of raisings generated and disposed in Malaysia. Data on waste minimization and recycling provided by local authorities is widely varied, not standardized, non-uniform, and inconsistent [19].

Disposal activities are very limited and the waste surface is estimated to be 38 m high above the mean sea level beside the Jelutong landfill [21]. With the shortage of open area in Penang Island [22], Penang has difficulty in locating new site for solid waste disposal because landfill are requires large area. Almost all construction wastes in Penang are sent to Jelutong landfill [21]. Residents at a high-rise condominium are up in arms over the worsening stench from the open burning at Jelutong landfill in Penang Island. They are infuriated by the delay of local authorities in relocating the landfill. The residents had complained to the authorities many times, but no action had been taken to relocate the landfill. Nearby residents want Jelutong Landfill be relocated due to the raging fire and the resulting smog which posed threat to health. They are worried as the landfill had been expanding and it is located a mere 500m from their condominium. The landfill has been getting bigger and moving closer to the sea. Fire has been spotted at the landfill especially during the hot and dry season [23].

*1.2. Negative impacts due to ineffective of implementation of construction waste reduction through 3R practice*

Rapid growth in construction activities increases construction waste problems around the world [24]. Construction wastes become serious environmental problems since large volumes of construction waste reduces landfill capacities and leads to environmental issues [8]. Construction waste has not only created environmental problem, but also poses negative impact on human health and hygiene [25]. Besides, expanding construction waste shows that the bulk dissipation of natural resources [26].

- **Environmental:** The environmental impact of the construction industry is substantial in developing countries than developed countries because developing countries are still under construction and going through rapid urbanization and industrialization [27]. Construction waste may cause environmental pollution on landfill. When it rains, construction waste may seep into and pollute the ground water, surface water and soil [28]. Contractors dump construction waste illegal at road side or into river will cause environment pollution [29].
- **Human health:** Traditional disposal method requires large of land which impose apprehensive and psychological fears due to the health and ecological risks [30]. Construction waste which is not properly managed will lead to human health problems [18].
- **Resource:** Higher construction waste being generated will result in higher consumption of natural resources [18]. 3R practice is an efficient and effective way to reduce waste going into landfills and decrease resource throughput, as using recycled material means reducing amount of new resources needed to manufacture a new product [31].

## 2. Research methodology

In this study, semi-structure interviews were carried out with experts from Penang Local Authority, CIDB in Penang and its Headquarters, National Solid Waste Management Department, and Headquarters of Solid Waste and Public Cleansing Management Corporation. A total 143 set of validated questionnaires were collected randomly from construction contractors of grade G7, G6 and G5 under CIDB accreditation located in Penang. These 143 responses are equivalent to 65.9% of response rate. According to [33], as a rule of thumb, a 30% return is seen as fairly satisfactory while more than 50 per cent is good. Collected questionnaire were analyzed using Statistical Packages for Social Sciences (SPSS) while interview data were analyzed through Content Analysis.

## 3. Current practice of construction waste reduction through 3R among contractors in Penang

Based on the questionnaire survey, it has been found that 67% of contractors have the knowledge on 3R concept. Meanwhile, 20% of the contractors are not sure and 13% of contractors have no knowledge whatsoever on 3R concept. Figure 1 summarises the finding. Majority of contractors get to know the about 3R through media, workshop and courses. Small portion of contractors obtained information on 3R through law and regulation and requirement of contract. Besides that, a few of the contractors also learn about 3R through university and reading materials.

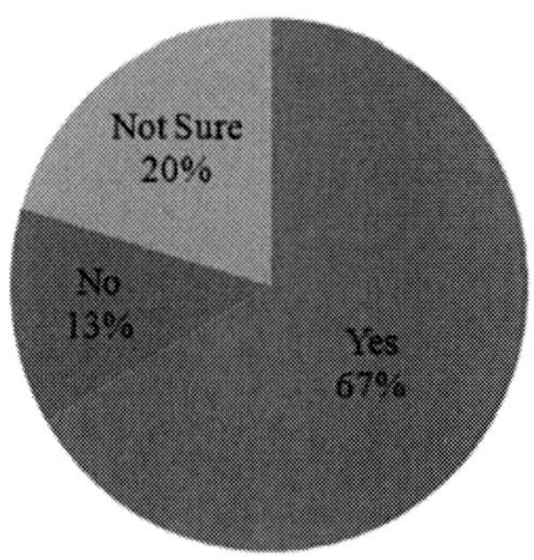


**Figure 1.** Level of knowledge on 3R concept among contractors in Penang.

Based on experts' interview, only an expert pointed out that effectiveness of the implementation of waste reduction through 3R in Penang has achieved 40% but this does not account for illegal dumping. Two experts highlighted that construction waste reduction through 3R among contractors in Penang is minimum since there is no ruling, technology, incentive, market and figure or record on it that exist. Meanwhile, three experts mentioned that waste management is under local authority and their agencies have no information on the current implementation of waste reduction through 3R in Penang.

Survey also shows that only 39.2% of contractors practiced 3R in managing construction waste, as depicted in figure 2. More than half of contractors (60.8%) do not practice 3R in managing construction waste. Most of the contractors sent their construction waste directly to landfill. This reveal that 3R implementation in construction waste reduction among contractors in Penang is still at its infancy. Table 1 shows percentage of waste being managed through 3R by contractors. From a total of 56 (39.2%) contractors who practice 3R, 37% of them managed only 20% or lower of their total construction waste through 3R. Only 11 contractors managed as much as 50% and more of their total construction waste through 3R.

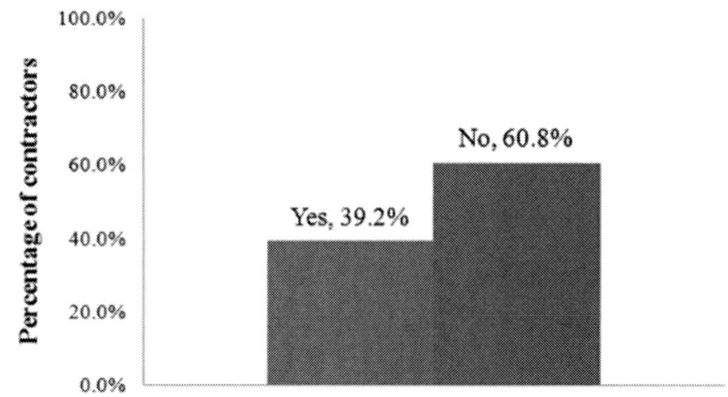


**Figure 2.** Percentage of contractors that practice 3R in managing construction waste at Penang.

**Table 1.** Total percentage of waste managed through 3R practice by contractors.

| Total percentage of waste managed through 3R | Number of contractors |
|---|---|
| 10% | 19 |
| 20% | 18 |
| 30% | 8 |
| 40% | 1 |
| 50% | 7 |
| 60% | 1 |
| 70% | 1 |
| 80% | 2 |
| 90% | 0 |
| 100% | 0 |
| Total | 56 (39.2%) |

The reasons why these 56 contractors (39.2%) practiced 3R in managing construction waste are shown in table 2. 45 of them practice 3R in construction waste management for environmental protection while 38 contractors are because of the high scrap value of recycled material. Monetary value of waste is the reason for 26 contractors to practice 3R. 12 others contractors practiced 3R because of contract requirement, while 2 contractors indicated that 3R practice may save cost.

Conversely, most of contractors do not practice 3R in managing construction waste, which is unsustainable. It has been identified that weak implementation of 3R practice is due to constraint in time and cost, lack of space, lack of enforcement, lack of awareness and knowledge, lack of coordination, and contractors' attitude and low participation. Therefore, governing bodies should put in efforts in encouraging, promoting and enforcing 3R practices. 3R practice is an appropriate approach for contractors in managing construction waste because it is sustainable and beneficial.

**Table 2.** Reason of contractors practiced 3R in managing construction waste.

| Reason | Number of contractors |
|---|---|
| Environmental Protection | 45/56 |
| High scrap value of recycled material waste | 38/56 |
| Profit | 26/56 |
| Contractor requirement | 12/56 |
| Others | 2/56 |

Survey also found that only 33.6% of contractors indicated that 3R is mandatory in construction waste management as shown in figure 3. Meanwhile, 37.1% contractors revealed that 3R is not mandatory and 29.4% of contractors not sure whether 3R is mandatory in construction waste management. However, based on the feedbacks provided by experts interviewed, one expert stated that 3R practice is mandatory to legal contractor because minimization and recycling construction waste are stated in Planning Permission Plan. Meanwhile, others experts stated that 3R practice are not mandated and not stated in project contract. However, some contractors do practice 3R in managing waste and stated in contract and in pursuing sustainable building rating like Green Building Index (GBI) and Malaysian Carbon Reduction & Environmental Sustainable Tool (MyCREST) where 3R practice is one of the criteria. Their agencies are in progress enforcing the law and regulation to mandatory 3R practices. Overall, this study revealed that 3R practice currently is not mandatory in construction waste management and it depends on contractor's initiative and pursuing the sustainable building rating. But, it will be mandatory when law and regulation has been enforced.

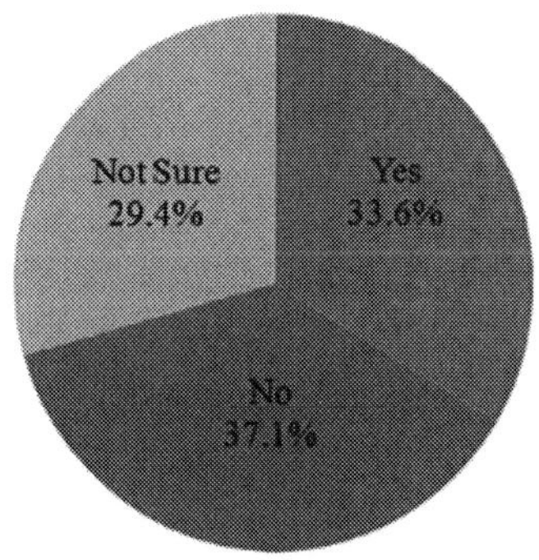


**Figure 3.** Practice 3R is mandatory in construction waste management.

All the experts being interviewed indicated that environmental pollution, human health risk and dissipation of resources are the definite negative impacts due to ineffective implementation of construction waste reduction through 3R. Environmental pollution is in existence due to improper dumping of construction waste and the carbon footprint of vehicle use to send waste to landfill. Further, improper segregation of construction waste also led to environmental pollution. Construction wastes that contain hazardous chemical will also pollutes the water and soil. Meanwhile, environmental pollution can also occur due to unavailable of very good inert landfill. All experts agreed that environmental pollution affected human health directly especially from polluted drinking water sources and air. They also stated that a lot of recyclable and reusable waste at landfill reflects the dissipation of resources. These unsustainable impacts will remain exist if new resources such as timber are continually used without practice 3R. Table 3 shows that majority of the contractors (86.7%) pointed out that environmental pollution is ranked first as the negative impacts that arises from ineffective implementation of construction waste reduction. Impact of human health risk was pointed out by 90 contractors while 37.1% of contractors indicated that dissipation of resources is also one of the impacts. Only 2.1% contractor indicated the others impacts such as excessive spending on waste cleaning.

**Table 3.** Negative impact arises of ineffective implementation of construction waste reduction through 3R practice.

| Negative impacts | Number of contractors (%) |
|---|---|
| Environmental Pollution | 124 (86.7) |
| Human health risk | 90 (62.9) |
| Dissipation of resources | 53 (37.1) |
| Others | 3 (2.1) |

## 4. Recommendation

In order to improve the effectiveness of waste reduction through 3R in construction industry, there are important elements which should be emphasized, i.e. law and regulation, scheme and incentive, awareness and knowledge on 3R, participation of contractors, and available technology. Moreover, top-down approach with enforcement of legislation and policies and good governance concept should be implemented towards sustainable construction waste management. Good governance is beneficial and leads a clear and legitimate process towards effective policy implementation. A framework encompassing these important elements with top-down approach and principle of good governance is needed to be developed in order to have effective construction waste reduction through 3R among contractor.

## 5. Conclusion

Rapid growth of construction industry has produced large amount of construction waste and increases the quantity of waste to be disposed on landfills. Landfills cannot accommodate the accumulating wastes for long term. Penang, which is an island area has scarce land for new landfills since landfill requires sizeable area. 3R implementation in construction waste reduction among contractors in Penang is still at its infancy. More than half of contractors do not practice 3R in managing construction waste. Many studies have shown that various types of construction waste have the potential to be reused or recycled. Therefore, contractor should fully reduce, reuse and recycle construction waste instead of sending the waste to landfill. All parties involved should be accountable and be responsible to achieve sustainable construction waste management.

## 6. References

[1] Poon C S 2007 Reducing construction waste, *Waste Management* **27** 1715–16
[2] Shen L Y, Tam W V Y, Tam C V and Drew D 2004 Mapping approach for examining waste management on construction sites, *J. of Construction Engineering and Management* **130(4)** 472-481
[3] Huang J F, Liang Q B, Wang H, Xu S Y, Zhang L S and Xing Y 2011 Comprehensive recycling of construction waste, *Applied Mechanics and Materials* **117-119** 385-389
[4] Minks W R 1994 The construction contractor's waste management plan: Optimizing control and cost *Proc. of the 1st Int. Conf. of CIB-TG16 Sustainable Construction* (Tampa, Florida, USA) pp 765-774
[5] Nitivattananon V and Borongan G 2008 *Report on Reduce, Reuse and Recycle (3R) Practices in Construction and Demolition Waste Management in Asia* Retrieved on April

22, 2012 from http://www.3rkh.net/3rkh/files/3RKH_C&D_waste_FinalReport.pdf

[6] National Solid Waste Management Department 2012 *National Solid Waste Management Policy* Retrieved on March 9, 2013 from National Solid Waste Management Department: http://www.kpkt.gov.my/jpspn/main.php?Content=articles&ArticleID=64&IID=

[7] National Solid Waste Management Department 2013 *Categories of controlled solid waste* Retrieved on March 9, 2013 from National Solid Waste Management Department: http://www.kpkt.gov.my/jpspn_en_2013/main.php?Content=articles&ArticleID=41&IID=

[8] Esin T and Cosgun N 2007 A study conducted to reduce construction waste generation in Turkey, *Building and Environment* **42(4)** 1667–74

[9] Tam V W and Tam C M 2008 Waste reduction through incentives: A case study, *Building Research & Information* **36(1)** 37-43

[10] Lu W and Yuan H 2011 A framework for understanding waste management studies in construction, *Waste Management* **31(6)** 1252-60

[11] Winkler G 2010 *Recycling Construction & Demolition Waste: A LEED-Based Toolkit* (USA: McGraw-Hill)

[12] Ling Y Y and Leo K C 2000 Reusing timber formwork: importance of workmen's efficiency and attitude, *Building and Environment* **35(2)** 135-143

[13] Seow T W and Mohamad A H 2007 *Construction Waste Management at site (Pengurusan Sisa Pembinaan Di Tapak Binaan)* Retrieved on March 20, 2012 from http://sssums.files.wordpress.com/2007/09/pengurusan- sisa-pembinaan-di-tapak-bina.pdf

[14] Yuan H, Shen L and Wang J 2011 Major obstacles to improving the performance of waste management in China's construction industry, *Facilities* **29(5/6)** 224-242

[15] Franchetti M J 2009 *Solid Waste Analysis and Minimization: A Systems Approach* (USA: McGraw-Hill)

[16] Hezri A A 2010 Toward 3R-based waste management: Policy change in Japan, Malaysia and the Philippines ed Kojima M in 3R Policies for Southeast and East Asia *ERIA Research Project Report 2009-2010* (Jakarta: ERIA) p 274-290

[17] Ministry of Housing and Local Government 2007 *Solid Waste and Public Cleansing Management Act (Act 672)* Retrieved on November 10, 2007 from http://www.kpkt.gov.my/jpspn/main.php?Content=vertsections&SubVertSertionID=6&VertSectionID=5&CurLocation=5&IID=&Page=1

[18] Abdul Rahman I Nagapan S 2015 *Causative Factors of Construction Waste Generation in Malaysia* (Johor: Penerbit UTHM)

[19] Moh Y C and Manaf L A 2014 Overview of household solid waste recycling policy status and challenges in Malaysia, *Resources, Conservation and Recycling* **82** 50-61

[20] Begum R A, Siwar C, Pereira J J and Jaafar A H 2009 Attitude and behavioral factors in waste management in the construction industry of Malaysia, *Resources, Conservation and Recycling* **53(6)** 321-328

[21] Ng L S, Seow T W and Goh K C 2015 Implementation on solid waste reduction through 3R (NSWM Policy) and elements to close gap between policy and contractors in construction industry in Penang, *International J. on Science and Development* **6(9)** 668-675

[22] William R *Penang should allow more reclamation to address land shortage: The Edge Malaysia* Retrieved on March 9, 2013 from http://www.theedgemalaysia.com/in-the-financial-daily/178363-penang-should-allow-more-reclamation-to-address-land-shortage.html

[23] New Straits Times 2016 *Nearby Residents Want Jelutong Landfill Relocated* (Febuary 29)

[24] Nagapan S, Abdul Rahman I and Asmi A 2011 A review of construction waste cause factors *Proc. of the Asian Conf. of Real Estate: Sustainable Growth Managing Challenges (ACRE)Malaysia)*

[25] Wei Z and Rotter S 2008 The current situation of construction & demolition waste management in China *Proc. of the 2nd Int. Conf. on Bioinformatics and Biomedical Engineering* (IEEE) pp 4747-4750

[26] Hu Y P 2011 Minimization management of construction waste *The Int. Symp. on Water Resource and Environmental Protection* (IEEE) pp 2769-2772
[27] Begum R A and Pereira J J 2011 C & D waste profile of the Malaysian construction industry: Need a centralized database *World Congress on Sustainable Technologies (IEEE)* pp 73-76
[28] Liu W and Huang W 2013 Research on construction waste resourcelization legislation *Proc. of the 5th Int. Conf. on Measuring Technology and Mechatronics Automation* (United States: IEEE) pp 893-897
[29] Ng L S, Seow T W and Goh K C 2013 Constraints of construction waste management in Penang Island *1st FPTP Postgraduate Seminar* (Batu Pahat, Malaysia)
[30] Chuen K P and Jamal O 2011 A choice experiment analysis for solid waste disposal option: A case study in Malaysia, *Environmental Management* **92(11)** 2993-3001
[31] Shareh Musa S M, Shafii H and Syed Muhammad Zubir S N 2009 Potensi kitar semula dalam industri pembinaan: Kajian kes daerah Batu Pahat, Johor *Proc. of the 3rd Malaysian Technical Universities Conf. on Engineering and Technology* (Kuantan, Pahang) pp 553-559
[33] Gillham B 2000 *The Research Interview* (London: Continuum) pp 14

**Acknowledgments**
This paper is produced with the contribution and assistance of important parties. The authors would like to thank for financial support by the Ministry of Higher Education, Malaysia (KPT) for the scholarship MyBrain 15 (MyPhD) and the UTHM Research and Innovation Centre; and the UTHM Center for Graduate Studies and faculty members for the guidance provided throughout the study. The authors would also like to thank all the construction contractors and experts involved in providing useful data and information that is reported in this paper.

IOP Conf. Series: Materials Science and Engineering **271** (2017) 012040 doi:10.1088/1757-899X/271/1/012040

# Construction waste generation in Malaysia construction industry: illegal dumping activities

**M H I A Rahim[1], N Kasim[1], I Mohamed[2], R Zainal[1], N Sarpin[1] and M Saikah[1]**

[1] Department of Construction Management, Faculty of Technology Management and Business, Universiti Tun Hussein Onn Malaysia, 86400 Parit Raja Batu Pahat, Malaysia.
[2] Department of Construction and Business, SWCorp Malaysia, Blok J, Bangunan MKN Embassy Techzone Jalan Teknokrat 2, 63000 Cyberjaya, Selangor.
Corresponding author: narimah@uthm.edu.my

**Abstract.** Nowadays development of construction in Malaysia has been effect to the increasing of construction waste. Additionally, the production of construction waste from construction projects has given negative impact to the environment especially in illegal dumping activities. The increasing number of illegal dumping activities from construction projects in Malaysia gives a sign that Malaysian construction waste management needs to be concerned. To date, a comprehensive criterion for construction waste management, particularly for a construction project in developing countries is still not clearly defined. Therefore, construction waste management in Malaysia needs further research. The objectives of this paper are to explore illegal dumping activities, and discuss the contributory factors of illegal dumping activities. Hence, this research conducted an interview with expertise in the area of construction waste management in order to scrutinise illegal dumping activities in Malaysia. The data from semi-structured interviews were analysed by content analysis. Findings from this research will help to find out the strategies to reduce the illegal dumping activities. The final result also expected to increase the awareness and better solution for reducing illegal dumping activities in construction projects among construction players.

## 1. Introduction

Construction waste is wastes that unused from construction activities, such as pre construction, construction and post construction. All the construction waste was generated by human activities will contribute to the environmental problems [1]. Previous research has been defined that construction waste as construction fragments, ruins, disaster, construction materials and building construction and demolition, site clearance and any kind of waste from construction projects [2]. Additionally, previous research has been mentioned that construction waste as useful, unwanted or discarded materials that arise from the human activities and are not free flowing [3]. However, there is need an appropriate of waste which is to avoid, reduce and reuse the waste generated by the human activities [1,4,5].

From the previous study has been found that illegal dumping of construction waste becoming depressing in Malaysia [2,4,5]. Illegal dumping activities from construction projects have contributed bad impacts to the environment, social and economy. Additionally, the average total of construction waste was generated 28.6 tons daily in the year 2015 [5]. The data shows the increasing number of construction waste generated and potential contribution of illegal dumping activities. Furthermore, several studies have been conducted at Penang, Johor, Melaka, and Selangor from 2004 to 2013, shows the increasing of illegal dumping activities from construction waste from year to year [5]. Thus, illegal dumping activities is an critical issue which 851 of illegal dumping area has been identified by SWCorp Malaysia and which has to be emphasized in order to reduce the construction waste generation in Malaysia [5].

## 2. Waste generation in construction industry

### *2.1. Construction waste generation and illegal dumping activities*

Malaysian construction projects have been shown in the value of projected new construction works as reported by CIDB in the year of 2017. The value of construction projects for 2017 is forecast at RM138.0 billion compare to RM 131.0 Billion in the year 2016. This projection and development of construction project in Malaysian construction industry show a view of an implementation of megastructure projects and a good investment for Malaysia. From the discussion above, increasing demand in new development and refurbishment construction project are significant to increasing of generation construction waste [6]. The development of infrastructure especially in a construction project has contributed to the environmental issues especially in the production of construction waste [6,7]. Additionally, by increasing number of demand in infrastructure, it was directly increasing the development of construction in Malaysia.

Growing number of waste generation is due to several factors which economy, demand, living standard, population growth and industrial and urbanisation. Due to the factors of construction waste generation has been contributed to illegal dumping activities [1-4, 6-8]. Figure 1 shows the factors generation of construction waste from Solid Waste Corporation Malaysia (SWCorp Malaysia) perspective. Figure 1 shows the generation factors of construction waste which are the development of economy growth, increasing of infrastructure projects demands, commercial buildings and housing projects, increasing of living standard, increasing of urbanisation and industrial and finally increasing of population growth in Malaysia [8]. The increasing number of waste generation directly has been effect to the illegal dumping activities. Furthermore, the illegal dumping activities are due to several factors which serve to environment [8,9].

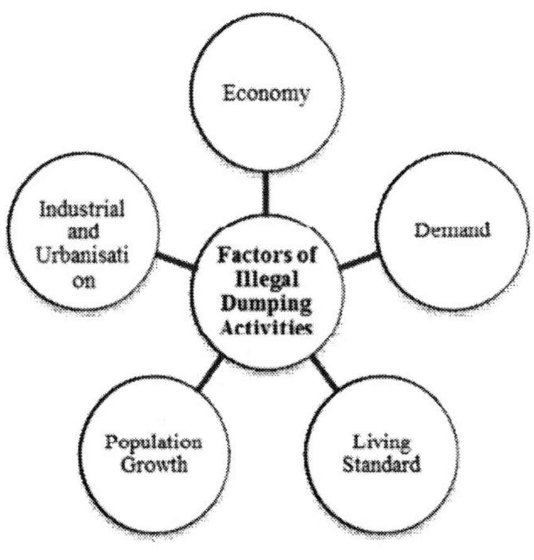


**Figure 1.** Factors Generation of Construction Waste [8,9].

SWCorp Malaysia has defined illegal dumping as a waste material dump into public and it can effect to the environmental issues [10]. In this case, waste materials from construction are not transported to the landfills can define as illegal dumping activities. Hence, the logistic factors are a major problem for developing countries. This is due to the development of the construction sector in the developing countries [11]. In Malaysia, normally construction waste dumped at municipal solid waste landfill. This is because this method is easy for responsible contractor and they do not need any of technology to manage this construction waste. However, the disposal of waste contributed to the potion of waste dump, cost running during the logistic and the environment [12]. Mostly construction waste has been thrown at landfill, there are still reported that more of waste disposed at illegal dumpsites [13]. Illegal dumping is a major problem in many areas which raises concerns with regard to safety, property value, and quality of life in the community [5] [13]. In Malaysia, dumping of waste at illegal dumping sites is a common practice, particularly industrial and construction wastes.

### *2.2. Impact of illegal dumping activities*

Generation of construction waste is one of the major factor by improper planning management during the construction phase [14]. Generally in every single phase of construction there are possibilities of construction waste generation if there are no concerns from construction parties [13]. Attention should be given emphasis to its environmental issues as the main impact is seen in public health over the last three decades [15]. The impacts of construction waste will stretch into the lives of the current and future generation. One of the major causes of construction waste is improper waste management. Lack of waste regulations and improper disposal facilities are the sources of unsuccessful waste management [16]. Construction waste management is important to avoid the negative impacts on the environment, social and economy [17]. It will become an issue if it is handled poorly. However, construction wastes which create issues such as illegal dumping require greater attention. According to the previous study the impacts of construction waste on the environment are unbalanced ecology, change of living environment, potential sewage, depletion of natural sources, energy consumption and generation waste [18].

Furthermore, a recent study of development squander impacts found that the ecological corruption, decimation of the biological system and dangers to general wellbeing are caused by poor administration of waste as far as transfer and waste taking care of handling waste management [18]. The previous research has found that the major factors contributed to an increasing of illegal dumping activities are the financial issues and logistic of the construction waste [19]. Additionally, the distance between the project location and landfill site is too far that has led the contractors to refuse disposing waste to the illegal dumping area. Continuity from that, it was reported that the contractors failed to pay for landfill charge and logistic cost in order to maximise their profit [19]. All parties in construction industry have together with the full commitment to ensure that construction waste can be organized and manage well. Thus, to archive the effort government should take an action by introducing new legislation and enforcement regime.

## 3. Methodology

This section explained the process of methodology adopted in this paper. Research methodology is a guide to achieve the aim and objectives of the research. Hence, this research will adapt qualitative approach as research approach because this approach is more suitable to capture information on construction waste management from respondents. Moreover, this approach allows researcher to go in-depth into the research topic. Apart from that, this research will use qualitative approach to gain an insight toward respondent perception and understanding toward on illegal dumping activities and construction waste management. The methodology divided into three phase which are literature review from the previous research, data collection by using qualitative method and data analysis by content analysis. Qualitative approach is an approach for examining respondent perceptions based on their work experiences [22]. For this paper, an illegal dumping activity has been explored by way of document review. Then semi-structured interview has been prepared to identify the contributory factors of illegal dumping activities. Interview protocol has been prepared to ensure the reliability and validity of qualitative data. Samples selected based on related and expert agencies in Malaysia construction waste management. Selection of SWCorp Malaysia from Cyberjaya, Kuala Lumpur, Johor, Melaka, Negeri Sembilan, Pahang, and Perlis and *Perbadanan Pengurusan Sisa Pepejal Negara* has been made based on the preliminary beforehand and there are only state have SWCorp Malaysia.

IOP Conf. Series: Materials Science and Engineering **271** (2017) 012040 doi:10.1088/1757-899X/271/1/012040

## 4. Result and discussion

### *4.1. Respondents background*

Ten respondents have been interviewed in this research. However, part of the rest of the respondents will be discussed in future publications. Working experience of respondents is at least 5 years and more. Table 1 shows two of respondents that have been involved in construction waste management are highly experienced in *Perbadanan Pengurusan Sisa Pepejal Negara*. Additionally eight of respondents from SWCorp Malaysia. Table 1 shows the background of respondents from the interviews.

**Table 1.** Respondents' background.

| Respondents | Organisation | Position | Experience |
|---|---|---|---|
| R1 | *Perbadanan Pengurusan Sisa Pepejal Negara* | Timbalan Pengarah | 10 years |
| R2 | *Perbadanan Pengurusan Sisa Pepejal Negara* | Timbalan Pengarah | 6 years |
| R3 | SWCorp Malaysia Cyberjaya | Pengarah Bahagian Binaan dan Perniagaan | 20 years |
| R4 | SWCorp Malaysia Cyberjaya | Engineer | 5 years |
| R5 | SWCorp Malaysia Johor | Engineer | 6 years |
| R6 | SWCorp Malaysia Melaka | Engineer | 7 years |
| R7 | SWCorp Malaysia Negeri Sembilan | Engineer | 5 years |
| R8 | SWCorp Malaysia Kuala Lumpur | Engineer | 8 years |
| R9 | SWCorp Malaysia Pahang | Pengarah Bahagian | 18 years |
| R10 | SWCorp Malaysia Perlis | Engineer | 5 years |

### *4.2. Illegal dumping activities*

Table 2 shows the percentages data of illegal dumping activities has been closed and reported in the year of 2015 by the several state in Malaysia.

**Table 2**. Illegal dumping activities year 2015 [8].

| State | Jan | Feb | Mac | Apr | May | June | Jul | Aug | Sept | Oct | Nov | Dec |
|---|---|---|---|---|---|---|---|---|---|---|---|---|
| **Johor** | 54% | 67% | 68% | 73% | 63% | 32% | 93% | 96% | 96% | 98% | 98% | 98% |
| **Kedah** | 90% | 93% | 100% | 100% | 100% | 100% | 100% | 100% | 100% | 100% | 100% | 100% |
| **Melaka** | 94% | 100% | 100% | 97% | 100% | 100% | 100% | 97% | 100% | 100% | 100% | 100% |
| **Negeri Sembilan** | 61% | 70% | 85% | 90% | 90% | 90% | 92% | 100% | 100% | 100% | 100% | 100% |
| **Pahang** | 100% | 70% | 62% | 90% | 92% | 100% | 14% | 100% | 100% | 100% | 100% | 100% |
| **Perlis** | 100% | 100% | 100% | 100% | 100% | 100% | 100% | 100% | 100% | 100% | 100% | 100% |
| **Kuala Lumpur** | 100% | 100% | 100% | 100% | 100% | 100% | 100% | 100% | 100% | 100% | 100% | 100% |

Table 2 shows the percentages of illegal dumping activities for the State of Johor, Kedah, Melaka, Negeri Sembilan, Pahang, Perlis and Kuala Lumpur. Data has been displayed in Table 3 are from January to December 2015. The percentage indicated is explained SWCorp Malaysia actions to implement the action plan and enforcement operations against illegal dumping areas for each state. SWCorp Malaysia aim to cover all the illegal dumping area (100%) for every month.

From the figure in Table 2 show that, every month illegal dumping activities exist and it will give negative impact to environment and public [15-18]. In overall of data, Johor are the only state are not archive 100% cover for illegal dumping site which only 32% illegal dumping sites has been covered. It shows that population of illegal dumping site in Johor are the highest compare to other state. Furthermore, from the data shown that Perlis and Kuala Lumpur are the only state archive 100% illegal dumping site has been covered from January to December 2015.

Finally, from Table 2 shows illegal dumping activities in every state are still active and need more action to reduce the illegal dumping activities and participated by all parties concern. Consistent with government need which to be focus in issues solution and have impact to nation.

*4.3. Contributory factors of illegal dumping activities*

Table 3 shows the respondents overview of contributory factors of illegal dumping activities based on the semi structured interview.

**Table 3.** Overview contributory factors.

| **Respondents** | Factors | | | | | **Additional Factors** |
|---|---|---|---|---|---|---|
| | **Economy** | **Demand** | **Living Standard** | **Population Growth** | **Industrial and Urbanisation** | |
| **R1** | ✓ | ✓ | ✓ | ✓ | ✓ | Facilities; Educations |
| **R2** | ✓ | ✓ | ✓ | ✓ | ✓ | Facilities; Attitude |
| **R3** | ✓ | ✓ | ✓ | ✓ | ✓ | Facilities; Educations; Renovations; Attitudes |
| **R4** | ✓ | ✓ | ✓ | ✓ | ✓ | Facilities; Educations; Renovations; Attitudes |
| **R5** | ✓ | ✓ | ✓ | ✓ | ✓ | Facilities; Educations; Attitudes |
| **R6** | ✓ | × | ✓ | ✓ | ✓ | Facilities; Renovations |
| **R7** | ✓ | × | ✓ | × | ✓ | Facilities; Renovations |
| **R8** | ✓ | ✓ | ✓ | ✓ | ✓ | Facilities; Educations; Renovations; Attitudes |
| **R9** | ✓ | ✓ | × | ✓ | ✓ | Facilities; Educations; Renovations; Attitudes |
| **R10** | ✓ | ✓ | × | ✓ | ✓ | Facilities; Educations; Renovations; Attitudes |

Description:
✓ : Respondents Agreed
× : Respondents Disagreed

The results of previous study (Table 3) identified that there are five factors has been contributed to the illegal dumping activities which are economy, demand, living standard, population growth, and industrial and urbanization. From the Table 3 it has shows that six out of ten respondents totally agreed with all contributory factors has been contributed to illegal dumping activities. Based on the data, R1, R2, R3, R4, R5, and R8 agreed that economy, demand, living standard, population growth and industrial and urbanisation has contributed to illegal dumping activities in fact, another four respondents are not agree with several factors. Respondent R6 mentioned that demand does not give a big contribution to the illegal dumping activities.

Besides that Respondent R7 were not agreed with demand and population growth has contributed to the illegal dumping activities. In fact Respondent R7 mentioned the renovation work has given big contribution to the illegal dumping activities. Additionally, Respondents R7 said the renovation work contributed to 70% of construction waste production. In Addition respondents R9 and R10 are not agree with the living standard can contributed to the illegal dumping activities. Hence R10 mentioned that Major contribution factor of illegal dumping activities is lack of facilities providing for construction waste management. In conclusion, there are several additional factors has been defined from industry expertise which is facilities, educations, renovations and attitudes which can give big contribution if not been seriously managed in Malaysia construction industry.

## 5. Findings

Illegal dumping activities generated throughout the construction waste from construction site to landfill. The statistic data of illegal dumping activities in Table 2 shows that there is illegal dumping occurs on a monthly basis. This clearly defined that the existence of the problem of illegal dumping still needs to be more emphasized by all parties in order to reduce illegal dumping issues. In addition, figure in Table 2 shows the illegal dumping activities are still not yet fully covered by government. Government agencies are needed to do an action plan and operation exterminate of illegal dumping site. The increasing of illegal dumping activities because of increasing demand in construction and refurbishment projects as mentioned in CIDB report [6,7,20]. Additionally, others factors of illegal dumping activities are increasing of industrial and urbanisation development in Malaysia [8,21].

Therefore, there are various factors that have contributed to the illegal dumping activities. The results of previous study confirm that there are numerous factors causes of illegal dumping activities. Therefore, finding from data collection has been found additional factors contributed to the increasing of illegal dumping activities which shows in figure 2.

**Figure 2.** Additional Factors of Illegal Dumping Activities

Factors that contribute to illegal dumping activities are facilities, educations, renovation, and attitude. Other than that, respondents have mentioned that one of the major factors related to the environmental is the illegal dumping activities from construction and demolition projects. According to respondent 1, in order to improve the current practices in the construction industry, SWCorp Malaysia have prepared the guidelines on construction waste management to assist the stakeholders in order to reduce the illegal dumping activities.

However, increasing number of illegal dumping has been reported shows the weaknesses of construction waste management in Malaysian construction industry and effect to environment [14,15]. The previous researcher also mentioned that the weaknesses of the guideline are not being enforced and translated in the form of the strong legal instrument [6]. Hence contractor in Malaysian construction industry can decide to adopt their own company initiatives for construction waste minimization during construction activities and it directly can contribute to the illegal dumping activities. As a result, respondents have found that new factors of illegal dumping activities and it were strongly proven the factors have increased the illegal dumping activities in Malaysia construction industry. Even though there are several programme handle by government to assist the construction parties to minimize an illegal dumping activities such as campaign, meeting and others programs. Then most of the respondents mentioned that illegal dumping activities are from the renovation works.

From the issues has been discussed the illegal dumping activities can effect to the unbalance ecology, environment pollution and others environment pollution [2,3,19]. Hence, the awareness among all parties especially in construction projects should be more concerned in order to reduce these issues. Other than that, the cooperation between all parties is needed to achieve successful construction waste management and reduce the illegal dumping activities. Several strategies for enhancing waste minimisation which are, all parties need to enhance the level of awareness on how to minimise the waste, then do an initiative to reduce, recycling and reused activities and strengthening the roles of government agencies to ensure the policies are effectively implemented [19,23].

## 6. Conclusion and recommendations

The overview of illegal dumping activities from construction and demolition projects reveals that construction waste in Malaysian construction industry are need to be more attention. From the studies has been prove that there is no proper construction waste management in Malaysia construction industry. This is evidenced by 21, 9000 tone daily or 8 million tone yearly estimated produced in Malaysia. Additionally, SWCorp Malaysia reported that in year of 2014/ 2015 there are 851 of illegal dumping (construction waste) has been detected. It shows that there are still lacks on construction waste management in Malaysia construction industry. Thus, a more holistic implementation is needed to ensure the illegal dumping activities can be reduced and protected. In conclusion, a strong support from the stakeholders, construction players, and government agencies are needed to more effective in managing construction waste in Malaysia. Otherwise, the sustainability and environmental problems will not be addressed effectively. Furthermore next stage will explore the requirement for controlling of illegal dumping activity from construction waste management in Malaysian construction industry.

## 7. References

[1] Seow T W 2016 *Senario Pengurusan Sisa Pepejal di Malaysia* (UTHM: Penerbit UniversitiTun Hussein Onn Malaysia)

[2] Shen L Y, Hao J L, Tam V W Y and Hou H 2007 A checklist for assessing sustainable performance of construction project, *J. Civil Eng. Manage.* **13**(4) 273-281

[3] Yahya K and Boussabaine A H 2006 Eco-costing of Construction Waste, *Manag. Env. Qual. Int. J.* **17**(1) 6-19

[4] The Waste and Resources Action Programme WRAP 2014 *Achieving Good Practice Waste Minimisation and Management. Practical solutions for sustainable construction* (United Kingdom: Waste & Resources Action Programme) pp 3-43

[5] SWCorp Malaysia 2009 *Pelan Strategik 2009-2013* (SWCorp Malaysia) pp 14-15

[6] Saadi N, Ismail Z and Alias Z 2016 A Review Of Construction Waste Management And Initiatives In Malaysia, *J. Sustainability Sci. Manag.* **11**(2) 101–14.

[7] Rahim M H I A and Kasim N 2017 Conceptual Model for Systematic Construction Waste

Management *MATEC Web of Conf.* vol. 87 (Paris: EDP Sciences) 1008
[8] SWCorp Malaysia 2014 *Pelan Strategik 2014 - 2020* (SWCorp Malaysia) pp 13-15
[9] Taha M P 2015 *What a Waste: Solid Waste Management and The Malaysian Perspective on Construction Waste Generation and Management* Retrieved on May 2, 2017 from http://www.swcorp.gov.my/index.php/en/
[10] SWCorp Malaysia 2016 *Progress Report 2016* (SWCorp Malaysia)
[11] Muhwezi L, Chamuriho L M and Lema N M 2012 An Investigation into Materials Wastes on Building Construction Projects in Kampala-Uganda, *Journal of Engineering Research* **1**(1) 11-18
[12] Osman W, Udin Z and Salleh D 2012 Adoption Level of Sustainable Construction Practices: A Study on Malaysia's Construction Stakeholders, *J. Southeast Asian Stud.* **2012**(2012) 270273
[13] Begum R A, Satari S K and Pereira J J 2010 Waste Generation and Recycling: Comparison of Conventional and Industrialized Building Systems, *American Journal of Environmental Sciences* **6**(4) 383-388
[14] Hassan S H, Ahzahar N, Fauzi M A and Eman J 2012 Waste Management Issues in the Northern Region of Malaysia *Procedia - Social and Behavioral Sciences* vol 42 (United Kingdom: Elsevier) pp 175-181
[15] Hossain M S, Santhanam A, Nik N N A and Omar A K M 2011 Clinical Solid Waste Management Practices and its Impact on Human Health and Environment - A Review, *Int. J. Integr. Waste Manage., Sci. & Technol.* **31** 754-766
[16] Njoroge G K 2012 *Environmental Pollution and Impacts on Public Health: Environmental pollution and impacts on public Health* Retrieved on May 2, 2017 from http://www.unep.org/urban_environment/PDFs/DandoraWasteDump-ReportSummary.pdf
[17] Nagapan S, Rahman I A and Asmi A 2011 A Review of Construction Waste Cause Factors *Asian Conf. on Real Estate: Sustainable Growth Managing Challenges (Terengganu)* vol 11 num 2 (Terengganu, Penerbit Universiti Malaysia Terengganu) pp 101-104
[18] Raja Mohd Noor R N H, Mohd Ridzuan A R, Endut I R, Noordin B, Shehu Z and Abdul Ghani A H 2013 The Quantification of Local Construction Waste for the Current Construction Waste Management Practices: A Case Study in Klang Valley *IEEE Xplore Digital Library* (United States: IEEE) pp 183-188
[19] Nagapan S, Rahman I A, Asmi A, Hameed A and Zin R M 2012 Identifying Causes of Construction Waste - Case of Central Region of Peninsula Malaysia, *International Journal of Integrated Engineering* **4**(2) 22-8
[20] Construction Industry Development Board 2017 *Construction Industry Prospect 2017 CIDB Malaysia* Retrieved on May 2, 2017 from http://www.cidb.gov.my/cidbv5/images/content/2017/bisnes/Chapter-4---Edited.pdf
[21] Sreenivasan J, Govindan M, Chinnasami M and Kadiresu, I 2012 *Solid Waste Management in Malaysia – A Move Towards Sustainability, Waste Management* Retrieved on October 15, 2016 from https://www.intechopen.com/books/waste-management-an-integratedvision/solid-wastemanagement -in-malaysia-a-move-towards-sustainability.
[22] Fellows R F and Lui A M M 2008 *Research Methods for Construction* (United States of America: Blackwell Publishing Ltd) pp 316
[23] Nizam M and Yusoff B I N 2010 *Waste Minimization by Recycling of Construction Waste* Retrived on March 21, 2016 from http://umpir.ump.edu.my/1296/1/MOHD_NIZAM_BIN_YUSOFF.PDF

**Acknowledgments**
The author would like to thank, Center for Graduate Studies Universiti Tun Hussein Onn Malaysia. This paper was partly sponsored bu the Centre for Graduate Studies UTHM. The author also likes to thank Department of Construction Management, Faculty of Technology Management and Business, UTHM, and SWCorp Malaysia, for supporting this research.

# Minimizing delays in the Jordanian construction industry by adopting BIM technology

**M Btoush[1] and A T Harun[1]**

[1] Faculty of Civil Engineering & Earth Resources, University Malaysia Pahang, Lebuhraya Tun Razak, 26300 Gambang, Pahang, Malaysia
Corresponding author: wardbtoush2012@yahoo.com

**Abstract.** The Jordanian construction industry plays a significant role and contributes immensely to the gross domestic product (GDP) of the economy. However, the Jordanian public work and housing ministry and most industry players including engineers and contractors have reported that most of the projects experience delays which lead time and cost overruns, and extra efforts. The main causes of delays identified by researchers include poor scheduling and planning, change orders, site conditions, weather, late deliveries, incompetent technical staff. To address these challenges, the implementation of building information modelling (BIM) is paramount. This paper presents BIM as a powerful tool for reducing delays in Jordan construction projects. The paper focuses on two main parts; the first part involves the identification of the major causes of delays, and the second part is to accurately outline the roles and responsibilities of BIM specialist in construction projects. Finally, the paper matches the roles and responsibilities of BIM specialist and the causes of delays, and how the delays are addressed through BIM specialist.

## 1. Introduction

The Jordanian construction industry plays a significant role and contributes immensely to the gross domestic product (GDP) of the economy [1]. However, the Jordanian public work and housing ministry and most industry players including engineers and contractors have reported that most of the projects experience delays which lead time and cost overruns, and extra efforts. The main causes of delays identified by researchers include poor scheduling and planning, frequent change orders, site conditions, late deliveries, poor designs, lack of competent technical staff, among others [1,2]. In order to address these challenges, the implementation of building information modelling (BIM) is paramount [3,4]. The use BIM in the Middle East is a recent phenomenon, especially, its implementation in Jordan is very scarce [5–7]. Nonetheless, BIM is emerging as a new tool, both in construction management and practical sphere of building construction. Researches have demonstrated that BIM has the capacity to improve methods and tools to control delays, fragmentation, risks and increase collaboration in construction projects. Likewise, the use of BIM facilitates the integration of disjointed practices, enhances sustainability, reduces poor quality and acts as a catalyst for changing construction process [8].

Several studies have shown potentials of BIM implementation to increase productivity, information integration, business process flow, and reduce uncertainties, complexities, fragmentations, conflicts, among others [3,4,9]. However, not much has carried out in Jordan to explore the effectiveness of

BIM to minimize the issues of delays in Jordan. This paper presents BIM as a powerful tool for reducing delays in Jordan construction projects. The paper focuses on two main parts; the first part involves the identification of the main causes of delays, and the second part is to accurately outline the roles and responsibilities of BIM specialist in construction projects. Finally, the paper matches the roles and responsibilities of BIM specialist and the causes of delays, and how the delays are addressed through BIM specialist. Thus, this will enable the industry towards more extensive BIM implementation to address delays and other wastes in the Jordanian building industry.

## 2. Related research

### *2.1. Delays in Jordan construction projects*

Delays are common in most projects [8,10–21] and the Jordanian construction industry is not an exception. For instance, in the Middle East, several delays have been reported in Saudi Arabia [12,22,23], Egypt [24–26], UAE [14], Kuwait [27–29], Qatar [28,29], Bahrain [30], Oman [26,31], Yemen [32,33], among others.

In Jordan, there are major public and private infrastructure projects that have experienced delays. The factors causing delays in the Jordanian construction industry include frequent changes made by clients, climate, site and weather conditions, delays in deliveries, economic conditions, financial problems, lack of technical know-how, poor designs, poor planning and management, and others [1,2,34,35]. Based on a comprehensive literature review, the delay factors have been summarized in table 1. The table presents the factors causing delays from some relevant studies conducted about the delays in Jordan construction projects. The three most common delays from the available literature are *poor design, lack of communication between parties and poor scheduling and planning* [1,2,29,31,34–41].

**Table 1.** Major causes of delays in Jordan.

| Researchers | Major causes of delay |
|---|---|
| [1,2,34–36,38,40] | Poor design |
| | Negligence of the owner |
| | Change orders |
| | Site condition |
| | Weather condition |
| | Late delivery |
| | Economic conditions |
| | Increase in quantities |
| [34,39–42] | Mistakes during construction |
| | Slow decision-making by clients |
| | Construction methods |
| | Improper planning |
| | Shortage of materials |
| | Documents |
| | Lack of communication between project parties |
| | Preparation and approval of drawings |
| [1,29,31,36–38] | Poor scheduling and planning of projects by contractors |
| | Frequent change orders by owners |
| | Shortage of manpower (skilled, semi-skilled, unskilled labor) |
| | Incompetent technical staff |
| | Financial difficulties faced by contractors |

### *2.2. Building Information Modeling (BIM)*

BIM is one of the most recent developments, an emerging procedural and technological shift within the architecture, engineering, and construction (AEC) sector [43–45]. Researches have shown that BIM, a three-dimensional (3D) parametric modelling software has the capacity to address a number of technical challenges in construction projects. For instance, BIM can reduce the costs of engineering and rework due to defects. It also enables uninterrupted flow of information, improves engineering productivity, improves customer services, reduces lead times and design errors. BIM is more reliable in delivering projects within the shortest possible time and in streamlining the procurement of component parts or materials for production [8,46]. Similarly, BIM can reduce the time, mistakes and errors caused by modifications [47,48]. Reports on BIM indicate that the adoption of BIM can save between 3% and 5% in costs, can boost GDP by 0.2 base point above the 'business as usual scenario' and has a remarkable effect design, construction and the economy [8,49].

BIM technology has increasingly been adopted by many AEC companies. These companies regularly seek to hire the most technically-skilled and creative design specialists available, especially with regard to BIM [50]. The motivation for these specialists is to address the issues of quality, schedule and cost overruns. According to Odeh and Battaineh [42], successful execution of construction projects and keeping them within the agreed schedule and estimated cost depend largely on the methodology which requires sound engineering judgment. In order to minimize delays through the use of BIM, the understanding of the roles and responsibilities of the BIM specialist is vital. Table 2 and table 3 presents the roles and responsibilities of BIM specialists [50–56]. The understanding of the roles and responsibilities could be used to gauge BIM potentials with regard to delay mitigation.

**Table 2.** Roles and responsibilities of BIM specialist.

| BIM Specialist | Roles and Responsibilities |
|---|---|
| **BIM Modeler** | • The roles of the BIM Modeler are to develop, create and extract 2D documentations from BIM models.<br>• May occupy the position of the Draftsperson |
| 3D Modeler | • Develops the geometry in the BIM models, working in groups to create different parts of the model. |
| Cost Modeler | • Embeds information about the processes and resources required. |
| Detailing Modeler | • Designing, drawing, sizing, specifying, detailing, verifying and documenting the design process. |
| Sequencing Modeler | • Adds phases to the resources develops building phasing files used for planning by the general contractors. |
| **BIM Analyst** | • Performs simulations and analyses based on BIM models, e.g., security analysis, building performance analysis, etc. |
| **Modelling Specialist** | • Contributes along with specialists in various areas of the industry, to the standards, from the software product's requirements to the final characteristics. |
| **BIM Application Developer** | • Creates and customizes the software to bolster integration and BIM processes, from plug-ins to servers, data repositories, and integrated project management tools. |
| **BIM Facilitator** | • Assists other specialists. Generally, he works with those that physically construct the building, helping the engineers to communicate with contractors or foremen.<br>• Extracts information from the BIM model for space planning, asset management, and maintenance schedule. |
| **BIM Consultant** | • Guides designers, builders and developers in BIM project implementation. |
| Functional | • Develops action plans according to strategies. |
| Operational | • Executes the process of implementation. |
| Strategic | • Develops strategies, typically, from medium to long-term based on a vision of accomplishment. |
| **BIM Researcher** | • BIM experts in research institutions, universities and government organizations who coordinate, teach or train and develop research on BIM. |

**Table 3.** Roles and responsibilities of BIM specialist.

| BIM Specialist | Roles and Responsibilities |
|---|---|
| **BIM Manager (General)** | • Oversees people in the implementation as well as maintenance of BIM processes.<br>• Coordinates the team, production, and implementation of the model.<br>• Inspects and evaluates the objectives of the BIM process and afterward designs a plan to meet the desires and demands of clients.<br>• Performs other functions including coordinating entity model's integration, setting design templates, etc.<br>• Guides the decision-making of the team. |
| Project Model Manager | • Concentrates on the production model, interacts with other project actors and the system.<br>• Coordinates information from various construction stakeholders, which guarantees non-dilution of the responsibilities between the contractor and the design team.<br>• Plans the mechanisms of maintenance and exchange of data<br>• Determines the necessary conventions for the management and reviewing of model versions.<br>• Holds meetings or calls with the modelers and the client to determine the required models, owners of the models and the responsibilities.<br>• Divides projects into teams of modelers.<br>• Verifies interference and keep the models accurate and updated.<br>• Ensures the alteration of models when there is a request for changes.<br>• transfers, access rights control, and information compilation from smaller models of different members.<br>• Avails model files to general contractors.<br>• Identifies communication errors, improves communication and information technologies and facilitates communication protocols.<br>• Manages and supports detailing teams for the implementation and establishment of BIM standards.<br>• Specifies detailing budgets for cost estimations. |
| **BIM Manager (Design Firm)** | • Responsible for BIM implementation<br>• Coordinates project teams and establishes connections for correspondence between various offices.<br>• Attends developmental plan meetings for BIM project implementation.<br>• Arranges training for workers to keep them updated about the trends and versions of BIM software.<br>• Attends and participates in industrial conferences.<br>• Designs materials to educate clients of the company's BIM capabilities, evaluates new products and verifies technical problems and deficiencies in the software, hardware and the net.<br>• Ensures teams' commitment by developing an effective procurement system so that every agent performs its responsibilities. |
| **BIM Manager (Contractor)** | • Assists in the estimation of time and cost for implementation<br>• Uses BIM software.<br>• Designs a plan for small teams that after training perform works with BIM.<br>• Responsible for resources management (people, software, and hardware) involved in model development.<br>• Select the right individuals for every project activity.<br>• Promotes credibility and confidence in the project team including clients, subcontractors and internal team members. |

## 3. Discussion

It has been indicated that delays have become an integral part of project's construction lifecycle. Even with project management understanding and modern advanced technology, construction projects still experience delays [57]. The question which arises is, "is BIM capable of resolving the delay issues?" To address this question, the major delays in the Jordanian construction projects are matched to the responsibilities of the BIM specialists in table 4 to explore BIM's potentials for minimizing delays.

**Table 4.** Delay cause and the required strategies.

| Delay Cause | Strategy to adopt by organization's top management |
|---|---|
| **Poor design** | Hiring BIM specialist to divide the phase into intervals (designing, drawing, verifying) which demand technological infrastructure (ex. 3D modelling software, valid account with Autodesk., well trained staff to deal with software). |
| **Lack of communication between project parties** | In this case, BIM specialist will act as a communication cycle between the projects partners by modelling the project and make it available to all the parts, update using integrated communication web site so any correction on the model will be available to all partners. The specialist' duty is to identify communication errors, improve communication and information technology and facility communication protocol. Also, he coordinates project teams and establishes connections for correspondence between various offices. |
| **Poor scheduling and planning** | To address this issue, the BIM specialist will develop the geometry in the BIM models, working in groups to create different parts of the model, embeds information about the processes and resources required, add phases to the resources develops building phasing files used for planning by the general contractors, extracts information from the BIM model for space planning, asset management, and maintenance scheduling, develops strategies, typically, from medium to long-term based on a vision of accomplishment. |

## 4. Conclusion

BIM implementation can lead to a successful construction of projects. BIM application can increase productivity, business process flow, and reduce wastes, uncertainties, complexities, fragmentations, conflicts, thus leading to quality construction projects. It also benefits construction projects by minimizing construction cost and improving information and communication between project parties, leading to delay mitigation. The use of BIM facilitates the integration of disjointed practices, enhances sustainability, reduces poor quality and acts as a catalyst for changing construction process. From the discussion, it was seen that BIM can resolve the major delay issues including poor design, lack of communication between project parties, and poor scheduling and planning in the Jordanian construction industry. With training, client and government support, it is expected that BIM technology would be used more widely in the Jordanian construction projects. This study opens the door for more realistic research's about adopting BIM as powerful tool to address delay issues in the Jordan construction industry.

## 5. References

[1] Sweis G, Sweis R, Abu Hammad A and Shboul A 2008 Delays in construction projects: The case of Jordan, *Int. J. Proj. Manag.* **26** 665–674

[2] Al-Momani A H 2000 Construction delay: a quantitative analysis *Int. J. Proj. Manag.* **18** 51–59

[3] Aranda-Mena G, Crawford J, Chevez A and Froese T 2009 Building information modelling demystified: does it make business sense to adopt BIM?, *Int. J. Manag. Proj. Bus.* **2** 419–434

[4] Mihindu S and Arayici Y 2008 Digital Construction through BIM Systems will drive the Re - engineering of Construction Business Practices IEEE *Xplore Digital Library* vol 308 (United Kingdom: IEEE) pp 9–11

[5] Gerges M, Austin S, Mayouf M, Ahiakwo O, Jaeger M and Saad A 2017 An Investigation into the Implementation of Building Information Modeling in the Middle East, *J. Inf. Technol. Constr.* **22** 1–15

[6] Awwad R and Ammoury M 2013 Surveying BIM in the Lebanese Construction Industry, *Int. Assoc. Autom. Robot. Constr.* 1–9

[7] Mehran D 2016 Exploring the Adoption of BIM in the UAE Construction Industry for AEC Firms *Procedia Engineering* vol 145 (United Kingdom: Elsevier) pp 1110–1118

[8] Ansah R H, Sorooshian S, Mustafa S Bin and Duvvuru G 2016 Advancing Towards Delay-Free Construction Project: A Review *Proc. of the 2016 Int. Conf. on Industrial Engineering and Operations Management (Detroit, USA)* (Detroit, USA: IEOM Society International) pp 744–751

[9] Khosrowshahi F and Arayici Y 2012 Roadmap for implementation of BIM in the UK construction industry, *Eng. Constr. Archit. Manag.* **19** 610–635

[10] Ansah R H, Sorooshian S, Bin Mustafa S and Oludapo O S 2017 Constructions project management risks' framework, *Calitatea* **18**(158) 90

[11] Ansah R H, Sorooshian S, Mustafa S Bin and Duvvuru G 2016 Lean Construction Tools *Proc. of the 2016 Int. Conf. on Industrial Engineering and Operations Management (Detroit, USA)* (Detroit, USA: IEOM Society International) pp 784–793

[12] Assaf S A and Al-Hejji S 2006 Causes of delay in large construction projects, *Int. J. Proj. Manag.* **24** 349–357

[13] Iyer K C and Jha K N 2006 Critical Factors Affecting Schedule Performance: Evidence from Indian Construction Projects, *J. Constr. Eng. Manag.* **132** 871–881

[14] Faridi A and El-Sayegh S 2006 Significant factors causing delay in the UAE construction industry, *Constr. Manag. Econ.* **24** 1167–1176

[15] Lowsley S and Linnett C 2006 *About time - : delay analysis in construction* (Coventry: RICS)

[16] Sambasivan M and Soon Y W 2007 Causes and effects of delays in Malaysian construction industry, *Int. J. Proj. Manag.* **25** 517–526

[17] Alaghbari W, Razali A. Kadir M, Salim A and Ernawati 2007 The significant factors causing delay of building construction projects in Malaysia, *Eng. Constr. Archit. Manag.* **14** 192–206

[18] Memon A H, Rahman I A, Abdullah M R, Asmi A and Azis A 2011 Assessing the Effects of Construction Delays on MARA Large Projects, *Int. J. Adv. Sci. Eng. Inf. Technol.* **1** 624–629

[19] Mahamid I, Bruland A and Dmaidi N 2012 Causes of Delay in Road Construction Projects, *J. Manag. Eng.* **28** 300–310

[20] Shehu Z, Endut I R and Akintoye A 2014 Factors contributing to project time and hence cost overrun in the Malaysian construction industry, *J. Financ. Manag. Prop. Constr.* **19** 55–75

[21] Ansah R H and Sorooshian S 2017 Effect of lean tools to control external environment risks of construction projects, *Sustain. Cities Soc.* **32** 348–356

[22] Al-Khalil M I and Al-Ghafly M A 1999 Delay in public utility projects in Saudi Arabia, *Int. J. Proj. Manag.* **17** 101–106

[23] Al-Kharashi A and Skitmore M 2009 Causes of delays in Saudi Arabian public sector construction projects, *Constr. Manag. Econ.* **27** 3–23

[24] Abd El-Razek M E, Bassioni H A and Mobarak A M 2008 Causes of Delay in Building Construction Projects in Egypt, *J. Constr. Eng. Manag.* **134** 831–841

[25] Aziz R F 2013 Ranking of delay factors in construction projects after Egyptian revolution, *Alexandria Eng. J.* **52** 387–406

[26] Aziz R F and Abdel-Hakam A A 2016 Exploring delay causes of road construction projects in Egypt, *Alexandria Eng. J.* **55**(2) 1515-1539

[27] Koushki P A, Al-Rashid K and Kartam N 2005 Delays and cost increases in the construction of private residential projects in Kuwait, *Constr. Manag. Econ.* **23** 285–294

[28] Ai-Tabtabai H M 2002 Causes for Delays in Construction Projects in Kuwait, *Eng. J. Univ. Qatar* **15** 19–37

[29] Emam H, Farrell P and Abdelaal M 2015 Causes of delay on infrastructure projects in Qatar *Procs 31st Annual ARCOM Conf. (Lincoln, UK)* (Lincoln, UK: Association of Researchers in Construction Management) pp 773–782

[30] Hasan R, Suliman S M A and Malki A Y 2014 An Investigation into the Delays in Road Projects in Bahrain - Technology, *Int. J. Res. Eng. Sci.* **2** 38–47
[31] Alnuaimi A S and Mohsin M A Al 2013 Causes of Delay in Completion of Construction Projects in Oman *Int. Conf. on Innovations in Engineering and Technology (Bangkok, Thailand)* (Bangkok: International Institute of Engineers) pp 267–270
[32] Sultan B and Kajewski S 2003 The Yemen construction industry: Readying the industry for the successful implementation of sustainability *International Conference on Smart and Sustainable Built Environment (Brisbane, Australia)* (Brisbane: Queensland University of Technology) pp 1–9
[33] Al-Fadhali N and Zainal R 2017 A Theoretical Framework on Factors Causing Delay of Construction Industries Projects, *Soc. Sci.* **12** 393–399
[34] Al-Hazim N, Salem Z A and Ahmad H 2017 Delay and Cost Overrun in Infrastructure Projects in Jordan *Procedia Engineering* vol 182 (United Kingdom: Elsevier) pp 18–24
[35] Ahmad H S, Bazlamit I M and Ayoush M D 2017 Investigation of Document Management Systems in Small Size Construction Companies in Jordan *Procedia Engineering* vol 182 (United Kingdom: Elsevier) pp 3–9
[36] Owalabi J D, Amusan L M, Oloke C O, Olusanya O, Tunji-Olayeni P, OwolabiDele, PeterJoy and OmuhIganatious 2014 Causes and effects of delay on project construction delivery time, *Int. J. Educ. Res.* **2** 197–208
[37] Afshari H, Khosravi S, Ghorbanali A, Borzabadi M and Valipour M 2011 Identification of Causes of Non-excusable Delays of Construction Projects, *Management* **3** 42–46
[38] Kikwasi G 2013 Causes and Effects of Delays and Disruptions in Construction Projects in Tanzania, *Australas. J. Constr. Econ. Build. - Conf. Ser.* **1** 52–59
[39] Al-Hazim N, Abusalem Z and abusalem zaydoun 2015 Delay and cost overrun in road construction projects in Jordan, *Int. J. Eng. Technol.* **4** 288
[40] Samarghandi H, Moosavi Tabatabaei S M, Taabayan P, Hashemi A M, Willoughby K, Mohammad S and Tabatabaei M 2016 Studying the Reasons for Delay and Cost Overrun in Construction Projects: The Case of Iran, *J. Constr. Dev. Ctries.* **21** 51–84
[41] Al-Werikat G 2017 Analysing the Role of Clients in the Jordanian Construction Supply Chain Delays, *Int. J. Eng. Res. Sci. Technol.* **6** 22–29
[42] Odeh A M and Battaineh H T 2002 Causes of construction delay: traditional contracts, *Int. J. Proj. Manag.* **20** 67–73
[43] Azhar S 2011 Building Information Modeling (BIM): Trends, Benefits, Risks, and Challenges for the AEC Industry, *Leadersh. Manag. Eng.* **11** 241–252
[44] Abdirad H 2015 Advancing in Building Information Modeling (BIM) Contracting: Trends in the AEC/FM Industry *AEI 2015* (Reston, VA: American Society of Civil Engineers) pp 1–12
[45] Azhar S, Khalfan M and Maqsood T 2012 Building information modelling (BIM): now and beyond, *Australas. J. Constr. Econ. Build.* **12** 15
[46] Sacks R, Eastman C M, Lee G and Orndorff D 2005 A Target Benchmark of the Impact of Three-Dimensional Parametric Modeling in Precast Construction, *PCI J.* **50** 126–139
[47] Ku K and Taiebat M 2011 BIM Experiences and Expectations: The Constructors' Perspective, *Int. J. Constr. Educ. Res.* **7** 175–197
[48] Andújar-Montoya M, Gilart-Iglesias V, Montoyo A and Marcos-Jorquera D 2015 A Construction Management Framework for Mass Customisation in Traditional Construction, *Sustainability* **7** 5182–5210
[49] Construction Training Fund 2014 The Impact of New Technologies on the Construction Industry, *Constr. Train. Fund, Aust.* 1–18
[50] Barison M B and Santos E T 2010 An overview of BIM specialists *Proc. of the Int. Conf. on Computing in Civil and Building Engineering* (Nottingham: Nottingham University Press) pp 1–6
[51] Van Berlo L A H M and Papadonikolaki E 2016 Facilitating the BIM coordinator and

empowering the suppliers with automated data compliance checking *11th Eur. Conf. Prod. Process Model (Limassol, Cyprus)* (United States: CPC Press) pp 145-153

[52] Gathercole M and Thurairajah N 2011 The influence of BIM on the responsibilities and skills of a project delivery team International Conf. on Construction in a Changing World (Heritance Kandalama, Sri Lanka) (Manchester: University of Salford) pp 1–15

[53] Papadonikolaki E and Papadonikolaki E 2016 Alignment of Partnering with Construction IT, *Archit. Built Environ.* **6** 1–328

[54] Barison M B and Santos E T 2011 The Competencies of BIM Specialists: A Comparative Analysis of the Literature Review and Job Ad Descriptions, *J. Comput. Civil Eng.* 594–602

[55] Uhm M, Lee G and Jeon B 2017 An analysis of BIM jobs and competencies based on the use of terms in the industry, *Autom. Constr.* **81** 67–98

[56] Abdulkader S E 2013 *Common BIM Roles and their Responsibilities* Retrieved on April 30, 2013 from http://www.bimuserday.com/wp-content/uploads/3rd-Qatar-BIM-User-DaySharaf -Abdulkader.pdf

[57] Ansah R H, Sorooshian S, Mustafa S Bin and Duvvuru G 2016 Assessment of Environmental Risks in Construction Projects: A Case of Malaysia *Proceedings of the 2016 International Conference on Industrial Engineering and Operations Management (Detroit Michigan, United States)* (Detroit, Michigan, USA: IEOM Society International) pp 752–763

IOP Conf. Series: Materials Science and Engineering **271** (2017) 012042 doi:10.1088/1757-899X/271/1/012042

# A review of risk management process in construction projects of developing countries

**R A Bahamid[1] and S I Doh[1]**

[1]Faculty of Civil Engineering and Earth Resources, Universiti Malaysia Pahang, 26300 Gambang Kuantan, Pahang, Malaysia
Corresponding author: ra2007mi2007@gmail.com

**Abstract.** In the construction industry, risk management concept is a less popular technique. There are three main stages in the systematic approach to risk management in construction industry. These stages include: a) risk response; b) risk analysis and evaluation; and c) risk identification. The high risk related to construction business affects each of its participants; while operational analysis and management of construction related risks remain an enormous task to practitioners of the industry. This paper tends towards reviewing the existing literature on construction project risk managements in developing countries specifically on risk management process. The literature lacks ample risk management process approach capable of capturing risk impact on diverse project objectives. This literature review aims at discovering the frequently used techniques in risk identification and analysis. It also attempts to identify response to clarifying the different classifications of risk sources in the existing literature of developing countries, and to identify the future research directions on project risks in the area of construction in developing countries.

## 1. Introduction

In order to have a better understanding of what we mean by risk, risk can be defined as exposure to loss/gain or the likelihood of an event of loss/gain multiplied by its corresponding magnitude [1]. Project risks are of two categories, known and unknown. Known risks are recognized and can be analyzed, which makes planning and preparing response possible. It's not possible to proactively unravel an Unknown risk. Therefore, to anticipate and provide a risk reserve becomes necessary [2]. Risk can be reduced, managed, transferred, shared or accepted but cannot be overlooked [3].

In construction perspective, risks are generally considered as incidences that influence the principal objectives of a particular project (time, cost, quality) [4]. As a result of its construction activities that are perceived to be unique, the construction industry is exposed to greater risks in comparison with other industries. It has unique features as: projects taken long periods, with processes that are complicated, financial intensity, environment that are abominable and organization structures that are dynamic [5]. Also, when dealing with risks, most construction industries are known for their poor reputation, this is because may projects do not meet their cost targets and proposed deadlines, which in turn adversely affects each of its participants (Contractors, clients the public and others) [6].

Risk management can be defined an organized and comprehensive method tailored toward "analysing", "identifying" and "responding" to risk factors in order to achieve the project goals[4]. Having a good understanding of risks allows the parties involved to take steps in order to reduce their negative impacts [7]. The lack of a project risk management approach that is effective has a lot of unpleasant effects on the project participants as a result of deficiency of actions to prevent the uncertainty and risks that are present in a project [8]. Risk identification and assessment is a crucial risk management process. It is not possible to manage all the risks in a construction project. However, it is important to focus on the vital risks. Attempting to identify all the risks can be a waste of time and counterproductive [7]. It was observed that, majority of decisions on construction risk management are done based on intuition, previous experience, and the manger's professional judgement. As a result of ignorance and doubts on the sustainability, formal methods available are not been applied for the activities of the construction [9].

In developing countries, construction projects are liable to a lot of unknown factors [10]. Numerous project challenges are prevented through the implementation of effective risk management in projects [11]. Various publications on risk management are available both online on the internet and in hard copies in books and journals, little information is known on the applications of risk management in practice, most especially in developing countries[12] and the process of risk management has not been adequately focused [11] . According to a study carried out by Bowers and Khorakian (2014), he is of the opinion that in most constructions industries in developing countries of the world, very little facts on successful implementation of risk management systems is known. Due to the minimal attention paid to risk management, the focus of researchers in developing countries has been on encouraging the use and application of risk management systems [13].

This paper shows extensively the result of the review of the published literature on risk management in construction industry, with a focus on developing countries. The focus of this research is on published peer reviewed academic journals, most especially in construction management. The primary goal of this paper is to discover the techniques in risk identification, analysis and respond, that is mostly in use, in other to clarify the different groups of risk sources in the existing literature of developing countries, and identify the directions for future research in construction project risks, in developing countries.

## 2. Risk management process in construction projects

Risk management can be defined as the systematic process of analyzing, identifying, and responding to projects risk. It consists of maximizing the chances and the impact of positive events while minimizing the probability and the impact of negative events, in other to meet the project objectives [14]. Risk management can be thought of as a decision-making process, and it entails having a full understanding of a known risk and/or necessary actions to reduce the effect and chances of the event of such risks, in other to reduce its complications and increase the chances of success [15]. There are three stages in the methods to managing risk in construction industry a) risk identification; b) risk analysis and evaluation; and c) risk response[16,17]. The term risk identification is refers to identifying and keeping records of the associated risks. Risk assessment on the other hand, means to examine the identified risks critically, refine the description of the risk, and making an estimate of their respective chances and effects on the project. The risk response refers to the identification, selection, evaluation, and the action to implement the project [18]. By making use of the risk management process, one can achieve a major improvement in the performance of the construction project management. The goal of risk management process is not to completely remove all project risks. Its aim is to produce an organized framework that will make management to manage project risks, most importantly the crucial ones, in a more efficient and effective way [16].

### *2.1. Risk identification*

Risk identification can be defined as the process of analytically and constantly identifying, assessing and categorizing the initial importance of the risks related to construction projects [19] and the interrelationships that exists among these risks[20]. The idea of risk identification seems to be very popular and practiced [21]. It is of substantial value as the process of response management and risk analysis may only be implemented on recognized potential risks [16]. It could have effects on project development and its success [2]. Failure in identifying potential risks can result into inadequacy in the whole process. This can in turn have critically effects on the resources available to the organization. Risk identification however helps organizations involved in risk management to: (a) identify the best and most crucial input data (b) have a better knowledge of the relevance of the process (c) identify risks and the effects they can have (d) provide information for those who make decisions [22].

With the aid of diverse tools and methods, risk identification process can be achieved [22]. These tools and techniques include: brainstorming, Delphi method, interview, cause analysis, SWOT analysis, and presumption analysis. The first four techniques relate with common techniques, but the last two techniques are used in particular to investigate larger scope of possible events [2]. Table1 shows the most four techniques are popularly being used to identify risks in developing countries when it comes to construction of projects [12,14,23,24]:

**Table 1.** Most techniques used to identify risks in developing countries.

| Techniques | Description |
|---|---|
| Checklists | Know potential points that can fail in previous projects and thus is very helpful in risk identification. This allows project managers to know the risks present and makes them to be involved in the process of risk identification, which will ultimately lead to greater acceptance of any means implemented to minimize the risks. |
| Interviews with experts | Historical data analysis for projects that appear similar and examine similar past or present projects, risk analysis, lessons learned or project evaluations are other methods available for getting feedback about risks involved in a project. |
| Past experience | Checking historical data of past projects that are similar can only be useful in a limited number of conditions. Such systems are most often restricted in terms of their usability or important data that are stored. |
| Brainstorming | Can be of use for projects involving new risks, new management arrangements or for developing initial checklists. This may be useful in risk management workshops. |

In previous studies, diverse sources of construction risk have being identified [25]. Various approaches have been proposed in the literature for classifying risk. The manager can better understand the nature of risks by categorizing the risks [26]. Risks can be categorized in various ways for different purposes as shown in table 2.

*2.2. Risk analysis*

Risk analysis is regarded as the procedure involving the critical evaluation of prospective risks, arranging them according to importance, and allowing the management team to select the important ones [7]. Risk analysis is the most tasking procedure in managing risk. This is due to the fact that it involves assessing the chances of the event of a risk and their outcomes on a project's objectives [36]. Its main aim is to evaluate risk by separating the unnecessary events, the chances of the unwanted event happening, and the size of such events [37]. Meaning that, it is the transitional process between identifying risk and its management. It includes uncertainty in a qualitative and quantitative manner to evaluate the potential effects of risk. The evaluation should largely focus on risks that have high chances or effects [16].

In risk analysis, two main approaches are broadly used. They are: qualitative risk analysis and quantitative risk analysis and sub-category semi-quantitative [24]. The choice of method depends on the following: the type and magnitude of the intended project, available information, the financial implication and time available, the experience of the analysts, the extent of innovation and the ultimate purpose of the results [15]. Quantitative approach is primarily based on probability spreading of risks.

However, if sufficient data are available it can provide objective results. Qualitative method on the other hand, is subject to personal experience, intuition and judgment. The outcomes can therefore significantly vary from one analyst to another. Consequently, the quantitative approach remains the preferred option by most practitioners [9].

**Table 2.** Risk identification categories in developing countries.

| Classification of risks | Author |
|---|---|
| project level, country level, market level | [16] |
| Resource, productivity, design, managerial, payment, client, technical, subcontractors | [27] |
| Client, Consultant, Labour, Contract, Contractor, Relationship, Materials | [28] |
| project management, Engineering, execution, suppliers | [29] |
| Construction safety-related, construction management; engineering design, natural hazards, socio and economic | [30] |
| Legal, Physical, design, political, fanatical, environmental, Logistics, Management, construction | [31] |
| Quality related, Environment related, Safety related, Cost related, Time related | [32] |
| Economic-financial and political, Technical and contractual, Managerial, External and site condition | [33] |
| Internal (Owners, Contractors, subcontractors, Designers, Suppliers), External (Social & Cult, Economic, Political, Natural, others). | [34] |
| Health and safety risks, Contractual risks, Financial risks, Design risks, Management risks, Construction risks, External risks | [35] |

The core qualitative analysis techniques are: brainstorming, expert judgment; cause and effect diagram; checklists; Delphi; Event Tree Analysis (ETA; Risk Breakdown Matrix (RBM); risk data quality assessment. While the core quantitative techniques are: decision tree analysis; expected monetary value; Fault Tree Analysis (FTA); fuzzy logic; probability distributions; sensitivity analysis/tornado diagram[4, 38]. Also, risk analysis techniques use computer-based simulation tools such as: Monte Carlo simulations and system dynamics applications for PRM [35].
For construction projects, wide research has been done in the area of risk management. Main summary of these attempts are the identification of the project goals related risks and the project phase related risks. While these recognized risks relates to diverse construction projects in the context of different countries, they are all of great importance in directing the risk management research and practice for the Chinese construction industry. Table3 shows the most risk analysis techniques used past research in developing countries.

*2.3.Risk response*
Acceptable mitigation steps of treating risk must be employed once the project risks have been known and analysed. These mitigation steps are based mostly on the nature and potential consequences involved in the risk. The main objective is to increase the level of control of risk, reduce the negative impact of the risk and remove as much as possible the potential impact. Measure becomes more effective when there is more control of one mitigation measure on one risk [16]. Six distinctive risk responses are retention, reduction, control, sharing, transfer and avoidance as it is being detailed in table 4. The choice of response must correspond to the importance of the risk; it should be financially cost effective and realistic with regard to the project timing; it also must be accepted by other parties involved [15].

**Table 3.** Most risk analysis techniques used in developing countries.

| Technique | Description | Author |
|---|---|---|
| Risk matrix | Risk classification into a limited number of categories | [39] |
| Monte Carlo simulation | Aggregate the combined effects resulted from uncertain parameters. Precise experimental statistical data. Computationally expensive | [35] |
| AHP | Risks events are ranked based on pairwise comparisons. Limited to a few number of pairwise comparisons | [17] |
| Fuzzy logic assessment | Useful where probabilistic data are absent. Not as precise as probabilistic methods | [40] |
| FMEA | Identify critical risk events. Only quantify one consequence in a particular time | [41] |
| Expected Monetary Value Analysis | This looks at the likelihood part of the system states and is based on a gain matrix | [12] |
| Expert judgment | Based on advice of a professional to analyse the failure rate and chances of success of the total project | [12] |

**Table 4.** Risk management responds [4, 15, 23].

| Techniques | Description |
|---|---|
| Risk retention | Involves considering that a particular risk situation exists and making a conscious steps to accept the level of risk, without engaging in any special efforts to control it. |
| Risk reduction | An approach adopted to bring the chances and effects of the risk down below a threshold that's acceptable. |
| Risk sharing | Principally obtained through a contractual mechanism to develop a sense of collective responsibility among the project stakeholders. |
| Risk control | Does not seek to stop completely the source of the risk, but takes steps to reduce the risk present |
| Risk avoidance | A refusal to accept the action taken or risk, to ensure that the risk is not going to continue. |
| Risk transfer | Shifts and changes, along with ownership, from one party to another third party, without changing the total amount of risk or reducing how crucial the risk sources are. |

In this phase available options and actions are developed to promote opportunities and to decrease threats to the project objectives. It appears that each party involved in the contract should accept some level of risks, be fully aware of its own share of risks and put into consideration the losses [43]. The significance of awareness of the allotted risks for each party involved and successive preparation necessary for dealing with the risks is imperative and contributes to the success of projects [32]. Based on the degree of severity, there are numerous paths that can be followed to respond to risks. [14].

In developing countries, the method of risk respond frequently takes preventive (avoidance) techniques and remedial techniques as [1, 44] Used in their studies:

- Preventive risk management technique: The best decision to make in managing risk is to eliminate it so that it can be avoided at the stage of planning. Early consideration of risks before the start of a project and making effective plan for it are preventive management approach used during the planning stage to avoid or minimize a necessary risk.

- Remedial risk management technique: Remedial risk management methods are required to minimize the effects of risk and totally stop them if possible. This is because it is obvious that all risk cannot be managed during the planning stage and some risks are going to happen during the operation phase. Therefore, remedial method becomes important.

According to Wang and Chou [45], data obtained of the findings, contractors usually use three means to transfer risk in project construction: through insurance to insurance companies, through modifying the contract terms and conditions to client or other parties, through subcontracting to subcontractor.

## 3. Conclusion

The method of risk management is sparingly applied because of fewer know-how and awareness among the people. The track record is also small when handling risks in projects, resulting in it affecting the project goals. This paper shows an elaborate review of the risk management process (analysis, identification and respond) of published literature. It is mainly focused on the development of risk process most especially in developing countries of the world. Diverse contributions towards investigating various techniques are also discussed.

In past studies, different sources of construction risk have being identified. Various approach for classifying risk has been recommended in the literature. Management can understand better the nature of risks by categorizing the risks. There are various ways for categorizing risk to achieve different goals. To some, in construction projects, risk can be categorized largely into external risks and internal risks while others classify risk in more elaborate categories. These categories depend on the situation of the project and the surrounding environment.

All the evaluated methods are important for making decisions, and all decision-making suggests considering the risks of the alternatives. Risk is better measured or more specifically measured than others in some of the methods, but they all have things in common: there is need for a medium to high level of experience, time resources and detailed data. Even though the quantitative approach uses more resources than the qualitative approach, they are also more complex. This paper has reviewed the actual practice of risk analysis as published in the literature. The findings refer to a weighty dependence on practical experience and professional judgment when assessing construction risk. Unfortunately, there is still a wide gap between theory and practice. However, the existing body of knowledge establishes a firm basis from which to investigate new options that can bridge the existing gap between theory and practice.

It is impossible to capture risk categories, the interdependencies between risks, the interaction with the complex project environment and the management team experience in improvement proposals. While project management literature is rich in papers addressing risk management, few papers have researched the actual practice of risk assessment and investigated the practitioners' points of view regarding the available tools. That has been established in the review that the literature lacks a comprehensive risk assessment framework which considers the different types of effects of risk on various project objectives concurrently. Such a framework is important for obtaining realistic risk assessments which is the first approach towards obtaining a realistic project risk level.

The current status of risk management approach of the construction industry of developing countries of the world, generally attempts to avoid or shift these risks, which results in the risk in the risk management practices of a large number of the industries being reactive and informal to deal with. Nevertheless, presently the awareness about risk management is on-going and a there is an intense desire to learn from past mistakes.

This research work is considered a fraction of a comprehensive research project that is aimed at re-evaluating construction risk management process and facilitating the elimination of the existing gap between theory and practice of construction risk in developing countries of the world.

## 4. References

[1] Iqbal S, Choudhry R M, Holschemacher K, Ali A and Tamošaitienė J 2015 Risk management in construction projects, *Technol. Econ. Dev. Eco.* **21** 65-78

[2] Crnković D and Vukomanović M 2016 Comparison of trends in risk management theory and practices within the construction industry, *e-GFOS* **7**(13) 1-11

[3] Taroun A 2014 Towards a better modelling and assessment of construction risk: Insights from a literature review, *Int. J. .Proj. Manag.* **32** 101-15

[4] PMI 2013 *A Guide to The Project Management Body of Knowledge (PMBOK guide)* (Newtown Square: Project Management Institute)

[5] Taylan O, Bafail A O, Abdulaal R M and Kabli M R 2014 Construction projects selection and risk assessment by fuzzy AHP and fuzzy TOPSIS methodologies, *Appl. Soft Comput.* **17** 105-16

[6] Tesfaye E, Berhan E and Kitaw D 2016 A Comprehensive Literature Review on Construction Project Risk Analysis, *Int. J. Risk Contingency Manag* **5** 1-15

[7] El-Sayegh S M and Mansour M H 2015 Risk assessment and allocation in highway construction projects in the UAE, *J. Manage. Eng.* **31**(6) 04015004

[8] Serpella A F, Ferrada X, Howard R and Rubio L 2014 Risk management in construction projects: a knowledge-based approach *Procedia - Social and Behavioral Sciences* vol 119 (Untied Kingdom: Elsevier) pp 653-62

[9] Jarkas A M and Haupt T C 2015 Major construction risk factors considered by general contractors in Qatar, *J. Eng. Des. Tech.* **13** 165-94

[10] Ebrahimnejad S, Mousavi S M and Seyrafianpour H 2010 Risk identification and assessment for build–operate–transfer projects: A fuzzy multi attribute decision making model, *Expert syst. .Appl.* **37** 575-86

[11] Tadayon M, Jaafar M and Nasri E 2012 An assessment of risk identification in large construction projects in Iran, *J. Constr. Dev. Countr.* **17** 57-69

[12] El-Sayegh S M 2014 Project risk management practices in the UAE construction industry, *Int. J. Pro. Org. Manag.* **6** 121-37

[13] Ghahramanzadeh M 2013 *Managing risk of construction projects:A case study of Iran* PhD Thesis (London: University of East London)

[14] Tipili L G and Ibrahim Y 2015 Identification and assessment of key risk factors affecting public construction projects in Nigeria: stakeholders' perspectives *Pro. of the 2nd Nigerian Institute of Quantity Surveyors Research Conf. Federal University of Technology (Akure)* (Nigeria: The Nigerian Institute of Quantity Surveyors) pp 707-21

[15] Goh C S and Abdul-Rahman H 2013 The identification and management of major risks in the Malaysian construction industry, *J. Constr. Dev. Countr.* **18** 19-32

[16] Wang S Q, Dulaimi M F and Aguria M Y 2004 Risk management framework for construction projects in developing countries, *Constr. Manag. Econ.* **22** 237-52

[17] Zayed T, Amer M and Pan J 2008 Assessing risk and uncertainty inherent in Chinese highway projects using AHP, *Int. J. .Proj. Manag* **26** 408-19

[18] Zhang Y and Fan Z-P 2014 An optimization method for selecting project risk response strategies, *Int. J. .Proj. Manag.* **32** 412-22

[19] Al-Bahar J F and Crandall K C 1990 Systematic risk management approach for construction projects, *J. Constr. Eng. M.* **116** 533-46

[20] Liu J, Zhao X and Yan P 2016 Risk paths in international construction projects: Case study from Chinese contractors, *J. Constr. Eng. M.* **142** 05016002

[21] Hassanein A A and Afify H M 2007 A risk identification procedure for construction contracts—a case study of power station projects in Egypt, *Civ Eng Environ Syst* **24** 3-14

[22] Rostami A 2016 Tools and Techniques in Risk Identification: A Research within SMEs in the UK Construction Industry, *Universal J. Manag.* **4**(4) 203-210

[23] Smith N J, Merna T and Jobling P 2009 *Managing Risk: in Construction Projects* (Oxford: John Wiley & Sons)
[24] Choudhry R M and Iqbal K 2012 Identification of risk management system in construction industry in Pakistan, *J. Manage. Eng.* **29** 42-9
[25] Odimabo O and Oduoza C 2013 Risk Assessment Framework for Building Construction Projects' in Developing Countries, *Int. J. Constr. Eng. M.* **2** 143-54
[26] Sharma S K 2013 Risk Management in Construction Projects Using Combined Analytic Hierarchy Process and Risk Map Framework, *IUP J. Oper. Manag.* **12** 23
[27] Dikmen I, Birgonul M T and Han S 2007 Using fuzzy risk assessment to rate cost overrun risk in international construction projects, *Int. J. Proj. Manag.* **25** 494-505
[28] Al-Kharashi A and Skitmore M 2009 Causes of delays in Saudi Arabian public sector construction projects, *Constr. Manag. Econ.* **27** 3-23
[29] Nieto-Morote A and Ruz-Vila F 2011 A fuzzy approach to construction project risk assessment, *Int. J. Proj. Manag.* **29** 220-31
[30] Kuo Y-C and Lu S-T 2013 Using fuzzy multiple criteria decision making approach to enhance risk assessment for metropolitan construction projects, *Int. J. Proj. Manag.* **31** 602-14
[31] Enshassi A and Mosa J A 2015 Risk management in building projects: owners' perspective, *IUG J. Nat. Stud.* **16**(1) 95-123
[32] Zou P X, Zhang G and Wang J 2007 Understanding the key risks in construction projects in China, *Int. J. Proj. Manag.* **25** 601-14
[33] Perera B, Dhanasinghe I and Rameezdeen R 2009 Risk management in road construction: the case of Sri Lanka, *Int. J. Strate. Prop. M.* **13** 87-102
[34] El-Sayegh S M 2008 Risk assessment and allocation in the UAE construction industry, *Int. J. Proj. Manag.* **26** 431-8
[35] Choudhry R M, Aslam M A, Hinze J W and Arain F M 2014 Cost and schedule risk analysis of bridge construction in Pakistan: Establishing risk guidelines, *J. Constr. Eng. M.* **140** 04014020
[36] Thomas D R 2006 A general inductive approach for analyzing qualitative evaluation data, *Am. J. Eval.* **27** 237-46
[37] KarimiAzari A, Mousavi N, Mousavi S F and Hosseini S 2011 Risk assessment model selection in construction industry, *Expert Syst. Appl.* **38** 9105-11
[38] Del Cano A and de la Cruz M P 2002 Integrated methodology for project risk management, *J. Constr. Eng. M.* **128** 473-85
[39] Mahamid I 2013 Common risks affecting time overrun in road construction projects in Palestine: Contractors' perspective, *Constr. Econ. Build.* **13** 45-53
[40] Naderi M 2008 *Fuzzy logic application in risk analysis of construction management* M. Sc. Thesis (Alberta: Unversity of Alberta) pp 93-96
[41] Ahmadi M, Behzadian K, Ardeshir A and Kapelan Z 2016 Comprehensive risk management using fuzzy FMEA and MCDA techniques in highway construction projects, *J. Civ. Eng. Manag.* 1-11
[42] Yafai K, Hassan J, Balubaid S, Zin R and Hainin M 2014 Development of a risk assessment model for Oman Construction industry, *Jurnal Teknologi* **70** 55-64
[43] Perera B, Rameezdeen R, Chileshe N and Hosseini M R 2014 Enhancing the effectiveness of risk management practices in Sri Lankan road construction projects: A Delphi approach, *Int. J. Constr. Manag.* **14** 1-14
[44] Kartam N A and Kartam S A 2001 Risk and its management in the Kuwaiti construction industry: a contractors' perspective, *Int. J. Proj. Manag.* **19** 325-35
[45] Wang M-T and Chou H-Y 2003 Risk allocation and risk handling of highway projects in Taiwan, *J. Manage. Eng.* **19** 60-8

# Adoption of Building Information Modelling in project planning risk management

**M M Mering[1], E Aminudin[1], C S Chai[2], R Zakaria[1],C S Tan[1],Y Y Lee[3] and A A Redzuan[1]**

[1]Faculty of Civil Engineering, Universiti Teknologi Malaysia, 81310 Johor Bahru, Johor, Malaysia.
[2]School of the Built Environemnt, University of Reading Malaysia, 79200 Iskandar Puteri, Johor, Malaysia.
[3]Jamilus Research Center, Faculty of Civil and Environment Engineering, Universiti Tun Hussein Onn, 86400 Parit Raja, Johor, Malaysia.
Corresponding author: eeydzah@utm.my

**Abstract.** An efficient and effective risk management required a systematic and proper methodology besides knowledge and experience. However, if the risk management is not discussed from the starting of the project, this duty is notably complicated and no longer efficient. This paper presents the adoption of Building Information Modelling (BIM) in project planning risk management. The objectives is to identify the traditional risk management practices and its function, besides, determine the best function of BIM in risk management and investigating the efficiency of adopting BIM-based risk management during the project planning phase. In order to obtain data, a quantitative approach is adopted in this research. Based on data analysis, the lack of compliance with project requirements and failure to recognise risk and develop responses to opportunity are the risks occurred when traditional risk management is implemented. When using BIM in project planning, it works as the tracking of cost control and cash flow give impact on the project cycle to be completed on time. 5D cost estimation or cash flow modeling benefit risk management in planning, controlling and managing budget and cost reasonably. There were two factors that mostly benefit a BIM-based technology which were formwork plan with integrated fall plan and design for safety model check. By adopting risk management, potential risks linked with a project and acknowledging to those risks can be identified to reduce them to an acceptable extent. This means recognizing potential risks and avoiding threat by reducing their negative effects. The BIM-based risk management can enhance the planning process of construction projects. It benefits the construction players in various aspects. It is important to know the application of BIM-based risk management as it can be a lesson learnt to others to implement BIM and increase the quality of the project.

## 1. Introduction

One of the major industries that contribute to Malaysian economy is the construction industry. Through provision of basic infrastructure, the industry continues to support social development. In everyday life in an organization and projects, there is always uncertainty [1], representing a clear danger to the business, but also in it significant opportunity that must be taken [2]. Risk may serves as opportunities, but the fact that most risks usually have negative results led individuals to only consider the negative side of the risk [2, 3]. Today risk management is important in project management, where one of the most difficult activities is determining what are the project's risks and how should they be prioritized [4].

Adoption of BIM can be implemented to all construction project phases, which are pre-construction phase, construction phase and post-construction phase [5, 6]. According to Furneaux and Kivit [5], the BIM application in construction projects helps in controlling the project efficiently. The design process decision is facilitates by the ability of the adoption of BIM to foster collaboration between construction players. Besides, time and construction cost can be reduced during the design stage with the detection of clash and clash analysis. Strategies and technologies as management tools is critical to minimize the dispute occurrence [7]. Hence, construction players play a big role in realizing the benefits of BIM in improving the implementation of construction processes. It is true that BIM is still in the early phase of introduction in Malaysian construction industry. Even though BIM is seen as an expensive technology to be implemented, but it has proven to provide a lot of solution [8].

## 2. Literature review

### *2.1 Adoption of risk management using BIM in the project planning phase*

Planning is a basic part of contracting and the construction process. In the projects that implement BIM, visualizations can be made from previously produced models, in contrast to the traditional projects, in which the building visualization had to be made from the very start. With the speedy development of theory and computer applications over the last few years, BIM has achieved a remarkable awareness in the construction industry. According to Volk *et al.* [9], there is a compelling increase of the implementation of BIM to support planning, design, construction operation and maintenance phases. According to Porwal and Hewage [10], BIM is systematic method and process that is changing the project delivery, designing [11] and the communication and organisational management of construction [12] instead considering technology. Table 1 presents the process of implementing BIM can be seen as a systematic way for managing risks in the project planning phase.

**Table 1.** Examples for applying or developing BIM-based risk management in the project planning stage [13]

| BIM-based Risk Management Functionality | Benefits for risk management | Research |
|---|---|---|
| **3D Visualisation** | Facilitating early risk identification and risk communication | Hartmann *et al.* [14] |
| **Clash detection** | Automation of detecting physical conflicts in model | Hartmann *et al.* [14], Tang *et al.* [15] |
| **4D construction scheduling/ planning** | Improving construction management level by facilitating early risk communication and risk identification | Hardin [12], Hartmann *et al.* [14], Whyte [16] |
| **5D cost estimation or cash flow modeling** | Planning, controlling and managing budget and cost reasonably | Hardin [12], Hartmann *et al.* [14], Whyte [16], Marzouk and Hisham [17] |
| **Construction progress tracking** | Quality, safety, time and budget management level can be improved | Bhatla *et al.* [18], Eastman *et al.* [19] |
| **Safety management** | Personnel safety hazards reduced | Whyte [16], Teizer [20] |
| **Space management** | The consideration of space distribution and management in design can be improved | Hartmann *et al.* [14], Kim *et al.* [21] |
| **Quality control** | Construction quality is improved | Chen and Luo [22] |
| **Structural analysis** | Structural safety is improved | Lee *et al.* [23], Sacks and Barak [24], Shim *et al.* [25] |
| **Operation & maintenance (O&M), facilities management (FM)** | Management level is improved and risks are reduced | Becerik-Gerber [27], Volk *et al.* [28] |
| **Urban planning and design** | Land-use planning, design and management can be facilitated by integrating planning and design of urban space and AEC projects | Kim *et al.* [21], Lee *et al.* [23] Rajabifard [29] |

From the beginning of the project, BIM can be used freely. BIM can be used as a strong and effective process in the intimate combination of the designer and the builder. Design Model and Construction Model can be identified when BIM is used. Design Model which is developed by the engineers and architect are expected to be completed at the level of detail of 2D construction documents. Construction Model is developed by the contractor and subcontractor where it consists of the modelling of shop drawings and related information. Developer from each distinct model can work and update their own files and are given the responsibilities for the dimensional accuracy of model. To form a federated model, distinct models can be linked to each other. The federated model can be used for various purposes. This includes clash detection, marketing and facilities maintenance. According to Westin and Sein [30], with the use of a 3D model, quantities, boards of objectives and production meetings, BIM should be part of the daily agenda. The reinforcement in the production phase of civil works, information of steel and concrete can be obtained by using BIM. Westin and Sein [30] said that the information of the stages of the concrete casting, quality and its characteristics can be found in the 3D model.

## 3. Methodology

Despite having all of the information from the literature review, a further research is still needed in order to achieve the study's aim and objectives. In order to obtain data, respondents selected are not only those who utilize BIM, but also those who understand and apply the concept of risk management in construction. 70 respondents are targeted to answer the questionnaires. A quantitative approach is adopted in this research as it is more reliable and objective. It reduces and restructures a complicated problem to a limited number of variables. The questionnaire is parted into four parts. Section A is the respondent's personal details, Section B is to identify the risk management practices, Section C is to determine the best functions of BIM in risk management and Section D is used to investigate the efficiency of adopting BIM-based risk management during the project planning phase. The data obtained will be analysed using Statistical Package for the Social Sciences (SPSS) software. Data collected from questionnaires are transformed from raw data into meaningful information that can be used to analyse and achieve research objectives. Once data have been collected and data file are formed, the next step is to analyse the data to make statistical estimates and reach conclusions.

## 4. Result and discussion

### *4.1. Respondents personal details*

According to the result collected, the questionnaires distributed showed 42.86% of response rate. Figure 1 shows the respondent's background. There were 60% respondents from civil and structure, 13% were quantity surveyor, 10% were mechanical and electrical, 7% were architect. Other than these, the other 10% were from turnaround engineering, construction management and safety, health and environment.

One's education level is usually related to their management level. Most diploma level entries are usually at the level of junior executive. However, this is also depending on their years of experience in the industry. Those who handle risk management are usually those who have bachelor degree and higher level. This is because risk management requires more attention and only be managed by the professionals. Figure 2 show the respondent's education level respectively. The respondents were mostly bachelor degree graduate in which they sit on the middle executive level.

IOP Conf. Series: Materials Science and Engineering **271** (2017) 012043 doi:10.1088/1757-899X/271/1/012043

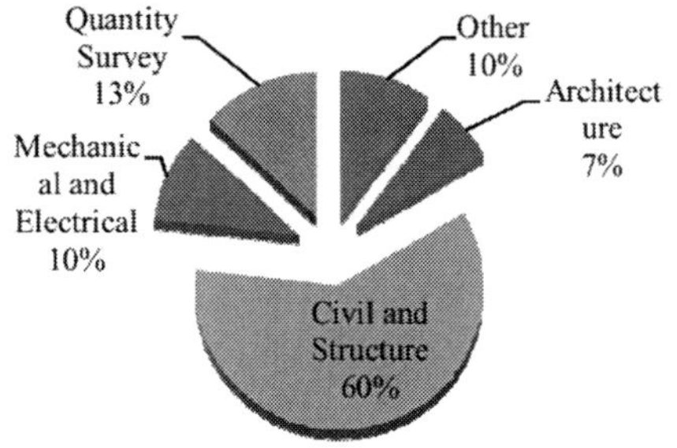


**Figure 1** Respondent's Professional Discipline.

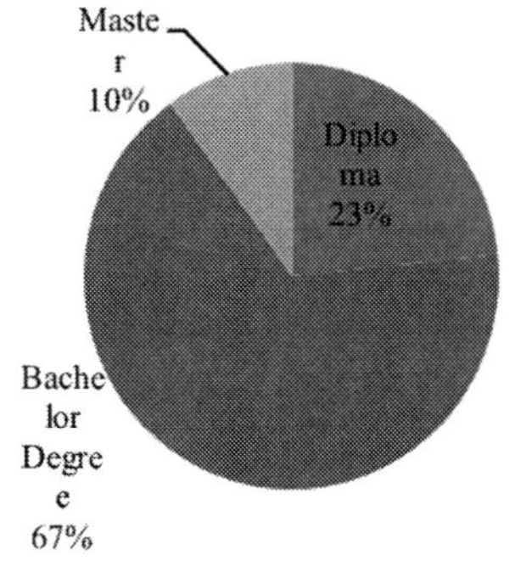


**Figure 2** Education Level.

*4.2. Risk management practice*

Traditional risk planning relies on gut-feelings, based on experience and manual observation using 2D drawings. There are some respondent who mentioned that it is based on analysis and five whys method. One of the five whys method is drawing based, depending on what type of activity that is going to be conducted. Almost majority agreed on practicing identifying risk based on experience, with 83.33% of response rate. From figure 3, many identified risk based on experience. However, there were some issues that they encountered during the implementation of traditional methods. This can be seen in figure 4.

Throughout the project lifecycle, risks can be changed. However, this can only happens when the controls are in match. Although risks are always treated as challenges, it poses opportunities as well. According to table 2, the descriptive statistics showed that most company preferred to reduce risks, with variance value of .478.

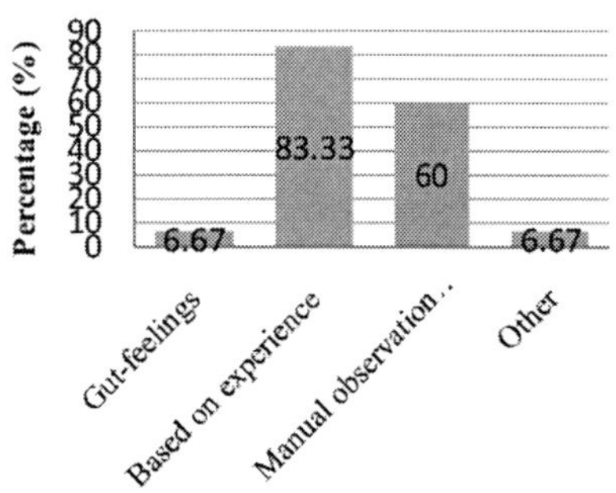


**Figure 3** Risk Management Traditional Practices.

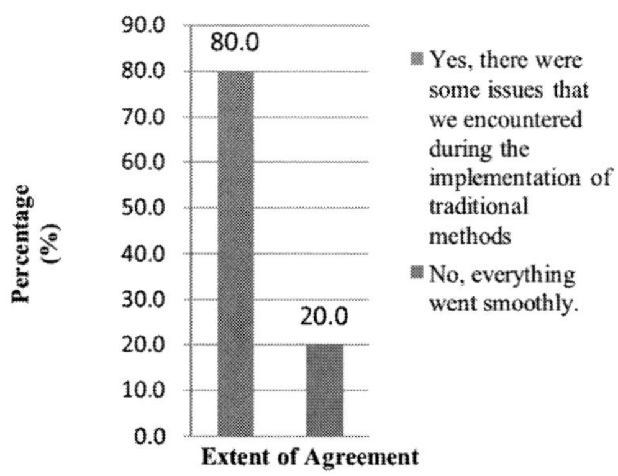


**Figure 4** Collaboration Issues Using Traditional Practices.

With variance value of 1.253, avoiding risks seem to be difficult to be done as it can caused quality risks, personnel risks, cost risks and dateline risks. Table 3 is the Kaiser-Meyer-Olkin and Bartlett's Test (KMO) in which it was recorded that the value is 0.677 where this means that the data collected were reliable.

**Table 2** Descriptive Statistics of Agreement in Position of the Company Handle Risks.

| Handling Risk | Mean | Std. Deviation | Variance |
|---|---|---|---|
| Tranferring risks | 3.0667 | 1.11211 | 1.237 |
| Avoiding risks | 3.3000 | 1.11880 | 1.252 |
| Accepting risks | 3.3333 | .84418 | .713 |
| Reducing risks | 4.0667 | .69149 | .478 |

**Table 3** KMO and Bartlett's Test of Agreement in Position of the Company Handle Risks.

| | | |
|---|---|---|
| Kaiser-Meyer-Olkin Measure of Sampling Adequacy. | | .677 |
| Bartlett's Test of Sphericity | Approx. Chi-Square | 10.985 |
| | df | 6 |
| | Sig. | .089 |

*4.3. Best function of BIM in risk management*

According to Furneaux *et al.* [18], BIM is seen as a concerted action to ensure collaboration between construction players such as architects, engineers and contractors. According to the data collected in Table 4, BIM was seen to be more useful in providing better tracking of cost control and cash flow with variance of .178. This can concludes that BIM has many uses in project planning as the tracking of cost control and cash flow give impact on the project cycle to be completed on time. As BIM is still considered new in Malaysia, it is still not up to the level of managing the facilities. This can be seen related to the variance value of .322 where BIM was considered also useful but with standard deviation of .56765, not many picked on this potential. Figure 5 is the answer of the respondents.

**Table 4.** Descriptive Statistics of the Agreement in relation to the Potential Benefits of using BIM.

| Code | Potential Benefits | Mean | Std. Deviation | Variance |
|---|---|---|---|---|
| P1 | Decreases costs of utility demand and demolition | 4.0000 | .47140 | .222 |
| P2 | Manages facilities proactively | 4.1000 | .56765 | .322 |
| P3 | Provides better tracking of cost control and cash flow | 4.2000 | .42164 | .178 |
| P4 | Identifies schedule phasing issues | 4.3000 | .48305 | .233 |
| P5 | Enables project manager and contractor to see construction work sequence, equipment and materials | 4.3000 | .48305 | .233 |
| P6 | Enables generation of takeoffs, counts and measurements directly from a 3D project model | 4.4000 | .51640 | .267 |
| P7 | Facilitates better communication and increase design effectiveness | 4.5000 | .52705 | .278 |

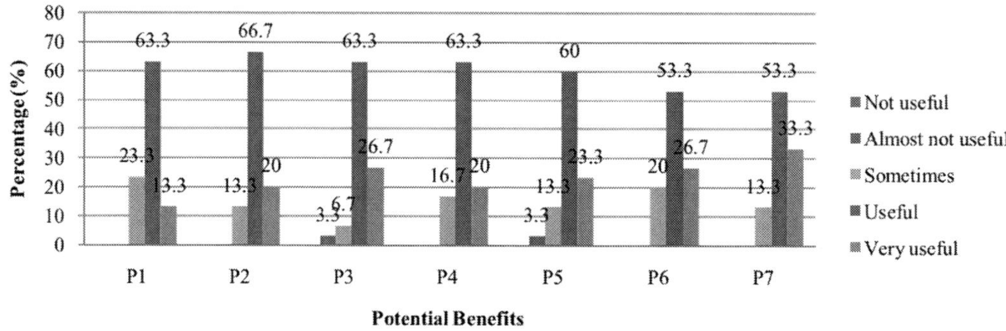


**Figure 5.** Percentage of the Extent of Agreement in Relation to the Potential Benefits of using BIM.

Table 5 is the Kaiser-Meyer-Olkin and Bartlett's Test (KMO) in which it was recorded that the value was 0.712 where this means that the data collected were reliable.

**Table 5.** KMO and Bartlett's Test of the Extent of Agreement in Relation to the Potential Benefits of using BIM.

| | | |
|---|---|---|
| Kaiser-Meyer-Olkin Measure of Sampling Adequacy. | | .712 |
| Bartlett's Test of Sphericity | Approx. Chi-Square | 138.768 |
| | df | 21 |
| | Sig. | .000 |

As regards to figure 6 and table 6, 5D cost estimation or cash flow modelling had been chosen as the almost always effective way for managing risk with the variance value of 0.178. 5D cost estimation or cash flow modelling benefit risk management in planning, controlling and managing budget and cost reasonably.

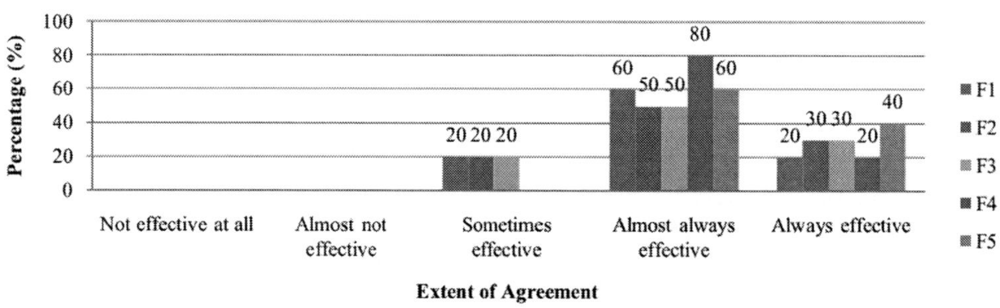


**Figure 6.** Percentage of the Extent of Agreement in Relation to BIM as a Systematic Way for Managing Risks.

*4.4. BIM-based risk management in project planning phase*

According to Zhou *et al.* [31], risk management had becomes proactive approaches rather than reactive. Although before the construction phase started, there is no project that challenges their risk issues [32], project manager could have predicted the risks [33] if appropriate risk identification were done. Without any doubt, risks exist in various stages of the project. The project lifecycle and the performance of the risk management have a forward impact on the project's successfulness. There are many factors that can lead to a successful BIM-based technology in risk management during the project planning phase.

In reference to table 6, there were two factors that mostly benefit a BIM-based technology which were formwork plan with integrated fall plan and design for safety model check.

**Table 6.** Descriptive statistics of the factors lead to a successful BIM-based technology in risk management.

| Code | Factors | Mean | Std. Deviation | Variance |
|---|---|---|---|---|
| L1 | Formwork plan with integrated fall protection | 4.1000 | .31623 | .100 |
| L2 | Design for safety model check | 4.1000 | .31623 | .100 |
| L3 | Hazard identification | 4.2000 | .42164 | .178 |
| L4 | Reducing construction fatalities | 4.2000 | .42164 | .178 |
| L5 | Safety protective equipment planning | 4.2000 | .42164 | .178 |
| L6 | Minimizing documentation errors | 4.3000 | .48305 | .233 |
| L7 | Replacing traditional delivery method with more integrated methods | 4.4000 | .51640 | .267 |
| L8 | Reduced rework | 4.5000 | .52705 | .278 |

Numerous investigations from Hartmann *et al.* [14], Shim *et al.* [25] and Zhang [34] have pointed out that there is a need to combine BIM-based and traditional risk management to improve practical applicability. Table 7 is the Kaiser-Meyer-Olkin and Bartlett's Test (KMO) in which it was recorded that the value was .768 where this means that the data collected were reliable.

**Table 7.** KMO and Bartlett's Test

| | | |
|---|---|---|
| Kaiser-Meyer-Olkin Measure of Sampling Adequacy. | | .768 |
| Bartlett's Test of Sphericity | Approx. Chi-Square | 117.005 |
| | df | 28 |
| | Sig. | .000 |

In conclusion, recognizing potential risks and turn it into chances is a strategy to avoid losses. Effective risk management must include in all areas, function and processes of the project.

## 5. Conclusion

This research addressed the topic of risk management and BIM, investigating the efficiency of adopting BIM-based risk management during the project planning phase. The objectives of this research were achieved throughout the process of literature review and questionnaires filled out by BIM experts and also those who understand and apply the concept of risk management in construction. The implementation of BIM in the AEC industry has been encouraged by the Malaysian Government to facilitate achievement of the vision of the CIDB Master Plan as well as the nation's 2020 vision through an increase in key performance indicators (KPIs) and productivity [6]. There are many benefits that each construction players can achieved with the adoption of BIM. Organization itself should educate their employees about the technology.

## 6. References

[1] Olsson R 2007 In search of opportunity management: Is the risk management process enough?, *Int. J. Proj. Manag.* **25** 745-752

[2] Hillson D 2011 *Dealing with Business Uncertainty* Retrieved on November 20, 2016 from http://www.risk-doctor.com/briefings.

[3] Baloi P and Price A 2003 Modelling global risk factors affecting construction cost performance, *Int. J. Proj. Manag.* **21** 261–269

[4] Anderson S 2009 *Risk Identification and Assessment* (United States: PMI Virtual Library) pp 1 - 3

[5] Furneaux C and Kivit R 2008 *BIM: Implications for Government* CRC for Construction Innovation Case Study No. 5 [2004-032-A + Case study no. 5] (Australia: Cooperative Research Centre for Construction Innovation) pp 10-31.

[6] Jabatan Kerja Raya (JKR) 2011 *Laporan Tahunan JKR 2011* Retrieved on March 12, 2013 from http://www.jkr.gov.my/var/files/File/dokumen/laporan_tahu nan_jkr_2011.pdf

[7] Charehzehi A, Yusof A M, Chai C and Chong H Y 2015 Conceptualizing building information modeling: In construction conflict management *In Proc. of the 26th Int. Business Information Management Association Conf. - Innovation Management and Sustainable Economic Competitive Advantage: From Regional Development to Global Growth, IBIMA 2015 (Madrid, Spain)* (United States: International Business Information Management Association (IBIMA)) pp 1337-1358

[8] Latiffi A A, Mohd S, Kasim N and Fathi M S 2013 Building Information Modeling (BIM) Application in Malaysian Construction Industry, *Int. J. Constr. Eng. Manag.* **2** 1-6

[9] Volk R, Stengel J and Schultmann F 2014 Building Information Modeling (BIM) for existing buildings — Literature review and future needs, *Automat. Constr.* **38** 109–127.

[10] Porwal A and Hewage K N 2013 Building Information Modeling (BIM) Partnering Framework for Public Construction Projects, *Automat. Constr.* **31** 204-214.

[11] Liu W P, Guo H L, Li H and Li Y 2014 Factors Constraining the Development of Professional Project Management in China's Construction Industry, *Int. J. Project Manage.* **22** 203-211.

[12] Hardin B 2011 *BIM and Construction Management: Proven Tools, Methods, and Workflows* (United States: John Wiley & Sons) pp 28-29

[13] Zou Y, Kiviniemi A and Jones S W 2016 A review of Risk Management through BIM and BIM-related Technologies, *Safety Sci.* **97** 88 - 98

[14] Hartmann T, Van Meerveld H, Vossebeld N and Adriaanse A 2012 Aligning building information model tools and construction management methods, *Automat. Constr.* **22** 605-613

[15] Tang P, Anil E, Akinci B and Huber D 2011 Efficient and effective quality assessment of as-is building information models and 3D laser-scanned data. *In: Proc. of the ASCE Int. Workshop on Computing in Civil Engineering (Miami, Florida, USA)* (United States: American Society of Civil Engineering) pp. 486-493.

[16] Whyte J 2003 Industrial Applications Ofvirtual Reality in Architecture, *ITcon* **8** 44-50

[17] Marzouk M and Hisham M 2014 Implementing earned value management using Bridge Information Modeling, *KSCE J. Civ. Eng.* **18** 1302-1313

[18] Bhatla A, Choe S Y, Fierro O and Leite F 2012 Evaluation of accuracy of as-built 3D modeling from photos taken by handled digital cameras, *Automat. Constr.* **28** 116-127

[19] Eastman C, Teicholz P, Sacks R and Liston K 2008 *BIM Handbook: A guide to building information modelling for owners, managers, designers, engineers and contractors* (Hoboken, New Jersey: John Willey & Son Inc) vol. 1 pp 23-25

[20] Teizer J 2008 3D range imaging camera sensing for active safety in construction, *ITcon* **13** 103-117.

[21] Kim T W, Kavousian A, Fischer M and Rajagopal R 2012 *Improving Facility Performance Prediction by Formalizing an Activity-Space-Performance Model* CIFE Technical Paper

TR210 (California: Stanford University - Center for Integrated Facility Engineering) pp 1 - 41
[22] Chen L and Luo H 2014 A BIM-based construction quality management model and its applications, *Automat. Constr.* **46** 64-73
[23] Lee S I, Bae J S and Cho Y S 2012 Efficiency analysis of set-based design with structural building information modeling (S-BIM) on high-rise building structures, *Autom. Constr.* **23** 20-32
[24] Sacks R and Barak R 2008 Impact of three-dimensional parametric modeling of buildings on productivity in structural engineering practice, *Autom, Constr.* **17** 439-449
[25] Shim C S, Lee K M, Kang L S, Hwang J and Kim Y 2012 Three-dimensional information model-based bridge engineering in Korea, *Struct. Eng. Int.* **22** 8-13
[26] Azhar S 2011 Building information modeling (BIM): trends, benefits, risks, and challenges for the AEC industry, Leadersh, *Manag. Eng.* **11** 241–252
[27] Becerik-Gerber B, Jazizadeh F, Li N and Calis G 2011 Application areas and data requirements for BIM-enabled facilities management, *J. Constr. Eng. Manage.* **138** 431-442
[28] Volk R, Stengel J and Schultmann F 2014 Building Information Modeling (BIM) for existing buildings — Literature review and future needs, *Automat. Constr.* **38** 109–127.
[29] Rajabifard A, Williamson L and Kalantari M 2012 *A National Infrastructure for Managing Land Information-Research Snapshot* (Australia: CSDILA, The University of Melbourne)
[30] Westin S and Sein M K 2014 Improving data quality in construction engineering projects: An action design research approach, *J. Manag. Eng.* **30** 1 - 17
[31] Zhou Z, Goh Y M and Li Q 2015 Overview and Analysis of Safety Management Studies in Construction Industry, *Safety Sci.* **77** 143-151
[32] Piperca S and Floricel S 2012 A Typology of Unexpected Events In Complex Projects, *Int. J. Managing Proj. Bus.* **5** 248-265
[33] Ramasesh R V and Browning T R 2014 A Conceptual Framework for Tackling Knowable Unknown Unknowns in Project Management, *J. Operations Manage.* **32** 190-204
[34] Zhang L and Li F 2014 Risk/Reward Compensation Model for Integrated Project Delivery, *Eng. Econ.* **25** 558-567

**Acknowledgements**
This present work benefited from financially supported by Universiti Teknologi Malaysia Flagship Grant, Vote: 03G55 and Universiti Teknologi Malaysia (UTM) Vot. Number of Q.J130000.2622.12J53 which directly supported by Ministry of Education Malaysia. The appreciation is also extended to Research Management Centre, Green PROPMT research team and members Universiti Tun Hussein Onn (UTHM)

# Understanding BIM Adoption in the AEC Industry: The Case of Jordan

**M Btoush**[1], **AT Haron**[1]

[1]Faculty of Civil Engineering & Earth Resources, University Malaysia Pahang, Lebuhraya Tun Razak, 26300 Gambang, Pahang.
Corresponding author: wardbtoush2012@yahoo.com

**Abstract.** Building information modelling (BIM) is a new and powerful technology implemented by many countries. The construction industry in Jordan plays a vital role and contributes immensely to the economic growth and development. In order to boost the industry and the economy, many industry players including engineers and contractors have recommended the implementation of BIM in Jordan. However, research demonstrates that successful BIM implementation is possible through the awareness of the different levels of BIM, which is a basic precondition for BIM implementation. Without a clear understanding of BIM, many companies would be unable to fully achieve BIM potentials or implement BIM in their building lifecycle. The objective of this study is to assess the current awareness of BIM technology in the Jordanian construction industry. A field interviews were conducted and 15 responses were collected and. The findings indicate that a significant proportion of respondents have little or no understanding of the concept of BIM. Also, the usage was found to be very low. Based on the results, a holistic roadmap was developed to spread the BIM adoption through the Jordanian construction industry. It is expected that this roadmap would lead to a better understanding and enable the industry towards more extensive implementation of BIM.

## 1. Introduction

Building information modelling (BIM) is a new and powerful technology implemented by many countries [1–6]. The construction industry in Jordan plays a vital role and contributes immensely to the economic growth and development [7]. In order to boost the construction industry and the economy, many industry players including engineers and contractors have recommended BIM's adoption in Jordan. However, research demonstrates that successful BIM implementation is possible through the awareness of the different levels of BIM, which is a basic precondition for BIM implementation [3,8–10]. That is, the first step in BIM technology adoption is to understand what exactly BIM is about [11]. Without a clear understanding of BIM, many companies are unable to fully achieve BIM potentials or implement BIM in their building lifecycle [3,8]. Similarly, despite several attempts by companies to interpret and apply the BIM, many are still struggling with design quality issues, high design and execution costs, long lead times, and these are mostly attributed to the lack of clear understanding and the absence of a comprehensive implementation roadmap. Moreover, several studies have been conducted on BIM implementation issues in other countries [1–3], conversely, not much have been carried out on BIM implementation issues in Jordan.

The objective of this article is to assess the current state of the BIM awareness within the architecture, engineering, and construction (AEC) industry in Jordan.A field study was conducted and

15 survey responses were collected and analyzed. The findings indicate that a significant proportion of respondents have little or no understanding of the concept of BIM. Also, the usage was found to be very low. Based on the results, a holistic roadmap was developed to spread the BIM adoption through the Jordanian construction industry. It is expected that this roadmap would lead to a better understanding and enable the industry towards more extensive implementation of BIM.

## 2. Literature review

BIM is an IT enabled tool with integral digital representation (data repository) for different project lifecycle phases [11]. BIM technology is emerging as one of the most promising tools, both in construction management and practical sphere of building construction [4]. The use of BIM improves methods and tools to control risks, fragmentation and increase collaboration in construction projects. BIM is a framework designed to ensure sustainability, facilitate the integration of disjointed practices, reduce poor quality and act as a catalyst for changing business process [1,2,12].

Several studies have been conducted on BIM's adoption in several countries including Norway, Finland, Sweden, Germany, Malaysia, Singapore, France, USA, Australia, and UK [1–3,10,13–16], and BIM has demonstrated the capacities to improve information integration, business process flow, productivity, and reduce complexities, uncertainties, conflicts, fragmentations, among others [3]. Explicitly, BIM can save between 3% and 5% in costs, boost gross domestic product by 0.2 basis points, and present an accurate virtual model to digitally construct a building [5,11]. BIM is a promising recent development, an emerging procedural and technological shift within the AEC sector [12,17,18]. It is more reliable in streamlining the procurement of component parts or materials for production and in delivering projects within the shortest possible time [5,19]. BIM can be used for design, planning, construction and operation of a building facility. Generally, BIM can improve scheduling and minimize the mistakes and errors caused by modifications [20,21].

BIM is spreading widely and quickly, yet it is still relatively new in the Middle East [8], specifically Jordan. BIM has been adopted in Dubai, Kuwait, among others, however, the implementation rate is very low [8]. According to van Berlo L A H M and Papadonikolak [16] and [9], the adoption of BIM in the Middle East region is low. The study indicated that only 24% of companies used BIM, about 43% did not know how to utilize BIM, 41% were interested in using BIM but did not have the technical capabilities to start. Also, the study by Gerges *et al.* [8] indicated that only 20% of the AEC organizations are using or in the processing of adopting BIM in the Middle East. Moreover, it is confirmed that the Middle East has the most reduced BIM implementation rate due to lack of awareness and understanding as professionals only see BIM as a tool for 3D modeling of the building [9]. Thus, BIM is mainly used in the Middle East for design authoring, 3D coordination, and clash detection [9].

Although the use BIM in the Middle East is a recent phenomenon, its adoption in Jordan is very scarce [8–10]. Moreover, through the synthesis of literature, it has been recommended that more research to identify and analyze BIM definitions, changes, awareness, challenges and how the challenges should be addressed are needed to facilitate BIM adoption, especially in Jordan. Thus, the need for a quantitative analysis that is focused on Jordan. Considering the varying understandings of BIM from the literature review, the subsections below address BIM capability stages, BIM maturity levels, and BIM competencies. These are further evaluated and analyzed through a survey in Jordan.

### *2.1. BIM capability stages*

BIM is an interactive set of processes, policies, and technologies producing an approach to manage essential building designs and project data in a digital format all through the building's life-cycle. However, according to Succar [22] in his "*Five Components of BIM Performance Measurement*," this definition delimits the BIM domain which continuously expands in connotation and coverage. To fully grasp the meaning of BIM, Succar suggested that professionals need to understand the BIM capability stages. The capability stages could be divided into three, which can further be subdivided into competency sets:

- **BIM Stage 1:** Object-Based Modelling of using virtually integrated design, construction and operation. For instance, the organizations considered at this stage should have employed an Object-Based Modelling software tool like Revit, ArchiCAD, Constructor or Tekla.
- **BIM Stage 2:** Model-Based Collaboration. For this stage, the organizations should have been part of a multidisciplinary Model-Based collaborative project.
- **BIM Stage 3:** Network-Based Integration. For this stage, the organizations should have deployed a Network-Based solution such as model servers to share Object-Based with not less than two other disciplines.

### *2.2. BIM maturity levels*
The BIM maturity is repeatability, quality, and excellence with the BIM Capabilities. There are three main BIM maturity levels [3,23–25]. The maturity levels are characterized as 2D/3D, BIMs and iBIM. The BIM maturity level in an organization affects their understanding of BIM and its definition. Similarly, an organization's level of maturity can be assessed through their general objectives. For instance, organizations conducting pilot projects and testing to determine BIM's benefits may be at stage 1 (Object-Based Modelling) and within that stage they are at a defined or an ad-hoc maturity level, striving to be more optimized through continuous testing. Barlish and Sullivan [25] provides a detailed description of the maturity levels. Figure 1 presents the BIM maturity map, indicating the levels, capabilities, and descriptions.

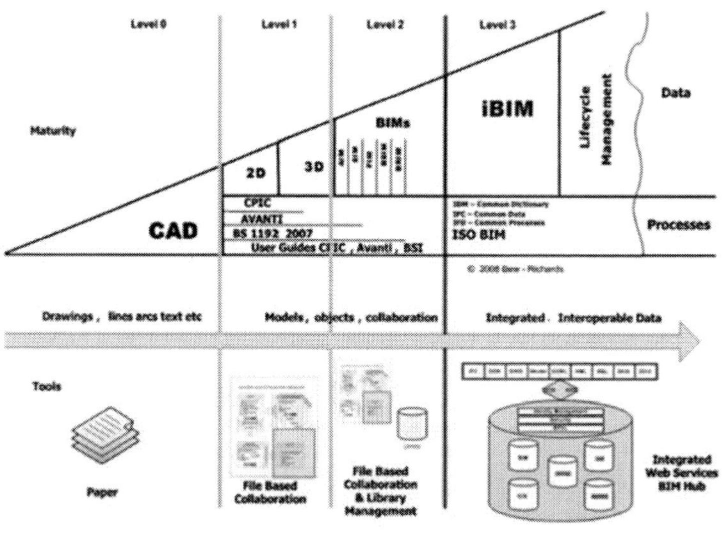


**Figure 1.** BIM Maturity Map [25].

### *2.3. BIM competencies*
A set of BIM competency is a hierarchical pool of individual sets of competencies recognized for BIM assessment and implementation purposes [22]. There are three major sets of BIM competencies:

- Technology sets in hardware, software, and networks. For instance, the readiness of BIM tools allows the transfer from drafting-based to Object-Based workflow (BIM Stage 1 requirement).
- Process sets in human resources, infrastructure, leadership and services/products. For instance, skills in database-sharing and collaboration processes are required to allow Model-Based Collaboration (BIM stage 2).
- Policy sets in regulations, contracts and education/research. For instance, contractual agreements are required for Network-Based Integration (BIM stage 3). Figure 2 presents the

structure of BIM competency sets, indicating the fields, stages, lenses, competency sets and their descriptions.

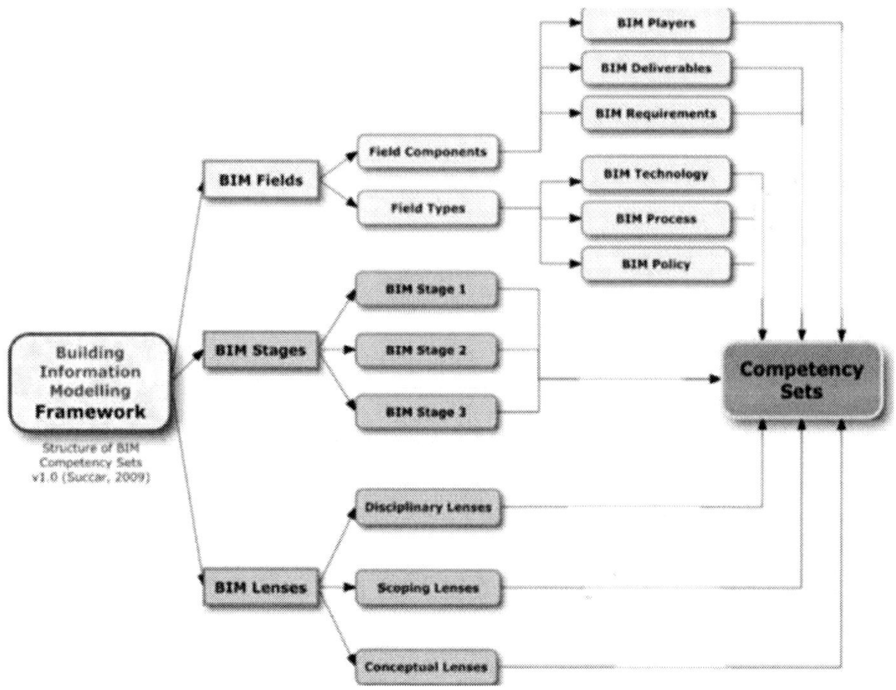


**Figure 2.** Structure of BIM Competency Sets [22].

### 3. Research methodology

The expert interview as a method of qualitative empirical research, designed to explore expert knowledge, has been developed considerably since the early 1990s. Furthermore, the expert interview has become very popular as a "streamlined" method [4]. The main focus of this article was to assess the current state of the BIM awareness within the architecture, engineering, and construction (AEC) industry in Jordan. Therefore, qualitative methods were applied in this research. Because it is a new phenomenon, the contextual analyses and qualitative methods seem appropriate as they better explain why and how [26]. The researcher demonstrated this by analyzing survey responses from the viewpoint of construction organizations representatives to measure the different levels of BIM awareness. First, an interview questionnaire structure was developed to assess the perceptions of the respondents of the relative awareness of BIM technology. The criteria for respondent's selection was based on the objectives of this study, and simplified as following: a) The respondent experience in the construction industry is more than 15 years b) The respondent work for a high-profile construction organization (Grade A based on the Jordanian Contractors Association) c) The respondent has excellent academic background (has published more than two papers in the Jordanian construction industry field).

Further, the responses were collected and analyzed. The analysis focused on measuring the different levels of BIM awareness. Table 1 illustrates the questions about BIM definition, BIM implementation, and BIM awareness responsibility.

**Table 1.** BIM Definition, BIM Implementation, and BIM Awareness.

| | | A | B | C | D | E |
|---|---|---|---|---|---|---|
| Q1 | BIM Definition | Modelling Software | Model-Based Collaboration | Network-Based Integration | BIM is an interactive set of processes, policies, and technologies producing an approach to manage essential building designs and project data in a digital format all through the building's life-cycle. | I am not sure |
| Ans. | | - | 2 | 8 | 5 | - |
| Q2 | BIM Implementation | BIM is well-known in Jordan and many firms implement it | BIM is well-known in Jordan but a little number of firms implement it | Some construction firms in the industry know what BIM is but don't implement it | No single firm knows what BIM is nor implement it | I am not sure |
| Ans. | | - | 3 | 9 | 2 | 1 |
| Q3 | Awareness Responsibility | It is the government's responsibility | Private sector's responsibility | Construction associations' responsibility | Educational institutes' responsibility | All the mentioned |
| Ans. | | 9 | 2 | 1 | 1 | 2 |

## 4. Results and discussion

The categories of the analysis of the data are based on the personal experiences and background of the respondents, their understanding, and awareness of BIM. The responses reflected a variation in terms of BIM definition. While 2 (13.3) individuals characterized BIM as a Model-Based Collaboration, 5 (33.3%) defined BIM as "an interactive set of processes, policies, and technologies producing an approach to manage essential building designs and project data in a digital format all through the building's life-cycle." Likewise, 8 (53.3%) believed that BIM is a Network-Based Integration. In terms of BIM implementation, 6.7% (1 respondent) was not sure about BIM implementation in Jordan. 13.3% (2 respondents) believed no single firm knows what BIM is nor implement it, 20% (3 respondents) indicated that BIM is well-known in Jordan but a little number of firms implement it, and 60% (9 respondents) believed that some construction firms in the industry know what BIM is but don't implement it. In terms of BIM awareness responsibility, the majority of the respondents (60%) agreed that it is the responsibility of the government to promote BIM. However, 13.3% attributed the responsibility to the private sector, 13.3% to all the mentioned, 6.7% to educational institutes and 6.7% to construction association.

With such a varying understanding, these definitions were evaluated to reflect the maturity levels of understanding and awareness. The general overview of the results was dominated by more respondents on the individual questions. Thus, the survey results were able to reflect the understanding of BIM across the Jordan construction industry. Furthermore, the results indicated that professionals have an understanding of BIM yet lack the capacity to implements it. Perhaps clear guidance, training, and financial and technical support would boost extensive BIM implementation across Jordan. Based on the results, a holistic roadmap has been developed to spread the BIM adoption through the

Jordanian construction industry. This is illustrated in figure 3 indicating the overall model for BIM awareness.

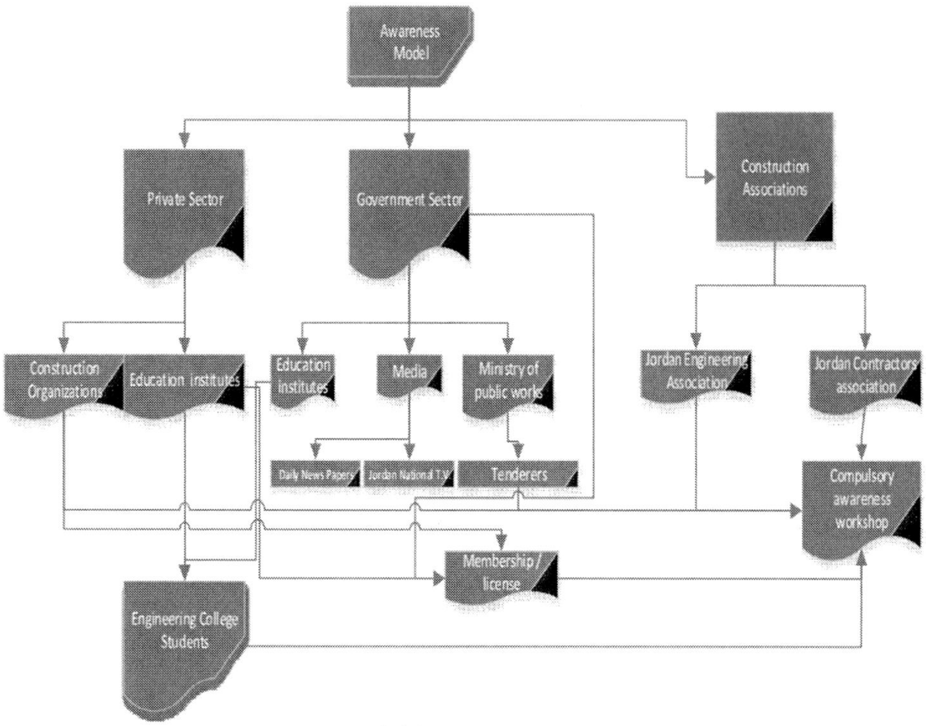


**Figure 3.** Model for BIM Awareness.

## 5. Conclusion

BIM adoption is increasingly becoming a must for the AEC industry to integrate the construction process and address challenges in the building lifecycles. In using BIM maturity gauge, the results demonstrated that professionals have a better understanding and are aware of BIM (53.3% and 60% respectively), but lack the capacity to implement it. This may be due to the lack of training staff on new process and workflow, new software and technology, the lack of familiarity with BIM use, client expectations, lack of the capital to invest, resistance to culture change, among others. Perhaps clear guidance, training, and financial and technical support would boost extensive BIM implementation across Jordan. More also, the government, the private sector, construction association and educational institutes are called upon to promote BIM adoption in the Jordanian construction industry. Based on the results, a comprehensive roadmap has been developed to spread the BIM adoption throughout the Jordanian construction industry. It is expected that this roadmap would lead to a better understanding and enable the industry towards more extensive implementation of BIM.

## 6. References

[1] Aranda Mena G, Crawford J, Chevez A and Froese T 2009 Building information modelling demystified: does it make business sense to adopt BIM?, *Int. J. Manag. Proj. Bus.* **2** 419–434
[2] Mihindu S and Arayici Y 2008 Digital Construction through BIM systems will drive the Re - engineering of construction business practices *12th Int. Conf. Information Visualisation (London)* (United States: IEEE Computer Society) pp 29 - 34
[3] Khosrowshahi F and Arayici Y 2012 Roadmap for implementation of BIM in the UK

construction industry, *Eng. Constr. Archit. Manag.* **19** 610–635
[4] Ansah R H and Sorooshian S 2017 Effect of lean tools to control external environment risks of construction projects, *Sustain. Cities Soc.* **32** 348–356
[5] Ansah R H, Sorooshian S, Mustafa S Bin and Duvvuru G 2016 Advancing towards delay-free construction project: a review *Proc. of the 2016 Int. Conf. on Industrial Engineering and Operations Management (Detroit, Michigan, USA)* (United States: IEOM Society International) pp 744–751
[6] Ansah R H, Sorooshian S, Mustafa S Bin and Duvvuru G 2016 Lean construction tools *Proc. of the 2016 Int. Conf. on Industrial Engineering and Operations Management (Detroit, Michigan, USA)* (United States: IEOM Society International) pp 784–793
[7] Sweis G, Sweis R, Abu Hammad A and Shboul A 2008 Delays in construction projects: The case of Jordan, *Int. J. Proj. Manag.* **25** 665–674
[8] Gerges M, Austin S, Mayouf M, Ahiakwo O, Jaeger M and Saad A 2017 An investigation into the implementation of building information modeling in the Middle East, *J. Inf. Technol. Constr.* **22** 1–15
[9] Awwad R and Ammoury M 2013 Surveying BIM in the Lebanese construction industry *30th Int. Symp. on Automation and Robotics in Construction and Mining (ISARC 2013) (Montreal, Quebec, Canada)* vol 1 (Slovakia: International Association for Automation and Robotics in Construction) pp 118 - 125
[10] Mehran D 2016 Exploring the adoption of BIM in the UAE construction industry for AEC firms, *Procedia Eng.* **145** 1110–1118
[11] Gu N and London K 2010 Understanding and facilitating BIM adoption in the AEC industry, *Autom. Constr.* **19** 988–999
[12] Azhar S 2011 Building Information Modeling (BIM): trends, benefits, risks, and challenges for the AEC industry, *Leadersh. Manag. Eng.* **11** 231–242
[13] Uhm M, Lee G and Jeon B 2017 An analysis of BIM jobs and competencies based on the use of terms in the industry, *Autom. Constr.* **81** 67–98
[14] Al-Werikat G 2017 Analysing the role of clients in the Jordanian construction supply chain delays, *Int. J. Eng. Res. Sci. Technol.* **6** 22–29
[15] Gerges M, Austin S, Mayouf M, Ahiakwo O, Jaeger M, Saad A and Gohary T-E 2017 An investigation into the implementation of Building Information Modeling in the Middle East, *J. Inf. Technol. Constr.* **22** 1–15
[16] van Berlo L A H M and Papadonikolaki E 2016 Facilitating the BIM coordinator and empowering the suppliers with automated data compliance checking *11th European Conf. Product and Process Modelling (ECPPM )2016 (Limassol, Cyprus)* (United States: CRC Press / Balkema - Taylor & Francis Group) pp 145 - 154
[17] Abdirad H 2015 Advancing in Building Information Modeling (BIM) contracting: Trends in the AEC/FM industry *AEI 2015: birth and life of the integrated buildin: Proc. of the AEI Conf. 2015) (Milwaukee, Wisconsin)* (Reston, Virginia: American Society of Civil Engineers) pp 1–12
[18] Azhar S, Khalfan M and Maqsood T 2012 Building information modelling (BIM): now and beyond, *Australas. J. Constr. Econ. Build.* **12** 15
[19] Sacks R, Eastman C M, Lee G and Orndorff D 2005 A target benchmark of the impact of three-dimensional parametric modeling in precast construction, *PCI J.* **50** 125–139
[20] Ku K and Taiebat M 2011 BIM experiences and expectations: the constructors' perspective, *Int. J. Constr. Educ. Res.* **7** 175–197
[21] Andújar-Montoya M, Gilart-Iglesias V, Montoyo A and Marcos-Jorquera D 2015 A construction management framework for mass customisation in traditional construction, *Sustainability* **7** 5182–5210
[22] Succar B 2010 The five components of BIM performance measurement *Proc. of CIB World Congress (Salford, United Kigdom)* (United Kingdom: University of Salford) pp 1–14

[23] Morlhon R, Pellerin R and Bourgault M 2014 Building Information Modeling implementation through maturity evaluation and critical success factors management *Procedia Technology* vol 16 (United Kingdom: Elsevier) pp 1125–1134
[24] Succar B, Sher W and Williams A 2012 Measuring BIM performance: Five metrics, *Archit. Eng. Des. Manag.* **8** 120–142
[25] Barlish K and Sullivan K 2012 How to measure the benefits of BIM ? A case study approach, *Autom. Constr.* **23** 149–159
[26] Li S, Wu X, Zhou Y and Liu X 2017 A study on the evaluation of implementation level of lean construction in two Chinese firms, *Renew. Sustain. Energy Rev.* **71** 846–851

IOP Conf. Series: Materials Science and Engineering 271 (2017) 012045 doi:10.1088/1757-899X/271/1/012045

# Baseline for food waste generation – A case study in Universiti Tun Hussein Onn Malaysia cafeterias

**A R Alias[1], N A Mohd Mokhlis[1] and N Y Zainun[1,]**

[1] Department of Building and Construction Engineering (JKBP)
[2] Faculty of Civil and Environmental Engineering, Universiti Tun Hussein Onn Malaysia, Batu Pahat, Malaysia
Corresponding author: nryasmin@uthm.edu.my

**Abstract.** Increasing population and economy status have contributed to the increasing volume of solid wastes produced in Malaysia and it creates problems on the existing solid waste management system. Ineffective waste management system was one of the issues that often discussed. The purpose of this study was to suggest the best method for managing food waste in Universiti Tun Hussein Onn Malaysia (UTHM) cafeterias. The scope of the study was to identify the type and quantity of waste generated in each cafeteria. The study area was carried out at six cafeteria in UTHM including residential college cafeteria which are Tun Dr.Ismail (TDI), Tun Fatimah (TF) and Tun Syed Nasir (TSN), G3's cafeteria, Arked, and Dr. Munie's cafeteria located at the Faculty of Civil and Environmental Engineering (FKAAS). In this study, food waste was quantified in unit of kilogram (kg). Results of the study showed that total food waste in selected UTHM's cafeterias was 6197.5 kg for two months. Food waste generated in G3's cafeteria was the highest value with 1823.5 kg among another cafeteria. This is due to strategic location for students and staff to take meals, the variety of food sold and reasonable price were major factors of generating food waste. Meanwhile, the Dr. Munie's Cafeteria located in FKAAS recorded the least total production of food waste as staffs and students take their meals at others cafeterias. Through literature review, there are list of methods on waste management were identified and composting method was suggested for food waste management in UTHM since the waste was produce in very large quantity.

## 1. Introduction

The amount of solid waste generated was directly proportional to the population growth [1]. Solid waste or garbage and all kinds of waste materials generated by human activity refers to materials that are not useful and no longer required [1]. Solid waste was one of three major environmental problems faced by most Local Authorities (PBT) in addition to the problem of air pollution and water pollution [2]. In Malaysia, food wastes make up 45 percent of the total 29,000 tons of solid waste generated within a day [3]. By 2020, an estimated 30,000 tons of solid waste will be generated within a day due to the increasing number of a population exceeding 28 million people [3]. This development resulted in curtailment lifespan of 176 landfills operating in Malaysia [3]. UTHM allocated about one million a year to pay contractors for picking up trash from campus to landfill and this cost was excluding the cost of waste management from cafeterias and residential colleges. Cafeteria operators pay another private contractor for the collection of solid waste generated in their cafeterias. Therefore, a suitable solid waste management method for food waste in UTHM will be proposed in this paper to reduce cost of waste management. This method hopefully will assist UTHM in sustainable waste management for food waste and generate new income for UTHM.

## 2. Literature review

Solid waste management was one of the main problems faced by society, particularly in urban areas [4]. The process of urbanization and rapid industrialization led to an increase in solid waste [4]. The cost of solid waste management from the stage of collecting, transportation, and disposal was very high [4]. Therefore, the quantity of solid waste, especially domestic solid wastes should be reduced in order to reduce government spending [4]. Municipal Solid Waste (MSW) was known as the waste generated from the area around the city, residential, commercial, institutional and industrial [5]. Municipal solid waste produced containing material which is durable and non-durable material that is deemed to be heterogeneous [5]. MSW can be classified into two major categories of organic waste and inorganic waste (table 1). Organic waste consists of plant and animal waste, where the waste was easily decomposed by soil microbial reactions produced a foul odor and harmful gasses such as methane. While, non-organic waste takes a long time to decompose [5].

**Table 1.** Sources of municipal solid waste (MSW) generation

| Classification | Composition | General |
|---|---|---|
| Organic waste | Food | Beef, chicken, vegetables |
| | Garden wastes | Dried leaves, twigs |
| | Textile / rubber | Leathers, clothes, carpet |
| | Paper product | Newspaper, box, paper cups |
| Non-organic waste | Plastic | *Polyethylene terephthalate* (1) *, *High density polyethylene* (2), *Polyvinyl chloride* (3), *Low-density polyethylene* (4), *Polypropylene* (5), *Polystyrene (6)*, *Multilayer plastic (7)* |
| | Glass | Glass plates, glass windows, light bulbs |
| | Metal | Ferrous, aluminium, zinc, chromium |

* Plastic coding system, Society of the Plastics Industry, Inc.[5]

The problem of solid waste management and disposal sites in Malaysia has increased its landfill management system and became inefficient. These include weaknesses in the system of waste collection and management of waste in landfills [6]. The daily amount of solid waste generated reaches approximately 30,000 tons and about 70 percent will be collected and reported about 95 percent of waste collected will be disposed of in a landfill with only about 5 percent recyclable [7]. Table 2 shows the number of existing landfills in Malaysia.

**Table 2.** Total disposal site in Malaysia

| Country | Landfills that operated | Landfills that not operated | Total |
|---|---|---|---|

| | | | |
|---|---|---|---|
| Johor | 14 | 23 | 37 |
| Kedah | 8 | 7 | 15 |
| Kelantan | 13 | 6 | 19 |
| Melaka | 2 | 5 | 7 |
| Negeri Sembilan | 7 | 11 | 18 |
| Pahang | 16 | 16 | 32 |
| Perak | 17 | 12 | 29 |
| Perlis | 1 | 1 | 2 |
| Pulau Pinang | 2 | 1 | 3 |
| Sabah | 19 | 2 | 21 |
| Sarawak | 49 | 14 | 63 |
| Selangor | 8 | 14 | 22 |
| Terengganu | 8 | 12 | 20 |
| WP Kuala Lumpur | 0 | 7 | 7 |
| WP Labuan | 1 | 0 | 1 |
| Total | 165 | 131 | 296 |

Based on table 2, total landfills were 296 but only 165 operated. In Johor, total landfills were 37 and only 14 landfills operated. Solid waste management continues to be a major challenge in urban areas around the world, especially in towns and villages where the increase of solid waste at an alarming rate, especially in underdeveloped countries [8]. The steady economic growth and low unemployment are driven by the political situation and there are a number of resources to make Malaysia a developed country on a par with [9]. Table 3 shows the methods of solid waste management.

Table 3 indicates the waste management practiced in Malaysia. According to Periathamby [10], in 2006, sanitary landfill produces 30.9% from total waste disposed. In 2020, landfill will target becoming high percentage of waste disposed with 44.1% follow by 22.0%, recycling, 16.8% for incineration, inert landfill, 9.1% and composting 8.0%. Referring to Jabatan Pengurusan Sisa Pepejal Negara [15], construction of the landfill will have an effect on greenhouse gas emissions to 50% by 2020. In Japan, from the 1960s and up to now, solid waste was beginning to dispose by incinerator. Japan has the world's leading incinerator facility. In 2011, a total of 1211 incinerator facilities is provided for managing the waste generated. Features of this high-technology combustion were promoted as a safe technology and risk of contamination of the population can be reduced [11].

In Singapore, 90% of solid waste was disposed by using incinerators and only 10% disposed in sanitary landfills. The resulting energy recovered from incineration was for electricity generation in which 980 million kWh of electricity could be generated each year [12]. While in India, Refuse Derived Fuel (RDF) is defined as the heat generated from the waste materials that have been processed to be used as a secondary or substitute fuel in industrial solid fuel. RDF is used as a replacement for coal (fossil fuels) in high-energy industrial processes such as power production, cement kilns, steel making where the use of RDF can be optimized to improve economic performance. In waste management, RDF placed under aerobic composting as waste-to-energy incineration to generate electricity. RDF method is different because the goal is to increase the thermal processing of fuels [13].

**Table 3.** The methods of solid waste management

| Country | Method | Article |
|---|---|---|

| | | |
|---|---|---|
| Malaysia | i. Landfill<br>ii. Recycling<br>iii. Composting<br>iv. Incineration<br>v. Inert landfill<br>vi. Sanitary landfill | [10] |
| Japan | i. Incinerator<br>ii. Technology biomass<br>iii. Landfill | [11] |
| Singapore | i. Landfill<br>ii. Incinerator | [12] |
| India | i. Composting<br>ii. Landfill<br>iii. Refuse Derived Fuel (RDF)<br>iv. Incinerator | [13] |
| Thailand | i. Landfill<br>ii. Incinerator<br>iii. Composting<br>iv. Open burning | [14] |

Furthermore, the methods used to manage solid waste in Thailand are sanitary landfill, composting, open burning, and incineration. There were 97 disposal facilities in which 91 sanitary landfills, 3 incinerators, and 3 integrated facility system that is designed to be placed. Open dumping of solid waste remains an option for some areas because of its low cost and performance effective in getting rid of a large amount of waste. However, this method is not appropriate as it may cause the area to become unsanitary and potential to environmental disaster [14].

## 3. Methodology

The research process was described in details which each level begins with the initial survey stage to analyze the data. In this study, the primary data collected by manual in each cafeteria. While, there are some obtained through literature review. This review process includes four phases that describe the flow of the study. The first phase refers to the beginning of the study and literature review, the second phase refers to the collection of data and information, third phase and fourth phase refers to data analysis and conclusion and recommendation respectively. Figure 1 shows a flow chart of the study.

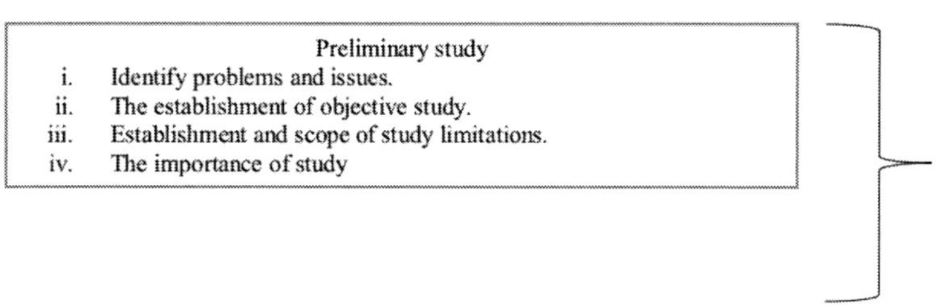

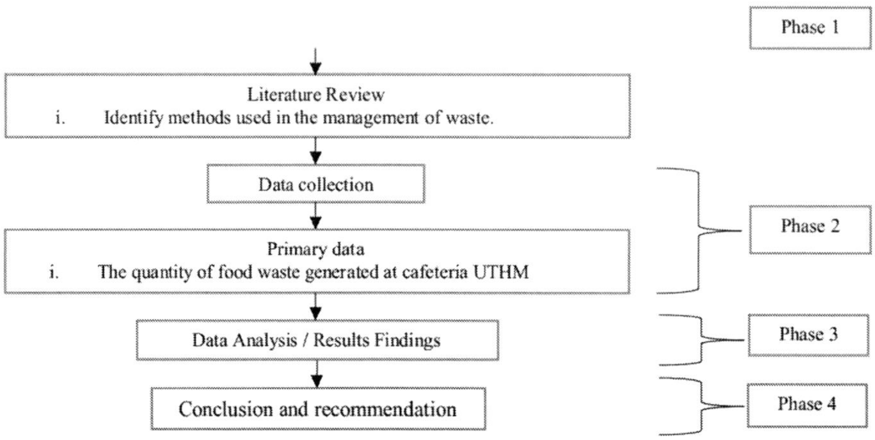


**Figure 1.** Flow chart of the study

In this paper, the method proposed was composting that seen suitable with high generation produced by all cafeteria in UTHM.

## 4. Data Analysis

### *4.1. Total of food waste generated in UTHM*

The analysis was performed to determine the quantity of food waste generated in each cafeteria located at the Universiti Tun Hussein Onn Malaysia (UTHM). From the data collected in each cafeteria located at UTHM, food waste generation schedule has been created as shown in table 4.

Table 4 shows the amount of food waste generated in each cafeteria located in the UTHM. It was found that 390.4 kilograms (kg) of food waste generated on Thursday (March 9, 2017) and was the highest weight of food waste collected. On the other hand, only 35.9 kg of food waste generated on Sunday (April 2, 2017). The drastic reduction of food waste was due to mid-sem holiday for students at the University. Total food waste collected can for two months was 6197.5 kg where 3738.8 kg in March and 2458.7 kg in April.

### *4.2. Comparison of food waste in March and April*

Figure 2 shows a comparison of food waste collected in each cafeteria in March and April respectively.

Referring to figure 2, more food waste was collected in March. Food waste collected at G3's cafeteria was the highest compare to others cafeteria because many students come to have meals. This was due to many types of food sold and students have many choices of food to eat. The strategic location, diversity of food sold and reasonable price were also among factors in the generation of food waste in the cafeteria. In contrast, the result shows the total food waste generated in the Cafeteria Dr. Munie's produced the least because most of the staff and students take meals in another cafeteria. This was due to food prices are less affordable and the location of the cafeteria was not strategic.

**Table 4:** Table of food waste generated in each cafeteria located at UTHM (Kg)

| Date | Day | Faculty | Total |
|---|---|---|---|

| | | FKAAS | TDI | TF | TSN | ARKED | G3 | |
|---|---|---|---|---|---|---|---|---|
| 5-March 17 | Sunday | 4.3 | 28 | 32.5 | 18.7 | 68 | 140.8 | 292.3 |
| 7-March 17 | Tuesday | 5.8 | 43.5 | 51.7 | 52.5 | 72.5 | 132.1 | 358.1 |
| 9-March 17 | Thursday | 0 | 54.4 | 80.4 | 63.2 | 107.2 | 85.2 | 390.4 |
| 12-March 17 | Sunday | 6.4 | 40 | 43.4 | 38 | 92 | 112 | 331.8 |
| 14-March 17 | Tuesday | 11.2 | 60 | 54.8 | 55.1 | 101.3 | 93.8 | 376.2 |
| 16-March 17 | Thursday | 4.2 | 45.9 | 23.2 | 47.1 | 73.2 | 68.8 | 262.4 |
| 19-March 17 | Sunday | 6.3 | 48.7 | 38.9 | 17.1 | 70.4 | 108 | 289.4 |
| 21-March 17 | Tuesday | 7.1 | 32.4 | 34 | 25 | 81 | 67.7 | 247.2 |
| 23-March 17 | Thursday | 5.2 | 43.2 | 45.3 | 41.3 | 82.7 | 78.4 | 296.1 |
| 26-March 17 | Sunday | 5.3 | 51.4 | 42.1 | 33 | 82 | 117.4 | 331.2 |
| 28-March 17 | Tuesday | 7.2 | 14 | 36 | 21 | 144 | 122 | 344.2 |
| 30-March 17 | Thursday | 6.3 | 17 | 21 | 15.9 | 68.3 | 91 | 219.5 |
| 2-April 17 | Sunday | 0 | 6.7 | 4.2 | 5.1 | 8.3 | 11.6 | 35.9 |
| 4-April 17 | Tuesday | 0 | 7.2 | 3.9 | 4.3 | 5.1 | 28.3 | 48.8 |
| 6-April 17 | Thursday | 0 | 10.8 | 8.2 | 9.8 | 10.1 | 15.8 | 54.7 |
| 9-April 17 | Sunday | 3 | 78 | 42 | 23.3 | 35.8 | 38.6 | 220.7 |
| 11-April 17 | Tuesday | 4 | 29 | 21 | 14 | 32 | 68 | 168 |
| 13-April 17 | Thursday | 3.2 | 24.5 | 33.1 | 35.5 | 41.3 | 55 | 192.6 |
| 16-April 17 | Sunday | 0 | 36 | 45 | 41.2 | 52.4 | 53 | 227.6 |
| 18-April 17 | Tuesday | 0 | 63.3 | 68.5 | 32 | 66 | 51 | 280.8 |
| 20-April 17 | Thursday | 5 | 46 | 32 | 64 | 86 | 68 | 301 |
| 23-April 17 | Sunday | 4.1 | 48 | 37.2 | 24.6 | 55 | 58 | 226.9 |
| 25-April 17 | Tuesday | 2.9 | 25.3 | 41.8 | 57 | 72.5 | 61 | 260.5 |
| 27-April 17 | Thursday | 5.3 | 32 | 33 | 43 | 56 | 51 | 220.3 |
| 30-April 17 | Sunday | 0 | 32 | 38 | 37.9 | 66 | 47 | 220.9 |
| TOTAL | | 96.8 | 917.3 | 911.2 | 819.6 | 1629.1 | 1823.5 | 6197.5 |

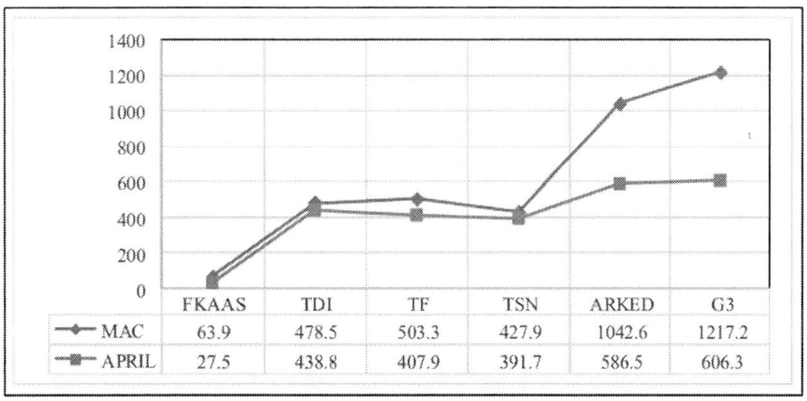


**Figure 2.** Comparison of food waste in March and April

*4.3. Amount of food waste generated in each cafeteria at UTHM*

Figure 3 shows amount of food waste generation in two month (March and April) at Dr. Munies, TDI, TF, TSN, ARKED and G3 respectively.

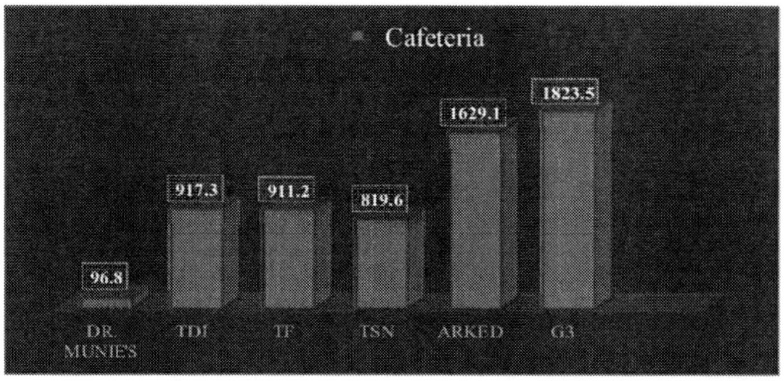


**Figure 3.** Amount of food waste generated in each cafeteria at UTHM

Figure 3 shows the total amount of food waste generated in March and April. 1823.5 kg of food waste collected in the G3's cafeteria was the highest in two months. The percentage of food waste collected at cafeteria Dr. Munie's was the least with only 96.8 kg. The study showed that students and staffs who come to take meals at the cafeteria Dr. Munie's was the least crowded. Based on the analysis carried out for two months, the overall amount of food waste generated in UTHM was 6197.5 kg. The total food waste generated at G3's cafeteria was the highest because a lot of students who come to take meals. Thus, much of the food produced in the cafeteria. Strategic places, diversity of food sold and reasonable price was also an among factors in generated of food waste in cafeterias. Next, the study also found that total amount of food waste generated in the Cafeteria Dr. Munie's produces was the least because most of the staff and students take meals at another cafeteria.

## 5. Conclusion

In conclusion, disposal of waste food in March was higher than in April. G3's cafeteria has recorded the most number of waste generated with the amount of 1823.5 kilograms (kg). While only 96.8 kilograms (kg) of waste generated in the cafeteria Dr. Munie's. Next, the data collected found that the overall amount of food waste generated by all six cafeterias for two months was 6197.5 kilograms (kg). With these data, other studies can be carried out in more detail and used as a reference. Based on previous studies, some methods of solid waste management have been identified including landfill, recycling, composting, incineration, inert landfill and sanitary landfill. According to waste generation waste produced in UTHM, composting method has identified the best and efficient method to manage food waste in six cafeterias in UTHM. This method was recommended as the best solid waste management for the recovery of organic waste. Further, this method cannot completely solve the problem of solid waste, but it will be able to move a number of waste mainly organic waste from landfill and thereby reduce the problem of organic waste. By using this propose a baseline figure for food waste generation in UTHM cafeterias, UTHM can save cost of waste management and also can generate new revenue.

## 6. References

[1] Perbadanan Pengurusan Sisa Pepejal dan Pembersihan Awam (PPSPPA) 2012 *Definisi Sisa Pepejal* Jabatan Pengurusan Sisa Pepejal Negara
[2] Sinar Harian 2009 *SWM Edar 587 Tong Sampah 1,100 Liter* (2009, Julai 20)
[3] Jabatan Pengurusan Sisa Pepejal Negara 2013 *Jenis Sisa yang Terkawal* Jabatan Pengurusan Sisa Pepejal Negara (March 2013)
[4] Hasnah Ali, Dody Dermawan, Noraziah Ali, Maznah Ibrahim, Sarifah Yaacob 2012 Masyarakat dan amalan pengurusan sisa pepejal ke arah kelestarian komuniti: Kes isi rumah

wanita di Bandar Baru Bangi Malaysia *Geografia Malaysia Journal of Society and Space* **8(54)** 64-75

[5] Pichtel, J. 2005 *Waste Management Practices: Municipal, Hazardous and Industrial.* (Boca Raton, Florida: Taylor and Francis Group)

[6] Zaini Sakawi 2005 *Pengurusan sisa pepejal domestik di Mukim Ampang Jaya oleh Alam Flora Sendirian Berhad, Selangor* (Universiti Kebangsaan Malaysia, Bangi) 281-301

[7] Agamuthu P, Fauziah S. H. 2011 Challenges and issues in moving towards sustainable landfilling in a transitory country- Malaysia *Waste Management and Resources* **29** 13-19

[8] Kibert C. J., Thiele. L., Peterson A., Monroe M., Ethics. Sustain. (2011) 266.

[9] Hamatschek, E. 2010 Current Practice of Municipal Solid Waste Management in Malaysia and the Potential for Waste-to-Energy Implementation *ISWA World Congress 2010* (Hamburg)

[10] Periathamby A 2009 Evolution of Solid Waste Management in Malaysia: Impacts and Implications of The Solid Waste *Journal Of Material Cycles And Waste Management* **11(2)** 96-103

[11] Japan Environmental Sanitation Center 2013 *Solid Waste Management and Recycling Technology of Japan (Toward a Sustainable Society)*

[12] National Environment Agency and Ministry of the Environment & Water Resources Singapore (2006) *Integrated Solid Waste Management in Singapore*

[13] Ranjith Kharvel Annepu 2012 Sustainable Solid Waste Management in India (University Columbia, New York)

[14] Nguyen Ngoc U and Schnitzer H 2009 Sustainable solutions for solid waste management in Southeast Asian countries *Waste Management* **29** 1982-1995

[15] Jabatan Pengurusan Sisa Pepejal Negara 2012 *Lab Pengurusan Sisa Pepejal* Jabatan Pengurusan Sisa Pepejal Negara

**Acknowledgement**

This research is supported by Universiti Tun Hussein Onn Malaysia (UTHM) under grant VOT B077 and MyBrain 15. The authors fully acknowledged to UTHM and Ministry of Higher Education for give the facilities and the approved fund for which makes this important research viable and effective.

# Rethinking construction: inclusion of slow learners as taker-off in quantity surveying practice

**Masidah Abdul Majid[1], Norul Izzati M. Ashaari[1], Suhaida @ Suhana Kamarudin Nurul Aini Osman[1] and Mohamad Saifulnizam Mohd Suhaimi[2]**

[1]Department of Quantity Surveying, INTI International University
2 Department of Quantity Surveying, International Islamic University Malaysia.
Corresponding author: masidah.majid@newinti.edu.my

**Abstract:** The objective of this paper is to present the preliminary findings regarding the participation of OKU with learning disability in Science Technology, Engineering and Mathematics (STEM) sectors. Review of the works of past researchers suggested that OKU is a potential workforce in STEM sectors but still under-represented due to lack of efforts from stakeholders and learning institutions in providing information on the opportunities that are available. A research has been initiated to explore the potential of slow learners to become workforce in the construction industry as a taker off – part of work of a Quantity Surveyor. Against the findings from the literature review, the modest attempt to attract slow learners to become taker off in the construction industry require the formulation of appropriate learning environment and strong support from the respective key players and stakeholders.

## 1. Introduction

According to government estimate, there are approximately 1.3 million people in Malaysia with various forms of disability [1]. In Malaysia, people with disability is referred as *Orang Kurang Upaya* (OKU). However, Malaysian society has yet to give serious thought on equipping the OKU with certain skills so that they can have more opportunities and be gainfully employed. There is a perception that assistive technologies and building modifications in the workplace to meet the needs of people with disabilities is difficult and expensive. This notion is subject to further argument as there are also evidences captured from the US Department of Labour which suggest that job accommodations to accommodate the OKU is cheap and up to 58 per cent came at no cost at all [2]

Malaysia has embarked the implementation of policy of 1% job quota for the OKU in the public sector [3-4]. Subsequently, the private sector is also expected to follow-suit and absorb OKU in the workforce [5-6]. All these efforts are made to help OKU to be more self-reliant and obtain decent jobs in the market.

Nonetheless, giving opportunities to the OKU or attempting to build a diverse team is a challenge for a lot of employers, and the STEM fields are certainly no exception. Persons with disabilities have been under-represented in the science, technology, engineering, and mathematics (STEM) fields for many years [7]. Reasons for this include low exposure for students with disabilities on STEM courses, and poor career advice, support and networking with people form scientific industries [7-9].

The present study is initiated since the authors believe that OKU can get a better job if there are efforts from various bodies to change the societies' perceptions on the ability of the OKU so that they can work in various fields. This on-going research is a novel attempt in investigating the ability of OKU under the category of slow learners to become taker-off; which is part of the roles of Quantity Surveyor in preparing Bills of Quantity [10]. The authors acknowledge that there are many categories of OKU. The explanation of some categories of OKU was discussed in the next section. Some researchers refers the slow learners as person with below average cognitive level [11]. Hence, in this document, the term slow learners and person with below average cognitive level will be used interchangeably in which both carries the same meaning.

This paper represents the report of the initial finding of this on-going research whether there is a possibility to include OKU in general to venture in STEM fields. The outcome from this literature review will become a meaningful guide on the possibility of including OKU with below cognitive level as assistant to quantity surveyor in the construction industry of Malaysia. In carrying out this research, past studies and literature with the key words of OKU, slow learners, handicapped person, and person with below average cognitive level and handicapped person in STEM fields were searched, reviewed and presented in this paper.

## 2. Overview of disability

There are various ways in categorizing disability or OKU. According to the United Nations [12], disability results from the interaction between persons with impairments and attitudinal and environmental barriers that hinder their full and effective participation in society on an equal basis with others. Oxford Dictionary [13] define the term disability as physical or mental condition and the state of not being able to use a part of the body completely of easily or mental disability means is not be able to learn easily. The Malaysian Disability Act 2008 and United Nations (2017) defined "persons with disabilities" include those who have long term physical, mental, intellectual or sensory impairments which in interaction with various barriers may hinder their full and effective participation in society and equal basis with others[12,14]. This definition further explained disability in relation to a person as a substantial restriction in the capacity of the person to carry on a profession, business or occupation in the State or to participate in social or cultural life in the State by reason of an enduring physical, sensory, mental health or intellectual impairment. The categories of disabilities as stated in Jabatan Kebajikan Masyarakat (JKM) [15] are as the followings.-

### *2.1 Hearing*

Hearing Disability means cannot hear clearly in both ears without the use of a hearing aid or not to hear directly even with the use of hearing aids.

### *2.2 Vision*

Disability Visibility means both blind eyes blind in one eye or vision in both eyes or etc. permanent vision disorders.

### *2.3 Speech*

Speech Disabilities means they cannot speak up cause interference to communicate properly and cannot be understood by those who interact with it. This condition is permanent or will not heal. For children must be based on the assessment at the age of five years.

### *2.4 Physical disability*

Physical Disability means the inability to remain members of either body caused by loss or inability absence of any member of the body that can affect their function in the fundamental activities. Activity is defined as basic as personal care, movement and body position changes. This condition can occur as a result of injury (trauma) or disease in any of the nervous system, cardiovascular, respiratory, haematology, immunology, urology, hepatic, musculoskeletal, gynaecology and others that cause malfunction

*2.5 Learning problems*
Learning problems means intelligence problems that are not consistent with biological his/her age. Those who fall into this category is Late Global Development, Down Syndrome and intellectual disabilities. This category also includes conditions that affect individual learning capabilities such as autism (Autistic Spectrum disorder), Attention Deficit Hyperactivity Disorder (ADHD) and specific learning difficulties such as (dyslexia, dyscalculia and Dysgraphia).

*2.6 Mental disability*
Mental Disability refers to a state of severe mental illness makes a person unable to function either partially or fully in matters pertaining to his or relationships in society.

*2.7 Multiple disabilities*
Multiple Disabilities has more than one type of disability, and is generally not appropriate to be classified in the above category in 2.1 to 2.7.

Upon synthesizing past literatures, the characteristics of slow learners conform with the definition of JKM under the category of 'learning problems'. Slow learners are the group of peoples with below cognitive capacity, poor memory, capability of having distraction, lack of concentration and inability to express ideas. The slow learners relatively unable to do work normally expected of their age group [16]. In Richard (2005) slow learners were described as group of people who are not able to do well in a particular project that sometime appear immature in interpersonal relationships, and often times do not understand simple skills while performing a task. In order to grasp new concepts, a slow learner may require more time, a number of repetitions, patience and often more resources from teacher or the trainer to be successful [17]. Slow learning young adults are capable of achieving further intellectual growth as such adolescents are become more sensitive to the felt needs of life and become more ready to accept help [16].

In psychometric terms, slow learners have a measured intelligence quotient (IQ) between 70 to 85%. Incidentally, IQ score should not be used as the sole indicator of diminished potential for the slow learner, as that determination is usually accompanied by myriad assessment evidence that include behavioral, academic, formal and informal measures [18]. Although their poor learning rate and limited intellectual capacity may not be a marked handicap in the employment, they will have realized that further education is essentials for promotions at works. This prove that they are the potential group of the people who are able to be guided with the instructional strategy to bring out the uniqueness in the individual to the surface.Variables of human ecological environment such as type of family, caste, education of family members, age, friends and neighboring hood exerted a powerful influence on the slow learners' multiple intelligence [19].

## 3. Background and prior work on OKU in STEM courses.

In general, the STEM field is hardly to be diversified and include people who are with disability in the workforce [8]. In Malaysia, the OKU are often portrayed by the local society as weak and could not afford to perform jobs as productive as other workers [20-21]. Nonetheless, findings from literature shows that the education system across countries either gives less attention or are still struggling and exploring the ways to increasing STEM workforce diversity that meets the needs of all concerned [8-9]. Hence, the scientific workforce remains primarily for people with a non-minority-dominated field, especially as technician or professional levels [8]. Against this background, the main issues concerning the inclusion of disable person in STEM courses or careers is how to provide appropriate information, learning environment and support system so that the OKU can obtain more opportunity to venture in STEM field. Such effort, although minimal will manage to open up opportunity for OKU in STEM field.

The inclusion of persons with disabilities in STEM career need a lot of effort especially from the stakeholders and education institution. In a study [8], the researchers highlighted that there is still lack of efforts from the stakeholders and education institution in providing information on the opportunities with regard to STEM courses for persons with disability although some platforms have been established to attract more people with different ability. For example, Institute for Accessible Science is certainly concerned and provides a specific channel for disabled person to be in the field of biomedical science. The establishment of organization such as the American Association for the Advancement of Science Entry Point is another channel that is encouraging STEM students with disability to become professional workforce [8].

However, reaching and encouraging people with different ability requires more effort than just providing a channel or platform. Mankin and Rohde [8] further highlighted that mentoring and networking programs need to be established and yet there is difficulties in performing collaboration with key players in STEM fields. Hence, this community will remain under-represented should there is lack of effort from respective key players.

Another important study with regard to people with disability in STEM is done by Rule Stefanich [7]. The study again highlighted the need of strong commitment between teachers, stakeholders and education institution in providing support and appropriate learning environment to the slow learners to be in STEM courses or careers. He proposed nine main ideas in response to the central problem of supporting students with disabilities in STEM courses or careers, i.e. 1) Self-advocacy and knowledge of one's disability, along with possible effective accommodations, will help students succeed. 2) Teachers need high expectations for students with disabilities and should encourage their participation in STEM subjects early so that students have appropriate knowledge bases for pursuing STEM. 3) Teachers need to be creative and willing to make accommodations so that students with disabilities can reach the same objectives as other students. 4) Universal design strategies, inquiry, hands-on learning, and real-world applications will benefit all students. 5) Laboratory assistants need to facilitate access and learning. 6) Assistive technology allows students to more fully participate in lessons. 7) Teamwork and sharing between faculty and support personnel of effective practices is needed. 8) Internships and mentorships build confidence, teach knowledge of the field, and help with transitions to work. 9) Transitioning to community colleges, four year institutions, or work can be facilitated by appropriately educated support personnel.

The study by Lam et al. [9], gives important indication that people with disability have strong interest in venturing STEM field. An academic program which initiated to stimulate interest towards STEM courses has received positive feedbacks from both parents to and students with learning disability. although students with learning disability faced some difficulties in understanding some of the scientific areas, but that does not stop them from giving encouraging comments, feedbacks and instilling interest in STEM field. This indicate that there is high possibility to include OKU or disable person in STEM field should they were given the exposure and chances.

In terms of teaching and training, in general, past studies have shown that there are constraints experienced both by the students and instructors or teachers in teaching disable person with below average cognitive level in the classroom. The teaching of OKU with below average cognitive level must be done with different approach than the normal students. Researchers suggested that the teaching of disable person must be done in different pace and in a controlled class room. In Wettasinghe and Mazlan [22], the researchers suggested the approach of teaching person with learning disability and can be enhanced with support of Information and Communication Technology (ICT). ICT is found to able to students' understanding and later increase self confidence in learning.

## 4. Quantity surveyor and taking off

It is the focus of this study to include OKU with slow learning ability to be part of the workforce in the construction industry of Malaysia. In the economy of Malaysia, the construction industry has been recognised as a major employer that employed more than 1 million people per annum [23–25]. There are various fields in the construction sector and some may be appropriate for the slow learners.

Quantity surveying is one of a team of professional advisers in the construction industry which concerned with contracts and costs of construction projects [10]. A quantity surveyor (QS) estimates and monitors construction costs, from the feasibility stage of a project through to the completion of the construction period. They work closely with architects, financiers, engineers, contractors, suppliers, project owners, accountants, insurance underwriters, solicitors and with all levels of government authorities. One of the tasks of the QS is to prepare bill of quantities which involve the measurement or "taking off" quantities of various items of work from design drawings prepared by the project architect and engineer [10].

In relation to the construction industry, the term taking off can be presumed to be specific to one of the roles of QS. It can be described as an operation which consists of taking the dimensions of the various items from the drawings and entering them with their proper descriptions on "dimension paper" or so called "taking off paper" [7].

## 5. Conclusion and the way forward

This paper represents the report of the first finding of this on-going research whether there is a possibility to include OKU in general to venture in STEM fields. Specifically, the present study undertakes to perform a preliminary study on the potential of OKU under the category of slow learners to perform part of the job of QS. While the job description of QS encompasses various dimensions, the present study is limited only on the role of QS to do taking off.

Based on the reviews which has been discussed above, it was found that the research pertaining to OKU in general is still at infancy stage. Nonetheless, the existing research posit that attracting OKU in STEM is a challenging task and require and in-depth study on the teaching module and establishing collaboration with respective stakeholders and learning institutions.

Some minimal attempts have been done by past researchers in providing opportunities for disabled persons to join STEM courses. These attempts are positive signs that disable persons can become part of workforce in STEM fields. The present research will need to establish the appropriate modules as well as in depth study on the capability and capacity of slow learners in capturing the modules so that they can have options to become part of the construction industry players. In the next phase of the study, the team of researchers will identify students who are slow learners and further explore methods to equipped them with the knowledge as taker - off.

## 6. References

[1] Khor H T 2010 *Turning Disability into a National Asset*, Penang Economic Monthly September 2010 pp16-20.

[2] Kurt D 2016 *6 steps to a More Inclusive Workforce* Retrieved on July 15 2017 from https://jobs.newscientist.com/article/6-steps-to-a-more-inclusive-workforce

[3] BERNAMA 2013 *1754 OKU Bekerja di Sektor Awam*Utusan Malaysia Kuala Lumpur pp 10 (October 10).

[4] Jabatan Perkhidmatan Awam 2010 *Pencapaian Dasar Satu Peratus Peluang Pekerjaan dalam Perkhidmatan Awam kepada Orang Kurang Upaya* Malaysia

[5] BERNAMA *Tengku Adnan Urges Private Sector to Absorb OKU into Workforce*The Sun DailyKuala Lumpur (October 12, 2014)

[6] Jabatan Perdana Menteri 2012 *Syarikat Korporat Perlu beri Peluang OKU Bekerja* Retrieved on 30 June 2017 from http://www.jpm.gov.my/userfiles/file/Syarikat korporat perlu beri peluang OKU bekerja.jpg.

[7] Rule A C and Stefanich G P 2012 Using a thinking skills system to guide discussions during a working conference on students with disabilities pursuing STEM fields *J. STEM Educ. Innov. Res.*13(1) pp 43–54

[8] Mankin R W and Rohde 2014 Differently Abled in Entomology pp 154–155

[9] Paul C M, Lam C, Doverspike D, Zhao J and Zhe J 2008 An Evaluation of a STEM Program for Middle School Students on Learning Disability Related IEPs *J. STEM Educ.*9(1-2) 21–29

[10] Ashworth A, Hogg K and Higgs C 2013 *Willis's Practice and Procedure for the Quantity Surveyor*, 13th Editi. (West Sussex: John Wiley & Sons)

[11] Carroll S 1998 Slow Learners' in the Regular Classroom: A Handout for Teachers *National Association of School Psychologists* Bethesda pp 205–206

[12] United Nations 2017 *What Is Disability And Who Are Persons With Disabilities?* Retrieved on

Jun 13, 2017 from http://www.un.org/esa/socdev/enable/faqs.htm
[13] Stevenson A 2007 *Oxford Dictionary* 2007 Internatio. New York (Oxford University Press)
[14] Laws of Malaysia Act 685 2008 *Laws of Malaysia Persons With Disabilities Act 2008*Pers. with Disabil. Act 2008, no. 24 January, p Act 685
[15] Jabatan Pembangunan Orang Kurang Upaya 2012 *Senarai Kemudahan dan Keistimewaan bagi Orang Kurang Upaya*
[16] Kusuma R R A and Reddy G L 2010 *Slow Learners: Their Psychology and Instruction* (1st Edition Delhi: Discovery Publishing House Pvt. Ltd)
[17] Hancock R S 2005 *The Wrong Direction: An Educator Speaks Out* (First Edit. Victoria: Trafford Publishing)
[18] Cooter K S and Cooter R B 2004 One Size Doesnt Fit All: Slow Learners in the Reading Classroom *Int. Read. Assoc.* **57(7)** pp 680–684
[19] Sangwan S and Duhan K 2010 Human Ecological Systems and Multiple Intelligence of Ecological Multiple Ow Learner Young Adolescents Slo *J. Dairy. Foods H.S* **29(3/4)** 216-221
[20] Yusoff N M and Wahat N W A 2013 *Sumbangan Keadilan Organisasi dan Persepsi Keterbatasan dalam Keboleh Upayaan Bekerja bagi Orang Kelainan Upaya (OKU) di Malaysia* (GREDUC 2013 no. Greduc) pp 162–165.
[21] Ariffin Z 2006 *Kerjaya untuk Orang Kurang Upaya* (Kuala Lumpur: PTS Professionals Publishing Sdn Bhd)
[22] Wettasinghe M and Hasan M 2008 Exploring the efficacy of IT with slow learners: case studies in primary schools *Proceedings of the 2nd International Convention on Rehabilitation Engineering & Assistive Technology* pp 254–257.
[23] Abdullah F 2004 *Construction Industry and Economic Development: The MAlaysian Scene*, (First Edit. Skudai: Universiti Teknologi Malaysia)
[24] Rashid K A 2002 *Construction Procurement in Malaysia, Process and Systems, Constraints and Strategies* (First Edit. Gombak: IIUM Press)
[25] Department of Statistics Malaysia 2017 *Labour Force Survey Report* Putrajaya

**Acknowledgement**
This research paper is truly contribution and guidance who belong to the list of authors. The authors pay homage to Jabatan Kebajikan Masyarakat for continuous encouragement, guidance and support throughout this work.

# The most common problem facing by the maintenance department: A case Study between Universiti Tun Hussein Onn Malaysia (UTHM) and Universiti Teknologi Malaysia (UTM)

**Mohd Norazam Yasin**[1,2], **Rosli Mohamad Zin**[2], **Abd Halid Abdullah**[1], **Muhammad Shafiq Mahmad**[1], **Muhammad Fikri Hasmori**[1]

[1] Faculty of Civil and Environmental Engineering, Universiti Tun Hussein Onn Malaysia, 86400 Parit Raja, Batu Pahat, Johor, Malaysia
[2] Faculty of Civil Engineering, Universiti Teknologi Malaysia, 81310 Johor Bahru, Johor, Malaysia
Corresponding author: norazam@uthm.edu.my

**Abstract.** From time to time, the maintenance works become more challenging due to construction of new building and also aging of the existing buildings. University buildings without any exception require proper maintenance services to support their function requirements and this can be considered as major responsibilities to be fulfilled by the maintenance department in the universities. Maintenance department specifically will face various kinds of problems in their operation works and thus this might influence the maintenance work operations itself. This study purposely to identify the common problem facing by the maintenance department and also to examine the current status of the maintenance department. In addition, this study would also propose any suitable approach that could be implemented to overcome the problem facing by the maintenance department. To achieve the objectives of this study, a combination of deep literature study and carrying out a survey is necessary. Literature study aimed to obtain deeper information about this study, meanwhile a survey aimed at identifying the common problem facing by the maintenance department and also to provide the information of the maintenance department's organization. Several methods will be used in analyzing the data obtained through the survey, including Microsoft Office Excel and also using mean index formula. This study has identified three categories of problem in the maintenance department, which are management problems, human resource problem, and technical problems. Following the findings, several solutions being proposed which can be implemented as the solution to the problem facing. These suggestions have the potential to improve the maintenance department work efficiency, thus could help to increase the department productivity.

## 1. Introduction

Nowadays, it's become a great challenge for a developing country to ensure that all the building's condition and quality are at the top so that their serviceability could be maintained and can fulfil their function requirements [1]. It is obviously necessary to adopt a good protection method or can be said as a building maintenance method. Building maintenance is a sector that most closely related to the building construction industry. Building maintenance practically been adopted after the completion of a construction project. In general, building maintenance can be defined as any practice that could assist in extending the lifespan of a building so that it could fulfil its function [2]. However the maintenance work for every building is not the same for each of the buildings has different kind of function. Determination of maintenance works is being influenced by the types of material used in construction and also the type of the structure itself. Obviously, the existence of buildings plays many important roles in human life nowadays. Regarding to this, it is vital to adopt the necessary maintenance for the building itself. It can be concluded that building maintenance as the 'support' for the construction industry. Good or bad condition of university building could describe the quality of maintenance service that being received. The responsibility of building maintenance is fully under the maintenance department. The effectiveness of maintenance management is closely related to the organization in the maintenance department itself [3]. From time to time, the maintenance work become more challenging since the public university keep expanding and developing rapidly. There would be problem or issues arising in the maintenance department, which could make the implementation of maintenance works become more difficult. This would likely influence the work performed by the department itself.

## 2. What is Maintenance?

Maintenance is defined as "the combination of all technical and administrative actions, including supervision actions, intended to retain an item, or restore it to, a state in which it can perform a required action [4]. Another definition of maintenance also being offered by the Chartered Institute of Building, which is "work undertaken in order to keep, restore or improve every facility, every part of the building, its services and surrounds, to an agreed standard, determined by the balance between need and available resources". Meanwhile The Committee of Building Maintenance recommended the adoption of the maintenance definition as follows, "Work undertaken in order to keep, restore or improve every facility, every part of the building, its services and surrounds to a currently acceptable standard and to sustain the utility and value of the facility" [5].

## 3. Maintenance Organization

Organization can be defined as a group of people who must coordinate their activities in order to meet organizational objectives [6]. Meanwhile, other definition for organizing is the process of arranging resources include peoples, materials and technology together to achieve the organisation's strategic and goals [7]. By the way, organization structure is referred as the way in which the various parts of an organization are formally arranged. The structure of the organization is determined by various factors such as technology and its rate of change, complexity, resource availability, product and/or services, competition, and decision making requirements.

## 4. Methodology

This study is continues with deeper information search through literature study. This information was gathered from two main sources. Firstly, through the external references such as internet, books, articles, and also journals. The questionnaire was built from two objectives of this study. Before that pilot survey and expert pilot was done to validate the questionnaire by interviewing selected industry professionals and academic staffs. A total of respondents were participated in this research. The questionnaire was developed and distributed among the targeted respondents.

### *4.1. Pilot Survey*

Pilot study was performed just before the final draft of questionnaire draft had been approved. It is important because the questionnaire need to be tested on how long it takes to complete the questionnaire, ensuring all the questions and instructions are clear and also to review the suitability of the question.

*4.2. Expert Pilot*
This part is necessary where the questionnaire draft is presented to the expert in the building maintenance field such as professional engineers and lecturer. This is to gain their opinion and advice in improving the questionnaire form. There were two professional engineers and two lecturer whom experts in building maintenance selected for the validation of questionnaire draft. This is vital in identifying any error and also any unsuitable parameter in the questionnaire form.

*4.3. Site Visit*
During this phase, it is crucial to gain as much information as can. Through the site visit, a brief description of the organization can be obtained besides interviews and questionnaire distribution could be done at the same time. In this study the site visit was carried out at two selected public university campuses in Johor which comprise of UTHM and UTM.

*4.4. Questionnaire*
A set of questionnaires to be prepared in advance, in which it aims is to make the collection of data necessary. This form is designed based on the information obtained from the study of literature and also based on the research objectives. Views from various urgently needed, especially supervisors who had guided so that the data obtained will help in realizing the objectives of the study. The questionnaire was created and is broken down into three parts which is section A, B and C. Before the questionnaires were distributed, a pilot study was carried out and also expert pilot. Two lecturers were asked to validate the questionnaire during the expert pilot. Then, the questionnaire was distributed to 80 respondents.

*4.5. Data Analysis*
The mean index will apply because it presents the frequency and percentage of results in tabular form of Likert scale.

## 5. Results and Discussions

*5.1. The current practice of maintenance department.*
This could be achieved by gaining the information through the observations to maintenance department. Among the information that need to be gathered are organization structure, standard operation procedure (SOP) in maintenance works, mission and vision, objective of department and also the scope of work. There are a few discussions that have been conducted with the top management from both universities maintenance department.

*5.2. The most common problem facing by the maintenance department.*
Figure 1 shows that maintenance department at UTHM facing a serious problem in human resource. The mean index value of elements in sub-section of the human resource problem are higher than other sub-section. Based on the figure, the highest mean index value is 3.76. The respondent agreed that the employees always been given the job/task that are not suitable with their skill and lack of training provided by management. These two element are closely related where the employee skill can be influenced by the training that being received. Due to lack of training provided, the employee's skill cannot be enhanced and developed.

IOP Conf. Series: Materials Science and Engineering **271** (2017) 012047 doi:10.1088/1757-899X/271/1/012047

**Table 1.** Comparison between maintenance departments at UTHM and UTM.

| Element | UTHM | UTM |
|---|---|---|
| **Organization Structure** | Of line and staff | Of line and staff |
| **Mission, vision, and objective** | Provide efficient maintenance management and best service in maintenance works | Manage facilities efficiently to provide excellent and sustainable facilities |
| **Staff Population** | 43 | 134 |
| **Scope of Work** | Civil<br>Electrical<br>Mechanical | Civil<br>Electrical<br>Mechanical<br>Renovation & space<br>Landscape<br>Furniture |
| **Work Area** | Approximate 600 acres | 2400 acres |
| **Contract** | In house & outsource ng | 100% In house |

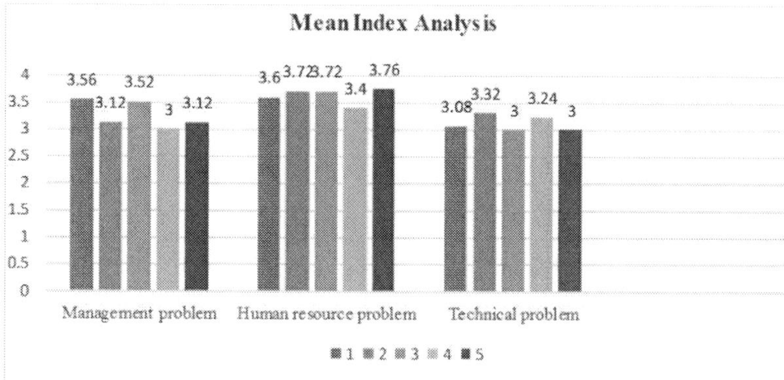


**Figure 1.** The most common problem facing by the maintenance department at UTHM.

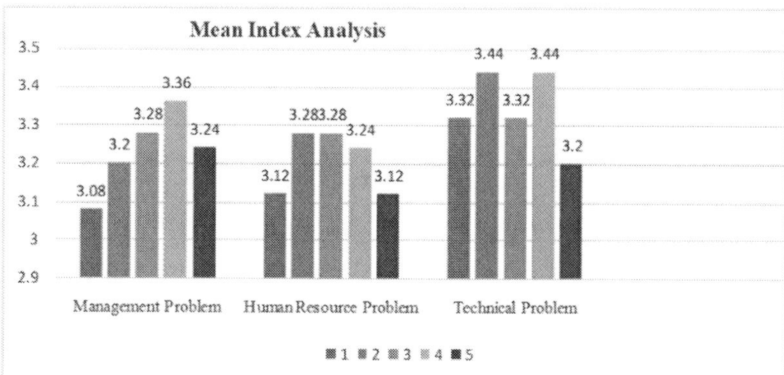


**Figure 2.** The most common problem facing by the maintenance department at UTM.

Based on figure 2, maintenance department at UTM facing a serious problem in technical problem. Elements in the sub-section of technical problem have the highest mean index value which is 3.44. The tools and materials were not proper and outdated and lack in quality performances of the maintenance work. Besides that, maintenance department at UTM also faces the management problem since the mean index value in the sub-section of management problem also high which is 3.36. Respondent agreed that the top management failed in planning the maintenance department budget wisely.

*5.3. The solution for the common problem facing by the maintenance department.*

From the figure 3, respondent at UTHM agreed that for the management problem, the organization must be led or managed by the expert in the maintenance operation. Meanwhile, in human resource problem, where the respondent agreed that the maintenance organization should provide the adequate facilities to give training suitable with the employee scope of work. Both elements obtained the highest mean index value which is 4.08. This two solutions will perceive the best to give a good work performance of maintenance at UTHM.

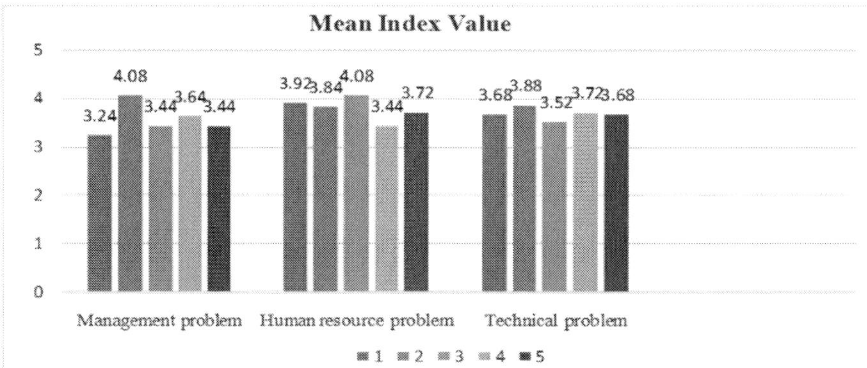


**Figure 3.** The solution for maintenance department at UTHM.

From figure 4, the respondent in UTM agreed that for managing problem, the maintenance organization should be led or managed by the expert in maintenance operation. For human resource problem, the respondent agreed that it is vital for the employees to be clearly explained about their scope of work and also prepared the work instructions for them. Meanwhile, for technical problems, the respondent agreed that in identifying any external resources of materials, proper planning needs to be done if emergency maintenance is highly recommended. Those elements obtained the highest mean index value which is 4.00.

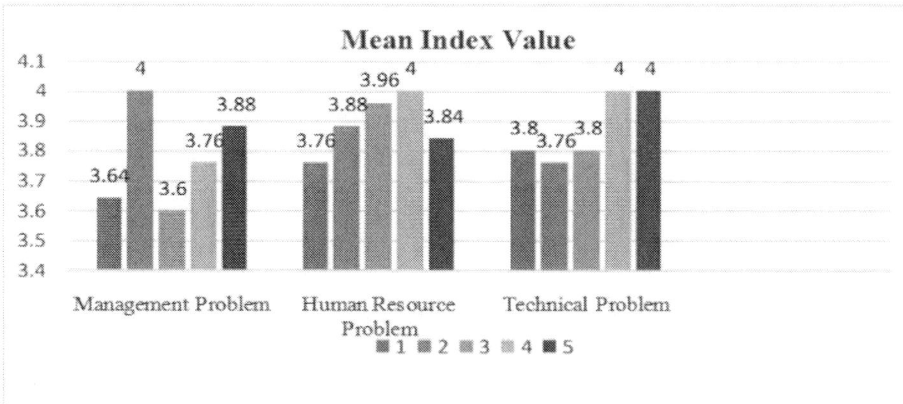


**Figure 4.** The solution for maintenance department at UTM.

## 6. Conclusion

A study was conducted through two public university maintenance department, which are UTHM and UTM to identify the common problem facing by the maintenance department. A site visit and observations were conducted to obtain the information about current practice of the maintenance department. A deeper literature study was also conducted to understand the building maintenance operation and also about the maintenance organization. Through the study the common problem facing by the maintenance department can be divided into management problem, human resource problem, and technical problem. Based on the data results obtained, the objectives of this study managed to be fulfilled. The common problem in maintenance department is important in identifying the critical factor of the problem arising in the maintenance department. The result of this study can be used by the concerned parties to identify any lacking in their organization.

## 7. References

[1] Hani Muftah 2010 Faulty *Design and Construction Factors In Building Maintenance* Master Thesis (Johor : Universiti Tun Hussein Onn Malaysia)

[2] Barrie Chanter & Peter Swallow 2007 *Building Maintenance Management Second Edition* (Oxford, UK: Blackwell Publishing Ltd)

[3] Norazam, M Abdullah A H Singh B and Abd R 2014 *Maintenance Organization at Universiti Tun Hussein Onn Malaysia* September 2016

[4] British Standard Institution: BS 3811 *1993 Glossary of Terms Used In Terotechnology* (London:United Kingdom)

[5] Great Britain. Commitee on Building Maintenance, Department of the Environment Directorate General of Development 1972 *Building Maintenance The Report Of The Committee* (London: H.M. Stationery Off)

[6] Harold Kerzner 2009 *Project Management: A systems approach to Planning, Scheduling, and Controlling 10th Edition* (New York : John Wiley & Sons, Inc)

[7] J Templer J Cattaneo D Cenzo and P Robbins 1999 *Human Resource Management. Canadian Edition* (Ontario, Canada: John Wiley & Sons, Ltd)

IOP Conf. Series: Materials Science and Engineering **271** (2017) 012048 doi:10.1088/1757-899X/271/1/012048

# The importance of cleanliness in a proper construction site management in malaysia: a contractor's perspective

**M A Shazwan[1], J V Quintin[1], N A Osman[1], S K Suhaida[1] and M I N Ma'arof[2]**

[1] Department of Quantity Surveying, Faculty of Engineering and Quantity Surveying, INTI International University, 71800 Nilai, Negeri Sembilan
[2]Department of Mechanical Engineering, Faculty of Engineering and Quantity Surveying, INTI International University, 71800 Nilai, Negeri Sembilan
Corresponding author: amirshazwan.hashim@newinti.edu.my

**Abstract.** Construction site's cleanliness and tidiness is one of top main concerns of construction site management. A good site management with adequate planning in regards to housekeeping will ensure safety to both the site's working personnel and the neighbouring environment. This is especially of priority today due to the rapid growth of construction projects in Malaysia. Nevertheless, to date, statistics had shown that housekeeping related accidents happened repeatedly despite the awareness on site's cleanliness and tidiness. The objective of this study was to explore constructor's perspective on site cleanliness and tidiness. A set of questionnaire was distributed to thirty-four (34) Grade 7 CIDB contractors' firms in Petaling Jaya. Petaling Jaya was chosen since this area are developed area covered with residential area and cleaning management an importance issues. The goals of the survey study were to identify the following items from the perception of the contractors: (i) the hierarchy of importance of several purposes in ensuring construction site's cleanliness and tidiness, and (ii) the risk factors that influence a construction site's level of cleanliness and tidiness. It was found that from the contractor's perspective, ensuring a site cleanliness and tidiness is of importance mainly due to the need in protecting the environment, whilst, the least in cost saving. In addition, poor personnel's working attitude was found to be the main risk factor that will influences a construction site's level of cleanliness and tidiness. Conclusively, construction site's cleanliness and tidiness is highly vital. Even so, for nationwide awareness on the topic matter will requires full cooperation from all related parties.

## 1. Introduction

According to Gurmeet [5] has predicted that in the construction industry in Malaysia will experience a considerable growth in the year 2017. This forecasting is based on the expectation that construction will be a strong pipeline to the country in achieving the National Transformation 2050 (TN50). Indeed, the construction industry covers a wealth amount of scopes, yet, this study is emphasizing specifically on construction site's cleanliness and tidiness. Aside from the obvious dire effects to the environment, this study strongly stress that construction site's cleanliness and tidiness is of high importance due to the fact that insufficient housekeeping practices in construction site may lead to potential hazards to workers.

Statistics by DOSH [14] has shown that major cases of mortality are considerably attributed to the construction industry in comparison to other occupational industries. DOSH [14] has indicated that from January to October 2016 alone, there are 58 deaths have been reported in the construction sector due to accidents. It was noted by the Health and Safety Department that the highest numbers of hazards in the construction sites are slipping and tripping – all related to housekeeping [17]. In addition, Datuk Lokman Hakim Sulaiman – the Health Deputy director-general (public health) noted on the increase of dengue cases in the construction sites as of late. This is also attributed to poor housekeeping [15] since stagnated rainwater in the construction sites has becomes the breeding ground for aedes. This is evidence when several construction sites were instructed to halt their operations until proven safe due to the infestation of aedes [15].

**Table 1:** Important of Site Cleanliness

| Descriptions | Authors |
|---|---|
| 1) Cost saving | Abdul Hamid and Singh [1] |
| 2) Improve work productivity | Roth [10], Federated [17], Bates [2], Watson (2014) & S.W.Poon et.al [11] |
| 3) Good company image | Bates [2]), Watson (2014) & Roth [10] |
| 4) Neighborhood safety | Roth [10] |
| 5) Protect the environment | Nagapan [7] & Bates [2] |

The literature such as shown in Table 1.0 has noted that by keeping the construction site to be clean, the particular company could also save on cost, improve work productivity and illustrating good company image, whilst simultaneously ensuring the neighbourhood's safety and protecting the environment. For instances, a clean working environment will facilitate in keeping the working personnel's level of health. Therefore, cases of the degradation or the drop of working performance and absentee due to medical conditions could be reduced, thus, maintaining the overall productivity of the organization. Ahmadon [6] further added that proper construction site cleanliness can improve the image of Malaysian construction industry. This study highly agreed to the suggestions and ideas which were presented by these studies.

### *1.1. Importance of Site Cleanliness*

The Social Security Organization (SOCSO) [18] has acknowledged that the fatality rate in the Malaysian construction industry is than three times higher in comparison to other industries. By reaching over than RM650 millions of compensation cost is a high number of value. Furthermore, this figures do not include the hidden cost of accidents which is usually thirty-three to eighty percent more than the direct cost, thus, summing into billions of ringgits [1] This compensation costs are indeed high, hence, actions need to be taken in overcoming the issue right from the root.

As mentioned by Roth [10], a construction project can be successful if housekeeping is made as one of their priorities in the construction site. Federated [17] commented that it is essential to have a good housekeeping program in the construction site to ensure a safe working environment. Bates [2] highlighted that when there are scraps or construction wastes being scattered at the construction sites un-attended, these items will create obstacles for the workers in being mobile. Suggestively, these obstacles may lead to case such as tripping and slipping as highlighted by Watson (2014). In timely clearing up construction site from trash and scraps, the overall working environment safety could be controlled. This notion is fully supported by Roth [10] and S.W. Poon [11]. The latter on their factor analysis on safety performance chapter 3.4 (d); which Tidy site: The tidiness and cleanliness of sites are important for safety. As mentioned by Nagapan [7], sorting out construction material to recycle and re-used certain material such as timber and metal will support on environmentally friendly construction works, which is good for sustainability. This can help promoting environmental friendliness scraps and supporting for the green building [10].

Moreover, [2] emphasized that a clean and tidy construction site will leave a good impression and protects the contractor's image. This is because, a clean working environment is a good tool for the company to illustrate their efficiency with the construction work to their potential customers (Watson, 2014). This is supported by Bates [2] upon pointing out that if the working tools and materials are not systematically kept, this will add to the need in relocating these items upon necessity. Therefore, as a whole, these unsystematic working environment will result in work deficiency and prolonging the overall time span for the project – which is of course, a negative setback.

Roth [10] also accentuated that contractors who keep their site clean will not only create a good working environment to their workers, but, also ensures that the community who lived close to construction site are not negatively affected by the construction works – physically, physiologically and psychologically.

### *1.2. Causes of Poor Site Cleanliness*

There are various factors that contribute to the occurrence of poor construction site cleanliness and tidiness. Firstly, is poor material usage and storage system. According to Wyatt [13], ordering error such as overestimation, is a factor which affects the site condition. The material which are not immediately utilized for construction will have the need for storage or allocation space. If this is not provided, the materials are left scattered around in construction site and this is a case of poor material handling [12]. Phu [9] has stated that the frequency of material being used also affects the flow component that provides the material movement and placement during the construction process. Hence, the management must have a systematic approach in controlling the in-flux and out-flux of construction materials during the entire course of the project. Else, the construction site may face issues in relation to the inventory.

Secondly, is poor waste management system. Managing the construction site cleanliness need a good waste management system. However, this involves a cost to hire a waste management specialist. This is especially true for waste material involving soil and oil material. As noted by Permana [8], the lack of site inspection and technology is among the factors that lead to overall poor construction site waste management. The lack of encouragement from construction related agencies to give incentive for contractors to support waste management and keep the construction site clean [3] may lead to the contractors to proceed with the waste management themselves as they see fit. According to Bukhari [4] however, based on the previous survey there is little attention to cleanliness on the construction site from the organization senior management in Malaysia. If little to no amount of commitment from the organization in managing construction waste, this issue will never be resolved. can cause on poor site cleanliness. Due to the lack of programs, campaigns and poor waste management technology implemented, this resulted in the increasing number of hazard in construction industry in Malaysia.

According to S.W. Poon [11] there several more factors that affect construction site's cleanliness and tidiness. The other factors are: (i) time schedule of projects, (ii) inefficient communication due to multi-level subcontracting system, and (iii) linear penalty for not implementing safety regulation. All of these factors will in a way influence the overall achievement of construction site's cleanliness and tidiness either independent or cooperatively. For instances, a very tight times schedule will require the organization to be fully committed to completing the project, thus, hindering the company from having adequate time for housekeeping. Communication breakdown will definitely be a hindrance for housekeeping since the messages in ensuring construction site's cleanliness and tidiness might not be delivered to the responsible individuals. Furthermore, the penalty given to the contractor for not complying to the health and safety with respect to cleanliness and tidiness may just be too linear or minimal for the contractors to take immediate actions in contrasting to the penalty for delay or incompletion of the project. Yet, this study strongly stressed that the main factor which contributed the most to construction site hazards in Malaysia is the overall poor working attitude of the construction site personnel. This particular factor will affect and infects the entire system of the organization's construction site management i.e. not only in regards to construction site cleanliness and tidiness. Poor working attitude could only be resolved via education/training and continuous monitoring.

## 2.Research Methodology

A survey study was conducted in gathering the data for this study. Questionnaire was distributed by emails to all the contractors whose name are listed in the CIDB Grade 7 contractor list. G7 was chosen since these contractors usually involve in big and high scale project. All of the contractor's main headquarters are also allocated in Petaling Jaya. A sampling size method had been used in order to determine the sample size for this research. As per listed in by CIDB, there are 159 contractor's registered in Petaling Jaya and based on random sampling size calculation, a number 60 respondent are needed in order to get 95% confident with 10% margin of error, however due to the time constraint, we only able to collect 34 samples out of 60.There questionnaire was divided to three (3) sections which were: (i) 'Part A - Background of Respondents'; (ii) 'Part B – The Important of Site Cleanliness and Causes of Poor Site Cleanliness', and (iii) 'Part C – Recommendation in Improving Cleanliness of Construction Site". Likert Scale was used for all of the questions in Part B and C.

## 3.Results and Findings

### *3.1. Respondents' Background*

A total of 34 respondents had responded in this research. Site Quantity Surveyor at 70% has the highest percentage of this survey participation. This is followed by Project Manager and Site Manager at 15% and 9% respectively, and finally by safety officer at 3%. In regards to working experience, only 3% out of the respondents are having more than 15 years of working experience, 26% having 10 to 15 years of working experience, 15% having 5 to 10 years working experience, whilst, mostly at 56% having less than 5 years of working experience. It was also noted that at 76%, most of the respondent knows about site housekeeping.

### *3.2. The Importance of Site Cleanliness in Construction Site*

The result of the importance of site cleanliness in construction site rated by the respondents are shown in figure 1. It was apparent that the respondents strongly agreed that site cleanliness is important at the utmost level in order to protect the environment, as our construction industry had been highlighted to adopted green building as stated in Malaysian Construction Industry master plan (2005-2015) [16]. Only then it is followed by neighbourhood safety, good company image, improve work productivity. However, the means are slightly lower in terms of keeping the site clean can help in cost saving. In short, for the respondents, maintaining the environment as the current state that it is, is more important than housekeeping cost and compensation cost that need to bear by the contractor due the injuries. This finding is such a great welcome.Line thickness

In general, try to avoid extremely fine lines (often called 'hairline' thickness) because such lines often do not reproduce well when printed out—your diagrams may lose vital information when downloaded and printed by other researchers. Try to ensure that lines are no thinner than 0.25 pt. Note that some illustrations may reduce line thickness when the graphic is imported and reduced in size (scaled down) inside Microsoft Word.

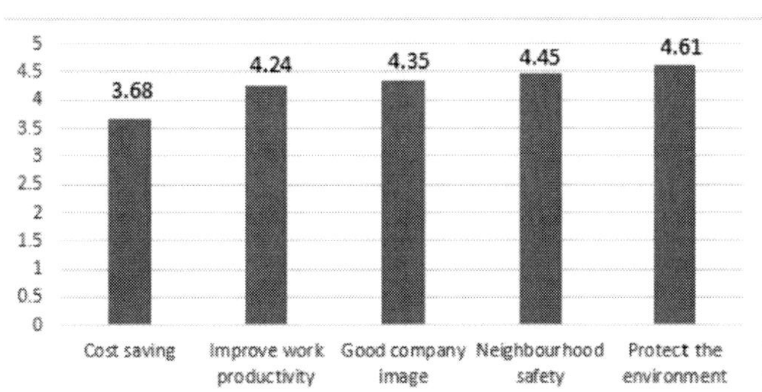


**Figure 1:** Important of Site Cleanliness

*3.3. Factors Affecting the Site Cleanliness Issues that Leads to Poor Site Cleanliness*

The result and analysis on the factors affecting the site cleanliness issues in Malaysian construction site can be referred to figure 2. From the respondents' feedback, top five highest mean responses of factors that caused poor site cleanliness are: (i) poor attitude of workers, (ii) poor supervision, (iii) no commitment from the organization, (iv) lack of construction waste management technology, and (v) lack of inspection on site. The rest of the factors which are ranked lower, yet, still agreed upon are: (i) ordering error, (ii) lack of experienced worker, (iii) language barrier between supervisor & workers, (iv) cost of managing construction waste, and (v) lack of staff for site cleanliness. Based on the five highest factors, most of the poor cleanliness affected by the individual factors due to the non-encouragement and guidance related to cleanliness on site . Therefore, by having adequate information on the causational factors that will influence construction site's cleanliness and tidiness, preventive measures could be taken by the contractors in overcoming the issue.

Captions should be below the figure and separated from it by a distance of 6 points—although to save space it is acceptable to put the caption next to the figure. Figures should be numbered sequentially through the text—‘

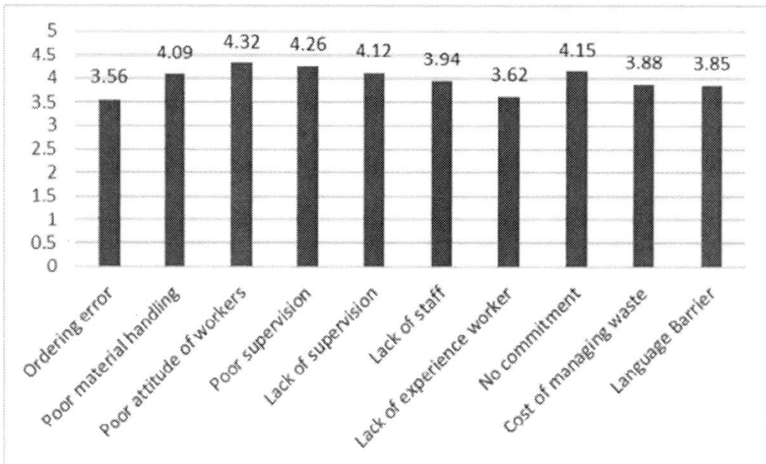


**Figure 2**: Factors that lead to poor site cleanliness in construction site.

*3.4. Suggestions and comments on how to reduce site cleanliness issues.*

Tables should be centred unless they occupy the full width of the text.

*3.5. Tables in parts*

The suggestions and comments were filtered and ranked accordingly by the frequencies of the responses. The strategies to reduce the site cleanliness problem at construction site were noted as follow:

1) Awareness campaign on site cleanliness
2) Enhance and incentive to contractor firms who keep site clean
3) Enhance and enforce 5S to site
4) Government enforces laws emphasizes on housekeeping
5) Housekeeping team should clear up the place after work
6) Proper waste disposal and effective recycling policy
7) Provide training course to worker
8) Regular inspection on site cleanliness
9) Worker should be briefed about site keeping before work
10) Impose penalty to contractors who offend the cleanliness policy

11) Language barrier between supervisors and workers
12) Assign person to in-charge site housekeeping

Overall, there were 12 listed suggestions given by the contractors in ensuring construction site's cleanliness and tidiness. Nonetheless, as good as these suggestions are, this study emphasize that full-on cooperation at every level of the organization is vital in making sure that the suggestions could actually be fruitful. Without cooperation and support from every level of the organization, the suggestions will only be mere ideas without actual impact.

## 4. Conclusion

Conclusively, construction site's cleanliness and tidiness is highly vital. From the result, it is concluded that most of the contractors will maintain their construction site cleanliness in order to ensure that they do not harm the environment – both in term of residents and nature. Nevertheless, the factors that are most affecting the site cleanliness problems on site are because of poor attitude exhibited by the construction working personnel, poor supervision, the lack of commitment from every level of the organization, the shortage of construction waste management technology, and finally is the deficiency of inspection. There are several of proper ways that were suggested to help reduce the site cleanliness problems in Malaysian construction site. Nevertheless, for nationwide awareness and implementation of construction site cleanliness and tidiness will requires full cooperation from all related parties.

## 5. Reference

[1] Hamid A R A and Singh B 2003 *Hazard of Construction Site* Retrieved on January 26, 2017 from https://www.researchgate.net/publication/264622908_HAZARDS_AT_ CONSTRUCTION_SITES
[2] Bates L R 2015 *Construction site housekeeping: Why you want a clean workplace* Retrieved on January 26, 2017 from http://www.commonwealthtourism.com/construction-site-housekeeping-want-clean-workplace/ (Accessed: 26 January 2017).
[3] Begum R A, Siwar C, Pereir JJ and Jaafar A H 2006 Factors and Values of Willingness to Pay for Improved Construction Waste Management – A Perspective of Malaysian Contractors, *Journal of Waste Management* **27** 1902-1906.
[4] Bukhari A, Hassan S H, Ridzuan A R 2009 *The Effectiveness of Industrialised Building System (IBS) with Respect to Waste Minimisation Compare to Conventional Method* Final Year Thesis (UiTM Malaysia)
[5] Kaur G 2015 *Construction sector to be busy in 2016 with projects worth RM83bil.* Retrieved on January 20,2017 from http://www.thestar.com.my/business/businessnews/2015/12/29/ jobs-flow-to-drive-construction-sector/
[6] Bakri A, Zin R M, Misnan M S and Mohammed A H 2006 *Occupational Safety And Health (Osh) Management Systems: Towards Development Of Safety And Health Culture* Thesis (Universiti Teknologi Malaysia)
[7] Nagapan S 2013 *Study of Site's Construction Waste in Batu Pahat, Johor* Research Management and Innovation Centre (Perlis:University Malaysia)
[8] Permana 2007 *Construction Safety Practices in Batam, Indonesia.* Master Thesis of Science Construction Management (University Teknologi Malaysia)
[9] Phu N L and Cho A M 2014 *Factors Affecting Material Management in Building Construction Projects*
[10] Roth S 2005 *Clean Work Sites Help Keep Projects' Reputations Sparkling* ( Kansas City Business Journal)
[11] Poon S W, Tang S L and Francis K W W 2008 *Management and Economics of Construction Safety in Hong Kong,*Construction and Real Estate Series (Hong Kong University Press).
[12] Tan C K and Razak N A 2014 *Case Studies on The Safety Management at Construction Site* pp 90–108.

[13] Wyatt D P 1978. *Material Management, Part I. Occasional Paper* (England:The Chartered Institute of Building, Berkshire)
[14] Department of Occupational Safety and Health Ministry of Human Resources 2016 *Guidelines on Occupational Safety and Health in Construction Industry (Management)*
[15] Rajendra E 2016 "*Youngsters at high risk of dengue" 26 august 2016* Retrieved on April 1, 2017 from https://goo.gl/U5XQoN
[16] Kamar K and Hamid Z 2012. *Sustainable construction and green building: the case of Malaysia* Sustainability Today 167 15
[17] Federated 2007 *Housekeeping during Construction Operations* (Malaysia: ISO Services Properties, Inc. SOCSO Annual Report Social Security Organization 2000)

**Acknowledgments**
The authors acknowledge INTI INTERNATIONAL UNIVERSITY in giving full support.

IOP Conf. Series: Materials Science and Engineering **271** (2017) 012049 doi:10.1088/1757-899X/271/1/012049

# The effectiveness of Microsoft Project in assessing extension of time under PAM 2006 standard form of contract

**S K Suhaida[1] andZ D Wong[1]**

[1]Faculty of Engineering and Quantity Surveying, INTI International University, Nilai, Negeri Sembilan, Malaysia.
Corresponding author:suhaida.kamarudin@newinti.edu.my

**Abstract.** Time is equal to money; and it is applies in the construction industry where time is very important. Most of the standard form of contracts provide contractual clauses to ascertain time and money related to the scenarios while Extension of Time (EOT) is one of them. Under circumstance and delays, contractor is allow to apply EOT in order to complete the works on a later completion date without Liquidated Damages (LD) imposed to the claimant. However, both claimants and assessors encountered problems in assessing the EOT. The aim of this research is to recommend the usage of Microsoft Project as a tool in assessing EOT associated with the standard form of contract, PAM 2006. A quantitative method is applied towards the respondents that consisted of architects and quantity surveyors (QS) in order to collect data on challenges in assessing EOT claims and the effectiveness of Microsoft Project as a tool. The finding of this research highlighted that Microsoft Project can serve as a basis to perform EOT tasks as this software can be used as a data bank to store handy information which crucial for preparing and evaluating EOT.

## 1. Introduction

Delays are common yet severe issues encountered by professions in the construction industry around the globe. Delays can be defined as a time period which exceeds the completion date stated in the contract initially [1].

Unfortunately, most of the standard form of contracts fail to deliver clear and precise outline on the preparations and assessment of Extension of Time (EOT) application[2]. As in PAM (Pertubuhan Arkitek Malaysia) Contract 2006 (With Quantities), provisions of EOT are given under Clause 23 whereby details and information regarding to EOT are stated such as submission of notice and particulars for EOT by the contractors (Clause 23.1), time period given to contractor to submit claims and architect to response to contractor's claim as well as "Relevant Events" that lead to delays in project (Clause 23.8). PAM 2006 Contract does not include standard guideline for the Architect or contract administrators to access EOT claims by the contractors[3]. Furthermore, there is no specific methods and procedures in preparing EOT claims stated for the contractors in the said PAM Contract 2006.

Hence, it fails to provide a fair basis for Architect and contract administrators to access EOT claims, instead, these claims are evaluated according to knowledge, practices, opinions and personal perspectives of the Architect and contract administrators. The absence of clear guideline has in fact induces vast number of issues while handling EOT application when contractors are not aware of the documents and supporting details required to claim for EOT effectively. Even if the contractors realize the type and extent of documents needed, they still fail to trace back the site records and submit the claim in time.

On the other hand, the utilization of project management software applications ease the management of the project by tracking and managing the records, data, critical path, schedules and plans; and other information [4]. Microsoft Project is one of the project management software tools available in the market and its useful features are somehow related to EOT and able to solve some of the issues. The main feature of Microsoft Project is acting as a record keeper, information such as project duration, on site progress, delays anticipated, resources, budget and other useful information are stored in the software and they can be traced and retrieved easily for contractual claims such as extension of time application.

## 2. Extension of Time (EOT) in PAM2006 standard form of contract

Delays are not favoured by parties involving in the construction projects including clients, stakeholders, architects, engineers, builders, contractors and other professions even labours. It is because delays are indirectly lead to overrun of time, budget, determination of contracts, abandonment of works and disputes [5]. At the same time, completion date stated in the contract does not solely represent the date when the contractor should complete the works, it also serves as a basis and provide rights to the employer to impose liquidated damages to the contractor upon non-completion works after the completion date [6].

According to Royal Institution of Chartered Surveyors (RICS) of United Kingdom, extension of time (EOT) is defined as additional timeframe provided for the contractor to complete his works after the stated completion date in the contract while the liquidated damages are not imposed to the contractor. However, the EOTwill only be granted for the contractor under certain circumstances and delays.

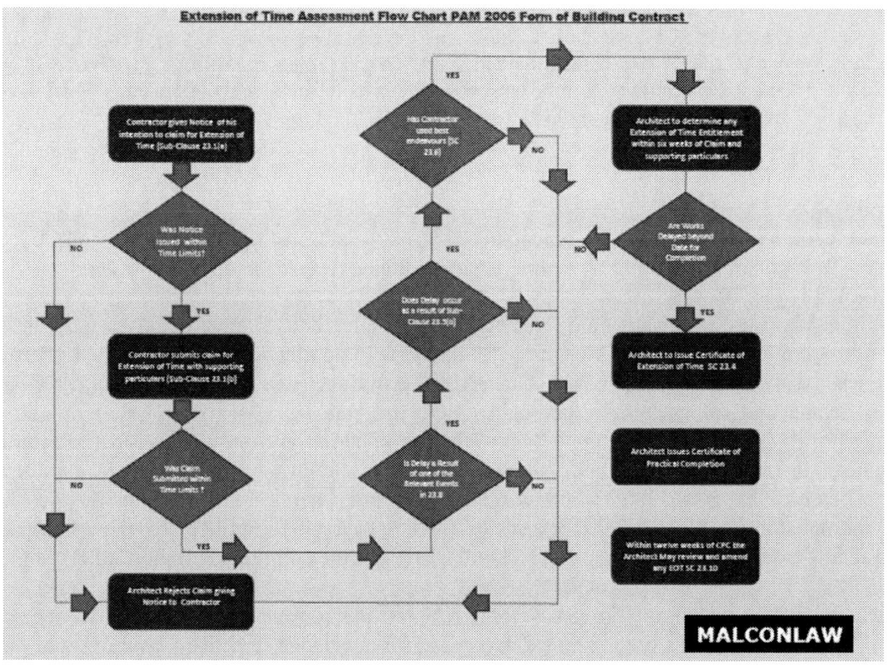


**Figure 1.** Extension of time assessment flow chart PAM 2006 form of building contract [7].

*2.1. Current practice of EOT assessment*

In the construction industry, the assessment of extension of time claims will be assisted with various type of delays analysis techniques. According to D.Aditi[8], there are five commonly used delay analysis techniques, as the followings:

**Table 1**. The delay analysis techniques.

| Techniques | Descriptions |
|---|---|
| As-Planned vs. As-Built | • An observational technique that compares the baseline or other planned programme to the as built programme[9] and among the preferable delay analysis method used [3].<br>• Does not scrutinize delay types and this makes it easy for it to be manipulated and distorted to reflect either the position of the claimant and defendant [2]. |
| Impacted As-Planned | • A method that measures the impact of the delays on the contractor's as-planned Critical Path Method (CPM) schedule, which can be used for analysis of delay during and after project completion. However, the original baseline programme may not be a realistic model on which to base the whole analysis and it has the potential of failing and lead to disputes [2]. |
| Collapse As-Built | • This technique consists of limitation as well. Similar with Impacted As-Planned technique, Collapse As-Built only emphasize on employer-caused delays while contractor-caused delays are excluded, hence, concurrent delays are not considered.<br>• It has limitation which does not consider the dynamic nature of the project's critical paths and highly subjective to manipulation [9]. |
| Window Analysis | • It is a dynamic delay analysis method that performed by using extracted schedule windows [9].<br>• However, it is time consuming and costly to operate and demands complete project records which often not available [2]. |
| Time Impact Analysis | • A technique that requires huge amount of information to be implemented and the as-planned schedule has to be prepared in critical path method (CPM) format andcausing the results to be inaccurate sometimes.<br>• Not practical or realistic to use if there are an overwhelming number of delay causing event, the analysis requires intensive effort and time consuming [2]. |

*2.2. Problems and challenges associating EOT claims*

EOT is equally important for all contracting parties, which to contractors a successful claim for EOT would absolve them from having to pay liquidated damages and to complete the project within the extended period; and to the employer, it would prevent time from being rendered at large [10].

Although there are provisions given in the standard form of contracts including PAM Contract 2006, however, disputes still occur commonly when dealing with extension of time issues. One of the main problems arise from the standard form of contract itself [3]. The issues associated with EOT claims also include the insufficient details submitted by the contractor in EOT claims, late submission of the claim and the failure of the claimants to collect relevant evidences on site. Table 2 shows the challenges and problems that associated with the EOT claims.

**Table 2**. Problems and challenges associated with EOT claims.

| Problems/ Challenges | Descriptions |
|---|---|
| Flaws in PAM2006 standard form of contract | • Absence of clear guidelines regarding the procedure of claiming and assessing EOT in the standard form of contract, as well as PAM 2006 Contract and it will lead to disputes[3],[11].<br>• Inexistence of agreed basis of EOT claiming and assessing procedures between contractor and the Architect/ contract administrator [10].<br>• Impropriate delay analysis methods. |
| Poor EOT claims' submission | • Failure in delays identification by contractor in order to provide sufficient supporting particulars[3], [12]. |
| Late EOT claims' submission | • Contractor failed to submit the EOT claim on time due to the time consuming process of compiling and keeping records[13]. |
| Failure in collecting relevant facts | • Contractor failed to retrace valid records and data regarding to the delays[13].<br>• Contractor failed to provide factual and valid evidences of the delays. |

### *2.3. Microsoft Project as project management tools*

Microsoft Project is a powerful project management tools designed for project managers and construction professions to supervise and manage the project effectively. It allows users to integrate information of data such as work resources, workers tasks, and visualizing reports of the project, supervise the budget and cost, create linking tasks and allow the planning and scheduling of the project in a single platform.

As the information and technology fields become advance these days, implementation of software applications in construction industry aids the project management routines and increase the successful project implementation rate. Meanwhile, software application such as Microsoft Project also reduce the time consumed to manage the project and the methods of managing the project are simplified into useful functions in the software[3].

Figure 2 shows the flow on how Microsoft Project act as a management tool which at the same time associates in assessing the EOT.

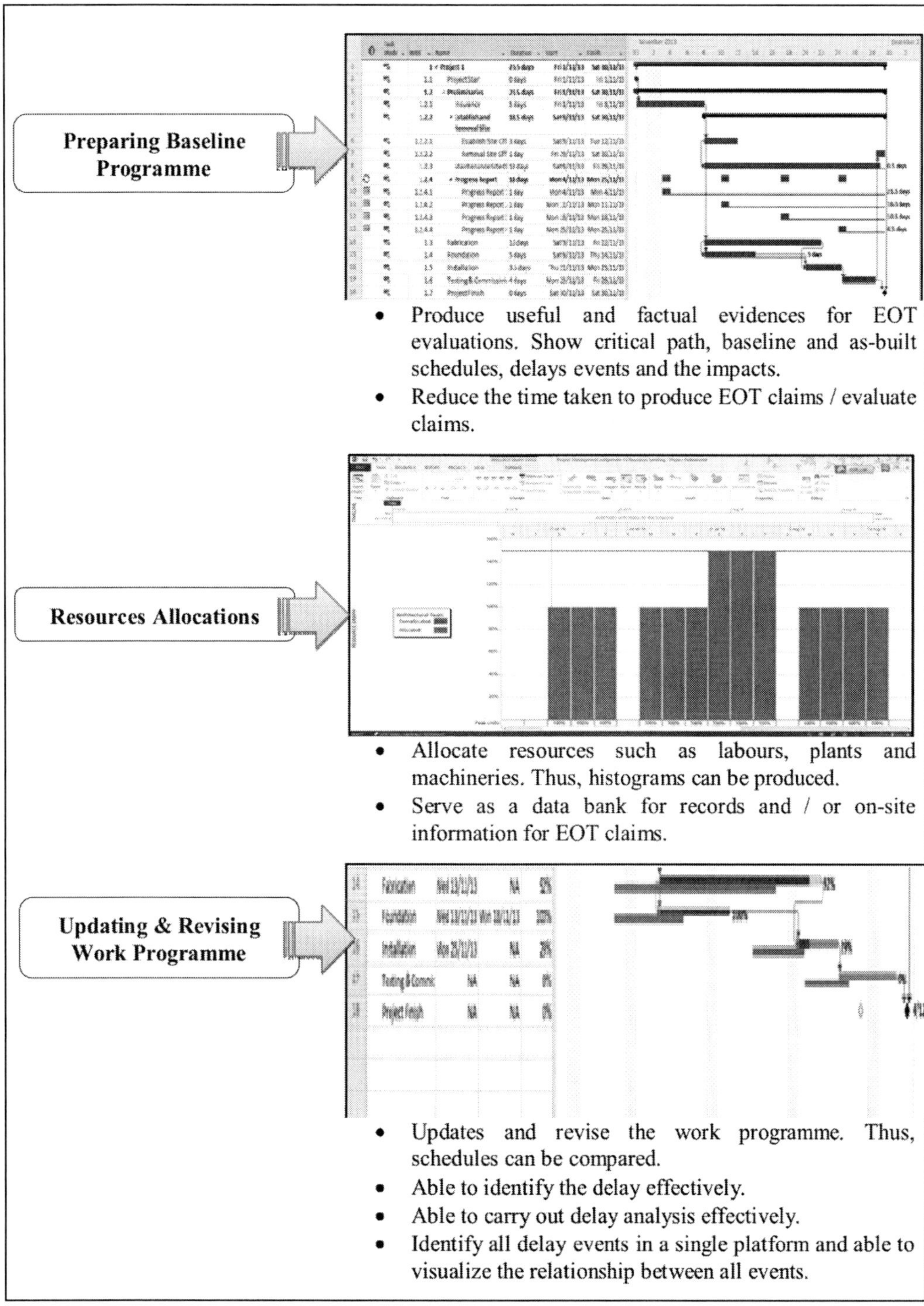


**Figure 2.** Microsoft Project as management tool in associating assessment of EOT.

## 3. Research methodology

This research applied both quantitative methods. The quantitative methods was used in order to determine the problems of the EOT in the current industry and the effectiveness of Microsoft Project in claiming and assessing EOT in forms of questionnaires. The respondents were architects and quantity surveyors in Kuala Lumpur Area.

The survey method was adopted and questionnaires were distributed to architect consultant firms and quantity surveyor consultant firms.The perspective of both sides were required to measure the effectiveness of Microsoft Project throughout the application process of EOT. On the other hand, the opinions of QS were also taken into accountability because QS also involve in the EOT associated tasks.

The questions in the questionnaire were designed in a multiple choices format. The multiple choices questions withFive-Points Likert Scale (1= highly not effective, 2= less effective, 3= moderate, 4= slightly effective and 5= highly effective)and dichotomous questions (Yes or No) was applied. Open-ended questions enabled a better explanation and relevant information regarding this research.

The respondents are made up of 100 respondents which consisted of fifty (50) numbers of architects and fifty (50) numbers of QS. Lists of architects and QS firms will be selected by "Simple and Random" method from Malaysian Institute of Architects (PAM) and Board of Quantity Surveyors Malaysia (BQSM).

## 4. Result and discussion

The absence of clear guideline on the procedures of claiming and assessing EOT is the major challenges faced by the consultants, especially the architect as the client's representative for a construction project. It is associated with the failure of contractors in retracing valid record and providing factual and valid evidence of the delays.

Inexistence of agreed basis of EOT claiming and assessing procedures between contractors and architect / contract administrators has been considered as the minor challenge faced by the architect or other consultants in assessing the EOT claims.

**Table3.**EOT Associated Problems and Challenges.

| | Problems and Challenges | RII Values | Ranks |
|---|---|---|---|
| Flaws in PAM2006 standard form of contract | • Absence clear guideline regarding to the procedures of claiming / assessing EOT application in standard form of contract. | 0.674 | 1 |
| | • Inexistence of agreed basis of EOT claiming and assessing procedures between contractors and Architect / contract administrators. | 0.587 | 7 |
| | • Impropriate delay analysis method. | 0.591 | 6 |
| Poor EOT claims' submission | • Failure in delays identification. | 0.600 | 5 |
| Late EOT claims' submission | • Late submission of claims by contractors. | 0.617 | 4 |
| Failure in collecting relevant facts | • Contractors fail to retrace valid records and data regarding to the delays. | 0.670 | 2 |
| | • Contractors fail to provide factual and valid evidences of the delays. | 0.665 | 3 |

Microsoft Project consists of numerous features that ease the management of the project. The respondents gave feedback that Microsoft Project is a tools that serve a data bank for records and on-site information for EOT claim. Therefore, it is considers as an effective tool which may completely assist the assessment of EOT claim submitted by the contractors, while at the same time it ability to carry out the delay analysis effectively.

**Table 4.** Effectiveness of Microsoft Project as a tool in assessing EOT claim.

| Effectiveness of Microsoft Project (Architects & QS) | | RII Values | Ranks |
|---|---|---|---|
| Preparing Baseline Programme | • Produce useful and factual evidences for EOT evaluations. Show critical path, baseline and as-built schedules, delays events and the impacts. | 0.639 | 4 |
| | • Reduce the time taken to produce EOT claims / Evaluate Claims. | 0.609 | 7 |
| Resources Allocations | • Allocate resources such as labours, plants and machineries. Thus, histograms can be produced. | 0.626 | 5 |
| | • Serve as a data bank for records and / or on-site information for EOT claims. | 0.691 | 1 |
| Updating & Revising Work Programme | • Updates and revise the work programme. Thus, schedules can be compared. | 0.665 | 3 |
| | • Able to identify the delays effectively. | 0.600 | 8 |
| | • Able to carry out delay analysis effectively. | 0.678 | 2 |
| | • Identify all delay events in a single platform and able to visualize the relationship between all events. | 0.622 | 6 |

It is agreeable by the architects/ contract administrator that Microsoft Project has the potential to be the reliable planning tool in assessing EOT. As it can be served as a data bank for records and / or on-site information for the EOT claims, it the potential to solve the difficulties or challenges of contractors in retrace and provide the factual and valid record of the delays.

## 5. Conclusion

It is recommended that, the project managers and contract administrators should update and revise the schedule regularly as some of the respondents (mainly architects and QS consultants) had mentioned that most of the contractors do not update the schedules unless problems are encountered during the construction process.On the other side, courses and trainings related to Microsoft Project and other similar project management software should be given to the employees by companies and firms to enhance the knowledge and encourage the usage of such software. These trainings can also begin in university level whereby lectures regarding to these software applications are given to construction students and therefore the students will be acquired with such knowledge before actual employments.

Furthermore, the industry and professions should be open-minded to accept the implementation of project management software such as Microsoft Project and other software as well as these software applications will keep on upgrading and improving with more features and functions to ease the flow of the projects in the future with lesser effort but greater efficiency and effectiveness. Last but not least, standard form of contracts such as PAM Contract 2006 do not provide sufficient information and instructions regarding to the EOT application and it is an undeniable fact and it is further proven in this study. Although enforcing the implementations of Microsoft Project and other similar software as mandatory in performing EOT tasks will be difficult, yet PAM should look into the problems of lacking clear guidance for preparing and evaluation EOT in the near future or upcoming edition of contract.

## 6. References

[1] Głuszak M and Leśniak A 2015 Construction delays in clients opinion - multivariate statistical analysis, *Procedia Eng.* **123**182–189

[2] Braimah N 2013 Construction delay analysis techniques—a review of application issues and improvement needs, *Buildings***3** 506–531

[3] Yusuwan N M and AdnanH 2013 Issues associated with extension of time (EOT) claim in malaysian construction industry*Procedia Technology*vol 9 (United Kingdom: Elsevier) pp 740–749

[4] KostalovaJ, TetrevovaL and SvedikJ 2015 Support of project management methods by project management information system*Procedia - Social and Behavioral Science*vol 210 (United Kingdom: Elsevier) pp 96–104

[5] Shehu A A,Masrom A N andMohammedM 2016 The effects of construction delayson the contractor's reputation in nigeria - the perception of professionals, *J. Multidiscip. Eng. Sci. Technol.***3**4154–4160

[6] ChappellD, Powell-SmithV andSimsJ 2008*Building Contract Claims: Fourth Edition* (United Kingdom: Blackwell Publishing Ltd) pp21-29

[7] Malaysian Construction and Contract Law 2012*PAM 2006 EOT Processes Flowchart* Retrieved on May 1, 2017 from https://simplymalaysia.wordpress.com/2012/01/11/pam-2006-eot-processes-flowchart/

[8] AditiD 2014 Delay analysis in construction contracts, *Int. J. Emerg. Technol. Adv. Eng.***4** 784–788

[9] MuhamadN H, MohammadM F, AhmadA C andIbrahimI H2016 Delay analysismethodologies (dams) in delivering quality projects: contractors and consultants' perceptions*Procedia - Social Behavioral Sciences*vol 222 (United Kingdom: Elservier) pp 121–131

[10] Chong H and LeongY 2012 Legal approach on assessment of contractors ' entitlement to extension of time,*African J. Bus. Manag.***6** 4815–4823

[11] Kumaraswamy M M andYogeswaranK 2003 Substantiation and assessment of claims for extension of time,*Int. J. Proj.Manag.***21** 27-38

[12] Gibson R 2008 *Construction Delays Extension of Time and Prolongation Claims*(USA: Taylor & Francis) pp 152-180

[13] Carmichael S and Murray M 2006 Record keeping for contemporaneous delay analysis: a model for effective event management,*Constr. Manage. Econ.***24**1007-1018

**Acknowledgments**
The authors would like to express gratitude to INTI International University for giving full support and encouragement towards this research.

IOP Conf. Series: Materials Science and Engineering **271** (2017) 012050 doi:10.1088/1757-899X/271/1/012050

# Performance of macro clay on the porous asphalt mixture properties

**C M Nurulain[1], P J Ramadhansyah[1], Y Haryati[1], A H Norhidayah[1], M E Abdullah[2] and M H Wan Ibrahim[2]**

[1]Faculty of Civil Engineering, Department of Geotechnics and Transportation, Universiti Teknologi Malaysia, 81310 Skudai, Johor Bahru, Malaysia
[2]Faculty of Civil and Environmental Engineering, Universiti Tun Hussein Onn Malaysia, 86400 Batu Pahat, Johor Bahru, Malaysia
Corresponding author: ramadhansyah@utm.my

**Abstract.** Porous asphalt pavement already well known as one type of pavement since many decades, however due to lack in strength it is not been widely used in road construction especially in high stresses areas. Many research had been conducted by researchers in order to improve the durability and increase the service life of the road. One of the methods that commonly used is through asphalt modification.Objectives of this study are to discuss influence of macro-clay in terms of physical and rheological properties and their effect on performance of porous asphalt mixture. Various percentage of macro-clay (starting from 2% to 8%) was blended in an asphalt binder and compacted using the Marshall compactor. The blended asphalt was characterized using penetration, softening point test and penetration index compared with unmodified binder. Performance tests of asphalt mixture were carried out such as Los Angeles Abrasion Value (LAAV) and Stability and Flow test in order to determine the strength and durability. The results show that the addition of macro-clay would increase in softening point but decrease in binder penetration. Based on the results, while macro-clay changes physical and rheological properties of bitumen and increase stiffness, it also improves strength and durability resistance, as well. Generally, the best improvements in the modified binders were obtained with 4% of macro-clay.

## 1. Introduction

Due to tropical climate in Malaysia where we receive excessive rainfall especially during monsoon season, proper management of surface water is a priority. This precipitation can prove severely harmful in areas with a limited or insufficient storm drain infrastructure in place. When water cannot penetrate into the ground, which resulting water splashing during vehicle run-off and may lead to hydroplaning. Hydroplaning occurs when a tire encounters more water than it can scatter which result in loss of steering, braking and power control. Due to this issues porous asphalt pavement will be consider as most suitable types of pavement that should be used in road construction. However, the main weakness found in porous asphalt binder are aging and stripping [1]. Porous asphalt has a possibility being water sensitive. Since rain water can infiltrate through the porous layer, it has chances to remain inside for long time. This will cause stripping of binder film from the aggregate to take place [2]. So the ideal asphalt should possess both: (1) high relative stiffness at high service temperatures (summer) to reduce rutting and shoving and; (2) increased adhesion between the asphalt and aggregate

IOP Conf. Series: Materials Science and Engineering **271** (2017) 012050 doi:10.1088/1757-899X/271/1/012050

in the presence of moisture to reduce stripping. In order to overcome this issue, modified binder has been introducing. There are many types of modifier that widely used in construction and one of them is clay. This study was focus on effect of Kaolin (macro- clay) on performance of porous asphalt mixture. Kaolin is mineral clay and normally found in white colour. Accordingly, an early assumption states that porous asphalt with modified asphalt possesses enhanced engineering properties. Researchers [3,4] suggested that clay modifications improve some characteristics of asphalt binders and asphalt mixtures, but more research is required before it can be applied on the large scale [5].

## 2. Materials and methods

### *2.1. Binder and aggregate*

This study used conventional bitumen with penetration grade (PEN) 80/100 supplied by Shell Malaysia. Kaolin as polymer modified and the granite aggregates supplied by the Ulu Choh quarry were used throughout the experiment. The basic bitumen tests for example penetration and softening point test were conducted to determine the bitumen properties.

### *2.2. Penetration test*

The consistency of the bitumen was measured by the penetration test according to the standard ASTM D5/D5M-13 [6], which is the simplest test to obtain the penetration value of the bitumen. A low penetration value indicates the hardness of the bitumen. The bitumen was heated and poured into a penetration cup prior to testing. The sample was then placed into a water bath for 1 h at 25°C after it was cooled. The penetration equipment was used to perform the test with the applied total load of 100g for 5 s at 25°C.

### *2.3. Softening point test*

Bitumen requires to definite temperature to soften. Therefore, determining the softening point of bitumen is important for result comparison. The softening point test was conducted according to ASTM D36/D36M-14e1 [7]. The bitumen was heated and poured into two rings and cooled for 30 min. The two rings and two balls were placed on each sample and heated. The temperature of the two samples was calculated.

### *2.4. Performance of porous asphalt*

The stability and flow value of porous asphalt samples were determined using the Marshall testing apparatus at a sample temperature of 60°C. Durability test was conducted to evaluate the performance of porous asphalt. Porous asphalt durability was evaluated in terms of resistance against abrasion loss. The compacted porous asphalt samples in this test were subjected to 300 rotations in the Los Angeles Abrasion Machine. The mass loss percent was used to determine the abrasion resistance of the porous asphalt samples.

## 3. Results and discussion

### *3.1. Effect of macro-clay on bitumen properties*

The effect of macro-clay as modified bitumen on penetration, softening point and penetration index value are given in table 1. The result shows that the penetration value was constantly decreased as the percentage of macro-clay increased. The results proved that macro-clay had chemically reduced the binder viscosity and increased the binder stiffness which result the decrease in the penetration value and would be useful to obtain stiffer asphaltic concrete [8]. The softening point of samples was meet ASTM D5/D5M-13 [6] specification which requires the softening point between 45°C to 52°C for bitumen PEN 80/100. According to table 1, when macro-clay content increases, the softening point value also increase. These were due to macro-clay reducing the binder viscosity and increasing the elasticity of the asphalt binder to resist deformation, thus resulting in the increased of softening point value. El-Shafie et al. [3] reported that modified asphalt were less sensitive to high temperature changes and may also be more resistant to plastic deformation (rutting) compared to unmodified asphalt. The Penetration Index (PI) is a measure bitumen susceptible to temperature and is calculated using equation (1) [9].

$$PI = \frac{1952 - 500\log pen - 20SP}{50\log pen - SP - 120} \quad (1)$$

Where pen is penetration (25°C) and SP is softening point. Table 1 shows the effects of macro-clay on PI of asphalt binder. The results indicated that penetration index had increased as percentage of macro-clay increased thus physically macro-clay modified binder become more harden. In addition the increase in PI value also reveals that macro-clay reduced the temperature susceptibility of the bitumen [10].

**Table 1.** Effect of macro-clay on bitumen properties.

| Kaolin Content (%) | Penetration (PEN) | Softening Point (°C) | Penetration Index (PI) |
|---|---|---|---|
| 0.0 | 89 | 46 | -0.84 |
| 2.0 | 70 | 52 | -0.59 |
| 4.0 | 60 | 56 | 0.66 |
| 6.0 | 54 | 58 | 0.81 |
| 8.0 | 45 | 62 | 1.15 |

*3.2. Effect of macro-clay on performance of porous asphalt mixture*

*3.2.1. Stability and flow.* From figure 1, the highest staility value occurred at the higher macro-clay content. The sample with a high content of macro-clay seems to improve the performance against permanent deformation. The result also show that the significant increasing in the stability value of the porous asphalt sample when the macro-clay content increase. The values were less varied in terms of flow. However, a slight reduction in the porous asphalt flow occurred when the macro-clay content increase. Previously, Zubeck et al. [11] found that polymer modified asphalt showed good workability and economically feasible while can meet other criteria which can improved the strength of asphalt mixture. The porous asphalt resistance against permanent deformation greatly depend on the binder, which provided stiffer and stronger mixture.

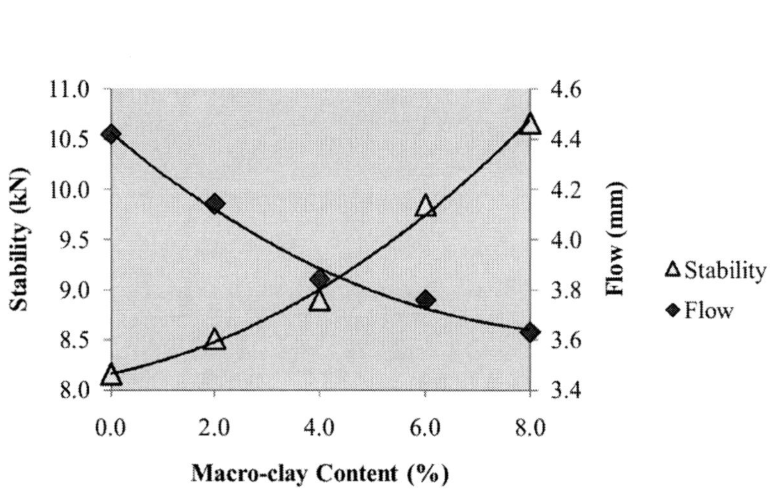


**Figure 1.** Relationship between stability and flow.

*3.2.2. Abrasion loss.* The relationship between the percentof abrasion loss and macro-clay content is illustrated in figure 2. The highest percentage of abrasion loss occurs at 8% of macro-clay. Vanessa and Jorge [12] found that porous asphalt mixtures require a modified asphalt binder in order to give greater cohesion bonding between aggregate and binder. However, high amount of bitumen will be reduce due to high macro-clay content which can cause coating work became harder and at the same times reduce the cohesion. The amount of abrasion loss increase significantly with the increases of macro-clay except for 4%. In term of cost view and ability to reduce the abrasion loss, this is possibly an indication that 4% of macro-clay is the optimum rate of macro-clay as poly modified asphalt.

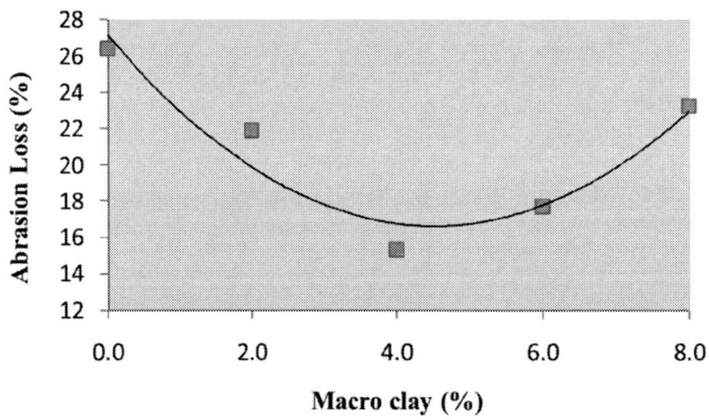


**Figure 2.** Relationship between abrasion loss and macro-clay content.

## 4. Conclusions

Based on the analysis and laboratory data, it is found that the addition of macro-clay has hardened the bitumen and reduces its temperature susceptibility as measured by the results from penetration and softening point tests. When penetration increase bitumen become stiff which is good to improve rutting however at the same time bitumen also become brittle which may increase cracking problem.This is also supported by the results obtained from the stability test where the additions of macro-clay increase the asphalt mixture stability compare to the control sample. However, the increase in the bitumen stiffness has reduced its ability to coat the aggregate particles and the bonding strength as shown in abrasion loss.In order to overcome this issue appropriate amount of macro-clay that use as modifier is is a main concern. The amount of macro-clay used should give optimum beneficial to improve the strength and durability of porous asphalt pavement but at the same time can maintain the main purpose of porous asphalt pavement as surface storm management. According to this study, 4% macro clay give the best improvement.

## 5. References

[1] Tarmuzi N A, Jaya R P, Yaacob H, Hassan N A and Aziz M M A 2015 Aggregate angularity effect on porous asphalt engineering properties and performance, *J. Teknologi* **73** 99-104

[2] Poulikakos L D, Pittet M, Arnaud L, Junod A, Gubler R, Simond E, Partl M and Dumont A G 2006 *Mechanical Properties of Porous Asphalt, Recommendations for Standardization* (Switzerland: Swiss Federal Laboratory for Materials Testing and Research, Empa) pp 70-76

[3] El-Shafie M, Ibrahim I M and Abd El Rahman A M M 2012 The additions effects of macro and nano clay on the performance of asphalt binder, *Egyptian J. of Petroleum* **21** 149-154

[4] Wan Azhar W N A, Bujang M, Jaya R P, Hainin M R, Aziz M M A and Ngadi N 2015 Application of nanotechnology in asphalt binder: a conspectus and overview, *J. Teknologi* **76** 85-89

[5] Jaya R P, Yusak M I M, Hainin M R, Warid M N M and Wan Ibrahim M H 2014 Porous concrete pavement containing nano-silica: Pre-review, *J. Adv. Mater. Res.* **911** 454-458

[6] American Society for Testing Materials ASTM D5/D5M-13 2013 *Standard Test Method for Penetration Of Bituminous Materials* (United States: ASTM Standard) pp 1-4

[7] American Society for Testing Materials ASTM D36/D36M-14e1. 2014 *Standard Test Method for Softening Point of Bitumen (Ring-and-Ball Apparatus)* (United States: ASTM Standard) pp 1-4

[8] Hainin M R, Jaya R P, Akbar N A A, Jayanti D S and Yusoff N I M 2014 Influence of palm oil fuel ash as a modifier on bitumen to improve aging resistance, *J. of Engineering Research* **2** 34-46

[9] Read J and Whiteoak 2003 *The Shell Bitumen Handbook* (Doncaster: Thomas Telford Publishing) pp 29-41

[10] Che'Mat N, Jaya R P, Hassan N A, Nor H M, Aziz M M A and Che Wan C M 2014 Properties of asphaltic concrete containing sasobit®, *J. Teknologi* **71** 27-31

[11] Zubeck H K, Lutfi R, Stephan S, George M and John Ryer P E 2003 Workability and performance of polymer-modified asphalt aggregate mixtures in cold regions, *Int. J. Pave* **4** 25-36

[12] Vanessa S A and Jorge E C M 2014 Resistance to degradation or cohesion loss in cantabro test on specimens of porous asphalt friction courses *Procedia – Social and Behavioral Sciences* vol 162 (United Kingdom: Elsevier) pp 290-299

**Acknowledgments**
The support provided by Malaysian Ministry of Higher Education and Universiti Teknologi Malaysia in the form of a research grant vote number Q.J130000.2522.18H05 and Q.J130000.2522.11H76 for this study is highly appreciated.

IOP Conf. Series: Materials Science and Engineering **271** (2017) 012051 doi:10.1088/1757-899X/271/1/012051

# Review of palm oil fuel ash and ceramic waste in the production of concrete

**Puteri Natasya Mazenan[1], Faisal Sheikh Khalid[1], Shahiron Shahidan[1], and Shamrul-mar Shamsuddin[1]**

[1] Jamilus Research Centre for Sustainable Construction, Department of Structural and Material, Faculty of Civil and Environmental Engineering, Universiti Tun Hussein Onn Malaysia, Batu Pahat, Johor.
Corresponding author:faisalsh@uthm.edu.my

**Abstract.** High demand for cement in the concrete production has been increased which become the problems in the industry. Thus, this problem will increase the production cost of construction material and the demand for affordable houses. Moreover, the production of Portland cement leads to the release of a significant amount of $CO_2$ and other gases leading to the effect on global warming. The need for a sustainable and green construction building material is required in the construction industry. Hence, this paper presents utilization of palm oil fuel ash and ceramic waste as partial cement replacement in the production of concrete. Using both of this waste in the concrete production would benefit in many ways. It is able to save cost and energy other than protecting the environment. In short, 20% usage of palm oil fuel ash and 30% replacement of ceramic waste as cement replacement show the acceptable and satisfactory strength of concrete.

## 1. Introduction

Palm oil fuel ash (POFA) is an agricultural waste produced from the combustion of oil palm plant residue in palm oil industry [1]. Both husks and shells are burnt in the boiler after the extraction of oil from palm oil fruit to produce steam for the turbine engine which generates electricity for use in the palm oil mills [2]. After the combustion in the boiler, 5% of ash produced which known as POFA is generally disposed of in open fields. This situation created the problems to the environmental and health problems [3-4].

Polishing process of ceramic tiles produced ceramic waste powder are dumped in landfills and can cause soil, air and groundwater pollution making a serious environmental problem [4]. The increases large piles occurs due to ceramic waste that disposed of factories producing construction industry materials and has been accumulating on frequently [5]. Ceramic industry has produced about 15% to 30% of waste material from the total production. However, this waste is not recycled in any form at present. Ceramic waste was characterized as durable, hard and highly resistant to biological, chemical and physical degradation forces [6].

## 2. POFA in concrete production

Mohammadhosseini et al. [7] present the study on the performance of concrete containing waste carpet fibers and POFA. This study only replaced cement with 20% POFA and 0%to 1.25% of carpet fiber used with OPC. In short, the replacement of cement with POFA resulted in lower workability. Moreover, the use of 0.75% carpet fibers and 20% POFA in concrete shows the decreases of drying shrinkage.

Ahamed and Siddiraju [8] study the strength of concrete with POFA as cement replacement. This study replaced the cement by 5%, 7.5%, 10%, 12.5%, 15% and 17.5% compared with conventional concrete of grade $M_{25}$ and the test is conducted for 3, 7 and 28 days. Researchers found that 12.5% is the optimum replacement of cement in the concrete production whereas it shows the increases in the strength. For the compressive strength test, it shows 34.3 MPa at age of 28 days while split tensile strength gave the result of 3.4 MPa.

For the experiment carried out by [9], the researchers study the effect of POFA in producing economical and eco-friendly self-compacting concrete. The study has replaced the cement by 0%, 50%, 60% and 70% with constant water/cement ratio of 0.35. The results found that fresh properties of SCC containing a high volume of treated POFA are better than control in terms of filling ability and passing ability. In the other hands, segregation index remains in the range specified by the EFNARC guidelines. In summary, the compressive strength of SCC containing 50% to 70% of treated POFA was higher than control at 28 days and beyond. Moreover, lower shrinkage occurred to all SCC specimens compared to control SCC after 12 months of drying shrinkage. The researchers conclude that the replacement of cement with high volume POFA can reduce the cost production of SCC with 8% to 12% compared with the normal SCC.

Research carried by Ranjbar et al. [10] used volume 10%, 15% and 20% of POFA as replacement of cement in the production of self-compacting concrete to study its durability and mechanical properties. Water to cement ratio used in this study was 0.35 for all specimens. In short, the researchers found that POFA is a great potential as the replacement of cement in SCC. However, POFA caused the reduction in workability of SCC but this can be improved by the additional of superplasticizer. The increases of POFA content lead to the reduction in early mechanical properties but the final strength of the SCC containing POFA was comparable with normal SCC. In addition, the SCC containing POFA showed the less surface water absorption and higher durability under acid and sulfate attack.

An experimental investigation has been carried out by Islam et al. [11] study the durability properties of sustainable concrete containing high volume palm oil waste materials namely oil palm shell (OPS) and POFA. POFA was used as partial cement replacement by 0%, 10%, 30% and 50% with water to binder ratio of 0.3. In the same way, POFA also replaced 0% and 70% of cement with water binder ratio of 0.4 and superplasticizer used was in the range 0.6% to 1.1% (by mass of binder) in order to maintain the slump within range. The compressive strength of POFA concrete was lower at 7 days and yet the strength of POFA concrete was higher at later ages. 10% POFA had higher strength gain at 90 days compared to the control concrete. The researchers also found that the water absorption and sorptivity of concrete containing a higher level of POFA were high.

Munir et al. [12] investigate the addition of POFA in producing the lightweight foamed concrete. The researchers carried out this study to evaluate the role of POFA in the foamed concrete production. This study has added the cement with 10%, 20%, 30%, 40% and 50% of POFA and used fixed water cement ratio which 0.5. With 50% addition of POFA in concrete, the compressive strength of concrete only decreased about 30 to 40% compared to control specimen. In the same way, 20% addition of POFA shows the acceptable strength of foamed concrete for non-structural purposes.

Mohammadhosseini et al. [13] studied the fresh and hardened state properties of SCC with 0%, 30% and 60% of POFA replacing cement in self-compacting concrete. Different water/cement ratios were used in this study which is 0.4, 0.45 and 0.5. Conclusion for fresh properties found that the increases of POFA as a cement replacement in SCC resulted in the lower workability with the acceptable range. The strength of the specimen was taken at 7, 28 and 90 days. The results found lower strength at an early age however the strength then increases at later ages. The strength of concrete at 30% replacement of cement almost achieved the same strength as control mix. Tensile and flexural strengths of SCC specimens show the similar trend with the compressive strength.

Awal et al. [14] determined the properties of concrete containing a high volume of POFA. This study had been done towards 50%, 60% and 70% replacement of cement with POFA with a variable water-cement ratio of 0.38, 0.42 and 0.48 and the addition of superplasticizer up to 2%. The researchers found that the workability of concrete is low with the high volume of POFA. However, the workability is improved to the satisfactory level with the addition of superplasticizer. In summary, the compressive and tensile strength of the concrete containing a high volume of POFA were found lower than control mix.

Ahmad et al. [15] present the compressive strength of POFA concrete. This study carried out with 5%, 10% and 15% of POFA as cement replacement with fixed water-cement ratio of 0.45. Compressive strength test was run on 3, 7, 28, 56 and 90 days. This research can be concluded that POFA concrete has higher workability than fly ash and quarry dust. However, the workability still decreases as the increases content of POFA. Moreover, 15% replacement of cement shows the optimum strength of concrete compared to others and the strength development of POFA has same patterns with OPC concrete. Table 1 shows the overall summary of previous study on POFA.

**Table 1.** Summary of previous work on POFA.

| Authors & Year | Percentage of replacement | Application of material | Findings |
|---|---|---|---|
| Mohammadhosseini, et al., 2017 | 0% and 20% | Concrete | The addition of POFA in concrete resulted in lower workability.<br>20% POFA in concrete composite resulted in the decrease of drying shrinkage. |
| Ahamed & Siddiraju, 2016 | 0%, 5%, 7.5%, 10%, 12.5%, 15% and 17.5% | Concrete | 12.5% cement replacement with POFA shows the increases in strength. |
| Alsubari et al., 2016 | 0%, 50%, 60% and 70% | Self-compacting concrete | Filling ability and passing ability are better than control.<br>70% of cement replacement was suggested to produce low-cost and sustainable concrete. |
| Ranjbar et al., 2016 | 0%, 10%, 15% and 20% | Self-compacting concrete | The increases of POFA content lead to the reduction in early mechanical properties but the final strength of SCC containing POFA was comparable. |
| Islam et al., 2016 | 0%, 10%, 30% and 50%<br>0% and 70% | Concrete | 10% of POFA shows the higher compressive strength than control at the age of 90 days. |
| Munir et al., 2015 | 10%, 20%, 30%, 40% and 50% (Addition) | Lightweight foamed concrete | 20% addition of POFA into foamed concrete shows the acceptable strength of foamed concrete for non-structural purposes. |
| Mohammadhosseini et al., 2015 | 0%, 30% and 60% | Self-compacting concrete | Replacement of POFA results in lower workability of concrete with acceptable range.<br>Higher volume replacement of POFA shows the reduction in strength. |
| Awal & Abubakar, 2011 | 0%, 50%, 60% and 70% | Concrete | The strength of concrete containing a high volume of POFA shows the lower strength.<br>Workability of high volume POFA low but improved with the addition of superplasticizer. |

| | | | |
|---|---|---|---|
| Ahmad et al., 2008 | 0%, 5%, 10% and 15% | Concrete | Workability shows a significant reduction with the increasing of POFA content. 15% of POFA replacement shows the optimum compressive strength and the same patterns of strength development with OPC concrete. |

## 3. Ceramic waste in concrete production

Rani's idea [5] was replaced the cement with ceramic waste powder was explored. The researcher had studied the effect of replacement level towards the strength of the concrete. This study replaced cement by 0%, 10%, 20, 30%, 40% and 50% of ceramic waste with water-cement ratio of 0.4 for M-40 grade concrete. Results were taken at days 7, 28 and 56. The conclusion made by author stated that 10% replacement of cement shows the increases in compressive strength and further replacement level shows the decreases in compressive strength.

The aim of Kanaan and El-deib [16] was to investigate the effect of the replacement of cement with ceramic waste powder as an ingredient to sustainable concrete. Ceramic waste powder has improved the workability retention for concrete mixture M25. Strength test was run on 7, 28, 56 and 90 days and the data demonstrated that the replacement from 20% to 40% have lower strength improvement. The researchers found the highest strength improvement with 10% replacement. In summary, researchers mentioned that 10% replacement is suitable for strength improvement while 40% replacement suitable for durability improvement.

Arthi [17] has compared the concrete containing ceramic waste powder as cement replacement with conventional concrete. Proportions used in this study are 0%, 15%, 30%, 35%, 40% and 45%. The specimens were curing and tested at age 7, 21 and 28 days. Replacement level less than 35% by weight of cement increases the compressive strength while further replacement leads to decreases in strength. Moreover, the cost of production can decrease up to 25% with 35% ceramic waste powder as cement replacement.

Anwar et al. [18] investigated the effect of substitution the cement with ceramic waste powder and fine aggregate with waste marble powder. Proportions used in this study were 0%, 5%, 10%, 15%, 20%, 25%, 30%, 35%, 40%, 45% and 50% for both ceramic waste powder and waste marble powder for the production of concrete grade M-20. The strength was only tested on age 28 days. Conclusions draw from this study is the strength of the concrete gradually decreased with the increases of the replacement level of ceramic waste powder. So, it was recommended to replace the cement less than 30% due to it is not affecting the characteristic strength of M20 grade concrete.

Patel, Arora and Vaniya [6] presents the result of the use of ceramic waste powder in cement concrete with 0%, 10%, 20%, 30%, 40%, 50% and 60% of replacement level. The concrete was cast with 0.46 of water-cement ratio and sodium silicate was added up to 2%. Workability of concrete without the addition of sodium silicate shows the decreases of slump value. However, with the 10% and 40% replacement of cement shows the increases in slump value. The strength test was carried out at age 7, 14 and 28 days. The researchers recommend that 30% replacement of cement with the 2% addition of sodium silicate due to its achieved the optimum compressive strength. Moreover, tensile strength test recorded only 1% loss compared to conventional concrete.

Raval et al. [19] have conducted experimental investigation by 0%, 10%, 20%, 30%, 40% and 50% of ceramic waste powder as cement replacement with water-cement ratio of 0.52. To evaluate the mechanical properties, tests were carried out at age 7, 14 and 28 days. Replacement of cement with ceramic waste powder less than 30% recorded the increases in strength while more than 30% replacement level shows the decreases in strength.

Since the concrete is widely used in construction industry many studies have been conducted to implement and improve the quality of the concrete. Reuse of agricultural waste and industry waste to replace the cement material appears to be a solution for the pollution problem and high cost of building materials [20-21]. Summary of previous work on ceramic waste powder are shown in Table 2.

**Table 2.** Summary of previous work on ceramic waste powder.

| Authors & year | Percentage of replacement | Application of material | Findings |
|---|---|---|---|
| Rani, 2016 | 0%, 10%, 20%, 30%, 40% and 50% | Concrete | 10% of ceramic waste as cement replacement shows the increases of compressive strength. |
| Kanaan & El-dieb, 2016 | 0%, 10%, 20%, 30% and 40% | Concrete | Use of ceramic waste powder in concrete improved its workability retention. Replacement more than 10% shows the decreases in strength. |
| Arthi, 2016 | 0%, 15%, 30%, 35%, 40% and 45% | Concrete | Less than 35% replacement level shows the increases in strength while further replacement caused the strength decreases. |
| Anwar et al., 2015 | 0%, 5%, 10%, 15%, 20%, 25%, 30%, 35%, 40%, 45% and 50% | Concrete | 30% of ceramic waste powder is suggested due to it was not affecting the characteristic strength. |
| Patel et al., 2015 | 0%, 10%, 20%, 30%, 40%, 50% and 60% | Concrete | Slump value of mix without the addition of sodium silicate increase with 10% and 40% replacement. 30% volume of cement waste powder with 2% addition of sodium silicate shows the optimum strength of concrete. |
| Raval et al., 2013 | 0%, 10%, 20%, 30%, 40% and 50% | Concrete | Replacement less than 30% of concrete waste powder lead to the increases in concrete strength. |

## 4. Conclusion
From the researchers discussed above, it is clearly shown that palm oil fuel ash and ceramic waste are suitable to use in the construction industry and more economic in the making of concrete. Palm oil fuel ash and ceramic waste are found to be suitable as a partial replacement of cement in concrete production. Most of the researchers suggested up to 20% replacement of palm oil fuel ash as a cement replacement in the concrete production. Replacement more than 20% leads to the decreases in the strength of concrete. For ceramic waste, the replacement suggested by the researchers is up to 30% and further replacement caused the decreases in the strength of concrete.

## 5. References
[1] Safiudin M, Jumaat M Z, Salam M A, Islam M S and Hashim R 2010 Utilization of solid wastes in construction materials *Int. J. Phys. Sci.* **5** (13) 1952–1963
[2] Mahlia T M I, Abdulmuin M Z, Alamsyah T M I and Mukhlishien D 2001 An alternative energy source from palm wastes industry for Malaysia and Indonesia *Energy Convers. Manag.* **42** 2109–2118
[3] Sheikh Khalid F, Herman H S and Azmi N B 2017 Properties of sugarcane fiber on the strength of the normal and lightweight concrete *Int. Symposium on Civil and Environmental Engineering* (Melaka, Malaysia) vol 103 01021
[4] Khalid F S, Azmi N B, Sumandi K A S M, and Mazenan P N 2017 Mechanical properties of concrete containing recycled concrete aggregate (RCA) and ceramic waste as coarse aggregate replacement *Int. Conf. on Applied Science and Technology (Langkawi)* vol 1891 (London: AIP Conference Proceedings) 020079

[5] Rani M S 2016 A study on ceramic waste powder *SSRG Int. J. Civ. Engineering* **3** (7) 1–6
[6] Patel H, Arora N K and Vaniya S R 2015 Use of ceramic waste powder in cement concrete *IJIRST - Int. J. Innov. Res. Sci. Technol.* **2** (1) 91–97
[7] Mohammadhosseini H, Mohamad J, Rahman A, Sam M and Awal A S M A 2017 Durability performance of green concrete composites containing waste carpet fibers and palm oil fuel ash *J. Clean. Prod.* **144** 448–458
[8] Ahamed V S and Siddiraju S 2016 Study of strength of concrete with palm oil fuel ash as cement replacement *Int. J. Civ. Eng. Technol.* **7** (3) 337–341
[9] Alsubari B, Shafigh P and Jumaat M Z 2016 Utilization of high-volume treated palm oil fuel ash to produce sustainable self-compacting concrete *J. Clean. Prod.* **137** 982–996
[10] Ranjbar N, Behnia A, Alsubari B, Birgani P M and Jumaat M Z 2016 Durability and mechanical properties of self-compacting concrete incorporating palm oil fuel ash *J. Clean. Prod.* **112** 723–730
[11] Islam M M U, Mo K H, Alengaram U J and Jumaat M Z 2016 Durability properties of sustainable concrete containing high volume palm oil waste materials *J. Clean. Prod.* **137** 167–177
[12] Munir A, Huzaim A, Irfandi S and Safwan 2015 Utilization of palm oil fuel ash (POFA) in producing lightweight foamed concrete for non-structural building material *Procedia Engineering* **125** 739–746
[13] Mohammadhosseini H, Awal A S M A and Ehsan A H A Q 2015 Influence of palm oil fuel ash on fresh and mechanical properties of self-compacting concrete *Sadhana* **40** 6 1989–1999
[14] Awal A S M A and Abubakar S I 2011 Properties of concrete containing high volume palm oil fuel ash: A short-term investigation *Malaysian J. Civ. Engineering* **23** (2) 54–66
[15] Ahmad M H, Omar R C, Malek M A, Md Noor N and Thiruselvam S 2008 Compressive strength of palm oil fuel ash concrete *Int. Conf. Constr. Build. Technology (ICCBT 2008)* vol 27 pp 297–306
[16] Kanaan D M and El-dieb A S 2016 Ceramic waste powder as an ingredient to sustainable concrete *4th Int. Conf. Sustain. Constr. Mater. Technology* vol 1
[17] Arthi A J J 2016 Effective replacement of cement by ceramic waste in concrete for sustainable development *Int. J. Res. Engineering Technology* **5** (11) pp 2319–2321
[18] Anwar A, Ahmad S, Husain S M A and Aqeel S 2015 Salvage of ceramic waste and marble dust for the refinement of sustainable concrete *Int. J. Civ. Eng. Technology* **6** (9) pp 79–92
[19] Raval A D, Patel I N and Pitroda P J 2013 Ceramic waste : Effective replacement of cement for establishing sustainable concrete *Int. J. Eng. Trends Technology* **4** (6) pp 2324–2329
[20] Khalid F S, Herman S H, Azmi N B and M I Juki 2017 Sand cement brick containing recycled concrete aggregate as fine-aggregate replacement *Int. Symp. on Civil and Environmental Engineering 2016 (Melaka)* vol 103 (Paris: MATEC Web of Conference) 01016
[21] Zamer M M, Irwan J M, Othman N, Faisal S K, Anneza L H, Teddy T and Alshalif A F 2017 Influence of ureolytic bacteria toward Interlocking Compressed Earth Blocks (ICEB) in improving durability of *ICEB Inter. Symp. on Civil and Environmental Engineering 2016 (Melaka)* vol 103 (Paris: MATEC Web of Conference) 01027

**Acknowledgment**

This work was financially supported by University Tun Hussein Onn Malaysia (UTHM) and Minister of Education Malaysia (KPM) through Dana Pembudayaan Penyelidikan (RAGS) Vot. No. R-058.

IOP Conf. Series: Materials Science and Engineering **271** (2017) 012052 doi:10.1088/1757-899X/271/1/012052

# The effect of sludge water treatment plant residuals on the properties of compressed brick

**Shamrul-Mar Shamsudin[1,3], S. Shahidan[1,3], M. A. M. Azmi[3], S. A. Ghaffar[1],M. B. Abdul Ghani[2], N. A. A. Saiful Bahari[1,3] and S. S. M. Zuki[1,3]**

[1]Jamilus Research Center, Faculty of Civil and Environmental Engineering, Universiti Tun Hussein Onn Malaysia, Batu Pahat

[2] PT392, Mahligai Idaman, Pusat Perniagaan Bandar BaruTunjong, Jalan Kuala Krai 16010 kota Bharu, Kelantan

[3] Faculty of Civil and Environmental Engineering, Universiti Tun Hussein Onn Malaysia, Batu Pahat

Corresponding author: shahiron@uthm.edu.my

**Abstract.**The focus of this study is on the production of compressed bricks which contains sludge water treatment plant (SWTP) residuals obtained from SAJ. The main objective of this study is to utilise and incorporate discarded material (SWTP) in the form of residual solution to produce compressed bricks. This serves as one of the recycling effortsto conservethe environment. This study determined the optimum mix based on a mix ratio of 1:2:4 (cement: sand: soil) in the production of compressed bricks where 5 different mixes were investigated i.e. 0%, 5%, 10%, 20%, and 30% of water treatment plant residue solution. The production of the compressed bricks is in accordance with the Malaysian Standard MS 7.6: 1972 and British Standard BS 3921: 1985 - Compressive Strength & Water Absorption. After being moulded and air dried,the cured bricks were subjected to compression tests and water absorption tests. Based on the tests conducted, it was found that 20% of water treatment plant residue solution which is equivalent to 50% of soil content replacement with a mix composition of [10: cement] [20: sand] [20: soil] [20: water treatment plant residue solution] is the optimum mix. It was also observed that the bricks containing SWTP residuals were lighter in weight compared to the control specimens

## 1. Introduction

In view of the wide usage of clay bricks especially in the construction industry today, numerous production techniques and processes have been investigated in order to reduce the cost of production without compromising the workability of bricks. Therefore, this research provides an alternative solution in improving the quality of compressed bricks and directly helps to reduce water treatment plant residue at SAJ residual disposal areas. The current production process of compressed bricks that is of interest is the production process of unburnt compressed bricks which are able to minimize the production cost [1].

With the accumulation of water treatment plant residue, a wide area is required to accommodate the safe storage and management of the residues. This in turn will increase the maintenance cost [6]. Apart from this, the disposal of polluted waste water treatment in accordance to SW 204 – Disposal category in First Schedule, "Peraturan-Peraturan Kualiti Alam Sekeliling

(Buangan Terjadual 2005)" [2] also increases the maintenance cost. Therefore, the main focus of this research is to overcome these issues and create improved construction material in civil engineering construction.

This research concentrates on the production of compressed bricks containing sludge water treatment plant residue in accordance to Malaysian Standard MS 7.6: 1972 and British Standard BS 3921: 1985) and takes into account the end users'/consumers' requirements [7]. The main objective of this research was to investigate the optimum mix ratio of the water treatment plant residual solution in the production of compressed bricks that complies with the requirements of the Malaysian Standard MS 7.6: 1972 and British Standard BS 3921: 1985 in terms of its compressive strength and water absorption rate [7].

## 2. Materials and methods

### *2.1. Sampling of Material*

The water treatment plant residuals were obtained from SAJ Holdings Sdn Bhd, Site 3, Loji Air Seri Gading, Lot 4704, Kawasan Perindustrian Sri Gading, Batu 6 Jalan Kluang, Batu Pahat, Johor Darul Takzim.

**Figure 1**.The process of collecting water treatment plant residuals at one of the dried lagoons.

### *2.2. Preparation of materials*

Mix ratios of 1:2:4 (1 Part : cement) : (2 Parts : sand) : (4 Parts : soil + WTP residual solution) were used for the evaluation of the compressive strength and water absorption rate of compressed bricks in accordance to the Malaysian Standard MS 7.6: 1972 and British Standard BS 3921: 1985 - Compressive Strength & Water Absorption [7].Sludge water treatment plant residuals were crushed and sieved before use as shown in Figure 2.

(a) (b) (c)

**Figure 2**. (a) WTP residuals before sieving, (b) WTP residuals after sieving (c) Laterite Soil.

*2.3 Production of compressed bricks*

The production of compressed bricks was doneaccording to the Malaysian Standard MS 7.6: 1972 and British Standard BS 3921: 1985 - Compressive Strength & Water Absorption. A total of 5 mix ratios were prepared where 10 bricks were produced for each mix ratio. Each compressed brick measured 222 mm in length, 114 mm in width and 75mm in height as shown in Figure 3. The compressed bricks producedwere of Class 1 Load Bearing brick. The mix ratios of the compressed bricks are shown in Table 1 [7].

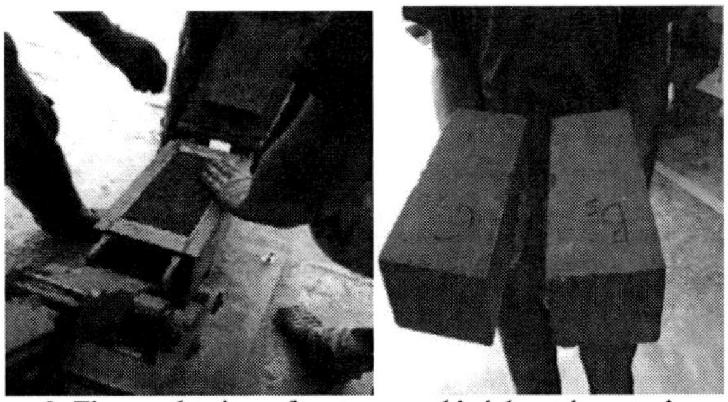

**Figure 3.** The production of compressed bricks using steel moulds.

**Table 1.** Compressed Bricks Mix Ratios 1:2:4 (1 Part: cement): (2 Parts: sand): (4 Parts: soil + WTP residual solution).

| Mix | Total Sample | Water (%) | Sand (%) | Cement (%) | Soil (%) | WTP Residual (%) |
|---|---|---|---|---|---|---|
| A | 10 | 10 | 20 | 10 | 40 | 0 |
| B | 10 | 10 | 20 | 10 | 35 | 5 |
| C | 10 | 10 | 20 | 10 | 30 | 10 |
| D | 10 | 10 | 20 | 10 | 20 | 20 |
| E | 10 | 10 | 20 | 10 | 10 | 30 |

*2.4. Brick testing*

The compressive strength test and the absorption rate test were conducted in accordance to the Malaysian Standard MS 7.6: 1972 and British Standard BS 3921: 1985 - Compressive Strength & Water Absorption [7] as shown in Figure 4.

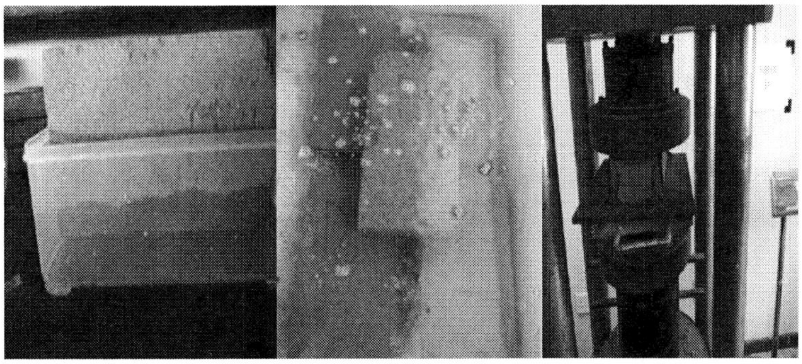

(a) (b) (c)
**Figure 4.** (a) Initial water absorption rate test (b) Absorption rate test, (c) Compressive strength test.

## 3. Results and discussion

### *3.1. X-Ray Fluorescence (XRF)*
Table 2 shows that the concentration levels of Aluminium Oxide (Al2O3), Silicon Oxide (SiO2) and Calcium Oxide (CaO) were the highest among all the materials used. Based on the XRF data, cement contains the highest element of Calcium Oxide (CaO) at 59.4%. In general, CaO is an important element in the production of cement. It reacts actively in the presence of water and produces heat to assist in the setting of cement. The XRF data also shows that laterite soil contains 36.1% of Silicon Dioxide (SiO2). SiO2 is one of the elements in soil that influences the strength and stiffness of a soil mass. Meanwhile for the WTP residual solution, XRF revealed that it contains 51.76% Silicon Dioxide (SiO2) and 36.70% of Aluminium Oxide (Al2O3). Al2O3 is known as Alum which is used in the coagulation and flocculation stage in the water treatment plant process [8].

**Table 2.** Concentration of elements in materials from XRF data.

| Element | Material Elements Concentration | | |
|---|---|---|---|
| | Cement (%) | Laterite Soil (%) | WTP Residual Solution (%) |
| CaO | 59.40 | 1.60 | 0.52 |
| SiO2 | 15.00 | 36.10 | 51.76 |
| SO3 | 3.56 | 0.85 | 0.38 |
| Fe2O3 | 3.11 | 18.90 | 6.67 |
| Al2O3 | 2.94 | 18.90 | 36.70 |
| MgO | 1.53 | 0.24 | 0.70 |

### *3.2. Moisture Content*
The Figure 5 shows the percentage of moisture content of the five (5) different mixtures. The moisture content of the mixture can be observed.Mixture A hasthe lowest moisture of 18.6% while the mixtures B, C and D contain a moisture content of 21.0%, 22.75% and 30.5% respectively. Mixture E contains the highest moisture content of 45.45%.

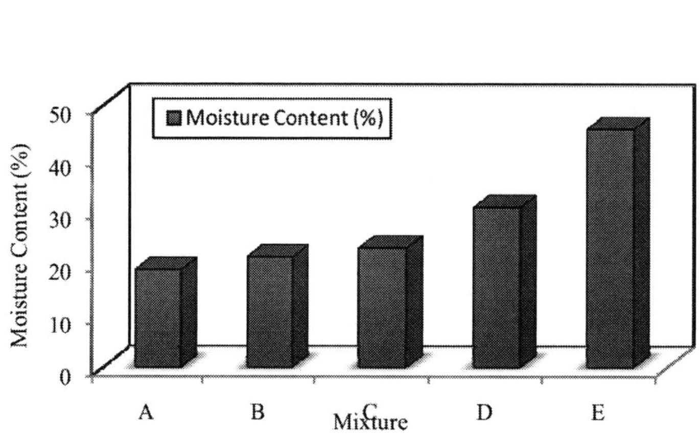


**Figure 5.** Percentage of moisture content (%).

From Figure 5, it was observed that the moisture content of the mixture increases as the ratio of the WTP residual solution increases. The weight of water in the mixture increases as the ratio of WTP residual solution increases in the mixture [8].

*3.3. Dimension test*

The value of shrinkage in terms of volume for the five (5) different mixes of compressed brick for a curing period of 7 days and 28 days respectively is shown in Figure 6. It was observed that mixtures A, B and C did not exhibit any change in volume. Mixture D exhibiteda volume shrinkage of 1.10 $mm^3$and 1.30 $mm^3$ after a curing period of 7 days and 28 days respectively. On the other hand, mixture E shows a volume shrinkage of 1.20 $mm^3$and 1.50 $mm^3$after a curing period of 7 days and 28 days respectively.

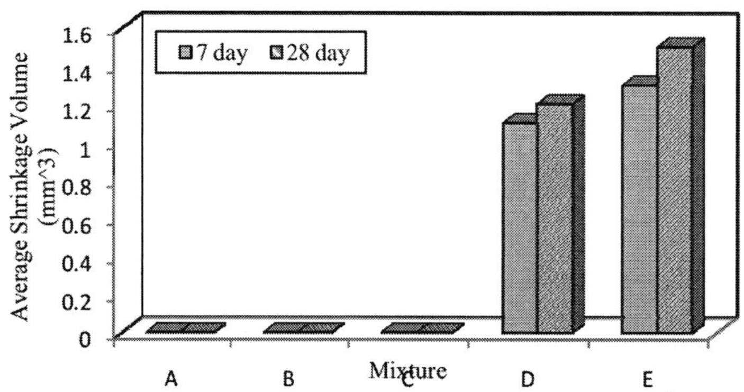


**Figure 6.**Shrinkage volume of compressed bricks ($mm^3$).

The occurrence of volume shrinkage in compressed bricks is due to the dissipation of water during the drying process. Improper drying process and curing practices increase the rate of water dissipation in the compressed brick mixture. The abrupt change in water content will increase the porosity of the resulting compressed brick. The rate of shrinkage and porosity structure in a compressed brick can influence its physical features and the rate of initial water absorption [8].

*3.4. Initial rate of absorption test*

Figure 7 shows that the rate of initial water absorption increases from mixture A to mixture E. The test results show that mixture A had the lowest average percentage rate of initial water absorption with a value of 1.0% for a 60-second test period. This is the followed by mixtures B, C and D with initial water absorption ratesof 3.5%, 5.3% and 8.0% respectively. Mixture E had the highest average percentage rate of initial water absorption (9.1%) for the same test period.

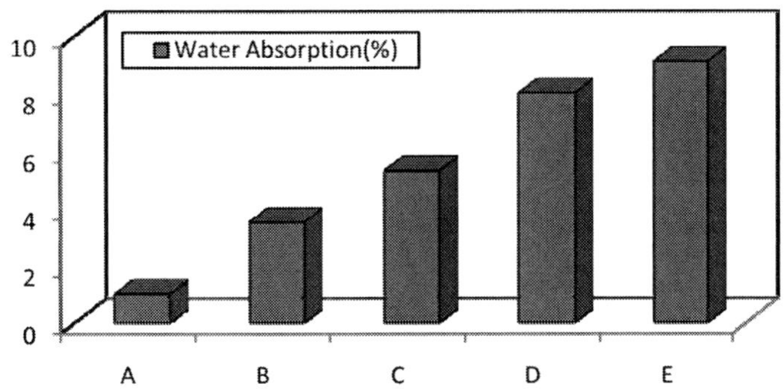


**Figure 7.** Initial rate of water absorption (%).

Based on the outcome of the test results, it was observed that the mixture with a higher ratio of WTP residual solution exhibiteda higher rate of initial water absorption in a test period of 60 seconds.The pattern of the obtained test results is consistent with the findings byAzizul[8] which stated that mixture E exhibited the highest rate of initial water absorption compared to other mixtures with a value of 11.20% in a 60-second test period. According to the research, mixture A had the lowest average rate of initial water absorption at 2.40%, Meanwhile, the average rate of initial water absorption for mixtures B, C and D were 4.9%, 7.9% and 8.6% respectively within the same test period. With the consistent pattern obtained from the test results, it can be concluded that as the ratio of WTP residual solution increases in a mixture, the rate of initial water absorption increases for a test duration of 60 seconds.

*3.5. Rate of absorption test*

Figure 8 shows the graphical presentation of the rate of water absorption in percentage for the five (5) mixtures. It was observed that the rate of water absorption increases from mixture A to mixture E. The value of rate of water absorption in percentage for mixture A, B, C, D and E are 13.22%, 14.65%, 15.51%, 17.61% and 27.81% respectively. The mixture A which is the control sample has the lowest rate of water absorption while mixture E has the highest rate of water absorption.

It was observed that as the WTP residual solution ratio increases in a mixture, the rate of water absorption also increases. Similar patterns were also observed from the analysis of the test results obtained from the moisture content test and the initial water absorption tests.

The pattern of results obtained from the rate of water absorption test was found to be consistent with the research conducted by Azizul[8] which stated that mixture A exhibited the lowest rate of water absorption compared to other mixtures with a value of 12.88% while mixture E has the highest rate of water absorption with a value of 52.31%. The research also indicated that the average rate of water absorption for mixtures B, C and D was 19.26%, 27.87% and 32.46% respectively. With the consistent pattern of test results obtained, it can be concluded that as the ratio of WTP residual solution increases in a mixture, the rate of water absorption increases.

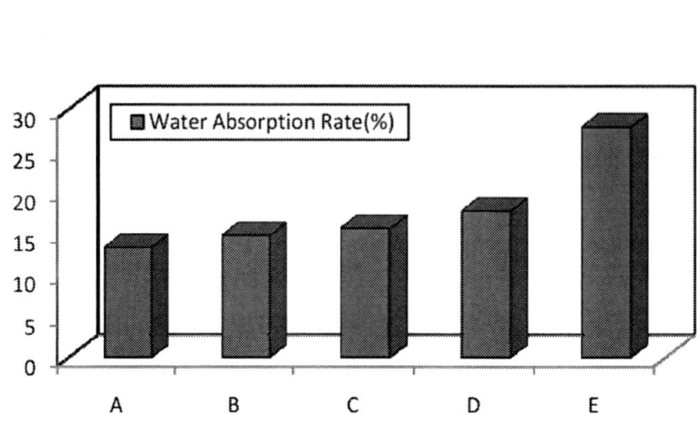


**Figure 8.** Water absorption rate (%).

*3.6. Compressive strength test*

The compressive strength test results for the five (5) mixtures are shown in Figure 9. It was observed that the compressive strength decreases from mixture A to mixture E. Mixture A, the control sample, has the highest average compressive strength with a value of 20.5 $N/mm^2$ and 22.6 $N/mm^2$after a curing period of 7 days and 28 days respectively. The average compressive strength of mixture B was 15.0 $N/mm^2$ and 16.7 $N/mm^2$for a curing period of 7 days and 28 days respectively. Meanwhile, mixture C showed an average compressive strength of 8.5 $N/mm^2$ and 11.8 $N/mm^2$after a curing period of 7 days and 28 days respectively. Mixture D has an average compressive strength of 7.0 N/mm2 and 7.2 $N/mm^2$ while Mixture E has the lowest average compressive strength of 5.2 $N/mm^2$ and 5.5 $N/mm^2$after a curing period of 7 days and 28 days respectively.

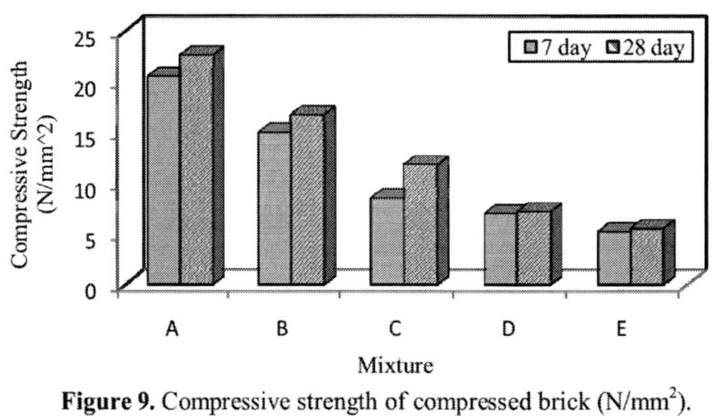


**Figure 9.** Compressive strength of compressed brick ($N/mm^2$).

Based on the compressive strength test results obtained, it was observed that the compressive strength of a compressed brick decreases with the increase of WTP residual solution ratio in a mixture. The pattern of results obtained from the compressive strength test was found to be consistent with the research conducted by Azizul [8] which indicated that mixture A exhibited the highest average compressive strength compared to other mixtures with a value of 6.45 $N/mm^2$ and 8.25 $N/mm^2$ after a curing period of 7 days and 28 days respectively. His research also shows that Mixture B had an average compressive strength of 2.87 $N/mm^2$ and 3.61 $N/mm^2$ while Mixture C had an average compressive strength of 2.45 $N/mm^2$ and 3.54 $N/mm^2$after a curing period of 7 days and 28 days respectively. The average compressive strength for Mixture D was at 1.30 $N/mm^2$(7 days) and 1.71 $N/mm^2$(28 days). The lowest average compressive strength was recorded for mixture E with value of

0.94 N/mm$^2$after a curing period of 7 days and 1.05 N/mm$^2$after a curing period of 28 days. Therefore, it can be concluded that with the increase of WTP residual solution ratio in a mixture, the compressive strength of a compressed brick will decrease.

## 4. Conclusion

Through the observations and analysis of the test results obtained in this study, it can be concluded that the ratio of 20% of WTP Residual Solution in a compressed brick mixture is the optimum value. In this study, the compressed brick produced via mixture D which represents a mix ratio of 10% cement, 20% sand, 20% soil and 20% WTP Residual Solution was found to be the optimum mixture. The compressive strength test shows that mixture D with a ratio of 20% soil and 20% WTP Residual Solution produced a strong, light and durable compressed brick that can be utilised in the construction of non-load bearing and load bearing brick walls.

## 5. References

[1] Mustaffa M 2001 *Kekuatan mampatan dan kebolehserapan bata bakar* Degree Thesis (Skudai, Malaysia:Universiti Teknologi Malaysia)
[2] AliN, Zainal N A,Burhanudin M K, Abdul Samad A A, Shahidan S 2016 Physical and mechanical properties of compressed earth brick (CEB) containing sugarcane bagasse ashvol 47 (Paris: MATEC Web Conference) pp 1–7
[3] RahimM. A. Ghazaly Z, Nurazira R, Mamat R, Azizan M A, Isa N F, Shahidan S 2016 Experimental study of slurry infiltrated fiber reinforced concrete *Mater. Sci. Forum* vol. 857 pp 363–366
[4] Ismail I, Jamaluddin N, Shahidan S 2016 Areview on performance of waste materials in self compacting concrete (SCC), *J. Teknologi* 5(78) pp 29–35
[5] Deraman S 2010 *Investigation of compressive strength foam brickwall panel with different bonding by using stretcher & flemish bond* Degree Thesis (Pahang, Malaysia: Universiti Malaysia Pahang)
[6] SAJ Holdings Sdn. Bhd. 2004Lukisanrekabentukprojekbekalan air Sri Gading (Johor: Ranhill Sri Gading Sdn. Bhd.)
[7] Malaysia Standard MS 76:1972 / British Standard BS 3921: 1985 - General brick specifications
[8] Aziz M A 2014 *Kajian terhadap ciri-ciri fizikal dan mekanikal bata termampat yang Mengandungi bahan enap cemar olahan air* Degree Thesis (Parit Raja, Malaysia:Universiti Tun Hussein Onn Malaysia)
[9] Leman A S, Shahidan S, Senin M S, Ramzi Hannan N I R 2016 Apreliminary study on chemical and physical properties of coconut shell powder as a filler in concrete vol 160 (London: IOP Conf. Ser. Mater. Sci. Eng.) p 12059
[10] Ramzi Hannan N I R, Shahidan S, Maarof M Z, Ali N 2016 Physical and chemical properties of coal bottom ash (CBA) from tanjung bin power plant vol 160(London: IOP Conf. Ser. Mater.) p 12056
[11] Azhar S, Tajudin A, Azim M, Azmi M, Shahidan S 2016 Relationship of physical parameters in Pb-contaminated by stabilization/solidification method vol 47(Paris: MATEC Web Conference) pp 3–8

## Acknowledgment

The authors acknowledge the technical and financial support (Vot U523) provided by Universiti Tun Hussein Onn Malaysia.

# Enhancing the compressive strength of landfill soil using cement and bagasse ash

**M A M Azim[1,2], A T S Azhar[1,2], A K A Tarmizi[2], S Shahidan[2] and A T A Nabila[1,2]**

[1] Research Center for Soft Soil, Universiti Tun Hussein Onn Malaysia, 86400 Batu Pahat Malaysia
[2]Faculty of Civil and Environmental Engineering, Universiti Tun Hussein Onn Malaysia, 86400 Batu Pahat, Malaysia
Corresponding author: mohamadazimazmi@gmail.com

**Abstract.** The stabilisation of contaminated soil with cement and agricultural waste is a widely applied method which contributes to the sustainability of the environment. Soil may be stabilised to increase strength and durability or to prevent erosion and other geotechnical failure. This study was carried out to evaluate the compressive strength of ex-landfill soil when cement and bagasse ash (BA) are added to it. Different proportions of cement (5%, 10%, 15% and 20%) was added to sample weights without BA. On the other hand, the cement in a different batch of sample weights was replaced by 2.5%, 5%, 7.5% and 10% of BA. All samples were allowed to harden and were cured at room temperature for 7, 14 and 28 days respectively. The strength of the contaminated soil was assessed using an unconfined compressive strength test (UCS). The laboratory tests also included the index properties of soil, cement and bagasse ash in raw form. The results indicated that the samples with cement achieved the highest compressive strength measuring 4.39 MPa. However, this study revealed that the use of bagasse ash produced low quality products with a reduction in strength. For example, when 5% of cement was replaced with 5% ash, the compressive strength decreased by about 54% from 0.72 MPa to 0.33 MPa. Similarly, the compressive strength of each sample after a curing period of 28 days was higher compared to samples cured for 7 and 14 days respectively. This is proved that a longer curing period is needed to increase the compressive strength of the samples.

## 1. Introduction

One of the most basic elements which make up our ecosystem is soil. It is necessary for the survival of mankind as well as the development of civilisations [1-2]. Unfortunately, soil is also extremely susceptible to contamination. John *et al.* [3] defines land contamination as areas with high concentrations (above normal background level) of substance which may have been left behind by previous land use. The contamination of soil, especially by heavy metals, may threaten the environment and human health. That is why there is a dire need to seek efficient and cost-effective solutions to curb soil contamination [4].

Soil contamination is mainly caused by industrial activities which generate waste and contaminants which seep into soil through direct disposal, emissions and other pathways [5]. Furthermore, past land use, human activities and the proximity to pollution sources may also have a negative impact on soil. Soil contamination can be caused by substances such as pesticides, fertilizers and other amendments to soil [6-7]. One of the important properties that is affected by contaminants is compressive strength which is vital for soil to withstand axial forces. The compressive strength of soil is important in nearly all geotechnical engineering designs because it is used for obtaining an estimate of the soil strength.

Similarly, unconfined compressive strength (UCS) is often used to provide a baseline comparison between unstable and stabilised wastes. In general, unstable waste materials are unable to provide good shear strength. Nevertheless, soil strength may increase significantly if it is stabilised using certain materials such as concrete [8-9]. This study aims to evaluate the compressive strength of contaminated soil mixed with cement and bagasse ash material.

## 2. Materials and methods

### *2.1. Collection of soil samples*

The contaminated soil samples were collected at the Bukit Bakri Landfill Site (BBLS), Muar, Johor. The top of the soil to a depth of 1 meter was removed in order to avoid taking the humus, waste and plant roots. Then, the soil was placed in polystyrene containers. After that, the soil samples were taken to the laboratory to be dried in the oven at 105°C for 24 hours. After being dried for 24 hours, the soil was crushed using a rubber hammer before being decimated into 2 mm in size using a grinder machine. The soil which passes the 2 mm sieve was stored in polyethylene plastic. Figure 1 shows the soil sample collection site at BBLS, Muar, Johor.

**Figure 1**. Soil sample collection site at BBLS, Muar, Johor.

### *2.2. Cement and Bagasse Ash (BA)*

Cement was used as the primary binder reagent in the stabilisation/solidification (S/S) method. The type of cement that was used in this study was Ordinary Portland Cement (OPC) type 1. Normally, when soft soil is mixed with cement, it can stabilise the soil because cement and water react to form cementitious calcium silicate and aluminate hydrates, which bind the soil particles together [2]. Sugarcane bagasse ash, a type of agricultural waste, was chosen to be incorporated with Ordinary Portland Cement (OPC) in the S/S technique. Tajudin et. al. [10] stated that sugarcane bagasse ash is known as a mineral additive in cementitious materials. Besides that, sugarcane bagasse ash can also reduce the remediation cost normally incurred by cement usage as it is relatively cheap. Sugarcane bagasse ash is also environmentally friendly [11]. The raw form of sugarcane bagasse was dried under the sun and burnt until a grayish-black ash was obtained. These materials are considered as uncontrolled burned bagasse because it is burnt under uncontrolled temperature. The bagasse ash (BA) was further burnt under a controlled temperature at 650° C for one hour. This burning process brought down the carbon content by at least 4.9% [11]. After cooling, the ash was grounded using grinder machines until a particle size of 90 μm was obtained.

*2.3. Production of soil samples*
In this study, the percentage of cement was partially replaced with bagasse ash (BA) as an additive to the soil. A study conducted by Ahmaruzzaman, [12] indicated that 5% to 20% cement content was the optimal percentage for the formation of cementitious product with clay. Another study by Voglar & Lestan [13] stated that 5% and 20% of ordinary Portland cement (by weight to soil) showed good results in the treatment of heavy metals using the S/S method. The weight of each sample was measured to avoid any material waste whereas the quantity of each component was calculated through its percentage in the sample according to Table 1. Sample batches were triplicated for 3 hydration durations namely 7, 14 and 28 days respectively. The sample were mixed in bulk to obtain relative homogeneity prior to packing and storage. The water to cement ratio used in this research has been determined from the optimum moisture content (OMC) obtained from the compaction test which ranged from 0.20 to 0.40 (w/c) depending on the quantity of bagasse ash (BA) added. Next, the raw material was mixed in using a small mixer to ensure the homogeneity of the S/S sample. Later on, a split mould was used to make S/S samples measuring 38 mm in diameter and 76 mm in height. The mixture was compacted in 4 layers using a custom-made hand compacting tool where 50 blows were applied to each layer [14]. The extruded samples were carefully wrapped and kept for 7, 14 and 28 days before the tests were carried out on them. The mix design of the S/S samples are shown in table 1.

**Table 1.** Mix design of S/S samples.

| No. | Mixing Type | Sample Name | Label | Percentage of Binder (%) | | |
|---|---|---|---|---|---|---|
| | | | | Soil | OPC | BA |
| 1. | Soil only (control) | Soil | A | 100 | 0 | 0 |
| 2. | Cement only | soil + 5% cement | B | 95 | 5 | 0 |
| 3. | | soil + 10% cement | C | 90 | 10 | 0 |
| 4. | | soil + 15% cement | D | 85 | 15 | 0 |
| 5. | | soil + 20% cement | E | 80 | 20 | 0 |
| 6. | Cement with BA | soil + 2.5% cement + 2.5% BA | F | 95 | 2.5 | 2.5 |
| 7. | | soil + 5% cement + 5% BA | G | 90 | 5 | 5 |
| 8. | | soil + 7.5% cement + 7.5% BA | H | 85 | 7.5 | 7.5 |
| 9. | | soil + 10% cement + 10% BA | I | 80 | 10 | 10 |
| 10. | BA only | soil + 5% BA | J | 95 | 0 | 5 |

*2.4. Soil properties*
Laboratory tests were conducted to determine the index properties of the natural ex-landfill soil in accordance to BS 1377. Figure 2 shows the particle size distribution curve of ex-landfill soil determined based on dry sieving and a hydrometer test. The index properties of ex-landfill soil are summarised in table 2.

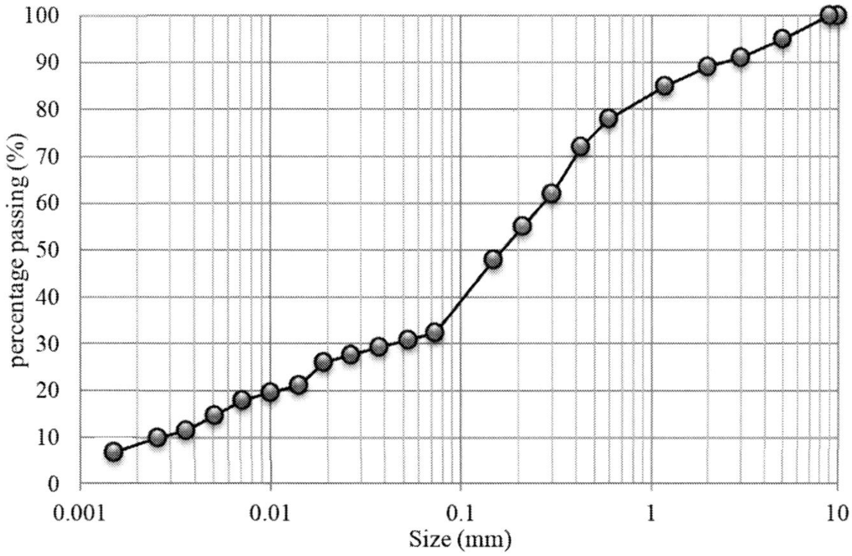


**Figure 2.** Particle size distribution of ex-landfill soil.

**Table 2.** Index properties of ex-landfill soil.

| Properties | Result |
|---|---|
| Moisture Content (%) | 24.25 |
| Particle Size analysis (%) | |
| Gravel (4.75 – 75mm) | 4.0 |
| Sand (4.75-0.075mm) | 63.5 |
| Silt & Clay (<0.075) | 32.5 |
| Atterberg Limit (%) | |
| Liquid limit | 47.15 |
| Plastic limit | 21.31 |
| Plasticity index (%) | 25.84 |
| Linear Shrinkage (%) | 21.4 |
| Specific Gravity | 2.49 |
| Maximum Dry Density (g/cm$^3$) | 1.60 |
| Optimum Moisture content (%) | 18.90 |
| Loss of Ignition (%) | 7.33 |
| Soil type | MH |
| Degree of plasticity | Low plasticity |

### *2.5. Unconfined compressive strength test*

The mixture was casted into cylindrical specimen (D = 38 mm, H = 76 mm) moulds of five layers, where each layer was compacted with 50 blows each. A further 7, 14 and 28 days were allowed for curing and the solidified samples were air dried in a cabinet under controlled conditions (temperature 25±2 ◦C, humidity > 80%) before the test was carried out. The UCS test was carried out using the Geocomp LoadTrac II system available in RECESS at UTHM. The test was conducted according to BS1377-7:1990. The samples were placed in a loading device and subjected to the uniaxial compression test (UCT) at a constant strain at of 1% per minute.

## 3. Results and discussion

### *3.1. Effect on compressive strength*

Compressive strength is an important factor because it determines the load that can be applied to soil [15]. For example, according to U.S. EPA, if materials are disposed of at landfills, they will have a minimum compressive strength of 0.35 MPa which is the minimum strength required for soil to bear heavy equipment that will be used to cut and fill the surface of the area.

Unconfined compression test results showed an increase in strength with the addition of cement and also the addition of ash mixed with cement as shown in figure 3. In addition, the figure also shows that there were 4 types of samples measuring less than 1.0 MPa beginning with soil sample G which consisted of 5% cement + 5% ash. However, the weakest sample was sample A which consisted of soil alone with a compressive strength of less than 0.35 MPa. This may be due to high organic carbon content (2.9%) and a low level of calcium content in the ash (<15%) as well as a relatively low density of 1.3 g/cm$^3$. This is a pozzolanic characteristic where it does not harden or solidify very slowly on its own but increases the strength of the material if it is mixed with cement or lime [7,14].

As expected, the samples with cement were found to achieve the highest compressive strength measuring 4 MPa. The figure shows that ash does not affect the strength of the samples significantly. For example, a soil sample mixed with 5% cement and 5% ash has an average strength of 0.63 MPa compared to the samples which were mixed with 10% of cement alone which achieved compressive strength up to 1.78 MPa. However, small quantities of alkaline mixture do not affect compressive strength. Soil samples mixed with 2.5% cement and 2.5% ash cement have an average compressive strength of 0.40 MPa while soil samples which were mixed with 5% cement only achieved an average compressive strength of 1.40 MPa. This has been proven through research conducted by Cordeiro *et al.* [14] who obtained a compressive strength ranging between 2 to 10 MPa for soil stabilised and solidified with cement after a dry curing period of 28 days depending on the type of binder used.

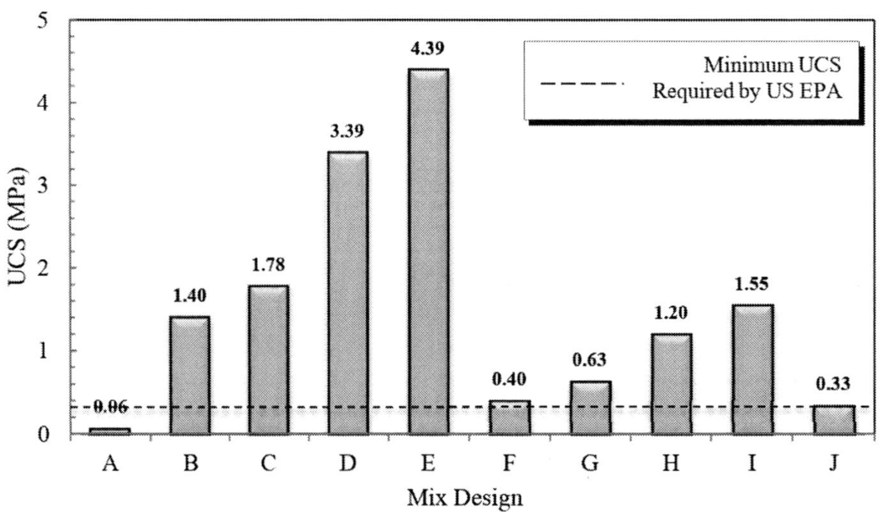


**Figure 3.** Compressive strength of stabilised samples.

### *3.2. Effect of UCS on curing period*

The effect of UCS on curing time is important to measure whether the treated soil gains the expected strength in a specified period. In this study, soil samples were prepared in several mix ratios and cured for a period of 7, 14 and 28 days respectively. Based on the results obtained, the compressive strength of the soil sample for all the mix ratios increases with the curing period. Figure 4 shows the changes in compressive strength for different curing periods. Soil sample E (20% OPC) has the highest compressive strength which is 4.39 MPa after a curing period of 28 days. Based on results obtained, it can be concluded that the increase in the number of curing days will also result in the increase of compressive strength.

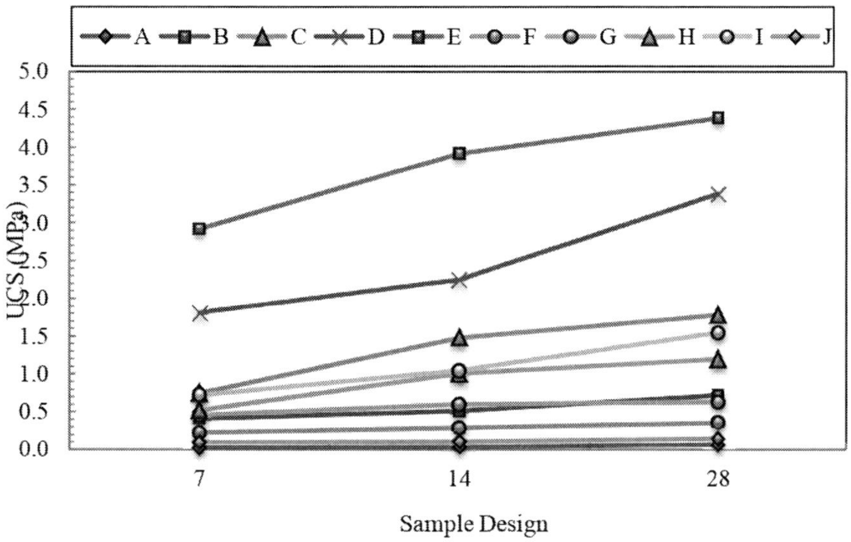


**Figure 4.** The variation in compressive strength for different curing periods.

In addition, the figure also shows that the compressive strength value for each sample after a curing period of 28 days was higher compared to samples cured for a period of 7 days and 14 days respectively. This may be caused by the formation of C-S-H (calcium silicate hydrate) which contributes to strength development in the S/S method [16]. However, the formation of C-S-H only occurs when the samples go through a certain period of hydration which allows water molecules to penetrate into the pores of cementitious products.

## 4. Conclusion

The findings revealed that highlight the stabilised soil by with cement and bagasse ash are successful on improve the compressive strength of samples. The samples containing cement alone show highest value in terms of compressive strength compared to samples containing cement with bagasse ash. However, even these two groups of samples (cement alone and cement with bagasse ash) were contradictory in terms of value. The samples containing bagasse ash still exceeded the minimum landfill disposal limit stated by US EPA of 0.35MPa after a curing period of 28 days. Thus, this implies that the two most important factors in strength development are cement content and curing period. On the other hand, the water/cement ratio is considered to be a secondary factor compared to matrix constituents [17]. This was concurred by a study by Basha *et al.* [18] where it was found that the compressive strength of cement structures is influenced by the quality of pore structures which depend on the type and quantity of constituents (cement hydration products and admixtures).

## 5. References

[1] Yao Z, Li J and Xie H, Yu C 2012 Review on remediation technologies of soil contaminated by heavy metals, *Procedia Environ. Sci.* **16** 722–729
[2] Azhar A T S, Azmi M A M and Nabila A T A 2016 Stabilization/solidification remediation method for contaminated soil: A review *IOP Conf. Series: Materials, Science and Engineering* vol 136 (London: IOP Publishing)

[3] John U E, Jefferson I, Boardman D I, Ghataora G S and Hills C D 2011 Leaching evaluation of cement stabilisation / solidification treated kaolin clay, *Eng. Geol.* **123** 315–323

[4] Azmi M A M, Tajudin S A A, Shahidan S, Masrom M A N and Nabila A T A 2016 Preliminary study on remediation of contaminated clay soil using cement and sugarcane bagasse *IOP Conf. Series: Materials Science and Engineering* vol 136 (London: IOP Publishing) 07001

[5] Voglar G E and Lestan D 2010 Solidification/stabilisation of metals contaminated industrial soil from former Zn smelter in Celje, Slovenia, using cement as a hydraulic binder, *J. of Hazard. Mater.* **178(1-3)** 926–33

[6] Du Y J, Wei M-L, Reddy K R, Liu Z –P and Jin F 2014 Effect of acid rain pH on leaching behavior of cement stabilized lead-contaminated soil *J. Hazard. Mater.* **271** 131–40

[7] Tajudin S A A, Marto A, Azmi M A M, Madun A and Abidin M H Z 2015 Utilization of sugarcane bagasse ash for stabilization/solidification of lead-contaminated soils, *J. Teknologi* **77 (11)** 119-125

[8] Gonzalez-Nunez R, Alba M D, Orta M M, Vidal M and Rigol A 2012 Remediation of metal-contaminated soils with the addition of materials - part II: leaching tests to evaluate the efficiency of materials in the remediation of contaminated soils, *Chemosphere* **87** 829–37

[9] Azhar A T S, Azmi M A M, Aziman M and Nabila A T A 2016 Leachability of arsenic (AS) contaminated landfill soil stabilised by cement and bagasse ash *IOP Conference Series: Materials Science and Engineering* vol 160 (London: IOP Publishing) pp 1–9

[10] Tajudin S A A, Mohammad Azmi M A, Shahidan S, Abidin M H Z and Madun A 2016 Relationship of physical parameters in Pb-contaminated by stabilization/solidification method *MATEC Web of Conf.* vol 47 (Paris: EDP Sciences) 03015

[11] Ganesan K, Rajagopal K and Thangavel K 2007 Evaluation of bagasse ash as supplementary cementitious material, *Cem. Concr. Compos.* **29** 515–524

[12] Ahmaruzzaman M 2010 A review on the utilization of fly ash, *Prog. Energy Combust. Sci.* **36** 327–363

[13] Voglar G E and Lestan D 2010 Solidification/stabilisation of metals contaminated industrial soil from former Zn smelter in Celje, Slovenia, using cement as a hydraulic binder, *J. of Hazard. Mater.* **178** 926–933

[14] Cordeiro G C, Toledo Filho R D, Tavares L M and Fairbairn E M R 2008 Pozzolanic activity and filler effect of sugar cane bagasse ash in Portland cement and lime mortars, *Cement and Concrete Composites* **30(5)** 410–418

[15] Chen Q Y, Tyrer M, Hills C D, Yang X M and Carey P 2009 Immobilisation of heavy metal in cement-based solidification/stabilisation: A review, *Waste Manag.* **29** 390–403

[16] Wang F, Wang H and Al-Tabbaa A 2014 Leachability and heavy metal speciation of 17-year old stabilised/solidified contaminated site soils, *J. Hazard. Mater.* **278** 144–51

[17] Bolan N, Kunhikrishnan A, Thangarajan R, Kumpiene J, Park J, Makino T and Scheckel K 2014 Remediation of heavy metal(loid)s contaminated soils to mobilize or to immobilize?, *J. of Hazard. Mater.* **266** 141–66

[18] Basha E A, Hashim R, Mahmud H B and Muntohar A S 2005 Stabilization of residual soil with rice husk ash and cement, *Constr. Build. Mater.* **19** 448–453

**Acknowledgment**

The authors would like to acknowledge the Center for Graduate Studies, Universiti Tun Hussein Onn Malaysia in providing a fund to produce this research paper.

IOP Conf. Series: Materials Science and Engineering **271** (2017) 012054 doi:10.1088/1757-899X/271/1/012054

# Preliminary evaluation of physical and chemical characterization of waste palm oil shell as cool material replaced in asphaltic concrete as fine aggregate

**M A M Anuar[1], N Anting, S Shahidan[1], Y Y Lee[1], M F Md Din[2], F S Khalid[1] and W M H W Nazri[1]**

[1]Jamilus Research Centre, Faculty of Civil and Environmental Engineering, Universiti Tun Hussein Onn Malaysia, 86400 Johor, Malaysia
[2]Centre for Environmental Engineering and Water Security (IPASA), Research Institute for Environmental Sustainability, Universiti Teknologi Malaysia, 81310 Johor, Malaysia
Corresponding author: nickholas@uthm.edu.my

**Abstract.** Malaysia is one of the biggest producer of palm oil product and currently as main source of economy for the country. During the production of crude palm oil, a large amount of waste material is generated, such as palm oil fibres, palm oil shells and empty fruit bunches. Palm oil shell aggregate (POSA) is identified as a material that shows good potential to be used as a fine aggregate replacement in asphaltic concrete. On other hand, the chemical compound that exist in the Palm Oil Shell (POS) have shown a good potential as reflective component in cool-material. The purpose of this study is to obtain the physical and chemical properties of palm oil shell. The result shows the apparent particle density of Palm Oil Shell is 1.6mg/m$^3$. The specific gravity of palm oil shell was obtained with the value 1.6 and the water absorption amount of palm oil shell recorded from this study was 25.1%. The X-Ray Fluorescence study shows that palm oil shell contains the highest amount of $SiO_2$ (46.412 wt%) and the second highest amount of $Fe_2O_3$ (34.016 wt%), both is the main output of relectivity compound. As a conclusion, waste palm oil shell has a potential to be used as alternative material for fine aggregate replacement. Besides that, the amount of chemical element that consist in palm oil shell which high in $SiO_2$ and $Fe_2O_3$, promising the benefit to mitigate urban heat island as a cooling material agent.

## 1. Introduction

The air temperature in the city is found warmer than rural region is called as Urban Heat Island (UHI) phenomenon. The properties of urban substantial are encourage more into solar heat absorption from solar energy through sun instead of influencing the reflectivity and emissivity [1,2]. Thus, the UHI can become worst due to application of this material. Pavement is one of the main contributors to the UHI phenomenon. About 20 - 39 % of the land surface in the city are covered by asphalt pavement[3,4].The heat generated by the pavement from the sunlight significantly increased the surrounding temperature in the city. Pavements with more reflective surface than traditional pavements, enable evaporative cooling, or other methods that allow the paved surface to remain cooler than traditional pavements [5,6]. This is because cool materials contain high solar reflectance and infrared emittance [7,8]. From the previous study the existence of near infrared reflective compound, such as $SiO_2$, $Al_2O_3$, $Fe_2O_3$ and $TiO_2$

is the main component that contributes to the cooling effects by study the chemical element of the material [9,10].

Oil palm industry refinement in Malaysia included 4.49 million hectares of land and producing 17.73 million tonnes of palm oil [11,12]. The industry becomes bigger , the fresh fruit bunch production is 75 million tons per annum and a significant amount of palm oil waste was generated problem of biomass waste overload, the problem was disposal difficulties and escalates the operating cost [13, 14]. Palm Oil Shell (POS) was partially a waste in 90s and early 2000 considering more than 350,000 tons were available for sale [15,16]. The waste material from oil palm production is also increases year by year [17,18]. Palm Oil shell is a by-product of palm oil processing, the properties of POS show that it can find useful application in light weight construction as material filler, as aggregate replacement and as sorbent material for industrial water treatment [17,19,20]. According to [21] POS can uses as an alternative material for more robust performance of asphalt mixes, specially tailored to the Malaysian condition.

Therefore, this research paper focuses mainly on the properties of waste POS sample to evaluate and investigate the physical and chemical characterization of the POS. Based on the study, the criteria of fine aggregate and cooling agent material for POS can be determined. The POS can be applied onto pavement as a cool material to develop cool pavement technology. On other hand, the utilization of POS waste can be varies and help to overcome the environmental problem.

## 2. Methodology

### *2.1. Material preparation*

The POS waste materials utilized for this research work were collected from palm oil industry company which is Syarikat Perusahaan Kelapa Sawit Yong Peng Sdn. Bhd., where located in Yong Peng, Johor. A material was collected with sufficient amount from the factory and stored in the container. The waste POS were washed to remove dirt, dust or any impurities and free from deleterious material. Furthermore, the POS waste was air-dried naturally under the sunlight for 24 hours. Then, the material was oven dried for 24 hours in a lab-scale oven at 110°C. Once the drying process was completed, the waste POS material was crushed into fine grain size. After that, the crushed palm oil shell was re-cleaning by filtering and segregate the palm oil fibre where stick with POS. Finally, the fine size wasted POS material was sieved to obtain the grading particle size from 1.18mm to 0.075mm for the purpose of fine aggregate replacement.

### *2.2. Physical properties of palm oil shell*

*2.2.1. Particle density and water absorption.* All the measurement is significant to presume the performance of asphaltic concrete when the waste POS replaced with fine aggregate. The determination of particle density and water absorption for fine aggregate were conducted and calculated accordance with the procedure [22–24] standard, which used small pycnometer method.

*2.2.2. Soundness of aggregate by using magnesium sulfate.* Soundness of aggregate by use magnesium sulfate test was performed and calculated based on ASSHTO Part 1: T 104-99 [25] standard. The soundness test determines an aggregate's resistance to disintegration by weathering and freeze-thaw cycles. Aggregates that are durable (resistant to weathering) are less likely to degrade in the field and cause premature hot mix asphalt (HMA) pavement distress and potentially failure.

### *2.3. Chemical properties of palm oil shell*

Chemical analyse was carried out to determine the major chemical component or minerals that are present in the experimental of palm oil shell. The chemical characteristic element of palm oil shell samples was determined by using X-Ray Fluorescence (XRF) analyser (X-Supreme Oxford Instrument). It is a non-destructive analytical technique that used the interaction of x-ray with a material to determine accurately the elemental composition and it has become widely used tool in the laboratory because of its diverse range of application. The rhodium X-ray tube with power as low as 3kW and voltage from 40 to 60kV was equipped in the XRF analyser [26]. Furthermore, to measure the main and trace element that present in the test sample, the sample of palm oil shell were crushed into fine powder form after sieve 63μm in size approximately. The pellet or thin disk with diameter 2.7cm and thickness 4mm were performed and compressed by using 10g the fine powder of palm oil shell sample [10]. The sample was sent to Kolej Kemahiran Tinggi Mara Masjid Tanah, Melaka, for XRF analysis. Figure 1 shows the crushed POS with 1.18 mm and 0.063 mm.

**Figure 1.** Crushed POS.

## 3. Results and Discussions

### *3.1. Particle density and water absorption determination.*

The result of particle density, water absorption and specific gravity of waste POS material obtained from the experimental works are shown in table 1.

**Table 1.** Particle density, water absorption and specific gravity of waste POS material.

| Physical Properties | POS Test Value | Comparison with Previous Studies | | | |
|---|---|---|---|---|---|
| | | [27] | [28] | [29] | [30] |
| Particle Density on oven dried basis | 1.2 | - | - | - | - |
| Particle Density on a saturated and surface dried basis (mg/m$^3$) | 1.4 | - | - | - | - |
| Apparent Particle Density, $\rho_s$ (mg/m$^3$) | 1.6 | 0.65 | 1.6 | - | - |
| Water Absorption (%) | 22.5 | 14 | 23.30 | 24.39 | 23.8 |
| Specific Gravity | 1.6 | 1.62 | 1.17 | 1.4 | 1.3 |

As seen in table 1, the result shows from the particle density and water absorption test of palm oil shell, from the table the value of the specific gravity of palm oil shell obtained is 1.6. Specific gravity of the palm oil shell is not in the category of common rock but its showed the same porosity as granite and limestones [27]. Other than that POS specific gravity is pass the standard of asphalt properties which the specific gravity value need to be exceed 1.0 [24]. As a result, by comparing with previous studies, the range of POS specific gravity value is 1.17 – 1.6. The apparent particle density recorded are much higher than particle density on saturated and surface dried basis and particle density on oven dried basis, which is 1.6mg/m$^3$, 1.4mg/m$^3$ and 1.2mg/m$^3$. From the result, the POS has a great number of pores so the density was changed due to condition applied. Besides that, based on the experimental conducted, the palm oil shell has an amount of water absorption which is 22.5%. Referring the previous study by depending on type of trees of POS obtained, it can be described the value of water absorption varies between range 14% – 25%. Thus, it shows palm oil shell absorbed more water to be compared with normal fine aggregate. This is because of POS is a high in fibre, so the presence of pores is great in that material, hence the material absorbed more water. Meanwhile, the studies indicate pumice aggregate absorbed more than 37% of water, pumice are categories as lightweight aggregate [27, 31].

### 3.2. *Soundness of aggregate by use magnesium sulfate*

The result of the soundness value of waste POS material obtained from the experimental works is shown in figure 2.

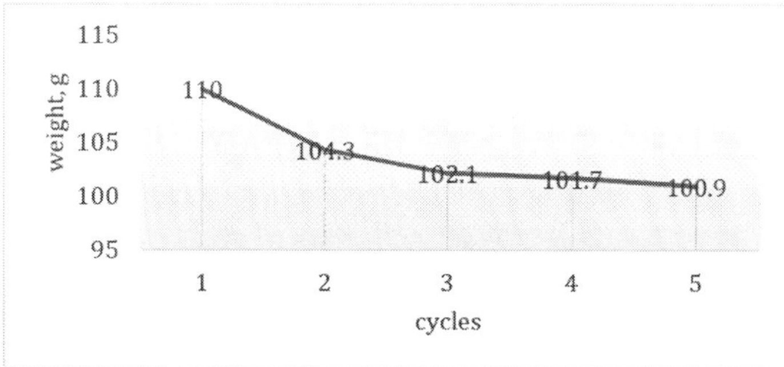


**Figure 2**. Soundness aggregate value (The erosion of POS).

Figure 2 shows the soundness test value of palm oil shell to analyse physical characteristic of the material. As shown, the palm oil shell was experience the erosion when the test was carried out. Initial weight of the sample before tested are 110g and after 5 cycles of immersion the POS in the magnesium sulphate, the result shows POS has loss 8.27% of the weight. Based on the specification by Public Road Department of Malaysia, the requirement of weighted average loss for fine aggregate that used in asphaltic concrete must not more than 20%, hence the POS are acceptable to be used as fine aggregate for replacement.

### 3.3. *Chemical properties of waste palm oil shell*

This section discusses the major pure element which is present in POS aggregate. The result of XRF analysis for the chemical compound for the waste POS is shown in table 2.

**Table 2.** Chemical composition of wasted palm oil shell.

| Compound | Concentration Wt % | | | | |
|---|---|---|---|---|---|
| | POS Concentration Value | Comparison with Previous Studies | | | |
| | | [32] | [14] | [33] | [29] |
| MgO | 0.662 | 6.11 | 4.8 | 3.19 | 3.7 |
| $Al_2O_3$ | 7.739 | 11.4 | 11.4 | 2.47 | 5.5 |
| $SiO_2$ | 46.412 | 54.81 | 43.6 | 63.72 | 63.4 |
| $P_2O_5$ | 2.448 | - | - | - | - |
| $SO_3$ | 2.276 | - | 2.8 | 0.77 | 0.9 |
| Cl | 0.767 | - | - | - | - |
| $K_2O$ | 1.928 | 6.25 | 0.87 | 6.20 | 6.3 |
| CaO | 3.027 | 8.79 | 3.5 | 7.16 | 4.3 |
| $TiO_2$ | 0.219 | - | - | - | - |
| $Cr_2O_3$ | 0.258 | - | - | - | - |
| $Mn_2O_3$ | 0.085 | - | - | - | - |
| $Fe_2O_3$ | 34.016 | 0.36 | 4.7 | 1.39 | 4.2 |
| ZnO | 0.165 | - | - | - | - |

The result XRF analysis for major chemical composition element that present in palm oil shell is presented in table 2. As can be seen, the test result generally show silica oxide ($SiO_2$) was most abundant in the samples followed by iron oxide ($Fe_2O_3$). The composition of $SiO_2$ and $Fe_2O_3$ which present in POS material are more than 40% and 30% each element and the weight percentage are dominantly higher than other elements. Other than that, Aluminium oxide ($Al_2O_3$) has a possibility to contribute to NIR solar reflectance although the element compound present in palm oil shell material nearly 10 wt% of its present. Based on previous study, Silica Dioxide, $SiO_2$ acquire a high reflectivity tint where it is suitable to become material for cool pavement [34]. Appreciable amount of CaO, $P_2O_5$, $SO_3$ and $K_2O$ were also present with traces less than 1% of each oxide is MgO, Cl, $TiO_2$, $Cr_2O_3$, $Mn_2O_3$ and ZnO. The value of MgO is different compared to previous studied because of the the type of palm oil trees and the presence of impurities inside POS samples. As a result, by comparing with previous studies and considered different species of palm oil trees, POS material contain high Silica dioxide with range 46wt% to 63wt%. Hence, these pure element are considered high NIR reflective agent, based on NIR reflectance value stated by previous study[9]. There are three type solar reflectance, ultraviolet (UV), visible region (VIS), and near infrared region (NIR). In addition, 50% of solar energy occurs in the VIS and 45% from NIR [6]. Thus, the higher VIS reflectivity the more effective to reduce heat penetration and surface temperature. However, it will affect human vision which cause by glare effect when the VIS reflectivity is high. Hence, higher NIR reflectivity can be considered to reduce the heat. Subsequently, palm oil shell has the potential as a reflective agent to perform solar reflectance due to the presence of high quantity of $SiO_2$ and $Fe_2O_3$. The increasing reflectance of pavement is important to reflect the solar radiation onto pavement to mitigate UHI, thus the surface temperature will reduce. The reflectivity characteristic in POS when used in pavement can be a combating technology of this problem.

## 4. Conclusion

This study has been carried out to evaluate the physical and chemical properties of waste palm oil shell. Palm oil shell as partial replacement of fine aggregate in asphaltic concrete has been used because of its possibility to reduce surface temperature of pavement. Finding show that palm oil shell specific gravity value exceeds the value of asphalt properties. The palm oil shell absorbed more water than normal aggregate which is the 22.5%. Moreover, the result also shows the amount of major chemical element existed in palm oil shell. The present of $SiO_2$ element in a palm oil shell waste material encourages the high amount of solar reflectance of NIR part in the solar spectrum. Other element including $SiO_2$, $Fe_2O_3$, $Al_2O_3$ and $T_iO_2$ are also possible for NIR reflectance contribution. In addition, based on finding, palm oil shell waste as fine aggregate in general have the potential as UHI mitigation when replaced with fine aggregate in asphaltic concrete.

## 5. References

[1] Gui J G, Phelan P E, Kaloush K E and Golden J S 2007 Impact of pavement thermophysical properties on surface temperatures, *J. Mater. Civ. Eng.* **19**(8) 683–690
[2] Barnes K B, Roberge J M M Iii M C and Roberge M 2002 *Impervious Surfaces and The Quality of Natural and Built Environment* (Maryland: Baltimore)
[3] Akbari H 2005 Energy Saving Potentials and Air Quality Benefits of Urban Heat Island Mitigation *Sol. Energy* 1–19
[4] Shahidan S, Ismail I, Zulkarnaian M S S and Rahman N A 2017 Influence of Asphalt Dust Waste Material in Mix Design for Self-Compacting Concrete, *Key Eng. Mater* **73** 473–478
[5] Rymer B and Levine K K 2011 *Cool Pavements Research and Technology* (United States: UC Berkeley Transportation Library)
[6] Anting N, Din M F Md, Iwao K, Ponraj M, Jungan K, Lee Y Y and Siang A J L M 2017 Experimental evaluation of thermal performance of cool pavement material using waste tiles in tropical climate, *Energy Build* **142** 211–219,.
[7] Santamouris M, Synnefa A and Karlessi T 2011 Using advanced cool materials in the urban built environment to mitigate heat islands and improve thermal comfort conditions, *Sol. Energy* **85**

(12) 3085–3102
[8] Anting N, Din M F Md, Ponraj M and Iwao K (2014) Thermal Performance of Developed Coating Material as Cool Pavement Material for Tropical Regions, *J. Mater. Civ. Eng* **26** 755–760
[9] Thongkanluang T, Chirakanphaisarn N and Limsuwan P 2012 Preparation of NIR reflective brown pigment *Procedia Engineering* vol 32 (United Kingdom: Elsevier) pp 895–901
[10] Guntor N A, Md Din M F, Ponraj M, Iwao K and Jungan K 2014 Physical And Chemical Charaterization Of Wasted Tiles As Coating Materials For Pavement Surface Temperature Reduction, *Malaysian J. Civ. Eng.* **26**(3) 446–455
[11] Malaysian Palm Oil Council 2016 *The Oil Palm Tree* Retrieve On January 1, 2017 from http://www.mpoc.org.my.
[12] Malaysian Palm Oil Council 2012 *The Oil Palm Tree* Retrieve On January 1, 2017 from http://www.mpoc.org.my.
[13] Awalludin M F, Sulaiman O, Hashim R and Nadhari W N A W 2015 An overview of the oil palm industry in Malaysia and its waste utilization through thermochemical conversion, specifically via liquefaction, *Renew. Sustain. Energy Rev.* **50** 1469–1484
[14] Borhan M N, Ismail A and Rahmat R A 2010 Evaluation of Palm Oil Fuel Ash ( POFA ) on Asphalt Mixtures, *Aust. J. Basic Appl. Sci.* **4**(10) 5456–5463
[15] Dit M Palm Kernel Shell (PKS) is More Than Biomass for Alternative Fuel After 2007 *Proc. of the Int. Palm Oil Congress (Chemistry & Technology) (Kuala Lumpur, Malaysia)* vol 35 (Kuala Lumpur: Malaysian Palm Oil Council) pp. 275–283
[16] Shahidan S, Leman A S, Senin M S and Hannan N I R R 2017 Suitability of Coconut Shell Concrete for Precast Cool Wall Panel-A Review *MATEC Web of Conf.* vol. 87 (Paris: EDP Sciences) 1005
[17] Putri E E, Adrian A, Hariadi M and Makinda J 2015 Using oil palm shell aggregate (OPSagg) for flexible pavements, *Electron. J. Geotech. Eng.* **20**(8) 2041–2054
[18] Leman A S, Shahidan S, Senin M S and Hannan N I R R 2016 A Preliminary Study On Chemical and Physical Properties of Coconut Shell Powder As A Filler In Concrete *IOP Conf. Series: Materials Science and Engineering* vol 160 (London: IOP Publishing) 12059
[19] Okoroigwe E C, Christopher M S and Pascal D K 2014 Characterization of palm kernel shell for materials reinforcement and water treatment, *J. Chem. Eng. Mater. Sci.* **5**(1) 1–6
[20] Leman A S, Shahidan S, Yusuf M Y, Zuki S S M and Misnon M A 2017 Workability and Compressive Strength for Concrete With Coconut Shell Aggregate 3 Behaviour Of Concrete With Coconut *MATEC Web of Conf.* vol. 87 (Paris: EDP Sciences) 1017
[21] Al Mansob R, Ismail A and Algorafi M 2013 Comparison between Mixtures of Asphalt with Palm Oil Shells and Coconut Shells as Additives, *J. Kejuruter.* **25** 25–31
[22] BS 812 1995 *Testing Aggregate - Part 2: Method of Determination of Density BS 812* (London: British Standard Institution) pp 3-14
[23] BS 812:Part2:1995 *Testing Aggregate Part2. Methods of determination of density* 3rd Issue (London: British Standard Institution) pp 4-12
[24] Jabatan Kerja Raya Malaysia 2008 *Section 4: Flexible Pavement JKR/SPJ/2008-S4 in Standard Specification for Road Work* (Kuala Lumpur: Jabatan Kerja Raya Malaysia) pp. 24–25
[25] AASHTO T 104 2004 Soundness of Aggregate by Use of Sodium or Magnesium Sulfate, Part 2A in *Standard Specification for Transportation Materials and Method of Sampling and Testing* 24th ed (Washington, DC: American Association of State Highway and Transportation Official) pp T104
[26] Kadir A A, Mohd Zahari N A and Azizi Mardi N 2013 Utilization of palm oil waste into fired clay brick, *Adv. Environ. Biol.* **7** (SPEC. ISSUE 12) 3826–3834
[27] Ndoke P N 2016 Performance of Palm Kernel Shells as a Partial replacement for Coarse Aggregate in Asphalt Concrete, *J. Pract. Technol.* 2–7
[28] Mannan M A and Ganapathy C 2004 Concrete from an agricultural waste-oil palm shell (OPS), *Build. Environ.* **39**(4) 441–448

[29] Liu M Y J, Alengaram U J, Jumaat M Z and Mo K H 2014 Evaluation of thermal conductivity, mechanical and transport properties of lightweight aggregate foamed geopolymer concrete, *Energy Build.* **72** 238–245
[30] Johnson Alengaram U, Al Muhit B A, Jumaat M Z and Jing M L Y 2013 A comparison of the thermal conductivity of oil palm shell foamed concrete with conventional materials, *Mater. Des.* **51** 522–529
[31] Alengaram U J, Al Muhit B A and Jumaat M Z 2013 Utilization of oil palm kernel shell as lightweight aggregate in concrete - A review, *Constr. Build. Mater.* **38** 161–172
[32] Nwaobakata C and Agunwamba J C 2014 Effect of palm kernel shells ash as filler on the mechanical properties of hot mix asphalt, *Arch. Appl. Sci. Res.* **6**(5) 42–49
[33] Mujah D 2016 Compressive strength and chloride resistance of grout containing ground palm oil fuel ash, *J. Clean. Prod.* **112** 712–722
[34] Lavat A E, Trezza M A and Poggi M 2009 Characterization of ceramic roof tile wastes as pozzolanic admixture, *Waste Manag.* **29**(5) 1666–1674

**Acknowledgments**
This paper was partly sponsored by the Centre for Graduate Studies UTHM. The authors express gratitude to Research, Innovation Commercialization and Consultancy Management Office UTHM, the Short-Term Grant Vot. No. U344, Graduate Research Assistant Grant Vot. No. U447, My Master Scholarship from MyBrain15 Program, Universiti Tun Hussein Malaysia for the facility and Ministry of Higher Education Malaysia.

IOP Conf. Series: Materials Science and Engineering **271** (2017) 012055 doi:10.1088/1757-899X/271/1/012055

# Compressive and tensile strength for concrete containing coal bottom ash

**A I F Ahmad Maliki [1], S Shahidan[1], N Ali[1], N I R Ramzi Hannan[1], S S Mohd Zuki[1], M H W Ibrahim[1], M A Mohammad Azmi[1] and M. Abdul Rahim[2]**

[1]Jamilus Research Center,Faculty of Civil and Environmental Engineering, Universiti Tun Hussein Onn Malaysia, Batu Pahat, Johor, Malaysia
[2]School of Environmental Engineering, Universiti Malaysia Perlis, Jejawi, Perlis, Malaysia
Corresponding author:shahiron@uthm.edu.my

**Abstract.** The increasing demand in the construction industry will lead to the depletion of materials used in construction sites such as sand. Due to this situation, coal bottom ash (CBA) was selected as a replacement for sand. CBA is a by-product of coal combustion from power plants. CBA has particles which are angular, irregular and porous with a rough surface texture. CBA also has the appearance and particle size distribution similar to river sand. Therefore, these properties of CBA make it attractive to be used as fine aggregate replacement in concrete. The objectives of this study were to determine the properties of CBA concrete and to evaluate the optimum percentage of CBA to be used in concrete as fine aggregate replacement. The CBA was collected at Tanjung Bin power plant. The mechanical experiment (compressive and tensile strength test) was conducted on CBA concrete. Before starting the mechanical experiment, cubic and cylindrical specimens with dimensions measuring 100x100x100mm and 150x300mm were produced based on the percentage of coal bottom ash in this study which is 0% as the control specimen. Meanwhile 10%, 20%, 30%, 40%, 50%, 60%, 70%, 80%, 90% and 100 % of CBA were used to replace the fine aggregates. The CBA concrete samples were cured for 7 days and 28 days respectively to maintain the rate of hydration and moisture. After the experimental work was done, it can be concluded that the optimum percentage of CBA as fine aggregate is 60% for a curing period of both 7 days and 28 days with the total compressive strength of 36.4Mpa and 46.2Mpa respectively. However, the optimum percentage for tensile strength is at 70% CBA for a curing period of both 7 days and 28 days with a tensile strength of 3.03MPa and 3.63MPa respectively.

## 1. Introduction

Concrete Coal bottom ash (CBA) is a by-product of coal combustion. The rock detritus filled in the fissures of coal becomes separated from the coal during pulverization. Swirling air carries the ash particles out of the hot zone where it cools down. The boiler flue gas carries away the finer and lighter particles of coal ash. The boiler flue gases pass through the electrostatic precipitators before reaching the environment. In the electrostatic precipitators, coal ash particles are extracted from the boiler flue gases [1-5]. The coal ash collected from the electrostatic precipitators is called fly ash. Fly ash constitutes about 80% of coal ash. During the combustion process some particles of the coal ash accumulate on the furnace walls and steam pipes in the furnace and form clinkers. These clinkers build up and fall to the bottom of furnace. In addition, the coarser particles, which remain in suspension with the flue gases, settle down at the base of the furnace. The ash collected at the bottom of furnace is called coal bottom ash. Coal bottom ash constitutes about 20% of coal ash and the rest is fly ash [6-9].

CBA is a coarse granular and incombustible by-product from coal-burning furnaces. It is composed of mainly silica, alumina and iron with small amounts of calcium, magnesium sulphate and etc. The appearance and particle size distribution of CBA is similar to that of river sand. These properties of coal bottom ash make it attractive to be used as fine aggregate in the production of concrete [8,10,11]. CBA is a potentially viable material to be used as fine aggregate in the production of concrete [12]. Bottom ash exhibits high shear strength and low compressibility [13-15]. These engineering properties make bottom ash an ideal material in the design construction of dams and other civil engineering applications. Bottom ash also exhibits a relatively high permeability and grain size distribution that allows the design engineer to use it in direct contact with impervious material. Bottom ash has proven to be an economical material and a good engineering property. The previous studies in which coal bottom ash has been targeted as a substitute for sand in the production of concrete demonstrated that the strength development pattern of CBA concrete is similar to that of conventional concrete [14-19].

A research was conducted by Yahya *et al.* [2] on the properties of concrete using Tanjung Bin power plant CBA and fly ash. CBA and fly ash were utilised to partically replace fine aggregates and cement respectively in varying percentages of 0%, 5%, 10%, 15% & 20%. The results showed that grade M35 concrete combined with CBA and fly ash can achieve 30MPa in strength after a curing period of 28 days. Kurama *et al.* [10] and Ramzi [11] concluded that the use of CBA and fly ash in concrete has the potential to produce durable and strong concrete. Singh and Siddique [1] studied the effects of coal bottom ash as sand replacement on the properties of concrete with different w/c ratio. Ramzi [11] concluded that the compressive strength, split tensile strength and flexural strength of concrete increased after a curing period of 7 days, 10 days, 14 days, 28 days, 56 days and 90 days when up to 30% of CBA is used to replace fine aggregrates. Besides that, the research work carried out by Ramzi [11] indicated that the compressive strength of concrete mixtures made of CBA as a substitute for sand was not strongly affected. However, the flexural strength and modulus of elasticity of concrete decreased with the increase in the content of CBA. Even though a lot of research has been conducted, there is still a lack of research on the use of CBA in concrete. Thus, this study will emphasise on the incorporation of CBA in concrete in varying percentages.

## 2. Material and procedure

### 2.1. *Materials*

The constituent materials used in this research were cement, coarse aggregate with a nominal size of (4.75-20) mm, water, fine aggregate and coal bottom ash (CBA).Coal bottom ash (CBA) was collected at the Tanjung Bin power plant. The coal bottom ash will be used in this test as a partial replacement of fine aggregate to produce concrete. Figure 1 show example of coal bottom ash.

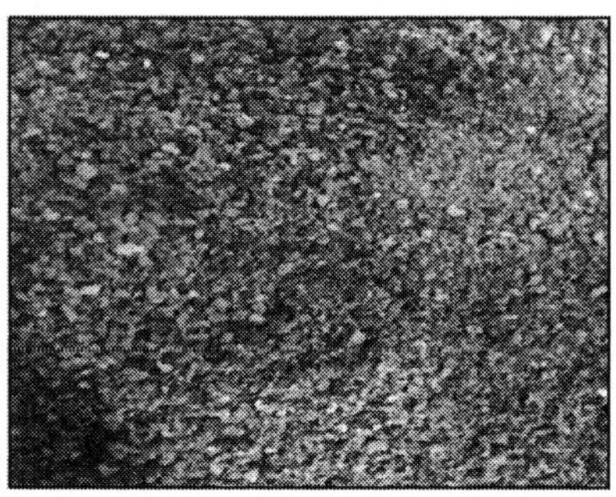

**Figure 1.** Coal Bottom Ash.

### *2.2. Mix design*
The cement content, aggregate content and water content for concrete was determined by using standard mix design of concrete (DOE Method). Table 1 presented the design of mixture used in this study.

**Table 1.** Design mix of concrete.

| Percentage (%) | Coal Bottom Ash (Kg) | Cement (kg) | Water (kg/L) | Fine aggregate (kg) | Coarse aggregate (kg) |
|---|---|---|---|---|---|
| 0 | 0 | 5.89 | 2.65 | 5.94 | 12.10 |
| 10 | 0.31 | 5.89 | 2.65 | 5.35 | 12.10 |
| 20 | 0.62 | 5.89 | 2.65 | 4.75 | 12.10 |
| 30 | 0.93 | 5.89 | 2.65 | 4.16 | 12.10 |
| 40 | 1.24 | 5.89 | 2.65 | 3.56 | 12.10 |
| 50 | 1.55 | 5.89 | 2.65 | 2.97 | 12.10 |
| 60 | 1.87 | 5.89 | 2.65 | 2.38 | 12.10 |
| 70 | 2.18 | 5.89 | 2.65 | 1.78 | 12.10 |
| 80 | 2.49 | 5.89 | 2.65 | 1.19 | 12.10 |
| 90 | 2.80 | 5.89 | 2.65 | 0.59 | 12.10 |
| 100 | 3.11 | 5.89 | 2.65 | 0.00 | 12.10 |

### *2.3. Specimens and tests*
Concrete cubes of 150 x 150 x 150 mm (66 numbers of Samples) were used to evaluate the compressive strength. However, concrete cylinders of 150 x 300 mm (66 numbers of sample) were used to evaluate the tensile strength. Compressive strength test was conducted based on BS EN 12390-3: 2009 [9]. Three cubes were tested for each percentage and age. However, tensile strength was carried out as specified in BS1881-117 (Testing Concrete - Method for determination of tensile splitting strength. British Standards Institute, London). Three cylindrical specimens were tested for each age and percentage. Figure 2 and figure 3 shows those specimens under compressive and tensile strength test.

**Figure 2.** Compressive strength test.

**Figure 3.** Tensile strength test.

## 3. Result and discussion
In this paper, all the results and analysis from the laboratory work that had been conducted are presented. The optimum percentages of CBA as a partial sand replacement and the suitability of CBA in concrete are explained in detail.

### *3.1. Slump test*
Figure 4, summarized the slump values of CBA concrete mixtures decreased with the increase of percentage of CBA in concrete. By replacing 10% up to 40% of CBA in the mixture, its indicate in range

of true slump compare to others. However the slump value for 100% containing of CBA, shown drastically reduce compared to control sample approximately 25 mm of slump. Consequently, the good percentages for replacing the CBA in concrete mixtures are between range 10%-40%. The range also supported by previous researcher with the range of 5% - 45% of CBA [10-11].

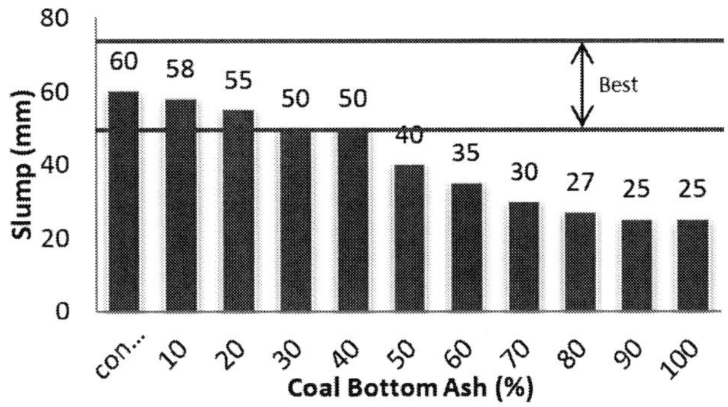


**Figure 4.** Relationship between slump and concrete containing different percentage of CBA.

### *3.2. Compressive strength*

Figure 5 and figure 6, summarized the values of compressive strengths for all mixes at 7 and 28 days of curing respectively. From the figures, it can be observed that the graph trend shows the strength decreased significantly at 10% of CBA in concrete but the strength increased slowly between 20% until 40% CBA concrete. At 50%, the strength of the concrete decreased again. At 60% of CBA concrete, the strength value increased significantly and gradually decreased until it reached 100% of CBA concrete.

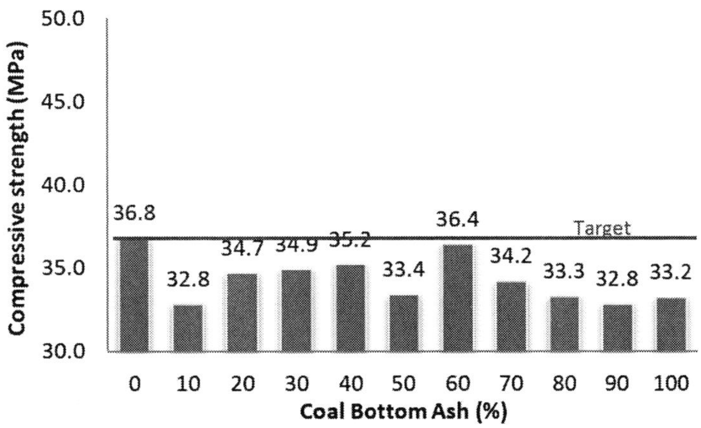


**Figure 5.** Compressive strength of CBA for 7 days.

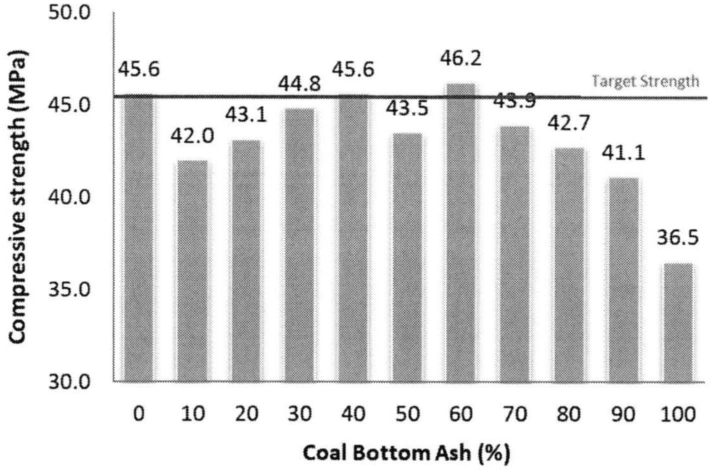


**Figure 6.** Compressive strength of CBA for 28 days.

The optimum percentage to achieve the maximum strength is 60% of CBA as sand replacement in concrete. However, according to Singh [1], the optimum strength was achieved by 30% CBA concrete cured for 7 and 28days. The delay in hydration due to the use of CBA as fine aggregate in concrete may be a possible explanation for the varying results achieved in terms of optimum strength. This may also be due to the strength of individual constituent materials of concrete which can also influence the strength of concrete as stated by Ramzi [11].

### *3.3. Tensile strength*

Figure 7 and figure 8 show the results of tensile strength. It can be summarised that all percentages of CBA in concrete as sand replacement for 7 days achieved lower strength than the target requirement of control concrete which is 32.9Mpa. However for 28days, the varying percentages of CBA as fine aggregate replacement significantly increased in strength compared to concrete cured for 7 days. The strength of 90%, 100% and 70% CBA concrete surpassed 3.0 MPa in strength. However, 70% CBA concrete achieved a value exceeding the tensile strength of control concrete which is 3.63MPa. Currently, the tensile strength of control concrete is 3.50MPa. Therefore, the optimum percentage of CBA in concrete to be used as fine aggregate replacement is 70%.

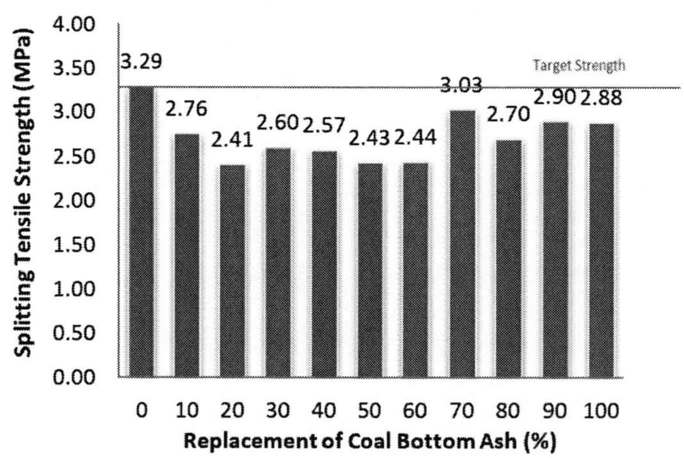


**Figure 7.** Tensile strength of CBA for 7 days.

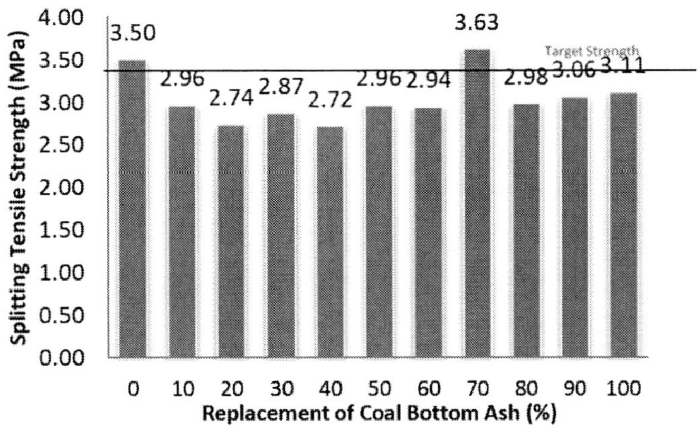


**Figure 8.** Tensile strength of CBA for 28 days.

There are varying results of tensile strength based on previous research studies. According to Singh [1], the optimum strength for 28 days is at 50%. However, according to Ramzi [11] the optimum tensile strength is up to 30%. The data obtained from this research shows that the optimum tensile strength achieved was at a higher percentage compared to other researchers.

## 4. Conclusion

In a nutshell, the optimum percentage of CBA as a fine aggregate replacement is range of 40% - 70%. This range is according to the testing of slump, compressive and tensile strength. The optimum percentage of CBA as a fine aggregate replacement from this paper indicated that slightly different compared to other researchers. Therefore its need more detail investigation on durability and detail percentage of CBA in mixture.

## 5. References

[1] Singh M and Siddique R 2015 Properties of concrete containing high volumes of coal bottom ash as fine aggregate, *J. Clean Prod.* **91** 269–278

[2] Yahya A A, Ali N, Mohd Kamal N L, Shahidan S, Beddu S, Nuruddin M F and Shafiq N 2017 Reducing Heavy Metal Element from Coal Bottom Ash by Using Citric Acid Leaching Treatment *MATEC Web of Conf.* vol 103 (Paris: EDP Sciences) 01004

[3] Ramzi H N I R, Shahidan S, Ali N and Maarof M Z 2017 A Comprehensive Review on the Properties of Coal Bottom Ash in Concrete as Sound Absorption Material *MATEC Web of Conf.* vol 103 (Paris: EDP Sciences) 01005

[4] Ismail I and Shahidan S 2017 A Preliminary Modified Volumetric Mix Design for Self-Compacting Concrete (SCC) by Utilizing Asphalt Dust Waste (ADW) *MATEC Web of Conf.* vol 103 (Paris: EDP Sciences) 01008

[5] Khalid F S, Herman H S and Azmi N B 2017 Properties of Sugarcane Fiber on the Strength of the Normal and Lightweight Concrete) *MATEC Web of Conf.* vol 103 (Paris: EDP Sciences) 01021

[6] Shahidan S, Aminuddin E, Hannan N I R R and Bahari N A S 2017 Potential of Hollow Glass Microsphere as Cement Replacement for Lightweight Foam Concrete on Thermal Insulation Performance *MATEC Web of Conf.* vol 103 (Paris: EDP Sciences) 01014

[7] Khalid F S, Herman S H, Azmi N B and Juki M I 2017 Sand Cement Brick Containing Recycled Concrete Aggregate as Fine-Aggregate Replacement *MATEC Web of Conf.* vol 103 (Paris: EDP Sciences) 01021

[8] Raju R, Paul M M and AboobackerK A 2014 Strength Performance of Concrete Using Bottom Ash As Fine, *Int. J. Res. Eng. Technol.* **2**(9) 111-122.

[9] Singh M and Siddique R 2013. Effect of coal bottom ash as partial replacement of sand on properties of concrete, *Resour. Conserv. Recy.* **72** 20–32
[10] Kurama H and Kaya M 2008 Usage of coal combustion bottom ash in concrete mixture, *Constr. Build. Mater.* **22**(9)1922–1928
[11] Ramzi N I R, Shahidan S, Maarof M Z and Ali N 2016 Physical and Chemical Properties of Coal Bottom Ash (CBA) from Tanjung Bin Power Plant *IOP Conf. Series Materials Science and Engineering* vol 160 (London: IOP Publishing) 12056
[12] Senin M S, Shahidan S, Leman A S and Hannan N I R R 2016 Analysis of Physical Properties and Mineralogical of Pyrolysis Tires Rubber Ash Compared Natural Sand in Concrete material *IOP Conf. Series Materials Science and Engineering* vol 160 (London: IOP Publishing) 12053
[13] Senin M S, Shahidan S, Leman A S and Hannan N I R R 2016 Properties of Cement Mortar Containing Rubber Ash as Sand Replacement *IOP Conf. Series Materials Science and Engineering* vol 160 (London: IOP Publishing) 12055
[14] Lynn CJ, Dhir OBE R K and Ghataora G S 2016 Municipal incinerated bottom ash characteristics and potential for use as aggregate in concrete, *Constr. Build. Mater.* **127** 504–517
[15] Shahidan S, Senin M S, Kadir A A, Yee L H and Ali N 2017 Properties of Concrete Mixes with Carwash Wastewater *MATEC Web of Conf.* vol 87 (Paris: EDP Sciences) 1018
[16] Shahidan S, Leman A S, Senin M S and Hannan N I R R 2017 Suitability of Coconut Shell Concrete for Precast Cool Wall Panel-A Review *MATEC Web of Conf.* vol 87 (Paris: EDP Sciences) 1005
[17] Leman A S, Shahidan S, Yusuf M Y, Mohd Zuki M M and Misnon N A 2017 Workability and Compressive Strength for Concrete With Coconut Shell Aggregate 3 Behaviour Of Concrete With Coconut *MATEC Web of Conf.* vol 87 (Paris: EDP Sciences) 1017
[18] Shahidan S, Ismail I, Zulkarnaian M S S and Abd Rahman N 2017 Influence of Asphalt Dust Waste Material in Mix Design for Self-Compacting Concrete, *Key Eng. Mater.* **730** 473-478
[19] Raju R, Paul M M and Aboobacker K A 2014 Strength Performance of Concrete Using Bottom Ash As Fine, *Int. J. Res. Eng. Technol.* **2**(9) 111-122

**Acknowledgements**
The author would like to thank to Jamilus Research Center, University Tun Hussein Onn Malaysia and Grant VOT U523, Ministry of Higher Education, Malaysia

# Rutting resistance of asphalt mixture with cup lumps modified binder

**E Shaffie[1], W M M Wan Hanif[1], A K Arshad[1,2] and W Hashim[1]**

[1] Faculty of Civil Engineering, Universiti Teknologi MARA, Uitm Shah Alam, Selangor, Malaysia.
[2] Institute for Infrastructure Engineering and Sustainability Management (IIESM), Universiti Teknologi MARA, Shah Alam, Selangor, Malaysia.
Corresponding author:eka@salam.uitm.edu.my

**Abstract.** Rutting is the most common pavement distress in pavement structures which occurs mainly due to several factors such as increasing of traffic volume, climatic conditions and also due to construction design errors. This failure reduced the service life of the pavement, reduced driver safety and increase cost of maintenance. Polymer Modified Binder has been observed for a long time in improving asphalt pavement performance. Research shows that the use of polymer in bituminous mix not only improve the resistance to rutting but also increase the life span of the pavement. This research evaluates the physical properties and rutting performance of dense graded Superpave-designed HMA mix. Two different types of dense graded Superpave HMA mix were developed consists of unmodified binder mix (UMB) and cup lumps rubber (liquid form) modified binder mix (CLMB). Natural rubber polymer modified binder was prepared from addition of 8 percent of cup lumps into binder. Results showed that all the mixes passed the Superpave volumetric properties criteria which indicate that these mixtures were good with respect to durability and flexibility. Furthermore, rutting results from APA rutting test was determined to evaluate the performance of these mixtures. The rutting result of CLMB demonstrates better resistance to rutting than those prepared using UMB mix. Addition of cup lumps rubber in asphalt mixture was found to be significant, where the cup lumps rubber has certainly improves the binder properties and enhanced its rutting resistance due to greater elasticity offered by the cup lumps rubber particles. It shows that the use of cup lumps rubber can significantly reduce the rut depth of asphalt mixture by 41% compared to the minimum rut depth obtained for the UMB mix. Therefore, it can be concluded that the cup lumps rubber is suitable to be used as a modifier to modified binder in order to enhance the properties of the binder and thus improves the performance of asphalt mixes.

## 1. Introduction

It is observed that the traffic volume on the highway keep increasing every year thus causing higher tire pressures, especially in the busy traffic areas such as tolls area, car park, intersections and at traffic light. The increases of traffic volume will lead to the permanent deformation of the pavements which is rutting. Rutting is the result of the progressive accumulation of permanent deformation of each layer of the pavement structure under repetitive loading [1]. Superpave mix design is the new method of mix design that is currently used in the US which is helping to improve the performance on the road. Superpave made its debut in Malaysia through the Kuala Lumpur International Airport (KLIA). Its directly correlates laboratory methods with pavement performance, which concern to the permanent deformation which are rutting and fatigue cracking. The tests are performed at the temperatures and aging conditions that more realistically represent the site conditions.

Previously, by using modified binder, the performance of asphalt mixtures can be improved and also the binder becomes more stable and stiffer at high temperature and more flexible at low temperatures. Some of the significant advantages of utilization of polymer modified binder are higher elastic recovery, a higher softening point, greater viscosity, greater cohesive strength and greater ductility [2]. Polymer modified binder also can increase the surface roughness of the aggregate and improved adhesion and degree of cohesion by creates an aggregate coating material thus produces superior asphalt mixtures [3]. Some previous studies had carried out to improve asphalt properties by using several kind of natural rubber latex such as field latex, concentrated latex and skim latex. They found that the the natural rubber latex is the best alternative for road making due to flexibility and stability improvement in asphalt pavement and extend service life of the road pavement. It was also found that the roads paving with natural rubber latex shall have more strength and durability than using unmodified binder [4 - 7]. Although there are many other materials that have been used as the modifiers in the bituminous mix such as crumb rubber, SBS, SBR latex, chemical additives and engineered binders but most of them are high cost materials.

Cup lump rubber or fresh latex is one of the modifiers that can be used in order to improve the binder properties. Cup lumps are obtained directly from rubber trees without going through any manufacturing process at the plant. This is why the researchers and engineers are looking for better materials in order to improve the bituminous mix. The uses of cup lumps rubber can be one of the solutions. It can be used as modifiers in the binder of bituminous mix in order to improve the binder mix properties. Therefore, in this study a new asphalt binder based on cup lumps rubber was investigated and characterized for possible improvements in the asphalt pavement mixes. It is believed that the cup lumps rubber modified bitumen would offer better performing asphalt mixes and such study is worthwhile for a sustainable road infrastructure that will provide a long lasting, safe and comfortable ride for the road users.

## 2. Experimental plan

The study focused on the evaluation volumetric properties and rutting performance of cup lumps rubber modified Binder (CLMB) of Hot Mix Asphalt (HMA) using Superpave methods. The Superpave mix design procedure involves careful material selection and volumetric proportioning as a first approach in producing a mix that will perform successfully. The aggregates were obtained from Blacktop Quarry, Rawang. The binder used in this study was 80/100 penetration grade supplied by SHELL. Superpave mix design was performed in order to prepare HMA Superpave mixtures for the mix design of SP 19 mm. The performance evaluation to be performed is rutting test. For rutting performance test, the laboratory designed mixtures was evaluated using the Asphalt Pavement Analyzer (APA).

### *2.1. Material properties*

Physical properties test were then conducted on the aggregates which conforms to the consensus and source properties as required in the Superpave system. The tests included Flat and Elongated particles (ASTM D4791), Abrasion Test (AASHTO T96)) and Aggregate Impact Value (AASHTO T96). Properties of aggregate were tested prior before preparing the asphalt mixture specimen. Table 1 and Table 2 shows the aggregate gradation and aggregate properties respectively. The aggregates meet the aggregate properties requirement. It can be concluded that Blacktop Quarry, Rawang, Selangor has a good quality aggregate. Thus, the aggregates are acceptable and suitable for use in road works.

**Table 1**. Aggregate Gradation.

| Sieve Size, mm | % Passing | % Retained |
|---|---|---|
| **25** | 100 | 0 |
| **19** | 96 | 4 |
| **12.5** | 81 | 15 |
| **9.5** | 75 | 6 |
| **4.75** | 55 | 20 |
| **2.36** | 43 | 12 |
| **1.18** | 32 | 11 |
| **0.6** | 23 | 9 |
| **0.3** | 13 | 10 |
| **0.15** | 8 | 5 |
| **0.0075** | 4 | 4 |
| **Pan** | - | 4 |
| **Total** | - | 100 |

**Table 2.** Aggregate Properties.

| Aggregate Test | Specification | Result |
|---|---|---|
| **Flakiness** | <20% | 3.1% |
| **Elongation** | <20% | 16.6% |
| **Aggregate Impact Value** | <45% | 21.75% |
| **LA Abrasion** | <45% | 25.35% |

*2.2. Binder modification*

The amount of binder used in the mixture is different, depending on the design. The binder used in this study was 80/100 PEN for control sample while PEN 80/100 + Cup lumps rubber for modified sample. The modified binder was prepared by mixing control sample with a proportion of cup lumps rubber. The most effective amount of cup lumps rubber was 8% by weight of binder, where it has great potential to improve the physical properties and the performance of bituminous mix [5]. The apparatus used to mix the polymer with the binder was hot plate and mechanical mixer. The modification process begun by heating and stirring the binder using the hot plate and mechanical stirrer at 500 rpm for 5 minutes. Then 8% of cup lump rubber by weight of binder was added slowly and mixed continuously for 60 minutes using mechanical stirring machine. The rotation speed of mechanical stirrer increases until the rotation reaches 1650 rpm.

*2.3. Optimum Binder Content (OBC)*

Superpave mix designs were conducted to determine design binder content. Superpave samples were prepared and tested according to AASHTO TP4 requirement and specification. Traffic level was selected based on common traffic level operating on most Malaysia highways. In this study, the traffic will be limited to medium to high roadway application. In Superpave system, medium to high traffic loadings is equivalent to between 3 to 30 million design equivalent single axle loads (ESALs) with 20 years design life. The design number of gyrations, Ninitial=8; Ndesign=100; and Nmax=160 was used to vary the compactive effort of the design mixture and was a function of traffic level. Selection of design asphalt binder content was obtained based on 4% air voids specimens. All other mixture properties were checked at the design binder content to verify that they meet the criteria. Table 3 shows the summary of the OBC determination process for both unmodified and modified samples. The results show that the modification binder with cup lumps rubber give more accurate result for the %VMA and %VFA which is 15.1 and 73.5 which is lesser than UMB sample. For the dust proportion the CLMB result is 0.9 which followed the criterion which is between 0.6 to1.2. While for percentage of theoretical maximum density (Gmm) is 85.9% lesser than 89% which is fulfill the requirement. Based on this result, it can be concluded that all the mixture properties meet the Superpave Criteria. The design binder content obtained for UMB is 5.7% and 5.1% for NRMB.

**Table 3.** OBC Determination.

| Mix Design Properties | UMB | CLMB | Criterion |
|---|---|---|---|
| **Air Void, %** | 4 | 0 | 4% |
| **VMA, %** | 15.8 | 15.1 | Min 13% |
| **VFA, %** | 74.7 | 73.5 | 65-76 |
| **Dust Proportion** | 0.8 | 0.9 | 0.6-1.2 |
| **$G_{mm}$, %** | 85.5 | 85.9 | <89 |
| **OBC, %** | 5.7 | 5.1 | - |

*2.4. Wheel tracking test*

The Asphalt Pavement Analyser (APA) equipment was used to evaluate the rutting resistance. The APA machine is the new version of the wheel tracker device. Generally, APA is a wheel tracking device that applies a vertical load to a steel wheel. The loaded wheels are applied to sample test of three pneumatic cylinders, where each of it is equipped with standard aluminium wheels. The total load from each of the moving wheel is shifted and transferred to a test sample through a pressurized rubber hose mounted along the top of the sample. The advantage of using APA is it can evaluate not only the rutting performance of an HMA mixture, but it also can determine the fatigue cracking and moisture susceptibility under certain condition of the service. Performance simulative tests have widely been used to compare the performance of a broad range of asphalt paving mixtures by modelling the anticipated traffic loads. Using these tests, stress conditions developed in the mixture under loading are not calculated, but similar to those applies to laboratory compacted samples and field cores to determine the performance figure 1 illustrates the Asphalt Pavement Analyser while figure 2 shows the samples after underwent the wheel tracking test.

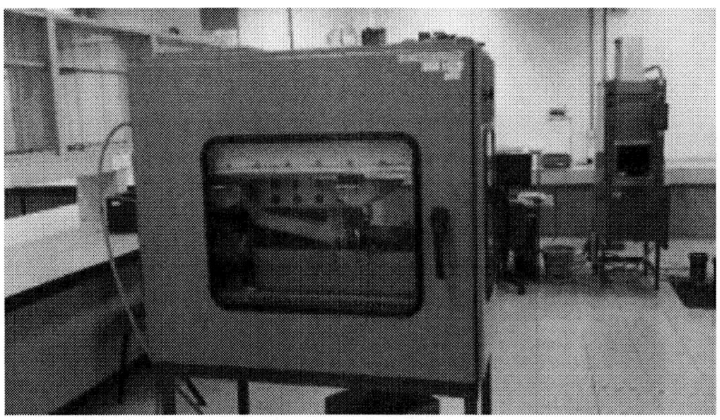

**Figure 1.** Asphalt Pavement Analyser.

**Figure 2.** Samples after Testing.

## 3. Result and discussion

The samples were prepared by using Superpave Gyratory Compactor in order to compact 150mm samples of 75mm height. The samples were tested for 8000 numbers of wheels which is 60 cycles per minute and the computer software measured and average the rut depths. The rate of rutting determined by referring to the rut depth of the sample in millimetre. The equipment allows more realistic real world conditions to be simulated in the laboratory including temperature control. The unmodified samples and modified samples were inserted into two different moulds. Then the mould was put into the APA machine side by side. Two tubes were put on top of the samples in the mould. The samples then were conditioned at the testing temperature of 60°C with 690 kPa for three hours and another two hours for running the test. This test was simulated to the wheel passes within a pavement and the procedure is described in BS 598. Table 4 shows the comparison of rut depth at difference cycle for unmodified and modified samples.

**Table 4.** Rut Depth Values.

| Stroke Count | Rut Depth, mm | |
|---|---|---|
| | UMB | CLMB |
| **0** | 0 | 0 |
| **25** | 0.273 | 0.155 |
| **4000** | 3.763 | 2.289 |
| **8000** | 4.764 | 2.801 |

From the result, the modified sample shows more resistance towards rutting compare to unmodified sample, where the average rut depth is 2.801 mm at 8000 cycles. For unmodified sample (PEN 80/100), the average rut depth is 4.764 mm at 8000 cycles. The difference between both values of rut depth is 1.963 mm, which shows a significant difference. Higher rut depth indicates that the mixture has lower resistance towards rutting. This showed that the used of cup lumps modified binder can significantly enhance the performance of asphalt mixture in terms of rutting resistance. Figure 3 shows the graph of wheel tracking test for both control (red) and modified binder (blue).

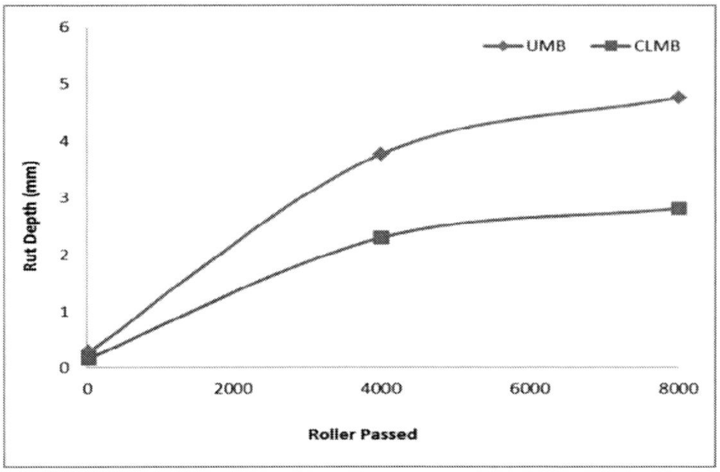


**Figure 3.** Rut Depth Versus Number of Cycle.

From the graph, it shows that modified sample has better rutting resistance compared to unmodified sample. At the beginning of the test, the rut depth values for both samples were zero. At 25, 4000 and 8000 stroke count, the rut depth value of the unmodified sample are 0.273 mm, 3.763 mm and 4.764 mm respectively. For the modified sample, the rut depth value at 25, 4000 and 8000 stroke count were 0.155 mm, 2.289 mm and 2.801 mm respectively. The differences in both samples at 25, 4000 and 8000 are 0.118 mm, 1.474 mm and 1.963 mm respectively. The differences increase as the stroke count increase. This shows that the increment of rut depth for modified sample is lower than unmodified sample. The properties of the unmodified sample mixture were not strong enough to provide the adequate bonding between binder and aggregate. The minimum rut depth obtained for the modified sample could reduce the rut depth by 41% compared to the minimum rut depth obtained for the unmodified sample. This higher resistance of modified sample to rutting compared to unmodified sample can be contributed by the greater elasticity offered by the cup lumps rubber particles which enhancing the rutting resistance. This finding is supported by previous studies [7] and [8] which indicates that the rut depths of modified sample are lower compared to the control sample. In addition, the natural characteristics of cup lump rubbers which is tends to glue itself strongly to the aggregates is also contribute to the better performance of the mixture. High amount of rut depth may be dangerous especially to the motorist. Road accident can happen when the tyre entering the rut area of the road.

## 4. Conclusion

In this study, the volumetric properties of Superpave mix design and rutting performance of the mix is evaluated to determine how resistance the control sample and modified sample is when used in asphaltic mixtures. Based on the findings, the following main conclusions can be drawn;

i. Local aggregates meet aggregate properties requirement. It can be concluded that Hanson quarry, Semenyih has a good quality aggregate. Thus, the aggregates are acceptable and suitable for use in road works.
ii. Locally aggregate gradations are suitable for Superpave mix design, since all mix design in accordance to Superpave system meet the Superpave requirements. The study also has proven the suitability of using Superpave mix design as the new system in our road pavement.
iii. The addition of cup lumps rubber in asphalt mixture was found to be significant, where cup lumps rubber has enhanced its rutting resistance due to greater elasticity offered by the cup lumps rubber particles. The minimum rut depth obtained for the modified sample could reduce the rut depth by 41% compared to the minimum rut depth obtained for the unmodified sample.

Cup lumps rubber also helps to increase the performance and the ability of asphalt mixture to improve its action toward traffic loading hence increase life span.

## 5. References

[1] Ahmad J, Rahman M Y A and Hainin M R 2011 Rutting evaluation of dense graded hot mix asphalt mixture, *Int. J. Eng. Technol.* **11** 56–60
[2] Yildirim Y 2007 Polymer modified asphalt binders, *Constr. Build. Mater.* **21** 66–72
[3] Zhang B, Xi M, Zhang D, Zhang H and Zhang B 2009 The effect of styrene–butadiene–rubber/montmorillonite modification on the characteristics and properties of asphalt, *Constr. Build. Mater.* **23** 3112–3117.
[4] Ruggles C 2005 *The Use of Natural Rubber Latex in Modified Asphalt Road Binders in the UK* AMAP 2005 Annual General Meeting report (Las Vegas: World Class Natural Rubber Solutions) pp 1 - 29
[5] Shaffie E, Ahmad J, Arshad A K, Kamarun D, Kamaruddin F and Mara U T 2015 Stripping Performance and Volumetric Properties Evaluation of Hot Mix Asphalt ( HMA ) Mix Design Using Natural Rubber Latex Polymer Modified Binder ( NRMB ) *Proc. of the Int. Civil and Infrastructure Engineering Conf. (InCIEC2014) (Sabah:Kota Kinabalu)* vol 1 (Singapore: Springer) pp 873 - 884
[6] Shaffie E, Ahmad J and Kamarun D 2015 Rutting performance of hot mix asphalt mixture using nanopolyacrylate polymer modifier, *Adv. Eng. Technol.* **753** 194–198
[7] Al-maamori M H and Hussen M M 2014 Use of Crumb Rubber As a Way To Improve Performance Grade for Asphalt Cement, *Acad. Res. Int.* **5** 66–78
[8] Celauro B, Celauro C, Lo Presti D and Bevilacqua A 2012 Definition of a laboratory optimization protocol for road bitumen improved with recycled tire rubber, *Constr. Build. Mater.* **37** 562–572

**Acknowledgments**
The authors expressed utmost gratitude to LESTARI Research Grant: 600-IRMI/DANA 5/3/LESTARI (0008/2016) from Ministry of Higher Education, Malaysia for financial support which enables this paper to be written.

# Effect of re-vibration on the compressive strength and surface hardness of concrete

**H B Koh[1,*], DYeoh[1] and S Shahidan[1]**

[1]Jamilus Research Centre, Faculty of Civil and Environmental Engineering, Universiti Tun Hussein Onn Malaysia
Corresponding author: koh@uthm.edu.my

**Abstract.** This paper reports the results of experimental investigations carried out on the effect of re-vibration on the compressive strength and surface hardness of normal weight concrete. Five different concrete mixes with varying water cement ratio ranging from 0.35 to 0.7 were prepared. The compacted concrete was subjected to re-vibration during the initial setting time period. During the initial setting time, the compacted concrete was re-vibrated at the time of 30 minutes up to 150 minutes of an interval of 30 minutes. The compressive strength and surface hardness of re-vibrated concrete were tested at the age of 35 days. The experimental results showed a significant increment of compressive strength and surface hardness in all re-vibrated concrete. In general, the maximum gain in compressive strength and surface hardness was when the re-vibration occurred at the initial setting time of 2 hours. The average increasing of the compressive strength of concrete by re-vibration is ranging from 3.5% to 21.8%.

## 1. Introduction

Entrapped air bubbles in the fresh concrete are needed to expel out. Blowholes or honeycomb will weaken the concrete strength and increase the porosity. Honeycombing will also reduce the bond between the concrete and reinforcement. Generally, every 1% of trapped air reduces the concrete strength by about 6%, therefore proper compaction is required in concrete construction [1].

Compaction is the process which expels entrapped air from freshly placed concrete and packs the aggregate particles together so as to increase the density of concrete. Compaction of concrete is a two-stage process as shown in figure 1. At the beginning stage concrete will liquefy and giving a level of top surface. In the second stage, entrapped air is expelled. The total time for both stages is around 10 to 20 seconds [2].

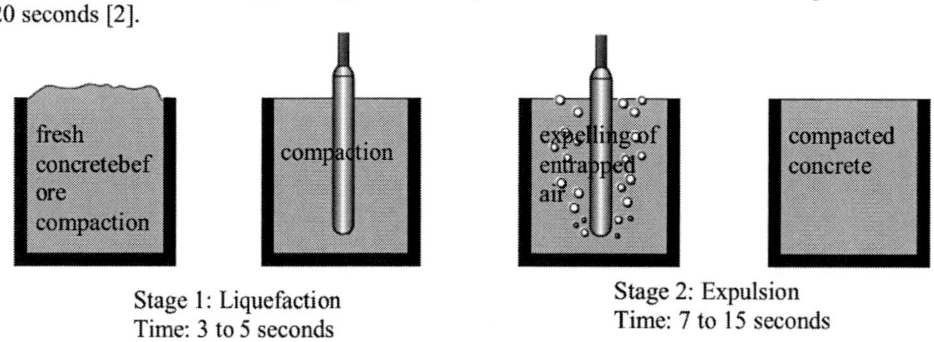


**Figure 1.** The process of compaction [2]

Re-vibration of concrete can be defined as the application or process of vibration to concrete after completion of placing and initial compaction, but preceding initial setting of the concrete. Many of people believe that partially set concrete should not be disturbed and re-vibration may cause a loss of bond between steel and concrete [3]. However, re-vibration at 1 to 2 hours after placing showed an increment of the compressive strength of concrete by up to 15%, but the actual values depend on the workability of the mix. The improvement in strength is more pronounced at earlier ages, and is greatest in concretes liable to high bleeding since the trapped water is expelled by re-vibration [4]. Concrete which subjected to re-vibration is gained higher strengths and bond, better impermeability, reduction in shrinkage and creep, reduction in surface and other voids as well as cracks [5]. Re-vibration can be done usually at any time as long as the running internal vibrator can sink by its weight in to the concrete or when the external vibrator or vibrating table can liquefy the concrete momentarily. The accepted stiffness limit for re-vibration is then the penetration resistance of the standard steel needle specified in ASTM C 403 [6] reaches the value of 3.5 N/mm$^2$.

An experimental work has been carried out to study the effect of re-vibration on the compressive strength, and also on the surface hardness of concrete which has been not studied in the previous concerning researches.

## 2. Experimental Program

### *2.1 Materials and Concrete Mix*

The materials for concrete mixes included Portland composite cement meeting MS EN 197-1:2007 [7] requirements, crushed granite aggregate with a nominal maximum size of 20 mm and fine aggregate was from river sand passing the 4.75-mm sieve and predominantly retained on the 75 μm sieve.

Five concrete mixes, based on the mix proportion used in previous study [5] are shown in table 1 were used in the study. The difference between the mixes was in the water cement ratio (w/c) and mix proportion between cement and total aggregate. M1 has the highest cement content with the lowest w/c. Meanwhile, M5 has the lowest cement content with the highest w/c. The Total aggregate is consists of fine aggregate and coarse aggregate. The w/c was 0.35, 0.5, 0.55, 0.6 and 0.7 for the five mixes. The cement to total aggregate ratio was 1:2.53, 1:4.3, 1:4.91, 1:5.51 and 1:7.01 for the five mixes. Meanwhile, the percentage of the fine aggregate over the total aggregate was fixed with 34.8%.

**Table 1.** Mix proportion

| Mix | Cement (kg) | Fine Aggregate (kg) | Coarse Aggregate (kg) | Water / Cement Ratio |
|---|---|---|---|---|
| M1 | 1 | 0.88 | 1.65 | 0.35 |
| M2 | 1 | 1.50 | 2.80 | 0.5 |
| M3 | 1 | 1.71 | 3.20 | 0.55 |
| M4 | 1 | 1.92 | 3.59 | 0.6 |
| M5 | 1 | 2.44 | 4.57 | 0.70 |

### *2.2 Specimens and Re-vibration Procedure*

The concrete was mixed using portable electric concrete drum mixer. The fresh concrete was fully poured inside the 150 mm steel moulds and immediately initial vibration was applied using poker vibrator for about 10 seconds.To prevent the forming of honeycomb on the hardened cubes, observation during re-vibration is important by ensuring entrapped air is expelled out to the top surface. A total of 90 cubes were prepared,18 cubes for each concrete mix. Three compacted concrete cubes were vibrated again at 30, 60, 90, 120 and 150 minutes. Previous study had shown that re-vibration before the final setting time at 1 to 2 hours will benefit compressive strength of concrete [4]. Therefore, re-vibration was done before the final setting time of concrete. All cubes were de-mould after 24 hours and subjected to water curing. Figure 2 and figure 3 shows the vibration process and curing of specimens, respectively. The cubes were cured until the age of testing.

**Figure 2.** Vibration using poker vibrator

**Figure 3.** Concrete cubes subjected to water curing

*2.3 Compressive Strength and Surface Hardness Tests*

Three concrete cubes from each group were used to conduct the surface hardness and compressive strength tests at 35 days. The concrete cubes were have been scheduled for testing on day 28 but due to the availability of the rebound hammer then all cubes were tested at 35 days. Compressive strength test was conducted using concrete compressive machine of the loading rate of 0.6 MPa/s. Concrete cubes used for surface hardness are the cubes used for the compressive strength test. Rebound hammer 'DIGI-SCHMIDT 2000' (figure 4) was used for the surface hardness test. Rebound hammer test was conducted on 3 surfaces of each cube. 9 rebound readings were recorded for each surface.

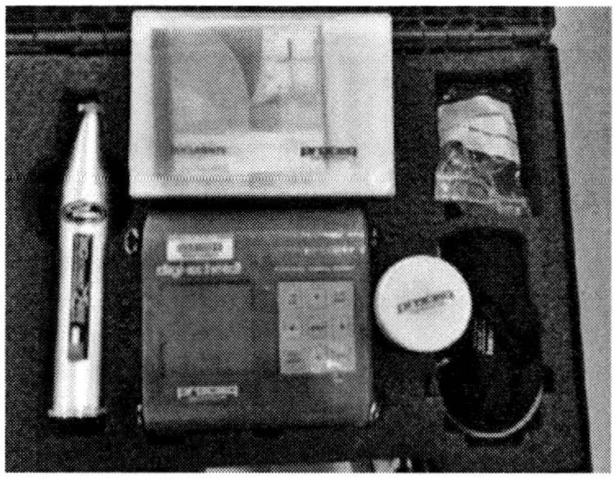

**Figure 4.** Rebound hammer 'DIGI-SCHMIDT 2000'

## 3. Results and Discussions

### *3.1 Density of Hardened Concrete*

Table 2 and figure 5 show the density of hardened concrete before and after re-vibration. It was found that concrete which has been vibrated again achieved higher density. The density of concrete was increased gradually after re-vibrated. All concrete achieved the highest density after re-vibrated at 120 minutes except M4 which achieved the highest density at 150 minutes. The highest density of concrete after re-vibrated was increased ranging from 0.3% to 3.0%.

**Table 2.** Average hardened density of concrete

| Mix | Density (kg/m³) | | | | | |
|---|---|---|---|---|---|---|
| | Re-vibration Time ( minutes) | | | | | |
| | 0 | 30 | 60 | 90 | 120 | 150 |
| M1 | 2326 | 2351 | 2354 | 2367 | 2381 | 2364 |
| M2 | 2350 | 2351 | 2353 | 2356 | 2356 | 2355 |
| M3 | 2282 | 2304 | 2319 | 2341 | 2350 | 2329 |
| M4 | 2329 | 2362 | 2346 | 2348 | 2365 | 2391 |
| M5 | 2334 | 2353 | 2363 | 2365 | 2378 | 2372 |

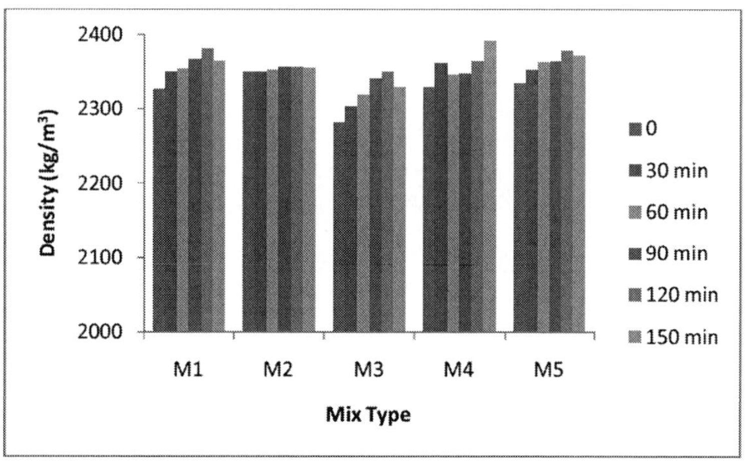


**Figure 5.** Hardened density of concrete

*3.2 Compressive Strength*

The 35-day compressive strength of 150 mm cubes of M1-M5 mixes are tabulated in table 3. In order to compare the relative gain of strength at different re-vibration time, the compressive strength of the mixes is also presented in figure. 6. As expected, M1 cubes with the lowest water/cement ratio and highest cement content has achieved the highest compressive strength. Compressive strength of M1 is approximately double of other mixes. All concrete have gained higher compressive strength after subjected to re-vibration. All concrete gained the maximum compressive strength resulting from the re-vibration applied at 2 hours after placing. An increment of compressive strength ranging from 3.5% to 21.9% was recorded for concrete subjected to re-vibration at 2 hours.

**Table 3.** Compressive Strength of Concrete

| Mix | Average 35-day Compressive Strength (MPa) | | | | | |
|---|---|---|---|---|---|---|
| | Re-Vibration Time (minutes) | | | | | |
| | 0 | 30 | 60 | 90 | 120 | 150 |
| M1 | 62.8 | 63.6 | 67.3 | 68.9 | 70.8 | 68.4 |
| M2 | 37.1 | 37.8 | 38 | 38.3 | 38.4 | 37.8 |
| M3 | 33.7 | 33.1 | 34.8 | 37.5 | 38.9 | 33.9 |
| M4 | 30.1 | 34.1 | 34.3 | 35.9 | 36.7 | 36.6 |
| M5 | 26.2 | 29 | 29.1 | 30 | 29.3 | 29.3 |

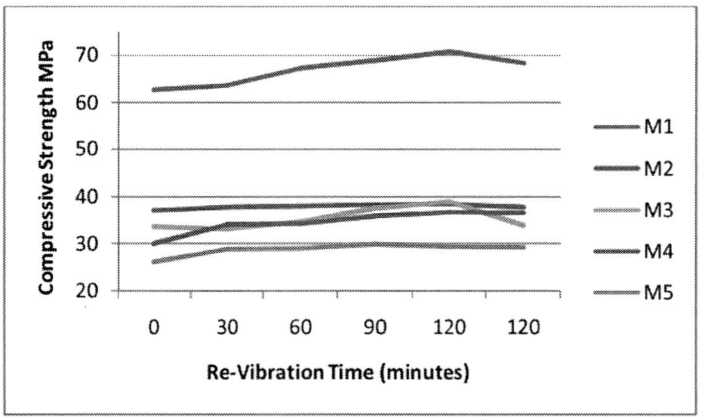


**Figure 6.** Compressive strength development of concrete subjected to re-vibration

*3.3 Surface Hardness*

Results of the average rebound hammer readings (27 rebound readings for each cube) are given in table 4. It can be noticed that re-vibration generally increases the surface hardness. All concretes except M4 gained the highest surface hardness resulting from the re-vibration applied at 2 hours.

**Table 4.** Average rebound hammer readings

| Mix | Average Rebound Reading | | | | | |
|---|---|---|---|---|---|---|
| | Re-Vibration Time (minutes) | | | | | |
| | 0 | 30 | 60 | 90 | 120 | 150 |
| M1 | 34.7 | 36.7 | 37.3 | 38.3 | 38.5 | 38 |
| M2 | 26.7 | 27.2 | 28.2 | 28.3 | 28.2 | 27.2 |
| M3 | 26.2 | 27.5 | 27.3 | 28.7 | 29.3 | 27.3 |
| M4 | 25.5 | 26.1 | 26.8 | 26.7 | 26.5 | 27.2 |
| M5 | 22.7 | 23.0 | 23.1 | 23.2 | 23.2 | 23.0 |

## 4. Conclusions

Based on the results of the experimental work carried out in the present work, the following conclusions could be made:

1. Re-vibration applied on the concrete within the initial setting time has caused an increment of hardened density.
2. Re-vibration improves the compressive strength and surface hardness of hardened concrete.
3. Re-vibration at 2 hours after placing gained the maximum compressive strength.
4. Generally, maximum compressive strength is gained at which the concrete has maximum density and surface hardness.

## 5. References

[1] British Ready Mixed Concrete Association 2008 *A practical guide for site personnel: Placing, compacting and curing concrete* (London, UK)1-2
[2] Cement Concrete and Aggregates Australia 2006 *Compaction of concrete* (NSW Australia) 1-7
[3] Gamal E. Khaleel. 1995. Effect of re-vibration on concrete quality. *Civil Engineering Research Magazine***17(7)** 119-131
[4] Neville A M, Brooks J J 1990 *Concrete Technology* (Longman. England) 137-140
[5] Krishna Rao M V, Rathish Kumar P, Bala Bhaskar N V R C 2008 Effect of re-vibration on compressive strength of concrete *Asian Journal of Civil Engineering (Building and Housing)* **9(3)** 291-301
[6] ASTM C 403 1999 *Standard Test Method for Time of Setting of Concrete Mixtures by Penetration Resistance* (United States of America) 1-6
[7] MS EN 197-1 2014 *Department of Standards Malaysia, Cement - Composition, specifications and conformity criteria for common cements* (Malaysia) 1-55

### Acknowledgements

The experiments were performed in the Laboratory of Materials Engineering at Universiti Tun Hussein Onn Malaysia. The concrete samples were prepared by Mr. Hafizan Mohd Saleh with the help of laboratory technicians. The authors wish to express the acknowledgements to their help.

IOP Conf. Series: Materials Science and Engineering **271** (2017) 012058 doi:10.1088/1757-899X/271/1/012058

# Influence of polypropylene fibres on the tensile strength and thermal properties of various densities of foamed concrete

**Ashfaque Ahmed Jhatial[1], Goh Wan Inn[1], Noridah Mohamad[1], U. Johnson Alengaram[2], Kim Hung Mo[2] and Redzuan Abdullah[3]**

[1] Jamilus Research Center, Faculty of Civil and Environmental Engineering, Universiti Tun Hussein Onn Malaysia, Batu Pahat 86400, Johor, Malaysia.
[2] Department of Civil Engineering, Faculty of Engineering, University of Malaya, 50603 Kuala Lumpur, Malaysia.
[3]Universiti Teknologi Malaysia, Skudai, Johor Malaysia.
Corresponding email: wigoh@uthm.edu.my

**Abstract.** As almost half of the world's population now lives in the urban areas, the raise in temperature in these areas has necessitated the development of thermal insulating material. Conventional concrete absorbs solar radiation during the daytime while releasing it at night causing raise in temperature in urban areas. The thermal conductivity of 2200 kg/m$^3$ density conventional concrete is 1.6 W/mK. Higher the thermal conductivity value, greater the heat flow through the material. To reduce this heat transfer, the construction industry has turned to lightweight foamed concrete. Foamed concrete, due to its air voids, gives excellent thermal properties and sound absorption apart from fire-resistance and self-leveling properties. But due to limited studies on different densities of foamed concrete, the thermal properties are not understood properly thus limiting its use as thermal insulating material. In this study, thermal conductivity is determined for 1400, 1600 and 1800 kg/m$^3$ densities of foamed concrete. 0.8% of Polypropylene fibres (PP) is used to reinforce the foamed concrete and improve the mechanical properties. Based upon the results, it was found that addition of PP fibres enhances the tensile strength and slightly reduced the thermal conductivity for lower densities, while the reverse affect was noticed in 1800 kg/m$^3$ density.

## 1. Introduction

The increase in migration of people from rural areas to urban cities due to economic benefits, has led almost half of the world's population [1] to dwell in the urban areas. Due to this urbanization, it has become necessary to manage the urban microclimate [2]. Building materials, such as concrete and asphalt used in the construction of pavements, roofs and other building components, have the tendency to absorb significant amount of solar radiation, ultimately causing concentrations of urban roads and buildings [3] to overheat compared to the rural areas during the day.

Urban overheating is a product of global warming, which effects the urban climate and the urban heat island (UHI) effect which occurs mainly in cities [4]. UHI is the phenomenon in which the temperature in the urban areas raises slightly compared to the surrounding rural areas, thus causing urban cities to become heat 'islands'. Though UHI varies from city to city in terms of intensity and magnitude, however it is the measure of difference between the urban and its surrounding rural temperatures [5,6].

The ability of any material to allow heat to flow through it is known as the thermal conductivity. Concrete and asphalt which are extensively used in urban construction, have profound impact on the urban thermal balance. These building materials absorb the solar radiation which is dissipated by convective and radiative heat transfer into the urban atmosphere and increases air temperature consequently [7]. Type, amount and form of moisture content present in the concrete has considerable influence on its thermal conductivity [8]. The conventional concrete of 2200 kg/m$^3$ has thermal conductivity of 1.6 W/mK [9]. The greater the value of thermal conductivity, the lesser resistance to heat transfer through the medium. It is therefore become necessity for the tropical countries, where constant hot climate is maintained throughout the year, to use thermal insulating building materials to counter the UHI effect. A good thermal insulating building material is one which has low thermal conductivity value.

Foamed concrete is a potentially thermal insulating material. It has become well-known material in the construction industry due to its ability to resistance the flow of heat through it. Dry density of foamed concrete ranges from 300 to 1840 kg/m$^3$ [10]. Homogenous pore voids are created with the introduction of air using the foaming agent [11,12] which is also used to control the density of the foamed concrete [13]. The higher the amount of foam used, reduced density of foamed concrete is achieved. The introduction of air voids makes the foamed concrete up to 87% lighter than conventional concrete. This light-weightiness achieved by foamed concrete allows it to reduce the self-weight which is another advantage over conventional concrete. Due to the presence of air voids in the foamed concrete, the heat flow is slowed down making foamed concrete, an excellent thermal insulating building material.

Due to reduced density, foamed concrete achieves lower compressive and tensile strengths while also developing micro-cracks. It has been observed that lower densities develop less mirco-cracks than higher densities of foamed concrete [14,15]. Therefore, nowadays concrete is being enhanced with the addition of Polypropylene fibres (PP fibres). PP fibres, which are synthetic organic fibres, are a product of polymerization process in which monomer polymers are turned into long chains of polypropylene. Over the years, various amount of PP fibres has been used in foamed concrete and it was observed that 0.8% of PP fibres are the optimum addition rate for enhancing the properties and performance of concrete [16].

Although the lightweight foamed concrete has been around for years, but it still lacks the study in terms of its thermal ability. Therefore, this study is done to reduce the energy consumption of the building and contribute to mitigation of UHI effect. To enhance the mechanical properties, 0.8% PP fibres are added into the three densities of foamed concrete, 1400 kg/m$^3$, 1600 kg/m$^3$ and 1800 kg/m$^3$. The thermal conductivity was also determined to understand the thermal performance of the different densities of foamed concrete with and without the addition of PP fibres.

## 2. Materials and Methodology

### *2.1 Materials*

Ordinary Portland cement (OPC), sand, water and foaming agent were used for the making of foamed concrete [17] while PP fibres were also added. The foaming agent is used to control the density and obtain the three different densities of foamed concrete used in this study. The foaming agent was used in a ratio of 1:20 with water. The mix proportion of foamed concrete was taken as 1:2 cement-sand ratio, with 0.55 water-cement ratio. Cylindrical specimens of 0.3 m length and 0.15 m diameter were cast to determine the splitting tensile strength. For the thermal properties, panels of 0.3 m x 0.3 m x 0.05 m were cast and thermal conductivity test was conducted.

### *2.2 Properties of Polypropylene fibres*

Fibrillated PP fibres were used in this study. The characteristics of the PP fibres are shown in table 1.

**Table 1.** Characteristics of polypropylene fibres [18]

| | |
|---|---|
| **Composition** | 100% Polypropylene fibre |
| **Configuration** | Fibrillated/Multi-Filament |
| **Fibre Length** | 19 mm |
| **Specific Gravity** | 0.9 |
| **Melting Point** | 160°C - 170°C |
| **Tensile Strength** | 45-60 ksi (0.31 – 0.42 kN/mm2) |
| **Thermal Conductivity** | Low |
| **Electrical Conductivity** | Low |
| **Absorption** | None |
| **Modulus of Elasticity** | 0.5 × 10 ksi (3.5 kN/mm2) |

The permeability as well as the plastic shrinkage is decreased when the fibrillated PP fibres are used in the concrete, while it is also known to enhance the impact resistance, cohesiveness and abrasion resistance fatigue [19]. In this study, dosage of 0.8% of PP fibres is added into each sample of foamed concrete with densities 1400 kg/m$^3$, 1600 kg/m$^3$ and 1800 kg/m$^3$.

*2.3 Experimental Procedures*

OPC, sand, water, foam and PP fibres were mixed together in the mixer and poured into moulds and kept for 24 hours. Afterwards these were unmoulded and kept for air curing for 28 days. The splitting tensile test was conducted using the Universal Testing Machine according to the BS EN 12390-6:2009 [20].

Panels of the three densities were used in the thermal conductivity test. The test was conducted according to BS EN 12664:2001 [21]. Four holes were drilled on the surface of sample with the depth of 10, 15, 20 and 30 mm to insert with the thermocouples. The panels were then placed on the container of the thermal conductivity apparatus that was attached with cold plate on top and hot plate on bottom. The initial temperature was controlled as for hot plate was set at 40°C by using power system while cold plate set at 18°C by using chiller system.The temperature of hot and cold plates was recorded at every 10 minutes interval for 24 hours. Thermal conductivity value can be determined based on the calculated data recorded from the computer.

## 3. Results and Discussion

*3.1 Splitting Tensile Strength*

The results of splitting tensile test conducted on cylinder specimens after 28 days curing is shown in table 2 and figure 1 shows the tensile strength vs. PP fibres for the three densities of foamed concrete.

**Table 2.** Average splitting tensile of various densities of foamed concrete

| Target Density (Kg/m$^3$) | % of PP Fibres | Splitting Tensile (MPa) |
|---|---|---|
| 1400 | 0% controlled | 0.69 |
| | 0.8% | 1.00 |
| 1600 | 0% controlled | 1.20 |
| | 0.8% | 1.54 |
| 1800 | 0% controlled | 2.60 |
| | 0.8% | 2.10 |

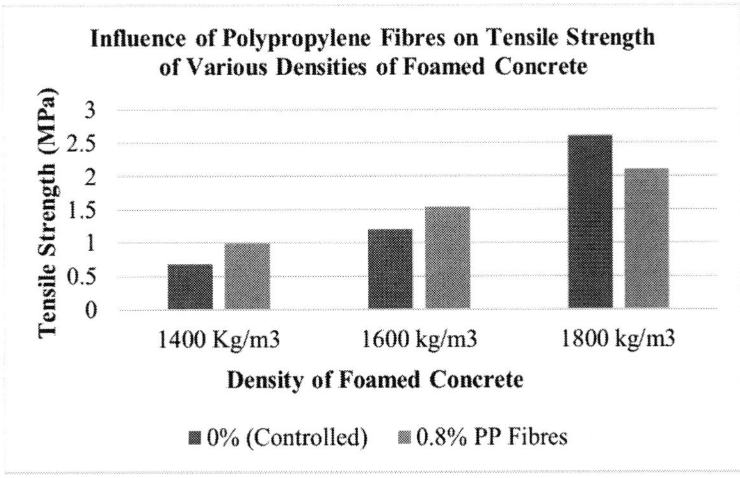


**Figure 1.** Average splitting tensile of various densities of foamed concrete

From the results, it can be observed that 0.8% PP fibres increases the tensile strength by 44.93% in 1400 kg/m$^3$ density foamed concrete while 28.33% increase was achieved in 1600 kg/m$^3$. But for the 1800 kg/m$^3$ density foamed the addition of PP fibres had adverse impact, it lost 19.23% of its tensile strength compared to the controlled sample.

The PP fibres can enhance the split tensile strength as the fibre can provide a bridging effect with the foamed concrete matrix. Shear stress acted can be transferred from the interface to the fibres and consequently the entire load can be supported. The transfer of stress helped in improving the tensile strain capacity which resulted in the increase of splitting tensile strength for foamed concrete added with PP fibres. Sudden reduction of split tensile strength at density of 1800 kg/m$^3$ can be due to the non-uniform distribution of the fibres that may occur during the mixing process.

### *3.2 Thermal Conductivity*

The thermal conductivity values of the three densities were obtained and are shown in table 3 while figure 2 shows thermal conductivity value vs. percentage of PP fibre added for various densities of foamed concrete.

**Table 3.** Average thermal conductivity of various densities of foamed concrete

| Target Density (Kg/m$^3$) | % of PP Fibres | Thermal Conductivity, k (W/mK) |
|---|---|---|
| **1400** | 0% controlled | 0.66 |
| | 0.8% | 0.64 |
| **1600** | 0% controlled | 0.71 |
| | 0.8% | 0.70 |
| **1800** | 0% controlled | 0.82 |
| | 0.8% | 0.84 |

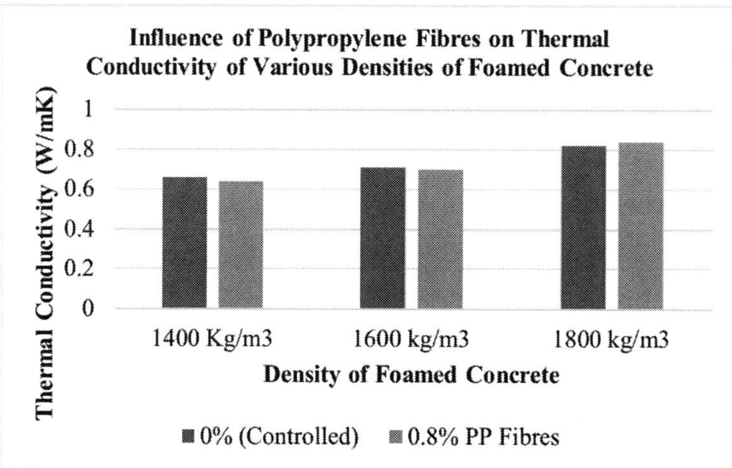


**Figure 2.** Thermal conductivity of various densities of foamed concrete

Lower the value of thermal conductivity makes the material to resist the heat transfer through it. The thermal conductivity for the controlled sample of 1400 kg/m$^3$ was recorded to be 0.66 W/mK but when the addition of 0.8% PP fibres were added, 0.64 W/mK thermal conductivity was achieved, which is approximately 3.03% reduction. The decrease in thermal conductivity value was also noticed with the addition of PP fibres in 1600 kg/m$^3$ density foamed concrete, reducing the thermal conductivity value from 0.71 W/mK (controlled sample) to 0.70 w/mK (0.8% PP fibres) with reduction of 1.41%. Interestingly the thermal conductivity increased 2.43% from 0.82 W/mK to 0.84 W/mK for the 1800 kg/m$^3$ density foamed concrete when PP fibres were added.

PP fibres being hydrophobic, retain water, during the thermal conductivity test, the water attained with PP fibres may dry out leaving the voids to be occupied by air. The air voids trap the heat and slow down the transfer. So theoretically, the thermal conductivity value should decrease with the addition of PP fibres. From the results, this is visible for both the lower densities of foamed concrete. The gain in 1800 kg/m$^3$ can be explained due to the non-uniform distribution of fibres or due to other factors such as variation in hole depth which are drilled to place the thermocouples to measure the internal temperatures.

## 4. Conclusion

Based upon the results obtained, the following conclusion are drawn:

1. Splitting tensile strength is higher when the density of foamed concrete increase as the amount of pore and voids decrease in higher density.
2. The addition of PP fibres enhances the tensile strength of 1400 kg/m$^3$ and 1600 kg/m$^3$ density foamed concrete, achieving significant increase strength. But the higher density of foamed concrete, 1800 kg/m$^3$, however showed decrease in tensile strength upon the addition of PP fibres.
3. The PP fibres also had a positive impact on thermal conductivity, reducing the value slightly for both 1400 kg/m$^3$ and 1600 kg/m$^3$ while the value was increased in 1800 kg/m$^3$ density foamed concrete when PP fibres were added.
4. The loss in splitting tensile, and slight gain in thermal conductivity can be associated to the non-uniform distribution of fibres in the 1800 kg/m$^3$.
5. The effect of PP fibres increases in lower densities of foamed concrete, which is evident from the splitting tensile and thermal conductivity of all the three densities of foamed concrete.

## 5. References

[1] UNPD 2011 World urbanization prospects: *The 2011 revision* (New York: United Nations) 1-302.

[2] Guan K K 2011 Surface and ambient air temperatures associated with different ground material: a case study at the University of California, Berkeley *Environmental Science* **196** 1–14

[3] Solecki W D, Rosenzweig C, Parshall L, Pope G, Clark M, Cox J and Wiencke M 2005 Mitigation of the heat island effect in urban New Jersey *Global Environmental Change Part B: Environmental Hazards* **6(1)** 39–49

[4] Santamouris M, Cartalis C, Synnefa A and Kolokotsac D 2015 On the impact of urban heat island and global warming on the power demand and electricity consumption of buildings—A review, *Energy and Buildings* **98** 119–124

[5] Elsayed I S M 2012 A Study on the Urban Heat Island of the City of Kuala Lumpur, Malaysia. *Journal of King Abdulaziz University: Metrology, Environment and Arid Land Agricultural Sciences* **23(2)** 121-134.

[6] Streutker D R 2003 *A Study of the Urban Heat Island of Houston, Texas,* PhD diss., (Rice University)

[7] Aflaki A, Mirnezhad M, Ghaffarianhoseini A, Ghaffarianhoseini A, Omrany H, Wang Z-H and Akbari H 2017 Urban heat island mitigation strategies: A state-of-the-art review on Kuala Lumpur, Singapore and Hong Kong Cities 131-145

[8] Kodur V 2014 Properties of Concrete at Elevated Temperatures *ISRN Civil Enginnering* **2014** 1–15.

[9] Zahari N M, Rahman I A and Zaidi A M A 2009 Foamed Concrete : Potential Application in Thermal Insulation Heat in Heat out Air- voids Foam Concrete, *Proceedings of MUCEET 2009 Malaysian Technical Universities Conference on Engineering and Technology June 20-22, 2009, MS Garden, Kuantan, Pahang, Malaysia,* 47–52.

[10] Mohamad N, Samad A A A, Ali N, Hadipramana J, Jamaluddin N 2015 Performance of connected precast lightweight sandwich foamed concrete panel under flexural load *Jurnal Teknologi (Sciences & Engineering)* **75(9)** 111–115

[11] Goh W I, Mohamad N, Abdullah R and Samad A A A 2014 Compression test and finite element analysis of foamed concrete cube *Journal of Engineering and Technology* **5(1)** 1-9.

[12] Tharkele M K H 2014 Experimental study on foam concrete *International Journal of Civil, Structural, Environmental and Infrastructure Engineering Research and Development (IJCSEIERD)* **4(1)** 145–158

[13] Abd A M and Abd S M 2017 Modelling the strength of lightweight foamed concrete using support vector machine (SVM), Case Study *Constr. Mater* **6** 8–15

[14] Wei S, Yiqiang C, Yunsheng Z and Jones M R 2013 Characterization and simulation of microstructure and thermal properties of foamed concrete, *Construction and Building Materials* **47(10)** 1278–1291

[15] Aldridge D 2005 Introduction to foamed concrete: what, why, how? *Use of Foamed Concrete in Construction* (London: Thomas Telford) 1-14.

[16] Chen B, Wu Z and Liu N 2012 Experimental research on properties of high strength foamed concrete, *J. Materials in Civil Engineering* **24** 113-118

[17] Mohamad N, Khalil A I, Samad A A A and Goh W I 2014 Structural Behavior of Precast Lightweight Foam Concrete Sandwich Panel with Double Shear Truss Connectors under Flexural Load *ISRN Civil Engineering* vol. **2014** 1-7.

[18] Awang H, Ahmad M H and Al-Mulali M Z 2015 Influence of Kenaf And Polypropylene Fibres on Mechanical and Durability Properties of Fibre Reinforced Lightweight Foamed Concrete, *Journal of Engineering Science and Technology* **10(4)** 496–508.

[19] Hazlin A R, Iman Afif, Mohamad N, Goh W I, Sia L M, Samad A A A and Noorwirdawati Ali 2017 Microstructure and Tensile Strength of Foamed Concrete with Added Polypropylene Fibres, *MATEC Web of Conferences **103**, 1-7.*

[20] BS EN 12390-6:2009, Testing Hardened Concrete Part 6: Tensile Splitting Strength of Test Specimens, *British Standard Institution.*

[21] BS EN 12664:2001, Thermal performance of building materials and products. Determination of thermal resistance by means of guarded hot plate and heat flow meter methods. Dry and moist products of medium and low thermal resistance, *British Standard Institution.*

**Acknowledgement**

The authors would like to thank the Universiti Tun Hussein Onn Malaysia (STG-528) for the financial support for this project and the University of Malaya for their service during the thermal conductivity test.

# A review on seashells ash as partial cement replacement

**Wan Ahmad Soffian Bin Wan Mohammad[1], Nor Hazurina Othman[1], Mohd Haziman Wan Ibrahim[1], Masazurah A Rahim[2], Shahiron Shahidan[1] and Raha Abd Rahman[1]**

[1]Faculty of Civil and Environmental Engineering, Universiti Tun Hussein Onn Malaysia, 86400 Parit Raja, Johor, Malaysia
[2]Fisheries Research Institute, Department of Fisheries Malaysia,11960 Batu Maung, Penang
Corresponding author: hazurina@uthm.edu.my

**Abstract.** This review paper emphasis on various sea shells ash such as cockle, clam, oyster, mollusc, periwinkle, snail, and green mussel shell ash as partial cement replacement and its objective is to create sustainable environment and reduce problems of global warming. Cement production give huge impact to environment in every stage of its production. These include air pollution in form of dust and, gases, sound and vibration during quarry crushing and milling. One of the solutions to solve this problem is by using modified cement. The modified cement is a cementitious material that meets or exceeds the Portland cement performance by combining and optimizes the recycle and wasted materials. This will indirectly reduce the use of raw materials and then, become a sustain construction materials. Therefore, the replacement of cement in concrete by various sea shell ash may create tremendous saving of energy and also leads to important environmental benefits. This study includes previous investigation done on the properties of chemical and mechanical such as specific gravity, chemical composition, compressive strength, tensile strength and flexural strength of concrete produced using partial replacement of cement by seashells ash. Results show that the optimum percentage of seashells as cement replacement is between 4 – 5%.

## 1. Introduction

Concrete is one of the two most used structure materials in construction [1]. Every year, the used of concrete increasing 2 tonne for each people [2]. In order to reduce reliance of raw material in concrete producing, the green concrete had been promoted. Green concrete is the concrete that had been produced using recycle or wasted natural materials [3]. One of the ways to produce green concrete is by using modified cement. Cement is the second largest volume materials used by human being after water [4]. Cement plays the role of a binder, a substance that sets and hardens and might bind alternative materials along. During production of cement and hydration process of cement, the amount of $CO_2$ emitted by the industry is nearly 900 kg of $CO_2$ for every 1000 kg of cement produced [5]. This $CO_2$ production causes serious environmental damages.

The sea shells are high potential materials to become partial cement replacement and filler in concrete. The calcium carbonate ($CaCO_3$) in the sea shells is more than 90% and is similar to the contain of calcium carbonate in the limestone dust that been used in the Portland cement production [6]. Impressively, the crystal structures of seashells are largely composed of calcite and aragonite, which have higher strengths and density than limestone powder [7]. Also, the particle size of seashells is between 36μm to 75μm and are similar to the particle size of Portland cement [8].

## 2. Literature Review

*2.1. Specific gravity*

**Table 1.** Specific gravity of the seashells

| Researcher | Type of seashell/Portland cement | Value |
|---|---|---|
| Noel D. Binag (2016) | Portland cement | 3.15 |
| | Oyster | 3.09 |
| | Mussel | 3.01 |
| | Mollusk | 3.01 |
| Lertwattanaruk Pusit et al. (2012) | Portland cement | 3.11 |
| | Oyster | 2.65 |
| | Mussel | 2.86 |
| | Cockle | 2.82 |
| | Clam | 2.71 |
| Adewuyi et al. (2015) | Periwinkle | 2.50 |
| | Oyster | 2.33 |
| | Snail shell | 2.44 |

Specific gravity is the ratio of the density of a substance to the density of a suggest substance; equivalently, it is the ratio of the mass of a substance to the mass of a suggest substance for the same given volume. In construction industry, the specific gravity is important to determine the suitability of materials that will be used in concrete production. The specific gravity also used to estimate the weight of the concrete. From Table 1, the specific gravity of the seashells was approximately similar to the Portland cement. Although the specific gravity among the seashells was different, the value was approximately similar to specific gravity of Portland cement. [9-10] found that the specific gravity of oyster is different which is 3.09, 2.65 and 2.33. But, this value is below than the Portland cement which was 3.15[9] and 3.11[8]. Also, from the table it can be concluded that the specific gravity value of oyster shell is 2.33 and it was the most suitable to be used as cement replacement compare to other seashells.

*2.2. Chemical composition*

Calcium carbonate has the common properties of other carbonate. It reacts with acids, releasing carbon dioxide.

$$CaCO_3 + 2H \rightarrow Ca + CO_2 + H_2O \quad (1)$$

It also produces carbon dioxide when heating above 840 °C to form calcium oxide, commonly called burnt lime or quicklime. Through thermal decomposition process known as calcination, $CaCO_3$ can be converted into calcium oxide (CaO) [22].

$$CaCO_3 \rightarrow CaO + CO_2 \quad (2)$$

Calcium carbonate also will react with water that is saturated with carbon dioxide to form the soluble calcium bicarbonate.

$$CaCO_3 + CO_2 + H_2O \rightarrow Ca(HCO_3)_2 \quad (3)$$

According to the Table 2, the highest content of CaO was found in the clam shell [19]. The quantity of CaO in the seashells is sufficient to form silicates and aluminates of calcium. The deficiency in CaO will reduce the strength of cement. Also, the deficiency in CaO causes cement to set quickly.

**Table 2.** Chemical composition of seashells and Portland cement

| Chemical composition (%) | Type of seashells/Portland cement | | | | | | | |
|---|---|---|---|---|---|---|---|---|
| | Portland Cement[a] | Clam[b] | Mussel[b] | Oyster[b] | Periwinkle[c] | Snail[c] | Cockle[d] | Clam[e] |
| $SiO_2$ | 21.20 | 0.84 | 0.73 | 1.01 | 26.26 | 10.20 | 1.6 | 0.39 |
| $Al_2O_3$ | 5.47 | 0.14 | 0.13 | 0.14 | 8.79 | 4.81 | 0.92 | 0.28 |
| $Fe_2O_3$ | 3.31 | 0.06 | 0.05 | 0.07 | 4.82 | 3.15 | - | 0.02 |
| CaO | 65.52 | 53.99 | 53.38 | 53.59 | 55.53 | 61.95 | 51.56 | 67.70 |
| MgO | 1.97 | 0.08 | 0.03 | 0.46 | 0.40 | 0.18 | 1.43 | - |
| $K_2O$ | 1.71 | 0.03 | 0.02 | 0.02 | 0.20 | 0.05 | 0.06 | - |
| $Na_2O$ | 0.46 | 0.39 | 0.44 | 0.23 | 0.25 | 0.04 | 0.08 | - |
| $SO_3$ | 1.90 | 0.16 | 0.34 | 0.75 | 0.18 | 0.03 | - | - |
| Cl | - | 0.02 | 0.02 | 0.01 | - | - | - | - |
| $SO_4$ | - | 0.06 | 0.11 | 0.43 | - | - | - | - |
| $CaCO_3$ | - | 96.8 | 95.6 | 96.8 | - | - | - | - |

Source: [a][16], [b][8], [c][17], [d][18], [e][19]

### *2.3. Compressive strength*

**Table 3.** Compressive strength of the seashells

| Type of seashell | Type of replacement | Grade | Control sample strength (MPa) | | Optimum percentage of replacement (%) | Compressive strength (MPa) | |
|---|---|---|---|---|---|---|---|
| | | | 7 days curing | 28 days curing | | 7 days curing | 28 days curing |
| Periwinkle[a] | concrete | - | - | 17.5 | 10 | - | 19 |
| snail[a] | | | - | 17.5 | 20 | - | 18 |
| oyster[a] | | | - | 17.5 | 15 | - | 16.5 |
| cockle[b] | | 35 | 38 | 45 | 5 | 34 | 36 |
| cockle[c] | | | 32 | 36.5 | 4 | 31.5 | 36 |
| clam[d] | | | 33.8 | 37 | 4 | 34 | 39.8 |
| oyster[e] | Masonry cement mortar | - | 31.6 | 43.04 | 5 | 28.59 | 37.59 |
| mussel[e] | | | 31.6 | 43.04 | 5 | 28.34 | 38.33 |
| mollusk[e] | | | 31.6 | 43.04 | 5 | 28.73 | 39.54 |
| clam[f] | | | 10.2 | 14.9 | 5 | 9.9 | 12.6 |
| oyster[f] | | | 10.2 | 14.9 | 5 | 9 | 13 |
| cockle[f] | | | 10.2 | 14.9 | 5 | 8.5 | 12.6 |
| mussel[f] | | | 10.2 | 14.9 | 5 | 9 | 10.5 |

Source: [a][20], [b][7], [c][18], [d][19], [e][9], [f][8]

For the compressive strength, there are two type of material that had been studied which were concrete and masonry cement mortar. Adekunle [20] stated that the concrete which contained periwinkle shell as partial cement replacement gained the most compressive strength compare to snail and oyster shell recorded at 19 MPa after 28 days curing which was higher than the control sample reading at 17.5 MPa. For the cockle, the concrete of grade 35 MPa had been made [7,18]. Although the material was the same, the optimum percentage of replacement was different which is 5% [7] and 4% [18]. According to Nor Hazurina *et al.* [7], the compressive strength of control sample for the 7 days curing is 38 MPa and it was higher than cockle shells concrete (34 MPa). While for the 28 days curing, the compressive strength of control sample was 45 MPa and the compressive strength for the concrete that use cockle shells as cement replacement was lower which is 36 MPa. But for the 4% of optimum percentage of replacement, the result was different. Olivia *et al.*[18] found that the compressive strength for the 7 and 28 days curing cockle shell concrete are 31.5 and 36 MPa respectively, while for the control sample, it was 32 and 36.5 MPa. From both result, it can be seen that the compressive strength of the cockle shell was lower than the control sample. But, its value is still exceeding the minimum strength of grade 35 MPa concrete. For the grade 35 MPa concrete using the clam shells, the result is different. According to *Olivia et al.* [19], the compressive strength of clam shell concrete was higher than the control sample which is 34 MPa (7 days curing) and 39.8 MPa (28 days curing) while for the control sample, the value was 33.8 MPa (7days curing) and 37 MPa (28 days curing). Also, based on the Table 3, the clam shell was the only seashell that gains higher compressive strength than control sample. It can be concluded that the clam shell is the most suitable material to be used as cement replacement in concrete and the optimum percentage of replacement is 4%.

There are also a few research had been done on the use of seashells as cement replacement in cement mortar for masonry. Noel D. Binag [9] had been used oyster, mussel and mollusk shell in cement mortar while Lertwattanaruk Pusit et al. [8] used clam, oyster, mussel and cockle shell as cement replacement materials. Both of the studies used 5% as an optimum percentage of cement replacement. Both studies also found that the compressive strength of 7 and 28 days curing cement mortar result is lower than the control sample. But the cement mortar result is still acceptable and can be used in construction industry.

### *2.4. Tensile strength*

**Table 4.** The tensile strength of the seashells

| Type of seashells | Researcher | Year | Control sample tensile strength (MPa) | | | Optimum percentage of replacement (%) | Tensile strength (MPa) | | |
|---|---|---|---|---|---|---|---|---|---|
| | | | 7 days curing | 28 days curing | 90 days curing | | 7 days curing | 28 days curing | 90 days curing |
| Cockle *(Anadara Granosa)* | Nor Hazurina et al.[7] | 2013 | 6.1 | 7.2 | 9.8 | 5 | 7.1 | 6.9 | 10.2 |
| Cockle *(Anadara granosa)* | Monita Olivia et al.[18] | 2015 | 4.5 | 5.5 | 6.1 | 4 | 4.6 | 6.1 | 6.7 |
| Clam *(Polymesoda expansa)* | Monita Olivia et al.[19] | 2017 | 2.7 | 3.3 | 3.6 | 4 | 2.5 | 3.15 | 3.25 |

From Table 4, Nor Hazurina *et al.* [7] found that the tensile strength of cockle shell concrete was higher than the control samples for 7 and 90 days curing. The strength of the concrete increases with the time when it is cured for longer duration [21]. The period of curing the concrete is important as it is very essential for keeping the hydration process of cement with water until the concrete attains the optimum strength. The longer curing duration also important to make sure the concrete strength was developed from the pozzolanic reaction [21]. Olivia Monita *et al.* [18,19] found that the tensile strength of the cockle and clam shells concrete are higher than the tensile strength of the control samples in 7, 28 and 90 days curing. It was reported that with an addition of 5% cockle ash, the tensile strength icreased significantly [7]. Those highly tensile strength was likely due to the improved bonding at the interface of the cement paste and aggregates. The addition of the ground seashell could increase the density of the concrete, thus could change the interfacial transition zone phase between the aggregate and cement paste.

*2.5. Flexural Strength*

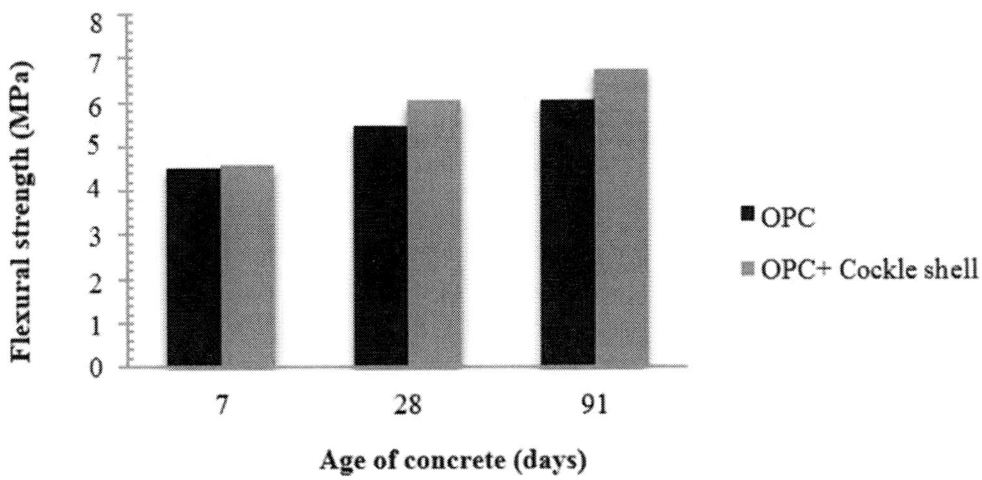


**Figure 1.** Flexural strength of OPC and cockle shell concrete at 7, 28 and 91 days by using 5% of cockle shell as cement replacement [18]

The average flexural strength of seashell concrete and ordinary Portland cement (OPC) was shown in Figure 1. Entirely, the values varied in the range of 4.5-6.75 MPa. It shown that the flexural strength of seashell concrete will rise at 28 and 91 days. After 28 days, a considerable strength was gained probably due to calcium content in the cement that produced a better bonding between aggregates and paste. The tension properties was encourage by the bonding. This additional increase in tension properties is worthwhile to improve composite action between concrete and steel reinforcement bars.

## 3. Type of seashells and it's usage in concrete

Table 5 shows a few types of seashell and how it was used in concrete mixture. [11,12,14] and [15] used different types of seashell to replace the coarse or fine aggregate in the concrete in range of 5% to 50% and found out that the compressive strength increased significantly. [13] had used oyster shells ashes to replace lime in production of cement based brick. It is showed that, at 15% replacement, the increment of the compressive strength was also significant.

**Table 5.** Another use of various seashells in concrete

| Researcher | Type of seashells | Application | Result |
|---|---|---|---|
| Martínez-García *et al.* (2017) | Mussel shells | Replace coarse aggregate in concrete | Tensile strength increase at 5% replacement |
| Muthusamy *et al.* (2012) | Cockle shells | Replace coarse aggregate in concrete | Compressive strength increase at 20% replacement |
| Gengying Li *et al.* (2015) | Oyster shells | Produce cement based-brick using oyster shell ash | Compressive strength increase at 15% replacement |

| | | | |
|---|---|---|---|
| Sugiyama (2004) | Scallops shells | Replace coarse aggregate in concrete | Compressive strength decrease at 50% replacement |
| Yusof (2011) | Clam shells | Replace fine aggregate in concrete | Compressive strength increase at 30% replacement |

## 4. Conclusion

From this review, we can say that the development of sea shell ash as partial cement replacement could be produce as a cement-like material where the particle size will be the same or finer than cement. Concrete with seashells as cement replacement will produce better concrete in term of chemical composition, specific gravity, compressive strength, flexural strength and tensile strength. It could be investigated by using difference types of sea shells to reduce environmental issues. Implicitly, this effort will create better benefit in future economic value to the local community and industries and also, provide better solution in concrete technology. Good values on developing the future of concrete industry should be expend through quality research among industry players and higher learning institutions to involve all parties into sustainable situation. Besides that, collaboration with local authorities through regulations and laws will create better opportunity to stake holders to come up with programs on waste minimization and utilization. Therefore, utilizing of waste materials such as sea shell in developing green concrete should be explored, expended and supported through focus researches. Development of green concrete should be proved through performance on hardened state and fresh state to understand the behaviour of each material either in long or short term effect.

## 5. References

[1] Neville A M 1996 *Properties of concrete 4th and final ed.* (Addison Wesley Logman) 43-44

[2] Harrison A W John 2003 New cements based on the addition of reactive magnesia to Portland cement with or without added pozzolan *CIA Conference: concrete in the third Millenium* CIA (Brisbane, Australia)

[3] Meyer Christian 2009 The greening of the concrete industry *Cement and concrete composites 31* **8** 601-605

[4] Patel A J, Patel V M and Patel M A 2015 Review on Partial Replacement of Cement in Concrete *UKIERI Concrete Congress – Concrete Research Driving Profit and Sustainability 1* **7** 831-837

[5] Mahasenan Natesan; Steve Smith; Kenneth Humphreys; Kaya Y 2003 The Cement Industry and Global Climate Change: Current and Potential Future Cement Industry CO2 Emissions *Greenhouse Gas Control Technologies – 6th International Conference* **2** 995-1000

[6] Mosher Shad, Gregory W Cope, Frank X Weber, Damian Shea, and Thomas J. Kwak 2012 Effects of lead on Na+, K+-ATPase and hemolymph ion concentrations in the freshwater mussel Elliptio complanata *Environmental toxicology 27* **5** 268-276

[7] Nor Hazurina Othman, Badorul Hisham Abu Bakar, Mashitah Mat Don and Megat Azmi Megat Johari 2013 Cockle Shell Ash Replacement for Cement and Filler in Concrete *Malaysian Journal of Civil Engineering* **25(2)** 201–211.

[8] Lertwattanaruk Pusit, Makul N and Siripattarapravat C 2012 Utilization of ground waste seashells in cement mortars for masonry and plastering *Journal of Environmental Management* **111** 133–141

[9] Noel D. Binag 2016 Powdered Shell Wastes as Partial Substitute for Masonry Cement Mortar in Binder, Tiles and Bricks Production *International Journal of Engineering Research & Technology 5* **7** 70-77

[10] Arroyo G Maria Antonia, Macias P Katrina Anna, Tayzon M Karlo Emir, Lorenzo D Maria Abigail 2005 The Feasibility of a Mollusk Shell – Based Adhesive as a Substitute for Mortar *Bato balani, Sophomore* **24(2)** 2004-2005

[11] Martínez-García Carolina, Belén González-Fonteboa, Fernando Martínez-Abella, and Diego Carro-López 2017 Performance of mussel shell as aggregate in plain concrete *Construction and Building Materials* **139** 570-583

[12] Muthusamy K and Sabri N A 2012 Cockle shell: a potential partial coarse aggregate replacement in concrete *International Journal of Science, Environment and Technology 1* **4** 260-267

[13] Gengying Li, Xiaoyang Xu, Chen E Jie Fan and Guangjing Xiong 2015 Properties of cement-based bricks with oyster-shells ash *Journal of Cleaner Production* **91** 279-287.

[14] Sugiyama M. 2004 The Compressive Strength of Concrete Containing Tile Chips, Crushed Scallop Shells & Crushed Roofing Tiles, edited by Limbachiya M C & Roberts J J in Sustainable Waste Management & Recycling: Construction Demolition Waste, London, Thomas Telford Publishing 165-172

[15] Yusof M, Ujai S J J, Sahari F, Taib S N L & Noor Mohamed N H 2011 Application of Clam (Lokan) Shell as Beach Retaining Wall *Proceeding of EnCon 2011: 4th Engineering Conference* (Kuching, Sarawak)

[16] Limbachiya M C 2009 Bulk engineering and durability properties of washed glass sand concrete. *Construction and building materials* **23** 1078-1083

[17] Etuk B R, Etuk I F and Asuquo L O 2012 Feasibility of using seashells ash as admixtures for concrete. *Journal of environmental science and engineering* **1** 121-127

[18] Olivia, M., Mifshella, A.A. and Darmayanti, L., 2015 Mechanical properties of seashell concrete.*Procedia Engineering,* **125** 760-764

[19] Olivia, M. and Oktaviani, R., 2017 Properties of Concrete Containing Ground Waste Cockle and Clam Seashells. *Procedia Engineering,* **171** 658-663

[20] Adewuyi, A.P., Franklin, S.O. and Ibrahim, K.A. 2015 Utilization of Mollusc Shells for Concrete Production for Sustainable Environment. *International Journal of Scientific & Engineering Research,* 6 **9** 201-208

[21] Mehta P K and Monteiro J M 1993 *Concrete: Structure, Properties and Methods, Second Edition* (New Jersey: Prentice-Hall International (UK) Limited) 207-208

[22] Nor Hazurina Othman, Badorul Hisham Abu Bakar, Mashitah Mat Don and Megat Azmi Megat Johari 2013 Potential Use of Cockle (*Anadara Granosa*) Shell Ash as Partial Cement Replacement in Concrete *Caspian Journal of Applied Sciences Research* **2** 369-376

**Acknowledgments**
This work was supported by Postgraduate Research Grant of Universiti Tun Hussien Onn Malaysia (U693 and U637). Also, this paper was partly sponsored by the Centre for Graduate Studies UTHM.

IOP Conf. Series: Materials Science and Engineering **271** (2017) 012060 doi:10.1088/1757-899X/271/1/012060

# Heat treatment on *keruing* and *light red meranti*: The effect of heat exposure at different levels of temperature on bending strength properties

**Nur Ilya Farhana Md Noh[1] and Zakiah Ahmad[2]**

[1] Faculty of Civil Engineering, Universiti Teknologi MARA (UITM), Malaysia
[2] Institute of Infrastructure Engineering and Sustainable Management (IIESM), UITM
Corresponding author: nurilyafarhana@yahoo.com

**Abstract.** Heat treatment on timbers is a process of applying heat to modify and equip the timbers with new improvised characteristics. It is environmental friendly compared to the common practice of treating timber by chemical preservatives. Malaysian hardwood timbers namely Keruing and Light Red Meranti which are in green condition were heat treated at temperature 150$^0$C, 170$^0$C, 190$^0$C and 210°C, in a specially designed electronic furnace within one hour duration. The objectives were to determine the effect of heat treatment on bending strength properties of heat treated timbers in terms of Modulus of Elasticity (MOE) and Modulus of Rupture (MOR) and to examine the significance changes at each temperature level. Untreated samples for each species were used as a control sample. The results indicated that the bending strength properties for both species of timbers were affected by the heat exposure. Both MOE and MOR values for heat treated Keruing were increased when subjected to the temperature levels at 150°C, 170°C and 190°C except at 210°C. Heat treated Light Red Meranti shows the same pattern of increment on its MOE and MOR values after exposure to heat at three temperature levels applied and the values dropped at 210°C. However, for both of species, even though there were decrement occurred at 210°C, the value is still higher compared to the control sample. The increments of MOE and MOR values are an indicator that heat treatment had successfully improvised the bending strength properties of these two species of hardwood timber.

## 1. Introduction

One of the oldest structural materials used by human to build their shelters and home is timber. Other than useful to build a building and house, timber also useful to build other structure such as bridge, boats, and the other civil engineering works in the construction industry [1]. Timber comes with good natural durability and characterization is good to be used structural members but this kind of timber is limited in numbers and high in price. Timber is unique as it is one kind of material which comes with non fix properties. It is depend on its species, surrounding and condition of the weather [2]. Due to the limitation of numbers of timbers with natural durability and good properties, an alternative way to solve the issue is by using the timber with non natural durability and low in properties [3]. The timber with low durability and properties need to be modified to equip the timber with new and improvise character before it can be used. Most of the timber in Malaysia which need to be modified had been treated by using chemical water based preservative known as Copper Chromium Arsenic (CCA) [4]. CCA has proven to improvise the lifespan and the properties of the treated timber [5]. This kind of treatment is world most used timber preservative. One of the big countries which applied CCA as timber preservative to modify the timber is United State (US), but in 2004 US Environmental Protection Agency (EPA) had announced that manufacturing of timber treated by CCA can no longer be produced [9]. Other big countries which applied CCA on timber earlier but had banned this application are Canada, Japan, Denmark, Switzerland, Vietnam and Indonesia. According to EC Scientific Committee on Toxicity, Eco- toxicity and the Environment (CSTEE), the chemical used in CCA may lead to the risk of cancer and genotoxic. Human may face lung, liver, kidney and bladder cancer. The ability of hearing may be reduced, decrease the immune system and can caused skin ulcer [6]. In Malaysia, the alternative method need to be applied in order to modified the properties of timber as Malaysia still allows CCA to be applied on the timber modification. An alternative treatment with no chemicals usage which is possible to make available is heat treatment. This treatment will modify the properties of timber with high level of temperature during the process of modification without any preservatives [10]. Timber treated with heat treatment may be equipped with new and improvised properties in term of physical, chemical and mechanical properties. Two species used in this study are Keruing which comes from strength group 5 (SG5) and Light Red Meranti from Strength Group 6 (SG6). Both of these species are local Malaysian hardwood timber species in Malaysia and specified as species of timber that need treatment to help in improving the durability and strength properties as stated in Malaysian Standard MS544: Part 2: 2001. The main objective in running this study is to determine the bending strength properties in term of Modulus of Elasticity (MOE) and Modulus of Rupture (MOR) for these two selected species after had been treated by four different level of temperature.

## 2. Heat Treatment

Modification of timber to improvised its durability and equips the timber with new properties by using high level of temperature in certain duration is known as heat treatment [11]. Heat treatment is using only temperature without any added preservative that make it known as one of environmental friendly way application to alter the properties of timber [12]. The timber treated by heat will be exposed to temperature ranging from 150°C to 260°C as at 150°C, the changes occur on timber properties will be permanent [13]. Commonly, heat treatment procedure is applying different range of temperature and duration. These two parameters are selected based on the species and properties of the timber been used for the treatment. The temperature of heat treatment is usually started at 180 to 280°C within 15 minutes or more [14]. The researcher also found that heat treatment on timber will not change its properties it applying low temperature even for a longer duration of treatment. [15] found that applying heat treatment on timber with temperature level above than 150°C will help to change the properties of timber in term of physical and chemical. The dimensional stability and the durability of timber are increases after treated by high level of temperature [16]. The movement stability of timber in term of shrinkage and swelling will be reduce as the equilibrium moisture content is low after the exposure of heat treatment. Previous researcher found that the heat treatment will lead to slow water absorption [17]. The high temperature level during heat treatment also will lead to reduction on ability of bound water migration through the treated timber as the chemical composition of the timber also had been change. The heat treatment had lead to cell wall modification [18].

## 3. Experimental Works

This study consists of few phases in its experimental works. The phases involved are samples preparation, heat treatment process, and bending strength measurement in order to gain the MOE and MOR values of the samples of timbers.

### 3.1*Sample Preparation*

There are two (2) species of Malaysian timbers in range of hardwood timbers had been utilized in this study. All of these species selected are the species that need to be treated before it can be used as stated in MS544: Part 2. The species are Keruing and Light Red Meranti. Both of these species of timber is in green state, means that the timber is wet and the moisture content level higher than 19%. The preparation of samples in term of size and number of control and heat treated samples is summarized in table 1 below.

**Table 1.** Samples Preparation

| Species | Number of Sample | | | Size (mm) | Strength Group (SG) |
|---|---|---|---|---|---|
| | **Control** | **Heat Treatment** | | | |
| | | **Temperature (°C)** | **Sample** | | |
| Keruing | 15 | 150 | 15 | 50 x 90 x 1800 | 5 |
| | | 170 | 15 | | |
| | | 190 | 15 | | |
| | | 210 | 15 | | |
| Light Red Meranti | 15 | 150 | 15 | 50 x 90 x 1800 | 6 |
| | | 170 | 15 | | |
| | | 190 | 15 | | |
| | | 210 | 15 | | |

Before the heat treatment, all of the samples of timber were placed in the conditioning room with a proper stack to let the samples conditioned properly. The relative humidity of that room and the temperature level will be adjusted to 65% and 24°C respectively.

After two (2) weeks, part of samples prepared went for the heat treatment while the others were treated as control samples. The timber samples were treated by heat at four ranges of temperature level; 150°C, 170°C, 190°C and 210°C, within one hour duration. Only ten numbers of samples were treated for each cycle of heat treatment to ensure that the heat applied on the timber is distributed uniformly among samples. Three wires were installed to each sample; one was in the middle part of the sample and the other two were each end of the sample. Those wires are connected to the temperature controller which can be monitored to ensure that the heat is uniformly supplied. The furnace was then closed properly before the treatment can be run.

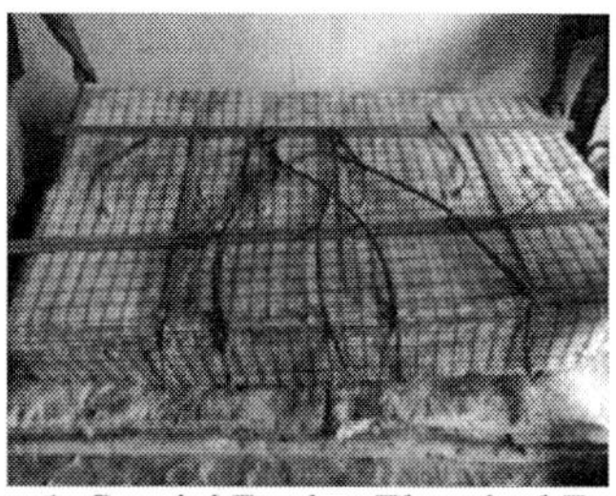

**Figure 1.** Special Design Electrical Furnace

**Figure 2.** The samples arrangement in the heat treatment electrical furnace

**4. Data Analysis and Discussion**

A four (4) point bending test was conducted to measure the mechanical properties in term of bending strength for the selected timber species in this study, in accordance to ASTM D198. The sample size of timber beam used to conduct this test is 1800mm of length x 50mm of width x 90mm of depth. This test was run after moisture content and density of the timber beams was determined. All of the samples prepared (control, heat treated and kiln dried) have been tested with this bending test. According to the clause in the standard, the half shear span to depth ratio (a/h) of the timber beams tested on four points loading must be in the range of five to twelve. The timber beams were arranged horizontally during the setup, to receive load at crosshead rate of 0.06mm/min, as shown in figure 3. In order to monitor the deflection occurred on the beam, there were four (4) linear variable differential transducers (LVDT) moving loading crosshead attached vertically during the test, shown in figure 4.

IOP Conf. Series: Materials Science and Engineering **271** (2017) 012060 doi:10.1088/1757-899X/271/1/012060

**Figure 3.** Timber Beam for Bending Test Set up

**Figure 4.** Arrangement of LVDT and Loading Crosshead

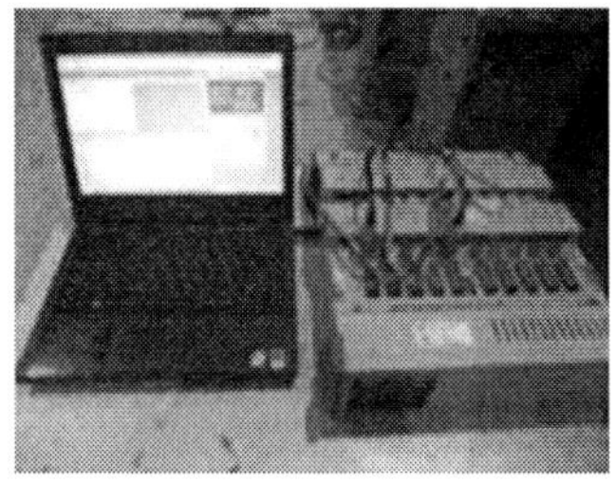

**Figure 5.** The Data Logger Connected to each LVDT

The data logger used to receive and record the reading of the data was connected to each of the LVDT attached on the timber beam, as shown in figure 5. The test had been conducted in Heavy Structure Laboratory, Faculty of Civil Engineering UiTM Malaysia using 250kN Universal Testing Machine (UTM). The value of MOE and MOR for each beam was then calculated based on the data from this bending test, using the equation shown in Equation 1 and Equation 2. The unit for both data is expressed in MPa.

$$MOR = \frac{pl}{bh^2} \tag{1}$$

$$MOE = \frac{23PL^3}{108bh^3\,\Delta} \tag{2}$$

where,
b= width of the timber beam (mm), h= depth of the timber beam (mm), P= load at the proportional limit (N), Δ= deflection at the proportional limit (mm), L= length of the span (mm), a= half shear span (mm)

Figure 6 shows the load versus deflection curve of structural size samples for Keruing. The curve shows different behavior of the slope between control samples and heat treated samples in four ranges of temperature. In elastic region of the curve, it can be seen that the linear slope of samples treated at temperature of 150°C, 170°C and 190°C are higher compared to the control samples. Treating Keruing at these three temperature level had improved the stiffness of the samples. According to the degree of slope in this figure, there is improvement on the ability of the heat treated samples to sustain load compared to the control sample. These three samples take the longest time to fail compared to control samples and the slope is drop abruptly after achieved the maximum load. The samples are samples treated at 150°C, 170°C and 190°C where the highest maximum load is performed by the sample treated at 190°C. Treated Keruing at temperature level of 210°C still improvised its bending capacity as both of the ability to sustain load and stiffness is is still higher than the control samples but low than other samples of temperature. It can be said that Keruing sample treated by heat at 190°C shows the highest improvement in bending behavior as it achieved highest load and take the longest time to fail compared to the other samples.

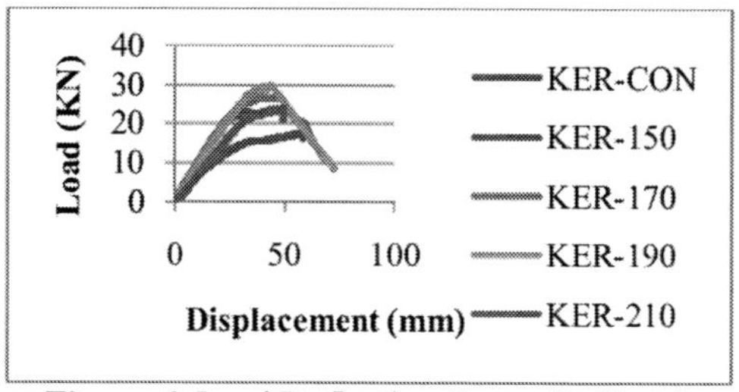


**Figure 6.** Load Deflection Curve for Keruing

Table 2 shows the MOE and MOR value of each sample for Keruing including control and heat treated samples at four different levels of temperatures. The MOE value indicated in the figure for control sample is 10.06GPa. The value then increasing to 17.24GPa for sample treated at temperature of 150°C. The figure shows that the MOE value is keep increasing with the increment of heat treatment temperature which is 19.55GPa at 170°C and 22.42GPa at 190°C. However, the value shows a slight reduction at temperature of 210°C which is 16.90GPa. It is learned that even there is reduction of value of MOE at the highest temperature of heat treatment on Keruing, the value is still higher compared to the control sample. The MOR value for control and heat treated samples for Keruing. The result shows that the value of MOR for this timber increase compared to the control sample when treated by heat. The MOR value for PAU-CON is 80.24MPa. The MOR value for sample treated at temperature of 150°C is 104.65MPa, 117.36MPa at 170°C and 135.0MPa at 190°C. The value decreased once the timber been treated by temperature of 210°C where the MOR value dropped to 104.87MPa.

**Table 2.** MOE and MOR Value for Keruing

| | MOE | | | | MOR | | | |
|---|---|---|---|---|---|---|---|---|
| **SAMPLE** | Mean (GPa) | S.D | COV (%) | Δ(%) | Mean (MPa) | S.D | COV (%) | Δ(%) |
| **KER-CON** | 10.06 | 0.62 | 6.20 | | 80.24 | 6.64 | 8.28 | |
| **KER-150** | 17.24 | 1.11 | 6.43 | -71.37 | 104.65 | 7.94 | 7.59 | -30.42 |
| **KER-170** | 19.55 | 1.39 | 7.11 | -94.38 | 117.36 | 7.92 | 6.75 | -46.26 |
| **KER-190** | 22.42 | 1.09 | 4.86 | -122.84 | 135.00 | 6.58 | 4.88 | -68.24 |
| **KER-210** | 16.90 | 1.40 | 8.29 | -68.04 | 104.87 | 7.57 | 7.22 | -30.69 |

During the bending test parallel to the grain, the load versus displacement had been monitored and recorded continuously for each of heat treated samples of Keruing. It was tested until the samples gained the maximum load and completely failed. The mode of failure shown by each samples tested was recorded. The samples failed at the tension zone with the different degree of tare on the samples beam. Figure 7 shows the pattern of failure occurred on the beam after been tested.

**Figure 7.** Mode of failure for Keruing

Figure 8 shows the behavior of load versus displacement curve under bending parallel to grain for structural size samples of Light Red Meranti. There are five numbers of slopes and each of the slopes represents the control and heat treated samples at four ranges of temperature. In this figure, it can be seen that the linear slopes are almost similar between all heat treated samples treated at different temperature level. Meanwhile, samples treated by heat at 150°C, 170°C and 190°C shown high improvement of linear slope which means these samples are stiffer compared to the control sample. The improvement of bending behavior on heat treated Light Red Meranti as the degree of slope had increasing from the lowest range of temperature to the highest range. It shows that the ability of Light Red Meranti to sustain the applied load is improving compared to control sample as the maximum load obtained keep increasing for heat treated samples from 150°C, 170°C to 190°C except for beam treated at 210°C. The maximum load achieved by LRM-210 is higher compared to the control beam, but it is low compared to maximum load achieved by LRM-190. It is an indicator that at temperature level above of 190°C, the bending behavior of Light Red Meranti started to decrease. In overall, heat treating Light Red Meranti had improved the bending behavior of the samples of each temperature applied but the best improvement occurred at 190°C temperature level where the beam is stiffer, sustain higher maximum load and take the longest time before completely fail.

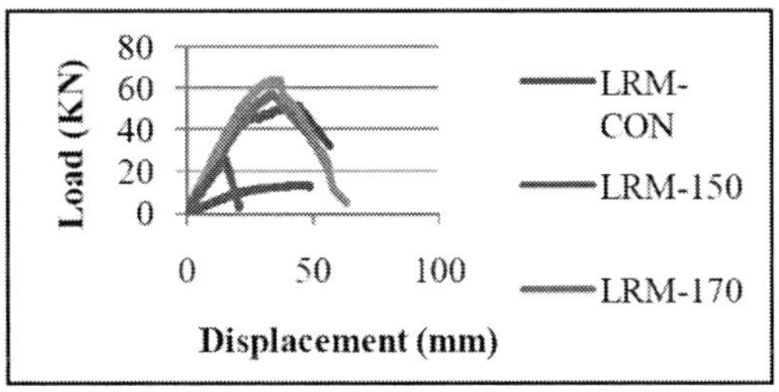


**Figure 8.** Load Deflection Curve for Kapur

Table 3 show the MOE and MOR value for all the samples of Light Red Meranti respectively. For MOE value, the result clearly indicates that the value increase for the heat treated timber of each level of temperature applied, compared to the control sample, except at 210°C. At 210°C, the MOE value started to decrease yet is still higher compared to the control and heat treated sample at 150°C. The MOE value of LRM-CON is 9.16GPa, then increased to 58.62Gpa at 190°C. At 210°C, MOE value had dropped to 55.60GPa. The value of MOE for Light Red Meranti for each sample; Table 3 shows the MOR value of Light Red Meranti before and after been treated by heat. The MOR value for LRM-CON is 56.87MPa. This value is then increasing once this timber been treated by heat as MOR value for LRM-150 is 227.92MPa, followed by LRM-170 with 251.30MPa. The value is keep increasing once the timber been treated by heat at temperature level of 190°C with 270.21MPa. The value is then dropped to 117.43MPa once temperature of heat treatment had been increased to 210°C. Even though the value at this temperature seems to be dropped compared to other three level of heat treatment temperature but if compared to the control samples, the value of MOR for LRM-210 is still shows improvement and better than the MOR value for control samples.

**Table 3.** MOE and MOR Value for Light Red Meranti

| | MOE | | | | MOR | | | |
|---|---|---|---|---|---|---|---|---|
| **SAMPLE** | Mean (GPa) | S.D | COV (%) | Δ(%) | Mean (MPa) | S.D | COV (%) | Δ(%) |
| **LRM-CON** | 9.16 | 1.55 | 16.97 | | 56.87 | 6.93 | 12.18 | |
| **LRM-150** | 37.67 | 2.08 | 5.51 | -311.25 | 227.92 | 7.96 | 3.49 | -300.81 |
| **LRM-170** | 55.55 | 3.56 | 6.41 | -506.42 | 251.30 | 9.97 | 3.97 | -341.93 |
| **LRM-190** | 58.62 | 3.00 | 5.12 | -540.02 | 270.21 | 8.09 | 2.99 | -375.18 |
| **LRM-210** | 55.60 | 1.67 | 3.01 | -507.04 | 117.43 | 5.91 | 5.03 | -106.51 |

Figure 9 shows the failure modes on Light Red Meranti's heat treated beam tested by bending parallel to the grain. Overall observation shows that all of the heat treated Kapur beams have same modes of failure. The beams failed at the tension zone except for the degree of tear shown by each species of sample beam tested. Light Red Meranti beam treated at heat of 210°C shows the highest degree of tear failure at tension zone compared to the other three beams treated at lower temperature.

**Figure 9.** Mode of failure for Light Red Meranti

Heat treatment had increased the bending strength properties of beech wood [19]. The increment of temperature level applied during heat treatment process had positively affected the bending strength properties of black pine [20].

## 5. Conclusion

According to the findings of this study, the value of MOE and MOR of heat treated Keruing and Light Red Meranti had been improvised and increased. It shows that, heat treatment is an alternative way in order to modify the properties and characterization of timber, which is better and environmental friendly compared to the chemical preservative. The heat treatment might give an acceptable effect on the properties of timber in term of physical, chemical or mechanical composition that will lead to the timber application in construction industry as structural materials without using chemical preservatives. To be concluded, bending strength properties of Keruing and Light Red Meranti had been improved by applying heat treatment. It can be said that, the timber with low strength properties can be modified and improvised by high level of temperature.

## 6. References

[1] Al-nagadi E M 2012 Saudi Arabia *Concrete Construction Industry* Cement Based Materials And Civil Infrastructure (CBM& CI)

[2] Patel, K. V. 2011 Construction Materials Management On Project Sites *National Conference on Recent Trends in Engineering & Technology*

[3] Hall N L and Beder S 2005 *Treated Timber, Ticking Time-bomb* (NSW, Australia)

[4] Tong H L 2005 *Treatment Of Timber For Trusses* Malaysian Wood Preserving Association (MWPA)

[5] Hardin R A and Beckermann C 2005 Simulation of Heat Treatment Distortion *Proceeding of the 59$^{th}$ SFSA Technical and Operating Conference* (Chicago, USA)
[6] Hedley M D 1997 An assessment of risks associated with use of CCA-treated timber in sensitive environments and options for its substitution with alternative timber materials *Conservation Advisory Science Notes No. 154* (Department of Conservation, Wellington, New Zealand)
[7] Barrie D E 2000 *Copper Chromium Arsenate (CCA) In The Environment*
[8] Read D 2003 *Report on Copper , Chromium and Arsenic ( CCA ) Treated Timber* (New Zealand)
[9] Arsenic Timber Treatments (CCA And Arsenic Trioxide) 2003 *Authority, Australian Pesticides and Veterinary Medicines*
[10] Korkut S, Alma M H and Elyildirim Y K 2009 *The effects of heat treatment on physical and technological properties and surface roughness of European Hophornbeam (OstryacarpinifoliaScop .) wood* **8(20)** 5316–5327
[11] Sundqvist B 2004 *Colour Changes and Acid Formation* (University of Technology, Sweden) 1402–1544
[12] Heat Treatment of Timber 2005 *Calu Technical Notes*, (July)
[13] Militz H, Biology W and Technology W 2002 *Heat Treatment Technologies in Europe : Scientific Background and Technological State-of-Art* (November 2000)
[14] Korkut D S, Korkut S, Bekar I, Budakç M and Dilik T 2008 *The Effects of Heat Treatment on the Physical Properties and Surface Roughness of Turkish Hazel (Coryluscolurna L .) Wood*, 1772–1783
[15] Syrjänen T, and Oy K 2000 *Production and Classification Of Heat Treated Wood In Finland*
[16] Yildiz S, Gümüşkaya E 2007 The effects of thermal modification on crystalline structure of cellulose in soft and hardwood *Building and Environment* **42(1)** 62–67
[17] Hakkou M, Pétrissans M, Zoulalian A, Gérardin P 2005 Investigation of wood wettability changes during heat treatment on the basis of chemical analysis *Polymer Degradation and Stability* **89(1)** 1–5
[18] Pot D, Chantre G, Rozenberg P, Rodrigues J C, Jones G L, Pereira H and Plomion C, 2002 Genetic control of pulp and timber properties in maritime pine (PinuspinasterAit.) *Annals of Forest Science* **59** 563–575
[19] Kortelainen S M, Antikainen T and Viitaniemi P 2005 *The water absorption of sapwood and heartwood of Scots pine and Norway spruce heat-treated at 170°C, 190°C and 230°C.* Journal of HolzalsRoh-und Wekstoff.
[20] Ordu M, Altinok M, Atilgan A, Ozalp M, and Peker H 2013 *The effects of heat treatment on some mechanical properties of laminated black pine ( Pinusnigra ).*

**Acknowledgments**
This research is financially supported by the Malaysian Public Work Department (MPWD), 100-RMI/GOV 16/6/2 (8/2013) and Universiti Teknologi MARA (UITM), Malaysia whom both are great acknowledged.

IOP Conf. Series: Materials Science and Engineering **271** (2017) 012061 doi:10.1088/1757-899X/271/1/012061

# Bacterial carbonate precipitation improves water absorption of interlocking compressed earth block (ICEB)

**M M Zamer[1], J M Irwan[1], N Othman[2], S K Faisal[1], L H Anneza[1], A F Alshalif[1] and T Teddy[1]**

[1]Jamilus Research Centre for Sustainable Construction (JRC), Faculty of Civil and Environmental Engineering, Universiti Tun Hussein Onn Malaysia, 86400, Parit Raja, Johor, Malaysia
[2]Micro-Pollutant Research Centre (MPRC), Faculty of Civil and Environmental Engineering, Universiti Tun Hussein Onn Malaysia, 86400, Parit Raja, Johor, Malaysia
Corresponding author: megatzamer@gmail.com

**Abstract.** Interlocking compressed earth blocks (ICEB) are soil based blocks that allows for mortarless construction. The addition of many alternative materials into interlocking block in order to improve the durability has been reported. However there are currently lack of report and evidence on the application of biocalcification or microbiologically induced calcite precipitation (MICP) in improving the engineering properties of ICEB. This paper evaluate the effect of UB in improving the water absorption properties of ICEB. This paper also provide the results on SEM analysis of addition of 1%, 3% and 5% UB in ICEB. The bacteria were added as partial replacement of limestone water in ICEB. The results showed the reduction of 14.72% with 5% UB on initial water absorption followed by the results for water absorption by 24-hour soaking which also indicates reduction of 14.68% with 5% UB on 28th days of testing compared to control specimen. It was expected that the reduction of water absorption was due to the plugging of pores by the bacterial calcite which prevent ingression of water in ICEB samples. Therefore this study hopes that the positive results from the UB as improving in water absorption of ICEB will lead to improve others ICEB properties and others construction materials.

## 1. Introduction

Interlocking compressed earth block (ICEB) is a common construction material which was used as the building blocks for many civil structures. It can be used for all types of building from wells, houses to schools and resort buildings. ICEB is an environmental friendly construction material, cost effective, dry stacked and sustainable construction material which has potential to bring durable and affordable homes to developing counties around the world [1]. Interlocking blocks can be divided into two types based on its raw materials: soil-cement block, a soil based block which was used in this study and a concrete block, in which the major component is cement. However, there are also others problems associated with ICEB namely low strength, higher water absorption, low fire resistance and high porosity [2].

Biocalcification also known as microbiologically induced calcite precipitation (MICP), is a normal biochemical process that naturally occurs in many environments such as sea water, fresh water and soil [3]. During this process calcium carbonate accumulates due to the activity of urease-producing microorganisms. The use of bacteria is one of new fundamental research in improving construction material in order to pursuit sustainable construction. Researcher such as Willem *et al.* [4] used *Bacillus sphaericus*, Navdeep *et al.* [5] used *Bacillus megaterium*, Abhjit [6] used *Bacillus megaterium* and Bernardi *et al.*[7] used *Sporosarcina paseurii*. All bacteria used by previous studies resulted in reduction on water absorption by comparing control sample and treated sample with bacteria. All previous reports mentioned above demonstrated that biocalcification could be used to improve the properties of construction material. Currently, lack information on locally isolated bacteria incorporated in ICEB from Malaysia has been documented. Therefore, a study on locally isolated strains incorporating in ICEB is necessary. This study was designed to investigate the influence of microbial induced calcite precipitation on the water absorption of ICEB.

## 2. Experimental works

Isolated strain UB from fresh urine was used in this studies. The standard Nutrient Broth (NB) media were used to demonstrate and compare the efficiency of microbial induced calcite precipitation (MICP) by UB in ICEB. Enrichment process was done follows previous researcher method [8]. In order to ensure the survival of UB in ICEB environment, enrichment process was done with addition of soil sample to acclimatize the ICEB condition. The composition for optimal growth was 300ml nutrient broth added with 120ml of 40% urea, 1mg of soil substances and lastly 1 cyrogenic bead of ureolytic bacteria. The differences between the compositions of control were the optimal growth was done without the soil sample.

Addition of UB as partial replacement of limestone water with 1%, 3% and 5% UB are added into ICEB mix. The initial rate of water absorption test was measured on how quickly the water were being absorbed through the ICEB in 1 minute. The water absorption test was measured for $7^{th}$, $14^{th}$ and $28^{th}$ days. The test was carried out according to the BS 3921:1985, [9] which specify for the testing of bricks. The scanning electron microscopy (SEM) analysis was done in determination of pore size and morphology precipitation calcium carbonate ($CaCO_3$).

## 3. Water absorption

### *3.1 Initial water absorption*

Figure 1 show the results of initial water absorption for $7^{th}$, $14^{th}$ and $28^{th}$ days of testing. The addition of UB as a liquid culture of 1%, 3% and 5% in ICEB reduced the initial water absorption within time compared to control specimen. The patterns of reduction water absorption are the same for all bacterial ICEB specimens. Addition of 5% UB indicates the highest reduction by 14.72% in water absorption at $28^{th}$ days compared to control sample follows with 3% of UB with 11.11% and 1% of UB with 7.5%. The highest reduction of initial water absorption recorded was 3.07kg/m$^2$.min at the $28^{th}$ days of testing for 5% addition of UB.

The positive result indicates the successfulness of UB surviving in ICEB environment and deposition of calcium carbonate, $CaCO_3$ for the water absorption improvement of ICEB. The reduction of initial water absorption with added bacteria agreed with Willem *et al.* [4] studies which also observed decrease in initial water absorption rate upon treatment with bacteria.

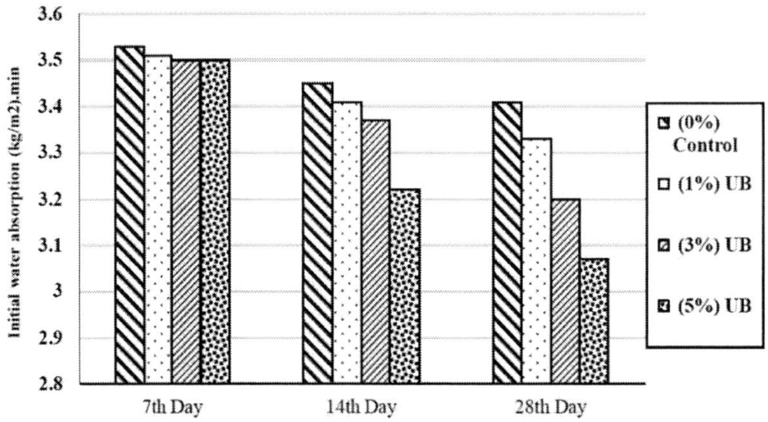


**Figure 1.** Initial water absorption of control and bacterial ICEB.

*3.2 Water absorption by 24-hour of soaking*

Figure 2 show the results of water absorption by 24hour soaking for 7[th], 14[th] and 28[th] days of testing. The addition of UB as a liquid culture of 1%, 3% and 5% in ICEB reduced the percentage of water absorption within time compared to control specimens. The patterns of reduction water absorption are the same for all bacterial ICEB specimen. Addition of 5% UB indicates the highest reduction by 14.68% in water absorption at 28[th] days compared to control sample follows with 3% of UB with 11.83% and 1% of UB with 10.73%.

According to Abhjit *et al.* [6], the calcite crystal acts as biosealent by filling the pores which leads to reduction in water absorption, porosity, and permeability. The presence of bacteria resulted in a significant decrease in the water absorption compared to control specimens. The deposition of a layer of calcium carbonate on the surface and inside pores of the ICEB specimens resulted in a decrease of water absorption and permeability. Hence, from the study it is clear that the presence of a layer of carbonate crystals by bacterial cells has the ability to improve the resistance of penetration of water absorption.

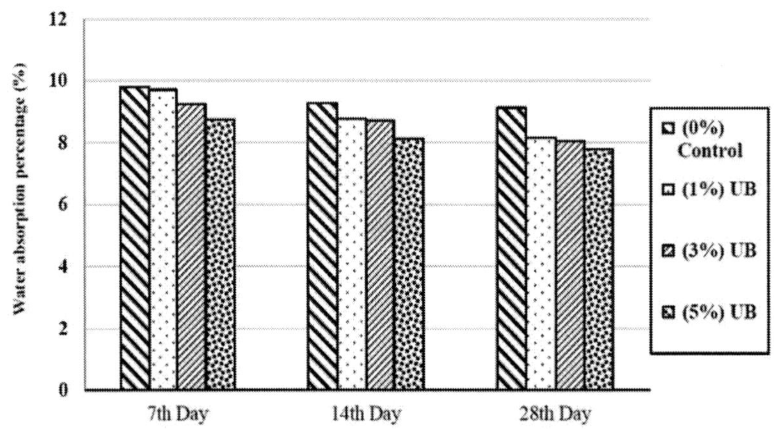


**Figure 2.** Results water absorption for 24-hour soaking of control and bacterial ICEB.

## 4. SEM analysis

Microstructure analysis using SEM image shows that UB successful to grow well inside the ICEB specimens in anaerobic condition and also capable to tolerate in alkaline environment. The pore size analysis was determine by using small size of crushed ICEB passing 64 micron sieve. Figure 3 shows the average results for pore size analyses conducted on control ICEB and bacterial ICEB. Control image generally showed rough and uneven surface. This porous surface from control samples led to increase of absorption capacity of ICEB especially water absorption and hence decrease in durability. In bacterial ICEB some part of the porous surface has been filled by calcite precipitation, thus reduction in pore size. The reduction of pore will promote to the reduction of water absorption, permeability and porosity. The appearance of calcite precipitation which can be seen as smooth surface of round sphere in bacterial ICEB microstructure image. The average size of calcite forms was evaluated and compared between control and bacterial ICEB. As shown in Figure 4, control sample which can be seen without precipitation of bacterial calcite due to 0% percentage of UB addition. Addition of 3% UB indicate highest average size of bacterial calcite form which is 12.469μm follows with 1% of UB with 9.212μm and 5% of UB by 7.815μm. Addition of 5% UB also resulted highest reduction in average pore size with 0.643μm follows with 3% UB by 0.650μm and 1% UB with 0.772μm compared to the control specimen which was 0.839μm. The calcium carbonate precipitation was to prove that addition of UB in ICEB promote to bacterial calcite precipitation.

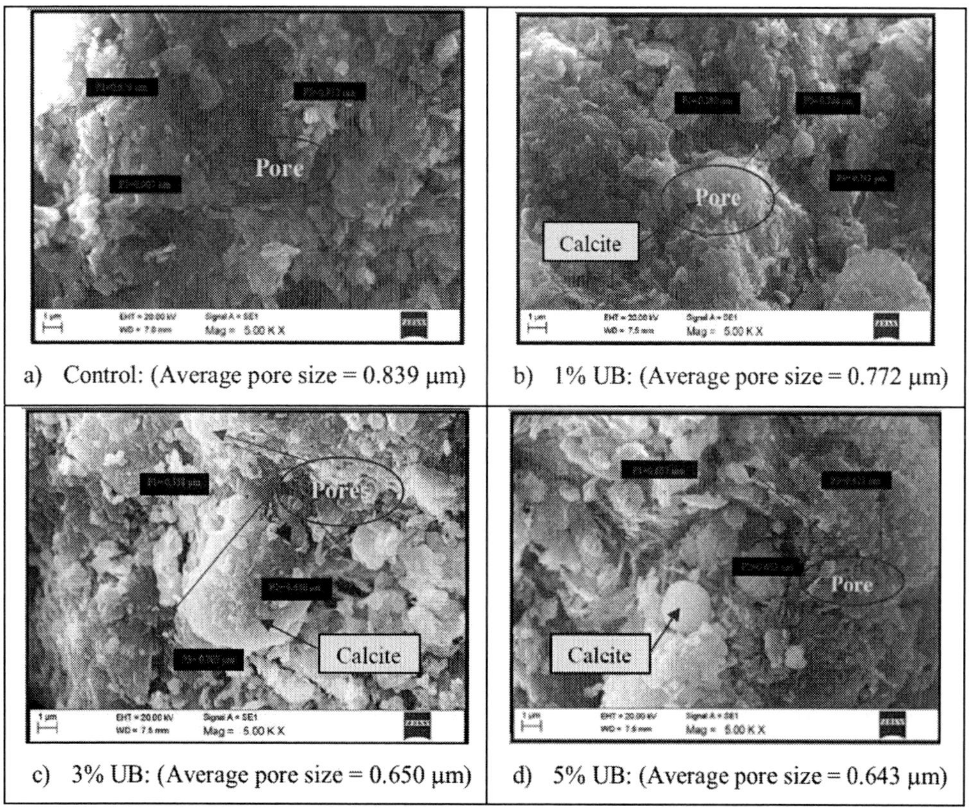


a) Control: (Average pore size = 0.839 μm) b) 1% UB: (Average pore size = 0.772 μm)

c) 3% UB: (Average pore size = 0.650 μm) d) 5% UB: (Average pore size = 0.643 μm)

**Figure 3.** Pore measurement for ICEB with and without addition of UB.

a) Control: (Average size = 0)

b) 1% UB: (Average size = 9.212 μm)

c) 3% UB: (Average size = 12.469 μm)

d) 5% UB: (Average size = 7.815 μm)

**Figure 4.** Precipitation of calcium carbonate for ICEB with and without addition of UB.

Based on the images obtained through Scanning electron microscope (SEM), the samples which contained bacteria has shown pores being filled and surface for these samples appear more compact and smooth compared to control which is rough and porous in nature. The appearance of calcite precipitation also can be seen clearly as smooth surface of round sphere in bacterial ICEB microstructure image. There are also decrement in pore size of bacterial ICEB compared to control ICEB. Addition of 5% UB in ICEB indicate the highest reduction of pore size compared to the control sample by 23.36%. Therefore, the results from microstructure analysis prove that by addition of UB in ICEB resulted in reducing the pores and bacterial calcite precipitation. Thus improve the water absorption of ICEB. This finding agreed with the previous studies [4-7] which indicated the improvement of material engineering properties due to deposition of calcium carbonate.

## 5. Conclusion

Based on the results of this study, it can be concluded that the addition of UB has positive effect on water absorption of ICEB. The present study suggests that addition of UB in ICEB resulted in improving the water absorption due to biocalcification activity of UB grown in ICEB compared to control specimens. SEM analysis shows that the successful of UB in depositing the calcium carbonate and reducing the pore size of ICEB. Hence this will further improve the properties of ICEB. This can favorably reduce the frequency of required maintenance for civil structures especially in bricks. However, further study on various parameter improvement for interlocking block manufacturing is needed. The development of a more effective process for large scale production is also necessary for practical applications in future.

## 6. References

[1] Laursen P T, Herskedal N A, Jansen D C and Qu B 2012 Out-of-plane structural response of interlocking compressed earth block walls, *Journal of Materials and Structures* **48**(1-2) 321-336

[2] Irwan J M, Zamer M M and Othman N 2016 A review on Interlocking Compressed Earth blocks (ICEB) with addition of bacteria *MATEC Web Conferences* vol 47(Paris:EDP Sciences) 01017

[3] Sada Y, Poon T and Wasu Pathom-aree 2015 The influence of biocalcification on soil-cement interlocking block compressive strength, *Journal biotechnology, Agronomy, and Society Environment* **19**(3) 262-269

[4] De Muynck W, Debrouwer D, De Belie N and Verstraete W 2008 Bacterial carbonate precipitation improves the durability of cementitious materials, *Journal of Cement and Concrete Research* **38**(7) 1005-1014

[5] Dhami N K, Reddy M S and Mukherjee A 2012 Improvement in strength properties of ash bricks by bacterial calcite, *Journal of Ecological Engineering* **39** 31-35

[6] Mukherjee A, Dhami N K, Reddy B V V and Reddy M S 2013 Bacterial Calcification for Enhancing Performance of Low Embodied Energy Soil-Cement Bricks, *Journal of Sustainable Construction Materials and Technology*, **1** 1-10

[7] Bernadi D, Dejong J T, Montoya B M and Mattinez B C 2014 Bio-bricks: Biologically cemented sandstone bricks, *Journal of Construction and Building Material* **55** 462-469

[8] Irwan J M, Anneza L H, Othman N, Husnul T and Alshalif A F 2016 Isolation and Identification of Concrete Environment bacteria, *IOP Conference Series: Materials Science and Engineering* vol 136 (1) (London:IOP Publishing) 012088

[9] BS 3921 1985 *Specification for Clay Bricks, Clay Product Standards Committee*, (London: British Standard) 8-20

**Acknowledgments**

This research was supported by University Tun Hussein Onn Malaysia (UTHM), Malaysia, The Ministry of Higher Education Malaysia through Fundamental Research Grant Scheme (FRGS), Vot. 1577 and collaboration with ICEB industry IRYAS INC. (M) Sdn Bhd.

IOP Conf. Series: Materials Science and Engineering **271** (2017) 012062 doi:10.1088/1757-899X/271/1/012062

# Green waste cooking oil-based rigid polyurethane foam

**N F Enderus[1] and S M Tahir[1]**

[1]Faculty of Applied Sciences, Universiti Teknologi MARA, 26400 Bandar Tun Abdul Razak Jengka, Pahang, Malaysia
Corresponding author:syuhadamohdtahir@uitm.edu.my

**Abstract.** Polyurethane is a versatile polymer traditionally prepared using petroleum-based raw material.Petroleum, however, is a non-renewable material and polyurethane produced was found to be non-biodegradable. In quest for a more environmentally friendly alternative, wastecooking oil, a highly abundant domestic waste with easily derivatized structure, is a viable candidate to replace petroleum. In this study,an investigation to determine physical and chemical properties of rigid polyurethane (PU) foam from waste cooking oil (WCO) was carried out. WCO was first adsorbed by using coconut husk activated carbon adsorbent prior to be used for polyol synthesis. The purified WCO was then used to synthesize polyol via transesterification reaction to yield alcohol groups in the WCO chains structure. Finally, the WCO-based polyol was used to prepare rigid PU foam. The optimum formulation for PU formation was found to be 90 polyol: 60 glycerol: 54 water: 40 diethanolamine: 23 diisocyanate. The rigid PU foam has density of 208.4 kg/m$^3$ with maximum compressive strength and capability to receive load at 0.03 MPa and 0.09 kN, respectively. WCO-based PU can potentially be used to replace petroleum-based PU as house construction materials such as insulation panels.

## 1. Introduction

Polyurethane (PU) is known as linear block copolymers prepared by reacting polyol and diisocyanate in the presence of surfactant, catalyst, blowing agent and other additives like fire retardants and smoke suppressants. Since variety of diisocyanates and polyols available to produce polyurethane, a broad spectrum of PU properties can be produced to meet the needs of specific applications. PU can be found in four main forms; 1) rigid, 2) flexible, 3) coatings, adhesives, sealants and elastomers, and 4) thermoplastic. Each type of PU has their own applications. For example, flexible PU are used as cushioning in bedding, furniture and in automotive industry, while rigid PU are used as insulation materials in house construction [1].

Currently, most PU products used petroleum as raw materials for polyol. Petroleum are non-renewable, not environmentally friendly and require high production costs with high technology processing system. PU made from vegetable oil based polyol such as rapeseed oil, palmeri oil, vernonia oil, castor oil and cardanol oil, are renewable andhas been found to be more biodegradable than petroleum-based [2-3]. However, virgin vegetable oil is also expensive. Recently, lots of effort has been devoted to synthesize polyol for rigid PU foam from waste cooking oil (WCO).

WCO is waste materials from food industry. During frying, oil will be degraded due to presence of air, heat and water via oxidation, polymerization and hydrolysis processes. Degradation reaction will reduce the quality of oil and will give adverse effect to human if consumed. Uncontrolled dumping of WCO to the drainage system without proper treatment can lead to severe environmental problems. WCO has potential as raw material for PU production due to its physical and chemical properties.However, an adsorbent is needed to remove the dissolved materials presence in WCO resulted from the degradation reaction[4-7].

Coconut husk, an agricultural waste material has been used as a low-cost adsorbent to adsorb dissolved materials in WCO. The bulk of coconut husk is composed of cellulose and lignin. The presence of hydroxyl groups in both polymeric substances provide sites for adsorption of dissolved materials in WCO [8].

In this study, WCO will be pretreated using coconut husk activated carbon, followed by transesterification reaction to synthesize polyol. The polyol was then used to prepare WCO-based PU rigid foam and the chemical and physical properties were studied.

## 2. Experimental

### *2.1. Materials*

The raw materials used for this study are WCO and coconut husk. For preparation of adsorbent, 1M potassium hydroxide solution (KOH) and 0.1M hydrochloric acid (HCl) were purchased from R&M Chemicals. For preparation of polyol, diethanolamine (DEA) and monoethyleneglycol (MEG) were purchased from R&M Chemicals while potassium acetate (PA) was purchased from HmbG Chemical. For synthesis of rigid PU foam, 2,4-diphenylmethane diisocyanate (MDI) was purchased from Merck, diethanolamine (DEA) from R&M Chemicals and glycerol from HmbG Chemicals.

### *2.2. Methodology*

Coconut husk used for preparation of activated carbon was obtained locally from coconut milk producer located in Kuala Terengganu. Coconut husk was first cut, washed with distilled water, oven dried for 12 hours at 85°C and grinded to 0.1 mm mesh size using Grinder Machine EFL2000. The powder was then sieved into 150 microns. 10 g of coconut husk powder mixed with 500 mL of 1 M KOH solution and placed into oven for 6 hours at 150 °C. The solution was filtered using filter paper. Next, the activation process was carried out at 800°C for 1 hour in a tubular furnace under nitrogen atmosphere. The activated product was then cooled to room temperature and washed with 0.1M HCl solution and deionized water repeatedly until the pH of the filtrate water became 7. Finally, the activated carbon was dehydrated in oven overnight at 105°C to remove moisture before being used to purify WCO [5].

Sample of WCO was obtained from a restaurant located in Kuala Terengganu, Malaysia. The WCO was first filtered using filter paper to remove any suspended materials. 100 mL of filtered WCO then heated at95-125°C.0.5 g of coconut husk activated carbon adsorbent was added to WCO and left for 24 hours. Next, the adsorbed WCO was stirred at 500 rpm and 110°C for 1 hour. Adsorbed WCO was then filtered by using filter paper and analyzed using FTIR and %FFA [5].

DEA, MEG and PA were mixed homogenously with a ratio of 90:7:3 to form polyhydric compound. The solution was then stirred to mix well and kept overnight. Transesterification reaction was carried out by mixing adsorbed WCO-based polyol with polyhydric solution with a ratio of 80:20 respectively. The solution mixture was heated and continuously stirred at 185-195°C with 200 rpm for 20 minutes to yield polyol. The polyol was then analyzed for FTIR spectroscopy, %FFA, acid value, hydroxyl value, iodine value and viscosity.

Resin was prepared in a plastic cup under elevated temperature. The resin was blended using an overhead mechanical stirrer at condition of 200 rpm rotor speed for 1 minute. MDI was added to the resin at a ratio of 1:1 at condition of 1000 rpm rotor speed for 10 second. The effect of amount of DEA as catalyst and glycerol as crosslinking agent on reaction behaviourof PU in terms of cream time, gel time, rise time and tack-free time of the foam was recorded. The rigid PU foam was then characterized for FTIR, densityand compressive strength [9].

### *2.3. Characterizations*

Fourier transformed infrared (FTIR) spectroscopy analysis was carried out by using Perkin Elmer FTIR Spectrum BX spectrophotometer to identify functional groups presence in WCO, polyol and PU. The samples were scanned in the range of 650-4000 $cm^{-1}$ wavenumbers. Free fatty acid percentage

(%FFA) was characterized using ASTM D5555–95 standard method to measure amount offree fatty acidscontained in WCO and polyol. Acid value was calculated from %FFA to measure mass of KOH in milligrams required to neutralize one gram of chemical substance in polyol. Hydroxyl value was analyzedusing Lubrizol Test Procedure TP-AATM-111A-02to measure the content of free hydroxyl groups in WCO and polyol. Iodine value was analysed using Lubrizol Test Procedure TP–AATM–112-01to determine the amount of unsaturation in fatty acids. Viscosity of WCO and polyol was determined by using Brookfield viscometer in the range of 10 to 100 000 mPa·s(cP) using ASTM standard method D4878-08. Compressive strength test for rigid PU foam was conducted on a Universal Testing Machine Model Instron. The specimen was cut into cubes of 50 mm × 50 mm × 50 mm in dimensions. The foam rise direction was marked and 50 mm/min crosshead speed was applied. The maximum stress and load of the foam was recorded.

## 3. Results and discussion

### *3.1. Synthesis of waste cooking oil-basedpolyol*

The properties of WCO and WCO-based polyol are summarized in Table 1. Adsorption process using coconut husk activated carbon had reduced %FFA and acid value in WCO sample.Transesterification reaction had been used to convert the purified WCO into polyol. Transesterification or alcoholysis is generally defined as a chemical reaction involving the displacement of alcohol from an ester by another.The viscosity of polyol produced was found to be 155 mPa.s which is quite high.Viscosity was measured to determine the resistance of polyol to uniformly continuous flow without other forces. High viscosity polyol is not favoured due to low stirring efficiency during PU processing.

**Table 1.** Properties of WCO and WCO-based polyol.

| Properties | Raw WCO | Adsorbed WCO | WCO-based polyol |
|---|---|---|---|
| Free Fatty Acid (%) | 4.4 | 1.4 | 1.5 |
| Acid Value (mgKOH/g) | 8.756 | 2.786 | 2.985 |
| Viscosity (mPa.s) | 106 | 97.8 | 155 |
| Colour | Dark brown | Light brown | Colourless |
| Hydroxyl value | - | - | 46.3 |

The structure of WCO-based polyol was confirmed by FTIR spectrometer. The polyol consisted of chains of fatty acid with carboxylic acid group. As shown in figure 1, the presence of peaks at 3303.13 $cm^{-1}$ and 1743.81 $cm^{-1}$of hydroxyl group and carbonyl stretching of ester group, respectively, confirmed the formation of polyester polyols. C-H stretching peaks can be seen at 2922.48 $cm^{-1}$ and 2853.29 $cm^{-1}$, C-O stretching and O-H bending near 1377.27-1237.44 $cm^{-1}$ and 1457.04-1395.60 $cm^{-1}$, respectively, and C-O-H bending peak was found at 1457.04-1395.60 $cm^{-1}$[9].

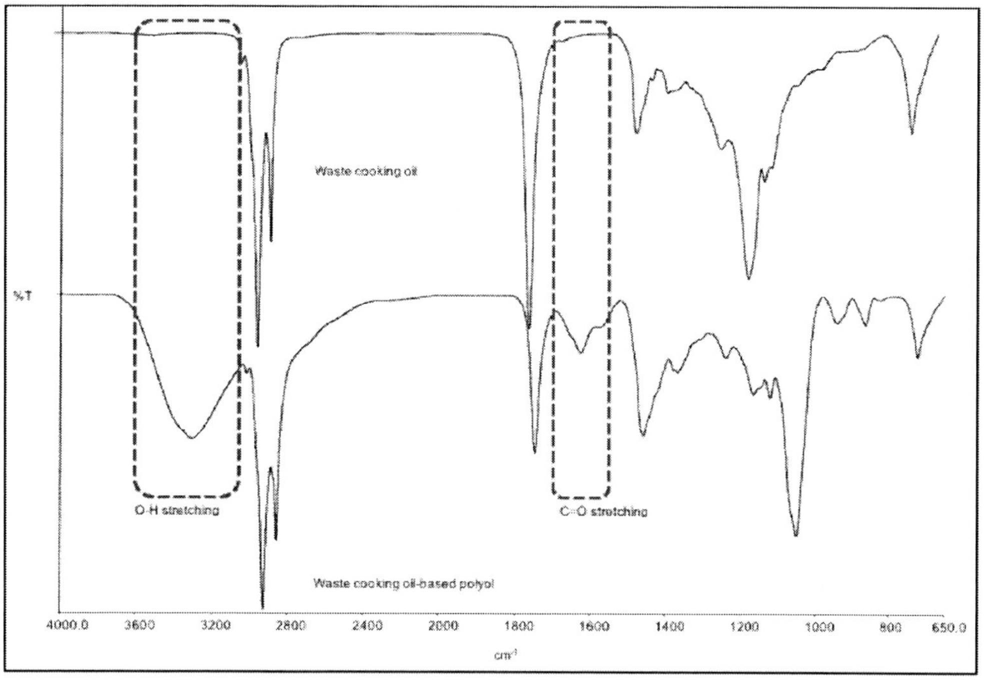


**Figure 1.** FTIR spectra of waste cooking oil and waste cooking oil-based polyol.

*3.2. Synthesis of waste cooking oil-based polyurethane foam*

Two main factors affecting the foam formation, which are amount of catalyst and glycerol were studied using various compositions. Table 2 shows the compositions used to prepare rigid PU foam in three different amount of DEA as catalyst. The function of DEA is to increase the rate of reaction involved in the foaming system. Cream time and rise time is defined as the rate of expansion, principally an exothermic effect, while the gel time and tack-free time could be used to determine the gelation rate taken to form rigid PU foam. The different composition of catalyst used could affect the cream time, gel time, rise time and tack-free time involved in the formation of rigid PU foam.From the result obtained, composition C showed faster expansion rate during foaming compare to A and B. The cream time, rise time, gel time taken were 4, 58 and 24 seconds, respectively. Composition C gave the optimum PU foam formation. The high amount of catalyst provided fast cream time, rise time and gel time of the PU foam [10]. Figure 2 shows the effect of catalyst on the reaction times of PU foam formation.

**Table 2.** Composition of PU foam formation with different amount of DEA as catalyst.

| **Raw Material** | **Composition** | | |
|---|---|---|---|
| | A | B | C |
| **Polyester polyol** | 90 | 90 | 90 |
| **Glycerol** | 60 | 60 | 60 |
| **$H_2O$** | 54 | 54 | 54 |
| **DEA** | 20 | 30 | 40 |
| **MDI** | 21 | 22 | 23 |

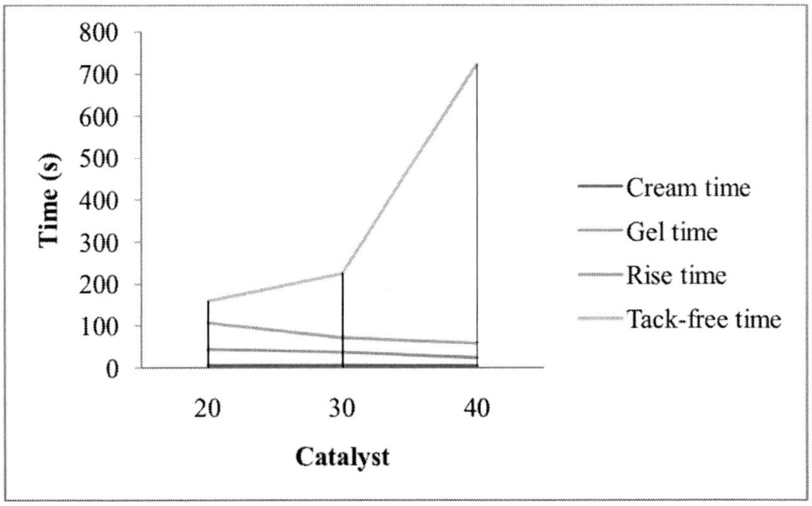


**Figure 2.** Effect of catalyst on reaction times during PU foam formation.

Glycerol acts as a crosslinking agent in PU foam formation. Three compositions; A, B and C, as shown in Table 3 were prepared to observe the effect of amount of glycerol.The amount of glycerol in composition C provided fast cream time for the PU foam to expand which was 4 seconds rather than 40 and 50. The time taken for the components in the resin to get cross-linked to each other was faster due to the high composition of glycerol used. Figure 3 shows the effect of glycerol on reaction times during PU foam formation.

**Table 3.** Composition of PU foam formation with different amount of glycerol.

| **Raw Material** | **Composition** | | |
|---|---|---|---|
| | A | B | C |
| **Polyester polyol** | 90 | 90 | 90 |
| **Glycerol** | 40 | 50 | 60 |
| **$H_2O$** | 54 | 54 | 54 |
| **DEA** | 40 | 40 | 40 |
| **MDI** | 21 | 22 | 23 |

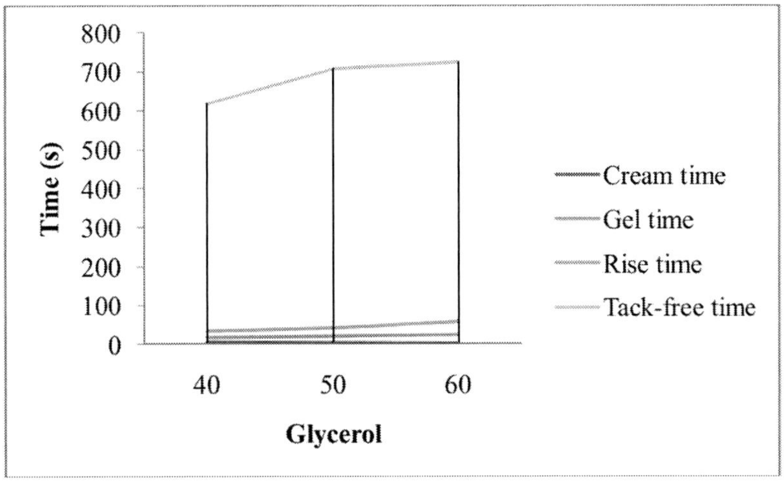


**Figure 3.** Effect of glycerol on reaction times during PU foam formation.

The composition of 90 polyester polyol: 60 glycerol: 54 $H_2O$: 40 DEA: 23 MDI was foundas the optimum formulation for rigid PU foam production, thus used to be characterized further, chemically and physically. Analysis by FTIR, as shown in Figure 4, proves the successful formation of polyurethane can be proven by the presence of amide (NH) at 3720.90 $cm^{-1}$, urethane carbonyl (C=O) at 1741.52 $cm^{-1}$, C-N-H at 1512.36 $cm^{-1}$, C-O-C at 1231.16 $cm^{-1}$ and the absence of hydroxyl (OH) absorption peak at around 3000 $cm^{-1}$. The stretching vibration of $-NH_2^-$ (amide) was located at 3294.19 $cm^{-1}$ overlapped with the OH peak as in PU from WCO based polyol structure. The synchronous reflection of asymmetric and symmetric of $CH_2$ bridges from the linkage of the urethane with DEA appeared at peak 2924.90 and 2854.37 $cm^{-1}$. At peak 1636.94 $cm^{-1}$, the absorption peak of N=C=O of urethane overlapped with ester linkage of the DEA was assigned.

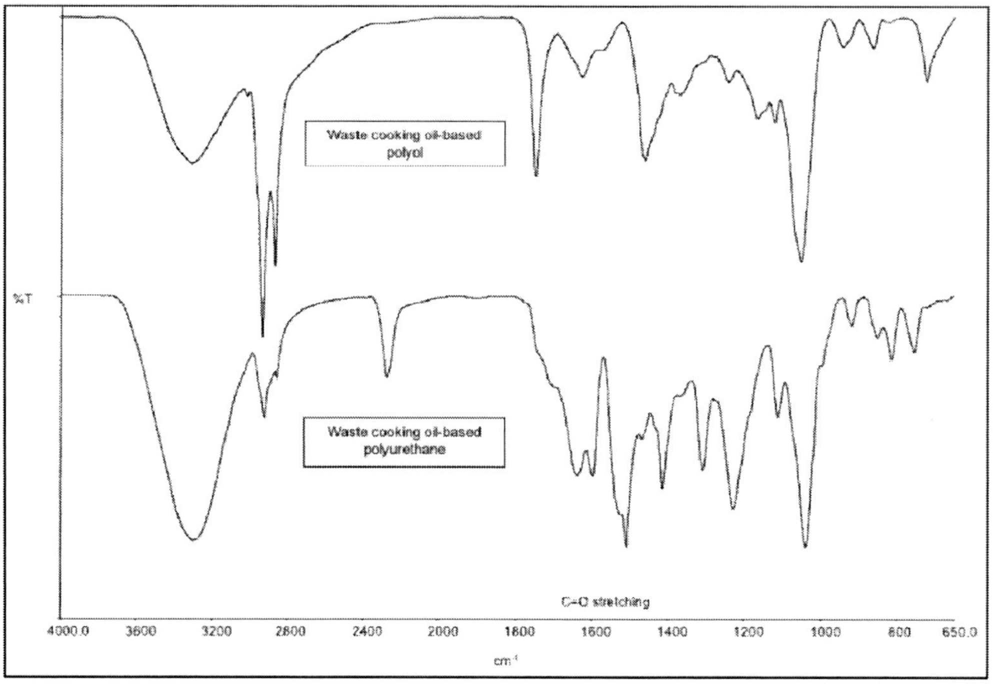


**Figure 4.** FTIR spectra of waste cooking oil-based polyol and polyurethane.

The rigid PU foam samples were cut into cubes of 1 cm × 1 cm × 1 cm in dimensions to obtain the volume of the foams. The mass of the cube foamwas weighed using an analytical balance. The volume is 1 $cm^3$ while the average mass of the cubes is found to be 0.2084 g. The density of the rigid PU foam was determined by using a simple mathematical equation, mass (kg)/volume ($m^3$). Thus, the density of the rigid PU foam is $2.084\times10^{-4}$ kg/$m^3$.

The compressive strength test of the rigid PU foam was carried out on a Universal Testing Machine Model Instron. The maximum stress of the foam was observed to be 0.03 MPa, considered the strength and elongation capability during the compressive strength test. The foam can receive a maximum load of 0.09 kN. The strength showed was affected by the segment content in the PU matrix [11].

## 4. Conclusion

In conclusion, WCO has great potential in replacing petroleum and vegetable oils as starting raw material for rigid PU foam. The use of the WCO also appears to be feasible and eco-friendly besides helping to reduce environmental problems. Pretreatment of WCO through adsorption process using coconut husk activated carbon as the adsorbent successfully removed the suspended materials and improved the colour and odour, which is important in producing commercial product. Viscosity of the WCO was decreased to 97.8 mPa·s after adsorption. Transesterification reaction was used to synthesize polyol from the adsorbed WCO in order to add O-H group in the structure. FTIR proved the WCO-based polyol was successfully synthesized as shown by the presence of O-H absorption peak. The rigid PU foam-based polyol has beensuccessfully synthesized as shown by the presence of urethane linkage in FTIR spectra. The effect of DEA as a catalyst and glycerol as a crosslinking agent on the PU foam formation was studied in terms of cream time, rise time, gel time and tack-free time during the foaming process. The composition of 90polyol: 60 glycerol: 54 $H_2O$: 40 DEA: 23 MDI foundas the optimum composition for rigid PU foam production. The compressive strength test revealed the maximum stress and load of the PU foam which was found to be 0.03 MPa and 0.09 kN. Thus, the properties shown by the rigid WCO-based PU foam is suitable to be used as alternative material forhouse constructionssuch as thermal insulation panels.

## 5. References

[1] Li Y, Luo X and Hu S 2015*Bio-Based Polyols and Polyurethanes*(Springer International Publishing)
[2] Zieleniewska M, Leszczynski M K, Kuranska M, Prociak A, Szczepkowski L, Krzyowska M and RyszkowskaJ2015 Biobased polyurethane foams modified with natural Fillers *Ind. Crops Prod.* **74** 887
[3] Arniza M Z , Hoong S S, Idris Z, Yeong S K, Hassan H A, Din A K and Choo Y M 2015 Synthesis of Transesterified Palm Olein-Based Polyol and Rigid Polyurethanes from this Polyol*J. Am. Oil Chem. Soc.* **92** 243
[4] Tahir S M, Salleh W N W, Hadid N S N and Enderus N F 2016Synthesis of Waste Cooking Oil-Based Polyol Using Sugarcane Bagasse Activated Carbon and Transesterification Reaction for Rigid Polyurethane Foam*Mater. Sci. Forum* **846** 690
[5] Wannahari R, Nordin M F M 2014 The recovery of used palm cooking oil using bagasse as adsorbent *Am. J. Eng. Appl. Sci.* **5**59-62
[6] Sipaut C S @ Nasri M, Sundang M, Saalah S, Hoon T C, Ibrahim M N M , Rahman I A and Abdullah A A 2012 The effect of different peroxide on ldpe foam properties in the presence of polyfunctional monomers *Cellular Polym.* **31** 19
[7] Rus A Z M, Salim N S M and Sapiee N H 2015Recycling of Cooking Oil Waste into Reactive Polyurethane for Blending with Thermoplastic Polyethylene *Int. J. Polym. Sci.***2015**
[8] Gupta V K, Jain R and Shrivastava M 2010 Adsorptive removal of dyes from aqueous solution onto carbon nanotubes: a review. *J. Colloid Interface Sci.* **347** 309
[9] Badri K H, MohdDawi L I and Abd Aziz N A 2013 Rigid polyurethane foam from glycolysed polyethylene terephthalate dissolved in palm-based polyol *Sains Malaysiana* **42** 449-457
[10] Dworakowska S, Bogdał D, Zaccheria F and Ravasio N 2014 The role of catalysis in the synthesis of polyurethane foams based on renewable raw materials*Catalysis Today* **223** 148-156
[11] Prociak A, Rojek P and Pawlik H 2012 Flexible polyurethane foams modified with natural oil based polyols *Journal of Cellular Plastics* **48** 489

**Acknowledgements**
The authors gratefully acknowledge the support from Universiti Teknologi Mara Pahang in terms of financial and facilities provided. A highly gratitude goes to the Ministry of Education, Malaysia for the financial support through Fundamental Research Grant Scheme, (FRGS/1/2016/STG07/UITM/03/1).

IOP Conf. Series: Materials Science and Engineering **271** (2017) 012063 doi:10.1088/1757-899X/271/1/012063

# Surface morphological properties of $Cd_xZn_{(1-x)}S$ thin films deposited by low-cost atmospheric pressure metal organic chemical vapour deposition technique (AP-MOCVD)

**Nurhafiza Kamaruddin[1,2], Yulisa Yusoff[3], Towhid Hossain[3], Nowshad Amin[1,3] and Md. Akhtaruzzaman[1*]**

[1]Solar Energy Research Institute (SERI), The National University of Malaysia, 43600, Bangi Selangor, Malaysia
[2]German-Malaysian Institute, Jalan Ilmiah, Taman Universiti, 43000, Kajang, Malaysia.
[3]Dept of Electrical, Electronic and Systems Engineering, The National University of Malaysia, 43600 Bangi, Selangor
Corresponding author: akhtar@ukm.edu.my

**Abstract.** The $Cd_xZn_{(1-x)}S$ films have been deposited by low cost atmospheric pressure metal organic chemical vapour deposition (AP-MOCVD) technique and the resulting morphological properties were successfully evaluated for solar cell applications. All morphological properties presented here were investigated by Field-Emission Scanning Microscopy (FESEM) and Energy Dispersive X-Ray (EDX) spectroscopy. It has been observed that the film thickness and grain size have been greatly influenced by the molar ratio of cadmium and zinc, deposition temperature and time. The average grain size of $Cd_{0.5}Zn_{0.5}S$ films (98.16 nm thick) was obtained with deposition temperature and time of 440°C and 2.5 minutes, respectively.

## 1. Introduction

Solar energy is one of green technologies that is rapidly evolve in the world. Second generation of solar cell known as thin film (TF) is made by depositing thin layer of photovoltaic material on a substrate such as glass or metal. The interfaces between each layer of solar cell has different application depends on the property of the layer of material. Window layer is one of the layer in solar cell which is essential to minimize recombination losses and affect the solar cell efficiency. Recombination loss might come from morphological properties of window layer such as grain size and pinholes at the layer. Deposition technique and parameters used can be optimized to minimize the problems occurred. Most of physical and chemical deposition techniques require high pressure and temperature which are costly. Therefore, in this work, ternary material cadmium zinc sulphide has been chosen as a window layer that has wide bandgap and low cost deposition technique has been implemented. Cadmium Zinc Sulphide (CdZnS) is a II-VI ternary compound with tunable band gap property [1]. It has been used widely in optoelectronics devices such as light emitting diodes (LED), blue lasers and optical devices [2]. CdZnS thin films have been a popular choice of material with a wide bandgap window layer material in heterojunction solar cell [3] and photoconductive devices [4]. The bandgap of CdZnS can be easily tuned and desirable lattice parameters can be achieved which attracts its implication in photo-conducting materials, non-linear optics, and other optoelectronic devices [5]. The band gap of CdS (2.42 eV) and ZnS (3.66 eV) can be tuned by varying the molar ratio of Cd and Zn via solvent engineered methods [6]. This higher energy band gap of these CdZnS films would prevent parasitic absorption losses and results in increased short-circuit current [7]. Recently, the numerical study of CIGS based solar cells utilizing the $Cd_xZn_{(1-x)}S$ showed better performance than the conventional CdS buffer layer, and subsequently boosted the photovoltaic performance of the CIGS solar cell [8]. In this area, many deposition techniques have been used to fabricate film based products such as chemical bath deposition (CBD), dip-coating, spray pyrolysis, proton exchange membrane, electro-active polymer, co-evaporation, thermal evaporation, solution growth technique, plasma polymerization, reactive diffusion, sol-gel, chemical vapor deposition, magnetron sputtering, successive ionic layer adsorption and reaction, and metal organic chemical vapor deposition (MOCVD) [9]. Among them, the MOCVD involves the formation of a thin solid film on a substrate material by a chemical reaction of vapour-phase precursors. It is a specific type of chemical vapour deposition (CVD) technique which utilizes metal-organic precursors. A metal-organic (or organometallic) compound contains a direct metal–carbon bond ($\sigma$ or $\pi$) (e.g. metal alkyls, metal carbonyls). However, the definition of MOCVD has broadened to include precursors containing metal–oxygen bonds (e.g. metal-alkoxides, metal-β-diketonates) or metal–nitrogen bonds (e.g. metal alkylamides), and even metal hydrides (e.g. trimethylamine alane) [10]. T.L. Chu et al. deposited $Cd_{1-x}Zn_xS$ films on glass substrates by MOCVD with the objective of controlling the composition and resistivity of these films for optoelectronic devices. They used 2 different temperatures for deposition process which are 375°C and 425°C [11]. Most of them used CVD technique in vacuum condition. CVD processes are extremely complex and involve a series of gas-phase and surface reactions which leads to increasing cost. For these reasons an alternative method for AP-MOCVD process is anticipated.

In this work, we have developed a low cost AP-MOCVD method in atmosphere condition without using complex instrument that lead to high cost to deposit $Cd_{0.5}Zn_{0.5}S$ thin films. Surface and cross sectional morphology as well as elemental composition analysis of $Cd_xZn_{(1-x)}S$ are studied with variation of deposition temperature and time. All material properties have been characterized by using Field-Emission Scanning Electron Microscopy (FESEM) and Energy Dispersive X-Ray (EDX) spectroscopy.

## 2. Methodology

### *2.1 Material preparation*

Soda lime glass substrates were cleaned in methanol-acetone-methanol-DI water using ultrasonic bath for 15 to 30 minutes in each solution followed by compressed nitrogen gas drying. After drying, the cleaned glass substrates were placed under the mask for screen printing process to prepare the source for AP-MOCVD. The precursor of $Cd_{0.5}Zn_{0.5}S$ was prepared using Cadmium Chloride ($CdCl_2$), Zinc Chloride ($ZnCl_2$) and Sulphur source in organic compound, which was dried overnight. Table 1 shows the molar ratio of each source used for the precursor preparation.

**Table 1.** $Cd_{0.5}Zn_{0.5}S$ precursor preparation.

| Material | Source | Ratio |
|---|---|---|
| Sulphur source in organic compound | Sulphur source | 1 |
| Cadmium Chloride ($CdCl_2$) | Cadmium source | 0.5 |
| Zinc Chloride ($ZnCl_2$) | Zinc source | 0.5 |

The dried precursor was mixed with three to four drops of 2-propanol for screen printing process. Then the AP-MOCVD source is dried in a convection oven at 100 °C for 10 minutes and then the samples are left in the oven for 1 day for natural cooling.

### *2.2 Experimental work*

Deposition of $Cd_{0.5}Zn_{0.5}S$ thin film is carried out by low cost AP-MOCVD technique by using two hot plates in atmospheric pressure. One hot plate is used to heat up the glass substrate while another hotplate is used for film deposition and post annealing process. For temperature variation, the temperature is varied up to 440 °C for certain time. After the deposition, air annealing was carried out. For deposition time variation, the time is varied with a constant post deposition annealing time. The temperature of deposition was kept constant at 440 °C.

## 3. Result and discussion

### *3.1 SEM images for temperature variation*

Surface morphology of $Cd_{0.5}Zn_{0.5}S$ films deposited by AP-MOCVD is studied by FESEM. Different temperature yielded variation in surface morphology of $Cd_{0.5}Zn_{0.5}S$. Figure 2(a) shows the FESEM surface morphology of $Cd_{0.5}Zn_{0.5}S$ film which has been deposited at 400°C. It shows that the grain sizes are small possibly due to insufficient thermal energy needed for film growth. The film also possesses pinholes which can lead to leakage current in a semiconductor device particularly in solar cell. Figure 2(b) shows the image of $Cd_{0.5}Zn_{0.5}S$ deposited at 420°C. The grain sizes are very uniform and the substrate is completely covered by $Cd_{0.5}Zn_{0.5}S$ film which could also act as a compact layer to prevent leakage current. Larger grain sizes have been observed in figure 2(c) and figure 2(d). The grain sizes are less uniform in terms of grain size as compared to the films deposited at 420°C but are more compact.

(a) (b) (c) (d)

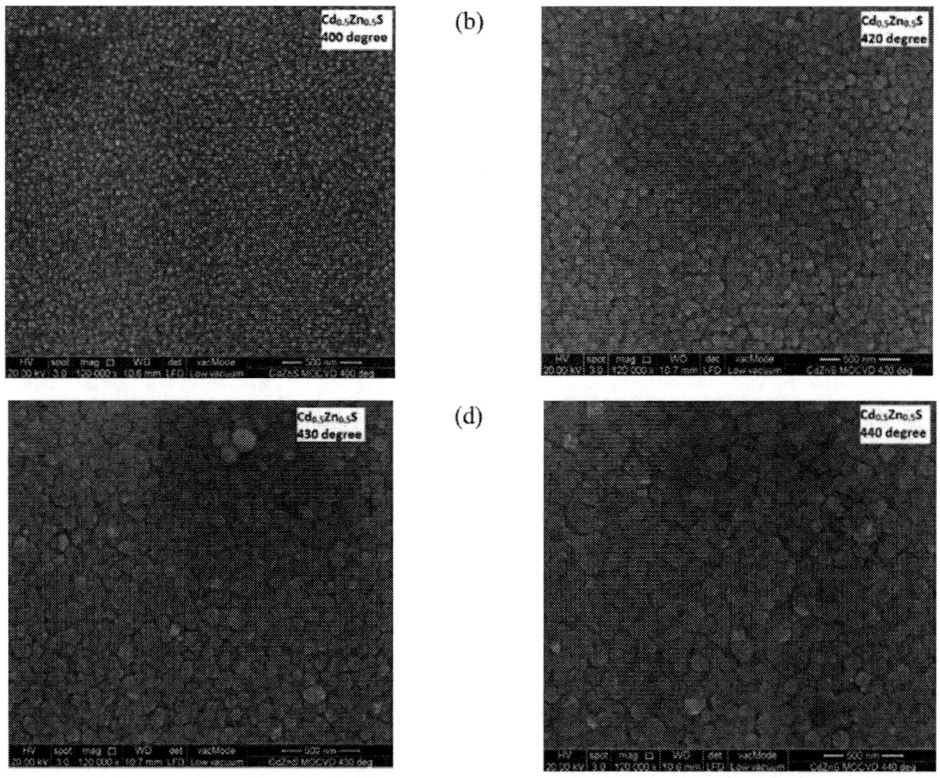


**Figure 1.** Surface FESEM image of CdZnS film on glass substrate at different temperature.

The grain sizes were also measured in respect to temperature variation as shown in table 2. Average grain sizes increased with deposition temperature. The deposited films were almost uniform with AP-MOCVD techniques and found that the largest grain size can be formed at 440°C temperature deposition. In figure 3, the line of the graph indicates that increase in average grain size with increasing deposition temperature.

**Table 2.** Variation of grain size with temperature.

| Sample | Temperature (°C) | Average grain size (nm) |
|---|---|---|
| 1 | 400 | 34.55 |
| 2 | 420 | 57.37 |
| 3 | 430 | 69.12 |
| 4 | 440 | 98.16 |

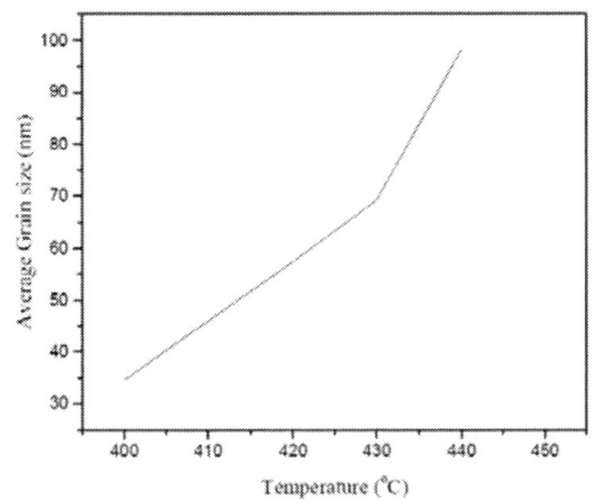


**Figure 2.** Graph of average grain size versus temperature of deposition.

*3.2 SEM images for variation of time*

Surface morphology properties of $Cd_{0.5}Zn_{0.5}S$ films deposited by AP-MOCVD is also studied by varying the deposition time and fixing the deposition temperature at 440°C. Figures 3(a) to 3(c), show that deposition of $Cd_{0.5}Zn_{0.5}S$ by this method produces films were compact, uniform and with good coverage. There are no significant changes of grain sizes with increased deposition time. The grain size varies from 28 nm to 72 nm for deposition time from 2.0 minutes to 3.0 minutes.

(a)
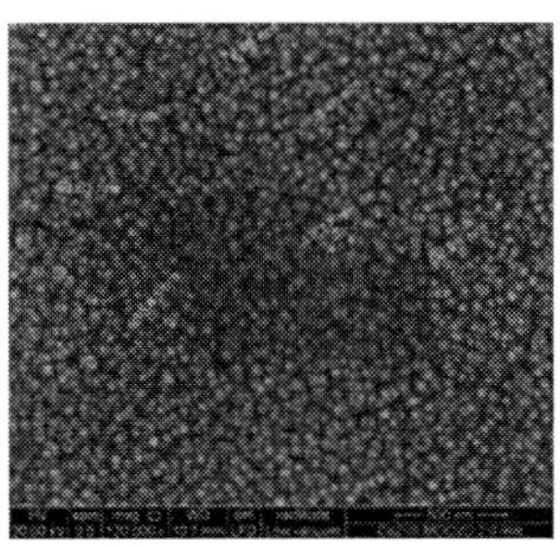
(b)
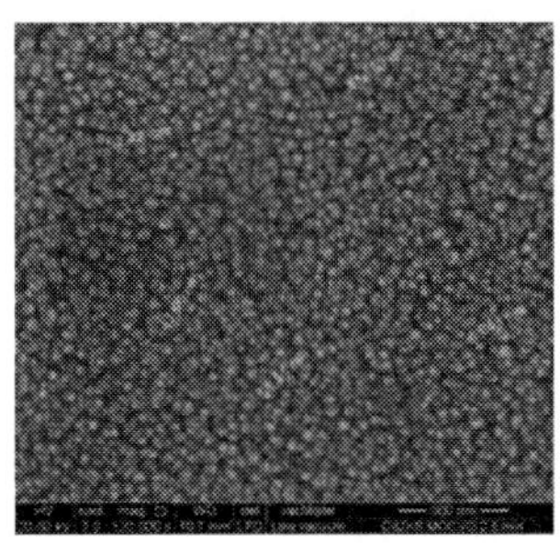
(c)
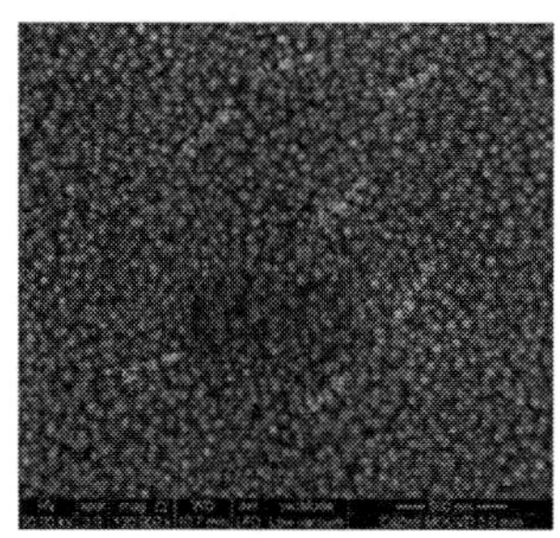

**Figure 3.** FESEM image of CdZnS (a) 2.0 min (b) 2.5 min and (c) 3.0 min deposition time.

The composition of deposited $Cd_{0.5}Zn_{0.5}S$ thin films was determined by Energy Dispersive X-ray (EDX) analysis and the results are provided in table 3. Figure 4 shows that the EDX spectrum of $Cd_{0.5}Zn_{0.5}S$ film with stoichiometric composition. The peak at 0.5 keV and 1.8 keV shows the presence of oxygen and silicon which most probably originate from the underlying glass substrates.

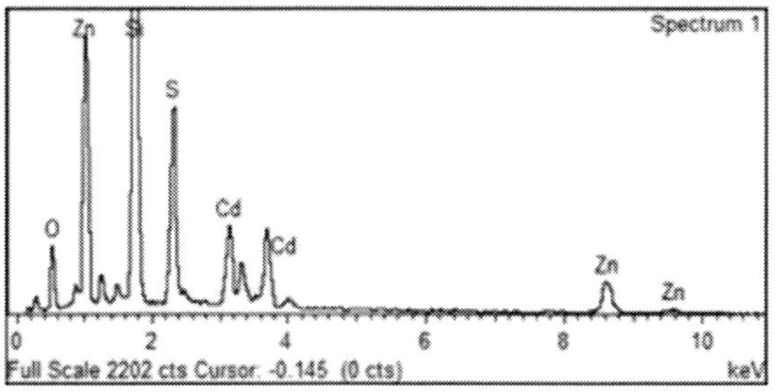


**Figure 4.** Composition of $Cd_xZn_{1-x}S$ from EDX result

**Table 3**. Composition of the elements on the substrate.

| Element | Atomic % |
|---|---|
| Silica (Si) | 37.31 |
| Sulphur (S) | 14.03 |
| Zinc (Zn) | 7.53 |
| Cadmium (Cd) | 4.57 |

## 4. Conclusion

Deposition of $Cd_xZn_{(1-x)}S$ by low cost AP-MOCVD has been investigated in this study. The $Cd_{0.5}Zn_{0.5}S$ thin films were deposited by this method under atmospheric pressure condition with various deposition time and temperature. From FESEM, it was found that by varying the temperature of deposition, the surface morphology and crystallinity of deposited film can be altered. By varying the deposition time, no significant difference in grain size and morphology was observed. These morphological results highlights the suitability of $Cd_xZn_{(1-x)}S$ in semiconductor devices particularly as buffer layer in thin film photovoltaic devices.

## 5. References

[1] Kariper İA Production Of $Hfo_2$ Thin Films Using Different Methods : Chemical Bath Deposition, SILAR And Sol – Gel Process 2014 *International Journal of Minerals, Metallurgy and Materials* **21(8)** 832–38.
[2] Abdolahzadeh Ziabari A and Ghodsi FE Growth, Characterization And Studying Of Sol-Gel Derived Cds Nanoscrystalline Thin Films Incorporated In Polyethyleneglycol: Effects Of Post-Heat Treatment 2012 *Sol Energy Mater Sol Cells Elsevier* **105** 249–62.
[3] Okamoto K, Yoshimi T and Miura S Tbof Complex Centers In Zns Thin-Film Electroluminescent 1988 *Devices Appl Phys Lett.* **53(8)** 678–80.
[4] Liu GK, Zhuang HZ and Chen XY Restricted Phonon Relaxation and Anomalous Thermalization of Rare Earth Ions in Nanocrystals 2002 *Nano Lett.* **2(5)** 535–39.
[5] Ashokkumar M and Muthukumaran S Size Dependent Structural and Optical Properties of $Cd_{0.9}Zn_{0.1}S$ Thin Films 2013 National Conf. on Spintronic Materials: Nanostructures and Devices (SMND-2011) *(Parundurai, India)* vol 49 (Amsterdam: Elsevier) pp 137–44.
[6] Deo SR, Singh AK, Deshmukh L and Abu Bin Hasan Susan M Metal Chalcogenide Nanocrystalline Solid Thin Films 2015 *J Electron Mater* **44(11)** 4098–127.
[7] Ezugwu SC, Ezema FI and Asogwa PU Synthesis and Characterization of Ternary $CuSbS_2$ Thin

Films: Effect Of Deposition Time 2010 *Chalcogenide Lett* **7(5)** 341–8.
[8] Moradi M, Teimouri R, Zahedifar M and Saadat M Optimization of $Cd_{1-y}Zn_yS$ Buffer Layer in $Cu(In,Ga)Se_2$ Based Thin Film Solar Cells *Optik (Stuttg)* *Elsevier GmbH* 2016 **127(8)** 4072–75.
[9] Asim Jilani, Mohamed Shaaban Abdel-Wahab and Hammad AH 2017 Advanced Deposition Techniques for Thin Film and Coating *Book chapter on Modern Technologies for Creating the Thin-film Systems and Coating* p138.
[10] Jones AC and Hitchman ML 2009 Overview of Chemical Vapour Deposition *Chemical Vapour Deposition: Precursors, Processes and Application* vol 1 (United Kingdom: Royal Society of Chemistry) chapter 1 pp 1–36.
[11] Chu TL, Chu SS, Britt J, Ferekides C and Wu CQ Cadmium Zinc Sulfide Films And Heterojunctions 1991 *J Appl Phys.* **70(5)** 2688–93.

**Acknowledgement**
The authors would also acknowledge the contribution by the research grant of Universiti Kebangsaan Malaysia (UKM) as coded DIP-2015-027 and Fundamental Research Grant Scheme (FRGS) with code FRGS/1/2014/TK06/UKM/02/3 of the Ministry Of Higher Education, Malaysia (MOHE).

IOP Conf. Series: Materials Science and Engineering **271** (2017) 012064 doi:10.1088/1757-899X/271/1/012064

# Strengthening and repair of RC beams with sugarcane bagasse fiber reinforced cement mortar

**Mohamad Syamir Senin[1], Shahiron Shahidan[1,], M Z Md Maarof[1], Alif Syazani Leman[1], S S Mohd Zuki[1], M A Mohammad Azmi[1]**

[1] Jamilus Research Center-Sustainable Construction, Faculty of Civil and Environmental Engineering, Universiti Tun Hussein Onn Malaysia.
Corresponding author: shahiron@uthm.edu.my

**Abstract.** The use of a jacket made of fiber reinforced cement mortar with tensile hardening behaviour for strengthening RC beams was investigated in this study. A full-scale test was conducted on beams measuring 1000mm in length. A 25mm jacket was directly applied to the surface of the beams to test its ability to repair and strengthen the beams. The beams were initially damaged and eventually repaired. Three types of beams which included unrepaired beams, beams repaired with normal mortar jacket and beams repaired with 10% sugarcane bagasse fiber mortar jacket were studied. The jacket containing 10% of sugarcane bagasse fiber enhanced the flexural strength of the beams.

## 1. Introduction

Concrete structures are prone to damage due to natural disasters such as earthquakes or other factors such as heavy traffic loading and aging. It is important to consider whether the damaged structure should be removed, repaired or replaced with a new one. In recent years, many research studies have focused on the repair and rehabilitation of deteriorated structures due to the magnitude of such problems on a global level [1-5]. The major portion of the infrastructure consists of reinforced concrete structures. The corrosion of steel which is normally used as reinforcement in concrete structures due to alkali-silica reactions or sulphate attacks is the principal cause of deterioration. In addition, deterioration could also be caused by impact loads such as the concrete bridge girders damaged by vehicle impact. Such deterioration results in loss of strength and stiffness of the concrete member. The need for rehabilitation and repair and the cost involved are matters of great concern among those responsible for the assessment and maintenance of affected structures [6-12].
Normally, designers prefer to use traditional techniques that utilise externally bonded steel plates or RC jacketing. However, it is now common to use externally bonded fiber reinforced polymer [13]. This technique has demonstrated a certain degree of success, but there are a number of limitations. This is because the R/C jacket can only be used by adding layers of concrete with a thickness of more than 60-70mm because the presence of rebar requires a minimum concrete cover [14-15]. The use of externally glued steel plates as FRP is also limited in terms of fire resistance. This technique may not satisfy the minimum requirement for serviceability limit states [16].

The use of concrete reinforced with fibers has increased in the past 15 years due to its enhanced properties after cracking of the cementitious matrix [17]. Some studies have demonstrated that fiber can be used to partially substitute shear reinforcement in concrete beams [18]. In addition, fiber reinforcement can be used as a total or partial replacement for conventional reinforcement such as rebar or welded mesh. The use of fiber reinforced cement mortar for strengthening or repairing

IOP Conf. Series: Materials Science and Engineering **271** (2017) 012064 doi:10.1088/1757-899X/271/1/012064

existing RC beams is proposed in this study. In order to verify the effectiveness of this application, experiments were conducted on full-scale beams measuring 1000mm in length. The proposed technique will consider the use of a thin sugarcane bagasse fiber cement mortar (SCBFCM) jacket which is 25mm thick. The main parameters and effects governing the behaviour of sugarcane bagasse fiber reinforced cement mortar jacketed beams were analysed and discussed in detail.

## 2. Experimental

The application of the SCBFCM jacketing technique was initially aimed at strengthening and repairing existing RC beams. It was applied to undamaged elements that required a higher load-carrying capacity.

### *2.1. Beam geometry and material properties*

Full-scale tests were performed on beams measuring 1000mm in length with a width of 150mm and a depth of 200mm as shown in figure 1. All the beams were reinforced with three bottom longitudinal rebars (Ø=12mm) and two top longitudinal rebars (Ø=12mm) as well as stirrups having a diameter of 8mm and a spacing of 125mm.

The beams were cast with concrete which has a nominal cube strength of 35MPa. Concrete was chosen in order to simulate the real case of a weak existing beam and to better highlight the strengthening effect.

One of the reinforced beams was used as a control specimen while SCBFCM jackets which were 25 mm thick were applied on two beams. Another two normal mortar jackets which were 25 mm thick were applied on another two beams as shown in figure 1.

The strengthening mortar material had a maximum aggregate size of 0.6mm and a water/binder ratio equal to 0.4 by volume. The mortar was reinforced with 10% (by volume) of sugarcane bagasse fiber. It had a maximum length of 12mm and a maximum diameter of 0.6mm. The compressive strength of the SCBFCM, as measured on the 50mm side cube after 28 days of curing was 11.7 MPa.

A direct tensile test called the three-point bending test was performed on small beams in order to characterise the material under tension.

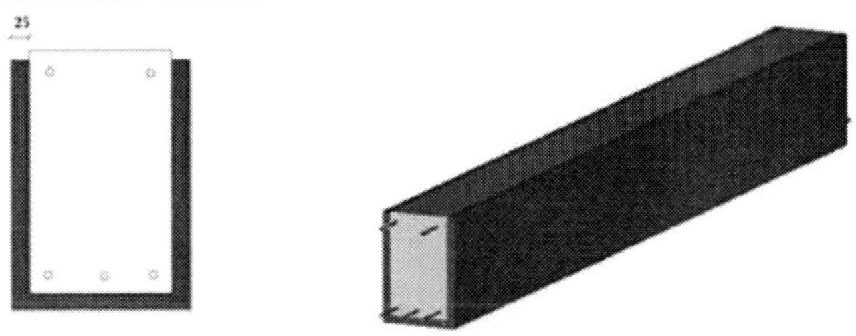


**Figure 1.** Strengthening scheme

### *2.2. Jacket application technology*

The surface of the full-scale beams was hacked in order to produce a roughness that was considered enough to avoid the use of bonding products. The SCBFCM material was directly cast on the beams without any vibration. Since the beams were cured under ambient temperature and humidity, a wet gunny was placed on the surface of the beams in order to limit water evaporation. The SCBFCM jacket was applied 28 days after casting the normal concrete beams while the test was performed 28 days after that.

## 3. Test Procedures

The full-scale beams were tested using a three-point bending scheme. The beams were placed on a 900mm span and loaded at one point located at a distance of 450mm from the support as shown in figure 2. The LVDT was adopted for measuring the vertical displacement.

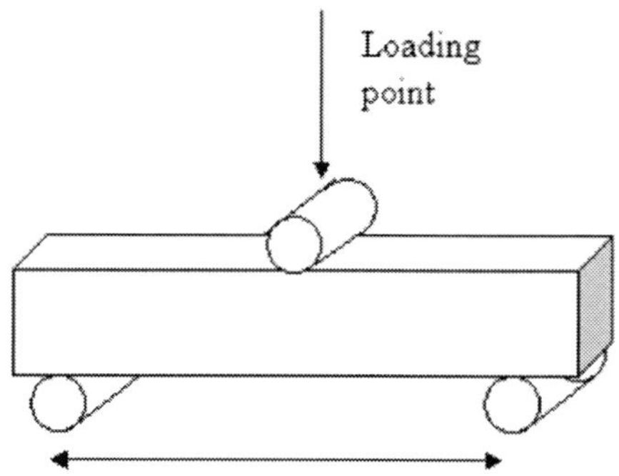


**Figure 2**. The test set-up of full-scale beams

**4. Results and Discussion**

The tests were conducted on the control beams (RC beam without the SCBFCM strengthening jacket), followed by beams strengthened with normal cement mortar (0%) and beams strengthened with SCBFCM jacketing (10%).

Figure 3 shows the curve of the total applied load versus the mid-span displacement. The main experimental observations are the following:

- The control specimens indicated the lowest value of deflection which was 5.03 mm. For specimens strengthened with (0%) SCBFCM, the value of deflection was 13.94 mm. The value of deflection for (10%) SCBFCM was 7.37 mm.
- The difference between (10%) and (0%) SCBFCM was 6.57mm. From the analysis, the SCBFCM improved by 47% compared to normal cement mortar and proved that mortar can be applied in concrete repair.
- The maximum load for repaired (10%) SCBF was 112 kN followed by repaired (0%) SCBF which was 109 kN. The flexural strength for repaired specimens with (10%) and (0%) SCBFCM improved by 13% and 10% respectively.

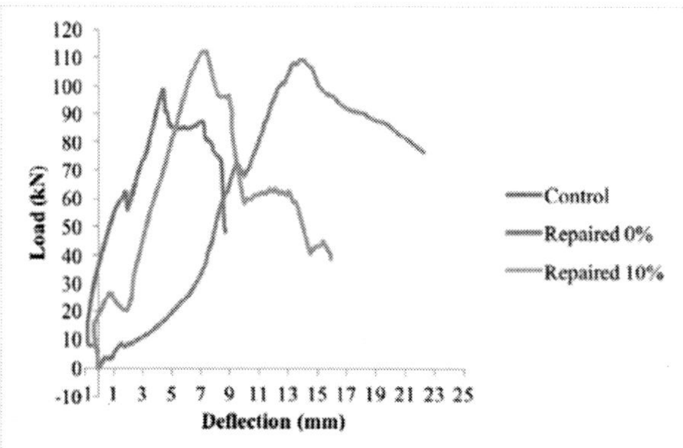


**Figure 3.** Load against deflection of the beams

**5. Conclusion**

This paper provides information on the structural behavior of beams strengthened and repaired using a thick layer of Sugarcane Bagasse Fiber Cement Mortar (SCBF). Experimental and numerical results demonstrated that this appears to be a promising technique. Based on the results discussed in this paper, the following remarks can be drawn:

- The application of a 25mm thick SCBFCM jacket on RC beams increased the ultimate load by 13%.
- The beams repaired with 10% SCBFCM improved by 47% compared to unrepaired beams.
- The proposed technique provides significant structural enhancement in terms of stiffness due the remarkable increase of the maximum load of the specimen.

Finally, it is worth mentioning it is possible to increase the durability of the structure by applying the SCBFCM jacket due to the improvement in flexural strength.

### 6. References

[1] Al-Jabri K S, Al-Saidy A H and Taha R 2011 Effect of copper slag as a fine aggregate on the properties of cement mortars and concrete *Constr. Build. Mater.* **25** (2) 933–938

[2] Shahidan S, Mohd Zuki S S, Rahim M A and Ali N 2016 Amplitude Distribution of Emission Wave for Cracking Process vol 47 (Paris: MATEC Web of Conference) pp 2–8

[3] Md Nor N, Bunnori N M, Ibrahim A, Shahidan S, and Mat Saliah S N 2011 Relationship between acoustic emission signal strength and damage evaluation of reinforced concrete structure: Case studies," in 2011 *IEEE Symposium on Industrial Electronics and Applications*, pp. 308–313

[4] Senin M S, Shahidan S, Leman A S and Hannan N I R R 2016 Properties of Cement Mortar Containing Rubber Ash as Sand Replacement vol. 160 (IOP Conf. Ser. Mater. Sci. Eng), p. 12055

[5] Shahidan S, Bunnori N M, Nor N, Rahim M A, Abdullah R and Zuki S S M 2016 Analysis Improvements of Acoustic Emission Data in Evaluation of Concrete Beam, *ARPN J. Eng. Appl. Sci.* **11** (20) 11804–11807

[6] Xie J and Hu R 2013 Experimental study on rehabilitation of corrosion-damaged reinforced concrete beams with carbon fiber reinforced polymer *Constr. Build. Mater.* **38** 708–716

[7] Ali N, N. Zainal A, Burhanudin M K, Abdul Samad A A, and Shahidan S 2016 Physical and Mechanical Properties of Compressed Earth Brick (CEB) Containing Sugarcane Bagasse Ash vol. 47 (Paris: MATEC Web of Conference) pp. 1–7

[8] Senin M S, Shahidan S, Leman A S, and Hannan N I R R 2016 Analysis of Physical Properties and Mineralogical of Pyrolysis Tires Rubber Ash Compared Natural Sand in Concrete material vol. 160 (IOP Conf. Ser. Mater. Sci. Eng) pp 12053

[9] Yahya, Ahmad Asyari, Ali, Noorwirdawati, Mohd Kamal, Nur Liyana, Shahidan, Shahiron, Beddu, Salmia, Nuruddin, Muhd. Fadhil and Shafiq, Nasir 2017 Reducing Heavy Metal Element from Coal Bottom Ash by Using Citric Acid Leaching Treatment vol. 103 (Paris: MATEC Web of Conference) pp 1004

[10] Ramzi Hannan, Nurul Izzati Raihan, Shahidan, Shahiron, Ali, Noorwirdawati, and Maarof, Mohamad Zulkhairi 2017 A Comprehensive Review on the Properties of Coal Bottom Ash in Concrete as Sound Absorption Material vol. 103 (Paris: MATEC Web Conf) p. 1005

[11] Shahidan, Shahiron, Maarof, Mohamad Zulkhairi Md, Raihan Ramzi Hannan, Nurul Izzati, and Ali, Noorwirdawati 2017 Existing Noise Level at Railway Stations in Malaysia vol. 103 (Paris: MATEC Web of Conference) p. 9012

[12] Senin M S, Shahidan S, Leman A S and Hannan N I R R 2016 Analysis of Physical Properties and Mineralogical of Pyrolysis Tires Rubber Ash Compared Natural Sand in Concrete material vol. 160 (IOP Conf. Ser. Mater. Sci. Eng) pp12053

[13] Aslam M, Shafigh P, Jumaat M Z and Shah S N R 2015 Strengthening of RC beams using prestressed fiber reinforced polymers: A review *Constr. Build. Mater.* **82** 235–256

[14] Chalioris C E, Thermou G E, and Pantazopoulou S J 2014 Behaviour of rehabilitated RC beams with self-compacting concrete jacketing – Analytical model and test results, *Constr. Build. Mater.* **55** 257–273

IOP Conf. Series: Materials Science and Engineering **271** (2017) 012064 doi:10.1088/1757-899X/271/1/012064

[15] Mohd Zuki, Sharifah Salwa, Shahidan, Shahiron, Keong, Choong Kok, Jayaprakash J and Ali, Noorwirdawati 2017 Concrete-Filled Double Skin Steel Tubular Columns Exposed to ASTM E-119 Fire Curve for 60 and 90 Minutes of Fire vol 103(Paris: MATEC Web of Conference) p 2009

[16] Martinola G, Meda A, Plizzari G A and Rinaldi Z 2010 Strengthening and repair of RC beams with fiber reinforced concrete, *Cem. Concr. Compos.* **32** (9) 731–739

[17] Zanotti C, Banthia N and Plizzari G 2014 A study of some factors affecting bond in cementitious fiber reinforced repairs, *Cem. Concr. Res.* **63** 117–126

[18] Ruano G, Isla F, Pedraza R I, Sfer D, and Luccioni B 2014 Shear retrofitting of reinforced concrete beams with steel fiber reinforced concrete,*Constr. Build. Mater.* **54** 646–658

**Acknowledgments**

The authors would like to thank Universiti of Tun Hussein Onn, Malaysia and the Ministry of Education, Malaysia for the generous grant awarded for this research (Short Term Grant Vot U114).

IOP Conf. Series: Materials Science and Engineering **271** (2017) 012065 doi:10.1088/1757-899X/271/1/012065

# Combination of pseudomonas putida and EK method to reduce the amount of mercury on landfill soil

**A T A Nabila[1,*], A T S Azhar[2], M S Nurshuhaila[3], M A M Azim[2] and S N Amirah[2]**
[1] Research Center for Soft Soil, Universiti Tun Hussein Onn Malaysia, 86400 Batu Pahat Johor, Malaysia
[2] Faculty of Civil and Environmental Engineering, Universiti Tun Hussein Onn Malaysia, 86400 Batu Pahat Johor, Malaysia
[3] Faculty of Engineering Technology, Universiti Tun Hussein Onn Malaysia, 86400 Batu Pahat Johor, Malaysia
*Corresponding author:ainnabila782@gmail.com

**Abstract.** Landfills usually lack of environment measures especially on soil. There are no guarantee that the landfill soil is free from being contaminated. It may cause harm for humans, animals and plants at surrounding area. In order to solve this problem, advance remediation technique is essential such as the electrokinetic combined with microorganisms known as electrokinetic bioremediation technique. The aim of this study is to investigate the performance of *P.putida* with 15 volt electric current supply (Ek-bio) and without electric current (Bio) in removal of mercury in landfill soil. Both treatments were running throughout 14 days. The *P.putida* was placed at anode compartment meanwhile sterile distilled water poured at cathode compartment. According to the both results, Ek-bio was removed mercury up to 48 % but by using standard bioremediation treatment, the removal only 32 %. Besides that, the migration of *P.putida* react more aggressively during the present of electric current compared with bioremediation. As the results, it was proven that by using Ek-bio technique can increase the activity of bacteria beside and the removal of mercury. Therefore, Ek-bio method can be commercialized to the parties concerned to solve the contaminated soil by mercury.

## 1. Introduction

Landfill is the most frequent waste disposal method worldwide. Since it is a place for waste disposal, so there may be a various contaminants have been spread around the landfill area [1-3]. Landfill usually lack of environment measures. There is no guarantee that landfills are free from being contaminated especially by heavy metals. Thus, it may cause harm to humans, animals and plants that live around that area. Then, this landfill soil needs to be treated. As we know, there are many technologies have been develop such as thermal, physical/chemical and biological [4-5]. The technique remediation of contaminated soil with electrokinetic will enhance the effectiveness of treatment at the landfill soil [6].

The application of electrokinetics helps to transport the ions where the electric field is applied along the soil, positive ions will be attracted to the cathode and negative ions to the anode. As well as objects such as pollutant molecules or bacteria in the soil will be affected by the process and electrokinetic forces that had been applied [7]. Electrokinetic separation is an emerging technology that relies on the application of a low-density, direct current through the soil to separate and extract heavy metal, radionuclides and organic contaminants from unsaturated soil, sludge and sediment [8,9]. This technology can be applied to the contaminant concentration range less than 100 ppm, but may not be effective for treating multiple contaminants that have significantly different concentration [10].

Other than that, the application of electrokinetic can be combined with microbial (electrokinetic bioremediation) to improve the performance of this method [11-13]. The microbial have a potential to reduce the amount of toxic in soil and can change the contaminants to less hazardous product. There are three different types of in situ bioremediation process which are bioattenuation, biostimulation and bioaugmentation [14]. Bioattenuation is greatly depends on the natural process of degradation while biostimulation is an intentional stimulation of degradation of chemicals is achieved by addition of water, nutrient, electron donors or acceptors. Subsequently followed by bioaugmentation where the microbial members with proven capabilities of degrading or transforming the chemical pollutants are added. In addition, there are several factors that contribute to the suitability of a particular bioremediation technology such as site conditions, indigenous population of microorganism, and the type, quantity and toxicity of pollutant chemical [15]. Besides that, the bacteria was used need to be select based on the capability itself to survive in the contamined soil [4].

Therefore, this study was used *pseudomonas putida* as an agent to reduce the amount of mercury in landfill soil. The selectrion of this bacteria based on their potential to survive in the contaminated soil that contained harmful compounds [18]. Azoddein *et* al. [16] used *pseudomonas putida* to removed the mercury up to 80 %. Besides, *pseudomonas putida* combined with electric current also was used to treat the soil contaminated by zinc and lead, and was removed up to 89 % respectively [11]. Based on the previous studies, it was proven that *pseudomonas putida* can be used to treat contaminated soil with existence low voltage of electric current to reduce the amount of mercury in landfill soil that convenient with the aim of this study.

## 2. Materials and Methods

### *2.1. Location of soil sample*

The soil samples were collected at coordinate 2.041317, 102.673008. In this study, standard used according to ASTM D5633-04 (2012) Standard Practice for Sampling with Scoop. The samples were collected from site by pushing the blade of a shovel into the soil to the 10-15 cm depth beneath the surface shown in the figure 1. Then, cut out a triangular wedge of soil by the shovel and set it aside (to be replaced after sampling). After that, push the blades into the soil again taking a thin (1.5 cm) slice from one side of the hole. With a knife, trim the slice to about 2.5 cm strip of soil down the center of the spade-top to bottom. Take the core as part of composite laboratory samples. The core samples were kept in flasks and subsequently transported to laboratory (USDA, 2009). Soil was oven-dried at 105 °C for 24 hours, grinded by a mortar grinder and then sieved to remove plants and debris.

**Figure 1.** Scope was pushed into the soil to a depth 10-15 cm.

### *2.2. Soil sample preparation*

Before the treatment was started, there were several step needed are taken. First, approximately 4 kg of soil was mixed with 4 kg of deionized water. The purpose of using deionized water was to prevent all the ions in the soil from attracted to the water before the testing started. Then, the soil sample was stirred using mechanical machine around 30 minutes and poured into transparent acrylic box. The slurry sample was kept overnight before undergoes consolidation process. Figure 2 shows the consolidation process to compact the soil and to make sure there are no excess water in soil. The consolidation process was running for three stage. First stage with loading 15 kg followed by 30 kg and the last stage is 60 kg. The selection of loading during consolidation was done according to the research by Azhar [17].

**Figure 2.** Consolidation process.

*2.3. Preparation of nutrient broth*
Approximately 30 g of soybean casein digest medium broth powder was mixed with 1000 ml in conical flask that had been filled with distilled water. Then, the mixture was mixed slowly until it is well mixed approximately around 45 min. Next, the mixture was placed into the autoclave machine to sterille at 121 °C for 15 min and the mixture solution was cooled down for a few minutes around 35 °C to 37 °C. After the mixture solution has cool down, pour a loop of *pseudomonas putida* into the nutrient broth. Then, it was leave into the incubator shaker at 32 °C and 200 rpm. After 24 to 48 h, the nutrient broth was ready to put in the tank at anode compartment.

*2.4. Preparation of nutrient media*
Approximately 24.2 g of agar powder of pseudomonas agar base mix with 500 ml of distilled water. The mixture was stired until it well mixed. After that, the mixture was placed in the autoclave at 121 °C for 15 min. The mixture was leave in a few minutes until it reach 48 °C then poured into petri dish. Agar media was keep in cooler at 4 °C before it been used for bacteria counting experiment.

*2.5. Preparation of electrokinetic*
Electrokinetic bioremediation tank was design by using transparent acrylic sized 20 cm with diameter of 22 cm as shown in figure 3. Contaminated soil were divided into three portions and paced in the electrokinetic cell. Filter paper and stainless steel plate was placed at the both end of soil. Cathode and anode part are considered as water reservoir. Anode part was filled with bacterial solution where the bacteria act as an agent to treat the contaminated soil. Whereas, distilled water was placed at the anode part. Liquid at both part was fixed at the highest level to prevent the differences in hydraulic gradient. 15 V of direct electric current was applied at the anode and cathode for 14 days. After 14 days, the electric current was stoped and the sample was collected for mercury testing using mercury analyzer model MA-3000 RD/SCC.

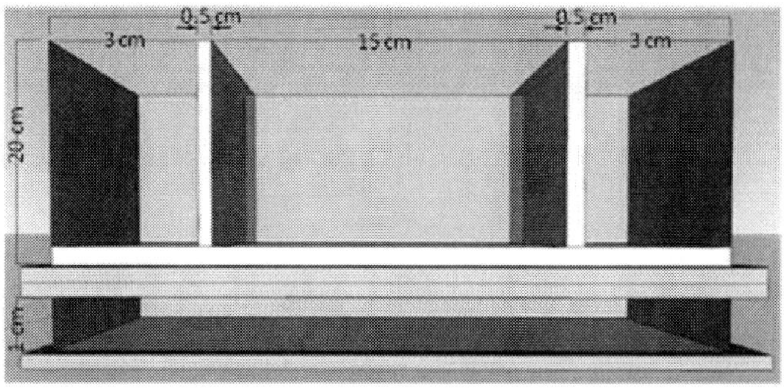


**Figure 3**. Dimension of electrokinetic bioremediation tank.

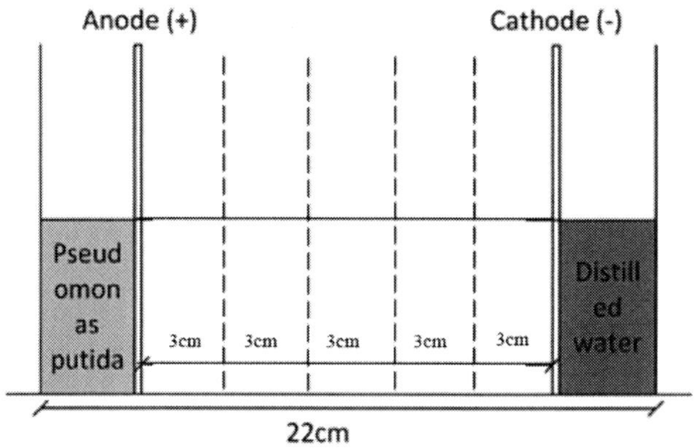


**Figure 4.** Cross-section of the physical modeling of the electrokinetic bioremediation tank.

## 3. Results and Discussions

### *3.1. Removal of mercury by Pseudomonas putida*

Both initial concentration of mercury before treatment is 16.615 ppb. Data collection for mercury was obtained based on the distances differences of soil between anode and cathode. Data obtained was classified into five section that was labelled as 3 cm, 6 cm, 9 cm, 12 cm and 15 cm from anode.

The reaction between bacteria was resulted the changes in landfill soil in term of differences contamination of mercury. At the same time, from the separation parts of soil, it can be identified whether mercury contaminants at soils has a changes or not. By referring to figure 5 for bioremediation, it was found that the lowest removal of mercury were at the middle section of the soils between anodlyte and cathodlyte (6 cm from anode) which is 7 % removal. Then for section 9 cm, the amount of mercury removal was increase to 32 % ppb. Meanwhile, at section 12 cm and 15 cm, the removal was decreased to 28 % and 26 % respectively. The results showed that the pattern removal of mercury is not sequence and slightly. It is effected because there are no element to transport *P.putida* from anode to cathode. However, it was showed *P.putida* can migrate itself without any supported from other sources. Therefore, it can be concluded that *P.putida* also can reduce the small amount of mercury in soil without using any combination method.

In the other hand, the combination of *P.putida* and electric current can influence in removal of mercury. With low voltage electric current can help the migration of *P.putida* through the soil from anode to the cathode to reduce the amount of mercury [12]. From figure 5 for Ek-bio, it shows that the trend removal is same with bioremediation but the removal of mercury is highest than bioremediation.

IOP Conf. Series: Materials Science and Engineering **271** (2017) 012065 doi:10.1088/1757-899X/271/1/012065

The lowest removal was occured at 6 cm which is 33 %. The removal was increase at 9 cm to 48 % and continue decrease at 12 cm and 15 cm. Both techniques shown the same pattern where the lowest removal was occurred at 6 cm. While, the highest removal was occurred at 9 cm for bioremediation while 12 cm for Ek-bio. The lowest removal for both treatment was accured at 6 cm caused by the activities of *P.putida* during the treatment. When the treatment has been stopped then caused the activities of *P.putida* was interrupted and cause the removal is low. Based on this result, it can be conclude with the addition of electric current through the soil can increase the removal of mercury [13]. So, it was proven that Ek-bio was more effective compared to bioremediation.

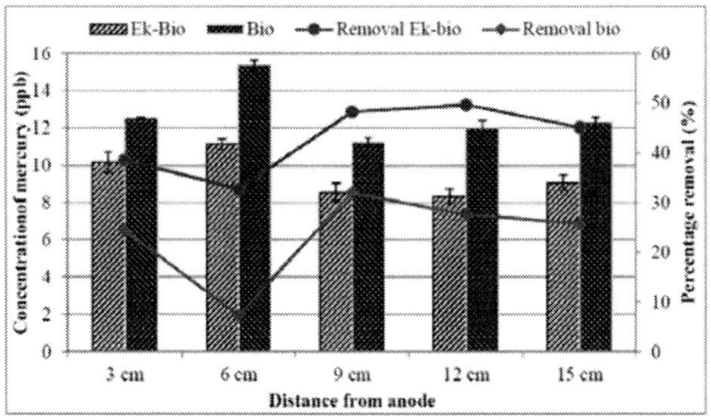


**Figure 5.** Concentration of mercury after Ek- bio and bioremediation treatment for 14 days.

### 3.2. *Migration of Pseudomonas putida through soil*

The migration of *P.putida* was recorded at three stages. First, the data was recorded before running the testing and was recorded as NB before treatment. Second, after 14 days treatment (NB after treatment) and lastly at the five different location of soil sample after treatment. Before that, the bacteria was count for soil before treatment to measure the amount of *P.putida* and the results is zero which there are no *P.putida* colonie was appear during counting. For the initial stage, the amount of nutrient broth *P.putida* was recorded. As shown in figure 6, both pattern show the reduction amount of *P.putida* occurred after treatment 14 days. Although, the pattern migration of *P.putida* from anode to cathode through soil was different for Ek-bio and bioremediation. The pattern migration for bioremediation is decreasing from 3 cm to 15 cm which are 42 x $10^5$ cfu/g, 31 x $10^5$ cfu/g, 31 x $10^5$ cfu/g, 24 x $10^5$ cfu/g and 23 x $10^5$ cfu/g, respectively. Then, the migration of *P.putida* for Ek-bio is increasing where the amount of *P.putida* is always increase start from 3 cm (104 x $10^5$ cfu/g), 6 cm (136 x $10^5$ cfu/g), 9 cm (143 x $10^5$ cfu/g), 12 cm (145 x $10^5$ cfu/g) and 15 cm (152 x $10^5$ cfu/g). Therefore, the result migration of *P.putida* was supported by Deflaun and Condee [12] which is by using low voltage of electic current can increase the migration of *P.putida* in soil and also can increase the amount of *P.putida* [17-18].

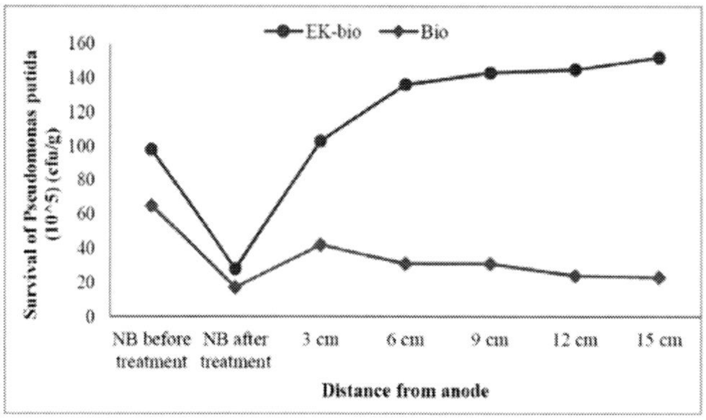


**Figure 6.** Bacteria migration through soil from anode to cathode.

## 4. Conclusion

The combination of *P.putida* and electrokinetic method used to reduce the amount of mercury on landfill soil was successfull and the method of research was presented. The *P.putida* and with the presence of low voltage of electric current do not give negative impact to lifespan of *P.putida.* By comparing the both method using only *P.putida* and *P.putida* with electrokinetic method on landfill soil, the combination technique give more improvement in removal of mercury. The results give meaningfull contribution to the environmental engineers and geotechnical engineers in solving the problem of mercury in soil. In addition, this method is the most recommended technology where it is also can be categorized as green remediation technology.

## 5. References

[1] Johari A, Alkali H, Hashim H, Ahmed S I, and Mat R 2014 Municipal Solid Waste Management and Potential Revenue from Recycling in Malaysia *Mod. Appl. Sci.* **8(4)** 37–49

[2] Ismail S N S and Manaf L A 2013 The challenge of future landfill : A case study of Malaysia *Journal of Toxicology and Environmental Health Sciences* **5** 86–96

[3] Raghab S M, Abd A M, Meguid E and Hegazi H A 2013 Treatment of leachate from municipal solid waste landfill *HBRC J.* **9(2)** 187–192

[4] United States Environmental Protection Agency 2006 "In Situ Treatment Technologies for Contaminated Soil-Report

[5] Yi Y M and Sung K 2015 Influence of washing treatment on the qualities of heavy metal–contaminated soil *Ecol. Eng.* **81** 89–92

[6] US Army Environmental Center 2000 *In-Situ Electrokinetic Remediation of Metal Contaminated Soils Technology* Status Report no SFIM-AEC-ET-CR-99022 pp 30

[7] Gill R T, Harbottle M J, Smith J W N and Thornton S F 2014 Electrokinetic-enhanced bioremediation of organic contaminants: A review of processes and environmental applications *Chemosphere* **107** 31–42

[8] Kim B K, Baek K, Ko S H and Yang J W 2011 Research and field experiences on electrokinetic remediation in South Korea *Sep. Purif. Technol.* **79(2)** 116–123

[9] Nordin N S 2015 Physicochemical Properties of Batu Pahat Clay using Electrokinetic Stabilisation Method Master Thesis (Malaysia: Universiti Tun Hussein Onn Malaysia) pp 28

[10] Dixit R, Malaviya D, Pandiyan K and Shu A 2015 Bioremediation of Heavy Metals from Soil and Aquatic Environment: An Overview of Principles and Criteria of Fundamental Processes *Sustainability* **7** 2189–2212

[11] Tajudin S A A, Talib A N A, Sunar N M, Zain N S M and Azmi M A M 2016 Electromigration of Contaminated Soil by Electro- Bioremediation Technique *IOP Conference Series: Materials Science and Engineering* (London: Institute of Physics Publishing (IOP) 160(**1**) pp 1-8

[12] Deflaun M F and Condee C W 1997 Electrokinetic transport of bacteria *J. Hazard. Mater.* **55(1–3)** 263–277
[13] Azhar A T S, Talib A N A, Sunar N M, Embong Z, Azmi M A M and Farhana S M S 2016 Assessment and Comparison of Electrokinetic and Electrokinetic-bioremediation Techniques for Mercury Contaminated Soil *Environmental Earth Sciences* **75(14)** 1-9
[14] Adams G O, Fufeyin P T, Okoro S E and Ehinomen I 2015 Bioremediation, Biostimulation and Bioaugmention: A Review *Int. J. Environ. Bioremediation Biodegrad.* **3(1)** 28–39.
[15] Megharaj M, Ramakrishnan B, Venkateswarlu K, Sethunathan N and Naidu R 2011 Bioremediation approaches for organic pollutants: A critical perspective *Environ. Int.* **37(8)** 1362–1375
[16] Azoddein A A M, Yunus R M, Sulaiman N M, Bustary A B and Sabar K 2015 Mercury Removal Using Pseudomonas putida ( ATTC 49128 ): Effect of Acclimatization Time , Speed and Temperature of Incubator Shaker *International Journal of Biological, Biomolecular, Agricultural, Food and Biotechnological Engineering* **9(2)** 204–209.
[17] Azhar A T S 2012 Electrokinetic Stabilisation of Soft Clay Ph.D. thesis (University of Birmingham) pp 120-123
[18] Azhar A T S, Sunar N M, Hanif M N S M, Azmi M A M, Majunit N I, Yusof M F and Yusof H N M 2015 Mobilization of Pseudomonas putida in Kaolin Soil by Electrokinetic Bioremediation Technique *Applied Mechanics and Material* vol 773-774 (Switzerland: Trans Tech Publications) pp 1486–1490

**Acknowledgment**
This paper was partly sponsored by the Centre for Graduate Studies UTHM.

IOP Conf. Series: Materials Science and Engineering **271** (2017) 012066 doi:10.1088/1757-899X/271/1/012066

# Compressive strength of marine material mixed concrete

**Adnan[1,*], H Parung[1], M W Tjaronge[1] and R Djamaluddin[1]**

[1] Department of Civil Engineering, Hasanuddin University, Indonesia.
*Corresponding author: adnan872015@gmail.com

**Abstract.** Many cement factories have been incorporated fly ash with clinker cement to produce blended cement. PCC is a type of blended cement incorporated fly ash that produced in Indonesia cement factories. To promote the sustainable development in the remote islands this present paper attempted to study the suitability of sea water, marine sand that available abundantly surround the remote island with Portland Composite Cement (PCC) and crushed river stone to produce concrete. Slump test was conducted to evaluate the workability of fresh concrete and also compressive strength with stress-strain relationship was carried out to evaluate the hardened concrete that cured with two curing condition (e.g. sea water curing, and tap water-wet burlap curing). Test result indicated that fresh concrete had proper workability and all hardened specimens appeared a good compaction result. Compressive strength of specimens cured which sea water was higher than the specimens which cured by tap water-wet burlap where stress-strain behavior of specimens made with sea water, marine sand, and PCC had similar behavior with specimens which made with PCC and tap water.

## 1. Introduction

In Indonesia, most of the coal used in this country is burned to generate electricity in the coal fired thermal power plant. Fly ash is the main residue from coal burning. The majority of fly ash is still disposed in landfills, which creates the environmental issue, where a landfill surcharge is needed in many areas. Hence, new recycling approaches are necessary to produce value-added products from fly ash instead of considering it as a waste material to be disposed. To obtain an environment friendly technology and to achieve material saving through the recycling of waste materials such as fly ash, then several cement factories produce and develop the blended cement containing fly ash such as Portland Composite Cement (PCC) [1-2]. Currently, most of the concrete constructions in Indonesia use PCC instead of Ordinary Portland Cement (OPC) or Portland Cement Type I. Therefore, this research attempted to use PCC containing fly ash as cementitous material. As the world's largest archipelagic state, Indonesia surrounded with 7.9 million square kilometres of sea water where many people live in the remote islands. In this concern, many serious works have been conducted enhance the sustainable infrastructure development in the remote islands. The utilization of sea water, and marine sand can reduce the consumption of fresh water and river sand, furthermore, will decrease the cost of concrete work in the low land areas and the distant-remote islands those lack of clean water and river sand. Several experimental studies have been conducted about the producing of concrete with sea water and marine sand [3-6].

This paper report for a part of the ongoing investigation which focus to study concrete made by PCC and marine material. The testing result on the workability of fresh concrete and compressive strength with stress-strain relationship of hardened concrete were discussed.

## 2. Materials and experimental methods

Table 1 shows the characteristics of sea water. The prominent compound are $Cl^-$, Na and Mg.

**Table 1.** Characteristics of sea water.

| Density ($g/cm^3$) | pH | Chemical composition (mg/L) | | | | | |
|---|---|---|---|---|---|---|---|
| | | Na | Ca | Mg | $Cl^-$ | $SO_4$ | $CO_3$ |
| 1.03 | 8.53 | 2085.22 | 348.35 | 1973.49 | 5303.70 | 134 | 576.58 |

The experiments were carried out using PCC with containing of fly ash and produced by Indonesia cement manufacture. Characteristics of PCC used in this research are shown in table 2 and meet with SNI 15-7064-2004 (Indonesia Standard for PCC).

**Table 2.** Characteristics of PCC.

| Physical properties | |
|---|---|
| Initial setting time (min) | 132 |
| Final setting time (min) | 224 |
| Specific surface ($m^2/kg$) | 341 |
| Specific gravity ($g/cm^3$) | 3.08 |
| Bulk density (kg/L) | 1.10 |
| Compressive strength: | |
| - 3 days ($kg/cm^2$) | 162 |
| - 7 days ($kg/cm^2$) | 231 |
| - 28 days ($kg/cm^2$) | 333 |
| Chemical properties | % |
| MgO | 0.99 |
| $SO_3$ | 1.81 |
| $SiO_2$ | 18.39 |
| $Al_2O_3$ | 5.15 |
| $Fe_2O_3$ | 3.41 |
| CaO | 61.79 |
| LOI | 4.61 |

Table 3 shows some physical properties of aggregates. Crushed river stone and marine sand were used as coarse aggregate and fine aggregate, respectively. Table 4 shows chemical properties of marine sand. All aggregates were used in saturated surface dry condition, and their were soaked into the sea water for 24 hours to produce saturated surface dry condition.

**Table 3.** Physical properties of aggregates.

| Property | | Crushed stone (diameter of 10-20 mm) | Marine sand |
|---|---|---|---|
| Specific gravity | Oven dry | 2.89 | 2.43 |
| | Saturated surface dry | 2.91 | 2.47 |
| Water absorption (%) | | 0.62 | 1.75 |

**Table 4.** Chemical composition of marine sand.

| Chemical Composition (%) | | | | | | |
|---|---|---|---|---|---|---|
| Mg | Fe | Ca | Al | Cl | $SiO_2$ | MgO |
| 1.1 | 3.6 | 1.8 | 12 | 0.04 | 51190 | 1.95 |

Table 5 shows proportion of concrete mix. The mixture was prepared in the laboratory. The slump design was 10 ± 2.5 cm. Fresh concrete was pouring into the cylindrical mold with diameter and heigth of 10 cm and 20 cm, respectively. All specimens were demolded after 24 hours of casting. After demolding, the specimens were divided into two sets the curing condition. Two curing condition, wet-burlap curing with tap water and sea water immersion curing were designed.

**Table 5.** Mix proportion of concrete (1 $m^3$).

| w/c | Water (kg) | Cement (kg) | (Fine aggregate) Marine sand (kg) | (Coarse aggregate) Crushed stone (kg) |
|---|---|---|---|---|
| 0.4 | 204 | 513 | 477 | 1099 |

Slump test was done according to SNI 1972-1990 (Method of test for slump test). The compressive strength and static modulus were tested with according to SNI 1974-2011 (Method of test for compressive strength of concrete). Two load variable transducers displacement (LVDT) used to measure displacement and to calculate vertical strain under compressive load. The compressive load and vertical strain recorded by a computerized data logging system.

## 3. Results and discussion

### *3.1. Slump test*

Fresh concrete had a slump value of 9 cm with met the slump design of 10±2.5 cm. Visual observation (figure 1) revealed that fresh paste made with sea water and PCC could maintain the workability and homogeneity of the mixture without segregation, bleeding did not occurr, and no coarse aggregates pilling at the bottom of specimen.

**Figure 1.** Slump Test.

### *3.2. Compressive strength and stress strain behavior*

The visual observation of the cylindrical specimens before compressive strength test shows that the surface of hardened concrete was smooth without any honey comb, and large air voids, also no pilling of coarse aggregates. This result showed that fresh concrete consisted of sea water, marine sand, PCC and crushed river stone can be poured with maintaining mixture homogeneity led to a good achievement of compaction.

**Figure 2.** Compressive strength test

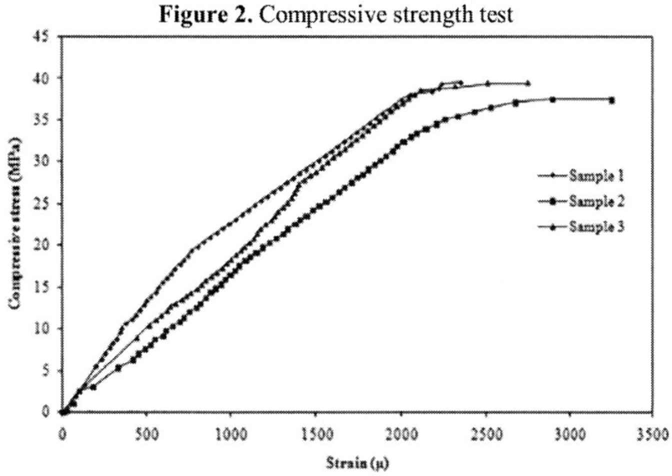


**Figure 3.** Stress-strain curve of concrete cured in the sea water

Figure 3 shows the stress-strain behavior curves of the three hardened concrete specimens with cured in sea water until 28 days. The concrete specimens attained the fracture strain from 2755 μ to 2900 μ. The average of compressive strength was 38.87 MPa.

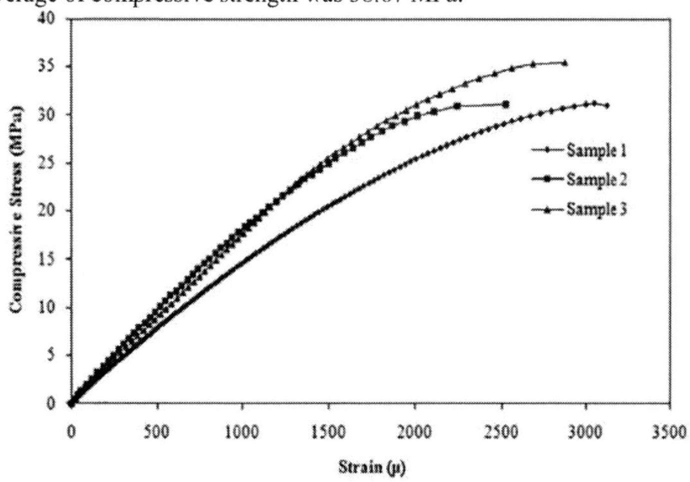


**Figure 4.** Stress-strain curve of cured with tap water-wet burlap

Figure 4 illustrates the stress-strain behavior curves of the three hardened concrete specimens that cured with tap water-wet burlap until 28 days. The normal concrete specimens attained the fracture strain from 2500 μ to 3100 μ. The average compressive strength of normal concrete specimen was 32.46 MPa.

Based on compressive strength test result, it can be noticed that an interaction between all materials through the bond between the paste made with sea water and PCC and the aggregates can achieve a good capability of concrete made with sea water, marine sand, PCC, and river coarse aggregate to load transfer mechanism from paste to aggregates and vice versa. Such attributes ensure the capability to withstand the compressive load.

Based on figure 3 and figure 4, it was observed that 19.74% higher in compressive strength by using sea water curing when compared with the concrete cured with tap water-wet burlap. This phenomenon showed that the presence of salt in sea water and marine sand did not affect hydration process and sea water can be used to cure concrete and the proper hydration process of concrete can take place to obtain a good compressive strength.

Furthermore, it can be seen that based on the literature review [7], the strain value of concrete made with sea water, marine sand, PCC and crushed river stone similar to the ordinary concrete made with OPC and fresh water.

## 4. Conclusions

1. The slump test result showed that fresh concrete had a proper workability while the hardened specimen exhibited that the mixture can maintain its homogeneity during pouring process into the mold and compaction process led to achieving a good compaction result without honeycombs, and large void appeared on the surface of specimens.
2. Compressive strength test showed that the specimens that cured in the sea water had a higher 28 days compressive strength as compared with the specimens cured with tap water-wet burlap while based on the literature review, the strain value of concrete made with sea water, marine sand, PCC and crushed river stone similar to the ordinary concrete made with OPC and fresh water.

## 5. References

[1] Antiohos S and Tsimas S 2005 Investigating the role of reactive silica in the hydration mechanism of high-calcium fly ash/cement systems, *Cement and Concrete Composites* **27(2)**171-181.

[2] Uddin M A, Jameel M, Sobuz H R, Islam M S and Hasan N M S 2012 Experimental study on strength gaining characteristics of concrete using PCC *KSCE Journal of Civil Engineering* **17(4)** 789-796.

[3] Tjaronge M W, Irmawaty R, Adisasmita S A 2014 Compressive strength and hydration process of self compacting Concrete (SCC) mixed with sea water, marine sand, and PCC, *Advance Material Research* **935** 242-246.

[4] Wegian F M 2010 Effect of sea water for mixing and curing on structural concrete, *The IES Journal Part A: Civil & Structural Engineering* **3(4)** 235-243.

[5] Otsuki N, Furuya D, Saito T and Tadokoro Y 2011 Possibility of sea water as mixing water in concrete, *Proceedings of 36th Conference on Our World in Concrete & Structures (Singapore)*
vol 1 pp 14-16.

[6] Katano K, Takeda N, Ishizeki Y and Iriya K 2013 Properties and application of concrete made with sea water and un-washed sea sand, *Third International Conference on Sustainable Construction Materials and Technologies (Japan)* vol 3 pp e172.

[7] Neville A M and Brooks J J 1987 *Concrete Technology* (New York: Longman Scientific & Technical).

**Acknowledgments**
Authors wishing to acknowledge all members of Eco-material laboratory Hasanuddin University.

# Permeability analysis of Asbuton material used as core layers of water resistance in the body of dam

**H Rahim[1], M W Tjaronge[2], A Thaha[3] and R Djamaluddin[3]**

[1] Doctoral student, Department of Civil Engineering, Hasanuddin University, Indonesia
[2] Professor, Department of Civil Engineering, Hasanuddin University, Indonesia
[3] Associated Professor, Department of Civil Engineering, Hasanuddin University, Indonesia
Corresponding author : hasrawatirahim@yahoo.com

**Abstract.** In order to increase consumption of the local materials and national products, large reserves of Asbuton material about 662.960 million tons in the Buton Islands became an alternative as a waterproof core layer in the body of dam. The Asbuton material was used in this research is Lawele Granular Asphalt (LGA). This study was an experimental study conducted in the laboratory by conducting density testing (content weight) and permeability on Asbuton material. Testing of the Asbuton material used Falling Head method to find out the permeability value of Asbuton material. The data of test result to be analyzed are the relation between compaction energy and density value also relation between density value and permeability value of Asbuton material. The result shows that increases the number of blow apply to the Asbuton material at each layer will increase the density of the Asbuton material. The density value of Asbuton material that satisfies the requirements for use as an impermeable core layer in the dam body is 1.53 grams/cm$^3$. The increase the density value (the weight of the contents) of the Asbuton material will reduce its permeability value of the Asbuton material.

## 1. Introduction

One of the main government programs is food sovereignty by construct 65 dams across Indonesia within 5 years, establishing a 1 million hectare irrigation network and rehabilitating an irrigation network of 3 million hectares, and encouraging the use of local products domestically. Serious research shall be conducted to find the dam type which more efficient, easier in operation maintenance, more resistance to disaster risk, easy in material availability. In Indonesia, the existing dams are generally embankment type. Embankment type dam is divided into three types, namely: homogeneous embankment, embankment type zonal and embankment membrane type [1]. Embankment Dam homogeneous, the former material of the dam body is the soil. Embankment Dam type zonal, material forming the body of the dam consists of several types of materials such as permeable materials and waterproof materials in the middle[2]. Material for Zonal-type embankment dam mostly using rock (big stone) as major component for dam stabilization. Waterproof layer will be installed on the upper stream or inside the body of the dam as core layer to preventing water penetrate through dam body. In general, the waterproof material used in Indonesia is clay, while abroad besides clay is also used as concrete asphalt mixture (hotmix)[3-4].

Indonesia reserve very large natural asphalt called *Aspal buton* (*Asbuton*) in Buton Island Southeast Sulawesi Province (figure 1). Based on assessment conducted by the Alberta Research Council on the 1980s and validated by The Pusjatan, Ministry of Public Works in 2010-2013 around

662.960 million tons [5]. Several studies showed that most of physical properties of bitumen within Asbuton similar with the petroleum bitumen [6-8].
The excess of asphalt-containing materials compared to clay soils used as a watertight core layer in a zonal type strain [9], is as follows:

- The risk of leakage is less likely
- The risk if there is a smaller earthquake.
- Shorter construction time.
- Maintenance is easier.

Based on the reasons mentioned above, the author intends to conduct research on "Permeability Analysis of Asbuton Material Used As A Waterproof Core Layer On The Body of Dam".

**Figure 1.** Asbuton mining in Buton Island, Inodonesia (natural rock asphalt)

## 2. Theoretical basis

*2.1. Asbuton granule for waterproof zone material*
Grain Asbuton is the result of the processing of solid-shaped Asbutons crushed by a stone crusher or other suitable cracking device having a specific grain size. The raw material to produce grain Asbuton can be obtained from solid Asbuton chunks.

*2.2. Permeability testing*
The permeability test is the most important test on material samples for dams. The permeability level of a material is usually characterized by a coefficient of permeability or filtration coefficient of cm/s. One of permeability testing in laboratory based on simple theory is relation between velocity of water through pores of the soil and hydraulic gradient, where flow of water through soil pores is laminar. Darcy formula raised up based on mentioned phenomenon, as follows;

$$Q = K.i.A \quad (1)$$

where:
Q is the discharge flowing through the cross-section unit per unit time ($cm^3/s$), I is hydraulic gradient, K is filtration coefficient (cm/s), A is cross section ($cm^2$).

In this formula, it is considered that water flows through the entire cross section (A), so not through the pores of the soil. In general, the level of permeability of soil materials has been grouped as follows:

- Pass water (permeable), $K>10^{-4}$ (cm/s)
- Semi-water pass (semi-permeable), $K=10^{-4}$ (cm/s)

- Waterproof (impermeable), $K<10^{-4}$ (cm/s)

The Darcy formula can only be used in slow laminar slow filtration flow, while for high-speed turbulent flow, the Darcy formula cannot be applied. Therefore, soil permeability testing should be carried out in laminar flow, so that the obtained filtration coefficient only gives identity to laminar flow because turbulent flow cannot occur on fine grained soils such as in clays and silt, so for the soil type, the Darcy formula is sufficient.

- Testing permeability with variable water level elevation. This test is carried out with examples of materials that are expected to have low permeability. The coefficient of filtration of the material sample can be calculated by using the following formula:

$$K = 2.3\left(\frac{a.L}{A.t}\right)\log_{10}\left(\frac{h_o}{h_t}\right) \tag{2}$$

where, K is coefficient of filtration (cm/s), T is duration of measurement (s), A is cross section of the material sample ($cm^2$), a is cross section of the pipe ($cm^2$), L is thickness of the material sample (cm), $H_0$ is the water level in the pipe at the start of the test (cm), $H_t$ is the water level in the pipe at the end of the test (cm)[10].

### *2.3. Core layer density*

One of the factors that support the strength and stability of the dam body is adequated compaction. The density of a material greatly affects the mechanical characteristics of the material, particularly the weight of the contents, permeability, stability and others. The most important factors affecting the level of compaction of the material is the water content, gradation and the amount of energy given to the compaction of the material. In essence the compaction of the material is essensial effort to remove the air from the gaps between the grains of material.

The parameters associated with the density are as follows:

$$W = \left(\frac{W_w - W_d}{W_d - W_c}\right)x100\% \tag{3}$$

$$\gamma_d = \frac{\gamma_w}{(w+100)}x100\% \tag{4}$$

$$\gamma_d = \frac{100G_s}{(100+wG_s)} \tag{5}$$

where, W is water content (%), w is moisture content (%), $W_w$ is weight of wet test object and container (gram), $W_d$ is weight of dry specimen and container (gram), $W_c$ is weight of container (gram), $\gamma_d$ is dry fill weight (gram/$cm^3$), $\gamma_w$ = wet content weight (gram/$cm^3$), $G_s$ is material density.

While the number of blow indicates the amount of energy given in the compaction process. If given energies to the material are increased, the weight of the dry content will be increased, while the optimum water content will move in a drier direction.

The amount of energy given when carrying out the compaction of the material, it can be calculated by the following formula.

$$E_c = \frac{W.H.N.L}{V} \tag{6}$$

Where, $E_c$ is amount of compaction energy (kg/$cm^2$), W is weight of hammer (kg), H is high falls hammer (cm), N = collision frequency at each layer, L = number of layers, V is print volume ($cm^3$)[10].

## 3. Research methodology

### *3.1. Research design*
The research flowchart can be seen on figure 2. The testing methods performed in this study are based on the Indonesian National Standard (SNI). Other standards such as the American Association of State Highways and Transportation Officials (AASHTO) and American Society for Testing and Materials (ASTM) are used if no available in SNI.

### *3.2. Testing of Asbuton Material Density*
The density testing procedure of Asbuton material refers to SNI 03-1743-1989 [11]. The pounder used weighs 4.54 kg with a falling height of 45.7 cm. Diameter of mold 152 mm with 116 mm height and volume is 2124 $cm^2$. The number of layers were 5, each layer was blowed 56 times according to the standard, but in this research the relationship between the density value and the blow energy per layer was considered.

### *3.3 Testing of Permeability of Asbuton Materials*
The testing procedure for permeability of Asbuton material refers to ASTM D 2434-68 [12]. The permeability test uses the falling head method, because the material to be tested is estimated to have low permeability.

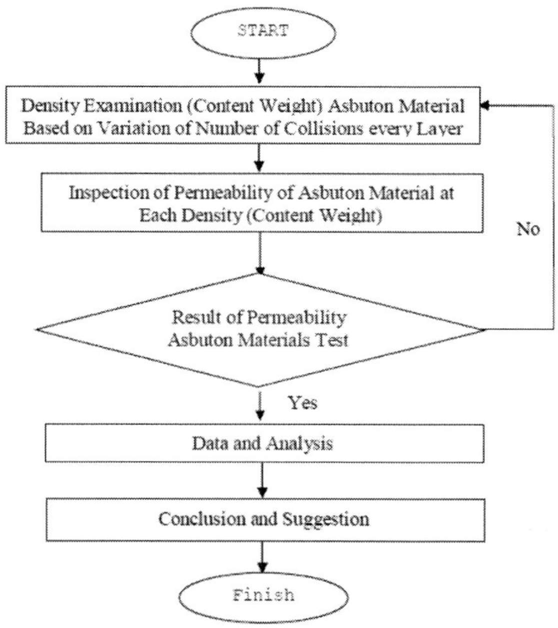


**Figure 2.** Flowchart of research design

## 4. Result and analysis

### *4.1. Data presentation*

#### *4.1.1. Test of result of Asbuton material density*
The results of density testing of Asbuton material to be used as a watertight core layer in the dam body can be seen in Table 1.

**Table 1.** Test result of Asbuton material density

| Energy of Compaction (kg/cm$^2$) | Density Value (gram/cm$^3$) |
|---|---|
| 27.35 | 1.24 |
| 34.19 | 1.40 |
| 41.52 | 1.51 |
| 48.84 | 1.53 |
| 67.17 | 1.55 |

*4.1.2. Testing result of Asbuton material permeability*

The results of the permeability testing of Asbuton material to be used as a watertight core layer in the dam body can be seen in Table 2.

**Table 2.** Test result of Asbuton material permeability

| Density Value (gram/cm$^3$) | Permeability Value (cm/s) (gram/s) |
|---|---|
| 1.24 | 2.12 x $10^{-3}$ |
| 1.40 | 1.86 x $10^{-4}$ |
| 1.51 | 4.19 x $10^{-5}$ |
| 1.53 | 1.12 x $10^{-6}$ |
| 1.55 | 7.33 x $10^{-6}$ |

*4.2. Data analysis*

The density test data the permeability testing of Asbuton material will be analyzed by correlating the number of compactions per layer with density value and the relationship between the density value and the permeability value.

*4.2.1. Relationship between density value energy compaction*

The relationship between compaction energy and the density value of Asbuton material can be seen in figure 3. Density value increase parallel with compaction energy given, indicating that increases number of blow will create denser and heavier Asbuton. The relationship between the compacting energy and the density value of the Asbuton material is illustrated in the equation:

$$\gamma = -0.0005E_c^{\ 2} + 0.0541E_c + 0.1579 \tag{7}$$

where, $\gamma$ is density value (gr/cm$^3$) and $E_c$ is energy compaction (kg/cm$^2$)

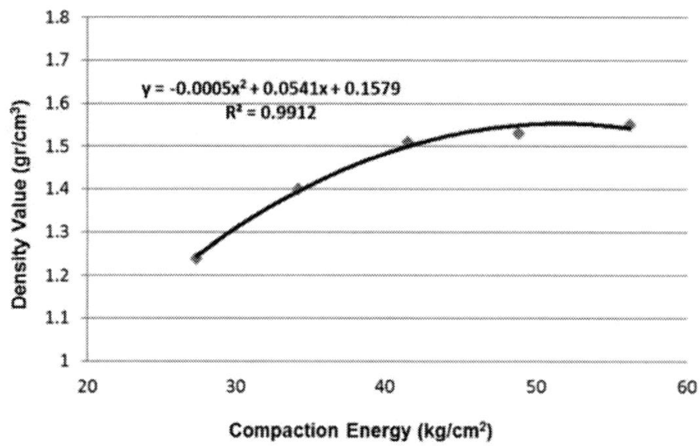


**Figure 3.** Relationship between compaction with density value

*4.2.2. Relationship between density value with permeability value*

The permeability value for the impermeable core coating material does not exceed $10^{-5}$ cm/s. Table 2 shows that the density value of the Asbuton material that meets the requirements for use as an impermeable core layer in the dam body is 1.53 gr/cm$^3$. The relationship between density value and the permeability value of Asbuton material can be seen on figure 4.

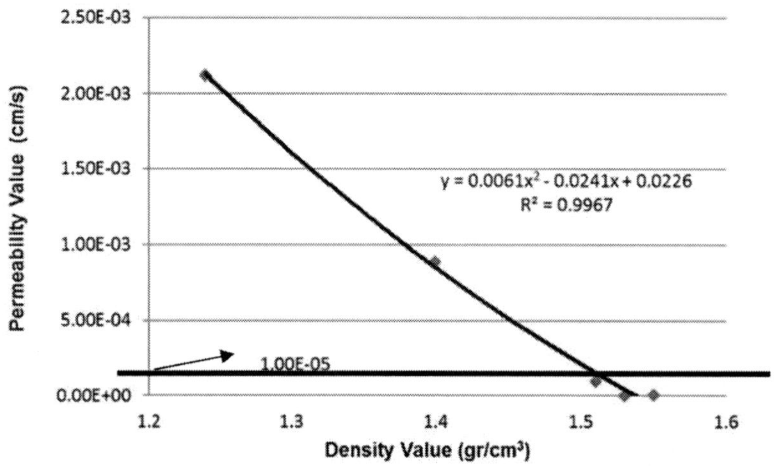


**Figure 4.** Relationship between density value and permeability value

Figure 4 shows that the permeability value of Asbuton will decrease when the value of density is increased, indicating that pores in the Asbuton will be less in the denser Asbuton. With the smaller pores in the Asbuton material, the volume of water passing through the pores is getting smaller. The relationship between the density value and the permeability value of the Asbuton material is illustrated in the equation:

$$K = 0.0061\gamma^2 - 0.0241\gamma + 0.0226 \quad (8)$$

where, K is permeability value, cm/s, $\gamma$ is density value, gram/cm$^3$

IOP Conf. Series: Materials Science and Engineering **271** (2017) 012067 doi:10.1088/1757-899X/271/1/012067

## 5. Conclusions and suggestions

### *5.1. Conclusions*
Based on the results of research and data analysis, it can be concluded as follows:

- Applying greater compacting energy on the Asbuton material in each layer will create higher density of the Asbuton material. Compacting energy applied to the Asbuton material in each layer, the higher the density (weight) of the Asbuton material.
- The density value (weight of contents) of Asbuton material that satisfies the requirements of impermeable core layer in the dam body is 1.53 gr/cm$^3$. The higher density value (weight of content) of the Asbuton material opposites with the permeability value of the Asbuton material.

### *5.2. Suggestions*
For further research development, we suggest some suggestions as follows:
- Still needed research to determine the durability of Asbuton material after submerged water.
- Make a modeling to find out the effective width of the Asbuton which can be used for impermeable core layer in the dam body.

## 6. References
[1] W Indradi 2013 *Material II Type of Urugan Dam.*
[2] Department of Public Works 2003 Guidance on general criteria of dam design. Office of the Secretariat of the Dam Security Commission *Directorate of Water Resources* (Indonesia).
[3] USS 2011 21st Century Dam Design – Advances and Adaptations *31st Annual USSD Conference.*
[4] Salemi S H, Baziar M H, Merrifield C M, and Heidari T 2008 Investigation on dynamic behavior of asphalt core dams *Sixth International Conference on Case Histories in Geotechnical Engineering* vol 13
[5] Ministry of Public Works 2011 *Utilization of Asbuton Directorate General of Highways* (Indonesia)
[6] A Gaus, Tjaronge M W, Ali N and Djamaluddin R 2014 Experimental study on characteristics of asphalt concrete bearing coarse (AC BC) mixture using Buton Granular Asphalt (BGA) *International Journal of Applied Engineering Research* **9(22)** 18037-18045.
[7] A Gaus, Tjaronge M W, Ali N, and Djamaluddin R 2015 Compressive strength of asphalt concrete binder course (AC-BC) mixture using buton granular asphalt (BGA) *Procedia Engineering* (Amsterdam: Elsevier) vol 125 pp 657-662.
[8] Budiamin, Tjaronge M W, S H Aly and Djamaluddin R 2015 Mechanical characteristics of hotmix cold laid containing buton granular asphalt (BGA) and flux oil as wearing course; *ARPN Journal of Engineering and Applied Sciences* **10(12)** pp 5200-5205.
[9] ICOLD 2015 Bulletin 84.
[10] Alicescu V, Tournier J P, Vannobel P and Moore V 2011 *Design and Construction of Nemiscau-1Dam The First Asphalt Core Rockfill Dam in North America*
[11] Standard National of Indonesia 1989 *Standard Method of Test for Moisture-Density Relation of Soils Using a 4.54 kg (10-lb) Rammer and a 457 mm (18 inch) Drop* SNI 03-1743-1989.
[12] ASTM D2434-68 2006 *Standard Test Method for Permeability of Granular Soil (Falling Head).*

IOP Conf. Series: Materials Science and Engineering **271** (2017) 012068 doi:10.1088/1757-899X/271/1/012068

# Effect of hydrated lime on compressive strength mortar of fly ash laterite soil geopolymer mortar

**F A Wangsa[1], M W Tjaronge[1], A R Djamaluddin[1] and A B Muhiddin[1]**

[1] Department of Civil Engineering, Hasanuddin University, Indonesia.
Corresponding author:fithriyahawz@gmail.com

**Abstract.** This paper explored the suitability of fly ash, hydrated lime, and laterite soil with several activator (sodium hydroxide and sodium tiosulfate) to produce geopolymer mortar. Furthermore, the heat that released by hydrated lime was used instead of oven curing. In order to produce geopolymer mortar without oven curing, three variations of curing condition has been applied. Based on the result, all the curing condition showed that the hardener mortar can be produced and exhibited the increasing of compressive strength of geopolymer mortar from 3 days to 7 days without oven curing.

## 1. Introduction

Portland cement has been widely used as a binding material in concrete and mortar production, where its demand increases along with rapid growth of infrastructure development. The problem emerges in Portland cement production because it releases much of carbon dioxide ($CO_2$) into the atmosphere and it could harm the environment. To reduce carbon dioxide emissions due to the Portland cement production, it is necessary to find other materials as Portland cement replacement materials. In this concern, fly ash geopolymer is one of the alternatives to Portland cement replacement as a binder on the concrete or mortar mixture. Geopolymer is a relatively new as a fastener material that can be a sustainable and economical because it was produced from industrial by-products such as fly ash. Number of studies have shown that fly ash geopolymer binders can form mortar and concrete, where geopolymer with fly ash has physical characteristics similar to cement mortar and cement concrete made from cement. Fly ash geopolymer mortar was made without the use of Portland cement, and instead used of fly ash which compound in Silica and Alumina that can react with alkaline liquids to produce binders. Sodium silicate and sodium hydroxide were used as an alkaline activator. Most previous studies that used fly ash geopolymer on concrete required curing ambient temperature of oven heat in hardening. From several studies indicated that geopolymer production requires heating ranging of 60-85°C at the curing period to provide the maximum strength of the test sample [1].

The geopolymer mortar using fly ash were investigated, where curing treatment was provided by using an oven curing with a temperature of 60°C [2]. The effect of time delay and $Na_2SiO_3$ concentrations was examined on the development of compressive strength of synthesized mortar geopolymers, with curing treatment of mortars provided by varied oven curing of 65°C, 75°C, and 85°C [3].

Laterite soil is quite widely available, it is attempted to be utilized and developed as a mixture of mortar geopolymer for replacing the sand. Using of laterite soil is one of the innovations providing alternative material geopolymer mortar. The laterite soil is formed in the tropics through a weathering process that supports the formation of iron, aluminum, manganese, and titanium oxide. This process breaks down silicate minerals into clay minerals such as kaolinite and illite. Iron and aluminum oxide stand out in the laterite soil, and with seasonal fluctuations in water tables, this oxide produces a reddish brown colour visible on the laterite soil.

Several studies have also used laterite soil as a material. Research on laterite soils and presented a global view of what has been done in the field of laterite soil repairs for construction purposes in tropical countries such as Cameroon as well as stabilizers. The laterite soil used is combined with Portland cement [4].

From previous research, producing of geopolymer requires curing temperature. Therefore several studies has begun to develop the use of geopolymer without using curing heat temperature. The presence of hydrated lime $Ca(OH)_2$ in the mortar mixture will be the activator that generates the heat required for the geopolymer reaction for giving maximum strength to the geopolymer mortar. Utilization of laterite soil and lime as a fly ash geopolymer mortar material are no longer using high temperatures with oven curing. The presence of lime in the mortar mixture will be the activator that generates the heat requirement for the geopolymer reaction. It was result that maximum strength to the geopolymer mortar. The amount of lime content has a considerable influence on the hardened geopolymer.

## 2. Experimental study

### *2.1. Materials*

Table 1 and table 2 show the chemical contents of fly ash and laterite soil, respectively. The material is a laterite and lime soil with a ratio of 1:1. Geopolymer used is a combination of fly ash with alkaline activator used is sodium hydroxide and sodium thiosulfate comparison of sodium hydroxide and sodium thiosulfate was given by a ratio of 1.5. While sodium hydroxide concentrate is 6 Molar. The mortar mixture can be seen in Table 3 below.

**Table 1.** Chemical contents fly ash

| | |
|---|---|
| MgO | 8.10% |
| $Al_2O_3$ | 19.16% |
| $SiO_2$ | 34.63% |
| $SO_3$ | 1.80% |
| $K_2O$ | 1.33% |
| CaO | 12.74% |
| $TiO_2$ | 1.26% |
| $Cr_2o_3$ | 0.07% |
| MnO | 0.25% |
| $Fe_2O_3$ | 19.96% |
| CoO | 0.05% |
| SrO | 0.13% |
| BaO | 0.21% |
| $Pr_6O_{11}$ | 0.05% |
| $Nd_2O_3$ | 0.07% |

**Table 2.** Chemical contents laterite soil

| | |
|---|---|
| $MgO$ | 1.90% |
| $Al_2O_3$ | 26.53% |
| $SiO_2$ | 40.12% |
| $SO_3$ | 0.30% |
| $K_2O$ | 1.61% |
| $CaO$ | 4.26% |
| $TiO_2$ | 2.26% |
| $Na_2O$ | 2.10% |
| $MnO$ | 0.20% |
| $Fe_2O_3$ | 19.66% |
| $Cl_2$ | 0.07% |
| $SrO$ | 0.11% |
| $BaO$ | 0.10% |
| $Pr_6O_{11}$ | 0.05% |
| $Nd_2O_3$ | 0.08% |
| $P_2O_5$ | 0.31% |
| $ZrO_2$ | 0.11 |

**Table 3.** The composition of a mixture of fly ash geopolymer mortar.

| No. | Material | Amount | Unit |
|---|---|---|---|
| 1 | Fly ash | 600 | gram |
| 2 | Laterite Soil | 900 | gram |
| 3 | $Ca(OH)_2$ | 900 | gram |
| 4 | Water | 915 | ml |
| 5 | NaOH | 146,4 | gram |
| 6 | $Na_2S_2O_3$ | 219,6 | gram |

*2.2. Preparation of geopolymer mortar specimens and cubicles*

The first stage is mixing a fly ash, laterite soil and lime for three minutes using the mixer to ensure homogeneity. Secondly, a solution of sodium hydroxide in the form of pellets is melted with a concentration of 6 Molar, then it was made the mix of sodium hydroxide and sodium thiosulfate by a ratio of 1.5. Thirdly, the solution of the ratio of sodium hydroxide and sodium silicate then it was stirring in 15 minutes. After a mixture was done then it was put in a cube mold with the size of 5 cm x 5 cm x 5 cm. Then the mold that has been filled then vibrated on the platform for 4 minutes to release the trapped air bubbles. After 24 hours the mold was opened cured to the test. The curing of sample test was performed in three conditions, i.e. air curing, where the sample is left in room temperature, then curing the water in which the sample is immersed in water and curing the lifting soap, wherein the test sample, soaked a day in water is then removed and dissipated for a day in air, then soaked again, done repeatedly until the time of testing.

*2.3. Stage of test*

This research is an experimental test and was done at Eco Material laboratory of Civil Engineering department, Hasanuddin University.

**Table 4.** Fine aggregate inspection (laterite soil).

| No. | Testing Material | Standard Method |
|---|---|---|
| 1 | Sieve Analysis | SNI 03-1968-1990 (Standard Test Method for Sieve Analysis of Fine and Coarse Aggregates) |
| 2 | Absorption and Specific Gravity | SNI-1970-2008 (Standard Test Method For Specific Gravity And Absorption of Rock For Erosion Control) |
| 3 | Volume Weight | SNI 03-1973-1990 (Standard Test Method for Calculation of Volume and Weight of Industrial Aromatic Hydrocarbons and Cyclohexane [Metric]) |
| 4 | Water Content | SNI 03-1971-1990 (Standard Test Method for Determination of Water Content of Soil By Direct Heating |
| 5 | Mud Content | SNI 03-4142-1996 |
| 6 | Organic Content | SNI 03-2816-1992 (Standard Test Methods for Moisture, Ash, and Organic Matter of Peat and Other Organic Soils |

**Table 5.** Lime stone inspection.

| Testing Material | Standard Method |
|---|---|
| Specific gravity | SNI 03-1964-2008 (Standard Test Methods for Specific Gravity of Soil Solids by Water Pycnometer) |

*2.4. Uniaxial compression strength (UCS) test*

There were three condition of curing used in this test (i.e. air curing, water curing, and dry-wet curing) and three samples prepared for each curing condition and tested by using compressive strength test used Universal Testing Machine (Tokyo Testing Machine Inc.). Strain of each test sample obtained by two Linear Variable Displacement Transducers (LVDTs). The compressive strength of specimen calculated by the average of compressive stress from three samples of each curing condition.

## 3. Result and discussion

*3.1. Compaction result*

Visual investigation was carried out to observe the compaction result of specimens prior to compressive strength test. All specimens showed that the hardened fly ash geopolymer mortar with laterite soil can be produce by using hydrated lime instead of using heat from oven to produce the specimen with ambident temperature. The heat that release from the hydration process between hydrated and water provide suitable temperatures to generate bonding between silica and activator alkaline 7 days curing in air.

*3.2 Effect of air curing on stress strain relationship of geopolymer mortar containing laterite soil*
Figure 1 shows the stress strain of relationship of specimen cured in air remain linear until reach the stress and compressive strength evolution to avoid environmental degradation.

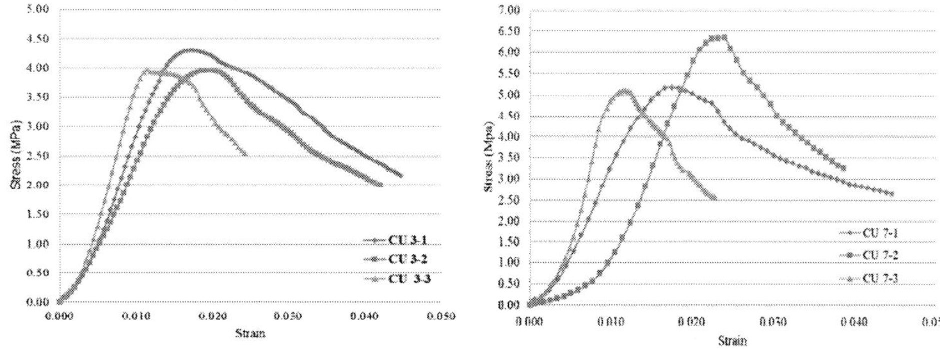


(a) 3 days air cured (b) 7 days air cured

**Figure 1.** Stress-strain relationship of geopolymer mortar with air curing.

Based on figure 1, it can be seen that sample of 3 days air curing has average compressive strength value of 4.07 MPa and the average strain of 0.0160 while 7 days air curing arised 5.55 MPa of average compressive strength corresponding to 0.018 of average strain. Compressive strength at 7 days air curing increase 26.67% as compared with the compressive strength at 3 days air curing. This result shows that the compressive strength developed without oven curing. Similar with the samples those cured in air and water, the presence of hydrated lime in mortar mixture, hence without the curing of oven temperature, the fly ash geopolymer mortar with this laterite soil material can still provide strength.

*3.3. Effect of water curing on stress strain relationship of geopolymer mortar containing laterite soil*
Figure 2 shows the stress strain of relationship of specimen cured in air remain linear until reach the stress and compressive strength evolution to avoid environmental degradation. Based on Figure 2, it can be seen that sample of 3 days water curing has average compressive strength value of 3.13 MPa and the average strain of 0.016 while 7 days water curing arised 4.62 MPa of average compressive strength corresponding to 0.016 of average strain . Compressive strength at 7 days water curing increase 32.29% compared with the compressive strength at 3 days air curing.

This result shows that the compressive strength developed without oven curing. As attribute by the presence of hydrated lime in this mortar mixture, which induced the heat, hence without the curing of oven temperature, the fly ash geopolymer mortar with this laterite soil material can still provide strength.

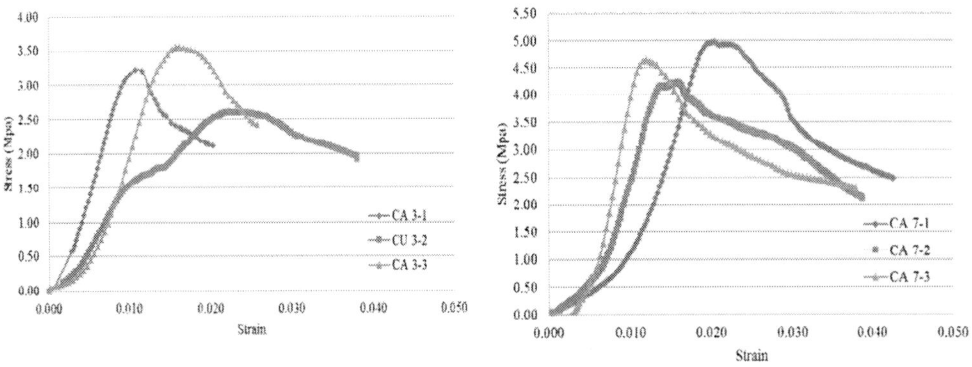


(a) 3 days air cured (b) 7 days air cured

**Figure 2.** Stress-strain relationship of geopolymer mortar with water curing.

*3.4. Effect of wet-dry curing on stress strain relationship of geopolymer mortar containing laterite soil*

Figure 3 shows the stress strain of relationship of specimen cured in wet-dry condition remain linear until reach the stress and compressive strength evolution to avoid environmental degradation. Based on Figure 3, it can be seen that sample of 3 days water curing has average compressive strength value of 3.57 MPa and the average strain of 0.015 while 7 days air curing arised 4.11 MPa of average compressive strength corresponding to 0.016 of average strain. Compressive strength at 7 days air curing increase 13.13% compared with the compressive strength at 3 days air curing. This result shows that the compressive strength developed without oven curing. As attribute by the presence of hydrated lime in mortar mixture without the curing of oven temperature, the fly ash geopolymer mortar with this laterite soil material can still provide strength.

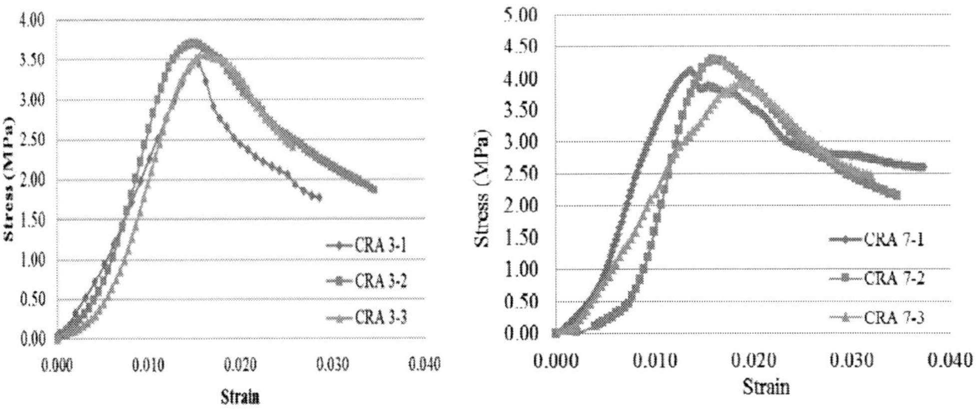


(a) 3 days air cured (b) 7 days air cured

**Figure 3.** Stress-strain relationship of geopolymer mortar with dry-wet curing.

IOP Conf. Series: Materials Science and Engineering **271** (2017) 012068 doi:10.1088/1757-899X/271/1/012068

## 4. Concluding remarks

Based on the experimental test, fly ash laterite soil geopolymer mortar can provide strength for all curing condition without heat which produced by oven. Three variations of curing for fly ash laterite soil geopolymer mortar has been conducted, the result shows that mortar with air curing had highest strength compared with two others curing (water curing and dry-wet curing).

## 5. References

[1] Shadnia R, Zhang L and Li P 2015 Experimental study of geopolymer mortar with incorporated PCM, *Construction and Building Materials* **84** 95 – 102.
[2] Djwantoro H, Wallah S E, Sumajouw D M J and Rangan B V 2004 Factor influencing the compressive strength of fly ash-based geopolymer concrete, *Civil Engineering Dimension* **6(2)** 88-93.
[3] Lemougna P N, MacKenzie K J D and Melo U F C 2011 Synthesis and thermal properties of inorganic polymer (geopolymers) for structural and refractory applications from volcanic ash, *Ceramics International* **37** 3011-3018.
[4] Hardjito D, Cheak C C and Ing C H L 2008 Strength and setting times of low calcium fly ash-based geopolymer mortar, *Modern Applied Science* **2(4)** 3-11.
[5] Mijarsh M J A, Johari M M.A and Ahmad Z A 2015 Effect of delay time and $Na_2SiO_3$ concentrations on compressive strength development of geopolymer mortar synthesized from TPOFA, *Construction and Building Materials* **86** 64–74.
[6] Lemougna P N, Melo U F Chinje, Kamseu E and Arlin B 2011 Laterite based stabilized products for sustainable building applications in tropical countries: review and prospects for the case of Cameroon, *Sustainability* **3** 293-305.

IOP Conf. Series: Materials Science and Engineering **271** (2017) 012069 doi:10.1088/1757-899X/271/1/012069

# Effect of Buton Granular Asphalt (BGA) on compressive stress-strain behavior of asphalt emulsion mixture

**Bulgis[1], M W Tjaronge[1], S A Adisasmita[1] and M Hustim[1]**

[1] Department of Civil Engineering Hasanuddin University, Indonesia..
Corresponding author:bulgisgis@gmail.com

**Abstract.** The natural rock asphalt which available in abundance in the southeast of Buton Island, Indonesia was attempted to use in production of cold mix asphalt. The purpose of this study is to replace of petroleum bitumen with local materials such as natural rock asphalt. Small granular size of natural rock asphalt that referred to Buton granular asphalt (BGA) was blended with petroleum bitumen as droplet phase in the asphalt emulsion mixture. Mixture with 5.5% of asphalt emulsion mixture made with cationic slow setting (CSS), that corresponded to the optimum asphalt emulsion content was used as a mixture control. At optimum asphalt emulsion content, bitumen of BGA was used for petroleum bitumen partial replacement in the asphalt emulsion mixture production. Test result revealed that utilizing BGA in asphalt emulsion mixture has proved for improving compressive strength where compressive strength of all mixture with BGA superpassed the compressive strength of mixture without BGA. Furthermore, mixture with BGA has less crack number as compared with the mixture without BGA.

## 1. Introduction

Annually, extensive development of the national road infrastructures need approximately 1.2 billion tonnes of bitumen material whereas half of the amount is imported from the other countries [1]. Hence, it is important to carry out proper works and inventions to reduce the import demand of the petroleum bitumen. Sedimentary rock with consists of high hydrocarbon compound with referred to the natural rock asphalt is available in abundance in the southeast of Buton Island, Indonesia. In recent years, a refinery process is conducted to produce Buton Granular Asphalt (BGA) in form of similar grains composes of water content, bitumen content and penetration value those are meet the requirement. Invention and production of asphalt pavement using BGA as partial replacement for petroleum bitumen in the mixture is being increasingly encouraged so as to reduce the petroleum bitumen consumption [2-5]. Recently, extensive technologies have been developed for decreasing the energy consumption related to producing and compacting asphalt pavement. Utilization of bitumen emulsion in production of cold mix asphalt mixture is an effective technology for decreasing the energy consumption [6-9]. Study of the stress strain relationship in static compressive testing of asphalt mixture can be considered as a good advancement for understanding the strength behavior of asphalt mixture. ASTM D1074-09 provides the standard method for measure the strength of asphalt mixture in compression [10]. In this paper, BGA is added into cold mix asphalt mixture using emulsified asphalt of cationic slow setting (CSS). This paper aims to report the experimental investigation on compressive strength test with stress strain relationship to determine the influence of partial replacement of petroleum bitumen by extracted bitumen from BGA on compressive strength behavior of the asphalt emulsion mixture containing BGA and mixture without BGA. Visual observation was conducted to investigate the crack and fracture pattern due to the compressive load.

## 2. Methodology

### *2.1. Material Preparation*

#### *2.1.1. Buton Granular Asphalt (BGA)*

BGA used in this research has a relatively uniform grain size with a maximum size of 1.18 mm. Figure 1 and table 1 show dimension of BGA and some properties of BGA, respectively.

**Figure 1.** Buton Granular Asphalt (BGA).

**Table 1.** Properties of BGA.

| Parameter | Value |
|---|---|
| Bitumen content of BGA (%) | 23.00 |
| Asphalt mineral content (%) | 77.00 |
| Water content (%) | 1.70 |
| Flash point before Extract ($^0$C) | 168 |
| Melting point of bitumen BGA Extract ($^0$C) | 86 |

#### *2.1.2. Characteristics of asphalt emulsion*

Table 2 shows characteristics of asphalt emulsion CSS-1h that used this research. Asphalt emulsion Cationic Slow Setting-1 hard (CSS-1h) is produced by Indonesian asphalt emulsion factory and available in national market. CSS-1h is produced according to SNI 4978:2011 [11]. Petroluem bitumen is used as droplet phase in CSS-1h.

**Table 2.** Characteristics of asphalt emulsion (CSS-1h).

| Kinds of Testing | Testing Result |
|---|---|
| Viscosity, Saybolt Furol 25°C, s | 39 |
| Storage stability, 24-h, % | 0.6 |
| Elementary charge | Positive |
| Sieve test number. 20, % | 0 |
| Distillation | |
| Water content, % | 36.65 |
| Oil content, % | 2.0 |
| Residue content, % | 62.35 |
| Residue penetration, 0.1 mm | 101 |
| Residue ductility, mm | 1030 |
| Solubility in trichloroethylene, % | 99.4 |

### *2.1.3. Characteristics of aggregates*

The physical properties of coarse aggregates, stone dust and filler are shown in table 3, table 4 and table 5, respectively. Stone dust and filler were obtained from stone crushed process. The aggregates used for material component in cold mixture were collected from Jeneberang river in Gowa. In this study, filler and stone dust were defined as material passing a 0.075 mm sieve and sieve no # 30, respectively. Crushed river stone with two fractions were used as coarse aggregate: one with aggregate diameter 5-10 mm and the other with crushed stone diameter 10-20 mm. At an optimum asphalt emulsion content of 5.5%, droplet phase made with petroleum bitumen in CSS-1h was partially replaced by bitumen of BGA where mineral of BGA become filler in the mixture.

**Table 3.** Properties of coarse aggregates.

| Properties | Crushed Stone | |
|---|---|---|
| | 5–10 mm | 10–20 mm |
| Water absorption, % | 2.07 | 2.08 |
| Bulk specific gravity | 2.62 | 2.63 |
| Saturated surface dry specific gravity | 2.68 | 2.68 |
| Apparent specific gravity | 2.77 | 2.78 |
| Flakiness index, % | 20.10 | 9.38 |
| Abrasion aggregate, % | 25.72 | 24.36 |

**Table 4.** Properties of stone dust (aggregate passing sieve no #30).

| | |
|---|---|
| Water absorption, % | 2.79 |
| Sand equivalent, % | 89.66 |
| Bulk specific gravity | 2.45 |
| Saturated surface dry specific gravity | 2.52 |
| Apparent specific gravity | 2.63 |

**Table 5.** Properties of filler.

| | |
|---|---|
| Water absorption, % | 2.28 |
| Sand equivalent, % | 69.57 |
| Bulk specific gravity | 2.60 |
| Saturated surface dry specific gravity | 2.65 |
| Apparent specific gravity | 2.76 |

### *2.2. Experimental works*

Figure 2 depicts the gradation and design limit of aggregate filler combination used in this study. The mid-point of control limits used in the cold asphalt mixture design was used as combined aggregate gradation. The combined aggregate gradation was kept for all mixture without BGA and with BGA. table 6 shows the composition of mix design of cold asphalt without and with 2.5% BGA. All mixtures were prepared in the laboratory. Mixing and compaction process were carried out in the laboratory at temperature room 27°C. Mixture without BGA and mixture with BGA were blended and compacted into the cylindrical mold with capacity of diameter of 101.6 mm and with capacity of 1,600 gram. All specimens were compacted by applying 50 blows to each side. After compaction, samples were subjected to the following curing regime prior to conducting compressive strength test: 24 hours ambient laboratory conditions in the sample mold, 24 hours in a forced draft oven at 38°C after demolding, and cooling at laboratory conditions.

IOP Conf. Series: Materials Science and Engineering **271** (2017) 012069 doi:10.1088/1757-899X/271/1/012069

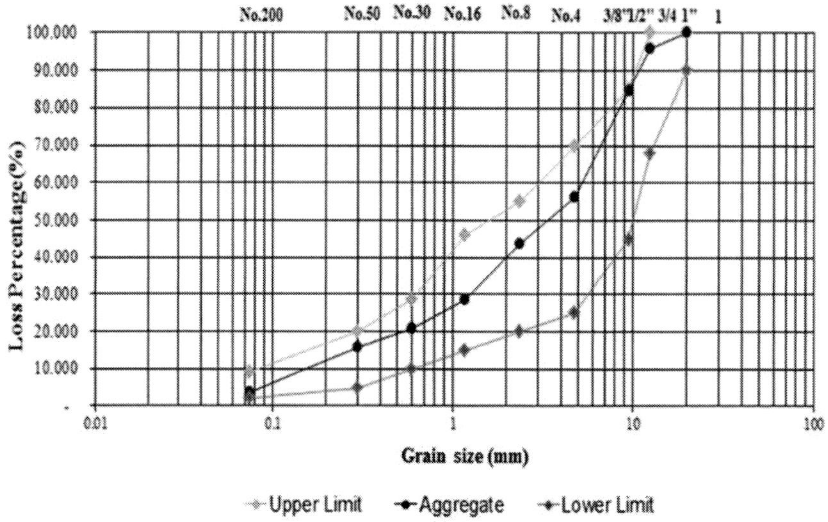


**Figure 2.** Combined Aggregate Gradation.

**Table 6.** Mixture design of cold asphalt mixture without BGA and with 2.5% of BGA.

| BGA | Asphalt content (5.5%) | | Aggregate (gram): Crushed stone 10-20 mm | Crushed stone 5-10 mm | Stone Dust (Passing sieve no. #30) | Filler (Passing Sieve 0,075 mm) | | Total |
|---|---|---|---|---|---|---|---|---|
| Without | Emulsion asphalt (88gr) / Residue (63%) 55.44 gr | Without BGA bitumen | 302.4 | 665.28 | 514.08 | Stone dust (30.24) | BGA minerals | 1600 gr |
| 2.5% (40 gram) | Emulsion asphalt (80.5 gram) / Residue (63%) 50.72gr | BGA bitumen (7.5gr) | 302.4 | 665.28 | 489.08 | Stone dust (30.24) | BGA minerals (32.5) | 1600 gr |

### *2.3. Compressive strength test*

Compressive strength test with stress strain relationship were carried out to study the performance of asphalt emulsion mixture containing BGA. Compressive strength test was conducted on asphalt emulsion-mixture specimens according to ASTM D1074-09. Figure 3 shows compressive strength test equipment. In order to determine the elasticity two linear variable differential transformers (LVDTs) were used to measure the average vertical deformation. The deformations of specimen and the load of Universal Testing Machine were monitored and recorded by a computerized data-logging system.

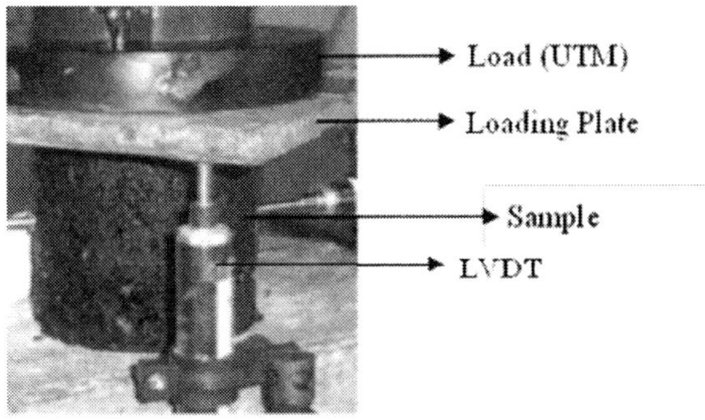


**Figure 3.** Compressive strength equipment.

## 3. Result and Discussions

### *3.1. Stress-strain relationship under compression*

Table 7 shows the result of compressive strength of the specimens. The results of compressive strength testing of emulsion asphalt without BGA obtained average compressive strength 0.215 MPa corresponding to peak strain average of 0.016 mm/mm and compressive strengths with 2.5% of BGA was obtained average compressive strength 0.292 MPa which corresponding to peak strain average of 0.021 mm/mm. It can be noticed that the addition of the 2.5% of BGA can increase the compressive strength and the peak strain 1.44 times and 1.105 times, respectively compared by the mixture without BGA. This result also indicated that within cold asphalt mixture with 2.5% of BGA can withstand the compressive load with slightly deform as compared to cold asphalt mixture without BGA.
The stiffness of specimens can be determined by the linear relationship of stress-strain curve The specimen of cold asphalt mixture without BGA still remain linear until the strain value reached approximately 0.007 mm/mm corresponding with stress of approximately 0.128 MPa, whilst the specimen of cold asphalt mixture with 2.5% of BGA have linear relationship up to strain of approximately 0.005 mm/mm associated with stress of approximately 0.207 MPa.

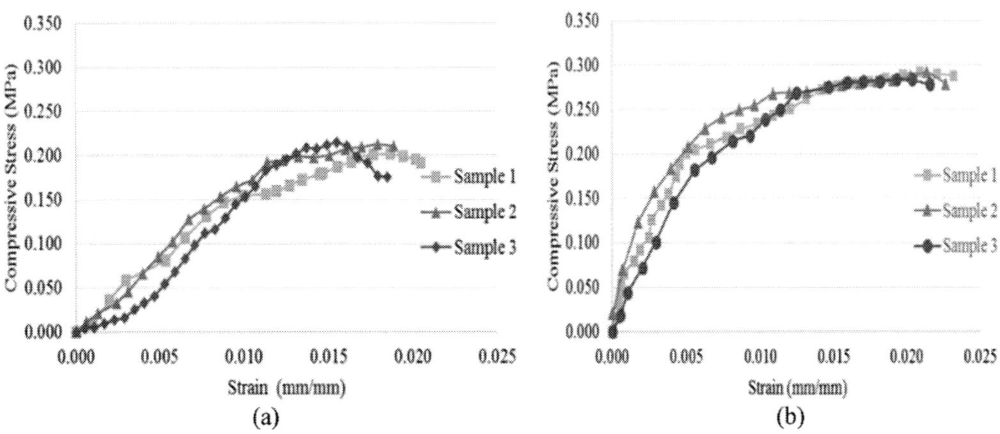


**Figure 4.** Stress –Strain relationship **(a)** without BGA, **(b)** with 2.5% BGA.

Elasticity was determined at 30% of peak stress of each specimen, where the elasticity for specimen without BGA and with 2.5% of BGA was 10.82 MPa and 13.95 MPa, respectively.

IOP Conf. Series: Materials Science and Engineering **271** (2017) 012069 doi:10.1088/1757-899X/271/1/012069

Based on the figure 4, it can be indicated that the addition of 2.5% of BGA resulting in higher elasticity and stiffness therefore cold asphalt mixture with 2.5% of BGA can withstand the compressive load with slightly deform as compared to cold asphalt mixture without BGA.

**Table 7.** Test result of compressive strength.

| Sample | Peak strain (mm/mm) | Compressive strength (MPa) | Elasticity (MPa) |
|---|---|---|---|
| without BGA | 0.016 | 0.215 | 13.43 |
| 2.5% BGA | 0.021 | 0.292 | 13.90 |

*3.2. Crack pattern*

Figure 5 shows that the crack and fracture pattern of the specimens after completion of compressive strength test. It can be seen that crack numbers of sample with 2.5% BGA was less than sample without BGA. This indicated that the addition of the BGA provided higher binding capacity of aggregate. The compressive strength result and crack pattern show that extracted bitumen of BGA can be combined with petroleum bitumen within cold asphalt mixture to create higher binding capacity as compared to cold asphalt mixture without BGA.

(a) (b)

**Figure 5. (a)** Crack pattern without BGA, **(b)** Crack pattern with 2.5% of BGA.

## 4. Concluding Remarks

The addition of 2.5% of BGA the in cold asphalt mixture resulting in higher compressive load capacity as compared to mixture without BGA. The combination between extracted bitumen of BGA and petroleum bitumen improve the binding capacity of mixture with BGA as compared to mixture without BGA.

## 5. References

[1] Furqon A 2009 Properties of bituminous mixes using Indonesian natural rock asphalt *Proceedings of 13th Conference of The Road Engineering Association of Asia and Australia (Incheon: South Korea)* vol. 13 (REAAA) pp 9–15.

[2] Gaus A, Tjaronge M W, Ali N and Djamaluddin R 2014 Experimental study on characteristics of asphalt concrete bearing coarse (AC BC) mixture using Buton Granular Asphalt (BGA) *International Journal of Applied Engineering Research* **9 (22)** 18037-18045.

[3] Gaus A, Tjaronge M W, Ali N and Djamaluddin R 2015 Compressive strength of asphalt concrete binder course (AC-BC) mixture using buton granular asphalt (BGA) *Procedia Engineering (Surabaya: Indonesia)* vol 125 (Amsterdam: Elsevier) pp 657-662.
[4] Budiamin, Tjaronge M W, S H Aly and Djamaluddin R 2015 Mechanical characteristics of hotmix cold laid containing buton granular asphalt (BGA) and flux oil as wearing course; *ARPN Journal of Engineering and Applied Sciences* **10(12)** 5200-5205.
[5] Mahyuddin A, Tjaronge M W, Ali N and Ramli M I 2017 Experimental analysis on stability and indirect tensile strength in asphalt emulsion mixture containing Buton Granular Asphalt. *International Journal of Applied Engineering Research (IJAER)* **12(12)** 3162-3169
[6] Stephen B 2015 *The Shell Bitumen Handbook* (University of Nottingham.)
[7] Shaowen D Interaction 2014 mechanism of cement and asphalt emulsion in asphalt emulsion mixtures *Materials and Structures* **47** 1149–1159.
[8] Salomon D R 2006 Asphalt emulsion technology *Transportation Research Board* (Washington-DC), Num E-C 102.
[9] Standard National of Indonesia 1991 *Standard Test Method of Asphalt Mix with Marshall Test.* SNI 06-2489-1991.
[10] ASTM D1074-09 2009 *Standard Test Method for Compressive Strength of Bituminous Mixtures.*
[11] SNI 4798:2011 2011 *Standard Specification for Emulsified Asphalt* (Indonesian National Standard SNI) Indonesia.

# Influence of seawater mixing and curing on strength characteristics and porosity of ground granulated blast-furnace slag concrete

**Adiwijaya[1], H Hamada[2], Y Sagawa[2] and D Yamamoto[2]**

[1]Civil Engineering Department, State Polytechnic of Ujung Pandang, Makassar 90245, Indonesia
[2]Faculty of Engineering, Kyushu University, Fukuoka 819-0395, Japan
Corresponding author: adiwijaya_ali@poliupg.ac.id

**Abstract.** Generally, in the concrete industry, several billion tons of fresh water are annually used for mixing water, curing water and cleaning water. Nevertheless, the utilization of seawater in the concrete industry is prohibited, because it increases the risk of corrosion of steel bars in concrete. This study presents strength characteristics and porosity of seawater mixed concrete and tap water mixed concrete incorporating Ground Granulated Blast-Furnace Slag (GGBS) with water-binder ratio (W/B) of 40%, 50% and 60%. The influence of seawater mixing, GGBS and curing conditions such as tap water curing (TC), seawater curing (SC) and air curing (AC) on the strength and porosity of concrete were evaluated. Based on investigation result, it was shown that there is no significant influence of seawater mixing in improving strength of GGBS concrete up to 365 days in TC and SC. Effectiveness of seawater-mixing on strength enhancement of GGBS concrete is larger in air curing than in water curing. Porosity of seawater-mixed concrete is decreased compared to tap water-mixed concrete in all curing conditions.

## 1. Introduction

Seawater contains about 35,000 ppm dissolved salt and total salinity is approximately 3.5%, of which 78% is sodium chloride [1]. Thus, seawater is avoided to use as mixing water for reinforced concrete, because the risk corrosion of reinforcement bars was induced by chloride in seawater compounds. However, in case of unavoidable circumstances, seawater may be used as mixing water, not only for plain concrete, but also for reinforced concrete [2-4].

Strength is a fundamental characteristic to assess durability of concrete in term quality of concrete. Durability and strength of concrete structure are interrelated, where a strength and durability relationship of concrete may always be estimated. For instance, abrasion resistance increases with increasing the compressive strength, as reported by Nurazuwa [5]. The investigation of concrete mixed with seawater on strength performance has been reported by several studies. Since 1974 to 2011, 68 papers have been published, which are related to concrete mixed with seawater. 42 papers among them revealed strength performance [6]. Nevertheless, the study on the strength of seawater mixed concrete is not still acquired the agreement among researcher, whether seawater-mixing improve the strength development of concrete is still unclear [7].

In this study, strength characteristics of seawater or tap water mixed concrete incorporating Ground Granulated Blast-Furnace Slag (GGBS) in tap water curing, seawater curing and air curing were investigated. The goal of this study is to evaluate the effectiveness of seawater mixing, GGBS and curing conditions on strength performance of concrete. By using GGBS in concrete, expected compressive strength of seawater mixed concrete are enhanced, as well mentioned by other references that mineral admixtures improve compressive strength and durability [8]. This paper also deals with porosity and the relation of compressive strength-porosity, as known that porosity and pore size distribution are one of important factor influencing compressive strength of concrete, as stated by some studies [9-11].

## 2. Research significance

There are plenty of archipelagic countries such as Indonesia, Japan and others countries, where lots of peoples live in the distant and isolated islands. Transportation of fresh water and others concrete materials for concrete production to distant and isolated islands will raise the cost of concrete work. This research was carried out in order to evaluate utilization of seawater in concrete production. By utilization of seawater in concrete production, it will not only benefit saving fresh water, but also reduce construction cost and $CO_2$ gas emissions in operational transportation, indirectly.

## 3. Experimental program

### *3.1. Materials*

Two types of binder were used in this study OPC and GGBS ($B_{4000}$), specified in Japan Industrial Standard (JIS). The physical properties of binders and concrete materials such as coarse aggregate (G), fine aggregate (S) and natural seawater as mixing water (W) can be obtained in the previous studies [12].

### *3.2 Concrete mixtures*

Table 1 presents concrete specimen series that were investigated in this research. Eight series of concrete mixtures with water-binder ratio (W/B) of 40%, 50% and 60% using tap water (T) and natural seawater (S) as mixing water were prepared by three type curing conditions such as tap water curing at 20°C (TC), seawater curing at 20°C (SC) and air curing at 20°C, 100% R.H. controlled room (AC) were used as curing conditions. The binder proportions of mixture series were OPC concrete (N) and 50% $GGBS_{4000}$ replacement (B4). Mix proportion of concrete specimen was designed according to Japan Society of Civil Engineering (JSCE). The results of mix proportion of concrete specimen are demonstrated in Table 2.

**Table 1.** Mixture series of concrete specimen.

| Mixture | W/B (%) | s/a (%) | Mixing Water | Binder Proportion (%) | |
|---|---|---|---|---|---|
| | | | | OPC | GGBS |
| T40-B4 | 40 | 45 | Tap water | 50 | 50 |
| S40-B4 | | | Seawater | 50 | 50 |
| T50-N | 50 | | Tap water | 100 | -- |
| S50-N | | | Seawater | 100 | -- |
| T50-B4 | | | Tap water | 50 | 50 |
| S50-B4 | | | Seawater | 50 | 50 |
| T60-B4 | 60 | | Tap water | 50 | 50 |
| S60-B4 | | | Seawater | 50 | 50 |

**Table 2.** Mix proportion of concrete specimen.

| Mixture | Unit Content ( kg/m³ ) | | | | | | | | Fresh Properties | | |
|---|---|---|---|---|---|---|---|---|---|---|---|
| | W | Binder | | S | G | WR | AE* (liter) | Sp | Slump (cm) | Air (%) | Temp. (°C) |
| | | OPC | $B_{4000}$ | | | | | | | | |
| T40-B4 | 160 | 200 | 200 | 764 | 1010 | 2.00 | 0.77 | | 10.5 | 4.0 | 24.0 |
| S40-B4 | 165 | 206 | 206 | 759 | 1004 | - | 0.79 | 3.71 | 8.0 | 5.1 | 21.0 |
| T50-N | 160 | 320 | - | 805 | 1087 | 1.00 | 2.00 | - | 8.0 | 6.2 | 16.0 |
| S50-N | 165 | 330 | - | 802 | 1083 | 1.03 | 1.27 | - | 8.0 | 6.0 | 18.0 |
| T50-B4 | 160 | 160 | 160 | 800 | 1081 | 1.4 | 1.08 | - | 12.0 | 4.5 | 19.0 |
| S50-B4 | 165 | 165 | 165 | 797 | 1076 | 1.24 | 1.11 | - | 5.0 | 4.0 | 20.0 |
| T60-B4 | 160 | 133 | 133 | 814 | 1077 | 0.83 | 1.41 | - | 6.5 | 5.0 | 23.0 |
| S60-B4 | 165 | 137 | 137 | 811 | 1073 | 0.94 | 1.45 | - | 7.0 | 4.8 | 21.0 |

WR: Water reducer (AE+WR), *100 times dilution

### *3.3. Test methods*

After at a certain curing period for 28 days and 365 days in TC, SC and AC, concrete cylinder specimens in size of ϕ100x200 mm were tested in compressive strength. Compressive strength test of concrete specimens was conducted in accordance with JIS A 1108:2006. The average compressive strength and elastic modulus of three specimens were determined for each concrete mixture in three curing conditions.

In regard with total pore volume (porosity) of concrete specimen were tested by Mercury Intrusion Porosimetry (MIP). Concrete cylinder specimens were cut into 5 mm-thick slice samples. Subsequently, the fragments were immersed in acetone for 15 minutes to stop further hydration of cement, and then dried in the vacuum desiccator for 2 days. In the MIP test, complete drying of the samples was required to obtain results without error (3.20). In this study, the maximum applied pressure of MIP test was 33.000 psi (227 MPa), and the surface tension and contact angle of mercury was 485 dynes/cm and 130°, respectively. Porosity of concrete specimens was carried out at 28 days and 365 days period, which was three times measurement in a specimen.

## 4. Result and Discussion

### *4.1. Strength characteristics*

Figure 1 represents the concrete strength development of seawater mixed (S) and tap water mixed (T) OPC concrete and GGBS concrete in TC, SC and AC with a curing period of 28 days and 365 days. It was observed that all concrete specimens were increased in compressive strength up to 365 days from 28 days in three W/B of 40%, 50% and 60%. When concrete specimens were cured in TC, the achieved strength development up to 365 days of OPC concrete is approximately 25% and 24% for tap water mixed and seawater mixed, respectively. It means that strength development up to 365 days of OPC concrete is not influenced by type of mixing water, tap water or seawater. It is also found that strength performance up to 365 days of seawater-mixed OPC concrete is higher than that of tap water-mixed OPC concrete in all curing conditions. This expresses seawater-mixing improves compressive strength up to 91-days compared with tap water-mixing.

In contrast, different trend on strength development up to 365-day from 28-day of GGBS concrete was obtained. When seawater-mixed GGBS concrete was maintained in TC, compressive strength up to 365-day of seawater-mixed GGBS concrete was lower than tap water mixed GGBS concrete. It indicates that strength enhancement up to 365-day from 28 days of tap water mixed GGBS concrete in TC is improved compared to seawater mixed GGBS concrete. However, in case of seawater-mixed GGBS concrete was cured in SC and AC, strength performance up to 365 days seawater-mixed GGBS concrete was higher compared to tap water mixed GGBS concrete. It indicates that strength performance of seawater mixed GGBS concrete is improved in SC and AC. Generally, it can be concluded that there is no significant effect of seawater mixing in improving strength of GGBS concrete up to 365 days in water curing, TC and SC.

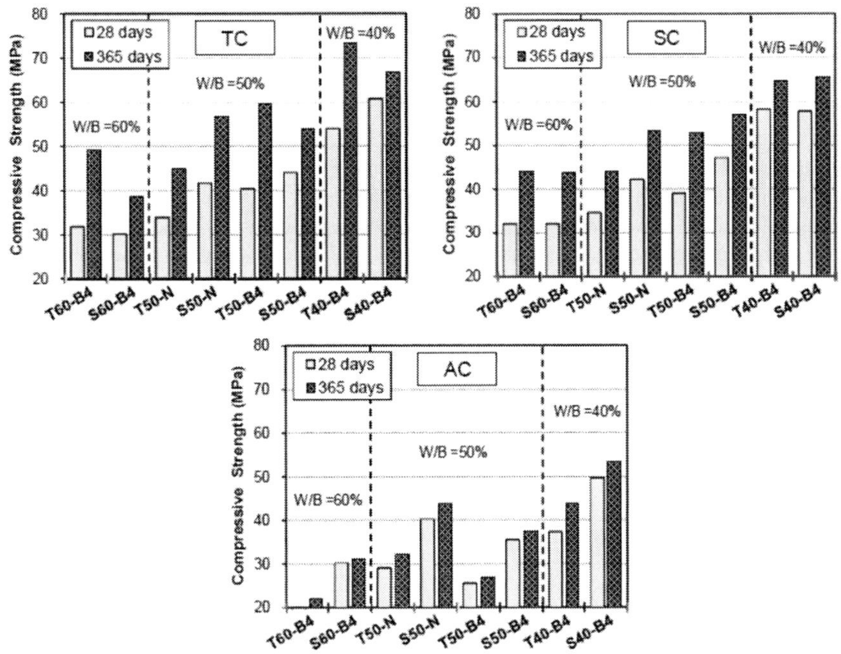


**Figure 1.** Strength development of concrete specimens.

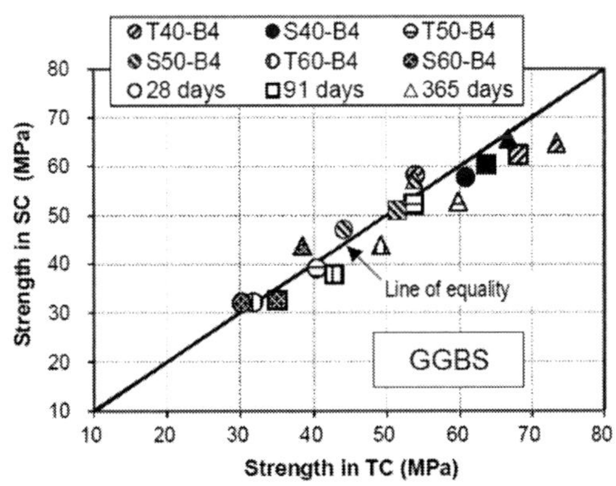


**Figure 2.** Strength in TC vs. Strength in SC.

The effect of curing water, TC and SC on strength performance of seawater mixed concrete and tap water mixed concrete is pointed out in figure 2. At 28-day curing period, compressive strength of GGBS concrete in SC were slightly higher than that in TC. Up to 365 days curing, compressive strength of concrete cured in TC slightly increased compared with SC. However, curing water (TC or

SC) does not have significant effect on compressive strength up to 365 days in all concrete mixture, in which the difference in compressive strength was less than 10 percent. It reveals that compressive strength of concrete mixed with tap water and mixed with seawater is not affected by type of curing water, tap water curing or seawater curing.

*4.1.1. Porosity* . Figure 3 shows the total volume of pores conducted by Mercury Intrusion Porosity (MIP) tests of concrete specimens at 28 days and 365 days. At 28 days, seawater mixed OPC concrete reduced in porosity of concrete compared with tap water mixed in all curing conditions, in TC, SC and AC. In contrast, 28-day porosity of tap water mixed GGBS concrete was lowered comparing to seawater mixed in water curing (TC and SC). When GGBS concrete specimens were cured in air, the porosity of concrete mixed with tap water was higher than that of concrete mixed with seawater. This phenomenon expresses that the porosity of concrete at 28 days is decreased by seawater mixing in AC. It is possibly inferred by accelerating the hydration of seawater mixed concrete, which densify the pore structure and reduces pore volume of concrete.

At 365-day curing period, the impact of GGBS on the porosity of seawater mixed and tap water mixed concrete could clearly be found. Porosity of seawater mixed GGBS concrete was lower than tap water mixed GGBS concrete. It suggests that porosity of seawater-mixed GGBS concrete up to 365 days is reduced in TC, SC and AC. The lowest concrete porosity in water curing, TC and SC were obtained for GGBS concrete in both seawater mixed and tap water mixed.

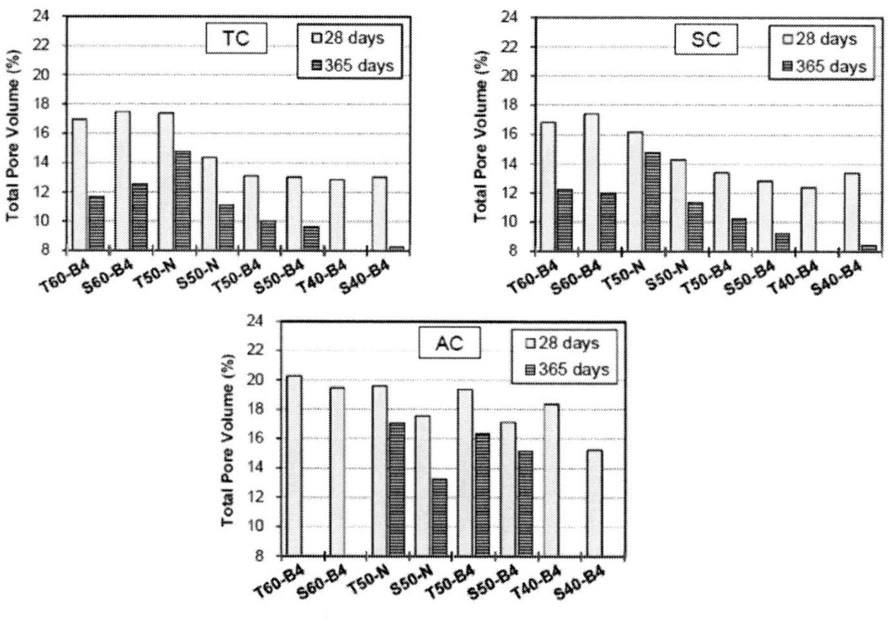


**Figure 3.** Total pore volume (porosity) of concrete specimens.

*4.2. Pore size distribution*

The typical pore size distribution at curing period of 28 days and 365 days of seawater mixed GGBS concrete and tap water mixed GGBS concrete in all curing conditions is shown in Figure 4. It was found that the pore distribution of tap water and seawater mixed are shown GGBS concrete in TC and SC is not significantly different. This indicates that pore size distribution of concrete is not affected by curing water, TC and SC. The maximum porosity and the large pore size of concrete specimen were achieved by specimens cured in air. This phenomenon is to be logical as hydration of cement paste in air curing, AC is not perfectly hydrated. No significantly distinct in pore size distribution of tap water mixing and seawater mixing was obtained. The trend of pore distribution of seawater mixed and tap water-mixed concrete were nearly similar. Nevertheless, the difference of various curing age (28 days and 365 days), the pore size distribution of concrete was altered for both seawater mixing and tap water mixing.

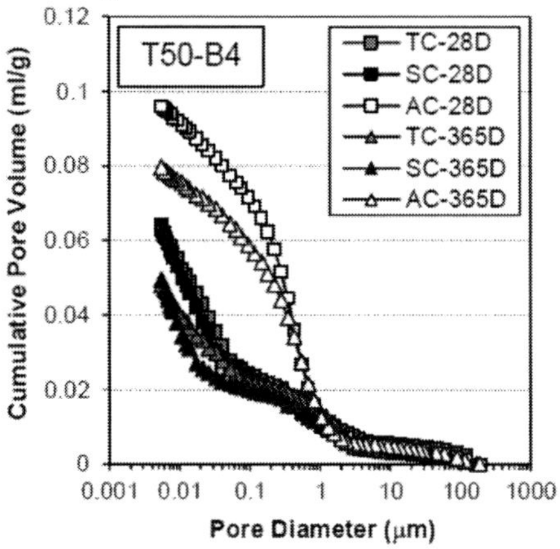


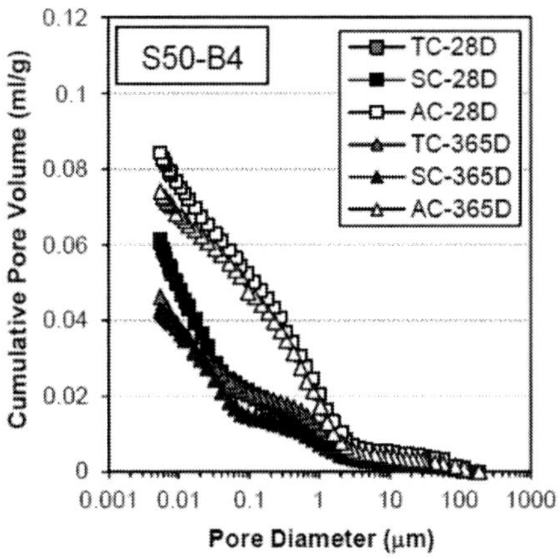


**Figure 4.** Pore size distribution of GGBS concrete specimens with 50% W/B.

### *4.3. Strength-porosity relationship*

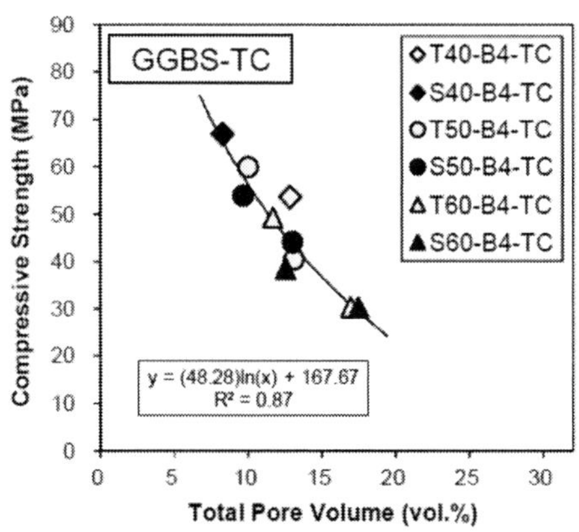


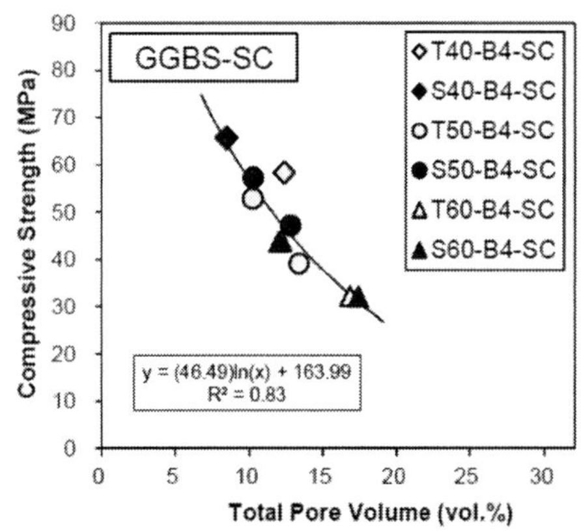


**Figure 5.** Strength-porosity relationship of GGBS concrete (at 28-day and 365-day).

Figure 5 demonstrates the relationship between compressive strength and total pore volume (porosity) of tap water mixed and seawater mixed GGBS concrete at 28 days and 365 days curing time in curing water. From this figure, it can be seen that, compressive strength was increased with a decreasing the porosity of concrete in logarithmic relation. A strong correlation is acquired in TC and SC, in which coefficient of correlation, $R^2$ value in TC and SC were more than 0.87 ad 0.83, respectively. It expresses that compressive strength of seawater mixed GGBS concrete and tap water mixed GGBS may be evaluated by its porosity. The higher compressive strength is, the lower porosity of concrete is, which is not influenced by mixing water, binder type and mix proportion of concrete. No significant difference of correlation coefficient, $R^2$ in TC and SC was obtained.

## 5. Conclusions

Based on the results of experimental study, some conclusion can be drawn:

1. There is no significant influence of seawater mixing in improving strength of GGBS concrete up to 365 days in water curing, TC and SC.
2. Compressive strength up to 365 days of seawater-mixed and tap water-mixed GGBS concrete is not affected by type of curing water, tap water or seawater.
3. Porosity up to 365 days of seawater mixed GGBS concrete is decreased compared to tap water mixed in TC, SC and AC.
4. Pore size distribution up to 365 days of seawater mixed concrete and tap water mixed GGBS concrete is not affected by type of curing water, TC or SC.

## 6. References

[1] Neville A 2001 Seawater in The Mixture *Magazine of Concrete International* **23**(1) pp 48-51
[2] Japan Society of Civil Engineers 2005 Standard Specifications for Concrete Structures-2002: *Material and Construction* (JSCE Japan) p 75
[3] Bureau of Indian Standards 2000 *Plain and Reinforced Practice, 4th revision,* (BIS India) p 14
[4] Bertolini L, Elsener B, Pedeferri P, Redaelli E, and Polder R 2013 *Corrosion of Steel in Concrete: Prevention, Diagnosis, Repair* (Germany: Wiley-VCH Verlag GmbH & Co. KGaA) p 205
[5] Noor N 2014 *Physical Performance and Durability Evaluation of Rubberized Concrete* Dissertation Thesis (Fukuoka: Kyushu University) p 82
[6] Nishida T, Otsuki N, Ohara H, Garba Z.M. and Nagata T 2015 Some considerations for the applicability of seawater as mixing water in concrete *Journal of Materials in Civil engineering* **27**(7) B4014004
[7] Adiwijaya, Yamamoto D, Dasar A, Hamada H and Sagawa Y 2013 Effects of seawater mixing and curing on strength and carbonation of fly ash concrete, *Journal of Structures and Materials in Civil Engineering* **29** 97-103
[8] Ramezanianpour A 2014 *Cement Replacement Materials: Properties, Durability, Sustainability* (Berlin: Springer Heidelberg) p 20
[9] Basheer L, Kropp J and Cleland D 2001 Assessment of the durability of concrete from its permeation properties: A Review, *Journal of Construction and Building Materials* **15** 93-103
[10] Kumar R and Bhattacharjee B 2003 Porosity, pore size distribution and in situ strength of concrete, *Journal of Cement and Concrete Research* **33** 155-164
[11] Kodraivendhan B, Divsholi B and Teng S 2013 Estimation of strength, permeability and hydraulic diffusivity of pozzolana blended concrete through pore size distribution, *Journal of Advanced Concrete Technology* **11** 230-237
[12] Adiwijaya, Hamada H, Sagawa Y and Yamamoto D 2014 Effects of Mineral Admixtures on Strength Characteristics of Concrete Mixed with Seawater *Proceeding of The 6th International Conference of Asian Concrete Federation (Seoul)* (Seoul: Asean Concrete Federation) pp 925-930

**Acknowledgments**

The first author would like to express his gratitude to Directorate General of Higher Education, Ministry of Research, Technology and Higher Education of Indonesian government for their financial support during my study at Kyushu University. Also, first author would like to address his highest appreciation and fully thanks to Professor Hidenori HAMADA for his priceless guidance, encouragement, and enthusiasm during the process of my three years research.

# Reuse of dredged marine soils as landfill liner: Effect of pH on *Escherichia coli* growth

**N M Anuar**[1] **and C M Chan**[1]

[1] Faculty of Engineering Technology, Universiti Tun Hussein Onn Malaysia, Parit Raja, 86400, Batu Pahat, Johor, Malaysia
Corresponding author: chan@uthm.edu.my

**Abstract.** A potential reuse area yet to be explored is the utilization of dredged marine soils (DMS) as geosorbent to retain pathogenic bacteria in landfill leachate. The use of DMS as geosorbent in landfill site could be considered as a new way of environmental friendly solid waste management. By laying DMS at the base of landfill like conventional clay liners, the geowaste could be simultaneously disposed of and act as passive geosorbent for microbes in leachate. DMS are known to serve as a hospitable environment for bacteria growth. Environmental factors such as soil's pH, salinity and particle size could affect the bacteria growth rate. This study investigated the effect range of pH value on the growth of indicator bacteria, *Escherichia coli* (*E. coli*) isolated from landfill leachate. The results showed that the number of *E. coli* grew higher in alkaline compared to acidic condition. Findings from this study will serve as a base for future studies for removing bacteria in leachate using DMS as geosorbent in a landfill site.

## 1. Introduction

Dredging operation is an important activity to maintain channel depth and expanding ports and harbours. For these purposes, a large volume of marine soils were dredged out yearly. According to the European Waste Catalogue, the Dredged Marine Soils (DMS) are classified as waste under section 17 05 05 (contaminated marine soils) and section 17 05 06 for other marine soils [1]. Therefore, most of the countries were used to dump back DMS at the disposal facilities (Figure 1). A high demand for dredging exist in the developing countries probably due to the growing maritime trade factor [2]. The project have been undertaken to maintain an adequate ferry and shipping channel depth in ports and harbours. In Malaysia, marine soils are dredged out once in a three years. Marine Department of Malaysia (Jabatan Laut Malaysia, JLM) have been authorized to perform the dredging process which took about six to ten months to be complete [3].

The dredging process in Malaysia commonly ends with the dredged material being disposed offshore in designated dumpsite. Nowadays the disposal of DMS into the sea is gradually being discontinued in developing countries due to concerns for the marine environment. Hence, such large volumes of DMS have created a greater challenge for sustainable disposal practice. Recently, a number of studies have reported the potential of DMS as a raw material for transformation into useful products of economic value. This is especially true for DMS which are marginally contaminated or treatable. For instance, [3] reported that the heavy metal concentrations in the DMS from Kuala Perlis, Malaysia were below the threshold values set by United State Environmental Protection Agency

(USEPA). Since marine soils such as them are considered harmless, the construction of landfill barriers can be potentially carried out by using this material.

**Figure 1**. Dredged marine soils from dredging activities.

As the minimization of the dredged material during dredging operation is impossible, extensive work has been done by researcher to develop various economically favorable for the reuse of DMS. It is a process where dredged material being reuse as raw material in obtaining productive materials such material for road construction, brick production, pavement base material and many more [4][5]. Other researchers have reported the use of marine soils to retain the migration of heavy metals in leachate. Du [6] and Du [7] provided the basis that marine clay soils could be used as landfill liner due to its low conductivity and high heavy metal sorption capability. A study by Chalermyanont [8] also propounded the potential of marine clay soils as landfill liner with better heavy metals retention than lateritic soils. However, the potential of DMS as a geosorbent material in landfill site to retain pathogen bacteria has yet to be thoroughly studied.

The role of marine soils as a reservoir for bacteria have been explored in previous study. Persistence of bacteria was found to be greater in marine soils than water column [9]. Marine soils are known to provide a hospitable environment for bacteria. The soils offer protection from predator, solar exposures, temperature, salinity, and pH that affect bacterial growth and survival. Removal of bacteria in landfill leachate would occur via adsorption when the leachate flow through the geosorbent. A group of researcher [10] investigated the soil properties affecting adsorption of *E. coli*. Thus, this paper aim to determine the influence of pH on the *E. coli* isolated from landfill leachate.

## 2. Materials and methods

### *2.1. Materials*

The dredged marine soils used in this study was obtained during dredging work from dredging site at Kuala Perlis, Perlis, Malaysia. The soil was packed in sampling bag and kept at 4°C until further analysis. The landfill leachate was collected from a conventional landfill (open dump) located in Johor, Malaysia. The landfill receive municipal solid waste as well as other non-hazardous waste. Prior to restricted issues, the exact location for the conventional landfill is not be mentioned in this paper. The samples were transport to the laboratory and kept at 4 °C until further analysis. Samples collection and preservation was conducted according to Standards Methods for the Examination of Water and Wastewater.

*2.2. Characteristic of DMS and landfill leachate*

The DMS were mixed before analysis to provide a homogeneity condition. The pH values of sediment were measured by mixing sediment and distilled water in 1:1 weight ratios. The pH value was taken after 60 minutes by inserting the probe in sediments solution. The pH of DMS and leachate was measured using the Eutech Instruments probe. The analysis of particle size, and moisture content of DMS were conducted according to the British Standard 1377:1990. Determination of bacteria from DMS and landfill leachate were conducted based on British Standards [11]. All the apparatus and distilled water used in this experiment underwent the sterilization process using an autoclave to avoid any contamination. Chromocult Coliform Agar (CCA) was used as media agar throughout the study for enumeration and isolation of *E. coli* in DMS. The use of CCA allows easy detection of *E. coli* from other coliform colonies.

*2.3. Isolation of Escherichia coli*

*E. coli* was isolated from the leachate sample using the streak plate method on the media agar namely Chromocult Agar (CCA). This method is rapid and simple in obtaining pure isolated colonies. Then the plate was incubated in inverted position for 24 hours at temperature 37 $^0$C. To avoid any contamination throughout the streaking process, the wire loop was sterilized by heating till red hot in the blue part of the Bunsen flame heat. Subculture of *E. coli* was obtained by transfer colonies from the plate culture contain CCA to a growth media. The growth media was sterilized before being inoculate with the bacteria colonies. The fresh inoculate medium allowed for bacteria growth as normal until such time the cells are used for experiments. After incubation period, single colony appeared on the CCA was then brought with sterilized loop into Nutrient Broth (NB). NB was prepared specified by manufacturer. Minimum 10 colonies were transferred into the NB. The incubation period was 18 to 72 hours. Bacteria growth in NB was indicated by turbidity.

*2.4. Effect of pH*

NB was used as a culture medium in pH effects experiment [12]. The pH of NB was adjusted by drop-wise addition of 1M hydrochloric acid (HCl) or sodium hydroxide (NaOH). The pH values obtained were 6, 7, 8 and 9. In order to measure pH effects on the survival of E. coli, the culture medium were inoculated with $10^6$ CFU per 1 ml of bacteria. The flasks were then sealed and incubate until Day 5 at 37$^0$C. The concentration of *E. coli* was counted after 24 h of incubation. Data obtained from the experiments was measured using optical density method and reported in Abs. The number of bacteria can also be counting in CFU/ml using the equation below:

$$y = 5 \, x \, 10^6 \, \chi^{3.0468} \tag{1}$$

where y = concentration of bacteria in Abs and $\chi$ = number of bacteria in CFU/ml. This equation was obtained from the relationship between Abs and colony counting using spread plate method. The relationship is as shown in figure 2.

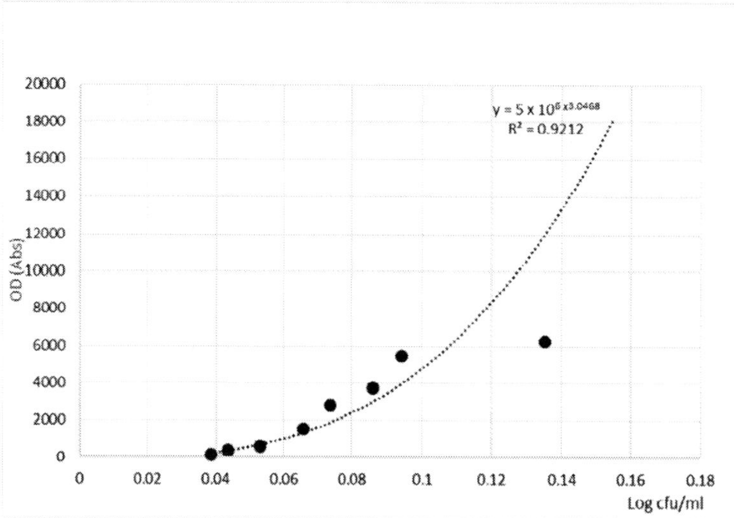


**Figure 2.** Standard curve for bacteria counting in OD and CFU.

## 3. Results and discussion

### *3.1. DMS and landfill leachate characteristic*

The marine soils used in this study contain a mixture of clay, silt and sand. Soil structure depends on the association between soil particles (sand, silt, and clay) in which aggregates of different sizes are formed. According to the Unified Soil Classification System (USCS), the fine-grained soil was classified according to the plasticity. The classification criteria are based on the relationship between Liquid Limit (LL) and Plasticity Index (PI). The relationship was determined from the plastic chart. Based on the USCS, the marine soil from Kuala Perlis was placed in highly plasticity silt categories. pH value of the DMS was found to be in the alkaline range with the moisture content of 218.1 %. The concentration of bacteria was 62 times higher in landfill leachate than in DMS. Lower concentration of bacteria was detected in DMS due to the disturbance of marine soils during dredging work. Due to this resuspension event, the bacteria mobilized into water column. The pH of landfill leachate was found to be in acidic while pH of DMS was in alkaline condition. The properties of DMS and landfill leachate was summarized in Table 1.

**Table 1.** Characteristic of DMS and leachate.

| **Properties** | **Kuala Perlis DMS** |
|---|---|
| Liquid Limit (%) | 66.5 |
| Plastic Limit (%) | 55.8 |
| Plasticity index | 10.7 |
| Soil classification | MH |
| pH | 8.0 |
| Moisture content (%) | 218.1 |
| *E. coli* (cfu/g) | $1.7 \times 10^2$ |
| **Properties** | **Landfill leachate** |
| pH | 6.5 |
| *E. coli* (cfu/ml) | $1.05 \times 10^4$ |

### *3.2. Effect of pH on E. coli growth*

In this study, pH of DMS appeared to be in alkaline while leachates' pH in acidic condition. The growth of *E. coli* normally range from pH 5 to 9.Despite offered protection from predator and solar exposures, alkalinity of DMS could provide a hospitable environment for the bacteria growth [14]. The survival of *E. coli* at low and high pH termed as "acid resistance" and "base resistance" respectively. There were differences in growth rate of *E. coli* among the six pH value tested range of pH 4 to 9. The range of pH were related with leachate and DMS (found in alkaline condition) characterisation. Time series (up to 28 days) results for range of pH effect on bacteria growth were shown in figure 3. In the experiment, the highest growth of *E. coli* was detected in alkaline compared to the acidic condition. More than 80 % of the bacteria was growth in the alkaline condition (pH of 8 and 9). For pH 6 and 7, the growth of bacteria was decreased within 50 % compared to the growth in alkaline condition. Lowest growth rate of bacteria was detected in acidic condition. These findings are consistent with the previous report where low pH shown to reduce the growth of *E. coli*. Ahmad [13] had found increasing bacteria concentration at neutral pH value and reducing bacteria concentration as pH levels dropped below 7. High concentration of bacteria in high pH levels indicated the bacteria are more survive in alkaline conditions as compared to the acidic conditions.

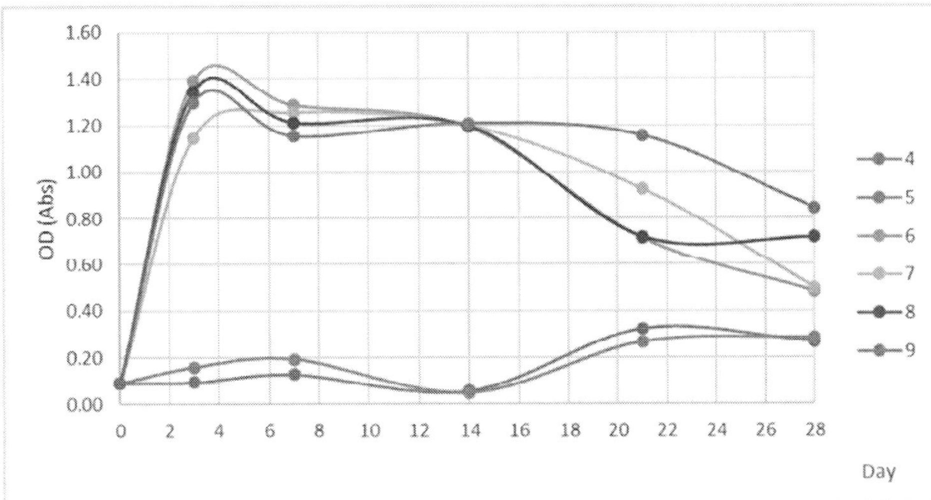


**Figure 3**. Number of bacteria growth in different pH condition.

## 4. Conclusions

The purpose of the current study was to determine the influence of pH on bacteria isolated from landfill leachate. The isolated bacteria were found to be capable to survive in the range of pH 4 to 9. The results showed that the number of bacteria tend to grew higher in alkaline condition (up to 80%) and vice versa. The findings from this study supported the idea that the bacteria can normally growth in alkaline condition. This will serve as a base for future studies for removing bacteria in leachate using DMS as geosorbent in landfill site.

## 5. References

[1] Dubois V, Abriak, N E, Zentar R and Balliy G 2009 The use of marine sediments as a pavement base material. *Waste management* **29** 774 – 782

[2] Manap N and Voulvoulis N 2016 Data analysis for environmental impact of analysis. *J. of Cleaner Production* **137** 392-404

[3] Ramli N, Haji Jumali M H and Wan Salim W S 2013 Fundamental characterization of dredged marine sediments from Kuala Perlis Jetty by XRF, XRD and FTIR *Advanced Materials Research* **620** 469-473.

[4] Siham K, Fabrice B, Edine A N and Patrick D 2008 Marine dredged sediments as new materials resource for road construction *Waste management* **28** 919 – 928

[5] Dubois V, Abriak N E, Zentar R and Balliy G 2009 The use of marine sediments as a pavement base material. *Waste management* **29** pp 774 – 782

[6] Du Y J, Hayashi S and Xu Y F 2004 Some factors controlling the adsorption of potassium ions on clayey soils *Applied Clay Science* **27** 209 – 231

[7] Du Y J, Hayashi S, Hino T and Tanaka K 2000 Contaminant sorption characteristics of Kyushu regional soils *Low land Technology Internationa.* **2**(2) 31-41

[8] Chalermyanont T, Arrykul S and Charoenthaisong N 2009 Potential use of lateritic and marine clay soils as landfill liners to retain heavy metals *Waste Management* **29** 117-127

[9] An Y J, Kampbell D H and Breidenbach G P 2002 Escherichia Coli and Total Coliforms in water and sediments at lake marinas *Environmental Pollution* **120** 771-778

[10] Kunito T, Ihyo Y, Miyahara H, Seta R, Yoshida S, Kubo H, Nagaoka K, Sakai M and Saeki K 2015 Soil properties affecting adsorption of plasmid DNA and its transformation efficiency in Escherichia coli *Biology and Fertility of Soils* **52(2)** pp 223-231

[11] British Standard BS 6068 1989 *Water Quality-Microbiological method (*London: British Standards Institution) pp 1-6

[12] Hrenovic J and Ivankovic T 2009 Survival of Escherichia coli and Acinetobacter junii at various concentration of sodium chloride, *Eurasian J. of Biosciences* **3** 144-151

[13] Ahmad A, Dada A C, Usup G and Lee Y H 2013 Validation of the Enterococci indicator for bacteriological quality monitoring of beaches in Malaysia using a multivariate approach *Earth and Environmental Science*, **2** p 425.

[14] Bradshaw J K, Snyder B J, Oladeinde A, Spidle D, Berrang M K, Meinersmann R J, Oakley B, Sidle R C, Sullivan K and Molina M 2016 Characterizing relationship among fecal indicator bacteria, microbial source tracking markers, and associated waterborne pathogen occurrence in stream water and sediments in a mixed land use watershed *Water research* **101** 498-509

**Acknowledgments**

The authors thanks to Universiti Tun Hussein Onn Malaysia for providing the laboratory facilities and also to the field and laboratory staff. This work was funded by Centre of Graduate Studies (CGS) and Office for Research, Innovation, Commercialization and Consultancy management (ORICC), UTHM.

# Axial compression behaviour of cross laminated wood-wool panel wallettes

**M S Md Noh[1], S Shahidan[1], N Ali[1], A F Kamarudin[1], S N Mokhatar[1], M F F A Razak[1], Z Ahmad[2] and A Ibrahim[3]**

[1]Jamilus Research Centre (JRC), Faculty of Civil and Environmental Engineering, Universiti Tun Hussein Onn Malaysia, 86400 Parit Raja, Batu Pahat, Johor, MALAYSIA
[2]Institute for Infrastructure Engineering and Sustainable Management (IIESM), Universiti Teknologi Mara, 40450 Shah Alam, Selangor, MALAYSIA
[3] Faculty of Civil Engineering, Universiti Teknologi Mara, 40450 Shah Alam, Selangor, MALAYSIA
Corresponding author: soffi@uthm.edu.my

**Abstract.** This study reported on the investigation of the structural behaviour of wallettes fabricated using wood-wool cement panel (WWCP) under axial compression load. WWCP is a wood based product, which produced by mixing a wood-wool, cement and water and compressed to form a stable panel with dimension of 600 mm x 2400 mm x 50 mm. In the fabrication of wallettes, two parameters were considered which are wallettes without surface plaster denoted as WNP and wallettes with 16 mm thickness of surface plaster on both sides denoted as WP. The 50 mm thickness of WWCP was used and cut into size 300 mm and 600 mm length. The cut panels were then horizontally laid in two layers (each layer consists of two panel strips) at different panel orientation to form a 600 mm x 600 mm wallettes. The front and rear layer were bonded together using 15 mm thick of normal mortar mix at a mixing ratio of 1: 3 (Cement : Fine sand). A total of five wallettes specimens were fabricated and tested under axial compression load test until failure. The testing results indicated that, the application of surface plaster on both side significantly increased the load carrying capacity of wallettes about 31 % compared to the wallettes without surface plaster.

## 1. Introduction

Every single year human tries to improve one by one the building construction. The improvements are made as to ensure better dream house for future generation [1]. From a sustainability point of view, the best improvement of the structure in a green building should balance between building development and the sustainable environment. In addition, the sustainable construction can help developing in creating greener structures. The aim of the environmental construction is to cut down the impact of built on human health and the natural surroundings [2-3].

One of the improvements that can be made is by replacing the less eco-friendly material to the material that more eco-friendly such as replacing the conventional brick to the wood-wool cement panel for wall construction [4]. Previously, the wood wool cement panel has been used for a long time ago, but this material not widely used in the building construction industry as a structural element [5]. There are several types of wood-wool that have been produced from different type of timber species throughout the world, but in Malaysia wood-wool is produced from timber species known as

"Kelampayan". The timber log is shredded from this species that also known as low density and fast grown timber species in Malaysia [6]. As a fast grown timber species, the tree can be harvested within five to six years and the replanting process will be conducted thereafter to ensure the availability of these resources for the next cycle.

Wood-wool cement panel is a composite material that built up from the combination of water, cement and wood-wool. The manufacturing of this panel undergoes the compression process of the mixture into the wooden mould to make a panel after it hardened. The wood-wool reacts as core material when mixing with cement paste and potential to be a lightweight material that essential for structural application. The previous studies [6-7] reported that, the strength properties of WWCP meet the minimum requirement stated in the standards of ISO 8335 and DIN 1101 and can be employed in the building industry.

The wall is the compulsory building element constructed in all housing construction. It provides shelter for the safety purposes and control the comfort level of the building occupants. The selection of the right material for wall construction is very important, especially for the country with hot and humid climate like Malaysia [8]. Due to this, the demand on the fired clay brick as a wall material is quite high due to its high strength and high thermal insulation properties [9]. However, the resources used in producing this brick are made from non-renewable resources, emit large quantities of carbon emission and heavy weight. The wall construction using this brick significantly labours intensive and consumed a lot of time. Therefore, the use of wood-wool cement panel as new material for wall construction is a great potential to improve the sustainability in the construction industry [3].

The acceptance in using WWCP as a wall element in Malaysia has been increased due to the advantages of this panel such as environmental friendly, low energy consumption, lightweight, low-cost and easy to process and fabricate. However, the construction technique and the behaviour of wall subjected to axial compression load are not well established [7]. Figure 1 shows the application of WWCP as wall element in the building construction in Malaysia.

**Figure 1**. Application of WWCP in low cost housing construction in Malaysia [4].

The compatibility between wood and cement is the main issue related to the production of wood-wool cement. There are only a few timber species that suitable to be used in producing this panel, since the organic materials released from the wood inhibit cement setting and reduced the bonding strength of cement [6, 7, 10]. To ensure the suitability of the timber species, the mechanical properties of wood-wool panel should be conducted prior being used in structural applications. The mechanical properties of wood-wool cement panel produced using Kelampayan timber species was conducted by several researchers and they were found that, the properties of WWCP achieved the requirement as specified in DIN 1101 standards and potential to be used as a structural element in building construction [6, 10, 13]. In terms of structural application of WWCP, the performance of reinforced concrete column using WWCP as embedded permanent formwork had been studied by Ahmad et al, [14]. The columns were tested under axial compression load until failure and the results revealed that the column using WWCP as permanent formwork significantly increased the load capacity about 30 % compared to the tested column without permanent formwork. Beside of that, the similar columns configuration was further evaluated under fire resistance performance test inside the fire furnace for two hours. The experimental observation shows that, the columns using WWCP as permanent formwork covered with surface plaster have capabilities to resist two hour fire rating without critical damage occurred [15].

The structural performance of wall constructed using WWCP under axial compression load was developed by Fathihah [16] via her dissertation project report. The construction technique used is similarly as currently practiced shown in figure 1. The results indicated that, the tested walls have low strength capabilities and the premature splitting failure between panel joint was observed. Due to this, the wall construction technique using WWCP has been improved by Md Noh et al, [10, 13]. In the previous studies, the construction technique, namely as a cross laminated technique has been proposed in the fabrication of 600 mm x 600 mm wallettes where two layers of 50 mm thickness of WWCP were horizontally laid at different panel orientation. The 15 mm thickness of EMACO R1 premix mortar was used as an adhesive to bond in between the two layers of panels. The wallettes with and without 16 mm thickness of surface plaster were tested under axial compression load until failure. The results indicated that, the arrangement of panel at different orientation significantly increased the load capacity of wallettes and the application of 16 mm plaster thickness (EMACO R1 premix mortar) gave additional restraint to the wallettes to resist 8 % higher loading capacity.

As a continuation from the previous study, the fabrication technique and size of wallettes was remained at 600 mm x 600 mm. However, in this study the mortar used for panel bonding agent and the surface plaster is a normal mortar with mixed cement to sand ratio of 1 : 3. The wallettes with and without surface plaster were fabricated accordingly and tested under axial compression load test in order to investigate its compression load capacity and behaviour. Through this experimental testing, the structural performance of the prefabricated wall has been identified and presented in this paper.

## 2. Materials and Methods

### *2.1. Materials*

WWCP is the main material used in fabrication of wallettes. The panel is a locally manufactured factory product which produced in standard panel size of 600 mm width and 2400 mm length and varies the panel thickness of 25, 50, 75 and 100 mm. However in this study, to fabricate 600 mm x 600 mm size of wallettes, the 50 mm thickness of WWCP has been used and cut into small panel size of 300 mm x 600 mm as shown in figure 2. The strength properties of WWCP used which obtained from the previous study is highlighted in Table 1.

The normal mortar mix was used as an adhesive and surface plaster where the ordinary Portland cement and fine sand were uniformly mixed at 1: 3 ratios. The strength properties in term of flexural and compressive strength of hardened normal mortar at 28 days were investigated throughout this study and the results obtained was presented in Table 2.

### *2.2. Fabrication of wallettes*

The fabrication of wallettes in this study was similarly as constructed in the previous study [13] where the two strips of 300 mm x 600 mm WWCP were horizontally laid in two layers at the different orientation. The front and rear panel strips were bonded together using 15 mm thickness of mortar to form the wallettes with the dimension of 600 mm x 600 mm x 115 mm. The two replicates of wallettes without surfaces plaster (W-NP) and three replicates of wallettes with 16 mm thickness of mortar (W-P) were prepared accordingly in Material Laboratory, UTHM. The configuration of wallettes and the fabrication process are shown in figure 3 and figure 4 respectively.

**Figure 2.** 50 mm thickness of WWCP used in this study (a) 600 mm x 2400 mm standard panel size (b) 300 mm x 600 mm cut panel size.

**Table 1.** Strength properties of wood-wool cement composite panel [10].

| WWCP Thickness | Density | Bending Properties | | Compressive Strength | | Tensile Strength |
|---|---|---|---|---|---|---|
| | | MOE | MOR | Perpendicular | Parallel | |
| (mm) | (kg/m$^3$) | (N/mm$^2$) | (N/mm$^2$) | (N/mm$^2$) | (N/mm$^2$) | (N/mm$^2$) |
| 50 | 328 | 444 | 1.15 | 0.84 | 1.00 | 0.060 |

**Table 2.** Strength properties of the hardened mortar at the age of 28 days.

| Application of mortar | Mix ratio (Sand : Cement : Water) | Flexural Strength (N/mm$^2$) | Compressive Strength (N/mm$^2$) |
|---|---|---|---|
| Adhesive | 1 : 3 : 1 | 41.68 | 37.88 |
| Plaster | 1 : 3 : 1 | 42.50 | 41.75 |

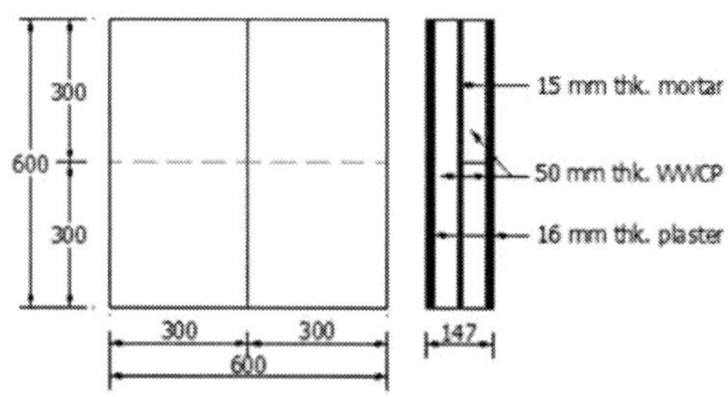


**Figure 3.** Configuration of wallettes constructed using a cross laminated technique (Units in milimeter).

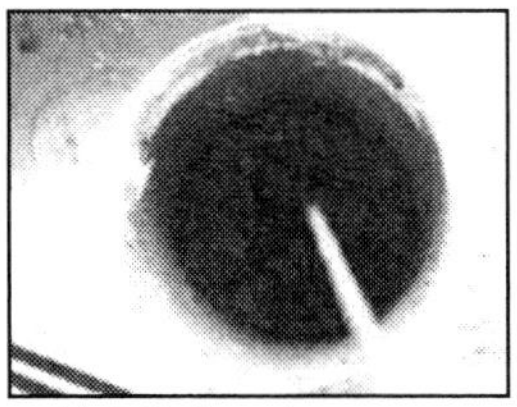
(a) Mixing the mortar

(b) Laying the rear panel strips in transverse direction

(c) Placing the stopper on top of rear panel

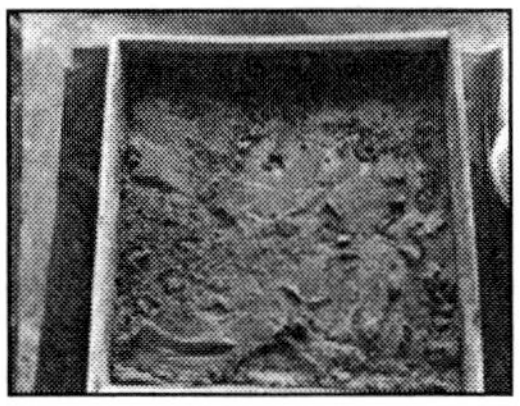
(d) Rendering the 15 mm thickness of mortar on top of rear panel

(e) Laying the front panel strips in longitudinal direction

(f) Plastering the wallettes on both surfaces

**Figure 4.** The Fabrication process of cross laminated wood-wool cement wallettes.

### *2.3. Axial compression load testing setup*

In the axial compression load test, a total of five wallettes specimens were prepared and undergoes curing process under air dried condition in the laboratory for 28 days. The wallettes specimens then vertically stand on top of steel beams with verticality adjustment using spirit level. To ensure the uniformity of the axial load application on top of wallettes specimens, the 20 mm thickness of steel plate was attached to steel spreader beam and directly connected to the 1000 kN load cell. The testing setup was instrumented with three LVDTs, where LVDT 1 was provided to measure the vertical displacement and LVDTs 2 and 3 were used to monitor the lateral displacement of the wallettes. The load was then applied at a uniform rate of 0.005 mm/sec up to failure and the axial compression load capacity, displacement and failure mode were recorded and observed [10]. The data logger that connected to the computer was used to record the load and displacement throughout the test conducted. The illustrated and actual test setup is shown in figure 5 below.

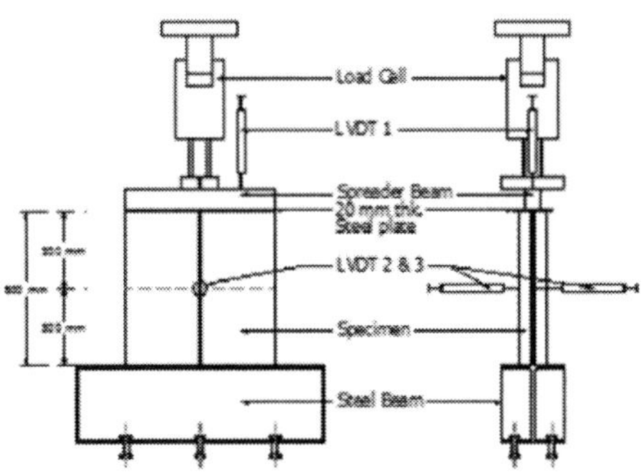

(a) Illustrated testing set-up

(b) Actual testing set-up

**Figure 5.** Axial compression load testing set-up of WWCP wallettes.

## 3. Results and Discussions

This section highlights the result of an axial compression load test of wallettes fabricated using wood-wool cement panel. The axial compression load capacity is a significant parameter in order to simulate the actual behaviour of wall under the action of permanent and variable gravity load of the building. The test conducted also can be used as an indicator to determine the maximum vertical load that can be carried by the walls. The wallettes without surface plaster (W-NP) and plastered with 16 mm thickness of mortar (W-P) have been observed in term of maximum load carrying capacity, load-vertical displacement behaviour and failure mode of each type of wallettes. A summary of maximum load and displacement at the maximum applied load of W-NP and W-P are shown in Table 3.

**Table 3.** Maximum load and vertical displacement of wallettes under axial compression load.

| Wall types | Wall reference | Maximum load (kN) | Vertical displacement (mm) | Maximum load (Mean) (kN) | Vertical displacement (Mean) (mm) |
|---|---|---|---|---|---|
| W-NP | W-NP1 | 236.50 | 6.27 | 275.00 | 12.17 |
| | W-NP2 | 313.50 | 18.07 | | |
| W-P | W-P1 | 285.50 | 12.57 | 360.30 | 14.83 |
| | W-P2 | 394.50 | 16.64 | | |
| | W-P3 | 401.00 | 15.27 | | |

From the Table 3, the results showed that the highest applied axial compression load of wallettes W-NP was recorded for W-NP2 which achieved its maximum load of 313.50 kN at a vertical displacement of 18.07 mm. Then the load significantly dropped to 236.50 kN at a small displacement of 6.27 mm for specimen W-NP1. Based on two replicates of tested wallettes type W-NP, the average maximum axial compression load recorded was 275.00 kN and average vertical displacement was 12.17 mm. Meanwhile, for wallettes type W-P, it can be seen that the highest maximum applied load was recorded on W-P3 with failure load of 401.00 kN at a vertical displacement of 15.27 mm. Then it was identically dropped by the W-P2 which reached its maximum load of 394.50 kN at maximum displacement of 16.64 mm. The maximum applied load was further decreased to 285.50 kN at a displacement of 12.57 mm for W-P1. The average maximum applied load and displacement of three replicates of wallettes W-P was 360.30 kN and 14.83 mm respectively.

The axial compression load capacity that was presented above provides information on the vertical load resistance of the prefabricated wall constructed using a cross laminated technique bonded with normal sand-cement mortar. From the test results, both of wallettes W-NP and W-P have their own capabilities to resist axial compression load. For wallettes without surface plaster W-NP the compressive strength calculated as high as 458 kN/m. The strength properties recorded shows the cross laminated technique used significantly increased the stiffness of the wallettes as well as increased the load carrying capacity of wallettes even tested without surface plaster. This is very beneficial since at the most condition in real construction practice the wall starts to support the vertical load after being constructed even without surface plaster. This performance will ensure that the wall able to safely resist an initial service before the full service is applied. For wallettes with surface plaster W-P, the calculated compressive strength increased about 31 % up to approximately 600 kN/m. This show that, an application of surface plaster on both surfaces of wallettes not just enhanced the appearance of wall surfaces but significantly increase the wall strength. Based on theoretical load transfer estimation, the minimum requirement of an axial compressive strength of the WWCP wall for two storey building is about 50 kN/m and the experimental results indicated that the wall strength recorded is 12 times higher. It is no doubt to classify that, the cross laminated WWCP wall system as a low weight load bearing wall system.

If compared against the previous study [10], the used of different mortar types as an adhesive and surface plaster had the greatest impact on the compressive strength of the wallettes. The used of normal mortar as panel bonding agent significantly increased the load carrying capacity of wallettes about 75 % compared to the wallettes that fabricated using EMACO R1 premix mortar. Further than that, the application of surface plaster using normal mortar also increase the load carrying capacity about approximately two times of wallettes using premix mortar. These shows that, the appropriate selection of mortar in the fabrication of WWCP wall, contributed to enhance the strength properties of wall as well as reduced the overall wall construction cost since the price of premix mortar is expensive compared to the normal mortar.

In terms of axial load - vertical displacement behaviour, figure 6(a) and 6(b) shows that, the vertical deformation of wallettes W-NP and W-P behave similar response where the applied load increase linearly with the vertical displacement up to the maximum failure load. After reaching the maximum load, the loads were observed drastically drop until final failure and the wallettes lost an ability to resist more load beyond this point. It can be seen that, the overall strength of the wallettes were governed by the strength of the normal mortar and once the mortar start to crush, the immediate failure will occur. Therefore, the precaution should be care taken as the actual imposed load should not exceed the maximum capacity of wallettes in order to avoid immediate fracture failure.

In terms of failure modes, figure 7(a) and 7(b) shows the failure behaviour of wallettes W-NP and W-P under axial compression load respectively. The failure mode of both wallettes shows the similar pattern where the vertical crack start to propagate along the cross section of the panels near to the panel and mortar bonding surface. As the applied load increase, the panel at each layer start to split out and at this stage load was totally carried by the mortar joint and the surface plaster up to the maximum load. The final failure was observed when the mortar reaches its ultimate strength and the load drastically drop due to crushing failure of the mortar joint.

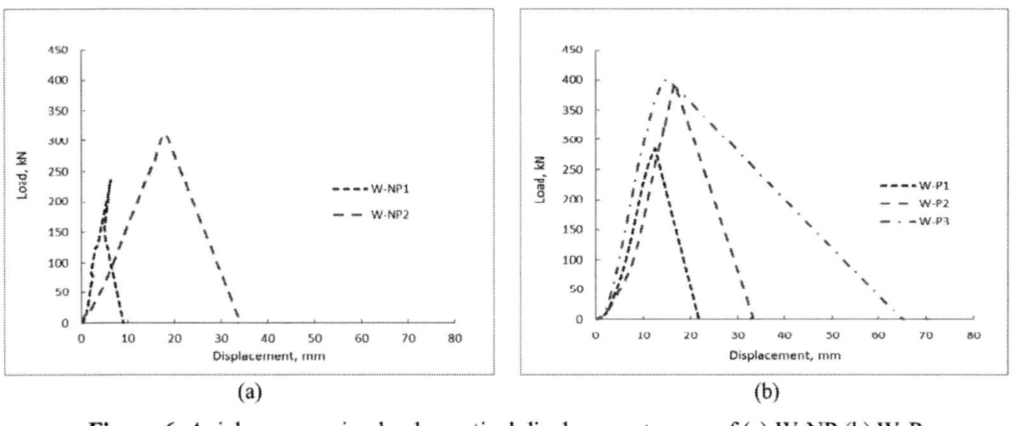


**Figure 6.** Axial compression load - vertical displacement curve of (a) W-NP (b) W-P.

*(a) Wallettes W-NP*

*(b) Wallettes W-P*

**Figure 7.** Failure mode behaviour of WWCP wallettes under axial compression load test.

## 4. Conclusions

Based on the experimental axial compression load test conducted on the wallettes, the conclusions can be drawn as follows;

- The new proposed fabrication technique, namely as cross laminated technique enhanced the wall construction process using wood-wool cement panel since it was easy to fabricate and install, lightweight and stable.
- The application of 16 mm thickness of surface plaster on both surfaces significantly increases the load carrying capacity of wallettes about 31 % up to 600 kN/m compressive strength.
- The compressive strength of wallettes is governed by the mortar used as a bonding agent and surface plaster.
- The used of normal mortar at mixed cement to sand ratio of 1:3 as a bonding agent in the fabrication of wallettes significantly 75 % increased the carrying capacity compared to wallettes fabricated using premix mortar.

## 5. References

[1] Goverse T, Hekkert M P, Groenewegen P, Worrell E and Smits R E H 2001 Wood innovation in the residential construction sector; opportunities and constraints *J. Resources, Conservation and Recycling* **34**(1) 53–74

[2] Okino E Y, Souza M R d, Santana M a, Alves M V d S, Sousa M E De and Teixeira D E 2004 Cement-bonded wood particleboard with a mixture of eucalypt and rubberwood *J. Cement and Concrete Composites* **26**(6) 729–734

[3] Idris N S, Boon K H, Kamarudin A F and Sooria S Z 2016 Ambient vibration test on reinforced concrete bridges *MATEC Web of Conference* **47** (Paris: EDP Sciences*)* 02012

[4] Md Noh M S, Ahmad Z, Ibrahim A and Walker P 2016 Development of new prefabricated wall constructed using wood-wool cement composite panel. *Proceedia Environmental Sciences* **34** (Netherland: Elsevier) pp 298 – 308

[5] Elten V 2006 Production of wood-wool cement board and wood strand cement board (Eltoboard) on one plant and application of product *Conference Proceeding of 10th International Inorganic-Bonded Fiber Composite Conference* IIBCC 2006 (Sao Paulo, Brazil) pp 206-217

[6] Ahmad Z, Wee L S and Fauzi M A 2011 Mechanical properties of wood-wool cement composite board manufactured using selected Malaysian fast grown timber species *ASM Science Journal* **5**(1) 27–35

[7] Ashori A, Tabarsa T, Azizi K and Mirzabeygi R 2011 Wood–wool cement board using mixture of eucalypt and poplar *J. Industrial Crops and Products* **34**(1) pp 1146–49

[8] Yusoff S M A M, Sooria S Z, Kamarudin A F and Boon K H 2016 Identification of dynamic characteristic of bridge crossing sungai simpang kiri using free vibration analysis *MATEC Web of Conference* **47** (Paris: EDP Scineces ) 02011
[9] Ali N, Zainal N A, Burhanudin M K, Samad A A A, and Shahidan S 2016 Physical and mechanical properties of compressed earth brick ( CEB ) containing sugarcane bagasse ash *MATEC Web of Conferences* **47** (Paris: EDP Sciences ) 01018
[10] Md Noh M S, Ahmad Z and Ibrahim A 2014 Axial compression behaviour of wallettes constructed using wood-wool cement composite panel *Advanced Materials Research* **1051** 671-677
[11] Hadipramana J, Mokhatar S N, Samad A A A and Hakim N F A 2016 An Exploratory compressive strength of concrete containing modified artificial polyethylene aggregate (MAPEA) *IOP Conference Series: Materials Science and Engineering* Vol. **160** (United Kingdom: IOP Publishing) 012065
[12] Mokhatar S N, Hadipramana J, Mohd Isa S N and Mustafa M M 2016 The Potential of artificial polyethylene aggregate (APECA) on compressive strength of concrete after exposed by temperature *MATEC Web of Conferences* **47** (Paris: EDP Sciences) 01005
[13] Md Noh M S, Ahmad Z, Ibrahim A, Walker P, 2016 Axial Compression Behaviour of Plastered Wood-Wool Cement Composite Panel Wallettes. *Proc. of Int. Civil and Infrastructure Engineering Conference* (Singapore: Springer) pp 483-495
[14] Ahmad N, Ahmad Z, Rahman A A, Hamid A H and Fauzi M A A 2013 . Increasing the capacity of concrete column with integrated permanent formwork using wood-wool cement board. *J. Applied Mechanics and Materials* **325-326** 1305-1309
[15] Ahmad Z, Ahmad N, Ali Rahman A, Abdul Hamid H and Md Noh M S 2014 Fire resistance performance of reinforced concrete column with embedded permanent formwork using wood-wool panel *Applied Mechanics and Materials* **661** 111-117
[16] Fatihah W W M 2011 Behaviour of Wall Constructed Using Wood-Wool Cement Board *Master Dissertation* (Malaysia: Universiti Teknologi Mara)

**Acknowledgements**
The authors would like to acknowledge the Duralite (M) Sdn. Bhd. for the WWCP sponsorship, Faculty of Civil and Environmental Engineering, Universiti Tun Hussein Onn Malaysia (UTHM) for the experimental equipment and testing. Many thanks to research members of the Jamilus Research Centre for their great co-operation.

IOP Conf. Series: Materials Science and Engineering **271** (2017) 012073 doi:10.1088/1757-899X/271/1/012073

# Compressive strength, flexural strength and water absorption of concrete containing palm oil kernel shell

**Nurazuwa Md Noor[1], Jun Xiang-ONG[1], Hamidun Mohd Noh[1,2], Noor Azlina Abdul Hamid[1], Salsabila Kuzaiman[1,3] and Adiwijaya Ali[1,3]**

[1] Faculty of Civil and Environmental Engineering, Universiti Tun Hussein Onn Malaysia, Parit Raja, Batu Pahat, 86400, Johor, Malaysia
[2]Faculty of Technology Management and Business, Universiti Tun Hussein Onn Malaysia, Parit Raja, Batu Pahat, 86400, Johor, Malaysia
[3]Jurusan Teknik Sipil, Politeknik Negeri Ujung Padang, Jalan Perintis Kemerdekaan, KM 10 , Tamalanrea, 90245 Makassar, Indonesia
Corresponding author: nurazuwa@uthm.edu.my

**Abstract.** Effect of inclusion of palm oil kernel shell (PKS) and palm oil fibre (POF) in concrete was investigated on the compressive strength and flexural strength. In addition, investigation of palm oil kernel shell on concrete water absorption was also conducted. Total of 48 concrete cubes and 24 concrete prisms with the size of 100mm x 100mm x 100mm and 100mm x 100mm x 500mm were prepared, respectively. Four (4) series of concrete mix consists of coarse aggregate was replaced by 0%, 25%, 50% and 75% palm kernel shell and each series were divided into two (2) main group. The first group is without POF, while the second group was mixed with the 5cm length of 0.25% of the POF volume fraction. All specimen were tested after 7 and 28 days of water curing for a compression test, and flexural test at 28 days of curing period. Water absorption test was conducted on concrete cube age 28 days. The results showed that the replacement of PKS achieves lower compressive and flexural strength in comparison with conventional concrete. However, the 25% replacement of PKS concrete showed acceptable compressive strength which within the range of requirement for structural concrete. Meanwhile, the POF which should act as matrix reinforcement showed no enhancement in flexural strength due to the balling effect in concrete. As expected, water absorption was increasing with the increasing of PKS in the concrete cause by the porous characteristics of PKS

## 1. Introduction

Concrete is construction materials which consist of cement, sand, gravel and water [1]. The concrete industry today is the largest consumer of limited natural resources caused by the high concrete demand and natural resources such as for gravel has reduced due to this tremendous demand [2]. It is crucial to find surrogate materials which have sustainable features. As a result, it motivates researchers to focus their investigation on the use of waste and by- product materials into possible construction material. Recently, the waste materials such as fly ash, clinker and expanded slag cinder have led for sustainable materials because it provides the purpose of both the structural stability and economic viability. Hence, the best alternative to achieve sustainable development of the concrete industry is the use of waste and by-product materials instead of natural resources in the concrete mixture. In this study, a by-product named palm oil kernel shell (PKS) was used as the replacement of coarse aggregate. It is irregularly shaped like an oval, circular, polygonal or flaky shaped with 0.15-8mm thickness [2]. The previous study showed that the PKS is suitable for light and dense concrete production and also as a road building material [3]. In addition, PKS concrete can achieve an acceptable compressive strength value which is among the design strength of 21 to 35MPa with the density of 800 to 2240kg/m3 [4] .

Besides adding the PKS, palm oil fibres (POF) is added to determine the flexural strength of concrete. POF has a high potential which can slightly improve the flexural strength of concrete in order to affect overall cement fibre properties [5]. Hence, it has a higher demand for production of concrete since Malaysia being the largest palm oil producer in the world [1] .

Another reason for this study is the suitability and effectiveness of PKS and POF which act as a construction material for concrete production. Based on the previous research, PKS is used as lightweight aggregate to build a one-storey building and a foot bridge which is being monitored for their structural behaviour [2]. While for the POF, it is considered as natural fibre which can behave like reinforcement as well as improve the mechanical properties of concrete composite [5]. Hence, this study more concentrates on the comparison of conventional concrete, PKS concrete and PKS concrete containing with POF related with their workability, water absorption and strength as well as the performance of these two materials (PKS and POF) when subject to compressive and flexural strength test. In a nutshell, four series of concrete mix consists of 0%, 25%, 50% and 75% PKS is replaced by volume according to coarse aggregate content. Besides, the concrete is divided into two categories; PKS concrete mixed with and without POF. The use of 5cm length and 0.25% of the volume fraction of POF was conducted in this study.Formatting the title, authors and affiliations

Please follow these instructions as carefully as possible so all articles within a conference have the same style to the title page. This paragraph follows a section title so it should not be indented.

## 2. Material Preparation and Testing

### *2.1. Materials*

Ordinary Portland cement of Grade 42.5 that conforming Malaysia Standard MS 197-7: 2007 was used as a binder. Fine aggregates origin from river sand was sieved through 5mm. Meanwhile, coarse aggregates, were sieved and minimum size of 10 mm and maximum size of 20 mm was selected. PKS in this study used as partially 0%, 25%, 50% and 75% coarse aggregate replacement. While, POF acted as an additive to determine its effect on the compressive and flexural strength of concrete when combined with PKS. The volume fraction of 0.25% with the 5cm length of POF was used in this study. Details of the density of each material are presented in table 1.

**Table 1.** Materials Density.

| Materials | Density ($kg/m^3$) |
|---|---|
| Ordinary Portland cement | 3160 |
| Water | 1000 |
| Sand | 2441 |
| Gravel | 3087 |
| Palm oil kernel shell (PKS) | 1203 |

**Figure 1.** Palm oil kernel shell (PKS)

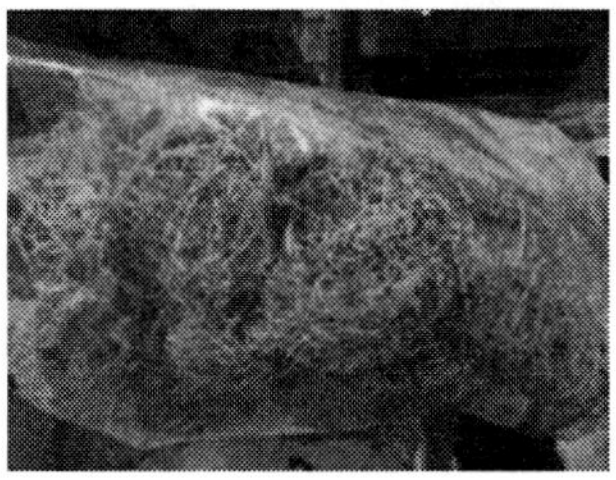

**Figure 2.** Palm oil fibre (POF)

*2.2. Experimental*

The compressive strength test was performed based on BS EN 12390-3:2009 Testing Hardened Concrete Part 3: Compressive Strength of Test Specimens [6]. Specimen size of 100 x 100 x 100 mm tested after 7 and 28 days of hydration in tap water. The compressive strength of concrete cube was calculated by the following formula:

$$f_c = F/A_c$$

where $f_c$ is compressive strength in MPa; $F$ is the maximum load at failure in N; $A_c$ is the concrete cross sectional area in mm$^2$.

As for flexural strength test, a specimen of size 100 x 100 x 500mm are tested on third-point loading test based on BS EN 12390-5:2009 Testing Hardened Concrete: Flexural Strength of Test Specimens [7]. It involves load at a rate of 0.06 ± 0.04MPa per second until failure. The flexural strength of beam prism was calculated by the following equation:

$$fcf = Fl/bd^2$$

where $ff$ is flexural strength in MPa; $F$ is the maximum load at the fracture point in N; $l$ is the length of the support in mm; $b$ is the width of the beam in mm; $d$ is the thickness of the beam in mm

## 3. Result and discussiom

*3.1. Workability*

Figure 3 showed the workability of the concrete mixtures. In concrete mixture without POF, the workability of 25%, 50% and 75%-PKS had achieved 90mm and 80mm slump which is lower than the slump of conventional; 120mm. It is because the porous feature of PKS than conventional aggregate, causing the concrete does not slump as much as conventional-weight concrete. However, the slump of all concrete mixtures without adding POF still achieve the design slump value ranging between 60 to 180mm. Four concrete mixtures with adding POF showed equal slump which is 30mm that not fulfilled the design slump value. Based on previous research conducted by Jayamani, POF is one of the types of natural fibre which has reduced water resistance, low durability, and poor fibre-matrix interactions. It can establish that composite materials are sensitive to humidity through absorption of water leading to differential swelling between the fibres and the matrix. Therefore, POF is good in water absorption. The dry mixture made the particles of aggregate difficult in the bonding process, resulting in the variation of a slump and concrete strength. In other words, POF absorbs the free water, lowering the water-cement ratio, causing the concrete mixture becomes dry.

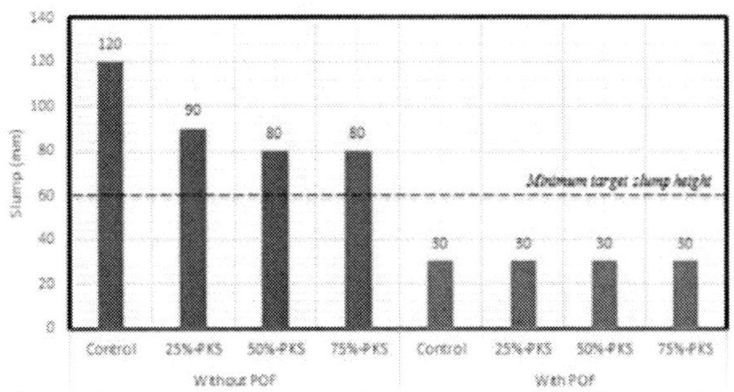


**Figure 3.** Workability performance of concrete with/without PKS and POF

*3.2. Physical Properties*

*3.2.1. Compressive Strength.* Results of compressive strength is presented in figure 4 and figure 5, where conventional concrete gave 28 days' average compressive strength 43.8MPa and 42.5MPa for concrete without POF and concrete with POF, respectively. The strength was decreasing with the inclusion of PKS. Concrete with 25% PKS (no POF) achieved 25.6MPa lower than target compressive strength but it still above the accepted strength for lightweight structural purposes. Based on the previous study, the minimum compressive strength of lightweight structural concrete for 28 days is 17.0MPa [1]. Therefore, the replacement of coarse aggregate to 50% and 75% of PKS without and with adding POF cannot be considered due to their low compressive strength which achieves below 17MPa at 28 days compared to another specimen. Hence, the replacement of coarse aggregate with PKS in volume fraction (50% and 75%) cannot be applied to structural design due to its failure to achieve above general concrete strength requirement.

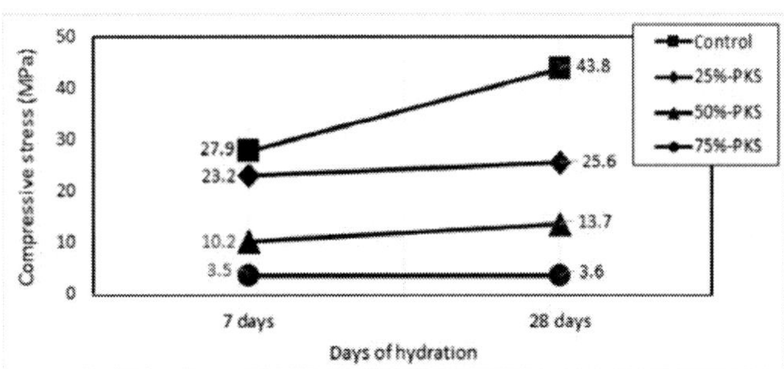


**Figure 4.** Concrete compressive stress without POF

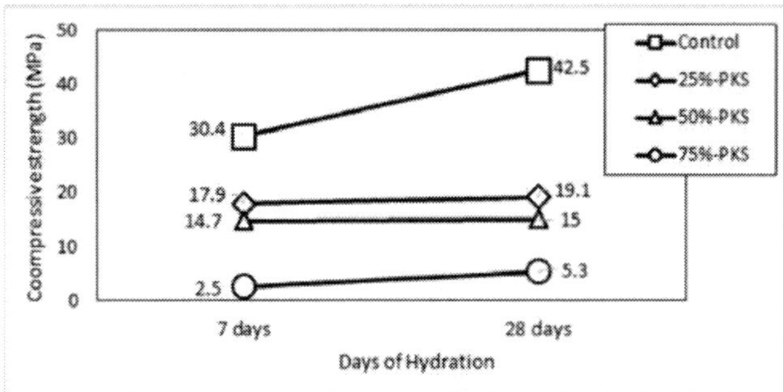


**Figure 5.** Concrete compressive stress with POF

*3.2.2. Flexural Strength.* Figure 6 showed that the addition of 5cm and 0.25% of the volume fraction of POF into each concrete mixture does not give any significant improvement in flexural strength. The 50%-PKS and 75%-PKS concrete show a slight increase in flexural strength when adding POF. While for conventional and 25%-PKS concrete with adding POF, the flexural strength drops to 6.8MPa and 5.2MPa, respectively. This was explained by Huzaifa (2008), who concluded that the reduction of the modulus of rupture might be due to the length of fibre used. A longer length of POF tends to give the small value of flexural strength due to too much of POF that contributes to weaker bonding between particles [8].

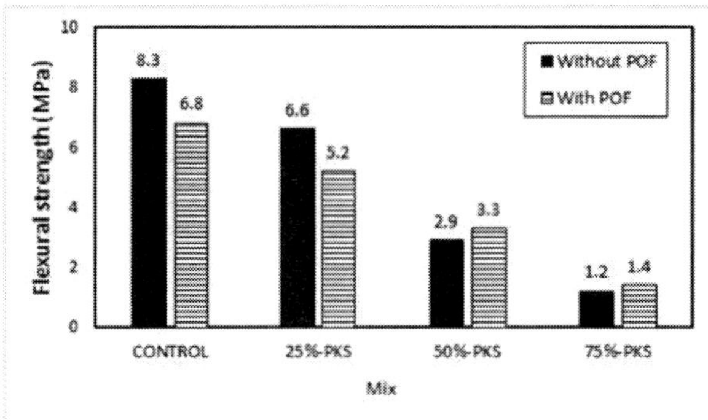


**Figure 6.** Flexural stress of concrete with and without POF

*3.2.3. Water Absorption.* Results on water absorption test is presented in figure 7. This evaluation was only on concrete with PKS, without any addition of POF. Conventional specimen showed slightly decrement when it reached 56 days. However, a similar pattern was observed on all specimen with PKS where water absorption kept increasing from day 7 to day 56. When inclusion of PKS in concrete increasing from 25% to 75%, water absorption was increasing due to the fact that PKS is a porous material which has the ability to absorbed water better than gravel. Thus, it is important to ensure PKS used in mixing was in the complete saturated surface dry state to avoid free water being absorbed by PKS.

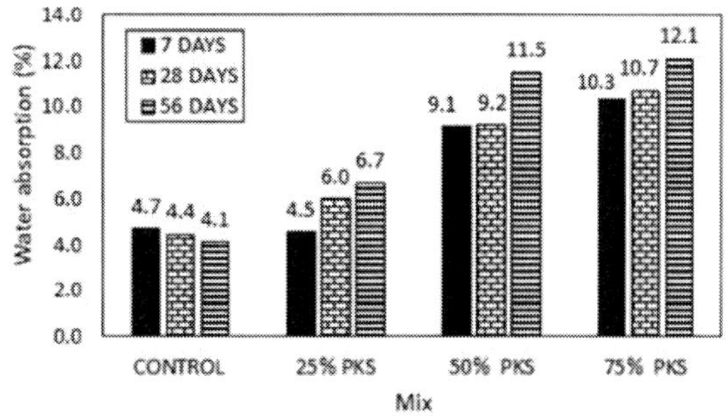


**Figure 7.** Water absorption of PKS concrete without POF

## 4. Conclusion

The following conclusions can be derived from the experimental results:

1. Workability of all PKS concrete without POF has a good workability. While for PKS concrete with POF, the slump was below than expected target.
2. The inclusion of PKS reduces the compressive strength of concrete. However, the replacement of 25% PKS gives acceptable values as it falls within the range of requirement for lightweight structural concrete. The 25%-PKS without and with POF achieve acceptable strength value which is 25.6MPa and 19.1MPa, respectively, at 28 days of age.
3. Adding POF into concrete mixture showed no improvement in flexural strength of PKS concrete. The congestion of POF may lead to reducing of bonding and disintegration. As a result, the 5cm length of POF in this study gave no significant result of flexural strength due to the inappropriate POF length which contributes to the possible balling effect.
4. Meanwhile, water absorption was increasing with the increasing percentage of PKS in concrete.

## 5. References

[1] Khan M, Deepak T and Nair S 2016 Use of palm shell as replacement of coarse aggregate for investigating properties of concrete *International Journal of Applied Engineering* Research **11** 2379–2383

[2] Alengaram U J, Muhit B A A and bin Jumaat M Z 2013 A comparison of the thermal conductivity of palm shell foamed concrete with conventional materials *Construction and Building Materials* **38** 161–172

[3] Olanipekun E, Olusola K and Ata O 2006 A comparative study of concrete properties using coconut shell and palm kernel shell as coarse aggregates *Building and Environment* **41** 297–301

[4] Itam Z, Beddu S, Kamal N L M, Alam M A and Ayash U I 2016 The feasibility of palm kernel shell as a replacement for coarse aggregate in lightweight concrete *IOP Conference Series: Earth and Environmental Science* vol 32 (London: IOP Publishing) 012040

[5] Raut A N and Gomez C P 2016 Thermal and mechanical performance of oil palm fiber reinforced mortar utilizing palm oil fly ash as a complementary binder *Construction and Building Materials* **126** 476–483

[6] EN BS 2009 *Testing hardened concrete–Part 3: Compressive strength of test specimens* (London: British Standard Institution )

[7] EN BS 12390-5 2009 *Testing hardened concrete. Flexural strength of test specimens* (London:British Standard Institution).

[8] Hashim H 2008 *The effect of palm oil fibre on concrete properties*

**Acknowledgments**

Authors would like to express their gratitude to Universiti Tun Hussein Onn Malaysia for supporting this study.

IOP Conf. Series: Materials Science and Engineering **271** (2017) 012074 doi:10.1088/1757-899X/271/1/012074

# A review on the suitability of rubberized concrete for concrete bridge decks

**Mohamad Syamir Senin**[1], **Shahiron Shahidan**[1,*], **Siti Radziah Abdullah**[1], **Nickholas Anting Guntor**[1], **Alif Syazani Leman**[1]

[1]Jamilus Research Center-Sustainable Construction, Faculty of Civil and Environmental Engineering, Universiti Tun Hussein Onn Malaysia.
Corresponding: shahiron@uthm.edu.my

**Abstract.** Road authorities manage a large population of ageing bridges, a substantial number of which fail to meet the current requirements either due to deterioration and other structural deficiencies or as a result of the escalating demands imposed by increased traffic.This problem is related to the dynamic load from vehicles. This problem can be solved by producing a type of concrete that can reduce the amplitude of oscillation or vibration such as rubberized concrete. Green construction has been a very important aspect in concrete production field in the last decade. One of the most problematic waste materials is scrap tires. The use of scrap tires in civil engineering is increasing by producing rubberized concrete. Rubberized concrete is a type of concrete that is mixed with rubber. The purpose of this review is to justify the suitability of rubberized concrete for concrete bridge decks. Several parameters named physical, chemical and mechanical properties were measured to ensure the suitability of rubberized concrete for concrete bridge decks.Rubberized concrete has similar workability to normal concrete. The rubber reduced the density and compressive strength of the concrete while increased the flexural strength, water absorption and damping ratio. The used of rubber in concrete beyond 20% is not recommended due to decreasing in compressive strength. Rubberized concrete recommended to be used in circumstances where vibration damping was required such as in bridge construction as shock-wave absorber.

## 1. Introduction

Today, many concrete bridge structures are ageing due to large impact load. As a result, the maintenance, upgrading and replacement of existing bridges have become a very challenging task for the construction industry. Complications associated with these tasks are even more critical in urban areas, where the overall cost of the project is often governed by indirect costs due to traffic disruption. New upgrading and refurbishment methods integrated with accurate urban planning minimize the traffic disruption and disturbance in highly populated areas and are therefore very useful for bridge authorities and owners. Engineers need to find a suitable material which has resistance towards the impact, vibration or damping such as rubber material. The United States generates approximately 300 million scrap tires annually, about 40% of which are used as fuel for generating energy, 26% ground into crumb rubber, 13% discarded in landfills and 5.5% used in civil engineering applications [1-4]. There are more that 300 million tires which are currently stockpiled throughout the United State [2, 5]. These stockpiled waste tires occupy large landfill spaces that contain nesting insects and rats. Stockpiles of tires destined for landfills are also known to be flammable. There are many waste materials which become problematic to the environment [6]. One option to reduce this environmental concern is for the construction industry to consume a high number of recycled tires accumulated in stockpiles. Utilizing waste materials in concrete provides significant contribution to the environment and indirectly helps to generate economic value for some other industries [7]. In the pavement industry, the use of crumb rubber has been initiated with asphalt mixes. However, some difficulties, such as the high viscosity of the rubberized bitumen and the requirement of higher temperature for production of rubberized asphalt have limited its application [8]. Consequently, the introduction of rubber to concrete pavements was the basis of this investigation.

The replacement of fine aggregates with waste tire rubber was done at a volume of 5% whereas the strength was increased up to 10% [9]. The addition of shredded scrap tires to concrete provides some favorable characteristics for concrete and alters some concrete properties. The ordinary cement-based concrete is generally brittle. However, the addition of rubber in the concrete material is able to increase the ductility and impact resistance [10]. The rubberized concrete was recommended to be used in circumstances where vibration damping was required, such as in bridge constructions as a shock-wave absorber [11]. Experimental studies on rubberized concrete materials have shown that using rubber in concrete as partial replacement of mineral aggregates enhances its ductility, toughness, impact resistance, energy dissipation, and damping ratio [12]. Using recycled tire rubber in cement concrete can provide an efficient way of utilizing rubber. In addition to environmental benefits, the use of tire rubber particles could provide a new type of concrete with unique mechanical and fracture characteristics. Although recycled tire rubber has been widely used in highway asphalt, there are limited studies on its application in cement concrete [13].

The current applications of recycling waste tires in civil engineering practice are mainly as follows:

- Used as modifiers for asphalt paving mixtures;
- Used as an additive for Portland cement concrete;
- Used as light weight fillers; and
- Used in whole tires as crash barriers, bumpers, artificial reefs and etc.

The purpose of this review is to justify the suitability of rubberized concrete for concrete bridge decks. There are several parameters which need to be measured such as physical, chemical and mechanical properties before determining the use of rubberized concrete for concrete bridge decks.

## 2. Waste tires

There are many ways to recycle scrap tires, but the preferred method for recycling tire rubber is by grinding it before converting it into various applications. There are several sizes of rubber particles as shown in Figure 1 after grinding. The different sizes of rubber particles will result in different effects to the concrete. Ganjian *et al.*[14] have classified scrap tires into two: car tires and truck tires. Car tires are different from truck tires in terms of its constituent materials, especially natural and synthetic rubber content as shown in Table 1. Considering the high production volume of car tires compared to truck tires, the former is usually of more interest to researchers. In most of the researches performed, usually three broad categories of discarded tire rubber are considered such as chipped, crumb and ground rubber:

- Shredded or chipped rubber replaces gravel. To produce this rubber, the tire has to be shredded in two stages. By the end of stage one, rubber has a length of 300-340 mm and a width of 100-230 mm. In the second stage, its dimension changes to 100-150 mm by cutting. If shredding is further continued, particles of about 13-76mm in dimension are produced. These are called "shredded particles".
- Crumb rubber that replaces sand is manufactured by a special mill in which big rubber is converted into smaller torn particles. In this procedure, different sizes of rubber particles may be produced depending on the kind of mill used and the temperature

generated. Using a simple method, particles are made with high irregularity in the range of 0.4250-4.75 mm.

- Ground rubber that may replace cement is dependent on the equipment for size reduction. The processed used tires are typically subjected to two stages of magnetic separation and screening. Various sizes of rubber are produced using more complex procedures. In the micro-milling process, the particles made are in the range of 0.075-0.475 mm.

Tires are made of rubber (60-65 wt%), carbon black (CB) (25-32 wt%) and the rest consists of accelerators and fillers, which are added during the manufacturing process. The rubbery materials are present in the form of $C_xH_y$ with some fibrous materials and they are considered as thermoset polymer [15].

**Table 1** Typical constituent materials of tires [16].

| Composition weight (%) | Car tire | Truck tire |
|---|---|---|
| Natural rubber | 14 | 27 |
| Synthetic rubber | 27 | 14 |
| Black carbon | 28 | 28 |
| Fabric, filler accelerators, and antiozonants | 16-17 | 16-17 |
| Steel | 14-15 | 14-15 |

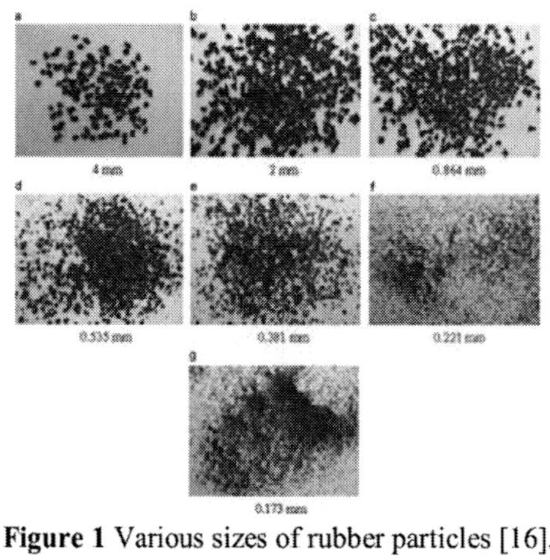

**Figure 1** Various sizes of rubber particles [16].

## 3. Properties of rubberized concrete

One of the possible solutions to reduce waste tire is to incorporate it into concrete by replace the natural aggregates. The properties of the fresh and hardened rubberized concrete had been reviewed.

### *3.1. Fresh concrete*

*3.1.1. Workability.*Fresh concrete is a plastic concrete that can be molded to any shape. Hundred percent compaction of fresh concrete is an important parameter to enable maximum strength for concrete. A highly workable concrete can ensure full compaction. Workability of concrete is the ease with which concrete can be mixed, handled and compacted. Trilok et al. [17] found that the workability of concrete is not affected significantly by adding rubber fibers. Zheng *et al.* [18] indicated that the mixture made with fine ground rubber is more workable than that with coarse tire chips. Ayman *et al.* [10] observed the replacement of up to 10% did not affect the workability much; however, an increase beyond 10% caused a severe reduction in workability. Youssf *et al.* [19] mentioned that the workability of rubberized concrete can be controlled by using an appropriate quantity of super plasticizer (1-3% by weight of cement). Elchalakani [20] explained that rubberized concrete having a combination of rubber powder and crumb rubber exhibited good workability in contrast to plain concrete when an adequate quantity of admixture was used. Rubber particles have acceptable workability when they are used to replace the total aggregate of up to 50% but the strength is reduced proportionally [21].

*3.1.2. Bulk density.* Gesog Lu *et al.* [22] prepared the lowest density of 1900 kg/m$^3$ while the highest density was 2240 kg/m$^3$ and found that the finer the rubber particles, the higher the density of the rubberized concrete. This is because the fine rubber played the role of fine aggregate and filled the small gaps among the concrete particles which improve the density and did not affect the permeability coefficient "K. Nadal Gisbert *et al.* [23] explained that the increase of the rubber content in the concrete reduces the density of the concrete due to the gaps surrounding the rubber. Gesog Lu *et al.* [22, 24] discovered that lighter weight concrete was produced by adding rubber to concrete. The rubberized concrete densities were 2-11% lighter than control concrete specimens. Pelisser *et al.* [25] explained that the density of the rubberized concrete was lesser by 13% compared to normal concrete but only 9% reduction was observed when silica fume was added to rubberized concrete due to the higher densification of the concrete structure. Parheco-Torgal *et al* [26] studied tire chips partially replacing coarse aggregates and crumb rubber replacing total mineral aggregate in rubberized concrete. It was observed that the density of the concrete with fine aggregate replacement was reduced by 34% while the coarse aggregate replacement had a reduction of 45%. The combination gave a decrease in density of 33% in total. Kardos *et al* [27] found that an increase in air content when crumb rubber was added into the concrete.

*3.1.3. Air content.*Richardson et al. [28] compared the percentage air entrainment for plain concrete which was 1.9% and the percentage value for <0.5 mm crumb rubber which was 3.3%. The additional air entrained in plain concrete was 74% compared to the <0.5 mm crumb rubber sample. This is a significant difference as a 3% air entrainment is adequate for providing freeze/thaw protection, especially as the mix has evenly dispersed crumb rubber particles that will permit pressure absorption, consisting of a particle size small enough to provide effective freeze/thaw protection. Al-Akhras et al. and Li et al. [29] discovered that the larger the particle of the tire fiber, the higher the air content in the concrete. Benazzouk *et al.* [30] indicated that increasing rubber volume ratio results in higher air content. Al-Akhras *et al.* [31] explained that the use of tire rubber ash in concrete can reduce the air content. The air content of tire rubber ash mortar decreased with increasing tire rubber ash content. The air content decreased from 2.6% for the control mortar to 1.5% for the mortar containing 10% tire rubber ash.

*3.2. Hardened concrete*

*3.2.1. Compressive strength.*The compressive strength of the rubberized concrete slightly decreased compared to normal concrete [10, 13, 17, 32-35]. However, Obinna Onuaguluchi *et al.* [36] found a significant improvement in compressive and tensile strength recorded in the mixture containing coated rubber crumb and silica fume. The synergy between limestone powder and silica fume caused this mixture to outperform the reference mixture at 5% and 10% fine aggregate replacement. Gupta *et al.* [17] explained that the compressive strength of waste rubber tire fiber concrete decreases with the increase in the replacement level of fine aggregates using rubber fiber. However, the strength of rubberized cement increases upon partial replacement of cement using silica fume. Al-Akhras and Smadi [31] observed an increase in compressive strength when the tire rubber ash replaced the fine aggregates up to 10%. The increase in compressive strength of mortar specimens at 90 days were 14%, 21%, 29%, and 45% where the tire rubber ash content were 2.5%, 5%, 7.5% and 10% respectively. Feng Lie *et al.* [35] explained that the increase in rubber content in concrete decreases the compressive strength of rubberized concrete. Larger rubber particles reduced the compressive strength of rubberized concrete. Therefore, the finer the rubber particles produced, the higher the compressive strength of rubberized concrete.

*3.2.2. Flexural tensile.*Matthew *et al.* [37] have found thatrubber particles in rubberized concrete reduce the fracture energy in external macro cracks. Farhan *et al.* [38] noted that the tensile strength of rubber is reduced due to the inclusion of rubber particles in cement stabilized aggregate mixture due to the weakness of these introduced particles. However, an increase was observed in the post peak behavior which caused an improvement in the toughness of the modified mixtures. In addition, the high stiffness of the original mixtures was mitigated after partial replacement with rubber particles. Gupta et al. [39] pointed out that the flexural strength of concrete containing rubber ash decreased with an increase in the percentage of rubber ash while the flexural strength of modified concrete (containing 10% rubber ash and a varying percentage of rubber fibers) increased with the increasing amount of rubber fiber (aspect ratio 8-10). Al-Akhras and Smadi [31] studied the properties of mortar containing tire rubber ash. An increase in flexural strength was observed when the tire rubber replaced fine aggregates up to 12%, 27%, 32%, and 43% whereas the tire rubber ash content was 2.5%, 5%, 7.5% and 10% respectively compared to the control mix mortar.

*3.2.3. Abrasion resistance. The* abrasion resistance of concrete may be defined as its ability to resist being worn away by rubbing. Concrete which is more resistant to abrasion can be used for the construction of pavements, floors and concrete highways; in hydraulic structures such as tunnels and dam spillways, or on other surfaces upon which abrasive forces are applied between surfaces and moving object during services. B. Thomas *et al.* [40] found that rubberized concrete is more resistant to abrasion compared to the control mix. Da Silva *et al.* [9] observed an improvement in abrasion resistance with the addition of rubber. In their study, the increase in the rubber content decreased the amount of abrasion whereas with 50% replacement, the amount of wear reduction was 18% compared to without rubber.

*3.2.4. Water absorption andpenetration.* Obinna Onuaguluchi *et al.* [36] discovered that crumb rubber increased the porosity of concrete. Matthew *et al.* [37] indicated that the rubber in concrete will increase water penetration depth and the water absorption coefficient. Thomas *et al.* [41] also found the water absorption of rubberized concrete was higher than the control mix concrete. However, Gupta *et al.* [17] also stated that water permeability increases with an increase in rubber fiber content. However, the permeability of rubberized concrete can be reduced by replacing cement with silica fume.

*3.2.5. Damping ratio.*Bravo and Brito [42] observed that the damping capacity of rubberized concrete was greater than normal concrete. Xue and Shinozuka [32] explained that the average damping ration of rubberized concrete columns was 7.70%. In contrast, the average damping ratio of normal concrete columns is only 4.75%. Adding rubber crumbs into concrete increases the damping ration by 62%.Zheng *et al.* [18] indicated that the damping ratios of rubberized concrete increased considerably with the increase of rubber content. However, the relationship between the damping ratio and rubber content is not linear. Since high rubber content would cause dramatic reduction in the modulus of elasticity of rubberized concrete, the optimal content of rubber would be less than 30% to obtain satisfactory static and dynamic properties. Youssf *et al.* [12] explained that replacing 20% of the concrete sand with rubber aggregates decreased the damping ratio by 49% compared to average conventional concrete specimens. Gisbert *et al.* [23] found that a rise in the proportion of rubber also increases the damping ratio of rubberized concrete. The bigger the grain size of rubber, the lower the damping.

## 4. Discussion and conclusion

Workable rubberized concrete mixture can be made with scrap tires [43]. The workability can be similar to normal concrete by using admixtures without any increase in the quantity of water [10]. The literature made it clear that the density of rubberized concrete decreases with increasing amount of rubber [44]. The loss in density would be severe when powdered rubber is used to replace aggregates. The rubber can also produce light weight concrete for special purposes.

The compressive strength of the rubberized concrete would be affected by the use of rubber in concrete. The reduction of compressive strength can be avoided if the replacement of rubber does not exceed 20% of the total aggregate content [8]. Rubber powder reduces the flexural strength but rubber fiber increases the flexural tensile strength. By using any coupling agent, severe loss in compressive strength could be avoided.

Most researchers mentioned that the abrasion resistance got improved with the addition of rubber in cement. Only few researchers found that rubberized concrete reduces the abrasion of concrete [45]. Water absorption and water penetration increase with the increase of rubber content in concrete. The addition of rubber increases the damping ratio of the concrete.

Rubberized concrete seems to be suitable to be used for concrete bridge deck as there were positive results. As recommendations for future work, a proper study on the application of rubberized concrete could be performed. Furthermore, a study on carbonation and damping effect of rubberized concrete could be performed. An in-depth study on the properties of tire rubber ash concrete could also be carried out to maximize its applications.

## 5. References

[1] Senin M S, Shahidan S, LemanA S and Hannan N I R R 2016Analysis of physical properties and mineralogical of pyrolysis tires rubber ash compared natural sand in concrete material *IOP Conf. Ser. Mater. Sci. Engineering* vol 160(London: IOP Publishing) 12053

[2] Shahidan S, Mohd Zuki S S, Rahim M A and Ali N 2016 Amplitude distribution of emission wave for cracking process vol 47(Paris:MATEC Web Conf.) pp 2–8

[3] Ali N, Zainal N A, Burhanudin M K, Abdul Samad A A and ShahidanS2016 Physical and mechanical properties of compressed earth brick (CEB) containing sugarcane bagasse ashvol MATEC Web Conf. vol **47** (Paris:EDP Sciences) pp 1–7

[4] Shu X and Huang B 2014 Recycling of waste tire rubber in asphalt and Portland cement concrete: An overview *Constr. Build. Materials* **67** 217–224

[5] Siddique R and Naik T R 2004 Properties of concrete containing scrap-tire rubber-an overview *Waste Management* **24**(6) 563–569

[6] Rahim M AGhazaly Z, Nurazira R, Mamat R, Azizan M A, Isa N F and Shahidan S 2016 Experimental study of slurry infiltrated fiber reinforced concrete *Mater. Sci. Forum* **857**3 63–366

[7] Shahidan S, Isham I and Jamaluddin N 2016 A review on waste minimization by adopting in self compacting concrete *MATEC Web Conf* vol 47(Paris:EDP Sciences)) pp1–7

[8] Yahya A A, Ali N, Kamal N L M, Shahidan S, Beddu S, Nuruddin M F and Shafiq N 2017 Reducing heavy metal element from coal bottom ash by using citric acid leaching treatment *MATEC Web Conf.* vol 103 (Paris: EDP Sciences) p 1004

[9] Senin M S, Shahidan S, Leman A S and Hannan N I R R2016 Properties of cement mortar containing rubber ash as sand replacement *IOP Conf. Ser. Mater. Sci. Engineering* vol 160 (London: IOP Publishing) p 12055

[10] Moustafa A and El Gawady M A 2015 Mechanical properties of high strength concrete with scrap tire rubber *Constr. Build. Mater.* **93** 249–256

[11] Chou L H, Yang C K, Lee M T and Shu C C 2010 Effects of partial oxidation of crumb rubber on properties of rubberized mortar *Compos Part B Eng.* **41**(8) 613–616
[12] Youssf O, El Gawady M A and Mills J E 2015 Experimental investigation of crumb rubber concrete columns under seismic loading*Structures* **3**13–27
[13] Md Nor N, Bunnori N M, Ibrahim A, Shahidan S and Mat Saliah S N 2011 Relationship between acoustic emission signal strength and damage evaluation of reinforced concrete structure: Case studies *2011 IEEE Symposium on Industrial Electronics and Applications* pp 308–313
[14] Ganjian E, Khorami M and Maghsoudi A A 2009 Scrap-tyre-rubber replacement for aggregate and filler in concrete *Constr. Build. Mater.* **23** (5) 1828–1836
[15] Martínez J D, Puy N, Murillo R, García T, Navarro M V and Mastral A M 2013 Waste tyre pyrolysis-A review *Renew. Sustain. Energy Rev.* **23**179–213
[16] Li L, Ruan S and Zeng L 2014 Mechanical properties and constitutive equations of concrete containing a low volume of tire rubber particles *Constr. Build. Mater.* **70** 291–308
[17] Gupta T, Chaudhary Sand Sharma R K 2015 Mechanical and durability properties of waste rubber fiber concrete with and without silica fume *J. Clean. Prod.* **112**702–711
[18] Zheng L, Sharon Huo X and Yuan Y 2008 Experimental investigation on dynamic properties of rubberized concrete *Constr. Build. Mater.* **22** 939–947
[19] Youssf O, El Gawady M A, Mills J E and Ma X 2014 An experimental investigation of crumb rubber concrete confined by fibre reinforced polymer tubes *Constr. Build. Mater.* **53**522–532
[20] Elchalakani M 2015 High strength rubberized concrete containing silica fume for the construction of sustainable road side barriers *Structures* **1** 20–38
[21] Abdullah S R, Wan Zainal Abidin W R, Shahidan S 2015 Strength of concrete containing rubber particle as partial cement replacement *3rd Int. Conf. Civ. Environ. Eng. Sustain.* vol 9 pp. 2–5
[22] Gesoğlu M, Güneyisi E, Khoshnaw G and İpek S 2014 Investigating properties of pervious concretes containing waste tire rubbers *Constr. Build. Mater.* **63**206–213
[23] Nadal GA 2014 Analysis behaviour of static and dynamic properties of Ethylene-Propylene-Diene-Methylene crumb rubber mortar *Constr. Build. Mater.* **50** 671–682
[24] Gesoğlu M, Güneyisi E, Khoshnaw Gand and İpek S 2014 Abrasion and freezing–thawing resistance of pervious concretes containing waste rubbers *Constr. Build. Mater.* **73** 19–24
[25] Pelisser F, Zavarise N, Longo T A and Bernardin A M 2011 Concrete made with recycled tire rubber: Effect of alkaline activation and silica fume addition *J. Clean. Prod.* **19** (6–7) 757–763
[26] Pacheco-Torgal F, Ding Y and Jalali S 2012 Properties and durability of concrete containing polymeric wastes (tyre rubber and polyethylene terephthalate bottles): An overview *Constr. Build. Mater.* **30** 714–724
[27] Kardos A J and Durham S A 2015 Strength, durability, and environmental properties of concrete utilizing recycled tire particles for pavement applications *Constr. Build. Mater.* **98** 832–845
[28] Richardson A, Coventry K, Edmondson V and Dias E 2016 Crumb rubber used in concrete to provide freeze–thaw protection (optimal particle size) *J. Clean. Prod.* **112** (1) 599-606
[29] Li G, Garrick G, Eggers J, Abadie C, Stubblefield M A and Pang S S 2004 Waste tire fiber modified concrete *Compos. Part B Engineering* **35** (4) 305–312
[30] Benazzouk A, Douzane O, Mezreb K, Laidoudi B and Quéneudec M 2008 Thermal conductivity of cement composites containing rubber waste particles: Experimental study and modelling *Constr. Build. Mater.* **22** (4) 573–579
[31] Al-Akhras N M and Smadi M M 2004 Properties of tire rubber ash mortar *Cem. Concr. Compos.* **26**(7) 821–826
[32] Xue J and Shinozuka M 2013 Rubberized concrete: A green structural material with enhanced energy-dissipation capability *Constr. Build. Mater.* **42** 196–204

[33] Atahan A O and Yücel A Ö 2012 Crumb rubber in concrete: Static and dynamic evaluation *Constr. Build. Mater.* **36** 617–622
[34] Khalil E, Abd-Elmohsen M and Anwar A M 2015 Impact resistance of rubberized self-compacting concrete *Water Sci.* **29** 45–53
[35] Liu F, Chen G, Li L and Guo Y 2012 Study of impact performance of rubber reinforced concrete *Constr. Build. Mater.***36** 604–616
[36] Onuaguluchi O and Panesar D K 2014 Hardened properties of concrete mixtures containing pre-coated crumb rubber and silica fume *J. Clean. Prod.* **82**125–131
[37] Hall M R and Najim K B 2014 Structural behaviour and durability of steel-reinforced structural Plain/Self-Compacting Rubberised Concrete (PRC/SCRC) *Constr. Build. Mater.* **73** 490–497
[38] Farhan A H, Dawson A R and Thom N H 2016 Characterization of rubberized cement bound aggregate mixtures using indirect tensile testing and fractal analysis *Constr. Build. Mater.* **105** 94–102
[39] Gupta T, Chaudhary Sand Sharma R K 2014 Assessment of mechanical and durability properties of concrete containing waste rubber tire as fine aggregate *Constr. Build. Mater.* **73**562–574
[40] Thomas BS, Gupta R C, Kalla Pand Cseteneyi L 2014 Strength, abrasion and permeation characteristics of cement concrete containing discarded rubber fine aggregates *Constr. Build. Mater.* **59** 204–212
[41] Thomas B S, Gupta R C and Panicker V J 2016 Recycling of waste tire rubber as aggregate in concrete: durability-related performance *J. Clean. Prod.* **112** 504–513
[42] Bravo M and de Brito J 2012 Concrete made with used tyre aggregate: Durability-related performance *J. Clean. Prod.* **25** 42–50
[43] Aiello M A and Leuzzi F 2010 Waste tyre rubberized concrete: Properties at fresh and hardened state *Waste Manag.* **30**(8–9) 1696–1704
[44] Meddah A, Beddar M and Bali A 2014 Use of shredded rubber tire aggregates for roller compacted concrete pavement *J. Clean. Prod.* **72** 187–192
[45] Thomas B S and Gupta R C 2016 A comprehensive review on the applications of waste tire rubber in cement concrete *Renew. Sustain. Energy Rev.* **54** 1323–1333

**Acknowledgments**
The authors would like to thank the University of Tun Hussein Onn, Malaysia and the Ministry of Education of Malaysia, for their generous grant awarded for this research, Short Term Grant Vot U114.

# The durability of concrete containing recycled tyres as a partial replacement of fine aggregate

**Mohamad Syamir Senin[1], Shahiron Shahidan[1,*], Alif Syazani Leman[1], Nurulain Othman[1], Shamrul-mar Shamsuddin[1], M. H.W. Ibrahim[1], S. S. Mohd Zuki[1]**

[1] Jamilus Research Center-Sustainable Construction, Faculty of Civil and Environmental Engineering, Universiti Tun Hussein Onn Malaysia.
*Corresponding author: shahiron@uthm.edu.my

**Abstract.** Nowadays, uncontrolled disposal of waste materials such as tyres can affect the environment. Therefore, careful management of waste disposal must be done in order to conserve the environment. Waste tyres can be use as a replacement for both fine aggregate and coarse aggregate in the production of concrete. This research was conducted to assess the durability of concrete containing recycled tyres which have been crushed into fine fragments to replace fine aggregate in the concrete mix. This study presents an overview of the use of waste rubber as a partial replacement of natural fine aggregate in a concrete mix. 36 concrete cubes measuring 100mm x 100mm x 100mm and 12 concrete cubes measuring 150mm x 150mm x 150mm were prepared and added with different percentages of rubber from recycled tyres (0%, 3%, 5% and 7%) as fine aggregate replacement. The results obtained show that the replacement of fine aggregate with 7% of rubber recorded a compressive strength of 43.7MPa while the addition of 3% of rubber in the concrete sample recorded a high compressive strength of 50.8MPa. This shows that there is a decrease in the strength and workability of concrete as the amount of rubber used a replacement for fine aggregate in concrete increases. On the other hand, the water absorption test indicated that concrete which contains rubber has better water absorption ability. In this study, 3% of rubber was found to be the optimal percentage as a partial replacement for fine aggregate in the production of concrete.

## 1. Introduction

Along with the demands of globalisation, the increasing demand for building materials as well as the increasing human population around the world have led to a shortage of construction materials [1-3]. Engineers have been challenged to convert industrial waste into useful construction materials [4-5]. This is because the construction sector is important in order to assist in the development of economic development in Malaysia to achieve the status of a developed nation. The use of rubber from waste tyres to replace fine aggregate may help to alleviate disposal problems [6]. Various studies have been conducted to replace the main components in concrete production with alternative materials in order to reduce cost and conserve the environment. In this research, rubber was used to replace fine aggregate in concrete. According to Blessen Skariah Thomas and Ramesh Chandra Gupta, by the year 2030, the number of discarded tyres would reach 1200 million tyres yearly (including the stockpiled tyres, there would be 5000 million tyres to be discarded on a regular basis) [7]. Besides that, 60% of the waste tyres are continuously being disposed at places which are categorised as hazardous [8]. The disposal of waste rubber is a threat the environment and there is no shortcut in solving this problem due to the lack of management of waste rubber [9]. The build-up of old rubber tyres in landfills is commonly consider a major threat to the environment, and it is a burden on landfill space [10]. Therefore, in the process of making concrete, rubber particles originating from recycled tyres have been chosen to replace fine aggregate in concrete. There are some features in the rubberised concrete to be compared with ordinary concrete in terms of strength and durability. This is done to make sure that the rubber concrete is safe for use in the construction industry. Concrete is an important building material in the production of building structures [11]. Concrete is widely used because it has durability and optimum compressive strength. The shape, strength, ability and value of the regular concrete are different from those of rubberised concrete. Additional material or materials used in the production of concrete should be tested in terms of effectiveness according to the standards set. As such, this study makes use of recycled waste tyres as a partial replacement for fine aggregate in the production of concrete. The main objective of this paper is to evaluate the level of durability of concrete containing rubber from recycled tyres as fine aggregate replacement.

## 2. Literature review

### *2.1. Rubber tyres*

In order to prevent environmental problems from worsening, recycling tyres is an innovative idea or way in this case [12-16]. Recycling tyres is the processes of recycling vehicle tyres that are no longer suitable for use on vehicles due to wear or irreparable damage (such as punctures) [17]. Previous studies have contributed greatly to the understanding of the behaviour of fibrous concrete, especially when tyre chips and waste plastic strips were used in concrete [18]. Early studies on the use of scrap tyres in asphalt mixes were very promising [19]. They showed that rubberised asphalt had better skid resistance, reduced fatigue cracking, and achieved longer pavement life than conventional asphalt. Large benefits can result from the use of scrap tyre rubber in Portland cement concrete (pcc) mixtures, especially in circumstances where properties like lower density, increased toughness and ductility, higher impact resistance, and more efficient heat and sound insulation are desired [20-23].

The use of waste tyres as a replacement for sand in concrete has great potential in the construction industry. Rubber is the principal element of tyres where both synthetic and natural rubbers may be used. Rubber consists of many polymeric repeated units (mainly polybutadiene) chained together [24]. Natural rubber is an elastic variety of plants. Synthetic rubber can be produced as a thermoset polymeric material in which individual monomer chains are chemically linked by covalent bonds during polymerisation [25]. The disposal of discarded tyres has been a major concern throughout the world because rubber is not easily biodegradable. The simplest way to get rid of waste tyres is by burning. However, this method generates toxic fumes that contaminate air, soil and water. The use of tyre rubber in the production of concrete is an alternative way to recycle this residue [26]. Most countries have to deal with the problem of waste rubber tyre disposal which greatly affects the environment [27]. Therefore, many research studies have been done to evaluate the influence of rubber tyres on cement. Many researchers have observed that the presence of rubber tyre in concrete reduces its mechanical properties.

## 3. Materials and methods

This part describes the methods used in this research. Basic materials such as cement, fine aggregate, coarse aggregate, water and rubber were used in this study.

### *3.1. Material preparation*

*3.1.1. Cement.* Ordinary Portland cement with characteristic strength was used in accordance with BS EN 197-1. This cement was taken into account during the mix design process. It was stored in airtight packages before use.

*3.1.2. Fine aggregate.* Natural river sand with a maximum particle size of 4.75 mm was used as fine aggregates. The procedures for measuring the durability of sand were the same as those for measuring the durability of gravel. A sieve analysis test was carried out in accordance with BS EN 933-1. As shown in Figure 3.3, the sand used in this study presented continuous grading.

*3.1.3. Coarse aggregate.* Crushed gravel with a nominal maximum size of 20 mm was used as coarse aggregate. The water absorption of the coarse aggregate used in this study under SSD (saturated surface dry) condition was measured by immersing them in water for 24 hours followed by the removal of excess surface water with a wet cloth after they were taken out of the water. At the time when there is no free water on the surface, the aggregates were assumed to be under the SSD condition.

*3.1.4. Water.* Tap water that is reasonably free from contamination in the laboratory was used to hydrate the cement in the mixtures.

*3.1.5. Rubber.* In this study, rubber obtained from recycled tyres was used as a partial replacement of fine aggregate in concrete. In order to obtain a suitable size, raw rubber tyres were placed in a machine to produce rubber particles measuring 0.16mm. This is important to ensure that the size and shape of rubber particles are similar to sand. In addition, the rubber particles must be free from impurities.

*3.2. Specimen preparation and testing*
48 concrete cube specimens were prepared for three separate tests. For the compressive strength test, 24 concrete cubes measuring 100mm x 100mm x 100mm were used. For the water absorption test, 12 concrete cubes measuring 150mm x 150mm x150mm were used whereas the carbonation test involved the use of 12 concrete cubes measuring 100mm x 100mm x 100mm. Table 1 show the mix proportion used in this study.

**Table 1**: Mix proportion.

| Type of concrete | Cement (kg) | Sand (kg) | Coarse aggregate (kg) | Water/cement ratio | Rubber ash (kg) | Rubber crumb (kg) |
|---|---|---|---|---|---|---|
| 0% | 430.00 | 530.00 | 1240.00 | 0.45 | 0.00 | 0.00 |
| 3% | 430.00 | 514.10 | 1240.00 | 0.45 | 0.00 | 4.56 |
| 5% | 430.00 | 503.50 | 1240.00 | 0.45 | 0.00 | 7.60 |
| 7% | 430.00 | 492.90 | 1240.00 | 0.45 | 0.00 | 10.64 |

## 4. Results and discussion

*4.1. Compressive strength*
This test was conducted using a compression machine as shown in figure 1. The machine was used to investigate the compressive strength of the composite concrete specimens shown in figure 2. Moreover, this test was done on composite concrete samples containing different compositions of rubber (0%, 3%, 5% and 7%). Recycled rubber particles were used to partially replace sand in the concrete samples. All samples including the control samples and the composite samples were tested after a curing period of 7 days and 28 days respectively.

**Figure 1.** Compression machine.

**Figure 2.** Cube specimens.

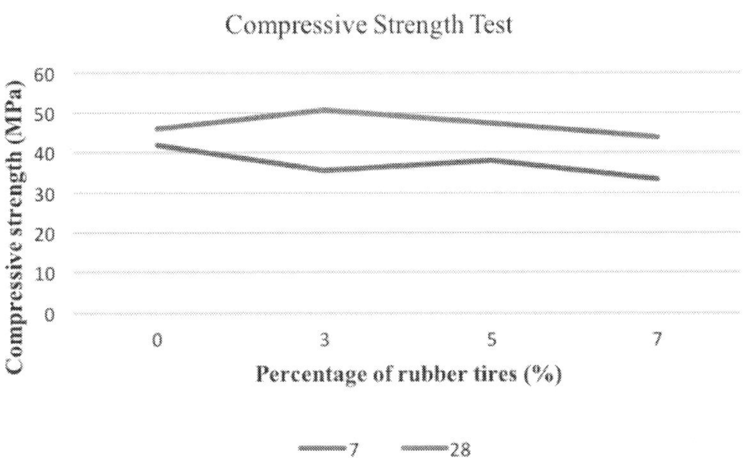


**Figure 3.** Compressive strength of cube specimens.

Figure 3 shows the compressive strength of the cube specimens after a curing period of 7 days and 28 days respectively. In the study, concrete samples containing 7% of rubber recorded the lowest compressive strength (43.7 MPa) while the highest compressive strength (50.8 MPa) was achieved by concrete samples containing 3% of rubber. Hence, it can be concluded that the optimum amount of rubber which should be used in concrete is 3%.

*4.2. Water absorption*

The purpose of this test was to investigate the water absorption ability of the concrete samples. The test was conducted on after a curing period of 28 days. The water absorption test was conducted on concrete samples containing 0%, 3%, 5% and 7% of rubber. All samples were then weighed before being submerged into a tub of water for 24 hours. The samples were weighed again after 24 hours. The data is shown in Figure 4.

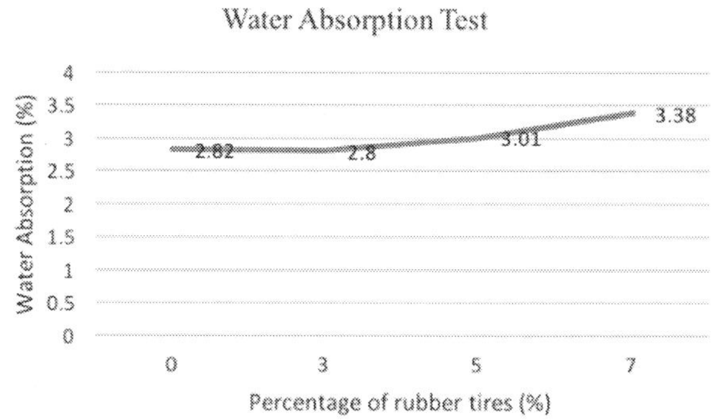


**Figure 4.** Water absorption (%) vs percentage of rubber (%) in concrete specimens.

It can be concluded that the percentage of rubber which should be used to replace fine aggregate is most suitable at 3% since it does not absorb too much water. Therefore, a high amount of rubber should not be used for building construction. A lower water absorption rate is not always an indicator of better frost resistance whilst a high one does not always mean poor frost resistance. Structural engineers can use materials with lower water absorption rates to improve the safety factors in their designs.

*4.3. Carbonation*

This test was carried out to determine the depth of concrete affected due to a combined attack of atmospheric carbon dioxide and moisture which can cause a reduction in the level of alkalinity of concrete. A solution containing 0.2% of phenolphthalein was used as the pH indicator of concrete. Figures 5,6,7 and 8 show the depth of carbonation for each of the concrete specimens (control, 3% rubber, 5% rubber and 7% rubber)

**Figure 5.** Control

**Figure 6.** 3%

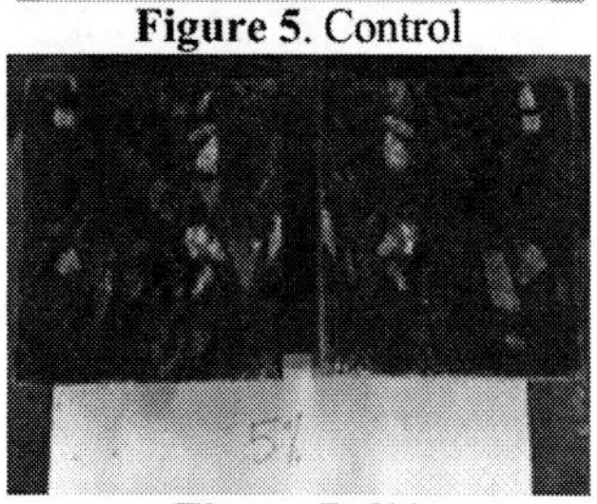

**Figure 7.** 5%

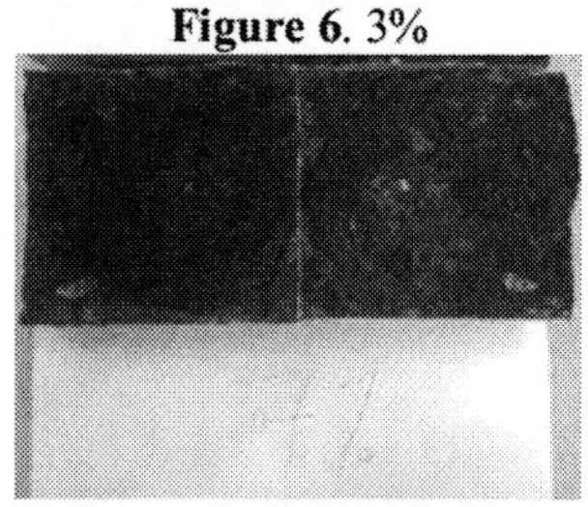

**Figure 8.** 7%

The carbonation in the concrete samples which were exposed to air for two weeks was not more than 1mm. After a year or so, it may reach a depth of 1 mm for dense concrete of low permeability (low water/cement ratio), or up to 5 mm or more for more porous and permeable concrete (high water/cement ratio).

## 5. Conclusion
The highest compressive strength of 50.8 MPa was achieved by concrete samples which contain 3% of rubber. The durability of concrete which can be assessed by testing the water absorption of the replacement tyre recycling as much as 3% low percentage compared to the other percentage of 2.8%. Meanwhile, the carbonation tests revealed that the carbonation of samples which were exposed to air for two weeks was not greater than 1mm.

## 6. References
[1] Md Nor N, Ibrahim A, Muhamad Bunnori N, Saman H M, Mat Saliah S N and Shahidan S 2014 Diagnostic of fatigue damage severity on reinforced concrete beam using acoustic emission technique *Engineering Failure Anal.* **41** 1–9

[2] Shahidan S, Isham I and Jamaluddin N 2016 A review on waste minimization by adopting in self compacting concrete vol 47 (Paris: MATEC Web Conf.) pp 1–7

[3] Shahidan S, Zuki S S M and Jamaluddin N 2016 Damage grading system for severity assessment on concrete structure *Case Stud. Constr. Mater.* **5** 79–86

[4] Ali N, Zainal N A, Burhanudin M K, Abdul Samad A A and Shahidan S 2016 Physical and mechanical properties of Compressed Earth Brick (CEB) containing sugarcane bagasse ash *MATEC Web Conf.* vol 47 (Paris: EDP Sicences) pp 1–7

[5] Abdullah S R, Wan Zainal Abidin W R and Shahidan S 2016 Strength of concrete containing rubber particle as partial cement replacement *MATEC Web Conf.* vol 47 (Paris: EDP Sciences) pp 2–5

[6] Yogender A, Vivek V and Bhupinder S 2014 Rubberized concrete with crumb rubber *Int. J. Sci. Res.* (IJSR) **3** (5) 1481-1483

[7] Thomas B S and Gupta R C 2016 Properties of high strength recycling of waste tyre rubber as aggregate in concrete: Durability-related performance *J. Clean. Prod.* **113** 86–92

[8] Miguel B and Jorge B 2012 Concrete made with used tyre aggregate: Durability-related performance *J. Clean. Prod.* **25** 42-50

[9] Ashraf F, Fouad A R, Abu-Lebdeh T and Ellie F 2014 Use of crumb rubber to improve thermal efficiency of cement-based materials *Am. J. Eng. Appl. Sci.* **7** (1) 1-11

[10] Mehmet G, Erhan G, Ganjeena K, Süleyman I P 2014 Investigating properties of concretes containing waste tyre rubbers *Constr. Build. Mater.* **63** 206-213

[11] Shahidan S, Mohd Zuki S S, Rahim M A and Ali N 2016 Amplitude distribution of emission wave for cracking process *MATEC Web Conf*. vol. 47 (Paris: EDP Sciences) pp 2–8

[12] Leman A S, Shahidan S, Senin M S and Ramzi Hannan N I R 2016 A Preliminary study on chemical and physical properties of coconut shell powder as a filler in concrete *IOP Conf. Ser. Mater. Sci. Engineering* vol 160 (London: IOP Publishing) p 12059

[13] Ramzi hannan N I R, Shahidan S, Maarof M Z and Ali N 2016 Physical and chemical properties of Coal Bottom Ash (CBA) from Tanjung Bin Power Plant *IOP Conf. Ser. Mater. Sci. Engineering* vol. 160 (London: IOP Publishing) p 12056

[14] Senin M S, Shahidan S, Leman A S and Ramzi Hannan N I R 2016 Analysis of physical properties and mineralogical of pyrolysis tires rubber ash compared natural sand in concrete material *IOP Conf. Ser. Mater. Sci. Engineering* vol. 160 (London: IOP Publishing) p 12053

[15] Senin M S, Shahidan S, Leman A S and Ramzi Hannan N I R 2016 Properties of cement mortar containing rubber ash as sand replacement *IOP Conf. Ser. Mater. Sci. Engineering* vol. 160 (London: IOP Publishing) p 12055

[16] Shahidan S, Ramzi Hannan N I R, Maarof M Z and Leman A S 2017 A comprehensive review on the effectiveness of existing noise barriers commonly used in the railway industry *MATEC Web Conf*. vol 87 (Paris: EDP sciences) pp 1–8
[17] Audrius G, Henrikas S and Mindaugas D 2012 Type rubber additive effect on concrete mixture strength *J. Civil Engineering Manag.* **18** (3) 393-401
[18] Erhan G 2010 Fresh properties of self-compacting rubberized concrete incorporated with fly ash *Mater. and Struct.* **43** (8) 1037-1048
[19] Yong-chang G, Jian-hong Z, Guang-ming C and Zhi-hong X 2014 Compressive behavior of concrete structures incorporating recycled concrete aggregates, rubber crumb and reinforced with steel fiber, subjected to elevated temperatures *J. Clean. Prod.* **72** 193-203
[20] Hall M R, Najim K B and Hopfe C J 2011 Transient thermal behaviour of crumb rubber-modified concrete and implications for thermal response and energy efficiency in buildings *Applied Thermal Engineering* **33-34** 77–85
[21] Hilal, Ameer Abdulrahman, 2011. Effect of crumb tyres rubber on some properties of foamed concrete *Anbar Journal for Engineering Sciences* **4** (2) 1- 17
[22] Anh Cuong H, Anaclet T, Rashid H and Duc Chinh V 2012 Effects of rubber aggregates from grinded used tyres on the concrete resistance to cracking *J. Clean. Prod.* **23** 209–215
[23] Huang X, Ranade R, Ni W and Li V C 2013 On the use of recycled tire rubber to develop low E-modulus ECC for durable concrete repairs *Constr. Build. Mater.* **46** 134–141
[24] Issa C A, Salem G 2013 Utilization of recycled crumb rubber as fine aggregates in concrete mix design *Constr. Build. Mater.* **42** 48-52
[25] Khatib Z K and Bayomy F M 1998 Rubberized Portland cement concrete *J. of Mater. in Civ. Eng.* 206-213
[26] Ling Tung-Chai 2011 Prediction of density and compressive strength for rubberized concrete blocks *Constr. Build. Mater.* **25** (11) 4303-4306
[27] Al-Tayeb M M , Abu Bakar B H, Ismail H and Hazizan M A 2013 Effect of partial replacement of sand by fine crumb rubber on impact load behavior of concrete beam: Experiment and nonlinear dynamic analysis. *Mater. and Struct.* **46** (8) 1299–1307
[28] Parveen, Dass S and Sharma A 2013 Rubberized concrete: Needs of good environment (overview) *Int. J. Emerg. Technol. Adv. Eng.* **3** (3) 192-196

**Acknowledgments**

The author would like to thank to Jamilus Research Center, Universiti Tun Hussien Onn Malaysia and Grant VOT U523 and VOT 572 (GPPS).

IOP Conf. Series: Materials Science and Engineering **271** (2017) 012076 doi:10.1088/1757-899X/271/1/012076

# Compatibility improvement method of empty fruit bunch fibre as a replacement material in cement bonded boards: A review

**Hayana Dullah[1], Zainal Abidin Akasah[1*], Nik Mohd Zaini Nik Soh[1], Sajjad Ali Mangi[1]**

[1] Faculty of Civil and Environmental Engineering, UTHM Malaysia
*Corresponding author: zainal59@uthm.edu.my

**Abstract.** The utilization of oil palm empty fruit bunch (OPEFB) fibre on bio-composite product has been introduced to replace current material mainly wood fibre. OPEFB is widely available as palm oil is one of the major agricultural crops in Malaysia. EFB fibre are lignocellulosic materials that could replace other natural fibre product especially cement bonded board. However, the contains of residual oil and sugar in EFB fibre has been detected to be the reason for incompatibility issue between EFB fibre and cement mixtures. Regarding on the issue, a study has been conducted widely on finding the suitable pre-treatment method for EFB fibre to remove carbohydrate contained in the said fibre that are known to inhibit cement hydration. Aside from that, cement accelerator was introduced to enhance the hydration of cement when it was mixed with natural fibre. Hence, this paper will summaries the previous findings and in-depth study on the use of EFB fibre as a replacement material in cement bonded fibre boards.

## 1. Introduction

The potential use of natural fibre as bio-composites has been accepted globally, because it is cheap, sustainable, biodegradable and reduces the carbon dioxide ($CO_2$) emissions [1]. The use of wood fibre has been increased rapidly as a natural composite in cement based products i.e. cement bonded board [2]. However, the utilization of wood fibre will lead to the increase demand of forest sources. Ultimately, the use of wood fibre as a bio-composites material will increase the rate of deforestation [3]. The growth in the use of palm oil production caused the environmental issue and created solid waste in terms of abundance quantity of empty fruit bunch (EFB) from oil palm mills [4]. Oil palm empty fruit bunch fibre (OPEFB) is the most important fibre that is readily available within Malaysia.
subsequently, the attention is required on the production of OPEFB-cement products and study on the incompatibility of cement and OPEFB fibre. The issue of compatibility between natural fibre and cement has been highlighted in many previous studies on application of bio-composites. Based on research by Yi, et al. [5], there is compatibility issue between cement and fibre due to the existing of hemi-cellulose, starch, sugar, tannins and lignin which appear to inhibit significantly the setting of cement hydration.
To resolve this issue, previous research suggested the most effective method is pre-treatment of the natural fibre using Sodium hydroxide (NaOH) and cement-curing accelerator to accelerate the curing time of cement which could improve the compatibility of natural fibre with the cement [6, 7]. This article summarises the methods of OPEFB pre-treatment and cement accelerator for cement bonded

fibre board (CBFB) used by previous researchers. It is expected that this information will provide an idea to utilize OPEFB as a main material for cement bonded composites.

## 2. Previous research work

This paper summarises the previous findings on the factors that affect compatibility, improving compatibility, pre-treatment and the use of chemical additive as cement accelerator by utilizing EFB fibre.

### *2.1. Factors affecting compatibility*

The compatibility affecting of EFB fibre on cement is consisted by a number of factors. One of the factors is of EFB fibre's wide range of carbohydrates such as hemi-cellulose, starch, sugar, tannins and lignin that are known to inhibit normal setting and delayed the setting time of the cement matrix [8]. The natural fibre cannot be used directly on cement matrix due to the existence of residual oil that interrupts the penetration of binding agent and thus affecting the properties of the final EFB-products [9, 10].

### *2.2. Method of compatibility improvement*

Many researchers carried a series of tests to improve the compatibility of fibre and cement by using some modification of fibre method mainly physical, chemical and thermal pre-treatments (extraction or soaking of wood particles in some solutions such as hot water and sodium hydroxide before its mixing with cement) [11, 12]. Natural fibre needs to be modified before being utilized on cement bonded board, thus reducing carbohydrate contained in EFB fibre is essential to improve the bonding between EFB fibre and cement.

On the other hand, some studies carried out a series of tests designed to improve the natural fibre-cement mixtures by addition of some chemicals such as magnesium chloride and calcium chloride [14, 15]. Chemical additive is one of the modification method commonly found to accelerate the setting time of cement matrix which is suitable for application of natural fibre cement composite products. It is reported by Nazerian and Sadeghiipanah [15] that fibre-cement compatibility and rapid hydration of cement paste could be improved by addition of an accelerator. There were different methods tried by researchers to treat the EFB fibre in order to enhance its compatibility with many types of product.

Table 1 shows the previous research focusing on methods of OPEFB treatment. It is clearly shown that the most common fibre pre-treatment has been used by other researcher is chemical treatment by using sodium hydroxide (NaOH).

Table 1: Previous study on EFB pre-treatment method.

| Ref. | Research | Fibre treatment | | Finding |
|---|---|---|---|---|
| | | Type | Method | |
| [9] | Oil Content Effect | 0.2, 0.4, 0.6 & 0.8% of; -NaOH -Acetic acid | Soaked 24 hours | Treated fibre using NaOH successfully removed more residual oil from the EFB surface compared to treatments using acetic acid. |
| [10] | Thermal Behavior | 10% NaOH<br>2% $H_2O_2$ | Soaked 48 hours<br>Stirred 3 hours (60 – 70 °C) | NaOH has removed a significant percentage of hemicellulose thus increasing the cellulose content compared to $H_2O_2$ treated fibre. |
| [12] | Effect of OPEFB Pre-treatments | Physical<br>Chemical (NaOH)<br>Thermal | Soaked 4 hours | Fibre treated by chemical followed by thermal treatment has produced the best reducing sugars as compared to the physical pre-treatment. |
| [16] | Medium Density Fibre Board | 2% NaOH<br>Water | Soaked 30 minutes<br>Boiled 30 minutes | Fibre treated with NaOH is more effective than water to remove the residual oil. |
| [17] | Biocomposite | 20g in 500ml Erlenmeyer flask of distilled water | Soaked 3 hours at temperature 75°C | Tensile and flexural modulus of treated EFB fibre is higher than untreated EFB fibre. |
| [18] | Medium Density Fibre Board | 2% NaOH<br>Water | Soaked 30 minutes<br>Boiled 30 minutes | Fibre treated with NaOH shows the highest MOR (31.4 MPa) and increased the mechanical and physical properties of fibreboards. |
| [19] | Biocomposite | 5% NaOH | Soaked 2 hours | Treated fibre biocomposite shows increasing values of mechanical properties compared to untreated fibre. |
| [20] | Corrugating Medium & Fibreboard | 0, 1, & 2 % NaOH | 121 °C for 1, 2 & 3 hours | Optimum conditions for chemical treatment was 2 hr with 2% NaOH. |
| [21] | Medium Density Fibre Board | 0.2, 0.4, 0.6 & 0.8% of; -NaOH -Acetic acid | Soaked 24 hours | Fibre treated with 0.4% of NaOH produced the lowest TS compared with other treatment concentrations. |
| [22] | Reinforced Epoxy Composite | 2% NaOH | Soaked 30 minutes | Fibre treatment improved the tensile strength of the OPEFB fibre reinforced epoxy composites. |

Sodium hydroxide (NaOH) is reportedly could modify the surface of EFB fibre and disrupts the crystalline region in the cellulose, thus undergo complete de-crystallization and weaken the energy bonding of cellulose [13, 14]. Furthermore, oil and impurities of EFB fibre could be removed by using chemical pre-treatment method. It was noted by Ndazi et al. [24] that the sodium hydroxide (NaOH) have good ability to break hydrogen bonding in network structure of fibre cellulose which can increase fibre surface roughness. From Table 1, Sodium hydroxide (NaOH) is the reagent that is widely used for EFB fibre pre-treatments, which can be conducted over a wide range of operating condition such as different concentration of NaOH and soak period of EFB fibre.

Apart from pre-treatment, the previous studies indicated that the chemical accelerator could affect the hydration setting and compressive strength of cement-bonded fibre board [24, 25] i.e. Modulus of Rupture (MOR) and Modulus of Elasticity (MOE). The percentage and type of chemical additives on compatibility of fibre-cement water mixtures are summarized in Table 2. Normally the percentage of chemical additive that have been used in range between 0.5 – 5% depending on the type of chemical. Based on the table, the favourable type of chemical used as cement accelerator is calcium chloride ($CaCl_2$) followed by magnesium chloride ($MgCL_2$).

Table 2: Previous study of chemical additive used as cement accelerator.

| **Ref.** | **Material** | **Chemical additive (%)** | | | | | **Remarks** |
|---|---|---|---|---|---|---|---|
| | | **$MgCL_2$** | **$AlCl_3$** | **$CaCl_2$** | **$Al_2(SO_4)_3$** | **Other** | |
| [3] | Wood (Afzelia Africana) | 2 | 2 | 2 | - | - | $CaCl_2$ was the best accelerator follow by $MgCl_2$ and $AlCl_3$. |
| [5] | Wood (Sugi) | 4 | 4 | 4 | 4 | 4 CaO, $Na_2SiO_3$ | $MgCl_2$ is most effective additive among other chemicals. |
| [14] | Waste Wood | 0.5, 1, 2.5 | 0.5, 1, 2.5 | 0.5, 1, 2.5 | 0.5, 1, 2.5 | 0.5, 1, 2.5 $FeCl_3$ | 2% $CaCl_2$ was selected for the best compatibility. |
| [25] | Pinus Wood Dust | 4 | - | 4 | - | 4 $Na_2SiO_3$, $Al_2SO_4$ | Admixture with a chloride basis had an excellent behaviour. |
| [26] | Wood (Tree Pruning Waste) | 3 | - | 3 | 3 | - | Three chemical additives were suitable as an accelerator. |
| [27] | Oil palm Frond | 0, 2.5, 5 | - | - | - | - | MOR & MOE increased when added 5-7 % of $MgCl_2$. |
| [28] | Bagasse, Cotton stalk, Sunt sawdust | - | - | 3 | - | - | The accelerator was compatible with cement. |
| [29] | Bamboo | 2 | - | 2 | 2 | 2 $Al_2(SO_4)_3$ + $Na_2SiO_3$ | Aluminium sulphate meets the requirements of MS 934. |
| [30] | 6 forest species | 1.5, 2.5, 3.5 | - | 1.5, 2.5, 3.5 | - | - | 1.5% of $CaCl_2$ was similar to 3.5% of $MgCl_2$. |
| [31] | Newsprint paper | - | - | 3, 5 | - | - | Optimum condition obtained when $CaCl_2$ is 5%. |
| [32] | Kenaf Fibre | - | 2, 4, 6 | 2, 4, 6 | - | 2, 4, 6 CaO | $CaCl_2$ and CaO were found to be effective accelerator. |
| [33] | Wood | 3 | - | 3 | - | - | Improved physical and mechanical properties of wood-cement composite. |
| [34] | Musa Paradisiaca | - | - | 3 | - | - | Additional chemical accelerator affects the properties of CBFB. |
| [35] | Eucalypt and Poplar | - | - | 5, 8 | - | - | 5% dose of $CaCl_2$ enhanced the MOE & MOR properties. |
| | **Total** | **9** | **4** | **13** | **4** | **5** | |

It was observed that additional chemical additive can modify the setting time of cement hydration for utilization of natural fibre into cement bonded fibre board products. Early study on utilization of oil palm fibre was conducted by Hermawan et al. [27] to cement bonded fibre board product which was reported to enhance the mechanical performance with addition of 5-7 % of $MgCl_2$. However, the additional of other accelerator need to be emphasized based on previous studies that applied several types of chemical accelerator on different type of natural fibre which have same structural composition with EFB fibre. Therefore, other types of chemical additive should also need to be considered to accelerate the setting time of EFB fibre-cement mixture.

## 3. Conclusion

Based on the extensive literature review as suggested by the several researchers, the following conclusions have been drawn;

(a) EFB fibre can be utilized as replacement of other natural fibre i.e. wood fibre by improvement method of chemical treatment with sodium hydroxide (NaOH) which could remove the part of hemicellulose and lignin contains present in EFB fibre.
(b) Amount of silica body removed in EFB fibre is depend to the concentration of Sodium Hydroxide (NaOH) and soaking period.
(c) It was perceived from the previous studies that the cement accelerator could affect the hydration setting and compressive strength of cement-bonded fibre boards.
(d) Chemical additive by using $CaCl_2$ and $MgCl_2$ seems to be a good potential as they increased the setting time of cement matrix.

As per conclusions above, it is recommended to modify the structure of EFB fibre by using Sodium Hydroxide (NaOH) as a pre-treament agent and additional of cement-curing accelerator i.e. calcium chloride ($CaCl_2$) and magnesium chloride ($MgCl_2$) to improve the compatibility between EFB fibre and cement for the cement bonded board product. However, it is essential to study on the different concentration of the chemicals to observe the optimum percentage for application of EFB fibre since there are very limited and rare to found published research finding about the effect of EFB fibre as a main material for the cement bonded board product.

## 4. References

[1] Brandt A M 2008 Fibre reinforced cement-based (FRC) composites after over 40 years of development in building and civil engineering *Compos. Struct.* **86** (1–3) 3–9
[2] Kochova K, Schollbach K and Brouwers H J H 2015 Use of alternative fibres in wood wool cement boards and their influence on cement hydration *19th International Conference on Building Materials* (London: ICABMMA 2017) pp 1375–1382
[3] Sotannde O A, Oluwadare A O, Ogedoh O and Adeogun P F 2012 Evaluation of cement-bonded particle board produced from *Afzelia africana* wood residues *J. Engineering Sci. Technolology* **7** (6) 732–743
[4] Sumathi S, Chai S P and Mohamed A R 2008 Utilization of oil palm as a source of renewable energy in Malaysia," *Renew. Sustain. Energy Rev.* **12** (9) 2404–2421
[5] Yi M W, Tomita B, Hiramatsu Y, Miyatake A and Fujii T 2002 Study of hydration behaviors of wood-cement mixtures: Compatibility of cement mixed with wood fiber strand obtained by the water-vapor explosion process *J. Wood Sci.* **48** (5) 365–373
[6] Bin N, Zhiqiang W, Haiqin W and Xiaoning L 2014 Wood-cement compatibility review *Wood Research* **59** (5) 813–826
[7] Karade S R 2010 Cement-bonded composites from lignocellulosic wastes *Constr. Build. Mater.* **24** (8) 1323–1330
[8] Pacheco Torgal F and Jalali S 2011 Cememtitious building materials reinfoced with vegetable fibres: A review *Constr. Build. Mater. J.* **25** (2) 575–581

[9] Ibrahim Z, Abdul Aziz A and Ramli R 2015 Effect of treatment on the oil content and surface morphology of oil palm (*Elaeis guineensis*) empty fruit bunches ( EFB ) fibres *Wood Research* **60** (1) 157–166
[10] Hassan N S and Badri K 2016 Thermal behaviors of oil palm empty fruit bunch fiber upon exposure to acid-base aqueous solutions *Malaysian J. Anal. Sci.* **20** (5) 1095–1103
[11] Aanifah F J M, Yee P L, Wasoh H and Abd-Aziz S 2014 Effect of different alkaline treatment on the release of ferulic acid from oil palm empty fruit bunch fibres *J. Oil Palm Res.* **26** (4) 321–331
[12] Ariffin H, Hassan M A, Md Shah U K, Abdullah N and Shirai Y 2008 Effect of physical, chemical and thermal pretreatments on the enzymatic hydrolysis of oil palm empty fruit bunch (OPEFB) (*Kesan prarawatan fizikal, kimia dan termal terhadap hidrolisis enzimitik tandan kosong kelapa sawit*) *J. Trop. Agric. and Fd. Sc.* **36** (2) 1–10
[13] Frybort S, Mauritz R., Teischinger A and Müller U 2008 Cement bonded composites - A mechanical review *BioResources* **3** (2) 602–626
[14] Wang L, Chen S S, Tsang D C W, Chi S P and Kaimin S 2016 Value-added recycling of construction waste wood into noise and thermal insulating cement-bonded particleboards *Constr. Build. Mater.* **125** 316–325
[15] Nazerian M and Sadeghiipanah V 2013 Cement-bonded particleboard with a mixture of wheat straw and poplar wood *J. For. Res.* **24** (2) 381–390
[16] Ramli R, Shaler S and Jamaludin M A 2002 Properties of Medium density fibreboard from oil palm empty fruit bunch *J. oil palm Res.* **14** (2) 34–40
[17] Suradi S S, Yunus R M, Beg M D H and Yusof Z A M 2009 Influence pre-treatment on the properties of lignocellulose based biocomposite *National Conference on Postgraduate Research (NCON-PGR) 2009* 67–78
[18] Norul Izani M A, Paridah M T, Mohd Nor M Y and Anwar U M K 2013 Properties of medium-density fibreboard (MDF) made from treated empty fruit bunch of oil palm *J. Trop. For. Sci.* **25** (2) 175–183
[19] Senawi R, Mohd Alauddin S, Mohd Saleh R and Shueb M I 2013 Polylactic acid/empty fruit bunch fiber biocomposite: Influence of alkaline and silane treatment on the mechanical properties *Int. J. Biosci. Biochem. Bioinforma* **3** (1) 59–61
[20] Mulyantara H L T, Rizaluddin A T, Nakagawa I A and Ohi H 2015 Properties of fibers prepared from oil palm empty fruit bunch for use as corrugating medium and fiberboard *J—STAGE Adv.* **1148** 1349–1159
[21] Ibrahim Z, Ahmad M, Aziz A A, Ramli R, Jamaludin M A, Muhammed S and Alias A H 2016 Dimensional Stability Properties of Medium Density Fibreboard (MDF) from Treated Oil Palm (Elaeis guineensis) Empty Fruit Bunches (EFB) Fibres, *Open J. Compos. Mater.*, **6** (4) 91–99
[22] Hassan S, Chellaiah N R, Sahari B, Salit M S and Abdul Aziz N 2016 Effect of chemical treatment on oil palm empty fruit bunch (OPEFB) fiber on water absorption and tensile properties of OPEFB fiber reinforced epoxy composite *Key Engineering Materials* **701** 295–299.
[23] Mat Soom R, Wan Hassan W H, Md Top A G and Hassan K 2006 Thermal properties of oil palm fibre *J. Palm Oil Res.***18** 272–277
[24] Ndazi B S, Karlsson S, Tesha J V and Nyahumwa C W 2007 Chemical and physical modifications of rice husks for use as composite panels *Compos. Part A Appl. Sci. Manuf.* **38** (3) 925–935
[25] Matoski A, Hara M M, Iwakiri S and Casali J M 2013 Influence of accelerating admixtures in wood-cement panels: Characteristics and properties *Acta Sci. Technology* **35** (4) 655–660
[26] Nasser R A, Salem M Z M, Al-Mefarrej H A and Aref I M 2016 Use of tree pruning wastes for manufacturing of wood reinforced cement composites *Cem. Concr. Compos.* **72** 246–256
[27] Hermawan D, Hata T, Kawai S, Nagadomi W and Kuroki Y 2002 Manufacturing oil palm fronds cement-bonded board cured by gaseous or supercritical carbon dioxide *J. Wood Sci.* **48** (1) 20–24

[28] Mohamed T E 2004 *Effects of mixing some wood and non-wood lignocellulosic materials on the properties of cement and resin- bonded particleboard* University of Khartoum Ph.D Thesis
[29] Sudin R and Swamy N 2006 Bamboo and wood fibre cement composites for sustainable infrastructure regeneration *Journal of Materials Science*, **41** (21) 6917–6924
[30] Noor Azrieda A R, Razali A K, Izran K, Rahim S and Adbul Aziz M 2009 Hydration permormance of cement-bonded wood composites: Compatibility assessment of six pioneer forest species *Borneo Sci.* **25** 47–58
[31] Ashori A, Tabarsa T and Valizadeh I 2011 Fiber reinforced cement boards made from recycled newsprint paper *Mater. Sci. Engineering A* **528** (25–26) 7801–7804
[32] Amel B A, Paridah M T, Rahim S, Osman Z, Zakiah A and Ahmed S H 2014 Effects of kenaf bast fibres on hydration behaviour of cement *J. Trop. For. Sci.* **26** (3) 340–346
[33] de Castro V G, Braz R L, Azambuja R R, Loiola P L, Iwakiri S and Matos J L M 2015 Wood-cement boards of Eucalyptus saligna with different chemical additives and methods of formation *Floresta* **45** (2) 349–360
[34] Ogunsile B O and Adepegba J A 2015 Cement bonded particle board from Musa paradisiaca stalk *Pacific J. Sci. Technol.* **16** (1) 12–20
[35] Ashori A, Tabarsa T, Azizi K and Mirzabeygi R 2011 Wood-wool cement board using mixture of eucalypt and poplar *Ind. Crops Prod.* 34 (1) 1146–1149

**Acknowledgments**
The authors are gratefully acknowledging the technical and financial support of Faculty of Civil and Environmental Engineering, and Centre of Graduate Studies, UTHM, Malaysia.

# Sound absorption coefficient of coal bottom ash concrete for railway application

**N I R Ramzi Hannan[1], S Shahidan[1], Z Maarof[1], N Ali[1], S R Abdullah[1] and M H Wan Ibrahim[1]**

[1]Jamilus Research Center, Faculty of Civil and Environmental Engineering, UniversitiTun Hussein Onn Malaysia.
Corresponding author: shahiron@uthm.edu.my

**Abstract.** A porous concrete able to reduce the sound wave that pass through it. When a sound waves strike a material, a portion of the sound energy was reflected back and another portion of the sound energy was absorbed by the material while the rest was transmitted. The larger portion of the sound wave being absorbed, the lower the noise level able to be lowered. This study is to investigate the sound absorption coefficient of coal bottom ash (CBA) concrete compared to the sound absorption coefficient of normal concrete by carried out the impedance tube test. Hence, this paper presents the result of the impedance tube test of the CBA concrete and normal concrete.

## 1. Introduction

Railway as one of main transportation mode, faces the demand to guarantee that the value of noise emissions does not exceed the permitted noise limits monitored in tracks for different speed zones which are divided into several categories such as standard, speed and high speed tracks [1-3]. Noise from railway transportation mode can come from two different sources either from construction activities or operational activities. Noise from operational activities is the main source of noise problem produced by railway transportation. It derives from several factors such as audible warning devices of all types, whether mounted on the train or near at grade road crossings, passing by of trains on tangential tracks, train movements on curved track sections which can generate wheel squeal and rail yard operations involving trains stopping and starting, assembling of trains, shunting of cars, retarders, use of Phisignalling devices and repair work [4].

EU (European Union) has funded research aimed at investigating human response to railway noise and vibration and has encouraged the development of engineering mitigation measures for example Cargo vibes, RIVAS (Railway Induced Vibration Abatement Solutions). Individual EU countries have also conducted research to evaluate the effects of the increase in railway noise and vibration. All these studies have concluded that an increase in railway noise or vibration exposure results in an increase in self-reported adverse effects and sleep disturbance [5].

Existing noise level in Malaysia for train operation has been recorded by ERE Consulting Group Sdn Bhd in preparation for the Klang Valley Mass Rapid Transit (KVMRT) project. This detailed environmental impact assessment (DEIA) report was approved by the Department Environment (DOE) on 19th November 2010. Noise barrier has been proposed to mitigate the high noise level due to the train operation [6]. This noise barrier will be built along the track especially at the critical area where the train curve will produce high friction between the wheels of the train and the track which resulted in high noise level.

Noise barriers are solid obstruction built between the highway and residential areas. Noise barrier can be built out of wood, concrete, masonry, metal and transparent materials [7].Noise barrier perform at its best if long enough and high enough to block the view of the road. However, the function of the noise barrier is only as a noise reducer but not completely blocking the sound annoyance [7].

Nowadays, noise barriers are usually made out of concrete, wood or steel. However, there are increasing demands for industries to use waste material as their main material to develop a new product made out of 100% from waste material or incorporation between natural resources and waste material.

In Malaysia, Solid Waste and Public Cleansing Management Act 2007 (Act 672)states that the controlled solid waste was defined as any solid waste that belong to any categories such as household solid waste, construction waste and industrial waste. Among these types of solid wastes, industrial solid waste was considered as the worst type of pollution towards the environment compared to the others. One of the examples of industrial waste is coal bottom ash, product of burning coal in electric generation industry that is located in Tanjung Bin, Jimah and Sultan Salahuddin Abdul Aziz/ Kapar.
Coal bottom ash (CBA) is simply disposed-off on open land which poses environmental danger to living beings [8]. The coal bottom ash has different composition of chemical composition depending on different source of coal and mainly composed of silica, alumina and iron with small amounts of calcium, magnesium and sulphate [9]. Large production of coal bottom ash from coal fired power plant has been a major factor to human health problems and environmental pollutions. This is due to the hazardous chemical composition contained in coal bottom ashes that potentially contaminate ground and sources of water near the power plant.

Therefore, due to the lack of research on the new noise barriers from waste material for railway application, this research will emphasize on the new noise barrier material for reducing noise level that currently exist in railway engineering application. The sound absorption coefficient of the CBA material will be measured and tested through impedance tube test and will be compared with the sound absorption coefficient of commonly used concrete for railway noise barrier. This is to improve the ability of noise barrier that is currently used in Malaysia and to reduce noise pollution problem especially for railway application. Besides, the purpose of this study is not limited to reduce the noise pollution in railway application only but may also solve other environmental problems that affect the ecological balance system in Malaysia causes by industrial waste such as coal bottom ash.

## 2. Material and experimental program

### *2.1. Material preparation*
In order to prepare mix concrete design, the mix concrete samples are prepared based on the calculation that had been analysed. It involves preparation of mold, concrete work and curing. Each processes are being explained in the following subtopics. Table 1 represents the properties of mix concrete design.

**Table1.** Mix design for 0.012 $m^3$ of concrete.

| Cement (kg) | Water (kg) | Fine aggregate (kg) | Coarse aggregate(kg) | Bottom ash (kg) |
|---|---|---|---|---|
| 6.42 | 2.892 | 6.48 | 13.2 | 3.394 |

Concrete is made up of three main ingredients which consist of Portland cements, water and aggregates. The ratio of the ingredients changes the properties of final product, which allows the engineer to design concrete that meets their specific needs. Admixtures are added to adjust the concrete mixture for specific performance criteria. Therefore, the selection of material is important in order to determine the require strength of the building and also the lifespan of a building.

However, it is a must to ensure the quality of material that being used in the construction. In this research, there are two type of concrete being casted which is conventional and previous concrete, with different percentage of power plant waste (CBA) as a replacement of fine aggregate in order to determine the best proportion of material that able to absorb sound. Previous concrete can be produce using conventional concrete-making materials, namely cement, all types of coarse and fine aggregates and water.

### *2.2. Concrete mix design*

Table 2 represents the details of concrete mix design proportion that being tested in this research for both samples which involve cube shape and cylinder shape. It can be seen from the table 2 that the percentage of coal bottom ash increase as the fine aggregate decreased. This is because the CBA act as a fine aggregate replacement to compare the acoustic and non-acoustic performance of the CBA concrete mixture with the control concrete which have zero CBA material. The other material such as cement, water and coarse aggregate were kept constant along this concrete preparation.

**Table 2.** Proportion of concrete mix design for 0.012 $m^3$ of concrete samples for both cube shape and cylinder shape sample.

| Percentage of CBA (%) | Cement (kg) | Water (kg) | Fine aggregate (kg) | Coarse aggregate (kg) | Coal bottom ash (CBA) (gram) |
|---|---|---|---|---|---|
| 0 (control) | 6.42 | 2.892 | 6.48 | 13.2 | 0 |
| 10 | 6.42 | 2.892 | 6.141 | 13.2 | 339.4 |
| 20 | 6.42 | 2.892 | 5.801 | 13.2 | 678.8 |
| 30 | 6.42 | 2.892 | 5.462 | 13.2 | 1018.2 |
| 40 | 6.42 | 2.892 | 5.122 | 13.2 | 1357.6 |
| 50 | 6.42 | 2.892 | 4.783 | 13.2 | 1697.0 |
| 75 | 6.42 | 2.892 | 3.935 | 13.2 | 2545.5 |
| 100 | 6.42 | 2.892 | 3.086 | 13.2 | 3394.0 |

Table 3 represents the volume and density of the specimens for impedance tube test purpose. The impedance tube test required two type of test which involve high frequency test and low frequency test. Thus, there are two size of the cylinder shape sample as shown in table 3. The cylinder 1 is for low frequency test and cylinder 2 for high frequency test. Then, the CBA material percentages were being increased from 10% to 100% as shown in table 2.

**Table 3.** Proportion of concrete mix design for 0.012 $m^3$ of concrete samples for both cube shape and cylinder shape sample.

| Type of specimen | Volume of specimen ($m^3$) | Density materials ($kg/m^3$) |
|---|---|---|
| Cylinder 1 | $\pi$ x 50mm$^2$ x 100mm | 0.064 |
| Cylinder 2 | $\pi$ x 14.5mm$^2$ x 100mm | 0.064 |

### *2.3. Impedance tube test*

Impedance tube used in this research as shown in figure 1 is accompanied with AFD 1001 software which has several steps to carry out the test.

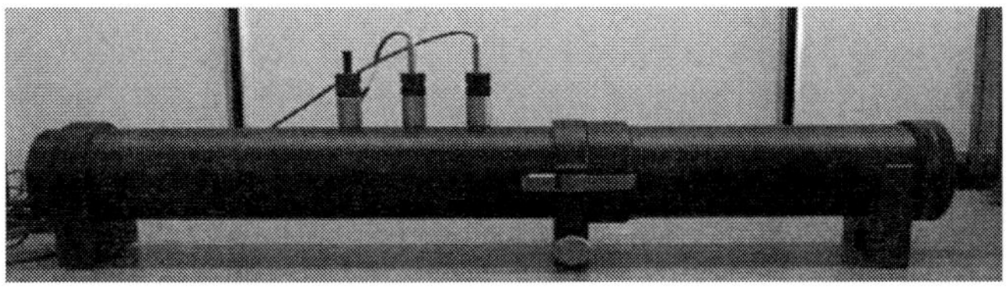

**Figure 1.** Impedance tube used in this research.

To determine the ability of the samples to absorb the noise and to determine the reflecting on the noise that emitted from the train with coal bottom ash concrete, the impedance tube testing had been conducted. The details of dimension of impedance tube that being used to test all the specimens were 29mm in diameter for high frequency while for low frequency will have 100mm in diameter.
The steps to use the impedance tube are divided into 8 steps as follows:

1. The data collected being processed by software named AFD 100. The software being opened and ready to run the data collected.
2. The distance between microphone and the distance between samples being inserted.
3. The calibration process with normal position being conducted at this stage.
4. The microphone being located at the above part of impedance tube.
5. The impedance tube being calibrated and the microphone being switched into opposite direction.
6. The position of the microphones switched between each other
7. The measurement of low frequency and high frequency are started and the results being recorded

**3. Results and discussions**

When a sound wave strikes a material, a portion of the sound energy was reflected back and another portion was absorbed by the material while the rest was transmitted. First, the influence of coal bottom ash percentage as a replacement for fine aggregate in concrete mixtures was tested with low frequency sound wave in impedance tube. The results were presented in table 4.

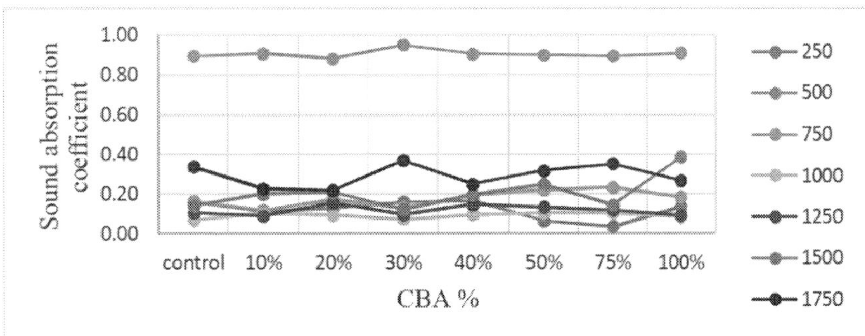


**Figure 2.** Sound absorption coefficient of the CBA concrete mixture at low frequency.

Figure 2 represents the value of sound absorption coefficient recorded by the CBA concrete mixture with CBA percentages ranges in between 10% to100%. All the sound absorption coefficients recorded were compared to the reference sample which was control sample where the CBA materials were absent. All the recorded values were summed up into table 4. The highest sound absorption coefficient recorded by all ranges of CBA percentages were at 500 Hz of sound frequency for low frequency of impedance tube test as can be seen in figure 2.

**Table 4.** The sound absorption coefficient of the coal bottom ash concrete mixture base on its percentages for low frequency.

| CBA % / Sound Frequency | Control | 10% | 20% | 30% | 40% | 50% | 75% | 100% | Mean |
|---|---|---|---|---|---|---|---|---|---|
| **250** | ***0.16*** | 0.11 | 0.13 | 0.16 | 0.16 | 0.06 | 0.04 | 0.14 | ***0.11*** |
| **500** | ***0.89*** | 0.91 | 0.88 | 0.95 | 0.91 | 0.90 | 0.89 | 0.91 | ***0.91*** |
| **750** | ***0.16*** | 0.12 | 0.17 | 0.13 | 0.19 | 0.22 | 0.23 | 0.18 | ***0.18*** |
| **1000** | ***0.07*** | 0.10 | 0.09 | 0.08 | 0.10 | 0.11 | 0.10 | 0.09 | ***0.09*** |
| **1250** | ***0.11*** | 0.09 | 0.16 | 0.10 | 0.15 | 0.13 | 0.12 | 0.09 | ***0.12*** |
| **1500** | ***0.14*** | 0.20 | 0.21 | 0.12 | 0.19 | 0.25 | 0.14 | 0.39 | ***0.21*** |
| **1750** | ***0.34*** | 0.23 | 0.22 | 0.37 | 0.25 | 0.32 | 0.35 | 0.27 | ***0.29*** |
| **Mean** | ***0.27*** | ***0.25*** | ***0.27*** | ***0.27*** | ***0.28*** | ***0.28*** | ***0.27*** | ***0.29*** | - |

Table 4 presents the recorded values of the sound absorption coefficient by CBA concrete mixture. It can be seen that all the recorded values were above the sound absorption coefficient of the control sample where the CBA materials were absent. This implies that, the CBA concrete mixture had a better acoustic performance compared to the control sample. Thus, the CBA concrete mixture was proven to have a better acoustic performance for low frequency of impedance tube test.

There were 3 samples of coal bottom ash concrete mixtures per batch that were tested with low frequency of noise using impedance tube. Table 4 shows the averages value of sound absorption coefficient for each batch and the value of sound wave being strike to the samples were between 250-1750 Hz. The results revealed that the sound absorption coefficient of the CBA concrete mixture samples with a yellow highlighted that from 10%-100% shows no significance difference compared to the control concrete mixture. The 10% CBA concrete mixtures was the only CBA concrete mixtures that recorded a lower sound absorption coefficient than the control concrete mixture. Hence, the performance of the coal bottom ash concrete mixtures almost same or at the same level to the control sample. It also specifies that the CBA did not deteriorate the acoustic performance of the control sample where the coal bottom ash fine aggregates were absent.

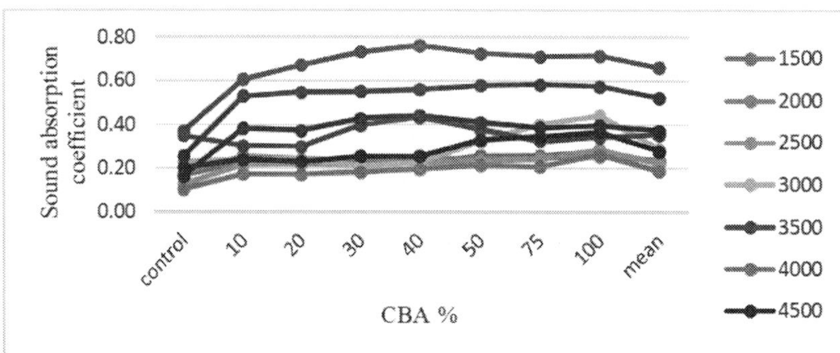


**Figure 3.** Sound absorption coefficient of the CBA concrete mixture at high frequency.

Figure 3 illustrates the recorded values of sound absorption coefficient for the high frequency impedance tube test. The results obtained as presented in figure 3 demonstrate almost a similar pattern with the result obtained for the low frequency impedance tube test. All the sound absorption coefficients recorded were compared with the control sample where the CBA materials were absent. However, the highest recorded of sound absorption coefficient of all ranges of CBA percentages were recorded at 6000 Hz frequency of sound for this high frequency impedance tube test. It can be disclosed that the highest sound absorption coefficient for both control samples and coal bottom ash concrete mixtures were recorded at 6000 Hz sound wave frequency. Figure 3 also reveals that the maximum sound absorption coefficient was recorded at 40% of coal bottom ash as a fine aggregate replacement of concrete in a range of 1500-6000 Hz of sound wave frequency.

**Table 5.** The sound absorption coefficient of the coal bottom ash concrete mixture base on its percentages for high frequency.

| CBA % / Sound Frequency | Control | 10% | 20% | 30% | 40% | 50% | 75% | 100% | *Mean* |
|---|---|---|---|---|---|---|---|---|---|
| **1500** | ***0.17*** | 0.23 | 0.22 | 0.22 | 0.22 | 0.26 | 0.26 | 0.28 | ***0.23*** |
| **2000** | ***0.18*** | 0.26 | 0.24 | 0.23 | 0.24 | 0.24 | 0.25 | 0.26 | ***0.24*** |
| **2500** | ***0.13*** | 0.22 | 0.22 | 0.22 | 0.22 | 0.23 | 0.25 | 0.29 | ***0.22*** |
| **3000** | ***0.23*** | 0.24 | 0.24 | 0.21 | 0.23 | 0.32 | 0.40 | 0.44 | ***0.29*** |
| **3500** | ***0.35*** | 0.30 | 0.30 | 0.40 | 0.43 | 0.38 | 0.33 | 0.34 | ***0.35*** |
| **4000** | ***0.10*** | 0.17 | 0.17 | 0.18 | 0.20 | 0.21 | 0.21 | 0.26 | ***0.19*** |
| **4500** | ***0.20*** | 0.24 | 0.23 | 0.26 | 0.26 | 0.33 | 0.35 | 0.37 | ***0.28*** |
| **5000** | ***0.16*** | 0.38 | 0.37 | 0.43 | 0.44 | 0.41 | 0.39 | 0.40 | ***0.37*** |
| **5500** | ***0.26*** | 0.53 | 0.55 | 0.55 | 0.56 | 0.58 | 0.58 | 0.58 | ***0.52*** |
| **6000** | ***0.37*** | 0.61 | 0.67 | 0.73 | 0.76 | 0.73 | 0.71 | 0.71 | ***0.66*** |
| *Mean* | ***0.21*** | ***0.32*** | ***0.32*** | ***0.34*** | ***0.36*** | ***0.37*** | ***0.37*** | ***0.39*** | - |

Table 5 reveals all the recorded values of the sound absorption coefficient for high frequency of impedance tube test. It indicates that all the recorded values were above than the control sample where the CBA materials were absent. Thus, it proves that the CBA concrete mixture demonstrates a better acoustic performance for both low and high frequency of impedance tube test compared to the control sample which was known as common concrete which was used as a noise barrier by KTMB.

Table 5 also represents the average value of sound absorption coefficient for each batch and the values of sound wave being strike to the concrete samples were ranged between 1500-6000 Hz to be considered as high frequency impedance tube test. Table 6 exhibitsa similar pattern of the result between low and high noise frequency. There was no significant difference of sound absorption coefficient from 10% of CBA concrete mixture to 100% of CBA concrete mixture. However, there was a significant difference in terms of optimal sound frequency being absorbed by the concrete samples as being shown in figure 3.

As presented in table 5 and figure 3, the sound absorption coefficient of all coal bottom ash concrete mixture were higher than the control or normal concrete mixture. Therefore, the development of coal bottom ash concrete mixture in this research was comparable with the normal or control concrete mixture. The acoustic performance of coal bottom ash concrete mixture was better than the acoustic performance of control or normal concrete mixture. The reference concrete presented in both table 4 and 5 recorded the lowest sound absorption coefficient compared to the CBA concrete mixture. The combination of CBA concrete mixture had been found to exhibit better sound absorption behaviour than the control concrete. According to the result obtained in figure 3, 40% of CBA fine aggregate replacement was the best combination for the acoustic concrete mixture giving the most effective sound absorption coefficient.

According to Ramzi Hannan *et al.*[10], a porous material is effective from the sound absorption point of view when its thickness is approximately one-tenth the wavelength of the incident sound. This could explain why the sound absorption coefficient recorded by all ranges of CBA percentages for both CBA concrete mixture and control samples were highest at certain frequency of sound. The previous research also mentioned that "The maximum absorption occurs at a resonance frequency of a quarter the wavelength of the incident sound" [11]. Thus, the ideal resonance frequency for CBA concrete mixture at all percentages were recorded at 500 Hz and 6000 Hz for low and high frequency of impedance tube test respectively as shown in figure 2 and 3.

The phenomenon where the acoustic performance of the CBA concrete mixture performs better than the control sample without the presence of the CBA materials could be explained by the previous research that stated "The increase in sound absorption coefficient of the coal bottom ash concrete mixture as the percentage of coal bottom ash concrete mixture increase was due to the increase of the porous structure of the concrete due to the vesicular texture, a characteristic of popcorn particles" [12]. A porous concrete creates lots of voids in the concrete and significantly increase the reflection of sound wave in the concrete structure itself and dissipates the sound energy into heat energy due to the porosity effects [13]. Thus, this explanation could prove why the CBA concrete mixture performs better than control sample where the CBA materials were absent in terms of acoustic performance.

**Table 6.** Sound absorption coefficient provided by akustik.ua.

| **Material** | *Frequency (Hz)* | | | | | |
|---|---|---|---|---|---|---|
| **Masonry wall** | **125** | **250** | **500** | **1000** | **2000** | **4000** |
| Rough concrete | 0.02 | 0.03 | 0.03 | 0.03 | 0.04 | 0.07 |
| Smooth unpainted concrete | 0.01 | 0.01 | 0.02 | 0.02 | 0.02 | 0.05 |
| Smooth concrete, painted or glazed | 0.01 | 0.01 | 0.01 | 0.02 | 0.02 | 0.02 |
| Breeze block | 0.2 | 0.45 | 0.45 | 0.4 | 0.45 | 0.4 |
| Plaster on solid wall | 0.04 | 0.05 | 0.05 | 0.08 | 0.04 | 0.06 |
| Plaster, lime or gypsum on solid backing | 0.03 | 0.03 | 0.03 | 0.03 | 0.04 | 0.05 |

Table 6 represents the details of the sound absorption coefficient of the commonly used noise barrier materials and the highlighted yellow colour were the commonly used material as a noise barrier in KTMB industry which were rough concrete, smooth unpainted concrete and plaster on solid wall. These types of materials were the materials that were found at the four railway stations involved in this research. All the materials recorded a lower sound absorption coefficient compared with the CBA concrete mixture at the same frequency of sound.

**Table 7.** Sound absorption coefficient comparison between CBA concrete mixture with non-CBA concrete mixture.

| **Material** | *Frequency (Hz)* | | | | | |
|---|---|---|---|---|---|---|
| **Masonry wall** | **125** | **250** | **500** | **1000** | **2000** | **4000** |
| Rough concrete | 0.02 | 0.03 | 0.03 | 0.03 | 0.04 | 0.07 |
| Smooth unpainted concrete | 0.01 | 0.01 | 0.02 | 0.02 | 0.02 | 0.05 |
| Plaster on solid wall | 0.04 | 0.05 | 0.05 | 0.08 | 0.04 | 0.06 |
| CBA concrete mixture | | 0.11 | 0.91 | 0.09 | 0.24 | 0.19 |

Table 7 indicates the comparison values of the sound absorption coefficient recorded by both CBA concrete mixture and non-CBA concrete mixture concrete that was commonly used as a noise barrier in KTMB industry. From table 8, it can be distinguishedthat the CBA concrete mixture perform significantly better acoustic performance by recorded significantly higher sound absorption coefficient than the commonly used concrete mixture without the presence of CBA materials.

## 4. Conclusion

The results obtained in this research proved that the CBA concrete mixture performed much better in terms of both acoustic and non-acoustic performance. Thus, the main objectives of this study was proven since the acoustic and non-acoustic performance of the CBA concrete mixture was evaluated and compared to the existing normal concrete mixture.

## 5. References

[1] Huang X, He P and Hua Z 2016 A cooperative differential game of transboundary industrial pollution between two regions *J. Clean. Prod.* **120** 43–52

[2] LidénT 2015 Railway Infrastructure Maintenance - A Survey of Planning Problems and Conducted Research, *Transp. Res. Procedia.* **10** 574–583

[3] Leman A S, Shahidan S, Senin M S and Hannan N I R R 2016 A preliminary study on chemical and physical properties of coconut shell powder as a filler in concrete *Int. Engineering Research and Innovation Symp. (Melaka, Malaysia)* vol 160 (London: IOP Conference Series Materials Science and Engineering) 12059

[4] Shahidan S, Hannan N I R R, Maarof M Z, Leman A S and Senin M S 2017 A comprehensive review on the effectiveness of existing noise barriers commonly used in the railway industry *MATEC Web of Conference* vol 87 (Paris:EDP Sciences) 01007

[5] Mcgrory M, Cirac D C, Gaussen O and Cabrera D 2012 Sound absorption coefficient measurement : Re-examining the relationship between impedance tube and reverberant room methods*Proc. of Acoustic 2012 – Fremantle (Fremantle, Australia)*(Australia: Australian Acoustic Society) pp 1–8

[6] Peris E., Woodcock J, Sica G, Sharp C, Moorhouse A. T, and Waddington D. C 2016 Guidance for new policy developments on railway noise and vibration. Transportation Research Part A, 85, 76–88.

[7] Halim H, Abdullah R, Abang Ali A A and Mohd Nor M J 2015 Effectiveness of existing noise barriers: comparison between vegetation, concrete hollow block, and panel concrete *Procedia Environmental Sciences* vol 30 (United Kingdom:Elsevier) pp 217–221

[8] Siddique R 2014 Utilization of industrial by-products in concrete *Procedia Engineering* vol 95 (United Kingdom:Elsevier) pp 335–347

[9] Singh M and Siddique R 2013 Effect of coal bottom ash as partial replacement of sand on properties of concrete, *J. Clean. Prod.***112** 620–630

[10] Ramzi N I R, Shahidan S, Maarof M Z and Ali N 2016 Physical and Chemical Properties of Coal Bottom Ash (CBA) from Tanjung Bin Power Plant *IOP Conference Series Materials Science and Engineering* vol 160 (London: IOP Publishing) 12056

[11] Abu bakar A U and Baharudin K S 2012 Potential Use of Malaysian Thermal Power Plants *Int. J. Sustain. Constr. Eng. Technol.* **3** 25–37

[12] Md No N, Ibrahim A, Muhamad Bunnori N, Saman H M, Mat Saliah S N and Shahidan S 2014 Diagnostic of fatigue damage severity on reinforced concrete beam using acoustic emission technique, *Eng. Fail. Anal.* **4 1** 1–9

[13] Khaiwal R, Singh T, Tripathy J P, Mor S, Munjal S, Patro B and Panda N 2016 Assessment of noise pollution in and around a sensitive zone in North India and its non-auditory impacts, *Sci. Total Environ.* **566–567** 981–987

## Acknowledgement

The authors acknowledge the technical and financial support provided by UniversitiTun Hussein Onn Malaysia Vot (U523) and FRGS Vot 1576.

IOP Conf. Series: Materials Science and Engineering **271** (2017) 012078 doi:10.1088/1757-899X/271/1/012078

# Marshall properties and rutting resistance of hot mix asphalt with variable reclaimed asphalt pavement (RAP) content

**A K Arshad[1,2], Z Abd Karim[2], E Shaffie[2], W Hashim[2] and Z Abd Rahman[2]**

[1] Associate Professor, Institute for Infrastructure Engineering and Sustainability Management (IIESM), Universiti Teknologi MARA, Shah Alam, Selangor, Malaysia
[2] Faculty of Civil Engineering, Universiti Teknologi MARA, Selangor Malaysia
Corresponding author:drahmadkamil@salam.uitm.edu.my

**Abstract.** Hot mix recycling is the process in which reclaimed asphalt pavement (RAP) materials are combined with new materials to produce hot mix asphalt mixtures. This paper details a laboratory study in which the effects of different RAP contents were evaluated on the performance of hot recycled mixes. The objective of this study is to evaluate the rutting resistance of asphalt mixes containing different percentages of RAP. The Marshall mix design method was adopted in this study to determine the OBC for the asphalt mixes containing four aggregate combinations with RAP contents of 0% (control), 15%, 25% and 35%. Volumetric analysis was performed to ensure that the result is compliance with specification requirements. The resilient modulus test was performed to measure the stiffness of the mixes while the Hamburg wheel tracking test was used to evaluate the rutting performance of these mixes. The results obtained showed that there were no substantial differences in volumetric properties, stability values and stiffness properties between the control mix and recycle mixes. It can be concluded that recycled mixes performed as good as the performance of conventional Hot Mix Asphalt (HMA) in terms of resilient modulus and rutting.

## 1. Introduction

The pavement industry has long emphasized the need to reuse RAP materials obtained from deteriorated roads as these materials still possess desirable properties to be used for the surfacing layers, subject to the limitations set in the specification used. With the dwindling supply of new resources and spiraling cost of materials, the use of RAP is a suitable way to conserve non-renewable resources, that is aggregates and bitumen used for asphalt mixes [1,2]. The increase in asphalt materials prices has led the road construction industry to strive for the use of higher percentages of RAP to reduce the cost of road projects. A higher addition of RAP should enhance the value of the recycled pavements and have the potential for reducing the quantity of waste materials [3].

The use of RAP in asphalt concrete also adheres to the requirement of sustainable solutions in pavements because it is both environmentally friendly and cost effective [4]. By adding RAP to asphalt mixtures, it is estimated that the environmental impact of production is reduced by 23%. Furthermore, RAP presents a significant material cost reduction. Quality virgin aggregate material is becoming increasingly difficult to find and purchase. Therefore, the use of RAP can offset costs and allow road authorities to build and rehabilitate more roadways with similar budget capacities.
Most road authorities allow the incorporation of not more than 30% of RAP in hot mix asphalt (wearing course) to avoid any detrimental effects on the mix properties [5,6]. For example, the Illinois

Department of Transportation allows up to only 30% RAP in binder and surface mixtures depending upon the traffic level present on a given roadway. This is because RAP is usually considered to be a stiff material primarily due to oxidative hardening and other aging mechanisms it undergoes while exposed to the environment during its service life. Consequently, the increased stiffness in RAP may lead to various forms of cracking failures which deter producers and road authorities from further increasing the RAP content [7].

The Asphalt Institute suggests that when 20 percent or less RAP is used in a mix, no change in asphalt grade is required [8]. However, for mixes with greater than 20 percent RAP, a drop in one grade (softer asphalt cement) is recommended to compensate for the higher viscosity of the oxidized binder. Many road authorities use the same grade of asphalt cement regardless of the RAP content.

The Asphalt Institute's manual on mix design methods for asphalt concrete provides a method to determine necessary mix design characteristics (such as stability, flow, and air voids content) for the Marshall mix design method. The final mix design proportions for the recycled hot mix paving mixture will be determined by completing mix design testing using standard procedures to satisfy the applicable mix design criteria. Additional virgin aggregates may be required to satisfy the gradation requirements to improve stability and to limit the RAP content in recycled hot mixes. The Marshall mix design method was adopted in this study to determine the optimum binder content (OBC) for the asphalt mixes containing four aggregate combinations with RAP contents of 0% (control), 15%, 25% and 35%. The objective of this study is to evaluate the rutting resistance of the asphalt mixes containing different percentages of RAP.

## 2. Methodology

The samples for control mix and recycled mixes were prepared based on the Marshall mix design method in accordance with ASTM D1559. The mix design used AC14 aggregate gradation in accordance with Public Works Department of Malaysia's (PWD Malaysia) Specification for Road Works [9], with the range of binder content from 4.0% to 6.0%. A total of 15 samples each were prepared for the control mix (100% virgin aggregates) and the asphaltic concrete mixes with RAP contents of 15% (RAP15), 25% (RAP25) and 35% (RAP35) by weight. All the samples were compacted with 75 blows on both surfaces of the samples using the Marshall compactor. Volumetric properties and analysis were carried out to determine the optimum binder content (OBC) for the control mix and recycled mixes. The optimum binder content value was used to prepare samples for the rutting resistance test. The Hamburg Wheel Tracking test was carried out in accordance with AASHTO T324 to evaluate the rutting susceptibility of mixtures under high temperature [11]. The equipment used for this test was the Asphalt Pavement Analyzer (APA) that is capable of running Hamburg Wheel Tracking tests. The test was run at a rate of 50 passes of a steel wheel per minute with a load of 705 + 22N. This test was conducted until the maximum number of 8,000 passes has occurred or until the maximum rut depth value of 20mm has been reached.

## 3. Result and discussion

Initial laboratory tests were conducted to determine the physical properties of the materials used. Marshall mix design was conducted to determine the Marshall properties and the optimum binder content (OBC) for control mix and RAP mixes. The result of resilient modulus test and rutting resistance test as performance tests were then compared between the control mix and RAP mixes.

### *3.1. Comparison of RAP Aggregate and Virgin Aggregate Properties*

The results of Aggregate Abrasion Value, Aggregate Impact Value, Aggregate Crushing Value and Specific Gravity for recycled aggregate and virgin aggregate is shown in table 1. The LA abrasion value, aggregate impact value, aggregate crushing value and specific gravity of recycled aggregate are higher than the virgin aggregate. This may be due to the fact that the recycled aggregate was exposed to traffic loadings for a long period of time, reducing the strength of the aggregate. However, the result

IOP Conf. Series: Materials Science and Engineering **271** (2017) 012078 doi:10.1088/1757-899X/271/1/012078

shows that the properties of recycled and virgin aggregate are within the specification requirements. This indicates that the aggregate is suitable to be used for the preparation of the mixtures.

**Table 1.** Basic Properties of virgin aggregate and RAP aggregate.

| Aggregate Properties | Recycled Aggregate | Virgin Aggregate | PWD Malaysia's Requirements |
|---|---|---|---|
| LA Abrasion Value (%) | 25.00 | 23.00 | <25 |
| Aggregate Impact Value (%) | 21.80 | 19.40 | <25 |
| Aggregate Crushing Value (%) | 24.31 | 22.24 | <25 |
| Bulk Specific Gravity, $G_{sb}$ (g/cm) | 2.625 | 2.648 | - |
| Apparent Specific Gravity, $G_{sa}$ (g/cm) | 2.711 | 2.705 | - |
| Water Absorption (%) | 1.063 | 0.626 | <2.0 |

### *3.2. Marshall properties*

Marshall tests were conducted to determine the volumetric properties of mixture. The volumetric properties determined included bulk specific gravity, air voids, voids filled with bitumen (VFB), stability and flow. The Optimum Binder Content (OBC) for the control and RAP mixes were determined from the individual plots of bulk density, air void, voids fill with asphalt (VFA), stability and flow versus percent asphalt content.

There are significant effects on the Marshall properties with the inclusion of RAP in the asphalt mixes. Marshall test results showed that the Marshall flow, VFA and Bulk SG increased as the RAP content increased. However, Marshall stability value and air voids in mix showed a decreasing trend with increasing RAP content. The results are shown in figures 1 to 5 below. These trends could be attributed to the existing bitumen in RAP which influences the properties as its percentage increases. However, the results and analysis showed that there is not much variation between conventional mixes with the RAP mixes and all the mixes satisfy the specification requirements for the Marshall tests.

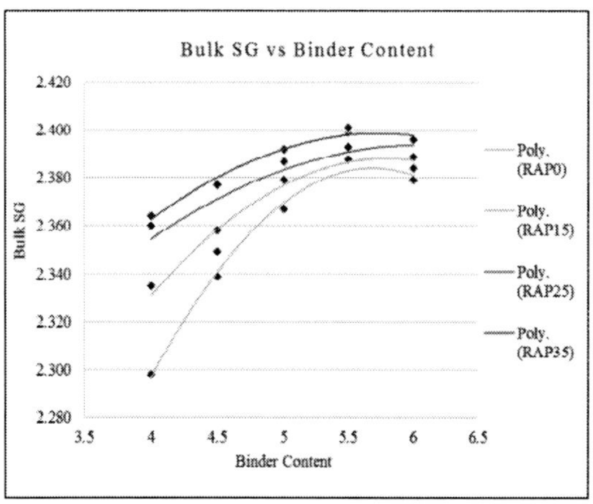


**Figure 1.** Bulk SG versus binder content for different RAP content.

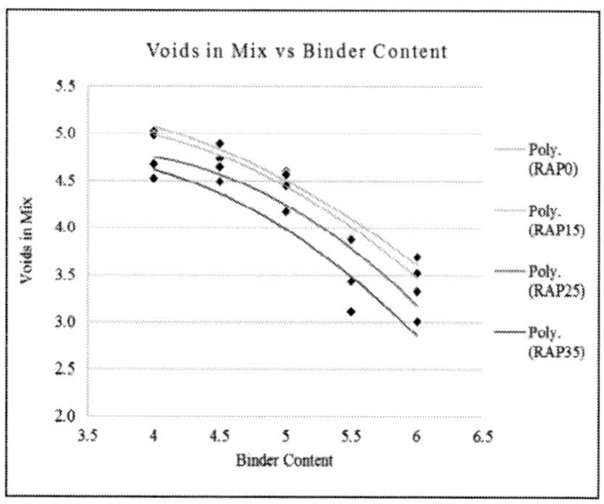


**Figure 2.** Air Voids versus binder content of different RAP content.

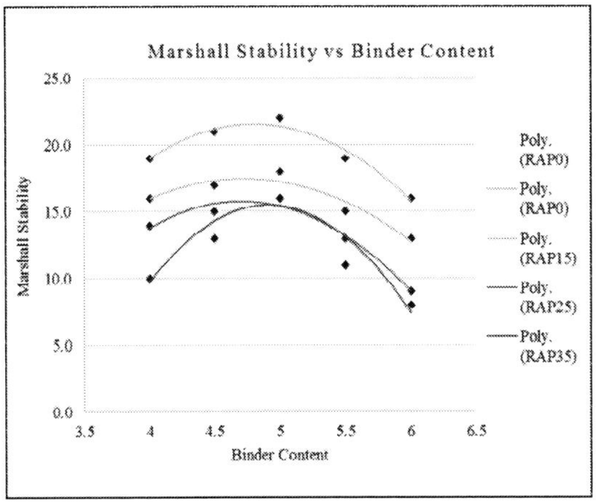


**Figure 3.** Stability versus binder content of different RAP content.

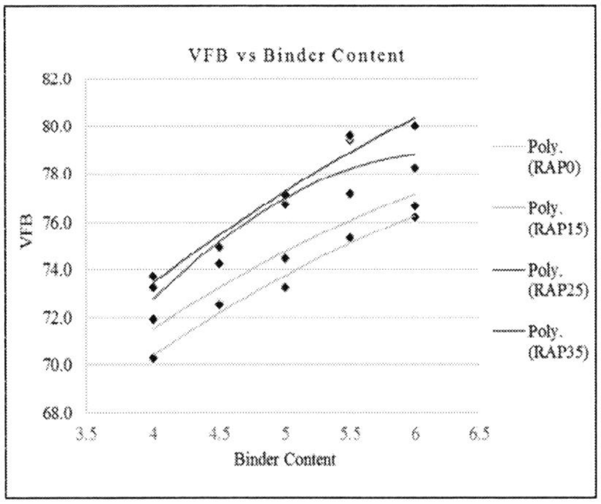


**Figure 4.** VFB versus binder content of different RAP content.

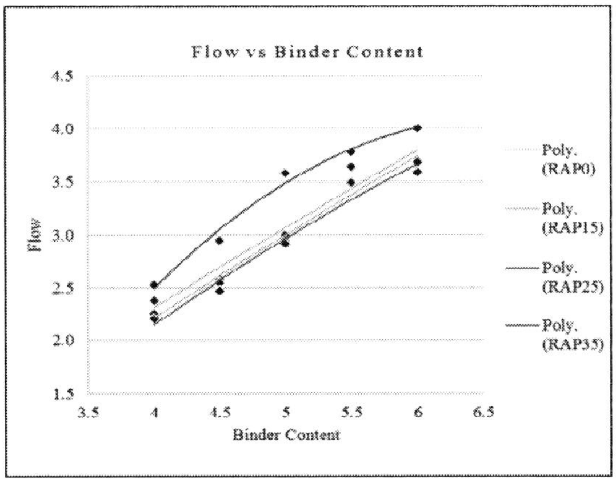


**Figure 5.** Flow versus binder content of different RAP content.

### *3.3. Marshall properties and determination of optimum binder content (OBC)*

The OBC for control mix and RAP mixes were taken as the arithmetic mean of the value of binder content from the five smooth curves plotted and the value was checked with the limits set in PWD Malaysia's specification. Table 2 shows the Optimum Binder Content (OBC) for control and RAP mixes. The OBC decreased as the RAP content increased. The OBC for control mix, RAP15, RAP25 and RAP 35 are 5.34%, 5.22%, 5.10% and 4.88% respectively. The result obtained showed that there were no substantial differences in Marshall Properties.

**Table 2**. Results for marshall mix design for different RAP contents.

| Properties | Control Mix | RAP 15 | RAP 25 | RAP 35 | PWD Malaysia's Specification |
|---|---|---|---|---|---|
| OBC (%) | 5.34 | 5.22 | 5.10 | 4.88 | - |
| Bulk S.G | 2.388 | 2.393 | 2.399 | 2.401 | - |
| Air voids (%) | 4 | 4 | 4 | 4 | 3-5 |
| VFB (%) | 75 | 75 | 75 | 75 | 70-80 |
| Stability (kN) | 22 | 18 | 16 | 18 | >8 |
| Flow (mm) | 3.0 | 2.95 | 2.92 | 2.95 | 2-4 |

*3.4. Rutting resistance test*

The result for rutting test for the asphalt mixes with different RAP contents is shown in figure 6. The rut depth increased with the increase in the RAP content of the asphalt mix. The control sample exhibit the highest resistance to rutting compared to mixes containing RAP with a depth of 3.4 mm at 8000 wheel passes. RAP15 reached a rut depth of 3.7 mm at 8000 wheel passes, followed by RAP25 and RAP35 with rut depths of 4.2 mm and 4.4 mm respectively. The higher rut depth indicates that the mixes with higher RAP content have lower resistance to rutting. This increase in rut depth may possibly be due to the effect of the existing binder in the RAP which increases the deformation of the mix and increases the rut depth.

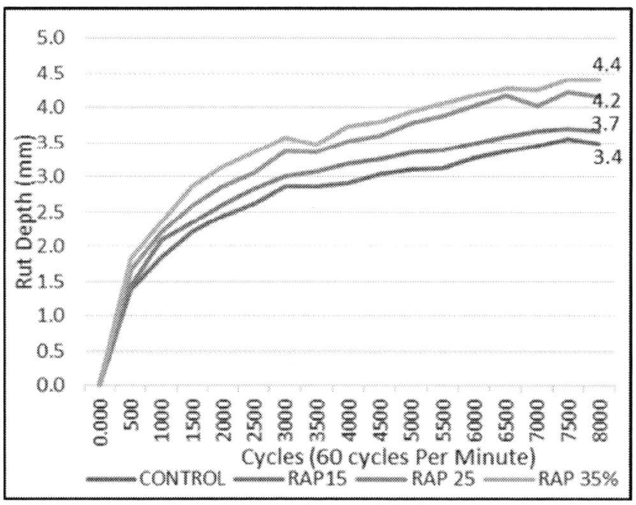


**Figure 6.** Rut depth for Hamburg wheel tracking test for different RAP content.

## 4. Conclusion

From this study, the following conclusions can be made:

- The laboratory result for the physical properties shows that the properties materials used in this study satisfy the specification requirements of the Public Works Department of Malaysia. However, from the tests carried out, the RAP aggregate was found to have lower strength than the virgin aggregate. This is due to the RAP material which was exposed to the environment and traffic load for a long time.

- From the Marshall test result, the OBC for control mix is 5.34%, RAP 15 is 5.22%, RAP25 is 5.10% and RAP35 is 4.88%. The OBC for mixes containing RAP reduced as the amount of RAP increased, possibly due to the influence of the existing binder in the RAP. It can be concluded that the use of high percentages of RAP could influence the properties of the Marshall mix design.

- The Hamburg wheel tracking test showed that the rutting resistance also decreased when the RAP content increased. The control mix sample shows the highest rutting resistance, followed with RAP15, RAP25 and RAP35. However, the difference in rut depth between the control sample and RAP mixes is minimal, indicating that the mixes with RAP have comparable rut resistance compared to control mix.

It can be concluded that the rutting performance of mixes containing RAP of up to 35% by weight is comparable to asphaltic concrete mixes containing virgin aggregates based on the Hamburg wheel tracking test carried out in the study. However, a detailed study and investigation is required for higher RAP contents to ensure compliance with specification requirements.

## 5. References

[1] ARRA 2001 *Basic Asphalt Recycling Manual* (Annapolis: Asphalt Recycling and Reclaiming Association)
[2] Abu Abdo A M 2016 Utilizing reclaimed asphalt Pavement (RAP) materials in new pavements - A Review, *Int. J. of Thermal & Env. Eng.* **12(1)** 61 - 66
[3] Michelle S 2007 *Materials Recycling and Reuse: Finding Opportunities* Report No. CDOT-2007-12. (Colorado: Colorado Department of Transportation)
[4] Chiu C, Hsu T and Yang W 2008 Life cycle assessment on using recycled materials for rehabilitating asphalt pavements, *Resources Conservation and Recycling* **52(3)** 545 - 556
[5] Tao M and Mallick R B 2009 Effects of warm-mix asphalt additives on workability and mechanical properties of reclaimed asphalt pavement Material, *Transportation Research Record* **2126** 151 – 160
[6] Kuehl R, Korzilus J and Marti M 2016 *Synopsis of Recycled Asphalt Pavement Material* (St Paul: Minnesota Local Road Research Board)
[7] Xiao F, Amirkhanian S and Juang C H 2007 Rutting resistance of rubberized asphalt concrete pavements containing reclaimed asphalt pavement mixtures, *J. Mater. Civ. Eng.* **19(6)** 475–483
[8] Asphalt Institute 1986 *Asphalt Hot-Mix Recycling* (Maryland: Asphalt Insitute)
[9] Public Works Department of Malaysia 2008 *Standard Specification for Road Works* (Kuala Lumpur: PWD Malaysia)
[10] ASTM 1997 *Standard Test Method for Indirect Tension Test for Resilient Modulus of Bituminous Mixtures D4123* (West Conshohocken: American Society for Testing and Materials)
[11] AASHTO 2011 *Standard Method of Test for Hamburg Wheel-Track Testing of Compacted Hot Mix Asphalt T324* (Washington DC: American Association of State Highway and Transportation Officials)

**Acknowledgments**
The authors expressed utmost gratitude to Universiti Teknologi MARA, Shah Alam and FRGS Research Grant: FRGS/1/2015/TK08/UITM/02/3 from Ministry of Higher Education, Malaysia for financial support which enables this paper to be written.

# Mechanical behaviour of fibre reinforced concrete using soft – drink can

**J Ilya[1] and C Cheow Chea[2]**

[1] School of Environmental, Kompleks Pusat Pengajian Jejawi 3, Universiti Malaysia Perlis, 02600 Arau, Perlis, Malaysia
[2] School of Environmental, Kompleks Pusat Pengajian Jejawi 3, Universiti Malaysia Perlis, 02600 Arau, Perlis, Malaysia
Corresponding author: ilya@unimap.edu.my

**Abstract.** This research was carried out to study the behaviour of concrete, specifically compressive and flexural strength, by incorporating recycled soft drink aluminium can as fibre reinforcement in the concrete. Another aim of the research is to determine the maximum proportion of fibres to be added in the concrete. By following standard mix design, Ordinary Portland Cement (OPC) concrete was made to have a target mean strength of 30 N/mm$^2$ with not more than 30 mm of slump. Having the same workability, OPC concrete with 0%, 1% and 2% of soft drink can aluminium fibre was prepared based on weight of cement. The specimens were tested for compressive strength and flexural strength. Laboratory test results based on short term investigation reveals that the compressive strength and flexural strength of concrete containing fibre are higher than of normal OPC concrete. Among two volume fractions, concrete with 1% of soft drink can fibre have performed better result in compressive strength and flexural strength compared with 2% amount of soft drink can fibre. The optimum proportion of aluminium fibre to be added in the concrete as fibre reinforcement is 1% fibre content by weight of cement which gave all the positive response from all the tests conducted.

## 1. Introduction

These Concrete is a Portland cement concrete when the aggregate is mixed together with dry cement and water, they form a liquid mass that is easily to mould in shape. Concrete is the most widely used construction material due to its high compressive strength, long service life and low cost. However, concrete is relatively strong in compression but weak in tension and low crack resistance. To improve such weakness of the concrete, fibre reinforced concrete have been developed. Fibre reinforced concrete has already found wide range of practical application and has proved as a reliable construction material that having good performance characteristic compared to conventional concrete [1]. Fibres in the concrete commonly are produced from a large variety of materials like steel, glass, organic, polyester and other materials various shapes and sizes. Considering the waste management system and global energy crisis, there are many research work have been carried out in recent years on waste materials to determine the characteristics and advantages [2].Therefore, this research study about one of the potential recycled materials which is aluminium soft drink can.

Considering the availability and environmental issues in term of waste management, a research was conducted to study the effect of aluminium soft drink can as fibre reinforcement in concrete construction. Standard concrete specimens were cast with different percentage of fibre based on weight of cement and tested for density and strength. The mechanical properties of the specimens were tested with experimental tests such as compressive strength test and flexural strength test.

### *1.1. Problem statement*
To overcome some disadvantages of concrete such as having relatively low tensile strength (compared to other building materials), low ductility, low strength-to-weight ratio and susceptible cracking, it is essential to find alternatives to improve the qualities of concrete. The use of fibre reinforcement is applied to concrete as fibre reinforced concrete. The application of fibre in concrete does not only change the properties of normal concrete but also will have some good improvement for the performance of concrete in building construction.

Recycling various kinds of waste materials undoubtedly will be the one of the problems that will most afflict the society in the future and that we must address and resolve in all possible ways. It is necessary that researchers find solutions that are original, imaginative and brilliant to reuse the waste. With the scarcity of space for landfilling and due to an ever increasing cost, the attention is towards the reuse of waste as an alternative to disposal. Research is always more interested in the use of such products in the concrete mix. This makes the concrete more economic and, at the same time, there is a reduction of problems regarding waste materials. For this research, by using waste material specifically aluminium soft drink cans as the fibre reinforcement material, it not only can decrease the number of dumping waste but also can strengthen the concrete in low cost.

### *1.2. Scope of study*
Please In this research, the mechanical properties of fibre reinforced concrete that containing soft drink cans aluminium fibre were studied. Important properties such as compressive strength and flexural strength were the main focus. To understand the development of fibre reinforced concrete adding aluminium fibre, few tests were conducted. The test conducted was at fresh state and hardened state. Slump test was conducted at fresh state of concrete mix to ensure the workability of concrete and also meets the required slump height in the range between 10mm to 30mm as followed concrete mix design. At hardened state, compressive strength test was conducted on the cubes and cylinder specimens. The cube specimens were 100mm x 100mm x 100mm size and cylinder specimens were 150mm x 300mm. Both types of specimens were cured with water curing method. The ages of curing for cubes were 7 days and 28 days respectively. Cylinder specimens were cured for 14 days. For flexural test, rectangular beam specimens with 100mm x 100mm x500mm were tested on the age of 14 days cured and also wet-cured.

To determine the maximum proportion of the fibre that can be added into the concrete, the percentage of fibre based on weight of cement was the variable for this research. Concrete with 1% and 2% fibre content were used to determine the effect of fibre reinforced concrete with different amount of fibre.

## 2. Literature review
The Concrete is a typical construction material that is commonly used due to its several advantages such as high concrete strength. High concrete strength can be produced by fulfilling the requirement of construction demand. However, concrete is relatively strong in compression but weak in tension and quite brittle [3].

The weakness of concrete can be overcome by the use of steel bars reinforcement or mixing with a sufficient amount of fibres. Fibres varying from size, material and geometry have been added into the concrete can produce a better construction material called Fibre Reinforced Concrete (FRC). Fibres added to the concrete are generally measured with the portion of total volume of concrete and known as volume fraction. The use of fibres also recalibrates the behaviour of the fibre-matrix composite after it has cracked through improving its toughness [4].

Solid waste is a major environmental problem in Malaysia and is significantly reducing our environment capacity to sustain life. There have been increasing construction wastes attributable from insufficient waste management practices in the construction projects since the last two decades [5]. The amount of waste generated continues to increase due to the increasing population and development and only small percentage of waste is being recycled. Most of the wastes which can be recycled include glass, paper, plastic and aluminium. There are many researches done by using waste material as fibre in the concrete. The investigation that have been made shows that the behaviour of fibre concrete by using the waste material such as lathe waste, soft drink bottle caps and beverage tin wastes and found that the strength of concrete for all the fibre has been increased significantly than ordinary concrete [6]. Besides, a research was conducted using recycled waste PET (polyethylene terephthalate) bottles fibres for the reinforcement of concrete. The results tested on concrete reinforced with circular PET fibres and long strips shows fibres give more ductile behaviour of concrete and a high concrete-PET adherence [7]. Therefore, this research study about one of the potential recycled materials which is aluminium soft drink can.

## 3. Methodology

To study the behaviour and characteristic of fibre reinforced concrete with aluminium fibre, the laboratory testing conducted were compressive strength test and flexural strength test.

### *3.1. Material*

The material used for this concrete is ordinary Portland cement (OPC), sand, aggregate and tap water. The ratio of concrete mix is 1:1:2 which are for cement, sand and aggregate respectively. The size of sand is according the 600μm sieve by passing 70% and the size of aggregate is 10 to 20mm. the concrete was design to be grade 20. For the preparation of soft drink can aluminium fibre, soft drink aluminium can were collected and cleaned. The top and bottom of the can is then cut off to obtain aluminium sheets. Then the sheets are cut into strips and finally the strips are cut into fibre form with constant dimension.

### *3.2. Concrete design mix*

The purpose of concrete mix design is to ensure the most optimum proportions of the constituent materials to fulfil the requirements. Table 1 shows the constituent material composition for the design mix.

**Table 1.** Constituent material composition.

| Material | Composition |
|---|---|
| Cementitious materials | 380 kg/m$^3$ |
| Fine Aggregate | 549 kg/m$^3$ |
| Coarse Aggregate | 1281 kg/m$^3$ |
| Water-cement ratio (by mass) | 0.5 |
| Fine-coarse aggregate ratio (by mass) | 1:2 |

## 4. Result and discussion

### *4.1. Density of specimens*

As shown in table 2 below, the density of the cube specimens for all fibre content increased from 7 days to 28 days. The result shows that for each types of specimens, concrete with 1% fibre have the highest density compared to others. On the other hand, concrete with 2% fibre have the lowest density. This is because excess amount of fibre may affect the moisture content or hydration of concrete during curing process so that the weight of concrete after cured become lighter compared to others. From the data obtained, this can be concluded that fibre content can affect the density of the concrete.

**Table 2.** Density of specimens.

| Type of Specimens | Fiber Content | Density ($kg/m^3$) | | |
|---|---|---|---|---|
| | | 7 days | 28 days | 14 days |
| Cube | Normal | 2409 | 2412 | - |
| | 1% | 2443 | 2450 | - |
| | 2% | 2377 | 2394 | - |
| | Average | 2409.67 | 2418.67 | |
| Cylinder | Normal | - | 2346.53 | - |
| | 1% | - | 2351.82 | - |
| | 2% | - | 2338.99 | - |
| | Average | - | 2345.78 | - |
| Rectangular Beam | Normal | - | - | 2426.78 |
| | 1% | - | - | 2456.00 |
| | 2% | - | - | 2391.80 |
| | Average | - | - | 2424.86 |

*4.2. Compressive strength test*

Referring to the table 3 below, the compressive strength increased corresponding with the age of curing. Specimens with no fibre content have the increasing strength from 18.59 $N/mm^2$ to 25.80 $N/mm^2$. For 1% fibre content of specimens, the strength increase from 21.36 $N/mm^2$ to 29.072 $N/mm^2$. Then, 2% fibre content of specimens also showed an increasing strength value from 18.13 $N/mm^2$ to 23.093 $N/mm^2$. The mixes also exceed the design grade of 25 and shows good compressive strength. The result show the increment of strength from curing period 7 days till 28 days because the specimens gain strength over the time during water curing in curing tank. The curing process is to maintaining the moisture level inside cast concrete, as long as free moisture exists inside the concrete, the strength, hardness and density increase gradually [8].

The addition of aluminium fibre among two different percentage of fibre had shown different effect on the specimens compared to the normal concrete. Concrete with 1% fibre shown increased strength compared to the normal concrete and had the highest strength among all the specimens. However, concrete with 2% fibre shown negative influence and had the lowest strength among all the specimens. The negative response in strength gain of concrete with 2% fibre is because the loss of bond inside the specimens due to excess amount of fibre.

It can be concluded that maximum or suitable proportion of fibre that can be added as improve the strength of concrete is 1% of fibre content by weight of cement.

**Table 3.** Compressive strength result

| Type of Specimen | Fiber Content | Day | Compressive Strength ($N/mm^2$) |
|---|---|---|---|
| Cube | Normal | 7 | 18.59 |
| | 1% | | 21.36 |
| | 2% | | 18.13 |
| | Normal | 28 | 25.80 |
| | 1% | | 29.07 |
| | 2% | | 23.09 |
| Cylinder | Normal | 28 | 18.84 |
| | 1% | | 22.00 |
| | 2% | | 11.75 |

### *4.3. Flexural strength test*

The flexural strength of normal concrete and fibre reinforced concrete were tested at the curing age of 14 days with rectangular beam specimens [9]. Table 4 show the result of flexural strength test. The results show an increasing value of flexural strength with the increasing amount of fibre in the specimens. The normal concrete has the value of 0.922 N/mm². However, when the strength reach the highest value at specimen with 1% of fibre which is 1.003 N/mm², the value decrease significantly at the 2% of fibre specimen with the value of 0.224 N/mm². These result shown that the increase in strength only up to 1% fibre content but decrease when the amount of fibre increased. Specimens with excess amount of fibre cannot carry more load because the cause of congestion of fibre. The congestion of fibre may cause the bonding of concrete become weak and the concrete system will become disintegration.

**Table 4.** Flexural strength of rectangular beam specimens.

| Type of Specimen | Fibre Content | Day | Modulus of Rupture ($N/mm^2$) |
|---|---|---|---|
| Rectangular Beam | Normal | 14 | 0.922 |
| | 1% | | 1.003 |
| | 2% | | 0.224 |

## 5. Conclusion

The values compressive strength increase with the increase amount of fibre in the concrete mix [10]. However, there are limits to which the excess amount of fibre will cause the strength to decrease. Therefore, it shows that if the fibre content higher it affects the strength of the concrete [11]. The negative response that concrete with 2% fibre have the least strength is because loss of bonding due to excess amount of fibre. Overall, the optimum percentage to obtain the highest compressive strength is 1%.

For flexural strength test, similar trend of results was observed when comparing to the compressive strength test. The strength increase when 1% fibre was added into the concrete but decrease in strength with the percentage of 2% fibre added into the concrete. This shows that increase in strength is only up to certain percentage of fibre content. It also can be deducted that specimens with higher amount of fibre cannot carry extra load due to congestion of fibre. The congestion of fibre may cause the bonding of concrete become weak and the concrete system will become disintegration.

The amount of fibre is corresponding to the density and strength of concrete was observed from each type of specimens. The highest density and strength is from concrete with 1% of fibre content. As the density is increased, the strength of concrete also increased. Concrete with 2% of fibre have the lowest density due to excess amount of fibre that may affect the moisture content or hydration of concrete during curing process so that the weight of concrete after cured become lighter compared to others.

From all the result, it can be concluded that maximum proportion of fibre to be added into concrete as fibre reinforced concrete is 1% fibre content by weight of cement which gave all the positive response from all the tests conducted.

## 6. References

[1] Johnston C D 1994 Fiber-Reinforced Cement and Concrete *Advances in Concrete Technology* CANMET, (Ottawa: National Resources Canada) pp 603-673

[2] Siddique R 2008 *Waste Material and By-Product in Concrete* (Berlin: Springer-Verlag Heidelberg)

[3] Banthia N 2012 Milestone in international research and development *Proc. of FIBCON 2012, ICI* (Nagpur, India) pp 48

[4] Nataraja M C 1998 Steel fiber reinforced concrete under compression, *The Indian Concrete Journal* 26(3) 353-356
[5] Nagapan S, Rahman I A, Asmi A and Adnan N F 2013 Study of site's construction waste in Batu Pahat, Johor *Procedia Engineering* vol 53 (United Kingdom: Elsevier) pp 99-103
[6] Murali G 2012 Experimental investigation on fibre reinforced concrete using waste materials, *Int. J. of Engineering Research and Applications* **2(2)** 278-283
[7] Foti D 2013 Use of recycled waste pet bottles fibers for the reinforcement of concrete, *Composite Structures* **96** 396-404
[8] BS 1881-Part 116 1993 *Method for Determination of Compression Strength of Concrete Cubes* (London: British Standard)
[9] BS 1881- Part 118 1993 *Method for Determination of Flexural Strength* (London: British Standard)
[10] Teychenné D C, Franklin R E and Erntroy H C 1992 *Design of Normal Concrete Mixes :Building Research* (United Kingdom: Establishment Ltd.)
[11] The Aluminium Association 2011 *Aluminum: The Element Of Sustainability A North American Aluminum Industry Sustainability Report* (Arlington: The Aluminium Association) pp1-70

**Acknowledgments**
We would like to acknowledge the assistance and guidance by the technical staff of Structural Laboratory of the School of Environmental Engineering, UniMAP, and also for financial support from UniMAP in undertaking this lab work.

# Effect of nylon fiber on mechanical properties of cement based mortar

**I M Hanif[1], M R Noor Syuhaili[2], M F Hasmori[1] and S M Shahmi[1]**

[1] Faculty of Civil and Environmental Engineering, Universiti Tun Hussein Onn Malaysia, Parit Raja, 86400, Batu Pahat, Johor, Malaysia
[2] Sekolah Menengah Kebangsaan Senggarang, Jalan Parit Jabar, 83200, Senggarang, Batu Pahat, Johor, Malaysia
Corresponding author: mohdhanif@uthm.edu.my

**Abstract.** An investigation has been carried out to study the effect of nylon fiber on the mechanical properties of cement based mortar after receiving large quantities of nylon waste. Subsequently, this research was conducted to compare the compressive, tensile and flexural strength of normal cement based mortar with nylon fiber cement based mortar. All samples using constant water-cement ratio of 0.63 and three different percentages of nylon fiber were added in the mixture during the samples preparation period which consists of 0.5%, 1.5% and 2.5% by total weight of cement based mortar. The results obtained with different nylon percentage marked an increases in compressive strength (up to 17%), tensile strength (up to 21%) and flexural strength (up to 13%) when compared with control cement based mortar samples. Therefore, the results obtained from this study shows that by using nylon fiber as additive material can improve the mechanical properties of the cement based mortar and at the same time produce a good sustainable product that can protects and conserve the marine environment.

## 1. Introduction

According to Duggal [1], mortar has been selected as building materials because of it special properties such as producing a strong bond between the mortars and stone and its whether and chemical resistant makes it long lasting and suitable as a building material. Thus, the quality improvement of mortar continues till today. In addition, the performance of mortar was improved by inclusion of fiber and was very popular as a method to enhance the strength. Fiber has several of types such as glass, carbon, plastic and nylon. The special characteristic of nylon is strong, elastic and fast drying [2]. Unfortunately, lot of waste nylon net were been dumped in the sea and it was reported that one of the whale's internal organs had ruptured abdominal area caused by the nylon net that was stuck in the whale stomach [3]. Hence, utilizing of waste nylon or nylon fiber as an additive in mortar will reduce the pollution and protect the marine environment. Habib *et al.* [4] prove that with additive of 2% of nylon fiber in cement based mortar shows higher compressive strength by 42.8% compared with normal cement based mortar. The compressive strength of 1.5% nylon fiber cement based mortar with 1.0 inch length was 36.3 MPa, while the compressive strength of 1.5% nylon fiber cement based mortar with 1.5 inch length was 39.6 MPa. This results shows that the length of nylon fiber affects the compressive strength of cement based mortar. The addition of shorter length of nylon fibers causes a decrease in the compressive strength of cement based mortar. The performance of the mortar can be improved with utilization of higher percentage and longer length of nylon fiber [5]. By adding higher percentage of nylon fiber up to 2%, the tensile and flexural strength of cement based mortar can be increase. This is because by adding fiber in cement based mortar, the plasticity of cement based mortar will increase thus delaying the formation of the cracks in the cement based mortar when load are applied[4]. The effect of using nylon fiber in cement based mortar should be carried out further in laboratory to gather additional prove regarding the effectiveness of nylon fiber in mortar in term of compressive strength, tensile and flexural strength. The addition of nylon fiber in the cement based mortar can improve the durability and mechanical properties of ordinary mortar [4]. Henceforth, the effect of cracking also can be improved [6].

## 2. Materials

Nylon fiber (figure 1) was used in this research. The fiber was collected from TCE Tackles Sdn Bhd, Lunas, Kedah and was prepared manually by cutting 0.33 mm diameter fishing line fiber into 1.5 inch length. The Ordinary Portland Cement (Tasek Cement) was used throughout this research. Meanwhile, river sand was collected from the nearest local river and was oven dried for 24 hours with $110^0$C. Constant water-cement ratio of 0.63 was used for all samples and three different percentages of nylon fiber were added in the cement based mortar consists of 0.5%, 1.5% and 2.5% by total weight of cement based mortar.

**Figure 1.** Nylon fiber.

## 3. Methodology

The compressive, tensile and flexural strength of nylon fiber cement based mortar were determined using BS EN 12390-3:2009 [7], BS EN 12390-6:2009 [8] and BS EN 1015-11:1999 [9]. Mortar mix design ratio of 1:3 was used in this research. Mortar mixture was prepared by mixing cement and sand. Then, water was added with constant water cement ratio of 0.63. When mortar mixture becomes uniformly mix, nylon fibers was added slowly to prevent bounding of the fibers. Cubical samples of 100 mm x 100 mm x 100mm, cylindrical samples of 150 mm diameter with 300 mm height and beam samples of 350 mm x 75 mm x 75 mm were prepared based on the mortar mix design. All the samples were stored under normal water at room temperature for 7 and 28 days for curing. Cubical and cylindrical samples were tested by compression machine while beam samples were tested using universal testing machine. Figure 2 shows the flexural test set-up with specimen dimension.

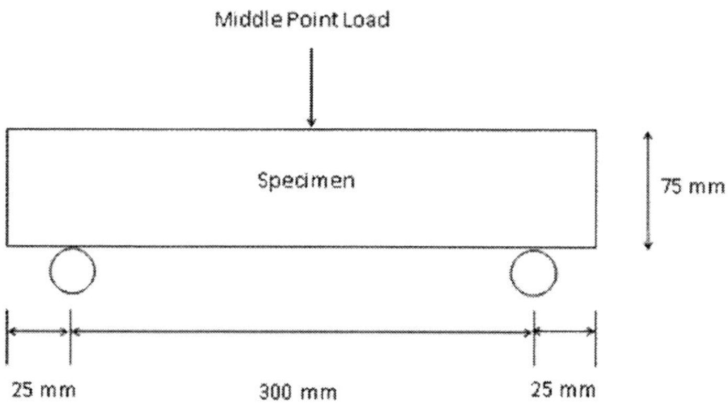


**Figure 2**. Flexural test set-up.

## 4. Results and discussions

### *4.1. Effect of nylon fiber on the compressive strength of cement based mortar*

The compressive strength results were presented in figure 3.

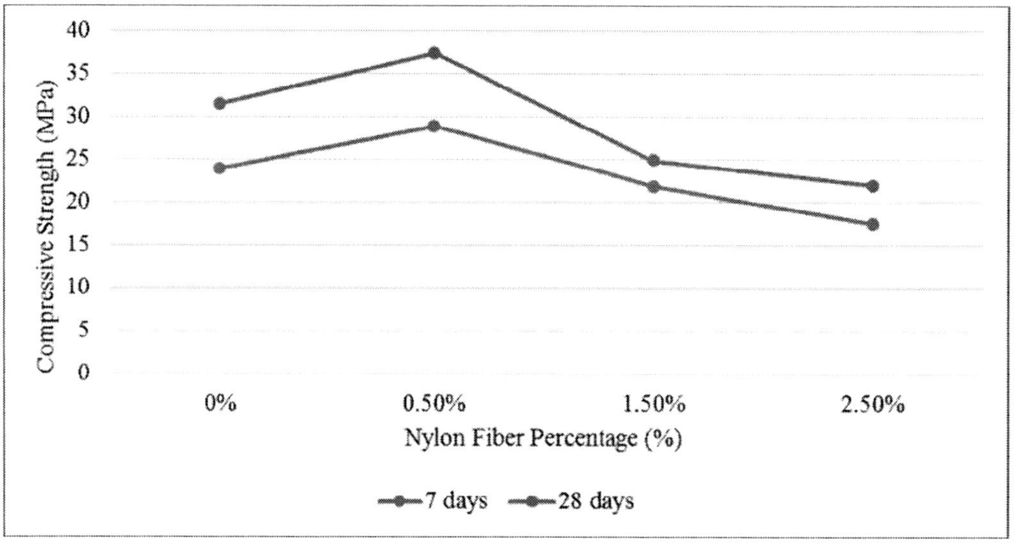


**Figure 3.** Compressive strength of cement based mortar with different percentage of nylon fiber.

Figure 3 showed that 0.5 % nylon fiber cement based mortar has the highest compressive strength compared to control cement based mortar and others nylon fiber cement based mortar. The compressive strength of 0.5% cement based mortar at the age of 7 and 28 days is 28.8 MPa and 37.3 MPa meanwhile the compressive strength of control cement based mortar at the age of 7 and 28 days is 23.8 MPa and 31.4 MPa. Hence, 0.5% nylon fiber cement mortar has higher compressive strength compared to control cement based mortar by 21% and 19% at 7 and 28 days of age. On the other hand, 1.5% and 2.5 % nylon fiber cement based mortar showed a reduction in the compressive strength compared to control cement based mortar by 7-30%. According to Spadea *et al.* [5] and Campello *et al.* [10] by adding more nylon fibre will created higher porosity or cavity between the mortar matrixes which causes compressive strength of cement based mortar to decrease. Another factor affecting the compressive strength of cement based mortar is the water-cement ratio. Additional of water in the mix mortar will produce higher porosity in hardened cement based mortar and led to lower compressive strength [11].

*4.2. Effect of nylon fiber on the tensile strength of cement based mortar*
The tensile strength results were presented in figure 4.

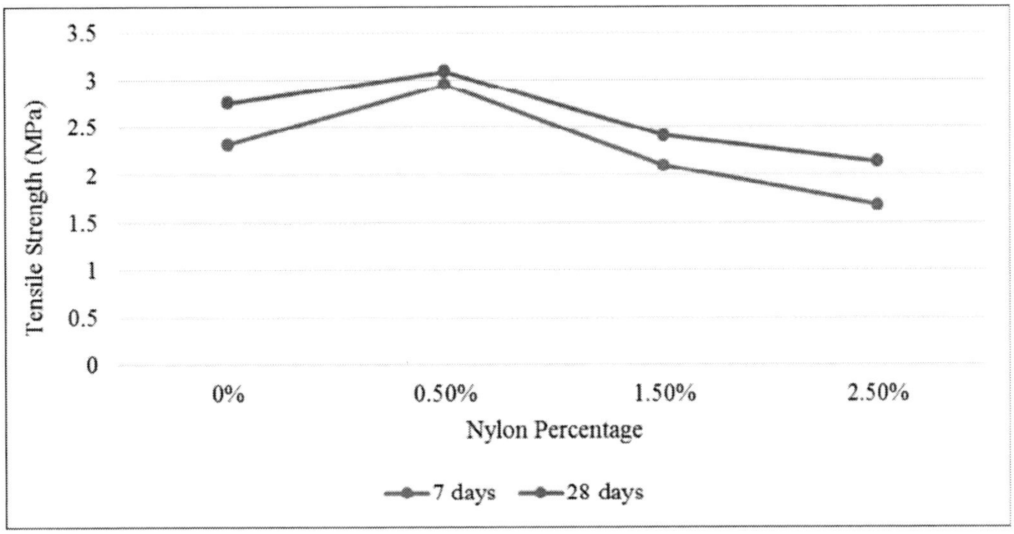


**Figure 4.** Tensile strength of cement based mortar with different percentage of nylon fiber.

Figure 4 showed that 0.5 % nylon fiber cement based mortar has the highest tensile strength compared to control cement based mortar and others nylon fiber cement based mortar. The tensile strength of 0.5% cement based mortar at the age of 7 and 28 days is 2.95 MPa and 3.09 MPa meanwhile the tensile strength of control cement based mortar at the age of 7 and 28 days is 2.32 MPa and 2.75 MPa. Hence, 0.5% nylon fiber cement mortar has higher tensile strength compared to control cement based mortar by 27% and 12% at 7 and 28 days of age. On the other hand, 1.5% and 2.5 % nylon fiber cement based mortar showed a reduction in the tensile strength compared to control cement based mortar by 9-28%. Previous studies by Orasutthikul *et al.* [12] demonstrated that increase of length and quantity of nylon fibers causes a decrease in tensile strength of mortar and high water-cement ratio will contributed to the reduction of the tensile strength of mortar [11].

*4.3. Effect of nylon fiber on the flexural strength of cement based mortar*
The flexural strength results were presented in figure 5.

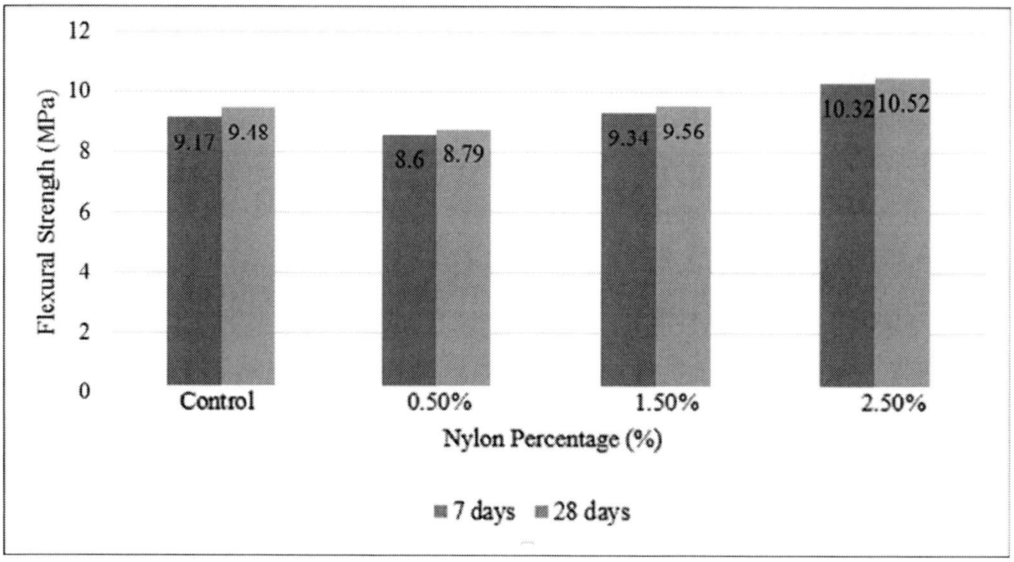


**Figure 5.** Flexural strength of cement based mortar with different percentage of nylon fiber.

Based on figure 5, by adding more nylon fiber up to 2.5% in cement based mortar increases the flexural strength of cement based mortar. 1.5 % and 2.5% nylon fibers cement based mortar have higher flexural strength by 1% - 13% at the age of 7 and 28 days compared to control cement based mortar. Meanwhile 0.5% nylon fiber cement based mortar showed slight reduction of flexural strength compared to control cement based mortar by 6% - 7%.

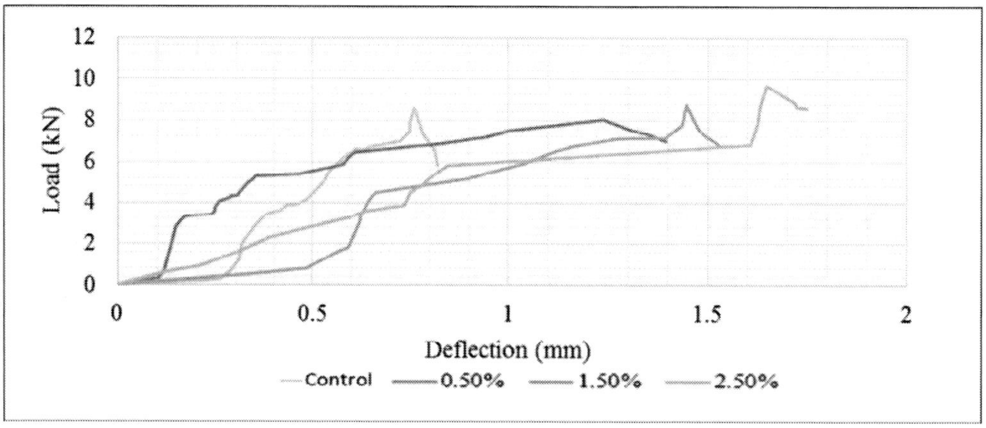


**Figure 6.** Load vs deflection of cement based mortar with different percentage of nylon fiber (7 days).

Figure 6 above shows the results of the maximum load and deflection of control and nylon fibers cement based mortar samples at the age of 7 days. 2.5% nylon fiber cement based mortar beam able to bear maximum load of 9.68 kN. Meanwhile, 1.5%, 0.5% nylon fiber and control cement based mortar beams able to bear maximum load of 8.75 kN, 8.06 kN and 8.60 kN. Higher deflection represents higher lifespan for the tested beam. Based on figure 6, 2.5% nylon fiber cement based mortar has the highest beam deflection value with 1.64 mm, followed by 1.5 %, 0.5 % nylon fiber and control cement based mortar with 1.44 mm, 1.24 mm and 0.76 mm.

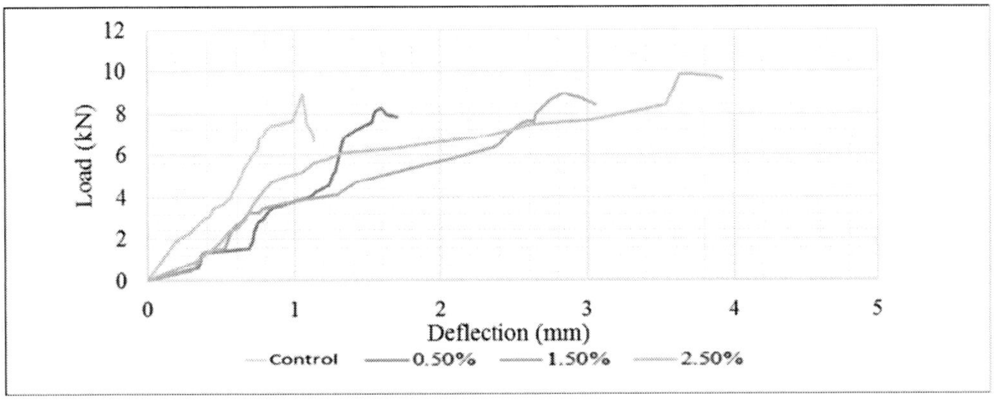


**Figure 7**. Load vs deflection of cement based mortar with different percentage of nylon fiber (28 days).

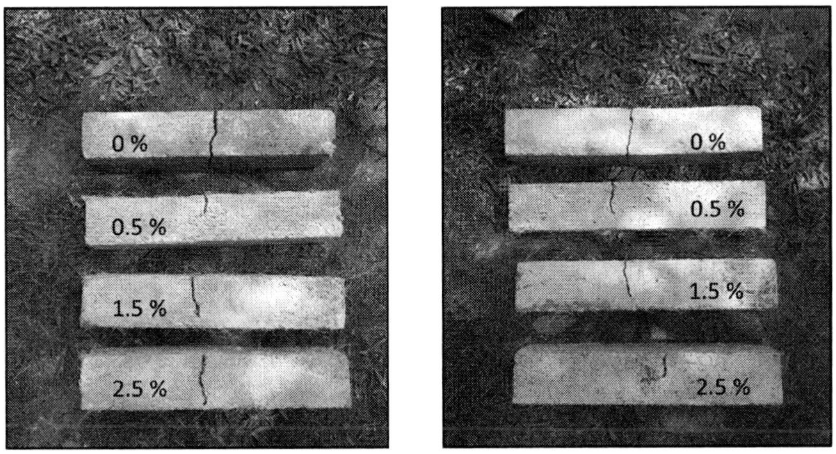


**Figure 8**. Cement based mortar specimen after undertaken flexural test.

Figure 7 displayed the results of the maximum load and deflection of control and nylon fibers cement based mortar beams at the age of 28 days. 2.5% nylon fiber cement based mortar beam able to bear maximum load of 9.86 kN. Meanwhile, 1.5%, 0.5% nylon fiber and control cement based mortar beams able to bear maximum load of 8.96 kN, 8.24 kN and 8.80 kN. 2.5% nylon fiber cement based mortar has the highest beam deflection value with 3.62 mm, followed by 1.5 %, 0.5 % nylon fiber and control cement based mortar with 2.84 mm, 1.59 mm and 1.05 mm. Hence, 2.5% nylon fiber cement based mortar was the most durable and has the longest lifespan compared to control and other nylon fibers cement based mortar beam.

According to Spadea *et al.* [5], the utilization of longer length and higher percentage of nylon fiber in cement based mortar causes the specimens able to bear higher load and able to extend the specimens lifespan. Additional percentage of nylon fiber will increase the plasticity of the mortar thus delaying the process of cracking and enhance the flexural strength of mortar [4]. De Oliveira and Castro-Gomes [13] stated that nylon fibers can increase the bending load to withstand the specimen capacity, especially at the stage after a failure occurs. Based on figure 6,7 and 8, the control cement based mortar beam was completely broken when subjected to the maximum load while nylon fiber cement based mortar beams did not fully broken as a result of bonding of nylon fibers with mortar.

## 5. Conclusion

The results showed a positive effect on the compressive and tensile strength of cement based mortar by utilizing 0.5% nylon fiber. By using higher quantity of nylon fiber, the compressive and tensile strength of cement based mortar decreases. Adding more water and percentage of nylon fiber in cement based mortar creates the development of higher porosity between mortars thus causing cement based mortar compressive strength to decrease. Utilizing nylon fiber in cement based mortar can increase the load capacity, especially at the stage after the failure occurred and also increase the plasticity of the mortar hence delaying the process of cracking and improved on the flexural strength of mortar. Overall, the addition of 0.5% of nylon fiber in the mortar mix able to enhance the compressive strength, tensile strength and flexural strength. Furthermore, the use of nylon fiber in the mortar can create a sustainable cement mortar and provide a good change to the environment by reusing the waste material thereby reducing pollution.

## 6. References

[1] Duggal S K 2008 *Building Materials Third Revised Edition* (New Delhi: New Age International (P) Limited)

[2] Thamizharasan K, Srinivasan S R, Varutharaju P and Sathishkumar V 2016 Study On Characteristics Of Textile Fibre Reinforced Concrete, *Issue of International Journal of Applied Sciences* **8**(1) 41-57

[3] Sigler M 2014 The effects of plastic pollution on aquatic wildlife: Current situations and future solutions, *Water. Air. Soil Pollut.* **225**(11) 2184

[4] Habib A, Begum R, Alam M M 2013 Mechanical properties of synthetic fibers Reinforced Mortars, *Int. J. Sci. Eng. Res.* **4**(4) 923–927

[5] Spadea S, Farina I, Carrafiello A and Fraternali F 2015 Recycled nylon fibers as cement mortar reinforcement, *Constr. Build. Mater* **80** 200–209

[6] Song P S, Hwang S and Sheu B C 2005 Strength properties of nylon- and polypropylene-fiber-reinforced concretes, *Cem. Concr. Res.* **35**(8) 1546–1550

[7] British Standard Institution 2009 *BS EN 12390-3 Testing Hardened Concrete: Compressive Strength Of Test Specimens* (London: British Standard Institution)

[8] British Standard Institution 2009 *BS EN 12390-6 Testing Hardened Concrete: Tensile Splitting Strength Of Test Specimens* (London: British Standard Institution)

[9] British Standard Institution 1999 *BS EN 1015-11 Methods of Test for Mortar for Masonry: Determination of Flexural and Compressive Strength of Hardened Mortar* (London: British Standard Institution)

[10] Campello E, Pereira M V and Darwish F 2014 The Effect of Short Metallic and Polymeric Fiber on the Fracture Behavior of Cement Mortar *Procedia Materials Science* vol 3 (United Kingdom: Elsevier) pp 1914–1921

[11] Singh S B, Munjal P and Thammishetti N 2015 Role of water/cement ratio on strength development cement mortar, *J. Build. Eng.* **4** 94-100

[12] Orasutthikul S, Yokota H, Hashimoto K and Unno D 2016 Effectiveness of Recycled Nylon Fibers in Mortar Comparing with Recycled PET and PVA Fibers *4th Int. Conf. on Sustainable Construction Materials and Technologies (Las Vegas)* (London: International Committee of the SCMT conferences) pp 1-10

[13] De Oliveira L A P and Castro-Gomes J P 2011 Physical and mechanical behaviour of recycled PET fibre reinforced mortar, *Constr. Build. Mater.* **25** 1712-1717

# A study on the performance of concrete containing recycled aggregates and ceramic as materials replacement

**N B Azmi[1], F S Khalid[1], J M Irwan[1], N Anting[1] and P N Mazenan[1]**

[1] Jamilus Research Center, Faculty of Civil and Environment Engineering, Universiti Tun Hussein Onn Malaysia, 86400 Parit Raja, Johor, Malaysia
Corresponding author:faisalsh@uthm.edu.my

**Abstract.** Natural fine aggregate materials are commonly used in development and commercial construction in Malaysia. In fact, concrete production was increased as linear with the growing Malaysia economy. However, an issue was production of concrete was to locate adequate sources of natural fine aggregates. There lot of studies have been conducted in order to replace the fine aggregate in which natural fine aggregate replace with the waste material in concrete preparation. Therefore, this study aims to utilize the Recycled Concrete Aggregate (RCA) and ceramic waste which has great potential to replace the natural aggregate in concrete mix with different type of method, admixture, and parameters. This research were focused on compressive strength and water absorption test to determine the optimum mix ratio of concrete mix. The concrete aggregate was chosen due to improvement capillary bonding mechanisms and ceramic presented similar strength compared to the conventional concrete using natural aggregate. Percent of replacement have been used in this study was at 25%, 35% and 45% of the RCA and 5%, 10% and 15% for ceramic, respectively. Furthermore, this research was conduct to find the optimum percentage of aggregate replacement, using water-cement ratio of 0.55 with concrete grade 25/30. The best percentage of replacement was the RCA35% C15% with the compressive strength of 34.72MPa and the water absorption was satisfied.

## 1. Introduction

Concrete is very familiar in civil engineering field since almost all the part of structure need to be concrete. This is because the properties of the concrete which have strong chemical bonding, high compressive strength and their durability can last for long life term very suitable to become part of structure. Many research have been done on concrete mixture which substitute the natural material with the waste material to counter environmental problems in construction engineering that increasing year by year due to the renovation and demolished of old building. The most popular study been done on the aggregate in average they study to improve the recycled waste material in building and replace it with natural source for cuts down cost of construction and achieve the development of sustainable construction method. To overcome this entire problem, research on recycling construction waste concrete and ceramic tile will be held to find the mechanical properties of the material as the substitution material in concrete replacing the natural aggregates.

This study aims to determine workability, compressive strength and water absorption of concrete containing RCA and ceramic wastes as fine aggregate and cement replacement materials, respectively. Based on the study of relationships between recycled concrete aggregates characteristics and recycled aggregates concretes properties, the open porosity is observed and that RCA are much porous than natural ones and both methods provide quite similar values [1]. Moreover, it can be pointed out that fine recycled aggregates are characterized by a higher porosity than coarse ones. This can be explained mainly by the higher amount of old cement mortar in fine recycled aggregates [2,3]. Water absorption coefficient of RCA whatever the used method although a slight increase is observed for recycled sand between 24 and 48 h. It can be concluded, that compared with NA, RCA are characterized with higher values of WA due to the amount of attached old cement mortar.

Based the study on mechanical performance of concrete made with aggregates from construction and demolition waste recycling plants, the sharp shape of the aggregates make the bulk density lower compare to NA [4]. Moreover, the high substitution of RCA in concrete mix gives high value of water absorption and lower density compare to NA. By comparison with the NA, RCA have lower particles density and higher water absorption mixing water than the NA mixes. This results from the nature and porosity of the RCA. It demonstrates that there is a propensity of the RCA mixes to require more mixture. Water absorption coefficient of RCA has a slight increase is observed for recycled sand between 24 and 48h [1]. RCA has higher values of water absorption due to the amount of attached old cement mortar because The old cement paste always give an effect to the concrete in term of water absorption. Hence, RCA are much porous than natural ones. This can be explained mainly by the higher amount of old cement mortar in fine recycled aggregates.

The previous study stated that the value of water absorption of concrete with RCA aggregates is higher compared to concrete with natural aggregate [5]. The durability of RCA seems less durable compared to the NA. It is due to the ratio coarse aggregate replacement and the concrete age itself, make the RA less durable compared to the NA because it has high porosity value. The replacement of natural aggregate from 5 to 100% with the recycled aggregate reduce their compressive strength by 5 to 25%. However, the workability of concrete present that more water is need for RCA mixture to gain similar workability Due to the present of impurities and old cement paste attached on it enough water is needed in order to get the satisfied workability [5,6].

Based on earlier research, the production of ceramic waste is lower compared to the others waste on construction. However, the ceramic waste that have been recycled only 1% from the total of its production. Many study have been made to the ceramic waste in order to determine it properties such as the density, strength, water absorption and workability. However, this studies are carried out to determine the characteristic of ceramic waste [7]. As we know the ceramic is durable, hard and highly resistant to biological, chemical and also physical degradation forces. Besides that, it seems very suitable in term of lead to sustainable concrete design and greener environment. Moreover, previous study stated that the studies is about possible reuse ceramic sanitary wastes as the aggregate (both fine and coarse) in concrete [8]. The procedure of aggregate production (crushing, dividing particles into two groups – fine and coarse particles and establishing their proportion) and designing the concrete mix.

The result stated that increased by 45%, in comparison to strength recorded immediately after heating, as opposed to compressive strength of specimens based on granite aggregate that decreased by 40% and specimens made of concrete based on gravel which were completely damaged [8]. Compressive strength of concrete based on ceramic aggregate, tested 30 days after heating was 65% of compressive strength of unheated concrete. From the study on application of waste ceramic tiles aggregates in concrete, the ceramic properties which has high water absorption value [9]. Besides that, failures modes toward compressive loading test for concrete sample is similar with normal concrete failure. Water absorption for an average of six sample that been conducted, the result of water absorption of concrete with ceramic waste aggregates is lower compared to the natural aggregate concrete mix. In addition, the ceramic waste concrete mixture is more cohesive and workable [7].

Besides, the compressive strength for 10 percent of ceramic that been substitute indicates the increase about 3.5 percent [9]. After 30% of coarse waste ceramic were added, the compressive strength value decrease. However, he stated that to gain the good strength of concrete, the replacement of the ceramic waste as aggregates must not more than 30%. Moreover, water absorption of ceramic is higher that the gravel whereas, the porosity of the ceramic is also higher than the gravel [10].

However, the most important factor that influent the properties of the concrete with ceramic is the chemical properties of ceramic itself which is has greater water absorption value. Besides that, the dry ceramic has lower density compare to the NA concrete mix. Furthermore, the result of compressive strength obtained for compression strength, at 7, 28 and 90 days reference concrete. The case of recycled concrete with 25% substitution, there was an increase from 12% to 25% in compressive and splitting tensile strength compared to the reference concrete (RC).

As the conclusion, the replacement of RCA and waste ceramic has a great properties and performance in order to achieve the objective of this study which to determine workability, compressive strength and water absorption of concrete containing RCA and ceramic wastes as fine aggregate and cement replacement materials.

## 2. Methodology

To complete the objective of this study, a few step need to be followed to ensure the study on the RCA and ceramic waste as aggregate substitution material and cement substitution material in manufacturing concrete. In this study, several laboratory testing have been conducted to get the mechanical properties of these material and suitable design of concrete mix containing the RCA and ceramic waste. Mechanical testing that had been held was compression and water absorption. Data have been recorded from this test and analysed so the best result can be observed. In this study, only the good quality of materials had been used such as Ordinary Portland Cement (OPC), coarse aggregates, sand, water and substitution material, which is RCA and ceramic. The mixtures were designed according to the British mix design or the Department of Environment mix design revised in 1998. The absolute volume method was used to calculate the absolute volume of each component to occupy 1m$^3$ of concrete. The water cement ratio that have been used in the mix design was 0.55 used to all 90 cube specimens. Table 1 shows the mixtures proportion of concrete.

**Table 1.** Mixtures proportion for 1m$^3$ of concrete.

| Mix design | | Cement (kg/m$^3$) | | Coarse Aggregate (kg/m$^3$) | Fine Aggregate (kg/m$^3$) | | Water (kg/m$^3$) | Super-plasticizer (kg/m$^3$) |
|---|---|---|---|---|---|---|---|---|
| | | N.C | Ceramic | (kg/m$^3$) | NA | RCA | | |
| Normal | 345 | 0 | 805 | 980 | 0 | 189.75 | 41.4 | Normal |
| RCA 25% | Ceramic 5% | 327.75 | 17.25 | 805 | 735 | 245 | 189.75 | 41.4 |
| | Ceramic 10% | 310.5 | 34.5 | 805 | 735 | 245 | 189.75 | 41.4 |
| | Ceramic 15% | 293.25 | 51.75 | 805 | 735 | 245 | 189.75 | 41.4 |
| RCA 35% | Ceramic 5% | 327.75 | 17.25 | 805 | 637 | 343 | 189.75 | 41.4 |
| | Ceramic 10% | 310.5 | 34.5 | 805 | 637 | 343 | 189.75 | 41.4 |
| | Ceramic 15% | 293.25 | 51.75 | 805 | 637 | 343 | 189.75 | 41.4 |
| RCA 45% | Ceramic 5% | 327.75 | 17.25 | 805 | 539 | 441 | 189.75 | 41.4 |
| | Ceramic10 % | 310.5 | 34.5 | 805 | 539 | 441 | 189.75 | 41.4 |
| | Ceramic 15% | 293.25 | 51.75 | 805 | 539 | 441 | 189.75 | 41.4 |

Slump test have been conducted to the concrete mixture according to BS EN 12350 - 2:2009.The mould for slump test was fill with concrete mix in three layers, each approximately one-third of the height of the mould when compacted. Each layer was compacted with 25 strokes of the tamping rod. After the concrete was poured, the concrete was been harden for 24 hours. After it was hardened, the concrete will be soaked in water for a certain period of time which is 28 days and this activity is called curing. All result of slump test, compressive strength and water absorption were recorded and analyse to determine the pattern of concrete performance as the percentage of aggregate replacement increased.

## 3. Results and discussion

This section discusses the properties of fresh state of concrete such as slump, compressive strength and water absorption of the concrete containing RCA and ceramic wastes as fine aggregate and cement replacement materials. The test was carried out to determine the strength and absorption of the cube sample.

### *3.1. Slump test*

Slump test have been conducted to the concrete mixture according to BS EN 12350 - 2:2009. The concrete mixture has been through the slump test before the mixture was put into cube mold for curing. The result for slump test was record as shown in table 2 and the data was summarized in figure 1.

**Table 2.** Slump test data.

| Types of mix | | Slump test result |
|---|---|---|
| Normal | | 75mm |
| RCA 25% | C5% | 71mm |
| | C10% | 70mm |
| | C15% | 70mm |
| RCA 35% | C5% | 68mm |
| | C10% | 63mm |
| | C15% | 58mm |
| RCA 45% | C5% | 57mm |
| | C10% | 54mm |
| | C15% | 53mm |

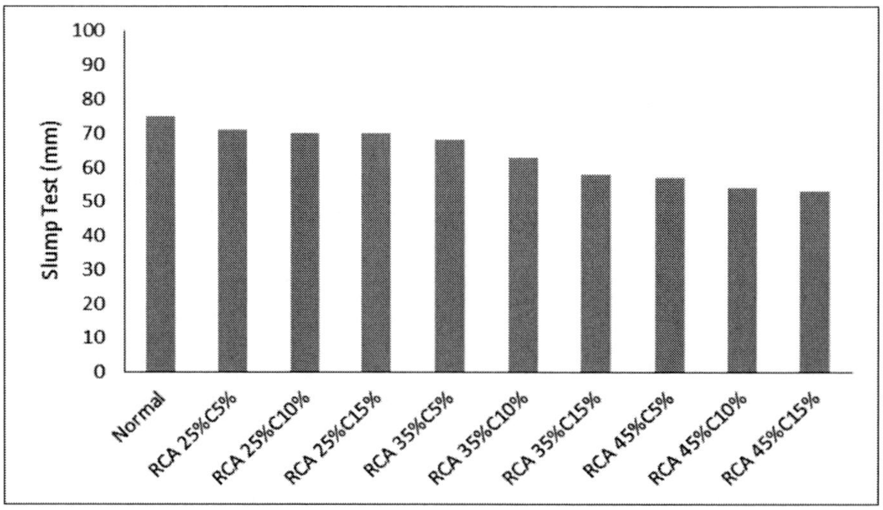


**Figure 1.** Slump test data (mm) of difference type of mixture.

By referring the figure 1, we can see that the slump for the normal concrete is 75 mm. It in the medium range of workability of concrete, meanwhile the concrete containing RCA and ceramic waste slump value is variant in mm but still have the workability. The pattern of the slump is decrease when the number of the RCA increase, the slump reading will be lower compared to the concrete mix which contain less RCA. Moreover, it can be concluded that the slump test result was decreasing as increasing the percentage replacement of both RCA and ceramic. This result was parallel to the study by Irwan *et al.* [2] where the value of slump test were increased as the percentages of RCA increased was obtained in their study.

### 3.2. Compressive strength test

The compressive strength of the concrete cube was tested at 14 and 28 days. The average of three test results for each batch has been taken. The details results of compressive strength against age for each samples were illustrate in figure 2.

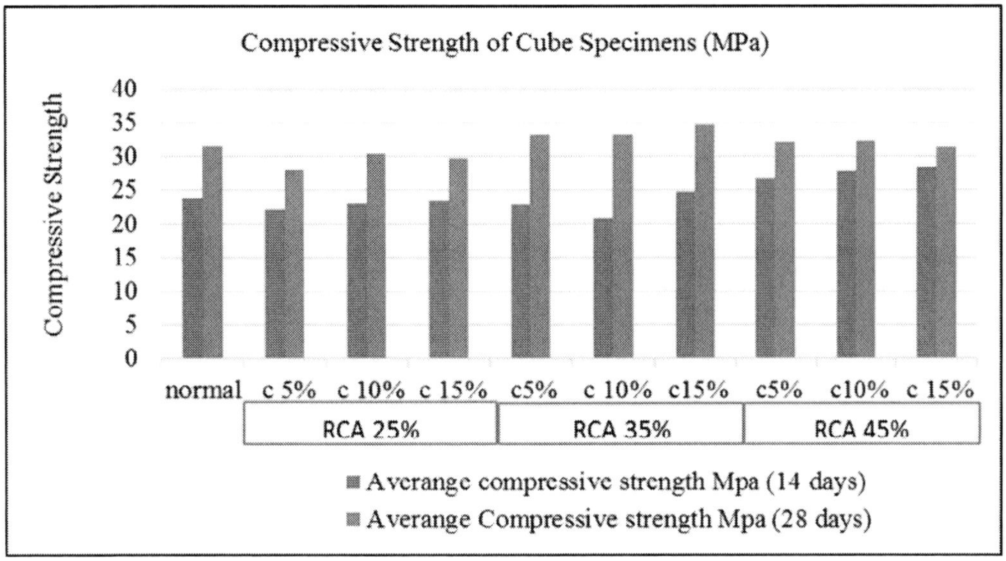


**Figure 2.** The average compressive strength of cube samples with RCA and ceramic versus ages.

By referring to figure 2, the result show that the maximum compressive strength at age 28 days for samples RCA35% C15% with the value 34.72 MPa. However, these sample show the lowest compressive strength compared to others mix and also control mix. Nevertheless, majority of the compressive strength values of RCA and Ceramic mixtures were found to be satisfactory since all samples almost exceed the target strength of 25MPa. It can also be observed that the compressive strength increased with the addition of RCA and ceramic up from RCA25%C10% to RCA45%C15% for both 14 days age and it was found that the compressive strength increased as the percentage of RCA and ceramic increase. The addition of RCA and ceramic at 28 days increase from RCA 25%C10% to RCA 35%C15%. The reduction in compressive strength might be due to poor bonding between cement pastes with RCA and Ceramic, therefore, the cube become more voided within the particles and affected the concrete strength [11,13]

This result is parallel to Yehia, as their experimental results obtained that with the addition up to 30% RCA the compressive strength increases [6]. However, the addition of RCA more than 50%, the strength starts to decrease [11]. The previous study demonstrated that the compressive strength between samples of concrete between 50%RCA replacement shown better results compared to larger percentages RCA replacement. This is because the RCA was more porous than the natural aggregates. In addition the old cement paste bond to its particles effected the concrete strength [12, 14].

The overall findings reveal that the addition of RCA and Ceramic were recommended that the replacement from 35%RCA to 45%RCA with 5% up to 15% ceramic content shows better results. Since the value of ceramic replacement was too small, the materials only gives minor effect to the concrete strength. So the addition RCA plays important role to the concrete strength. At this level of replacement, the mixtures providing good interlocking structure. But after 50%RCA of replacement, the compressive strength of the sample start decrease. Therefore too much of RCA replacement was not recommended.

### *3.3. Water absorption test*

Water absorption was carried out to determine the percentage of water absorption of the cube sample. The detailed of these result were illustrated in figure 3.

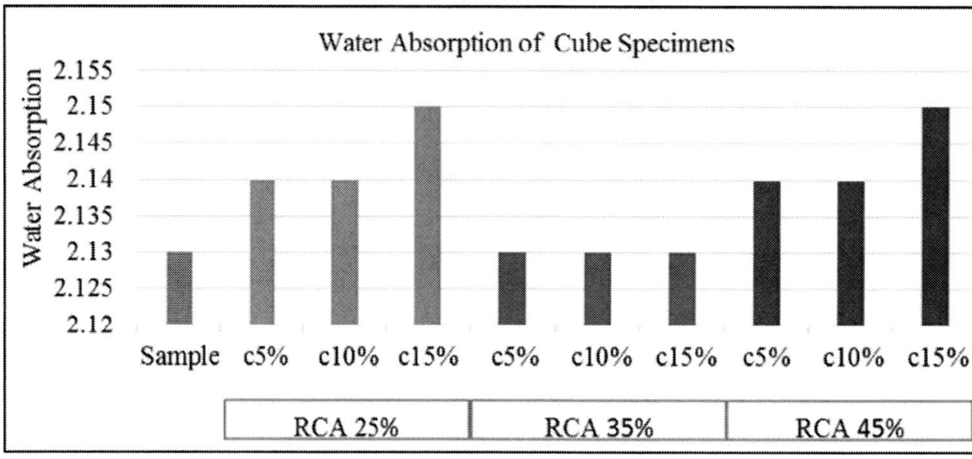


**Figure 3.** Average of water absorption on cube sample on 28 days.

Figure 3 above illustrated the pattern of average water absorption based on experiment that has been conducted. By refer the chart, it can be observe that the water absorption on 28days were gradually increase at RCA25%C5% to RCA25%C15% as the percentage of the RCA increase. However, it is start decrease at RCA35%c5% as the percentage of RCA increase. The value of water absorption at RCA 45%C5% to RCA45%C15% start increase as the percentage of RCA and ceramic increase. From the previous study [7], it obtained that the water absorption result for concrete aggregate with Ceramic aggregate was higher compared to conventional ones. Besides that, based on previous study of RCA incorporating with concrete [6], it got the result for water absorption of concrete with RCA aggregates was higher compared to conventional ones. Referring the figure 3, the replacement of 25% RCA and 45%RCA with 5%C, 10%C and 15%C recorded the highest percentage of water absorption among the samples. It has been proved by previous study for both RCA and ceramic incorporating with concrete stated that these two material had high result in water absorption compared to the conventional concrete. The presence of old cement paste makes the concrete become more effective in absorb the water had been mentioned in their study [15]. It can be concluded that the water absorption of concrete will increase as the increase the replacement of these material.

## 4. Conclusion

Based on the study that have been done, the conclusion by observing the compressive strength of cube specimens at 28 days, it can be concluded as the increase of RCA and ceramic replacement will increased the compressive strength of specimens. RCA35%C15% shows the highest value of compressive strength at 28 days with 34.72 MPa. It was also observed whenever the replacements of RCA and ceramic reach the optimum content, the compressive strength of the specimens decreased. The water absorption of cube specimens at 28days, it can be concluded that the increase of RCA and ceramic content will increased the value of water absorption of specimen. This is parallel to previous study stated, regarding the study on ceramic obtained the result for water absorption value increased as the percentage of ceramic content increase [9].

However, variety of replacements have been done with RCA and ceramic. Then the specimens through several test to determine the best performance of replacements. Compressive strength and water absorption of the specimens were evaluated and based on the test the best mixture of cube specimens were obtained. Based on the compression and water absorption test result, it can be conclude that the best percentages of replacement was RCA35%C15% with the value for both compressive strength and water absorption were 34.72 MPa and 2.13% respectively.

## 5. References

[1] Omary S, Ghorbel E and Wardeh G 2016a Relationships between recycled concrete aggregates characteristics and recycled aggregates concretes properties, *Constr. Build. Mater.* **108** 163–174

[2] Irwan J M, Asyraf R M, Othman N, Koh H B, Aeslina A K, Annas M M K and Faisal S K 2014 Deflection behaviour of irregular-shaped polyethylene terephthalate fibre reinforced concrete beam *Advanced Materials Research* vol 911 (Switzerland: Trans Tech Publications) pp 438-442

[3] Khalid F S, Herman S H and Azmi N B 2017 Properties of Sugarcane Fiber on the Strength of the Normal and Lightweight Concrete *MATEC Web of Conf.* vol 103 (Paris: EDP Sciences) 01021

[4] Bravo M, De Brito J, Pontes J and Evangelista L 2015 Mechanical performance of concrete made with aggregates from construction and demolition waste recycling plants, *J. Clean. Prod.* **99** 59–74

[5] Khalid F S, Herman S H, Azmi N B and M I Juki 2017 Sand Cement Brick Containing Recycled Concrete Aggregate as Fine-Aggregate Replacement *MATEC Web of Conf.* vol 103 (Paris: EDP Sciences 01016

[6] Yehia S, Helal K, Abusharkh A, Zaher A and Istaitiyeh H 2015 Strength and Durability Evaluation of Recycled Aggregate Concrete, *Int. J. Concr. Struct. M.* **9**(2) 219–239

[7] Senthamarai R, Manoharan P D and Gobinath D 2011 Concrete made from ceramic industry waste: Durability properties, *Constr. Build. Mater.* **25**(5) 2413–2419

[8] Halicka A, Ogrodnik P and Zegardlo B 2013 Using ceramic sanitary ware waste as concrete aggregate, *Constr. Build. Mater.* **48** 295–305

[9] Daniyal M and Ahmad S 2007 Application of Waste Ceramic Tile Aggregates in Concrete, *Int. J. of Innovative Research in Science, Engineering and Technology* **4**(12) 12808-12815

[10] Medina C, Sánchez De Rojas M I and Frías M 2012 Reuse of sanitary wastes as coarse aggregate in eco-efficient concretes, *Cement and Concrete Composites* **34**(1) 48–54

[11] Grdic Z J, Toplicic-Curcic G A, Despotovic I M and Ristic N S 2010 Properties of self-compacting concrete prepared with coarse recycled concrete aggregate, *Constr. Build. Mater.* **24**(7) 1129–1133

[12] Khalid F S, Azmi N B, Sumandi K A S M, and Mazenan P N 2017 Mechanical properties of concrete containing recycled concrete aggregate (RCA) and ceramic waste as coarse aggregate replacement *AIP Conference Proc.* vol 1891 (United States: AIP Publishing) 02007

[13] Irwan J M, Annas M M K, Aeslina A K, Othman N, Koh H B, Asyraf R M and Faisal S K 2014 Cracking propagation of reinforced concrete using polyethylene terephtalate (PET) bottles as fine aggregate, *Adv. Mat. Res.* **911** 474-478

[14] Faisal S K, Irwan J M, Othman N, Wan Ibrahim M H 2017 Flexural toughness of ring-shaped waste bottle fiber concrete *MATEC Web of Conf.* vol 47 (Paris: EDP Sciences) 01002
[15] Zamer M M, Irwan J M, Othman N, Faisal S K, Anneza L H, Teddy T and Alshalif A F 2017 Influence of Ureolytic Bacteria Toward Interlocking Compressed Earth Blocks (ICEB) in Improving Durability of ICEB *MATEC Web of Conf.* vol 103 (Paris: EDP Sciences) 01027

**Acknowledgment**
This work was financially supported by University Tun Hussein Onn Malaysia (UTHM) and Minister of Education Malaysia (KPM) through Dana Pembudayaan penyelidikan (RAGS) Vot. No. R-058.

# Investigation on dynamic performance of concrete column crumb rubber steel and fiber concrete

**M Z Siti Nurul Nureda[1],A K Mariyana[1],M Iqbal Khiyon[1],M S Abdul Rahman[1]and Z Nurizaty[1]**

[1]Faculty of Civil Engineering, Universiti Teknologi Malaysia, 81310 UTM Skudai, Johor, Malaysia.
Corresponding author: snnureda2@live.utm.my

**Abstract.** In general the Normal Concrete (NC) are by quasi-brittle failure, where, the nearly complete loss of loading capacity, once failure is initiated especially under dynamic loadings. The significance of this study is to improve the damping properties of concrete structure by utilization of the recycled materials from waste tires to be used in concrete as structural materials that improve seismic performance. In this study, the concrete containing 10% of fine crumb rubber and 1 % volume fraction of steel fiber from waste tires is use to investigate the dynamic performance (natural frequency and damping ratio).A small scale column were fabricated from Treated Crumb Rubber and Steel Fiber Concrete (TCRSFC) and NC were cast and cured for 28 days to investigate the dynamic performance. Based on analysis, dynamic modulus, damping ratio and natural frequency of TCRSFC has improved considerably by 5.18%, 109% and 10.94% when compared with NC. The TCRSFC producing concrete with the desired properties as well as to introduce the huge potential as dynamic resistance structure from severe damage especially prevention on catastrophic failure.

## 1. Introduction

Concreteis a composite material, consisting of Portland cement, aggregates and water. This material has been utilized as construction materials since 1824. Characteristic of the concrete depends on the aggregates or cement used in the mixture. Now a days, there are many innovation has be initiating to improve the characteristic of the concrete. One of the ongoing research is concrete properties can be modified to perform in a more ductile manner by replacing fine aggregates using crumb rubber and the addition of randomly distributed discrete fibers in the concrete matrix, which prevent and limit initiation, propagation and integrate of cracks. Therefore, many researchers have been carried out extensive studies on the utilization of the crumb rubber and steel fiber derived from waste tires into concrete mixes. The introduction of steel reinforcement in concrete structure has proved its improvement in tensile strength but when exposing to dynamic loading such as earthquake event, it is cause severe damage due to less energy dissipation [1]. Previous research has shown that the utilization of crumb rubber has improve its damping ratio which is low in seismic response but give reduction in compressive strength and elastic modulus [1 – 3].It is suggested that the replacement of crumb rubber in range of 0%-20% by aggregate replacement [4,5]. The reduction in compressive strength can be tackle by modification in water cement ratio [6 – 8]. Therefore, the problematic reduction in strength with inclusion of crumb rubber can be improved by modification in water cement ratio. Besides, the addition of steel fiber from waste tire will helps in improving the concrete properties. Based on past studies, the inclusion of steel fiber in concrete mixtures improve the tensile capacity, toughness and reduce surface cracking and it is suggested that the addition is in range of 0.5%-2.5% [9]. Meanwhile, production of waste tire is one of the main problem faced by most of the country all over the world [10 – 12]. These production cause a lot of environmental pollution especially when stored in landfills or stockpile [13]. Besides, burning activities of waste tires causes health hazard from excessive smoke and toxic during burning process [14,15].

In this research, the mechanical and dynamic property of concrete has been investigated and compared with those derived given Treated Crumb Rubber Steel Fiber Concrete (TCRSFC) and Normal Concrete (NC).The optimum replacement of crumb rubber and of steel fiber was examined. The effect of combination between crumb rubber with steel fiber with at the replacement ratios obtained, modified water cement ratio on the mechanical and dynamic properties of concrete column is determined as well.

## 2.Materials and test methods

### *2.1.Materials*

All concrete mixtures in the test are using a same binder materials which is Ordinary Portland Cement (OPC) which is design for concrete grade 30. Fine aggregates and crumb rubber used in concrete mixture is passing through 4.75mm with specific gravity of 2.60 and 1.21 while maximum size of coarse aggregates is 10mm. Crumb rubber was treated with Sodium Hydroxide (NaOH) solution for 20 minutes for removing unnecessary impurities on crumb rubber surface. Then, crumb rubber was rinsed with water and drying for 24 hours at ambient temperature. Meanwhile, the average length of steel fiber is 2.35cm, diameter of 0.30mm with tensile strength and tensile strength of 897 MPa were used in the mixture. Figure 1 shows the (a) crumb rubber and (b) steel fiber used in concrete mixtures.

(a) (b)

**Figure 1.** (a) Crumb rubber and (b) Steel fiber.

### *2.2.Mix design*

There are two types of concrete mixtures that have been used in this research with different mix proportions. The design was done according to Building Research Establishment:1988[16]. Treated Crumb Rubber Steel Fiber Concrete (TCRSFC) (0.50 water cement ratio) were using lower water cement ratio compared to Normal Concrete (NC) (0.55 water cement ratio) because previous studies has shown that concrete mixtures exhibits significant strength reduction when using same water cement ratio [1, 2] .Table 1 denotes the mix proportion for concrete specimen.

**Table 1.**Mix design of concrete specimen.

| | C, (kg/m$^3$) | W, (kg/m$^3$) | FA, (kg/m$^3$) | CA, (kg/m$^3$) | W/C | CR, (%) | CR, (kg/m$^3$) | SF (1%) | SF, (kg/m$^3$) | SP (%) |
|---|---|---|---|---|---|---|---|---|---|---|
| NC | 425 | 233 | 750 | 661 | 0.55 | - | - | - | - | - |
| TCRFC | 466 | 233 | 920 | 897 | 0.50 | 10 | 75 | 1.0 | 53.40 | 1.0 |

Note: C: Cement, W: Water, FA: Fine Aggregate, CA: Coarse Aggregate, W/C: Water/Cement ratio, CR: Crumb Rubber, SF: Steel Fiber, SP: Super plasticizer

*2.3.Fabrication of column specimen*

Column specimen consist of three parts, lump mass, column, and foundation, that were fabricated by using plywood and column specimen consist of two reinforcement bars that have 500 N/mm$^2$ characteristic strength, fy. The diameter (d) of reinforcement bar is 3 mm with length of 724 mm. The curtailment of reinforcement at lump mass is one-quarter of column length and the anchor is 5d from reinforcement which is designed according to BS EN 1992-1-1 (2004)[17].

**3.Sample preparation**

Modulus of elasticity test are determined according to BS 1881-121:1983 [18].Normal Concrete (NC) and Treated Crumb Rubber Steel Fiber Concrete (TCRSFC) mixture were cast for six cylinders with 100 mm diameter and 200 mm height (modulus of elasticity)and small scale column with two reinforcement (3mm diameter) which is designed by 1:6 ratio according to Malaysian Standard MS1064: Part 10: 2001 [19]. There are three parts for small scale column which is lump mass (120mm×120mm×190mm), column (40mm×40mm×500mm) and foundation (275mm×275mm×80mm). Small scale column was fixed with bolts and nuts on shaking table and accelerometer was glued on lump mass, column and shaking table before being tested under free vibration and seismic test as shown in figure 2.

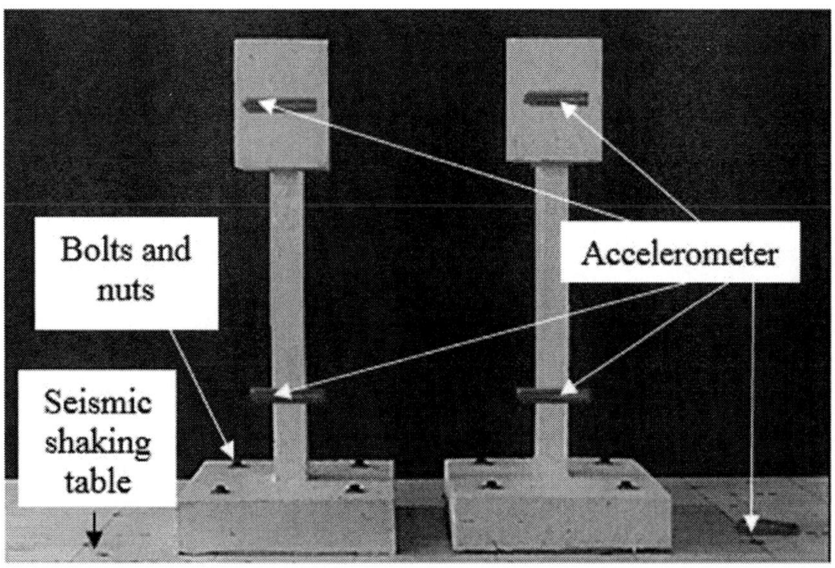


**Figure 2.** Illustration of small scale column fixed on shaking table.

For free vibration test, an impact load was applied at lump mass by a hammer with 2.439 kg to induce the vibration, and the vibration was recorded by accelerometer. After free vibration test, column specimen was tested by exiting the shaking table test with modified north-south Ranau earthquake ground motion which is 1.0g. The ground motion was scale up from 0.126g (6.1 magnitude) to 1.0g in order to examine the effectiveness of combination 10% crumb rubber and 1% steel fiber in concrete column (TCRSFC) during earthquake event.

## 4.Results and discussion

### *4.1.Modulus of elasticity*

The modulus of elasticity for both Normal Concrete (NC) and Treated Crumb Rubber Steel Fiber Concrete (TCRSFC) was tested and determined by using British Standards 1881-Part 121[18] which is defined as secant modulus (A) at one-third of compressive strength cube at age 28 days. Meanwhile the dynamic modulus (B) is defined as small instantaneous strain. Generally, the dynamic modulus is 20%-40% higher than secant modulus. Besides, dynamic modulus are more suitable for stress analysis of structure that subjected to earthquake or impact loading[20]. Figure 3 illustrate the stress-strain curve from tested specimen.

Based on analysis, secant modulus, Es and dynamic modulus, $E_D$ of TCRSFC cylinder shows an increment by 5.60% (27.82 GPa) and 5.18% (31.05GPa)NC cylinder which is 26.28 GPa and 29.52 GPa. Figure 4 shows the mode of failure for concrete specimen under modulus of elasticity. Based on observation, NC exhibited a brittle failure which is small pieces of concrete are shattered meanwhile TCRSFC are more ductile and less brittle because it was confined by 1% of steel fiber in concrete specimen thus improve the modulus of elasticity.

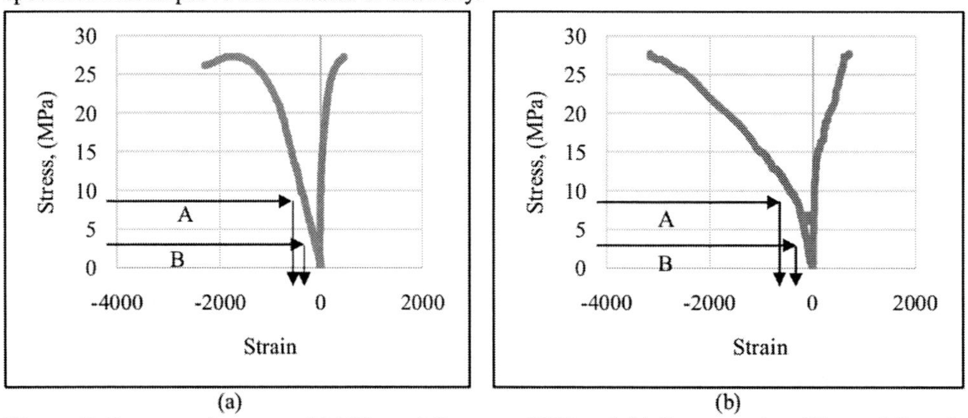


**Figure 3.** Stress-strain curve of (a) Normal Concrete (NC) and (b) Stress-strain of Treated Crumb Rubber Steel Fiber Concrete (TCRSFC), A is secant modulus and B is dynamic modulus.

(a) (b)

**Figure 4.** Mode of failure for (a) Normal Concrete (NC) and (b) Treated Crumb Rubber Steel Fiber Concrete (TCRSFC).

*4.2. Free vibration test*

Damping ratio is defined as energy dissipation which mean can be measured by free vibration test. Four concrete column consist of two Normal Concrete (NC) and Treated Crumb Rubber Steel Fiber Concrete (TCRSFC) was tested under free vibration test. Figure 5 and 6 shows the Time-History and Power Spectrum Distribution (PSD) that was extract from free vibration test.

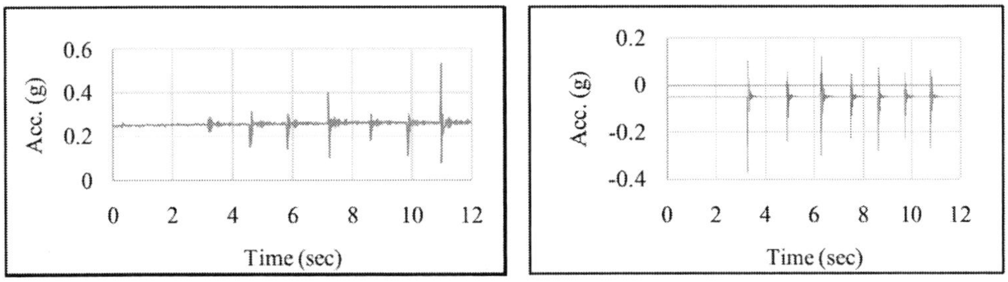


(a) (b)

**Figure 5.**Free vibration decay Time-History of (a) Normal Concrete (NC) and (b) Treated Crumb Rubber Steel Fiber Concrete (TCRSFC).

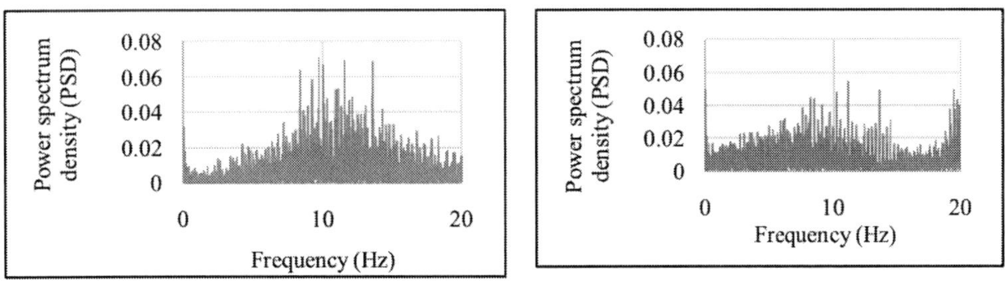


(a) (b)

**Figure 6.** Power Spectrum Distribution of (a) Normal Concrete (NC) and (b) Treated Crumb Rubber Steel Fiber Concrete (TCRSFC).

Damping ratio was computed by using logarithm decrement equation (1), which A1 denotes as the first amplitude of acceleration, an is the amplitude after next cycle and n is next cycle of oscillation.

IOP Conf. Series: Materials Science and Engineering **271** (2017) 012082 doi:10.1088/1757-899X/271/1/012082

$$\text{Damping, } \zeta = {}^{1}/_{2n\pi} \text{ x In } {}^{A1}/_{An} \tag{1}$$

Meanwhile, the natural frequency was determined from PSD curve. Table 2 denotes the value of damping ratio and natural frequency for NC and TCRSFC. Based on analysis, TCRSFC give 109% increment in damping ratio, where is the damping ratio for TCRSFC give 9.6% with respect to NC with 4.6%. According to the past research, the range of damping ratio for concrete are varied from 4%-7% [1]. Meanwhile, the natural frequency of TCRSFC has increase by 10.96% (11.14 Hz) compared with NC (10.04Hz). According to resonance method, the natural frequency is directly proportional to dynamic modulus, $E_D$ [21]. Therefore, TCRSCFC column has the highest in natural frequency as compared to NC due to the highest in stiffness (dynamic modulus).

**Table 2.** Damping ratio and natural frequency of Normal Concrete (NC) and Treated Crumb Rubber Steel Fiber Concrete (TCRSFC).

| | NC | TCRSFC |
|---|---|---|
| Mass (kg) | 21.60 | 22.60 |
| Percentage different (%) | - | +4.60 |
| Amplitude, (A1) | 0.5341 | 0.3735 |
| Amplitude, (An) | 0.2987 | 0.1111 |
| Damping, ζ (%) | 4.6 | 9.6 |
| Percentage different (%) | - | +109 |
| Natural frequency (Hz) | 10.04 | 11.14 |
| Percentage different (%) | - | +10.96 |

*4.3.Seismic shaking table test*

By using the same column specimen, seismic shaking table were tested after free vibration test. Figure 7 shows the crack on column specimen surface after exited with North-South Ranau earthquake 1.0g ground motion. Based on observation, the crack lines on Normal Concrete (NC) and Treated Crumb Rubber Concrete (TCRSFC) appear at connection between foundation and column, and at the column surface but crack on TCRSFC relatively fewer than NC as shown in figure 7.

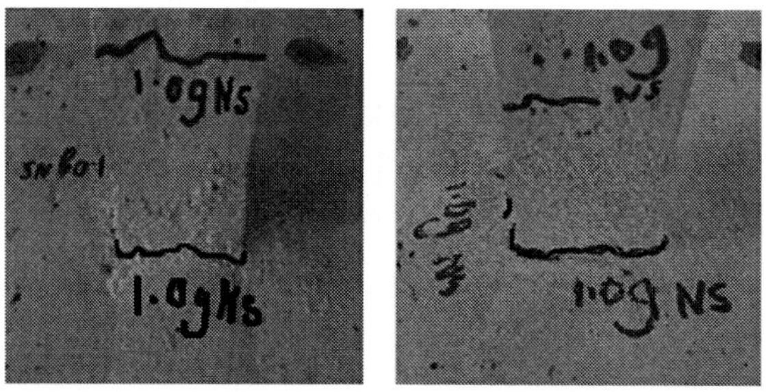


(a) (b)

**Figure 7.** Damage of column specimen after seismic shaking table test. (a) Normal Concrete (NC)(b) Treated Crumb Rubber Concrete (TCRSFC).

The response of acceleration was recorded by accelerometer and the result was plotted as a Time-History graph. Figure 8 (a) shows the input of 1.0g ground motion while figure 8 (b) shows the response of acceleration on column specimen. Based on analysis, the seismic response of TCRSFC column demonstrated smaller acceleration response which is 1.21g as compared to NC column (1.91g). Smaller acceleration response on TCRSFC is due to inclusion of 10% treated crumb rubber and 1% of steel fiber because crumb rubber itself has high resistance in impact loading while steel fiber improved the concrete stiffness [1, 22] which is proved by an improvement in dynamic modulus and damping ratio. On average, the inclusion of 10% treated crumb rubber and 1% steel fiber in concrete mixtures reduce the acceleration response approximately 36%.

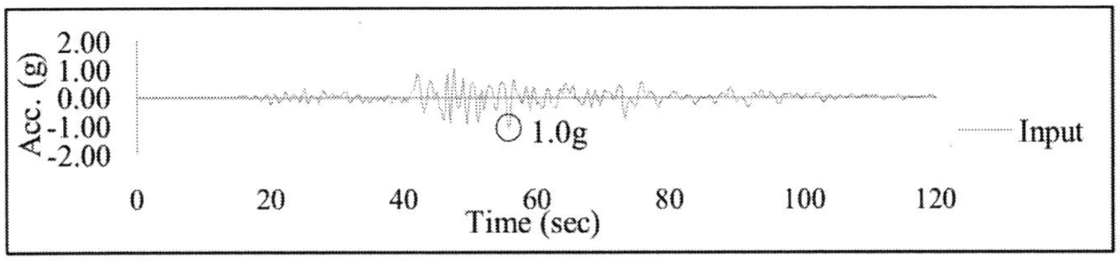


(a)

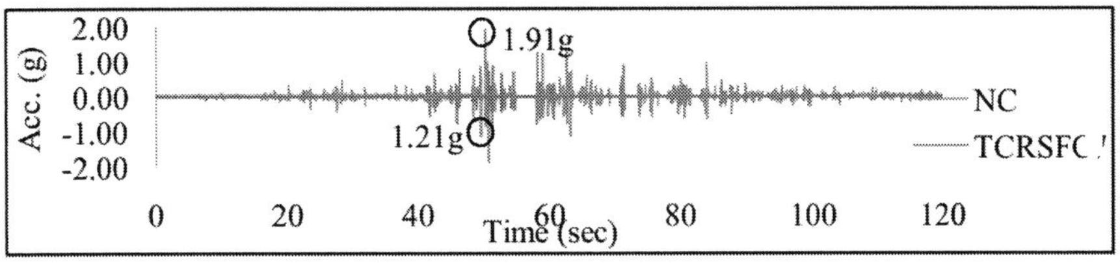


(b)

**Figure 8.**(a) Seismic input and (b) seismic response for column specimen.

## 5.Conclusions

This research focused on investigation of dynamic performance of concrete with inclusion of 10% crumb rubber as sand replacement and 1% of steel fiber as addition in concrete. The conclusion are presumed:

(a) Dynamic modulus for Treated Crumb Rubber Steel Fiber Concrete (TCRSFC) has increased by 5.18% with respect to Normal Concrete (NC). This increment consistent with increment of natural frequency from free vibration test.

(b) Based on free vibration analysis, damping ratio and natural frequency of Treated Crumb Rubber Steel Fiber Concrete(TCRSFC) column is 9.6% and 11.14Hz while Normal Concrete (NC) column is 4.6%and 10.04 Hz. Inclusion of treated crumb rubber and steel fiber has improve 109% of damping ratio and 10.96% natural frequency.

(c) The peak response acceleration of Treated Crumb Rubber Steel Fiber Concrete(TCRSFC) column (1.21g) is reduce by 37% which is less than Normal Concrete (NC) column (1.91g). This is showing that the presence of treated crumb rubber and steel fiber in concrete mixture helps in reducing seismic force due to an increment in damping ratio.

## 6. Reference

[1] Xue J and Shinozuka M 2013 Rubberized concrete: A green structural material with enhanced energy-dissipation capability,*Constr. Build. Mater.* **42** 196–204.

[2] Youssf O, EI Gawady M A, Mills J E and Ma X 2014 An experimental investigation of crumb rubber concrete confined by fibre reinforced polymer tubes,*Constr. Build. Mater.***53** 522 -

532.
[3] Naji Hilal N 2017 Hardened properties of self-compacting concrete with different crumb rubber size and content,*Int. J. Sustain. Built Environt.* **6** 191–206.
[4] Khatib Z K andBayomy F M 1999 Rubberized portland cement concrete, *J. Mater.Civil Eng.* **11** 206–213.
[5] Noaman T A, Abu Bakar B H and Md Akil H 2015The effect of combination between crumb rubber and steel fiber on impact energy of concrete beams *Procedia Engineering* vol 125(United Kingdom: Elsevier) pp 825–831.
[6] Alawode O and Idowu O 2011 Effects of water-cement ratios on the compressive strength and workability of concrete and lateritic concrete mixes, *Pac. j. sci. technol.* **12**99–105.
[7] Zivica V 2009 Effects of the very low water/cement infrastructures in developing countries,*Constr. Build. Mater.* **23** 3579–3582.
[8] Rahmani K, Shamsai A, Saghafian B and Peroti S 2012 Effect of water and cement ratio on compressive strength and abrasion of microsilica concrete, *Middle-East J. Sci. Res.* **12** 1056–1061.
[9] Atiş C D and Karahan O 2009 Properties of steel fiber reinforced fly ash concrete, *Constr. Build. Mater.* **23**392–399.
[10] Eldin N andSenouci A 1993 Rubber-Tire particles as concrete aggregates,*J. Mater. Civil Eng.* **5** 0899–1561.
[11] Zheng L. Sharon Huo X and Yuan Y 2008 Strength, modulus of elasticity, and brittleness index of rubberized concrete,*J. Mater. Civil Eng.* **20** 692–699.
[12] Waris M B, Ali N N and Al-Jabri K S 2016 Use of recycled tire in concrete for partial aggregate replacement, *Int. J. Struct. Civ. Eng. Res.* **5** 273 - 276.
[13] Herman K and Bisesi M S 2002 *Handbook of Environmental Health, Forth edition, Volume II: Pollutant Interactions in Air, Water and Soil*(United States: CRC press LLC) pp1-83.
[14] Issa C A andSalem G 2013 Utilization of recycled crumb rubber as fine aggregates in concrete mix design,*Constr. Build. Mater.* **42** 48–52.
[15] Moustafa A and Elgawady M A 2015 Mechanical properties of high strength concrete with scrap tire rubber,*Constr.Build. Mater.* **93** 249–256.
[16] Teychenné D C, Nicholls J C, Franklin R E and Hobbs D W 1988*Design of Normal Concrete Mixes, second edition*(London: CRC Ltd - Building Research Establishment) 9–25.
[17] BS EN 1992-1-1 Eurocode 22004 *Design of concrete structures -Part 1-1: General Rules and Rules of Buildings* (London: British Standard) 153–155.
[18] BS 1881-121 1983 *Testing Concrete - Part 121: Method for Determination of Static Modulus of Elasticity in Compression*(London: British Standard) 3–5.
[19] MS 1064-10 2009 *Guide to Modular Coordination in Buildings: Part 10: Coordinating Size and Preferred Size for Reinforced Concrete Component* (Malaysia: Malaysian Standard) 5–8.
[20] Metha P K andMonteiroar P J M 2006 *Concrete Microstructure, Properties, and Materials, 3rd edition* (United States: McGraw-Hill) pp 89–90.
[21] Inman D J 1994 *Engineering Vibration* (United States: Prentice-Hall) pp 390–463.
[22] Minnetyan L and Batson G B 1984 *Steel Fibrous Concrete under Seismic Loading* Technical Report AD-P003 691 (United States: Department of Civil and Environmental Engineering Clarkson College of Technology) pp 589–597.

**Acknowledgement**
This work was supported by Ministry of Education Malaysia and Universiti Teknologi Malaysia (UTM) for providing the financial support under FRGS grant Q.J130000.2522.12H44 and partially financed by Smart IBS Sdn. Bhd.

# Compressive strength and initial water absorption rate for cement brick containing high-density polyethylene (HDPE) as a substitutional material for sand

**Noorwirdawati Ali[1], Norhasmiza Din[1], Faisal Sheikh Khalid[1], Shahiron Shahidan[1], Siti Radziah Abdullah[1], Abdul Aziz Abdul Samad[1] and Noridah Mohamad[1]**

[1]Jamilus Research Centre, Faculty of Civil and Environmental Engineering, Universiti Tun Hussein Onn Malaysia, 86400 Parit Raja, Batu Pahat, Johor
Corresponding author: wirda@uthm.edu.my

**Abstract.** The rapid growth of today's construction sector requires high amount of building materials. Bricks, known to have solid properties and easy to handle, which leads to the variety of materials added or replaced in its mixture. In this study, high density polyethylene (HDPE) was selected as the substitute materials in the making of bricks. The reason behind the use of HDPE is because of its recyclable properties and the recycling process that do not emit hazardous gases to the atmosphere. Other than that, the use of HDPE will help reducing the source of pollution by avoiding the millions of accumulated plastic waste in the disposal sites. Furthermore, the material has high endurance level and is weatherproof. This study was carried out on experimenting the substitute materials in the mixture of cement bricks, a component of building materials which is normally manufactured using the mixture of cement, sand and water, following a certain ratios, and left dried to produce blocks of bricks. A series of three different percentages of HDPE were used, which were 2.5%, 3.0% and 3.5%. Tests were done on the bricks, to study its compressive strength and the initial water absorption rate. Both tests were conducted on the seventh and 28th day. Based on the results acquired, for compressive strength tests on the 28th day, the use of 2.5% of HDPE shown values of 12.6 $N/mm^2$ while the use of 3.0% of HDPE shown values of 12.5 $N/mm^2$. Onto the next percentage, 3.5% of HDPE shown values of 12.5 $N/mm^2$.

## 1. Introduction

There are many building materials used in construction. One of them is bricks. Usually, the use of bricks in the sector is widely practiced such as its use in structures, walls and etc. Among the types of bricks manufactured and normally used in construction are clay bricks, cement bricks and cement blocks. Cement brick is a chunk made out of a mixture of cement, sand and water. It is also known as unfired bricks where it need not to be heated up with fire in its production, but only compressed in a mould and left to dry under the sunlight. The production of unfired bricks helps to avoid the incurred cost of burning with fire and reduce air pollution out of smoke and haze. Every bricks are manufactured for different uses in terms of shapes, strength and as well as its capabilities. The properties of bricks itself are not known for supporting loads, hence, the elements of contributing to sustainable construction nowadays consists of the use of other materials to replace sand.

A huge number of building materials are made out of recyclable materials to reduce cost and save the environment. Many efforts were done to combine various kinds of waste products in the production of bricks and concrete such as natural fibres, processed tea waste, sewage sludge, structured glass waste from computers and televisions, sugarcane dregs, organic waste, grass ashes, marbles and granite waste [1- 5]. Recycling the waste produced by industrial and agricultural activities to form building materials is a viable resolution to not only pollution problems, but also to designing buildings economically [6].

Many researchers have conducted experimental investigation on the use or addition of plastic material (polyethylene terephthalate (PET), polypropylene (PP) and high density polyethylene (HDPE)) as concrete materials or replacement [1, 3, 7, 8, 9]. From their findings, plastic have shown to give promising results as partial concrete materials as addition or replacement. More studies need to be conducted on various parameters to investigate the performance of plastic in other parameters and testing.

This study focuses on the compressive strength and initial water absorption rate involving the use of different HDPE ratios. The produced samples are made out of HDPE, sand, water and Ordinary Portland Cement, where the size of the bricks are 215mm in length, 102.5mm of width and 65mm of thickness based on BS 3921: 1985 [10]. As for the tests on the compressive strength and water absorption rate, the samples were tested on the seventh and 28th day. The purpose of test done on the seventh day is to observe the condition of the samples on its premature state, as on the 28th day, the sample bricks can be classified as matured.

All data acquired were recorded. 24 samples of sand cement bricks were produced, where 12 of them were used in compressive strength test and the 12 rest were for initial water absorption rate test. HDPE ratios used to replace sand is at 2.5%, 3.0% and 3.5%. As for the mixture, the ratio of water-cement used is at 0:4.

## 2. Materials and methodology

Materials used to produce bricks are water, cement, sand and HDPE. The cement used for this particular study is Ordinary Portland Cement. This cement need to be stored in an air tight container, or sealed properly to prevent air from getting in. The reason for this is because cement when exposed to air, will lack in quality which makes it hardened and affect the strength of bricks produced later.

River sand is used in the production of the bricks. In order to get a suitable size of sand grains to make bricks, a sieve analysis is done. According to British Standards BS EN 1052:1998 [11] the sand to be used are not more than 10% passing 75micron sieve, and the retained amount found on the 2.36mm sieve are not more than 1%. The percentage of HDPE used is at 2.5%, 3.0% and 3.5%. These ratios are to be used to replace sand in the mixture to make bricks. Recycled HDPE materials shown in figure 1 are acquired from a recycle factory.

**Figure 1.** HDPE used in this study.

*2.1. Compressive strength test*

This test is conducted to study the compressive strength of bricks that contains a certain percentage of HDPE as a replacement of sand. It was conducted based on BS EN 772-1:2011 [12]. The surface of the machine is cleaned from any kinds of dusts and dirt that can affect the test results. Specimens are sorted out in a uniform placement for uniform results to be obtained. Calculation of the compressive strength is done by dividing the maximum strength obtained with the area of which loads are forced to, and rounded off to the nearest 0.1 N/mm$^2$.

*2.2. Initial water absorption rate test*

Specimens were placed in an oven where the temperature is near to 105°C ± 5° for 24 hours ($m_{dry,s}$). They were then left to cool down to room temperature. When cooled down, dimensional measurements were done to obtain the bricks surfaces and calculation of the area, $A_s$. The height need to be sustained until the end of test. If the surfaces of the bricks are irregular, the level of water for soaking purpose need to be increased so that the whole bricks surfaces are in contact with water. The time taken for this particular test ($t_{so}$) is around (1 ± 0.2) min. Once the time is right, specimens were removed from the water and then bricks were weighed ($m_{so,s}$).

## 3. Results and discussion

This test study upon samples of cement bricks are done on the seventh and 28th day. The sum of bricks used in both test is 24 bricks. Every samples mixture use the same ratios of water and different percentages of High Density Polyethylene which are 0% (controlled), 2.5%, 3.0% and 3.5%.

*3.1. Compressive strength test*

Based on figure 2, it is found that controlled sample that used 0% of HDPE reached 11.7 N/mm$^2$ of compressive strength on the seventh day, and increased by a small portion on the 28th day to 14.4 N/mm$^2$. As for the sample with 2.5% of HDPE, it showed a drop in value of compressive strength which was 9.7 N/mm$^2$ on the seventh day and 12.6 N/mm$^2$ on the 28th day. Meanwhile, samples with 3.0% of HDPE showed almost similar value of compressive strength with samples that used 0% of HDPE, which were 11.1 N/mm$^2$ on the seventh day and 15.9 N/mm$^2$ on the 28th day. Next up, samples with 3.5% of HDPE showed an increment of compressive strength values on the seventh day which was 12.4 N/mm$^2$ compared to samples with 0% and 2.5 % of HDPE. Whereas, it showed a significant drop of value on the 28th day, which was only 12.5 N/mm$^2$, making it the lowest value when compared to samples with 0%, 2.5% and 3.0% of HDPE. This value almost matched up to the one on the seventh day.

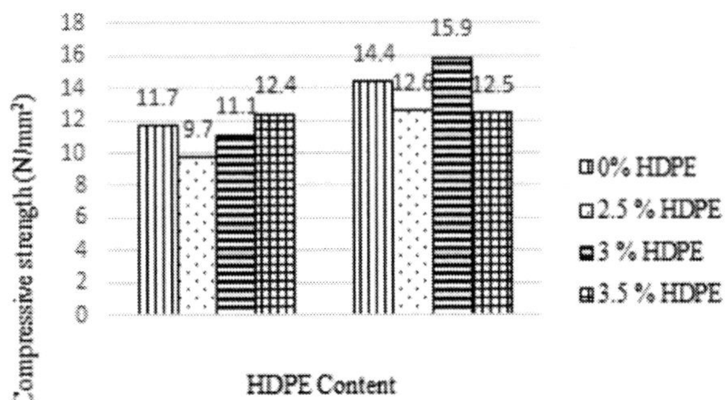


**Figure 2.** Results of compressive strength.

Bricks samples with 3.5% of HDPE was going through a drop of values of compressive strength on the 28th day, and this might be caused by the imperfect condition of bricks placement, which in turn cause it to not drying out properly, and this indirectly affected the compressive strength value of the samples. Based on the analyzed data, it is found that compressive strength of bricks containing HDPE meets the standards set in MS 76: 1972 [13]. It also proves that compressive strength of bricks containing 3.0% of HDPE matches up to the ones with 0% (controlled). This in a whole, signifies that the optimum percentage of HDPE is at 3.0%.

As for the difference in compressive strength value for bricks that contains certain percentage of HDPE composition as shown in Figure 2 above, it showed that HDPE percentage of 2.5% gives out the highest difference which was 17.09%, as compared to others. This is caused by the lower value of compressive strength than the ones of controlled bricks. While the 3.0% HDPE percentage shows the lowest difference (5.13%) compared to all readings, caused by the almost similar values of compressive strength with the controlled bricks.

### *3.2. Initial rate water absorption test*

Figure 3 shows a percentage graph of initial water absorption rate on the seventh and 28th day. Based on the data obtained, a high rate of initial water absorption recorded for samples with 3.5% of HDPE on the seventh day, at 3.33 kg/m$^2$.min while it showed lower in values on the 28th day, at 2.36 kg/m$^2$.min. On the 28th day, the initial water absorption rate recorded for all samples are lower than their seventh day.

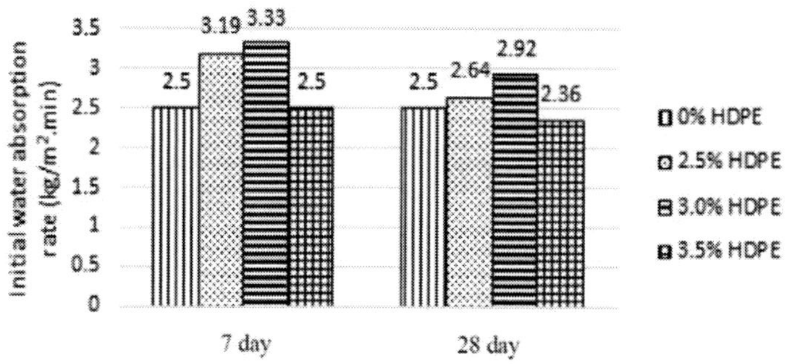


**Figure 3.** Results of initial water absorption rate.

The good rate of initial water absorption is between 0.25 kg/m$^2$.min to 2.00 kg/m$^2$.min. This value can be referred to evaluate the performance of initial water absorption rate on certain bricks. As shown in Figure 3, it is found that each bricks showed different rates of initial water absorption. However, on the seventh day, bricks samples with 0% and 3.5% of HDPE gives out the similar values. The graph also shows that all sample bricks exceed the values of 2.00kg/m$^2$.min.

### *3.3. Relationship between compressive strength and initial water absorption rate*

On the relationship of both tests done on seventh day as shown in figure 4, low values of compressive strength recorded on samples with 2.5% of HDPE which was 9.4N/mm$^2$ and the rate of initial water absorption recorded was 3.19 kg/m$^2$.min. Compressive strength of 2.5% HDPE bricks samples is affected by the higher rate of initial water absorption where results of compressive strength is lower than those samples with 0% HDPE (controlled). It was recorded that bricks samples with 3.5% HDPE was denser than other samples and this value is affected by the lower rate of initial water absorption which was at 2.50 kg/m$^2$.min, similar to 0% HDPE bricks (controlled).

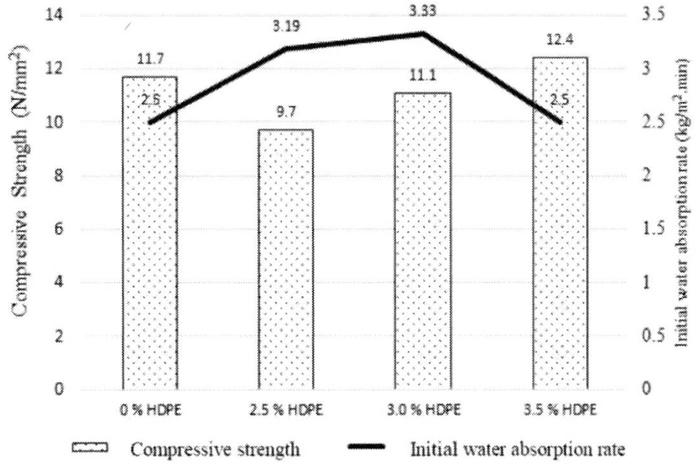


**Figure 4.** Relationship between compressive strength and initial water absorption rate on 7$^{th}$ day.

Meanwhile, for the results of test done on 28th day, samples with 3.0% HDPE gives out highest value of compressive strength which was 15.9 N/mm$^2$ compared to samples with 2.5% and 3.5%. However, the rate of initial water absorption for this set of samples was higher than others as shown in figure 5. This shows that its rate of initial water absorption did not affect the compressive strength when it reaches its maturity state.

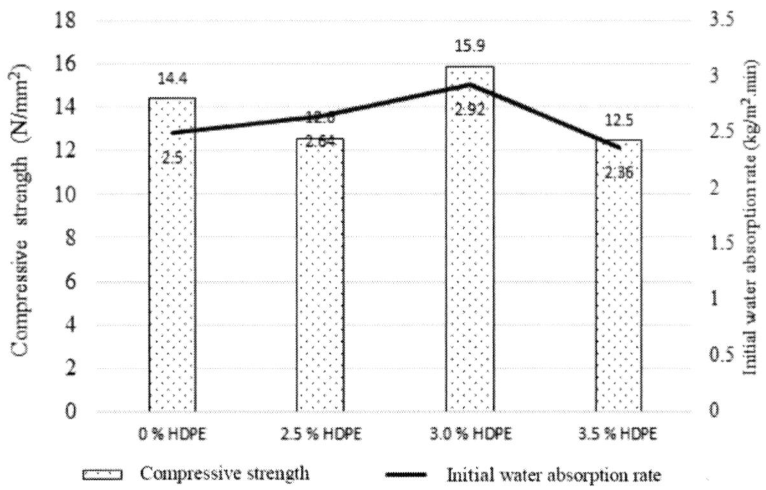


**Figure 5.** Relationship between compressive strength and initial water absorption rate at 28$^{th}$ day.

## 4. Conclusion

The use of HDPE in making cement bricks can affect the physical properties of the bricks in various ways. Among the visible effect is the sign of the presents of HDPE in the form of small black beads. Other than that, it is concluded that:

- The values obtained for compressive strength recorded the highest at 3% replacement of HDPE (15.9 $N/mm^2$) while the lowest initial water absorption rate recorded for 3.5% HDPE replacement of 2.35kg/$m^2$.min.
- The initial water absorption rate shows an increasing trend up to 3% replacement. Beyond that, at 3.5% replacement, both values on seventh and $28^{th}$ day recorded a decreased initial water absorption rate.

As a whole, the higher amount of HDPE used will cause higher compressive strength obtained up to a certain amount. However, this study found that 3.0% use of HDPE gives higher value of compressive strength than the controlled bricks, and at the same time reduce the use of sand in the production of bricks, as well as reducing pollution to the environment caused by plastics. As to improve the study carried out, a few considerations and suggestion could be applied in future studies. Among the suggestions for future studies are:

- The study was done with consistent ratios of water on all samples. Various ratios of water-cement can be used for future studies, to evaluate the optimum strength of bricks using different ratios of water and different percentage of HDPE.
- Substituting sand with HDPE into other kinds of bricks such as clay bricks in future studies, as HDPE has high potential to be used as replacement of sand, and not specific to only cement bricks.
- Smaller size of HDPE beads may be used. Theoretically, the smaller the aggregate size used, the higher the values of compressive strength of the bricks.

## 5. References

[1] Hadipramana J, Samad A A A, Zaidi A M A, Mohammad N and Wirdawati W 2012 Influence of Polypropylene Fiber in strength of foamed concrete, *Advanced Materials Research*, **488-489**, 253-257

[2] Ali N, Yaacob K Y, Burhanudin M K, Shahidan S and Abdullah S R 2016 Investigation of compressed earth brick containing ceramic waste, *ARPN Journal of Engineering and Applied Sciences* **11** (8) 5459-5462

[3] Khalid F S, Juki M I, Othman N and Wan Ibrahim M H 2017 Pull-out strength of polyethylene terephthalate bottle fibre in concrete matrix, *Malaysian Construction Research Journal* **21**(1) 75-85

[4] Abdullah S R, Wan Zainal Abidin W R and Shahidan S 2016 Strength of concrete containing rubber particle as partial cement replacement *3rd Int. Conf. on Civil and Environmental Engineering for Sustainability, IConCEES 2015 (Malacca, Malaysia)* vol 47 (Paris: MATEC Web Conference) 01009

[5] Chee-Ming C 2011 Effect of natural fibres inclusion in clay bricks: physico-mechanical properties, *Geotechnical and Geological Engineering* **73**(51) 1-8

[6] Perez J A, Terradas R, Manent M R, Seijas M and Martinez S 1996 Inertization of industrial wastes in ceramic materials, *Journal of Industrial Ceramics* **1** 1-7

[7] M Frigione 2010 Recycling of PET bottles as fine aggregate in concrete, *Waste Management* **30** 1101-1106

[8] Saikia N and de Brito J 2014 Mechanical properties and abrasion behaviour of concrete containing shredded PET bottle waste as a partial substitution of natural aggregate, *Construction and Building Materials* **52** 236-244

[9] Fathi H, Lameie T, Maleki M and Yazdani R 2017 Simultaneous effects of fiber and glass on the mechanical properties of self-compacting concrete, *Construction and Building Materials* **133** 443-449
[10] BS EN 771-1:2011+A1:2015 2011 *Specification for Masonry Units. Clay Masonry Units* (London: British Standard) pp 1-74
[11] BS EN 1052-1 1999 *Methods of Test for Masonry. Determination of Compressive Strength.* (London: British Standard) pp 1-12
[12] BS EN 772-1 2011 *Methods of Test for Masonry Units. Determination of Compressive Strength.* (London: British Standard) pp 1-18
[13] MS 76 1972 *Specification for Bricks and Blocks of Fired Brick Earth, Clay or Shale* (Selangor: Malaysian Standard) pp 1-74

**Acknowledgments**
This research was funded under Fundamental Research Grant Scheme (1576) by Ministry of Higher Education. Authors also wishing to acknowledge the assistance from Mr. Noorfairul Faizal Ali for proofreading this paper.

IOP Conf. Series: Materials Science and Engineering **271** (2017) 012084 doi:10.1088/1757-899X/271/1/012084

# Properties of concrete containing foamed concrete block waste as fine aggregate replacement

**K Muthusamy [1], A M A Budiea[2], A L F Zaidan[1], M H Rasid[1], D S Hazimmah[3]**

[1] Faculty of Civil Engineering and Earth Resources Universiti Malaysia Pahang, Lebuhraya Tun Razak, 26300, Gambang, Pahang
[2] Galactic Bridge –Research and Technology Centre, Bandar Sri Damansara, 52200 Kuala Lumpur
[3] School of Civil Engineering and Technology, University College of Technology Sarawak, 868 Persiaran Brooke, 96000 Sibu, Sarawak
Corresponding author: khairunisa@ump.edu.my

**Abstract.** Environmental degradation due to excessive sand mining dumping at certain places and disposal of foamed concrete block waste from lightweight concrete producing industry are issues that should be resolved for a better and cleaner environment of the community. Thus, the main intention of this study is to investigate the potential of foamed concrete block waste as partial sand replacement in concrete production. The foamed concrete waste (FCW) used in this research that were supplied by a local lightweight concrete producing industry. The workability and compressive strength of concrete containing various percentage of foamed concrete waste as partial sand replacement has been investigated. Prior to the use, the foamed concrete waste were crushed to produce finer particles. Six concrete mixes containing various content of crushed foamed concrete waste that are 0%, 10%, 20%, 30%, 40% and 50% were used in this experimental work. Then the prepared specimens were placed in water curing until the testing age. Compressive strength test and flexural strength tests were conducted at 7, 14 and 28 days. The result shows that integration of crushed foamed concrete waste as partial sand replacement in concrete reduces the mix workability. It is interesting to note that both compressive strength and flexural strength of concrete improves when 30% crushed foamed concrete waste is added as partial sand replacement.

## 1. Introduction

Foamed concrete is manufactured by entraining relatively large volumes of air into the cement paste by the use of a chemical foaming agent [1]. Currently, foamed concrete which is classified as lightweight concrete are increasingly used in construction projects mainly due to its lightness, more economic structural design and also ease of handling at construction site. Foamed concrete is available in various form of product ranging from fresh foam concrete mix, precast wall panel and precast concrete block. Foamed concrete block is one of the lightweight concrete favoured by the contractors owing to its lower density and ease of handling in comparison to conventional concrete. Increase in demand for this product by local contractors has results in factories to manufacture this lightweight concrete product. The production of this blocks in huge quantity which then checked for is quality has caused some of the products are rejected due to defects resulting from mishandling or other factors. Those foamed concrete blocks are usually discarded as profitless waste near the factory. Being a concrete material which does not degrade as organic plant, continuous disposal of these foamed concrete block waste consume larger dumping area as it piled up higher with the time. Utilizing this construction material waste in production of any products would be one of the solutions to turn the waste into profit making element. This would also lead to benefits in terms of reduction on cost, energy savings, promoting ecological balance and conservation of natural resources [2].

At the same time, sand that is one of the important mixing ingredients in concrete is used in increasing quantity for construction industries. The local sand used is mined from the rivers is also an income generating industry in the country. Excessive sand mining would cause negative impacts to the environment both to the river as well as water quality. This happens when the rate of extraction of sand, gravel and other minerals exceeds the rate at which natural processes generate these materials [3]. Sand mining activity cannot be stopped as it contributes to the development of other local industries. However, the quantity of sand harvested from the nature is possible to be controlled by introducing other waste materials that able to function as partial sand replacement material. Realization on the importance of managing the resources smartly for sustainable environment, has inspired local researchers [4-8] to investigate the potential existing waste materials as partial fine aggregate in concrete. So far, no literature review is available on the use of foamed concrete waste as partial sand replacement in concrete. Thus, the present research looks into the potential of the freely available foamed concrete block waste to be transformed into environmental friendly sand replacement material for the use in concrete production. The current research looks into the workability and mechanical properties of concrete containing crushed foamed concrete block waste as partial sand replacement.

## 2. Methodology

### *2.1. Materials*

The materials used in this study were cement, fine aggregate, foamed concrete block waste, coarse aggregate and water. The type of cement used throughout this research was Portland Composite Cement (PCC) of Pheonix brand. Local river sand and granite aggregate has been used as fine aggregate and coarse aggregate respectively. Tap water is used for concrete mixing work and curing of the specimen. No superplasticizer is employed during the mixing work. The foamed concrete block chunks illustrated in figure 1 is obtained from Binaan Desjaya Sdn Bhd, Kuantan, Pahang. The foamed concrete wastes were collected from the factory and transported to laboratory. Then, the foamed concrete is crushed using jaw crusher shown in Figure 2 to produce fine size aggregate. The finely crushed foamed concrete block wastes which are ready to be used as partial fine aggregate replacement in this research were kept in closed container.

**Figure 1.** Foamed concrete block chunks.

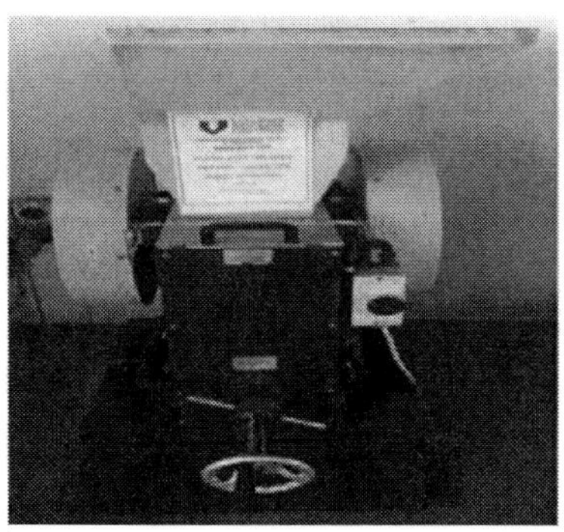

**Figure 2.** Jaw crusher used to produce fine sand.

### 2.2. *Mix proportions*
Six mixes were used in this experimental work. The control mix used is grade 30. The control mix consist 100% river sand as fine aggregate. The rest of the mixes contain various percentage of crushed foamed concrete waste (FCW) that is 10%, 20%, 30%, 40% and 50% as partial sand replacement. The mix proportion used to produce specimens presented in Table 1.

**Table 1**. Mix proportion of concrete mixes.

| Mix Design | Cement ($kg/m^3$) | Sand ($kg/m^3$) | Coarse Aggregate | Foamed Concrete Waste ($kg/m^3$) | Water ($kg/m^3$) |
|---|---|---|---|---|---|
| FCW 0 | 330 | 1020 | 905 | 0 | 190 |
| FCW-10 | 330 | 918 | 905 | 102 | 190 |
| FCW-20 | 330 | 816 | 905 | 204 | 190 |
| FCW-30 | 330 | 714 | 905 | 306 | 190 |
| RFC-40 | 330 | 612 | 905 | 408 | 190 |
| RFC-50 | 330 | 510 | 905 | 510 | 190 |

### 2.3. *Specimen preparation and testing method*
All the mixes were prepared in form of cubes (100x100x100mm) and beam (100x100x500) for determination of compressive strength and flexural strength respectively. Specimens were subjected to water curing until the hardened properties testing date. The effect of crushed foamed concrete waste (FCW) content when used as partial sand replacement on concrete mix workability were determined through slump test. Slump test was carried out adhering to the steps stated in BS EN 12350:2 [9]. Compressive strength test conducted following the procedures in BS EN 12390:3 [10]. Flexural strength test was conducted by referring to BS 1881 – 118 [11] to determine the flexural strength for concrete specimens. Both types of destructive test were conducted when the specimens reach the curing age of 7, 14 and 28 days.

## 3. Result and discussion

### 3.1. *Workability*
Figure 3 illustrates the effect of foamed concrete waste content on the slump of concrete. Generally, inclusion of crushed foamed concrete waste as partial fine aggregate replacement effects the concrete mix workability. The slump reading of the concrete increases as the percentage of foamed concrete waste used as partial sand replacement increased. All slump readings are classified as medium workability. The highest slump value recorded at 50% replacement level, which still considered medium value. The concrete slump value at 10%, 20% and 30% of RFC replacement had given low workability with 58mm, 63mm and 66mm respectively, which are higher as compared to control specimen with value of 53mm.

The highest slump value observed at 50% of sand replacement that is 72mm. Increment in the workability of mixture when FCW replacement increasing is may be due to the fineness of FCW as compare to normal sand. The increase in slump value is due to high porosity of FCW particles, which cause the absorption of more water. This is due to large voids in the foamed concrete. In addition, the high demand of water as the FCW content increases could cause slump to collapse due to loss of adhesion between the mixing ingredients. This is an additional function of water in the mixture to prevent the slump from collapse. That means, more water is required in addition to water required during hydration of cement [12].

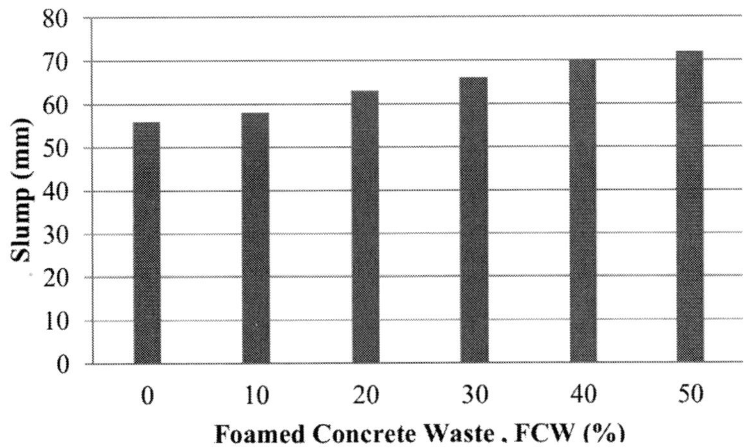


**Figure 3.** Effect of foamed concrete waste content on concrete workability.

*3.2. Compressive strength and flexural strength*

Figure 4 and 5 illustrates the compressive strength and flexural strength development of the concrete containing various percentage of foamed concrete waste as partial sand replacement. On overall, all mixes exhibit increment in the strength value as the curing age increases from 7 to 14 and 28 days. Application of water curing ensures continuous hydration process, which is vital for production of larger amount of calcium silicate hydrate that is responsible for compressive strength development. Looking at the effect of crushed foamed concrete waste on concrete strength, it is evident that integration of this waste material at suitable percentage able to increase the strength of concrete.

The result indicates that utilization of 30% crushed foamed concrete waste as partial fine aggregate replacement able to produce concrete with the highest strength as compared to all mixes. The compressive strength of concrete mix containing 30% of sand is almost 4.8 % higher than control specimens. This positive contribution is possibly due to the role of this fine waste material as filler that tend to occupy existing voids in the concrete making it denser and stronger. Similar observation has been made by [13-15] who reported that when solid waste in form of fine particles were integrated as partial fine aggregate replacement in concrete at suitable proportion the concrete exhibit strength increment. However, decrease in both compressive strength and flexural strength of concrete was observed as the percentage of sand is replaced with by foamed concrete waste 40% and 50%. That could be explained as the content of FCW increased above 30%, the water content became insufficient for the hydration process to take place. This causes the concrete to experience loss of adhesion between the aggregates owing lower amount of C-S-H gel generated finally resulting in drop of strength.

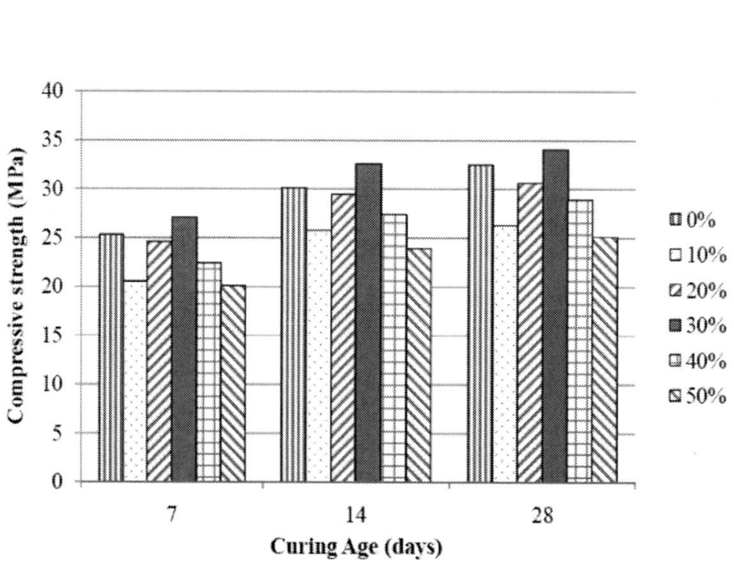


**Figure 4.** Effect of foamed concrete waste content on compressive strength of concrete.

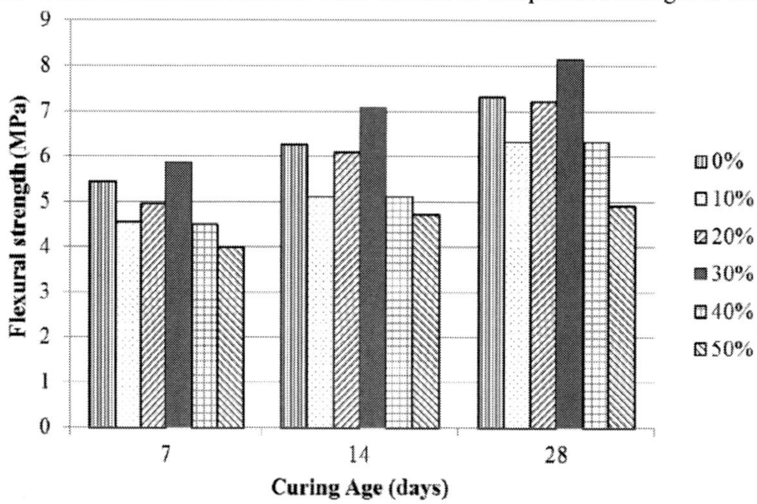


**Figure 5.** Effect of foamed concrete waste content on flexural strength of concrete.

## 4. Conclusion

Compressive strength of concrete reaches its optimum at 30% replacement level of foamed concrete waste (FCW) with the sand. However, integration of FCBW beyond 30% decreases the compressive strength of concrete gradually. Success in integrating crushed foamed concrete block waste as partial fine aggregate in concrete able to reduce amount of waste disposed by foamed concrete block manufacturing industry and save the use of natural river sand resulting in cleaner environment.

## 5. References

[1] Kearsley E P and Wainwright P J 2001 Porosity and Permeability of Foamed Concrete *Cement and Concrete Research* **31** 805 – 812

[2] Kaminskas N, Subramaniam K and Mustapha N 2006 Development of bricks using thermal power plant bottom ash and fly ash *Asian Journal of Civil Engineering* **13** 275-287

[3] Ashraf M A, Maah M J, Yusoff I, Wajid A and Mahmood K 2011 Sand mining effects, causes and : A case study of Bestari Jaya, Selangor, Peninsular Malaysia *Scientific Research and Essays.* **6**(6) pp 1216-1231

[4] Muthusamy K, Zulkepli N A and Mat Yahaya F 2013 Exploratory study on oil palm shell as partial sand replacement in concrete. *Research J. of Applied Sciences and Engineering Technology* **5**(7) 2372-2375

[5] Shafigh P, Mahmud H B, Jumaat M Z, Ahmmad R and Bahri S 2014 Structural lightweight aggregate concrete using two types of waste from the palm oil industry as aggregate *J. of Cleaner Production.* **80** pp 187-196

[6] Mohamed Idris M H, Muthusamy K, Othman R, Zamri N A, Ahmad Zawawi M N A and Mohd Ali N 2016 Workability and strength performance of foamed palm oil waste concrete *Proc. of Putrajaya International Built Environment, Technology and Engineering 2016* (Bangi, Selangor) p 177

[7] Ahmad Zawawi M N A, Muthusamy K, Nordin N, Zamri N A, Mohamed Idris M H, and Mohd Ali N 2016 Compressive strength palm oil waste lightweight aggregate concrete containing palm oil fuel ash as partial cement replacement. *Proc. of Putrajaya International Built Environment, Technology and Engineering 2016* (Bangi, Selangor) p 194

[8] Mat Yahaya F, Muthusamy K and Hussin M W 2016 Long term investigation on sulphate resistance of aerated concrete containing palm oil fuel ash *ARPN J. of Engineering and Applied Sciences* **11**(4) pp 2406-2411

[9] British Standard 2000 *BS EN 12350-2:Testing fresh concrete: Slump test* (London: British Standard Institution)

[10] British Standard 2002 *BS EN 12390-3 :Testing Hardened Concrete: Compressive strength of test specimens* (London: British Standard Institution)

[11] British Standard 1983 *BS 1881-118:1983: Testing concrete: Method for determination of flexural strength* (London: British Standard Institution)

[12] Adesanya D A and Raheem A A 2009 A study of the workability and compressive strength characteristics of corn cob ash blended cement concrete, *Construction and Building Materials* **23** 311–317

[13] Shettima A U, Hussin M W, Ahmad Y and Mirza J 2016 Evaluation of iron ore tailing as replacement for fine aggregate in concrete *Construction and Building Materials* **120** 72-79

[14] Wan Ahmad S, Muthusamy K, Hashim H and Yaacob M A 2017 Properties of concrete containing unground palm oil fuel ash as partial sand replacement *Applied Mechanics and Materials* **854** 278-283

[15] Muthusamy K, Hashim H and Ariffin N F 2017 The effect of unground palm oil fuel ash content as partial sand replacement on compressive strength of lightweight aggregate concrete *Proc. of Intervarsities of Multidiciplinaries International Conference 2017* (Kuantan,Pahang)

IOP Conf. Series: Materials Science and Engineering **271** (2017) 012085 doi:10.1088/1757-899X/271/1/012085

# Selection of suitable NDT methods for building inspection

**Mohamad Pauzi Ismail**[1]

[1]Industrial Technology Division, Agensi Nuklear Malaysia, Bangi.
Corresponding author: pauzi@nm.gov.my

**Abstract.** Construction of modern structures requires good quality concrete with adequate strength and durability. Several accidents occurred in the civil constructions and were reported in the media. Such accidents were due to poor workmanship and lack of systematic monitoring during the constructions. In addition, water leaking and cracking in residential houses was commonly reported too. Based on these facts, monitoring the quality of concrete in structures is becoming more and more important subject. This paper describes major Non-destructive Testing (NDT) methods for evaluating structural integrity of concrete building. Some interesting findings during actual NDT inspections on site are presented. The NDT methods used are explained, compared and discussed. The suitable methods are suggested as minimum NDT methods to cover parameters required in the inspection.

## 1. Introduction

British Standard defined non-destructive testing (NDT) as a test that does not impair the intended performance of the element or member under investigation [1]. NDT has been used widely in industrial sectors such as petroleum, petrochemical, ship construction, aircraft, railway, power station, gas pipeline etc. Materials used in these industries are mostly made from steel and aluminum, which involved processes such as welding, casting, forging or machining during construction. The quality of the product was checked and evaluated by NDT methods. Similarly, in-service inspections are also performed by this method to detect so-called in-service defects such as corrosion or fatigue cracks.

The use of NDT in inspecting concrete is not as common as in the metallic construction. This may be due to lack of demand or no specific requirement by code or standard from the related sectors as compared to metallic construction. No criteria for acceptance except for visual inspection and no personnel certification scheme available for NDT inspectors. However, research and development activities in various NDT methods indicate that NDT has a great potential to be applied on concrete structures. Most materials in buildings, bridges, dam, tunnels, etc. are made of concrete. This construction requires concrete of high quality in terms of strength and durability. NDT has the ability to estimate the strength and durability of critical construction without damaging them and the test can be carried out on-site.

Following the collapse of the New World Hotel in Singapore, the Singaporean government has imposed an act to ensure that all buildings of more than 2 floors must be inspected for their quality at an interval of every 5 years. Similarly in Malaysia, The Street, Drainage and Building Act (Amendment 1994, Act A903) was amended as a follow up of the Highland Tower tragedy. The Act stipulates the requirements for the periodical inspection of buildings. These requirements apply to buildings exceeding five storeys and all buildings must be checked for their safety and quality in every 10 years before certificate of fitness (COF) can be issued.

In mechanical construction, there are existing codes or standards available to be used as a guideline. For example ASME code is commonly used for construction of boiler and pressure vessel, API code for constructions in petroleum industry and AWS code for constructions of steel structures. In such cases the application of NDT is well specified. However in the case of civil construction, there is no specific code or standard currently available that can be used as a guideline for the selection and application of suitable NDT methods. British Standard only specifies visual inspection for criteria of acceptance, e.g. based on crack width.

Since there are so many NDT methods available including new and conventional which make us confuse which method to be chosen, the author would like to recommend and discuss the most effective, reliable and cheaper method based on practical experience on site.

## 2. Non-destructive testing methods

There are several NDT methods applicable to concrete structures. Table 1 lists the suitable NDT methods for checking specific properties of concrete quality [2]. The importance of checking certain properties according to the type of structures is listed in Table 2 [2]. Table 1 indicates that there are 11 NDT methods applicable for concrete inspection. They are rebound hammer, penetration resistance, pull out, ultrasonic, radar, thermography, radiography, acoustic emission, magnetic or eddy current, half-cell potential and visual (photography) methods.

The Schmidt rebound hammer is basically a surface hardness test with little apparent theoretical relationship between the strength of concrete and the rebound number of the hammer. However, within certain constraints, empirical correlation has been established between strength properties and the rebound number. The correlation between rebound index *N* and compressive strength *K* varies from one researcher to another. The author [3] proposed a polynomial correlation which agrees with the manufacturer's calibration curve.

$$K = 10.8 + 0.95N + 0.017N^2 \quad (1)$$

The estimated error is 30%.

The Windsor probe, like the rebound hammer, is a hardness tester, and its inventors claim that the penetration of the probe reflects the precise compressive strength in a localized area is not strictly true. However, the probe penetration does relate to some property of the concrete below the surface, and, within limits, it has been possible to develop empirical correlations between strength properties and the penetration of the probe. It can be used for estimating compressive strength, uniformity and quality of concrete. Expertise required is low, however license from Police Department is required for operation of explosive materials. The equipment is easy to use and does not require surface preparation prior to testing. It is good for determining in-situ quality of concrete. The results are not subject to surface conditions, moisture content or ambient temperature. It requires minimum edge distance and member thickness. It slightly damages small area. Calibration by manufacturers does not give precise prediction of strength for concrete older than 5 years and where surface is affected by carbonation or cracking [4].

**Table 1.** List of items can be evaluated by non-destructive testing (NDT).

| NDT methods | | | | | | | | | | | | |
|---|---|---|---|---|---|---|---|---|---|---|---|---|
| 1- Rebound hammer | ● | | | | | | | | | | | |
| 2- Penetration resistance | ● | | | | | | | | | | | |
| 3- Pull-out | ● | | | | | | | | | | | |
| 4- Ultrasonic | ● | ● | ● | ● | | | | ● | ● | ● | | |
| 5- Radar | | | | | | | | ● | ● | ● | | |
| 6- Thermagraphy | | | | | | ● | | ● | ● | | | |
| 7- Radiography | | | ● | | | | | ● | | ● | ● | ● |
| 8- Acoustic emission | | | | | | | ● | | | | | |
| 9- Magnetic or eddy current | | | | | | | | | | ● | ● | ● |
| 10- Half-cell Potential | | | | | | | | | | | | ● |
| 11- Photography | | | | | ● | ● | | | | | | |

**Table 2.** Inspection of structure.

| Items | | | | | | | |
|---|---|---|---|---|---|---|---|
| Strength | | ● | ● | ● | ◉ | ◉ | ● |
| Elastic Modulus | | | ● | | ● | | ● |
| Deformation | | ● | ● | | | ● | ◉ |
| Crack | Distribution | ● | ● | ● | ● | ● | ● |
| | Width | ● | ● | ● | ● | ● | ● |
| | Depth | ● | | ● | | ● | ◉ |
| Defect | Delamination | ◉ | ● | ● | | ● | ◉ |
| | Honeycombing, voids | ● | ● | | | ● | ● |
| | Backwall voids | | ◉ | | | | |
| Rebar | Location, size | ◉ | | ● | | ● | ◉ |
| | Corrosion | ◉ | ● | ◉ | | ◉ | ◉ |
| Thickness | | ● | ● | | | | ● |

◉ very important ● important

Pullout Testing is the test involves drilling a hole in which a standard threaded or wedge anchor is placed. This is then pulled until the concrete raptures. With a calibration charts the maximum force gives an indication of the strength of concrete. Pullout devices can be inserted during casting of concrete. It provides an estimation of the compressive and tensile strengths of hardened concrete. User expertise is low and can be used in the field. In-place strength of concrete can be measured quickly and appears to give good prediction of concrete strength. Pullout devices must be preplanned and inserted during the construction stage, or inserted in hole drilled in hardened concrete. A cone of concrete may be pulled out, necessitating minor repairs. It can only test a limited depth of material. As it is a surface method, and in reinforced concrete could only be used to assess the concrete cover quality [4]. As indicated in Table I, the ultrasonic method can be used to estimate strength, elastic modulus, slab thickness, crack depth, and to detect voids, lamination and bar location. It is recognized as the only NDT method available at this time, which is capable of determining the crack depth to a certain degree of reliability. The technique was also used to monitor the mixing materials during construction, determine the concrete uniformity, thickness measurement and to estimate the depth of damage caused by fire [5]. The ultrasonic pulse velocity method can only estimate strength with an accuracy of 20% [6] at best. Better and reliable strength estimation is by combination with rebound hammer. Although there are many other combination, combination between rebound hammer and pulse velocity is the most popular. Figure 1 show typical strength estimation curve by combination between rebound number and pulse velocity for Malaysian concrete design based on DoE method.

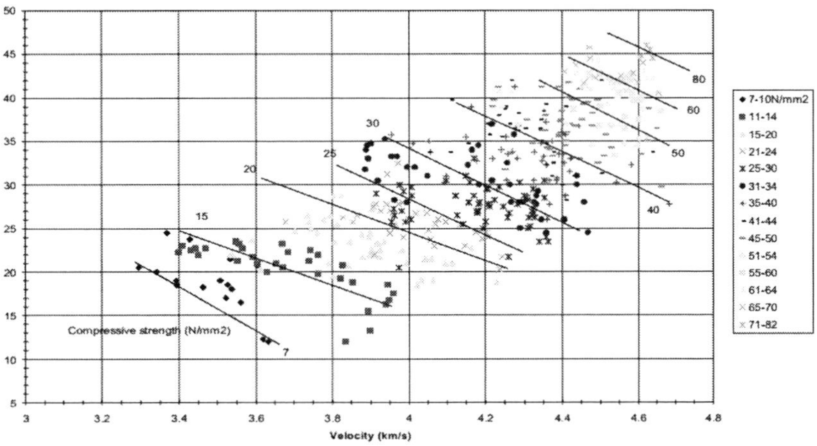


**Figure 1.** Strength estimation by combined rebound no. and pulse velocity [7].

Reinforcement bar in concrete can be located using radar technique or magnetic method (cover meter). Radar method is an easy, fast and accurate technique for measuring cover thickness and arrangement of the bar. On the other hand, cover meter can also perform the same as the radar method but the method is rather slow and needs some experiences to understand/interpret the signal. It is worth to mention that both methods are not so accurate for measuring the diameter of the bar. Figure 2 and 3 show radar profile for rebar detection.

An infrared camera is used to detect variations in infrared radiation output of a surface. Thermal gradients arise because of different in surface temperature between sound and unsound concrete. The temperature gradients are displayed on a monitor screen in the form of colour thermal contours. It can be used for detecting delamination, heat loss and moisture movement through concrete elements especially for flat surfaces. User expertise is not high but interpretation of results requires understanding of thermal behaviour and patterns. It is portable and permanent records can be made. Testing can be done without direct access to surface and large areas can be rapidly inspected using infrared cameras. It is very sensitive to thermal interference from other heat sources [4]. Figure 4 shows typical photographs and thermographs of wet area. The thermographs were taken a few minutes after sunset [5].

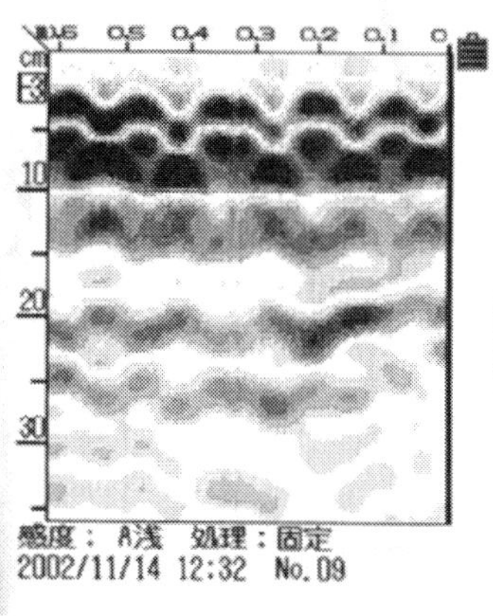


**Figure 2.** Radar profile of different bar size.

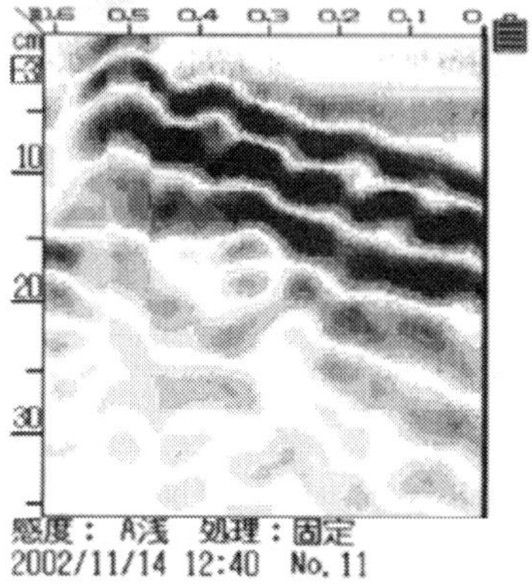


**Figure 3.** Radar profile of 20mm diameter bar at different depth

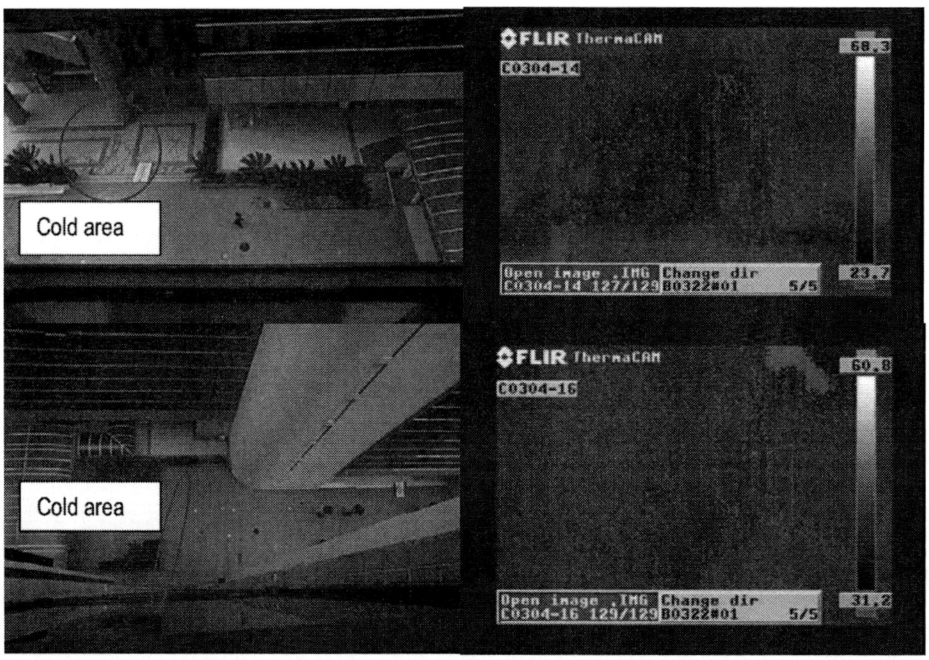


**Figure 4.** Photographs (left) and thermographs (right).

Radiography is the best method to check the condition of the bar (including sizing) and to detect void in concrete. Typical sources are Iridium-192 and Cobalt-60. Although Cobalt-60 has better radiation penetration in concrete, the equipment is bulky and difficult to handle in order to meet on-site safety requirement. It requires a very large safe operational area which is not practical in most of the construction site. Modern equipment such as Linac or betatron produce high energy X-ray, although the energy is high but dose rate is considerably low since X-rays are produced not continuously like gamma equipment but in the form of pulses and they use imaging plate as recording media instead of film. Figure 5 shows typical radiographic image on concrete sample using Iridium 192 gamma source [5].

**Figure 5.** Example of radiographic image of void and bar in concrete. SFD: 500mm, concrete thickness: 150mm, source: Ir-192, exposure: 4 Ci-hr, film: Agfa D7.

Acoustic emission is produced when a material is loaded beyond their elastic limit. Kinetic energy is released. They are inaudible but can be detected by sensors attached to the surface of a test object. This method has also been used in recent years to study the initiation and growth of cracks in concrete under stress. Extensive knowledge is required to plan the test and to interpret results. It monitors response of existing structure to applied load it is capable of detecting onset of failure and locating source of possible failure. Since acoustical signals come from defects throughout the structures a few transducers are enough to detect and locate defects over large areas [4]. The method is suitable for structures with dynamic loading such as bridges. The equipment is expensive on the market; however Nuclear Malaysia has developed a low cost acoustic emission system. A laboratory study has been carried out to evaluate the feasibility of detecting corrosion damage in rebars by the eddy current and magnetic flux leakage techniques. Many practical low frequency systems are capable of locating and estimating the size of the cross-section of the rebars via calibration charts [4].

Electrical potential of steel reinforcement is measured relatively to a reference electrode (half-cell). This enables potential contour maps to be plotted. The electrode potential of steel in concrete indicates the probability of corrosion. The half-cell provides a relatively quick method of assessing reinforcement corrosion over a wide area without the need for wholesale removal of the concrete cover. Quantitative measurements are made so that a structure can be monitored over a period of time and any deterioration can be noted. It is portable equipment. Field measurements can be readily made and results can be plotted in the form of equipotential contour diagram, which can indicate likely areas of corrosion [4].

Visual inspection is the method applied during the first survey of the structures. The purpose is to observe for any surface discontinuities. Visual features may be related to workmanship, structural serviceability and material deterioration. Figure 6 to 11 show a few of these in their typical form. Typical equipment used during the visual inspection includes special tapping hammer, crack gauge, magnifying glass, binocular and camera. For buildings, survey may be started from the outside of the building and followed by the inside survey with particular attention given to areas at the corner or any tension area of the building.

**Figure 6.** Efflrouresence caused by water leakage.

**Figure 7.** Structural cracks at beam.

**Figure 8.** Water leakage causes cracking.

**Figure 9.** Damage timber pile.

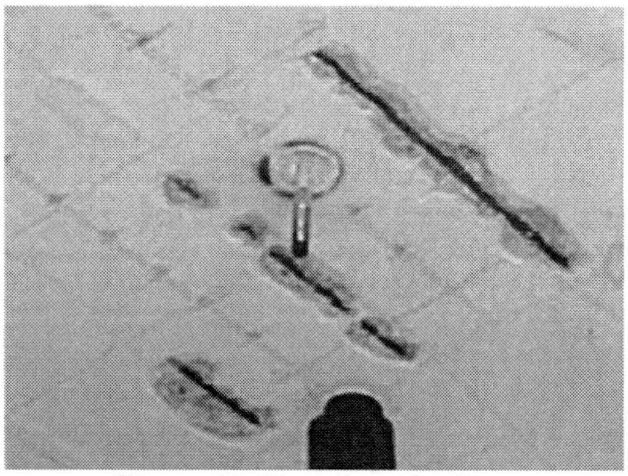

**Figure 10.** Corrosion of rebars.

**Figure 11.** Honeycomb.

## 3. Selection of test method

Rebound hammer, penetration resistance, pull out and ultrasonic methods can be used to estimate concrete strength. So far no NDT methods can give the best estimation. However the popular combination for strength estimation is rebound hammer and ultrasonic testing. Ultrasonic is among the widest applications. It can be used for estimation of strength, elastic modulus, thickness and crack depth using conventional ultrasonic pulse velocity (UPV) method. Using advanced ultrasonic method such as multi array transducer, it can detect void, honeycomb, lamination and bar location. Ultrasonic is the only NDT method for crack depth measurement.

Radar is very good for bar location, plastic or PVC conduit and to some extent can be used for detection of lamination and void detection. Thermography is applicable for imaging crack distribution, wet area (water leakage), near surface void, honeycomb and lamination. Radiography is the most effective and easier method for detection of void, honeycomb, bar location, size and corrosion. Acoustic emission for concrete inspection is still under development it was used in steel or metallic construction for monitoring crack growth. It has potential application in structure with dynamic loading such as bridges. Magnetic and eddy current method can also be used to detect reinforcement bar and to some extent can be used to estimate the bar size.

However advanced ultrasonic with multi array probes probably can do the same function for radar, radiography and magnetic and eddy current method methods. Half-cell potential was the old method for bar corrosion but the reliability of the method is questionable. Visual inspection is the basic inspection method. It is used during first survey to decide whether or not there is a need for further investigation.

## 4. Conclusion

The cheapest NDT method is by visual inspection. However the valuable information can only be gathered by the well-trained eye. Like in mechanical construction, four or five NDT methods may be adequate for evaluating concrete structural integrity. It is recommended that Visual, rebound hammer, advanced ultrasonic and thermography methods are the minimum NDT methods to cover the required parameter for assessing the structural integrity of concrete building.

## 5. References

[1] British Standard Institution 1986 Guide to the use of non-destructive methods of test for hardened concrete Part 201 (London: British Standard)

[2] *The summary of National Seminar on NDE of Non-Metallic Materials (Concrete)* (Singapore) pp 25-26

[3] Ismail M P, Yusof KM and Ibrahim A N 1996 *Reliability of the rebound hammer method of estimating concrete strength* Bulletin Institute of Engineers Malaysia (Malaysia: Institute of Engineer Malaysia) pp 5-7

[4] Training course series no. 17 2002 *Guidebook on non-destructive testing of concrete structures* (Vienna: International Atomic Energy Agency)

[5] Mohamad Pauzi Ismail 2015 *Kaedah Ujian Tanpa Musnah (UTM) terhadap konkrit struktur* (Kuala Lumpur: Dewan Bahasa dan Pustaka).

[6] Bungey J H 1984 The use of ultrasonics for NDT of concrete, *British Journal of NDT* **26(5)** 366-369

[7] Ismail M P, Yusof K M and Ibrahim A N 1996 A Combined ultrasonic method on the estimation of compressive concrete strength, *INSIGHT* **38(11)** 781-785

[8] Devices for non-destructive testing of metals, plastics and concrete 2017 *Development and production* Retrived on July 6, from http://www.acsys.ru/eng/production/detail/a1040-mira/

**Acknowledgement**

The author would like to thank to NDT team of Nuclear Malaysia for their helps in collecting NDT data. They are Suhairy Sani (UT2 & RT 2), Mohd Zaki Umar (IR 2), and Abdul Bakhri Muhamad (UT 2).

IOP Conf. Series: Materials Science and Engineering **271** (2017) 012086 doi:10.1088/1757-899X/271/1/012086

# A review of the application Acoustic Emission (AE) incorporating mechanical approach to monitor Reinforced concrete (RC) strengthened with Fiber Reinforced Polymer (FRP) properties under fracture

**S M S Syed Mazlan[1], S R Abdullah[1], S Shahidan[1] and S R Mohd Noor[1]**

[1]Jamilus Research Centre, Faculty of Civil and Environmental, Department of Structural and Material Engineering, Universiti Tun Hussein Onn Malaysia, Batu Pahat, Johor, Malaysia.
Corresponding author: syedmsyahir@gmail.com

**Abstract.** Concrete durability may be affected by so many factors such as chemical attack and weathering action that reduce the performance and the service life of concrete structures. Low durability Reinforced concrete (RC) can be greatly improved by using Fiber Reinforce Polymer (FRP). FRP is a commonly used composite material for repairing and strengthening RC structures. A review on application of Acoustic Emission (AE) techniques of real time monitoring for various mechanical tests for RC strengthened with FRP involving four-point bending, three-point bending and cyclic loading was carried out and discussed in this paper. Correlations between each AE analyses namely b-value, sentry and intensity analysis on damage characterization also been critically reviewed. From the review, AE monitoring involving RC strengthened with FRP using b-value, sentry and intensity analysis are proven to be successful and efficient method in determining damage characterization. However, application of AE analysis using sentry analysis is still limited compared to b-value and intensity analysis in characterizing damages especially for RC strengthened with FRP specimen.

## 1. Introduction

Fiber Reinforced Polymer (FRP) strengthening system was first developed in Europe since 1980 as a replacement for steel strengthening. This technique was widely used to upgrade many bridges and other structures around the world. Nowadays, FRP material shows an increased number of application in the construction industry and it is successfully being used in strengthening deficient concrete structures [1]. FRP is more effective as it provides a better solution in terms of properties and application. These materials are lightweight, noncorrosive, and exhibit high tensile strength [2-3]. FRP strengthening systems can be used to rehabilitate or restore the strength of the deteriorated structural member, retrofit or strengthen the sound structural member to resist increased loads due to change in the use of structure, or address design or construction errors found in a structure [4]. These FRP materials are readily available in several forms such as plates, rodes and sheets. Application of a Non-Destructive Technique (NDT) is reliable and robust to provide real time information and important to monitor and assess the condition of buildings. There are many types of NDT testing and one of the promising methods of NDT testing commonly used to monitor the damage or deterioration in structures is known as AE technique [5]. AE technique method was first introduced by a German scientist, Kaiser in year 1950 [6]. Meanwhile in civil engineering, AE application has started since the late 1970s with an investigation on concrete material response towards AE carried out by Fetis and Cabe [7]. AE has the ability for real time process monitoring applications. It can monitor the occurring defects in concrete structure such as concrete fracture starting from the initial micro crack to macro crack formation and finally, structural failure [8]. Nowadays, many researchers are motivated to improve the use of AE instrumentation, AE source localization and work extensively in the AE signal processing system [6]. In this paper, an overview of the application of AE for RC strengthened with FRP under mechanical tests such as bending test (four-point bending and three-point bending) and cyclic loading test are discussed. Numerical method and statistical method from AE are usually performed to evaluate b-value, sentry and intensity analysis to study the effect of loading rate on variation, understand the behavior of the material, and quantify and evaluate the damage severity on samples. Studies on AE monitoring of application of AE for FRP in the form of sheets laminated onto RC specimen being put under mechanical tests have been reported in extensive literatures. The formation of cracks in FRP laminated onto RC structures within fracture process can be noticed by recording AE signals during the process and data obtained from such studies could be used to assess the amplitude (stress-strain), delamination (crack initiation and its internal microevents) and level of deterioration (quantification).

## 2. Acoustic emission based on b-values

B-value method was historically introduced by Gutenberg and Richer in year 1949 [9]. This method can be used to evaluate structural damage [10]. The b-value is functioning in the form of log-linear slope representing the cumulative amplitude distribution of AE. Previously, study of b-value on RC structures has proved that it is useful in providing details of early crack formation up to the failure fracture with the support of quantitative b-value ranges and damage levels descriptions as shown in table 1 [10].

### *2.1. Formulation of b-value*

B-values may change systematically due to the different stages of fracture growth [10]. The most commonly used equation is known as Gutenberg-Ritcher formula presented in equation (1).

$$\log_{10} N = a - bM \tag{1}$$

According to equation (1), M is the peak amplitude of AE hit in decibel unit obtained from AE, N is number of AE hits with magnitude greater than M while α is an empirical constant. The cumulative frequency distributions of the linear gradient descend, b is taken as AE b-values. Value of M can be determined by calculating the magnitude of waveform using equation (2) while conversely, the coefficient of 20 can also be divided with the slope to obtain the b-values [11-14].

$$\log_{10} N = a - b\left(\frac{M}{20}\right) \tag{2}$$

The b-value is known as the log-linear slope of the frequency–magnitude distribution of AE. From this formula, Colombo et., al. [10] found that larger b-values are obtained when the distributed micro crack are occurring in an early stage of damage, meanwhile b-value is low when the macro cracks begin to localize. The b-value is one of the AE functions which experienced continuous improvement accordance to findings from case studies. To date, the function of b-value is improved with Ib-value to evaluate the slope failure and process of fracture [9-12]. The calculation of Ib-value is based on the slope of the amplitude distribution from AE signal. The Ib-value formula is shown in equation (3) below:

$$Ib = \frac{\log N(\mu - \alpha_1\sigma) - \log N(\mu - \alpha_2\sigma)}{(\alpha_1 + \alpha_2)\sigma} \tag{3}$$

Where σ is a standard deviation, μ is the mean value of amplitude distribution; $\alpha_1$ and $\alpha_2$ are the coefficients related to the smaller amplitude and fracture level, respectively. This formula varies depending on the damage level. The function of the Ib-value is to evaluate the fracture process in concrete.

*2.2. Applications of b-value*

There are numerous studies on damage characterization using b-value contributing deeper information for damage descriptions. Therefore, marks b-value analysis as one of the most promising method to be used for quantitative damage classification. In the four-bending test, AE monitoring of b-value has been reported by Guzman *et al.* [13]. In the study, AE event represented by accumulated frequency of the AE and maximum amplitude was characterized in a form of semi-logarithmic scale and lineal correlation scale. The b-value ranges were characterized based on the four-bending test final load to define severity of damages as two stages as shown in table 1. The minimum and maximum b-values were considered for each cycle load failure phase. All the cycles and channels were plotted to expose the value variation and the development of micro and macro cracks. The maximum b-value shows the trend of the micro crack growth and the minimum b-value shows the trend of the formation of macro cracks as shown in figure 1(a). Table 1 shows the quantitative b-value. Previously, Colombo *et al.* [10] has established quantitative results for b-value. The research showed that the b-value is very useful for understanding the data obtained from all the monitoring work of concrete structure. To date, classification of damage particular for concrete specimens strengthened with FRP was reported in Ma and Li [15]. The characteristics of damage level were made based on the distribution of AE hits during the test and the observed damage states of the FRP-strengthened column of AE hits can be related to four different damage levels of the FRP-strengthened column qualitatively as shown in table 2. Therefore, it can be concluded that the crack process of FRP-strengthened columns can be monitored by AE techniques and the damage levels can also be determined qualitatively and conveniently based on the distribution of AE hits. In the study, AE event represented by accumulated frequency of the AE and maximum amplitude was characterized in a form of semi-logarithmic scale and lineal correlation scale. Figure 1(b) indicates the inhomogeneous mechanical properties of concrete, the crack process of the concrete in the plastic region was essentially non-uniform, resulting in that the entire b-value curve along the time axis was not smooth.

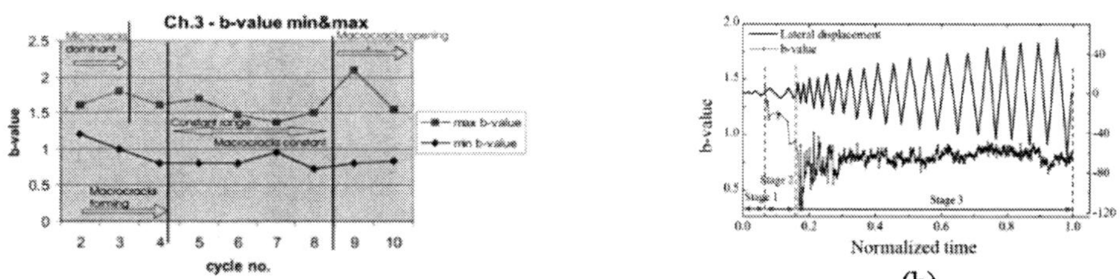


(a) (b)

**Figure 1.** (a) b-value analysis pattern for RC structure [10], (b) b-values correlative to the number of AE hits in a certain range of time for RC structure strengthened with FRP [15].

Investigations for the assessment of signal characteristics for RC members strengthened with Carbon Fiber Reinforced Polymer (CFRP) sheets have only begun [16]. In three-point bending test conducted by Do Hyun *et al.* [16], AE monitoring based on b-value is correlated to the fracture process of the RC beams bonded with CFRP sheets and the degree of localization of damage. The evolution of acoustic activity caused by micro-fractures within concrete is often quantified using the concise framework originated by Gutenberg and Richter in their analysis of earthquake magnitudes, which is a reflection of the view that large-scale (geological) and small-scale (micro-fracture) acoustic events share a common origin in cascades of strain energy release events. Choi and Yun [14] claim the AE cumulative amplitude distribution emitted was used as a statistical data to look at parametric AE data. The b-value formula indicated three ranges of damage levels or fracture levels as shown in table

3. In the development stage of micro cracking, the b-values show a value higher than 1.25. The b-value range in the crack propagation and crack widening stage is between 1.25 and 1.15. At the micro crack level, the b-value varies between 1.15 and 0.08.

**Table 1.** Quantitative damage progressions by Colombo *et al.* [10].

| B-value ranges | Types of damage progressions |
|---|---|
| 1.0 < b-value < 1.2 | Indicates that the channel is in the closest position towards a large crack such as macrocracks forming |
| 1.2 < b-value < 1.7 | Cracking is uniformly distributed suchlike the macrocracks are constant |
| b-value > 1.7 | Microcracks are prevailing or macrocracks are opening |

**Table 2.** Quantitative damage levels based on the distribution of AE hits by Ma and Li [15].

| Damage levels | AE hits stage | Damage states FRP-strengthened column |
|---|---|---|
| Slight | Stage 1 | No crack |
| Minor | Stage 2 | No crack |
| Moderate | Early part of stage 3 | Some horizontal cracks above FRP sheets, cracks between column and pedestal |
| Severe | Latter part of stage 3 | Horizontal cracks on FRP sheets, interface failure between column and pedestal. |

**Table 3.** B-value for damage level by Choi and Yun [14].

| B value ranges | Damage description |
|---|---|
| b-values > 1.25 | Formation of micro-cracks |
| 1.25 > b-values > 1.15 | Propagation of micro-cracks and crack-widening |
| 1.15 > b-values > 0.80 | Formation of macro-cracks, de-bonding of CFRP, rupture |

The most recent study conducted by Ma and Li [15] involving AE monitoring and damage assessment of FRP strengthened reinforced concrete columns. Analysis of b-value reported that the accumulative AE energy exhibited a good correlation with the accumulative hysteretic energy dissipated during the cyclic test for the FRP-strengthened column. By rule of thumb, high b-value density represented a large number of AE hits (high degree of crack activity) meanwhile low b-value density meant a small number of AE hits (low degree of crack activity).

## 3. Acoustic emission based on sentry function

Sentry method was first discovered and introduced by Minak [17]. By definition, sentry method is known as combination of mechanical and acoustic energy information in AE parameters [16]. The sentry values are calculated by combining the logarithm ratio of Strain Energy (Es) and the acoustic energy ($E_{AE}$) using particular formula. This method has been extensively used in the study of composite material and steel [16]. The rule of thumb for sentry method, f is divided into four levels demonstrating increasing of sentry function f(x). The four levels were labelled from PI(x) until PIV(x). Figure 2 illustrate the schematic diagram of PI(x), PII(x), PIII(x) and PIV(x) which used to describe the whole function of sentry.

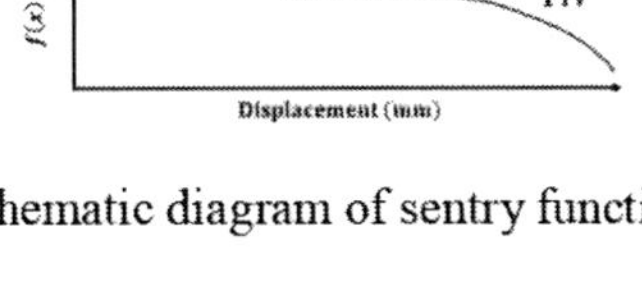


**Figure 2.** Schematic diagram of sentry function (f) [18].

**Table 4.** Standard sentry function sentry function (f) behaviors [14,18].

| Type | Symbols | Descriptions |
|---|---|---|
| Type I | PI(x) | An increasing trend, which represents the strain energy storing phases |
| Type II | PII(x) | Illustrated by the sudden drops of the function f, which may be related to a significant internal material failure occurrence |
| Type III | PIII(x) | There is an equilibrium state between the mechanical and AE energy |
| Type IV | PIV(x) | The decreasing behavior of f, related to the fact that the AE activity is greater than the material strain energy storing capability and the damage has reached a maximum |

### *3.1. Formulation of sentry function*

Sentry function requires vital combinations of both mechanical and acoustic energy information to perform deeper quantitative analysis of damages occurs in structure members. The formulation of sentry function is represented in the equation (4):

$$f(x) = Ln\left[\frac{E_S(x)}{E_{AE}(x)}\right] \quad (4)$$

According to equation (4), $E_s$ is expressed as the strain energy, $E_{AE}$ refers to AE energy while $x$ is the test driving variable (displacement or strain).

### *3.2. Applications of sentry function*

Application of sentry method was widely used for composite material and steel. Previously, sentry method was used to detect important events conforming to sudden drops and occurrences of delamination for various specimens such as glass fiber or epoxy composite materials [17], glass fiber reinforced plastics (GFRP), CFRP composite laminate, polyester resin reinforced with glass fiber [18] and many more. However, for specimen involving RC strengthened with FRP, the number of study is still limited. Selman *et al.* [19] conducted a study on a RC beam strengthened with CFRP undergoing a cyclic load test. The damage characterization was made by calculating sentry function. AE parameters such as cumulative acoustic emission energy and acoustic emission signals analysis was taken into account to further study the fracture progression in the specimen. Outcomes was summarized and test driving ($x$=1) was chosen to be presented in table 5.

**Table 5.** Damage characterization based on sentry function (f) behaviors [19].

| Type | Symbols | Sentry value (f) | Damage characterization |
|---|---|---|---|
| Type I | PI | $12.5 \leq f \leq 12.3$ | No significant event detected |
| Type II | PII | $12.8 \leq f \leq 12.4$ | Sudden released of elastic waves. Initial flexural crack developed at CFRP joint and grew rapidly. Flexural crack started to propagate towards RC sub-layer. Shear cracking occurs at CFRP joint. Drop of stiffness in both flexure and shear. |
| Type III | PIII | 13.9 | Specimen is in equilibrium state |
| Type IV | PIV | $13.8 \leq f \leq 13.7$ | Opening of shear crack increased and peeling of CFRP. Rebars yields. Shear damaged. |

By considering both acoustic emission and strain energy simultaneously using sentry function, the trends variation of sentry can be accurately presents the phenomenon occurred whilst loading. Therefore, the researcher concluded that the sentry function is an effective tool which can be used to investigate the fracture mechanism of CFRP strengthening reinforce concrete. It also clearly demonstrates the damage progression.

## 4. Acoustic emission based on intensity analysis

One of the AE analyses used to categorize the level of damage is known as intensity analysis [11]. The intensity analysis also known as a graphical method to assess structural performance using AE. The origin of this analysis is found in the structural evaluation of FRP structures by Fowler since year 1989 [20]. This intensity analysis technique had already been successfully applied to FRP and metal piping system evaluations [21]. Figure 3 illustrates a typical chart on which maximum values of historic index and severity are plotted to determine the intensity of a source. Based on figure 3, the established intensity chart by Gostautas [22] was classified in five zones and written more detailed as shown in table 6. This typical intensity chart has been widely used for miscellaneous systems.

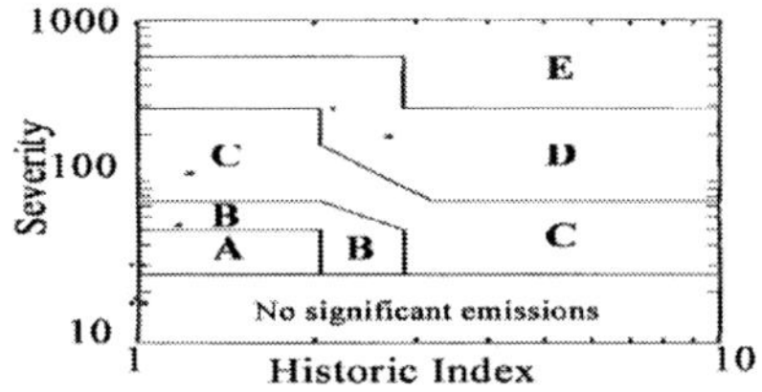


**Figure 3.** Representative intensity chart of FRP material [21].

**Table 6.** The general intensity zone [22].

| Zone Intensity | Recommended Action |
|---|---|
| Zone A – Insignificant | Nonsignificant of AE |
| Zone B – Minor | Note for reference in future tests. Normally represents minor surface defects for example corrosion, pitting, gouges or crack by weld attachment |
| Zone C - Intermediate | Defects considering follow-up evaluation. Evaluation may be made based on further data analysis or complementary non-destructive analysis |
| Zone D – Follow up | Significant defect considering supplementary follow up or inspection task |
| Zone E - Major | Major defects require immediate closure and supplementary inspection task to be carried out |

*4.1. Formulation of intensity analysis*

Intensity analysis is well known as a statistical approach that calculates two values which are historical index (HI) and severity (Sr) [11,12]. The HI is an analytical quantity that traces the slope change of the cumulative signal strength parameter measured during a test. Meanwhile Sr is the value is obtained by averaging the strongest signal strength values and helps normalize the collected AE data making it independent of the location of the AE source. The equation of the Historical Index (HI) and Severity ($S_r$) [11,20-22] are shown below as equation (5) and (6) respectively:

$$H(I) = \frac{N}{N-K}\left(\frac{\sum_{i=K+1}^{N} S_{oi}}{\sum_{i=1}^{N} S_{oi}}\right) \tag{5}$$

According to equation (5), H (I) is indicates as historical index, N is the number of hits up to time t and $S_{oi}$ is the signal strength of the *i*th hit. K is a parameter that depends on the number of AE hits and the types of material. K value for concrete: K = 0, N ≤ 50; K = N – 30, 51≤ N ≤ 200 and K = 0.85N, 201≤ N≤ 500.

$$S_r = \frac{1}{J}\left(\sum_{m=1}^{J} S_{om}\right) \tag{6}$$

Based on equation (6), $S_r$ indicates as the Severity index; J is an empirically derived constant based on material type. J value for concrete material: J = 0, N < 50 and J = 50, J ≥ 50, $S_{om}$ is the signal strength of the *m*th hit where the *m* is based on the magnitude of the signal strength.

*4.2. Applications of intensity analysis*

Damage characterization using intensity analysis approach involving correlations between the properties of the AE such amplitude, duration and signal strength. Degala *et al.* [23] presented an Acoustic Emission (AE) study performed during a four-point bending test on concrete slab specimens strengthened with CFRP strips. The findings for all slabs on parametric analysis reported that flexural cracks near the laminates and debonding create AE activities much higher in terms of signal amplitude and frequency range than eventual shear cracking was shown in figure 4. The peak value of HI was associated with the variation in slope. The specimen experienced lost in energy and generates a significant increase in the slope of the severity line. The significant increase in slope notifying presence of CFRP strip debonding. The maximum values of HI and severity indicated that specimen experienced CFRP debonding together with shear failure. Thus, the correlation between HI and Sr are able to demonstrate the progression of the intensity values moving from lower-left of the chart to the upper-right can be used in real-time to flag combinations of HI and Sr which enter an intensity zone associated with severe damage.

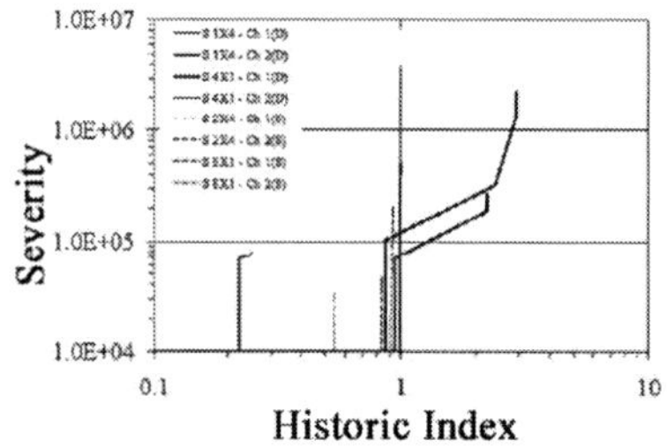


**Figure 4.** The intensity charts for all slabs strengthened with CFRP strips [23].

## 5. Conclusion

In a nutshell, these three types of AE analyses are available for identifying cracks and detecting the damage mechanism level of FRP concrete structures. This is because both of the AE analyses have their own advantages where the best application of the significant AE analysis can provide the damage mechanism and details of the fracture failure in concrete structures. To sum up, by utilizing the b-value, sentry and intensity analysis as parameter analysis, all AE analyses are able to deliver the damage progression and classification of damage in structures as their final output values.

## 6. References

[1] Mikami C, Wu H C and Elarbi A 2015 Effect of hot temperature on pull-off strength of FRP bonded concrete, *Constr. Build. Mater.* **91** 180–186

[2] Wang X, Shi J, Wu Z, and Zhu Z 2016 Creep strain control by pretension for basalt fiber-reinforced polymer tendon in civil applications, *Mater. Design.* **89** 1270–1277

[3] Syed Mazlan S M S, Abdullah S R, and Shahidan S 2017 Failure Behaviour of Concrete Prisms Strengthened by Various Bond Widths of Carbon Fibre Reinforced Polymer (CFRP) *MATEC Web of Conf.* vol 103 (Paris: EDP Sciences) 02015

[4] De Domenico D, Fuschi P, Pardo S, and Pisano A 2014 Strengthening of steel-reinforced concrete structural elements by externally bonded FRP sheets and evaluation of their load carrying capacity, *Compos. Struct.* **118** 377–384

[5] S. Shahidan, S. R. Abdullah and I. Ismail Relationship between AE Signal Strength and Absolute Energy in Determining Damage Classification of Concrete Structures, *J. Teknol.* **78** 1–6

[6] Beattie A G 2013 *Acoustic Emission Non-Destructive Testing of Structures using Source Location Techniques* (California: Sandia National Laboratories) pp 1-128

[7] Ohtsu M 1996 The history and development of acoustic emission in concrete engineering, *Mag. Concr. Res.* **48** 321–30

[8] Hu S, Lu J and Xiao F 2013 Evaluation of concrete fracture procedure based on acoustic emission parameters, *Constr. Build. Mater.* **47** 1249–1256

[9] Behnia A, Chai H K and Shiotani T 2014 Advanced structural health monitoring of concrete structures with the aid of acoustic emission, *Constr. Build. Mater.* **65** 282–302

[10] Colombo S, Main I G and Forde M C 2003 Assessing Damage of Reinforced Concrete Beam Using "b-value", *Analysis of Acoustic Emission Signals Managing* **15** 280–286

[11] Shahidan S, Md Nor N and Bunnori N M 2011 Analysis methods of acoustic emission signal for monitoring of reinforced concrete structure: A review *IEEE Xplore Digital Library* (United States: IEEE) pp 480–484
[12] N. Md Nor, N. M. Bunnori, A. Ibrahim, S. Shahidan and S. N. Mat Saliah Relationship between acoustic emission signal strength and damage evaluation of reinforced concrete structure: Case studies *IEEE Xplore Digital Library* (United States: IEEE) pp 308–313
[13] Guzmán C, Torres D, Hucailuk C and Filipussi D 2015 Analysis of the Acoustic Emission in a Reinforced Concrete Beam Using a Four Points Bending Test *Procedia Materials Science* vol 8 (United Kingdom: Elsevier) pp 148–154
[14] Choi W and Yun H 2015 Cold Regions Science and Technology Acoustic emission activity of CFRP-strengthened reinforced concrete beams after freeze – thaw cycling, *Cold Reg. Sci. Technol.* **110** 47–58
[15] Ma G and Li H 2017 Acoustic emission monitoring and damage assessment of FRP-strengthened reinforced concrete columns under cyclic loading, *Constr. Build. Mater.* **144** 86–98.
[16] Do Yun H, Choi W C and Seo S Y 2010 Acoustic emission activities and damage evaluation of reinforced concrete beams strengthened with CFRP sheets, *NDT E Int.* **43** 615–628
[17] Minak G and Zucchelli A 2008 *Composite Materials Research Progress* (New York: Nova Science Publishers Inc) pp 165–207
[18] Bakhtiary Davijani AA, Hajikhani M and Ahmadi M 2011 Acoustic Emission based on sentry function to monitor the initiation of delamination in composite materials, *Mater. Des.* **32** 3059–65.
[19] Selman E, Ghiami, A and Alver N 2015 Study of fracture evolution in FRP-strengthened reinforced concrete beam under cyclic load by acoustic emission technique: An integrated mechanical-acoustic energy approach, *Constr. Build. Mater.* **95** 832–841
[20] Fowler, Blessing T J, Conlisk J A and Swanson T L 1989 The MONPAC system, *J. Acoust. Emis* **8** 1–8
[21] Nair A, Cai C S, Pan F and Kong X 2014 Acoustic emission monitoring of damage progression in CFRP retrofitted RC beams, *J. Struct. Monit. Maint* **1** 111–130
[22] Gostautas R S, Ramirez G, Peterman R J and Meggers D 2005 Acoustic emission monitoring and analysis of glass fiber-reinforced composites bridge decks, *ASCE J. of Bridge Eng.* **10** 713-721
[23] Degala S, Rizzo P, Ramanathan K and Harries K A 2009 Acoustic emission monitoring of CFRP reinforced concrete slabs, *Constr. Build. Mater.* **23** 2016-2026

**Acknowledgement**
This work was supported by the Faculty of Civil and Environmental Engineering, University Tun Hussein Onn Malaysia, Center for Graduate Studies, UTHM and the Research Acculturation Grant Scheme (RAGS) Vot. 053.

# Nuclear gauge application in road industry

**Mohd Azmi Ismail**[1]

[1]Industrial Technology Division, Malaysian Nuclear Agency,Bangi, 43000 Kajang, Selangor, Malaysia
Corresponding author: azmi@nm.gov.my

**Abstract.** Soil compaction is essential in road construction. The evaluation of the degree of compaction relies on the knowledge of density and moisture of the compacted layers is very important to the performance of the pavement structure. Among the various tests used for making these determinations, the sand replacement density test and the moisture content determination by oven drying are perhaps the most widely used. However, these methods are not only time consuming and need wearisome procedures to obtain the results but also destructive and the number of measurements that can be taken at any time is limited. The test can on be fed back to the construction site the next day. To solve these problems, a nuclear technique has been introduced as a quicker and easier way of measuring the density and moisture of construction materials. Nuclear moisture density gauges have been used for many years in pavement construction as a method of non-destructive density testing The technique which can determine both wet density and moisture content offers an in situ method for construction control at the work site. The simplicity, the speed, and non-destructive nature offer a great advantage for quality control. This paper provides an overview of nuclear gauge application in road construction and presents a case study of monitoring compaction status of in Sedenak - Skudai, Johor rehabilitation projects.

## 1. Introduction

Effective control of compaction of soil and stone layers is an important factor in the construction of roads and other types of foundations for civil engineering structures [1]. Compaction is a process of increasing soil density or unit weight, accompanied by a decrease in air volume, with the intention to improve strength and stiffness of foundation materials. The failure to displace air from between soil particles may lead to unwanted soil movement and water penetration into the earth beneath the foundation.

To achieve a uniformly compacted base and sub-grade, one needs to be able to measure it. In the traditional content, the roller operator feels the level of compaction through their seat and makes adjustments according to this. Every operator has a different learned experience for what is acceptable or not [2]. The degree of compaction is measured by dry unit weight and depends on the water content. The maximum dry unit weight occurs at an optimum water content. There are several techniques that can measure the dry density and moisture content of the soil. Conventional methods such as the sand replacement density test the moisture content determination by oven drying are perhaps the most popular [3]. However, these methods are not only time consuming and need a wearisome procedure to obtain results but they are also destructive and the number of measurement that can be taken at any one time is limited. The results of the test can only be fed back to the construction site the next day [4].

To solve this problem, a quicker and easier way of measuring soil density and other properties of construction materials is required. Various methods have been studied and among them was nuclear technique. Nuclear moisture density gauges are testing devices that use low-level radiation to measure the wet density, dry density, and moisture content of the soil and granular construction materials. This technique which can determine both wet density and moisture contents offers an in-situ method for compaction control at the work site. Its simplicity, speed, and non-destructive nature offer a great advantage for quality control compared to other conventional methods.

## 2. Radiation interaction with matter

### *2.1. Interaction of gamma rays*

When a gamma ray passes through medium, successive interactions occur which the constituent atoms and energy of the gamma ray decrease gradually. Interaction takes place between the gamma ray and the atoms and the following three kinds of interactions are known:

*2.1.1. Photoelectric absorption.* This is the ionization of gamma ray in which the difference between the initial gamma ray energy and the binding energy of the orbital electron will be taken away by the released electron as its kinetic energy (figure 1). The differential cross-section ( $_p$: a probability of interaction) depends on the initial gamma ray energy (E) and atomic number (Z), and can be written approximately by the following equation:

— .. .. .. .. (1)

where,n = 3 (for low energy gamma ray) and n = 5 (for high energy gamma ray)

Contribution of photoelectric absorption to the attenuation of gamma ray will become important when the low energy gamma rays pass through a medium of large atomic number.

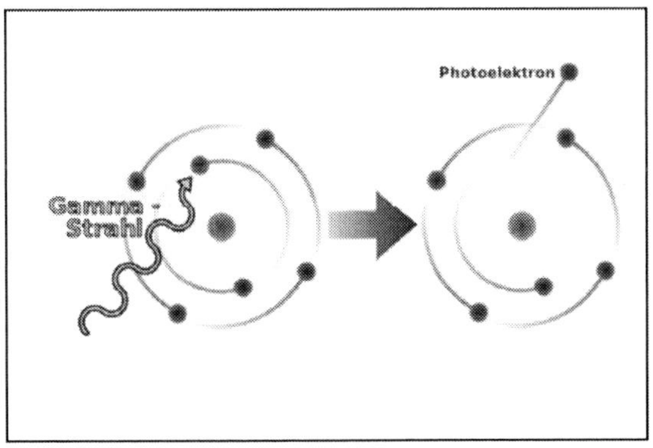


**Figure 1.** Schematic presentation of gamma ray interaction with matter - photoelectric absorption.

*2.1.2. Compton scattering.* It is an elastic scattering between the gamma ray and the orbital electrons in which total kinetic energy, as well as total momentum, are conserved throughout the scattering (figure 1). The energy of gamma ray after scattering can be written as a function of its initial energy ($E_0$) and scattering angle ( ) in the centre of the mass system.

$$= \frac{}{(\quad)} \quad .. \quad .. \quad .. \quad (2)$$

where, $E_0$ = initial energy gamma ray, $m_0$ = rest mass of electron, C= velocity of light

Cross section of the Compton scattering ( $_c$) depends on the energy of gamma ray (E) and atomic number of the medium (Z) as follows,

$$- \quad .. \quad .. \quad .. \quad .. \quad (3)$$

In Compton scattering, part of the initial energy of gamma ray will be lost during collision with the electrons (figure 2), but not all as in the case of photoelectric absorption. Therefore, after several successions of Compton scattering, the probability of the photoelectric absorption becomes greater until finally gamma ray disappears.

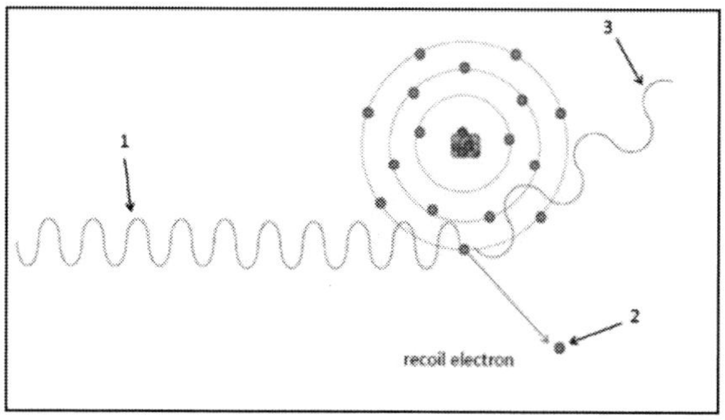


**Figure 2.** Schematic presentation of gamma ray interaction with matter - Compton scattering.

*2.1.3. Pair production.* In this interaction, a gamma ray with an energy greater than twice of electron mass energy (nearly 1.02MeV) disappears near the nucleus of the material and a pair of electron and positron is created (figure 3). The dependency of the cross section of the interaction on E and Z can be written as follows:

$$(\quad 2 \quad) \quad .. \quad .. \quad .. \quad (4)$$

where, $m_0$ = rest mass of electron C = velocity of light

The energies of the gamma rays emitted from $^{60}$Co are 1.17 and 1.33 MeV, the difference of which from the threshold of pair production is very small so that we can actually neglect the effect.

*2.2. Interaction of neutron*

A neutron is an electrically neutral particulate radiation. Being uncharged, the neutron can go through any material with little interaction with surrounding atoms. The high transparency of neutron makes it very different in character as compared with an electron or any other electrically charged particles. Neutron interaction with the material is only by nuclear force. In another word, only the nuclei of atoms have the capability to stop or slow down the neutrons.

IOP Conf. Series: Materials Science and Engineering **271** (2017) 012087 doi:10.1088/1757-899X/271/1/012087

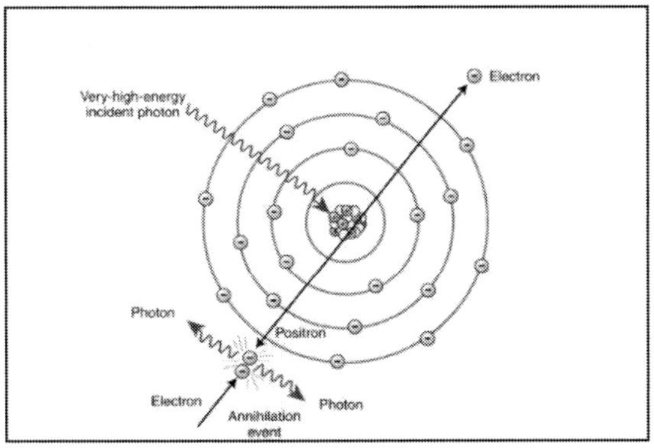


**Figure 3.** Schematic presentation of gamma ray interaction with matter - pair production.

Neutrons emitted from a neutron source are part of the product of nuclear interaction taking place in the source. The possible energy of the produced neutron varies one type of neutron source to another, ranging from near 0 to 11 MeV and the mean energy is in the order of several MeV. The significant of the neutron energy in the interaction of the neutron with matter makes it necessary to classify neutron according to their energies. The terms thermal, epithermal and fast each represents a broad, rather indefinite range of neutron energies. Thermal neutrons are those which have reached thermal equilibrium with their surroundings (less than 0.5 eV). Epithermal neutrons are within the group of thermal and fast neutrons. The lower boundary of fast neutron group is quite arbitrary, perhaps about 100 KeV.

The nature of interactions between neutron and atomic nuclei will depend both on the energy and nature of the nucleus. Practically, these are classified as absorption or capture and scattering reactions. Only those interactions with an appreciable probability of occurrence within the energy range from 11 MeV down to thermal energies are of interest in moisture measurement.

## 3.Nuclear moisture density gauge

*3.1. Density measurement*

Density measurement is based on the fact that the intensity of the transmitted gamma ray depends strongly on the density of the measured material. The flux of directly transmitted gamma ray can be given by the following well known empirical equation [5].

.. .. .. .. (5)

where, = flux, $S_0$ = initial source intensity, x = distance between the source and the centre of the detector, = density of the material, and = mass absorption coefficient.

Under the normal circumstances of measurement, a considerable amount of secondary gamma ray which is scattered in the medium will also be included in the detected flux. The contribution is signified by a term known as Build-up Factor, *B*. Therefore, in terms of count rate, (6) can be written as the following:

N = . .. .. .. .. (6)

where N is the number of atoms.

In general, $B$ is not only a complex function of the decay length ( ) but also depends on the type and dimensions of the detector. However, in the design of density gauge, the following simple equation is adopted, applicable for a specific and appropriate range of material density.

*3.2. Moisture measurement*

In the measurement of water content (moisture) of soft ground, such as embankment, direct transmission gauge is more widely used. Transmitted fast neutrons from the source are slowed-down during interaction with the nuclei of the constituent atoms. As mentioned earlier, fast neutrons can lose maximum energy when they collide with hydrogen nucleus. The number of fast neutrons will decrease as more hydrogen is present in the soil. Therefore, fast neutron counting rates are roughly expressed as an exponential function of moisture content (water density),

$$R = e \quad .. \quad .. \quad .. \quad .. \quad (7)$$

where,

$R_m$ = neutron count rate, A, $B$ = calibration coefficient, and $_m$ = moisture content

Hydrogen presents in the soil in many forms. It may exist as absorbed water (water content), crystalline water and in organic materials. In civil engineering works, we are only interested in the water content which is evaporated at 110 C in the oven. Other types of waters (so called ignition loss) remain in the soil after the drying process. Unfortunately, neutrons interact with all types of hydrogen in the soil and cannot distinguish between them. Therefore, when using the nuclear gauge, correction of ignition loss has to be made. This is done by introducing a moisture corrective coefficient, . From (8),

$$R = e \quad .. \quad .. \quad .. \quad .. \quad (8)$$

where,

$_m''$ = equivalent neutron content ( $_m''$ = $_m$ + . $_d$), = moisture content corrective coefficient, $_d$ = dry density, and $_m$= moisture content

*3.3. Field measurement*

The instrument introduced in this paper is a compact surface gauge of direct transmission type which employs a radiation source with a strength of 8 mCi of $^{137}Cs$ and 40mCi of $^{241}Am$:Be. Both the density and moisture gauges are incorporated into the instrument hence it is often called a combined gauge. Modern gauges offer optional features such as GPS location, remote start keypad near the handle, backlit keypad and external beeper [6].

*3.4. Calibration procedure*

*3.4.1. Calibration equation for different types of soil.* To compute density and moisture content from a measured amount of transmitted radiation, it is necessary to derive a conversion formula (calibration curve) beforehand. In the early design, only a single calibration curve of a particular type of soil can be stored in the gauge at one time. Since various kinds of soil are normally dealt with during construction of road earthwork, a different calibration curve has to be drawn every time the soil type changes. To overcome this problem, a single calibration curve which is applicable to various types of soils was studied.

A calibration test on 17 types of soil sample ranges from coarse- to fine-grained soils under 144 different conditions of densities and moisture contents were conducted. The results of the test for

IOP Conf. Series: Materials Science and Engineering **271** (2017) 012087 doi:10.1088/1757-899X/271/1/012087

density and moisture is plotted in figures 4 and 5, respectively, before and after corrections for interference by neutron-capture gamma ray to density measurement and effect of dry density and ignition loss to moisture measurement are made.

Although there are some scattered points in the calibrated data a indicated in these figures, we can safely conclude that a single calibration curve could be applied to all kinds of soil tested. Scattered data could be due to the heterogeneity of the soil samples.

*3.4.2. Calibration for individual gauge.* In actual measurement, a calibration curve must be programmed in the gauge beforehand. There are, however, some differences in the mechanical characteristics of each gauge and each gauge must have its own calibration curve.

Calibration for each gauge is carried out using a material of constant density and moisture content. The response characteristics of the gauge are recorded as standard values. Once a calibration curve is obtained, it is usable for any site unless the radiation source is replaced.

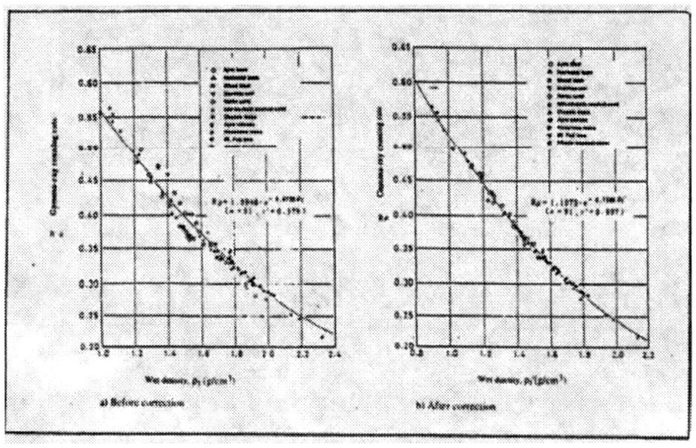

**Figure 4.** Influence of thermal neutron in density measurement [5].

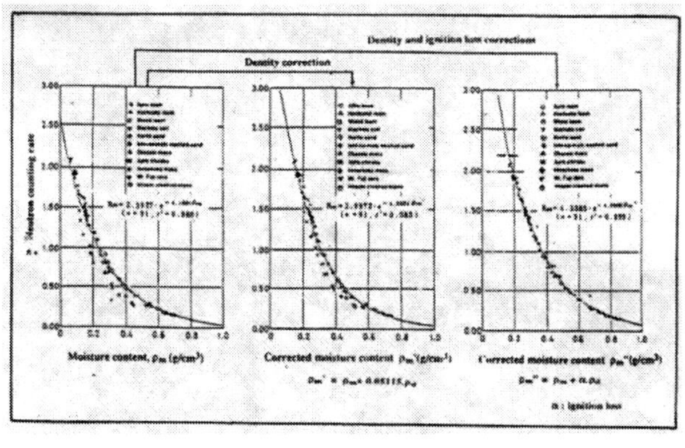

**Figure 5.** Influence of dry density and ignition loss in moisture measurement [5].

## 4. A Case Study

PLUS Malaysia Berhad has conducted a rehabilitation project on North - South Expressway at KM9.57, south bound of Kempas - Skudai Expressway. The road surface was deteriorated and needs to be replaced. PLUS has decided to reconstruct the road base using Cement Bound Material (CBM), consisting of 10% cement and 6% water. Cement was used to stabilize the in-situ material and placed prior to placement of a wearing course [7].

The status of compaction of the CBM need to be evaluated within hours after placement because it will harden. Conventional techniques require a longer time to evaluate the dry density and moisture content of the CBM material, for instance, the determination of water content using oven will require at least 12 hours. The nuclear gauge was used for assessment of compaction of the CBM was used with throughout the project and took only one-minute measuring time per each spot.

**Figure 6.** The problem of the deteriorating road surface.

**Figure 7.** The status of compaction was finally checked by nuclear gauge.

## 5.Conclusion

In conclusion, compaction is one of the most important processes in roadway construction. A uniform sub-base and sub-grade is a key ingredient in maximizing pavement performance and the stiffness of the sub-grade is a major contributor to this uniformity, whether it is in pavements, road embankments or bridge abutments. The determination of the status of compaction by nuclear technique has the following advantages:

i. The method is non-destructive and repeatable.
ii. Results for moisture and density at a single point can be obtained within minutes.
iii. The method is accurate and easy to perform, minimizing human error.
iv. The collection of as many sampling data as required can be done without any difficulty.
v. The status of compaction can be fed to the construction work instantly with confidence.

## 6. References

[1] Huet J 1965 Field determination of moisture and density in soils by nuclear methods, in A. Wexler ed., *Proceedings International Symposium on Humidity and Moisture, Van Reinfold and Co., NY***4** 185-193
[2] IAEA Technical Report Series 1970 *Neutron moisture gauges*
[3] Toyama M 1990 Compaction control by nuclear method in highway embankment. *Paper presented at the NEMS on the use of Nucleonic Control System and Nuclear Instruments in Civil Engineering. Kuala Lumpur, Malaysia*
[4] Power R 2013 The changing face of Compaction and Compaction Control for road building – a Global Review
[5] Nobuyama M 1988 Density measurement and its application to compaction control of the embankment. Paper presented at the NEMS on the use of Nucleonic Control System and Nuclear Instruments in Civil Engineering. Kuala Lumpur, Malaysia
[6] Troxler Electronic Laboratories, 2017. Surface nuclear gauges

[7] Mohd Azmi Ismail and Kanesan Sinnakarupan. 1993. The in-situ density measurement for CBM at km 9.57 south bound of the Kempas - Skudai Expressway. A report submitted to Projek Penyelenggaraan Lebuhraya Berhad (PROPEL)

**Acknowledgement**
The author is grateful to Malaysian Nuclear Agency for support and for sponsoring the training in this field. Thanks also are due to Japan Highway Public Corporation, Soil and Rock Engineering Co. Ltd, Kajima Corporation for supervision and discussion during the author's stay in Japan.

# Integrity inspection of main access tunnel using ground penetrating radar

**M A Ismail[1], A A Abas[1], M H Arifin[2], M N Ismail[3], N A Othman[3], A Setu[3], M R Ahmad[1], M K Shah[1], S Amin[1], and TSarah[1]**

[1]Industrial Technology Division, Malaysian Nuclear Agency,Bangi,43000, Kajang,Selangor, Malaysia,
[2] Geology ProgrammeSchool of Environmental Sciences and Natural Resources,Faculty of Science and Technology, Universiti Kebangsaan Malaysia, Bangi 43600, Kajang, Selangor,Malaysia
[3] TNB Research Sdn. Bhd.,Lorong Ayer Itam, Kawasan Industri Penyelidikan,43000 Kajang,Selangor, Malaysia
Corresponding author: azmi@nm.gov.my

**Abstract.** This paper discusses the Ground Penetrating Radar (GPR) survey performed to determine the integrity of wall of tunnel at a hydroelectric power generation facility. GPR utilises electromagnetic waves that are transmitted into the medium of survey. Any reflectors in the medium will reflect the transmitted waves and picked up by the GPR antenna. The survey was done using MALA GeoScience RAMAC CUII with 250MHz antenna. Survey was done on the left, the crown and the right walls of the underground tunnels. Distance was measured using wheel encoders. The results of the survey is discussed in this paper.

## 1. Introduction

Ground Penetrating Radar (GPR) is becoming a common non-destructive method in detection of pipes and utilities, tank and drum location, concrete and rebar inspection, sinkholes, geology and geologic hazards, landfill and burial trenches, cemetery and grave sites and archaeological studies. It produces a continuous cross-sectional profile of different subsurface features. In civil engineering, parameters such as thickness of asphalt layer, bar location and void detection are made possible. On the other hand, the essential information in geotechnical studies such as the depth to bedrock, different soil layers and subsurface voids can be obtained from GPR survey.

The GPR technique is similar in principle to sonar methods. The radar transmitter produces a short pulse of high frequency (25 - 1000 MHz) electromagnetic energy, which is transmitted into the ground through an antenna. Variations in the electrical impedance within the ground generate reflections that are detected at the ground surface by the same or another antenna attached to a receiver unit. Variations in electrical impedance are largely due to variations in the relative permittivity or dielectric constant of the ground. The greater the difference of dielectric constants at an interface between two materials, the greater the amount of electromagnetic energy reflected at the interface. Consider the behavior of an incident beam of electromagnetic wave (EM) energy as it strikes an interface, or boundary, between two materials of different dielectric constants (Figure 1). A portion of the energy is reflected, and the remainder penetrates through the interface into the second material. The reflection coefficient at the interface, $R$is given by Equation (1),

$$R = \frac{\sqrt{K_1} - \sqrt{K_2}}{\sqrt{K_1} + \sqrt{K_2}} \tag{1}$$

where $K_1$ and $K_2$ are the dielectric constant of materials 1 and 2, respectively.

Equation 1 indicates that when a beam of microwave energy strikes the interface between two materials, the amount of reflection, $R$ is dictated by the values of the relative dielectric permittivity of the two materials. If material 2 has a larger relative dielectric permittivity than material 1, then $R$ would have a negative value; i.e., with the absolute value indicating the relative strength of the reflected energy and the negative sign indicating that the polarity of the reflected energy is the opposite of that arbitrarily set for the incident energy. Water has a dielectric constant of 80 (compared, for example, to 5 for dry soil and 8 for rock) and hence there are high reflection coefficients between dry and wet materials. In addition, the water table is a strong reflector.

After penetrating the interface and entering into material 2, the wave propagates through material 2 with a speed, $V_2$, given by

$$V_2 = \frac{C}{\sqrt{\varepsilon_2}} \tag{2}$$

Where$C$ is the propagation speed of EM waves through air, which is equivalent to the speed of light, or 0.3 m/ns).

As the wave propagates through material 2, its energy is attenuated as follows:

$$\alpha = 12.863 \text{ x } 10^{-8} \text{ f } \sqrt{\varepsilon_2}\left(\sqrt{1+\tan^2\delta} - 1\right)^{1/2} \tag{3}$$

Where$\alpha$= attenuation, in decibel/meter, $f$ = wave frequency, in Hz, and $\delta$ = the loss tangent (or dissipation factor) is related to$\sigma$, the electrical conductivity (in mho/meter) of the material by:

$$\tan\delta = 1.80 \text{ x } 10^{\prime\circ} \frac{\sigma}{f\varepsilon_2} \tag{4}$$

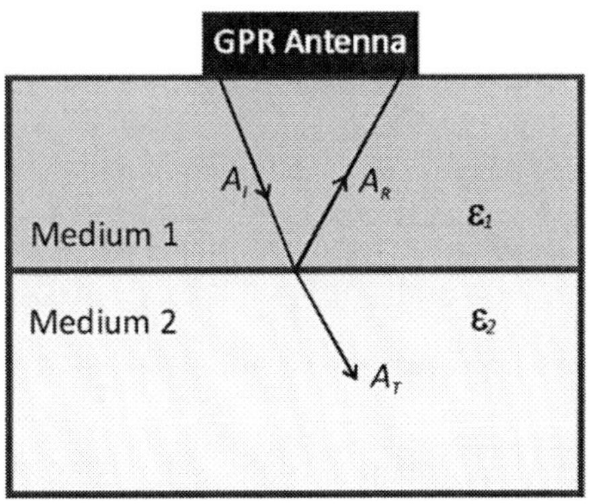


**Figure 1.** Propagation of radar energy through dielectric boundaries.

When the remaining microwave energy reaches another interface, a portion will be reflected back through material 2 as given by Equation 1. The resulting two-way transit time ($t_2$) of the microwave energy through material 2 can be expressed as

$$t_2 = \frac{2D_2}{V_2} = \frac{2D_2\sqrt{\varepsilon_2}}{C} \quad (5)$$

where $D_2$ is the thickness of material 2.

## 2. GPR survey methodology

The survey was performed on an underground tunnel of a hydroelectric power generation plant. The tunnel functions as a Main Access Tunnel (MAT) which stretches at 1km long. Both of the tunnels leads to a cavern which houses the turbines and control room. The tunnels were constructed in the mid 1990's. This survey is intended to determine the integrity of the tunnels.

GPR system is a real time technique making it a rapid and reliable method for tunnel wall integrity survey. The results are displayed on a radargram that represents the cross sectional view of the survey line. The x-axis is the distance travelled by the GPR antenna while the y-axis is the time of wave travel. Meanwhile, the colour scale represents the polarity of wave. The purpose of GPR survey is to map anomalies in the tunnel wall such as fracture lines, voids and determine water content relatively. Areas with relatively higher water content was identified. The tunnels experience areas of higher water content which can be seen visually with water dripping and seeping through the wall tunnel.

The GPR system used in this survey is MALA Geoscience RAMAC CUII. The antenna used is a shielded directional with 250MHz frequency. The choice of frequency is to ensure adequate penetration balanced to a reasonable resolution and sensitivity. Location was measured by a wheel encoder. The distance of survey was compared to the chainage labels of the tunnel and was found the variation was about 5%. This is due to the wheel experiencing bumps from the pavement which increases the distance and it accumulated over the total length of survey distance. The height of the crown of the tunnel was about 10 metres. Ideally the position of the antenna should be as close as possible to the wall surface. However, this is not possible since the use of a combustion engine is not allowed in the tunnels due to safety reasons. Therefore, the survey was done with the antenna placed on a battery operated carts with a antenna spacing of approximately one meter. It was found that the large gap did not produce any significant effects on the quality of results. Furthermore, the footprint of the wave is large, making the area of coverage larger. The gap was large enough for the crosstalk between the transmitter and receiver which appears as ringing to diminish and makes the reflection from the front surface of the tunnel wall 'clean'. Three surveys on the tunnel were done at the crown and two at angles on both sides of wall, respectively (Figure 2).

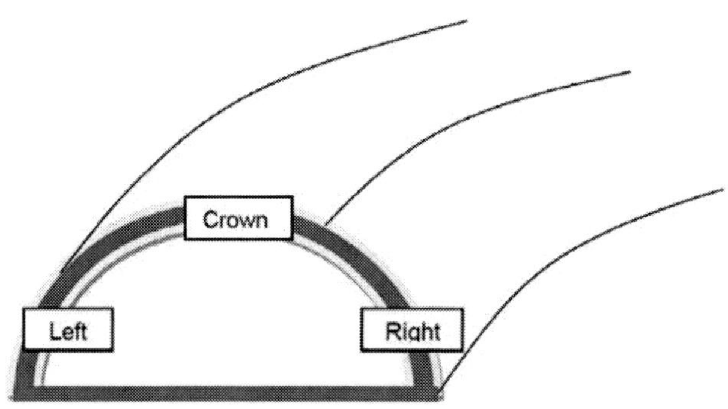


**Figure 2.** Survey lines of GPR survey (view from the tunnel entrance).

(a).The actual setup of the antenna for the left wall survey.

(b).The actual setup of the antenna for the crown wall survey.

(c). The actual setup of the antenna for the right wall survey.

**Figure 3.** GPR survey being done on three sides of the tunnel walls.

The radargrams were enhanced using post processing filters. A DC removal filter was added to normalize the voltage and remove effects of voltage drift due to electrical noise. Then a time corrected gain was applied which consist of a certain amount of linear and exponential amplitude time gain. The colour of radargram was selected to blue-white-red for easy viewing and interpretation. The radargrams were interpreted with minimal filters and image post processing to ensure minimal loss of data.

### 3. GPR survey results

The survey results for the left, the crown and the right walls of the tunnel are displayed in figures 4, 5 and 6, respectively. The findings are categorized into two categories which are saturated fracture and unsaturated fracture. The areas with saturated fracture are areas with higher intensity of reflection which indicates that the area is wet and highly contaminated with fine clayey particles. Areas marked with unsaturated fracture contains less fractures and it is relatively much drier. It is recommended that areas marked with saturated fracture to be investigated thoroughly. Indications from MAT's left wall shows several locations that are saturated fracture and much more significant. The indication is summarised in Table 1.

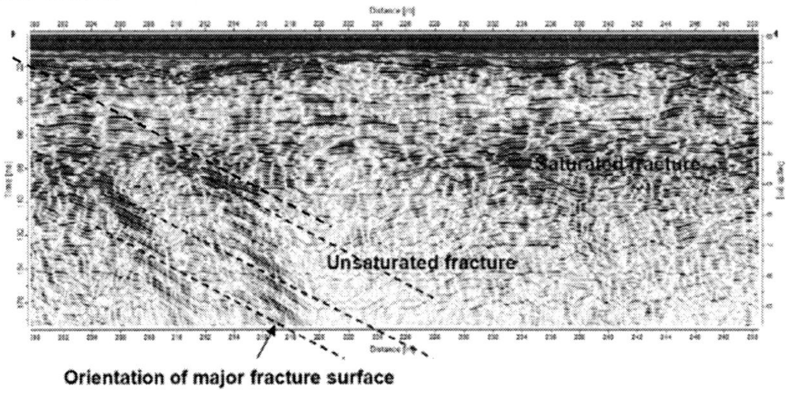


**Figure 4.** Typical radargram for the left wall.

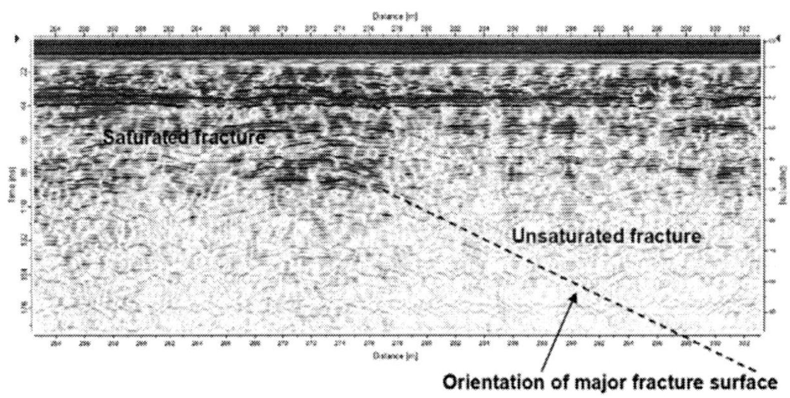


**Figure 5.** Typical radargram for the right wall.

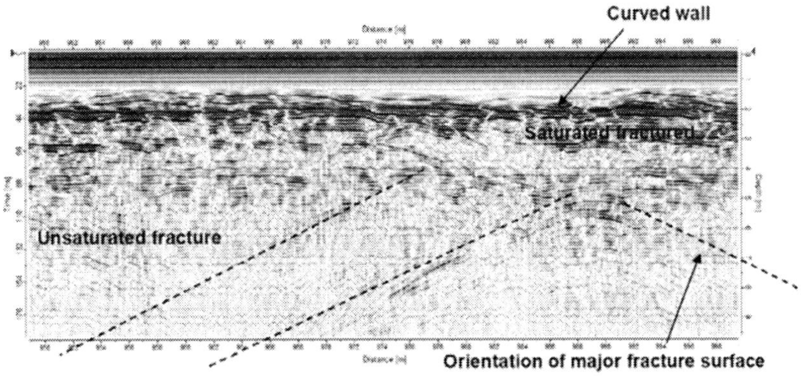


**Figure 6.** Typical radargram for the crown.

**Table 1.** Summary of GPR findings for MAT.

| **Relative Degree of Fractures** | | |
|---|---|---|
| **Left Wall** | **Crown** | **Right Wall** |
| 20 – 289 m saturated | 20 – 388 m saturated | 20 to 280 m saturated |
| 289 – 430 m unsaturated | 388 – 430 m unsaturated | 280 - 350 m unsaturated |
| 430 – 566 m saturated | 430 – 700 m saturated | 350 - 388 m saturated |
| 566 – 710 m moderate | 700 – 1036 m unsaturated | 388 - 434 m unsaturated |
| 710 – 780 m unsaturated | | 434 - 530 m saturated |
| 780 – 930 m saturated | | 530 - 768 m unsaturated |
| 930 – 1020 m unsaturated | | 768 - 902 m saturated |
| 1020 – 1046 m saturated | | 902 - 1034 m unsaturated |

## 4. Conclusions
This report highlights finding from GPR surveys conducted on the Main Access Tunnel of a hydroelectric power station. Areas of relatively saturated (wet) and unsaturated (dry) are identified. The wet areas are representing by fractures that is filled with water or clay particle, whereas the dry areas are representing by fractures that are dry and probably smaller in size. GPR 'sees' only fractures that is perpendicular to the radar beam as the reflected energy will be detected by the receiver. Reflected radar energy from fracture planes which are not perpendicular to the radar beam will be reflected towards the receiver.

## 5. Reference
[1] Vienna 2002 Guidebook on non-destructive testing of concrete structures, Training course series No. 17, International Atomic Energy Agency

[2] Parkinson G 2008 Ground Penetrating Radar Evaluation of Concrete Tunnel Linings, *12$^{th}$ International Conference on Ground Penetrating Radar* (Birmingham, UK) pp16-19

[3] Rister B P E,Dinger J Ph.D 2011Use of Ground Penetrating Radar for Void Detection and Hydro-Geochemical Water Testing Results at the Cumberland Gap Tunnel, *62nd Highway Geology Symposium*

[4] Ismail M P, Ismail M A, Abas A A, Sani S 2004 Application of GPR in Concrete inspection and Geotechnical studies, Seminar on Geophysics 2004, 2nd August 2004, ESSET, Bangi, organised by GSM and MINT.
[5] RAMAC/GPR 2004 3-day training course by MALA Geoscience, MINT

**Acknowledgments**
The authors would like to thank TNB Research Sdn. Bhd., Geo Technology Resources Sdn. Bhd, and Malaysia Nuclear Agency for funding this research project.

IOP Conf. Series: Materials Science and Engineering **271** (2017) 012089 doi:10.1088/1757-899X/271/1/012089

# GPR application on construction foundation study

**T S T Amran [1], M P Ismail[1], M A Ismail[1], M S M Amin[1], M R Ahmad[1] and N S M Basri[2]**

[1]Non Destructive Testing-Material Structure Integrity Group (NDT-MSI), Industrial Technology Division, Malaysian Nuclear Agency, 43000 Kajang, Selangor, Malaysia
[2]The Faculty of Science, Technology and Human Development, Universiti Tun Hussein Onn Malaysia (UTHM), 86400 Parit Raja, BatuPahat, Johor, Malaysia
Corresponding author: sarah@nm.gov.my

**Abstract.** Extensive researches and studies have been carried on radar system for commercialisation of ground penetrating radar (GPR) technology pioneered in construction, and thus claimed its rightful place in the vision of future. The application of ground penetrating radar in construction study is briefly reviewed. Based on previous experimentation and studies, this paper is focus on reinforcement bar (rebar) investigation on construction. The various data through previous references used to discuss and analyse the capability of ground penetrating radar for further improvement in construction projects especially in rebar placement in works.

## 1.Introduction

In year 1904, Huelsmeyer, a German inventor, introduced radar to detect metallic objects [1]. Radar is a detection system that uses radio waves to detect things.Ground Penetrating Radar (GPR) is one of the radar devices that use electromagnetic reflection method. GPR is one of the non-destructive testing (NDT) technique that can imaging at subsurface depthin high resolution.GPR has already established itself in various field such as archaeology, military and construction. GPR is applied in construction to evaluate the concrete depth, detect and locate of the reinforcing bars and metallic ducts andalso calculate the localization of crack and dimension of voids.Construction foundation study using GPR mostly emphasis on reinforcement bar (rebar). Rebar refers as a tensioning devise to reinforce concrete to help hold the concrete in a compressed state as depicted in figure 1. GPR is a very effective device to locate and map steel rebar in concrete structures and has been utilize to investigate rebar-related issues such as rebar location and rebar spacing. The evaluation of rebar using GPR has been a choice as non-destructive assessment of condition for engineering structures.

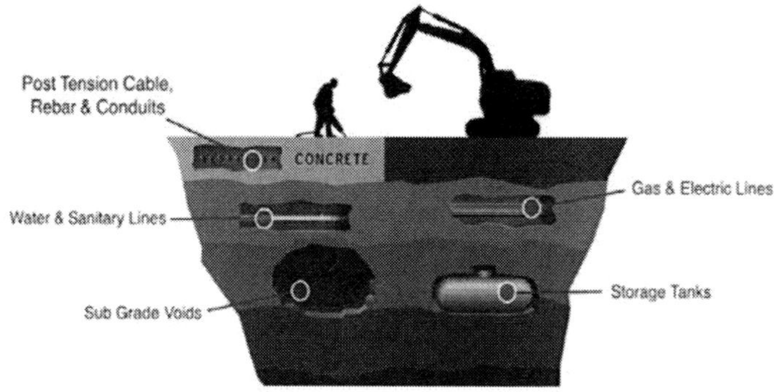


**Figure 1**. The location of rebar in concrete.

GPR used electromagnetic reflection strength approaches as the reflection strength of electromagnetic waves is examined in between two different dielectric components. It was proven that the reflection strength of a rebar decreasing with size of rebar, but increasing with depth of rebar through previous experiments using GPR. The reflection strength also refer to amplitude, where the amplitude decreasing with the presence of rebar corrosion. Rebar diameter cannot be measured directly through GPR scanning, and antenna polarization is used for determination of rebar diameter. The dipole and co-pole antenna are the types of antenna polarization.

This paper is briefly reviewed on various radar equipment for locating rebar as well asrebar detection, estimation of rebar diameter and rebar spacing using GPR technique was discussed in this paper.

## 2.Radar Equipment

Various radar devices are applicable and can be used for the detection of reinforcement bar, rebar. By using these radar equipment that utilize non-destructive testing, the evaluation of rebar does not require any grinding or drilling that causes any damages to the surface and the building during the inspection. Cost effective is also one of the advantages on using radar for rebar detection. Other than GPR, radar devices such as cover meter and rebar locator can be employed on construction study to detect rebar.

### *2.1. Cover Meter*

One of the radar equipment that useful for rebar locating is cover meter. Cover meter refers to magnetic device, where delivers information on cover of concrete to reinforcement steel bar by utilizing magnetic field. The cover meter is operated on electromagnetic pulse induction, where the eddy current reside two sections. The two sections are locating probe and meter. For that matter, the locating probe is a cover unit that have a search coil, while the meter is conveying the dimension of the cover over a strength of signal. And so, the meter displays a stronger magnetic field as the larger rebar size detected by the cover meter. To getan adequate result, a very detailed calibration of cover meter is necessary.

Following the detailed information on cover meter, an experimental investigation under laboratory condition was performed by Sivasubramanian[2]. Experiments were ran to observe the capability of cover meter in verifying cover depth. Two different cover thickness probes were used, small and large thickness probe, where the small probe used for smaller cover depths and the large probe employed for larger cover depths. For detecting clear covers depth larger than 70 mm, larger probe is more effective compared to smaller probe. Figure 2 shows, asthe cover depth increase, the error of detection the diameter of rebar also increasing. The results then prove that cover meter is not very accurate in detecting rebar based on error percentage that had been observed throughout the study.

The difference between cover meter and GPR is cover meter has a very limited detection range compared to GPR, because cover meter can onlydetermine the cover depth of the rebar while GPR is implemented to determine the cover depth, rebar spacing and rebar size.

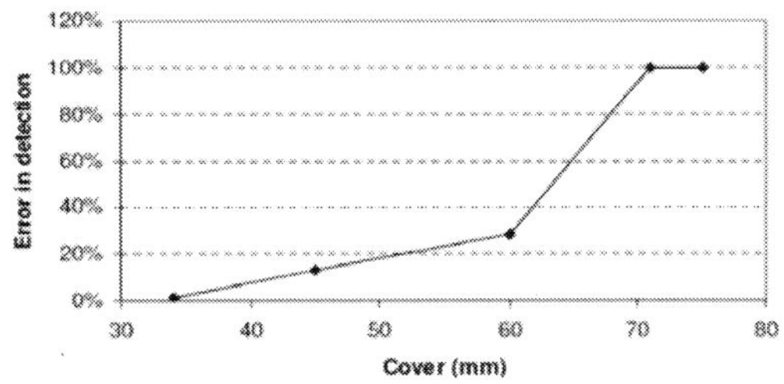


**Figure 2.** A graph of error in detection for rebar diameter vs cover depth [2].

*2.2. Rebar Locator*
Rebar locator is a device that specifically made on detecting rebar. The rebar locator can give an accurate location of rebar in building, where it massively can be mark the exact areas without causing any damages to the building. Other than that, rebar locator can efficiently scan concrete and discover any internal rebar, thus allows the interpreter to find the rebar position and its size precisely. Rebar locator can be greatly differentiate from GPR, where rebar locator can only locate rebar at shallow depth such as in beam, rather than GPR, which the GPR can locate rebar at deeper depth such as in concrete slab of floor area.

**3.Rebar Detection**
This section reviews the literatures that revised from previous experiment on rebar using GPR. Laboratory experiment was conducted by Einsenmann[3], where the group of researchers are doing experiment on the effect of position and orientation of GPR from structural rebar of highway bridge.The hypothesis that proposes on thin rebar would produce significant difference in signal amplitudes as between a flush-with-road-deck measurement and an elevated measurement is proven on the study [3].

Several attempts to locate rebar by recreating laboratory experiment was ran by Lakshmi [4].2.6 GHz antenna was used to collect data on locating rebar, where the scanning depth was 0.40 m. Figure 3 depicts the raw radar collected for the study conducted. The hyperbola peak as marked in figure 3 indicates the position of rebar in concrete slab.

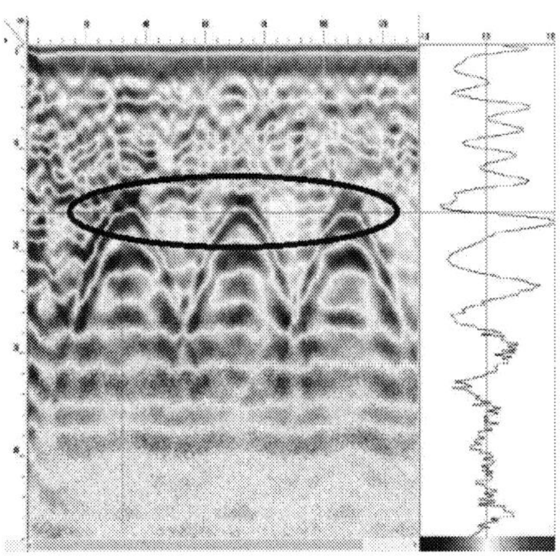

**Figure 3**. Radar profile of concrete slab, represents rebar position [4].

GPR scanning was also run by Einsenman [5] to detect rebar. Radar wave can detect rebar through different antenna positions. The line scan along X-axis as depicted in figure 4 (a) below resulting on one-dimensional scan of antenna. Figure 4 (b) shows vertical axis that represent arrival time, while the horizontal axis illustrates scan distance or the position of antenna. A-scan was presented in figure 4 (c), where the signal response of rebar is depicted in the same figure. It was concluded that the distance between the rebar and the antenna plays crucial part on determining the arrival time of the rebar echo.

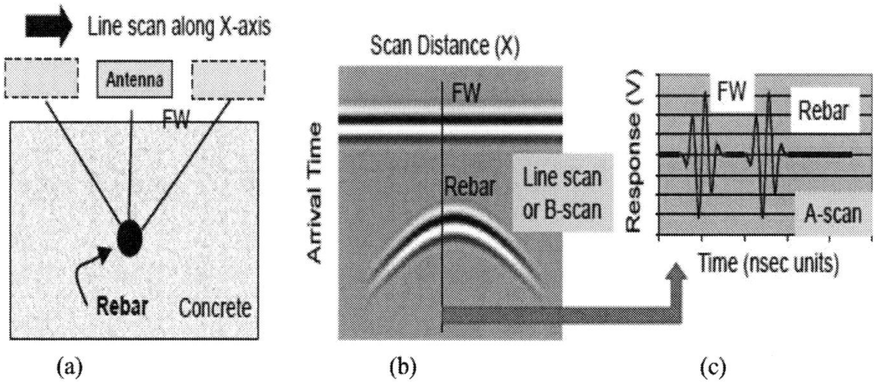


**Figure 4.** One-dimensional scan of GPR on concrete slab above an embedded rebar [5].

Hugenschmidt performed an inspection of retaining wallsusing different frequency antenna [7]. Figure 5 illustrates the radargram data obtained, where in figure 5(a) three rebar are obviously detected by 1.5 GHz antenna, while figure 5 (b) presents the corresponding three rebar using 900 MHz, and figure 5 (c) depicts another corresponding three rebar using 400 MHz antenna. It showed that radargram with lower frequency exhibit clear images of rebar compared to others two frequency antenna. Low resolution images wasproduced due to low frequency antenna but can penetrate more in term of depth of penetration.

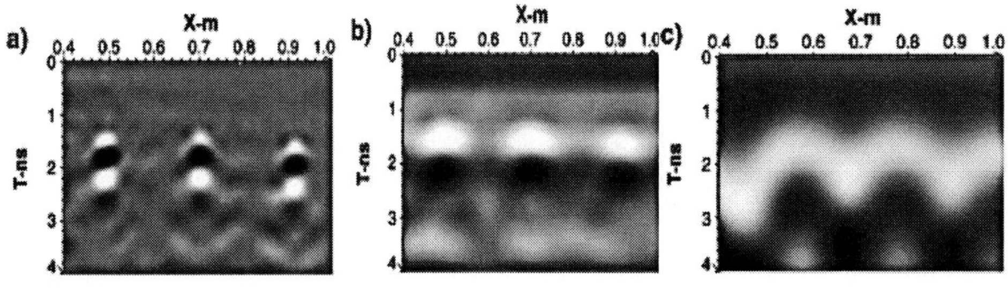


**Figure 5.** Radargram of three rebar located in retaining walls using different frequency antenna (a) 1.5 GHz (b) 900 MHz (c) 400 MHz [7].

We also carry out an inspection of concrete floor using different MALA GPR antenna equipment as shown in figure 6. The inspection was done using 800 MHz and 500 MHz frequency antenna. It is interesting to note that the 800 MHz frequency antenna (figure 7) demonstrate an obvious rebar images compare to 500 MHz frequency antenna (figure 8). It concluded that, high frequency antenna produce high resolution data of rebar through GPR scanning.

IOP Conf. Series: Materials Science and Engineering **271** (2017) 012089 doi:10.1088/1757-899X/271/1/012089

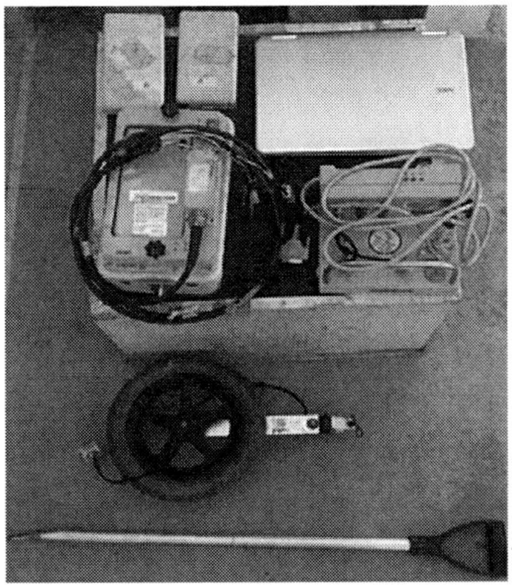

**Figure 6.** MALA ground penetrating radarequipment set

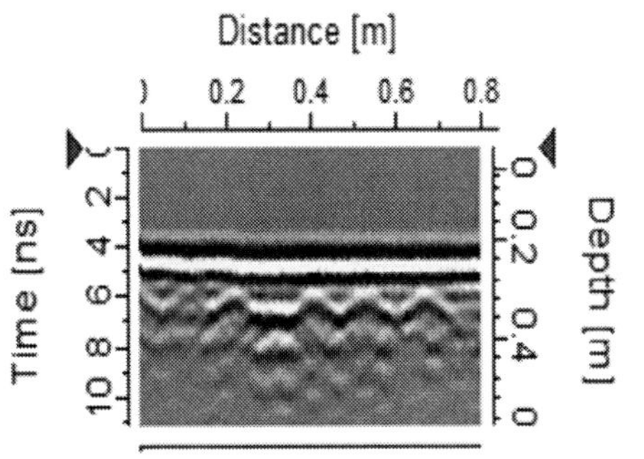


**Figure 7.** Radargram of rebar located at concrete floor using 800 MHz frequency antenna

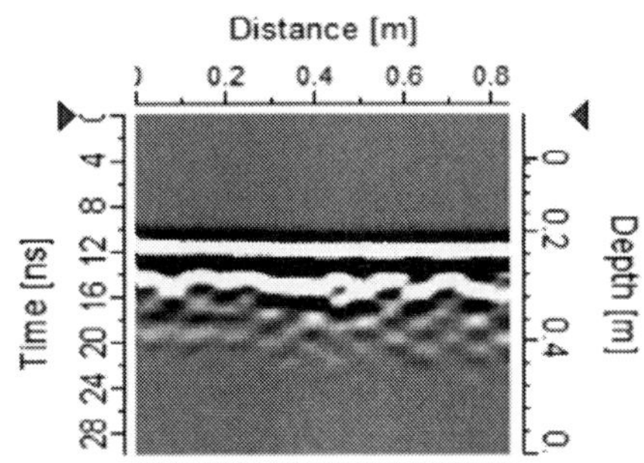


**Figure 8.** Radargram of rebar located at concrete floor using 500 MHz frequency antenna

**4.Rebar Diameter**

A study had been carried out, the research suggest to use different antenna polarization in order to measure the diameter of rebar. Special and specific conditions are required to estimate diameter of rebar, due to GPR do not measure the diameter of rebar directly from the signal [6]. The GPR is capable to measure the rebar diameter by the strength of reflection signal that come from small, medium or large of target (rebar) sizes.

Measuring rebar diameter of reinforced concrete is quite demanding task and its plays vital role to appraise the load bearing capacity of structures for construction. Following the successful studies from previous researches by Utsi [12], it was deduced that the method using different antenna polarization reach up to 20% of accuracy for rebar diameter evaluation [8]. Typical GPR antenna during scanning is the co-pole configuration as shown in figure 9. The co-pole arrangement is the alignment of transmitting and receiving antennas are aligned as parallel to each other. Thetransmitter and receiver antenna represent electric and magnetic field respectively [9]. The co-polarized and cross polarized antenna configurations as illustrated in figure 9 (a) and (b) respectively, are important for measurement of rebar diameter since difference method of antenna polarization are required.

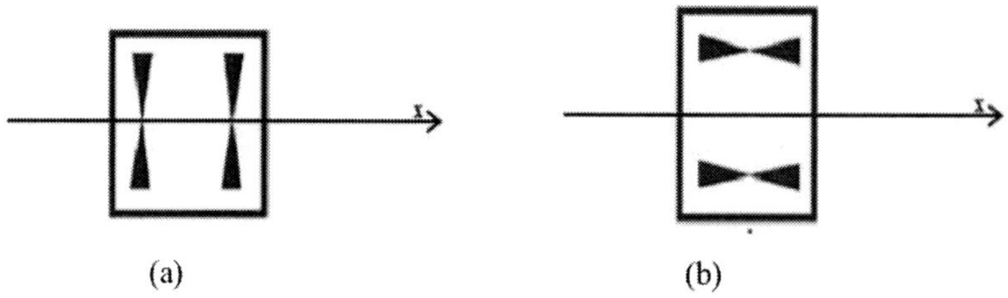


**Figure 9.** Co-pole arrangement of GPR (a) co-polarized (b) cross polarized [9].

However, the method approach is easy to be distorted. Analysing data taken with transducer and signal axis parallel and then orthogonal to the bar using Multilayer Perceptron (MLP) neutral network has been established by Shaw as to get the automatic estimation of rebar diameter [10].A cylinder shape target can be contemplated as rebar for GPR detector. The typical reflection signal presented in radar profile for rebar is hyperbola fitting as depicted in figure 10 [9].

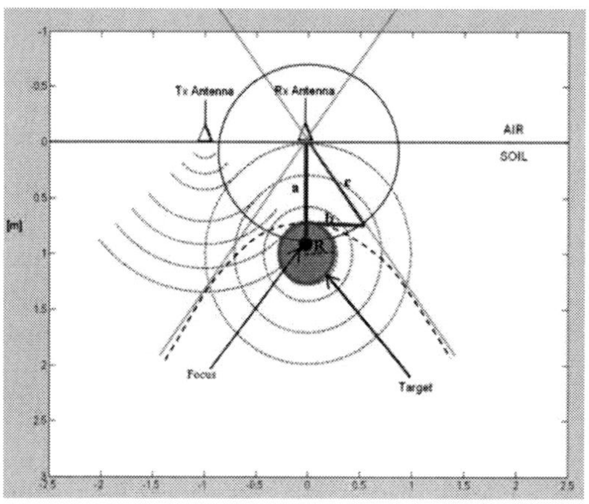


**Figure 10**. Hyperbola produced from cylinder object [9].

The connection between the rebar radius and the geometry of the hyperbola is verified as the presence of the reflection signal as mentioned. In year 2005, Al-Nuaimy has created an equation that illustrates cylinders can be detected and remarked from a single radar profile. He also demonstrated a model, that describes two-way travel time, t to the horizontal position, x, and the propagation velocity, v [11]. The model presented to characterize the signatures of rebar, and the model depends on hyperbolic signatures, which resulted in point reflectors.

A series of tests and experiments were performed and inspired by Utsi's previous research and findings to estimate the diameter of rebar. It was revealed from the study, the rebar yielded weak reflections when oriented orthogonal to the transmit antenna and produced strong reflections when oriented parallel to the long axis of a dipole transmit antenna [12]. Figure 11 illustrates the difference on radar wave for signal along EM field and signal across EM field [9].

IOP Conf. Series: Materials Science and Engineering **271** (2017) 012089 doi:10.1088/1757-899X/271/1/012089

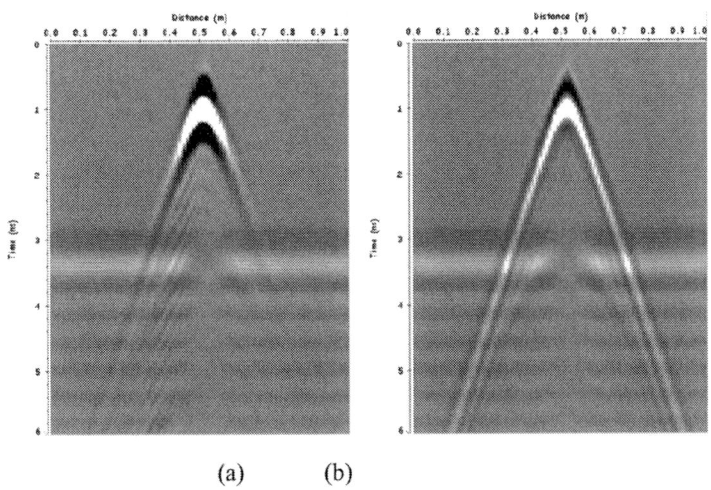


**Figure 11.** Radar profile on rebar signal (a) along electromagnetic field (b) across electromagnetic field [9].

## 5. Rebar Spacing

Rebar spacing also can be determined using ground penetrating radar (GPR). The integrity of structure for every reinforced concrete is also depends on size and spacing of reinforcement steel bar. A bridge girder was scanned using GPR by Ekes as shown in figure 12[13]. The inspection was operated by using a 1GHz antenna, where high metallic object such as rebar was detected as rebar has a huge conductivity contrast. Rebar spacing was detected through the GPR scanning as presented in figure 13 [13].

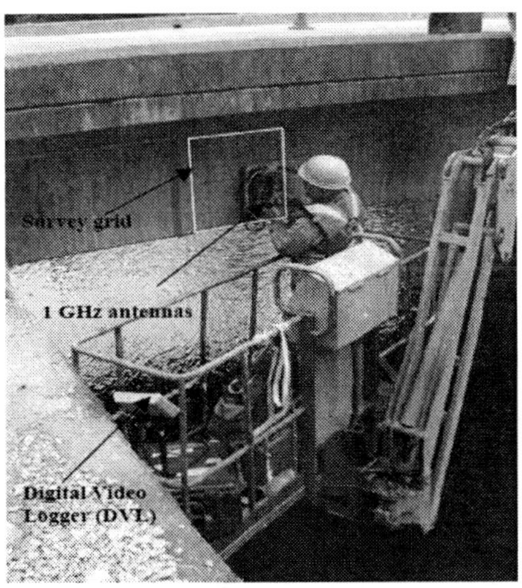


**Figure 12.** Inspection of bridge girder using GPR [13].

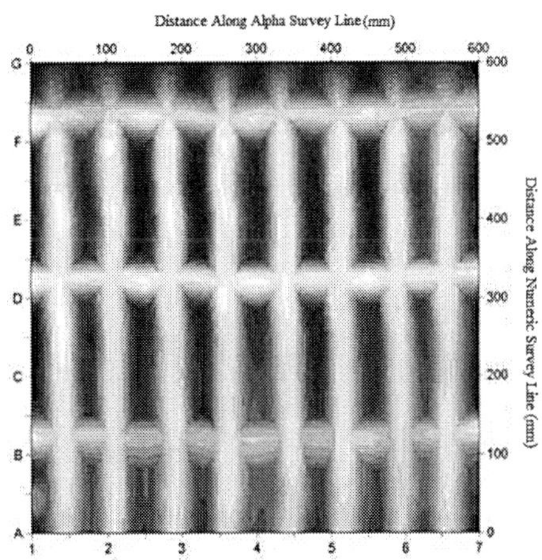


**Figure 13**. Radar profile of rebar spacing as detection of bridge girder through GPR scanning [13].

## 6.Conclusion

Ground penetrating radar (GPR) can be applied on construction foundation study where GPR can be used to identifying and detecting the location of rebar. Rebar diameter can be estimated by using different antenna polarization and configuration of GPR however further researches are needed to measure the diameter of rebar precisely.

## 7. References

[1] Ulricksen C P F 1982 *Application of Impulse Radar to Civil Engineering* Doctoral Thesis (Department Of Engineering Geology, Lund University of Technology, Sweden)

[2] Sivasubramanian K, Jaya K P and Neelemegam M 2013Covermeter for Identifying Cover Depth and Rebar Diameter in High Strength Concrete *Int J. Civil and Struct. Engr.* **3**(3) 557-563

[3] Eisenmann D, Margetan J F, Chiou P, Ellis S, Huang T and Tan Y J 2017 Effects of Position, Orientation, and Metal Loss on GPR Signals from Structural Rebar *AIP Conference Proceeding* vol 1806 (United States: AIP Publishing) 080005

[4] Lakshmi A K, Sangoju B, Vasanthakumar, and Rahamath A 2016 Estimation of Rebar Radius Using Ground Penetrating Radar. *Int J. of Emerging in CS & Electronics (IJETCSE)* **22** 2.

[5] Einsenmann D, Margetan J F, Koester L and Clayton D 2016 Inspection of a Large Concrete Block Containing Embedded Defects using Ground Penetrating Radar *AIP Conference Proceeding* vol 1706 (United States: AIP Publishing) 020014

[6] Geophysical Survey Systems, Inc. 2006 GSSI Handbook for RADAR Inspection of Concrete (New Hampshire USA: Geophysical Survey Systems Inc. Salem)

[7] Hugenschmidt J and Kalogeropoulos A 2009 The Inspection of Retaining Walls using GPR. *J. of ApplGeophy.* **67** 335-344.

[8] Hugenschmidt J and Wenk F 2016 Rebar Diameter and Rebar Orientation Using Different Polarizations. *16$^{th}$IntConf of Ground Penetrating Radar (GPR)* (Hong Kong:IEEE Explore)

[9] Leucci G 2012 Ground Penetrating Radar: An Application to Estimate Volumetric Water Content and Reinforced Bar Diameter in Concrete Structures. *J of Adv. Concr Tech* **10**(12) 411-422

[10] Shaw M R, Molyneaux T K, Millard S G, Taylor M J and Bungey J H 2003 Assessing Bar Size of Steel Reinforcement in Concrete using Ground Penetrating Radar and Neural Networks *Insight: Non-Destructive Testing and Cond MNT* **45**(12) 813-816

[11] Nuaimy A, Huang Y, Nakhkash M, Fang M., Nguyen V and Eriksen A 2000 Automatic Detection of Buried Utilities and Solid Objects using Neural Networks and Pattern Recognition. *J. Appl.Geophys.* **43**(2-4) 157-165

[12] Utsi V and Utsi E 2004 Measurements of Reinforcement Bar Depths and Diameters in Concrete. *Tenth IntConf on GPR.* (Delft: IEEE Xplore)

[13] Ekes C 2011 GPR: A New Tool for Structural Health Monitoring of Infrastructure *Proc. of 3rd International Conference on Structural Health Monitoring of Intelligent Infrastructure (SHMII-3)* (Vancouver, BC Canada)

## Acknowledgments

Authors wishing to acknowledge assistance and encouragement from colleagues, technical staff as wellas the Non Destructive Testing - Material and Structure Integrity Group (NDT-MSI) for the support. Lastly, authors would like to acknowledge the financial support from MOSTI and AgensiNuklear Malaysia.

# Crack classification in concrete beams using AE parameters

**N A A S Bahari[1], S Shahidan[1], S R Abdullah[1], N Ali[1], S S Mohd Zuki[1], M H W Ibrahim[1] and M A Rahim[2]**

[1] Jamilus Research Center, Faculty of Civil and Environmental Engineering, Universiti Tun Hussein Onn Malaysia, 86400 Batu Pahat, Johor.
[2] School of Environmental Engineering, University Malaysia Perlis, 02600, Arau, Perlis, Malaysia
Corresponding author: shahironshahidan@gmail.com

**Abstract.** The acoustic emission (AE) technique is an effective tool for the evaluation of crack growth. The aim of this study is to evaluate crack classification in reinforced concrete beams using statistical analysis. AE has been applied for the early monitoring of reinforced concrete structures using AE parameters such as average frequency, rise time, amplitude counts and duration. This experimental study focuses on the utilisation of this method in evaluating reinforced concrete beams. Beam specimens measuring 150 mm x 250 mm x 1200 mm were tested using a three-point load flexural test using Universal Testing Machines (UTM) together with an AE monitoring system. The results indicated that RA value can be used to determine the relationship between tensile crack and shear movement in reinforced concrete beams.

## 1. Introduction

The deterioration and damage in existing concrete structures could be caused by ageing, fatigue, corrosion, environmental hazards and environmental effects [1]. Thus, it is a priority to detect structural damage through regular inspection and maintenance. In addition, it is very important to understand the phenomenon associated with cracking in order to ensure safety and cost efficiency [2]. Traditionally, visual inspection has been used by trained instructors to monitor bridge structures. Due to the limitations associated with visual inspection, it is impossible to detect all sources of damage especially for internal structures [3]. Therefore, it is important to facilitate asset management by improving or using more efficient techniques in the evaluation of reinforced structures.

Nowadays, the monitoring of concrete structures is carried using destructive testing or non-destructive testing (NDT) methods. One of the NDT techniques currently being investigated for reinforce concrete beams and bridge management systems is acoustic emission (AE). The acoustic emission (AE) technique is widely accepted in order to create non-invasive and passive NDT and is being used in the field of civil engineering for structural health monitoring [4]. Using this technique, elastic waves are released during the propagation of cracks [5,6]. The elastic waves propagating inside the material are detected by a piezoelectric transducer as shown in figure 1. A piezoelectric transducer is an AE sensor which is also known as the heart of the AE system as it converts stress waves into electric signals [7]. Prior to testing, the AE sensor network is mounted on a flat surface of the RC structure to detect the propagating waves. Using timing strategies, the source of the emission could be identified [8]. The sensor has to be mounted on the flat surface of an RC structure due to the flat frequency response. A rough surface would cause the AE sensor to be less sensitive and make it difficult to detect AE signals [7]. In addition, grease can be used in order to improve the sensitivity between the concrete surface and that of the AE sensor. According to [9], grease is able to reduce the duration of the test by improving the surface contact between concrete and the AE sensor.

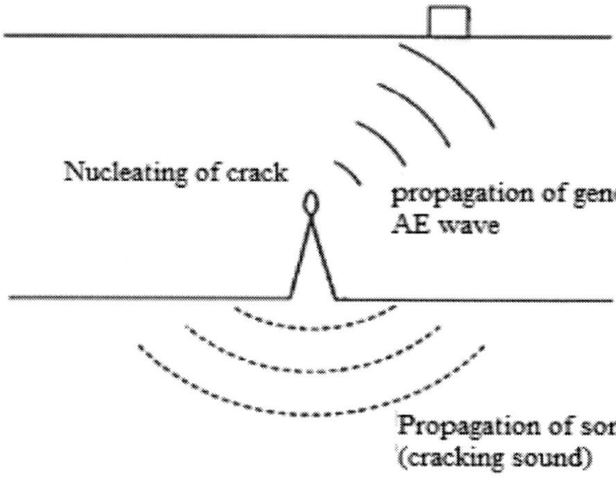


**Figure 1.** Detection of acoustic emission wave [10].

Even though the AE technique has been extensively utilised in damage monitoring for RC structures, there has been a lack of research on damage evaluation and correlating AE parameters. AE parameters are useful for better characterisation of AE sources such as peak amplitude, duration, frequency, counts and rise time. From the AE parameters, the rise time and amplitude can be used to define the RA value. The RA value can be calculated using equation (1) and equation (2):

$$RA = {}^{Rise\ time}/_{Amplitude} \quad (1)$$

$$Average\ frequency = {}^{Counts}/_{Duration} \quad (2)$$

The cracking behaviour of AE parameters could be classified into two types of fracture mode namely tensile crack and shear movement. Tensile cracks (Mode I) are nucleated during the early stages of the damage process. On the other hand, shear movement (Mode II) is generated when fretting or sliding occurs to existing cracks [7]. These types of crack classifications are indicated in figure 2. According to Aggelis *et al.* [11], the emission during the early stage has shown that the result corresponds to tensile cracks whereas the pattern of emission exhibits a high average frequency and low RA values. In contrast, during the later stages, shear cracks occur close to failure which indicate the value of average frequency. [2]. Hence, classifying the type of cracking behaviour in RC structures requires a combination of the average frequency against the RA value as shown in figure 3.

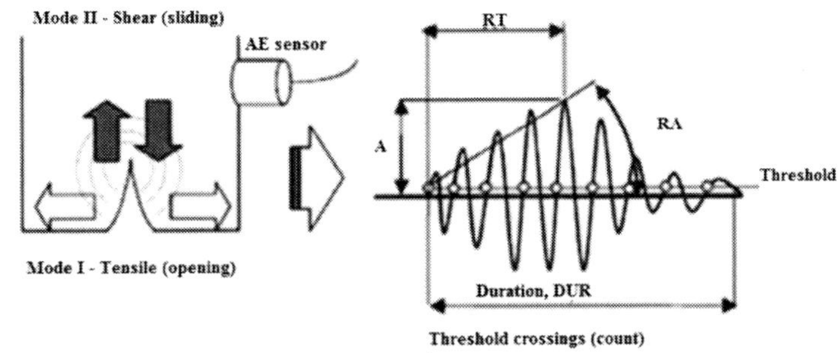


**Figure 2.** Cracking mode due to AE signals [12].

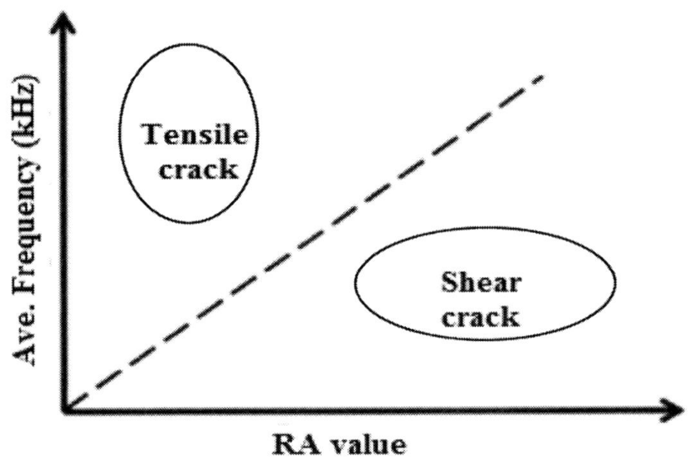


**Figure 3.** Classification of cracks [5].

## 2. Experimental program

The experimental program in this study consists of the preparation of beam specimens, acoustic emission measurement and flexural testing.

### *2.1. Preparation of beam specimens*

The specimens were reinforced concrete beams with a dimension of 150 mm x 250 mm x 1200 mm each. They were designed in accordance with Eurocode 2 (BS EN 1992-1-1:2004) as shown in figure 4. The beams were reinforced with steel bars with high tensile strength. Cement, water, fine aggregate and coarse aggregate were used to make the beam specimens. The concrete mix was then casted into moulds. The vibration of fresh concrete was done in order to avoid the formation of voids or pores within the concrete specimen. In order to ensure that the design fulfilled the required strength, three cube specimens measuring 150 mm x 150 mm x150 mm were prepared with grade G30 cement. These specimens had to undergo a compression test after a curing period of 7 days and 28 days respectively to gain uniform strength. Prior to the test, the surface of the beam specimens was smoothen to ease the installation of AE sensors.

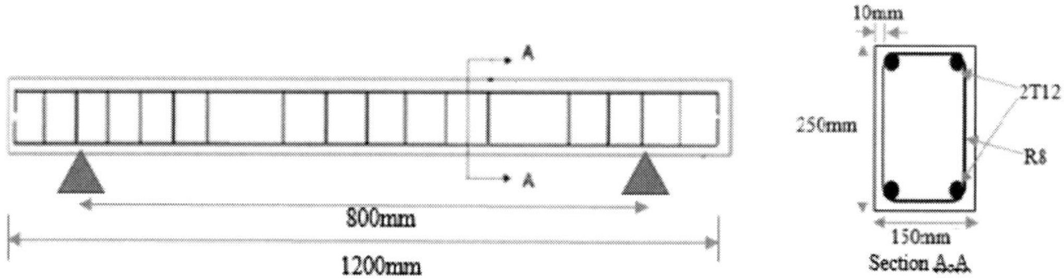


**Figure 4.** Beam cross section dimension and detail reinforcement.

### *2.2. Acoustic emission measurement*

An acoustic emission system board (8 channels) x 16 (hubs) bit acoustic emission channel were used for the AE measurement. During the test, AE was monitored using a MicroSAMOS (μSAMOS) supplied by Physical Acoustic Corporation (PAC). Four sensors (R6I-AST) for concrete materials were mounted on the beam surface at selected points shown in figure 5 by using a coupling agent [13]. Magnetic clamps were used to safeguard the sensor where the sensors were held in place by magnetic clamps [13,14]. The sensitivity of the AE sensor installation was verified using the Hsu-Nielson method. The sensitivity of the sensors were checked before the specimens were tested. The sensors are significantly coupled if three or more replicates of pencil lead fracture (PLF) produce an amplitude of 99 dB or the sensitivity is within ±3 dB in difference [15,16]. This is done to ensure that the sensor and the beam have good contact in order to provide accurate results throughout the test. If these criteria are not met, the sensor on the beam surface will be remounted and the sensitivity check has to be carried out again until the amplitude fulfills the requirement. The threshold level was set to 45 dB to eliminate noise from the surrounding area [10,17,18].

### *2.3. Flexural testing*

The simply supported specimens were tested under three-point load flexural testing by using a Universal Testing Machine (UTM) loaded statically until failure, as shown in figure 5. Flexural testing was monitored using an AE monitoring system [19,20]. Six different load ranges used during the text, LR1 to LR6, are shown in figure 6.

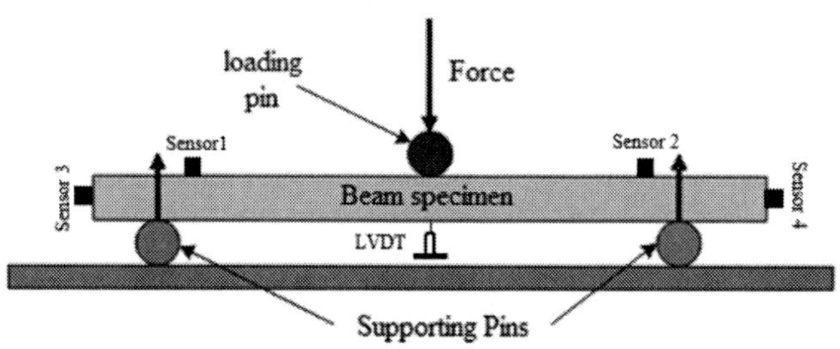


**Figure 5.** Experimental setup with three-point flexural testing.

## 3. Results and discussion

The AE data parameters were analysed post-test using RA value methods to classify the type of cracking in the RC beam.

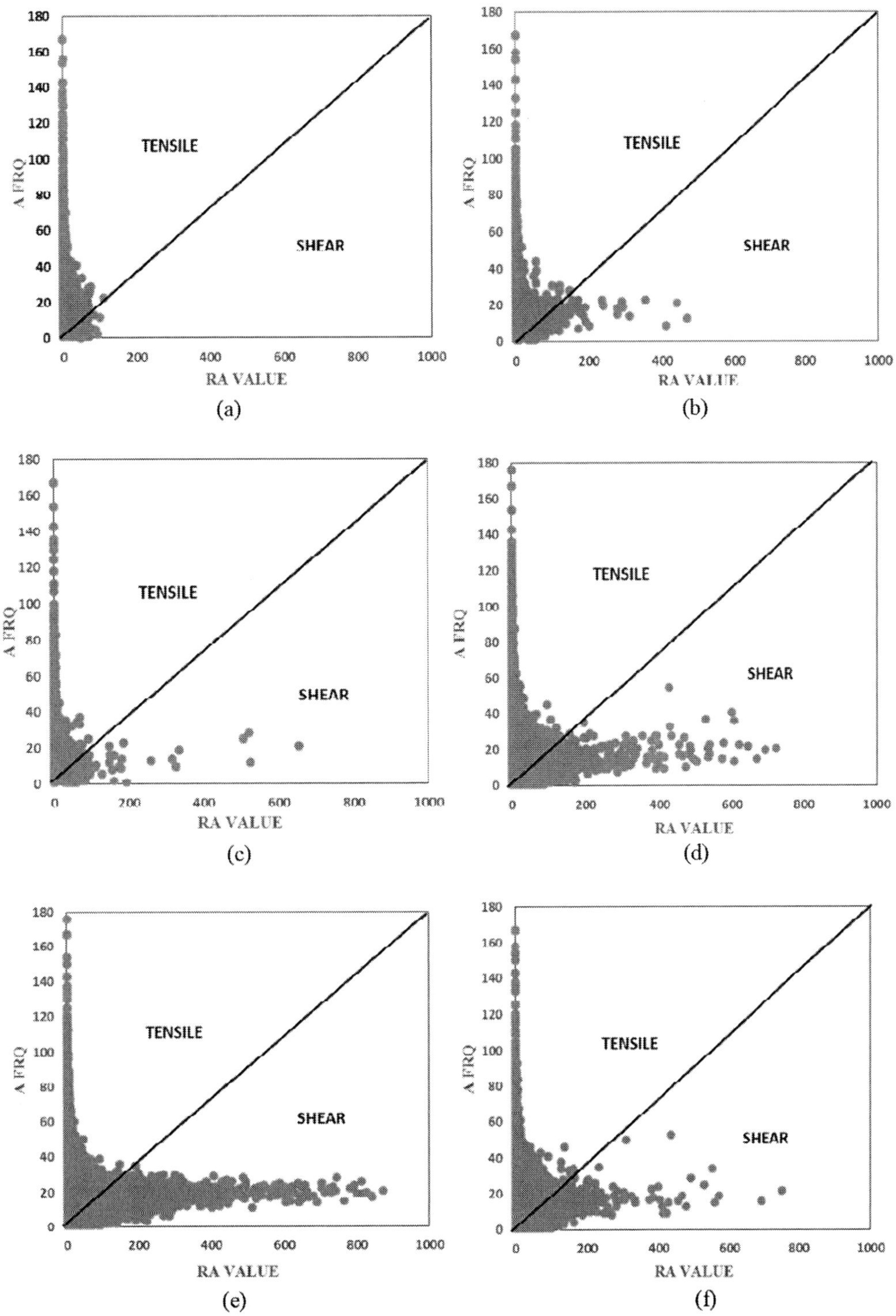


**Figure 6.** RA value analysis for each load range (a) LR1(0-21 kN) (b) LR2(21-40 kN) (c) LR3(41-60 kN) (d) LR4(61-80 kN) (e) LR5(81-100 kN) (f) LR6(100 until failure).

The RA value was assessed by using equation (1) and equation (2) to classsify the type of cracking mode present in concrete beams. Thus, the analysis was done for each specimen using RA values. In figure 6, the results of the RA value analysis are presented according to the load range (LR). Figures 6(a) to 6(f) present the typical development of the cracking process from LR1 to LR6.

During the initial stage, as shown in figure 6(a), micro-cracking appeared at the middle section of the reinforced beam when loads ranging between 0 kN to 20 kN (LR1) was applied. This type of cracking is classified as tensile cracking. At this time, the damage mechanism is not too obvious as the steel bars begin to bear the load [2]. Next, figure 6(b) shows that when a loads ranging between 21 kN to 40 kN was applied to the specimens, tensile cracks continue to grow. At the same time, some diagonal cracks were formed between the support and point of loading.

After that, when LR3 was applied as shown in figure 6(c), the tensile cracks grew slowly between the two points of loading and the cracks propagated to the compression zone [12]. As shown in figures 6(d) and 6(e) where LR4 and LR5 were applied respectively, diagonal shear cracks spread upward when fretting or sliding occurs to existing tensile cracks. The tensile cracks continue to grow along with some diagonal shear cracks.

In figure 6(f), the tensile cracks did not spread to the top when LR6 was applied. Instead, diagonal shear cracks were formed at the beam support which spread upwards until the specimen failed.

## 4. Conclusion

This experiment has successfully identified the crack classification using AE parameters such as amplitude, rise time and average frequency. RA value analysis was employed. It can be concluded that the acoustic emission (AE) technique can be applied to monitor the progress of deterioration progress in reinforced concrete beams under loading in laboratory tests. Thus, AE parameters can be used for crack classification.

The RA value method analysis is suitable and effective for the typical crack classification process from initial damage to failure. On the other hand, it was confirmed that when the average frequency increases and RA value decreases, damage mode I (tensile cracking) occurs. Damage mode II (shear cracking) occurs when the average frequency decreases whereas the RA value increases.

From the RA value analysis method, the initial phase of loading causes tensile cracks. Nevertheless, when the loading increases, shear cracking occurs.

In conclusion, this experimental study provides a promising method for the classification of cracks in reinforced concrete beams using AE parameters.

## 5. References

[1] Nair A and Cai C S 2010 Acoustic emission monitoring of bridges: Review and case studies *Eng. Struct.* **32(6)** 1704–1714
[2] Shahidan S, Pullin R, Bunnori N M and Zuki S S M 2017 Active crack evaluation in concrete beams using statistical analysis of acoustic emission data *Insight - Non-Destructive Test, Cond. Monit.* **59**(1) 24–31
[3] Carnì D L, Scuro C, Lamonaca F, Olivito R S and Grimaldi D 2016 Damage analysis of concrete structures by means of acoustic emissions technique, *Compos. Part B Eng.* 1–8
[4] Prem P R and Murthy A R 2017 Acoustic emission monitoring of reinforced concrete beams subjected to four-point-bending, *Appl. Acoust.* **117** 28–38
[5] Zhou J 2011 *A Study of Acoustic Emission Technique for Concrete Damage Detection* Master Thesis (United States: Michigan Technological University)
[6] Noorsuhada M N 2016 An overview on fatigue damage assessment of reinforced concrete structures with the aid of acoustic emission technique, *Constr. Build. Mater.* **112** 424–439
[7] Ono K 2011 Acoustic emission in materials research: A review, *J. Acoustic Emission* **29** 284–308
[8] Choi W C and Do Yun H 2015 Acoustic emission activity of CFRP-strengthened reinforced concrete beams after freeze-thaw cycling, *Cold Reg. Sci. Technol.* **110** 47–58

[9] Ziehl P H 2000 *Development of A Damage Based Design Criterion for Fiber Reinforced* vessels PhD Thesis (Austin: University of Texas)
[10] Shahidan S, Pullin R, Holford K M and Nor N 2013 Quantitative evaluation of the relationship between tensile crack and shear movement in concrete beams, *Adv. Mater. Res.* **626** 355–359
[11] Aggelis D G, Tsimpris N, Chai H K, Shiotani T and Kobayashi Y 2011 Numerical simulation of elastic waves for visualization of defects, *Constr. Build. Mater.* **25** (4) pp 1503–1512
[12] Shahidan S, Pulin R, Muhamad N and Holford K M 2013 Damage classification in reinforced concrete beam by acoustic emission signal analysis, *Constr. Build. Mater.* **45** 78–86
[13] Shahidan S, Bunnori N M, Md Nor N and Basri S R 2011 Damage severity evaluation on reinforced concrete beam by means of acoustic emission signal and intensity analysis *Symp. Ind. Electron. Appl.* (United States: IEEE) pp 337–341
[14] Md Nor N, Ibrahim A, Muhamad Bunnori N, Saman H M, Mat Saliah S N and Shahidan S 2014 Diagnostic of fatigue damage severity on reinforced concrete beam using acoustic emission technique, *Eng. Fail. Anal.* **41** 1–9
[15] Behnia A, Kian H and Shiotani T 2014 Advanced structural health monitoring of concrete structures with the aid of acoustic emission, *Constr. Build. Mater.* **65** 282–302
[16] Md Nor N, Muhamad Bunnori N, Ibrahim A, Shahidan S, and Saliah S N M 2011 An investigation on acoustic wave velocity of reinforced concrete beam in-plane source *7th Int. Colloquium on Signal Processing and Its Applications (Penang, Malaysia)* (United States: IEEE) pp 19–22
[17] Shahidan S, Zuki S S M and Jamaluddin N 2016 Damage grading system for severity assessment on concrete structure, *Case Stud. Constr. Mater.* **5** 79–86
[18] Shahidan S, Mohd Zuki S S, Rahim M A and Ali N 2016 Amplitude distribution of emission wave for cracking process MATEC Web Conference vol 47 (Paris: EDP Sciences) pp 2–8
[19] Shahidan S, Bunnori N M, Nor N, Rahim M A, Abdullah R and Zuki S S M 2016 Analysis improvements of acoustic emission data in evaluation of concrete beam *ARPN J. Eng. Appl. Sci.* **11**(20) 11804–11807
[20] Bohse J and Brunner J 2008 Acoustic emission in delamination investigation, *Delamination Behav. Compos.* 217–277
[21] Huguet S, Godin N, Gaertner R, Salmon L and Villard D 2002 Use of acoustic emission to identify damage modes in glass fibre reinforced polyester, *Composite Science Technology* **62** 1433–1444

**Acknowledgments**
The author would like to thank Center Graduate Studies Universiti Tun Hussein Onn Malaysia, Grant Kontrak U523 for making this important research viable and effective.

IOP Conf. Series: Materials Science and Engineering **271** (2017) 012091 doi:10.1088/1757-899X/271/1/012091

# Damage Identification Based on Curvature Mode Shape using Cubic Polynomial Regression and Chebyshev Filters

**C M Hasrizam[1*] and Noor Fawazi[1]**

[1] Mechanical Precision Engineering, Malaysia-Japan International Institute of Technology (MJIIT), Universiti Teknologi Malaysia, Jalan Semarak, 54100 Kuala Lumpur, Malaysia.
Corresponding author: mhasrizam2@graduate.utm.my

**Abstract.** Structure Health Monitoring (SHM) has been applied in various application such as aerospace, machinery and civil structures to maintain structure's safety and integrity. Gapped smoothing method (GSM) is most popular non-destructive identification (NDI) method due to its simplicity and did not require baseline data for comparisons. However, GSM is less accurate to detect wide size of damage in structure and cause false detection. Objective of this study is to propose a method to detect damage in structure using curvature mode shape data estimated from damaged structure and did not require data from undamaged structure. Finite element analysis (FEA) on a free-free boundary condition steel beam was carried out to demonstrate the feasibility of the proposed method that estimate undamaged curvature mode shape data using cubic polynomial regression (CPR) and Chebyshev filters (CF) methods. The results shows proposed method that used Chebyshev filters has better accuracy damage detection on wide notch compared to GSM. Although application of an interpolation and Chebyshev filters showed results with a high potential for overcoming the issue of false detection due to different notch size, however the proposed method still need refinement to better detection of different damage cases.

## 1. Introduction

Structure Health Monitoring (SHM) is one of important tools to maintain safety and integrity of structure such as aircraft, automotive, machinery, multi-storey building and bridges [1-4]. A reliable non-destructive identification (NDI) is important to ensure effectiveness of such monitoring system. Because undetected damage may grow and reduce structure integrity which subsequently lead to catastrophic failure.

Vibration-based damage detection method has been studied extensively in the past few decades [5]. This method was based on physical changes in structures which manifest by changes in modal parameters (i.e. natural frequencies, mode shape and damping coefficient). Therefore, the changes of modal characteristics in structures can be treated as the damage indicator.

Study by Ratcliffe [6] shows curvature mode shape as parameter to detect damage in structure by using the mode shape data from undamaged and damaged structures, and they are found curvature mode shape can be better indicators for damage identification compared to natural frequency. However this method requires data from the undamaged structures which may impossible to be obtained in real life structure. Wahab and Roeck [7] has extended the curvature mode shape method by averaging the damage index over all modes and applied their technique to a concrete bridge. Gapped Smoothing Method (GSM) was introduced by Pandey *et.al* [8], advantage of the GSM compare to other methods it did not require data from the undamaged structures. However, this method not accurate to detect big size crack because the damage index will create 2 peaks which indicate presence of the two small size cracksinstead of one big size crack [9].

This paper will propose a method to detect damage in structure using curvature mode shape data from damaged structure and did not require data from undamaged structure. This technique will used small number of measurement data to estimate undamaged curvature mode shape data, $\omega_u$ using cubic polynomial regression (CPR) and Chebyshev filters (CF) methods. The data subsequently used to calculate damage index in order to determine location and size of the damage in structures. This method is expected to enable detection of wide size of crack without data from undamaged structures.

## 2. Methodology

### *2.1. Finite Element Analysis*

For this study, geometry of the beam is same as used by Pandey *et.al* [8]. Figure 1 shows a diagram of a steel beam model with notch to represent damage that was used in the study. The model has dimension of 914.4 mm in length, 6.35 mm in height and 50.8 mm in width (36 in x 2 in x ¼ in). The mode shape data are obtained at a total of 36 grid points located from 12.7 mm to 901.7 mm by a grid spacing ($\Delta x$) of 25.4mm along the distance. 12.7 mm narrow notch with node 12 at the centre of the notch and the node 12 was located 292.1 mm from one end are introduced at bottom of the beam to represent the narrow notch damage. In addition, 127 mm wide notch with node 13 at the centre of the notch and the node 13 was located 317.5 mm from one end are introduced at bottom of the beam to represent the wide notch damage. The depth for both notch type is 0.1905 mm or 3% from total beam thickness.

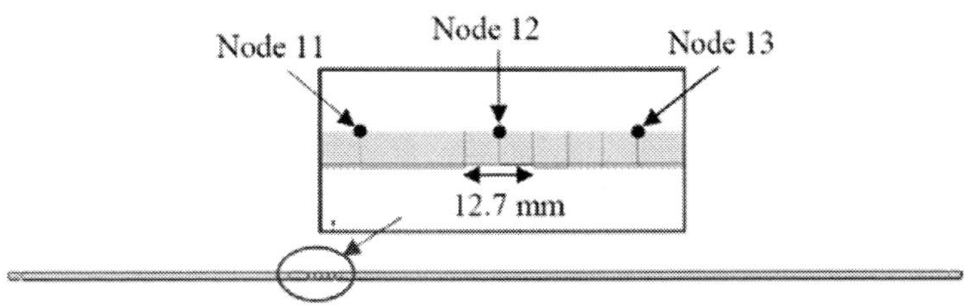


**Figure 1.** Diagram of the Beam and Notch Location (Narrow Notch).

Finite element analysis (FEA) was performed using commercial code ABAQUS to obtain the vibration responses of the steel beam models. Three-dimensional continuum solid elements with 20 nodes with reduced integration and second order accuracy (C3D20R) are used to discretize the model. The model has isotropic material properties with Young Modulus, density and Poisson's Ratio are 204.9 GPa, 7850 kg/m$^3$ and 0.3 respectively. Both cases have free-free boundary condition and the first bending mode shape was used for data processing.

### *2.2 Data Analysis Procedures*

Data analysis process for this study is presented in this section. Two beam with different damage cases were considered in this study, namely narrow notch and wide notch. Firstly, displacement mode shape data, $U_2$from 14 nodes was obtained from FEA performed as explained in section 2.1. Later, the displacement mode shape data, $U_2$ was interpolated into 36 nodes using Barycentric interpolation (n=2) equation as follows:

$$f(x) = \frac{(x-x_2)(x-x_3)\dots(x-x_n)}{(x_1-x_2)(x_1-x_3)\dots(x_1-x_n)}y_1 + \frac{(x-x_1)(x-x_3)\dots(x-x_n)}{(x_2-x_1)(x_2-x_3)\dots(x_2-x_n)}y_2 + \cdots + \frac{(x-x_1)(x-x_2)\dots(x-x_n)}{(x_n-x_1)(x_n-x_2)\dots(x_n-x_{n-1})}y_n \quad (1)$$

Secondly, the curvature mode shape data, $\omega_u$ was estimated using interpolated displacement mode shape data, $U_2$throughcentral finite different equation [8]:

$$curvature\ mode\ shape, \omega_i = \frac{u_{i-1} - 2u_i + u_{i+1}}{\Delta x^2} \tag{2}$$

where $u_i$ is mode shape at the grid point $i$ and $\Delta x$ is the distance between two successive grid points.

Thirdly, the interpolated curvature mode shape data fitted using two method, cubic polynomial regression and Chebyshev filters. Through this process, fitted curvature mode shape will be used as undamaged curvature mode shape data. The undamaged curvature mode shape data fitted with cubic polynomial regression is called $\omega_{u(\mathrm{CPR})}$ and the undamaged curvature mode shape data fitted with Chebyshev filters is called $\omega_{u(\mathrm{CF})}$. Subsequently, modal curvature difference or damage index was by subtracting the undamaged curvature mode shape, $\omega_u$ from the damaged curvature mode shape, $\omega_d$ (calculated using displacement mode shape data, $U_2$from 36 nodes obtained from FEA performed as explained in section 2.1), given by:

$$Damage\ Index, \delta_i^m = |\omega_d]_i - \omega_u]_i| \tag{3}$$

where $\omega_d]_i$ and $\omega_u]_i$ are curvature mode shape dataat the grid point $i$of the damaged and undamaged structures respectively.

## 3.Result and Discussions

### *3.1. Natural Frequency*

Table 1 shows natural frequency data from FEA and experimental data [9] for narrow notch beam. Only the fundamental bending mode shapes are considered for data processing. The difference between experiment and FEA is less than 3% for the first four bending mode shapes, it indicates that the data from FEA has good correlation with experimental data.

**Table 1.** Natural frequency data from Present FEA and experimental data by Wahab and Roeck [9].

| Mode Shape | Natural Frequency, $\omega_n$ (Hz) | | Difference, $\Delta\omega_n$ (%) |
|---|---|---|---|
| | Experiment (Ratcliffe, 2009) | Present FEA | |
| 1 | 39.092 | 39.884 | 2.03 |
| 2 | 108.640 | 109.96 | 1.22 |
| 3 | 210.108 | 215.62 | 2.62 |
| 4 | 352.940 | 356.55 | 1.02 |

### *3.2 Curvature Mode Shape*

Curvature mode shape is the data that was obtained by differentiating the displacement mode shape data twice, for this study the curvature mode shape was approximated using central different finite different of the displacement mode shape data.

Figure 2 shows plot of curvature mode shape, it was noted that curvature mode shape for undamaged beam has clean shape from any irregularity while the narrow notch has very small irregularity that not clearly visible, and the wide notch has big irregularity that clearly observable.

Thus, differences in curvature mode shape between damage and undamaged can show damage location in the structures.

Table 2 shows standard deviation for narrow notch and wide narrow that use cubic polynomial regression and Chebyshev filters fitting methods. It was noted that Chebyshev filters fitting method has better approximation undamaged curvature mode shapes compared to cubic polynomial regression. Better approximation for Chebyshev filters fitting method because the method remove unwanted components before fitting the data, while cubic polynomial regression only fitting the data. Thus, Chebyshev filters fitting method can detect damage location more accurate compared to cubic polynomial regression.

### *3.3. Damage Index*

Damage index plot is an indicator to show presence of damage in the structure. It was calculated by subtracting the undamaged curvature mode shape, $\omega_u$ from the damaged curvature mode shape, $\omega_d$. In addition, size and location of crack in the structure can be estimated from damage index plot. Figure 4 shows damage index for narrow notch cases. It was noted, GSM and Chebyshev method able to determine damage location accurately while Cubic polynomial regression unable to detect damage location.

Figure 5 shows damage index plot for narrow notch case. From this figure, GSM indicates presence of two nearby small notch at node 11 and node 15 which is false detection. Cubic polynomial regression indicates presence of wide notch in between node 11 to 15, however this method indicate presence of another wide notch in between node 5 to 10 which is false detection. Among all method, only Chebyshev filters method can predict presence of wide notch in between node 11 to 15 accurately.

Different in damage detection capability between GSM and proposed method (cubic polynomial regression and Chebyshev filters) is due to lack of curvature change at node 12, 13 and 14. It caused GSM to only detect damage at sudden changes (node 11 and 15) in curvature mode shape plot which later indicated as two small damage in damage index plot.

On the other hand, since Chebyshev is a filter that removes unwanted components in interpolated undamaged curvature mode shape data, it also capable to create more smooth approximation of undamaged curvature mode shape compared cubic polynomial regression. Lack of filter capability caused cubic polynomial regression to have deviation in curvature mode shape data than damaged curvature mode shape and the different is indicated in damage index as false damage detection. Hence, Chebyshev method capable to smooth the interpolated curvature mode shape data to estimate the undamaged curvature mode shape data more precisely allows it to detect wide notch type of damage in structure.

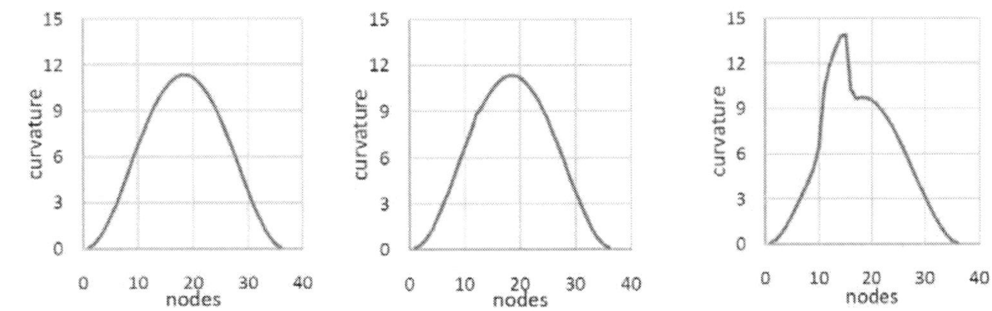


a) undamaged b) narrow notch c) wide notch

**Figure 2.** Curvature mode shape plot for a) undamaged, b) narrow notch and c) wide notch.

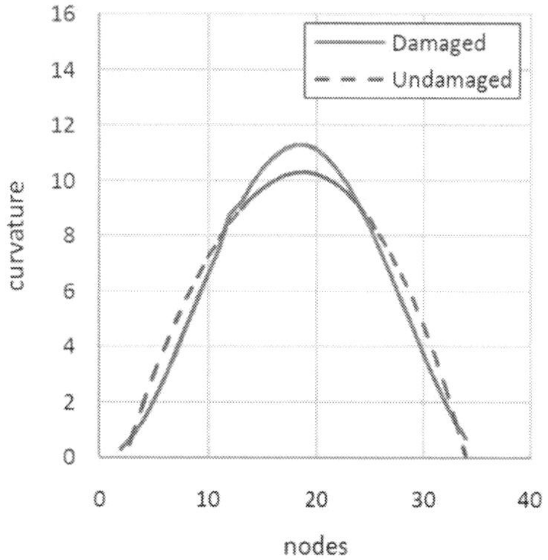


a) narrow notch - CPR

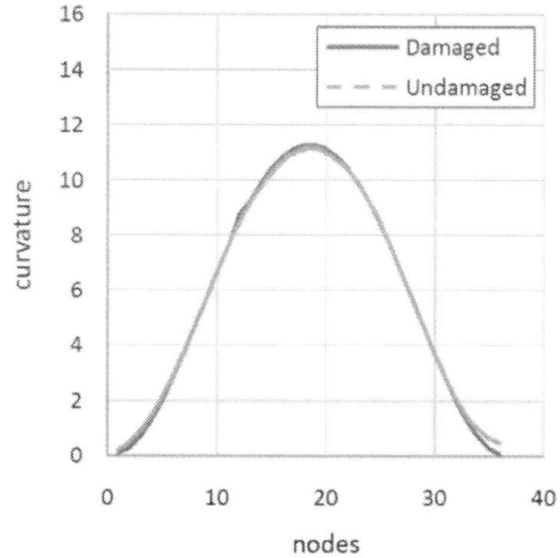


b) narrow notch - CF

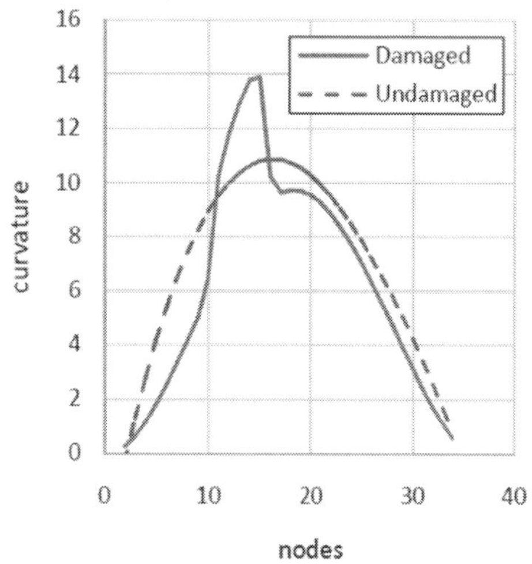


c) wide notch - CPR

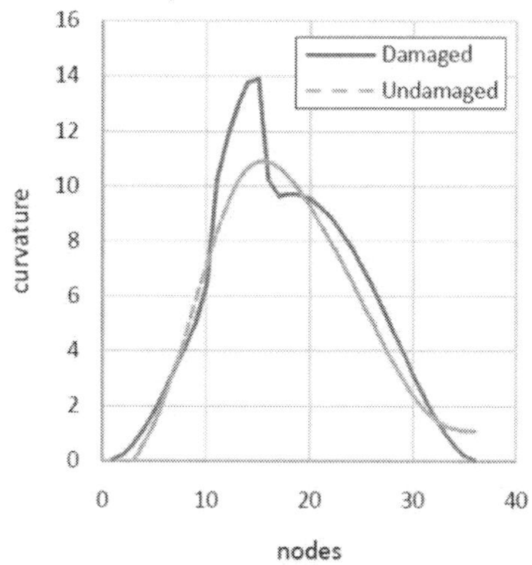


d) wide notch - CF

**Figure 3.** Curvature mode shape plot for narrow notch and wide notch.

**Table 2.** Standard Deviation for different Fitting Methods.

| Standard Deviation | | |
|---|---|---|
| Fitting Method | Narrow Notch | Wide Notch |
| Cubic Polynomial Regression (CPR) | 6.7210 | 6.1449 |
| Chebyshev Filters (CF) | 0.1718 | 1.2576 |

IOP Conf. Series: Materials Science and Engineering **271** (2017) 012091 doi:10.1088/1757-899X/271/1/012091

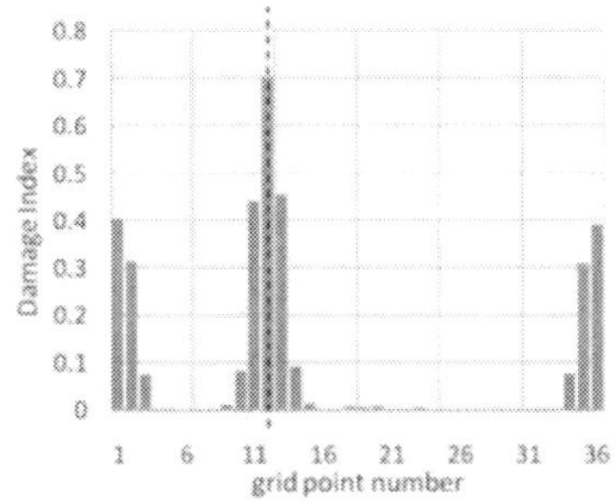


a) GSM

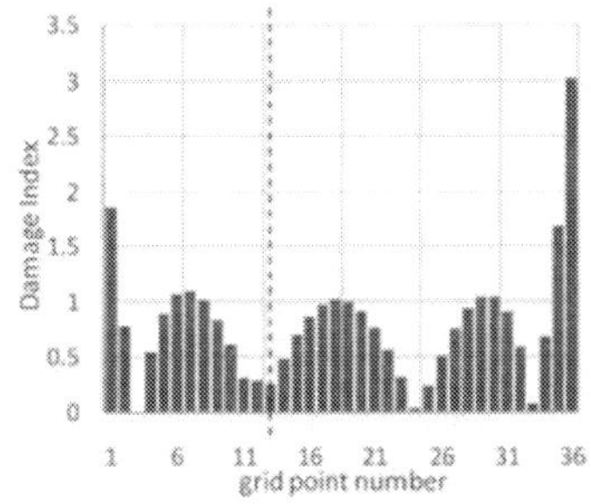


b) CPR

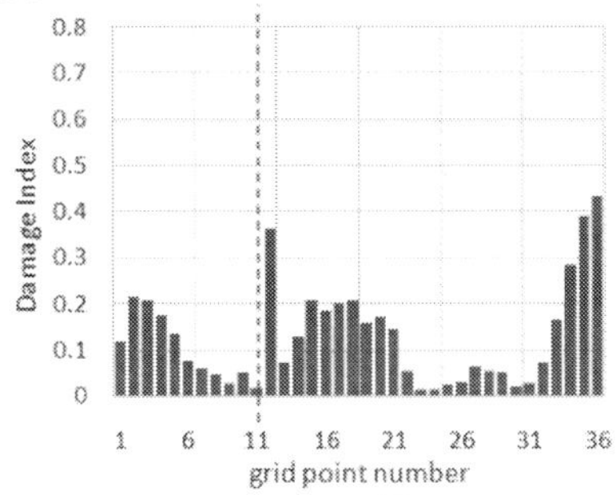


c) CF

**Figure 4.** Damage Index plot for Narrow Notch for a) GSM, b) CPR and c) CF.

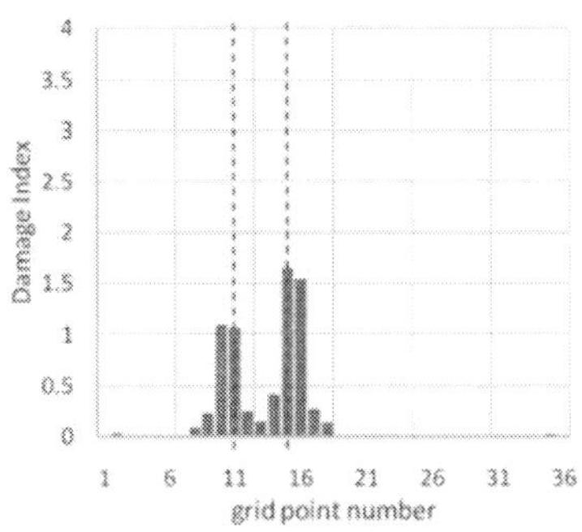


a) GSM

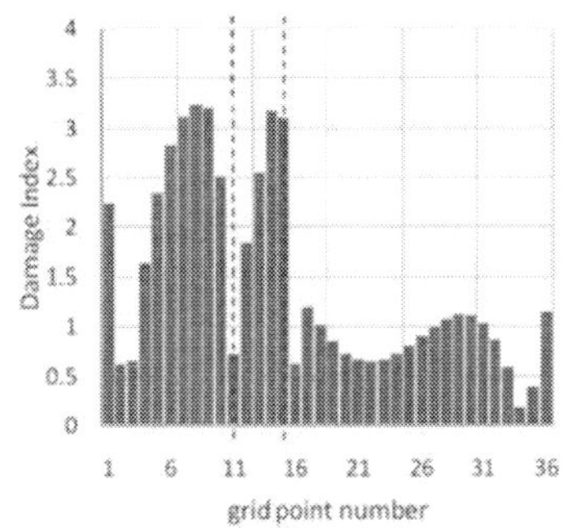


b) CPR

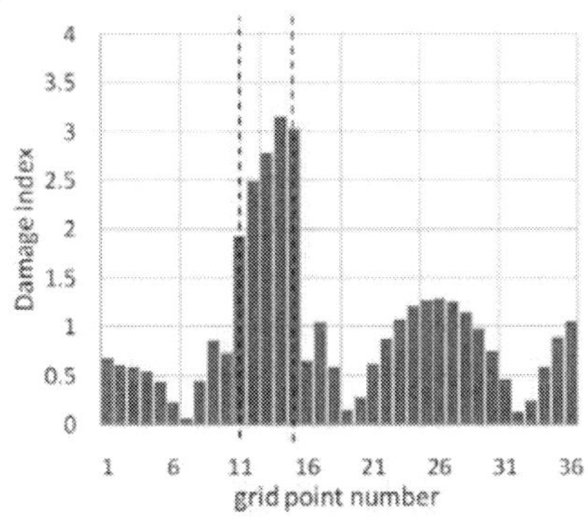


c) CF

**Figure 5.** Damage Index plot for Wide Notch for a) GSM, b) CPR and c) CF.

## 4. Conclusion and future work

This paper proposed new method to estimate undamaged curvature mode shape data using damaged curvature mode shape data using cubic regression and Chebyshev filter. Numerical analysis has been conducted using steel beam with 3% damage and narrow and wide notch to demonstrate the feasibility of proposed method.

For narrow type notch, GSM and Chebyshev filters able to determine damage location accurately while Cubic polynomial regression unable to detect the damage location. However, Chebyshev filters create some noise that can lead to fault detection. For wide type notch GSM and cubic polynomial regression can predict location however not the size of the crack in structure while only Chebyshev filters method can predict presence of wide notch in between node 11 to 15 accurately. Unlike GSM and cubic polynomial regression method, Chebyshev filter is able to estimate undamaged curvature mode shape near to actual undamaged data. This is because Chebyshev filter only remove unwanted component in the data whilst maintain the shape of curvature mode shape.

However, this study is only limited to one type of filter. Hence, damage detection using other filter or other smoothing formulation to estimate undamaged curvature mode shape that closer to actual undamaged data is recommended for future study.

## 5. References

[1] Meo M and Polimeno U2009 Detecting barely visible impact damage detection on aircraft composites structures *J. Composite Structures* **91** 398

[2] Medeiros R, Ribeiro M and Tita V 2014 Computational methodology of damage detection on composite cylinders: structural health monitoring for automotive components *Int. J. of Automotive Composites* **1** 112

[3] Dilena M, Limongelli M P and Morassi A2015 Damage localization in bridges via FRF interpolation method *J. Mechanical Systems and Signal Processing* **52** 162

[4] Wang Y 2010 *A Non-destructive Damage Detection Method for Reinforced Concrete Structures Based on Modal Strain Energy* P hD Thesis (Sydney: University of Technology Sydney) pp1-13

[5] Lu Z R, Liu J K and Xu W H 2009 Identification of local damages in coupled beam systems from measured dynamic responses *J. of Sound and Vibration* **177-189** 362

[6] Ratcliffe C 1997 Damage Detection Using a Modified Laplacian Operator on Mode Shape Data *J. of Sound and Vibration* **204** 505

[7] Yoon M, Heider D, Gillespie J, Ratcliffe C and Crane R 2009 Local Damage Detection with the Global Fitting Method Using Mode Shape Data in Notched Beams *J. of Nondestructive Evaluation* **28** 63

[8] Pandey A, Biswas M and Samman M 1991 Damage detection from changes in curvature mode shapes *J. of Sound and Vibration* **145** 321

[9] Wahab M and Roeck G 1999 Damage detection in bridges using modal curvatures: application to a real damage scenario *J. of Sound and Vibration* **226** 217

## Acknowledgements

The authors would like to acknowledge the financial support in this study from Malaysia-Japan International Institute of Technology (MJIIT) and Universiti Teknologi Malaysia (UTM) through MJIIT Incentive and Research University Grant (Tier 2), PY/2017/00468.

IOP Conf. Series: Materials Science and Engineering **271** (2017) 012092 doi:10.1088/1757-899X/271/1/012092

# Comparative analysis of different weight matrices in subspace system identification for structural health monitoring

**H Shokravi**[1], **NH Bakhary**[1]

[1] Faculty of Civil Engineering, Universiti Teknologi Malaysia,81310 Skudai, Johor Bahru, Malaysia
Corresponding author: hf.shokravi@gmail.com

**Abstract.** Subspace System Identification (SSI) is considered as one of the most reliable tools for identification of system parameters. Performance of a SSI scheme is considerably affected by the structure of the associated identification algorithm. Weight matrix is a variable in SSI that is used to reduce the dimensionality of the state-space equation. Generally one of the weight matrices of Principle Component (PC), Unweighted Principle Component (UPC) and Canonical Variate Analysis (CVA) are used in the structure of a SSI algorithm. An increasing number of studies in the field of structural health monitoring are using SSI for damage identification. However, studies that evaluate the performance of the weight matrices particularly in association with accuracy, noise resistance, and time complexity properties are very limited. In this study, the accuracy, noise-robustness, and time-efficiency of the weight matrices are compared using different qualitative and quantitative metrics. Three evaluation metrics of pole analysis, fit values and elapsed time are used in the assessment process. A numerical model of a mass-spring-dashpot and operational data is used in this research paper. It is observed that the principal components obtained using PC algorithms are more robust against noise uncertainty and give more stable results for the pole distribution. Furthermore, higher estimation accuracy is achieved using UPC algorithm. CVA had the worst performance for pole analysis and time efficiency analysis. The superior performance of the UPC algorithm in the elapsed time is attributed to using unit weight matrices. The obtained results demonstrated that the process of reducing dimensionality in CVA and PC has not enhanced the time efficiency but yield an improved modal identification in PC.

## 1. Introduction

Civil engineering structures are generally designed to serve for the lifetime of the occupants or facilities and their failure may lead to catastrophic economic or human losses. Hence, it is important to monitor the health state of these structures during their service life. Structural Health Monitoring (SHM) is an effective solution introduced for the need of a safer and more efficient condition assessment of structures. SHM is widely accepted tool for damage diagnosis in civil engineering communities and has been subject of various studies for the past three decades. SHM systems are used in many civil engineering structures including buildings [1-2], bridges [3-4] or dams [5- 6]. Vibration-based Damage Detection (VDD) is an area of significant interest in SHM and considerable effort have been dedicated toward developing novel approaches and improving existing strategies [7-8]. VDD methods rely on change of dynamic properties as an indicator of damage existence. These methods exploit the observable variation in modal parameters such as resonant frequency, damping and mode shape or their derivative to identify changes in physical properties of a structure. VDD Variation in environmental and operational condition of structures alters the modal parameters of structures that pose negative effect in detectability of the VDD algorithm. The influence of environmental and operational condition and discrimination of their effects are discussed in [9-10]. Time-domain methods are the simplest VDD methods to perform due to direct use of time series in analysis process. Time-domain VDD methods generally use acceleration response due to its higher sensitivity and richer dynamic contents compared to velocity and displacement signals. SSI method is reported as one of the most reliable time-domain methods in VDD and most of the studies are specially concentrated on subspace method in the recent years [11]. The outstanding performance of SSI algorithm is particularly due to its capability in global noise rejection [12]. The most researched subspace methods in the field of system identification are i) Canonical Variate Analysis (CVA), ii) Multivariable Output Error State-sPace (MOESP) and iii) Numerical algorithms for State-Space Subspace System Identification (N4SID) [3]. In order to reduce the complexity of implementing the aforementioned algorithms, Overschee, Moor [12] introduced an unified stochastic, deterministic and combined subspace schemes which put CVA, MOESP and N4SID into a pragmatic approach. Peeters and De Roeck [13] for the first time adapted SSI algorithm to deal with output-only measurements of ambient input. Output-only system identification approaches are the most appropriate methods for damage detection of civil engineering structures where the ambient method is used as excitation source [14-15].

Several research studies have been conducted to improve the performance of the SSI to deal with low quality of the input data including short length of measurement data [16], gluing the extracted data from multiple sets of sensors [17], noise inclusion [18], bias errors [19], complexity of structures [20], non-structural elements [21], Hammerstein systems [22], unexcited modes [23] and superiors modes [15]. Selecting of the appropriate weight matrix is a factor that must be considered in implementation of the SSI algorithm. Weight matrix was first introduced by Overschee, Moor [12] to reduce the dimensionality of data space through using principle parameters in PC, UPC and CVA methods. PC analysis is a multivariate statistics technique for reducing dimensionality of a data space. CVA is linear regression method for data analysis that is applied to quantify the relation between expectation and the extracted normalized variables. UPC algorithm is simpler than PC and CVA algorithms and use matrices of unit weight in the approximation process [24]. Some studies are conducted to compare the performance of PC, UPC and CVA as below. Cismaşiu, Narciso [25] studied optimization routine for implementation of FE updating techniques based on identified dynamic response of a real footbridge structure. It was stated that the SSI–UPC was chosen due to its simplicity and superior performance dealing with modes having comparable energy levels. Nguyen [26] proposed a unified sensing configuration for VDD of complex civil engineering structures. It was demonstrated that the SSI-UPC algorithm was selected due to powerful estimation capabilities and its application in most of modal analysis used in civil structures. Pioldi, Pansieri [27] investigated modal dynamic properties of buildings under earthquake base-excitations. It is stated that the CVA was found to be the most stable weighting option to achieve a reliable estimation at seismic excitation. It is claimed that CVA returns less noise or mathematical poles and higher capabilities to separate true physical modes from spurious earthquake harmonics. However, no direct comparison between these three subspace algorithms has been reported in the aforementioned studies and no evidence was given to prove the advantage of the CVA approach. Kompalka, Reese [28] presented a monitoring framework to deal with progressive damage using SSI-DATA algorithm together with model updating. In the paper it is demonstrated that the use of different weighting matrices of PC, CVA or UPC yield similar results. Miguel, Lopez [29] presented a hybrid stochastic/deterministic optimization algorithm to provide a starting point for optimizer. It is stated that the performance of three different variants of CVA, PC and UPC is quite similar, thus the variant PC was chosen for system identification. Herlufsen, Andersen [30] presented a damage detection technique including SSI technique together with a recently developed projection channel technique. It was mentioned that the analysis was performed on UPC, CVA and PC algorithms for different channels scenarios and all three algorithms gave almost identical results. However just the evaluation result of the SSI-PC was presented in the paper. As mentioned above, several researches have reported the effect of selecting weight matrices in SSI algorithm. However, there are limited studies, which comprehensively investigate on the influence of weight factor in efficiency of the SSI algorithm.

The aim in the present paper is to evaluate the influence of weight matrix on estimation accuracy, noise robustness and time efficiency of the SSI-PC, SSI-UPC and SSI-CVA algorithm. Three different metrics of fit values, poles analysis and the elapsed computation time are introduced to compare the performance of the SSI algorithms. The obtained results show that UPC algorithm had the best record in computation time analysis. The superior performance of the UPC algorithm in the elapsed time is attributed to using unit weight matrices. The use of unit weight matrix reduces the computational burden significantly and improves the time efficiency of the algorithm. The obtained results demonstrated that the process of reducing dimensionality in CVA and PC has not enhanced the time efficiency but yield an improved modal identification in PC. It is observed that the principal components obtained using PC algorithms are more robust against noise uncertainty and give more stable results for the pole distribution. Furthermore, higher estimation accuracy is achieved using UPC algorithm. CVA had the worst performance for pole analysis and time efficiency analysis.

## 2. Subspace algorithm and implementation of weight matrices

SSI is a time-domain identification method to extract the parameter of dynamic system. The subspace algorithm is divided into four steps of i) QR decomposition, ii) state sequence determination, iii) least square estimation and iv) Kalman filter. In the first step, the extracted response signal is cast into the form of block Hankel matrix. Then, the Hankel matrix is decomposed into triangular matrix and block matrix where the extracted triangular matrix is subspace representation of the Hankel matrix. The oblique projection of the past and future output data are used to determine the weighting matrices. The extracted oblique projection is pre and post multiplied by appropriate weight matrices to infer the system order and state sequence. In the second step, a geometrical projection is adapted to eliminate dependence of the SSI algorithm onto future output. The oblique projection of the past and future output data are used to determine the state-sequence of the system. In the third step, the LS is deployed to drive system matrices $(A \text{ and } C)$. Finally, the Kalman predictor is used to estimate the system model by inferring the Kalman gain $K$ of the state-space model. High resistance against noise in SSI algorithm to a great extent is achieved by adopting Singular Value Decomposition (SVD). SVD provide a simple way to improve the performance of SSI algorithm and reduce dimensionality with minimal loss of information. The principle angle and direction between subspaces can be determined using singular values of the obtained oblique projection. The cosines of the principle angles $(U$ and $V)$ are denoted by the singular values $(S)$.

$$W_1 O_i W_2 = USV^T \tag{1}$$

Where $W_1$ and $W_2$ are the weighting matrices of the oblique projection. Weighting matrices of $W_1$ and $W_2$ allows to draw the most proper state-space basis of an identified model. Three weighting algorithm are defined for implementation of the SSI algorithm including PC, UPC and CVA. The PC method incorporates right weight matrix to determine singular values. The output covariance matrix of the past $\Phi_{[Y_p, Y_p]}$ is used to determine the block Toeplitz matrices in PC algorithm whereas the output covariance matrix of the future data $(\Phi_{[Y_p, Y_p]})$ is used for CVA algorithm. UPC method is special case of PC analysis that gives the first principle component index of a system with equal weight factors to each set of data. The CVA algorithm selects equal weights for the all incorporated system. The weighting matrices in CVA are obtained from singular values.

IOP Conf. Series: Materials Science and Engineering **271** (2017) 012092 doi:10.1088/1757-899X/271/1/012092

**Table 1.**The $W_1$ and $W_2$ weighting matrices in PC, UPC and CVA algorithms.

| | $W_1$ | $W_2$ |
|---|---|---|
| PC | $I_{li}$ | $Y_p^T \Phi^{-1/2}_{[Y_p,Y_p]} Y_p$ |
| UPC | $I_{li}$ | $I_j$ |
| CVA | $\Phi^{-1/2}_{[Y_f,Y_f]}$ | $I_j$ |

Table 1 presented the weighting matrices for PC, UPC and CVA algorithms. The weights matrices of PC, UPC and CVA algorithms are used to determine the singular values. PC analysis is a left side weighting obtained from the covariance matrix of the past and future data. UPC method use identity matrix for weighting and CVA incorporates right-hand weight matrix.

## 3. Numerical simulation case study

Mass-spring-damper (MSD) system is the most common reduced-order engineering model in structural dynamics. A 6-DOF, Mass-Spring-Damper (MSD) simulation model is used in this study to compare the performance of PC-SSI, UPC-SSI and CVA-SSI algorithms.

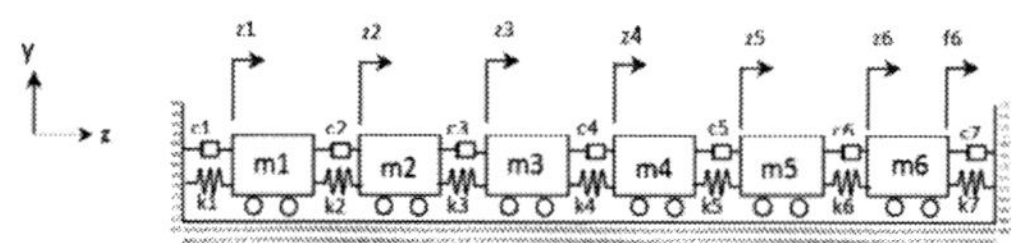


**Figure 1.**The schematic of the simulation system.

Figure 2 schematically shows the simulation model used in this case study. The proposed numerical model includes six mass elements with seven massless components, which are excited in z direction at node 6. The dynamic simulation example was defined by mass $C\,(M)$, stiffness $(K)$ and damping $(C)$ matrices. The mass matrix $M$ is an identity matrix of order six. The stiffness of each mass to the adjacent elements equals $Ak = 2000N/m$ and the Rayleigh damping was given by a combination of damping matrix proportion to stiffness and mass matrices as: $A$, equal to:

$$C = \begin{bmatrix} 1.377 & -0.3486 & 0 & 0 & 0 & 0 \\ -0.3486 & 1.377 & -0.3486 & 0 & 0 & 0 \\ 0 & -0.3486 & 1.377 & -0.3486 & 0 & 0 \\ 0 & 0 & -0.3486 & 1.377 & -0.3486 & 0 \\ 0 & 0 & 0 & -0.3486 & 1.377 & -0.3486 \\ 0 & 0 & 0 & 0 & -0.3486 & 1.377 \end{bmatrix}$$

An impulsive load, with duration of 0.01 second and excitation force of 1 $kN$ is applied to the MSD structure in node 3. The corresponding acceleration time-history is calculated with sampling frequency of 500 Hz and 1000 Hz. The imposed noise uncertainity in a real structure is simulated by adding random values into the extracted response signal.

### *3.1. Fit analysis*

Fit-value is a similarity measure between two signals, which is described with Variance Accounted For (VAF). VAF is an evaluation metric associated to the fit quality of the estimation. Higher VAF values suggest a better fit of the estimation result into a particular signal and thus better identification performance[31]. The VAF value is used in this study to appraise the performance of the CVA, PC and UPC algorithms. VAF criterion is defined as:

$$VAF = \left(1 - \frac{Variance(y - y_1)}{Variance(y)}\right) \times 100 \quad (14)$$

Where ($y$) is response signal of the simulation model and ( $y_1$ ) is the predicted values of the dynamic system.

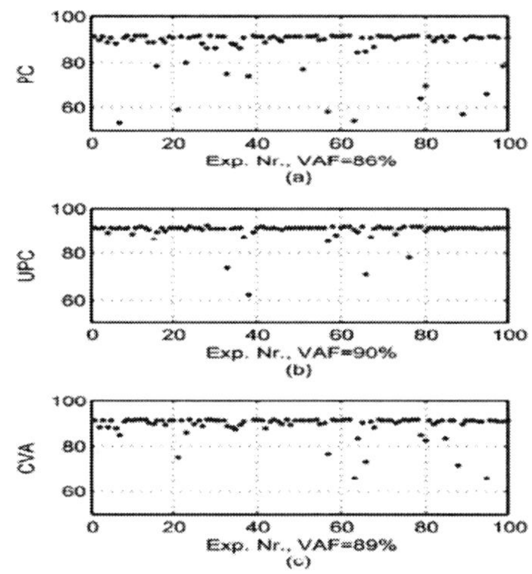


**Figure 2.**Distribution of the VAF values for the subspace algorithms using (a) PC algorithm (b) UPC algorithm and (c) CVA.

Acceleration response of the simulation FE model with 100 sets of noise patterns was used for evaluation of the SSI weight algorithms. The noise ratio of 30% is used in fit analysis experiment. Figure 6 displays the fit analysis for three SSI algorithms. In Figure 6 (a,b and c) the oscillation pattern of PC, UPC and CVA algorithm are plotted. UPC algorithm has the best prediction capability among all, and PC is only slightly worse, while the CVA performs the worst in obtained results.

### *3.2. Analysis of the system poles*

Poles are dynamic parameters of a system and are depend on distribution of mass, stiffness and damping within a system. Complex poles are a common phenomenon in modal identification of damped structures however, there is not any unified procedure to quantify the poles complexity [32- 33]. Poles can be plotted in a complex plane of real and imaginary components. In an undamped structure, the poles lie on the imaginary axis whereas the real portions of the complex values are zero.

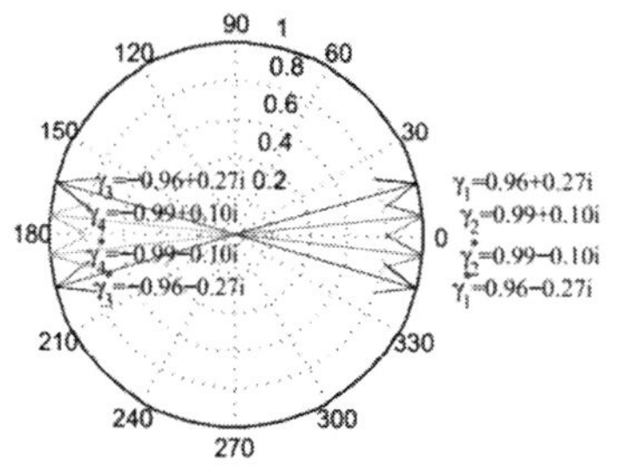


**Figure 3.**Complex poles nomenclature of the simulation numerical system in complex plane.

Figure 7 illustrates the extracted poles of the four first orders in the noise-free simulation model. The poles correspond to each of the four orders are complex conjugate of each other. Hence, $\lambda_1, \lambda_2, \lambda_3$ $\lambda_1^*, \lambda_2^*, \lambda_3^*$ and $\lambda_4^*$ are the first order of the numerical model and $\lambda_1^*, \lambda_2^*, \lambda_3^*$ and, $\lambda_4^*$ $\lambda_1^*, \lambda_2^*, \lambda_3^*$ are the conjugates. The poles are plotted on real and imaginary axes of a complex plane. In a separate numerical experiment, various noise ratios ranged from zero to 30% was used to determine the optimum ratio, which provides the best resolution for assessing the inherent oscillatory patterns of system poles.

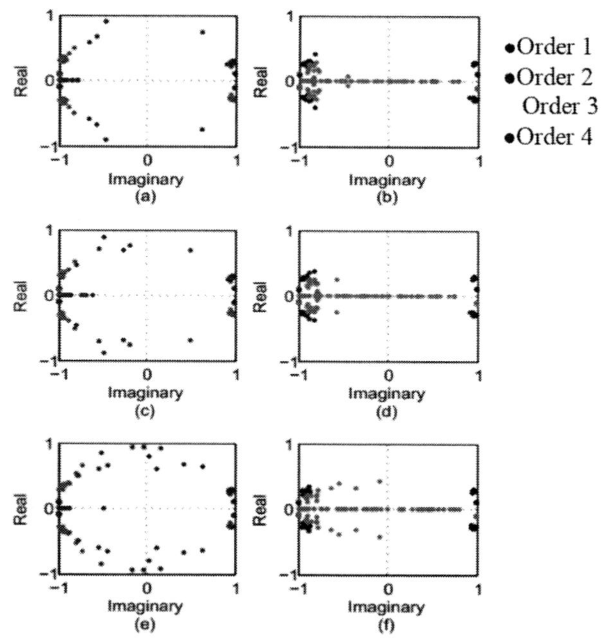


**Figure 4.** Plot of the estimated poles of the simulation case study obtained from: (a) PC algorithm for order 1 and 2, (b) PC algorithm for order 3 and 4, (c) UPC algorithm for order 1 and 2, (d) UPC algorithm for order 3 and 4, (e), CVA for order 1 and 2, and (f) CVA for order 3 and 4.

The 5% noise ratio was found to be the optimum ratio, which shows a clear insight into the transition pattern of the system's complex poles from the concentrated into the oscillated phase. The oscillation pattern of the numerical model under various noise distributions is used to evaluate the noise-robustness of the identification algorithms [34]. The results obtained from different SSI algorithms of the simulation model are illustrated in complex plane considering the four order system. 100 sets of output signal with different noise patterns are used as the input of the SSI algorithms. Sampling frequency of 500Hz for 2048 data point was used for poles analysis. Figure 8 shows the comparisons of the oscillation patterns for complex poles correspond to PC, UPC and CVA algorithms. As it could be seen the poles of a dynamic system lie inside a circle centered at zero with radius of one. The obtained values are in the form of conjugate pair symmetric about the real axes. The poles values are heavily distributed near the real axes than imaginary one. For the purpose of brevity and informativeness the zeros are not plotted in the figures. In Figure 8(a) and (b), the oscillation pattern of the four-order dynamic parameters of the numerical system obtained from the PC algorithm is plotted. The extracted results represent a relatively lower oscillation of the pole values. The plots show that the poles in the $4^{th}$ order are heavily distributed about the real axes. The complex pole values of the Figure 7 still could be traced through the distribution density. Figure 8 (c) and (d) illustrates the oscillation pattern of the poles associated with UPC algorithm for the conducted experiments. The scattering increases to approach the imaginary axis. Figure 8 (e) and (f) depicted the oscillation pattern for the CVA algorithm. The CVA algorithm provides high scattering compared to other SSI algorithms. The PC algorithm has the best prediction capability among all, and UPC is only slightly worse, while the CVA performs the worst.

### *3.3. Analysis of the elapsed time*

In a continuous monitoring process, a large repetition cycle is carried out to obtain the system parameters. The large number of repetition time executed in a highly populated dataset may have a prohibitive computational cost in structural systems. Accordingly, it is important to reduce the elapsed analysis time by striking a reasonable balance within the required accuracy level. Therefore the best SSI algorithm is the one that achieves higher accuracy in an optimal time.

In this subsection, the elapsed time for the repetitive running of the subspace algorithms of PC, UPC and CVA is analyzed to verify the effectiveness of each algorithm. The comparison scheme of the time-efficiency was implemented in a desktop computer of a *msi* Dual Core CPU with two 3GHz cores and 2 GB RAM running windows 7 having Matlab 2014a installed. 100 different noise patterns was extracted and used as input data for the three subspace algorithms. The elapsed computation time for each algorithm is measured and recorded in a separate data-base.

**Table 2.**Elapsed computation time for subspace algorithms.

| Algorithm | Computation time (s) |
|---|---|
| PC | 10.8 |
| UPC | 5.2 |
| CVA | 11.3 |

In Table 2, the elapsed computation time for SSI algorithms is presented. The mean values of the computation time for PC, UPC and CVA algorithms are 10.8, 5.2 and 11.3, respectively. The UPC algorithm has the best performance in these three techniques with nearly two times better in term of computation efficiency. CVA algorithm has the worst performance while PC algorithm has resulted slightly lower time for the same set of estimation data.

## 4. Remarks

This study present comparison of three different subspace implementation of PC, UPC and CVA. Numerical simulation was used for evaluation process. The subspace algorithms were evaluated based on the fit values, poles variances and the elapsed time. Table 3 outlines brief remarks of this study. The best VAF value among three algorithms was observed in UPC algorithm whereas the lowest performance was for PC algorithm. The highest efficiency was in UPC algorithm while the recorded time for the PC and CVA algorithms were nearly two times of UPC's. The least oscillation of poles for the first four orders were achieved in PC algorithm for both numerical and field test data and the smallest variances was mapped in CVA.

**Table 3.** Evaluation of the performance in PC, UPC and CVA subspace algorithms.

| | | PC | UPC | CVA |
|---|---|---|---|---|
| Case study I | VAF analysis | +* | +++ | ++ |
| | Pole analysis | +++ | ++ | + |
| | Elapsed time analysis | ++ | +++ | + |

* The notion '+', '++' and '+++' represent the 'poor', 'medium' and 'strong' identification results in the incorporated analysis technique, respectively.

## 5. Summary

The PC and CVA algorithms are designed to reduce the dimensionality of the identification result using orthogonal transformation and linear regression, respectively. Since the UPC algorithm doesn't use any weighting matrix (unit weight is used), the computation time is significantly low compared to the other counterparts. As it can be seen by the computation time analysis, the process of reducing dimensionality has not enhanced the time efficiency of the PC and CVA algorithms but yield an improved modal identification in PC. The principal components obtained using PC algorithms are more robust against noise uncertainty and give more stable results for the pole distribution. Higher estimation accuracy is achieved using UPC algorithm however it is not as good as PC for discrimination of the poles in the first four orders. CVA had the worst performance for pole analysis and time efficiency analysis.

## 6. References

[1] E P Carden, and J M Brownjohn 2008 Fuzzy Clustering of Stability Diagrams for Vibration-Based Structural Health Monitoring *Computer-Aided Civil and Infrastructure Engineering* **23**(5) pp 360-372

[2] Y Guo, D K Kwon and A Kareem 2015 Near-Real-Time Hybrid System Identification Framework for Civil Structures with Application to Burj Khalifa *Journal of Structural Engineering* **142**(2) 1-15

[3] H C Gomez 2012 *System Identification of Highway Bridges using Long-Term Vibration Monitoring Data*, PhDThesis (University of California, Irvine) pp. 1-350

[4] A Cunha, E Caetano, F. Magalhaes, C. Moutinho 2013 Recent perspectives in dynamic testing and monitoring of bridges *Structural control & health monitoring***20**(6) pp. 853-877

[5] C Loh2014Sensing solutions for assessing and monitoring dams *Sensor Technologies for Civil Infrastructures: Applications in Structural Health Monitoring* **2** pp 275

[6] R Tarinejad and M. Pourgholi 2016 Modal identification of arch dams using balanced stochastic subspace identification *J. of Vibration and Control* **333**(3) 1024-1045

[7] W Fan, and P Qiao 2011 Vibration-based damage identification methods: a review and comparative study *Structural Health Monitoring* **10**(1) 83-111

[8] C H Loh, W T Hsu, S F Chen,C K Chan 2015 Comparison on Identification and Damage Detection Methods Using Output-only Measurement: Application to Bridge Monitoring During Scouring Test*Structural Health Monitoring 2015: System Reliability for Verification and Implementation***12** pp. 1395-1402

[9] R D Nayeri, S. F. Masri, R G Ghanem,R L Nigbor 2008 A novel approach for the structural identification and monitoring of a full-scale 17-story building based on ambient vibration measurements," *Smart Materials and Structures* **17**(2) 025006

[10] H Zhou, Y Ni, and J Ko 2010 Constructing input to neural networks for modeling temperature-caused modal variability: mean temperatures, effective temperatures, and principal components of temperatures *Engineering structures* **32(**6) 1747-1759

[11] M Gevers 2006 A personal view of the development of system identification: A 30-year journey through an exciting field *IEEE Control Systems* **26**(6) 93-105

[12] P V Overschee, B Moor 2012 *Subspace Identification for the Linear Systems: Theory–Implementation* (Springer Science & Business Media) pp. 1-116
[13] B Peeters and G De Roeck 1999 Reference-based stochastic subspace identification for output-only modal analysis *Mechanical systems and signal processing* **13**(6) 855-878
[14] M Brehm, V Zabel and C Bucher 2013 Optimal reference sensor positions using output-only vibration test data *Mechanical Systems and Signal Processing* **41**(1–2) 196-225
[15] D Li, W-X. Ren, Y-D Huand D Yang 2016 Operational modal analysis of structures by stochastic subspace identification with a delay index *Structural Engineering and Mechanics* **59**(1) 187-207
[16] S Marchesiello, S Bedaoui, L Garibaldi andP Argoul 2009 Time-dependent identification of a bridge-like structure with crossing loads *Mechanical systems and signal processing* **23**(6) 2019-2028
[17] M Döhler, E Reynders, F Magalhaes and L Mevel 2011 Pre-and post-identification merging for multi-setup OMA with covariance-driven SSI *Dynamics of Bridges* Vol 5 (Germany : Springer) pp 57-70
[18] A Benveniste and L Mevel 2007 Nonstationary consistency of subspace methods *IEEE Transactions on Automatic Control* **52**(6) 974-984
[19] E Reynders, R Pintelon and G De Roeck 2008 Uncertainty bounds on modal parameters obtained from stochastic subspace identification *Mechanical systems and signal processing* **22**(4) 948-969
[20] B Alıcıoğlu, and H. Luş 2008 Ambient vibration analysis with subspace methods: case studies *J. of Structural Engineering* **134**(6) 1016-1029
[21] A Brasiliano, G Doz, J L V Brito and R Pimentel 2008 Role of non-metallic components on the dynamic behavior of composite footbridges *Proc. Of the 3rd Int. Conference Footbridge 2008*
[22] J Wang, A. Sano, T. Chen and B Huang 2009 Identification of Hammerstein systems without explicit parameterisation of non-linearity *Int. J. of control* **82**(5) 937-952
[23] A E Ashari and L Mevel 2013 Auxiliary input design for stochastic subspace-based structural damage detection *Mechanical systems and signal processing* **34**(1) 241-258
[24] D Foti, S Ivorra, D Bru and G Dimaggio 2012Dynamic Identification of a Pedestrian Bridge using Operational Modal Analysis*Proc. Of the eleventh Int. Conf. on Computational Structures Technology* paper 180
[25] C Cismaşiu, A Narciso and F Amarante dos Santos 2014 Experimental dynamic characterization and finite-element updating of a footbridge structure *J. of Performance of Constructed Facilities* **29**(4) 04014116
[26] T Nguyen 2014 *SHM through flexible vibration sensing technologies and robust safety evaluation paradigm* PhD Thesis (Queensland University of Technology) pp1-342
[27] F Pioldi, S Pansieri and E Rizzi 2016 On the processing of earthquake-induced structural response signals by suitable Operational Modal Analysis identification techniques *Proc. of the 27th Int. Conf. on Noise and Vibration engineering ISMA2016* pp 2873-2883
[28] A S Kompalka, S Reese and O T Bruhns 2007 Experimental investigation of damage evolution by subspace identification *Archive of Applied Mechanics* **77**(8) 559-573
[29] L F F Miguel, R H Lopez and L F F Miguel 2013 A hybrid approach for damage detection of structures *J. of Sound and Vibration* **332(18)** pp 4241-4260
[30] H Herlufsen, P. Andersen, S Gade and N Moller 2005 Identification techniques for operational modal analysis–an overview and practical experiences *Proc. Of the 1st Int. Operational Modal Analysis Conf.(IOMAC)* (Copenhagen, Denmark)
[31] B Peeters, S Kanev, M Verhaegen H Van Der Auwerarer2007 Vibration-Based Damage Assessment for Controller Reconfiguration: Application to an Oilpan *Key engineering Materials* **347** 645-650
[32] U Fuellekrug 2008 Computation of real normal modes from complex eigenvectors *Mechanical systems and signal processing* **22**(1) 57-65, 2008

[33] H Koruk, and K Y Sanliturk 2013 A novel definition for quantification of mode shape complexity *J. of Sound and Vibration* **332**(14) 3390-3403
[34] P Van Overschee, and B De Moor 1995 unifying theorem for three subspace system identification algorithms *Automatica* **31**(12) 1853-1864

**Acknowledgement**
The authors would like to thank the Ministry of Higher Education, Malaysia, and Universiti Teknologi Malaysia (UTM) for their financial support through the Fundamental Research Grant Scheme (4F308 and 4F800) and Research University Grant Scheme (05H21).

IOP Conf. Series: Materials Science and Engineering **271** (2017) 012093 doi:10.1088/1757-899X/271/1/012093

# Finite element modelling of concrete beams reinforced with hybrid fiber reinforced bars

**Santa binti Sinring**[1,a*], **Norhafizah Salleh**[1,b], **NoorAzlina Abdul Hamid**[1,c], **Masni A.Majid**[1,e]

[1]Department of Structural and Material Engineering, Faculty of Civil Engineering and Environmental, Universiti Tun Hussein Onn Malaysia
Corresponding author: santasinring@gmail.com

**Abstract.** Concrete is a heterogeneous composite material made up of cement, sand, coarse aggregate and water mixed in a desired proportion to obtain the required strength. Plain concrete does not with stand tension as compared to compression. In order to compensate this drawback steel reinforcement are provided in concrete. Now a day, for improving the properties of concrete and also to take up tension combination of steel and glass fibre-reinforced polymer (GFRP) bars promises favourable strength, serviceability, and durability. To verify its promise and support design concrete structures with hybrid type of reinforcement, this study have investigated the load-deflection behaviour of concrete beams reinforced with hybrid GFRP and steel bars by using ATENA software. Fourteen beams, including six control beams reinforced with only steel or only GFRP bars, were analysed. The ratio and the ordinate of GFRP to steel were the main parameters investigated. The behaviour of these beams was investigated via the load-deflection characteristics, cracking behaviour and mode of failure. Hybrid GFRP-Steel reinforced concrete beam showed the improvement in both ultimate capacity and deflection concomitant to the steel reinforced concrete beam. On the other hand, finite element (FE) modelling which is ATENA were validated with previous experiment and promising the good result to be used for further analyses and development in the field of present study.

## 1. Introduction

Fibre reinforced polymer (FRP) composites were first developed during the 1940's for the military and aerospace applications [1]. The characteristics of FRP include high resistance to corrosion, lightweight, high strength-to-weight ratio, high tensile and impact strength, fatigue resistance, non-conductive and magnetically neutral [2]. As a result, FRP have been successfully used in many construction applications such as chemical and waste-water treatment plant, underwater structures, bridges, substation reactor bases, airport runways, laboratories and water tanks [3].

The other application is to strengthen the structurally deficient beams or columns with FRP sheets or plates [4]. Today, applications of FRP in the building industry have realized the increasing popularity in Malaysia. As proven, many construction projects are using FRP water tank to replace the conventional water tank that facing steel corrosion problems. In fact, cost of manufacturing these products is quite high compared to the conventional materials.

Hybrid composite beam (HCB) is a structural member, utilizing several different building materials with dissimilar attributes. Furthermore, the hybrid composite beam is a sustainable technology that combines the strength and stiffness of conventional concrete and steel. In this report, we investigated the flexural performance of hybrid GFRP-steel reinforced concrete beams by using finite element modelling, ATENA. Our analysis model is to predict the load-deflection relationship of the beams. Design models for predicting flexural strength, deflection, and crack behaviour are presented. We analysed beams with different reinforcement ratios and with different ordinate of GFRP and steel, and compared to the control beams including only GFRP and steel reinforcement bar respectively.

## 2. Literature Review

Conventional reinforced concrete is widely applied in the construction industry owing to the availability and low cost of steel and concrete, the knowledge regarding design, and the vast experience of its use in practise. Due to their different mechanical properties, the behaviour of FRP RC members is quite different to that of traditional steel reinforced concrete [5].

Recently, there has been a rapid increase in the use of Fibre Reinforced Polymer (FRP) bars substituting for conventional steel bars for concrete structures, due to the advantages of noncorrosive characteristic, high strength, and light weight of FRP bars [6]. According to [7] the elastic modulus of FRP bars is much less than that of steel bars. This low elastic modulus leads to higher deflection and larger crack width in FRP bar- reinforced concrete beam that have an equivalent reinforcement ratio to steel-reinforced concrete beams. In addition, while steel bars behave in elastically after yield strength, FRP bars show perfect elastic behaviour up to failure, and fail in brittle manner. In order to overcome this problems in term of deformability and ductility of concrete beams reinforced with FRP bars, alternative solutions of hybrid reinforcing with FRP and steel bars, and using fibre reinforced concrete (FRC) were proposed by [8,9].

Based on [7,10] again, they conclude that the lower stiffness and higher deflection of FRP bar-reinforced beams were controlled and improved by hybrid reinforcing with steel bars. Because of the increased ultimate concrete compressive strain, the GFRP bar reinforced beams with steel and synthetic fibres displayed inelastic and ductile behaviour near the failure, and higher ultimate flexural strength than the beam with no fibres. The steel fibres in concrete and hybrid reinforcing with steel controlled the propagation of cracks and the crack width of FRP-bar reinforced beams. And last but not least, addition of fibres and hybrid reinforcing with steel bars can be possible methods to overcome the low ductility of FRP bar-reinforced beams.

Furthermore, the ductility of RC structure and brittle failure of FRP structure can be resolve by laying the FRP tendons in the corner area where concrete easy corrupted, and it still need to keep on loading to up to the ultimate load after the steel bars yield and shows certain safety factor and good ductility. After all, it is goodish reinforced format state by [11].

Paper studied by [10,12] was investigated accuracy of deflection prediction made by the finite element package ATENA and the design code methods ACI and EC2. Finally G. Kaklauskas stated that deflection was calculated of the beam within load interval ranging from 30 to 90% of the theoretical ultimate load by using the ATENA FE.

## 3. ATENA Modelling

### *3.1. Validation of ATENA*

Three sets of data RC in direct tension tests have been selected from the literature. The results gathered from the previous experimental stated at methodology which satisfactory demonstrates the accuracy of the FE modelling. Figure 1, 2 and 3 show the trend of the validation of finite element modelling with experimental data.

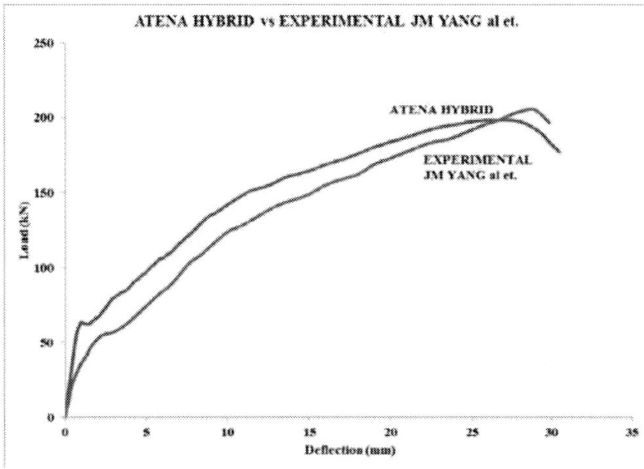


**Figure 1**. Validation of hybrid beam

**Figure 2.** Validation of control steel beam (C-3S)

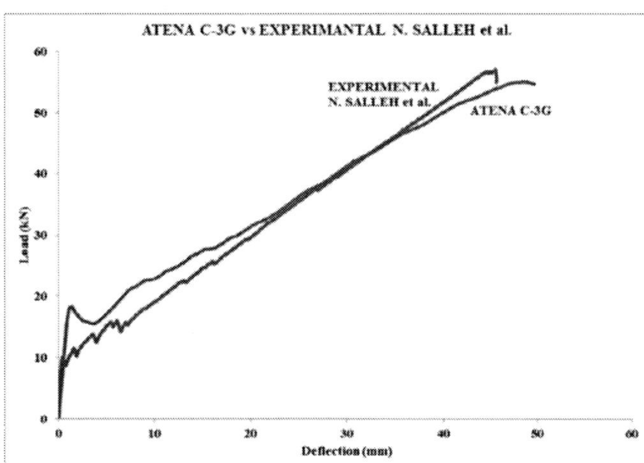


**Figure 3**. Validation of control GFRP beam (C-3G)

### *3.2. Details of specimen*

Fourteen high-performance concrete beam specimens reinforced with different types of flexural reinforcement and fibre including six control beams were constructed and analysed. Figure 1 shows the details of fourteen beams specimens. All specimens were 2300 mm long with a rectangular cross section of 200 x 250 mm. Group A were reinforced with a layer of reinforcement and Group B and C were reinforced with two layers of reinforcement, and the effective depths of the outer layer (d1) and the inner layer (d2) were 213.5 mm and 200.5 mm, respectively. Each specimen had different combinations of flexural reinforcement of conventional steel bars and GFRP bars.

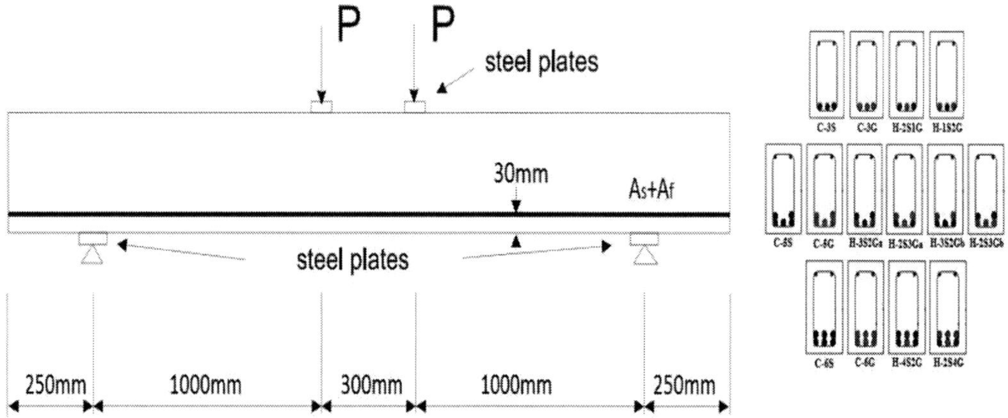


**Figure 4.** Details of specimens and analyse setup

The first letter in the specimen names indicate the control beam and hybrid beam respectively. All reinforcement used the same diameter which is 12 mm for steel bar and 13 mm for GFRP bar. A 30 mm concrete cover was used, and 8 mm steel bars were used as closed stirrups at 100 mm spacing and as longitudinal compression reinforcements for all specimens. All specimens were designed to fail by concrete crushing to avoid brittle failure. The mechanical properties of concrete, steel and GFRP reinforcement are shown in table 1 below.

**Table 1.** Mechanical properties of concrete, steel and GFRP reinforcement

| Material | Elastic Modulus (GPa) | Compressive Strength (MPa) | Yield Strength (MPa) | Tensile Strength (MPa) |
|---|---|---|---|---|
| High Performance Concrete | 36.95 | 61.80 | - | 6.18 |
| Steel bars (12 mm) | 200 | - | 550 | - |
| Steel bars (8 mm) | 200 | - | 250 | - |
| GFRP bars (13 mm) | 44.1 | - | - | 920 |

### *3.3. ATENA Setup*

Modelling by ATENA was completed by two stages including pre-processing and post processing. In pre-processing stage, there are geometrical model, material selection and boundary condition. The model for finite element analysis will be created during the pre-processing with help of the fully automated mesh generator.

First step before start geometrical modelling is to set the material properties used including High Performance Concrete (HPC), reinforcement bars and steel plate that use for support and reaction. For HPC, SBeta material selected that made a perfect bond between concrete and reinforcement is assumed within the smeared concept. Reinforcement bar will include the steel, stirrup and GFRP respectively. For steel plate, dimension 100 mm x 200 mm x 30 mm will be used and the material use is Plane Elastic Isotropic by use Hooke's Law. All properties of material are entered by information given above.

Next step is to form the geometrical model including the geometrical joint that connected to the boundary lines. Then the model will mesh with 0.005 m size of element. Boundary condition that used in this analysis is force and prescribed deformation which located at the support and reaction respectively. Three monitoring point also included in the modelling as indicator to monitor the analysis while running by the Newton-Raphson Method chosen. The overall review of modelling is shown at figure 5 below.

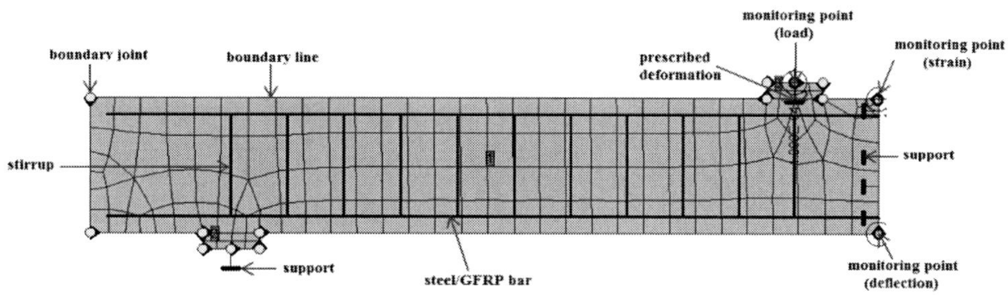


**Figure 5.** Overview the model setup in ATENA

## 4. Results and Discussion

*4.1. Load-Deflection behaviour of specimens*

Figure 6 (a) to (c) shows the load versus mid-span deflection analysis, while table 2 summarizes the loads and mid-span deflection of the theoretical and ATENA value as well as the failure mode for all beam specimens. Initially all beams show relatively linear elastic behaviour up to the cracking load when the concrete cracked at the tension face. Due to the low elastic modulus of GFRP bars, the stiffness of the beam was reduced faster with larger deflection.

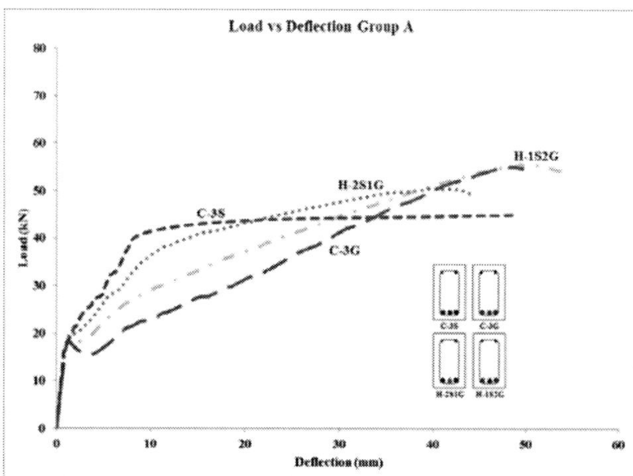


**(a)**

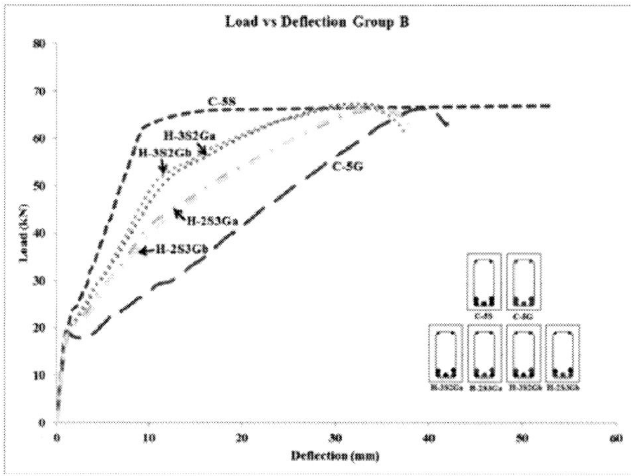


**(b)**

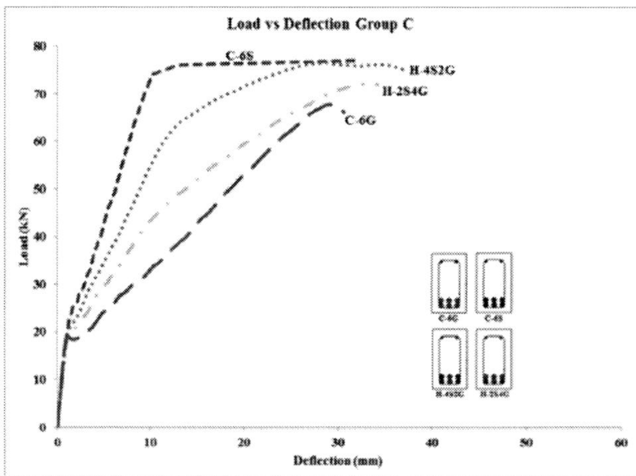


**(c)**

**Figure 6(a) – (c ).** Load versus deflection analysis

**Table 2.** Summarize of analysis results

| Group of Specimen | | $P_{theo}$ (kN) | $\delta_{theo}$ (mm) | $P_{ATE}$ (kN) | $\delta_{ATE}$ (mm) | Δ P (%) (kN) | Δ δ (%) (mm) | ρ (%) |
|---|---|---|---|---|---|---|---|---|
| A | C-3S | 32.77 | 22.62 | 45.00 | 48.76 | 0.73 | 0.46 | 0.79 |
| | C-3G | 60.78 | 41.95 | 55.14 | 40.51 | 1.10 | 1.04 | 0.93 |
| | H-2S1G | 42.49 | 29.33 | 50.47 | 33.36 | 0.84 | 0.88 | 0.84 |
| | H-1S2G | 51.83 | 25.77 | 49.70 | 29.74 | 1.04 | 0.87 | 0.89 |
| B | C-5S | 50.77 | 35.04 | 67.00 | 42.10 | 0.76 | 0.83 | 1.37 |
| | C-5G | 89.81 | 61.99 | 65.25 | 34.02 | 1.38 | 1.82 | 1.60 |
| | H-3S2Ga | 67.55 | 46.62 | 67.38 | 32.23 | 1.00 | 1.45 | 1.46 |
| | H-2S3Ga | 75.35 | 52.01 | 65.96 | 35.21 | 1.14 | 1.48 | 1.51 |
| | H-3S2Gb | 67.55 | 46.62 | 66.01 | 34.33 | 1.02 | 1.36 | 1.46 |

| | | | | | | | | |
|---|---|---|---|---|---|---|---|---|
| | H-2S3Gb | 75.35 | 52.01 | 65.94 | 36.10 | 1.14 | 1.44 | 1.51 |
| C | C-6S | 59.67 | 41.19 | 77.00 | 32.27 | 0.88 | 1.41 | 1.64 |
| | C-6G | 102.95 | 71.06 | 67.78 | 29.22 | 1.34 | 2.20 | 1.92 |
| | H-4S2G | 75.64 | 52.21 | 76.11 | 33.52 | 1.05 | 1.56 | 1.73 |
| | H-2S4G | 90.07 | 62.17 | 72.60 | 33.52 | 1.24 | 1.85 | 1.83 |

Although the effect of plastic strains is less significant for under-reinforced members, deflection are underestimated by the method at large loads [12]. Thus, the accuracies of the theoretical analysis and the FEA model for beams with low or high effective reinforcement ratios need further verification according to the [13] that have the same results.
After first flexural cracking, the stiffness of the cracked was reduced, and each specimen started to behave differently depending on the combination of the flexural reinforcement. In term of the control beam, we can see they react differently when the increasing of the GFRP bars. In the time the GFRP bars increasing to five and six bars the strength of the beam same and reduce with the strength of the steel bars respectively.
Hybrid beams with different reinforcement ratio and ordinate of reinforcement show a high potential in the comparison in the group A instead of the group B and C. Meanwhile, the GFRP bars react as good bond with concrete when it is set at the outer of the tension [14]. Furthermore, the bonding of the GFRP bars getting low when the quantity is increase. The load carried by beam reinforced with more GFRP bars are getting low than the beam reinforced with more steel bars. This conclude that the beam reinforced with GFRP bars already achieve the maximum strength according to the increasing number of the bar in group B. Hybrid beams showed a significant improvement in terms of beams deformability under service load conditions, since the beams were over-reinforced, the contribution of steel bars to increase the flexural capacity was less than 15% [14].

### *4.2. Cracking behaviour*
Sections
The depth of cracks in beam reinforced with steel bar was very short, while the cracks in beams reinforced with GFRP bar propagated quite deeply into the compression zone. This indicates that immediately after the formation of the first cracks the neutral axis shifted up very near the top compression of the GFRP bar-reinforced beams. Cracks in hybrid reinforced specimens with steel bars and in GFRP reinforced specimens were shallower than those in beam reinforced with GFRP bar only. This indicates that the deep propagation of cracks in the GFRP bar-reinforced beams can be restrained by hybrid with steel bars. Figure 7 shows the pattern of cracking for hybrid beam that perform well than other beams.

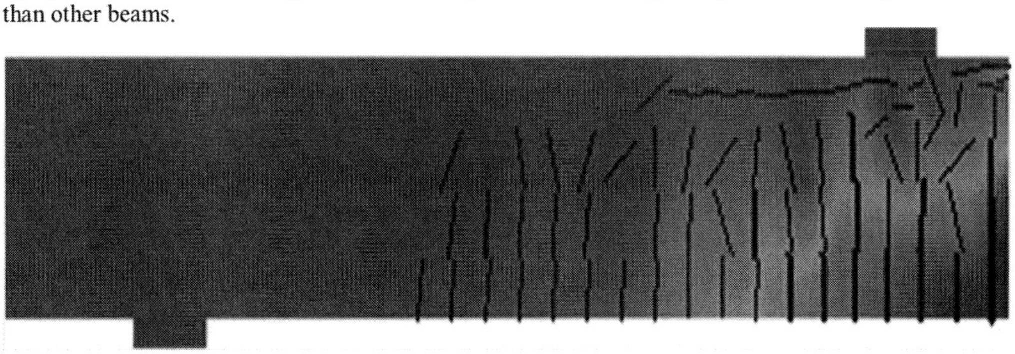

**Figure 7**. Crack Pattern of Hybrid Beam

Crack behaviour is summarizes in the table 3. As expected, the beam contains GFRP bar will crack before the beam contain steel bar because of the lowest modulus of elasticity of the GFRP reinforcement bar.

**Table 3.** Cracking Behaviour and Mode of Failure

| Group of Specimen | | $P_{cr}$ (kN) | Failure Mode |
|---|---|---|---|
| A | C-3S | 17.00 | Flexural |
| | C-3G | 15.90 | Flexural |
| | H-2S1G | 16.37 | Flexural |
| | H-1S2G | 16.14 | Flexural |
| B | C-5S | 18.00 | Flexural |
| | C-5G | 16.03 | Flexural |
| | H-3S2Ga | 16.62 | Flexural |
| | H-2S3Ga | 16.50 | Flexural |
| | H-3S2Gb | 16.74 | Flexural |
| | H-2S3Gb | 16.38 | Flexural |
| C | C-6S | 18.00 | Flexural |
| | C-6G | 16.09 | Flexural |
| | H-4S2G | 16.93 | Flexural |
| | H-2S4G | 16.59 | Flexural |

All beams failed in the flexural tension mode besides the major factor for the failure of beam is the concrete failure in the compression zone. Cracking pattern indicated that the steel reinforced concrete beam was more ductile compared to the GFRP reinforced concrete beam.

## 5. Conclusion

i. Hybrid GFRP-steel reinforced concrete beam had achieved ultimate load at the same time higher compared to the conventional concrete beam.
ii. The lower stiffness and higher deflection of FRP bar-reinforced beams were controlled and improved by hybrid reinforcing with steel bars.
iii. ATENA 2D was employed to simulate the structural behaviour of the hybrid reinforced GFRP-steel beams.
iv. Hybrid GFRP-steel reinforced concrete beams had good bond performance when installing the GFRP bars as near as possible to the outer surface of the concrete element, which is beneficial in term of structural performance of the reinforcing bar.
v. The steel bars increased the ductility of hybrid GFRP-steel reinforced concrete beams.

## 6. References

[1] R. Weatherhead 1980 *FRP Technology: Fibre Reinforced Resin Systems* Applied Science Publishers
[2] M. Thériault and B. Benmokrane 1998 Effects of FRP Reinforcement Ratio and Concrete Strength on Flexural Behavior of Concrete Beams *Journal of Composites for Construction* **2(1)** 7–16
[3] V L B and C L Bartholomew 1993 FRP Reinforcing Bars in Reinforced Concrete Members *Mater. J.* **90(1)**
[4] Nanni A 2003 North American design guidelines for concrete reinforcement and strengthening using FRP: principles, applications and unresolved issues *Constr. Build. Mater* **17** 439–446.
[5] Barris C, Torres L, Turon A, Baena M, Catalan A 2009An experimental study of the flexural behaviour of GFRP RC beams and comparison with prediction models *Compos. Struct.* **91(3)** 286–295
[6] Nanni A 2005 Guide for the Design and Construction of Concrete Reinforced with FRP Bars (ACI 440.1R-03) *Struct. Congr.* **88** 1–6
[7] Yang J M, Min K H, Shin H O, and Yoon Y S 2011 Behavior of high-strength concrete beams reinforced with different types of flexural reinforcement and fiber **L** 27–30

IOP Conf. Series: Materials Science and Engineering **271** (2017) 012093 doi:10.1088/1757-899X/271/1/012093

[8] J.-M. Yang, K.-H. Min, H.-O. Shin, and Y.-S. Yoon 2012 Effect of steel and synthetic fibers on flexural behavior of high-strength concrete beams reinforced with FRP bars *Compos. Part B Eng.* **43(3)** 1077–1086

[9] Salleh N, Mohd S A R, Mohd Y J, and Osman M F 2014 Flexural behaviour of reinforced concrete beam with glass fiber reinforced polymer (GFRP) bar strengthened with carbon fiber reinforced polymer (CFRP) plate *Adv. Mater. Res.* **1051** 748–751

[10] Salleh N, Azlina N, Hamid A, Rahman A, Sam M, Yatim J M, Thamrin R, Hafizah N, Khalid A, Majid M A and Jamellodin Z 2017 Finite Element Simulation of GFRP Reinforced Concrete Beam Externally Strengthened With CFRP Plates *MATEC Web of Conf* **103**

[11] Ge W, Zhang J, Dai H and Tu Y *2010* Experimental Study on the Flexural Behavior of Concrete Beam Hybrid Reinforced with FRP Bars and Steel Bars *Advances in FRP Composites in Civil Engineering: Proceedings of the 5th International Conference on FRP Composites in Civil Engineering (CICE 2010)* 301–303

[12] Kaklauskas G, Cervenka V and Cervenka J 2004 Deflection Calculation of RC Beams : Finite Element Software Versus Design Code Methods in *International Conference on Computing in Civil and Building Engineering - ICCCBE* 1–8

[13] Du J, Wang C, Qiao M, Chang X and Chen H 2010 Flexural behavior of concrete beams reinforced by CFRP bars *Mech. Autom. Control Eng. (MACE), 2010 Int. Conf.* **13(5)** 350–359.

[14] Aiello M A and Ombres L 2002 Structural Performances of Concrete Beams with Hybrid Fiber-Reinforced Polymer-Steel Reinforcements **6** 133–140

IOP Conf. Series: Materials Science and Engineering **271** (2017) 012094 doi:10.1088/1757-899X/271/1/012094

# Literature review of the benefits and obstacle of horizontal directional drilling

**Mohd Norizam M S[1], Nuzul Azam H[2], Helmi Zulhaidi S[3], Abdul Aziz A[4] and Nadzrol Fadzilah A[5]**

[1,2,3] Department of Civil Engineering, Faculty of Engineering, Universiti Putra Malaysia, 43400 UPM Serdang, Selangor Darul Ehsan, Malaysia

[4] Centre For Continuing Education (CCE), Universi Sultan Zainal Abidin, Gong Badak, 21300 - Kuala Terengganu, Terengganu, Malaysia

[5] Malaysian Highway Authority, Jalan Serdang - Kajang, Kawasan Institusi Bangi, 43000 Bandar Baru Bangi, Selangor, Malaysia

Corresponding author: norizam32@yahoo.com.my

**Abstract.** In this new era the construction industry not only need to be completed within budget, timely, at acceptable quality and safety but the stakeholders especially the local authorities and the public realises for the important need of sustainable construction method to be used for our younger generation to heritage if not better a safer world for them to live and raise up their children's. Horizontal Directional Drilling method is the most commonly recognised trenchless utilities method as a preferred construction method in this age. Among the reasons HDD method offers less disturbance on traffic, the public, business activities and neighbourhood, lower restoration cost, less noise, dust and minimum import/export of the construction materials. In addition HDD method can drill through congested utilities areas with minimum cutting and shorter time. This paper aims to appraise the benefits and obstacle of HDD method in construction industry. It is an endeavour to fulfil the local authorities cry for alternative method that less damages to the roads, road furniture's and public complaints compared to the conventional open cut method. In addition HDD method is seem to be in line with sustainable development requirements e.g. reduce, reuse, recycle and etc. Hence, it is important to determine the benefits and obstacle factors of HDD implementation. The factors are based on the literature review conducted by the author on the subject matters gathered from previous studies, journals, text books, guidelines, magazine articles, newspaper cutting and etc.

## 1. Introduction

"Horizontal Directional Drilling" which its acronym is HDD method is not a new born utilities' installation method but was introduced since early 1970s [1]. Horizontally directional drilling was developed in California in the 1970s and was introduced in Europe in 1986 [2]. Although the first HDD river crossing across Pajaro river near Watsonville, California was carried out in 19971 only 36 successful crossings were done up to year 1979 [3]. For the first two decades the technology which was formally very difficult to operate/manage had evolved significantly to what we have today which is user friendly, well equip with hydraulics system, power source, drill frame, drilling fluids and detection guidance systems [4].

IOP Conf. Series: Materials Science and Engineering **271** (2017) 012094 doi:10.1088/1757-899X/271/1/012094

HDD method had gone through three (3) phases, in the early days it was used only in the oil and gas industry in the installation/laying of oil and gas pipelines but mostly in major crossing e.g. crossing rivers, railways and roads. Later this method was also adopted and had been very successfully accepted by the construction industries to lay/install utilities e.g. water pipes, telecommunication ducts, power cables, gas pipes, sewer pipes and etc. But its usage were limited for doing the crossing as this construction method cost is still higher compared to conventional open cut method [5]. Lastly HDD method was widely used in congested utilities area as the room for using other construction/installation method is almost impossible [6].

In the past three decades the demand in this techniques has increased tremendously and mostly was contributed by the environmentally awareness among the local authorities and the public. Among the reasons HDD method offers less disturbance on traffic, to the business activities and public, lower restoration cost in congested urban areas, less noise, dust and minimum import/export of the construction materials [7].

With the capacity to install utilities for varies sizes from 50m to 900mm in diameter and up to 1,700m in length allow HDD method to offer a clear alternative to the conventional methods with minimum surface excavation [6]. Since year 1992 up to 2016 there are about 40 thousand unit HDD machine were sold worldwide [8-10].

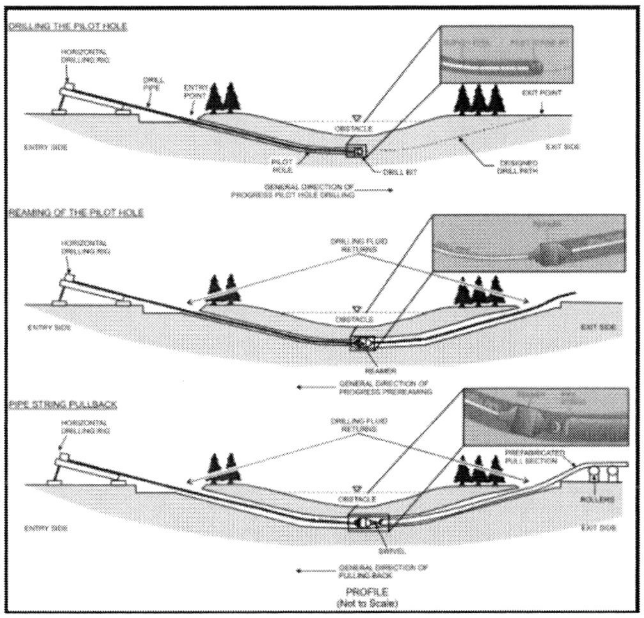


**Figure 1.** HDD process [11]

As illustrated in figure 1, HDD construction process starting with piloting a small pilot hole using a pilot head along the desired centre line of the proposed profile, followed with back reaming until the required drilled trench size to house the utility pipe, preparation of the product pipes and finally the pulling of pipes into the bored hole. Its major components are the drilling rig, drill pipes, slurry/bentonite, slurry recycling, survey equipment, drill bits, reamers and pipeline materials [11]. A transducer sonde is attached at behind of the cutter head during piloting and the signal is collecting by a locator carry by a person (tracker) that walk on top of the HDD alignment. The HDD operator will communicate with the tracker in order to control the pilot head drilling path.

## 2. Methodology

The factors determined are based on the literature review compilation from international and national journal papers, conference papers, proceedings, web pages, government's reports and bulletins. The literature review references cover the time frame from 1998 to 2016. The review comprises the history of HDD, all types of HDD machine, its benefits, obstacles and authorities requirements.

The author used both the exploratory and refines review. At the initial exploratory review the author used the terminology or keyword searches for the review. Once sufficient literatures on topic matter are obtained the author progressed with refine review. At this stage the review involves more focused observations, analysis, specific paper references and more importantly the bibliography.

## 3. HDD in Malaysia

Back here in Malaysia, it was believed the Horizontal Directional Drilling (HDD) method was introduced in late 80s for some critical crossing in the construction/installation of Peninsular Gas Utilization Projects/PGU II. This project was successfully completed in 1991[12-13]. On the other hand from the information gathered from the market players, the HDD works in telecommunication and power industries were started back in 1993. Mersing Construction Sdn Bhd, one of the trenchless leader in Malaysia founded in 1980s, started their HDD division in 1995 [14].

Since then and as almost happen almost everywhere in the world, HDD method becoming one of the prominent trenchless methods used to lay/install in Malaysia not only gas pipes but also for other utilities e.g. water pipes, telecommunication cables, power cables and etc.

Some local authorities in Malaysia e.g. Public of Works, DBKL (Kuala Lumpur City Hall), Malaysia Highway Authorities, highway concessionaires, railway authorities and others town council had restricted only trenchless methods can be used for road crossing especially for protocol road.

The usage of HDD in Malaysia is normally limited to obstacle crossing (e.g. highway, water ways, railways and etc.), relocation works due to road expansion or installation of new infrastructures (e.g. MRT, LRT, bridges and etc.), crossing protocol road and in congested urban areas. Among the reason the installation using HDD method is still more expensive compared to the use of conventional open cut method.

Every year many of local authorities/town councils cries for the poor repair works done by the utilities contractors due to poor reinstatement works that were carried out after the utilities installation works were carried out via conventional open cut method. Due to this and in order to reduce complaint from the road users and the public some of the local authorities had imposing for the repair works to be carried out by their panel contractors and some preferred for the utilities installation works to be carried out using HDD.

Ir Tan the former Director of DBKL's Civil Engineering and Drainage Department had stated that DBKL would favour on HDD method due to its minimum impact to the road and road furniture's damages but the setback is this method is 3-5 times more expensive from the open cut [5].

Figure 2, shows the photograph of a project site, plan view and the longitudinal plan where HDD works have been carried out to cross the Federal highway, railway track and river at Bt 3, Shah Alam

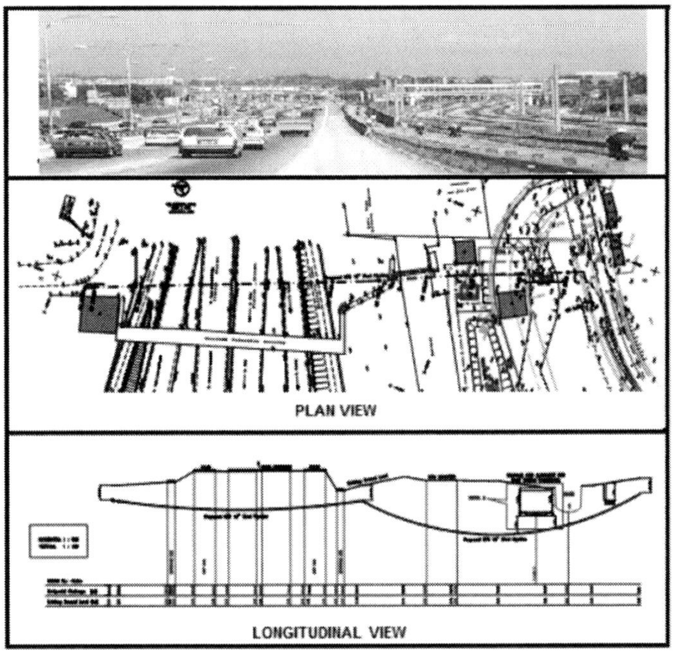


**Figure 2.** HDD crossing Federal highway, railway track and river at Bt 3, Shah Alam

*3.1. The Advantages of Using HDD Method*

In HDD method the open excavation works are minimal. The excavation are only required at the launching and receiving pits only. For example if the HDD profile is 200m distance the excavation works are only about 2m (width) x 4m (long) x 3m (depth) X 2 locations for its launching/receiving pits. In addition these pits can be located at the suitable location with the option of avoiding the road pavement. Even if it is required to be located on the road pavement the distance on the pavement is only about 8m out of the total distance. In contrarily in conventional open cut method the total length shall be required open excavation. The following are the HDD advantages;

1. No vertical shafts required as drilling launches from the surface, small entry and reception pits [15].
2. Steering ability to avoid existing utilities or other obstacles either horizontally, vertically and even pulled back [16]
3. Requires relatively little auxiliary equipment and short set up time [15].
4. HDD provides an efficient method for crossing obstacles such as rivers, highways, rail tracks, or airfield runways [17].
5. It minimizes the negative impact on residents and businesses and eliminates the need for the removal and repositioning of expensive restoration and landscaping [7].
6. Less environmentally-damaging compared to open cut method [18]
7. The HDD method eliminates the cost and time associated with installing dewatering facilities for operations carried out below the groundwater table level [7].
8. Able to be used in congested utilities area where other construction/installation method is almost impossible [19].
9. Minimal traffic disruption and associated social costs to the public [15].
10. Ability to accommodate large diameters [19] .
11. Ability to install pipes of different materials, including HDPE, PVC, steel and ductile iron [19].

12. Compatible with a variety of soil conditions, including loose sand and solid rock [19].
13. Satisfies environmental guidelines (especially in wet lands) [19].
14. The longest single line drive installation for trenchless method up to 3km distance [21].

### *3.2. Obstacle of HDD Works in Malaysia*

1. Poor drilling practices by some contractors may cause utility strikes that will resulted in major legal ramifications [21].
2. Lack of established good practices [22].
3. Insufficient experienced drillers [22].
4. Availability of training for HDD operators [22].
5. A through site and subsurface investigation is very important [16].
6. To establish through soil conditions [6].
7. Need to consider disposal dump area if there is no fluid recycling method. But U.S. Protection Agency (EPA) not considered bentonite and certain polymers as toxic materials [16].

## 4. Results and Discussion

Based on the literature review, there are fourteen (14) advantages of using HDD method and only seven (7) obstacles that that had been found may hindered the success of HDD implementation. But if these obstacles are being identified and taking care during planning and construction stages it will never be an issue. With a lot of advantages identified no doubt why HDD method becoming very famous everywhere in the world including in Malaysia. However some action need to be taken by Malaysian local authorities to avoid what have been faced by HDD market in USA in the 90s, where accidents and failures related to HDD method becoming a big issues. Back then, HDD market was exploding due to rapid growth in constructing fibre optic backbone across the United States. Unfortunately, this fast grown method caused insufficient experienced HDD drillers, inadequate geotechnical investigations and a lack of established good practices. In 1998, concern about potential damages from HDD installations beneath its roadways the California Department of Transportation (Caltrans) imposed regulations requiring training for HDD contractors when drilling under the state lands [23].

Similar concerns were being raised around USA by other transportation and environmental agencies. Agencies, such as Santa Clara County, considered declaring a moratorium on all drilling operations until a standardized set of practices was established and training of operators was implemented [23]. In addition poor drilling practices by some contractors have caused utility strikes that have resulted in major legal ramifications and subsequent negative image of the technique. The damage claims are normally huge as it is not only the utility repairing cost but also include "loss of revenue" where insurance company are normally not covered [22]. Table 1, summarised the advantages/disadvantages between HDD method and conventional open cut method.

**Table 1.** Comparison summary between HDD and conventional Open Cut method.

| No | HDD Method | Traditional Open Cut |
|---|---|---|
| 1 | • **Rapid Deployment**<br>✓ Easy setup & minimum machineries<br>✓ Open trench only at launching and receiving pits | • **Slow Deployment**<br>✓ Involve a lot of machineries for excavation & import/ export materials<br>✓ Open Excavation all the way |
| 2 | • **Flexible Installation**<br>✓ Drill from surface<br>✓ Can re-engineer the alignment vertically/horizontally to avoid utilities/obstacles<br>✓ Installation not necessary straight | • **Rigid Installation**<br>✓ Open Excavation all the way<br>✓ Difficult to change direction once meet utilities and obstacles.<br>✓ Difficult to change alignment |
| 3 | • **Environmental Friendly Process**<br>✓ Minimum open excavation<br>✓ Less noise and dust. Fewer amounts of debris & waste generation.<br>✓ Reduce construction materials (For road repairs).<br>✓ Bentonite can be recycle and chemically not harmful | • **Non-environmental Friendly Process**<br>✓ 100% open excavation<br>✓ Large dumping of excavated materials, import/export at site, large noise and dust pollution.<br>✓ Huge quantity of waste, required heavy machinery to export/import materials<br>✓ A lot import/export sand, crusher run requires for road repairs |
| 4 | • **Better Quality reinstatement**<br>✓ Less excavation easier to repair and control reinstatement quality | • **Poor Quality reinstatement**<br>✓ 100% Open excavation difficult to repair and control reinstatement quality |
| 5 | • **Avoid Traffic Congestion**<br>✓ No road closure<br>✓ Less construction machineries<br>✓ Less import/export construction material | • **Causes Traffic Congestion**<br>✓ Road closure<br>✓ Many Construction machineries<br>✓ Major import/export construction material |

HDD method comes with some sustainable construction criteria's e.g. reduce, reuse and recycle of the construction materials and bentonite, less noise, less impact to the environment and the public however the disposal of bentonite upon completion if not being managed properly may attract ecological issues e.g. turbidity and contamination to our water source [24].

## 5. Conclusions

No doubt HDD method is a good alternative method when compared to the conventional open cut method but it also a better option to most other trenchless method. Among the reasons the HDD method offers less disturbance on traffic, the public, business activities and neighbourhood, lower restoration cost, less noise, dust and minimum import/export of the construction materials, easy to setup, minimum support machineries/truck, can manoeuvre in between utilities/obstacles, not restricted to straight alignment drilling, reduce, reuse, recycle and etc.

However HDD method requires well trained operators, good soil/geological information, accurate utilities mapping, and good drilling practices. Therefore with the intensity of utilities are getting congested the local authorities/clients requires to ensure only trained and experience contractors are allowed to undertake the job. To safe guard the facilities/asset from get damage by poor HDD works, the time has come for the local authorities to enhance their technical condition for HDD works e.g. to ensure the HDD operator are well trained, HDD machines are in good shape, soil investigation report and utilities mapping were carried out, a good drilling practices is adopted in contractors' drilling procedure and etc.

With the present of cheaper HDD Machine from China although this can reduce the construction cost but it's also invites HDD sub-contractors with lacking of HDD knowledge and training to join the market. This can lead to the above problems. Moreover the rapid growth in using HDD drills method in the last two decade make some areas are almost saturated with utilities and any activities by inexperienced HDD contractors can invite catastrophic. To proof the concern on this issue is valid the damage to fibre optic cables of Custom Malaysia Information System in Klang in May 2016, that had affected Westport operation [25] and the power outage in Menggelebu district in Ipoh, Perak where 138 sub-station were down [26] both were strike by HDD activities. The numbers of accidents are actually getting bigger but the question now is how to get the statistic? Therefore more interviews to the stakeholders' e.g. local authorities, insurances, utilities companies, design consultant, project owners and etc. need be conducted to establish information and data.

## 6. References

[1] Arends G Trenchless technology research in the Netherlands 1998 *Tunnelling and Underground Space Technology* **12(Supplement 1)** 53–57

[2] Bayer H J 2005 *HDD-Practice Handbook.* (Ed.) (Huyssenallee Essen, Germany: Vulkan-verlag GmbH) p 13

[3] Sarireh M, Najafi M, Slavin L 2012 Usage and Applications of Horizontal Directional Drilling. *ICPTT 2012: Better Pipeline Infrastructure for a Better Life: Proc. of the Int. Conf. on Pipelines and Trenchless Technology* American Society of Civil Engineers (Wuhan, China) p. 1835–1847

[4] Pipeline and Gas Technology 2005 History of horizontal directional drilling. *Pipeline and Gas Technology* **4**(6 (September)): pp HDD3-HDD6

[5] Pek Mei W DBKL 2014 faces various challenges in building, maintaining and cleaning the city's streets The Star Retrived on Dec 7, 2014 from: http://www.thestar.com.my/news/community/2014/10/18/roads-much-travelled-dbkl-department-faces-various-challenges-in-building-maintaining-and- cleaning/

[6] Ariaratnam S and Allouche E. 2000 Suggested Practices for Installations Using Horizontal Directional Drilling *Practice Periodical on Structural Design and Construction* **5**(4) pp142–149

[7] Willoughby D A 2005 *Horizontal Directional Drilling, Utilities and Pipeline Application.* (New York: McGraw-Hill) pp 1-9

[8] Carpenter R. 2011 13th Annual Directional Drilling Survey-Mixed Market Recovery For HDD. *Underground Construction* **66**(6) 24–28

[9] Carpenter R 2014 16th Annual Directional Drilling Survey-HDD Evolves. *Underground Construction* **69**(6) pp 34–38

[10] Carpenter R 2016 18th Annual HDD Survey: Diverse Markets Drive HDD Forword. *Underground Construction.* **71**(6) 18–26

[11] Canadian Association of Petroleum Producers 2009 *Guideline; Planning Horizontal Directional Drilling for Pipeline Construction* (Alberta, Canada) pp 1-3

[12] Petroliam National Berhad 2014 *Peninsular Gas Utilisation Project* Retrieved on Dec 7, 2014 from http://www.petronas.com.my/our-business/gas-power/gas-processing-transmission/Pages/gas-processing-transmission/peninsular-gas-utilisation.aspx

[13] Hasan AF. Natural Gas Utilisation in Malaysia. Meeting Today's Challenges to Supply Tomorrow's Energy Clean Fossil Energy Technical and Policy Seminar *Asia Pacific Economic Cooperation (APEC) Energy Working Group's Expert Group on Clean Fossil Energy (EGCFE)*; 2005 p. (1-47).
[14] Mersing Construction & Engineering Sdn Bhd 2015 Retrieved on Jun 22, 2017 from: http://www.mersing-group.com/index.php/mersings-history/
[15] Allouche E N, Ariaratnam S T, Lueke JS 2000 Horizontal Directional Drilling-Profile of an Emerging Industry. *J. of Constr. Engineering & Management* **126**(1) 68–76
[16] Najafi M 2010 *Trenchless technology piping: Installation and Inspection* (New York, NY: McGraw-Hill)
[17] Woodroffe N J, Ariaratnam S T 2009 Contractor Perspective on Factors for Evaluating Installation Options for Small-Diameter Utilities. *J. of Constr. Engineering and Management* **135** 75-87
[18] Tammi C E, Desilets A M, Skonberg E R, Srivastava V, Evans J M 2008 A Horizontal Directional Drilling Best Management Practices Manual *Environment Concerns in Rights-of-Way Management 8th International Symposium* (Saratoga Springs, New York, USA: Elsevier B.V) pp 603–612
[19] Puckett J 2011 Guidelines for Preventing Underground Utility Damage as a Result of Horizontal Directional Drilling *18th BIENNIAL PRCI, AFIA, EPRG Joint Technical Meeting on Pipeline Research*. San Francisco: Pipeline Research Council International Retrieved on Jun 20, 2017 from: https://s3.amazonaws.com/media.cloversites.com/de/de188978-4510-4da7-bc09-2f39a73619e7/documents/Guidelines_for_Preventing_Underground_Utility_Damage_as_a_Result of HDD.pdf
[20] Bennett D, Ariaratnam S T, HDD Consortium 2008 *Horizontal Directional Drilling Good Practices Guidelines* 3rd ed. North American Society for Trenchless Technology; (Cleveland, Ohio) pp 2-45
[21] Najafi M 2013 *Trenchless Technology: Planning, Equipment, and Methods* (New York: McGraw-Hill)
[22] Ariaratnam S T and Proszek J J 2006 Legal Consequences of Damages to Underground Facilities by Horizontal Directional Drilling. *J. of Professional Issues In Engineering Education & Practice* **132**(4) pp 342–355
[23] Bennett D and Ariaratnam S 2011The HDD Industry's Comprehensive Guidebook Is Now in Its Third Edition. *Horizontal Directional Drilling Guide* pp 20–23
[24] Matar M, Osman H, Georgy M and El-said M 2017 A systems engineering approach for realizing sustainability in infrastructure projects *Housing and Building National Research Center Journal* - Elsevier **13**(2) 190-201
[25] Gangguan talian smk, internet di pelabuhan barat berpunca kerja pembinaan 2016 Utusan Online Retrieved on June 20, 2017 from http://www.utusan.com.my/berita/nasional/gangguan-talian-smk-internet-di-pelabuhan-barat-berpunca-kerja-pembinaan-1.335300
[26] TNB: Temporary power outage in Klang and Ipoh due to faulty cables 2017 *The Malay Mail Online* Retrieved on June 20, 2017 from http://www.themalaymailonline.com/malaysia/article/tnb-temporary-power-outage-in-klang-and-ipoh-due-to-faulty-cables#sthash.KHeT8wNz.dpuf

IOP Conf. Series: Materials Science and Engineering **271** (2017) 012095 doi:10.1088/1757-899X/271/1/012095

# Experimental study on the use of steel-decks for prefabricated reinforced concrete beams

**Y A Priastiwi[1], A L Han[2], A Maryoto[3] and E S Noor[4]**

[1]Department of Civil Engineering, Diponegoro University, Semarang-Indonesia
[2]Department of Civil Engineering, Jenderal Soedirman University, Purwokerto-Indonesia
[3]Master Program in Structural Engineering, Diponegoro University, Semarang-Indonesia
Corresponding author: yulita_tiwi@ymail.com

**Abstract.** This paper presents an experimental study on the use of steel-decks for concrete beams. The purpose of this research is to determine the beam's capacity, and the load-displacement relationships due to the use of steel-decks. The failure mechanism was also studied, since the behavior differs significantly from conventional concrete members. For analysis purposes, two beam prototypes with steel-decks (GB1 and GB2), and two conventional concrete beams having the exact same material properties and dimensions (NB1 and NB2) functioning as control elements, were tested. Load was applied by a two-point loading system, creating a pure bending state. To monitor vertical deflections, two LVDTs were used. All precision instruments were connected to a data logger, and a computer. The results showed that the beams GB had a significant ultimate moment capacity increase, which is 2,3 times the control element NB. The main enhancement contribution is originated from the presence of the bottom steel-deck, which due to bonding to the concrete, functioned as additional tensile reinforcement. The deck also increased the member's ductility performance by 1.3 times. Specimen GB2 underwent bond loss in the transition zone between the deck and the concrete, reducing the initial stiffness of the member.

## 1. Introduction

The revolutions and evolutions in the concrete industry today demand for innovations in productivity, effectivity and efficiency. Optimization leading to more effective and efficient construction methods in terms of time, execution, and cost will provide benefits for all parties.

The use of steel-decks as formwork for slabs and concrete floors have been widely used. The deck functions, besides as formwork, also as tensile reinforcement in the tension zone of the plate. The installation of steel deck formworks that do not require dismantling, and demand a minimum in scaffolding, will speed up the construction process and make the space below the construction operational faster. The steel deck formwork offers a very attractive alternative to the traditional scaffolding and framework that needs to be removed. In terms of cost efficiency, the addition of the permanent formwork will reduce the overall cost, and makes it an interesting choice. The research of Uji [1] demonstrated that the use of steel-decks as formworks can provide cost savings up till 28.12% when compared to traditional construction methods. In terms of time savings, the steel-decks can contribute to a 41.67% in time reduction.

The financial benefits of steel-decks forms would become more sophisticated when the structural improvements are analysed clearly. Kan *et al.* [2] conducted a study on the mechanical behavior and responses of steel-decks forms for light weight concrete beams. Their research concluded that the optimum load was achieved as a result of de-bonding in the interface between the steel-decks and the

concrete. The load-displacement curve then dropped abruptly, resulting in failure of the specimen. The presence of screws functioning as shear connectors influenced the ductility performance of the member, since the post peak curve was more pronounced for the specimens with screws. The use of screws is of crucial importance in maintaining the bond between the steel-decks and the concrete. A good bond will result in a compatibility mode, and then the steel and concrete will act as a composite member in carrying the load. Merryfield *et al.* [3] declared that the use of puddle-welds as shear connectors ensured a good bond, and the composite performance of the combined material improved significantly. At service loading, the actual vertical deflection of the beam was reduced to 70% of its allowed value, while the strain of reinforcing steel was lessened to only 20% of its yield strain. The main focus of the study was to determine the contribution of the steel-decks to load carrying capacity increase and the stiffness of the member, while the moment-curvature behavior was evaluated through the strain in the longitudinal tensile reinforcement and concrete in compression.
This study focused on the potentials, and evaluation in behavior of a steel-decks formwork for a beam element, since up till now the beams of a steel-decks formwork plate or slab were constructed using a conventional casting system.

## 2. Methods

The response and behavior of the steel-decks formwork concrete beams GB1 and GB2 was studied by method of comparison to the control elements. These control elements were identical to the GB specimens in terms of their dimension, reinforcing steel configuration and concrete strength aspects. The GB1 and GB 2 specimens were 200 mm by 300 mm. At the bottom, a corrugated sheet 1 mm thick was used as formwork, as well as tensile reinforcement [4]. The side panels were flat sheets having the same thickness as the bottom formwork. The panels were connected with a strip having a distance of 500 mm apart to ensure sufficient stiffness to maintain the design configuration of the beams' cross section (figure 1). Connections between elements was provided by the use of self-drilling screws having a diameter of 4 mm as can be seen in figure 1b. These screws were positioned at a distance of 200 mm apart. The steel deck has a bilinear stress-strain behavior with a yield stress ($f_y$) of 300 MPa in combination with a yield strain ($\varepsilon_y$) of 0.0015. The Young's modulus was recorded to be 200 GPa.

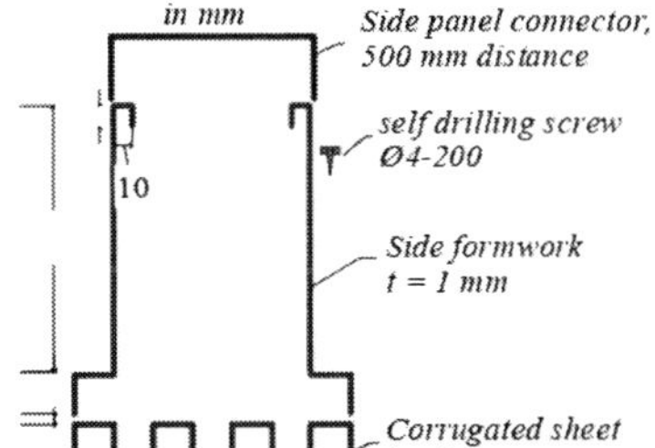


(a) Formwork elements.

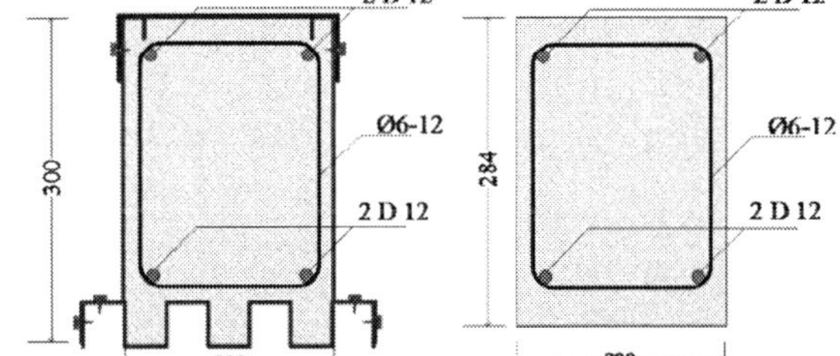


(b) Assembled formwork and convensional beam.

**Figure 1.** Details of steel-decks formwork. [4]

The control members NB1 and NB2 were 200 mm by 284 mm reinforced concrete beams, having 2D12 mm reinforcing steel in the tension as well as the compression zone. The steel had a yield stress ($f_y$) of 400 MPa in combination with a yield strain ($\varepsilon_y$) of 0.002. Further, the stress-strain behavior of these steel bars followed the standard yield pattern and a non-linear curve at strain hardening. The ultimate strength was 400 MPa and the Young's modulus was 192 GPa. The mechanical properties of the steel-decks thus closely approached the properties of the reinforcing steel.

The 284 mm height of the NB cross section was a conversion of the 300 mm GB beam's height, targeting a constant section area.The conversion accommodated the depth of the corrugated, part measuring40 mm (figure 1). The concrete had a volumetric material mix proportion of 1 : 2 : 3 for the cement, fine and coarse aggregates respectively, while the water-cement-ratio was set at 0.5. The cylindrical compression strength at 28 days was measured to be 19.8 MPa. This simplified mix design approach was favored since for low-rise industrial buildings this method is widely used, and has proven satisfactory during the lifetime of these buildings.

The beam was simply supported and had a length of 2.440 mm, the supports were distanced 2.280 mm apart. The two-points loading system had a distance of 760 mm as can be seen in figure 2. The load was recorded by a *load cell* type CLC-500kN with a capacity of 500 kN while the vertical displacements were monitored by two *Linear Variable Displacement Transducer* (LVDT) type SDP-100C. A acrylic base plate with sufficient stiffness was attached at the base of the beam at mid-span, as reading point for the LVDT's (figure 2a). To obtain the strain behavior of reinforcing steel and steel-decks, strain gauges type FLA 6-11 were attached to the reinforcing steel bars in tension, and the steel-decks at mid-point. The compression strain of the concrete was recorded using the PL 60-11 strain gauges situated at the top of the section in the longitudinal direction of the beam (figure 2b).

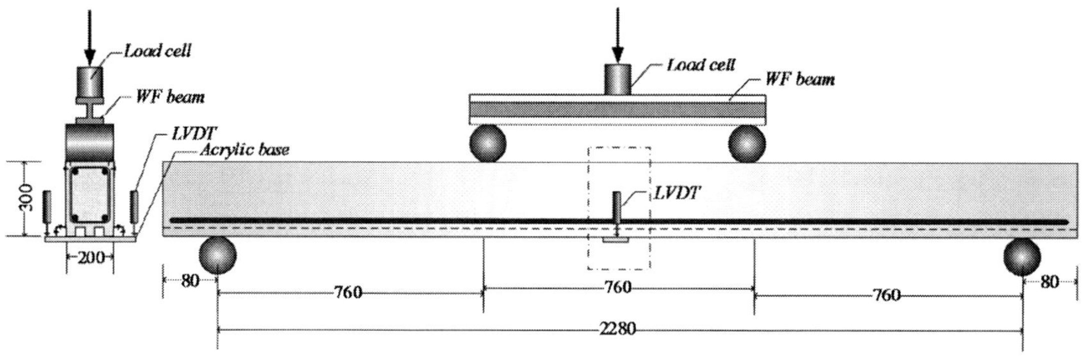


(a) Set up and precision apparatus placements.

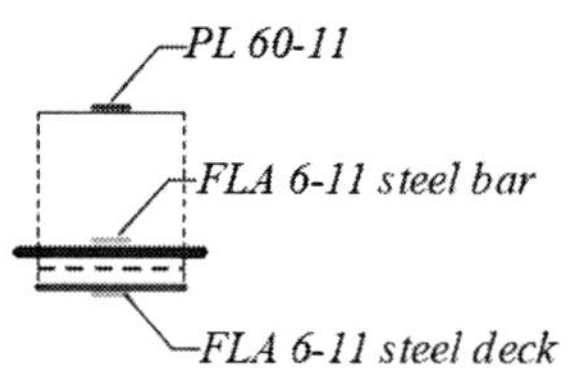


(b) Strain gauges at mid-point.

**Figure 2.** Test set up.

A monotonic increment load was applied to the beam till failure, and all the data was stored. The load data was further analyzed and transferred to the moment-deflection relationship, while the strain information was used to generate the curvature development as a function of the load increase. The bending moment area underneath the loads had a constant positive bending moment.

$$M = \left(\frac{1}{2}P\right)\left(\frac{1}{3}L\right) = \frac{1}{6}PL \quad (1)$$

where P is the load recorded by the load cell in kN and L is the beam's clear span 2.28 meters in length. This two-point loading system ensures a pure state of bending. As for the curvature determination, the strain in the concrete in compression $\varepsilon_c$ and the strain in the tension reinforcement $\varepsilon_s$ were used (figure 3). The strain in the steel deck $\varepsilon_D$ functioned as controlling strain. The curvature was calculated using simple mathematics as: $\varphi = \frac{\varepsilon_C+\varepsilon_S}{d}$ (2)

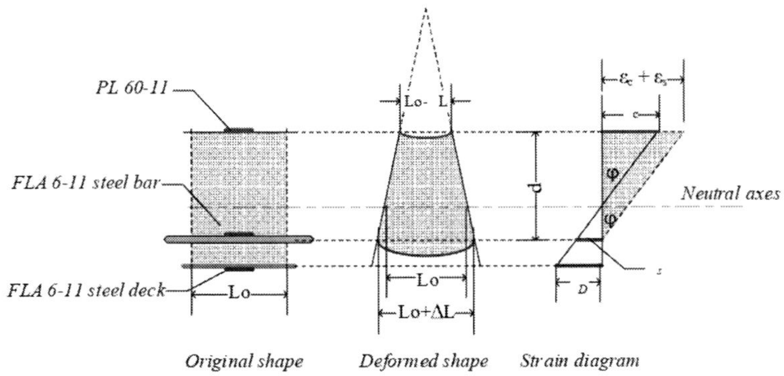


**Figure 3.** The bending deformation and corresponding strains.

## 3. Data analyses and discussion

### *3.1 Moment-displacement response and curvature*

The response of the load in kN was converted to the bending moment capacity of the section, since this provided a better understanding to the overall strength of the beam under consideration (figure 4). The control elements NB1 and NB2 showed a typical under-reinforced concrete member behavior in bending. The initial stiffness of the member was high and was calculated to be 11 kN/mm.

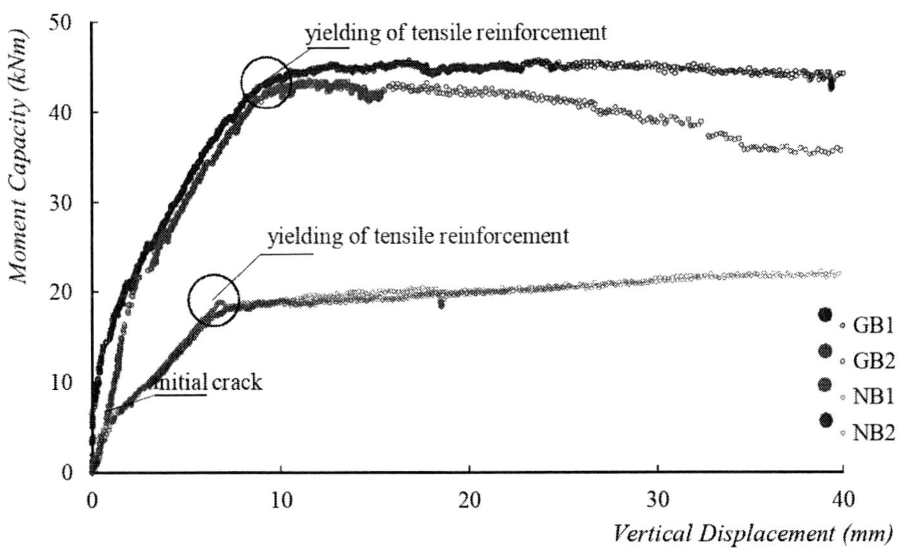


**Figure 4.** Moment-displacement response of specimens.

The initial stages exhibited a linear response, and the moment-displacement trajectory followed a straight path. The first crack in the concrete tension zone was identified at 6 kN-m, the curve underwent an unmistakable deviation in stiffness. This decrease in member stiffness was due to a reduction in moment of inertia of the section. Visual observation to the behavior of the beam showed that this first crack occurred at mid-span. The moment-displacement pattern tends to become non-linear. The load was increased, and at 18 kN-m the tensile reinforcement yielded. The curve diverged significantly and was distinguished by a yield plateau. The recordings of strain gauges on the reinforcing steel confirmed that the steel yielded. The strain hardening of these steel bars resulted in a slight capacity increase, and the average of ultimate moment capacity of the member was determined at 20 kN-m.

The composite beams demonstrated a different moment-displacement response. The initial stiffness of the member GB1 and GB2 were remarkably higher when compared to the NB beams. The GB1 specimen had an initial stiffness of 35 kN/mm, while specimen GB2 had a stiffness of 24 kN/mm. The steel-decks formwork increased the initial stiffness of the member 2.1 ~ 3.1 times. Closer analysis revealed that the major contribution to this stiffness enhance was originated from the side steel panels, especially during the period that the bond between these side-panels and the concrete was intact. As load increased, this bond was lost, and the concrete element underwent substantial slipping with respect to the steel-decks formwork, as can be seen in figures 5a and 5b. Prior to this, the stiffness dropped gradually, resulting in the non-linear behavior at later stages of loading.

(a) Initial bond loss. (b) Specimen bond-slip.

**Figure 5**. Side-panel bond loss and bond-slip.

The path of the curves then deviated substantially, and this was diagnosed as the stage at which the tensile reinforcement yielded. The moment capacity was calculated to be 45 kN-m and 44 kN-m for specimens GB1 and GB2 respectively. The use of the steel deck formwork enhanced the load carrying capacity 2.5 times. The first crack could not be detected, not numerically nor visually, since the concrete view was blocked by the presence of the steel-decks. Table 1 shows the values as obtained by the experimental test results. The control elements NB1 and NB2 had almost identical outcomes, and the average of these were taken for evaluation purposes.

**Table 1.** Moment capacity and stiffness data.

| Spesimen | $M_y$ (kN-m) | GB/NB ratio | $M_u$ (kN-m) | GB/NB ratio | Stiffness K (kN/mm) | GB/NB ratio |
|---|---|---|---|---|---|---|
| GB1 | 45 | 2.5 | 46 | 2.3 | 35 | 3.1 |
| GB2 | 44 | 2.4 | 44 | 2.2 | 24 | 2.1 |
| $NB_{av}$ | 18 | 1.0 | 20 | 1.0 | 11 | 1.0 |

When examining the data at failure it was also noticed that the elements GB1 and GB2 had a much greater ductility when compared to the conventional beam. The data are presented in table 2. Since ductility of a member is customary expressed in the ratio between the deflection at ultimate $\delta u$ and the yield deflection $\delta_y$, the ductility can be calculated. The GB specimens had ductility coefficient of 1.3 times the conventional sections.

**Table 2.** Ductility behavior.

| Spesimen | $\delta_y$ (mm) | $\delta_u$ (mm) | $\mu_\delta$ $\delta_u/\delta_y$ |
|---|---|---|---|
| GB1 | 11 | 58 | 5.3 |
| GB2 | 10 | 52 | 5.2 |
| $NB_{av}$ | 7 | 27 | 3.9 |

The moment-curvature data underlined the conclusions obtained from the moment-displacement curves. The specimens GB had a much stiffer section behavior as compared to the conventional concrete elements NB. These conventional concrete beams also demonstrated a clear yield plateau prior to failure of the element. The first cracking of the members was distinguished by an unmistakably increment in curvature.

*3.2 Failure mechanism*

The conventional specimens NB failed due to extensive cracking in the tension area of the member. The cracks were initiated at the extreme concrete fibers in tension, and propagated vertically towards the neutral axes. At advanced loading stages, the cracks multiplied along the area in between the loading points, and widened significantly. No shear failure was detected, and prior to yielding, the concrete in compression collapsed due to crushing of the material (figure 6).

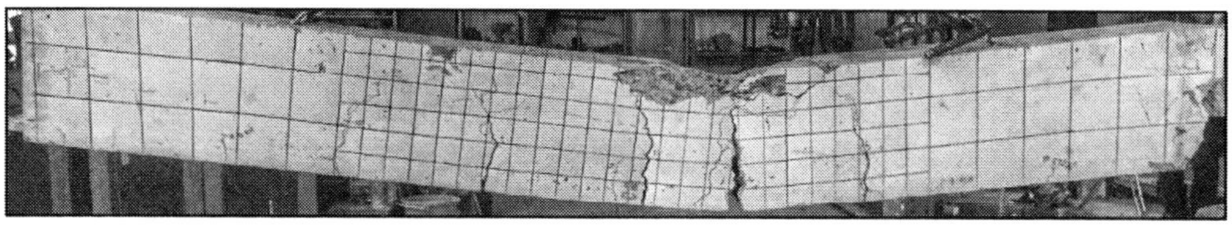

**Figure 6.** Failure mechanism of controlling elements.

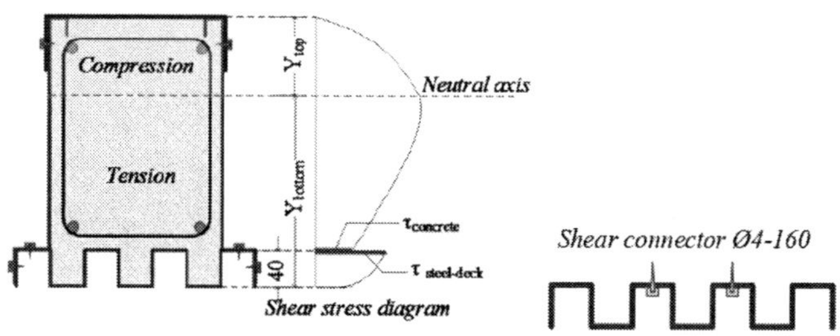


**Figure 7.** Bond-slip mechanism. [4]

The failure mechanism of the steel-decks beams differed from the failure mode of the conventional beams. The most distinguishing factor was the bond deterioration between the steel-decks and the concrete. The shear stresses disparities in the interface of the corrugated sheet and the concrete were carried by the shear connectors, situated at a distance of 160 mm apart (figure 7).

There were no shear connectors place at the side-panels, and the loss in bond resulted in a decrease in moment inertia of the section. The corrugated sheet acted as additional tensile reinforcement, and the capacity of the members increased accordingly [4,5]. To evaluate the failure behavior of the GB specimens, the formwork was stripped and the members visually inspected (figure 8). The crack pattern deviates from the pattern of the conventional beams NB. All cracks were situated beneath the point of impact of the load, suggesting that the concrete in this area suffered extensive bearing stresses that

IOP Conf. Series: Materials Science and Engineering **271** (2017) 012095 doi:10.1088/1757-899X/271/1/012095

weaken the section. A decrease in moment of inertia then initiated the propagation of tensile cracks towards the neutral axis.

**Figure 8.** Failure mechanism of steel-decks specimens.

When the steel bars in tension yielded, cracks in the concrete widened, and the shear stresses in the interface reached a maximum [6]. The shear connectors in this area at a distance of 1/3 of the span from the support collapsed, and the beam failed. Close examination to the shear connectors underlined this outcome. The shear connector in the mid-section were straight while the ones in the high shear stress area were bent outward to the direction of the beam (figure 9).

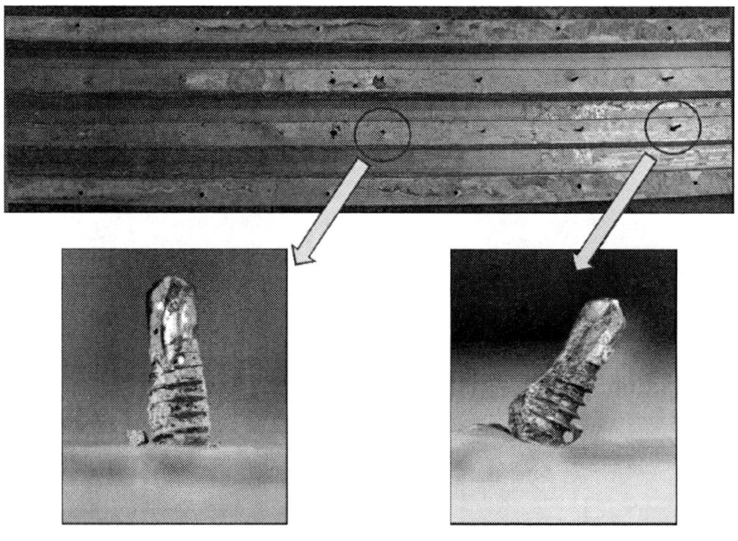

**Figure 9.** Shear connector failure mode

**4. Conclusion and remarks**

The steel-deck formwork has an advantage in that it provides a speed and clean construction environment. The formwork doesn't require dismantling so that the structure can be directly used upon reaching of the concrete design strength. The system also doesn't demand scaffoldings so that even during construction work, the space underneath the beam can be used.

The corrugated sheet when utilized with shear connectors, contribute to the composite action of the section. In positive bending mode, this sheet provides additional tensile reinforcement to the beam. Test results demonstrated a 2.3 times ultimate moment capacity increase when compared to an identical conventional concrete member with the exact same material properties and reinforcing configurations.

The initial stiffness of the member was enhanced significantly, thus grating a better serviceability level at service loading conditions. At ultimate, the steel-decks formwork element exhibits a much higher ductility ratio compared to the conventional beam.

The performance of shear connectors is determining to the composite action between the steel-decks and the concrete. Care should be taken in designing of these shear connectors especially in the areas subjected to high shear forces. Also, the side-panels do not contribute to the load carrying capacity of the member, so in design these sheets can be reduced in thickness. However, the bond between these side-panels and the concrete should be improved to ensure the high stiffness of the member under service loading conditions. These panel should also be sufficiently stiff to guarantee the final dimensions, so that no deformations occur.

## 5. References

[1] Uji A T 2012 *Perbandingan Biaya Pelaksanaan Pelat Beton Menggunakan Bondeck dan Pelat Konvensional Pada Gedung Graha Suraco Makassar* (Universitas Hasanuddin, Indonesia)

[2] Kan Y C, Chen L H, Yen T 2013 Mechanical Behaviour of Lightweight Concrete Steel Deck *Construction and Building Materials* **42** 78-86

[3] Merryfield G, El-Ragaby A and Ghrib F 2016 New Shear Connector For Open Web Steel Joist With Metal Deck And Concrete Slab Floor System *Construction and Building Materials* **125** 1-11

[4] Han A L 2017 *Report on the Experimental Test Result of Corrugated Steel Deck Formwork for Beams* (Material and Structural Laboratory Diponegoro University, Indonesia)

[5] McComb C, Tehran F M 2015 Enhancement Of Shear Transfer In Composite Deck With Mechanical Fasteners *Engineering Structures* **88** 251-261

[6] Elamary A, Ahmed M M, Mohmoud A M 2017 Flexural Behaviour And Capacity Of Reinforced Concrete–Steel Composite Beams With Corrugated Web And Top Steel Flange *Engineering Structures* **135** 136-148

**Acknowledgments**

Special gratitude is expressed to *Ir. Bambang Susilo*, the patent holder of the corrugated steel-decks formwork system for concrete beams, who made this study possible due to his initiation, cooperation and support.

IOP Conf. Series: Materials Science and Engineering **271** (2017) 012096 doi:10.1088/1757-899X/271/1/012096

# Modelling of concrete topping thickness effects on the vibration behaviour for lvl-concrete composite floor (LCC)

**NH Abd Ghafar[1,2] and N M Sahban[1]**

[1] Fakulti Kejuruteraan Awam dan Alam Sekitar, Universiti Tun Hussien Onn Malaysia, Parit Raja, Batu Pahat, Johor.
[2] Jamilus Research Centre, Universiti Tun Hussien Onn Malaysia, Parit Raja, Batu Pahat, Johor.
Corresponding author: noryati@uthm.edu.my

**Abstract.** This research was conducted on 2 m LVL - concrete composite (LCC) floor consisting of two parts between concrete floor and laminated veneer lumber (LVL) timber joist. The floor system was model using SAP 2000 software package. The aim of this research to study the vibration behaviour of the LCC floor with different concrete topping thickness which 25 mm, 65 mm and from 20 mm until 200 mm in every 20 mm interval. Natural frequency decision produced through SAP 2000 in thickness 25 mm and 65 mm is 57.45 Hz and 57.19 Hz. In thickness from 20 mm until 200 mm in every 20 mm interval, optimum value which found is during thickness reach 65 mm. For concrete topping below 65 mm thickness, the mass will be domain the behavior of the floor. When concrete topping increased more than 65 mm, the behavior of the floor will be domain by floor stiffness.

## 1. Introduction

Floor vibration is an important element that should be consider in a construction structure. Floor vibrations of a building are caused by human activity such as running, dancing, jumping or even walking and mechanical equipment excitation that are affect personal comfort. The force from walking transferred from one location to another with each step. This may create the serviceability issues which human feel annoy and not comfortable.

Traditional timber floor had an issues in vibration serviceability due the flexibility of the timber especially for the long span floor. Thus, the vibrations problem of traditional timber floors can be solved by applying a thin layer of concrete on top of the timber joists. The concrete slab can increases the mass, stiffness and strength of the timber floor. This means that the strength and stiffness are utilized to the greatest advantage giving the timber-concrete composite floor system the potential to improve the dynamic and vibration performance of traditional timber floors [1].

Timber-concrete composite (TCC) floor is a construction technique used to upgrading the traditional timber floor [2]. This system is the combination of timber beam with concrete topping slab using a joist system to transfer the loads between the timber and concrete. TCC floor system offers larger strength and stiffness, higher fire resistance and smaller effect of vibration compare with traditional timber floors [3]. The benefits of TCC systems caused it attractive use for new construction [4]. TCC structure provided more advantages as compared to traditional timber structure in term of strength and durability. TCC has higher

IOP Conf. Series: Materials Science and Engineering **271** (2017) 012096 doi:10.1088/1757-899X/271/1/012096

stiffness and strength, reduces susceptibility to vibration, more resist to fire and seismic and good in thermal mass and acoustic separation. Besides that, the overall weight of a floor can be reduced due to introduce of timber in construction of floor that can reduce the volume of reinforcement concrete usage, which directly can reduce the imposed load act on foundation [5].

In this research, the laminated veneer lumber (LVL) was used as timber joist to increase the strength of the TCC floor. LVL is engineered timber which has almost three times the strength of the sawn timber, more reliable and high modulus of elasticity (Abd Ghafar, 2008). The LCC floor was model using SAP 2000 finite element software package to study the effect of concrete topping thickness and to determine the optimum of concrete topping.

## 2. Materials and Methods

The material properties for LVL and concrete as illustrated in table 1. The properties of the LVL was adopted from H'ng [6]. The material properties is the first element need to insert in SAP 2000 software package.

**Table 1.** Material Properties of LVL and Concrete.

| **Material Properties** | **LVL** | | **CONCRETE** |
|---|---|---|---|
| Grade | SG3 | | C30 |
| Unit Weight (kN/m$^3$) | 9.0 | | 25.0 |
| Young Modulus, E (N/mm$^2$) | Mean | Minimum | 30000 |
| | 11300 | 499 | |
| Poisson's Ratio, U | 0.5 | | 0.2 |
| Shear Modulus, G (N/mm$^2$) | 753 | | 12500 |

The second step in SAP 2000 software package is to insert the element of TCC. Three types of element that been used in this research which (1) beam element, (2) shell element and (3) link element. Beam element represented the 400 mm thickness of LVL joist. While the shell element represented the concrete topping with varies concrete topping thickness were model form 20 mm to 200 m with 20 mm interval, including 25 mm and 65 mm thickness. The selection of 25 mm and 65 mm based on Abd Ghafar [7]. The link element represented the connector of the TCC. The link element was used to connect two beam elements and transfer the loads from concrete topping to the LVL joist. Two types of link were defined in link properties which are shear connector and rigid connector. The finite element modelling and cross-section of LCC as shown in figure 1.

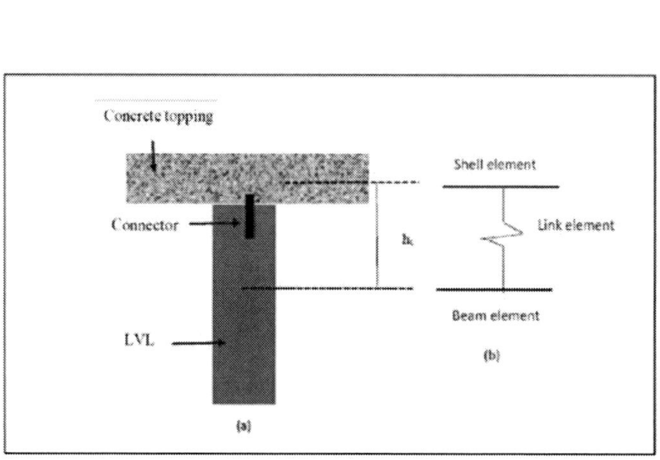


**Figure 1.** (a) Cross-section of LCC and (b) Finite element modelling of LCC floor.

## 3. Result and Discussions

The important of the mode shape is to make sure that the natural frequencies of the LCC floor models were correct. Table 2 shows the natural frequencies and mode shapes of 2m LCC floor with 25 mm and 65 mm concrete topping thickness. All floors model had the similar behavior of mode shapes as for models in table 2.

The first mode of LCC floor was presented as half sine curve with the natural frequency of 57.45 Hz and 57.19 Hz for thickness 25 mm and 65 mm. The first mode shape was similar to the deflection shape caused by uniform gravity loading LCC floor system. The highest displacement located in the mid span of LCC floor due to self- weight of span turn into stress focus at the mid-span where it is most far from the support. The second mode shape of LCC floor corresponds to a predicted natural frequency of 121.90 Hz and 121.38 Hz where it presented as full sine curve. The third mode shape of LCC corresponds to a natural frequency of 193.72 Hz and 188.57 Hz which is also not significant for floor vibration. The third mode of LCC floor was presented as one and the half sine curve. The natural frequencies of the floor were a bit higher due to the short span of the LCC floor. The shorter the span of the floor will increased the stiffness and the natural frequency of the floor.

The natural frequencies of different concrete topping thickness were obtained by using SAP 2000 software and theoretical analysis. The natural frequencies between theory and SAP 2000 software results as illustrated in figure 2. The results shown that the natural frequencies were decrease from 20 mm until 60 mm concrete topping thickness. While, the results increased starting from 65 mm concrete topping until 200 mm concrete topping. This behavior present that the natural frequencies of the LCC were dominant by mass for LCC with 65 mm concrete topping or less. LCC with 65 mm concrete topping or more will be dominated by stiffness of the LCC. This is due to the changing of the mass and the stiffness when changing the thickness of concrete topping as mention by Rijal et. Al [7], theoretically the natural frequency depends on the system properties, mass and properties.

Comparison between theoretical and modelling show that the theoretical gave the higher value of natural frequency due to no stiffness of the connector included during the theoretical calculation. The average percent difference between the theoretical calculation and SAP 2000 is in the range of 4.25% to 11.56%. Wherein, the lowest percentage difference is 4.25% when the thickness of the concrete is 200 mm. The highest percentage difference is when the concrete thickness 60 mm where it was 11.56%. While, on the thickness of 25 mm and 65 mm, the percentage difference that occurred was 11.46% and 10.78%.

IOP Conf. Series: Materials Science and Engineering **271** (2017) 012096 doi:10.1088/1757-899X/271/1/012096

**Table 2.** Mode Shape.

| Concrete Thickness, $h_s$ (mm) | Frequency (Hz) | | |
|---|---|---|---|
| | **Mode 1** | **Mode 2** | **Mode 3** |
| 25 | 57.45 Hz 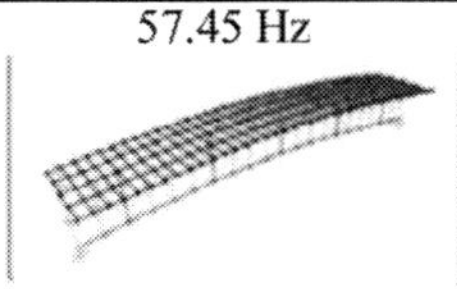 | 121.90 Hz 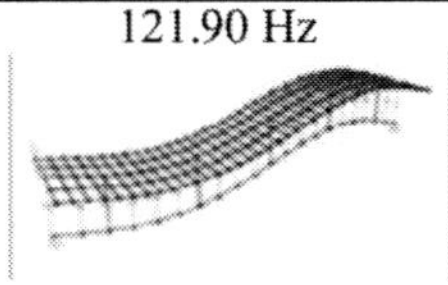 | 193.72 Hz 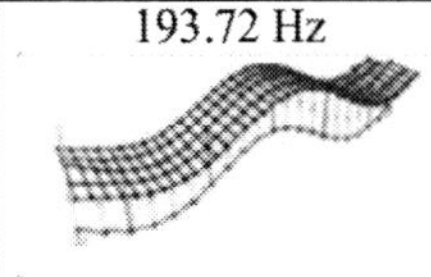 |
| 65 | 57.19 Hz 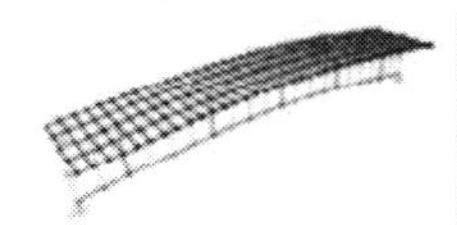 | 121.38 Hz  | 188.57 Hz 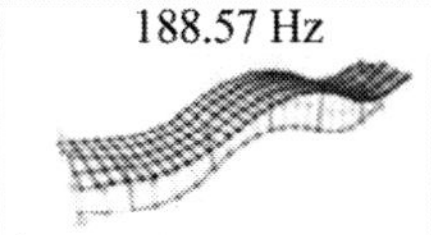 |

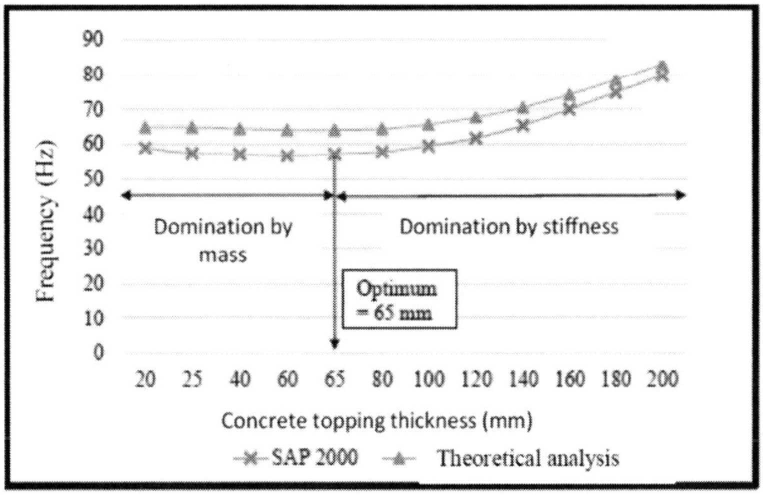


**Figure 2.** Comparison of natural frequency between finite element modelling and theoretical analysis.

## 4. Conclusions

The 2 m span LCC floor was model in Sap 2000 software package with different thickness of concrete topping. This study proof that the thickness of concrete topping influence the natural frequency of the LCC floor. The optimum concrete topping was determined at 65 mm thickness. For concrete topping below 65 mm thickness, the mass will be domain the behavior of the floor. When concrete topping increased more than 65 mm, the behavior of the floor will be domain by floor stiffness.

## 5. References

[1] Smith I and Chui Y H 1988 Design of lightweight wooden floors to avoid human discomfort. *Can J Civil Engineering* **15(2)** pp 254-262.

[2] Ceccotti A 1995 *Timber-concrete Composite Structures* (Timber Engineering, Step 1, First Edition, Centrum Hout, The Netherlands) pp E13/1-E13/12.

[3] Buchanan A, Deam B, Fragiacomo M, Pampanin S, and Palermo A 2008 *Multi-storey Pre-stressed Timber Buildings in New Zealand* Structural Engineering International. Special Edition on Tall Timber Buildings

[4] Lukaszewska E, Johnsson H, and Fragiacomo M 2008 Performance of connections for prefabricated timber-concrete composite floors *RILEM Journal of Materials and Structures* **41(9)** 1533–1550

[5] Yeoh D, Fragiacomo M and Deam B 2010 Experimental Behaviour of LVL-concrete Composite Floor Beams at Strength Limit State **33(9)** 2697-2707

[6] H'ng P S, Ahmad Z and Tahir, P M 2012 *Laminated Veneer Lumber from Malaysian Tropical Timber* (UiTM Press)

[7] Abd Ghafar, N H 2008 *Forced vibration testing on LVL-Concrete Composite floor systems* 7th fib PhD Symposiumin (Stuttgart, Germany)

[8] Rijal R, Samali, B and Crews K 2010 *Dynamic performance of timber concrete composite flooring systems* Sam Fragomeni and Srikanth Venkatesan (CRC Press) chapter 5 pp 315-319.

# The influence of main bar corrosion on bond strength in self-compacting concrete

**S S Ayop[1], A N K Emhemed[1], N Jamaluddin[1] and A Sadikin[1]**

[1]Jamilus Research Centre, Faculty of Civil and Environmental Engineering, Universiti Tun Hussein Onn Malaysia, 86400 Batu Pahat, Johor, Malaysia
Corresponding author: sallehs@uthm.edu.my

**Abstract.** The experimental study was conducted to determine the influence of main bar corrosion on bond strength in self-compacting concrete (SCC). A total 16 tension pullout tests specimens reinforced with 10 mm and 14 mm diameter bar were used for the bond strength test. The properties of SCC were determined from the slump flow, $T_{50cm}$, V-funnel and L box test. Reinforcing bars in the concrete were submitted to impressed current to accelerate the corrosion of the bar. It was found that the relationship between bond strength and concrete strength in un-corroded specimens differed from that of corroded specimens set in high-strength concrete because of brittleness in the corroded specimens, which caused a sudden loss of bond strength. The results revealed that specimens of un-corroded and corroded showed a higher percentage of bond strength degradation during the pullout tests.

## 1. Introduction

Corrosion of steel reinforcement continues to pose threats to the performance and integrity of reinforced concrete structures. Steel bar corrosion affects the reinforcement itself, the surrounding concrete and the composite action between the steel and concrete. The most obvious effect is the reduction in cross sectional area of the affected steel bars [1]. The corrosion of embedded reinforcement is one of the major problems that contribute to the deterioration of structural concrete. The direct effects of the reinforced corrosion are a loss of bar cross section; an increase in bar diameter resulting from the volumetric expansion of the corrosion products; a change in the characteristics of the bar/concrete interface upon the formation of corrosion product; and reduction of the concrete section [2]. The loss of bar section may reduce the ductility of the steel reinforcing bar. With insignificant ductility, a structure might fail and collapse in a brittle fashion without warning.

Self compacting concrete it a latest innovation in concrete technology is being regarded as one of the most promising developments in the construction industry due to numerous advantages of it over conventional concrete [3]. In construction industry, SCC is easier to handle than the normal concrete due its ability to consolidate or self-levelled without any internal or external compaction.

Many experimental works have been conducted previously to study the effects of steel bar corrosion in normal concrete [4, 5, 6] while less available data in self compacting concrete. In normal concrete, the general trend of changes in the bond strength after corrosion has been initiated is presented in figure 1 [7], although there are exceptions in certain circumstances that will be described later. Initially, the bond strength is increased by a small amount of corrosion but starts to decrease with a further increase in the corrosion level.

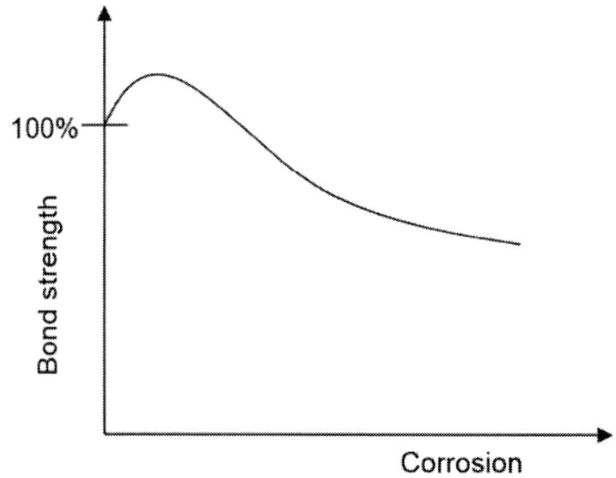


**Figure 1.** Bond strength versus corrosion [6].

## 2. Experimental program

*2.1 Specimen*

A total of 16 specimens were tested in the study. The specimens were subdivided into four groups that reinforced with 10 mm and 14 mm diameter of the bar with two different concrete cover thicknesses (figure 2). Each sample was then subjected to corrosion using an accelerated corrosion method. The crack width ($W_{cr}$) of the specimens was visually observed and measured with "Crack Detection Pocket Microscope". After the corrosion conditioning, tension pullout test was conducted to determine its bond strength. The length of embedment was limited to 12$Ø_b$. Transverse reinforcement was provided to study the influence of confinement on the bond strength.

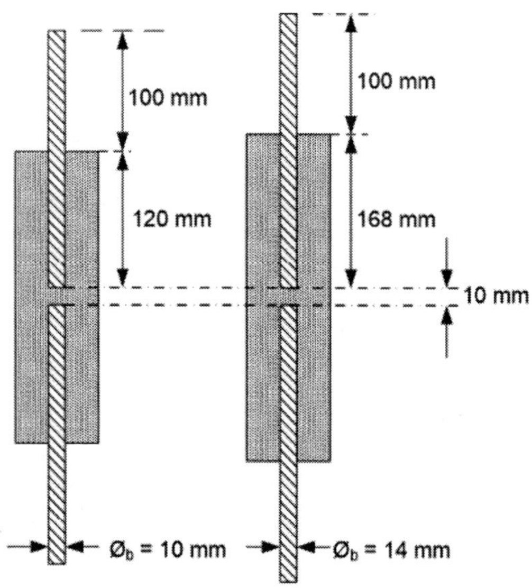


**Figure 2.** Tension pull out test specimen.

*2.2 Materials properties*

Table 1 shows materials the mixture proportions of concrete. Table 2 shows workability, compressive and tensile results. After pouring the concrete specimens were kept outside in normal temperature for 28 days. Before mixing the concrete and casting the reinforcement bars were carefully cleaned then mass of the reinforcement bar in each specimen were recorded.

**Table 1.** Mix proportions for self-compacting concrete.

| Material | Water ($kg/m^3$) | Cement ($kg/m^3$) | Fly ash ($kg/m^3$) | Fine aggregate ($kg/m^3$) | Coarse aggregate ($kg/m^3$) | Admixtures- 20% (SP 800) (ml) | Water-cement ratio |
|---|---|---|---|---|---|---|---|
| Quantity | 180 | 420 | 180 | 853 | 698 | 8400 | 0.43 |

**Table 2.** Workability, compressive and tensile results.

| Mix | Slump flow (mm) | $T_{50cm}$[a] (sec) | V-funnel $T_f$[b] (sec) | L-box bloking ratio ($H_2/H_1$)[d] | 7 day (MPa) | 28 day (MPa) | Tensile (MPa) |
|---|---|---|---|---|---|---|---|
| 1 | 710 | 3 | 8 | 0.85 | - | - | - |
| Cubic | - | | - | - | 23 | 36 | - |
| Cylinder | - | | - | - | 21 | 33 | 2.67 |

$T_{50cm}$[a]: Time taken for concrete to reach the 500 mm spread circle.
$T_f$[b]: V-funnel flow after keeping the concrete in funnel.
$H_1/H_2$: Heights of the concrete at both ends of horizontal section of L-box after allowing the concrete to flow.

*2.3 Accelerated corrosion method*

Acceleration corrosion was carried out by impressing an electric current using "ATTEN APS3005" power supply as in figure 3. Where the reinforcement bars were served as an anode and stainless steel plate at the concrete surface served as a cathodes. A total of 360 mA current, which corresponds to approximate current density ($i_{corr}$) of 0.09 $\mu A/cm^2$ was impressed to the main bar. The wet sponges were placed between the stainless steel plates and the concrete surface to provide an adequate current contact during corrosion conditioning process.

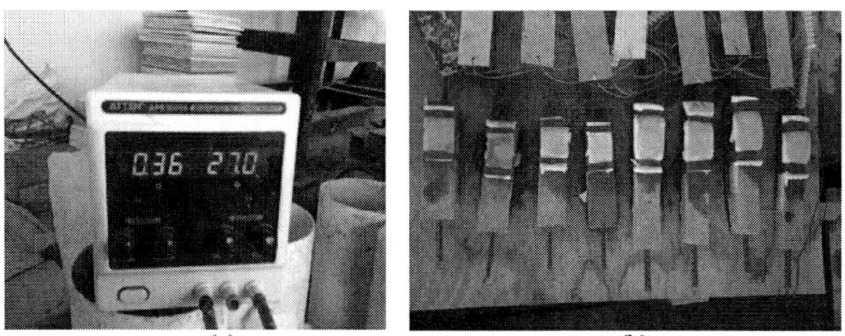

(a) (b)

**Figure 3.** Connection setup for the accelerated corrosion process.

*2.4 Calculation corrosion level and mass loss*

The gravimetric approach is used to determine the percentage of mass loss on each corroded bar using equation (1) based on the weight of the steel bar before and after corrosion conditioning [8].

$$G_R = \frac{G_0 - G_F}{g_0 l} \times 100\% \quad (1)$$

where $G_0$ is the initial weight of the steel bar before corrosion (gram), $G_F$ is the final weight of the steel bar after removal of the corrosion products (gram), $g_0$ is the weight per unit length of the steel bar (gram/mm) and $l$ is the corroded length (mm).

*2.5 Tension pull-out test*

Pull-out tests were conducted for both corroded and un-corroded specimens. The bond tests were performed using an Instron universal testing machine with a capacity of 50kN. The applied loads were controlled via computer using displacement control at a rate of 10 mm/min. The setup of the pull-out testing system is shown in figure 4. The maximum pull-out force was recorded and used to calculate the ultimate bond strength ($f_b$) according to equation (2).

$$f_b = \frac{P_{max}}{\pi \varphi_b L} \quad (2)$$

where $P_{max}$ is the maximum pull-out load, $\phi_b$ is diameter of the reinforcing bar and $L$ is the bonded length of the reinforcement bar in concrete.

**Figure 4.** Setup of the pull-out test system.

## 3. Test results and discussion

*3.1 Control bond strength*

The average bond strengths for control un-corroded specimens are plotted in figure 5. The trend shows that the linear increment on bond strength with increasing $C/Ø_b$ ratio. Furthermore higher bond strength is observed on the specimen with stirrup compared to non-stirrup specimens. This condition should be expected since the presence of stirrups has increased the confinement around the steel reinforcing bars, thus increased the bond strength. The difference between these two measurements ranges from 3% to 32%. Based on the regression line, the following equation (3) and (4) are developed to predict the bond strength ($f_b$) of un-corroded specimen:

$f_b = -1.85 + 1.9\ C/Ø_b$ for specimen with stirrup (3)

$f_b = -0.96 + 1.23\ C/Ø_b$, for specimen without stirrup (4)

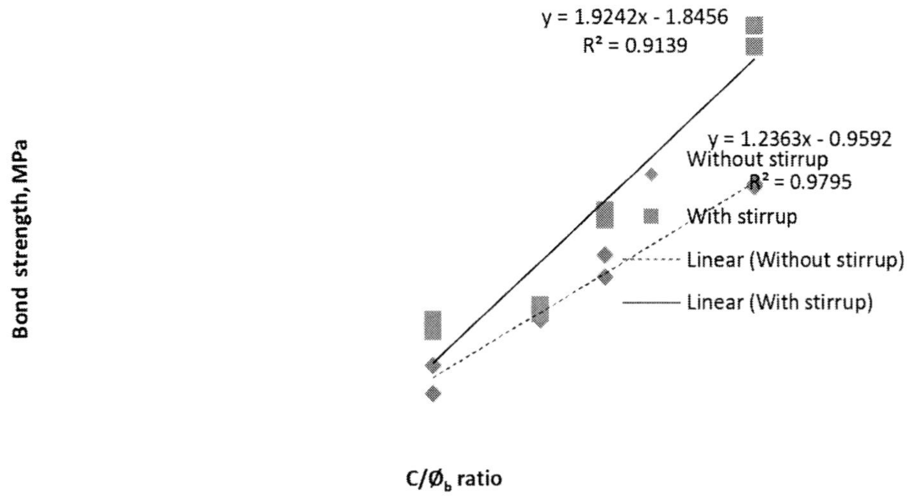


**Figure 5.** Control un-corroded bond strength.

*3.2 Corroded bond strength*

The influence of corrosion on bond strength is shown in figure 6. The level of corrosion in this study is up to 2% mass loss and for both bar diameters. The trend shows that the bond strength is decreased with increasing mass loss. This trend is consistence for both bar diameters and confinement conditions. Since thin concrete cover was used in this study, the bond reduction is more critical on bigger 14 mm bar diameter where its corrosion penetration is higher than the 10 mm bar diameter.

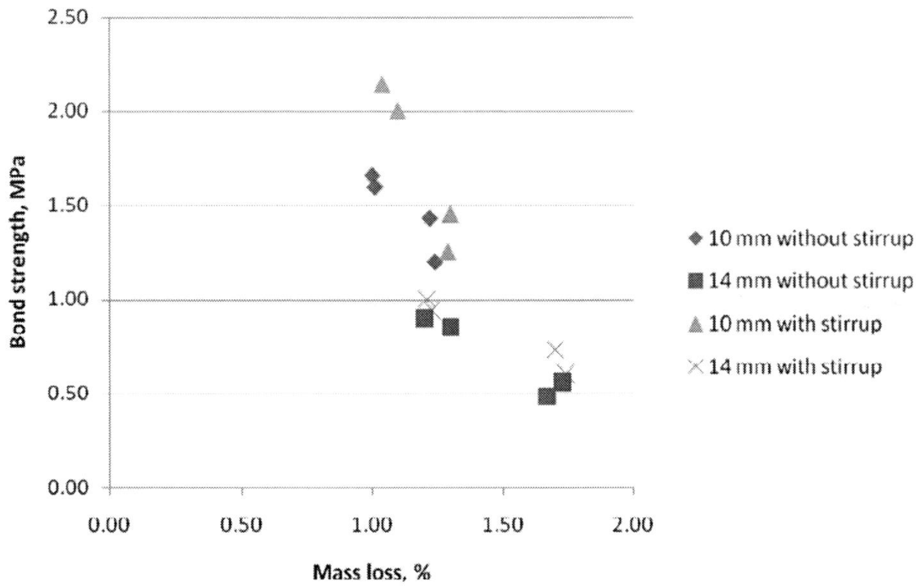


**Figure 6.** Bond strength versus mass loss.

The effects of corrosion on bond strength are depicted in figure 7 and 8 for specimens with and without stirrups. The presence of corrosion on both bar diameters had significantly reduced the bond strength where the ratio between corroded to control specimens between 0.6 to 0.8 for specimens with and without stirrups and 10 mm bar diameter had higher ratio than 14 mm bar diameter.

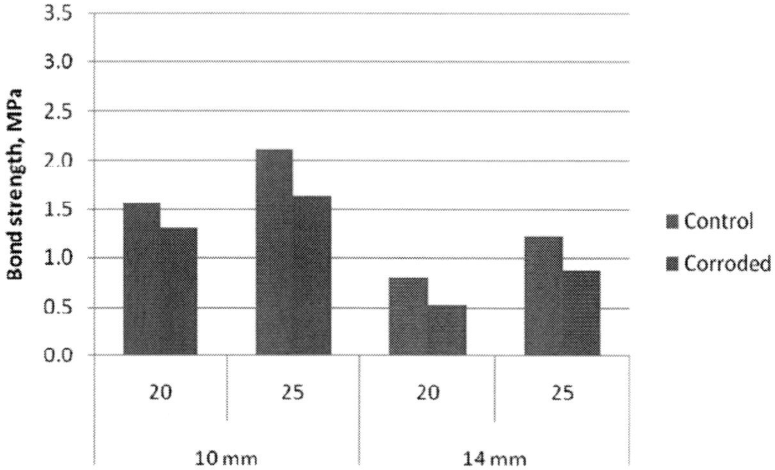


**Figure 7.** Control and corroded bond strength of specimens with stirrups.

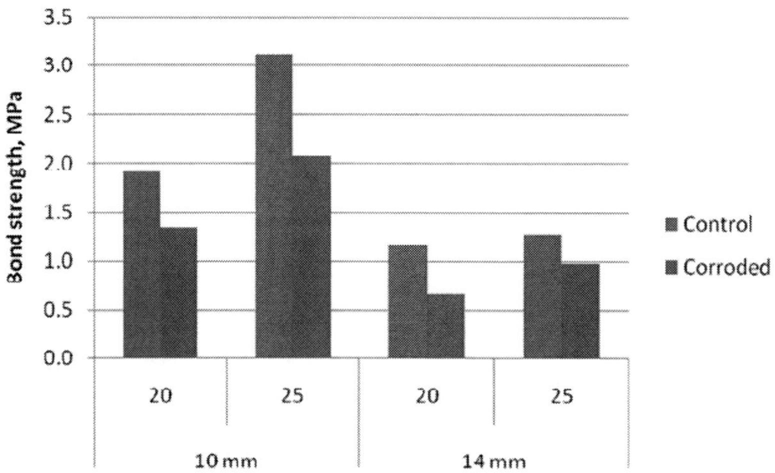


**Figure 8.** Control and corroded bond strength of specimens without stirrups.

## 4. Conclusion

Tension pull-out test method was used to study the influence of corrosion on self compacting concrete. The results showed a linear relationship was found between bond strength and cover-to-diameter ratio for both confined and unconfined specimens. Furthermore, higher bond strength was observed on confined specimen compared to unconfined ones as similar as observed in normal concrete. Corroded specimens had lower bond strength and this value was decreased with an increasing in mass loss. This trend was similar as observed in normal concrete.

## 5. References

[1] Sather I 2011 Bond deterioration of corroded steel bars in concrete, *Structure and Infrastructure Engineering* **7** 415–429
[2] Cairns J, Du Y and Law D W 2006 Residual bond strength of corroded plain round bars, *Magazine of Concrete Research* **58** 221- 231
[3] Naik T R, Kumar R, Ramme B W and Canpolat F 2012 Development of high-strength, economical self-consolidating concrete, *Construction and Building Materials* **30** 463–469
[4] Dahou Z, Castel A and Noushini A 2016 Prediction of the steel-concrete bond strength from the compressive strength of Portland cement and geopolymer concretes, *Construction and Building Materials* **119** 329–342
[5] Ayop S S and Cairns J 2013 Critical study of corrosion damaged concrete structures, *International Journal of Integrated Engineering* **5** 984- 989
[6] Rasheeduzzafar S, Al-Saadoun S and Gahtani A S 1992 Corrosion cracking in relation to bar diameter, cover and concrete quality, *Journal of Materials in Civil Engineering* **4** 327-342
[7] *fib* 2000 *Bond of Reinforcement in Concrete fib* Bulletin 10 (Lausanne: The International Federation for Structural Concrete) pp 187-212
[8] Fang C, Lundgren K, Plos M, and Gylltoft K 2006 Bond behaviour of corroded reinforcing steel bars in concrete, *Cement and Concrete Research* **36** 1931- 1938

IOP Conf. Series: Materials Science and Engineering 271 (2017) 012098 doi:10.1088/1757-899X/271/1/012098

# Design of reinforced areas of concrete column using quadratic polynomials

**Tjiang Arif Gunadi[1], Herman Parung[2], Abd Rachman Djamaluddin[3] and A Arwin Amiruddin[4]**

[1] Student of Civil Engineering program S3 Universitas Hasanuddin
Jln. Poros Malino Km. 6 Gowa
[2]Lecturer of Engineering Faculty of Universitas Hasanuddin
Corresponding email: parungherman@yahoo.com

**Abstract.** Designing of reinforced concrete columns mostly carried out by a simple planning method which uses column interaction diagram. However, the application of this method is limited because it valids only for certain compressive strenght of the concrete and yield strength of the reinforcement. Thus, a more applicable method is still in need. Another method is the use of quadratic polynomials as a basis for the approach in designing reinforced concrete columns, where the ratio of neutral lines to the effective height of a cross section ($\xi$) if associated with $\xi$ in the same cross-section with different reinforcement ratios is assumed to form a quadratic polynomial . This is identical to the basic principle used in the Simpson rule for numerical integral using quadratic polynomials and had a sufficiently accurate level of accuracy. The basis of this approach to be used both the normal force equilibrium and the moment equilibrium. The abscissa of the intersection of the two curves is the ratio that had been mentioned, since it fulfill both of the equilibrium. The application of this method is relatively more complicated than the existing method but provided with tables and graphs ($N$ vs $\xi_N$) and ($M$ vs $\xi_M$) so that its used could be simplified. The uniqueness of these tables are only distinguished based on the compresssive strength of the concrete, so in application it could be combined with various yield strenght of the reinforcement available in the market. This method could be solved by using programming languages such as Fortran.

## 1. Introduction

If a normal force acted on a column then it could be described as a normal force relationship with the moment in a diagram as below. The column interaction diagram was divided into 3 regions:

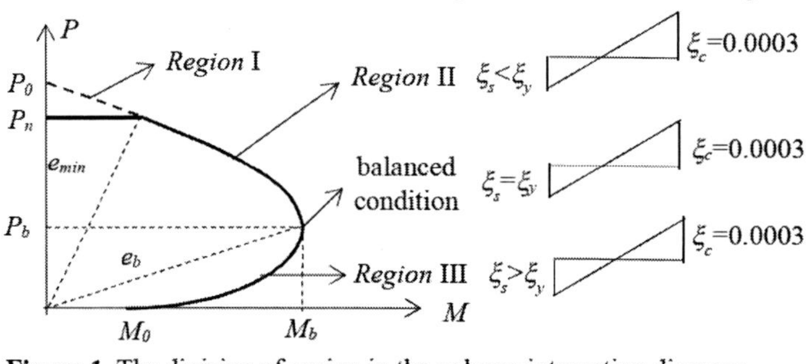


**Figure 1**. The division of region in the column interaction diagram

*1.1 The assumption of the stresses of the concrete and steel*

*1.1.a Assumption of concrete stresses*

1. Strain on reinforcement and concrete should be assumed to be proportional to the distance from the neutral axis. (Bernoulli - Navier)
2. The maximum strain that could be utilized in the outermost compression concrete was assumed to be 0.003.

3. Strenght at the reinforcing value smaller than the strength of $f_y$ should be taken by $E_s$ multiplied by the steel strain (Law of Hook). For strains of greater than $\varepsilon_y$ associated with $f_y$, the stress on the reinforcement should be taken equal to $f_y$.
4. In axial and bending calculations, concrete tensile strength should be ignored.
5. The relationship between the distribution of compressive stresses of concrete and concrete strain may be assumed to be square, trapezoidal, parabolic or other shaped, resulting in a good reasonably when compared to the compression test results.
6. The above conditions could be satisfied by an equivalent square distribution defined as follows:
   a. A stress of concrete average of 0.85 $f_c$', was assumed to be uniformly distributed over an equivalent compressive area bounded by a cross-sectional edge and a straight line parallel to a neutral axis of a = $\beta_1$ c of fiber with maximum compressive strain.
   b. The distance from the fiber with maximum strain to the neutral axis c should be measured in the direction perpendicular to the neutral axis.
   c. For $f_c$' between 17 and 28 MPa, $\beta_1$ must be taken at 0.85. For $f_c$' above 28 Mpa, $\beta_1$ must be reduced by 0.05 for any 7 MPa over 28 Mpa, but $\beta_1$ should not be taken less than 0.65 or could be written in mathematical form as:

$$\beta_1 = 0.85 - \frac{1}{140}(f_c' - 28) \geq 0.65 \quad (1)$$

*1.1.b The asumption of steel stress*

The idealization of steel stress with strain to be used as commonly used in structural analysis was bi-linear as shown below :

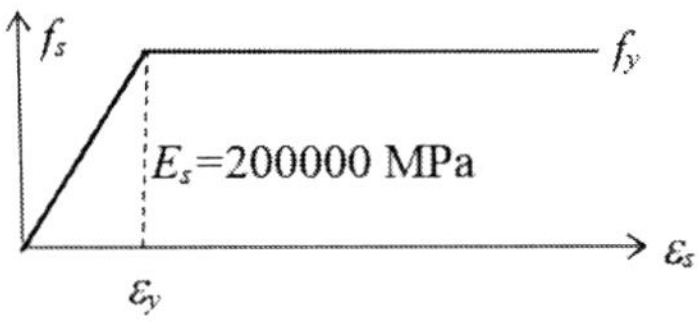


**Figure 2**. Stainless-steel strain idealization

*1.2 Balanced condition*

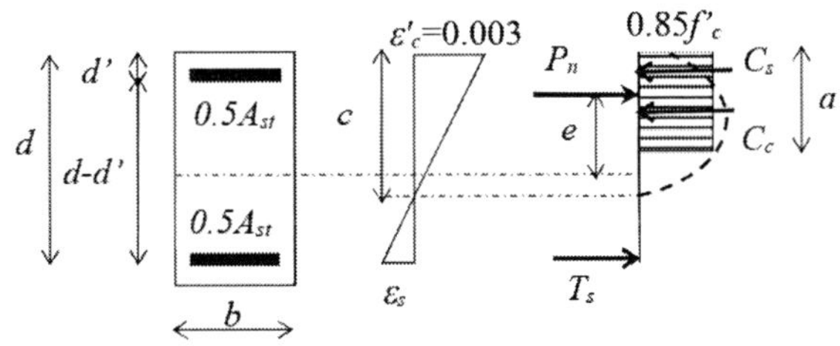


Figure 3. Cross section of columns and the stresses

From the drawing, obtained the relationship:

$$\frac{c}{d} = \frac{\varepsilon'_c}{\varepsilon'_c + \varepsilon_s} \rightarrow \xi = \frac{\varepsilon'_c}{\varepsilon'_c + \varepsilon_s} \quad (2)$$

For the balanced condition :

$$\xi_b = \frac{\varepsilon'_c}{\varepsilon'_c + \varepsilon_y} = \frac{0.003}{0.003 + \frac{f_y}{E_s}} = \frac{0.003}{0.003 + \frac{f_y}{200000}} = \frac{600}{600 + f_y} \quad (3)$$

$$\sum H = 0$$

$$C_c + C_s - T_s - P_b = 0 \quad (4)$$

Where : $C_s = T_s \rightarrow P_b = C_c = 0.85\beta_1 f_c' b c_b$ (5)

*1. 3 Region II*

$P_b < P < P_n$

Compression forces :

$$C_c = 0.85 f_c' ab = 0.85 \beta_1 f_c' bc \quad (6)$$

$$C_s = 0.5 A_{st} f_y = 0.5 \rho_t bd f_y \quad (7)$$

Tensile forces :

$$T_s = 0.5 A_{st} f_s = 0.5 \rho_t bd \varepsilon_s E = 300 \rho_t bd \left(\frac{d}{c} - 1\right) \quad (8)$$

$$\sum H = 0$$

$$C_c + C_s - T_s - P_n = 0$$

$$0.85 \beta_1 f_c' bc^2 + \{\rho_t bd(0.5 f_y + 300) - P_n\} c - 300 \rho_t bd^2 = 0 \quad (9)$$

$$\sum M = 0$$

$$-C_c\left(d - \frac{a}{2}\right) - C_s(d - d') + P_n\left(e + \frac{d-d'}{2}\right) = 0 \quad (10)$$

$$0.85 \beta_1^2 f_c' bc^2 - 1.7 \beta_1 f_c' bdc - \rho_t bd f_y (d - d') + P_n(2e + d - d') = 0 \quad (11)$$

From the above two squares equations (9 &11) it is seen that the value of c depended on $\rho_t$ and vice versa so it was difficult to solve.

*1.4 Region III*

$P < P_b$

$$C_c = 0.85 f_c' ab$$

$$C_s = 0.5 A_{st} f_y = 0.5 \rho_t bd f_y$$

$$T_s = 0.5 A_{st} f_y = 0.5 \rho_t bd f_y$$

$$\sum H = 0$$

$$C_c + C_s - T_s - P_n = 0$$

$$P_n = C_c = 0.85 f_c' ab \rightarrow a = \frac{P_n}{0.85 f_c' b} \quad (12)$$

$$\sum M = 0$$

$$-P_n\left(d - \frac{a}{2}\right) - C_s(d - d') + P_n\left(e + \frac{d-d'}{2}\right) = 0$$

$$A_{st} = \frac{P_n\left(2e - h + \frac{P_n}{0.85 f_c' b}\right)}{f_y(d - d')} \quad (13)$$

From this equation (13) it could be seen that the designed of reinforced concrete columns for region III could be calculated directly so that the discussion needed not be elaborated further.

*1.5 Region II for 4 sided reinforcement*

With each 0.25 $A_{st}$ reinforcement on each side. ($P_b < P < P_n$)

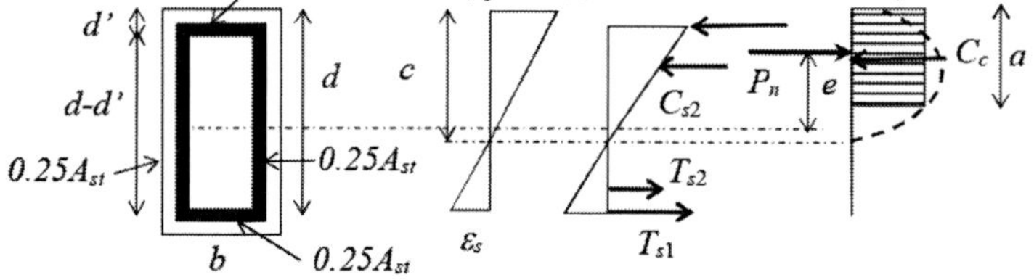


**Figure 4**. Cross section for 4 sided reinforement

The forces acting on the cross section :

Compression forces :

$$C_c = 0.85 f_c' ab = 0.85\beta_1 f_c' bc$$

$$C_{s1} = 0.25\rho_t bdf_y \quad (14)$$

$$C_{s2} = \frac{c-d'}{c} 0.5A_{st}\tfrac{2}{3} f_{s'} = 200\rho_t \frac{bd}{c^2}(c-d')^2 \quad (15)$$

Tensile forces :

$$T_{s1} = 0.25\rho_t bd\varepsilon_s E_s = 150\rho_t bd\left(\frac{d}{c}-1\right) \quad (16)$$

$$T_{s2} = \frac{d-c}{c} 0.5A_{st}\tfrac{2}{3} f_s = 200\rho_t \frac{bd}{c^2}(d-c)^2 \quad (17)$$

$$\sum H = 0$$

$$C_c + C_{s1} + C_{s2} - T_{s1} - T_{s2} - P_n = 0 \quad (18)$$

$$\text{If}: \frac{d'}{d} = R \quad (19)$$

$$0.85\beta_1 f_c' c^3 + \left\{\rho_t d\left(0.25f_y + 150\right) - \frac{P_n}{b}\right\}c^2 + 250\rho_t d^2(1-1.6R)c - 200\rho_t d^3 = 0 \quad (20)$$

$$\sum M = 0$$

$$-C_c\left(d - \frac{a}{2}\right) - C_{s1}(d-d') - C_{s2}\left\{\tfrac{2}{3}(c-d') + d - c\right\} + T_{s2}\tfrac{1}{3}(d-c) + P_n\left(e + \frac{d-d'}{2}\right) = 0 \quad (21)$$

$$0.85\beta_1^2 f_c' c^4 - 1.7\beta_1 f_c' dc^3 - \left\{0.5\rho_t df_y(d-d') - \frac{P_n}{b}(2e + d - d')\right\}c^2 - \frac{400}{3}\rho_t d\{(c - d')^2\{3d(1-\tfrac{2}{3}R) - c\} - (d-c)^3\} = 0 \quad (22)$$

## 2. Research methods

In this research the normal force equilibrium equation, the form was converted to equation with $\xi_N = \left(\frac{c}{d}\right)$ as a variable :

$$0.85\beta_1 f_c' \xi_N^2 + \left\{\rho_t\left(0.5f_y + 300\right) - \frac{P_n}{bd}\right\}\xi_N - 300\rho_t = 0 \quad (23)$$

This equation could not be solved because the value $\xi_N$ depended on $\rho t$ and vice versa so that there was a solution it was determined $\rho t$ first so that $\xi_N$ could be determined. While the other value was a known factor of a column.

Therefore this equation would be solved by taking the value of the column reinforcement ratio $\rho_{t1}$, $\rho_{t2}$ and $\rho_{t3}$ which had the same difference it would be obtained three pairs of numbers $\{\rho_{t1}, \xi_{N1}\}$, $\{\rho_{t2}, \xi_{N2}\}$ and $\{\rho_{t3}, \xi_{N3}\}$ which were assumed to satisfy **quadratic polynomial**.

Similarly, the moment equilibrium equation, the form was converted to equation with $\xi_M = \left(\frac{c}{d}\right)$ as variable :

$$0.85\beta_1^2 f_c' \xi_M^2 - 1.7\beta_1 f_c' \xi_M - \rho_t f_y(1-R) + \frac{P_n}{bd^2}(2e + d - d') = 0 \quad (24)$$

In the same way above, taking the values of the column ratios of $\rho_{t1}$, $\rho_{t2}$ dan $\rho_{t3}$ that had the same difference would be obtained three pairs of numbers $\{\rho_{t1}, \xi_{M1}\}$, $\{\rho_{t2}, \xi_{M2}\}$ dan $\{\rho_{t3}, \xi_{M3}\}$ which were assumed to satisfy **quadratic polynomial.**

This was identical to the assumption used by Simpson in numerical integration that using three points as the basis of calculation assuming it meeted the quadratic polynomial and given quite accurate results. The value of $\{\rho_t, \xi\}$ occurring was that which satisfies the normal force equilibrium curve and the moment equilibrium curve or, in other words, the point of intersection of the two curves.

*2.1 Determining the value of ξ from the normal force and moment equilibrium equations*

The normal force equilibrium equation could be written as:

$0.85\beta_1 f_c' \xi_H^2 + N\xi_H - 300\rho_t = 0$ (25)

where : $N = \rho_t(0.5f_y + 300) - \frac{P_n}{bd}$ (26)

By solving the quadratic equation, it was obtained :

$\xi_N = \frac{1}{1.7\beta_1 f_c'}\left(-N + \sqrt{N^2 + 1020\rho_t\beta_1 f_c'}\right)$ (27)

From the above equation it can be seen that $\xi_H$ depended on $N$ and the compressive strenght of theconcrete. Of the three values of $N_{1.0}$ , $N_{1.5}$ dan $N_{2.0}$ were determined by three variation $\rho_{1.0} = 1.0$ %, $\rho_{1.5} = 1.5$ % and $\rho_{2.0} = 2.0$ % so that the value $\xi_{N1.0}$, $\xi_{N1.5}$ and $\xi_{N2.0}$ could be obtained.

So that the three points obtained $\{(0.01, \xi_{N1.0}), (0.15, \xi_{N1.5})$ and $(0.02, \xi_{N2.0})\}$

The moment equilibrium equation could be written as :

$0.85\beta_1^2 f_c' \xi_M^2 - 1.7\beta_1 f_c' \xi_M + M = 0$ (28)

where : $M = -\rho_t f_y(1 - R) + \frac{P_n}{bd^2}(2e + d - d')$ (29)

By solving the quadratic equation, it was obtained :

$\xi_M = \frac{1}{\beta_1}\left(1 - \sqrt{1 - \frac{M}{0.85f_c'}}\right)$ (30)

From the equation above showed that $\xi_M$ depended on $M$ and the compressive strenght of the concrete. Of the three values of $M_{1.0}$, $M_{1.5}$ dan $M_{2.0}$ could be determined with three variations $\rho_{1.0} = 1.0$ %, $\rho_{1.5} = 1.5$ % dan $\rho_{2.0} = 2.0$ % so the value $\xi_{M1.0}$, $\xi_{M1.5}$ and $\xi_{M2.0}$ could be obtained.

Thus, there were three points $\{(0.01, \xi_{M1.0}), (0.15, \xi_{M1.5})$ and $(0.02, \xi_{M2.0})\}$

*2.2 Determining the reinforcement ratio on the basis of $\rho_{1.0}$, $\rho_{1.5}$ and $\rho_{2.0}$*

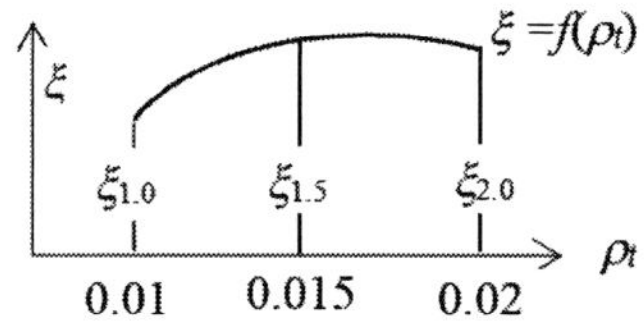


**Figure 5**. Quadratic equation curve

$$\xi = 20000(\xi_{2.0} - 2\xi_{1.5} + \xi_{1.0})\rho_t^2 + 100(-5\xi_{2.0} + 12\xi_{1.5} - 7\xi_{1.0})\rho_t + 3\xi_{2.0} - 8\xi_{1.5} + 6\xi_{1.0} \quad (31)$$

Forms of quadratic equations on the basis of $\xi_{N1.0}$, $\xi_{N1.5}$ and $\xi_{N2.0}$ or $\xi_{M1.0}$, $\xi_{M1.5}$ and $\xi_{M2.0}$

The $\rho_t$ value that occurs was the abscissa value of the intersection of the curve of normal force and the curve moment equilibrium and as an illustration saw the picture below in figure 6.

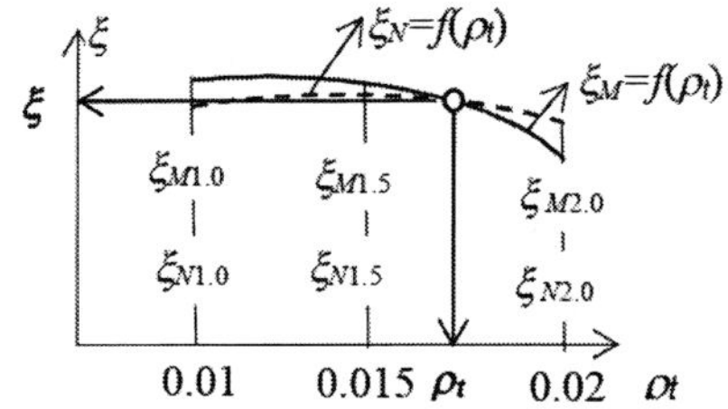


**Figure 6**. Illustration of intersection of two curves normal force and moment

For easy writing, next

$[\xi_{N2.0} - \xi_{M2.0}] = Y_{2.0}$ (32)

$[\xi_{N1.5} - \xi_{M1.5}] = Y_{1.5}$ (33)

$[\xi_{N1.0} - \xi_{M1.0}] = Y_{1.0}$ (34)

So the quadratic equation to be solved was:

$$20000(Y_{2.0} - 2Y_{1.5} + Y_{1.0})\rho_t^2 + 100(-5Y_{2.0} + 12Y_{1.5} - 7Y_{1.0})\rho_t + 3Y_{2.0} - 8Y_{1.5} + 6Y_{1.0} = 0 \quad (35)$$

Or : $20000A_{1.5}\rho_t^2 + 100B_{1.5}\rho_t + C_{1.5} = 0$ (36)

$A_{1.5} = Y_{2.0} - 2Y_{1.5} + Y_{1.0}$ (37)

$B_{1.5} = -5Y_{2.0} + 12Y_{1.5} - 7Y_{1.0}$ (38)

$C_{1.5} = 3Y_{2.0} - 8Y_{1.5} + 6Y_{1.0}$ (39)

$$\rho_t = \frac{B_{1.5}}{400A_{1.5}}\left(-1 + \sqrt{1 - 8\frac{A_{1.5}C_{1.5}}{B_{1.5}^2}}\right) \quad (40)$$

*2.3 Determining the reinforcement ratio on the basis of $\rho_{2.0}$, $\rho_{2.5}$ and $\rho_{3.0}$*

In case $\rho_t > 0.02$ then it was better to use $\rho_{2.0}$ = 2,0 %, $\rho_{2.5}$ = 2.5 % dan $\rho_{3.0}$ = 3.0 %, as the based of the analysis.

$$\xi = 20000(\xi_{3.0} - 2\xi_{2.5} + \xi_{2.0})\rho_t^2 + 100(-9\xi_{3.0} + 20\xi_{2.5} - 11\xi_{2.0})\rho_t + 10\xi_{3.0} - 24\xi_{2.5} + 15\xi_{2.0} \quad (41)$$

$[\xi_{N2.5} - \xi_{M2.5}] = Y_{2.5}$ (42)

$[\xi_{N3.0} - \xi_{M3.0}] = Y_{3.0}$ (43)

So the quadratic equation to be solved was :

$$20000(Y_{3.0} - 2Y_{2.5} + Y_{2.0})\rho_t^2 + 100(-9Y_{3.0} + 20Y_{2.5} - 11Y_{2.0})\rho_t + 10Y_{3.0} - 24Y_{2.5} + 5Y_{2.0} = 0 \quad (44)$$

Or : $20000A_{2.5}\rho_t^2 + 100B_{2.5}\rho_t + C_{2.5} = 0$ (45)

$A_{2.5} = Y_{3.0} - 2Y_{2.5} + Y_{2.0}$ (46)

$B_{2.5} = -9Y_{3.0} + 20Y_{2.5} - 11Y_{2.0}$ (47)

$C_{2.5} = 10Y_{3.0} - 24Y_{2.5} + 15Y_{2.0}$ (48)

$$\rho_t = \frac{B_{2.5}}{400A_{2.5}}\left(-1 + \sqrt{1 - 8\frac{A_{2.5}C_{2.5}}{B_{2.5}^2}}\right) \quad (49)$$

*2.4 Determining the reinforcement ratio on the basis of $\rho_{3.0}$, $\rho_{3.5}$ and $\rho_{4.0}$*

In case $\rho_t > 0.03$ then it was better to use $\rho_{3.0}$ = 3,0 %, $\rho_{3.5}$ = 3.5 % and $\rho_{4.0}$ = 4.0 %, as the base of the analysis.

$$\xi = 20000(\xi_{4.0} - 2\xi_{3.5} + \xi_{3.0})\rho_t^2 + 100(-13\xi_{4.0} + 28\xi_{3.5} - 15\xi_{3.0})\rho_t + 21\xi_{4.0} - 48\xi_{3.5} + 28\xi_{3.0} \quad (50)$$

$[\xi_{N3.5} - \xi_{M3.5}] = Y_{3.5}$ (51)

$[\xi_{N4.0} - \xi_{M4.0}] = Y_{4.0}$ (52)

So the quadratic equation to be solved was :

$$20000(Y_{4.0} - 2Y_{3.5} + Y_{3.0})\rho_t^2 + 100(-13Y_{4.0} + 28Y_{3.5} - 15Y_{3.0})\rho_t + 21Y_{4.0} - 48Y_{3.5} + 28Y_{3.0} = 0 \quad (53)$$

Or : $20000A_{3.5}\rho_t^2 + 100B_{3.5}\rho_t + C_{3.5} = 0$ (54)

$A_{3.5} = Y_{4.0} - 2Y_{3.5} + Y_{3.0}$ (55)

$B_{3.5} = -13Y_{4.0} + 28Y_{3.5} - 15Y_{3.0}$ (56)

$C_{3.5} = 21Y_{4.0} - 48Y_{3.5} + 28Y_{3.0}$ (57)

$$\rho_t = \frac{B_{3.5}}{400A_{3.5}}\left(-1 + \sqrt{1 - 8\frac{A_{3.5}C_{3.5}}{B_{3.5}^2}}\right) \quad (58)$$

*2.5 Research of region II for 4-sided reinforcement*

The normal force equilibrium equation :

As in the above of this research the normal force equilibrium equation, changed to the equation $\xi_N = \left(\frac{c}{d}\right)$ as a variable :

$$0.85\beta_1 f_c'\xi_N^3 + \left\{\rho_t(0.25f_y + 150) - \frac{P_n}{bd}\right\}\xi_N^2 + 250\rho_t(1 - 1.6R)\xi_N - 200\rho_t = 0 \quad (59)$$

Where : $N_4 = \rho_t(0.25f_y + 150) - \frac{P_n}{bd}$ (60)

The above equation (59) was in the form of a cubic polynomial so that $\xi_N$ was obtained with the aid of the Newton-Raphson method.

This equation would be solved by taking the value of the column reinforcement ratio $\rho_{t1}$, $\rho_{t2}$ and $\rho_{t3}$ which have the same difference it would be obtained three pairs of numbers $\{\rho_{t1}, \xi_{N1}\}$, $\{\rho_{t2}, \xi_{N2}\}$ and $\{\rho_{t3}, \xi_{N3}\}$ assumed to satisfy **quadratic polynomial.**

The moment equilibrium equation :

As in the above section of this research the the moment equilibrium equition, changed to the equation $\xi_N = \left(\frac{c}{d}\right)$ as a variable:

$$0.85\beta_1^2 f_c' \xi_M^4 - 1.7\beta_1 f_c' \xi_M^3 - \left\{0.5\rho_t f_y(1-R) - \frac{P_n}{bd^2}(2e + d - d')\right\}\xi_M^2 - 400(1-R)^2\rho_t\xi_M + \frac{400}{3}\rho_t(2R^3 - 3R^2 + 1) = 0 \quad (61)$$

Where : $M_4 = -\left\{0.5\rho_t f_y(1-R) - \frac{P_n}{bd^2}(2e + d - d')\right\}$ (62)

The above equation (61) was in the form of a quartic polynomial so that $\xi_M$ was obtained with the aid of the Newton-Raphson method.

This equation was solved by taking the values of the column ratios of $\rho_{t1}$, $\rho_{t2}$ and $\rho_{t3}$ which had the same difference then there would be three pairs of numbers $\{\rho_{t1}, \xi_{M1}\}$, $\{\rho_{t2}, \xi_{M2}\}$ and $\{\rho_{t3}, \xi_{M3}\}$ which is assumed to satisfy **quadratic polynomial.**

The value of $\{\rho_t, \xi\}$ occurring was that which satisfies the equilibrium the normal force curve and equilibrium the moment curve or, in other words, the point of intersection of the two curves.

## 3. Result

The case of double-sided reinforcement with the following data : $f_c'$ = 40 MPa, $f_y$ = 420 MPa, $b$ = 0.8 m, $d$ = 0.85 m, $d'$ = 0.15 m and $e$ = 0.3 m

Estimate $A_{st} = 2\%bd$ (Tied column)

$P_0 = 0.85f_c'(A_g - A_{st}) + f_y A_{st} = 26.207$ MN

$P_n = 0.8P_0 = 28.37$MN

$P = \frac{P_u}{0.65} = 16.2$ MN

$$c_b = \frac{600}{600 + 420}0.72 = 0.424$$

$P_b = 0.85\beta_1 f_c' b c_b = 9.36$ MN

$P_b < P < P_n$ → *region* II

| | | | |
|---|---|---|---|
| $f_c'$ = | 40 | MPa | |
| $f_y$ = | 420 | MPa | |
| $P_u$ = | 10,53 | MN | |
| $b$ = | 0,8 | m | |
| $d$ = | 0,85 | m | |
| $d'$ = | 0,15 | m | |
| $e$ = | 0,3 | m | |
| $\beta_1$ = | 0,764 | | |
| $\rho$ | 1,00 | 1,50 | 2,00 |
| $N$ | -18,724 | -16,174 | -13,624 |
| $\xi_N$ | 0,856 | 0,831 | 0,81 |
| $M$ | 32,977 | 31,248 | 29,518 |
| $\xi_M$ | 1,082 | 0,937 | 0,834 |
| $Y$ | -0,226 | -0,106 | -0,024 |
| $A_{1.5}$= | -0,038 | | |
| $B_{1.5}$= | 0,43 | | |

$C_{1.5}$ = -0,58
$\rho_t$ = 0,022196

That mean not cut between the two curves were at $0.01 \leq \rho_t \leq 0.02$ then recalculated with $\rho_{2.0}$, $\rho_{2.5}$ and $\rho_{3.0}$ as the basis :

| $\rho$ | 2,00 | 2,50 | 3,00 |
|---|---|---|---|
| $N$ | -13,624 | -11,074 | -8,524 |
| $\xi_N$ | 0,81 | 0,791 | 0,775 |
| $M$ | 29,518 | 27,789 | 26,060 |
| $\xi_M$ | 0,834 | 0,749 | 0,676 |
| $Y$ | -0,024 | 0,042 | 0,099 |

$A_{2.5}$ = -0,009
$B_{2.5}$ = 0,213
$C_{2.5}$ = -0,378
$\rho_t$ = 0,021741

Results from the fortran program : $\rho_t$ = 0.0217 and validation by the iteration method : $\rho_t$ = 0,0219 (taken from the analysis but not shown)

| Analysis method | $\rho_t$ | Ratio to the iteration method |
|---|---|---|
| Manual | 0.021741 | 99.27% |
| Fortran program | 0.0217 | 99.09% |
| Iteration Gauss-Seidel | 0.0219 | 100% |

## 4. Conclusion
The most unique of these approaches was that the tables or graphs ($N$ vs $\xi_N$) and ($N$ vs $\xi_N$) were only distinguished on the basis of the compressive strenght of the concrete alone while the yield strenght of the reinforcement could be adjusted to the available market. To ensure the accuracy of this method used $\rho_{1.0}$ , $\rho_{1.5}$ and $\rho_{2.0}$ or $\rho_{2.0}$, $\rho_{2.5}$ and $\rho_{3.0}$ or $\rho_{3.0}$, $\rho_{3.5}$ and $\rho_{4.0}$ as a basis for calculation two sides and calculation with 4 sided of reinforcement. If with the help of fortran program all combination of concrete and reinforcement could be completed so that the application was very flexible.

## 5. References
[1] Ayuddin, Parung H, Tjaronge W and Djamaluddin R 2013 *Behavior of Concrete Coloms Externally Confined By Carbon Fiber Reinforce Polymer* Master Thesis, (Indonesia: Universitas Hasanuddin) pp 36-38
[2] Badan Standarisasi Nasional (BSN) 2013 SNI 2487-2013 *Persyaratan beton struktural untuk bangunan gedung* Jakarta pp 74
[3] Biazar J and Ghanbary B 2008 A new technique for solving systems of nonlinear equations, *Applied Mathematical Sciences* **2** 1-5
[4] Chapra, Steven C, Canale and Raymond P 1996 *Metode numerik*, jilid 1, edisi kedua (Indonesia:Penerbit Erlangga, Jakarta) pp 135-328
[5] Mac Gregor and James G 1988 *Reinforced concrete (Mechanics and design) Third edition* (New Jersey: Prentice-Hall International Inc.) pp 446-448
[6] Muthumalai and Ramesh Kumar 2012 Note on Newton interpolation formula, *Int. Journal of Math. Analysis* **6** pp 1-4
[7] Suryoatmono and Bambang 1997 Bahasa Fortran : *Dari fortran IV hingga fortran powerstation* (Bandung: Penerbit Eresco) pp 5-67

# Behaviour of reinforced concrete slabs with steel fibers

**A O Baarimah[1] and S M Syed Mohsin[1]**

[1] Faculty of Civil Engineering and Earth Resources, Universiti Malaysia Pahang, Lebuhraya Tun Razak, 26300 Gambang, Kuantan, Pahang, Malaysia
Corresponding author: eng.baarmh@gmail.com

**Abstract.** This paper investigates the potential effect of steel fiber added into reinforced concrete slabs. Four-point bending test is conducted on six slabs to investigate the structural behaviour of the slabs by considering two different parameters; (i) thickness of slab (ii) volume fraction of steel fiber. The experimental work consists of six slabs, in which three slabs are designed in accordance to Eurocode 2 to fulfil shear capacity characteristic, whereas, the other three slabs are designed with 17% less thickness, intended to fail in shear. Both series of slabs are added with steel fiber with a volume fraction of $V_f = 0\%$, $V_f = 1\%$ and $V_f = 2\%$ in order to study the effect and potential of fiber to compensate the loss in shear capacity. The slab with $V_f = 0\%$ steel fiber and no reduction in thickness is taken as the control slab. The experimental result suggests promising improvement of the load carrying capacity (up to 32%) and ductility (up to 87%) as well as delayed in crack propagation for the slabs with $V_f = 2\%$. In addition, it is observed that addition of fibers compensates the reduction in the slab thickness as well as changes the failure mode of the slab from brittle to a more ductile manner.

## 1. Introduction

Concrete has become one of the most important construction materials commonly used in many types of engineering structures. Concrete is a material which is strong in compression and weak in tension, thus causes cracking in the tension zone. Reinforced concrete is a combination of concrete and steel where the steel reinforcement improves the tensile strength lacking in the concrete. In the past decades, fiber reinforced concrete (FRC) has been gaining more attention in the development and are used in numerous types of civil engineering application such as shotcrete, pavement slabs, precast products, tunnel linings, seismic structures, bridge deck slab repairs, marine and refractory applications [1-5]. There are many advantages of adding fiber into reinforced concrete, such as improving the load carrying capacity and ductility of the structural members, controlling crack propagation, increasing energy absorption and altering the mode of failure [6-11].

Steel fiber reinforced concrete (SFRC) is a composite material containing Portland cement, water, aggregate and adding discrete discontinuous steel fiber. Steel fiber has been demonstrated its efficiency in enhancing the structural behaviour of reinforced concrete structural members [12-17].

Based on the literature, promising results were observed in the potential of steel fibers to serve as part of shear reinforcement in reinforced concrete beams, columns and beam-column joints. However, limited research is carried out to study this potential in reinforced concrete slab, especially for the case in which the thickness of the slab is controlled or reduced. Therefore, this study aims to investigate the effect of steel fiber added into reinforced concrete slabs as well as its potential to serve as part of shear reinforcement through the decrease in slab thickness.

## 2. Materials and methods

The concrete mixture was designed in accordance with British Standards (BS EN 206-1, 2000) for 20 MPa of concrete compressive strength. Three concrete mixtures were produced for SFRC using different volume fraction ($V_f$) as listed in table 1. The first mixture was a reference mixture (control) without adding any fiber, $V_f = 0\%$. Hooked-end steel fiber was added into the concrete mixtures. Table 2 presents the properties of the steel fiber. In addition, super plasticizer (SP) was used in the mixtures to improve workability and achieve the required slump.

**Table 1.** Concrete mix design.

| Materials | Mix 1 ($V_f = 0\%$) | Mix 2 ($V_f = 1\%$) | Mix 3 ($V_f = 2\%$) |
|---|---|---|---|
| Cement (kg/m$^3$) | 280 | 280 | 280 |
| Coarse Aggregate (kg/m$^3$) | 1292 | 1292 | 1292 |
| Fine Aggregate (kg/m$^3$) | 556 | 556 | 556 |
| Water (Liter/m$^3$) | 162 | 162 | 162 |
| W/C ratio | 0.58 | 0.58 | 0.58 |
| SP (Liter/m$^3$) | 5.6 | 5.6 | 5.6 |
| Steel fiber (kg/m$^3$) | 0 | 75 | 150 |

**Table 2.** Steel fiber properties.

| Properties | Steel Fiber |
|---|---|
| Length, L (mm) | 60 |
| Diameter, D (mm) | 0.75 |
| Aspect Ratio, L/D | 80 |
| Tensile Strength (MPa) | 1100 |
| Unit Weight (kg/m3) | 7500 |

In order to measure the compressive and flexural stress of the concrete mixtures, compression test and flexural test were conducted in this study using compression (cube) test and four-point bending test, respectively. Three cubes as well as three prisms were prepared for each mixture. A total number of nine cubes with a standard size of 150 x 150 x 150 mm as well as nine prisms with a dimension of 100 x 100 x 500 mm were tested on 28$^{th}$ day as recommended in British Standards BS EN 12390-3, 2009 and BS EN 12390-5, 2009, respectively.

In this study, two thicknesses of reinforced concrete slabs were prepared. The dimension of the first series of slab (slab size 1) was 1000 x 500 x 120 mm, while the dimension of the second series (slab size 2) was 1000 x 500 x 100 mm in length, width and depth, respectively. The difference in the slab thickness is to cater for the potential of the fibers to serve as part of shear reinforcement in the reinforced concrete slab. The second series of slab was designed with the thickness of slab less than the required so that the slab would be failed in shear. The reinforcement of slab was provided as mesh with the diameter of 10 mm, the main steel bar was H10 – 150 mm (4H10) and secondary steel bar was H10 – 320 mm (4H10).

The loading arrangement and details of the slab are shown in figure 1. Six samples of SFRC slabs were prepared for casting and were tested on the 28$^{th}$ day. For the first series, the slab S1-0, S1-1, S1-2 were added with fibers by a volume fraction of $V_f = 0\%$, $V_f = 1\%$ and $V_f = 2\%$, respectively, whereas the second series, slab S2-0, S2-1, S2-2 were added with fibers by a volume fraction of $V_f = 0\%$, $V_f = 1\%$ and $V_f = 2\%$, respectively. Slab S1-0 is considered as the control slab.

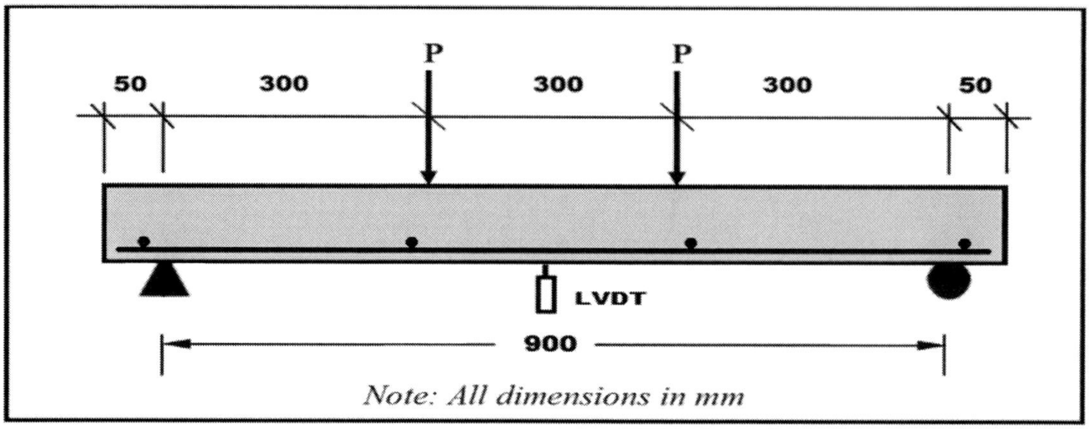


**Figure 1.** Load arrangement and details of the slab.

All specimens of the slab were tested under four-points bending test by using a hydraulic machine (Magnus Frame) with a capacity of 300 kN under static monotonic loading until failure over a clear span of 900 mm with a shear span of 300 mm as well as the distance between the two loading points of 300 mm. The mid-span deflection was measured by using linear variable differential transducer (LVDT) which was located at the center of the slab whereas the load cell indicated the applied load test set up for the slabs as shown in the figure above. During the loading, the crack propagation of the slabs sides was marked and identified their location.

### 3. Results and discussion

Table 3 shows the average compressive and flexural strength of the SFRC cubes and prisms. The observation shows that the compressive and flexural strength of the concrete with added fibers were higher than that of the concrete without fiber. The enhancement of compressive and flexural strength increase with increasing fiber volume fraction. The compressive strength of 1% and 2 % steel fiber specimens were compared to the control concrete increased by 6.3% and 24.7%, respectively. Furthermore, by adding fibers of 1% and 2 %, the flexural strength was increased by 114% and 197 %, respectively. These findings are in agreement with previous studies [18, 19].

**Table 3.** Compressive and flexural strength results.

| $V_f$% | Compressive Strength (MPa) | Flexural Strength (MPa) |
|---|---|---|
| 0 | 22.30 | 3.50 |
| 1 | 23.70 | 7.50 |
| 2 | 27.8 | 10.40 |

The relationship between load and deflection for first series and second series slabs are illustrated in figure 2 and figure 3, respectively. From figure 2, it can be seen that the inclusion of fibers has a moderate influence on the structural performance of the SFRC slabs. The strength of the SFRC slab with $V_f$ = 1% and 2% are significantly higher in comparison to the control slab. There was a sudden drop in the load-deflection curve for S1-2 which could be due to the over reinforcement which in turn leads to the brittle behaviour as compared to the control slab. In addition, the slab becomes stiffer and less deflection (this is largely attributable to the fibers' role in bridging cracks and limiting their opening) [14].

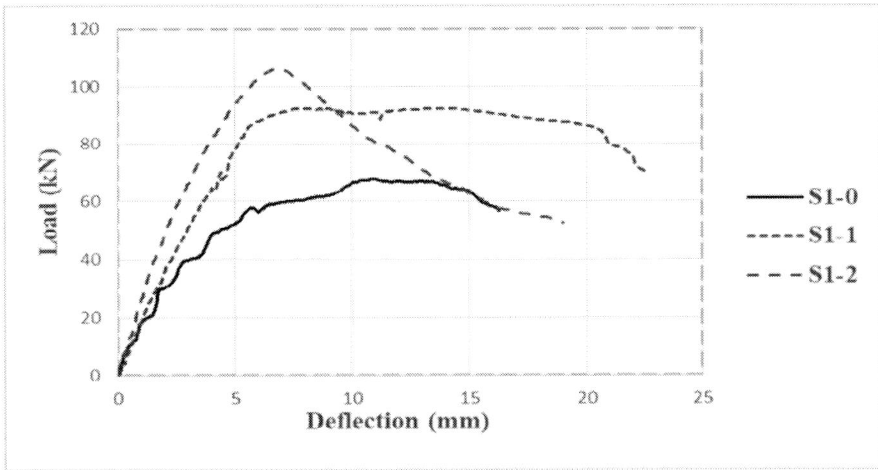


**Figure 2.** Load deflection curves for the first series of SFRC slabs.

For the second series slab, it can be seen that addition of fibers in the slab with reduced thickness resulted in better strength as compared to the slab without fiber (refer to figure 2 and 3). The results indicated that the steel fiber demonstrates its capability to compensate the loss of the shear capacity (from the slab thickness) and improves the slab structural behaviour while serving as part of shear reinforcement in the SFRC slabs. Furthermore, there was a proportional relationship between the ductility and the fiber volume fraction where the ductility of the SFRC slabs increases when fiber volume fraction was increased. This confirms the fact that adding fiber enhances the ductility of the brittle characteristics for concrete. It is apparent that the deflection of the S2-1 and S2-2 slabs was higher as compared to the slab S2-0. Moreover, the higher load was required to produce the deflection, suggesting that the slab is ductile and can sustain higher load carrying capacity.

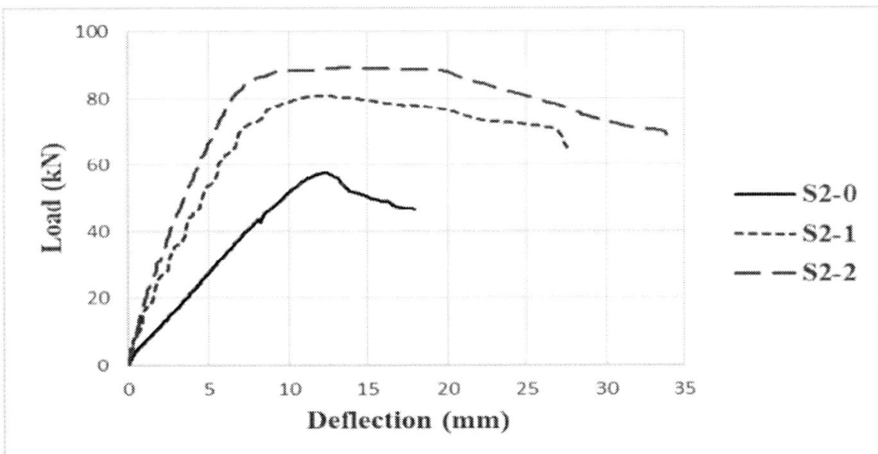


**Figure 3.** Load deflection curves for the second series of SFRC slabs.

The key parameters of strength and ductility from the load-deflection curves in Figure 2 and Figure 3 were summarized in table 4 and table 5, respectively. The key parameters involve the load at yield (*Py*) and its respective deflection ($\delta y$), the ultimate load (*Pu*) which represents the ultimate load at failure (taken as 85% of the maximum load) and its respective deflection ($\delta u$) and the maximum load (*Pmax*) and its respective deflection ($\delta max$). Ductility ratio ($\mu$) was computed by dividing the deflection at ultimate load to the deflection at yield ($\mu = \delta u / \delta y$).

It can be seen from the tables that the maximum strength (*Pmax*) and yield load (*Py*) for the second series of SFRC slabs are higher than that of slab without fiber. Added fibers are acting to keep the concrete matrix together. Subsequently, higher loading was needed to initiate the crack propagation. Moreover, fibers also serve as part of shear reinforcement to enhance the shear capacity of the slab as well as compensate the loss in concrete shear capacity of slab due to a reduction in the slab thickness.

**Table 4.** Result for the first series of SFRC slabs.

| Specimen | *Py* (kN) | *δy* (mm) | *Pu* (kN) | *δu* (mm) | *Pmax* (kN) | *δmax* (mm) | *μ= δu / δy* |
|---|---|---|---|---|---|---|---|
| S1-0 | 57.75 | 5.83 | 57.53 | 16.15 | 67.68 | 10.87 | 2.77 |
| S1-1 | 70.32 | 4.38 | 78.67 | 21.36 | 92.49 | 7.86 | 4.88 |
| S1-2 | 80.81 | 3.94 | 90.23 | 9.40 | 106.16 | 6.86 | 2.40 |

**Table 5.** Result for the second series of SFRC slabs.

| Specimen | *Py* (kN) | *δy* (mm) | *Pu* (kN) | *δu* (mm) | *Pmax* (kN) | *δmax* (mm) | *μ= δu / δy* |
|---|---|---|---|---|---|---|---|
| S2-0 | 43.53 | 8.12 | 49.21 | 15.93 | 57.50 | 12.46 | 1.96 |
| S2-1 | 64.79 | 6.59 | 68.77 | 27.13 | 80.90 | 12.31 | 4.12 |
| S2-2 | 70.47 | 5.43 | 75.93 | 28.10 | 89.33 | 13.80 | 5.17 |

In term of ductility, it can be seen that the ductility ratio ($\mu$) of SFRC slabs continue to increase with the increase in the fiber volume fraction. However, there is an optimum amount of fibers depending on the amount of reinforcement or concrete shear capacity of the slab. For instance, in the first series of slab, the optimum amount of fibers can be taken as 1%, whereas for the second series of slab, the optimum amount of fibers increased to 2%. This because over reinforced of reinforced concrete structures tends to show a more brittle behaviour and should be avoided. Thus, it can be concluded that the inclusion of fibers introduces a ductile property into the concrete material, however, there is a certain limitation depending on the amount of the initial reinforcement presents in the structure.

Figure 4 to figure 7 illustrate the ratio of strength, ductility and energy absorption of the SFRC slabs normalized to control slab (thickness of 120 mm and $V_f$ = 0%) against fiber volume fraction. Figure 4 and Figure 5 show the ratio of the maximum load and the yield load to control slab (*Pmax/ Pmax,0*) and (*Py/Py,0*), respectively.

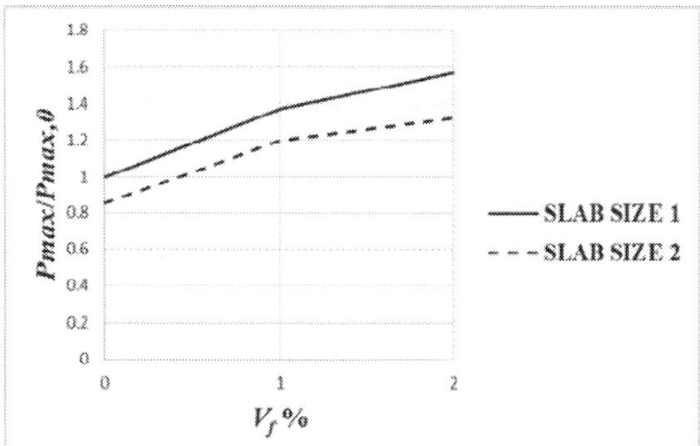


**Figure 4.** Graph *Pmax/Pmax,0* versus $V_f$.

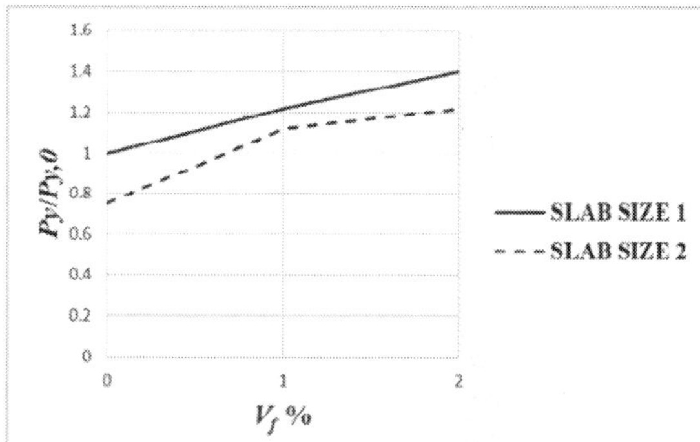


**Figure 5.** Graph *Py/Py,0* versus $V_f$.

A similar pattern was observed from the ratio of the maximum load (*Pmax/Pmax,0*) and load yield (*Py/Py,0*) as compared to the control slab. Slab S2-0 demonstrated a decrease in both maximum and yield ratios as compared to the control slab. On the other hand, the slabs with fibers show consistent enhancement the addition of fibers.

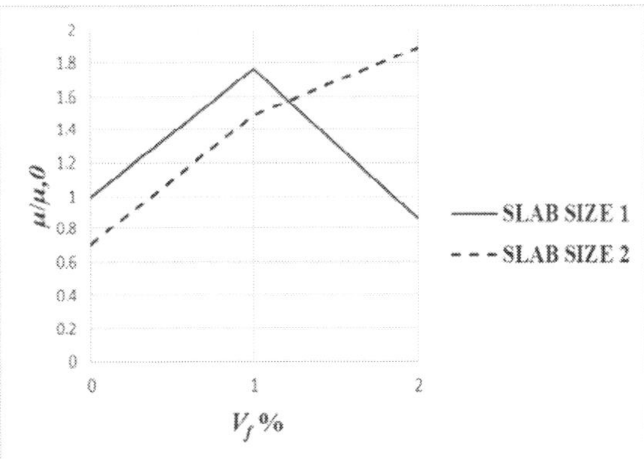


**Figure 6.** Graph $\mu/\mu,0$ versus $V_f$.

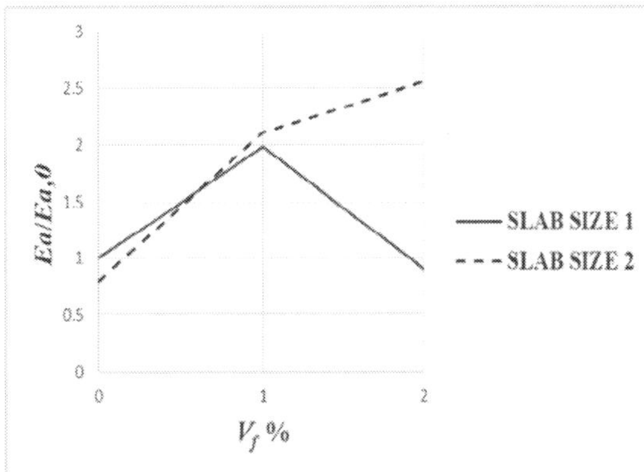


**Figure 7.** Graph *Ea/Ea,0* versus $V_f$.

Figure 6 illustrates the results for the ratio between ductility ratio for each slab ($\mu$) and that of the control slab specimen ($\mu,0$) plotted against volume fraction of steel fiber. The performance of ductility of the first series slab was an upward trend until $V_f$=1%, then followed by a sudden drop with $V_f$=2% due to over reinforcement which indicates to the early failure compared to the control slab. On the other hand, the ductility in second series slab was more promising with a continuous upward pattern. Moreover, at 1% of steel fiber volume fraction, the ductility ratio was higher than that obtained from the control slab. One of the key indicators of the structure's ability to absorb deformations which is energy absorption (*Ea*). Practically, energy absorption was calculated using the area under the load-deflection curve. The ratio between the energy absorption capacity of each slab (*Ea*) to the control slab (*Ea,0*) is given in figure 7. It is noticeable from the figures 6 and figures 7 that they have almost same pattern ratio graphs which confirmed that the actual trend and energy absorption ratio is similar to the ductility ratio which ensures its findings. These results are in agreement with previous studies reported by [3, 14]. It can be concluded that for significant improvement by inclusion steel fiber, the energy absorption was enhanced as compared with the control slab.

Figures 8, 9 and 10 represent the cracking pattern of the SFRC for the slabs S1-0, S1-1 and S1-2, respectively, whilst, the cracking pattern of the SFRC for the slabs S2-0, S2-1 and S2-2, are shown in figures 11, 12 and 13, respectively.

**Figure 8.** Cracking pattern at the failure of S1-0.

**Figure 9.** Cracking pattern at the failure of S1-1.

**Figure 10.** Cracking pattern at the failure of S1-2.

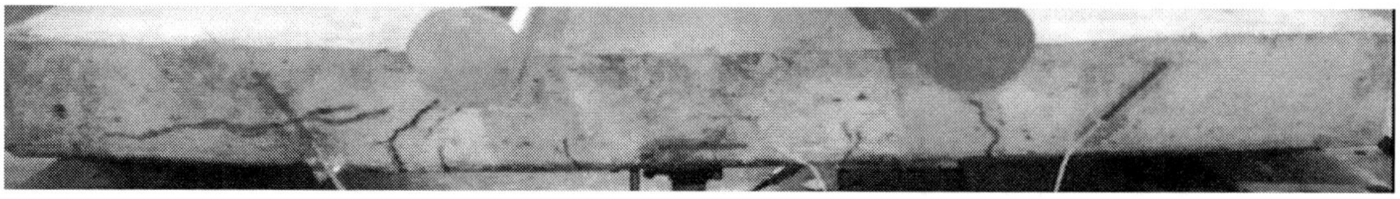

**Figure 11.** Cracking pattern at the failure of S2-0.

**Figure 12.** Cracking pattern at the failure of S2-1.

**Figure 13.** Cracking pattern at the failure of S2-2.

From the previous figures, most of the slabs showed cracking propagation along the mid-span and between support point and the loading. During testing, it was observed that S1-0 and S1-1 failed in bending, whilst, S1-2 failed in shear mode. As for the slabs with reduced in thickness, S2-0, without adding fibers was observed to fail in shear. Furthermore, as the fibers were added to the slabs, the mode of failure of the slabs change from shear to bending in S2-1 and S2-2.

## 4. Conclusion

Based on the results presented and discussed, it can be seen that the steel fibers have the potential to serve as part of shear reinforcement in reinforced concrete slabs as well as compensate the loss in concrete shear capacity of slab due to the reduction in the slab thickness.

The addition of steel fibers improves the load carrying capacity of the slabs consistently. In terms of ductility performance, the inclusion of fibers improved the ductility, delayed the crack propagation and managed to change the mode of failure of the slab from brittle to a more ductile manner. However, the optimum amount of fibers will reduce if the amount of initial steel reinforcement is higher. Reducing the shear capacity (through reducing the thickness) of the reinforced concrete slab causes higher optimum fiber volume fraction required to compensate the loss in the shear reinforcement.

## 5. References

[1] Swamy R N and Lankard D R 1974 Some practical applications of steel fibre reinforced Concrete *Proc. of the Institution of Civil Engineers* vol 56 (United Kingdom: ICE Publishing) pp 235-256

[2] Brandt A M 2008 Fibre reinforced cement-based (FRC) composites after over 40 years of development in building and civil engineering, *Compos. Struct.* **86** 3-9

[3] Syed Mohsin S M 2012 *Behaviour of Fibre-Reinforced Concrete Structures under Seismic Loading* Doctoral dissertation (London: Imperial College London) pp 46-50
[4] Mansur M A and Aziz M A 1982 A study of jute fibre reinforced cement composites, *Int. J. of Cement Compos. and Lightweight Concrete* **4** 75-82
[5] Abbas A A Syed Mohsin S M and Cotsovos D M 2016 A simplified finite element model for assessing steel fibre reinforced concrete structural performance, *Comput. and Struct.* **173** 31-49
[6] Hannant D J 2003 Fibre-reinforced concrete *Advanced Concrete Technology –processes* ed. J Newman and B S Choo (Oxford: Butterworth-Heinemann, Elsevier) chapter 6 pp 1 - 17
[7] Cucchiara C, La Mendola L and Papia M 2004 Effectiveness of stirrups and steel fibres as shear reinforcement, *Cement and Concrete Compos.* **26** 777-786.
[8] Khaloo A R and Afshari M 2005 Flexural behaviour of small steel fibre reinforced concrete slabs, *Cement and Concrete Compos.* **27** 141-149.
[9] Juárez C, Valdez P, Durán, A and Sobolev K 2007 The diagonal tension behavior of fiber reinforced concrete beams, *Cement and Concrete Compos.* **29** 402-408
[10] Ding Y, You Z and Jalali S 2011 The composite effect of steel fibres and stirrups on the shear behaviour of beams using self-consolidating concrete, *Eng. Struct.* **33** 107-117
[11] Minelli F, Conforti A, Cuenca E and Plizzari G 2014 Are steel fibres able to mitigate or eliminate size effect in shear?, *Mater. Struct.* **47** 459-473
[12] Kwak Y K, Eberhard M O, Kim W S and Kim J 2002 Shear strength of steel fiber-reinforced concrete beams without stirrups, *ACI Struct. J.* **99** 530-538
[13] Conforti A, Minelli F and Plizzari G A 2013 Wide-shallow beams with and without steel fibres: a peculiar behaviour in shear and flexure, *Compos. Part B: Eng.* **51** 282-290
[14] Abbas A A, Syed Mohsin S M, Cotsovos D M and Ruiz-Teran A M 2014 Shear behaviour of steel-fibre-reinforced concrete simply supported beams, *Struct. Build.* **167** 544-558.
[15] Gouveia N D, Fernandes N A, Faria D M, Ramos A M and Lúcio V J 2014 SFRC flat slabs punching behaviour–Experimental research, *Compos. Part B: Eng.* **63** 161-171
[16] Baarimah A O and Syed Mohsin S M 2016 An overview of using steel fibers in reinforced concrete structural elements to improve shear reinforcement *In: Proc. of the National Conf. for Postgraduate Research 2016 (Malaysia: Pahang)* (Malaysia: Universiti Malaysia Pahang) pp 260-265
[17] Marar K, Eren Ö and Roughani H 2017 The influence of amount and aspect ratio of fibers on shear behaviour of steel fiber reinforced concrete, *KSCE J. Civil Eng.* **21** 1393-1399
[18] Song P S and Hwang S 2004 Mechanical properties of high-strength steel fiber-reinforced concrete, *Constr. and Build. Mater* **18** 669-673
[19] Wang H T and Wang L C 2013 Experimental study on static and dynamic mechanical properties of steel fiber reinforced lightweight aggregate concrete, *Constr. and Build. Mater.* **38** 1146-1151

IOP Conf. Series: Materials Science and Engineering **271** (2017) 012100 doi:10.1088/1757-899X/271/1/012100

# A case study on the structural assessment of fire damaged building

**M H Osman[1], N N Sarbini[2], I S Ibrahim[1], C K Ma[2],M Ismail[2] and M F Mohd[3]**

[1]Forensic Engineering Centre, Institute for Smart Infrastructure and Innovative Construction, Universiti Teknologi Malaysia, 81310 Johor Bahru, Johor, MALAYSIA
[2]Department of Structures & Materials, Faculty of Civil Engineering, Universiti Teknologi Malaysia, 81310 Johor Bahru, Johor, MALAYSIA
[3]Malaysia Public Work Department, JKR Head Quarters, 50582 Kuala Lumpur, MALAYSIA
Corresponding author:noornabilah@utm.my

**Abstract.** This paper presents a case study on the structural assessment of building damaged by fire and discussed on the site investigations and test results prior to determine the existing condition of the building. The building was on fire for about one hour before it was extinguished. In order to ascertain the integrity of the building, a visual inspection was conducted for all elements (truss, beam, column and wall), followed by non-destructive, load and material tests. The load test was conducted to determine the ability of truss to resist service load, while the material test to determine the residual strength of the material. At the end of the investigation, a structural analysis was carried out to determine the new factor of safety by considering the residual strength. The highlighted was on the truss element due to steel behaviour that is hardly been predicted. Meanwhile, reinforced concrete elements (beam, column and wall) were found externally affected and caused its strength to be considered as sufficient for further used of building. The new factor of safety is equal to 2, considered as the minimum calculated value for the truss member. Therefore, this fire damaged building was found safe and can be used for further application.

## 1. Introduction

This paper discussed on the case study of a fire damaged building. The building is made up from reinforced concrete for the beam, column and wall, while have steel trusses as a main structural element. In general, the condition inside the office was affected by the fire (figure 1). There were a lot of things that were affected such as partitions, windows, electrical circuit, electrical sockets and insulations. All the things mentioned were found broken and damaged due to the fire. Effect of fire to the structural components can be seen from the shaded colour of walls, columns and beams, at all over the surfaces. On the other hand, the trusses components were also affected by showing rusted steel appearance all over the surfaces.The structural assessment was aimed to investigate the effect of fire to the elements inside the building, to investigate the condition of the main structural element due to the fire (steel trusses) and to recommend overall condition of the building.

In general, reinforced concrete structure performed well under fire [1-3]. Its behaviour as fire resisting element caused most of fire damaged structure being reused after the fire, provided structural assessment has been carried out. With sufficient information on the level of fire during the event will assist the assessment to investigate the existing condition of a building. Effect of fire level to the structural element inside the fire damaged building need to be investigated in order to ascertain the existing capacity and condition.

Several important information in carried out the structural assessment of a fire damaged building are; types of building elements and materials, point of fire initiated, exposure condition, predicted fire temperature, duration of fire and other possible data. As structure itself is made up by different other materials, it made each fire damaged building investigation to become more difficult. In other word, each investigation will be unique between different fire damaged building incidents. Therefore, it is very important for investigator to share their experienced wherever they facing such of fire damaged building investigation.

On the other hand, in reinforced concrete structure, colour changes are normally used to investigate the temperature of the material exposed to [4-7]. Table 1 is used to determine the level of temperature exposed to the building elements. Meanwhile, figure 1 shows the flowchart of current structural assessment of a fire damaged building. The flowchart is determine based on review of important data that need to be collected in order to ascertain the existing condition of the building element after exposed to fire.

**Figure 1.** Interior view of the building damaged by fire shows most of the elements inside the building were affected.

**Table 1.**Colour changes in concrete at specified range of temperature, T.

| Range of temperature, °C | Colour | Appearance | Condition |
|---|---|---|---|
| $T < 300$ | Normal | Normal | Normal |
| $300 \leq T < 600$ | Pink to red | Surface crazing, cracking and aggregates pop outs | Sound but strength may be reduced |
| $600 \leq T < 950$ | Whitish grey | Spalling, exposed of steel reinforcement and powdered existence | Weak |
| $T \geq 950$ | Buff | Extreme spalling | Extreme/severe |

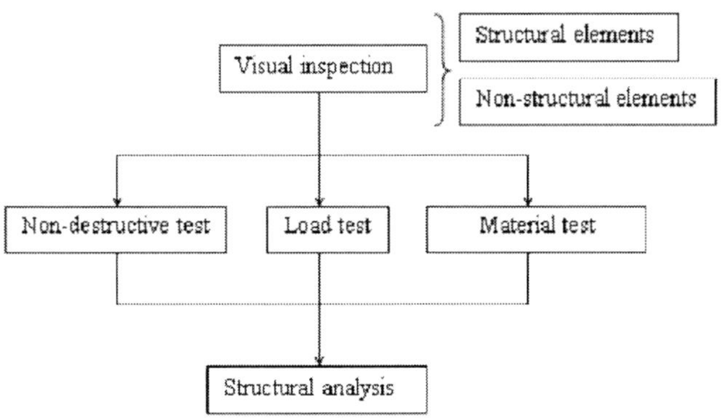


**Figure 2.**Flowchart of a structural assessment.

## 2. Site Investigation

### *2.1. Visual inspection*

Visual inspection has been conducted for all elements inside building (figure 3). Through visual inspection, it was found that there were many damages to the roof due to the fire. There was a few damages found at the metal sheets which made it not fit for roofing system. The same observation were found for the foil insulations. It was found that all of the foil insulations were damaged and was not suitable for use.

The condition of steel purlins was found most affected. These conditions were expected due to the material properties of steel purlin of which have a strength that is lower compared to the steel truss. Therefore, the effect of fire will create so much impact to the steel purlin compared to the steel trusses. General overview on the steel purlins that were corroded on-site was shown in figure 4. Most of the steel purlins were corroded along their lengths. The most affected steel purlin was found for a purlin located above the area of fire initiated. The steel purlin was heavily buckled and twisted along their length.

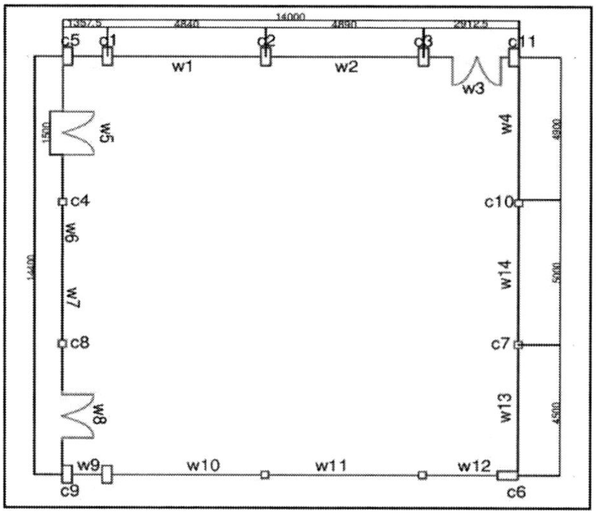


**Figure 3.** Plan view of a fire damaged building shows ten columns (cl to c10) and fourteen walls (wl to w14). Other elements that were not visible inside the plan view are two beams spanning from c5 to c9 and from c11 to c6; and six steel trusses spanning in between the beams.

**Figure 4.** Observation on steel purlins that were affected by fire.

The steel trusses are the main structural elements on-site, as the members carry the roof loads. Visual inspection shows that all members are affected by the fire. Other than the original colour of black was brownish in color which was found to be colour of corrosion.In determining the state of the corrosion; the corroded part was polished using a sandpaper and stiff wire brush to investigate the extent of the depth of the corrosion. After the process, it was found that the condition of the welding at most of the connections were not satisfactory. An example of the less satisfactory condition that was found on-site is shown in figure 5(a) is considered as poor quality.

On the other hand, the visual inspection shows that there are many spalling of the plaster on the surface of the beam. However, the concrete surface beneath the plaster layer is not affected. The concrete was found not affected due to the appearance and condition that still intact and shows the original grey in their colour (figure 6). The observation is also shown by column and wall elements.

In general, it can be summarized that reinforced concrete elements inside the building were just externally affected by the fire. Although comparison of the elements colour and texture to table 1 shows that the elements were experiencing about 600 to 800 °C temperature, nevertheless, only plastering of the RC elements were affected. In this case, only RC beams were found affected showing by spalling of the plastering. This proved on the capacity of RC elements in resisting fire as per mentioned by previous works [1-3]. Meanwhile, the steel purlins and trusses needs to be investigated in detail due to the confirmation on its behaviour after fire.

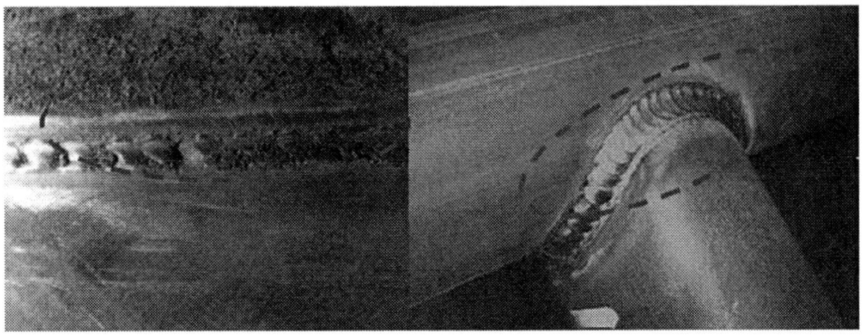

(a) Unsatisfactory (b) Satisfactory

**Figure 5.**Appearance of welding quality at connection.

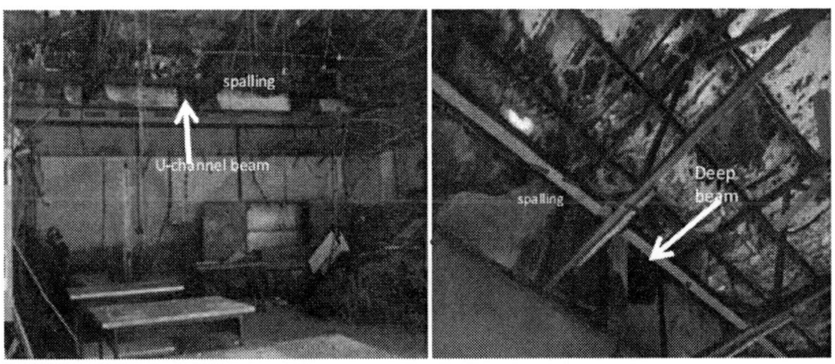


**Figure 6.**Physical appearance of the beams

*2.2 .Non-Destructive Test*

Although the observation from the visual inspection has shown that all RC elements (wall, column and beam) were just externally affected by the fire, a "Schmidt" rebound hammer test is still been carried out to ascertain the existing concrete uniformity of the elements. The test is in accordance to ASTM C805/C805M-13a [8] and was carried out for beam elements on-site due to the occurrence of concrete spalling. A few points on the beams were selected to carry out the test. In general, the test results show that the compressive strength for concrete beams are in between 40 $\pm$ 4 N/mm$^2$, which is considered as more than adequate.

IOP Conf. Series: Materials Science and Engineering **271** (2017) 012100 doi:10.1088/1757-899X/271/1/012100

### *2.3.Load Test*

The objective of load test is to ascertain current capacity of trusses on-site (figure 7). Previously, RC elements have been found externally affected; however, the truss is made up from steel material. Therefore, further test is needed for the truss elements. Two trusses are selected based on its condition as the most (Truss 1) and least (Truss 2) affected truss. Full load test was carried out to determine the truss behaviour due to the loading and unloading process at service load. Loading was applied in 1 kN increment; and up to maximum service load of 14 kN at point D and E (at the middle of the truss span). The results for load-deflection relationship for two LVDTs location is shown in figure 8(a).

In general, the deflection relationship have the same curve pattern at LVDTs position (point D and E) indicating that the loading system and test setup are closely symmetrical and aligned. The same pattern was observed for both trusses, during the loading and unloading process. The recorded deflection at LVDT 1 (point D) and LVDT 2 (point E) was 4.24 mm and 4.79 mm for Truss 1, respectively. Meanwhile, Truss 2 recorded deflection of 4.06 mm and 4.49 mm at LVDT 1 (point D) and LVDT 2 (point E), respectively.

Meanwhile, the sustained load test (for service load at 14 kN) was carried out for 24 hours and at the same time, deflection was recorded at every 30 minutes. Truss 1, the most affected truss was selected to undergo this sustained load test. The deflection relationship at points D and E is shown in figure 8(b). The deflection of the truss at LVDT 1 (point D) ranged between 3.51 mm and 4.24 mm, while LVDT 2 (point E) ranged between 3.96 mm and 4.85 mm. The graph pattern indicates a bit increasing trend as the load remained sustained on the slabs; at early 8 hours. However, the deflection became constant after 8 hours from the start of the sustained load.

The full load test and sustained load test results shows that the trusses are still stable and capable in resisting service load for further used.

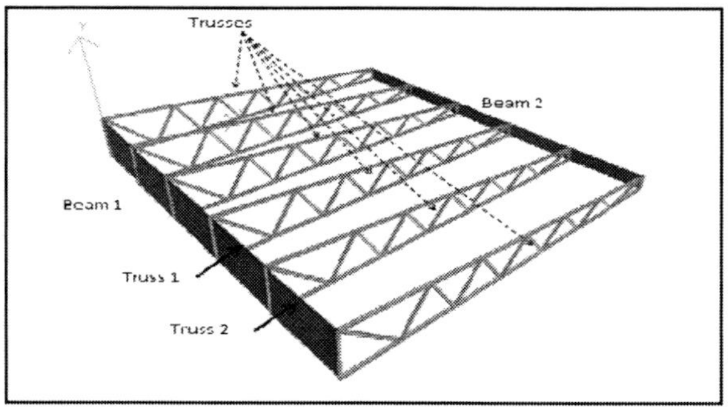


**Figure 7.**Steel trusses on-site

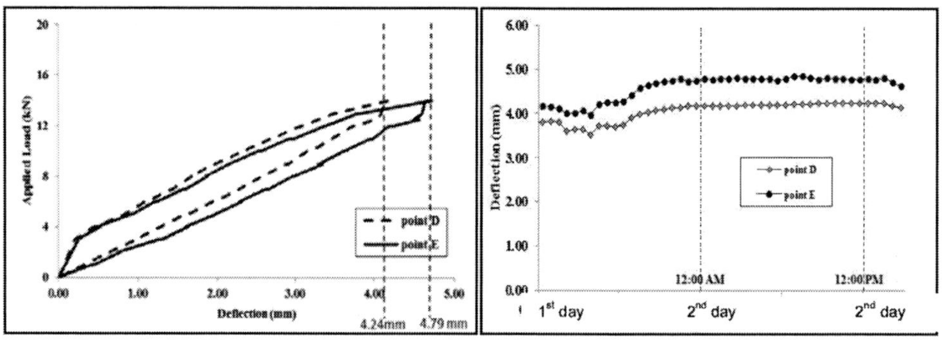


(a) Full load test results (b) Sustained load test results

**Figure 8.** Deflection relationship during the test

*2.4Material Test*

The coupon tensile test was carried out on the steel purlins [9]. The test is conducted for the steel purlin due the observation during visual inspection which found that steel purlins were the most affected element on-site. This is due to the appearance of the element that was heavily buckled and twisted compared to the steel trusses. A total of 6 samples that consisted of:3 samples from the most affected area and 3 samples from less affected area.The average tensile capacity is 341 N/mm$^2$ for least affected samples, which is 15% less than the design requirements of 400 N/mm$^2$. Meanwhile, the average tensile capacity is 324 N/mm$^2$ for the most affected samples, which is 19% less than the design requirements of 400 N/mm$^2$. Therefore, the test results show that the steel purlin was severely affected with it yield capacity reduced at maximum of 19% from the actual capacity.

## 3.Structural Analysis

In predicting the current capacity of the steel truss, a structural analysis has been carried out using STAAD.Pro v8i Analysis Software and in accordance to BS 5950 [10].The values under the design load and un-factored load were gained from the software analysis. Meanwhile, the design capacity was calculated based on the actual capacity of the member. Factor of safety, *FOS* was determined and it was calculated using equations (1) and (2). The reduce capacity was calculated by considering the higher temperature that the steel truss was exposed; in a range of 600 to 800 °C.

Based on an assumption that was made from the literature and investigation, the exposed temperature caused the capacity of the truss to drop by 30% from the actual. It means that, about 30% of the actual capacity of the truss members were reduced due to the exposure to the fire. This is also similar to the previous work reported [11]. From their findings, the steel with characteristic strength of 400 to 460 N/mm$^2$ will have a strength loss of about 30% when exposed to a temperature of between 800 to 1000 °C. Subsequently, the coupon tensile strength test of which the steel purlin was made from the steel with 400 N/mm$^2$ design strength dropped at about 20% from the strength. Therefore, assumption on 30% of strength dropped for the steel truss was make sense, although the load test has already giving unaffected fire condition of a steel truss .After calculation, the $FOS_{new}$ that was considering the reduced capacity of each truss members lies between a range of 2 and 64. To the best knowledge, this will conclude that the steel trusses have enough residual capacity for reused and further application.

$$FOS = \frac{Design\ capacity}{Design\ load} \tag{1}$$

$$FOS_{new} = \frac{Reduce\ capacity}{Design\ load} \tag{2}$$

## 4.Conclusion and Recommendation

The findings of the investigation are:

1) There is no appearance of severe damage for reinforced concrete elements, except for spalling of mortar screeding found appeared in certain parts of columns and beams. No crack found on the concrete surface. Strength of concrete beams and columns have been estimated by using Schmidt rebound hammer test. From the results of the test, it indicates that the strength of concrete is still unaffected by the fire. It proves the finding of the physical inspection on the appearance of the concrete surface, i.e. no sign of degradation of strength of concrete in the building structure.

2) From the comparison of colour change on mortar and concrete surface, we can assumed the building experienced temperature in the range of 600℃ to 800℃. This is confirmed with the occurrence of corrosion on steel truss members which is similar to the occurrence of corrosion on steel coupon specimen when exposed to fire. From the full-scale loading test, it was found that the steel trusses are still strong. The maximum deflection of the trusses was very small, i.e in the range of 4 mm, compared to the span of the truss which is 14 m (deflection/span ratio of 1/3500, as compared to the maximum allowable which is 1/200). The small deflection is also an indication that there is no structural degradation of the truss members and connections, and proved that they are over-designed.

## 5. References

[1] Stanislav Kmet, Michal T and Ivo D 2013 Experimental Diagnostics and Static Assessment of Fire Damaged Steel Structure *Lviv Polytechnic National University Institutional Repository*

[2] Emmanuel V R A, Luc R T 2013 Assessment techniques for the evaluation of concrete structures after fire *Journal of Structural Fire Engineering* **4(2)** 123-129

[3] Narendra K Gosain, Ray F Drexler and Dilip C 2008*Evaluation and Repair of Fire-Damaged Buildings* (STRUCTURE Magazine)

[4] Tide R H R 1998 Integrity of structural steel after exposure to fire*Engineering Journal* **35(1)** 26–38

[5] Awoyera P O, Akinwumi I I, Ede A N and Olofinnade M O 2014 Forensic Investigation of Fire-affected Reinforced Concrete Buildings*IOSR Journal of Mechanical and Civil Engineering* **11(4)** 17-23

[6] Technical Note - No. 102 ; April 2011. Fire Damaged Reinforced Concrete- Investigation, Assessment and Repair.

[7] Jiro T and Gregory G Deierlein 2007 Collapse Performance Assessment of Steel-Framed Buildings under Fires. Report No. 163

[8] ASTM C805/C805M - 13a Standard Test Method for Rebound Number of Hardened Concrete

[9] British Standard Institute. *Metallic Materials - Tensile Testing - Part 1: Method of Test at Room Temperature.* London, BS ISO 6892-1. 2009

[10] British Standard Institute. *Structural used of Steelwork in Building.* London, BS 5950. 2000

[11] Kigha F, Sadeeq J A, Abejide O S 2015 Effects of temperature levels and concrete cover thickness on residual strength characteristics of fire exposed reinforced concrete beams *Nigerian Journal of Technology* **34(3)** 429-437

## Acknowledgments

An appreciation goes to all parties that involved directly or indirectly to the completion of this work, and sponsorship through cost centre no. 14J31.

# Modelling the behaviour of steel fibre reinforced precast beam-to-column connection

**C E Chai[1], NN Sarbini[1], I S Ibrahim[2], C K Ma[1] and M Z Tajol Anuar[1]**

[1] Department of Structures & Materials, Faculty of Civil Engineering, Universiti Teknologi Malaysia
[2] Forensic Engineering Centre, Institute for Smart Infrastructure and Innovative Construction, Universiti Teknologi Malaysia
Corresponding author: noornabilah@utm.my

**Abstract.** The numerical behaviour of steel fibre reinforced concrete (SFRC) corbels reinforced with different fibre volume ratio subjected to vertical incremental load is presented in this paper. Precast concrete structures had become popular in the construction field, which offer a faster, neater, safer, easier and cheaper construction work. The construction components are prefabricated in controlled environment under strict supervision before being erected on site. However, precast beam-column connections are prone to failure due to the brittle properties of concrete. Finite element analysis (FEA) is adopted due to the nonlinear behaviour of concrete and SFRC. The key objective of this research is to develop a reliable nonlinear FEA model to represent the behaviour of reinforced concrete corbel. The developed model is validated with experimental data from previous researches. Then, the validated FEA model is used to predict the behaviour of SFRC corbel reinforced with different fibre volume ratio by changing the material parameters. The results show that the addition of steel fibre (SF) increases the load carrying capacity, ductility, stiffness, and changed the failure mode of corbel from brittle bending-shear to flexural ductile. On the other hand, the increasing of SF volume ratio also leads to increased load carrying capacity, ductility, and stiffness of corbel.

## 1. Introduction

Reinforced concrete (RC) is one of the most commonly used materials in the construction field since long ago. This material had been used for decades to build houses, bridges and even monuments all around the world [1, 8]. However, the trend had changed and the use of precast reinforced concrete (PC) had become popular in the construction field. It brings a lot of benefits compared to conventional cast in-situ RC structures. The PC components are fabricated in controlled environment which ensure the quality of the construction components [7, 10]. This increased building quality of a construction project. Besides, PC allows the use of smaller structural members which can enhance the appearance of a building [7]. In short, there are still many advantages of using PC compared to RC which it would makes PC as the future trend of construction, this includes the used of new composite high strength material and waste utilization as part of a component material. The invention is due to benefits of other materials to the precast components for a building. The benefits includes long lasting material, high strength and durable material, thermal and other important properties [18, 26].

The use of PC had been the future of construction field [18]. In a PC design, the connection of the precast member is the most important and critical component. Most of failure in PC structures is caused by the failure of connection [10]. To date, there are many researches have been conducted on PC [1-18]. However, there is limited research focused on a connection of PC using SFRC [1]. In this research, behaviour of precast concrete corbel reinforced with different volume ratio of SF is investigated through modelling of a component in LUSAS analysis software. FEA model is developed by implementing constitutive models for the component. Experimental data from previous research is adopted to validate the prediction of the model. The validated model is used to model the behaviour of SFRC corbels reinforced with different SF volume ratio.

## 2. Literature Review

### *2.1. Reinforced Concrete*

The ability of RC to provide a sufficient strength, stiffness and cost effective have made the RC as a first choice in construction industry [11]. Researches which aim to improve the ductility of plain concrete had been done throughout the time and finally found the approach to introduce the ductility into concrete is by adding a steel reinforcement which suggests the reinforced concrete [21-22]. To some extent, the steel reinforcement included both bars reinforcement and fibres reinforcement can provide sufficient tensile strength to improve the ductility of concrete to work under tension [23]. Besides, it also provides confinement effects in concrete [15]. The confinement effect is very important in a compression region of a member as it leads to improve the energy absorption capacity. Thus, steel fibre reinforcement is one of the alternatives approached in improved the ductility of a precast concrete component as it allows member to perform better under both tension and compression.

Bouafia *et al.* modified constitutive confined concrete model proposed by Mander *et al.* and the reliability of the modified model is supported with experimental data [2, 15]. Based on the comparison, the modified model shows a good agreement with experimental data for both circular and rectangular sections. To reduce the error in calculation, the equations proposed by Mander *et al.* is simplified by replacing unknown with constant values [15].Concrete in tension is highly nonlinear as concrete in compression [16]. Due to this property, the model proposed by Massicotte and MacGregor assumed concrete as a linear-elastic brittle material with strain softening and suggested a post-cracking stress-strain relationship for the tensile part of the curve [16].

### *2.2. Precast Reinforced Concrete*

Just like in a conventional RC structure, PC components such as beams and columns are constructed by using concrete material and steel reinforcements. The main difference between the conventional RC and PC structures are; method of construction and material used. In some application, a pre-stressed strand will be used to reinforce the precast products that cannot be found in conventional RC.

Corbel is a short cantilever projecting from column or wall and is commonly used in connection between beam and column PC. The main functions of corbel are to support other parts of a structure such as beams and transmit vertical and horizontal loads to the principal members [4, 6]. Since corbel is a part of precast component, researches also focused to the behaviour and capacity of corbel. Fattuhi [4] found out that the arrangement and amount of reinforcement in column has no significant effect on the strength of corbel. At most of the case, the shear span-to-depth ratio is the main factor that governs the capacity of corbel [6].

### *2.3.Steel Fibre Reinforced Concrete*

SFRC was introduced to reduce the brittleness of concrete by introducing SF into concrete to act as the reinforcement. With the addition of SF, the properties of concrete will improve from brittle to ductile [9] as well as the failure mode. Concrete will losses its tensile resistance after the formation of macro cracks [5]. It is proven that SF could prevent the forming of macro cracks which leads to concrete spalling in concrete members [9] by forming more microcracks. All the microcracks are linked through the bridging effect of SF in concrete matrix [14] where it allows the transfer of stress throughout the concrete member [17]. Therefore, SF could improve the post-cracking behaviour of concrete [9]. In some cases, adding of SF can reduce the steel reinforcement bars needed in a concrete structure [23].

Previous work proposed a compressive model based of concrete reinforced with hooked-end type of SF only [12]. The model can predict the strain at the compressive strength, elastic modulus of SFRC and applicable on a wide range of concrete strength. Actual compressive strength of SFRC is used in the model instead of compressive strength of plain concrete.

## 3.Methodology

2D model is used to predict the behaviours of corbel. Since the selected corbel specimen has a symmetrical shape, only half of the corbel was modelled to save the processing time without affecting the outcomes of analysis [3]. The boundary conditions at the top, bottom, and a side of column were assumed fixed. On the other hand, the reinforcement steel bars in column are not as the assigned parameters had considering the properties of such composite [4]. Next, the load was applied incrementally on a bearing plate made of an undefined strong material to prevent local crushing of concrete. Table 1 provides the model description.

**Table 1.**Model description

| Model Definition | $V_f$(%) | $f_{ct}$ (MPa) | $E$ (GPa) |
|---|---|---|---|
| Case Study Model | 0 | 6.97 | 44.27 |
| Reference Model | 0 | 6.97 | 44.27 |
| Model 1 | 0.5 | 8.24 | 32.80 |
| Model 2 | 1 | 8.42 | 31.92 |
| Model 3 | 1.5 | 9.89 | 31.04 |
| Model 4 | 2 | 11.38 | 30.15 |

Reasonable assumptions were made for the properties which are not given in the reference paper. Steel was assumed to behave as an elastic-perfectly plastic material in both tension and compression. Material properties of steel were specified using the elastic and the plasticity of metal with plastic options. From the selected experimental data, yield stress of steel was given as 400 MPa. Young's Modulus and Poisson's Ratio were assumed as 210 GPa and 0.3, respectively.

To predict the nonlinear behaviour of concrete, compressive and tensile stress-strain curve were developed from [2] and [16] models, respectively. In LUSAS, concrete was attributed as an isotropic material. The material properties of concrete are specified as elastic and the plasticity of concrete with plastic options. Besides, the values of Young's Modulus, compressive and tensile stress strain relationship were derived from the selected models and assigned to define the concrete material. Poisson's Ratio for concrete was assumed as 0.2.

After the nonlinear FEA performed by LUSAS, a load-displacement curve is generated. The predicted load-displacement curve was compared to the curve from [6] experimental data. Several trials had been run to determine the suitable meshing shape and size that exhibit the same behaviour as experiment. Once both graphs show a good agreement, then the model is considered validated and can be used to predict the effects of SFRC on corbels [25].

## 4. Results and Discussions

Figure 1 shows the comparison of load-displacement curves. The maximum deviation of base model from experimental model is 13.2%, which is in between 10-15%. This indicates that the prediction of base model has a reasonable good agreement with experimental data [25]. Thus, this confirmed the validity of the FEA model to predict the effects of SFRC on corbels.

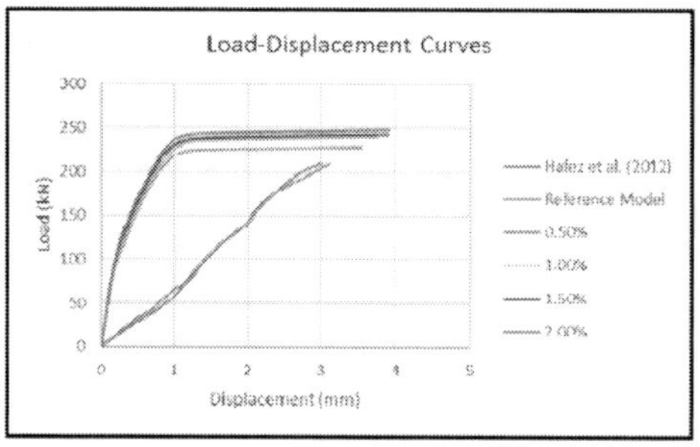


**Figure 1.** Comparison of load-displacement curves.

The study of the numerical effects of SFRC on corbels was carried out by increasing the SF volume ratio by $V_f$ = 0%, 0.5%, 1.0%, 1.5%, and 2.0%. The compressive and tensile stress-strain relations for different SF volume ratio is developed from [12] and [13] models, respectively.The results of analysis are summarized in table 2. From the results, it shows that the ultimate shear load is increased drastically by 9.25% when SF is added into the concrete. The increase of ultimate shear capacity might due to the shear stress induced by increasing loads which absorbed by the fibres. The stress is distributed all over the structure by SF and reduced the stress acting on the critical parts of corbels. Besides, the ultimate shear load increases with an average value of 3.07% when the volume of SF increases by 0.5%. Thus, the increase of SF volume leads to the increase of energy needed to pull out the SF from concrete matrix [20], and results in greater load carrying capacity of corbels.

The ductility of corbel is assessed by ductility ratio. Values of ductility ratio of corbels is calculated by dividing the displacement at ultimate load with displacement at yield and the values are as summarized in table 2. From the values, it shows an increasing trend when SF is added. The ductility ratio is increasing with an average value of 8.95% when the SF volume increases by 0.5%. Next, the model illustrates a decreasing trend in displacement of corbels at the loading of 208.73kN when the volume of SF increases. The point of interest was set at 208.73kN because that is the ultimate shear capacity of Reference Model and displacement data at this point for all models are available. The decrease of displacement indicates that larger loading is required to cause the same displacement in corbel with the increase of SF volume. Hence, it can be concluded that the addition of SF could improve the ductility of corbel.

Stiffness of corbel is derived from the gradient of load-displacement curves (elastic part). The stiffness indexes for all the models are shown in table 2. The significant increase in the stiffness index from plain concrete to SFRC corbel indicates the stiffness of corbel improved with the addition of SF. Besides, the stiffness indexes are increasing at an average value of 11.80% when the SF volume increases by 0.5%. Therefore, increasing SF volume will improve the stiffness of corbel.

The numerical cracking and crushing pattern of the corbel models are illustrated in figure 2. The compressive stress is indicated by dark orange contour, while tensile stress is indicated by dark yellow contour. From figure 2, it is observed that all models exhibited tensile cracking at the region around bottom column-corbel interface and compressive crushing at the area around top column-corbel interface.

Besides, it can also be observed that the tensile cracks zone becomes larger when the amount of SF increased. By observing the crack and crush zone of base model, it shows that the corbel failed in bending-shear as there is an obvious crack that slices through column-corbel interface. Besides, the failure was also accompanied by brittle failure mode as the concrete at the region of top column-corbel crushed. However, by comparing the base model with other SFRC models, it clearly shows that the addition of SF reduced the area of compressive crushing zone significantly and completely removed the crack that slices through the column-corbel interface. Hence, it can be concluded that addition of SF altered the failure mode of corbel from bending-shear accompanied with brittle failure to flexural ductile failure mode.

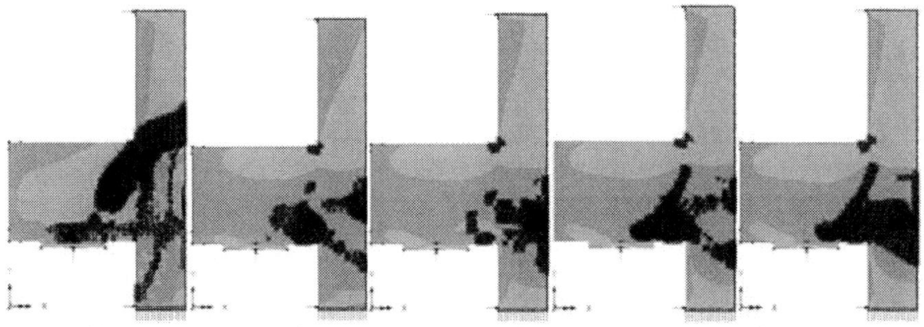

**Figure 2.** Stress distribution and cracking zone: (a) $V_f$ = 0%, (b) $V_f$ = 0.5%, (c) $V_f$ = 1.0%, (d) $V_f$ = 1.5%, (e) $V_f$ = 2.0%.

**Table 2.** Summary results for models.

| Model Definition | $P_{ult}$ (kN) | $\Delta P_{ult}$ (%) | $\delta_{ult}$ (mm) | $\Delta\delta_{ult}$ (%) | μ | Δμ (%) | S.I | ΔS.I (%) |
|---|---|---|---|---|---|---|---|---|
| Case Study Model | 210.00 | - | 3.01 | - | 1.03 | - | 75.13 | - |
| Reference Model | 208.73 | 0 | 3.10 | 0 | 1.08 | 0 | 71.77 | 0 |
| Model 1 | 228.03 | 9.25 | 3.55 | 14.52 | 1.46 | 35.19 | 202.78 | 182.54 |
| Model 2 | 239.92 | 14.94 | 3.76 | 21.29 | 1.57 | 45.37 | 212.24 | 195.72 |
| Model 3 | 243.19 | 16.51 | 3.89 | 25.48 | 1.66 | 53.70 | 216.19 | 201.18 |
| Model 4 | 247.25 | 18.45 | 3.91 | 26.13 | 1.75 | 62.04 | 228.19 | 217.95 |

## 5. Conclusion

Based on the results and discussions, conclusions can be drawn as summarized below:

i. A reliable nonlinear FEA model to represent the behaviours of concrete corbels was developed by implementing models from [2] and [16]. The capability of the model to predict the behaviours of concrete corbels subjected to vertical load only is validated with experimental data from [6].

ii. Addition of SF caused the load-carrying capacity of corbels to increase drastically by 9.25% due to the distribution of excess stresses by discrete SF away from critical parts of the corbels. Besides, the load-carrying capacity of SFRC corbels increase by an average of 3.07% when SF volume increases by 0.5%. This is contributed by the pull-out strength of SF.

iii. Corbels with SFRC can deform greater before failure compared to plain concrete corbel. This indicates the ductility of corbel is improved with the addition of SF and it is supported by the increasing values of ductility ratio. The ductility ratio increased with and average value of 895% when the SF volume increased by 0.5%. This indicates that SFRC corbels with higher volume of SF can deform greater before failure. This shows that the increasing volume of SF improved the ductility of SFRC corbels.

iv. Addition of SF caused an increase of 182.54% in stiffness. Besides, the stiffness index is increasing with and average value of 11.80% when the SF volume increases by 0.5%. This indicates the SF can improve the stiffness of corbel.

v. Increase of SF volume enlarges the tensile crack zone in corbels due to the improved ability to propagate cracks by the bridging effect of SF and more microcracks are formed. Failure mode of corbels changed from bending-shear with brittle failure to flexural ductile failure mode with the presence of SF.

## 6. References

[1] Alcocer S M, Carranza R, Perez-Navarrete D 2000 Behaviour of a precast concrete beam-column connection *12th World Conference on Earthquake Engineering* **52** 1-8

[2] Bouafia Y, Iddin A, Kachi M S and Dumontet H 2014 Barcelona. Stress-strain relationship for the confined concrete *5th European Conference on Computational Mechanics (ECCM V)*

[3] Canha R M F, Kuchma D A, Debs M K E and Souza R A D 2014 Numerical analysis of reinforced high strength concrete corbels *Engineering Structures* **74** 130-144

[4] Fattuhi N I 1990 Column-load effect on reinforced concrete corbels*Journal of Structural Engineering* **116(1)** 188-197

[5] Ganesan N, Indira P V and Abraham R 2007 Steel fibre reinforced high performance concrete beam-column joints subjected to cyclic loading *ISET Journal of Earthquake Technology* **44(3-4)** 445-456

[6] Hafez A M A, Ahmed M M, Diab H and Drar A A M 2012Shear behaviour of high strength fiber reinforced concrete corbels *Journal of Engineering Sciences* **40(4)** 969-987

[7] Hawileh R A, Rahman A and Tabatabai H 2010 Nonlinear finite element analysis and modeling of a precast hybrid beam-column connection subjected to cyclic loads *Applied Mathematical Modelling* **34** 2562-2583

[8] Izni S I, Khairul H P, Hamid M A B, Abdul A S and Noor N S 2014 Ultimate shear capacity and failure of shear key connection in precast concrete construction *Malaysian Journal of Civil Engineering* **26(3)** 414-430

[9] Kim K S, Lee D H, Hwang J H and Kuchma D A 2012Shear behaviour model for steel fiber-reinforced concrete members without transverse reinforcements *Composites Part B: Engineering* **43(5)** 2324-2334

[10] Korkmaz H H and Tankut T 2005 Performance of a precast concrete beam-to-beam connection subject to reversed cyclic loading *Engineering Structures* **27** 1392-1407

[11] Kwak H G and Filippou F C 1990 *Finite Element Analysis of Reinforced Concrete Structures Under Monotonic Loads* (Berkeley, California)

[12] Lee S C, Oh J H and Cho J Y 2015 Compressive behaviour of fiber-reinforced concrete with end-hooked steel fibers*Materials***8(4)** 1442-1482

[13] Lok T S and Xiao J R 1999 Flexural Strength assessment of steel fiber-reinforced concrete *Journal of Materials in Civil Engineering* **11(3)** 188-196

[14] Madandoust R, Ranjbar M M, Ghavidel R and Shahabi S F 2015 Assessment of factors influencing mechanical properties of steel fiber reinforced self-compacting concrete *Materials & Design* **83** 284-294

[15] Mander J B, Priestley M J N and Park R 1984 Theoritical Stress-Strain Model for Confined Concrete *Journal of Structural Engineering* **114(3)** 1804-1826

[16] Massicotte A E E B and MacGregor J G 1990 A Study of Combined Bending and Axial Load in Reinforced Concrete Members*ASCE Journal of Structural Engineering,* **116(11)**3039-3058

[17] Meskenas A, Gelazius V, Klauslaus G, Gribniak V and Rimkus A 2013 A New Technique for Constitutive Modeling of SFRC *Procedia Engineering* vol 57 (United Kingdom: Elsevier) pp 762-766

[18] National Precast Concrete Association 2008 *The Little Green Book of Concrete - Precast Concrete for Sustainable Construction* (United States: Carmel, IN)

[19] National Research Council 2007 *Guide for the Design and Construction of Concrete Structures Reinforced with Fiber-Reinforced Polymer Bars* (Rome: Italy)

[20] Pajak M and Ponikiewski T 2013Flexural behaviour of self-compacting concrete reinforced with different types of steel fibers *Construction and Building Materials***47** 397-408

[21] Rudnov V, Belyakov V and Moskovsky S 2016 Properties and design characteristics of the fiber concrete *Procedia Engineering* vol 150 (United Kingdom: Elsevier) pp 1536-1540

[22] Sermet F, Ozdemir A 2016 Investigation of punching shear behaviour of steel and polypropylene fibre reinforced concrete slabs under normal load *Procedia Engineering* vol 161 (United Kingdom: Elsevier) pp 458-465

[23] Soutsos M, Le T, Lampropoulos A 2012Flexural performance of fiber reinforced concrete made with steel and synthetic fibers *Construction and Building Materials***36** 704-710

[24] Thomas J, Ramaswamy A 2007Mechanical properties of steel fiber-reinforced concrete *Journals of Materials in Civil Engineering* **19(5)** 385-392

[25] Yassin A Y M, Noor M A A M, Kasiman E H and Jamil M Z 2013 Linear and Nonlinear Finite Element Analysis for Solids and Fluids [Unpublished]

[26] Aminudin E, Khalid N H A, Azman N A, Bakri K, Din M F M, Zakaria R and Zainuddin N A 2017 Utilization of Baggase Waste Based Materials as Improvement for Thermal Insulation of Cement Brick *Matec Web of Conferences* vol 103 (Paris: EDP Sciences)

**Acknowledgments**
An appreciation goes to all parties that involved directly or indirectly to the completion of this work and UniversitiTeknologi Malaysia grant sponsorship through cost centre no. 14J31.

IOP Conf. Series: Materials Science and Engineering **271** (2017) 012102 doi:10.1088/1757-899X/271/1/012102

# A Preliminary Experimental Study on Vibration Responses of Foamed Concrete Composite Slabs

**R H M Rum[1], Z M Jaini[2,*], N H Abd Ghaffar[2] and N Abd Rahman[2]**

[1]Faculty of Civil and Environmental Engineering, Universiti Tun Hussein Onn Malaysia, 86400 Parit Raja, Johor, Malaysia
[2]Jamilus Research Center, Universiti Tun Hussein Onn Malaysia, 86400 Parit Raja, Johor, Malaysia
Corresponding author: rizuan@uthm.edu.my

**Abstract.** In recent years, composite slab has received utmost demand as a floor system in the construction industry. The composite slab is an economical type of structure and able to accelerate the construction process. Basically, the composite slab can be casting by using a combination of corrugated steel deck and normal concrete in which selfweight represents a very large proportion of the total action. Therefore, foamed concrete become an attractive alternative to be utilized as a replacement of normal concrete. However, foamed concrete has high flexibility due to the presence of large amount of air-void and low modulus elasticity. It may result in vibration responses being greater. Hence, this experimental study investigates the vibration responses of composite slab made of corrugated steel deck and foamed concrete. The specimens were prepared with dimension of 750mm width, 1600mm length and 125mm thickness. The hammer-impact test was conducted to obtain the acceleration-time history. The analysis revealed that the first natural frequency is around 27.97Hz to 40.94Hz, while the maximum acceleration reaches 1.31m/s$^2$ to 1.88m/s$^2$. The first mode shape depicts normal pattern and favourable agreement of deformation.

## 1. Introduction

Composite slab is widely recognized as durable and efficient replacement to the conventional reinforced concrete slab. The genius preference of composite slab is lay on the fact that it able to speed the construction process, safe method of construction and generally economical for the leisure, commercial, industrial and residential buildings [1]. In addition, composite slab is also regarded as one of the best diaphragm strengthening method that allow higher load bearing capacity with lower thickness and reduction size of structural element [3]. Therefore, it is not surprising that composite slab has received high demand in the construction industry. Code of practice BS EN1994-1-1 [2] defined the composite slab as a structure in which corrugated steel deck is used initially as permanent shuttering, subsequently combine structurally with the concrete and act as tensile reinforcement. The presence of corrugated steel deck not only acts as permeant formwork where it remains for the whole service life, but also provide sufficient shear bond that contribute to the compositeness behaviour.

In the current practise, composite slab is basically constructed using a combination of corrugated steel deck and normal concrete in which the selfweight of topping become a major concern. A recent study by Hulimka *et al.* [4] revealed huge interest to utilize lightweight concrete as main material in structural element. Zulkarnain and Ramli [5] suggested that the dead load of structural element can be reduced by using lightweight concrete. In many choices of lightweight concrete to be associated with composite slab, foamed concrete is found more practical and sustainable due to its characteristic of low selfweight, excellent acoustic and fire resistance as well as good in shock absorption and thermal properties. Although foamed concrete exhibits drawback in strength but the latest studies by Hadipratama *et al.* [6] and Jaini *et al.* [7] revealed that the mechanical properties of foamed concrete can be improved through modification approach. However, the truncated value of elastic modulus may result in vibration responses being greater.

Many studies were conducted on the design aspect, static performance and structural behaviour of composite slab. Sanchez *et al.* [8], Khalaf *et al.* [9] and Chaudhari *et al.* [10] investigated the structural behaviour of composite slab in term of load bearing capacity, deflection and failure mode. Meanwhile, Abbas *et al.* [3], Cifuentes and Medina [11], Johnson & Shepherd [12] and Lakshmikandhan *et al.* [13] investigated the interaction and shear resistance. Johnson & Li [14] investigated the bearing capacity of foamed concrete composite slab under four-point bending test and found the promising performance. However, the vibration responses of composite slab also need a grave concern. Vibration not just cause discomfort but may lead to undesirable serviceability problems and outright structural failure. Reviews of past studies have shown that the knowledge on the vibration responses of foamed concrete composite slab are currently not readily available. Therefore, this study is necessary to assure the comfortability and applicability of foamed concrete composite slab.

## 2. Experimental study

### *2.1. Material preparation*

The material preparation involves the provision of mix proportion and required amount of cement, sand, gravel, water and foam agent to produce good quality of foamed concrete and normal concrete. DOE method was employed in the calculation of mix proportion. The targeted compressive strength of foamed concrete and normal concrete is setup at18MPa. Foamed concrete was designed with density of 1600kg/m$^3$ using cement to sand ratio of 0.50 and water to cement ratio of 0.55. The preformed stable foam was produced by diluting the water and foam agent with concentration of 0.05. On the other hand, the mix proportion of normal concrete follows the cement to sand ratio of 0.67 and sand to gravel ratio of 0.50, while 0.45 for water to cement ratio. Based on the mix proportion, the required quantity was obtained as presented in table 1.

**Table 1.** Mix design for foamed and normal concrete.

| Material | Quantity of FC for 1m$^3$ (kg) | Quantity of NC for 1m$^3$ (kg) | Required quantity FC (kg) | Required quantity NC (kg) | Remarks |
|---|---|---|---|---|---|
| **Cement** | 400 | 320 | 080 | 064 | OPC Type I based on BS EN 197-1 |
| **Sand** | 800 | 480 | 160 | 096 | Size less than 3mm based on BS 882:1992 |
| **Gravel** | - | 960 | - | 192 | Size less than 20 mm based on BS 882:1992 |
| **Water** | 220 | 144 | 044 | 28.8 | Tap water based on BS EN 1008:2002 |

*Note: FC = Foamed concrete, NC = Normal concrete, OPC = Ordinary Portland cement

### *2.2. Specimen preparation*

A total of nine cube specimens of foamed concrete were prepared using mould of 100mm x 100mm x 100mm. Meanwhile, another nine cube specimens of normal concrete were casting using mould of 150mm x 150mm x 150mm. Cube specimens of foamed concrete were undergone air curing at the ambient temperature, whilst cube specimens of normal concrete were submerged in water. It should be noted here that the curing process was conducted at 7, 14 and 28 days. The slab specimens were prepared using the galvanized corrugated steel deck. The composite slab has dimension of 750mm width, 1600mm length and 125mm thickness. The schematic design of composite slab can be seen in figure 1. Bolts were installed with an interval of 250mm center-to-center to provide perfect bond between corrugated steel deck and concrete topping, while wire mesh was utilized as reinforcement. The preparation of mould and hardened composite slab can be seen in figure 2.

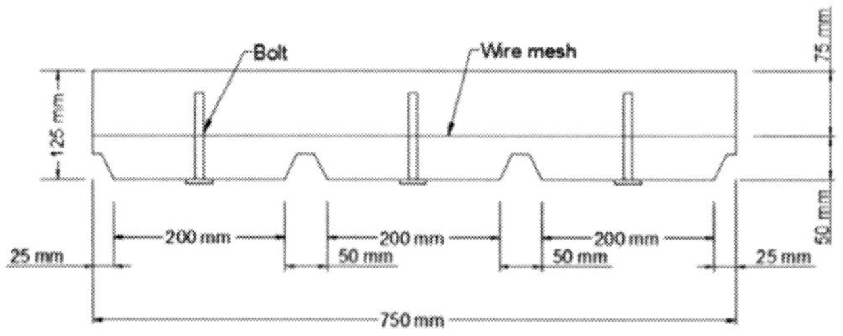


**Figure 1.** Schematic design of composite slab.

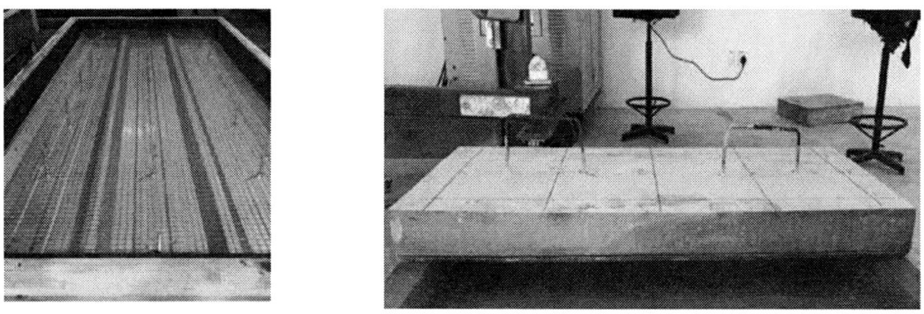

(a) (b)

**Figure 2.** Specimen preparation of composite slab: (a) mould and (b) hardened composite slab.

*2.3. Test programme*

The compression test on cube specimens of foamed concrete and normal concrete was conducted accordance to BS EN12390-3:2009 [15]. The Ele Compact Machine 1500 was used in the compression test. Before the compression test take place, the cube specimens were visually inspected, labelled, sized and weighted. Meanwhile, slab specimens were tested under the hammer-impact test after 28 days of curing process. The hammer-impact test requires special instruments of hammer, accelerometers and data logger as can be seen in figure 3. The hammer was designed for use with delicate structure. The force transducer on hammer measures the impact force that applied to the surface of composite slab. Meanwhile, the accelerometers were used to measure the wave vibration due to the impacted of hammer. The wave vibration was recorded as acceleration-time history (in relation to location) by data logger and can be displayed in both plot and digital using QuickDAQ. The set-up of hammer-impact test can be seen in figure 4.

Fifteen (15) points for accelerometers were drawn on top surface of composite slab as shown in Figure 5. Points of A1, A2, A3, A4 and A5 are classified as Series A, while Series B consists of B1, B2, B3, B4 and B5. Points for Series C are dictated by C1, C2, C3, C4 and C5. Hammer was stroked at point X which located at one third span of slab specimens. The hammer was stroked for ten times at ponint X for the accuracy of data and the wave vibration at A1, B1 and B2 were simultaneously captured until A5, B5 and C5.

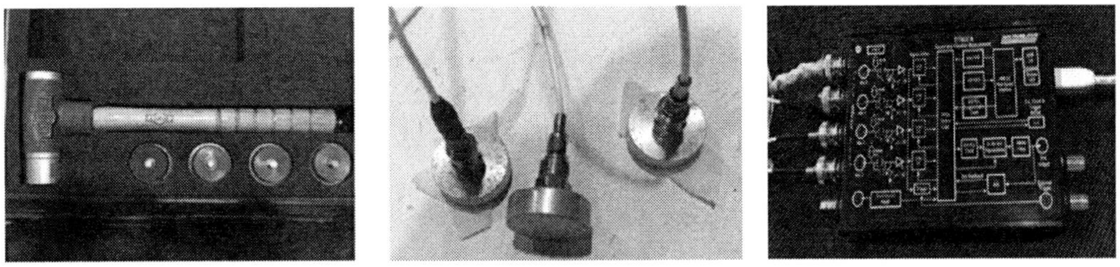

a) Hammer b) Accelerometer c) Data logger

**Figure 3.** Instrument of hammer-impact test.

**Figure 4.** Test set-up for hammer-impact test.

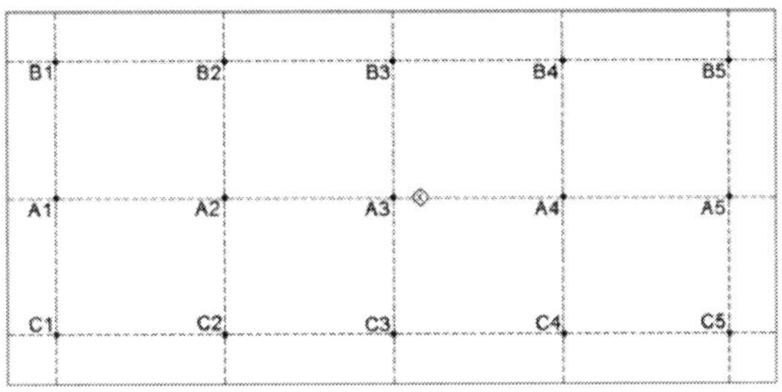


**Figure 5.** The position of the accelerometer during the test.

The raw data obtained from the impact hammer test was examined to get the graph of wave propagation and acceleration-time history, consequently the vibration responses can be determined. The vibration responses are referred to the first natural frequency, maximum acceleration and mode shape. In order to determine the first natural frequency, the average time of natural period need to be calculated using the following equation:

$$T = (T_f - T_i) / n \tag{1}$$

where, $T$ is the average time of the chosen circles period, $T_f$ is the final time of last completed sine cycle, $T_i$ is the initial time of first completed sine cycle and $n$ is the number of completed sine cycle. Therefore, the natural frequency, $f$ of composite slab can be calculated by the following equation:

$$f = 1 / T \tag{2}$$

On the other hand, analytical approach can be used to determine the first natural frequency of composite slab using the following equation:

$$f = \frac{1}{2\pi}\sqrt{\frac{k}{m}} \tag{3}$$

where $k$ is stiffness ratio related by properties of material of composite slab that that can be expressed as $EI$. Where $E$ is the young modulus of concrete ($kN/mm^2$) and I is the second moment of inertia of the composite slab for critical axis ($mm^4$). On the other hand, $m$ is the total mass of composite slab($kg$).

## 3. Results and Discussions

### *3.1. Compressive strength*

Figure 6 shows the compression strength of foamed concrete and normal concrete at 7, 14 and 28 days. It was obtained that foamed concrete and normal concrete achieved the targeted compressive strength. At 28 days of curing process, the compressive strength of foamed concrete is 17.5MPa, while normal concrete achieved 18.2MPa. The difference of compressive strength between foamed concrete and normal concrete is approximately 4%. The compressive strength foamed concrete indicated close compliance with British Cement Association [16]. In addition, investigations by Jaini *et al.* [7] and Abd Rahman *et al.* [17] confirmed that the compressive strength of foamed concrete with density 1600kg/m$^3$ should be in the range of 12MPa to 18MPa.

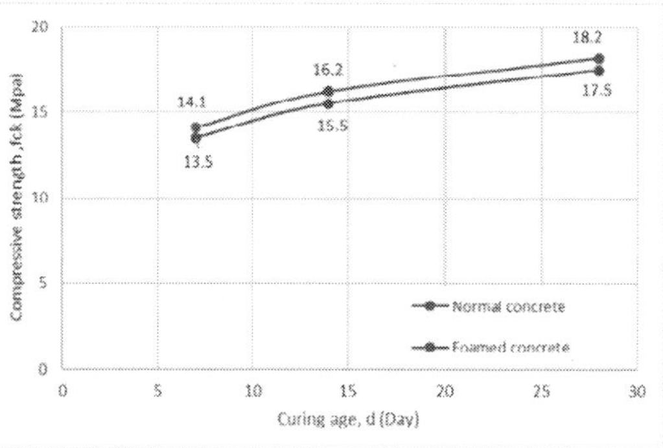


**Figure 6.** Comparison for compressive strength of foamed concrete and normal concrete.

### *3.2. Natural frequency*

The raw data from the hammer-impact test was examined to plot the acceleration-time history. Figure 7 shows the acceleration time-history of foamed concrete composite slab while the acceleration time-history of normal concrete composite slab can be seen in figure 8. It was identified that there are more completed sine cycles in the similar time interval for the wooden support condition. This is due to the wooden material can absorb the wave energy generated by the forced applied and dissipated it easily. This situation causes it had to take short time to complete one sine cycle. However, steel material will reflect the wave and mix with generated wave. Thus, it need longer time to complete one sine cycle of the wave. Natural frequency value is inversely proportional to the time. Therefore, the longer the time to complete a sine cycle, the smaller the fundamental natural frequency. Table 2 shows the summaries of fundamental natural frequency for foamed concrete and normal concrete composites slabs due to wooden and steel supports. It can be observed that the composite slabs with steel support have the lower fundamental natural frequency if compared to that supported by wood, which is 31.7% for foamed concrete composite slab while 13.9% for the normal concrete composite slabs. Moreover, an investigation by Emad El-Dardiry & Tianjian [18] showed that the fundamental natural frequency profiled composite floors are less than 40Hz.

A comparison of fundamental natural frequency discovered that foamed concrete composite slab has the highest value, which is 12.3% than normal concrete composite slab for wooden support. Despite that normal concrete slightly strong in term of compressive strength. However, under the influence of steel support, foamed concrete composite slab shows contradict behaviour. The design density of foamed concrete used was 1600 kg/m3, while the density of normal concrete used was 2400 kg/m3. There was 33% increment of mass of the specimen. Thus, this would explain the lower natural frequency occur on normal concrete composite slab.

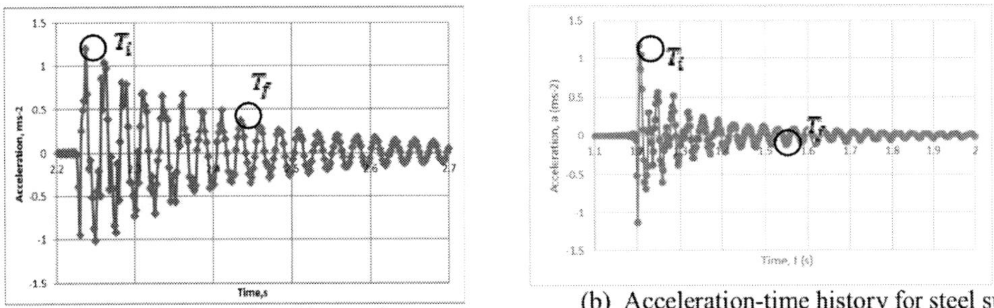

(a) Acceleration-time history for wooden support (b) Acceleration-time history for steel support

**Figure 7.** Acceleration-time history of foamed concrete composite slab at mid span (A3) for different type of support.

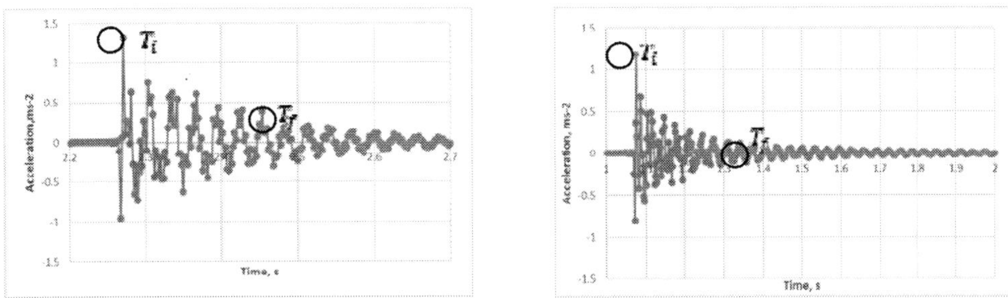

(a) Acceleration-time history for wooden support (b) Acceleration-time history for steel support

**Figure 8.** Acceleration-time history of normal concrete composite slab at mid span (A3).

**Table 2.** First natural frequency of composite slabs.

| Type | Wooden Support | | | | Steel Support | | | |
|---|---|---|---|---|---|---|---|---|
| | $T_i$ (s) | $T_f$(s) | $n$ | $f$ (Hz) | $T_i$ (s) | $T_f$(s) | $n$ | $f$ (Hz) |
| Foamed concrete composite slab | 2.241 | 2.412 | 7 | 40.94 | 1.212 | 1.498 | 8 | 27.97 |
| Normal concrete composite slab | 2.272 | 2.467 | 7 | 35.89 | 1.076 | 1.335 | 8 | 30.89 |

### *3.3. Mode shape*

The peak acceleration of each point in series A, B and C were plotted to form the first mode shape as can be seen in figures 9 and 10. In addition, it was also found that the series B and C show the nearest of acceleration at each side of the composite slab during the first mode of composite slab. This may be the reason of first natural frequency of composite slab was synchronized. Table 3 shows the highest of maximum acceleration of the slab with its location. The highest maximum acceleration of the composite slab with wooden support is approximately same which is 1.881 $m/s^{-2}$ for foamed concrete composite slab while 1.867 $m/s^{-2}$ for normal concrete composite slab. For the steel support, the highest maximum acceleration of the foamed concrete composite is 1.886 $m/s^{-2}$ which approximately while acceleration of the normal concrete composite slab of 1.305 $m/s^{-2}$. This indicates that the foamed concrete composite slab has the higher acceleration compare to normal concrete.

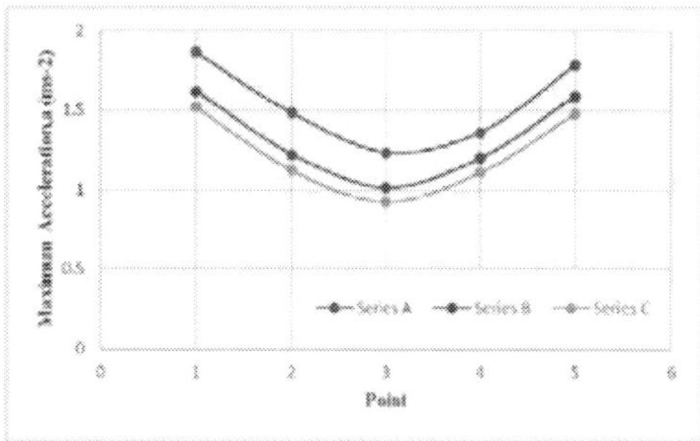


(a)

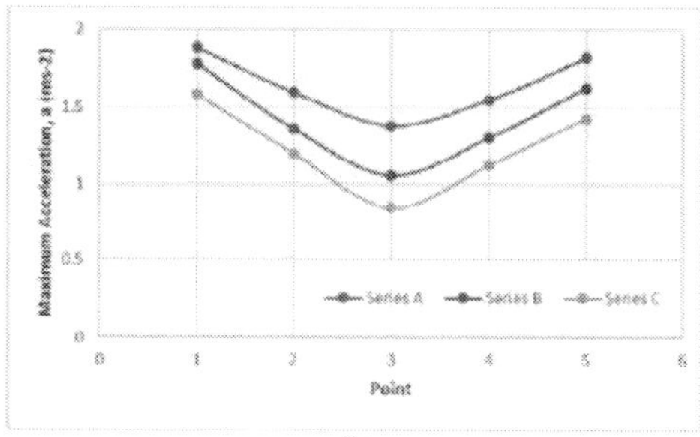


(b)

**Figure 9.** First mode shape of foamed concrete composite slab under: (a) wooden support (b) steel support.

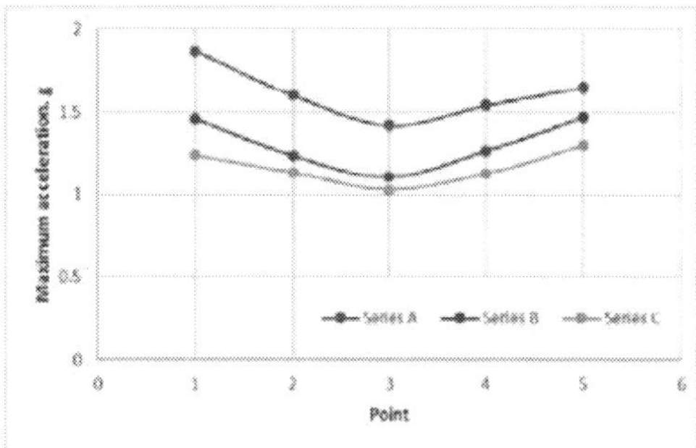


(a)

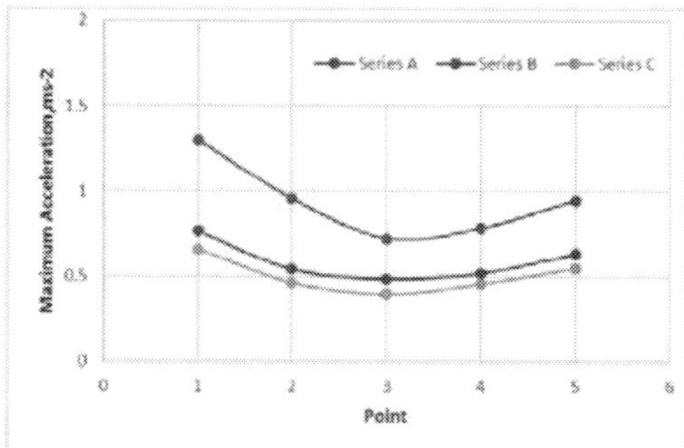


(b)

**Figure 10.** First mode shape of normal concrete composite slab under: (a) wooden support (b) steel support.

**Table 3.** Maximum acceleration of composite slabs.

| Type | Wooden Support | | Steel Support | |
|---|---|---|---|---|
| | Maximum Acceleration ($m/s^2$) | Location | Maximum Acceleration ($m/s^2$) | Location |
| Foamed concrete composite slab | 1.881 | A1 | 1.886 | A1 |
| Normal concrete composite slab | 1.867 | A1 | 1.305 | A1 |

## 4. Conclusions

An investigation on the vibration responses of foamed concrete composite slab with different type of support were compare with normal concrete composite slab. From the result, it showed that the lowest fundamental natural frequency of corrugated steel-foamed concrete slab was 40.94 Hz with wooden support while 27.97 Hz with steel roller support. According to the BS EN 1994-1-1 [2], the floor may be designed to have a fundamental frequency of at least 8.4 Hz vertically and at least 4 Hz horizontally, in which case the resonant effects need not be evaluated. The lowest fundamental natural frequency of the composite slab has fulfilled the minimum requirement of the design of floor based on the standard code of practice. Thus, it can conclude that the natural frequency of composite slab was reduced by replacing the wooden support with the steel roller support. Furthermore, the maximum acceleration obtained, the corrugated steel-foamed concrete slab has a highest maximum acceleration recorded. Hence, corrugated steel-foamed concrete slab has a higher vibration compared to corrugated steel-normal concrete slab.

## 5. References

[1] Rackham J W, Couchman, G H and Hicks S J 2009 Composite Slabs and Beams Using Steel Decking: Best Practice for Design and Construction *MCRMA Technical Paper SCI Publication* **13** 3550-3560

[2] BS EN 1994-1-1 2004 Design of composite steel and concrete structures Part 1-1: *General Rules and Rules for Buildings*, (British Standard Institution, London)

[3] Abbas H S, Bakar S A, Ahmadi M and Haron Z 2014 Experimental studied on corrugated steel-concrete composite slab *Journal Civil Engineering* **67(3)** 225-233

[4] Hulimka J, Krzywo Ĕ, and Agnieszka J Ĉ 2017 Laboratory tests of foam concrete slabs reinforced with composite grid *Procedia Eng.* **193** 337-344

[5]` Zulkarnain F and Ramli M 2008 Durability performance of lightweight aggregate concrete for housing construction *(Icbedc)* 541–551

[6] Hadipramana J, Samad A A A, Zaidi A M A, Mohammad N and Riza F V 2013 Effect of uncontrolled burning rice husk ash in foamed concrete *Adv. Mater. Res.* **626** 769-775

[7] Jaini Z M, Mokhatar S. N, Mohd Yusof A S, Zulkiply S and Abd Rahman M H 2016 Effect of pelletized coconut fiber on the compressive strength of foamed concrete *MATEC Web of Conf.* **47** 01013

[8] Sanchez T A, Davis B and Murray,T M 2011 Floor vibration characteristics of long span composite slab systems *Proceeding of Structure Congress* Las Vegas, United States 360-370

[9] Khalaf M, El-Shihy A, Shehab H and Mustafa S 2013 Structural behaviour analysis of two-ways composite slab. *Journal of Engineering and Innovative Technology* **2(12)** 47-54

[10] Chaudhari T, Macrae G, Bull D, Chase G and Hobbs M 2014 Composite slab effects on beam-column subassemblies *Proceeding of the International Conference on Structural and Earthquake Engineering* (Chrischurch, New Zealand) 1-9

[11] Cifuentes H and Medina F 2013 Experimental study on shear bond behaviour of composite slabs according to eurocode 4 *Journal of Construction Steel Research* **82(3)** 99-110

[12] Johnson R P and Shepherd A J 2013 Resistance to longitudinal shear of composite slabs with longitudinal reinforcement *Journal of Construction Steel Research* **82(3)** 190-194

[13] Lakshimikandhan K N, Sivakumar P, Ravichandran R and Jayachandran S A 2013 Investigations on efficiently interfaced steel concrete composite deck slabs *Journal of Structures* **13(6)** 1-10

[14] Johnson E A F and Li Q M 2012 Structural behaviour of composite sandwich panels with plain and fibre-reinforced foamed concrete cores and corrugates steel faces *Composite Structures* **94(5)** 1555-1563

[15] BS EN 12390-3:2009 *Testing Hardened Concrete - Compressive Strength of Test Specimens* (British Standard Institution, London, 2011)

[16] British Cement Association 1994 *Foamed concrete: Composition and properties*, Report 46.042 (Slough, United Kingdom)

[17] Abd Rahman N, Jaini Z M and Mohd Zahir N N 2015 Fracture energy of foamed concrete by means of the three-point bending test on notched beam specimen *ARPN J. Eng. Appl. Sci.* **10** 6562-6570

[18] Emad El-Dardiry and Tianjian Ji 2006 Modelling of the dynamic behaviour of profiled composite floors *Engineering Structures* **28** 567-579

**Acknowledgments**
Authors would like to thank Universiti Tun Hussein Onn Malaysia (UTHM) and Ministry of Education Malaysia for the continuous supports in term of facilities and financial. This experimental study was conducted under Research Acculturation Grant Scheme (RAGS), Vot. No. R068.

IOP Conf. Series: Materials Science and Engineering **271** (2017) 012103 doi:10.1088/1757-899X/271/1/012103

# Ultimate strength capacity of a square hollow section filled with fibrous foamed concrete

**Siti Amirah Azra Khairuddin[1], Norashidah Abd Rahman[2,*], Norwati Jamaluddin[2], Zainorizuan Mohd Jaini[2], and Noorwirdawati Ali[2]**

[1]Faculty of Civil and Environmental Engineering, Universiti Tun Hussein Onn Malaysia, 86400 Parit Raja, Johor, Malaysia
[2]Jamilus Research Center, Universiti Tun Hussein Onn Malaysia, 86400 Parit Raja, Johor, Malaysia
Corresponding author: nrashida@uthm.edu.my

**Abstract.** Concrete-filled sections used as building columns have become popular due to their architectural and structural elements. In recent years, there has been a renewed call for the improvement of materials used as concrete to fill the composite columns. Among these materials, foamed concrete has received great attention due to its structural characteristics and its potential as a construction material used in hollow sections. However, its behaviors as infill material in a hollow section, such as its strength and failure mode, should be investigated. In this study, experimental research was conducted to compare the experimental and theoretical values of its ultimate strength capacity. Eight specimens of hollow steel sections with two different thicknesses were filled with fibrous foamed concrete and then subjected to compression load. The obtained results were compared with those obtained from a hollow section with the same thicknesses, but were filled with normal foamed concrete. Results show that the ultimate strength capacity of the experimental value is the same as that of the theoretical value based on Eurocode 4. The largest percentage values between theoretical and experimental results for thicknesses of 2 and 4 mm are 58% and 55%, respectively.

## 1. Introduction

The use of a concrete-filled hollow section (CFHS) as a structural component has increased in recent years due to its aesthetically pleasing appearance and its ability to significantly improve axial capacity without increasing the cross-sectional area [1]. In other words, the CFHS offers basic advantages in aesthetic appeal and structural efficiency. Moreover, the CFHS is a structural member that efficiently combines the tensile strength and ductility of steel with the compressive strength of concrete. Chan, Huai, and Wang [2] reported that concrete infill contributed not only to the strength but also to the ductility of CFHS.

Studies on normal concrete and high-strength concrete as infill in CFHS have focused on the dead weight of the structure and the increased construction cost. Therefore, lightweight concrete has been proposed to promote the use of CFHS and minimize the weight of the structure. Extensive research on the use of lightweight aggregate concrete as infill in hollow steel sections have shown that lightweight aggregate concrete provides good compressive and bond strengths to CFHS compared with normal concrete [3–5]. Considerable research on lightweight concrete used as infill in CFHS has been conducted [6,7], and results from past studies indicated that CFHS with lightweight concrete is more ductile than that with normal and high-strength concrete [8]. The use of foamed concrete as infill material also increases axial load capacity due to the confining effect of the steel tube with concrete core [5,9].

Foamed concrete is a green concrete that is produced using a low amount of water; it optimizes energy, conserves natural resources, generates minimal water, and provides healthy spaces for occupants [10]. Foamed concrete is a type of concrete that includes an expanding agent, which increases the volume of concrete mixture while reducing the dead weight due to the minimal use of coarse aggregate. Despite these advantages, foamed concrete is also known for its brittleness. Thus, the inclusion of fiber has been proposed to improve the brittle properties of foamed concrete. Further research on additional fiber in concrete have contributed to increasing its tensile, flexural, and compressive strengths [11–14]. Awang, Ahmad, and Al-Mulali [15] proved that the compressive strength of foamed concrete increased with the addition of up to 0.4% polypropylene fiber. Subair, Fayas, and Paul [11] conducted an experimental study to determine the optimum percentage of steel fiber. They reported that the compressive strength increased as the percentage of steel fiber increased. However, the compressive strength of lightweight concrete reduced when the steel fiber comprised1% volume of the concrete. Therefore, the aim of the current research is to investigate the strength of CFHS with fibrous foamed concrete.

## 2. Analytical consideration

In this study, the stability of a short column structure is referred to as the ultimate capacity of the compression member to reach the maximum load without buckling, also known as squash load. The column must be perfectly straight and subjected only to axial load [4]. Meanwhile, Eurocode 4 [16] is an international standard code of practice in composite structure, which covers not only concrete-encased and partially encased steel section, but also concrete-filled section with or without reinforcement. The ultimate axial load of a composite column can be analyzed using an equation based on Eurocode 4, known as plastic resistance to compression (squash load). The equation also includes the limit state concept to achieve serviceability and safety by including the partial safety factor of material properties

$$N_u = A_s f_{sd} + A_c f_{cd}, \tag{1}$$

where $f_{sd}$ is the design value of the yield strength of steel, $f_{cd}$ is the design value of the compressive strength of concrete, and $f_c$ is the 28-day strength of concrete. The value of $0.83 f_c$ is recommended by the European Convention for Constructional Steelwork for experimental work.

The strength index (SI) is defined as the ratio of the ultimate strength capacity of the CFHS column to the sum of the strengths of the composite column. Values of SI that are higher than 1.0 suggest a positive interaction between the steel tube and the concrete core, and good concrete confinement [17] SI is described as

$$SI = {}^{N_e}\!/_{N_u} \tag{2}$$

where $N_e$ is the experimental value of ultimate strength capacity of CFHS and the summation of the load-bearing capacity of the steel tube and core concrete, $N_u$.

## 3. Experimental program

The aims of the experimental study were to determine the maximum load capacity of specimens and compare the result with theoretical value analyses based on the code of practice. In this study, 12 specimens were cast with different types of concrete and thicknesses.

### *3.1. Material properties*

The strength properties of mild steel hollow steel section were determined for each side of the steel section. The geometry of the coupon specimen was specified according to BS EN ISO 6892-1-2016

[18]. Table 1 shows the detail of the dimension and value of yield strength steel section. The cross section of the hollow steel section is shown in Figure 1.

**Table 1.** Details of section properties.

| Dimension (h × b × t) (mm) | Length (mm) | $f_{yk}$ (MPa) |
|---|---|---|
| **100 × 100 × 2** | 350 | 349 |
| **100 × 100 × 4** | 350 | 387 |

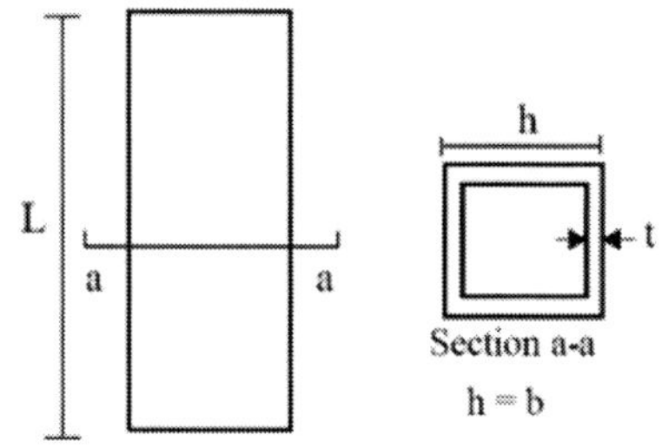


**Figure 1.** Cross section of the specimen.

The material components of the foamed concrete included used in this study included ordinary Portland cement, sand, water, preforming agent, and fiber. The mix design of foamed concrete was calculated based on the design of environmental method. The proposed density was 1600 kg/m$^3$. The optimum percentages of steel and polypropylene fibers used in foamed concrete were 0.8% and 0.4%, respectively. With the addition of steel fiber, a slight increase in density (~4%) was achieved. To achieve the targeted density, the quantities of cement and sand proportions in fibrous foamed concrete for 1 m$^3$ were 550 and 756 kg, respectively. Furthermore The ratios of water–cement, foam–water, and foam–cement were 0.55, 0.05, and 0.07, respectively.

*3.2. Testing*

Twelve specimens with a size of 100 mm × 100 mm × 350 mm and thicknesses of 2 and 4 mm were tested under static load test. The displacement of the specimens was measured by two linear variable differential transformers. Each specimen was filled with various types of foamed concrete, which were polypropylene fiber foamed concrete (PF), steel fiber foamed concrete (SF), and normal foamed concrete (FC). A strain gauge with a size of 10 mm was used to capture the deformation value of the specimens. The universal testing machine was set up to the minimum speed of 1 mm/min to provide sufficient time for specimens to fail. The loading rate was maintained, and the strain and displacement reading was recorded at each load increment. Figure 2 shows the equipment setup for the CFHS specimens.

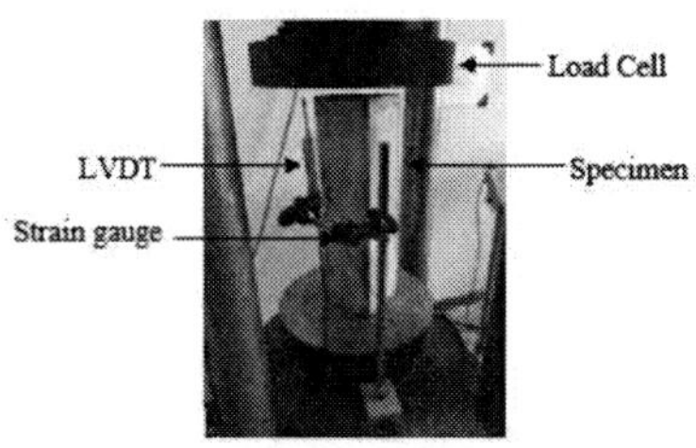


**Figure 2.** Testing specimen setup.

Cube specimens with a size of 100 mm × 100 mm × 100 mm were tested according to BS EN12390-3:2009 [19] in order to determine the compressive strength of fibrous foamed concrete. The cube specimens were tested at 7, 14, and 28 days of air-curing age. Before testing, the cube specimens were weighed to determine the density of each specimen.

## 4. Result and analysis

Figure 3 shows the compressive strengths of various types of foamed concrete. As can be seen, the compressive strength of FC is 14.5 MPa, which complies with the compressive strength of foamed concrete determined by the British Cement Association [20]. Rahman, Jaini, and Zahir [21] reported a relatively high fracture energy for the compressive strength of foamed concrete at 14.7 MPa with a density of 1600 kg/m$^3$. The compressive strengths of foamed concrete with the addition of polypropylene and steel fibers increase by up to 17.2 and 17.9 MPa, respectively. Furthermore, with the addition of fiber in foamed concrete, the strength of foamed concrete compared with that of FC increases at the range of 19% to 23%. Hence, the inclusion of fiber improves the mechanical properties of foamed concrete. This finding agrees with the results reported by Subair et al. [11], Mydin et al. [12], and Awang et al. [13], who concluded that the addition of fiber improved the structural strength and the impact resistance of foamed concrete.

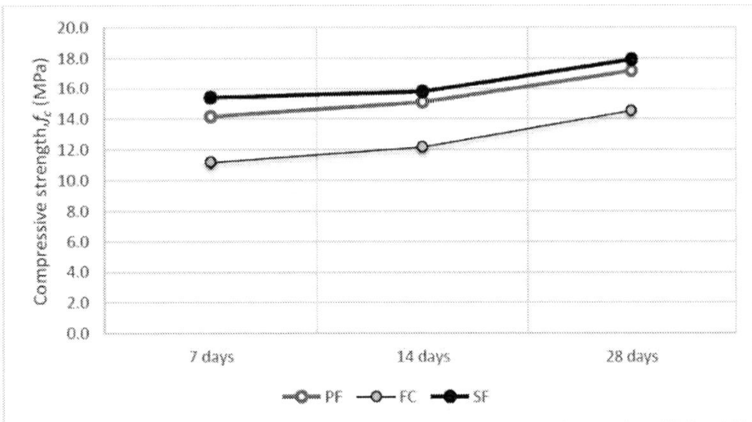


**Figure 3.** Compressive strength of various types of foamed concrete.

Table 2 shows the comparison of the theoretical and experimental results. As can be seen, the CFHS with fibrous foamed concrete with a density of 1600 kg/m$^3$ achieved the theoretical value. The maximum ultimate strength of the CFHS with additional fiber in foamed concrete for a thickness of 2 mm is 542 kN, and that of CFHS with FC is 520 kN. Moreover, the ultimate strength capacity for 4 mm thickness increases to a maximum (as much as 971 kN) compared with the CFHS with FC, which yields a value of 910 kN. The experimental result indicates that the value of ultimate strength capacity with additional fiber increases by up to 6% for 2 mm thickness, and that for 4 mm thickness increases in the range of 7% to 9%. Both thicknesses are compared with CFHS with FC. The result proves that the addition of fiber contributes to the increment of the ultimate strength capacity of CFHS.

The inclusion of steel fiber in CFHS with foamed concrete also shows higher experimental value due to the higher compressive strength of concrete with FC. Further, the inclusion of steel fiber in concrete core also significantly increases the ultimate strength of capacity of CFHS. Hence, the addition of steel fiber definitely improves the ductility and deformation behavior of CFHS specimen [22]. Meanwhile, a minimal increment in the ultimate strength capacity of CFHS with PF is observed compared with that of CFHS with FC. The experimental value of CFHS with polypropylene fiber shows an unstable result, because polypropylene fiber is a hydrophobic type of fiber that tends to absorb water and cause the generation of numerous pores [15,23].

The maximum value for the ultimate strength capacity of CFHS with fibrous foamed concrete for 4 mm thickness is 971 kN. The percentage difference between the experimental and theoretical values for 4 mm thickness for each type of foamed concrete as an infill ranges between 46% and 55%. For 2 mm thickness, the high value of experimental result is 535 kN, which ranges between 57% and 58%. The two thicknesses are compared with CFHS with FC due to the increasing thickness of the steel section. As previously mentioned, the ultimate strength capacity is influenced not only by concrete strength, but also by the thickness of steel section. The ultimate strength capacity of CFHS for 4 mm thickness increases in the range of 75% to 81% compared with that for 2 mm thickness. This result is due the decreased b/t ratio. An investigation conducted by Geetha and Swedha [24] showed that the percentage difference in ultimate strength capacity between 2 and 3 mm thicknesses was 48%. The ultimate strength capacity is also affected by b/t ratio, which agrees with the previous work conducted by Al-Mazini and Chkhewier [25]. Theoretically, SI value is affected by concrete strength and b/t ratio. The experimental result also shows that SI value is influenced by concrete strength, thus proving that fibrous foamed concrete ensures good confinement. Kassoul et al. [26] reported that the parameters influencing confinement can significantly improve the strength and ductility of local sections.

**Table 2.** Comparison of CFHS test results.

| Specimen | Thickness (mm) | b/t | Compressive strength, $f_c$ (MPa) | Theoretical value, $N_u$ (kN) | Experimental value, $N_e$ (kN) | Strength index, SI |
|---|---|---|---|---|---|---|
| FC21 | 2 | 50 | 14.5 | 323 | 495 | 1.53 |
| FC22 | | 50 | 14.5 | 323 | 520 | 1.61 |
| FC41 | 4 | 25 | 14.5 | 608 | 870 | 1.44 |
| FC42 | | 25 | 14.5 | 608 | 910 | 1.50 |
| PF21 | 2 | 50 | 17.2 | 336 | 513 | 1.52 |
| PF22 | | 50 | 17.2 | 336 | 538 | 1.60 |
| PF41 | 4 | 25 | 17.2 | 621 | 954 | 1.54 |
| PF42 | | 25 | 17.2 | 621 | 949 | 1.53 |
| SF21 | 2 | 50 | 17.9 | 340 | 542 | 1.59 |
| SF22 | | 50 | 17.9 | 340 | 535 | 1.57 |
| SF41 | 4 | 25 | 17.9 | 624 | 971 | 1.56 |
| SF42 | | 25 | 17.9 | 624 | 964 | 1.54 |

***FC: CFHS with normal foamed concrete***

***PF: CFHS with polypropylene fiber foamed concrete***

***SF: CFHS with steel fiber foamed concrete***

Figure 4 shows the typical modes of failure experienced by short-column CFHS with fibrous foamed concrete subjected to compression load. The failure modes of specimens show an outward buckling at the top and at the bottom of the stub specimens. According to Jayaganesh et al. [27], the steel section in the outer limit directly carries the applied load and ensures confinement to the inner concrete core, thus avoiding damage to the inner concrete core that can lead to the buckling of the steel section. However, the buckling effect cannot be seen clearly because the fiber can postpone the buckling effect of CFHS. Tao [28] reported that SI values greater than 1 effectively postpone the local buckling effect of specimens.

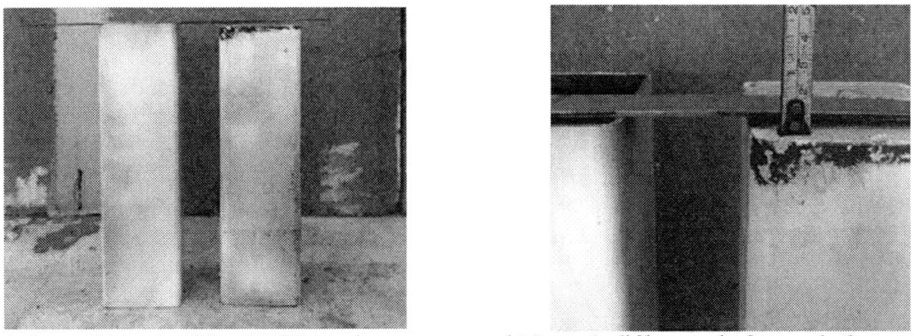

a) Overall length of the CFHS specimen. b) Length difference before and after testing.

**Figure 4.** Outward buckling of CFHS with SF.

## 5. Conclusion

This paper presents the experimental results of testing CFHS with fibrous foamed concrete, from which the following conclusions can be derived:

1) Normal concrete can be replaced with fibrous foamed concrete as infill material in steel hollow sections in order to reduce the dead weight of the concrete infill.
2) The confinement of the fibrous concrete infill improves its strength and delays the local buckling of thc steel section.
3) The ultimate strength capacity is influenced by b/t. When b/t is reduce, the ultimate strength capacity increases.
4) CFHS with fibrous foamed concrete can achieve the theoretical value based on Eurocode 4, in which 58% is the largest percentage difference between the experimental and theoretical results of ultimate strength capacity due to the thickness of wall section.

## 6. References

[1] Testo N and Lam D 2011 *Composite Construction in Steel and Concrete VI* (USA: American Society of Civil Engineers) pp 252-262

[2] Chan T M, Huai Y M, and Wang W 2015 Experimental investigation on lightweight concrete filled cold-formed elliptical hollow section stub columns *Journal Construction Steel Res*earch **115** 434–444

[3] Roseline K J and Tensing D 2013 State of The Art Report on Steel- Concrete Infilled Composite Column *International Journal of Scientific & Engineering Research* **4** 677–684

[4] Mouli M and Khelafi H 2007 Strength of short composite rectangular hollow section columns filled with lightweight aggregate concrete *Eng. Struct.* **29 (8)** 1791–1797

[5] Hunaiti Y M 1997 Strength of Composite Sections with Foamed and Lightweight Aggregate Concrete *J. Mater Civil Eng.* **9** 58–61

[6] Ghannam S, Al-Rawi O and El-Khatieb 2011 Experimental Study on Light Weight Concrete-Filled Steel Tubes *Jordan Journal of Civil Engineering* **5** 1-9

[7] Fu Z Q, Ji B H, Lei L, and Zhou W J 2011 Behavior of lightweight aggregate concrete filled steel tubular slender columns under axial compression *Advanced Steel Construction* **7** 144–156

[8] Chu K 2014 *Axial load behaviour of steel tube columns in-filled with various high-performance concretes* Thesis (Ryerson University, Toronto, Ontario)

[9] Hunaiti M Y, Nabil M F and Issam M A 2002 Evaluation of the Concrete Contribution Factor for Composite Section with Lightweight Concrete under Axial Compression *Journal of Applied Science* **10** 990–999

[10] Moon A S, Varghese V., Waghmare S S and Moon A S 2015 Foam Concrete as A Green Building Material *International Journal For Research In Emerging Science And Technology* **2(9)** 25–32

[11] Subair M D, Fayas C and Paul 2016 An experimental study on the mechanical properties of steel fibered sand replaced of foamed concrete *Int. Res. J. Engenieering Technol.* **3** 14
[12] Mydin N M, Othuman M A, Sahidun, N S, Yusof M. and Noordin, 2016 Compressive, Flexural and Splitting Tensile Strength of Lightweight Foamed Concrete with Inclusion of Steel Fibre *Journal Teknologi* **5** 45–50
[13] Awang M H, Hanizam and Ahmad 2014 Durability Properties of Foamed Concrete with Fiber Inclusion *International Journal of Civil, Environmental, Structural, Construction and Architectural Engineering* **8** 273–276
[14] Mydin A O and Soleimanzadeh S 2012 Effect of polypropylene fiber content on flexural strength of lightweight foamed concrete at ambient and elevated temperatures *Advances in Applied Science Research* **3** 2837–2846
[15] Awang H, Ahmad M H, and Al-Mulali M Z 2015 Influence of kenaf and polypropylene fibres on mechanical and durability properties of fibre reinforced lightweight foamed concrete *J. Eng. Sci. Technol.* **10(4)** 496–508
[16] EN 1994 1-1 2009 Design of composite steel and concrete structures *British Standard Institution*
[17] Jamaluddin N, Lam D, Dai X H and Ye J 2013 An experimental study on elliptical concrete filled columns under axial compression *Journal of Constructional Steel Research* **87** 6–16.
[18] EN ISO 6892-1 2016 Metallic Materials-Tensile Testing Part 1: Method of test at room temperature *British Standard Institution*
[19] EN 12390-3 2009 Testing hardened concrete (compression) *British Standards Institution* **vol. 3** pp. 420–457
[20] British Cement Association 1994 *Foamed Concrete: Composite and properties* Report ref46042 Slough:BCA.
[21] Rahman N A, Jaini Z M, and Zahir N N M 2015 Fracture Energy of Foamed concrete by means of the three-point bending tests on notched beam specimens *ARPN J. Eng. Appl. Sci.* **10** 6562–6570
[22] Tokgoz S and Dundar C 2010 Experimental Study on Steel Tubular Columns In-filled with Plain and Steel Fiber Reinforced Concrete *Thin-Walled Struct.* **48** 414–422
[23] Roslan H, Ahmad F and Awang H 2014 Effects of Various Additives on Drying Shrinkage, Compressive and Flexural Strength of Lightweight Foamed Concrete (LFC*)Advanced Material Research (Switzerland)* vol 2013(Trans Tech Publication) pp. 594-604
[24] Geetha H and Swedha T 2015 An Experimental study on Concrete Filled Tubular Columns Using Varying Steel Materials *International Journal of Innovative Science, Engineering & Technology* **2(5)** 737–742
[25] Al-mazini M A and Chkhewier A H 2017 Behavior of Concrete Filled Aluminum Square and Rectangular Hollow Section Columns under Axial Loads: Experimental and Analytical Study *Journal of Babylon University* **25(2)** 712–726
[26] Kassoul A, Bougara A, and Abdesselam H 2014 Analysis of Confinement Effect on Strength and Ductility in Reinforced Concrete Structure *Earthquake Engineering And Seismology. European Conference (Istanbul)* vol 3 (European Association for Earthquake Engineering (EAEE ))pp. 2584-2593
[27] Jayaganesh S, Murugadoss J R, Prabhu G G, and Jegan J 2015 Effects of Concentrical Partial (Local) Compression on the Structural Behavior of Concrete Filled Steel Tubular Column
[28] Tao Z, Han L, and Wang D 2008 Strength and Ductility of stiffened thin-walled hollow steel structural stub columns filled with concrete *Thin-walled Structure* **46** 1113-1128

**Acknowledgments**
The author would like to thank the Universiti Tun Hussein Onn Malaysia (UTHM) for the continuous support in terms of facilities and the Center for Graduate Studies UTHM as a sponsor for this paper. This study also was conducted under FRGS Vot 1625.

# Analysis of castellated steel beam with oval openings

**S Tudjono[1], Sunarto[2] and A L Han[1]**

[1]Department of Civil Engineering, Diponegoro University, Semarang-Indonesia
[2]Master program in Civil Engineering, Diponegoro University, Semarang-Indonesia
Corresponding author: sunarto.1392@gmail.com

**Abstract.** A castellated steel beam is per definition a wide flange (WF) or I shaped steel profile with openings, to reduce self-weight and improve the effectiveness in terms of material use. Recently, extensive study on these castellated steel beams has been conducted, involving different shapes in web openings. The main goal of these research works was to evaluate and analyze its optimum opening sizes and shapes configuration. More in-depth research work to the behavior and the influence of holes to WF beams need to be conducted. In this paper, an oval shaped web opening is chosen as alternate. The study involves a modification in the variation of oval web openings both in the horizontally and vertically direction. An experimental and numerical study based on the finite element method conducted with the Abaqus/CAE 6.12 software is used to analyze the buckling behavior of the web. The obtained results from the experimental test specimens are in good agreement with the obtained results from the finite element analysis. Furthermore, the numerical model can be expanded to be used as analyzing tool in evaluating and studying the effect and influencing factors of a variation in opening's parameters.

## 1. Introduction

A castellated steel beam is an I-shaped beam section with a variety in shape opening in the web. The opening can be hexagonal, rectangular, circular, diamond or oval in shape. The origin of the name "castellated" is derived from the pattern of holes in the web, because castellated means "built like a castle, or regular holes in the walls, like a castle". The castellated steel beam is made by expanding a standard rolled shape in a manner which creates a regular pattern of holes in the web. At first, the chosen pattern is made along the web on a path that will be cut. The cut halve beams are separated and then welded together based on the chosen opening shape.

The use of castellated steel beams nowadays has been rapidly catching attention due to its advantages. A castellated steel-beam in a structure gives the advantage of its lighter weight. Castellated steel-beams can also utilize the placement of installations. Basically, steel sections that satisfy strength requirement have difficulty in satisfying serviceability requirements. Castellated steel beams can be the way to overcome that problem by providing a greater depth, and thus a greater moment of inertia. Furthermore, the modification of web openings affects the failure modes and stability of castellated steel beams itself. Numerous researchers have been dedicating their work to studying the effect of modification of web shaped opening. In this paper, a deepen research on the behavior of castellated steel beams with oval shaped opening was conducted.

## 2. Review of previous studies

Extensive studies have been carried out in the field of design and analysis of castellated steel beams, starting from hexagonal until circular opening shapes. The aim of those studies was to explain the structural behavior of castellated steel beam with different opening shapes. The complexity in geometry and failure mode of castellated steel beams leads to a statement that there is no universally accepted design method for castellated steel beam. There are six possible failure modes for castellated steel beams, they are the formation of a Vierendeel mechanism, lateral torsional buckling of web post, rupture of a welded joint in a web post, lateral torsional buckling of an entire span, formation of a flexure mechanism, and the local buckling of web post [1].

A study concerning the horizontal shear strength of castellated steel beams with and without additional plates was conducted by Suharjanto [2]. In this study, the effects of the additional plate toward buckling are negligible. The result showed that maximum load on castellated steel beam with additional plate gave 26.16% higher values than without additional plates. It was concluded that additional plates on castellated steel beams resulted in a stronger section toward shear force and thus increased the load capacity and horizontal shear strength of circular castellated steel beam.

On the other hand, a research work about the vertical shear buckling capacity of web-posts in castellated steel beams with fillet corner for hexagonal web openings was investigated by Wang et al. [3]. This research was based on the finite element method. The authors compared the structural castellated steel beam performance of web post of fillet corner openings to circular and elongated circular openings. It was found that castellated steel beams with fillet corner opening have a better structural performance as compared to circular openings in terms of their failure mode, stress distribution, shear capacity, and out-of-plane displacement.

The recent study about the castellated steel beam with circular and diamond shaped openings was highlighted by Jamadar and Khumbar [4]. The aim of this research was to obtain the optimization of its size by considering the ratio of overall depth of the castellated steel beam to the depth of the opening, and the ratio of spacing between the opening to the depth of the opening. Results showed that the castellated steel beams with diamond shaped openings with an opening sized of 0.67 times the overall depth of the beam provided a better strength. It was also found that the castellated steel beams mostly tend to fail in their local failure modes.

## 3. Methodology

In this paper, the analysis was conducted experimentally, and numerically, using the finite element analysis. Finite element analysis is chosen due to the complexity of geometry of the castellated steel beam, and was executed with the Abaqus/CAE 6.12 software. Furthermore, the analysis result from the experimental tests were utilized to validate the model, analyzed by the Abaqus software.

In this research, an IWF 150 x 75 x 7 x 5 was chosen as original beam. The original beam was modified into a castellated steel beam with oval shaped openings, both in the vertical and horizontal direction. The modification can be shown in figure 1 and the detail of test set up is seen in figure 2.

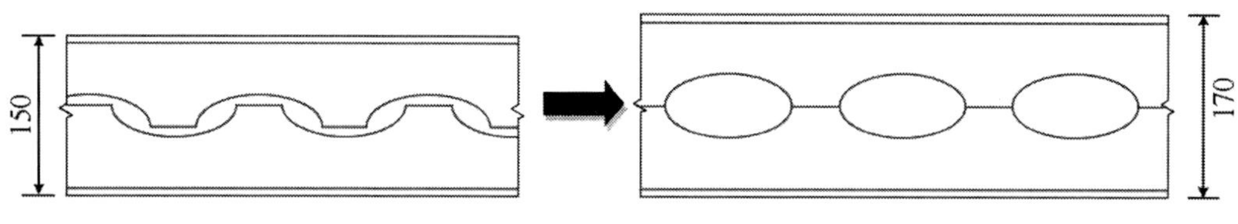


**Figure 1.** The modification of original beam into castellated steel beam

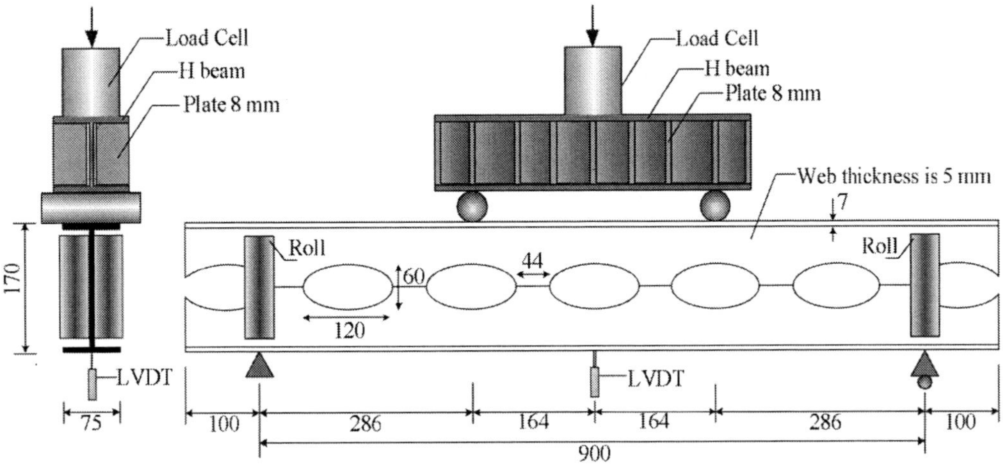


**Figure 2.** Test set up

The modification of castellated steel beams divided into three kind of specimens, they were CB1, CB2, and CB3. CB1 and CB2 are the modification of castellated steel beams with horizontal oval shaped web opening, having exact the same configurations but with a different loading position. This differentiation in loading position was used to study the effect of the position of the load. Meanwhile, the openings for CB3 are positioned vertically. When the developed FE model was proven accurate through a validation process, the model was further used as analyzing tool in evaluating and studying the effect of parameter-variations in the openings.

In the experimental test, the castellated steel beams were tested using a load cell with a capacity of 50 tons. As shown in figure 2., the load was applied to the beam using a two-points loading system, four rolls for each castellated steel beam were located at the center of web, right above the supports, to prevent the ends from out-of-plane buckling and sections above the supports to rotate. To measure the displacement, one LVDT (Linear Variable Differential Transducer) SDP 100-C with a capacity of 100 mm, was used.

## 4. Finite element analysis of castellated steel beam with oval openings

The first step in the development of the finite element model was the construction of a model identical to the experimental specimens CB1, CB2, and CB3. The finite element model was an exact replica of the laboratory specimens in terms of their dimensions, openings, boundary conditions, and loading system. To simulate the physical supports, restrains in the direction of the degree of freedom were applied, as can be seen in figure 3. A non linier material and geometry analysis together with isotropic material were chosen in this research work.

To evaluate the meshing techniques and configuration of the model, a sensitivity analysis was run to obtain the best meshing, by roughly comparing the ultimate load and the pattern of the load-displacement curves. The variations in meshing have to reach a convergence point to make sure that the chosen mesh-configuration and finesse is at optimum.

**Figure 3.** The model of castellated steel beam

From the obtained result of finite element analysis, it was shown that the horizontal loading position didn't influence the load carrying capacity of the beam substantially. On the other hand, the modification of castellated steel beam with horizontal oval openings gave a higher value for the maximum load, as compared to the vertically one. In the finite element analysis, the maximum load and displacement of CB1 reached 169.17 kN with a vertical displacement of 12.32 mm, as for CB2 and CB3 the loads were 164.34 kN and 128.75 kN, and the displacements 8.29 mm and 13.03 mm respectively.

## 5. Experimental Validation

The results from the numeric solution should be validated to the experimental tests, so that the model can further be used as analyzing tool in evaluating modification of castellated steel beam with various opening shapes. The set-up for the castellated steel beams CB1, CB2 and CB3 are shown in figure 4. Furthermore, the obtained load-displacement curves from the finite element analysis were validated to the curves from the experimental tests, which are shown in figure 5.

(a)

(b)

(c)

**Figure 4.** The set up for experimental test in laboratory (a) CB1, (b) CB2, and (c) CB3

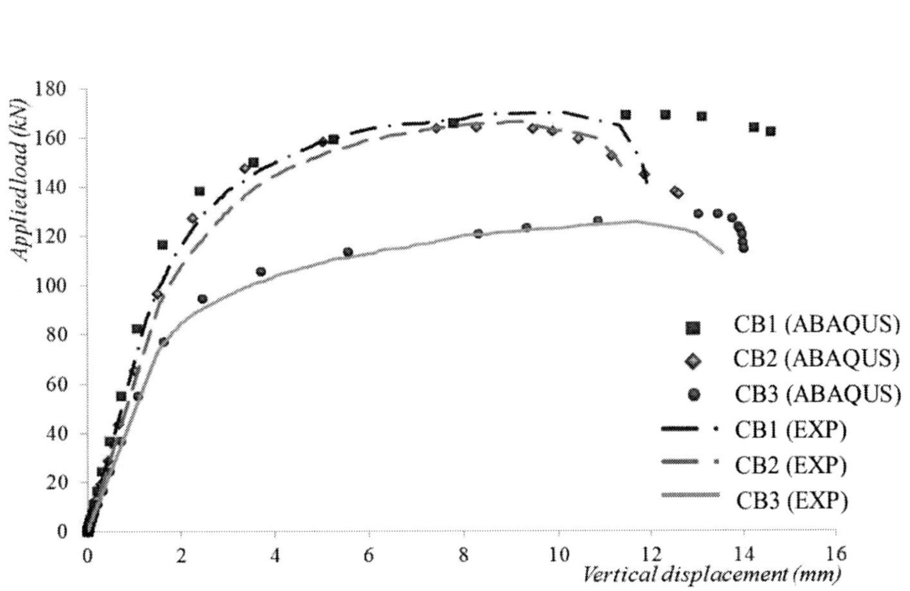


**Figure 5.** The experimental validation towards Abaqus

In the experimental test results, CB1 reached a maximum load of 170.17 kN in combination with the vertical displacement of 10.11 mm. CB2 reached the maximum load at 166.5 kN with a vertical displacement of 9.23 mm. Meanwhile for CB3, its maximum load was 125.584 kN and a vertical displacement of 11.71 mm. Based on the curves, it can be concluded that the horizontally modifications (CB1 and CB2) yielded in a better performance when compared to the section with vertically openings (CB3) in terms of load carrying capacity. It was also found that the different loading position on CB1 and CB2 had negligible effect on the load carrying capacity.

The results of the finite element model closely approach the experimental tests results. The differentiation between the load-displacement results from the model and the experimental specimens are shown in table 1. The origin from these deviations are most probably due to the fact that in the finite element analysis the welding of the joints was neglected. In general, the results from the finite element model are in good agreement with the experimental results. Thus, the results can be compared to the original beam in terms of its maximum load. The visual comparison between the numerical and experimental castellated steel beams' behavior in terms of deformation are shown in figure 6, 7, and 8.

**Table 1.** The obtained results for the experimental test and Abaqus

| | Experimental Test | | Abaqus | | Ratio Max. Load |
|---|---|---|---|---|---|
| | Maximum Load (kN) | Vertical Displacement (mm) | Maximum Load (kN) | Vertical Displacement (mm) | |
| Control Beam | - | - | 165.56 | 3.54 | - |
| CB1 | 170.17 | 10.11 | 169.17 | 12.32 | 1.01 |
| CB2 | 166.50 | 9.23 | 164.34 | 8.29 | 1.01 |
| CB3 | 125.58 | 11.71 | 128.75 | 13.03 | 0.98 |

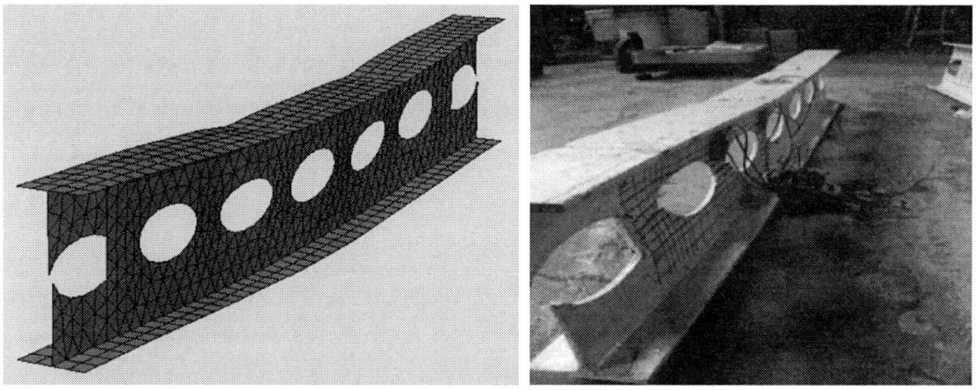

**Figure 6.** Deformation condition of CB1 (left) Abaqus; (right) Experimental Test

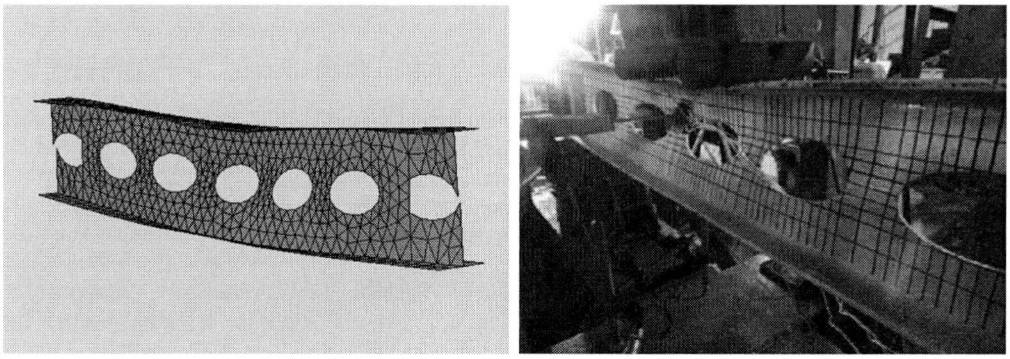

**Figure 7.** Deformation condition of CB2 (left) Abaqus; (right) Experimental Test

**Figure 8.** Deformation condition of CB3 (left) FEM; (right) Experimental Test

It is remarkably clear that the model represented the failure behavior in great detail. The local buckling of the flanges for CB1 are accentuated in the finite element model, while the web buckling in CB3 are the exact replica of the actual behavior in the laboratory. Based on the pictures above, it can be concluded that the results of finite element analysis provide an excellent representation to the behavior of castellated steel beams. Furthermore, the self-weight reduction of castellated steel beam towards the original beam can be analyzed by dividing the net web area of castellated steel beam with the web area of original beam. It was found the reduction ratio in self-weight for both horizontal and vertical modification toward the original beam were 0.92 and 0.90. Proven to be valid, the model can be used as analyzing tool in evaluating and studying the effect of a parameter variation in openings.

## 6. Conclusion

The castellated steel beam with horizontal oval shaped web openings resulted in a load ratio of 1.02 to the original beam. A small increase of 2.2% in load carrying capacity was recorded. The reduction ratio in self-weight was 0.92. Meanwhile, the castellated steel beam with vertical oval shaped web openings yielded in a strength capacity decrease, the ultimate load ratio to the control element was 0.78. The self-weight reduction ratio of this specimen was 0.90.

The member with the horizontal and vertical, oval shaped openings was substantially feasible, since no load-carrying capacity occur, but the self-weight of the member was optimized noticeably. Both of the horizontal and vertical modification almost gave the same weight reduction to the original beam, but the horizontal modification was satisfying better than the vertical one in terms of its load carrying capacity. Thus, it can be concluded that the horizontal modification gave a greater impact on its load carrying capacity rather than the vertical one. A lighter section has generally an encouraging impact on the overall construction cost, since a reduction in self-weight of beam members will, in turn positively influence the load acting on columns and foundation.

The results of finite element analysis using Abaqus are in a good agreement with the experimental results. The model was proven to provide a good estimation not only for the not load-displacement curves, but in predicting the failure mode of the element. Continuing research will be conducted using this basic model as analyzing tool in evaluating and studying the effect of a variation in opening's parameters.

## 7. References

[1] Kerdal D, Nethercot D A 1984 Failure modes for castellated beams *J. Const. Steel Res.* **4** 295-315

[2] Suharjanto 2011 Kajian kuat geser horisontal circular castellated steel beam with and without adding plates *Jurnal Teknik Sipil* **1** 65-71

[3] Wang P, Wang X and Ma N 2014 Vertical shear buckling capacity of web-posts in castellated steel beams with filler corner hexagonal web openings *J. Engin. Struct.* **75** 315-26

[4] Jamadar A M and Khumbar P D 2015 Parametric study of castellated beam with circular and diamond shaped openings *Int'l. Res. J. Engin. Technol.* **2** 715-22

IOP Conf. Series: Materials Science and Engineering **271** (2017) 012105 doi:10.1088/1757-899X/271/1/012105

# Green and sustainable median on a divided multilane highway

**K Ambak[1], F A A Bajuri[1], N A Yusoff[1], A Madun [1] and J Prasetijo[1]**

[1]Department of Infrastructure and Geomatic Engineering, Faculty of Civil and Environmental Engineering, Universiti Tun Hussein Onn Malaysia, 86400 Batu Pahat, Johor, Malaysia.
Corresponding author: kamardin@uthm.edu.my

**Abstract.** The road system in Malaysia showed good growth with increasing number of vehicles on the road. However, statistic of road accident is still at an alarming rate. There are many factor can contribute to the occurrence of an accident. If the accident was caused by road facilities such as street furniture, it is shown that the street furniture was failed to improve the safety of the road users. In this study, a new concept of materials for the construction of road dividers was purpose. The sustainable materials such as laterite soil, Cow Grass, tires and PVC pipes was used to replace the cement and iron guardrails in road divider construction. The main objectives of this study is to design characteristics of a road divider using natural material such as laterite soil, sand, grass and recycle material like used a tyre. A series of experiments such as test liquid limit, plastic limit test, standard proctor test and observations of root growth in 30 days were conducted. The result shows that the plastic index value of 35% was obtained. From the proctor compaction test, the optimum water content was 23% and maximum dry density at 15.3 kN/m$^3$. This value was used to build a prototype of green and sustainable raised road median. The suitability of sand used as the medium in irrigation system was approved. Permeability of sand is 10$^{-2}$ mm/s to 10 mm/s which is having a high rate of flow. The growth of Cow Grass roots was increased by 4.9 cm in 30 days. This shows that, the grass is suitable to grow in laterite soil and can be used as vegetation material. Through the impact test conducted on a model with a scale of 1:7.5, indicate the initial damage occurred when a force reaches 10N. However, the condition of the model is still stable. From the impact test conducted, the prototype of green and sustainable raised road median has a good potential to be used as existing road divider as it can absorb the impact of an accident.

## 1. Introduction

There are a many causes of accidents such as negligence of drivers, slippery road surfaces, weather factors and there are also some accidents due to the roadblocks. Road barrier function of protecting the driver is in an unsatisfactory level [1]. In the event of road accidents occur on the existing road divider, it will help reduce the acceleration of the vehicle is either slowly (rubbing) or dramatically, which will result in greater impact [2]. Therefore, the study of innovation on green and sustainable raised road median at multi-lane highway is conducted to determine the effectiveness in replacing the existing road divider.

Innovation on the material used to build a road median is to reduce the level of accidents and injuries. The use of cement and guardrails are replaced with laterite soil. A rigid surface such as concrete dissipate almost no kinetic energy than a soft surface such as plastic is ground to dissipate the kinetic energy [3]. The use of concrete and steel in building a road divider rail will not only have a negative impact on the environment [4], but it does not have the aesthetics. Green landscape on the road would reduce pressure on road users [5].

This study is expected to help reduce maintenance costs. The material is made up of laterite soil. Laterite soil easily found in Malaysia and the tropical region. Furthermore, the use of laterite soil could also reduce the use of cement and steel. In the process of cement preparation, cement dust and gas that contribute by cement factories that cause pollution and ultimately affect human health [4].

## 2. Materials and methods

In this study, the existence of road divider was studied in terms of material and structure. This information is obtained from the Public Work Department Guidelines *Arahan Teknik (Jalan)* 1/85. There were no such type of sustainable for road median and barrier to be referred. Therefore, the green and sustainable raised road median is proposed and to be designed.

### *2.1. Green median conceptual design*

Height of guardrails that commonly applied in Malaysia is about 600mm to 800mm according to *Arahan Teknik (Jalan)* 1/85. This range of height is applied in the design of green and sustainable raised road median. The vertical drainage system is 12 feet help to channel the water directly into the subgrade. The most suitable diameter is between 150 mm to 200 mm. The pipe is placed at a distance of 900 mm between each other (see figure 1). The use of sand in irrigation systems is a good choice because sand can be a separation medium by eliminating unwanted particle or material with good flow rate. The permeability of the sand is $10^{-2}$ mm/s to 10 mm/s. This value indicates that the sand has good water flow properties [6].

The focus of this study is to investigate suitable design to build raised road median using laterite soil. Some of the tests carried out on the laterite soil characteristics and the extent of soil suitability for the construction of this road divider. Type of grass that used was a Cow Grass and it has strong root suitability. The Cow Grass as vegetative material is durable and has a strong grip to avoid the grass from out rooted if an accident occurs at the raised road median. The function of grass is also to help maintaining the shape of the raised road median.

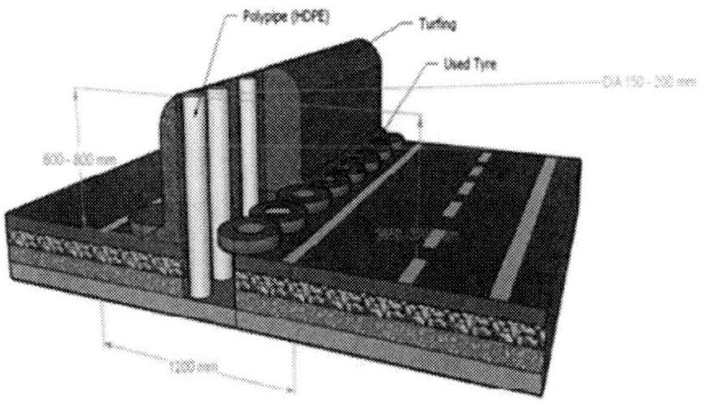


**Figure 1.** The design concept of the green and sustainable raised road median.

### *2.2. Testing the suitability of laterite soils*

In order to determine the suitability of laterite soil that are used to construct the green median, a few experiments are conducted. To know the plasticity characteristic of the laterite soil, atterberg test was conducted according to the British Standard 1377: Part 2: 1990. The value of plastic index will indicate the characteristic of laterite soil. Other than that, standard compacted are conducted based on

BS 1377: Part 4: 1990 to know the optimum water content. The optimum water content will be used in the construction of green and sustainable raised road median prototype.

*2.3. Impact test on the median*

To determine the effectiveness of green and sustainable raised median, an impact test is designed by using the pendulum swing theory as applied by Virginia Technology [7] as shown in figure 2. In the impact test, the median strip of the green model and sustainable was built in scale 1: 7.5 to facilitate the impact test are conducted. A small scale model was built due to lack of equipment to conduct impact tests on prototypes and sustainable green median strip in figure 3. Heavy pendulum that is released is equivalent to the power provided by a train at a certain velocity, in this test which used the pendulum weight was calculated using the following formula.

**Figure 2.** Impact test conducted by Virginia Technology [7].

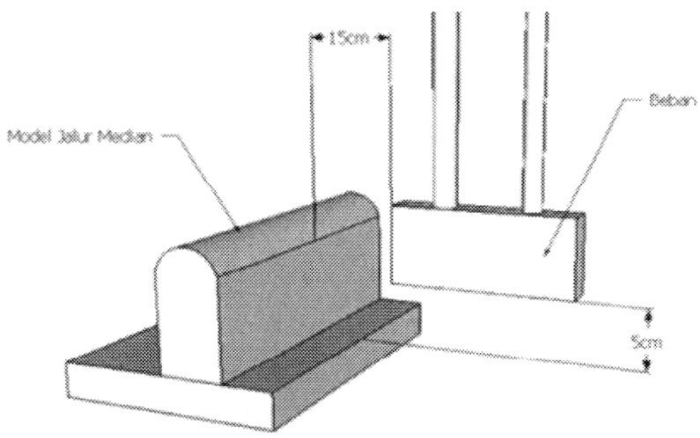


**Figure 3.** Impact test on small scale model.

*2.4. Observation on the growth of cow grass*

Suitability of grass that used as vegetative in the construction of green and sustainable raised road median are determined by observing the growth of the Cow Grass's root in 30 days. Previous studies done by planting grass type Axonopus Compressus in a cylindrical container which has a diameter of 18cm and height 40cm. In this study, the grass is planted in sandy loam soil (see figure 4). The length of the root measure and observe in 360 days. The growth of root at the end of this experiment of the previous study is 35cm [8].

The same observation is made for the development of grass roots within 30 days. The grass was planted in the laterite soil, in a cylinder container which has a diameter of 7.5cm and a height of 15cm. According to previous studies, the length of the Cow Grass root within 30 days can reach 5cm to 10cm. At the end of this experiment, the pattern of growth of grass can be observed. In addition, the level of suitability of Cow Grass grown using laterite soil can be determined.

**Figure 4.** Cow Grass is planted in a container.

## 3. Results and discussions

The results of laboratory analysis have been derived from the liquid limit test, plastic limit test and soil compaction test. Each experiment was carried out to determine the properties of laterite soil that will be used to build a prototype of a sustainable road divider and sustainable. In addition, the results of crash tests conducted on prototypes divider are discussed. Subsequently, the observation of the Cow Grass root growth is conducted.

### *3.1 Atterberg limit test*

This study is to design a raised road median using laterite soil. Therefore, plastic limit and liquid limit test are conducted to determine the laterite soil plastic index. The cone penetration method, the water content in the cone penetration depth of 20 mm was taken as the liquid limit. Based on the graph that has been plotted, the liquid limit obtained is 60%. From the plastic limit test, the water content of each sample was recorded and the value of plastic is the limit is 25%. Then, the value of the Plasticity index obtained is 35%. In addition, the proctor test on soil is conducted to obtain the optimum moisture content of 23% and maximum dry density at 15.3 kN/m$^3$. The percentage content of water is used in the construction of this prototype green and sustainable raised road median.

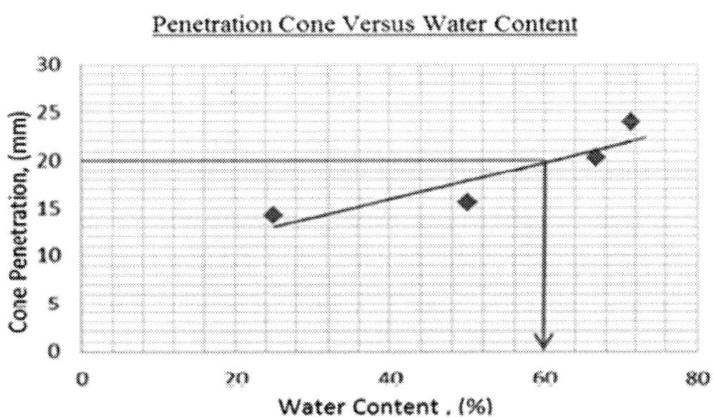


**Figure 5.** Graph of penetration versus water content.

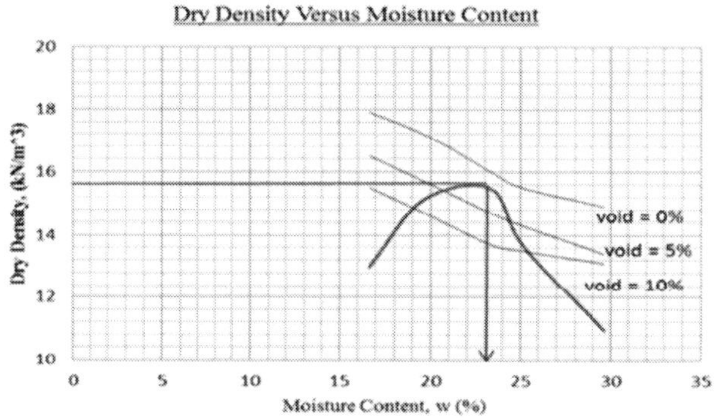


**Figure 6.** Graph of dry density versus moisture content.

*3.2 Impact test on model (1:7.5)*

From observation of the impact test, the smallest impact of 2.5kN did not show any damage to the surface of the model. Table 1 shows the damage occurs when the force of 10N imposed on the model (1:7.5). Damage that occurs at the third layer which is the layer compacted at a percentage of 19% water. Weight 20 kg, which is used on a small scale model (1:7.5) is equivalent to the weight of the pendulum 150kg levied on the real size raised road median (1:1). Therefore, it can be estimated at 150 kg weight resulting is 10N. The initial damage occurs at the green and sustainable raised road median when the force of 10N imposed on the model (1:7.5) in Figure 7. This damage is not serious and can be treated with appropriate because planting grass vegetative material, can help to protect the soil surface.

**Table 1.** Result of impact test.

| Weight, (kg) | Accelaration ($m/s^2$) | Force (kN) | Condition of Model |
|---|---|---|---|
| 5 | 0.5 | 2.5 | Stable |
| 10 | 0.5 | 5 | Stable |
| 15 | 0.5 | 7.5 | Stable |
| 20 | 0.5 | 10 | Small Damage |

**Figure 7.** Damage on the model (1:7.5).

*3.3 Observation on cow grass growth*

The grass is allowed to grow in a plastic container. After 14 days, the lawn is fertilized in order to speed up the grass growth. On day 30, the grass is removed and the length of the root length is recorded. Figure 8 shows the grass roots have grown up to 12.3 cm on the 30th day. The grass roots growth increased by 4.8cm from the initial day.

(a) Initial day (b) After 30 days

**Figure 8.** The root pattern on first day (a) and after 30 days (b).

Referring to previous studies, on day 30, the length of the roots of Cow Grass is 5cm [8]. The length of Cow Grass roots that planted in sandy loam soil in this previous study is similar to the growth of the grass on laterite soils. Therefore, the length of the roots of the Cow Grass planted in the laterite soil has potential to grow up to 35cm in a year as has been stated in previous studies. Furthermore, after 30 days, the roots patterns of the Cow Grass (b) have extended compare with the roots pattern on the initial day (a). This helps the grass to the ground with a stronger grip [8].

## 4. Conclusion

In conclusion, the objective of this study was achieved. Material such as laterite soil is a suitable material to build Green and sustainable raised road median. The structure of the green and sustainable raised road median is stronger by using PVC pipe and tire as additional material. In addition, Cow Grass is a suitable for as covering material to maintain the shape of the raised road median this kind of grass has a good pattern of growth. Based on the impact test was conducted on the model, this study has a high potential in achieving the objectives of the study. While medium irrigation bleak sand can serve to isolate unwanted substances.

## 5. References

[1] Suman M T 2009 *Kajian Terhadap Penghadang Jalan dari Segi Impak Kemalangan Jalan Raya* Final Year Project Report (Johor : Universiti Teknologi Malaysia).
[2] Udayakumar G and Chandralekha S 2014 Devising a new technique to reduce highway barrier accidents, *International Journal of Research in Engineering and Technology* **3** 773-780
[3] Sareen A K, Fasanella E, Sparks C, Jackson K and Mullins B R 2005 Comparions of hard surface and soft soil impact performance of a crashworthy composite fuselage concept, *American Helicopter Society International* **42** 116-125
[4] Mehraj S S and Bhat G A 2013 Cement Factories, Air Pollution and Consequences (New York : Marsland Press) pp 10-11
[5] Singh V 2013 Road Development and Management With Reference to Pollution, *International Journal of Geology, Earth and Environmental Sciences* **3** 23-28
[6] Daud R 2005 *Penentuan Indeks Penyusupan Tanah Di Kelantan* Final Year Project Report (Johor : Universiti Teknologi Malaysia).
[7] Hampton C E and Gabler H C 2013 Development of a missing post guideline for longitudinal barrier crash, *Safety Journal of Transportation Engineering* **139** 549-555

[8] Bordoloil S, Basumatary B, Saikia R and Chandra H 2012 Axonopus compressus (Sw.) P. Beauv. A native grass species for phytoremediation of hydrocarbon-contaminated soil in Assam, India, *Journal of Chemical Technology and Biotechnology* **8** 1219–1365

**Acknowledgments**
Authors would like to thanks Office for Research, Innovation, Commercialization and Consultancy Management (ORICC) providing Research & Innovation Fund. Also, Soft Soils Research Center (RECESS) and, Faculty of Civil and Environmental Engineering that gave full support by committed technical staff to ensure all the laboratories equipments in good and functional condition.

IOP Conf. Series: Materials Science and Engineering **271** (2017) 012106 doi:10.1088/1757-899X/271/1/012106

# Potential implementation of light steel housing system for affordable housing project in Malaysia

**M Saikah[1], N Kasim[1], R Zainal[1], N Sarpin[1] and M H I A Rahim[1]**

[1] Department of Construction Management, Faculty of Technology Management and Business Universiti Tun Hussein Onn Malaysia, 86400 Batu Pahat, Johor, Malaysia
Correspondence author: narimah@uthm.edu.my

**Abstract.** An unparalleled number between housing demand and housing supply in Malaysia has increased the housing prices, which gives consequences to the homeownership issue. One way to reduce the housing price is by faster increase the number of affordable housing, but the construction sector faces difficulties in delivering as expected number by using conventional and current industrialised building system (IBS) due to the issue related high project cost, time and labour. Therefore, light steel housing (LSH) system as one of another type of IBS method can be utilised in housing construction project. This method can replace the conventional method that was currently used in the construction of affordable housing project. The objectives of this study are to identify the potential of LSH and influencing factors of system implementation. This is an initial stage to review the previous study related to LSH implementation in developed and developing countries. The previous study will be analysed regarding advantages and disadvantages of LSH and factors that influence the implementation of the system. Based on the literature review it is expected to define the potential and influencing factors of the LSH system. The findings are meaningful in framing and enhance construction housing method of an affordable housing project in Malaysia.

## 1. Introduction

Nowadays, Malaysia is facing a great challenge towards living a happy life. More worried when based on the Annual Report 2015 by Bank Negara Malaysia, the demographic research analysis demonstrates that the cost of housing in all major cities in Malaysia is at the 'severely unaffordable' stage [1]. The main factor for this problem is when the housing price in Malaysia was increased drastically due to the high housing demand and unparalleled housing supply [2]. Various actions have been taken by the government to settle down the homeownership issue through the implementation of the Malaysian plan each year. However, the record under the PR1MA housing program is worsened when to date just only 560 units of houses have been completed from the total number 158,807 units' plan, while 41,187 are under construction [3,4]. The number of the affordable house completed is too low compared to the total of 102 million of scheme registration by household [3].

Thus, one way to reduce house prices is by fastest increase the quantity of housing supply for the middle household income. However, conventional construction methods are still widely used in the construction of affordable housing in Malaysia while it is high cost and unable to respond to this huge demand in a short time with standard quality [5]. The method is based on-site works include building the reinforced concrete frame and brick, beam, column, wall and roof are cast in-situ using timber framework while steel reinforcement is fabricated off-site [6]. But, the contractor generally faces other problem involving capital turnover, rising raw material prices, low margin and source degradation which will cause project delayed. Therefore, IBS system as one of the forms of off-site manufacturing (OSM) technique was introduced by the Malaysian Government since 1966 but until now just a few developers and contractor had used the method because of high cost, high risk and high level in setting up the programs [7]. For reducing the complexity of the system, the government should introduce the simple construction methods but reliable to be applied in all projects rates in Malaysia including affordable housing projects in the rural areas. Therefore, this study will establish the framework for improving residential construction system by utilisation of light steel housing system as one of the types of IBS system.

The light steel housing (LSH) can be built quickly and economically due to the installation method carried out in pre-manufactured which has saved construction time and labour [8]. The installation process starts from connecting the wall and roof framing components in a factory. The framing components then brought to the site and the labour starts joining the components using thousands of self-drilling screw. These erection steps implemented with the simple procedure and will shorten the projections time. LSH systems found extensive usage, due from that this system is extensively used for the construction of commercial and residential buildings with the low and moderate rise in United States, Canada, Australia, Japan and many other countries [9]. Therefore, to have a deep understanding about LSH system this study will focus on the influencing factors and advantages of LSH implementation for affordable housing projects in Malaysia.

## 2. Overview of affordable housing

Life will be more meaningful when basic needs have been met. Normally, every individual will allocate a portion of their income for housing expenses. In many countries, affordability issue is regard between salary and housing budget imbalance and synonyms with the affordable housing problem. A household is considered to face unaffordability problem when they spend more than 30% of their income on rent and utilities, and severely cost burdened when it spends more than 50% [10]. In Malaysia, cost burden situation has turned them to the difficulties to the homeownership. Most of them do not have the ability to buy their own homes because of lower salary rate and high housing prices. High housing price particularly in large cities caused by an imbalance between a number of housing supplies and housing demands. It is based on assumptions that have mentioned the house prices function as a trigger, inducing rapid movement towards the equality of housing supply and demand [11]. The lower number of housing supply is caused by several factors including high construction cost and high premium rates need to pay to the government for every housing project. Two factor has given burden to the costs incurred, and consequence to the developer or contractor reluctant to build an affordable housing.

In order to meet current housing needs, especially for the low and middle income, the government and the private sector has played their respective role to social responsibility in agenda to build the shelter for people. However, by utilising current constructions method the stakeholder face difficulties to push up the number of housing supply exceeds the housing demand. Construction skills by using conventional methods are a technique that has been widely used in the field of residential construction for centuries of years. The conventional method can be defined as fabricating components for the building on site through the process of installation timber or plywood formwork and steel reinforcement [12]. The cast in-situ process is mainly conducted by labour and overall process done on site work.

There are mainly a lot of advantages and disadvantages of the conventional housing system. Advantages of the conventional system include; first, the construction work can be carried out at any time because it is easy to get the skilled labour in the construction field. It is not surprising when a lot of contractors using too many labour crews to expedite the construction work [13]. Second, the structural cost of the conventional building including material, labour and transportation cost are not significantly influenced builder decision to choose others constructions method [13]. The structural cost is almost same compared other system, the increasing cost currently occurs due from other reason such as high labour cost and delay problem. Third, Conventional methods of construction are capable of produce a high level of stability concrete structure [14]. The stable concrete structure makes residential home looks sturdier and shape. Fourth, the concrete structure has a high resistance to the fire [14]. Lastly, it is easy to get a supplier of building materials for the conventional concrete house because it is based on 100% natural products such as sand, cement, and timber [14].

However, by using the conventional construction method, it will cost more in the whole construction project in such cost for the labour, raw material and transportation besides slow down the overall time duration of the project [12]. The conventional construction method is related to work done on-site, so it involves more on cost and resources inefficient [15]. In this case, the contractor needs to make a good estimation and high control a number of materials involved in the housing project. Besides that, delay problem always occurs when using unskilled labours consequence of the poor quality of work [16]. This is a major issue to be solved by contractors because almost part of the project costs is influenced by labour cost. Generally, conventional construction method involves major components of construction wastes such as wood, concrete, bricks, metals and others. The waste area always generated from finishing works, such as packaging of materials, ceramic tiles and insulation [17]. Thus, conventional housing construction system does not meet the sustainability housing standard which has been actively promoting by the government.

Admittedly, another alternative system that can replace the current method of construction work is by utilising IBS system. The government has encouraged utilisation of IBS system which mainly promotes precast concrete material. Since the year 2005, Construction Industry Development Board (CIDB) has given exception full levy imposed to developers that have used IBS component exceeding 50% [18]. Because of that, many of the contractors have applied the IBS precast concrete material in a housing project to earn IBS score by using the precast components [19]. By using IBS system it gives a lot of advantages include; overall work process has been done in the factory, so it will reduce the labour incentive on-site construction work. Less labour used will further reduce the construction cost [20]. In addition, building the house by using off-site construction will enable faster work processes. Work can quickly complete, will reduce labour costs and enable them to move to other projects [21]. Additionally, the system's workflow has good features include less construction site wastage, less environmental impacts, and reducing the consumption of electricity and water become an important medium which can bring country towards sustainability [21]. This step will automatically solve the problem of waste disposal site of landfills as one of the important issues in Malaysia.

However, there are some hinder to further utilisation of current IBS system which particularly it's involved a higher structural cost than conventional methods. The higher cost is mainly caused by the lack of supplier, purchasing expensive moulds and high transport cost [22]. Moreover, due to the complexity of the system and the needs to buy materials in volumetric it increases the initial capital to start the projects which can only bear the cost by the contractor that have strong financial position. Due to the elements of precast concrete that usually big, bulky and heavy, it needs to use a high cost of cranes for hoisting especially by installing the components [20]. Other critical issues related the system is by using the wrong technique to joining the prefabricated elements have caused the leakage problem [20]. Hence, based on the advantages and disadvantages of current IBS system, it is less suitable to be used as the system that can help to increase the number of affordable housing. In Malaysia, the authorities need to allocate different funding sources and ready to uptake more systematic delivery models that encourage the involvement of the various players in the supply chain and marketplace [2].

### *2.1. Potential implementation of light steel housing*

Light steel housing system as feeling under IBS component is one of the building concepts that can enhance the current system of construction for affordable housing in Malaysia. Light steel housing system was made by galvanised sheet steel which readily transformed onto shapes used for framing are has been utilised in the housing for about 70 years [23]. The extensive use of light steel housing was given a confidence to the several countries including United State, Canada, Australia and Japan in utilising the technology for construction of commercial and residential buildings with the low and moderate rise [24]. Malaysia also needs to recheck the current building code to allow the utilisation of LSH system in housing construction industry. There are three types of the assembly method of light steel housing as shown in figure 1 [25].

(a) (b) (c)

**Figure 1.** Three type of assembly method of light steel housing. (a) Stick-built construction system (b) Panelized system [33] (c) Pre-engineered system [34].

Stick-built construction usually suitable for small-scale construction projects such as build houses under the program of Perumahan Bantuan Rakyat (PBR) and MESRA RAKYAT, which refers to the house that is built on applicant own land. While the panelized and pre-engineered system is commonly for the medium and large-scale housing project. There are many advantages of light steel housing over to the other structures including higher in factory production quality and more simple, more reliable, much more consolidated and rapid connections system [8]. Lightweight house allowed high degrees of prefabrication, which major part of houses was made at the manufacturing firm under control production environment [26]. In addition, speed constructions work and precise steel material has further given a greater economic performance [27]. Moreover, the system can be applied whether for small and large scale housing projects without the use of heavy lift cranes [8]. Besides that, LSH system as a simple construction method, in which the installation technique same as wooden framing system given much easier to trained the new builder [25]. LSH also fulfil the green building specification which there are less scrap and waste from constructions process and high recycle ability [27].

Although light steel housing system has present many of advantages to the construction industry, there is a lot of disadvantages of this system. First, the framing labour needs to enter a phase of learning the techniques of installation and it will extend for the project's completion date. The costs incurred for the initial application of light steel housing system quite high due to the labour training process [28]. Second, there is very less or neither theoretical analysis and nor relevant experimental data of the structure exist as a guide for local designer and architecture to develop the systematic lightweight steel specification [28]. Third, the new housing system is very strange to the people in term of structure and they still have low confidence about its security, stability, comfort and etc. Fourth, fewer technicalities test regarding sound modes forces exerted by the wind and determines the effect due from the earthquake [29]. However, the disadvantages of LSH can be solved by several actions such as labour training, further experimental research and promotion. Therefore, several important requirements need to be aware to ensure LSH system can be implemented without fail; it includes the availability of capital, supplier, technology and labour.

### *2.2. Influencing factor of implementation of light steel housing system*

To bring the new systems successfully, needs to concern regarding several influencing factors includes the regulation requirement, sustainable issues, market opportunities, social demand, economic assessment and industrial capability. Firstly, regulation requirement needs authorities to build partnering relationship amongst stakeholder under the industry, such as high promotion action have been taken in China since there is a positive growth of steel in the construction industry [30]. Secondly, the utilisation of light steel housing system will ensure that Malaysians can enjoy the housing affordability standard. The best step is to articulate that the housing development carried out has meet sustainability standard in wide-ranging criteria including economic, environmental and social aspects [31]. Thirdly, in order to promote green building practices, authorities must expose the public about the factors and impacts of human activities on the environment once stimulate market demand for environmentally friendly products [27]. Sustainable products indeed meet the standards for economic, environmental and social.

Fourth, one of the factors that make a country being developed is to have people open-minded, creative and analytical. Thus, as long as they have facing great increasing housing prices and in addition to the burden of high debt relative to income, it will cause a great stress on the residents [32]. Hence, one way to increase the quality of life is in faster increase number of housing especially for affordable housing in the target to reduce the housing price. Fifth, all the metal stud come in precise measurement let the easy maintenance operations without great material loss, contribute to fewer expenses, faster constructions time, less waste, less water bill and hence to obtain a greater economic performance in the life-cycle costs indicators [27]. Lastly, the manufacturer of steel product should be involved in the initial process for introducing light steel housing in the constructions market. This is in order to strange industry capability to fulfil the future huge demand for metal product consequence from the implementation of light steel housing method in the delivery of affordable housing [32].

## 3. Methodology

This paper is written based on an overview of literature from books, journals, articles and conference papers related to this topic. The review focus about local and other country experience regarding housing affordability issue, and the way to solve the problem by emphasising on the need for improving current construction method. From this, the key point is in recognising the advantages of light steel housing, how it has covered the disadvantages of conventional and IBS (precast concrete) method. The important look for the potential of the LSH system is to faster increase the quantity supply of affordable housing. So that, many people can own the house without cost burden. To ensure the light steel housing can be implemented and accepted by clients and contractor the requirement of light steel housing system is identified. Requirement focus on the factors that enable the utilisation of the system in the construction field, it includes capital, technology, supplier and labour. While identification about influencing factor of implementation of LSH system also needed as a guideline for authorities to accepting the implementation of the system.

## 4. Results and discussion

After reviewing the literature related to light steel housing system, it is obvious that the utilisation of light steel housing system of the construction industry in Malaysia is sparse. The majority of studies focused on industrialised building system (IBS) as an alternative to conventional construction method in term of concept, system performance, improvement, promotion and barriers of adoption to the system. The literature suggest that if construction industry utilised the light steel housing system in affordable housing project there will found it can give a great economic performance to the contractor [27]. The economic performance is depends on three main aspects which are oriented on time, labour and cost. Firstly, the contractors get to reduce the constructions time due to the majorities of work is mainly involve metals panel installation by using the easy step of assembly works. The simple erection step has eased contractors to train the construction workers for utilising the system [25]. It is because the erection step is almost same, the difference is in which the steel house use screw system rather than a nail. Finally, if the time and labour have been reduced it will further lessen the housing cost. In addition, LSH implementation is suitable for small and large scale housing project. This because major part of metal component was made by bending machine and the production performs under control environment [26]. Generally, manufacturing firm will fulfil the supply order for the LSH component within one week whether for small scale or large scale housing project. In fact, with the light weight characteristic have enabled the workers to easily move the component material without using heavy crane [8]. In the other hand, since the material ship according to the requested size it results to very less scrap and waste from constructions process [27]. The interesting is, all the scrap and waste will send to the recycle centre to enter to the recycling process.

Based on the previous study, there are six influencing factors that are very important to be as guidelines for the authorities to implement the new approach that can improve the current constructions method, refer table 2. The literature have review that one of the way to reach affordable housing standard is by applying sustainable concept in housing development [31]. The government need set up the regulation regarding the sustainable issue to encourage the demand of LSH. This is a chance to the authorities to introduce to the society about green house building concept [27]. Government have the wide right to change the people mindset through a training program and promote the new housing concept from time to time [30]. In the other hand, as one of promotion strategy, the authorities need introduce the utilisation of LSH in the project of subsidies house such as PBR house, home placement of flood victims and others in early stage of method utilisation. Through this way government and stakeholders will obtain great economic and environment performance, also at the same time will fulfil the gap of affordable housing delivery. As a result, when people have met their needs to own an affordable housing they will less stress and will live in happy life. In fact, the difficulties to own the house due from high housing price and cost burden will cause to the stress life amongst the resident [32]. Therefore, the government needs to take an immediate step to reduce the gap of housing supply and demands, in order to have open-minded and creative society.

**Table 1.** Potential of implementation of light steel housing system for affordable housing project.

| No | Potential of Light Steel Housing System | Author |
|---|---|---|
| **1.** | Greater economic performance | [27] |
| **2.** | Faster construction period | [27] |
| **3.** | Ease training labour | [25] |
| **4.** | Production system under controlled production | [26] |
| **5.** | Reduced usage of heavy lift cranes | [8] |
| **6.** | Less waste | [27] |
| **7.** | Long life span | [27] |

**Table 2.** Influencing factor of implementation of light steel housing system for affordable housing project.

| No | Influencing Factor of Light Steel Housing System | Author |
|---|---|---|
| **1.** | Regulation requirement | [30] |
| **2.** | Sustainable issues | [31] |
| **3.** | Market opportunities | [27] |
| **4.** | Social Demand | [31] |
| **5.** | Economic Assessment | [27] |
| **6.** | Industrial Capability | [32] |

## 5. Conclusion

Mass housing targets can be achieved by implementation of light steel housing system in replacing the conventional methods in the construction industry. This is in advance to fulfil the needs of affordable housing supply. When a number of affordable housing supplies have fulfilled, it also will reduce the housing price. Therefore, it's put under authorities' decision to uptake the LSH system and promotes it to the public.

## 6. References

[1] Bank Negara Malaysia 2015 *Financial stability and payment systems report 2014* (Kuala Lumpur: Bank Negara Malaysia)

[2] Olanrewaju A, Seong Y and Lee L 2016 Rethinking Affordable Housing Delivery: an Analytical Insight *MATEC Web of Conf.* vol 47 (Paris: EDP Sciences) 04002

[3] Abas A 2015 *PR1MA Homes Of High Quality And Below Market Value* New Straits Times pp 15 (August 16)

[4] Mutalib A 2015 *Rent-to-own your home: scheme will help PR1MA applicants who cannot get a loan* The Star pp 12 (August 22).

[5] Nawi M N M, Lee A, Azman M N A and Kamar K A M 2014 Fragmentation Issue in Malaysian Industrialised Building System (IBS) Projects, *Journal of Engineering Science and Technology* **9**(1) 98

[6] Ramli M Z, Hanipah M H, Zawawi M H, Abidin M Z Z, Zainal N A and Halim N S A 2016 Cost Comparison on Industrialized Building System (IBS) and Conventional Method for School Construction Project, *Journal of Scientific Research and Development* **3**(4) 95-101

[7] Fathi, I M S, Abedi M and Mirasa A K 2012 Construction Industry Experience of Industralised Building System in Malaysia *9th Int. Congress on Civil Engineering (Isfahan)* (Iran: Isfahan University Of Technology IUT) pp 1

[8] Mashhadifarahani S 2015 Light Weight Steel Frames vurses Common Building Structures - Structural Performance Evaluation, *American Scientific Research Journal for Engineering, Technology, and Sciences* **2**(1) 229.

[9] Mehdizadeh H and Ravanshadniya M 2017 Technical and Economic Assessment of Building Performance Through Light metal Frame (LSF), *International Academic Journal Of Science And Engineering* **4**(2) 113-123

[10] Aurand A, Emmanuel D, Crowley S, Errico C E, Leong G M and Rodrigues K 2016 *The Affordable Housing Gap Analysis 2016* (Washington, Dc: The National Low Income Housing)

[11] Vries P D and Boelhouwer P 2005 Local House Price Developments And Housing Supply, *Property Management* **23**(2) 80 – 96

[12] Haron N A, Hassim I S, Kadir M R A and Jaafar M S 2005 Building Cost Comparison Between Conventional and Formwork System: A Case Study Of Four-Storey School Buildings In Malaysia, *American Journal Of Applied Sciences* **2**(4) 819-823

[13] Kadir M R A, Lee W P, Jaafar M S, Sapuan S M and Ali A A A 2006 Construction Performance Comparison Between Conventional and Industrialised Building Systems in Malaysia, *Structural Survey* **24**(5) 412-424

[14] Hontus A C 2014 Comparative Study on The Choice of Building Materials for Constructing a House, *Scientific Papers Series Management, Economic Engineering in Agriculture and Rural Development* **14**(4) 117-125

[15] Dave M, Bilbao J, Dickson J, Watson B, Prasad D and Sproul A 2014 Towards Sustainable Modular Housing: A Case Study of Thermal Performance Optimisation for Australia *Int. Plea Conf. (Ahmedabad)* (Ahmedabad: Cept University) pp 1-8

[16] Hamid A R A, Singh B S B J and Mazlan M S 2013 The Construction Labour Shortage in Johor Bahru, Malaysia, *International Journal of Research in Engineering and Technology* **2** (10) 508-512

[17] Lau H H, Whyte A and Law P L 2008 Composition and Characteristics of Construction Waste Generated by Residential Housing Project, *Int. J. Environ. Res.* **2**(3) 261-268

[18] Badawi A A 2004 *The 2005 Budget Speech* M.O.F. (MOF) (Kuala Lumpur: Dewan Rakyat) pp 1-43

[19] Azman MNA, Ahamad M S S and Hilmi N D 2012 The Perspective View of Malaysian Industrialized Building System (IBS) Under IBS Precast Manufacturing *Int. Engineering Conf. (Gaza)* (Gaza: Islamic University of Gaza) pp 1-13

[20] Tam V W Y, Tam C M, Zeng S X and Ng W C Y 2007 Towards Adoption of Prefabrication in Construction, *Build. Environ.* **42** 3642-3654
[21] Chen Y, Okudan G E and Riley D R 2010 Sustainable Performance Criteria for Construction Method Selection in Concrete Buildings, *Automat. Constr.* **19**(2) 235-244
[22] Nawi M N M, Lee A, Azman M N A and Kamar K A M 2014 Fragmentation Issue in Malaysian Industrialised Building System (IBS) Projects, *Journal of Engineering Science and Technology* **9**(1) 97-106
[23] Lawson R M and Ogden R G 2008 'Hybrid' Light Steel Panel and Modular Systems, *Thin. Wall. Struct.* **46** (7-9) 720–730
[24] Mirzaaghabeik H and Vosoughifar H R 2015 Evaluation HSE of a LSF System Subject to Near And Far-Field Earthquakes, *Natural Science Review: Natural Science And Engineering* **17**(3) 69-78
[25] Hacker J H and Gorges J A 1998 *Residential Steel Design and Construction* (New York: McGraw-Hill Companies)
[26] Eren O 2013 A Comparison with Light Steel Frame Constructional Building Systems for Housing, *World Applied Sciences Journal* **25** (3) 354-368
[27] Andrade J B, Bragança L and Camões A 2016 Steel Sustainability Assessment - Do BSA Tools Really Assess Steel Properties?, *Journal of Constructional Steel Research* **120** 106–116
[28] Li X, Wang J, Meng X P and Wang J 2014 Comparison and Analysis of Lightweight Steel Structure Residential Housing *4$^{th}$ Int. Conf. on Mechatronics Control and Electronic Engineering (Shenyang, China)* (Paris: Atlantis Press) pp 718-722
[29] Vallipour M and Omidinasab F 2015 Assessment of Seismic Behavior of Light Steel Frames (LSF) with use by Nonlinear Static, *Journal of Renewable Natural Resources Bhutan* **3**(1) 60-85
[30] Cheng X, Zhao X Z and Chen Y Y 2011 Overall Investigation of Affordable Residential Housing in China *IEEE Xplore Digital Library* (United States: IEEE) pp 539-543
[31] Zuo J, Pullen S, Rameezdeen R, Bennetts H and Wang Y 2017 Green Building Evaluation From A Life-Cycle Perspective In Australia, *Renewable And Sustainable Energy Reviews* **70** 358–368
[31] Corman H, Curtis M A, Noonan K and Reichman N E 2016 Maternal Depression as a Risk Factor for Children's Inadequate Housing Conditions, *Social Science & Medicine* **149** 76-83.
[32] World steel 2013 *World steel prices in profile 2013* Retrieved on June 22, 2017 from http://www.worldsteelprices.com
[33] Veljkovic M and Johansson B 2016 Light steel framing for residential buildings, *Thin. Wall. Struct.* **44** (2006) 1272–1279
[34] Rogan A L, Lawson R M and Brkljac N B 2000 *Value and Benefits Assessment of Modular Construction* (United Kingdom: The Steel Construction Institute)

**Acknowledgement**
The author would like to thank the Universiti Tun Hussein Onn Malaysia (UTHM) for supporting financial aid for this research. This paper was partly sponsored by the Centre for Graduate Studies UTHM.

# Carbon footprint of construction using industrialised building system

**P Y Lim[1], K Yahya[1], E Aminudin[1], R Zakaria[1], Z Haron[1], R Mohamad Zin[1], A A H Redzuan[1]**

[1] Faculty of Civil Engineering, Universiti Teknologi Malaysia, 81310 Johor Bahru, Johor, Malaysia.
Corresponding author: eeydzah@utm.my

**Abstract.** Industrialised Building System (IBS) is more sustainable to the environment as compared to the conventional construction methods. However, the construction industry in Malaysia has low acceptance towards IBS due to the resistance to change and also lack of awareness towards sustainability development. Therefore, it is important to study the amount carbon footprint produced by IBS during its manufacturing and construction stage, and also the amount of carbon footprint produced by one meter square of gross floor area of IBS construction using Life Cycle Assessment (LCA) to ease future research through the comparison of the carbon footprint of IBS with the conventional building system. As a result, a case study on a residential type of construction in the vicinity of Johor Bahru, Malaysia was carried out to obtain the necessary data and result. From the data analysis, the amount of greenhouse gases (GHG) for a residential type IBS construction based on the raw materials and resources involved to manufacture and construct IBS components is 0.127 tonnes fossil $CO_2Eq$ per meter square. Raw material that contributed to the most amount of carbon footprint is Ordinary Portland Cement (OPC), followed by steel bars, autoclaved aerated blocks and diesel. The LCA data acquired will be very useful in implementing IBS in the residential type construction. As a result, the awareness towards sustainable construction using IBS can be improved.

## 1. Introduction

As a developing country, construction sector represents a significant percentage of Gross Domestic Product (GDP) per annum in Malaysia. Despite all the advantages brought by the construction industry to the country in terms of healthy economic growth and job opportunities, the industry also causes alarming environmental issues such as construction waste, depletion of natural resources due to raw material extraction for building materials, emission of greenhouse gases (GHG) to the atmosphere during the manufacturing of building materials and the construction stage and etc [1]. Even though Industrialised Building System (IBS) is more sustainable to the environment as compared to the conventional construction methods, the construction industry in Malaysia has low acceptance towards IBS due to the resistance to change and also lack of awareness towards sustainability development in this area [2]. Moreover, Malaysia still does not have reliable statistics to showcase the amount carbon footprint in IBS determined via LCA to further convince the decision making in adopting this system.
Majority of the IBS product in Malaysia are originated from the United States, Germany and Australia with market share of 25%, 17% and 17% respectively. Malaysian's produced systems only account for

12% [3]. This indicates that there is a considerable room for improvement in the area of research and development of IBS in Malaysia. Most of the industries still holds preference towards conventional building system even though IBS was introduced for more than 50 years in Malaysia [4].

As a result, it is important to study the amount GHG produced by IBS during its manufacturing and construction stage using Life Cycle Assessment (LCA) to ease future research through the comparison of carbon footprint in IBS construction and the conventional building system. LCA is the most adopted framework in the world to evaluate the environmental load of processes and products during their life cycle from cradle to grave [5]. A case study on a residential type of construction in the vicinity of Johor Bahru, Malaysia was carried out to obtain the necessary data and result.

## 2. Literature review

### *2.1. IBS in Malaysia*

Early 1960s marks the beginning of IBS in Malaysia when the Ministry of Housing and Local Government visited several European countries and evaluate their housing development program [6]. After their successful visit in the year 1964, the Government had launched pilot project on IBS to speed up the delivery time, built affordable and quality houses. This was important as there were no short in the increasing demand for construction due to the increase of population and improvement in the quality of life. For instance, the demand for residential buildings alone in Malaysia between the years 1995 and 2020 has been projected to be around 8,850,554 units, which includes 4,964,560 units of new housing units [7].

According to a research done by Abdul Kadir in year 2006 [3], 55 projects out of 100 residential projects used conventional building system as the main structural system. This is followed by various types of IBS including cast in-situ table form system with 16 projects and precast concrete wall and precast half slab system with 15 projects [4]. The frequency distribution on the type of structural building systems is shown in table 1. Table 1 also shows the type of project of the 100 residential projects. Ibrahim *et al.* [8] added, IBS system such as precast concrete hollow core slabs with concrete topping may come in different joint. This may reduce the life cycle for one productConventional account for the most maybe due to the flexibility of conventional construction methods which can suit all types of construction work. Cast in-situ table and tunnel form systems were only used for apartment (21 projects) and condominium (four projects) because the steel moulds are costly, and are only suitable to construct large number of houses. Apartments were the major share for full precast concrete systems with 11 projects [4]. Nowadays the number of IBS project has obviously increased in Malaysia construction industry, opening a demand for whole new sets of data up-to-date.

**Table 1.** Distribution on the type of residential project and structural building systems [4].

| Type of residential project | Conventional | Cast in-situ table form | Cast in-situ tunnel form | Full precast concrete | IBS Composite | Block | Timber | Total |
|---|---|---|---|---|---|---|---|---|
| Apartment | 21 | 13 | 8 | 11 | 1 | Nil | Nil | 54 |
| Condominium | 5 | 3 | 1 | 1 | Nil | Nil | Nil | 10 |
| Terrace house | 19 | Nil | Nil | 2 | 2 | 1 | 1 | 25 |
| Bungalow | 4 | Nil | Nil | 1 | Nil | Nil | Nil | 5 |
| Semi-detached | 6 | Nil | Nil | Nil | Nil | Nil | Nil | 6 |
| Total | 55 | 16 | 9 | 15 | 3 | 1 | 1 | 100 |

Today, the use of IBS as a method of construction in the Malaysian construction industry is evolving. The obligation to implement IBS serves both to improve performance and quality in construction, as well as to minimize dependency on unskilled foreign labour in the construction market [2]. Previously in 2002, it was reported that at least 21 suppliers and manufacturers are actively involved in the dissemination of IBS in Malaysia [3]. As of December 2013, it was reported that there are 171 suppliers and manufacturers actively involved in the dissemination of IBS, of which only 5 of them are bumiputera. Selangor contributed the most with total of 67 suppliers and manufacturers [9]. Although members of the industry are open to the idea of IBS, a major portion of the industry stakeholders are indifferent, perhaps due to resistance towards change, insufficient information and lack of technology transfer method to support feasibility of change to IBS [10]. In this case, it has been proven that it is difficult to introduce new method and technologies in the construction sector. Construction sector is known as a traditional sector that can be characterised as reluctant and even resistant to change [11].

### *2.2. LCA tools related to construction industry*
There are many types of tools that can be used for environmental assessment. As summarised by Ortiz *et al.* [5], these tools have been classified according to three levels. Level 3 is called "Whole building assessment framework or systems" and consists of tools such as BREEAM, LEED, GBI and MyCREST. Level 2 is titled "Whole building design decision or decision support tools" and uses tools such as LISA, Ecoquantum, Envest, ATHENA and BEE. Finally level 1 is for product comparison tools and includes Gabi, SimaPro, TEAM and LCAiT. Some databases used for environmental evaluation are CML, DEAM TM, Ecoinvent Data, GaBi 4 Professional, IO-database for Denmark 1999, Simapro database, the Boustead Model 5.0 and US Life cycle inventory database [5]. Instead of analysing an environmental assessment tools, this study will focus on Life Cycle Assessment (LCA), a much more detailed process compared to rating tools. LCA is a tool to investigate environmental burdens of a product or process, considering the whole life cycle, from cradle to grave [12]. All aspects considering natural environment, human health and resource depletion are taken into account and together with the life cycle perspective, LCA avoids problem-shifting between different life cycle stages, between regions and between environmental problems [13]. Previous tools and databases vary according to users, application, data, geographical location and scope [5]. The data represents conditions in industrialized countries. Data from developing and emerging countries, however, is still lacking [14]. For example the use of European and American database may not lead to correct decisions in developing countries [5].

### *2.3. LCA studies of building materials*
An LCA study done in Spain by Cuchi (2007), shows the amount of energy invested in manufacturing some specific materials for one square metre (considering the gross floor area) in a standard building equals the amount of energy produced from the combustion of more than 150 L of petrol. Figure 1 and figure 2 show the relative contribution of the main building materials to the primary energy demand and CO2 emissions associated with a square metre in a Spanish standard block of flats. The high impact of commonly used materials such as steel, cement and ceramics is notable [15].

Another LCA study did by Bribian *et al.* [16] which to evaluate certain energy and environmental specifications of different building materials, analysing their possibilities for improvement and providing guidelines for materials selection. In this study, one kg of material is the selected functional unit and the stages considered are the material manufacturing, the transportation from production plant to building site, the construction and demolition of the building, and the final disposal of the product. The study was carried out according to a static focus, so the life cycle inventories include intermediate values of the current processes within the system analysed, without analysing their variation over time. The software tool used in the study is SimaPro v7.1.8. In the manufacture stage, the supply of starting materials, the associated transport needs and the factory manufacturing processes of the different construction materials analysed are considered. Regarding transport from the production plant to the building site, a 20 to 28-tonne lorry covering an average distance of 100 km has been considered. A sensitivity assessment for other means of transport has also been developed [17].

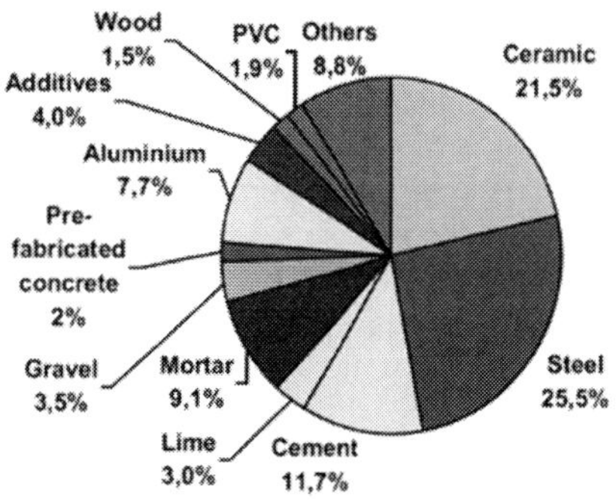


**Figure 1.** Contribution of energy demand for the manufacturing of construction materials per square meter of floor [14].

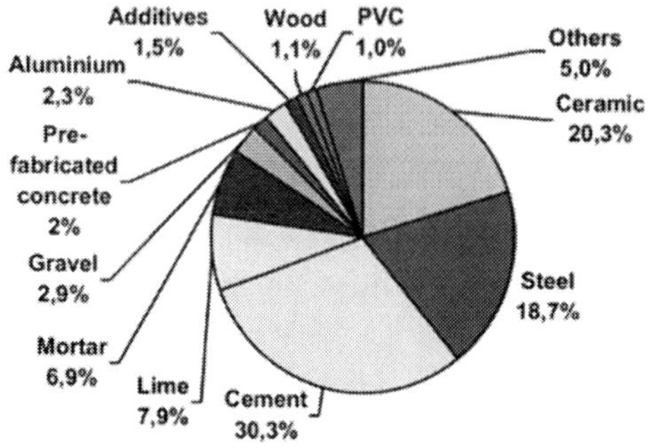


**Figure 2.** Contribution of carbon dioxide emissions for the manufacturing of construction materials per square meter of floor [14].

**Table 2.** Impact calculation coefficients for transport stage from production plant to building site of 1 tonne [16].

| Impact Category | Lorry, Road (m1) | Freight Rail (m2) | Transoceanic Freight Ship (m3) |
|---|---|---|---|
| Primary Energy Demand (MJ-Eq/km) | 3.266 | 0.751 | 0.170 |
| Global Warming Potential (kg CO2-Eq/km) | 0.193 | 0.039 | 0.011 |
| Water Demand (l/km) | 1.466 | 1.115 | 0.097 |

**Table 3.** LCA results for cement and concrete [16].

| Building Product | Density (kg/m3) | Thermal Conductivity (W/mK) | Primary Energy Demand (MJ-Eq/kg) | Global Warming Potential (kg CO2-Eq/kg) | Water Demand (l/kg) |
|---|---|---|---|---|---|
| Cement | 3150 | 1.4 | 4.235 | 0.819 | 3.937 |
| Cement Mortar | 1525 | 0.7 | 2.171 | 0.241 | 3.329 |
| Reinforced Concrete | 2546 | 2.3 | 1.802 | 0.179 | 2.768 |
| Concrete | 2380 | 1.65 | 1.105 | 0.137 | 2.045 |

## 3. Methodology

The information gathered for this study were obtained through site visits and through interviews with people who are related to the study such as the site engineer, senior site supervisor, casting plant production manager and etc. The interviewees have at least 5 years of experience in a similar nature of construction works from the selected site. During site visits, observation was done thoroughly to identify the type of IBS components utilised in the structure, the manufacturing process of the IBS components and also the construction process of the structure. These data leads to the amount of carbon footprint released by the IBS components of the structure. The site was selected based on several criteria, which are:

i. More than 80% of the structure was constructed using IBS system.

ii. The developer follows the IBS criteria set by CIDB strictly, proven by the accreditation obtained from various local and international accreditation bodies.
iii. The site is located in Johor Bahru vicinity and the permission to enter he site was granted by the developer.
iv. The type of project was residential type.
v. The construction stage and duration reached the practical completion of at least 90% during the research period.

After obtaining the data and information, the "Carbon Calculator" produced by Environmental Agency, UK was used to analyze the amount of carbon footprint, also known as the carbon dioxide equivalency ($CO_2Eq$) emitted during the manufacturing and construction of housing using IBS. $CO_2Eq$ is calculated through the multiplication of materials (plus unit of distance travelled) by the emission factor associate with that material. Estimates of mobile plant fuel mass consumption is derived by multiplying the hours of operation by the kW rating of the engine and by the appropriate fuel mass consumption rates.

## 4. Result and discussion

The selected residential type construction project is situated at Taman Skudai Indah 2, about 4.8 km from Universiti Teknologi Malaysia. The project consists of sixty (60) units of 3-storey cluster residential house. The total built up area of one (1) triple-storey cluster residential house is 315 meter squares. Data was taken up to 90% of completion. The IBS construction site is named as Site-A. The manufacturing processes of precast concrete components used in Site-A was observed in Site-B, an IBS components casting plant situated at Lima Kedai, Skudai, Johor, which is about 6.0 km away from Site-A.

### *4.1. Types of IBS component at Site-A*

Types of IBS component which were identified in Site-A were precast concrete beams, precast concrete wall panels, precast concrete slabs, in-situ steel formwork shear wall, and lightweight concrete blocks wall. The structure has more than 80% of its components constructed using IBS system.

Precast concrete beams, precast concrete suspended slabs and precast wall panels were produced in the casting plant owned by the developer, which is named as Site-B. The components were transported to Site-A based on the construction timeline to avoid overcrowding at Site-A. The 13 meter long shear wall was constructed via reusable steel formwork instead of being precast at Site-B because it is not practical to transport such a large panel from the casting plant to the construction site. Lightweight concrete blocks which replaced the usage of bricks in conventional construction were not produced in Site-B. The blocks were purchased from other manufacturers and transported to the site.

Figure 3 shows that precast concrete slabs have the highest percentage by volume, which is 36.6% among all other IBS components, equivalent to 29.21 $m^3$. This is because the entire first floor and second floor of the structure was constructed using precast concrete slabs which consist of a large area of 162.3 $m^2$.

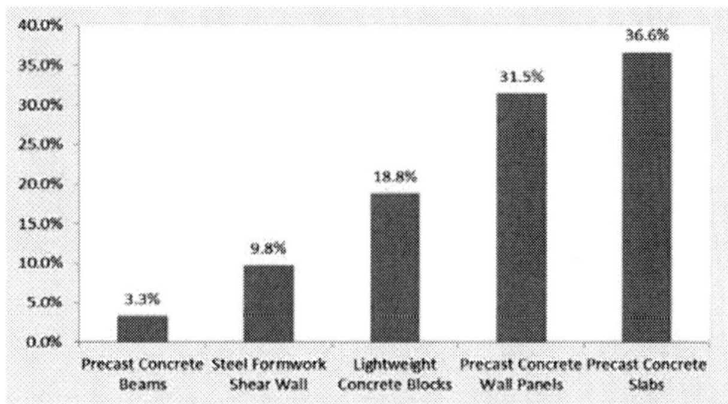


**Figure 3.** The composition of IBS components by percentage of volume.

*4.2. Manufacturing and construction processes of IBS*

The manufacturing processes began with the weighing of the raw materials, namely Ordinary Portland Cement, fine aggregates, coarse aggregates and water for the precast concrete components. The raw materials were mixed in the batching plant and poured into the concrete truck mixer to be transported to a location where steel moulds customised to produce the precast concrete components were located. The concrete mix was poured into the steel moulds where steel bars and wire mesh were fixed prior to concrete pouring. After the concrete hardened, the steel moulds were removed and the concrete components were cured. Prior to delivery to Site-A, the concrete components are refined to ensure that they adhere to the quality measurements. From the manufacturing processes, elements that could contribute to carbon footprint were raw materials used to produce precast concrete components, diesel and electricity used to operate batching plant and diesel to operate concrete truck mixer.

In the construction stage, the concrete mix was transported to Site-A to construct in-situ steel formwork shear wall. The lightweight concrete blocks were bonded using rendering mortar to form wall panels. Precast concrete components were transported to the site using trucks and were located into place using a crane. Elements that contributed to carbon footprint during the construction stage were raw materials to construct steel formwork shear wall, lightweight concrete blocks, rendering mortar for lightweight concrete blocks, diesel used to transport concrete mix and precast concrete components from Site-B to Site-A and diesel used to operate heavy cranes.

*4.3. Carbon footprint analysis*

The quantifying of carbon footprint was done by the method discussed in the previous section. In table 4, it is obvious that Ordinary Portland Cement (OPC) has the highest amount of carbon footprint among all other raw materials, which is 23.965 tonnes of fossil $CO_2Eq$ even though its quantity is lower than the quantity of 20mm aggregates and fine aggregates. This is highly related to the high amount of embodied carbon in the manufacturing process of cement that requires heating up to 1400 degree Celsius. Steel bar has the second highest amount of carbon footprint, which is 7.254 tonnes of fossil $CO_2Eq$. Despite having a relatively small quantity, which is about 3.1% out of the total tonnage of raw materials, steel bars contributed to 17.72% of the total carbon footprint in the IBS manufacturing and construction processes. This phenomenon is related to the hot-rolled process in steel bar manufacturing process, normally with coal, which is a considerable $CO_2$ source. Autoclaved aerated blocks, which are cured using steams, produced a carbon footprint of 3.686 tonnes fossil $CO_2Eq$, is ranked the third in table 4. Therefore, we can observe that raw materials that require complex manufacturing processes, such as OPC, steel bars and autoclaved aerated blocks produce more carbon footprint compared natural materials such as coarse and fine aggregates.

From table 5, diesel produced the highest amount of carbon footprint in the plant and equipment category, which is 4.841 tonnes fossil $CO_2Eq$. Diesel, in this study, is consumed to operate batching plant, concrete truck mixer, crawler crane and trucks. In the table, we can observe that water has no carbon footprint due to its low embodied energy. According to figure 4 when the carbon footprint in raw materials category and emissions from plant and equipment category is compared, OPC has the highest percentage of carbon footprint, which is 58.53%. This is followed by steel bar (17.72%), diesel (11.82%) and autoclaved aerated blocks (9.0%) respectively.

**Table 4.** Carbon footprint in IBS raw materials.

| Raw Material | Embodied $tCO_2Eq$ Per Tonne of Material | Quantity (Tonnes) | Carbon Footprint (Tonnes Fossil $CO_2Eq$) |
|---|---|---|---|
| **Ordinary Portland Cement** | 0.74 | 32.3850 | 23.965 |
| **20 mm Aggregates** | 0.005 | 64.7052 | 0.324 |
| **Fine Aggregates** | 0.005 | 48.5775 | 0.248 |
| **Steel Bars** | 1.40 | 5.1816 | 7.254 |
| **Steel Wire Mesh** | 0.269 | 1.9431 | 0.575 |
| **Autoclaved Aerated Blocks** | 0.31 | 11.988 | 3.686 |
| **Mortar (1:1:6 cement: lime: sand mix)** | 0.17 | 0.200 | 0.035 |
| **Total** | | | **36.087** |

**Table 5.** Carbon footprint from the emissions from plant and equipment.

| Emissions | Embodied $tCO_2Eq$ Per Unit of Material | Quantity (Unit) | Carbon Footprint (Tonnes Fossil $CO_2Eq$) |
|---|---|---|---|
| Diesel (litres) | 0.0031761 | 1524.15 | 4.841 |
| Grid Electricity (kWh) | 0.0005937 | 30 | 0.018 |
| Water (litres) | 0.00000034 | 200 | 0 |
| **Total** | | | **4.859** |

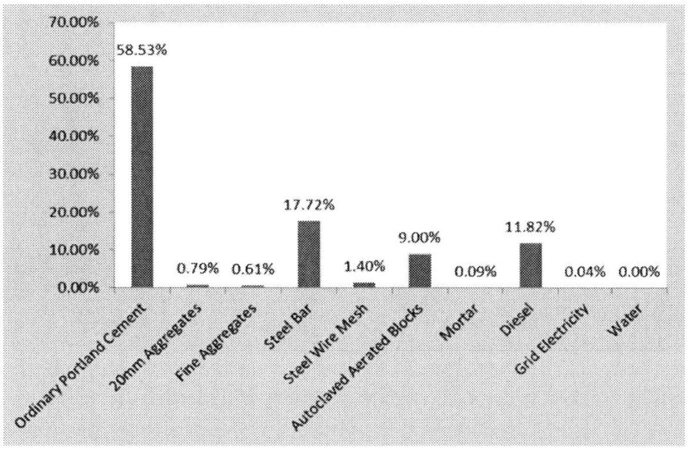


**Figure 4.** Tonnes fossil $CO_2Eq$ classified according to raw material and plant and equipment.

### *4.4. Amount of carbon footprint by 1 meter square of gross floor area*

The total amount of carbon footprint for the structure is 41 tonnes fossil $CO_2Eq$. Divided by its total build up area, which is equivalent to 315 $m^2$, we can obtain the amount of carbon footprint per meter square equals to 0.127 tonnes fossil $CO_2Eq$.

## 5. Conclusion

From this study, it can be concluded that OPC, steel bar, diesel and autoclaved aerated blocks contributed to the most significant amount of carbon footprint. The amount of carbon footprint can be minimized by replacing the raw materials used partly or entirely with materials with lower embodied energy. For example, the amount of OPC can be reduced by replacing it with a portion of ground granulated furnace slag or fly ash which has lower embodied carbon content. The estimated amount of carbon footprint generated per floor area for the IBS structure in this study is 0.127 tonnes fossil CO2Eq per square meter. This data is useful to estimate the amount of carbon footprint for other IBS structures with similar properties by multiplying the total build up area with the factor of 0.127 tonnes fossil CO2Eq.

## 6. References

[1] Thong J W, Ho C S and Noor Z Z 2014 Assessment of embodied energy and global warming potential of building construction using life cycle analysis approach: Case studies of residential buildings in Iskandar Malaysia, *Energ. Buildings* **93** 295–302

[2] Tamrin N, Nawi M N M and Nifa F A A 2017 Industrialised Building System in Construction Projects: A Study on Readiness, *Soc. Sci.* **12** 839-844

[3] Badir Y F, Abdul Kadir M R and Hashim A H 2002 Industrialised building system construction in Malaysia, *J. Archit. Engi.* **8** 19 - 23

[4] Abd Kadir M R, Lee W P, Jaafar M S, Sapuan S M and Ali A A A 2006 Construction Performance Comparison Between Conventional & IBS in Malaysia, *Structural Surv.* **24** 412 – 424

[5] Ortiz O, Castells F and Sonnemannn G 2009 Sustainability in the construction industry: A review of recent developments based on LCA, *Constr. Build. Mater.* **23** 28-39
[6] Thanoon W A M, Peng L W, Abdul Kadir M R, Jaafar M.S and Salit M S 2003 The experiences of Malaysia and other countries in industrialised and automated building system in Malaysia *Proc. on IBS Seminar (Selangor, Malaysia)* (Malaysia: Universiti Putra Malaysia)
[7] Nawi M N M, Lee A and Nor K M 2011 Barriers to Implementation of the Industrialised Building System (IBS) in Malaysia, *Built Human Environ. Rev.* **4** 22 - 35
[8] Ibrahim I S, Elliott K S, Abdullah R, Kueh A B H and Sarbini N N 2016 Experimental study on the shear behaviour of precast concrete hollow core slabs with concrete topping, *Eng. Struct.* **125** 80-90.
[9] MIDF Research 2014 *Construction IBS: Practical Solution to Rising Costs* Malaysia Equity Research Report (Malaysia: MIDF Amanah Investment Bank Berhad) pp 3 - 23
[10] Kamar K A M, Hamid Z A, Ghani M K and Rahim A H 2007 *Industrialised Building System: Current Shortcomings and The Vital Role of R&D* Master Builders 2nd Quarter article (Malaysia: Master Builders Association Malaysia) pp 62-65
[11] Hervas F C and Ruiz M I V 2007 Stakeholders requirement for open building manufacturing *Open Building Manufacturing: Core Concept and Industrial Requirement* ed Kazi A S, Hannus M, Boudjabeur S, Malone (Finland: Manubuild and VTT –Technical Research Centre of Finland) chapter 8 pp 133 - 152
[12] ISO 14040 2006 *Environmental management—life cycle assessment—principles and framework* (Geneva, Switzerland: International Organisation for Standardization) Retrieved on April 20, 2015 from https://www.iso.org/obp/ui/#iso:std:iso:14040:ed-2:v1:en
[13] Buyle M, Braet J and Audenaert A 2013 Life cycle assessment in the construction sector: A review, *Renew. Sustainable Energy Rev.***26** 379–388
[14] Hertwich E G 2005 Life cycle approaches to sustainable consumption: a critical review, *Environ. Sci. Technol.* **39** 4673 - 4684
[15] Cuchí A, Wadel G, Lopez F and Sagrera A 2007 *Guía de la Eficiencia Energética Para Los Administradores de Fincas* (Spain: Fundación Gas Natural)
[16] Bribian I Z, Capilla A V and Uson A A 2011 Life cycle assessment of building materials: comparative analysis of energy and environmental impacts and evaluation of the eco-efficiency improvement potential, *Build. Environ.* **46** 1133-1140

**Acknowledgments**
This work was financially supported by Vot no QJ130000.2622.12J53, Research Management Centre, and research team members Green PROPMT and Department of Construction Management, UTM. Thus, directly it is supported by Ministry of Education Malaysia.

IOP Conf. Series: Materials Science and Engineering **271** (2017) 012108 doi:10.1088/1757-899X/271/1/012108

# Traditional formwork system sustainability performance: experts' opinion

**Mohammed Taher Al-ashwal[1*], Redzuan Abdullah[2] and Rozana Zakaria[2]**

[1] School of Graduate Studies, University Technology Malaysia, 81310, Skudai, Malaysia.
[2]Department of Structure and Materials, Faculty of Civil Engineering, University Technology Malaysia, 81310 Skudai, Malaysia
*Corresponding author: alashwal52@yahoo.com

**Abstract.** The traditional formwork system is one of the commonly used systems in concrete construction. It is considered as one of the least observed activities in term of sustainability performance. In this paper, the sustainability performance of the traditional formwork has been assessed by using a multi-criteria assessment tool to facilitate the decision on the sustainability performance measurement. A quantitative five Likert scale survey study using judgemental sampling is employed in this study. A sample of 93 of engineering construction experts, with different fields including contractors, developers, and consultants in the Malaysian context has made the body of the collected primary data. The results show variety in the distribution of the respondents' working experience. The sustainability performance is considered moderately sustainable by the experts with only given 40.24 % of the overall total score for the three sustainable categories namely environmental, social and economic. Despite the finding that shows that the economic pillar was rated as the most sustainable aspect in comparison to the environmental and social pillars the traditional formwork system sustainability still needs enhancement. Further incorporation of the social and environmental pillars into the concrete construction the sustainability performance of traditional formwork system could be improved.

## 1. Introduction

Formworks are the moulds and dies that hold the concrete and carry the weight of materials, workers and equipment and for construction of reinforced concrete superstructure in building projects. [1, 2]. The conventional (Traditional) formwork system usually consists of standard plywood panels tied together with timber frame over their backs with horizontal members called walling to resisting the weight and horizontal force of wet concrete. A careful handling of the wall forms is needed as it is considered susceptible to edge and corner damage [3]. There is a need to choose a formwork type as there are different systems because they have their advantages and disadvantages. Cheap conventional formwork is too time consuming to use for larger structures. It has the tendency to damage and would seriously influence the cost, time, and quality of project delivery [4].

Conventional formwork can last longer if treated correctly. It is flexible and could be used in conjunction with engineered formwork for unique sections that require custom formwork [5]. However, conventional formwork also having disadvantages including requires extensive manpower, such as skill dependant and semi-skilled workers dependent would result in problems of cost, time, waste, poor finishes, leakages, and corrosion of structures. In term of cost and time, the formwork contributes approximately up to 20% of concrete cost and 60% of the concrete time [6]. The conventional timber formwork system is economical but highly environmentally undesirable [4].

Concerning the sustainability in formwork concrete construction, it is encouraged to follow most, if not all, sustainable construction criteria. They must be managed and assessed using multi-criteria means to facilitate the decision on the sustainability performance measurement [7]. Sustainable development is "the Development that meets the needs of the present without compromising the ability of future generations to meet their own needs" [8]. Sustainability includes social, economic and environmental pillars (Figure1.1).

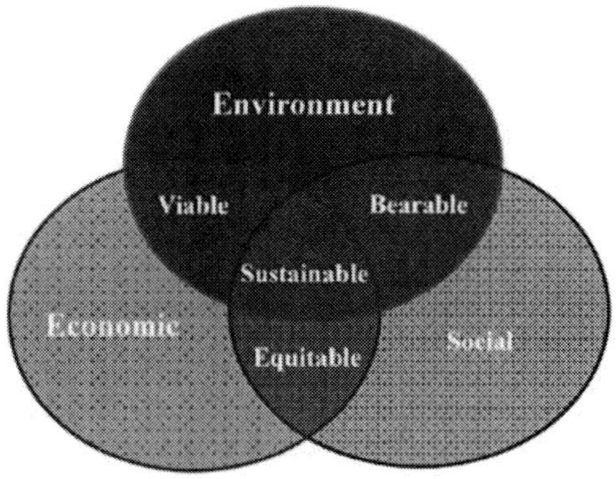


**Figure 1.** Sustainability three main pillars [9].

The construction should have a holistic design to meet the green building specifications to include materials used, energy consumption, air quality and adequate space [10]. However, in the current practice, the interest on the temporary work system is more due to its cost effectiveness ignoring other sustainability's factors [11]. This ignorance may happen due to several factors such as the focus on direct cost effectiveness and the level of awareness of the influence of the formwork type on the overall sustainability performance of the construction project [2]. The formwork system, in general, shows the least discussed aspect of concrete construction in civil engineering. Therefore, in this paper, we will present the result of the experts' opinion on sustainability requirement of the traditional formwork system using predeveloped multi-criteria indicators. The view of the experts could enlighten this area of research and provide the solution when knowing their views on the how far the existing traditional formwork system is fulfilling the sustainability requirement.

## 2. Methodology

Quantitative study design using five Likert survey designs has been used with the indicators as 1 = Strongly not responsive; 2 = less responsive; 3 = moderately responsive; 4 = responsive and 5 = strongly responsive in the study design. Online Google form has been used in a format of choice, and the questionnaire has been reviewed by the experts to check its face validity before the distribution. The survey consisted of two main sections, A for demographics data and B for sustainability rating. The results have been recruited and cleaned then analysed using the SPSS software version 23 (IBM, US).

### *2.1 Element pooling and rating*

Collection for the indicators has been used to evaluate the formwork system sustainability performance. The indicators have been grouped under the environment, social and economic pillars. The sustainability assessment model represents the relation between the indicators and the category using the factor loading as shown in Table 1. The sustainability score had been calculated using the mean and the factor loading as shown in equation (1) and (2).

$$\text{Indicator score} = \text{Mean} \times \text{Factor Loading} \quad (1)$$

$$\text{Category Score} = \sum \text{Indicator score} \quad (2)$$

The result of the model has been classified into four levels as 0-25% are basic, 26-50% moderate, 51-75% advanced and 76-100% optimum. The experts were asked to rate the fulfillment of the traditional formwork to the sustainability for each indicator.

**Table 1.** Formwork sustainability indicators.

| | Element Code | Element | Factor loading |
|---|---|---|---|
| ECO | ECO 1-1 | Labor Cost | .567 |
| | ECO 1-2 | Formwork Serviceability | .377 |
| | ECO 2-1 | Live-Cycle Cost | .595 |
| | ECO 2-2 | Installation Cost | .363 |
| | ECO 3-1 | Simplicity of Technology Use | .743 |
| | ECO 4-1 | Cost in Use | .669 |
| | ECO 4-2 | Material & Equipment Cost | .347 |
| ENV | ENV 1-1 | Energy and Resource Consumption | .996 |
| | ENV 1-2 | Waste Generation | .726 |
| | ENV 2-1 | Formwork Reusable | .979 |
| | ENV 2-2 | Using Renewable Material | .450 |
| | ENV 3-1 | Impact on Local Environment | .940 |
| | ENV 3-2 | Waste Efficiency | .367 |
| SOC | SOC 1-1 | Fire Safety | .993 |
| | SOC 1-2 | Safety Design of Formwork | .820 |
| | SOC 2-1 | Safety Measures | .559 |

## 3. Results and discussion

A total of 93 responses are collected with a variety of experience background from different types of companies as shown in Table 2 and Figure 2.

**Table 2.** Respondents' working experience and field of expertise.

| | Frequency | % |
|---|---|---|
| **Working Experience** | | |
| Less than 5 yrs | 22 | 23.7 |
| 5 – 10 yrs | 31 | 33.3 |
| 10 – 15 yrs | 21 | 22.6 |
| more than 15 yrs | 19 | 20.4 |
| Total | 93 | 100.0 |
| Total | 93 | 100 |
| **Field of Expertise** | | |
| Design | 27 | 29.0 |
| Management | 25 | 26.9 |
| Construction | 28 | 30.1 |
| Others | 13 | 14.0 |
| Total | 93 | 100 |

IOP Conf. Series: Materials Science and Engineering **271** (2017) 012108 doi:10.1088/1757-899X/271/1/012108

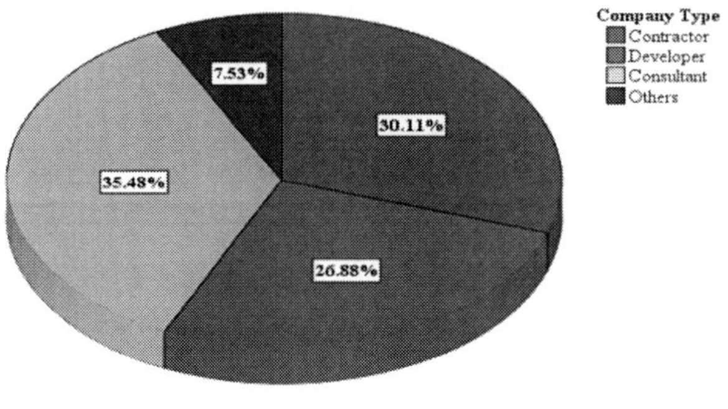


**Figure 2.** Respondents' company Type.

Regarding the respondent's knowledge on sustainable construction, 39.8 % of the respondents consider themselves as moderate experts in sustainable construction while 26.9 % of them just heard about it. Meanwhile; 23.7 % of the respondents are considered as experts in sustainability, and only 9.7 % had known nothing about sustainable construction (Figure 3). Despite the highest percentage of the respondents are moderately experts in sustainability, the results obtained from this survey are consistent as a total of 63.5 % of the respondents had enough knowledge to respond to the questionnaire.

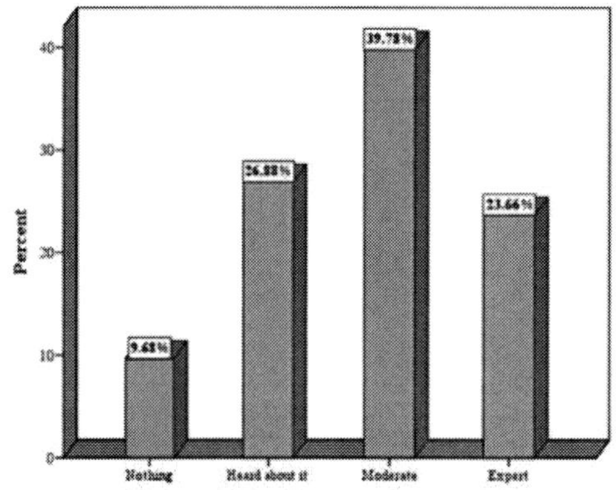


**Figure 3.** Respondents' knowledge level on sustainable construction.

As shown in Table 3 the element Eco 1-2 "Formwork Serviceability" had the highest mean average with 2.47 among the economic sustainability and followed closely by Eco 1-1 "Labour Cost" and Eco 4-1 "Cost in Use" with mean 2.45 and 2.42 respectively. As for the environmental category, the highest mean average is 2.18 for the element Env 1-1 "Energy and Resource Consumption" and following by to it both elements Env 2-1 "Formwork Reusable" and Env 3-2 "Waste Efficiency" with mean value 2.10. By looking at the social sustainability category, the highest mean average was for the element Soc 1-2 "Safety Design of Formwork" with mean 2.17. In general, these responses indicate that the respondent are not satisfied with the sustainability performance of the conventional formwork system as all assessment criteria achieve less than 2.5 in average. A quick glance on the survey responses the lowest performance was for installation cost, then to safety measures, which can be expected referring to the nature of the conventional formwork activities and work environment.

**Table 3.** Respondents' evaluation for the conventional formwork system fulfillment to the sustainability criteria.

| | Element Code | Element | Min | Max | Mean | Std. Dev |
|---|---|---|---|---|---|---|
| ECO | ECO 1-1 | Labor Cost | 2 | 4 | 2.45 | .648 |
| | ECO 1-2 | Formwork Serviceability | 1 | 4 | 2.47 | 1.077 |
| | ECO 2-1 | Live-Cycle Cost | 1 | 3 | 1.62 | .639 |
| | ECO 2-2 | Installation Cost | 1 | 3 | 1.52 | .649 |
| | ECO 3-1 | Simplicity of Technology Use | 1 | 4 | 2.12 | .980 |
| | ECO 4-1 | Cost in Use | 1 | 4 | 2.42 | .814 |
| | ECO 4-2 | Material & Equipment Cost | 1 | 4 | 1.99 | .916 |
| ENV | ENV 1-1 | Energy and Resource Consumption | 1 | 3 | 2.18 | .776 |
| | ENV 1-2 | Waste Generation | 1 | 3 | 1.97 | .583 |
| | ENV 2-1 | Formwork Reusable | 1 | 3 | 2.10 | .630 |
| | ENV 2-2 | Using Renewable Material | 1 | 3 | 1.67 | .738 |
| | ENV 3-1 | Impact on Local Environment | 1 | 3 | 1.89 | .700 |
| | ENV 3-2 | Waste Efficiency | 1 | 4 | 2.10 | 1.043 |
| SOC | SOC 1-1 | Fire Safety | 1 | 3 | 1.99 | .788 |
| | SOC 1-2 | Safety Design of Formwork | 1 | 3 | 2.17 | .556 |
| | SOC 2-1 | Safety Measures | 1 | 3 | 1.55 | .697 |

As the standard deviation for the responses were in the acceptable range "±2", the mean average was used to key in the values of the sustainability elements into the assessment model and calculate the final sustainability score of the conventional formwork system. The results are as shown in Figure 4, 5, 6, and

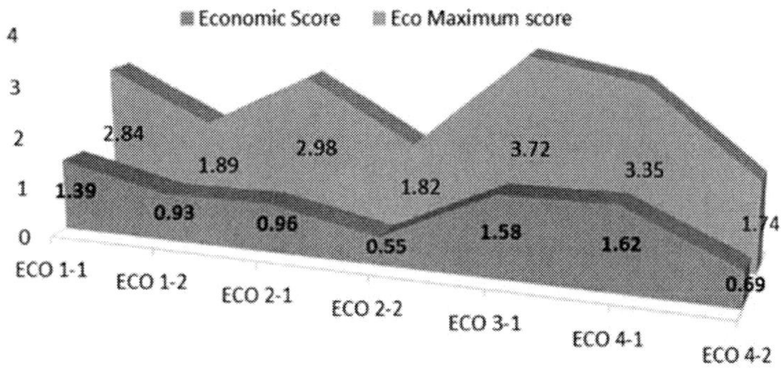


**Figure 4.** Economic Sustainability Score for conventional formwork.

The total sustainability score of the economic category was 7.7 from 18.3 total category scores i.e., 42.2%. The highest score in the economic categoy is for the element Eco 4-1 " Cost in Use" with value 1.62. While the lowest economic sustainability performance score was for the element Eco 4-2 "Material and Equipment Cost" with score only 0.69 (Figure 4).

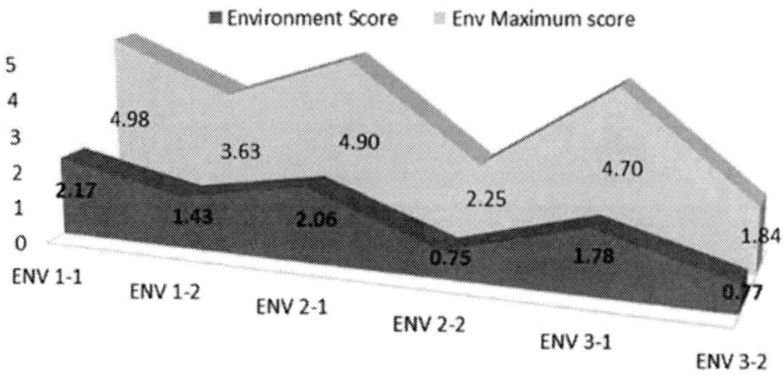


**Figure 5.** Environmental Sustainability Score for conventional formwork.

Referring to the environmental category (Figure 5), the total score is 8.96 of 22.3 i.e 40.2%. The highest score was for the element Env 1-1 "Energy and Resource Consumption" with 2.17, and the lowest score was for the elements Env 2-2 "Using Renewable Material" and Env 3-2 "Waste Efficiency" with the score 0.75 and 0.77 respectivly.

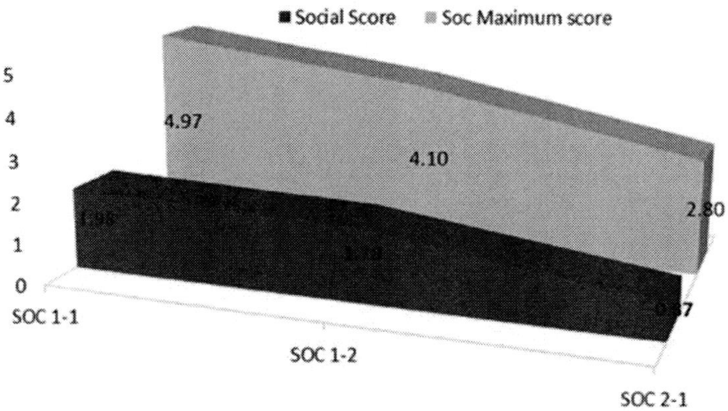


**Figure 6.** Social Sustainability Score for conventional formwork.

The social category had achieved total score 4.7 of 11.9 i.e., 38.9%, with the highest value for the elements Soc 1-1 "Fire Safety" and with score 1.98(Figure 6).

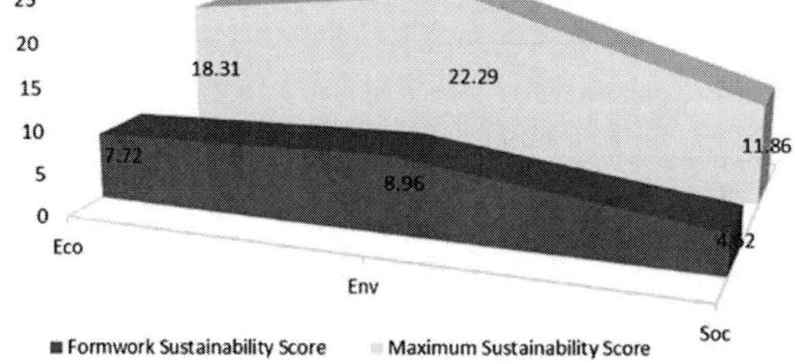


**Figure 7.** Sustainability Performance Score for conventional formwork.

The overall sustainability performance of the conventional formwork system is shown in Figure 7. The result is considered moderate for the three categories and also for the overall performance with the total score 40.6% as the results fall in the range of 26-50%.

## 4. Conclusion

As sustainability shows a trending growth for building development and safety, the sustainability measurement mode is not well developed and still in infancy stage. The findings of this study show that traditional formwork is performing well in term of the economy but not in social or environmental. This finding is tally with the reality in construction projects, as the use of conventional formwork system is mainly for its cost effectiveness. The sustainability pillars are not considered in the traditional formwork that calls for enhancement and focus in considering the environmental and social indicators to improve its sustainability performance.

## 5. References

[1] Hurd, M.K., 2005, *Formwork for Concrete*. 7 ed. Vol. 1. United States of America, Michigan American Concrete Institute.

[2] Naik, M.B. and Rathod, H.A. 2015, *A Review on Innovating Formwork Systems*. International Journal of Advanced Research in Engineering, Science & Management. **1**(6): p. 8.

[3] Karke, S.M. and Kumathekar, M.B 2014, *Comparison of the use of Traditional and Modern Formwork Systems*, in *Civil Engineering Systems and Sustainable Innovations*, G.C. Mishra, Editor., EXCELLENT PUBLISHING HOUSE: New Delhi, India. p. 348-351.

[4] Poon, C.S. and Yip, R., 2005, *Comparison of The Use of Traditional and Low Waste Formwork Systems in Hong Kong*, in *The 2005 World Sustainable Building Conference*, Tokyo: SB05Tokyo:. p. 2741-2748.

[5] Shen, L.-Y., Jorge Ochoa, J., Shah, M.N., and Zhang, X. 2011, *The application of urban sustainability indicators – A comparison between various practices.* Habitat International. **35**(1): p. 17-29.

[6] Nemati, K.M., 2007, *Temporary Structures: Formwork for Concrete*, in *Construction Management*. University of Washington: USA. p. 36.

[7] Yip, R. and Poon, C.S. 2008, *Comparison of timber and metal formwork systems.* Proceedings of Institution of Civil Engineers: Waste and Resource Management. **161**(1): p. 29-36.

[8] Concrete society U. K., 2012, *Formwork: A Guide to Good Practice*. 3rd ed., UK: Concrete society, UK. 300.

[9] UN General Assembly 1987, *Report of the World Commission on Environment and Development: Our Common Future*, UN General Assembly, Editor. UN: USA. p. 28.

[10] Strong, A.W. and Hemphill, L.A., 2006, *Sustainable Development Policy Directory*. USA: W.A. Strong & L.A. Hemphill Blackwell Publishing Ltd. 669.

[11] Kadir, M.R.A., Lee, W.P., Jaafar, M.S., Sapuan, S.M., and Ali, A.A.A. 2006, *Construction performance comparison between conventional and industrialised building systems in Malaysia.* Structural Survey. **24**(5): p. 412-424.

## Acknowledgments

We wish to acknowledge the assistance of The CIDB and GBI organisations for in distributing the survey and providing knowledge for this study success.

IOP Conf. Series: Materials Science and Engineering **271** (2017) 012109 doi:10.1088/1757-899X/271/1/012109

# Effect of cement dosage and early curing towards Kuala Perlis dredged marine sediments: a $\varepsilon_v - \sigma_v$ and SEM-EDX approach

**Nurul Syakeera Nordin**[1] **and Chee-Ming Chan**[1,*]

[1] Faculty of Engineering Technology, Universiti Tun Hussein Onn Malaysia, Batu Pahat, Johor, Malaysia
Corresponding author:chan@uthm.edu.my

**Abstract.** Cement is the primary material used in solidifying the soft soils. This material was applied in solidifying Kuala Perlis dredged marine sediments (DMS). These unwanted sediments are classified as high plasticity silt, *MH* with 3.36 *LL* of *wc/LL* value. At dosage 10 and 20 % of cemented-DMS and 3 days curing time, compression curve results shows the settlement criteria were enhanced than the natural DMS. Unfortunately, the settlement criteria are not complies with the permissible settlement limit and applicable pressure. The formation of cementing compounds appears in the SEM micrograph for 10 and 20 % of cemented-DMS. EDX analysis shows the Ca:Si ratio were increased for cemented-DMS due to the formation of C-S-H gel.

## 1. Introduction

It is noted that water will increasingly shallow over time due to the sedimentation and accumulation of fine particle. Hence dredging works in offshore area are required to maintain the water depth in shipping ports. The other purpose of dredging works are to allow the maintenance of waterways and construction works in harbour, reclamation of foreshores, and purification of water by the removal of contaminated sediment [1-2]. A large amount of sediments or known as DMS are generated due to the dredging works. The DMS are not appropriate to be used in construction activities because of its poor geophysical properties such as high water content, low shear strength, high compressibility and presence of organic matter and salt [3].

The way to manage DMS is either by disposed it offshore in designated dumping sites or by treating it with an adequate pre-treatment [4]. The offshore dumping of the DMS aggravate the situation with leave a negative impact to the marine ecosystem and environment. Furthermore, it causes the long term damaging pollution due to the contamination of heavy metals [3] and it is now prohibited by the international covenants [1]. While, the treated DMS can be reuse as a geomaterial in various civil engineering applications and it has also become a specific new environmental issue recently. For example, it can be reuse as a backfill material by create a new land bases or rehabilitate the eroded land in near-shore areas [4]. Thus, those applications of treated DMS could meet client demand and moreover it could reduce the construction cost by using the recycle material.

Solidification is the process of utilizing the chemically reactive formulation together with water, other components in sludge and other aqueous hazardous waste to form a stable solids. The material used also insolubilizes, immobilizes, encapsulates, destroys, sorbs or otherwise interacts with selected components. These interactions form a solids which are non-hazardous or less hazardous than the original soil. The effectiveness of solidified products are often investigate through their strength, settlement and leach resistance [5]. Solidification technique towards the soft soil will alter in primarily its physical properties to produce a solidified soil hence it will no longer contain free liquids and also improved the strength. The treated soil are much easier to handle and more easily transported to the environment. The higher strength, lower settlement and lower porosity of treated soil are caused by the hydration reactions. The binder will bind together with the pozzolanic material to reduce the rate of hydration kinetics as well as the effect of carbonation. Where the carbonated solidified product develops higher strength material [5-6]. A major factor that will affecting the solidification is the mix design of additive material (binder) and waste (soil/sediment). It will effect on its strength, settlement, hydration, carbonation, pore structure, pore volume, and environmental factors [5].

## 2. Experimental work

### 2.1 Materials

The studied materials was dredged from Kuala Perlis Jetty, Perlis of Peninsular Malaysia in year 2016 (figure 1). Selection of dredging site was made based on Malaysian Marine Department (Jabatan Laut Malaysia, JLM) dredging schedule. The sediments was dredged at depth 4 to 6 m from sea level using backhoe dredger (figure 2). The properties results of the DMS are reported in table 1. The natural water content is about 240.74 % (3.36 *LL*), dried in an oven at 100 °C (BS 1377-2:1990:3.2). The particle density value of DMS is 2.36 Mg/m$^3$, measured using a small pyknometer method (BS 1377-2:1990:8.3). The liquid limit value is 71.70 % by using a cone penetrometer method (BS 1377-2:1990:4.3). The plastic limit and plastic index value for DMS are 40.06 % and 31.64 %, respectively (BS 1377-2:1990:5). The DMS consist of 2.5 % of sand, 80.5 % of silt and 17 % of clay (figure 3). The DMS was classified in high plasticity silt, *MH* for its soil classification (Unified Soil Classification System).

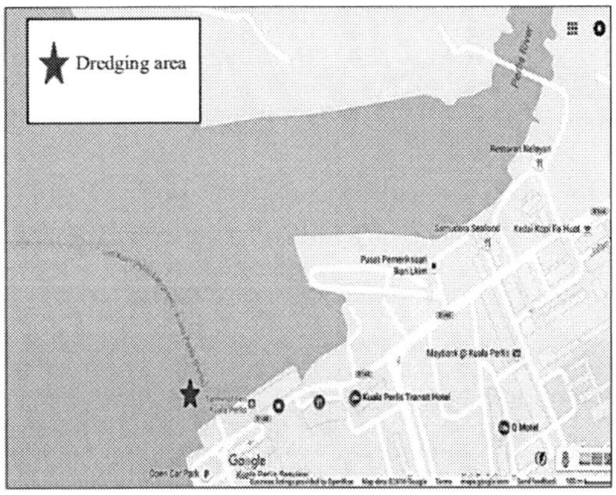


**Figure 1.** Location of dredging site at Kuala Perlis Jetty, Perlis, Malaysia.

**Figure 2.** Backhoe dredger.

**Table 1.** Properties of dredged marine sediments.

| | |
|---|---|
| **Natural water content (%)** | 240.74 |
| **Liquid limit (%)** | 71.70 |
| **Plastic limit (%)** | 40.06 |
| **Plastic index (%)** | 31.64 |
| **WC/LL** | 3.36 LL |
| **Specific gravity** | 2.36 |
| **Particle size distribution (%)** | Sand: 2.5 Silt: 80.5 Clay: 17 |
| **Soil classification** | High plasticity silt, MH |

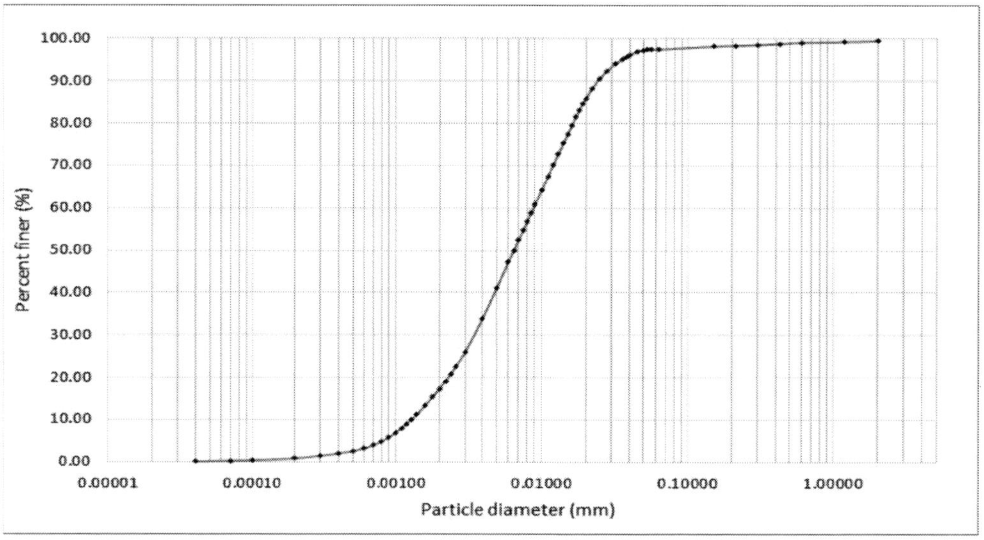


**Figure 3.** Particle size distribution of Kuala Perlis DMS.

*2.2 Preparation of specimens*

Distilled water was added to the DMS to achieve the consistency of 3.36 *LL* for each sample to achieve the same as natural water content of the DMS [7]. Cement was added to the DMS at a dosage of 10 and 20 %, corresponding to the *w/c* ratio as shown in Table 2. DMS and cement was thoroughly mixed by using kitchen mixer at a lower speed. The mixture was then transferred into a consolidation ring (H: 20 mm, ⌀: 75 mm) and left to cure for 3 days before a 1-dimensional compressibility test were performed. The sample was then dried in the oven and materials passing 63 μm are used in performing the SEM-EDX analysis.

**Table 2.** Specimen list and mix ratio.

| Specimen | Natural Water Content % | Mix Proportions | | | | Cement Dosage (%) | w/c |
|---|---|---|---|---|---|---|---|
| | | Wet Soil (g) | Dry Soil (g) | Water, W (g) | Cement, C (g) | | |
| 10C | 240.74 | 250 | 73.37 | 176.63 | 7.34 | 10 | 24.06 |
| 20C | | | | | 14.67 | 20 | 12.04 |

## 3. Results & Discussion

### *3.1 $\varepsilon_v - \sigma_v$ curves*

The compression curves of natural Kuala Perlis DMS, 10 (10C3D) and 20 (20C3D) % cemented-solidified DMS are tabulated in figure 4. The natural Kuala Perlis DMS was used as the control sample and for comparison purpose. Sample 10C3D displayed slightly lower settlement reduction than the 20C3D. But both dosages displays a quite similar trend of compressibility.

The 6 % of axial strain value denoted as the permissible settlement limit of embankment. Where it was simulated in a 1-dimensional compressibility test with maximum settlement of 1.2 mm from 20 mm specimen's height. National Cooperative Highway Research Program (1975), specified the allowable maximum settlement of 4.5 m height embankments are between 300 to 600 mm. Thus the 300 mm was taken as the permissible settlement limit. The 300 mm settlement from 4.5 m height of embankment are equal to the 1.2 mm from 20 mm of compressibility test. It shows, both cemented-solidified DMS start yielding after 6 % of axial strain. The applicable pressure for 4.5 m height embankment are 100 kPa. Those pressure were used and applied as the applicable pressure in compressibility test [8]. From the results, 10C3D and 20C3D shows both compression curves are not complies with the settlement criteria and applicable pressure.

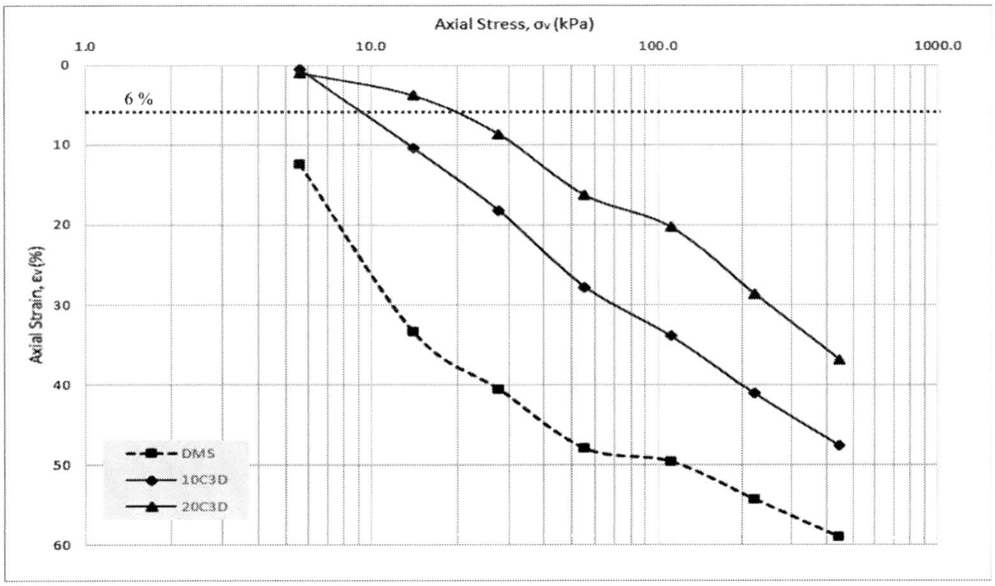


**Figure 4.** $\varepsilon_v - \sigma_v$ curves of DMS and cemented-DMS.

### *3.2 SEM-EDX*

Figure 5, 6 and 7 represents the SEM micrographs and EDX spectra of natural Kuala Perlis DMS, 10 and 20 % cement-solidified DMS, respectively. The microstructure of cemented DMS are quite different compare to natural DMS. Specimens with 10 and 20 % of cement illustrates the flocculated nature of the fabric and the formation of new cementing compounds in the matrix. The flocculated nature occurred due to the cation-exchange process. Hence, resulting in replacement of calcium ions with readily exchangeable ions during cement hydration. For specimen with 20 % of cement, it demonstrate a transform of soil fabric from a particle-based form to an integrated composition. Moreover, the existence of plate-shaped crystals of hydroxide calcium (cementing compounds) are clearly seen in 20 % cemented DMS [9-11].

Table 3 presents the Ca:Si ratio for natural Kuala Perlis DMS, 10 and 20 % cement-solidified DMS. It shows the Ca:Si ratio increased upon the cement application towards the DMS. The further increase in Ca:Si ratio of cemented DMS are probably due to the formation of C-S-H gel. Where the depositing of higher C-S-H gel on the surface of the particles lead to the increment in Ca:Si ratio [9,12].

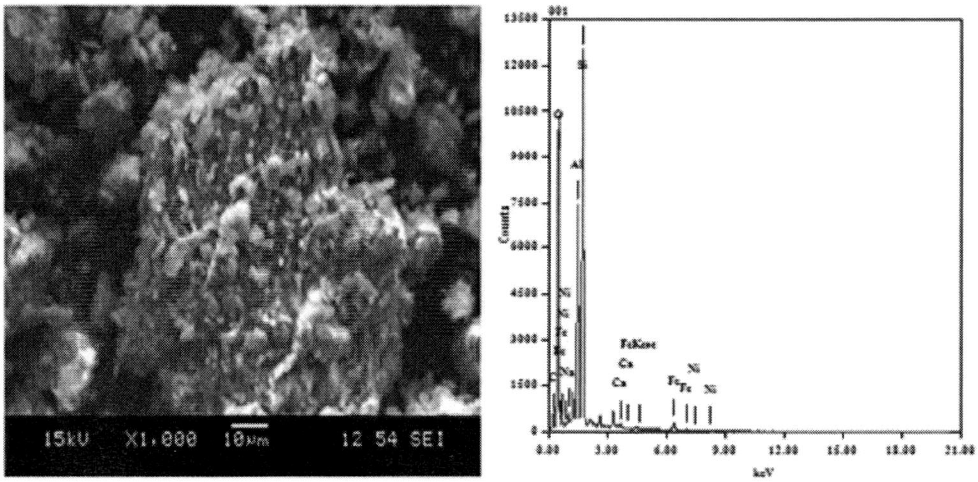


**Figure 5.** SEM image and EDX spectra of Kuala Perlis DMS.

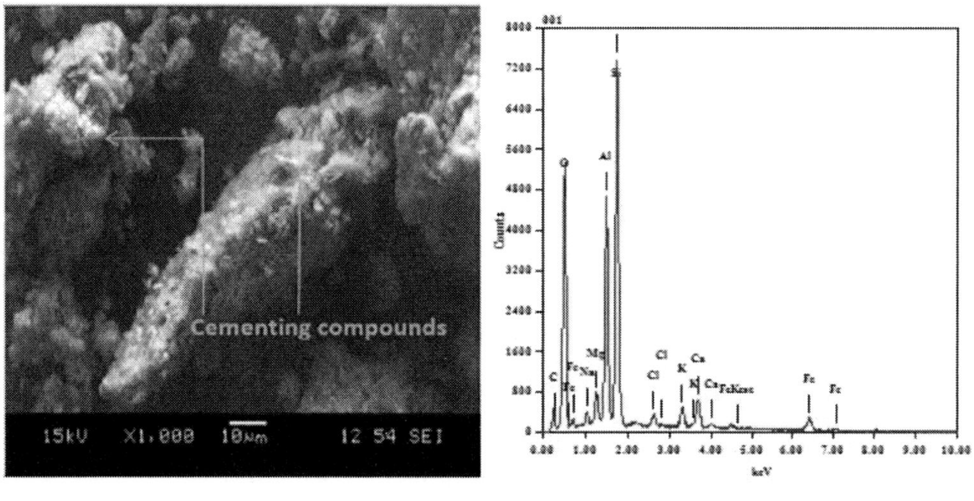


**Figure 6.** SEM image and EDX spectra of 10 % cemented-DMS.

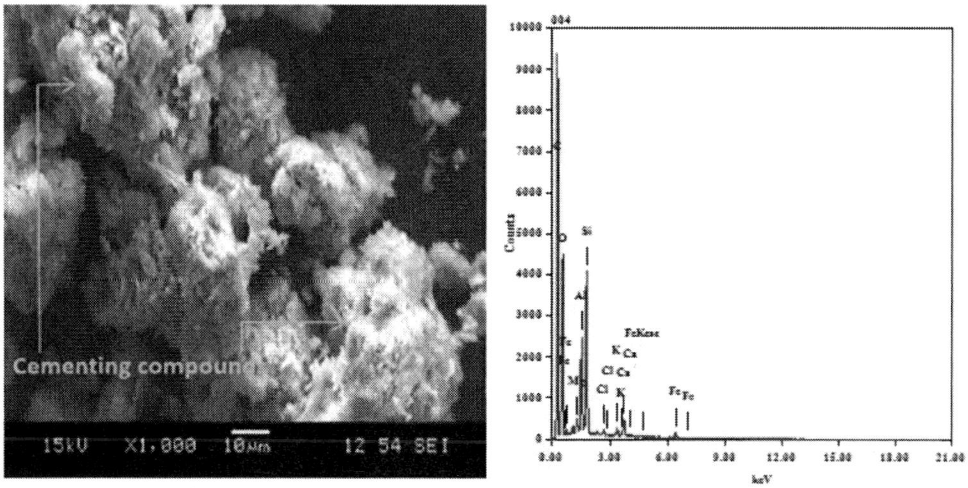


**Figure 7.** SEM image and EDX spectra of 20 % cemented-DMS.

**Table 3.** Ca:Si ratio from EDX analysis.

| **Specimen** | Kuala Perlis DMS | 10C DMS | 20C DMS |
|---|---|---|---|
| **Ca:Si (%)** | 1.63 | 17.78 | 26.48 |

## 4. Conclusions

From the compression curves, it can be conclude that in order to enhance the settlement criteria of cement-solidified DMS, a further studies are required in decrease the *w/c* ratio and increase curing time [7]. The prolonged curing are believe could enhance the settlement criteria of cemented-DMS even though at 10 and 20 % of cement dosage. Otherwise, the cement dosage need to be increase if the specimen will be left cured at a short times (ie. 3 days). SEM micrograph proves the formation of the cementing compounds in cement treated specimens. EDX results revealed the increment of Ca:Si ratio towards the specimens. The specimen's final leachate pH observations are recommended to support the explanation of the C-S-H gel formation. Where, the lower pH of specimens may imply the more cementing reactions. Hence, depositing higher C-S-H gel on the particles and increased the Ca:Si ratio.

## 5. References

[1] Kim KJ, Kim DH, Yoo JC and Baek K 2011 Electrokinetic extraction of heavy metals from dredged marine sediment *Separations and Purification Technology* **79** 164-9

[2] Hwang KY, Park SY, Baek WS, Jung JH, Shin WS, Lee NJ and Hwang IS 2007 Speciation and leaching potential of heavy metals in sediments of nakdong river *J. Korean Soc. Water Wastewater* **21** 113-22

[3] Prakash A and Paul A 2016 A study on stabilization of marine dredged soil using quarry dust *International Journal of Scientific Engineering and Research (IJSER)* **4(3)** 83-6

[4] Chan CM 2014 On the reuse of dredged marine sediments: no dumping, reuse please *Proc. of the Intl. Conf. on Advances in Civil, Structural, Environmental & Bio-Technology* (Institute of Research Engineers and Doctors) pp 1-9

[5] Malviya R and Chaudhary R 2006 Factors affecting hazardous waste solidification/stabilization: a review *Journal of Hazardous Materials* **137** 267-76

[6] Batchelor B 2006 Overview of waste stabilization with cement *Waste Management* **26** 689-98

[7] Nordin NS and Chan CM 2017 Undrained shear strength of low dosage cement-solidified dredged marine soils (DMS) for reclamation works *International Journal of GEOMATE* **13(35)** 180-6

[8] Kaliannan S 2016 *Light Solidification of Kuala Perlis Dredged Marine Soils via Admixtures of GGBS-Cement and Sand: 1-D Compressibility Study* Master's Thesis (Malaysia :Universiti Tun Hussein Onn Malaysia) chapter 4 pp 64

[9] Goodarzi AR and Zandi MH 2016 Assessing geo-mechanical and leaching behavior of cement-silica-fume-stabilized heavy metal-contaminated clayey soil *Environ Earth Sci* **75(911)** 333

[10] Du YJ, Jiang NJ, Liu SY, Jin F, Singh DN and Puppala AJ 2013 Engineering properties and microstructural characteristics of cement-stabilized zinc-contaminated kaolin *Can Geotech J* **51** 289-302

[11] Shen SL, Han J, Du YJ 2008 Deep mixing induced property changes in surrounding sensitive marine clays *J Geotech Geoenviron Eng* **134** 845-54

[12] Eisazadeh A, Kassim KA and Nur H 2012 Stabilization of tropical kaolin soil with phosphoric acid and lime *Nat Hazards* **61** 931-42

## Acknowledgments

The authors would like to acknowledge the Centre for Graduate Studies, Universiti Tun Hussein Onn Malaysia in providing a fund to produce this research paper.